PROCEEDINGS OF ICOSSAR'93 – THE 6TH INTERNATIONAL CONFERENCE ON STRUCTURAL SAFETY AND RELIABILITY / INNSBRUCK / AUSTRIA / 9-13 AUGUST 1993

Structural Safety & Reliability

Edited by

G.I.SCHUËLLER
University of Innsbruck, Austria
M.SHINOZUKA
Princeton University, New Jersey, USA
J.T.P.YAO
Texas A & M University, College Station, Texas, USA

VOLUME 1

A.A.BALKEMA / ROTTERDAM / BROOKFIELD / 1994

The texts of the various papers in this volume were set individually by typists under the supervision of each of the authors concerned.

Published by
A.A. Balkema, P.O. Box 1675, 3000 BR Rotterdam, Netherlands
A.A. Balkema Publishers, Old Post Road, Brookfield, VT 05036, USA

For the complete set of three volumes, ISBN 90 5410 357 4
For Volume 1, ISBN 90 5410 377 9
For Volume 2, ISBN 90 5410 378 7
For Volume 3, ISBN 90 5410 379 5

Printed in the Netherlands

Structural Safety & Reliability, Schuëller, Shinozuka & Yao (eds) © 1994 Balkema, Rotterdam, ISBN 90 5410 357 4

Preface

The 6th International Conference on Structural Safety and Reliability (ICOSSAR'93) was held in Innsbruck on August 9-13, 1993. More than 400 university researchers, representatives from all sectors of industry and officials from various government agencies gathered from 43 countries to discuss problems and solutions of stochastic mechanics, reliability analysis and risk assessment.

During the Conference, 326 papers were presented representing the work of the authors from 33 countries.

At this Conference the International Association of Structural Safety and Reliability (IASSAR) – which serves as the organizing institution of the ICOSSAR series – inaugurated the awarding of Research and Junior Research Prizes in various areas of structural reliability and stochastic mechanics.

ICOSSAR'93 covered all major aspects of structural safety, reliability analysis and risk assessment. The technical program contains new developments as well as practical and novel applications of reliability principles to all types of structural systems and mechanical components. In view of product liability requirements, moreover, safety, reliability performance and quality assurance aspects of major technological systems are covered with emphasis on related issues of risk analysis and risk acceptance pertaining to such systems. Most importantly, industrial applications of the principles are emphasized.

The papers presented at the Conference are contained in these volumes, which represent a permanent record of the proceedings of the ICOSSAR'93. Included herein are keynote lectures, a Plenary Panel discussion addressing the timely issue of 'How safe is safe enough?', technical and ongoing research papers. These volumes should serve as a valuable reference on the recent development in structural safety/reliability and stochastic mechanics. We believe that the material in these volumes will help further advance the state of science and practice of structural reliability in the safety assurance and design of structures.

The Editors
G.I.Schuëller
M.Shinozuka
J.T.P.Yao

Structural Safety & Reliability, Schuëller, Shinozuka & Yao (eds) © 1994 Balkema, Rotterdam, ISBN 90 5410 357 4

Conference organization

ORGANIZING INSTITUTION
International Association for Structural Safety and Reliability (IASSAR)

CONFERENCE CO-CHAIRMEN
G. I. Schuëller, University of Innsbruck, Innsbruck, Austria
M. Shinozuka, Princeton University, Princeton, N.J., USA

CONFERENCE SCIENTIFIC COMMITTEE
J.T.P. Yao, Texas A & M University, College Station, Tex., USA (Chairman)
N. Shiraishi, Kyoto University, Kyoto, Japan (Vice-Chairman)

A. H-S. Ang, University of California, Irvine, Calif., USA
G. Apostolakis, University of California, Los Angeles, Calif., USA
S.T. Ariaratnam, University of Waterloo, Ontario, Canada
G. Augusti, University di Roma La Sapienza, Rome, Italy
B. M. Ayyub, University of Maryland, College Park, Md., USA
R. G. Bea, University of California, Berkeley, Calif., USA
D. Blockley, University of Bristol, Bristol, UK
N. H. Cho, Han Yang University, Seoul, Korea
R. B. Corotis, The Johns Hopkins University, Baltimore, Md., USA
A. Der Kiureghian, University of California, Berkeley, Calif., USA
O. Ditlevsen, Technical University Denmark, Lyngby, Denmark
I. Elishakoff, Florida Atlantic University, Boca Raton, Fla., USA
D. M. Frangopol, University of Colorado, Boulder, Colo., USA
C. Guedes Soares, Technical University Lisboa, Lisboa, Portugal
M. Hoshiya, Musashi Institute of Technology, Tokyo, Japan
H. Ishikawa, Kagawa University, Kagawa, Japan
H. Itagaki, Yokohama National University, Yokohama, Japan
M. Ito, Saitama University, Saitama, Japan
P. Kafka, GRS, Garching, Germany
H. Kameda, Kyoto University, Kyoto, Japan
W. B. Krätzig, Ruhr University Bochum, Bochum, Germany
S. Krenk, University of Aalborg, Aalborg, Denmark
Y. K. Lin, Florida Atlantic University, Boca Raton, Fla., USA
L. D. Lutes, Texas A & M University, College Station, Tex., USA

CONFERENCE SECRETARIAT

ICOSSAR'93 Secretariat
c/o Institute of Engineering Mechanics
University of Innsbruck
Technikerstr. 13
A-6020 Innsbruck, Austria
Tel.: +43 512 218 4181
Fax: +43 512 218 4185
E-Mail: G.I.Schueller@uibk.ac.at
Telex: 75312720=ifm a

SPONSORING INSTITUTIONS
American Concrete Institute (ACI)
American Society of Civil Engineering (ASCE)
European Safety and Reliability Association (ESRA)
International Association for Structural Mechanics in Reactor Technology (IASMiRT)
International Association of Wind Engineering (IAWE)
Österreichischer Ingenieur- und Architekten-Verein (ÖIAV)
Réunion Internationale des Laboratories d'Essais et de Recherches sur les Matériaux et les Constructions (RILEM)
Society of Material Science, Japan (JSMS)
Verein Deutscher Ingenieure, VDI-Gemeinschaftsausschuß 'Industrielle Systemtechnik' (VDI-GIS), Germany

This Conference was held under the auspices of:
The Minister of Research and Science and Vice-Chancellor of Austria, Dr E. Busek
The Governor of the State Tyrol, Dr A. Partl
The Mayor of the City of Innsbruck, Mr R. Niescher

Special thanks for financial support to:
US National Science Foundation, NSF, Washington D.C., USA
Ministry of Research and Science, Vienna, Austria
University of Innsbruck, Innsbruck, Austria
US National Center of Earthquake Engineering, NCEER, Buffalo, USA
Tourismusverband, Innsbruck-Igls, Austria
State of Tyrol, Innsbruck, Austria
City of Innsbruck, Innsbruck, Austria
Österreichische Forschungsgemeinschaft, Wien, Austria
Liebherr Werke Nenzing GmbH, Nenzing, Austria
VOEST-Alpine Bergtechnik, GmbH, Zeltweg, Austria
Plasser & Theurer, Linz, Austria
Österreichische Vereinigung für Qualitätssicherung, Wien, Austria
Vorarlberger Illwerke, Bregenz, Austria
Hilti AG, Schaan, Liechtenstein
Commission of the European Communities, Brussels, Belgium
Hypobank, Innsbruck, Austria
Mayreder, Kraus & Co., Innsbruck, Austria
STRABAG, Linz, Austria
STUAG, Vienna, Austria
PORR AG, Vienna, Austria
Vorarlberger Kraftwerke, Bregenz, Austria
Germanischer Lloyd AG, Hamburg, Germany
Österreichische Raumfahrt- und Systemtechnik, GmbH, Wien, Austria
Gleitbau GmbH, Salzburg, Austria
Imatran Vouma Oy, Vantaa, Finland

FREUDENTHAL AWARD LECTURE 1993

Professor Jann N. Yang
University of California, Irvine, USA
'Applications of reliability methods to fatigue, quality assurance and maintenance'

IASSAR AWARD LECTURES 1993

Professor Torgeir Moan
University of Trondheim, Trondheim, Norway
'Reliability and risk analysis for design and operations planning of offshore structures'

Professor Naruhito Shiraishi
Kyoto University, Kyoto, Japan
'Frontiers of constructions and reliability of Honshu-Shikoku bridges'

IASSAR RESEARCH PRIZES 1993

Professor Mircea Dan Grigoriu
Cornell University, Ithaca, USA
'Contributions to modeling of non-Gaussian processes and the analysis of systems subjected thereto'

Professor Yu-Kweng Michael Lin
Florida Atlantic University, Boca Raton, USA
'Contributions to the advancement of methods of nonlinear stochastic dynamics'

Professor Alfredo H-S. Ang
University of California, Irvine, USA
'Contributions to development of methods for structural systems reliability'

Professor V.V. Bolotin
Russian Academy of Sciences, Moscow, Russia
'Contributions to the advancement of the field of stochastic fracture mechanics and fatigue'

Professor Masaru Hoshiya
Musashi Institute of Technology, Tokyo, Japan
'Contributions to the extension and practical applications of the Kalman Filtering Technique'

Professor Frank Kozin
Special Research Prize Posthumous
'Contributions to the advancement of the field of systems identification'

IASSAR JUNIOR RESEARCH PRIZES 1993

Dr Tsuyoshi Takada
Ohsaki Research Institute, Tokyo, Japan
'Contributions to the industrial applications of stochastic finite element methods'

Dr Christian G. Bucher
University of Innsbruck, Innsbruck, Austria
'Contributions to applying stochastic stability theories to practical engineering structures'

Dr Karl W. Breitung
Munich, Germany
'Contributions to the advancement of analytical procedures to calculate structural reliability by using asymptotic analysis'

Professor Bill F. Spencer Jr
University of Notre Dame, Notre Dame, USA
'Contributions to stochastic modeling of fatigue crack growth'

Professor Roger Ghanem
State University of New York, Buffalo, USA
'Contributions to the application of recursive system identification techniques in civil engineering'

Standing left to right: M.D.Grigoriu, K.W.Breitung, A.H-S.Ang, G.I.Schuëller (Conference Co-Chairman), M.Shinozuka (Freudenthal Award 1989), B.F.Spencer Jr, Mrs M.A.Freudenthal, J.N.Yang (Freudenthal Award 1993), R.Ghanem, N.Shiraishi, T.Takada, T.Moan, V.V.Bolotin, Y.-K.Lin, C.G.Bucher.

Not on this picture: M.Hoshiya

Structural Safety & Reliability, Schuëller, Shinozuka & Yao (eds) © 1994 Balkema, Rotterdam, ISBN 90 5410 357 4

Table of contents

Random vibration (ongoing research)

Offshore structures

Wood structures

Materials

Optimum design

Design codes

Design (ongoing research)

Freudenthal lecture

Structural Safety & Reliability, Schuëller, Shinozuka & Yao (eds) © 1994 Balkema, Rotterdam, ISBN 90 5410 357 4

Application of reliability methods to fatigue, quality assurance and maintenance

J.N.Yang
University of California, Irvine, Calif., USA

ABSTRACT: Early developments of the applications of fracture mechanics-based reliability methods to fatigue, quality assurance and maintenance are reviewed. Applications for the area of aeronautical engineering systems, such as airframes, gas turbine engines, pressure vessels, composite materials, etc. are emphasized. Whenever possible, experimental verifications of the theoretical approaches will be demonstrated. The reliability method is shown to be a rational approach not only for the analysis and design of fatigue-critical components but also for the assessment of the impacts of the quality assurance program and scheduled inspection/repair or replacement maintenance. Proof testing and scheduled inspection maintenance are important methods for ensuring the safety and integrity of structural systems in and beyond their intended design service lives. Issues for gaining wider acceptance of the reliability methods in industry are discussed.

1 INTRODUCTION

In his introductory remarks during the first ICOSSAR, late Professor A. M. Freudenthal (1969) outlined the general reliability analysis of structures, in which the distribution of the residual strength decreases as a function of the service time, whereas the distribution of the maximum load increases as the service time increases. The fact that the distribution of the maximum load increases is well-known in the theory of random process; however, the degradation of the residual strength is still a matter of research to date, for instance due to corrosion, erosion, etc. From the fracture mechanics standpoint, one form of the residual strength degradation for metals or superalloys is related to the crack propagation resulting from fatigue loading, sustained loading or stress-corrosion environments. Early works on reliability analyses of structures, based on the fracture mechanics principle, were conducted considering the strength degradation due to (i) fatigue crack propagation [e.g., Whittaker & Saunders 1973, Yang & Trapp 1974, Shinozuka 1976], and (ii) stress corrosion cracking [Heer & Yang 1971]. Stochastic crack propagation for a superalloy due to sustained loads was investigated later [e.g., Yang & Donath 1984a].

The purpose of this article is to describe some early developments of the applications of fracture mechanics-based reliability methods to fatigue, quality assurance and maintenance. Emphasis is placed on scheduled inspection maintenance and proof testings. One important subject in the scheduled inspection and proof testing maintenance is the stochastic fatigue crack propagation. Reliability methods using S-N fatigue data have been applied extensively [Committee Report 1982, ASCE]. However, these will not be discussed herein, because of their irrelevance to quality assurance and maintenance. Another form of the residual strength degradation is that of composite materials subjected to cyclic loadings, where fracture mechanics is not applicable. Investigations for reliability analyses and quality assurance of composite laminates have been conducted and will be described. In particular, the experimental verifications of the statistical theory of proof tests using composite laminates will be demonstrated.

Whenever possible, experimental verifications for the reliability approaches will be emphasized. A brief summary of an early systematic reliability method, based on fatigue crack propagation, residual strength degradation, random service loads and periodic inspection/repair maintenance will be made. Stochastic crack propagation will be reviewed in Section 2, including the experimental verifications using a jet engine superalloy under high temperature environments. Section 3 describes the inspection/replacement maintenance of gas turbine engine disks. Section 4 summarizes the reliability methods for durability analysis of airframes and points out the importance of reliability methods. Section 5 describes the impact of the proof test and scheduled periodic proof tests on the reliability and quality assurance of structural components. Section 6 summarizes the statistical approach to fatigue of composite materials and the experimental verifications of the theory of proof tests. Section 7 presents some observations for practical applications of reliability methods in industry, based on the author's point of view.

1.1 *Reliability Method for Scheduled Inspection Maintenance*

Consider a critical component (or location) of a transport-type aircraft subjected to the flight-by-flight service loads as shown in Fig. 1 [Yang & Trapp 1974b]. The service loads consist of ground loads, ground-air-ground loads and gust loads. The ground loads resulting from take-off and landing have been modeled as a random process. The ground-air-ground loads, representing one g loading, depend on the payload of each flight and it is a random variable. The gust loads consist of a series of turbulence patches that can be modeled as a series of stationary Gaussian random process, referred to as the composite Gaussian random process. Each of the stationary random process has a zero mean and a different standard deviation σ_i (i,=1,2,...), whereas these standard deviations follow a half-normal distribution. Such a gust turbulence model results in a load exceedance curve that consists of the clear air turbulence and thunderstorm turbulence.

The ultimate strength of the component is assumed to be a Weibull random variable with a scale parameter β as shown in Fig. 1. The ultimate strength distribution remains the same until a crack is initiated. The time to crack initiation is assumed to follow the two-parameter Weibull distribution. After a crack is initiated, the fracture mechanics crack growth rate equation can be used to predict the crack propagation under random service loads [e.g., Yang 1974a]. As the crack length increases, the residual strength decreases. The component residual strength is related to the crack length using the fracture mechanics formula for non-redundant structures or using testing and analysis for redundant structures.

Meanwhile, a periodic inspection/repair maintenance can be implemented. When a crack is detected during inspection, it is repaired so that the residual strength and the crack length are renewed, thus enhancing the structural reliability in service. However, the detection capability of a particular nondestructive evaluation (NDE) system involves considerable uncertainties and it is defined by the probability of detection (POD) curve. Accounting for all the uncertainties described above and the renewal process due to periodic inspection maintenance, the cumulative probability of first failure for a fleet of 50 aircraft components is shown in Fig. 2 [Yang & Trapp 1974b] as a function of service time. In Fig. 2, N denotes the number of inspections performed in the service life of 15,000 flight hours. The first segment in Fig. 2 with smaller failure rate in the early service life is due to the load exceedance over the ultimate strength (before crack initiation), whereas the failure probability increases rapidly in the later service time (after crack initiation).

With the information of failure probabilities, it is possible to determine the optimal inspection interval by minimizing the total expected cost [Yang & Trapp 1975]. Extensive sensitivity studies and new contributions in this regard were carried out by

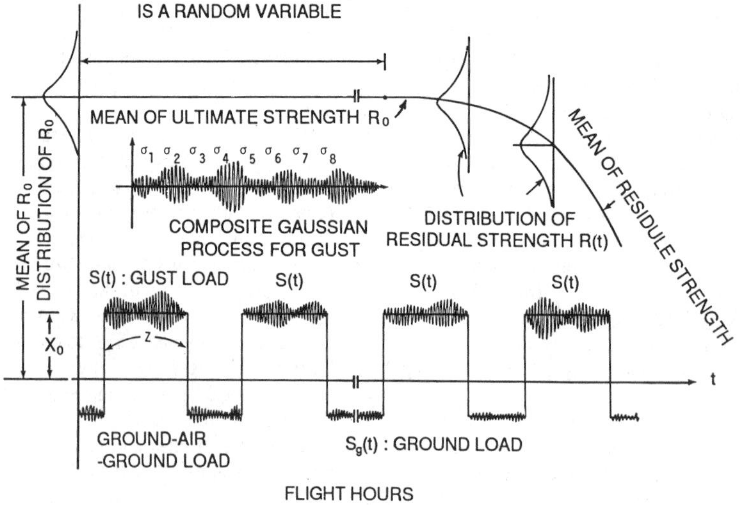

Fig. 1: Profile of flight-by-flight load spectrum, ultimate strength and residual strength

Shinozuka [1976]. Other contributions have been made in this subject area by Lincoln [1985], Berens [1989, 1993], etc. Practical applications of reliability methods to fleet maintenance scheduling have been made recently [Manning, Yang et al 1992a,b, Yang & Manning 1994].

The application of reliability methods to scheduled inspection/repair maintenance has received considerable attention after 1984 in offshore industry. Many contributions have been published in the literature, for instance, Itagaki & Shinozuka [1989], Madsen [1989], etc., using the Baysian approach, and more recent contributions by Deodatis et al [1992], Ito et al [1992], Rocha & Schueller [1993], Fujita & Rackwitz [1993], Moan [1993], etc., just to mention a few.

2 STOCHASTIC FATIGUE CRACK GROWTH MODELS

One important subject in the reliability analysis of fatigue-critical components under scheduled inspection maintenance is the prediction of stochastic crack growth. Experimental test results and field data indicate that the crack propagation involves considerable statistical variability. As a result, various stochastic crack growth models have been proposed and investigated in the literature for metallic materials and superalloys. In particular, a book edited by Provan [1987] contains extensive literature on such models. Early works on statistical crack growth were described in a committee report, ASCE [1982]. Bogdanoff and Kozin [1985] proposed a model that the evolution of the crack size a(t) is a discrete Markov chain. Most of the stochastic models investigated are based on the randomization of the well-known deterministic crack growth rate function $g(\Delta K,R)$ [Yang, et al 1982c, 1983b,c], i.e.,

$$da(t)/dt = X(\eta)g(\Delta K,R) \tag{1}$$

in which a(t) is the crack size at time t, t denotes either cycles or flights or flight hours, ΔK is the stress intensity range, R is the stress ratio, and $X(\eta)$ is a positive random process. Yang & Donath [1983b] suggested that $X(\eta)$ can be a random process of time, i.e., $X(\eta)=X(t)$. The random process model X(t) has been investigated extensively. Lin & Yang [1983] considered X(t) a Poisson pulse train and used the Markov approximation to solve the statistical moments of the time to reach any crack length a(t) from the Fokker-Plank equation. Later, Lin & Yang [1985a] solved the Fokker-Plank equation directly using the Stratonovic stochastic averaging method. Many investigations using the Markov approximation and Fokker-Plank equation have been conducted in the literature, e.g., Spencer [1988, 1989], Tang & Spencer [1989], Tsurui & Ishikawa [1986], Tsurui et al [1989], , Sobczyk [1986], Sobczyk & Spencer [1992], Ishikawa et al [1993], Itagaki et al [1993], etc., just to mention a few. X(t) was considered a lognormal random process by Yang et al [1985a, 1987] using (i) Monte Carlo simulation, and (ii) the second order approximation in which the autocorrelation function of X(t) was accounted for in a simple manner. Lin, Wu & Yang [1985b] used the method of Gaussian closure approximation. Yang et al [1982c, 1983c] suggested that $X(\eta)$ can be a random process of the stress intensity range ΔK, i.e. $X(\eta)=X(\Delta K)$; however, the analysis was carried out only for the completely correlated process, i.e., $X(\Delta K)$ is a random variable. The study for the random process $X(\Delta K)$ was carried out by Ortiz & Kiremedjian [1988]. Recently, $X(\eta)$ has been considered a random function of the crack size a(t), i.e., $X(\eta)=X(a)$, by Berens [1991] and Yang & Manning [1990a]. Finally, a different stochastic crack growth model for composite solid propellants has been studied [Liu & Yang 1993].

The positive random process $X(\eta)$ involves two extreme cases [Yang, et al 1982c, Lin & Yang 1983]: (i) when $X(\eta)$ is completely correlated, i.e., $X(\eta)=X$ is a random variable, the resulting crack growth behavior of a(t) has the largest statistical dispersion and hence the crack growth life prediction is most conservative, referred to as the random variable model, and (ii) when $X(\eta_1)$ and $X(\eta_2)$ are completely independent at any two time instants t_1 and t_2, i.e., X(t) is a white noise process, the resulting crack growth behavior has the smallest statistical variability, leading to the most unconservative life prediction due to the central limit

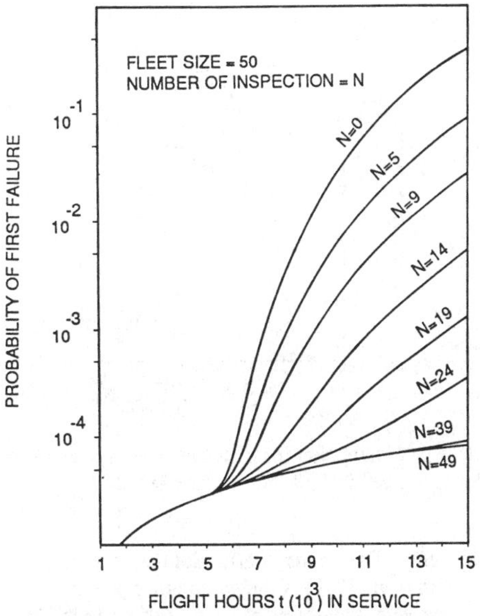

Fig. 2: Cumulative probability of first failure for a fleet of 50 aircraft components vs. service time and number of inspections

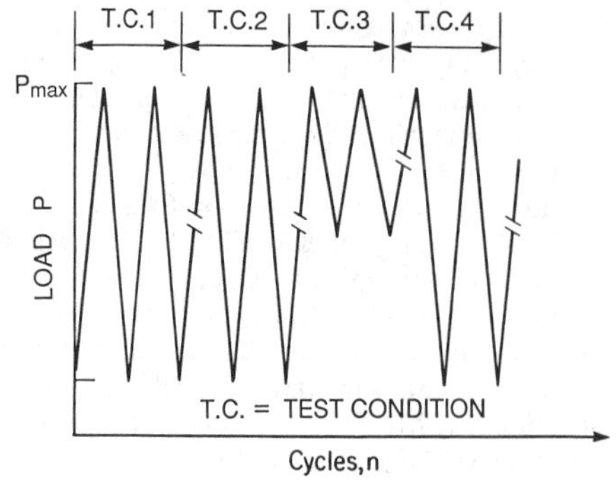

Fig. 3: Laboratory test block spectrum

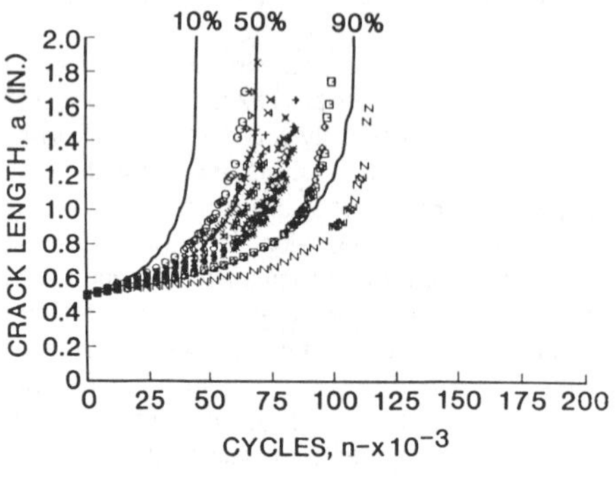

Fig. 4: Comparison of test results and predictions

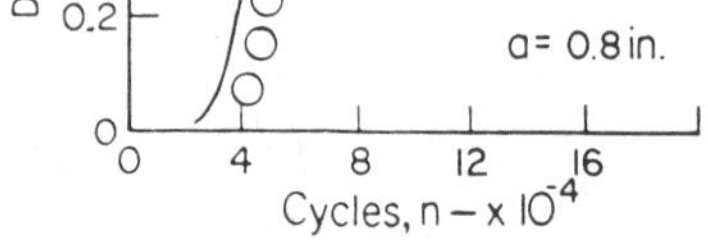

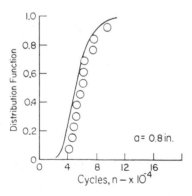

Fig. 5: Correlation for distribution of cycles to reach 0.8 in. (20.2mm) crack length

theorem. The Monte Carlo simulation performed by Virkler et al [1979] is the white noise process.

Extensive investigations have been made aiming at a simple stochastic crack growth approach for practical applications [e.g., Yang et al 1982c, 1983b,c, 1984b, 1985a,b]. The random variable model, i.e., the extreme case in which $X(\eta)$ is completely correlated, is found to be quite reasonable for practical applications

$$da(t)/dt = Xg(\Delta K,R) \qquad (2)$$

in which X is a lognormal random variable with a median value of 1.0. The reasons are as follows: (i) the solution for Eq. (2) is the simplest and it can be understood easily by practicing engineers, (ii) the prediction of the crack growth damage is slightly conservative and the accuracy is quite reasonable for most data sets of practical significance, (iii) since the model does not require the autocorrelation function, it is suitable for applications to baseline data with a small sample size, and (iv) various uncertainties, such as service loads and others, can easily be accounted for in the model. Reasonably accurate predictions for the random variable model have been demonstrated using extensive fastener hole data sets [e.g., Yang et al 1983b, 1984c, 1985a, 1987].

The verifications of the lognormal random variable model using experimental test results for a superalloy by Yang et al [1982c] and Salivar et al [1988] are summarized in the following. A spectrum loading for gas turbine engine consists of repeated identical cycle blocks. Each block consists of m different segments or test conditions. Each test condition, say jth condition, is defined by a constant value of temperature T_j, loading frequency v_j, stress ratio R_j, maximum load P_{jmax} and number of load cycles n_j. Thus, one cycle block consists of (T_j, v_j, R_j, P_{jmax}, n_j) for j=1,2...,m as shown in Fig. 3 where m=4.

For the jth test condition in one block cycle, the lognormal crack growth rate equation is given by

$$da/dn = X_j(10) ** \{C_{1j}\sinh[C_{2j}\log \Delta K + C_{3j}] + C_{4j}\} \qquad (3)$$

in which the hyperbolic sine function has been used for the crack growth rate and $Z_j = \log X_j$ = a normal random variable with zero mean and standard deviation σ_{zj}. Equation (3) can be used to predict the crack growth behavior under spectrum loading, where X_j for j=1,2,...,m are completely correlated.

Baseline data for each test condition shown in Fig. 3 for a superalloy IN100 have been compiled over a long period of time. However, not only the sample size (no. of specimens) available is very small as shown in Table 1, but also these data sets are highly nonhomogeneous, in the sense that each specimen was tested at different time for different purpose using a different maximum load, initial flaw size, final flaw size, specimen dimension, and maximum loading. As a result, the log (da/dn) vs. log ΔK data for each specimen do not cover the entire range of the variable logΔK. This type of highly nonhomogeneous data base is typical for gas turbine engine materials. Parameters c_{j1}, c_{j2}, c_{j3}, c_{j4} and σ_{zj} for each test condition using available data base were determined.

Experimental verification tests were conducted using twelve IN100 ASTM compact tension specimens under the block type spectrum shown in Fig. 3, and the results are displayed in Fig. 4. The solid curves shown in Fig.

4 are 10%, 50% and 90% crack growth trajectories, respectively, based on Eq. (3) and the baseline data for each test condition.

Predictions for the distributions of the number of load cycles to reach the crack length 0.8 and 1.4 inches (20.3 and 35.6mm) are presented as solid curves in Figs. 5 and 6, respectively. These curves are obtained by drawing a horizontal line through the respective crack lengths in Fig. 4. The circles in these figures represent the test data for the twelve specimens. The model (solid line) predicts a slightly smaller number of load cycles to reach a given crack length and a slightly larger dispersion, which is conservative.

The probability of crack exceedance, denoted by $p(x;n)= P[a(n)\geq x]$, is the probability that the crack length at n load cycles will exceed any value x. The solid curves in Fig. 7 shows the predicted crack exceedance probability at n=24,000 cycles. Again, the circles in these figures represent the test data. Figure 7 indicates that the model predicts a slightly larger crack size at a specified number of cycles with a slightly larger dispersion, which is conservative.

The slightly conservative prediction in dispersion comes from the nature of the lognormal random variable model (completely correlated random process). The slightly conservative prediction in the mean value comes from the possible retardation effect under block type loading spectrum as explained in Salivar, Yang & Schwartz [1988]. Nevertheless, correlations between experimental results and the theoretical predictions are very reasonable. It is noted that most of the literature on stochastic crack propagation are rarely verified by independent experimental results. Most of the studies in the literature used the same data set to calibrate the model parameters and to correlate the predictions.

Table 1. Test conditions (T.C.) for one cycle block

T.C.	T_j	R_j	σ_{zj}	Sample Size
1	1200 OF	0.1	0.103	9
2	1350 OF	0.1	0.169	4
3	1200 OF	0.5	0.124	4
4	1000 OF	0.1	0.167	5

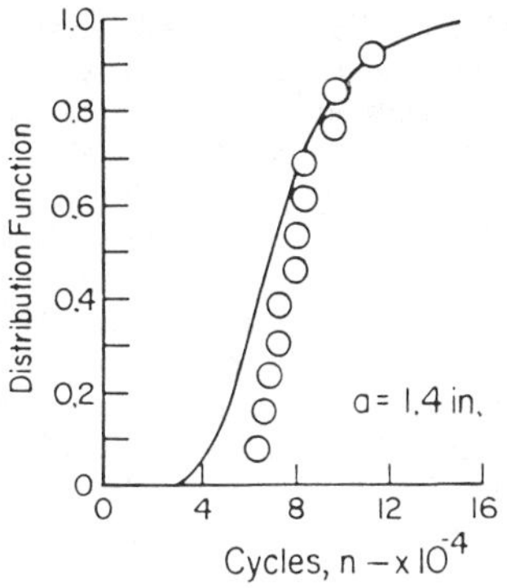

Fig. 6: Correlation for distribution of cycles to reach 1.4 in.(35.6mm) crack length

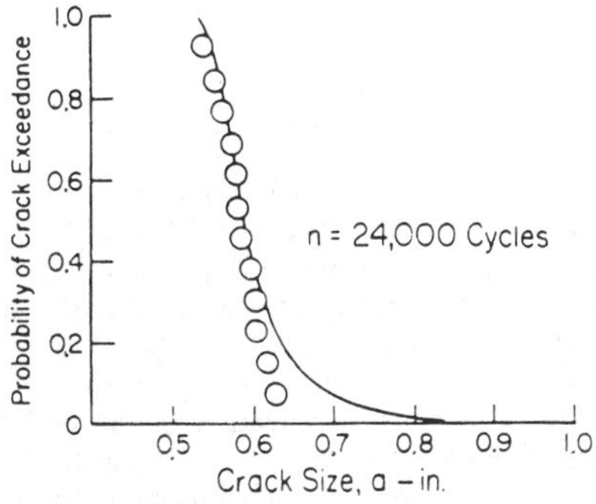

Fig. 7: Correlation for crack exceedance probability at 24,000 cycles

3 INSPECTION/REPLACEMENT MAINTENANCE FOR GAS TURBINE ENGINE COMPONENTS (RETIREMENT-FOR-CAUSE MANAGEMENT)

Most rotor disks of gas turbine engine are limited by low-cycle fatigue (LCF) life, generally expressed in terms of mission equivalent cycles. The low-cycle fatigue (LCF) life usually is defined as the number of cycles necessary to initiate a crack approximately 0.03 in (0.76mm) long at the critical locations, such as bolt holes, cooling air holes, rim holes. etc. The distribution of LCF life is obtained for a given set of loading conditions (stress/strain, time, temperature). Traditionally, the design life of most rotor disks of gas turbine engines, which are limited by LCF life, is determined from the distribution of the crack initiation life at an occurrence rate of 1 in 1000 disks (i.e., probability of 10^{-3}). It is at this design life that all LCF-limited disks are retired from service. This procedure has been very successful in preventing catastrophic failures in service. However, in retiring 1000 disks because one may fail, the remaining 999 unfailed disks are not fully utilized. Hence, the concept of retirement-for-cause (RFC) has been investigated [e.g., Hill 1981, Annis 1981, Besuner 1982, 1983, Yang 1984b,c] for screening one bad part and for certifying the remaining 99.9% parts for safe utilization of the life capacities of engine disks.

The retirement-for-cause procedure is based on the applications of probabilistic fracture mechanics and nondestructive evaluation (NDE), where each rotor disk is inspected periodically. When a crack is detected during inspection, the disk is retired; otherwise it is returned to service until the next inspection maintenance. This procedure could be repeated until the disk has incurred detectable damage (crack), at which time it is retired for that reason (cause). Retirement-for-cause is, then, a methodology for retiring an engine component from service when quantifiable damage occurs, rather than when a predicted minimum design life had been exceeded. Not only is the entire RFC life management quite complex, but also many physical quantities involve considerable statistical variabilities. These include the time to crack initiation, initial flaw size, crack growth damage accumulation, service loads, stress concentration factor, nondestructive evaluation, and others. As a result, reliability methods have been used to develop RFC analysis methodologies, in which the method of Monte Carlo simulation [e.g., Annis 1981, Besuner 1982, 1983] and analytical method [e.g., Yang & Chen 1984b,c] have been used. The probabilistic RFC methodologies developed are capable of estimating the probability of component failure, the percentage of replacement during each inspection maintenance, life-cycle-cost, etc. Hence, the optimal inspection interval can be determined for minimizing the life-cycle-cost of engine components while maintaining a high level of reliability in service.

The major components of the probabilistic RFC analysis methodology and an example are summarized briefly in the following. The detailed technical approaches are given by Annis [1981], Besuner [1982, 1983] and Yang & Chen [1984b,c].

3.1 *Initial Fatigue Quality*

The durability of engine materials is determined by the initial fatigue quality (IFQ). The IFQ defines the initially manufacturing state of a component or details prior to service. For some engine materials, such as titanium, Waspaloy, Astroloy, etc., the IFQ is represented by the distribution of the time to crack initiation, i.e., the time (cycles) required to initiate a reference crack size a_0 (e.g., 0.03 inches). The distribution of the time to crack initiation is determined from laboratory test results. For some materials, the initial fatigue quality may be defined by the distribution of the intrinsic defect size [Annis 1981] or both the distributions of the intrinsic defect size and the time to crack initiation at a smaller reference crack size.

3.2 *Fatigue Crack Propagation*

After a crack is initiated at the reference crack size, a_0, the Paris model for a corner crack has been modified for convenience of analysis

$$da/dn = QS^V a^b \tag{4}$$

in which Q, V and b are constants, and S is the "normalized" design stress spectrum, in the sense that S=1 represents the nominal design spectrum. The values of Q, V and b are determined from three or more crack growth curves, i.e., crack size a(n) versus cycles, n, using least-squares best fit procedures. The crack growth curves are obtained from the results of either spin pit tests, or field experience, or numerical integrations using a general computer program under spectrum loading. Experience indicates that Eq. (4) is quite versatile and it is capable of fitting crack growth curves under spectrum loading very well [Yang 1984b,c]. The deterministic crack growth rate given by Eq. (4) is randomized using the lognormal random variable model as follows

$$da/dt = XQa^b \tag{5}$$

in which X is a random variable accounting for the following contributions to the scatter of the crack growth rate,

$$X = H_1 H_2 H_3 H_4 S^V \tag{6}$$

In Eq. (6), H_1, H_2, H_3, H_4 and S are statistically independent random variables assumed to follow the lognormal distribution with the median value equal to unity, and H_1 = the material crack growth resistance variability; H_2 = the variability of crack geometry, such as aspect ratio, stress intensity factor, etc.; H_3 = the variability of crack modeling, e.g., two-dimensional crack model, etc.; and H_4 = the variability of crack growth damage due to each equivalent cycle. For instance, two stress records may be modeled by the same number of equivalent cycles, but the corresponding crack growth damages are not identical. Finally, S denotes the variability of the service loads, temperature profile, etc. from the nominal spectrum loading.

It follows from Eq. (6) that the random variable X is lognormal with the median value equal to unity, and the log standard deviation can be obtained analytically from the log standard deviations of H_1, H_2, H_3, H_4 and S.

3.3 *Nondestructive Evaluation (NDE)*

Current nondestructive evaluation (NDE) systems are not capable of repeatedly producing correct indications when applied to defects of the same size. The probability of detection (POD) for all defects of a given size has been used in the literature to define the capability of a particular NDE system in a given environment. The probability of detection as a function of the defect size "a" is referred to as the POD(a) curve [e.g., Berens & Hovey 1984, 1988]. An NDE system involves two types of errors: (i) Type I error allows components containing a large flaw length to remain in service, thus greatly increasing the failure probability; and (ii) Type II error rejects good components and adversely affects the life-cycle-cost. For the RFC life management, both Type I and Type II errors should be

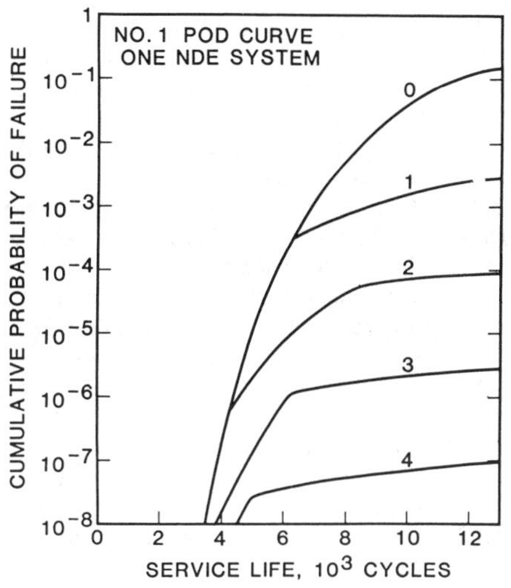

Fig. 8: Cumulative probability of failure as function of service life using No. 1 POD curve

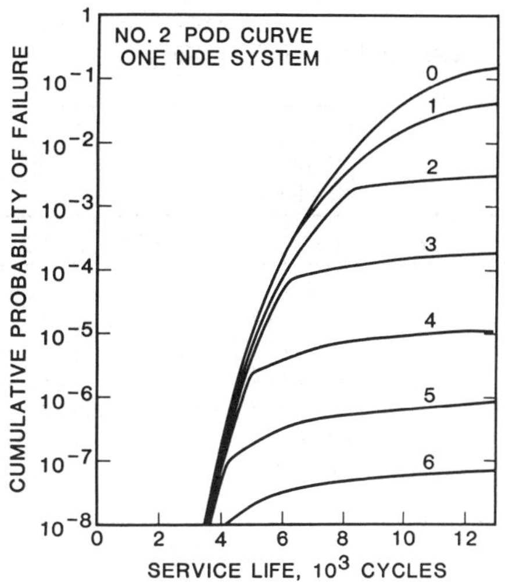

Fig. 9: Cumulative probability of failure as function of service life using No. 2 POD curve

reduced to minimize the life-cycle-cost. Methodologies to reduce either or both types of errors using multiple inspection procedures have been investigated by Yang & Donath [1982d, 1983d]. To reduce the rejection of good disks, rejected disks are reinspected using the second NDE system. Before the second inspection, some actions may be taken for surface preparations, such as cleaning, polishing or replicating of the rejected critical locations. Consequently, the detection capability of the second NDE system (inspection) is much better than that of the first inspection. The disk is replaced only when it is rejected by the second NDE system; otherwise it is returned to service. The resulting POD curve using multiple inspection procedures has been derived [Yang & Donath 1982d, 1983d].

3.4 *Demonstrative Examples*

The third stage turbine disk of the TF-33 jet engine is considered for demonstrative examples. The material is Incoloy 901 and 10 bolt holes are the critical locations. The conventional design life is 2,500 cycles, i.e., all disks are retired after 2,500 cycles. Under the RFC life management, the disk life is extended 5 times to 12,500 cycles. The cumulative probabilty of failure for a bolt hole is shown in Fig. 8 as a function of service life, in which Curve 0 denotes the failure probability without scheduled inspection. With scheduled inspection/ replacement maintenance, the failure probabilities are designated as Curves 1, 2, 3, and 4. The number shown in each curve indicates the number of inspection maintenances performed in 12,500 cycles. For instance, Curve 4 represents the failure probability for a bolt hole in which four service inspection maintenances are performed, i.e., the inspection interval is 2,500 cycles. Figure 8 clearly indicates that the cumulative probability of failure decreases significantly as the number of inspections in 12,500 cycles increases.

The results presented in Fig. 8 are based on a good NDE system, referred to as No. 1 POD curve [Yang & Chen 1984b]. For an NDE system with a less detection capability, referred to as No. 2 POD curve, the cumulative probabilities of failure are shown in Fig. 9. As observed from Fig. 9, the enhancement of disk reliability is not as drastic as that in Fig. 8. However, six (6) inspections in service could achieve the same structural reliability as that using four (4) inspections for No. 1 POD curve.

In addition to the failure probability, the percentage of disk replacement during each inspection can also be computed. In Fig. 8 for instance, with 4 inspections (inspection interval at 2,500 cycles), the percentages of disk replacement are 9.91%, 40.86%, 34.87% and 34.95%, respectively. Hence, the total replacement rate is 120.6%. With the conventional design, the replacement rate is 400%. The retirement-for-cause life management reduces the number of disk replacements and spare parts requirements, while minimizing the life-cycle-cost.

Based on the information obtained above as well as the information on the cost of failure, costs of inspections and replacements, etc., a life-cycle-cost savings analysis as compared with the conventional design can be performed [e.g., Besuner 1983, Yang & Chen 1984c]. The life-cycle-cost savings for 1,000 disks in $ millions, and the corresponding failure probability in 12,500 cycles are shown in Table 2 for two NDE systems. It is observed from Table 2, that four inspections can be chosen for No. 1 NDE system

Table 2: Life-cycle cost savings for 1000 TF-33 engine disks (in $million); N = No. of inspections

N	LCC Savings $M	Probability of Failure	LCC Savings $M	Probability of Failure
	POD Curve 1		POD Curve 2	
1	19.07	$3.0x10^{-3}$	-49.03	$3.86x10^{-2}$
2	22.27	$9.40x10^{-5}$	19.97	$2.91x10^{-3}$
3	21.22	$2.81x10^{-6}$	24.31	$1.73x10^{-4}$
4	20.16	$9.51x10^{-8}$	23.94	$1.08x10^{-5}$
5	19.32	$4.14x10^{-9}$	23.47	$7.89x10^{-7}$
6	18.99	$2.31x10^{-10}$	23.06	$6.75x10^{-8}$
7	18.43	$1.65x10^{-11}$	22.70	$6.71x10^{-9}$

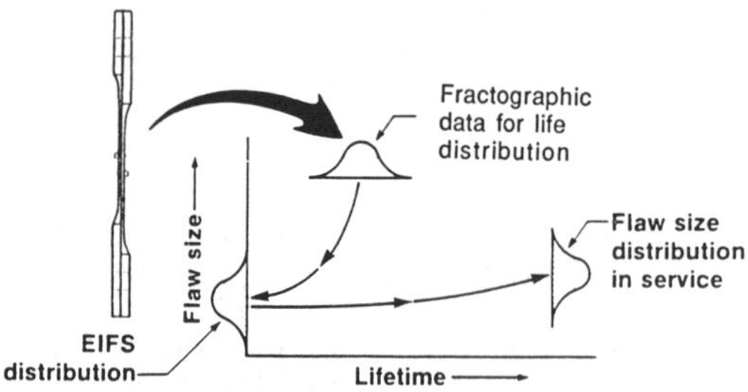

Fig. 10: Probabilistic durability analysis procedures

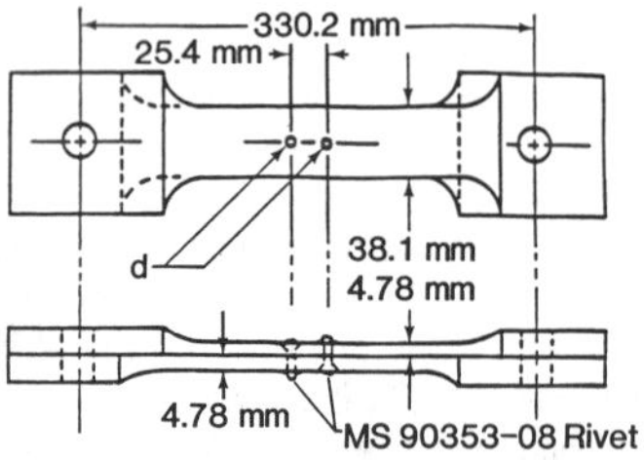

Fig. 11: Double reversed dog-bone specimen

and six inspections can be used for No. 2 NDE system, simply to maintain an acceptable level of reliability. Note that for the No. 2 NDE system, the life-cycle-cost saving is negative, if one inspection is considered. This is because the probability of failure and, hence, the cost of failure are too high.

4 DURABILITY ANALYSIS METHODS FOR METALLIC AIRFRAMES

The U.S. Air Force has damage tolerance and durability design requirements for metallic airframes [MIL-A-87221,1985]. Damage tolerance requirements are concerned with structural safety, whereas durability requirements are concerned with minimizing functional impairment, such as excessive cracking (e.g., cracks < 2.54mm), fuel leaks, and ligament breakage. Airframe durability affects structural maintenance requirements, economic life, and operational readiness. A metallic airframe contains thousands of structural details, and all such details are susceptible to fatigue cracking in service. Therefore, a statistical approach is essential for durability analysis to quantitatively estimate the extent of damage (i.e., number of details expected to have cracks exceeding specified crack size limits) at any service time.

Probabilistic damage tolerance analysis has been proposed by Palmberg et al (1987). Probabilistic-based durabilty analysis "design tools" have been developed recently for ensuring the Air Force's durability design requirements for metallic airframes. These methodologies are documented in the Air Force's durability design handbook [Manning & Yang 1984, 1989] and elsewhere [e.g., Manning & Yang 1987, Yang & Manning 1990b]. The methodologies have been demonstrated and verified for fastener holes in coupon specimens and full-scale aircraft structures for a wide range of crack sizes covering functional impairment. A probabilistic durability analysis methodology is summarized briefly along with the experimental verifications by coupon specimens and full-scale lower wing skins. The necessity of using the reliability method in the durability analysis will be emphasized later.

4.1 *Durability Analysis Procedures*

The reliability analysis procedures for predicting the extent of damage for any part of an airframe are schematically shown in Fig. 10. The procedures include (i) the determination of the equivalent initial flaw size (EIFS) distribution, and (ii) the crack growth damage analysis using the EIFS distribution. The initial fatigue quality defines the initially manufactured state of a structural detail or details with respect to initial flaws in a part, component, or airframe prior to service. For a group of replicate details (e.g., fastener holes), it is represented by an EIFS distribution. An equivalent initial flaw size is an artificial initial crack size, which results in an actual crack size at an actual point in time when the initial flaw is grown under service conditions [e.g., Rudd & Gray 1977, Manning & Yang 1984].

The Weibull compatible distribution function was found to be reasonable for representing the EIFS distribution [Yang & Manning 1980a]

$$F_{a(0)}(x) = exp\left\{-\left[\frac{ln(x_u/x)}{\phi}\right]^{\alpha}\right\} \quad ; 0 \le x \le x_u \qquad (7)$$

$$= 1.0 \quad ; \quad x > x_u$$

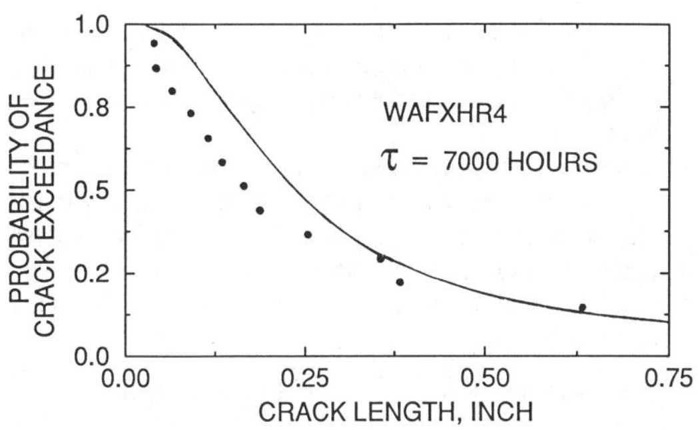

Fig. 12: Correlation between predicted crack exceedance probability and test results at 7000 flight hours

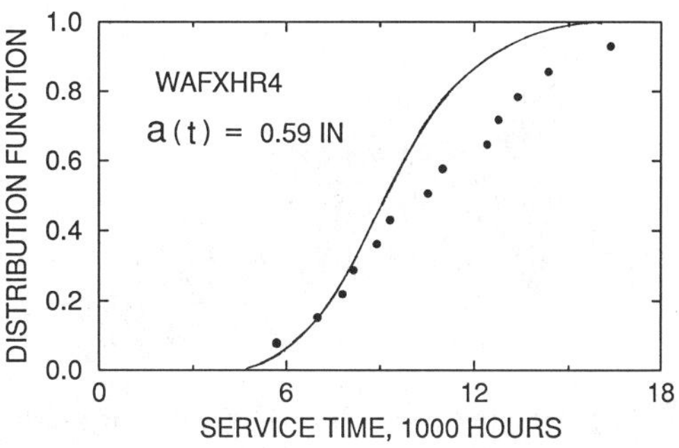

Fig. 13: Correlation between predicted distribution of time to reach 0.59 in crack length and test results

in which x_u = EIFS upper bound limit; α and ϕ are distribution parameters. Equation (7) is derived based on the assumption that the time to crack initiation follows the three-parameter Weibull distribution [Yang & Manning 1980a]. Other types of distribution functions have been investigated by Shinozuka [1979].

Fastener hole specimens, such as the one shown in Fig. 11, were fatigue tested under a fighter spectrum loading and fractographic data were acquired. An EIFS value for a fastener hole is determined by back-extrapolating fractographic data in a selected crack size range AL-AU (e.g., 0.01"-0.05"). After EIFS values a(0) are obtained from all available fractographic data in the crack size range AL-AU, they are fitted by Eq. (7) to determine the EIFS distribution parameters x_u, α, and ϕ. To predict the extent of cracking in service, the initial flaw size distribution, Eq. (7), is grown forward to derive the statistical distribution of the crack size a(t) at any service time t, Fig. 10. Then, the following quantities can be predicted: (1) the probability that a crack in the ith stress region at any service time τ will exceed any given crack size x_1, denoted by $p(i,\tau)$, and (2) the distribution of service time for a crack in the ith stress region to reach any given crack size x_1. The EIFS distribution given by Eq. (7) is grown forward using a service crack growth curve (SCGC) for a given stress region. The SCGC in each stress region is determined using a general fracture mechanic crack growth analysis computer code. To simplify the durability analysis procedures, the SCGC can be fitted by analytical crack growth rate equations, such as Eq. (5). Detailed approaches were described by Manning & Yang [1984, 1989]. Likewise, the statistical estimation of the economic life was studied [Yang 1980d].

4.2 *Experimental Verifications*

Experimental verifications of the theoretical durability analysis methodology were conducted for clearance-fit countersunk fastener holes for coupon specimens and for full-scale lower wing skins of a fighter aircraft. The EIFS distribution of clearance-fit countersunk fasteners in 7475-T7351 aluminum was determined using fractographic results for double-reversed dog-bone specimens with a 15% bolt load transfer, as shown in Fig. 11, fatigue tested under a fighter spectrum loading. Three different fractographic data sets were used to estimate the EIFS distribution parameters for the Weibull compatible distribution function, Eq. (7).

A set of 13 replicated reversed dog-bone specimens (twice the width of that in Fig. 11) was fatigue tested under the same fighter spectrum loading but with a higher stress level (40.8 ksi). The resulting fractographic data set is referred to as WAFXHR4. The theoretical predictions based on the EIFS distribution obtained above for (i) the probability of crack exceedance $p(i,\tau)$ at τ=7,000 flight hours, and (ii) the distribution of time to reach a crack size of 0.59 inches are shown in Figs. 12 and 13, respectively, as solid curves. Also shown in these figures as solid circles are the experimental results for WAFXHR4. As observed from Figs. 12 and 13, the correlation between the experimental results and the theoretical predictions is very reasonable. Reasonable correlations have been shown for other verification data sets [Manning & Yang, 1989].

Fractographic results are available for two lower wing skins from a fighter durability test article that was fatigue tested under spectrum loading to 16,000 flight hours (2 lifetimes). Teardown inspection results for cracks in each fastener hole are given in Tables 3 and 4. To predict the extent of damage, the lower wing skin is divided into ten stress regions as shown in Fig. 14. The maximum stress in each stress region is approximately the same and the number of fastener holes, N_i, in each stress region is shown in Table 3. Based on the crack growth damage prediction using the EIFS distribution, the probability of crack exceedance $p(i,\tau)$, i.e., the probability that the crack size at t=16,000FH will exceed any crack size x_1, was predicted for 10 stress regions. Using the binomial

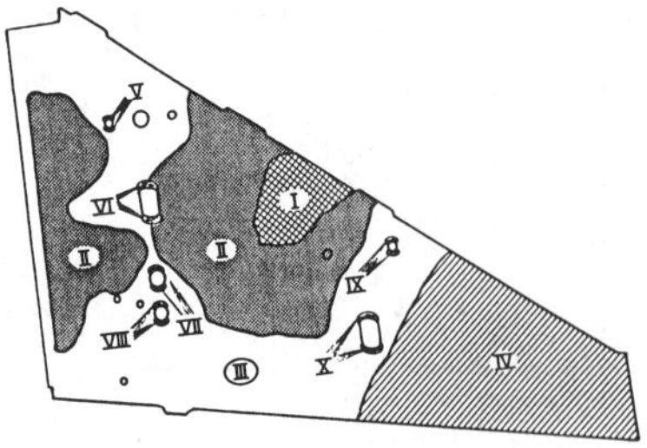

Fig. 14: Stress region for fighter lower wing skin

Table 3: Average number of fastener holes with crack size exceeding x_1 = 1.27mm at τ=16,000 flight hours in each stress region

Stress Region	No. of Fastener Holes, N_i	$p(i,\tau)$	$\bar{N}(i,\tau)$
1	59	0.0350	2.07
2	320	0.0145	4.64
3	680	0.00007	0.05
4	469	0.00	0.00
5	8	0.0371	0.29
6	30	0.0577	1.73
7	8	0.225	1.80
8	8	0.00714	0.06
9	12	0.00714	0.09
10	20	0.00403	0.08
Total	1614		10.81
Experimental	Results		9.5

Table 4: Statistics for number of fastener holes with crack size exceeding x_1 in a fighter lower wing skin at τ=16,000 flight hours

x_1(mm)	$\bar{L}(\tau)$	$\sigma_L(\tau)$	Experimental Results (Ave.)
0.762	35.80	5.800	14.5
1.27	10.81	3.185	9.5
2.54	5.38	2.262	7.0
5.08	2.19	1.450	1.0
7.62	1.24	1.097	0.5

distribution [Yang 1976b], the average number, $\bar{N}(i,\tau)$, and the standard deviation, $\sigma(i,\tau)$, of cracks for the ith stress region can be computed as $\bar{N}(i,\tau)=N_i p(i,\tau)$, $\sigma(i,\tau)=\{N_i p(i,\tau)[1-p(i,\tau)]\}^{1/2}$. Only the results for $p(i,\tau)$ and $\bar{N}(i,\tau)$, at x_1=1.27mm and τ=16,000FH, are presented in Table 3. As observed from Table 3, the theoretical prediction for the total average number of cracks exceeding 1.27mm in the entire lower wing skin is 10.81, whereas the average experimental result for both wing skins is 9.5.

Furthermore, predictions for the average number of fastener holes in the entire lower wing skin with a crack size > x_1 at 16,000 flight hours, $\bar{L}(\tau)$, and its standard deviation, $\sigma_L(\tau)$, are shown in Table 4. The $\bar{L}(\tau)$ and $\sigma_L(\tau)$ values are computed based on the binomial distribution as

$$\bar{L}(\tau) = \Sigma \bar{N}(i,\tau) \quad ; \quad \sigma_L^2(\tau) = \Sigma \sigma^2(i,\tau) \tag{8}$$

The teardown inspection results are shown in Tables 3 and 4 for comparison. These results reflect the average extent of damage for a lower wing skin based on the total extent of damage for left and right lower wing skins combined. It is observed that the durability analysis predictions based on specimen test results correlate well with the test results of the full-scale lower wing skins. Using $\bar{L}(\tau)$ and $\sigma_L(\tau)$, the extent of damage for the lower wing skin can be estimated for selected probabilities. Such results can be used to determine the mean and upper/lower bound limits for the extent of damage.

It is important to mention that the reliability method is essential for the durability analysis in predicting the extent of damage of a component as shown in Tables 3 and 4. Should a deterministic analysis be used, the probability of cracking would have been zero for $p(i,\tau)<0.5$ and 1.0 for $p(i,\tau)>0.5$. Consequently, all the prediction results for $\bar{N}_i$, $\bar{L}(\tau)$ and $\sigma_L(\tau)$ in Tables 3 and 4 would have been zero. As observed from these two tables, the full-scale experimental results are not zero.

Traditionally, the initial fatigue quality (IFQ) has been characterized by the distribution of time to crack initiation. The distribution of the equivalent initial flaw size (EIFS) approach for defining the IFQ is quite significant [Yang & Manning 1980a]. The main advantage of such an approach is that the EIFS distribution can be grown forward under different stress levels using a general crack growth computer code for theoretical predictions. Recently, such an approach has been used with modifications by ALCOA researchers [e.g., Magnusen, Bucci & Hinkle 1989, 1992, Owen et al 1988, etc.] for characterizing and comparing the initial fatigue quality of new materials and old materials. The EIFS istribution approach has also been used by some industrial companies and it is expected to gain more acceptance in industry in the future.

5 PROOF TESTING FOR QUALITY ASSURANCE AND MAINTENANCE

Proof testing has been used extensively in industry for quality assurance of selected structural components, such as pressure vessels, rocket motor cases, etc. [e.g., Barnett & Hermann 1965, Tiffany 1968, 1970]. The purpose of proof testing is to screen out weak components or components containing unacceptable flaw length or defects to ensure the safety and integrity of the structure in its design service life. It is well-known, however, that the structural strength, defect or flaw length, service loads, etc., involve statistical variabilities. As a result, the effect of proof tests can be interpreted rationally in terms of the improvement of the structural reliability [e.g., Shinozuka & Yang 1969].

Let $f_R(x)$ and $F_R(x)$ be the probability density function and the distribution function, respectively, of the resisting strength, R, of a component prior to the proof test, and $f_{R0}(x)$ be the probability density function of the strength after the component passing the proof test at a proof load level r_0. Assuming no structural damage is introduced during proof testing, it has been derived by Shinozuka and Yang (1969) that

$$f_{R_0}(x) = f_R(x)H(x-r_0)/[1-F_R(r_0)] \qquad (9)$$

in which $H(x-r_0)$ is the unit step function starting at r_0. Hence, proof testing truncates the lower tail portion of the strength distribution as shown in Fig. 15, thus improving the structural reliability in service. Appropriate procedures for performing proof testing to avoid structural damages have been described by Tiffany (1968, 1970). Note that for the general case in which the strength, R, of a component is a function of many random variables, the derivation above and Fig. 15 are valid [Yang et al 1993].

Corresponding to a proof load level r_0, there is a flaw length a_0 through the fracture mechanics relationship $K_{IC}=r_0\,\beta(a_0)\,\sqrt{\pi a_0}$ in which K_{IC} is the critical stress intensity factor and $\beta(a_0)$ is a geometric factor. Let $f_a(x)$ and $F_a(x)$ be the probability density function and the distribution function, respectively, of the maximum flaw length in a component prior to the proof test, and $f_{a*}(x)$ be the probability density function of the maximum flaw length after proof testing. Then, it can be shown that [Heer & Yang 1971]

$$f_{a^*}(x) = f_a(x)H(a_0-x)/F_a(a_0) \qquad (10)$$

Hence, proof testing truncates the upper tail portion of the distribution of the maximum flaw length in a component as shown in Fig. 16, thus enhancing the structural reliability in service. The effect of proof testing on the structural reliability and its relation to optimal structural design have been investigated [e.g., Shinozuka & Yang 1969, Heer & Yang 1971].

Proof testing also provides a maintenance option for certain structural components with critical details or locations that are not accessible for inspections. The effect of scheduled periodic proof tests is the truncation of the lower tail portion of the residual strength distribution, that is subjected to degradations in service, at the periodic service interval, thus enhancing the structural reliability in service. Structural reliability under periodic proof test maintenance and the determination of the optimal proof-test maintenance interval have been studied [Yang 1976a, 1977a,b] for both metallic and composite components.

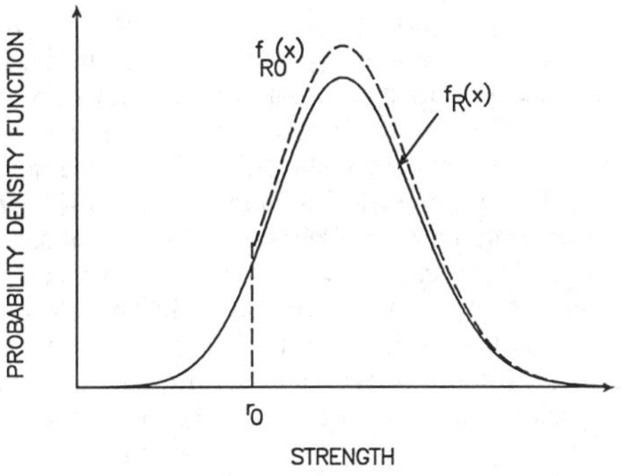

Fig. 15. Truncation of strength distribution by proof test

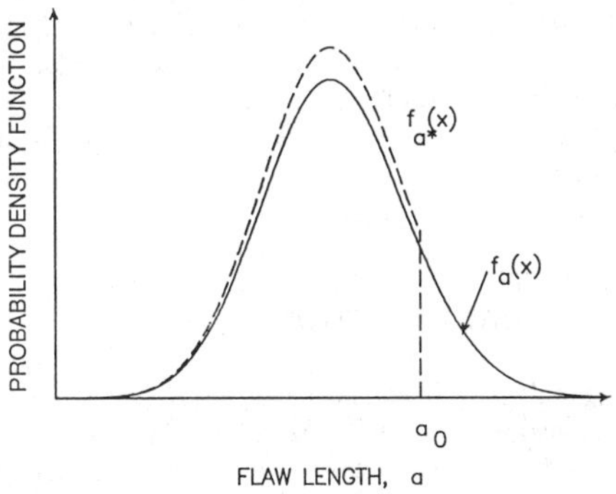

Fig. 16. Truncation of flaw length distribution by proof test

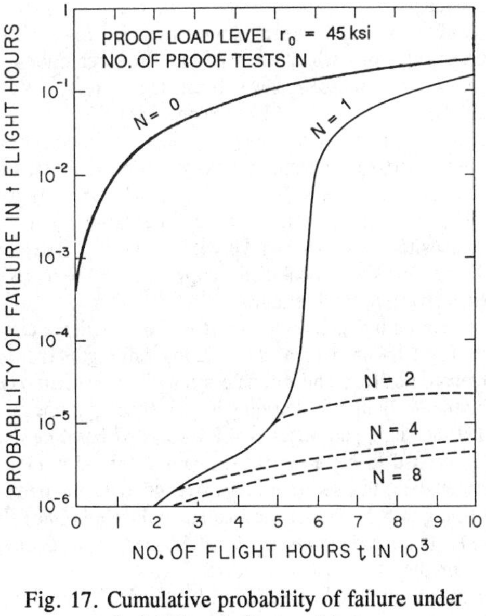

Fig. 17. Cumulative probability of failure under scheduled periodic proof tests

Consider a glass/epoxy $\pi/4$ composite laminate subjected to a transport-type aircraft spectrum loading and scheduled periodic proof tests in service [Yang 1977b]. The ultimate strength follows the two-parameter Weibull distribution with the shape parameter $\alpha=12.3$ and the scale parameter $\beta=53$ ksi. The cumulative probability of failure as a function of service time is shown in Fig. 17, in which N denotes the number of proof tests in 10,000 flight hours and the proof load level is $r_0=45$ ksi. Based on experimental results, a proof load level at 45 ksi will not damage the composite laminate. In Fig. 17, the curve for N=0 reflects no proof testing, whereas the curve for N=1 indicates that the component is proof-tested prior to service. Figure 17 shows that a single proof test prior to service is quite effective if the service life is 5,000 flight hours; however, it is not enough when the service life is 10,000 flight hours. As a result, scheduled periodic proof tests, such as N=2, are needed for a service life of 10,000 flight hours. As observed from Fig. 17, periodic proof test maintenance is quite effective.

6 STATISTICAL FATIGUE OF COMPOSITE MATERIALS AND PROOF TESTS

Fatigue damage in composite laminates is a complicated phenomenon involving different mechanisms, such as fiber breakages, matrix cracking, fiber-matrix debonding, delamination, etc. These fatigue damage mechanisms cause changes in the strength, stiffness and other material properties of composite laminates. From the phenomenological standpoint, fatigue damage can be evaluated, in the global sense, by the degradation of the residual strength and stiffness. In the past, various residual strength degradation models have been proposed from which the fatigue life predictions of composite laminates have been made [e.g., Yang 1978a,b, Whitney 1981, Sendeckyj 1981, etc.]. Likewise, residual stiffness degradation models have been proposed for statistical fatigue life predictions of composite laminates [e.g., Yang et al 1989, 1990c, 1992, Lee et al 1993]. Since the fatigue data of composite materials involve a large statistical dispersion, the probabilistic approach has been used extensively in the literature.

A strength degradation model is shown schematically in Fig. 18 in which the residual strength reduces monotonically. The ultimate strength is characterized by the Weibull distribution with the shape parameter α and the scale parameter β. Two sets of baseline data are needed to calibrate the strength degradation model parameters; one set of ultimate strength data and one set of fatigue S-N data at various stress levels. Once the strength degradation model is established, it can be used to predict the distribution of fatigue life under (i) constant amplitude loadings, (ii) variable amplitude loadings [e.g., Yang & Jones 1980b, 1982a,b], (iii) spectrum loadings [Yang & Du 1983a], and (iv) proof tests [e.g., Yang & Liu 1977c, Yang & Sun 1980c]. As observed from Fig. 18, a composite specimen passing a proof stress level r_0 will survive a certain minimum fatigue life.

6.1 *Experimental Verifications of Statistical Theory of Proof Tests*

Based on a residual strength degradation model, Fig. 18, experimental verifications for the statistical theory of proof tests were conducted [e.g., Yang & Liu 1977c, Yang & Sun 1980c]. These results are summarized briefly in the following. Fatigue test results for graphite/epoxy 5208/300 [90,+45,-45,0]s laminates were conducted. The baseline data included a set of 15 ultimate strength data, with a scale parameter $\beta=88.78$ ksi (621.1 MPa), and a set of 39 fatigue S-N data at various stress levels with a stress ratio R=0.1. Parameters in the strength degradation model were determined from the baseline data. For verification purposes, a set of 19 specimens which survived an initial proof test at $r_0=82.64$ ksi (569.8 MPa) was fatigue tested at a maximum stress $\sigma=64.6$ ksi (447.5 MPa) with R=0.1 until failure. The resulting fatigue life data are shown in Fig. 19 as circles. Note that none of the baseline S-N data was subjected to the stress level $\sigma=64.6$ ksi used for the verification tests. From the theoretical prediction, a specimen surviving the proof stress $r_0=82.64$ ksi will sustain at least $N_0=6{,}289$ cycles. The experimental results, Fig. 19, confirm that the minimum life is 7,750 cycles. The distribution functions for the fatigue life based on two residual strength degradation models are shown in Fig. 19 as a solid curve and a dashed curve, respectively. It is observed from Fig. 19 that the correlations between the experimental results and the theoretical predictions are reasonable.

To verify the statistical theory of periodic proof tests, fatigue test results for graphite/epoxy AS3501 [0,90, +45,-45]s laminates were obtained. The baseline data included a set of 12 ultimate strength data with a mean value $\mu=75.39$ ksi and a set of 21 fatigue S-N data at various stress levels with R=0.1. For verification purposes, a set of 14 specimens was subjected to an initial proof stress at $r_0=0.9\mu=67.8$ ksi and followed by 26,000 fatigue cycles at $\sigma=0.7\ \mu=52.8$ ksi. Then, a second proof test at r_0 was performed and followed by another 26,000 stress cycle at $\sigma=0.7\mu$. At this time, fatigue tests were terminated and the residual strengths of the specimens were measured. Out of 14 secimens, 3 specimens failed under proof tests and no specimen failed during cyclic loading. Note that none of the baseline S-N data was subjected to the same stress level $\sigma=0.7\mu$ conducted for the verification tests. The residual strength data for eleven (11) surviving specimens are shown in Fig. 20 as circles. Also shown in Fig. 20 as a solid curve is the theoretical prediction for the distribution of the residual strength.

Based on the theoretical prediction, a specimen surviving a proof load $r_0=0.9\mu$ will sustain at least 26,750 cycles at $\sigma=0.7\mu$. This was confirmed by the experimental results that none of the 14 specimens failed during cyclic loadings. The probability of failure under two (periodic) proof tests was computed as 0.278 and, hence, the average number of specimens expected to fail was 14x0.278=3.9. The experimental results showed 3 failures under proof tests. Considering the small sample size, the correlation is quite reasonable. Finally, as observed from Fig. 20, the correlation between the experimental results (circles) and the predicted distributuion of the residual strength (solid curve) is excellent. More verification tests were available in Yang & Liu [1977c].

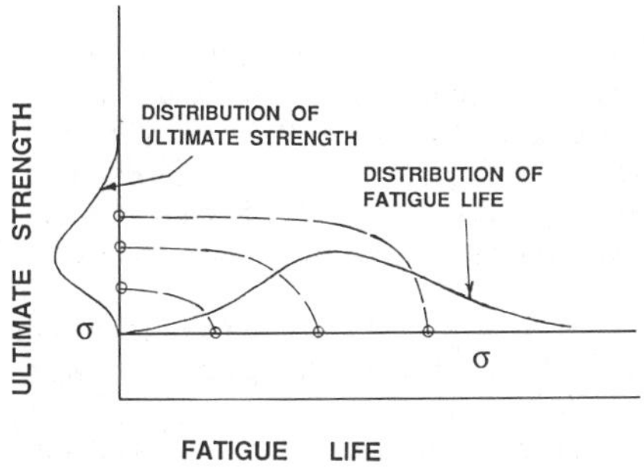

Fig. 18. Residual strength degradation of composite laminates

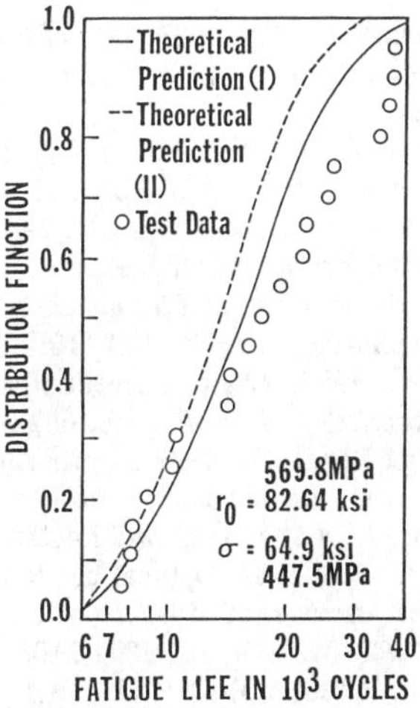

Fig. 19. Correlation between experimental results and theoretical prediction of fatigue life (single initial proof test)

7 CONCLUSIONS

Selected applications of reliability methods to fatigue, quality assurance and maintenance have been presented. Emphasis is placed on applications to aeronautical engineering areas based on the author's experience. Considerable progress has been made for practical applications of the reliability methods in the following specific areas: (i) the inspection/repair/replacement maintenance, durability analysis, fleet management, NDE, supportability, etc., (ii) statistical interpretations of proof tests and inspection for quality assurance and maintenance, and (iii) the definition of the initial fatigue quality for a component in terms of the equivalent initial flaw size distribution.

To date, the reliability method has not been widely used as an analysis or design tool in industry. To gain a wider acceptance of reliability methods for practical applications in industry, the following issues, based on the author's experience, are noted.

(1) Technical approaches for reliability methods should be user friendly. The simpler the methodology is, the easier the practicing engineers will accept it.

(2) While theoretical investigations of the reliability methods are very important, experimental verifications and demonstrations using either laboratory specimen data or full-scale test results or field data are extremely important.

(3) Reliability methodologies that can utilize deterministic design tools and practices are very desirable.

(4) Reliability methods require more data; however, the associated benefit and pay-offs are very significant.

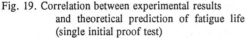

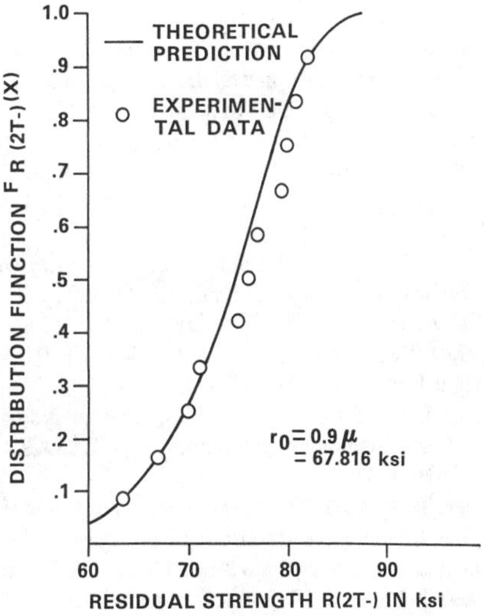

Fig. 20. Correlation between experimental results and theoretical prediction of residual strength (periodic proof tests)

Industrial companies usually are not willing to generate more data unless it is required by customers or government regulatory agencies.

(5) Most aerospace systems and subsystems are subjected to qualification tests or certification tests, such as static or dynamic tests, inspections, proof tests, etc. However, the probabilistic interpretations of the significance of these tests have not been widely used, partly because of a lack of meaningful statistical data. Future efforts are needed in this area.

ACKNOWLEDGEMENT

The author is most grateful to late Professor A. M. Freudenthal for his introduction to the subject area and to work with experts in various government laboratories. The author is indebted to Professor M. Shinozuka, Princeton University, for his constant guidance and encouragement. Most of the works discussed herein were carried out while the author was at the George Washington University. He would like to thank Professor H. Liebowitz, former Dean of the Engineering School, GWU, for his constant support and encouragement. Finally, the author would like to thank Dr. S. D. Manning of Lockheed Fort Worth Company for his collaboration in the subject area for more than 14 years.

REFERENCES

Annis, C.G., Jr. et al 1981. Gas turbine engine disk retirement-for-cause; an application of fracture mechanics and NDE, *J. of Eng. for Power*, ASME, 103: 198-200.

Barnett, R.L. and Hermann, P.C. 1965. Proof testing in design with brittle materials, *J. of Spacecraft and Rockets*, AIAA, 2, 956-961.

Berens, A.P. and Hovey, P.W. 1984. Flaw detection reliability criteria, vol. 1 - method and results, Air Force Wright Aero. Lab., AFWAL-TR-84-4022, WPAFB.

Berens, A.P. 1988. NDE reliablity data analysis. *Metals Handbook*, ASM International, 17:19th ed., 680-701.

Berens, A. P. et al 1991. Risk analysis for aging aircraft fleets, vol. I - analysis, WL-TR-91-3066, Wright Laboratory, WPAFB, Ohio.

Berens, A.P. and Burns, J.G. 1993. Risk analysis input for fleet maintenance planning, ICOSSAR '93, IASSAR.

Besuner, P.M. et al 1982. Analysis methods for F100 engine component retirement-for-cause, Failure Analysis Associates, Palo Alto, CA, FAA-81-12-5.

Besuner, P.M. et al 1983. Cost/risk analysis for disk retirement: vol. I, Failure Analysis Associates, Palo Alto, CA, FAA-82-3-8.

Bogdanoff, J.L. and Kozin, F. 1985. *Probabilistic Models of Cumulative Damage*, John Wiley & Sons, New York.

Committee on fatigue and fracture reliability, 1982. *J.of Struc. Div.*, ASCE, 108:ST1, 1-88.

Deodatis, G. 1992. Non-periodic inspection by Baysian method I, *J. of Prob. Eng. Mech.*, 7:4, 191-204.

Freudenthal, A.M. 1969. Introductory remarks, *Proc. 1st Inter. Conf. on Struc. Safety and Reliability*, 5-9, Pergammon Press, New York, 1972.

Fujita and Rackwitz, R. 1993. Influence of repair of components on reliability and the availability of redundant systems, ICOSSAR '93, IASSAR.

Heer, E. and Yang, J.N. 1971. Structural Optimization Based on Fracture Mechanics and Reliability Criteria", AIAA Journal, 9:5, 621-628.

Hill, R.J., et al, 1981. A retirement-for-cause study of an engine turbine disk, AFWAL-TR-81-2094, Air Force Wright Aero. Lab., WPAFB.

Ishikawa, H. et al 1993. Reliability assessment of structures based upon probabilistic fracture mechanics, *J. Prob. Eng. Mech.*, 8, 43-56.

Itagaki, H., Shinozuka, M., et al 1989. Baysian reliability analysis of structures with multiple components, *Proc. ICOSSAR '89*, IASSAR, 2143-2146.

Itagaki, H. et al 1993. Experimental estimation of probability distribution of fatigue crack growth model, *J. Prob. Eng. Mech.*, 8, 25-34.

Ito, S., Deodatis, G., Shinozuka, M. et al (1992), Non-periodic inspection by baysian method II: structures with elements subjected to different stress levels, *J. Prob. Eng. Mech.*, 7:4, 205-215.

Lee, L.J., Yang, J.N. and Sheu, D.Y. 1993. Prediction of fatigue life for matrix dominated composite laminates, *J. of Composite Science and Technology*, 46, 21-28.

Lin, Y.K. and Yang, J.N. 1983. On statistical moments of fatigue crack propagation, *J. Eng. Frac. Mech.*, 18, 243-262.

Lin, Y.K. and Yang, J.N. 1985a. A stochastic theory of fatigue crack propagation, *AIAA*, 23, 117-124.

Lin, Y.K., Wu, W.F. and Yang, J.N. 1985b. Stochastic modeling of fatigue crack propagation, *Probabilistic Methods in Mechanics of Solids and Structures*, Springer, Berlin, 103-110.

Lincoln, J.W., 1985, Risk assessment of an aging military aircraft, *J. Aircraft*, 22:8, 687-691.

Liu, C.T. and Yang, J.N. 1993. A probabilistic crack growth model for application to composite solid propellants, *J. of Spacecraft and Rocket*, AIAA.

Madsen, H.O. et al 1989. Optimal inspection planning for fatigue damage of offshore structures, *Proc. ICOSSAR '89*, IASSAR, III, 2099-2106.

Magnusen, P.E., Bucci, R.J., Hinkle, A.J., et al, 1989. The influence of material quality of an airframe structural durability, *Advances in Fracture Research*,

2; *Proc. 7th ICF*, 999-1007.

Magnusen, P.E., Bucci, R.J., Hinkle, A.J., Burns, J.G., Rudd, J.L. 1992. Effect of microporosity on fatigue durability of thick 7050 aluminum plate, *USAF Structural Integrity Program Conf.*

Manning, S.D. and Yang, J.N. 1984. USAF durability design handbook: guidelines for the analysis and design of durable aircraft structures, Air Force Wright Aeronautical Laboratories, WPAFB, AFWAL-TR-83-3027, 1st ed. 1984, 2nd ed. 1989.

Manning, S.D., Yang, J.N. and Rudd, J.L. 1987. Durability of aircraft structures, *Probabilistic Fracture Mechanics*, Martinus Nijhoff Publisher, Chapter 5, 213-267.

Manning, S.D., Yang, J.N, Pretzer, F.L. and Marler, J.E. 1992a. Reliability-centered maintenance for metallic airframes based on a stochastic crack growth approach, *Advances in Fatigue Life Predictive Techniques*, ASTM-STP-1122, 422-434.

Manning, S.D., Yang, J.N. and Welch, K.M. 1992b. Aircraft structural maintenance scheduling based on risk and individual aircraft tracking, Proc. Conf. on Theoretical Concepts and Numerical Analysis of Fatigue, Birmingham, England.

Military specification, aircraft structures, 1985. Air Force ASD, Military Standard MIL-A-87221, WPAFB.

Moan, T. 1993. Reliability and risk analysis for design and operation planning of offshore structures, Keynote lecture, ICOSSAR '93, IASSAR.

Ortiz, K. and Kiremedjian, A.S. 1988. Stochastic modeling of fatigue crack growth, *J. Eng. Frac. Mech.*, 29, 317-330.

Owen, C.R., Bucci, R.J. and Kegarise, R.J. 1988. Aluminum quality breakthrough for aircraft structural reliability, *J. of Aircraft*, 26:2, 178-184.

Palmberg, B., Blom, A.F. and Eggwertz, S. 1987. Probabilistic damage tolerance analysis of aircraft structures, *Probabilistic Fracture Mechanics*, Martinus Nijhoff Publisher, Chapter 2, 47-128.

Provan, J.W. (Ed.) 1987. *Probabilistic Fracture Mechanics and Reliability*, ISBN 90-247-3334-0, Martinus Nijhoff, The Hague.

Rocha, M.M. and Schueller, G.I. 1993. Markov chain modeling of non-destructive in-service inspection, ICOSSAR '93, IASSAR.

Rudd, J.L. and Gray, T.D. 1977. Quantification of fastener hole quality, *Proc. 18th AIAA/ASME/SAE SDM Conf.*,

Salivar, G.C., Yang, J.N. and Schwartz, B.J. 1988. A statistical model for the prediction of fatigue crack propagation under a block type spectrum loading, *J. Eng. Frac. Mech.*, 31, 371-380.

Sendeckyj, G.P. 1981. Fitting models to composite materials fatigue data, in *Test Methods and Design Allowable for Fibrous Composites*, ASTM-STP-734, 245-260.

Shinozuka, M. and Yang, J.N. 1969. Optimum structural design based on reliability and proof-load test, *Annals of Assurance Science*, Proc. of the Reliability and Maintainability Conf., 8, 375-391.

Shinozuka, M., 1976. Development of reliability-based aircraft safety criteria: an impact analysis, AFFDL-TR-76-36, Air Force Flight Dynamics Lab., WPAFB.

Shinozuka, M. 1979. Durability methods development, vol. iv: initial fatigue quality representation, Air Force Flight Dynamic Lab., AFFDL-TR-79-3118.

Sobczyk, K. 1986. Modeling of random fatigue crack growth, *J. Eng. Frac. Mech.*, 24, 609-620.

Sobczyk, K. and Spencer, B.F. 1992. *Random Fatigue: From Data to Theory*, Academic Press.

Spencer, B.F., Jr. and Tang, J. 1988. A Markov process model for fatigue crack growth, *J. Eng. Mech.*, ASCE, 114, 2134-2157.

Spencer, B.F. et al 1989. A stochastic approach to modeling fatigue crack growth, *AIAA Journal*, 27:11, 1628-1635.

Tang, J. and Spencer, B.J. 1989. Reliability solution for stochastic crack growth problem, *J. Eng. Fract. Mech.*, 34:2, 419-433.

Tsurui, A. and Ishikawa, H. 1986. Application of Fokker-Plank equation to a stochastic fatigue crack growth model, *J. Struc. Safety*, 4, 15-29.

Tsurui, A., Nienstedt, J., Schueller, G.I. and Tanaka, H. 1989. Time variant structural reliability analysis using diffusive crack growth models, *J. Eng. Frac. Mech.*, 34:1, 153-167.

Tiffany, C.F., Master, J.N. and Paul, F.A. 1968, Some fracture considerations in the design and analysis of spacecraft pressure vessels, American Society of Metals, TR-C6-2.3.

Tiffany, C.F. 1970. On the prevention of delayed time failures of aerospace pressure vessels, *J. of the Franklin Institute*, 290, 567-582.

Virkler, D.A. et al, 1979. The statistical nature of fatigue crack propagation, *J. of Eng. Mat. & Technology*, ASME, 101, 148-152.

Whitney, J.M. 1981. Fatigue characterization of composite materials, in *Fatigue of Fibrous Composite Mater.*, ASTM-STP-723, 133-151.

Whittaker, I.C. and Saunders, S.C. 1973. Application of reliability analysis to aircraft structures subjected to fatigue crack growth and periodic inspection, Air Force Materials Lab., AFML-TR-73-92, WPAFB.

Yang, J.N. 1974a. Statistics of random loading relevant to fatigue, *J. Eng. Mech.*, ASCE, 100:EM3, 469-475.

Yang, J.N. and Trapp, W.J. 1974b. Reliability analysis of aircraft structures under random loading and periodic inspection, *AIAA Journal*, 12:12, 1623-1630.

Yang, J.N. and Trapp, W.J. 1975. Inspection frequency optimization for aircraft structures based on reliability analysis, *J. Aircraft*, AIAA, 12:5, 494-496.

Yang, J.N. 1976a. Reliability analysis of structures under periodic proof test in service, *AIAA Journal*, 14:9, 1225-1234.

Yang, J.N. 1976b. Statistical estimation of service cracks and maintenance cost for aircraft structures, *J. of Aircraft*, AIAA, 13:12, 929-937.

Yang. J.N. 1977a. Optimal periodic proof test based on cost-effective and reliability criteria, *AIAA Journal*, 15:3, 402-409.

Yang, J.N. 1977b, Reliability prediction for composites under periodic proof test in service, *Composite Materials, Testing and Design*, ASTM, ASTM-STP 617, 272-295.

Yang, J.N. and Liu, M.D. 1977c, Residual strength degradation model and theory of periodic proof test for graphite/epoxy laminates, *J. of Composite Materials*, 11, 176-203.

Yang, J.N. 1978a. Fatigue and residual strength degradation for graphite/epoxy composites under tension-compression cyclic loading, *J. of Composite Materials*, 12, 19-29.

Yang, J.N. and Jones, D.L. 1978b. Statistical fatigue of graphite/epoxy angle-ply laminates in shear, *J. of Composite Materials*, 12, 371-389.

Yang, J.N. and Manning, S.D. 1980a. Distribution of equivalent initial flaw size, *1980 Proc. of Annual Reliability and Maintainability Symp.*, 112-120.

Yang, J.N. and Jones, D.L. 1980b. The effect of load sequence on statistical fatigue of composite, *AIAA Journal*, 18:12, 1525-1531.

Yang, J.N. and Sun, C.T. 1980c. Proof test and fatigue of composite laminates, *J. of Composite Materials*, 14, 801-817.

Yang, J.N. 1980d. Statistical estimation of economic life for aircraft structures, *J. of Aircraft*, AIAA, 17:7, 528-535.

Yang, J. N. and Jones, D.L. 1982a. Fatigue of graphite/epoxy $\pi/4$ laminates under dual loads, *Composite Technology Review*, 4:3, 63-70.

Yang, J.N. and Cole, R.T. 1982b. Fatigue of composite bolted joints under dual load levels, *Progress in Science and Engineering of Composites*, Proc. ICCM-IV, 333-340.

Yang, J.N., Salivar, G.C. and Annis, C.G. 1982c. Statistics of crack growth in engine materials - volume 1: constant amplitude fatigue crack growth at elevated temperatures; volume 2 spectrum loading and advanced techniques, Air Force Wright Aero. Lab., AFWAL-TR-82-4040, WPAFB.

Yang, J.N. and Donath, R.C. 1982d. Improving NDE capability through multiple inspections with applications to gas turbine engine, Air Force Wright Aero. Lab., AFWAL-TR-82-4111, WPAFB.

Yang, J.N. and Du, S. 1983a. Exploratory study into fatigue of composites under spectrum loading, *J. of Composite Materials*, 17, 511-526.

Yang, J.N. and Donath, R.C. 1983b. Statistical fatigue crack propagation in fastener hole under spectrum loading, *J. Aircraft*, AIAA, 20, 1028-1032.

Yang, J.N., Salivar, G.C. and Annis, C.G. 1983c. Statistical modeling of fatigue crack growth in a nickel-based superalloy, *J. Eng. Frac. Mech.*, 18, 257-270.

Yang, J.N. and Donath. 1983d. Improving NDE reliability through multiple inspections, *Review of Progress in Quantitative NDE*, Plenum Press, NY, 1, 69-78.

Yang, J.N. and Donath, R.C. 1984a. Statistics of crack growth of a superalloy under sustained load, *J. of Eng. Materials and Technology*, ASME, 106, 79-83.

Yang, J.N. and Chen, S. 1984b. Fatigue reliability of gas turbine engine components under scheduled inspection maintenance, AIAA paper no. 84-0850, April; *J. of Aircraft*, AIAA, 22: 415-422, 1985.

Yang, J.N. and Chen, S. 1984c. An exploratory study of retirement-for-cause for gas turbine engine components, AIAA paper no. 84-1220; *J. of Propulsion and Power*, AIAA, 2:1:38-49, Jan. 1986.

Yang, J.N., Hsi, W.H. and Manning, S.D. 1985a. Stochastic crack propagation with applications to durability and damage tolerance analysis, Air Force Wright Aero. Lab., AFWAL-TR-85-3062, WPAFB.

Yang, J. N., Manning, S.D., Rudd, J.L. and Hsi, W.H. 1985b. Stochastic crack propagation in fastener holes, *J. Aircraft*, AIAA, 22, 820-817.

Yang, J.N., et al 1987. Stochastic crack growth models for applications to aircraft structures, *Probabilistic Fracture Mechanics*, (Provan, Ed.), Martinus Nijhoff Publisher, Chapter 4, 171-209.

Yang, J.N., Yang, S.H. and Jones, D.L. 1989. A stiffness-based statistical model for predicting fatigue life of graphite/epoxy laminates, *J. of Composite Technology and Research*, ASTM, 129-134.

Yang, J.N. and Manning, S.D. 1990a. Stochastic crack growth analysis methodologies for metallic structures, *J. of Eng. Frac. Mech.*, 37:5, 1105-1124.

Yang, J. N. and Manning, S.D. 1990b. Demonstration of probabilistic-based durability analysis method for metallic airframes, *J. of Aircraft*, AIAA, 27:2, 169-175.

Yang, J.N., Yang, S.H. and Jones, D.L. 1990c. A stiffness degradation model for graphite/epoxy laminates, *J. of Composite Materials*, 24:7, 753-769.

Yang, J. N., Lee, L.J. and Sheu, D.Y. 1992. Modulus reduction and fatigue damage of matrix-dominated composite laminates, *J. of Composite Structures*, 21:2, 91-100.

Yang, J.N., et al 1993. Reliability analysis and quality assurance of rocket motor case considering proof testing, AIAA-93-1382-CP, La Jolla, CA.

Yang, J.N. and Manning, S.D. 1994. Aircraft fleet maintenance based on structural reliability analysis, *J. of Aircraft*, AIAA, 31:1.

Keynote lectures

Structural Safety & Reliability, Schuëller, Shinozuka & Yao (eds) © *1994 Balkema, Rotterdam, ISBN 90 5410 357 4*

Reliability and risk analysis for design and operations planning of offshore structures

T. Moan
Department of Marine Structures, The Norwegian Institute of Technology, Trondheim, Norway

ABSTRACT: A brief overview of the variety of offshore structures used for exploitation of hydrocarbons and the criteria used for their design and operation is given. Recent developments and applications of reliability and risk methodology for design and operations planning of offshore structures are reviewed. Particular attention is paid to developments of time-variant reliability analysis of fracture failure under fatigue degradation and systems reliability analysis considering overload and fatigue failure of components. Applications include code calibration and inspection planning, as well as design based on quantitative assessment of operational risks.

1 INTRODUCTION

Our time is the hydrocarbon age. Oil and gas are the dominant sources of energy in our society. Some of these hydrocarbons are recovered from reservoirs beneath the seabed. Various kinds of platforms are used to support exploratory drilling equipment, and the chemical (production) plants required to process the hydrocarbons. Pipelines or tankers are used to transport the hydrocarbons to shore.

The hazards of hydrocarbons, and harsh environmental conditions, and the potential consequences of failure in terms of fatalities, pollution or significant economic losses, make safety an important consideration for offshore structures.

This paper addresses the use of reliability and risk analysis to make optimal decisions regarding safety and life cycle costs of offshore structures. The emphasis is placed upon recent developments of component and systems reliability analysis relating to fatigue and fracture, including Bayesian updating. Applications include code calibration of ultimate and fatigue limit states and inspection planning. The broader aspects of structural risk assessment, considering human errors, are discussed with reference to accidental loads resulting from gross operational errors.

2 OFFSHORE STRUCTURES

2.1 *General*

The platforms used in connection with exploratory drilling are mobile, i.e., the units are easy to move to another site after completing their task in 1 to 3 months at one site. Typically, mobile platforms are floaters and anchored by an easily disconnectable mooring system, as illustrated by the two units in Figs. 1a and b. However, both of these structures are also used as production platforms.

Figs. 1c-e show the variety of permanent production platforms, mostly made of steel, but some, notably the one in Fig. 1d, are concrete structures. The majority of the 5000 production platforms in the world are steel truss/framework platforms, often denoted as jackets, as exemplified in Fig. 1c. They are constructed of fairly slender tubular members and are normally piled to the seafloor. A fairly novel platform concept is the so-called tension leg platform (TLP) shown in Fig. 1e. The hull provides excessive buoyancy which introduces high pretension in the mooring legs, or tethers. When subjected to waves and wind forces, it moves like an inverted pendulum. Usually there are 3 to 4 tethers in each of the four platform corners. A tether is a tubular made of 12-15 m long sections connected by welds or threaded joints. The tether foundation is by gravity, skirts, or piles.

Jackets and other fixed platforms in deep waters have to resist a significant overturning moment and shear force due to waves and current loads, and possibly seismic loads. They, therefore, become huge structures. The Bullwinkle jacket platform in the Gulf of Mexico has a total height of 492 m; a base of 122 m by 146 m; and the steel weight of the load carrying structure is 45000 tons. The heaviest fixed concrete platform to date is 380 m high and has a dead weight of 750000 tons, see, e.g., Clauss et al. (1992).

The deck structure of production platforms is a complex and compact chemical process plant. Fig. 1d shows the five to seven story deck structure of the Gullfaks-C production platform. The significant latent energy associated with hydrocarbons flowing through, or stored in production platforms, obviously represents a significant hazard.

The environmental and hydrocarbon hazards together make safety an important issue for offshore platforms. The location far off shore makes evacuation and rescue difficult, but on the other hand accidents on offshore plants affect the general public less than accidents on similar facilities on land.

Offshore platforms are designed for a service life of 20 years or more, with criteria for serviceability and safety against structural failure and overall instability or possible sinking. For buoyant structures, capsizing due to wind/waves, especially associated with some accidental flooding of the buoyancy compartments, is the most crucial failure mode. There has for a long time been robustness criteria for floaters in connection with instability caused by flooding (ISSC 1988-94b).

Adequate structural integrity is achieved by applying ultimate and fatigue limit state criteria for components. Fatigue is an important consideration for structures in areas with more or less continuous storm loading (such as the North Sea) and especially for dynamically sensitive structures. Progressive limit state criteria for structural integrity were first implemented in codes for the Norwegian Continental Shelf in the 1970's (NPD 1977). The purpose of such criteria is to avoid catastrophic accidents, i.e., system failure due to a small damage, e.g., caused by ship impacts, fires and explosions, and other accidental loads.

Inspection and monitoring, and repair, if necessary, are important measures for maintaining safety, especially with respect to fatigue, wear and other deterioration phenomena. But their effect on the reliability depends upon the quality of inspection, e.g., in terms of detectability vs. size of the damage. Hence, an inspection and repair measure can contribute to the safety only when there is a certain damage tolerance. This implies that there is an interrelation between design criteria (fatigue life, damage tolerance) and the inspection and repair criteria. Up to now, however, this interrelation has not been explicitly considered, due to lack of methods to deal with this problem in a rational way.

The quality of visual or non-destructive inspection (NDI) methods depends very much upon the conditions during inspection. Large volume offshore structures are accessible from the inside, while, e.g., jackets and tethers in TLPs, which are made of relatively slender tubulars, are not. Underwater inspections, e.g., of jackets, are carried out under difficult conditions, due to the presence of marine growth, bad visibility, wave motions, etc. Before an underwater NDI can be carried out, a time-consuming cleaning of the welded joints is needed (MTD 1992). For this reason and the high costs of diver operations, underwater inspections become expensive, yet with a quality deficient to that of dry inspections (Barnouin et al. 1993).

Tethers in so-called tension-leg platforms are tubulars with an inside diameter of 0.2 to 1.0 m, and are surveyed by a remotely operated special tool running inside the tether, commonly based on ultra sound.

Leak-before-break monitoring has been an issue for tethers as well as members in other offshore structures (MTD 1992).

The largest production platforms in, e.g., the North Sea, represent an investment of 8 to 10 billion U.S. dollars and significant maintenance costs.

2.2 *Wave loading and response*

The loading unique to offshore structures is the forces from waves and current. While the current is described by the variation of velocity in time and space, the description of wave loading is more complex; see, e.g., Sarpkaya and Isaacson (1981).

The wave process over a long-term period can be displayed as shown in Fig. 2. In a short-term period of 3 to 6 hours the wave elevation can be well approximated by a stationary, ergodic Gaussian process, described by its power spectrum, which is represented by an analytical function with, for instance, the significant wave height, H_s; peak period, T_p, as parameters. The ocean environment in a short-term period is further characterized by the mean wave direction, θ; and current velocity, v_c; see, e.g., ISSC (1988-94a). The parameters describing the ocean environment may be collected in a time dependent vector $\underline{Y}(t) = \{H_s, T_p, \theta, \ldots v_c \ldots\}$. The long-term variation of $\underline{Y}(t)$ can then be approximated by a sequence of random vectors $\underline{Y}$: $\{\underline{Y}_1, \underline{Y}_2, \underline{Y}_3 \ldots \underline{Y}_n\}$, each $\underline{Y}_i$ characterizing a short-term state. With about 3000 3 hour periods in one year, the dimension n may be of the order 100,000. When the period T considered is long compared to the scale of variation of Y(t), it may be represented by its joint density function, $f_Y(\underline{y})$. The long-term variation of wave forces and responses can then be determined by considering the short-term wave process represented by a superposition of regular waves of different frequencies and random phase angles. The wave pressure, particle velocity, v and acceleration, a are given by kinematic theories, and wave forces F are expressed by the kinematic variables; see, e.g., Sarpkaya and Isaacson (1981). For a slender tubular structure with diameter D, the wave force per unit length may be given by the Morison equation as:

$$F = F_I + F_D = C_M \rho \frac{\pi}{4} D^2 a + \frac{1}{2} C_D \rho D v|v| \qquad (1)$$

where ρ is the density of water and C_M and C_D are coefficients. While a and v are proportional to the wave height (for a given wave length), F will be a nonlinear function of the wave height.

This means that the probabilistic character of F depends on whether the first (inertia) or second (drag) term of Eq. (1) dominate. For instance, the individual maxima of F for a narrow-banded Gaussian wave process follow a Rayleigh and an exponential distribution when the first and second term predominate, respectively (Brouwers and Verbeek 1983). For a given diameter, D, the ratio F_D/F_I increases with wave height. Hence, F_I may dominate in moderate seas, while F_D may be most important for high seas. This fact should be observed when the model uncertainty of extreme loads is estimated based on measurements in moderate seas. This fact would also imply large samples when using Monte Carlo simulations to achieve extreme dynamic response.

The overall forces on a jacket or jack-up platform are obtained by integrating forces, F over the wet surface of the structure, which may be above or below the mean water level (Fig. 3). This fact contributes to an increased nonlinear character of drag forces; see, e.g., Karunakaran (1993).

Faltinsen (1993) reviews other hydrodynamic load mechanisms that cause nonlinear (non-Gaussian) response processes, which are treated, e.g., by Næss (1989, 1993) and Winterstein (1988).

The short-term dynamic response of a fixed platform subjected to nonlinear wave loads is achieved by using a finite element model of the structure and possibly the soil, and a time domain simulation of the resulting MDOF system; see recent review, e.g., by Karunakaran (1993). Particular care is then required when fitting an analytical distribution to the cumulative probability, $F_{x|\underline{y}}(x\,|\,\underline{y})$ of the calculated response maxima, x, especially when extrapolation to extreme values is needed. In general, the Weibull distribution provides a good fit, see Farnes (1990). Fig. 4 illustrates the non-Gaussian character of short-term response maxima in a North Sea jacket. The long-term probability distribution of individual maxima may be obtained by

$$F_x(x) = \int_{\underline{Y}} w(\underline{y})\, F_{x|\underline{y}}(x|\underline{y})\, f_{\underline{Y}}(\underline{y})\, d\underline{y} \qquad (2)$$

where $w(\underline{y})$ is a weighting function relating the response frequency in the actual seastate to the the long-term mean zero-crossing rate. Calculation of $F_x(x)$ involves significant computational efforts, especially for MDOF systems with nonlinear behavior, e.g., due to drag dominant loads, because $F_{x|\underline{y}}(x|\underline{y})$ must be determined for several representative sea-states. The extreme response, say, the expected maximum in 100 years, may be obtained from Eq. (2) using extreme value statistics.

In practical design, the extreme response is commonly predicted considering a regular wave with a 100 year wave height and an appropriate wave period, wave kinematics and hydrodynamic theory. Such, so-called design wave approaches have been outlined and validated by Tromans et al. (1991), Heideman and Weaver (1992) and Fu et al. (1993). Also, Olufsen and Bea (1990) and Gudmestad (1993) provide estimates of the inherent model uncertainty of hydrodynamic loads on platforms with quasistatic behavior. Farnes (1990) and Kanunakaran (1993) demonstrate good agreement between the stochastic long-term and the design wave approach for such structures. For dynamically sensitive platforms a stochastic approach should be applied to check possible simplified approaches.

3 HISTORICAL NOTES ON RELIABILITY AND RISK ANALYSIS OF OFFSHORE STRUCTURES

The evolution of offshore engineering has been influenced by civil engineering as well as naval architecture. The important role played by civil engineers in developing structural reliability theory, in general, has also lead to contributions to reliability analysis of offshore structures. While the naval architecture community already in the early 1960's advocated using the reliability approach to ships (ISSC 1964), the actual application to merchant ships has been very limited. The offshore applications of reliability methodology started in the mid-sixties, when a significantly increasing interest in such methods in general could be observed. The initial focus in offshore engineering was on risk-based storm load criteria (Marshall 1969, Bea 1973), but also on reliability engineering (Stahl 1977, Fjeld 1977).

The first offshore probability based limit state code was issued by the Norwegian Petroleum Directorate (NPD 1977, Fjeld 1977), three years after the CSA code appeared as the first one of its kind.

In the late 1970's research on Probabilistic Risk Assessment (PRA) in the offshore industry started (Moan and Holand 1981) based on experiences in the nuclear and aeronautic industries. The first guidelines on PRA were issued by NPD (1981).

Significant developments of reliability and risk methodology, including Bayesian updating techniques, took place in the 1980's. Techniques for dealing with the effect of inspection on the reliability were developed, with due account of previous research in, e.g., the aeronautical field. Some of the research relating to systems reliability and Bayesian updating, were directly motivated by needs in the offshore industry, as discussed subsequently.

The most serious offshore accident to date in terms of fatalities, the Piper Alpha Disaster (Cullen 1990), was the direct reason for introducing PRA, or QRA, in the UK in 1992, which certainly was a milestone.

The fact that structural reliability analysis now is becoming mature as an engineering discipline, is demonstrated by the current trend of establishing consulting companies, which are not necessarily based in research organizations or universities. An important factor contributing to this trend is the general availability of reliable reliability software (Bourgund and Bucher 1986, DnV 1989, RCP 1991, and W. S. Atkins 1992) accurate and efficient structural tools and computer capability. Also, the confidence in reliability methods has increased due to the recent substantiation of reliability predictions by in-service experiences (Vugts and Edwards 1992).

The amount of information published on risk and reliability of offshore structures has increased exponentially, as evidenced by the number of papers presented at this and other conferences, such as BOSS, ISOPE and OMAE, confined to offshore structures. Early developments and applications of risk and reliability analysis to offshore platforms are reviewed, e.g., by Moan (1983), Baker (1985) and ISSC (1988-94b). Here, only the main, recent developments and applications in decision making are dealt with.

4 COMPONENT RELIABILITY

4.1 *General*

For a time-invariant problem, which is described by a set, $\underline{Z}$ of random variables, the failure probability may be calculated as

$$P_F = \int_{g(\underline{z})\leq 0} f_{\underline{z}}(\underline{z})\, d\underline{z} \qquad (3)$$

where $f_{\underline{z}}(\underline{z})$ is the probability density and $g(\underline{z})$ the failure function. The integral of Eq. (3) may be calculated by direct integration, simulation or FORM/SORM methods as described in, e.g., textbooks by Madsen et al. (1986) and Melchers (1987). While reliability problems involving a single, time variant load effect, $X(t)$ and otherwise time independent variables ($\underline{Z}$), can be readily formulated in terms of Eq. (3), this is not the case when the load effects are a vector process, $\underline{X}(t)$. Under such circumstances, the conditional failure probability, $P_F(\underline{z})$ for a given $\underline{Z} = \underline{z}$, can be determined by estimating the rate of crossing from the safe to failure domain; see, e.g., Melchers (1987). P_F is then obtained by taking the expectation of $P_F(\underline{Z})$ over $\underline{Z}$. Leira (1987) studied vector processes relating to wave-induced load effects to establish reliability-based design criteria.

Time-invariant reliability methods have been extensively used to calibrate ultimate strength code checks based on partial safety factors, to comply with a certain target reliability level. This application is centered around current design practice, in the sense that the g( )-function can be based on the relevant design equations in an existing code. The main result of the calibration is improved safety factors.

Fatigue and fracture are important failure modes for offshore structures as described, e.g., in Chapter 1 of Almar-Næss (ed.) (1985). This fact and the significant inherent uncertainties associated with fatigue imply that a probabilistic treatment is important (Ang and Munse 1975 and Wirsching 1979). A reliability formulation for fatigue can be centered around design practice by using SN-curves and the Miner-Palmgren approach. However, a fracture mechanics model is required, especially to account for the effect of inspection; see, e.g., Almar-Næss (ed.) (1985), Rhee and Kanninen (1988), and Kirkemo (1988). For these reasons, much attention has recently been paid to developing fatigue and fracture reliability methods. Furthermore, because fatigue and fracture control relies on design as well as inspection and repair, much work has also been devoted to Bayesian updating of component reliability, partly based on approaches used in other fields, as reviewed by Yang (1993).

The following subsections are devoted to calibration of offshore codes and recent research efforts on fatigue and fracture reliability. Application of reliability methods in design and inspection planning is dealt with in Chapter 6.

4.2 *Code calibration and reliability-based design*

Code calibration involves defining the scope (class of structures, load cases, etc.) and code format – including definition of characteristic values; selecting reliability measure; assessing uncertainties in loads and resistances; establishing target reliability level; and accomplishing the calibration itself to determine the partial safety factors; see, e.g., Madsen et al. (1986) or Melchers (1987). The reliability analysis is a relatively simple task; most work is devoted to the uncertainty assessment. The significant effort by API (Moses 1986, 1987; Lloyd and Karsan 1988) to establish a load and resistance factor design code for steel jackets in U.S. waters is especially noteworthy. This code was put into effect as of July 1, 1993. In parallel with the API work, the Canadian Standards Assoc. conducted a study for fixed steel and concrete platforms (Jordaan and Maes 1991) which was implemented in a draft version in 1989. Moses (1991), Theophanatos et al. (1992), and Turner et al. (1992) have investigated the modifications of the API code which are necessary if it is to be implemented for platforms in other geographical areas. The main concern then has been the difference in uncertainties, especially in the environmental loads. However, it may also be necessary to differentiate the target level.

The calibration of the API Code was undertaken using the same average target reliability as implied by the previous API Code and, hence, primarily aimed at a more uniform reliability level. A similar approach has been used in most calibrations so far. However, in the Canadian effort mentioned above, the target level was explicitly assessed (Jordaan and Maes 1991) based on cost analysis (minimum total costs, including expected damage costs) and acceptable fatality risk for individuals. Jordaan and Maes only found the latter approach appropriate and proposed, assuming that a component failure implies fatalities, that the annual target component failure probability should be approximately 10% of the annual fatality rate in the society. However, whether a component failure will cause fatalities depends upon the system redundancy and the likelihood of accomplishing a safe evacuation before a possible system failure. For instance, the likelihood of fatality caused by storm overload in the Gulf of Mexico (GOM) will be less than in the North Sea (NS) because platforms in GOM can be evacuated in face of a storm (hurricane), while this is not the case in the NS. Bea (1993) favors using cost analysis to set the target level for requalification of platform systems.

Also, significant studies have been conducted and are going on to calibrate codes for floating platforms, especially for TLP's (Faulkner 1991; Banon et al. 1994), as well as for subsea pipelines (Sotberg and Bruschi 1992). Much of the ongoing work is directed towards implementation in a new world-wide offshore code, being developed under the auspices of ISO (1993b). The new offshore code is established within the framework of ISO 2394 (1993a), which is currently being revised, e.g., to better reflect the new developments in reliability technology, and the need for reassessment of existing structures.

All code calibrations described above refer to ultimate strength criteria. Because fatigue in offshore structures is mainly caused by one type of loading, there is limited merit of calibrating fatigue criteria in the same sense as ULS criteria relating to multiple loads. However, to achieve consistent design and inspection criteria, fatigue design criteria should be calibrated to reflect the consequences of failure and inspection plan. Wirsching (1983) reports an extensive study of reliability-based fatigue design criteria. Moan et al. (1993) focused on calibration of design criteria dependent upon the inspection plan. It is shown that the allowable cumulative damage (D) in design can be relaxed when inspection is carried out. With an inspection every four years and a mean detectable crack depth of 1.5 mm, the allowable D can be increased by a factor of 2 to 2.5 and 2.5 to 3, respectively, for a tether joint, and tubular joint in a jacket.

In certain situations when a new design falls outside the scope of existing codes, reliability analysis has been applied ad-hoc to establish design criteria. This was the case when the first offshore production ship was designed some years ago. It then became clear that application of the ship rules for merchant vessels and existing offshore codes differed significantly, implying a difference in steel weight of the order of 20-30%. Moan (1988) conducted a study to establish ultimate strength criteria for this type of vessel, which complied with the inherent safety level in the existing NPD code for offshore structures. Fig. 5 shows the reliability indices implied, and proposed target reliability indices for different components in various types of structures. Two sets of results are given for tubular joints in jackets because a revised code formulation for such components had been proposed at the time of the study. Actually, an additional benefit of the study was to improve the tubular joint criteria. But it is seen from Fig. 5 that offshore codes and ship rules implied very different reliability indices. The actual safety factors established for the production vessel was based on a target level of 3.7, and resulted in a 15% lower load factor for environmental loads than otherwise used for offshore structures.

4.3 *Fatigue and fracture reliability of welded steel structures*

By adopting a fracture mechanics model, the fatigue phenomenon may be described by the

a) crack occurrence in space

b) a probabilistic distribution for the time to crack initiation T_o, which is often neglected for welded structures

c) a probability model for crack propagation, e.g., based on a deterministic Paris-Erdogan's law for the crack increment per cycle, da/dN:

$$da/dN = C(\Delta K)^m \tag{4}$$

where the material parameters C and m, and the stress intensity factor ΔK are taken as random variables. Alternative models are reviewed, e.g., by Langley (1989).

d) recognition of the fact that inspection methods are not perfect by characterizing their quality by the probability of crack detection, POD, typically taken to be an exponential function; see, e.g., Barnouin et al. (1993).

Let $\Delta K = S \cdot Y(a,\underline{Z})\sqrt{\pi a}$, where S is the nominal stress range and $Y(\ )$ is a function of a and a set of random variables $\underline{Z}$. The time from an initial crack size, a_0 to failure (crack size, a_f) can then be obtained by integrating Eq. (4) as

$$T = \frac{1}{C \cdot \nu_o \cdot E(S^m)} \int_{a_o}^{a_f} \frac{da}{\left(\gamma_Y \cdot Y(a,\underline{Z})\sqrt{\pi a}\right)^m} \tag{5}$$

where ν_0 is the mean zero upcrossing frequency of S, and $E(S^m)$ is the expectation of S^m. γ_Y is the model uncertainty of $Y(\ )$.

The limit state function, $g(\ )$, for fatigue failure in the time τ is then:

$$g(\ ) = T_o + T - \tau \tag{6}$$

Three aspects relating to the Paris-Erdogan's law should be observed in this connection. First, Fig. 6 shows an example of crack increment da/dN vs. the stress intensity factor ΔK as obtained under different environmental and stress conditions. Clearly modeling da/dN vs. ΔK by a single straight line in a logarithmic plot, may give large errors especially when the crack is small. Often the largest contribution to the fatigue life stems from this stress range regime. An appropriate failure function is given by Kam and Dover (1989) for the general case when da/dN vs. ΔK is approximated by several line segments (in the log plot); and Hovde and Moan (1994) extended this approach by also taking the effect of stress ratio into account.

The second issue is concerned with modeling crack growth in complex joints. Fig. 7 shows how the crack front propagates at the intersection between two tubulars, which is the basic joint in, e.g., jacket structures. In principle, a multiparameter crack propagation law should be applied to describe the crack propagation of large cracks. Approximate formulations can be based on the assumption of a semi-elliptic crack with two parameters or a one parameter formulation in terms of crack depth, but with the aspect ratio as a function of crack depth; see Berge (1993). The Paris equation for the two-parameter formulation is expressed by two simultaneous differential equations in terms of crack increments, see, e.g., Shetty (1992).

The third point deals with the calculation of the "fatigue loading", $E(S^m)$ induced by random wave loading. In narrow band Gaussian response the stress ranges S are simply related to peaks (amplitudes, a) as $S = 2 \cdot a$. Experiments (Dowling 1972) show that stress ranges in a general response process should be determined by Rainflow or Local Range Counting. Many simplified theoretical and empirical formulations have been suggested to calculate the fatigue loading. Several of these approaches have been compared with the results of extensive simulations for broad-banded Gaussian processes by Boussy et al. (1993). They conclude that an empirical approach by Dirlik generally performs best over the range of band width variation of interest. In the case of non-Gaussian response processes, the counting normally needs to be accomplished on a case basis.

4.4 *Time variant vs. time invariant formulation of fatigue-induced fracture*

Steel structures subjected to dynamic loading may fail due to ductile overload, fatigue or fracture. In particular, fracture failure may occur after deterioration of the strength caused by fatigue crack growth. Failure then occurs when the load effect $X(t)$ first upcrosses the resistance $\xi(t)$ as indicated in Fig. 8. Calculation of fracture probability under fatigue degradation was first addressed by Guers and Rackwitz (1987). Marley and Moan (1992) deal with this time variant reliability problem for marine structures when the sea loads are described by the long-term approach; hence, extending the work of Bjerager et al. (1988), which is concerned with time invariant reliability under long-term load processes. In the following, the time variant long-term approach (Marley and Moan 1992) will be briefly outlined. In a short-term period, the load effect $X(t|\underline{y}, \underline{z})$ is conditional upon the value of the parameters $\underline{y} = \{H_s, T_p, \theta, ..., v_c\}$, and the time invariant parameters $\underline{z}$ relating to loading, geometry and material properties. If the crossings of X into the failure region is Poissonian; see Cramer and Leadbetter (1966), the conditional failure probability can be obtained by:

$$P_F(\underline{y},\underline{z}) = P\left[\min_{0\le t\le\tau}\left\{(\xi(t) - X(t))\big|(\underline{y},\underline{z})\right\} \le 0\right]$$
$$\approx 1-\exp\left[-\int_0^\tau \nu_X^+(t;\underline{y},\underline{z})\,dt\right] \tag{7}$$

where $\nu_X^+(t;\underline{y},\underline{z})$ is the time dependent frequency of upcrossing into the failure region, and is given by, e.g., Cramer and Leadbetter (1966) for Gaussian response. Results for some non-Gaussian processes are given by, e.g., Grigoriu (1984) and Winterstein (1988). The first significant study of outcrossing rates for vector-processes was made by Veneziano et al. (1977). Recently, Hagen and Tvedt (1991) reported an interesting method to calculate outcrossing rates using time independent reliability considerations.

The unconditional failure probability P_F can then be obtained by

$$P_F = E_{\underline{Z}}\left[E_{\underline{Y}}\left\{1-\exp\left(-\int_0^\tau \nu_X^+(t;\underline{y},\underline{z})\,dt\right)\right\}\right] \tag{8}$$

Eq. (8) may be calculated by a Monte Carlo step simulation. Random samples $\{\underline{y},\underline{z}\}^i = \{\underline{y}_1,\underline{y}_2,...\underline{y}_n, \underline{z}\}^i$ of $\underline{Y}$ and $\underline{Z}$ are generated according to $f_{\underline{YZ}}(\underline{y},\underline{z})$. The threshold deterioration is calculated for each seastate y_j^i by using a fracture mechanics approach. The failure probability in the j'th sea state of sample i, given survival at the beginning of the sea state, is

$$(P_F)_j^i = 1-\exp\left[-\int_{t_{j-1}}^{t_j}\left\{\nu_X^+(t)\big|(\underline{y}_1 ... \underline{y}_j, \underline{z})^i\right\}dt\right] \tag{9}$$

With a sea state safety margin $P[M_j^i > 0] = 1-(P_F)_j^i$ the sample failure probability may be written as

$$P_F^i = 1 - P\left[\bigcap_{j=1}^{n}(M_j^i > 0)\right] \tag{10}$$

and P_F may be estimated by a sample of m

$$P_F \approx \hat{E}[P_F] = \frac{1}{m}\sum_{i=1}^{m} P_F^i \tag{11}$$

The Monte Carlo approach is feasible for this problem. However, with a dimension of $\underline{Y}$ of 100,000 and the large sample of m required to estimate a small P_F, simplified methods are required. It is then important to recognize that fatigue crack growth occurs in moderate, relatively frequent seastates followed by fracture in a more extreme seastate. The "mean crack driving stress" $E(\overline{S^m})$ (see Chapter 4.2) and the extreme load giving rise to final fracture are the relevant response quantities. If the period τ considered, is long compared to the scale of variation of $\underline{Y}(t)$, the load effects induced by the environmental processes may be considered stationary in a long-term period. For this case, an equivalent long-term stress range, $E(S^m)$ and the exceedance probability crossing rate ν_X^+ may be determined which takes the variation of $\underline{Y}$ into account. If τ is long, $\underline{Y}_j$ may be assumed independent and identically distributed, and

$$E\left(\overline{S^m}\right) = \int_{\underline{Y}} \int_0^{\infty} w\left(\underline{y}\right) f_{S|\underline{Y}}\left(s|\underline{y}\right) f_{\underline{Y}}\left(\underline{y}\right) ds\, d\underline{y} \tag{12}$$

When τ is large, ν_X^+ may be approximated by

$$\nu_X^+ = \nu_0^+\left[-\ell n\left(F_X(x)\right)\right] \tag{13}$$

where $F_X(x)$ is the long term distribution of maxima calculated by Eq. (2). In this case, the failure probability P_F can be determined by Eq. (8) by omitting the expectation with respect to $\underline{Y}$ because it has already been incorporated in $X(t)$ and ν_X^+, and using either simulation or FORM/SORM based on the Wen and Chen (1987) approach. This simplified long-term method compares very well with results obtained by extensive step simulations (Marley and Moan, 1992).

A further simplification would be to replace the time variant (TV) approach by a time invariant (TIV) one, conservatively using the threshold corresponding to the fracture resistance at the end of the time period, τ. The failure probability $P_F(\underline{z})$ for this case is calculated by $P[\xi - X' \leq 0]$ where ξ is the deterministic threshold level and X' is the maximum value of $X(t)$ on τ. The TV and TIV methods have been compared for different cases by Marley and Moan (1992, 1993). Here, only one example relating to the welded joints of TLP tethers is briefly described. Fatigue is of main concern for these critical components.

The leak-before-break (LBB) principle may be applied for crack monitoring of tethers. In this connection it is of interest to calculate the failure probability in a repair period given that the initial crack is a through thickness crack. Here, 3 and 12 months repair periods are considered.

Fig. 9 shows the conditional reliability index for the repair period conditioned upon an initial through thickness crack; as a function of the residual welding stress in the joint and the duration of the repair period. The residual stress affects the fracture. It is seen that the time invariant approach yields only slightly conservative reliability indices for this fairly severe test of the TIV approach. This is of course satisfactory from a practical point of view, as the computational effort for the TV approach is 100 times larger. More detailed criteria for the applicability of the TIV approach are given by Marley and Moan (1993).

Another observation in this example is, of course, that the failure probability is unacceptably high in the repair period. The benefit of the LBB monitoring of tethers is, hence, very limited if failure of one tether is unacceptable! However, further analyses show that the LBB principle has some merit if the initial cracks are accidentally large, for instance, due to fabrication defects. Design and inspection planning are discussed more in detail in Chapter 6.

4.5 *Guidelines for reliability analysis*

Up to now, reliability analysis has been used in decision making primarily by experts in code committees. However, reliability analysis is increasingly being applied on a case basis, in design, inspection planning and reassessment of existing structures. It is, therefore, necessary to ensure a consistent practice by standardizing methods, uncertainty modeling and the approaches to set target levels. This issue is discussed in general terms by Ditlevsen and Madsen (1989). The first practical guidelines for reliability analysis of marine structures were issued by DnV (1992), NGI (1992) and Department of Energy and Mines (1993). This is an important step, but further work is clearly necessary to achieve internationally accepted standards.

5 SYSTEMS RELIABILITY ANALYSIS

5.1 *General*

Component reliability analyses have been applied in decision making especially in cases when they could be centered around deterministic design approaches. This has not been possible for systems analyses, because even deterministic systems analyses are scarce in current design practice. However, some codes, e.g., NPD (1992), require explicit robustness design checks of the system associated with accidental loads. Also, systems analysis is attractive in connection with reassessment of the ultimate capacity of existing platforms; see, e.g., Vugts and Edwards (1992), and Bea (1993).

While systems failure of fixed platforms occurs due to a progressive failure of the structure or foundation, systems failure of buoyant structures may take place in a sequence of structural failures or flooding events or by a combination of such events. The event chain for buoyant platforms may involve structural failure, equipment failure (valves, pipes) and operational errors, as illustrated by the (simplified) event tree in Fig. 10.

Much research on systems reliability analyses of fixed steel platforms (jackets) and the tether system of the TLPs, has been accomplished over the last 15 years. Certainly, practical applications will follow, for the obvious reason that the main safety issue is the integrity of the system. System failure represents a catastrophic event, for instance in the sense that fatalities caused by structural failure primarily occur due to failure to support the deck, where all people are located, or overall overturning or capsizing. Use of a systems approach for some platforms is also motivated by the significant reserve capacity beyond

a) Ship

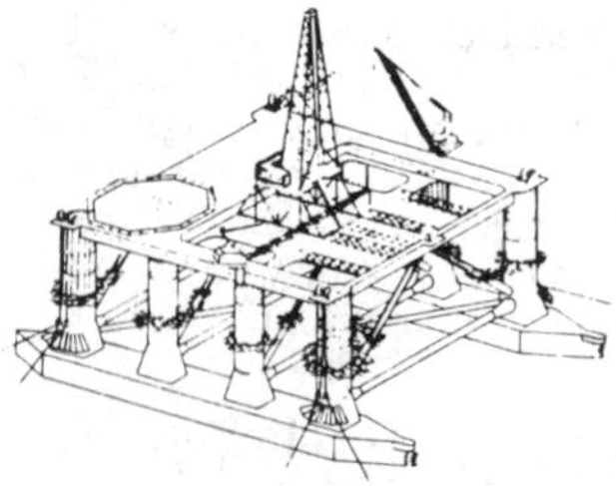

b) Semi-submersible

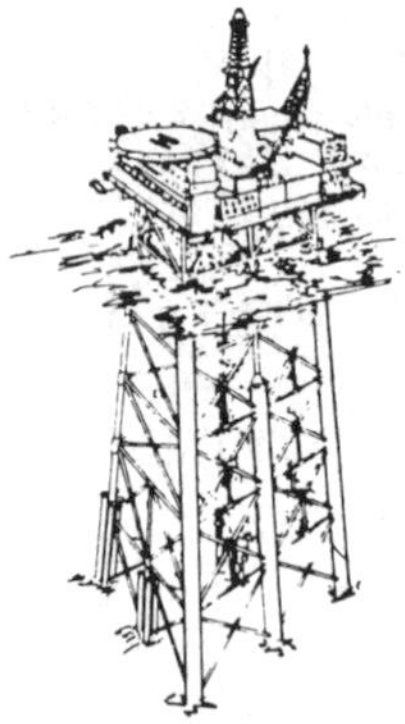

c) Jacket

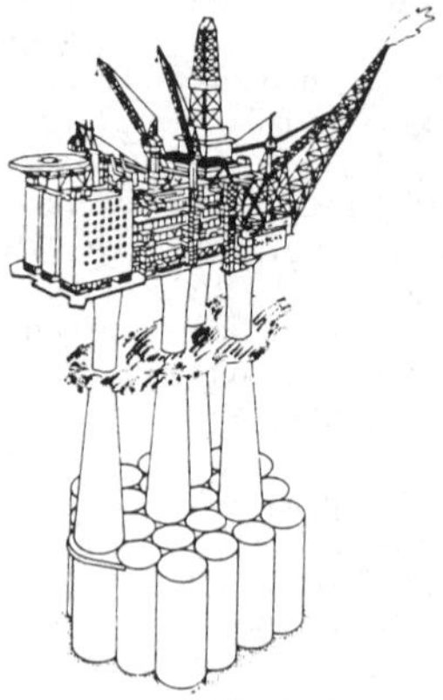

d) Gravity platform

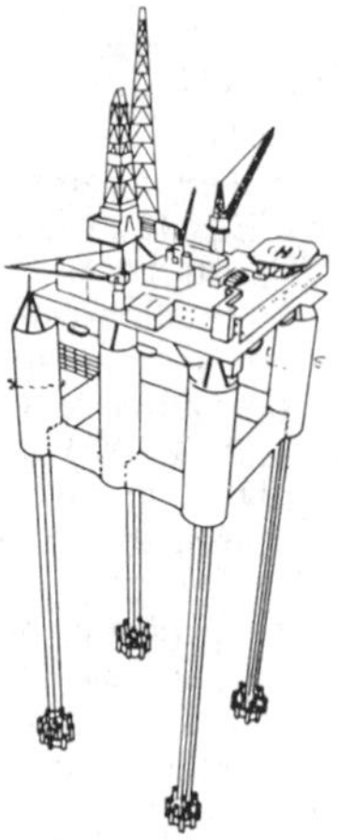

e) Tension-leg platform

Fig. 1 Offshore structures.

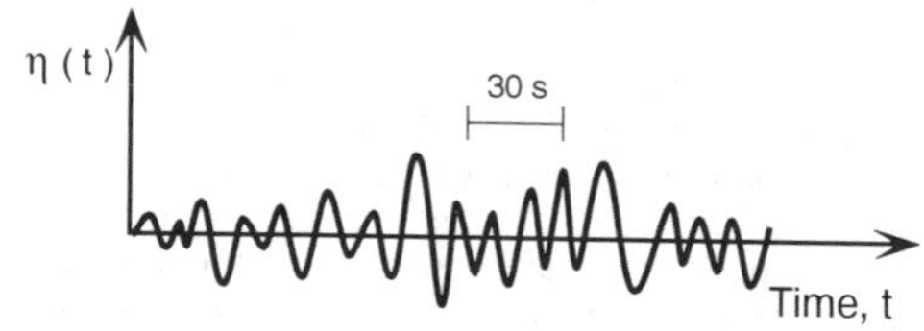

a) Short-term sea state (stationary, ergodic, Gaussian)

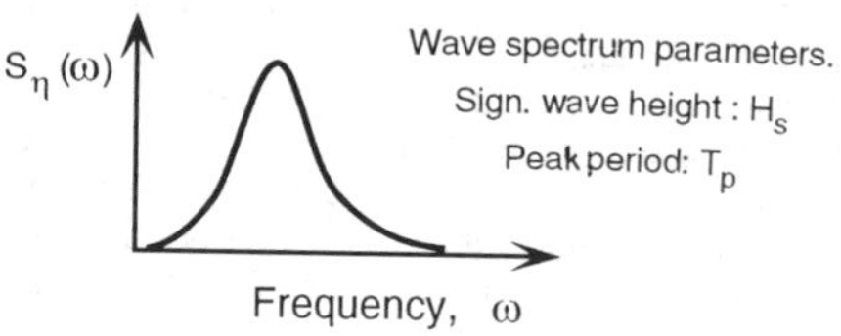

c) Wave power spectrum

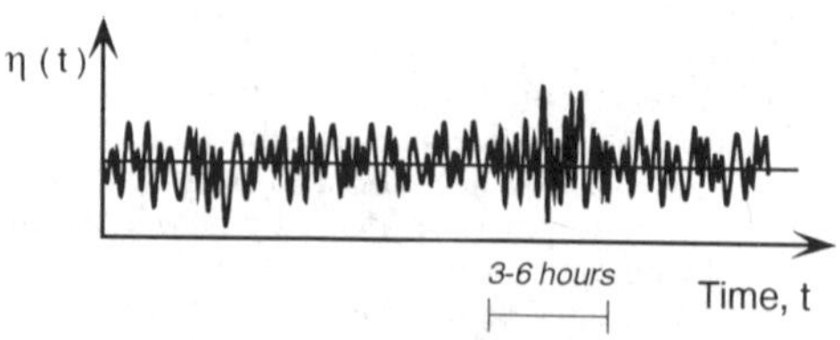

b) Long-term variation of wave elevation

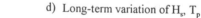

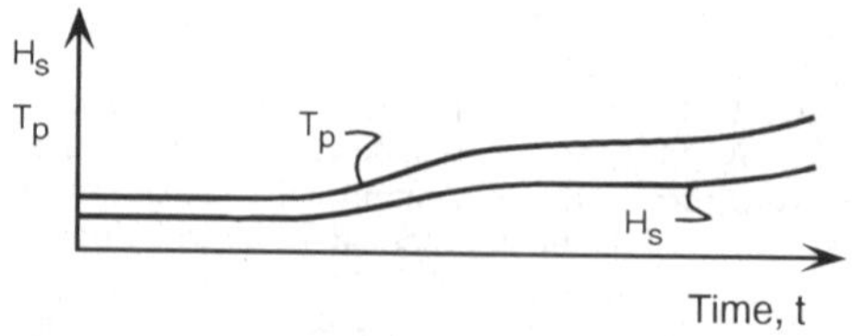

d) Long-term variation of H_s, T_p

Fig. 2 Long-term model of the wave elevation process.

first component failure which is not accounted for by the conventional component design methods (Lloyd and Clawson 1983).

5.2 *Fixed steel platforms with component overload failure*

Significant progress has been made over the last one to two decades; on the structural systems reliability analysis of fixed platforms subjected to extreme sea loads, as reviewed, e.g., by Moses and Liu (1992), and Baker and Vrouwenvelder (1992). Some of the general developments of systems reliability methodology have been motivated by applications to offshore steel truss/framed towers, which consist of easily identifiable components of slender tubular members and joints. The development of realistic mechanical models for truss/frame systems have made the reliability developments possible (see, e.g., Moan and Amdahl 1989, Ueda and Rashed 1991, and Stewart et al. 1993). Most efforts have been devoted to jackets subjected to design wave loads and functional deck loads, with a static behavior. Virtually no research on systems reliability analysis of dynamically sensitive structures has been made.

A truss structure collapses in a sequence of failures of members, joints or piles. Member failures occur in a beam-column mode with a semibrittle behavior. Tubular joints may fail due to collapse or fracture. The purpose of the systems reliability analysis is to express system failure mathematically by load and resistance parameters relating to the components and calculate the systems failure probability by the probabilistic properties of these parameters. Broadly speaking, this may be achieved by a failure mode (or survival mode) analysis, or direct-simulation methods (Karamchandani 1990, and Moses and Liu 1992).

The failure mode analysis consists in

- identifying the sequences of component failures (system failure modes), E_i, considering members, joints and piles.
- establishing a mathematical expression for the events of each sequence, E_i, based on structural mechanics. The event sequence (No. i) may involve failure of (n_i) components such that

$$E_i : E^0_{i_1} \cap E^1_{i_2} \cap \quad \cap E^{1,\dots n_i-1}_{i_{n_i}}$$

where $E^{1,\dots j-1}_{i_j}$ is the event that component no. j fails given that j-1 components already have failed. Neglecting the ordering of events, it is assumed that the events are failure events, as discussed, e.g., by Cornell (1987).

- establishing probabilistic measures for the random variables involved.
- calculating the failure probability of the system.

The first two steps in this approach are particularly easily accomplished when ideal plastic behavior can be assumed, see, e.g., Thoft-Christensen and Murotsu (1986). However, since buckling or collapse of members and fracture of joints are of concern for jackets (and other offshore structures), the more versatile member replacement or incremental approaches should be applied. The resulting failure function for the j-th failure given that j-1 components already have failed in sequence i, may be written as, see, e.g., Wu and Moan (1991):

$$g^{1,2\dots(j-1)}_{i_j} = u_j - \bar{u}_j - \sum_n c^i_{jn}\left(R_n - \bar{R}_n\right) - \sum f^i_{j\ell}\left(L_\ell - \bar{L}_\ell\right) \tag{14}$$

where u_j is the limit state displacement corresponding to the jth failure and R_n and L_ℓ are the n-th resistance and ℓ-th load parameter, respectively. The bar denotes mean value. c^i_{jn} and $f^i_{j\ell}$ are denoted strength and load influence coefficients, respectively, and are determined by an incremental load formulation.

Modeling of the uncertainties in loads and resistances of members, joints and pile foundation represents a crucial task (Baker and Vrouwenvelder 1992, and Shetty 1992). Typically, resistance parameters for members include yield stress, initial deflection and post collapse model-uncertainty. The initial deflection should have a fictious value to cover the effect of residual stress, which cannot be explicitly represented by the beam models commonly adopted in reliability analysis. The resistance of tubular joints is based on parametric formulas validated by experimental data (Marshall 1992). The associated uncertainty is included by uncertainty in the model and yield stress. The pile properties in reliability analysis are also represented by a simple model and the associated model and material uncertainty. Also, a measure of correlation between resistance parameters for different components needs to be included. Finally, a model uncertainty for global behavior should be associated with the system resistance and not the component resistances. Since the forces in the components depend upon the integrated effects – tower shear and bending moment at the relevant section, they actually relate to global load characteristics. Hence the loads in individual components are strongly correlated. The generic uncertainty level for jackets is indicated in Table 1.

Having established the limit states and uncertainty measures, the failure probability may be calculated by

$$P_{FSYS} = P[FSYS] = P\left[\bigcup_{i=1}^{N} \bigcap_{i=1}^{n_i} \left(g^{1,\dots j-1}_{i_j}() \le 0\right)\right] \tag{15}$$

by FORM/SORM, bounding techniques or simulation methods. Due to the effort involved, it is important to limit the number (N) of failure modes. To identify the stochastically dominant failure sequences the incremental method by Moses (1982), the branch and bound method of Murotsu et al. (1984), truncated

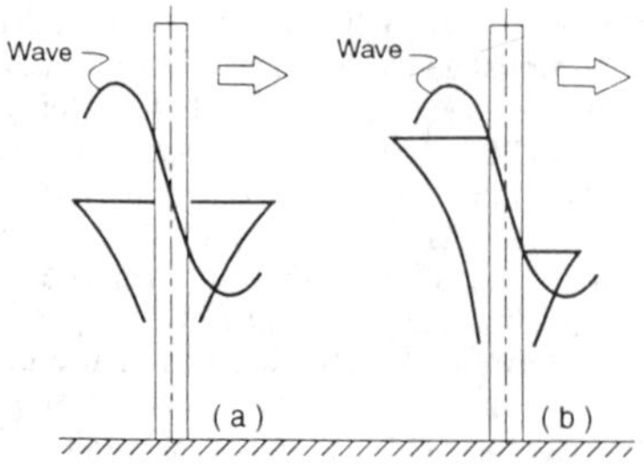

Fig. 3 Wave forces act on "wet surface".

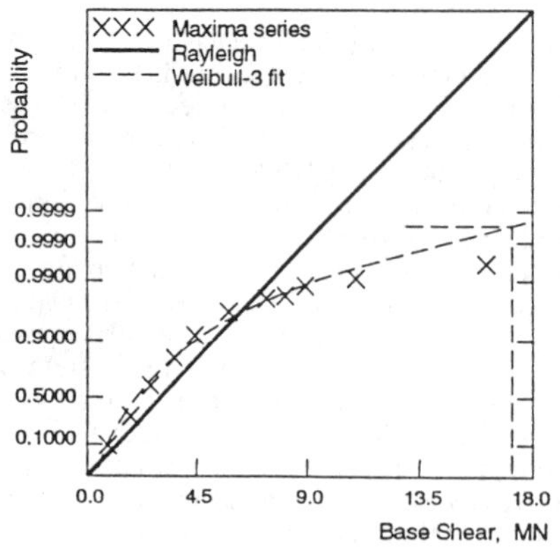

Fig. 4 Short-term cumulative distribution of base shear in the Veslefrikk jacket (Karunakaran 1993).

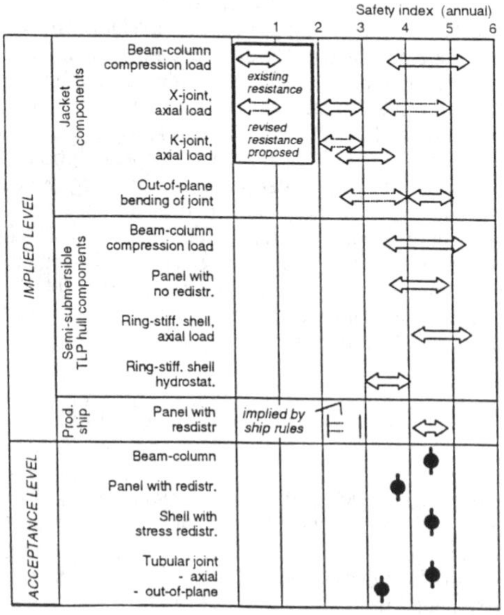

Fig. 5 Comparison of reliability levels implied by the 1985 NPD regulations; and target reliabilities (Moan 1988).

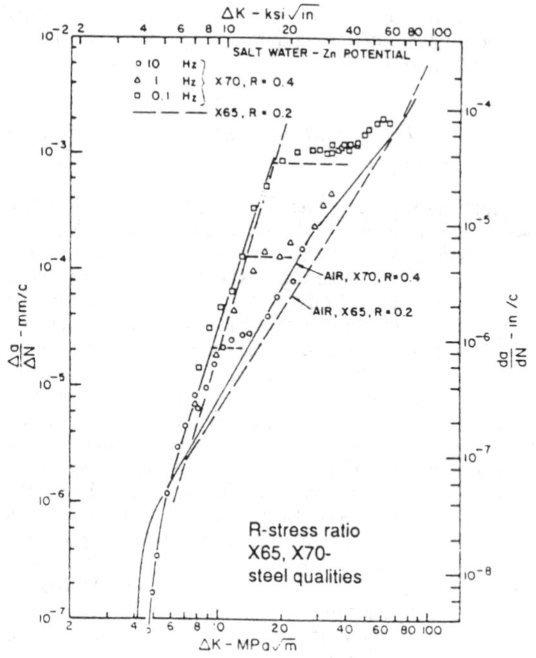

Fig. 6 Fatigue crack growth rates in salt water at zinc potential. Stress ratio (R) of 0.4, and three load frequencies (Vosikovsky 1980).

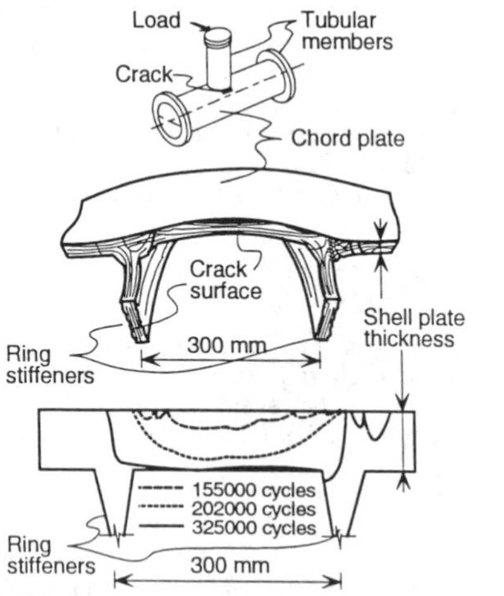

Fig. 7 Crack propagation in ring-stiffened T-joint subjected to out-of-plane bending load (Forbes 1991).

enumeration method of Melchers and Tang (1984) and the method of Guenard (1984) are applied.

Recent efforts have been made to extend the failure mode method to include members with more realistic post-collapse behavior (Karamchandani 1990, and Wu and Moan 1989) and efficient generation of failure sequences for large offshore structures (Cornell et al. 1988). Structural mechanics approaches currently used in the context of the failure mode analysis of systems, are simplified. The large deflection and plasticity/hardening effects which occur before failure, are not properly accounted for. On the other hand, a complete incremental finite element analysis is difficult to implement in a failure mode approach. The recent improvements of computer capability and simulation methods make it feasible to accomplish "direct systems reliability analyses" by simulation (Sigurdsson et al. 1993). Such methods are especially useful to achieve benchmark references for other approaches based on further approximations in the mechanics and probabilistic models.

The research on systems analysis of jackets has also shown that accurate estimates of the systems failure probability under extreme sea loading can be achieved with a very simple model, i.e., by referring both the load and resistance to a given load pattern and using the (overall) base shear as variable. Such a model is easily understood by designers and regulators and may hence contribute to a wider acceptance of systems reliability analysis. This model has been validated for cases where the load uncertainties are dominant and the component forces are highly correlated (Wu and Moan 1989, and De et al. 1990). With uncertainties typically as shown in Table 1, this condition is fulfilled when the overload failure sequences involve only members and the effect of joint flexibility on the systems behavior is negligible. The simplified model needs further justification for systems with failure sequences involving collapse or fracture of tubular joints; or pile foundation failure, which are associated with a higher uncertainty and smaller correlation between components. The failure probability for the simplified model may be obtained by using a Monte Carlo simulation to determine the distribution of the resistance. However, even estimating the mean system resistance by using mean values of all component resistance variables, and taking the random uncertainty (COV) as somewhere between zero and the maximum component COV often yields a sufficiently accurate resistance model (Dalane 1993, and Sigurdsson et al. 1993). This simplified approach can be easily extended to the calculation of conditional failure probabilities for platforms with damage caused by accidental loads or fatigue, for use in a more general risk assessment.

The computationally efficient "system approach" based on loads and resistances expressed by base shear, is especially attractive for probabilistic requalification of existing platforms (Bea 1993, and Vugts and Edwards 1992).

Table 1. Typical general uncertainty levels for jacket platforms (Moses 1986, 1987; Moan 1988; Baker and Vrouwenwelder 1992).

Random Variable		Component Uncertainty[1)] Bias	COV	Corr. between comp.
Extreme (design wave) loading	North Sea	0.8-1.0	0.20-0.30	high
	Gulf of Mexico	0.7-1.0	0.30-0.40	"
Fatigue hot spot loading	North Sea	0.8-1.0	0.30-0.40	0.5-0.9
	Gulf of Mexico	0.7-1.0	0.30-0.50	
Stiffness (elastic)	Member	1.0	0.02-0.05	0.2-0.9
	Tub. Joint	0.8-1.0	0.10-0.30	
	Pile	0.8-1.0	0.20-0.30	
Ultimate strength	Member	1.0-1.3	0.10-0.15	
	Tub. Joint	1.0-1.5	0.20-0.30	
	Pile	1.0-1.5	0.20-0.40	
Fatigue strength	Tubular joints	1.0	0.40-0.60	

[1)] For a given method the bais and COV will have more specific values within the given ranges of variation given.

5.3 *Fixed steel platforms with component fatigue or overload failure*

Failure of redundant offshore structures may also be initiated by a fatigue failure or a fatigue induced fracture. If repair is not accomplished, a second fatigue or overload failure may occur. Even if the damage is detected, it may not be repaired until some time later, depending upon weather conditions. The increased stress in the remaining members will contribute to this second failure. While progressive overload failures are assumed to take place instantaneously (during the 18-20 sec period of a storm wave), the failures in the sequences of concern here, occur at different points in time.

Until recently, relatively few studies on systems fatigue reliability of offshore structures had been accomplished. The first one is due to Martindale and Wirsching (1983). Their simplified approach was based on components behavior modeled by SN-Miner Palmgren method and the assumption of a log normal time to failure, evaluated as the sum of time intervals between successive fatigue failures. This approach was later extended to a parallel system with components with equal resistance and equal load (Stahl and Geyer 1984), and unequal resistance and load (Kumar and Karsan 1990). Recently, Karamchandani et al. (1991), Shetty (1992), Dalane (1993) developed fairly general approaches based on the failure mode approach.

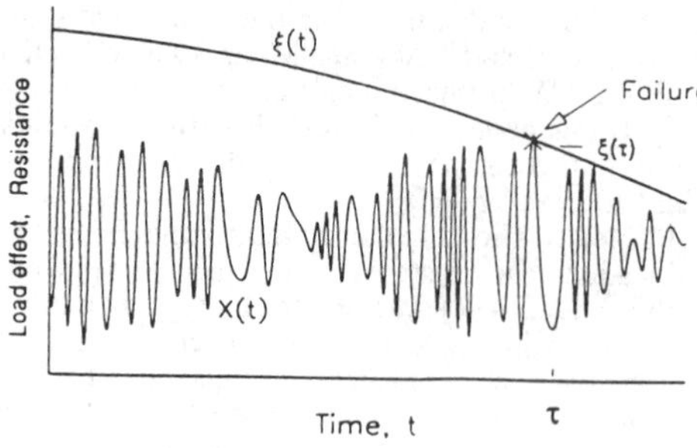

Fig. 8 Time dependent load and resistance.

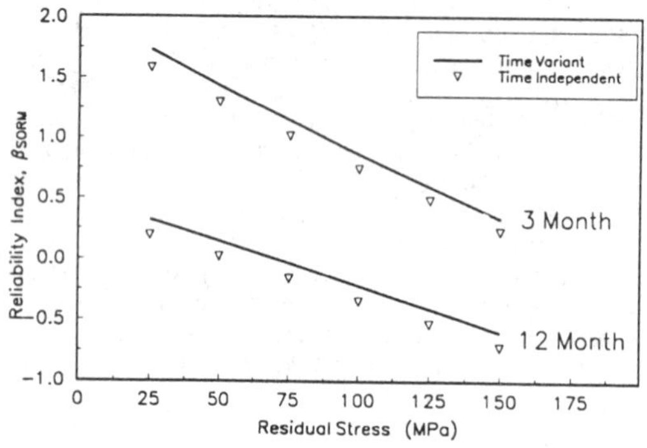

Fig. 9 Conditional reliability index for TLP tether versus residual stress for two different repair periods (Marley 1991).

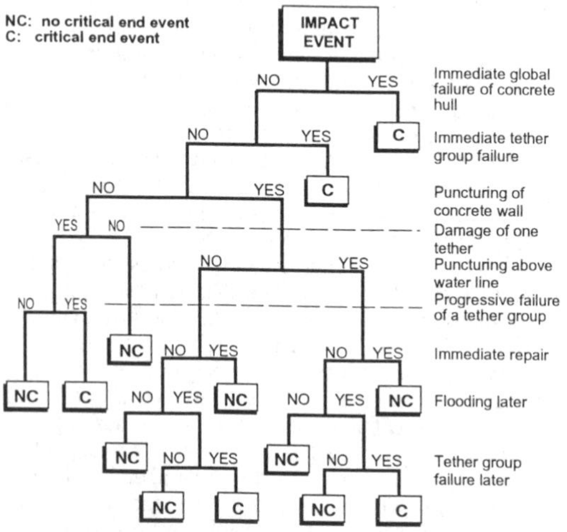

Fig. 10 Event tree for impairment of TLP (Moan et al. 1993).

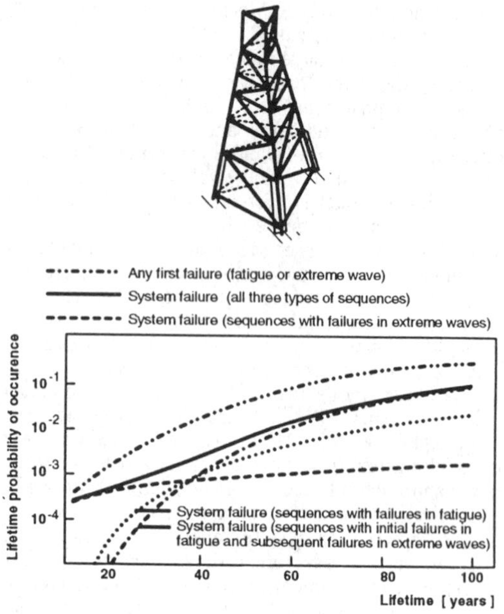

Fig. 11 Failure probability of North Sea tripod jacket in 70 m waterdepth, as a function of design life (Dalane 1983).

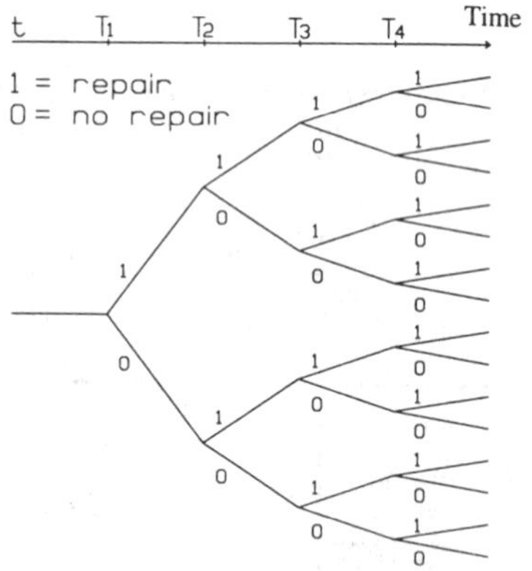

Fig. 12 Inspection event tree for a simple joint.

Fatigue failures are assumed to be brittle in nature so that components once failed remain in a failed state in the remaining life of the structure. The main difference from the case with the successive overload failures is the establishment of the failure functions, $g_{i_j}^{1,2...j-1}(\ )$.

In the failure mode approach, different (important) failure sequences are identified. For each failure sequence (i) the failure function, $g_{i_j}^{1,2...i_n}(\ )$ needs to be established. Consider a sequence with the first fatigue failure in section i_1, the second failure in joint i_2, etc. The first failure occurs when the time to failure, $T_{\text{sec}i_1}$, is less than the service period, τ, i.e.

$$g_{i_1}^0 = T_{\text{sec}i_1} - \tau \leq 0 \tag{16}$$

where $T_{\text{sec}i_1}$ is given by Eq. (5), and the initiation period is neglected. Similarly, the second failure occurs when

$$g_{i_2}^1 = T_{\text{sec}i_2/i_1} - \tau \leq 0 \tag{17}$$

By assuming linear cumulative damage, it may be shown (Shetty 1992, Dalane 1993), that the conditional time $T_{\text{sec}i_2/i_1}$ to the second failure, in section i_2, given failure in section i_1, may be written as

$$T_{\text{sec}i_2/i_1} = T_{\text{sec}i_1} + \left(T_{\text{sec}i_2} - T_{\text{sec}i_1}\right)\frac{d_2}{d_{2/1}} \tag{18}$$

where $T_{\text{sec}i_2}$ is the time to failure of section i_2 in the intact structure and

$$d_2 = \nu_{i02}\, E\left(\{S_{i2}(\underline{Z})\}^m\right) \text{ and } d_{2/1} = \nu_{i02/1}\, E\left(\{S_{i2/1}(\underline{Z})\}^m\right) \tag{18a,b}$$

ν_{i02} and $\nu_{i02/1}$ being the zero-upcrossing frequency of the response at section i_2 for the intact structure and the structure with failure of section i_1; and S_{i2} and $S_{i2/1}$ the corresponding stress ranges. $\underline{Z}$ is a set of time-invariant load and resistance parameters. The time to fatigue failure no. $k : T_{i_k|i_1 i_2 i_{k-1}}$ can be expressed in an analogous way as the time to the second failure (Shetty 1992, and Dalane 1993).

The probability of a given sequence of fatigue failures in the period τ may then be calculated by the intersection of events of the type $\left(g_{i_1 i_2 ... i_k}^{1,2...k-1}(\ ) \leq 0\right)$. The occurrence of an overload failure in the period τ, following the fatigue failures, may be included by referring the failure function for the overload to the period $\tau - T_{i_k/i_1,i_2...i_{k-1}}$.

In computing the time to failure for the intact and damaged structure by using Eqs. (16-18a,b), the main effort is associated with calculating d_k and $d_{k/k-1}$. If the corresponding times $T_{\text{sec}i_k}$ and $T_{\text{sec}i_k} - T_{\text{sec}i_{k-1}}$ exceed one year, S_k^i and $S_{k/k-1}^i$ should be determined by a long-term response analysis, which involves stochastic, dynamic response analyses of multi-degree-of-freedom systems for several sea states (Karunakaran 1993). In reliability analysis the effect of uncertainties inherent in sea-state, hydrodynamic, soil and structural models and parameters $\underline{Z}$ on the response (S), should be included. Finally, in a system reliability analysis it is necessary to do all these response analyses for many different configurations of damage. Clearly, the calculation of d_k and $d_{k|k-1}$ can be simplified by adopting a Weibull distribution with a scale and shape parameter A and B, respectively, for the long-term distribution of S and choosing A and B appropriately (Shetty 1992, and Dalane 1993).

The uncertainties in loads and resistances of components in fatigue reliability differ from those in overload reliability. While the load uncertainties predominate in the calculation of the probability of wave overload, load and resistance uncertainties are of the same order of magnitude in fatigue problems. Also, the correlation between component failure modes is less in fatigue. For fatigue failure modes the probability of first failure to the system failure probability will, hence, be large. Similarly, the small correlation between fatigue and overload failure events implies large systems effects for failure modes comprising fatigue events and a single overload failure.

The failure probability of different platforms has been determined by Shetty (1992) and Dalane (1993), considering various failure sequences: fatigue-fatigue- , and fatigue-overload. Here, some results for a tripod jacket located in 70 m waterdepth in the North Sea, are presented. The design fatigue lives of different joints are 80 years and more for this structure. System failure is considered to occur after failure of two components. Fig. 11 shows the lifetime probability as a function of the service life for different failure sequences. The upper curve shows the probability of any first failure, and the second, fully drawn curve refers to the system failure probability. For short service lives the (total) system probability is obviously dominated by the overload probability. For long lives, system failure caused by fatigue followed by overload, is most important.

By a closer look at the results obtained for this case (Dalane 1993), it is observed that the conditional probability of a second failure, given the first failure, is 0.6 for the dominant overload-overload sequence while it is only 0.02 and 0.07 for fatigue-fatigue and fatigue-overload, respectively. This is an example of the much greater systems effects in fatigue reliability compared to overload reliability. Clearly, these results depend critically on the correlation between components. In the base case of fatigue analysis the member forces are assumed to have a correlation coefficient of 0.5 to 0.9, while hot spot stresses and fracture mechanics resistance parameters are considered independent. An increase of the correlation does not reduce the probability of first failure (series systems effect) significantly, but increases the systems failure probability ("parallel system effect"). If the

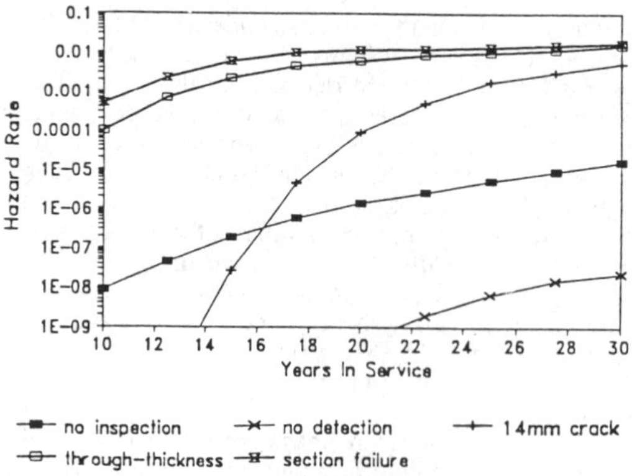

Fig. 13 Updated hazard rate for the second most dominant failure sequence (fatigue-fatigue) based on inspection after 10 years in service (Dalane 1993).

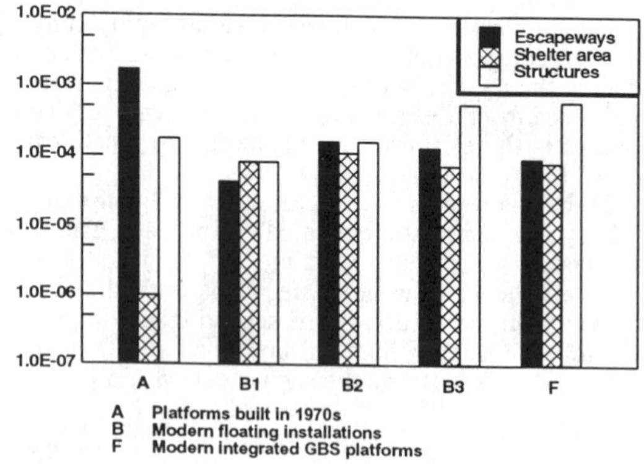

Fig. 16 Overview of estimated impairment frequencies for typical platform concepts (Vinnem 1992).

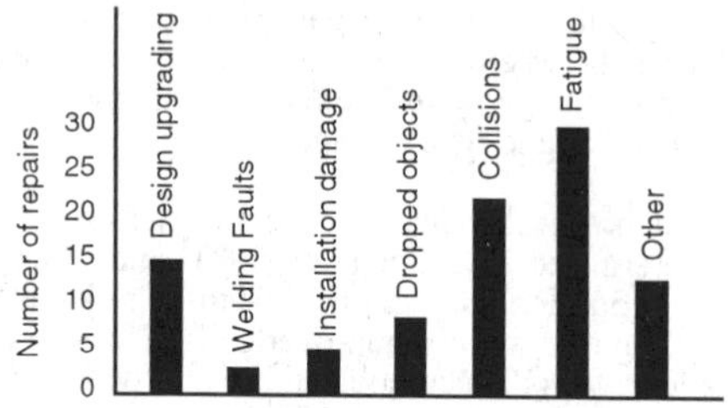

Fig. 14 Causes of repair for North Sea platforms, up to 1987 (Scharp et al. 1993).

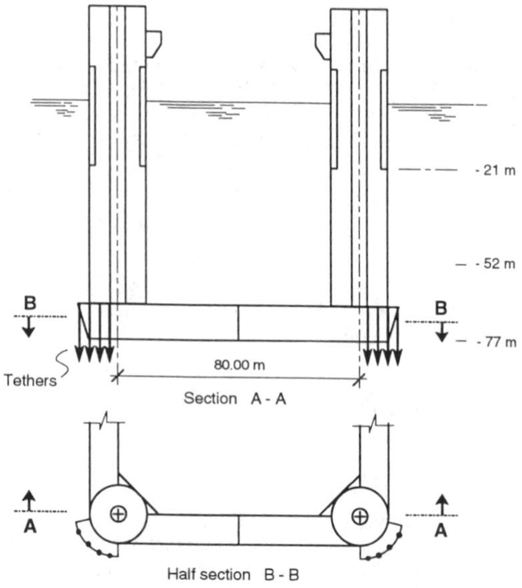

Fig. 17 Compartmentation of TLP hull (Moan et al. (1993b).

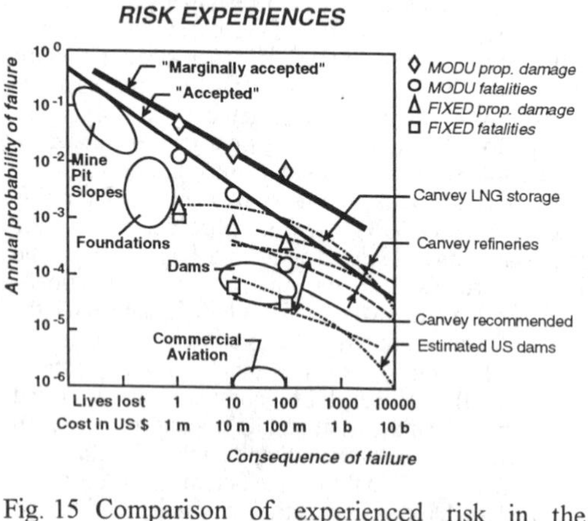

Fig. 15 Comparison of experienced risk in the offshore and other industries (ISSC 1991b).

correlation coefficient of ℓn C is taken to be 0.8 (instead of 0.0), the probability of fatigue-fatigue failure, in a lifetime of 30 years, increases by a factor of 5 and becomes approximately the same as that of overload.

5.4 *Tension leg systems*

A tension leg platform system consists of three subsystems, namely the hull, tension-leg and foundation systems. Here, the focus is on the tension-leg system, which is crucial to the safety of the TLP. The tethers are subjected to pretension, wave, current and wave loads and possible earthquake loads. Failure of all tethers in one of the four tether groups is considered to be catastrophic. A tether failure may occur due to tension or bending overload; or by going slack with excessive bending or subsequently experiencing a snatch load; or by fatigue. All these limit states are considered in the design. The tethers are normally designed with equal strength in all members, and in all joints.

Two features associated with a component failure of the tether system need to be observed. Firstly, the load shedding, following failure, involves dynamic effects. In this connection it is noted that the total load present when an overload occurs, will be larger than in the case of a fatigue failure (Jiao et al., 1990). Secondly, there may be a possible effect of failed components (tethers) on the system. A broken tether may hit other tethers and cause damage or even another tether failure as discussed by Moan and Amdahl (1989).

The tether system is an ideal parallel system of series subsystems and is hence well suited for systems analysis. The tension in tethers in a group is highly correlated, while the possible bending in different tethers is less correlated. Compared to tubular joints in jackets, the uncertainty in SCF's is negligible. The ultimate capacity is described by an interaction equation involving tension, bending and external pressure. However, tension is the dominant loading, especially for small waterdepths. Fatigue resistance is described by crack growth parameters, C and m, or SN curves.

The high correlation between component ultimate resistances makes the system analysis of overloads very simple and the systems effects relatively small (Guenard 1984). Moreover, fatigue is often the most critical failure mode for components and also for the system.

While the failure mode approach obviously can be applied to the tether system, simple and efficient alternative formulations have also been developed and applied.

Jiao et al. (1990) studied the reliability of a tether system by a simple model, considering both fatigue and overload component failures. The overload was assumed to take place in a one year period corresponding to the maximum period the system possibly could remain unrepaired after a first failure. Ximenes (1990) and Ximenes and Mansour (1991) applied a model based on extreme value statistics as well as a failure mode approach to determine the system fatigue reliability. Hovde and Moan (1994) extended the first approach by considering unequal loads and more general correlation properties between components.

Kung and Wirsching (1992) investigated the systems reliability in view of fatigue and overload in the tether system, by replacing the series system of each tether by an equivalent single component, using Monte Carlo simulation.

6 BAYESIAN RELIABILITY ANALYSIS

6.1 *General*

Adequate structural safety is ensured by design; but also load or response monitoring, or inspection; and by taking the necessary actions to reduce loads directly or indirectly by, e.g., removal of marine growth, or to repair, when necessary. It is important to establish an optimal balance of these different safety measures at the design stage, when the costs of increasing the strength are small compared to what they are for an existing structure. However, the planning at the design stage can only be based on predicted resistances and loads as well as outcomes of load monitoring or inspections. The effect of the possible outcomes on the reliability can then be estimated by Bayesian updating.

During operation a reassessment of the safety may be necessary due to
- required change of live loads
- occurrence of deterioration or overload damage
- extension of service life
- reduction of operational costs

Such assessments should take into account information about the geometry and material properties acquired during fabrication and observed response or damages during previous inspection, or condition assessments or survival of proof-loads or service loads. Observations both for the actual structure as well as for other relevant structures should be considered. In particular, it is essential to sequentially update the inspection plan made at the design stage as information is achieved during operation.

6.2 *Bayesian updating*

Updating of reliability can be made using the Bayesian approach in two ways, namely by updating of:
- the multivariant probability distributions and then recalculating the failure probability
- the probability of events, such as failure events, directly

Significant developments of the first type of approach were made by Shinozuka, Yang, Itagaki and others as reviewed by Yang (1993). Over the last decade FORM/SORM techniques have been used to develop general and efficient techniques for event updating, see Rackwitz and Schrupp (1985) and Madsen (1986, 1987).

Due to the significant computational efforts involved, it has not until recently been possible to update system reliabilities. When the failure mode approach is applied, the updated failure probability for a system may be written as:

$$P_{FSYS} = P[FSYS|I] = \frac{P[FSYS \cap I]}{P[I]} \qquad (19)$$

where FSYS is defined by Eq. (15) and

$$I: \bigcap_k (I_{\ell k} \leq 0) \qquad (19a)$$

with ℓ and k referring to the inspection number and the type of inspection outcome, respectively. For a multi-component system and an inspection event tree as shown in Fig. 12, the computational efforts to calculate Eq. (19) are significant. Therefore, simplifications are necessary. For instance, simplified component limit state functions may be achieved by using time-invariant rather than time-variant approaches. The T given by Eq. (5) in the failure function of Eq. (6) may be replaced by an equivalent one of the type $T = X_1X_2X_3$, where the X_i's are appropriately chosen (Hovde and Moan 1994). Using the simple single load-resistance variable model for jackets represents a particularly simple system. So does the model for tether systems used by Kung and Wirsching (1992) where the multicomponent series system of a tether is modeled by an equivalent component. The inspection event tree may be approximated by considering only repair events corresponding to no-repair. This approximation may be appropriate if the reliability level is high and only a few repairs are expected (Sørensen 1993).

In connection with FORM/SORM analysis, particular care needs to be exercised to eliminate non-active events (i.e., covered by other events). Also, the accuracy of the method needs to be validated, e.g., by simulation.

6.3 *Example of updating of systems reliability*

To illustrate the effect of system reliability updating during operation, consider again the simple three-legged jacket shown in Fig. 11.

Fig. 13 displays the hazard rate (annual P_F given survival up to the considered year) for the dominant failure sequence (fatigue-overload) as a function of time, given an inspection after 10 years of operation. Compared to the hazard rate, when no inspection, is performed, detection of a through-thickness crack or complete failure of a cross-section, imply very high hazard rates. On the other hand, no detection of damage during the inspection implies a very low hazard rate. In the latter case, the time to the next inspection can be significantly increased, and the inspection costs reduced.

In the approach described above, inspection results are used to update the initial estimates of fatigue loading, crack growth parameters and initial crack size based on an assumed probabilistic nature of these variables. Since only fundamental randomness and "normal" uncertainties are considered, the updated hazard rate will be very conservative if, e.g., the initial crack size had been accidentally large due to a fabrication defect. Additional information about loads, may be a second measurement of the crack size, etc., would then be necessary to achieve a better estimate.

6.4 *Requalification of jacket platforms*

There is an increasing need for reassessment of offshore structures. For instance, one third of the world's 5000 fixed platforms are 15 years or more (ILO 1993), and an extended service life or reuse is called for in many cases. ISO codes (ISO 1993a,b) and other codes currently under development, explicitly address reassessment issues.

Extension of the service life of jackets in the Gulf of Mexico has been of main concern recently (Bea and Smith 1987). A special condition assessment and determination of the overall safety have been the main basis for making decisions about service life extension. The reassessment is made recognizing the fact that there is no reason to apply more refined methods than necessary. Hence, advanced nonlinear structural analysis and reliability analysis are only applied when it is not obvious that demolition or life extension can be concluded based on simple linear and deterministic methods. Bea (1993) proposes to use reliability methods on the most advanced level in a requalification approach with four levels of analysis refinement.

Both sea and seismic loads are considered in the analysis. As mentioned in Chapter 4.2, the target level used by Bea (1993) is based on cost analysis. This approach implies a willingness to accept lower reliabilities for old systems compared to new ones. Iwan et al. (1993) proposed a target level based on separate considerations of the fatality rate and environmental damage, and arrive at a different target reliability level, however, also implying slightly lower target reliabilities for existing compared to new platforms. An example of a reliability-based requalification is reported by Bea et al. (1992). Several oil companies and consultants are still investigating the use of systems reliability methods in reassessment of offshore structures. Marioni et al. (1993) report such a study, but most of the work in this area is not published as yet.

6.5 *Inspection planning relating to crack damage*

Inspection and maintenance are important in ensuring continued integrity of offshore structures in the

North Sea. While such safety measures should be directed at all types of damages (MTD 1989), particular attention should be paid to cracks. This is partly because one third of repairs to date (see Fig. 14) have resulted from fatigue cracking and partly because of the significant efforts required in inspection and maintenance of cracks. Traditionally, inspection planning relating to cracks has been based on design fatigue life, although recently redundancy has become another parameter used in the planning process.

A number of limitations of inspection practice were identified in MTD (1989), especially the significant uncertainties in fatigue prediction, including design deficiencies, fabrication defects and crack detection probability, as well as the inadequacy of examining only a limited amount, typically 10-15%, of the joints in 4-5 year periods. The usefulness of probabilistic methods to deal with uncertainties, and especially Bayesian methods, are, hence, increasingly recognized in inspection planning.

Carr et al. (1986), Madsen and Sørensen (1990), Skjong and Torhaug (1991), Peers et al. (1992) and Faber et al. (1992) discuss reliability-based design and inspection criteria to minimize life-cycle costs on a component basis. The failure and repair probabilities necessary for calculating the expected costs can be expressed in terms of intersections of the events of inspection repair and failure. Information achieved by inspection or monitoring is accounted for by updating probabilities by using Bayesian methods as outlined in Chapter 6.2.

Due to the conservatism in some initial inspection plans, such methods may potentially lead to cost reduction. This can be achieved through sequential inspection planning, see Fujita et al. (1989), Fujimoto et al (1991). Carstens et al. (1992) report that the time between inspections could be significantly increased for some North Sea jackets as a result of component reliability updating based on inspection. The use of systems reliability models may still yield further reduction of costs, especially for structures with member redundancy, since the whole range of inspection methods, also those which aim at detecting member failure rather than "shallow" cracks, can then be considered. The simplest system approach is to let the target reliability level for components depend upon a deterministic or probabilistic measure of redundancy, see, e.g., Lotsberg and Kirkemo (1989). While the component target probability levels were based on simple considerations in Carstens (1992), improvement can be achieved by using the simple system model described in Chapter 5.2. Optimal inspection procedures based on a systems reliability model is discussed in general terms by Faber et al. (1992), and Sørensen (1993). Clearly, a practical implementation of optimal design and inspection planning for systems would require significant efforts since it will imply calculating P_{FSYS} in Eq. (19) and, hence, e.g., the structural response, repeatedly.

7 STRUCTURAL RISK ANALYSIS

7.1 *General*

Structural reliability analysis in the classical sense deals with fundamental random variability and natural or man-made uncertainties due to lack of knowledge. Human actions which influence design, fabrication and operation of structures normally follow defined procedures, which ensure that the corresponding uncertainties in structural resistances, loads and load effects follow certain patterns, which make it possible to measure and predict such uncertainties. However, sometimes the human actions depart from acceptable practice and result in so-called human errors. Design errors may be intentional or unintentional. They materialize as a deficient (or excessive) resistance, which cannot be derived from the parameters affecting the "normal" variability of resistance. Fabrication imperfections (such as cracks, plate misalignment, etc.), which also affect the resistance, are influenced by human actions. The "normal" variability of welders performance, environmental conditions, etc. lead to a "normal" variability in the imperfection size, characterized by a smooth variation of the relevant imperfection parameter. Sometimes an abnormal deviation from this behavior occurs, e.g., caused by using a wet electrode, etc. or another gross fabrication error. Man-made live loads also have a "normal" and an abnormal component, while some loads, notably fires and explosions, ship collisions, etc. do not have a normal counterpart. They are simply caused by operational errors.

Experiences with building structures (Matousek 1977), and offshore platforms (Moan 1993, WOAD 1990, Visser 1992, Bea and Moore 1993) show that gross errors is the dominant cause of failures. This is also illustrated in Fig. 14, which relates to moderate damages, which are repaired. In this connection, it is noted that abnormal fabrication defects contribute much to fatigue failures, and that catastrophic (system) failures are even more caused by gross errors. (On the other hand, gross errors may sometimes have resulted in excessively safe and expensive structures.)

Clearly, the nature of human errors differs from that of natural phenomena and "normal" man-made variability and uncertainty, and different safety measures are required to control error-induced risks. Primarily, quality assurance and control procedures should be used. This implies for instance inspection and repair of the structure during fabrication and operation. In addition, some accidental scenarios develop over time and they can be controlled to some extent (Event Control). For instance, fires and explosions may be controlled by detecting leakage of combustibles (hydrocarbons) and activating shut-down, fire extinguishing, etc. Finally, the risk especially associated with fabrication and operational human errors can be controlled by introducing a progressive or accidental limit state for the system, see, e.g., Moan (1983), NPD (1992) and ISO

(1993a,b). In this way there will be no conflict with considering randomness and "normal" uncertainties for the usual component ultimate and fatigue limit states. Table 2 summarizes the causes of structural failure from a risk management point of view, and how the associated risks may be ameliorated.

Table 2. Causes of structural failures and risk reduction measures, see, e.g., Moan (1983)

Cause	Risk Reduction Measure
• Less than adequate safety margin to cover "normal" uncertainties in loads and resistances	- Increased safety margins in (SLS), ULS, FLS; or inspection (FLS)
• Gross errors in - design - fabrication - operation	- Design QA/QC - Event control - Inspection/repair - Design for damage tolerance (PLS)
• Unknown phenomena	- R&D

In Fig. 15, the risk level experienced in the offshore oil industry worldwide in the period 1980-89 is compared with the risk inherent in other systems (ISSC 1991b). The risk of fixed platforms seems to be reasonable while the risk associated with mobile platforms is higher because such platforms are floaters, which are susceptible to buoyancy loss, especially during temporary modes of operation. Fig. 15 may also serve as a basis for setting target risk levels, relating to fatalities and property loss. In addition, target levels for environmental damage need to be addressed (Vinnem 1992; and Iwan et al. 1993).

7.2 *Quantitative risk assessment*

Quantitative Risk Assessment (QRA) is a tool for estimating the risk and, hence, for managing risk control measures. It is now performed on a regular basis in the North Sea Offshore Industry. The aim is especially to compare the risks associated with alternative platform concepts as a basis for improving decisions regarding design and operation (DnV-Technica 1992, Vinnem 1992). Special QRA studies are performed, for instance, to determine design accidental loads for structural design. As outlined by Vinnem (1992), QRA should include assessment of all subsystems of an offshore installation, including the load-bearing structure. Fig. 16 may serve to illustrate the main outcome of a conceptual risk assessment. The vertical axis represents the annual impairment frequency, that is the frequency of impairment of important safety functions such as the operability of escapeways, and shelter area (for personnel until they can escape from the platform) as well as integrity of the load-carrying structure. Clearly, the impairment frequency for the modern platforms B through F is quite close, simply reflecting the fact that the design ensured a risk level adhering to the target level. But, there is a noticeable difference in the impairment frequencies for escapeways and shelter area for platforms built in the 70's and recent ones.

QRA is a challenging task and the scientific basis is still limited. A particular difficulty is the treatment of human errors. Currently, operational errors are indirectly considered, by using generic data for events, i.e., accidental loads caused by operational errors (Veritec/SINTEF 1988; Moan and Amdahl 1989; Technica 1992). A databank for failure rates of equipment etc. has also been established (OREDA 1991). To some extent the effect of fabrication errors are also implicit in failure data. However, since the effect of both design and fabrication errors depends upon the QA/QC procedures, their treatment is more difficult than operational errors. Because of the especially complex nature of design errors, the approach currently used in offshore engineering is to plan QA/QC of design based on a qualitative assessment of error proneness and the consequences of a deficient resistance that may result from design errors. To estimate the likelihood of various types of design errors, accounting for the effect of QA/QC, is still far-fetched.

QRA is particularly useful for novel systems such as the TLP, which consists of a chemical process plant, risers and conductors for oil and gas, hull (to provide support of the deck and pretension), tethers (for mooring) and foundations (Vinnem 1992, Stahl et al. 1992).

7.3 *Design of a TLP based on an impact risk assessment*

The use of structural risk assessment to aid design decisions will here be illustrated by considering design of the reinforced concrete hull of the Heidrun TLP, currently under construction. The example is limited to the operational hazards relating to impacts caused by ships and objects accidentally dropped from cranes (Moan et al. 1993).

A spectrum of impact events on various parts of the structure may be envisaged. Fig. 10 shows different possible event sequences, starting with the impact event and leading to system impairment. Clearly, the impact, if sufficiently strong, may cause total hull or tether failure. If not, systems impairment may develop progressively after a relatively small damage of the hull that result in flooding, tether slackening and tether failure.

The probability of impairment is then estimated by assigning probabilities to each branch of the event tree. The probabilities of frequent impact events are based on historical data, while risk assessments involving the effect of human errors during ship operations, are used to quantify probabilities of rare impact events. The method developed (Haugen and Moan 1992) for predicting the frequency of rare events is found to agree reasonably well with experiences when applied collectively to all platforms over all years operating in the North Sea.

The conditional probabilities of wall penetration given an impact load and location of flooding given a wall penetration and of tether failure given flooding, are estimated by using structural reliability methodology.

Now, the probability of impairment can be influenced by the structural design in two ways (see, e.g., Fig. 17):

- sizing of the outer wall — which will affect the conditional probability of penetration
- subdivision of the hull to limit the amount of flooding and thereby the probability of tether failure, given penetration and flooding.

It turns out that the costs of providing internal bulkheads according to current deterministic compartmentation criteria, are very high. Since other design criteria imply quite robust scantlings of the column walls, it was thought that some further strengthening of the outer wall may give sufficient safety against impacts — without subdividing the columns into compartments. It was found that a limited compartmentation, with a compartment in the still water zone (Fig. 17), was sufficient to comply with the target probability for annual total loss of 10^{-6}. However, large safety factors were necessary for the frequent and hard local impact loads caused by service vessels attending the platform. But anyway, because of the significant wall scantlings implied by other design criteria, only a moderate strengthening was required. The chosen design was significantly cheaper than the best alternative design based on conventional deterministic subdivision criteria.

8 CONCLUDING REMARKS

Recent developments and applications of structural reliability and risk analysis to existing and new offshore structures, have been described. It is demonstrated that such methods are useful in achieving the desired safety level through design and adequate inspection/repair programs with minimum costs. Because of the benefits of reliability and risk analysis, as also recognized in new model codes, increasing use of such methods, also on a case basis, is anticipated in the future. Standardization of methods and data to ensure consistent results then becomes crucial. The first guidelines for structural reliability analysis have been issued, and should be further developed. Further work is also required regarding systems reliability methodology for structures with fatigue and overload component failures, especially for dynamically sensitive structures; updating of systems reliability, and integration of structural reliability methods in QRA. The latter issue is especially important for buoyant structures which may fail due to loss of structural integrity or buoyancy. There is always a need for more data. Priority should be given to establishing information about correlation between components and on the quality of inspection. Finally, further work should be devoted to establishing target reliability and risk levels.

REFERENCES

Almar-Næss, A. (ed.) 1985. *Fatigue Handbook for Offshore Steel Structures*, Trondheim: Tapir Publ.

Ang, A. H.S. & W.H. Munse 1975. Practical Reliability Basis for Structural Fatigue. Preprint 2394, *ASCE Natl. Struct. Engng. Conf.*

Baker, M.J. 1985. The Reliability Concept as an Aid to Decision Making in Offshore Structures. *Proc. 4th BOSS*: 75-94. Amsterdam: Elsevier.

Baker, M.J. & A.C.M. Vrouwenvelder 1992. Reliability Methods for the Design and Operation of Offshore Structures. *Proc. 11th OMAE*, Vol. II, 123-132. New York: ASME.

Banon, H. et al. 1994. Development of Reliability-Based Global Design Equations for TLPs. *Proc. 13 OMAE*, to appear.

Barnouin, B. et al. 1993. Underwater Inspection Reliability Trials for Offshore Structures. *Proc. 12th OMAE*, Vol. III-B: 883-890. New York: ASME.

Bea, R.G. 1973. Selection of Environmental Criteria for Offshore Platform Design. *Proc. 5th OTC*, Vol. 1: 185-196, OTC 1839. Houston: Offshore Technology Conf.

Bea, R.G. 1993. Reliability-based Requalification Criteria for Offshore Platforms. *Proc. 12th OMAE*, Vol. II: 351-361. New York: ASME.

Bea, R.G. & C.E. Smith 1987. AIM (Assessment, Inspection, Maintenance) and Reliability of Offshore Platforms. *Marine Structural Reliability Symp.*: 257-289. Arlington, Virginia: SNAME.

Bea, R.G. et al. 1992. Requalification of a Platform in Cook Inlet, Alaska. *Proc. 24th OTC*, Vol. 2: 551-562, OTC 6935. Houston: Offshore Technology Conference.

Bea, R.G. & W. H. Moore 1993. Operational Reliability and Marine Systems. In: *New Challenges to Understanding Organizations*, Roberts, K.H. (ed.). New York: Macmillan Publ. Co.

Berge, S. 1993. Fatigue Design of Tubular Joints – Developments and Problems. In: *Fatigue in Steel Structures Behaviour under Spectrum Loading and in Corrosive Environment.* Copenhagen.

Bjerager, P. et al. 1988. Reliability Method for Marine Structures under Multiple Environmental Load Processes. *Proc. 5th BOSS*: 1239-1253. Trondheim: Tapir Publ.

Bourgund, U. & C.G. Bucher 1986. Importance Sampling Procedures using Design Points – A User Manual. Innsbruck: Inst. of Engng. Mech., Univ. of Innsbruck.

Boussy, V. et al. 1993. Comparison of Analytical Counting Methods for Gaussian Processes. *Struct. Safety*, to appear.

Brouwers, J.J.H. & P.H.J. Verbeek 1983. Expected Fatigue Damage and Expected Response for Morison-type Wave Loading. *Applied Ocean Research* 15: 129-133.

Carr, P. et al. 1986. A Probabilistic Strategy for Subsea Inspection of Steel Structures. *Proc. European Petroleum Conf.* Paper No. SPE 15868. London: Society of Petroleum Engineers.

Carstens, H. et al. 1992. Reliability Based Inspection Planning – The Implementation. *Proc. 6th BOSS*, Suppl. Vol. London: Imperial College.

Clauss, G. et al. 1992. *Offshore Structures*. Vol. 1. Berlin: Springer Verlag.

Cornell, C.A. 1987. Offshore Structural System Reliability. A JIP-Report to Amoco Production Company, C. Allin Cornell Inc., Portola Valley, California.

Cornell, C.A. et al. 1988. Enhanced Failure-Path Based Structural System Reliability. A JIP-Report to Amoco Production Company, C. Allin Cornell Inc., Portola Valley, California.

Cramer, H. & M.R. Leadbetter 1966. *Stationary and Related Stochastic Processes*. New York: John Wiley & Sons.

Cullen, The Hon. Lord 1990. The Public Inquiry into the Piper Alpha Disaster. London: HMSO.

Dalane, J.I. 1993. System Reliability in Design and Maintenance of Fixed Offshore Structures. Dr.Ing. Thesis. Trondheim: The Norwegian Institute of Technology.

Dalane, J.I. et al. 1991. Updating in Structural System Reliability: An Application to Offshore Structures under Fatigue Loads. *Proc. 6th ICASP*. Mexico: CERRA-ICASP 6.

De, R. et al. 1989. Study of Redundancy in Near-Ideal Parallel Structural Systems. *Proc. 5th ICOSSAR*: 975-982. San Francisco. New York: ASCE.

Dept. of Energy and Mines 1993. Guidelines for the Reliability Analysis of Offshore Structures. Draft. Canada.

Ditlevsen, O. &H.O. Madsen 1989. Proposal for a Code for the Direct Use of Reliability Methods in Structural Design. Working Doc., Joint Committee on Structural Safety.

Dharmavasan, W. & W.D. Dover 1988. Nondestructive Evaluation of Offshore Structures using Fracture Mechanics. *Applied Mechanics Reviews* 41: 36-49.

DnV 1992. Structural Reliability Analysis of Marine Structures. Class Notes No. 306. Oslo: Det norske Veritas. (First ed. June 1991.)

DnV 1989. Probability Analysis Software Package (PROBAN). Oslo: Det norske Veritas.

DnV Technica 1992. OHRA-Offshore Hazard and Risk Analysis Toolkit. Oslo: Det norske Veritas.

Dowling, N.E. 1972. Fatigue Failure Prediction Methods for Complicated Stress-Strain Histories. *J. Materials* 7: 71-84.

Faber, M.H. et al. 1992. Optimal Inspection Strategies for Offshore Structural Systems. *Proc. 11th OMAE*: Vol. II: 145-151. New York: ASME.

Faltinsen, O.M. 1993. Sea Loads on Floating Offshore Structures. *Proc. 25th OTC*: Vol. 2: 73-83, OTC 7142. Houston: Offshore Technology Conference.

Farnes, K.A. 1990. Long-Term Statistics of Response in Non-linear Marine Structures. Dr.-Ing. Thesis, Report 1990-60. Trondheim: The Norwegian Institute of Technology.

Faulkner, D. 1991. Unified Design Codes for Floating Systems. *Proc. Integrity of Offshore Structures*, Glasgow, 1990. London: Elsevier.

Fjeld, S. 1977. Reliability of Offshore Structures. *Proc. 9th OTC*: Vol. 4: 459-472, OTC 3027. Houston: Offshore Technology Conf.

Forbes, J. 1991. Fatigue in Stiffened T-Tubular Joints for Offshore Structures. Ph.D. Thesis. Waterloo: University of Waterloo, Dept. of Mech. Engng.

Fu, S.L. et al. 1992. Evaluation of Environmental Criteria for Gulf of Mexico Platform Design. *Proc. 25th OTC*: Vol. 1: 243-252, OTC 6834. Houston: Offshore Technology Conf.

Fujita, M. et al. 1989. Adaptive Reliability Based Inspection Strategies for Structures Subjected to Fatigue. *Proc. 5th ICOSSAR*: 1619-1626, San Francisco. New York: ASCE.

Fujimoto, Y. et al. 1991. Inspection Planning for Deteriorating Structures Based on Sequential Cost Minimization Method. *Proc. 11th OMAE*: Vol. II: 219-226. New York: ASME.

Grigoriu, M. 1984. Crossings of Non-Gaussian Translation Processes. *ASCE Engng. Mech.* 110(4): 610-620.

Gudmestad, O.T. 1993. Measured and Predicted Deep Water Wave Kinematics in Regular and Irregular Seas. *Marine Structures* 6: 1-73.

Guenard, Y.F. 1984. Application of System Reliability Analysis to Offshore Structures. Report No. 71. Stanford, CA: John Blume Earthq. Engng. Center, Stanford University.

Guers, F. & R. Rackwitz 1987. Time-Variant Reliability of Structural Systems Subject to Fatigue. *Proc. 5th ICASP*: Vol. 1: 497-505. Vancouver.

Hagen, Ø. & L. Tvedt. Vector Process Outcrossing as a Parallel System Sensitivity Measure. *ASCE Engrng. Mech.* 117(10).

Hanna, S.Y. & D.I. Karsan 1989. Fatigue Modelling for Reliability Based Inspection and Repair of Welded Tubular Offshore Structures. *Proc. 8th OMAE*: 657-666. The Hague.

Haugen, S. & T. Moan 1992. Frequency of Collision between Ships and Platforms. *Proc. 11th OMAE*: Vol. II: 359-365. New York: ASME.

Heideman, J.C. & T.O. Weaver 1992. Static Wave Force Procedure for Platform Design. *Proc. Civil Engng. in the Oceans V*: 496-517. College Station, Texas: ASCE.

Hovde, G.O. & T. Moan 1994. Fatigue Reliability of TLP Tether Systems. *Proc. 13th OMAE*, to appear.

ILO 1993. Safety and Related Issues Pertaining to Work on Offshore Petroleum Installations. Geneva: Int. Labour Organization.

ISO 1993a. General Principles on Reliability for Structures, Revision of IS 2394. Draft No. 10.

ISO 1993b. Standard for Offshore Structures for the Petroleum and Natural Gas Industries. ISO Draft No. 10.

ISSC 1964. Proc. of the Second Int. Ship Structures Congress. Delft.

ISSC 1988-94a. Report of ISSC Committee I.1. Environmental Conditions. 10th ISSC, Copenhagen (1988), 11th ISSC, Wuxi (1991) and 12th ISSC, St. John's, Newfoundland (1994).

ISSC 1988-94b. Report of ISSC Committee IV.1. Design Philosophy. ibid.

Iwan, W.D. et al. 1993. A Reliability-Based Approach to Seismic Reassessment of Offshore Platforms. *This Conference*.

Jiao, G. et al. 1990. Reliability Analysis of TLP Tether Systems. *Proc. 9th OMAE*: Vol. II: 65-71. New York: ASME.

Jiao, G. & T. Moan 1990. Methods of Reliability Model Updating through Additional Events. *Struct. Safety* 9: 139-153.

Jordaan, I.J. & M.A. Maes 1991. Rationale for Load Specifications and Load Factors in the New CSA Code for Fixed Offshore Structures. *C. J. Civil Engng.* 18(3): 454-464.

Kam, J.C.P. & W.D. Dover 1989. Advanced Tool for Fast Assessment of Fatigue under Random Wave Stress Histories. *Proc. Inst. Civ. Eng.*: Paper 9425, Part 2, Vol. 8. London: ICE.

Karamchandani, A. 1990. New Methods in Systems Reliability. Ph.D. Thesis. Stanford, CA: Department of Civil Engineering, Stanford University.

Karamchandani, A. et al. 1991. System Reliability of Offshore Structures including Fatigue and Extreme Wave Loading. *Marine Structures* 4: 353-379.

Karsan, D.I. & A. Kumar 1990. Fatigue Failure Paths for Offshore Platform Inspection. *ASCE Struct. Engng.* 116(6): 1679-1695.

Karunakaran, D. 1993. Nonlinear Dynamic Response and Reliability Analysis of Drag-dominated Offshore Platforms. Dr.-Ing. Thesis. Trondheim: The Norwegian Institute of Technology.

Kirkemo, F. 1988. Applications of Probabilistic Fracture Mechanics to Offshore Structures. *Applied Mechanics Review* 41: 61-84.

Kumar, A. & D.I. Karsan 1990. Fatigue Reliability of Parallel Systems. *ASCE Struct. Eng.* 116(3): 719-729.

Kung, C.J. & P.H. Wirsching 1992. Fatigue and Fracture Reliability and Maintainability of TLP Tendons. *Proc. 11th OMAE*: Vol. II: 15-21. New York: ASME.

Langley, R.S. 1989. Stochastic Models for Crack Growth. *Engng. Fracture Mechanics* 32(1).

Leira, B.J. 1987. Gaussian Vector-Processes for Reliability Analysis Involving Wave-Induced Load Effects. Dr.-Ing. Thesis. Trondheim: The Norwegian Institute of Technology.

Lloyd, J.R. & W.C. Clawson 1983. Reserve and Residual Strength of Pile Founded Offshore Platforms. *Design – Inspection-Redundancy. Int. Symp. on the Role of Design, Inspection and Redundancy in Marine Structural Reliability*, Williamsburg, Virginia.

Lloyd, J.R. & D.I. Karsan 1988. Development of a Reliability-based Alternative to API RP2A. *Proc. 20th OTC*: Vol.4: 593-600, OTC 5882. Houston: Offshore Technology Conf.

Lotsberg, I. & F. Kirkemo 1989. A Systematic Method for Planning In-service Inspection of Steel Offshore Structures. *Proc. 8th OMAE*: Vol. I: 275-284. The Hague.

Madsen, H.O. 1985. Random Fatigue Crack Growth and Inspection. *Proc. 4th ICOSSAR*, Int. Assoc. for Struct. Safety and Reliability: 1: 475-484.

Madsen, H.O. 1987. Model Updating in Reliability Theory. *Proc. 5th ICASP*: 564-577. Vancouver.

Madsen, H.O., S. Krenk & N.C. Lind 1986. *Methods of Structural Safety*. Englewood Cliffs, NJ: Prentice-Hall Inc.

Madsen, H.O. et al. 1987. Probabilistic Fatigue Crack Growth of Offshore Structures with Reliability Updating Through Inspection. *Marine Struct. Reliability Symp.*: Vol. SY-23: 45-56. New York: SNAME.

Madsen, H.O. et al. 1990. Optimal Inspection Planning for Fatigue Damage of Offshore Structures. *Structural Safety and Reliability* 3: 2099-2106. New York: ASCE.

Marioni, A. et al. 1993. A Reliability based Methodology of Requalification for Existing Jacket Platforms. *Proc. 12th OMAE*: Vol. II: 335-349. New York: ASME.

Marley, M.J. & T. Moan 1992. Time Variant Formulation for Fatigue Reliability. *Proc. 11th OMAE*: Vol. II: 161-168. New York: ASME.

Marley, M.J. & T. Moan 1993. Approximate Time Variant Analysis for Fatigue. *This Conference*.

Marshall, P.W. 1969. Risk Evaluations for Offshore Structures. *ASCE St. Div.* 95(12).

Marshall, P.W. 1992. *Design of Tubular Connection – Basis and Use of AWS Code Prov.* Amsterdam: Elsevier.

Martindale, S.G. & P.H. Wirsching 1983. Reliability-Based Progressive Fatigue Collapse. *ASCE Struct. Engng.* 109(8): 1792-1811.

Matousek, M. 1977. Outcome of a Survey on 800 Construction Failures. *Proc. IABSE Colloq. on Inspection and Quality Control*. Zürich: Inst. of

Struct. Engng., Swiss Federal Inst. of Technology.
Melchers, R.E. 1987. *Structural Reliability Analysis and Prediction*. Chichester, UK: Ellis Horwood Limited.
Melchers, R.E. & L.K. Tang 1984. Dominant Failure Modes in Stochastic Structural Systems. *Structural Safety* 2: 127-143.
Moan, T. 1983. Safety of Offshore Structures. *Proc. 4th ICASP*: 41-85. Univ. of Florence. Bologna: Pitagora Editrice.
Moan, T. 1988. The Inherent Safety of Structures Designed According to the NPD Regulations. Report No. F88043. Trondheim: Div. Structural Engineering, SINTEF.
Moan, T. & I. Holand 1981. Risk Assessment of Offshore Structures Experiences and Principles. *Proc. 3rd ICOSSAR*: 803-820. Amsterdam: Elsevier.
Moan, T. & J. Amdahl 1989. Catastrophic Failure Modes of Marine Structures. In: T. Wierzbicki, & N. Jones (eds.), *Structural Failures*: 463-510. New York: John Wiley
Moan, T. et al. 1993. Reliability-Based Fatigue Design Criteria for Offshore Structures Considering the Effect of Inspection and Repair. *Proc. 25th OTC*: Vol. 2: 591-599, OTC 7189. Houston: Offshore Technology Conf.
Moan, T. et al. 1993. Analytical Risk Assessment and Risk Control of Floating Platforms Subjected to Ship Collisions and Dropped Objects. *Proc. 25th OTC*: Vol. 1: 407-418. OTC 7123. Houston: Offshore Technology Conf.
Moses, F. 1986. Load and Resistance Factor Design-Recommended Practice for Approval. Final Report, API PRAC 86-22. API.
Moses, F. 1987. Load and Resistance Factor Design-Recalibration LRFD Draft Report. API PRAC 87-22. API.
Moses, F. 1990. New Directions and Research Needs in System Reliability Research. *Structural Safety* 7(2-4): 93-100.
Moses, F. 1991. A Global Approach for Reliability-based Offshore Platform Codes. *Proc. 4th Integrity of Offshore Struct. Symp.*, Glasgow. London: Elsevier.
Moses, F. & Y. W. Liu 1992. Methods of Redundancy Analysis for Offshore Platforms. *Proc. 11th OMAE*: Vol. II: 411-416. New York: ASME.
Murotsu, Y. et al. 1981. Reliability Assessment of Redundant Structure. *Proc. 3rd ICOSSAR*. Amsterdam: Elsevier.
Murotsu, Y. et al. 1984. Automatic Generation of Stochastically Dominant Modes of Structural Failure in Frame Structures. *Structural Safety* 2: 17-25.
MTD 1992. Probability-based Fatigue Inspection Planning. London: Marine Technology Directorate Limited. Publication 92/100.
Næss, A. 1989. Prediction of Extremes of Combined First-order and Slow-draft Motions of Offshore Structures. *Applied Ocean Research* 11(2): 100-110.
Næss, A. 1993. Statistics of Combined Linear and Quadratic Springing Response of a TLP in Random Waves. *Proc. 12th OMAE*: Vol. II: 201-210. New York: ASME.
NGI 1992. Geotechnical Reliability Analyses for Offshore Structures. Report 514165-6. Oslo: Norwegian Geotechnical Institute.
NKB 1987. Recommendations for Loading and Safety Regulations for Structural Design. Report No. 55E. Copenhagen: The Nordic Committee on Building Regulations.
NPD 1977. Regulation for the Design of Fixed Structures on the Norwegian Continental Shelf. Stavanger: Norwegian Petroleum Directorate.
NPD 1981. Guidelines for Safety Evaluation of Platform Conceptual Design. Stavanger: Norwegian Petroleum Directorate.
NPD 1990. Regulations Concerning Implementation and use of Risk Analysis in the Petroleum Activity with Guidelines. Stavanger: Norwegian Petroleum Directorate.
NPD 1992. Regulations Concerning Loadbearing Structures in the Petroleum Activity. Stavanger: Norwegian Petroleum Directorate.
Olufsen, A. & R.G. Bea 1990. Loading Uncertainties in Extreme Waves. *Marine Structures* 3: 237-260.
OREDA 1991: Offshore Reliability Databank. Stavanger: Statoil.
Peers, S.M.C. et al. 1992. A Rational Inspection Scheduling Philosophy for Fixed Offshore Platforms. *Proc. 11th OMAE*: Vol. II: 133-139. New York: ASME.
Rackwitz, R. & K. Schrupp 1985. Quality Control Proof Testing and Structural Reliability. *Structural Safety* 2: 239-244.
RCP 1991. Structural Reliability Analysis Package (STRUREL). Munich: RCP, Gmbh.
Rhee, H.C. & M. Kanninen 1988. Opportunities for Application of Fracture Mechanics for Offshore Structures. *Applied Mechanics Reviews* 41: 23-35.
Rodenbusch, G. 1986. Random Directional Wave Forces on Template Platforms. *Proc. 18th OTC*: Vol. 1: 147-156, OTC 5098. Houston: Offshore Technology Conf.
Sarpkaya, T. & M. Issacson 1981. *Mechanics of Wave Forces on Offshore Structures*. New York: Van Nostrand.
Sharp, J.V. et al. 1993. Review of Criteria for Inspection and Maintenance of North Sea Structures. *Proc. 12th OMAE*: Vol. II: 363-368. New York: ASME.
Shetty, N.K. 1992. System Reliability of Fixed Offshore Structures under Fatigue Deterioration. Ph.D. Thesis. London: Imperial College, Univ. of London.
Shetty, N.K. & M.J. Baker 1990. Fatigue Reliability of Tubular Joints in Offshore Structures: Crack Propagation Model. *Proc. 9th OMAE*: Vol. II: 223-230. New York: ASME.

Sigurdsson, G. et al. 1993. Probabilistic Collapse Analysis of Jackets. *This Conference*.
Skjong, R. 1985. Reliability-based Optimization of Inspection Strategies. *Proc. 4th ICOSSAR*: Vol. III: 614-618. Kobe: IASSAR.
Skjong, R. & R. Torhaug 1991. Rational Methods for Fatigue Design and Inspection Planning of Offshore Structures. *Marine Structures* 4: 381-406.
Sørensen, J.D. 1993. Reliability-Based Inspection Planning for Structural Systems. *Proc. IFIP WG 7.5*. Takamatsu, Japan: Elsevier.
Sotberg, T. & R. Bruschi 1992. Future Pipeline Design Philosophy – Framework. *Proc. 11th OMAE*: Vol. V-A: 239-248. New York: ASME.
Stahl, B. 1977. Offshore Structure Reliability Engng. *Petroleum Engineer* 49: 76-86.
Stahl, B. & J.F. Geyer 1984. Fatigue Reliability of Parallel Members Systems. *ASCE Str. Engng*. 109(8): 2307-2323.
Stahl, B. et al. 1992. Methodology for Comparison of Alternative Production Systems (MCAPS). *Proc. 24th OTC*: Vol. 1: 389-398. OTC 6915. Houston: Offshore Technology Conf.
Stewart, G. et al. 1993. Nonlinear Re-assessment of Jacket Structures under Extreme Storm Cyclic Loading. *Proc. 12th OMAE*: Vol. II: 491-541, Glasgow. New York: ASME.
Technica 1992. Hydrocarbon Leak and Ignition Database. London: E&P Forum.
Theophanatos, A. et al. 1992. Adaption of API RP2A-LRFD to the Mediterranean Sea. *Proc. 24th OTC*: Vol. 2: 529-538, OTC 6932. Houston: Offshore Technology Conf.
Thoft-Christensen, P. & Y. Murotsu 1986. *Application of Structural System Reliability Theory*. Berlin: Springer-Verlag.
Tromans, P.S. et al. 1991. A New Model for the Kinematics of Large Ocean Waves – Application as a Design Wave. *Proc. ISOPE*: 3: 64-71. Edinburgh.
Turner, R.C. et al. 1992. Towards the Worldwide Calibration of API RP2A-LRFD. *Proc. 24th OTC*: Vol. 2: 513-520, OTC 6930. Houston: Offshore Technology Conf.
Ueda, Y. & S.M.H. Rashed 1991. Modern Method of Ultimate Strength Analysis of Offshore Structures. *Int. J. Offshore and Polar Engineering* 1(1): 27-41.
Veneziano, D. et al. Vector-Process Models for System Reliability. *ASCE Engr. Mech*. 103(3).
Veritec/SINTEF 1988. Design Guidance for Offshore Steel Structures Exposed to Accidental Loads. Report 88-3172. Oslo: Veritec.
Vinnem, J.E. 1992. Offshore QRA Where are We, Where are We Heading? *Proc. ESRA Workshop on Perspectives and Trends in Safety and Reliability*, Lisbon. 19 pp.
Visser, R.C. 1992. Offshore Platform Accidents: Their Effect on Regulations and Industry Standards. *Proc. 11th OMAE*: Vol. II: 97-103. New York: ASME.
Vosikovsky, O. 1980. Effects of Stress Ratio on Fatigue Crack Growth Rates in X70 Pipeline Steel in Air and Saltwater. *J. Testing and Evaluation* 8(2): 68-73.
Vugts, J.H. & G. Edwards 1992. Offshore Structural Reliability Assessment – From Research to Reality? *Proc. 6th BOSS*. London: Imperial College, Univ. of London.
Wen, Y.K. & H.C. Chen 1987. On Fast Integration for Time Variant Structural Reliability. *Prob. Engng. Mech*. 2(3): 156-162.
Winterstein, S.R. 1988. Nonlinear Vibration Models for Extremes and Fatigue. *Engng. Mechanics, ASCE* 114(10): 1772-1790.
Wirsching, P.H. 1979. Fatigue Reliability of Welded Joints in Offshore Structures. *Proc. 11th OTC*: Vol. 1: 197-206, OTC 3380. Houston: Offshore Technology Conf.
Wirsching, P.H. 1983. Probability Based Fatigue Design Criteria for Offshore Structures. Final Report, API-PRAC Project 81.15. Dallas: API.
WOAD 1990. World-Wide Offshore Accident Databank. Oslo: Veritec.
Wu, Y.-L. & T. Moan 1989. A Structural System Reliability Analysis of Jacket using an Improved Truss Model. *Proc. 5th ICOSSAR*: 887-894, San Francisco. New York: ASCE.
Wu, Y.-L. & T. Moan 1991. An Incremental Load Formulation for Limit States in the Reliability Analysis of Nonlinear Systems. *Structural Safety* 10: 307-325.
W.S. Atkins 1992. Reliability Analysis System for Offshore Structures (RASOS). W.S. Atkins Science and Technology.
Ximenes, M.C.C. 1990. System Fatigue Reliability – A Study of Tension Leg Platform Tendon System. Ph.D. Thesis. Berkeley, CA: University of California, Berkeley.
Ximenes, M.C.C. & A.E. Mansour 1991. Fatigue System Reliability of TLP Tendons Including Inspection Updating. *Proc. 10th OMAE*: Vol. II: 203-211. New York: ASME.
Yang, Y.N. 1993. Application of Reliability Methods to Fatigue, Quality Assurance and Maintenance. The Freudenthal Lecture. *This Conference*.

Structural Safety & Reliability, Schuëller, Shinozuka & Yao (eds) © 1994 Balkema, Rotterdam, ISBN 90 5410 357 4

Frontiers of constructions and reliability of Honshu-Shikoku Bridge

Naruhito Shiraishi
Kyoto University, Japan

ABSTRACT: The Honshu-Shikoku Bridge Authority was established in 1970 to construct 3 routes of connecting bridges between Honshu (Main Island of Japan) and Shikoku (Shikoku Island) to cross over the Seto Inland Sea. The central route, called as the Seto Bridge, was opened for traffic in April, 1988. The two other remaining routes are now under construction. The eastern route is particularly high-lightened by construction of the Akashi Strait Bridge, the longest suspension bridge in the world of 1990 m long center span. The two towers of the Akashi Strait Bridge have been constructed in the summer of 1993 to continue for cable constructions now. Frontiers of construction technologies, materials, methods of structural analysis and reliability evaluation method developed in connection with the Honshu-Shikoku Bridge Construction Project are briefly described in this paper.

KEYWORDS: Construction Machinery, Prepact Concrete, PWS (Parallel Wire Strand), Aseismic Design, Wind Resistant Design, TMD (Tuned Mass Damper), Quality Control, Fatigue Test, Vibration Test

INTRODUCTION: The Honshu-Shikoku Bridge Construction was a dream of more than 100 years. As well known, there are numerous small islands scattered in the beautiful Seto Inland Sea to separate Honshu-Island and Shikoku Island. For long time people could cross this inland sea only by ferry boat. In early morning of May 11, 1955, a tragedy of ship-collision of " Shi-un Maru " and " Daisan Utaka Maru " occurred and 168 lives lost. Most of fatalities in this accident were school boys and girls who were en route for school excursion. A year before this accident, the " Toya Maru " of ferry boat for Honshu (Main Island of Japan)-Hokkaido (Northern Island of Japan) sank by typhoon and 1,155 lives lost. These two accidents became a crucial impact to plan the connecting routes for both Honshu-Hokkaido and Honshu-Shikoku by either tunnel or bridges. Soon later the Honshu-Hokkaido route was resolved to plan to construct a tunnel, while the Honshu-Shikoku route was resolved to plan long span bridges. The Seikan Tunnel between Honshu and Hokkaido (Tunnel between Aomori of Honshu side and Hakodate of Honshu side) was opened for railway traffic in March, 1988 and the Seto Bridge between Honshu and Shikoku was opened for traffic in April, 1988 so that all major four islands of Japan are now connected with each other by either tunnels or bridges.

In order to clarify engineering problems associate with construction of the Honshu-Shikoku bridges, the ministry of construction and the national railway corporation requested an inves-

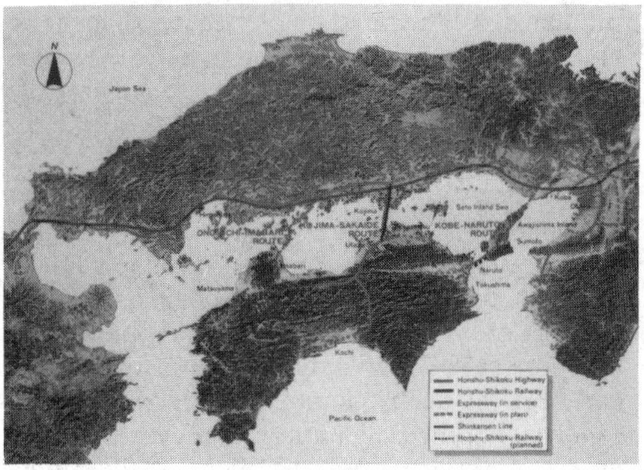

tigation for feasibility and technical possibility of Honshu-Shikoku Connecting Bridge for the Japan Society for Civil Engineers in 1963 and the final report was published in 1967 indicating that the planned Honshu-Shikoku Bridges for both roadway and railway would be possibly constructed under specially prepared design and material specifications with the design guide lines for earthquake and strong winds. And at almost same time the technical survey office of ministry of constructions was opened in Kobe in 1963 to start to investigate ground conditions as well as various environmental problems including not only natural problems but social problems. In 1970 the Honshu-Shikoku Bridge Authority was established as the public corporation responsible for construction of bridges and its maintenance of constructed bridges and affiliated facilities.

Major technical problems to be vanquished in connection with the Honshu-Shikoku Bridge Construction can be illustrated as follows;

1. To assure aerodynamic stability and safety against action of so strong winds as typhoon and to assure aseismic safety against earthquake

2. To assure travelling safety of high speed train on suspended bridges and fatigue resistant design of welded parts

3. To develop safe and assured method of construction of foundations in deep strait under high speed tidal current

4. To develop safe assured erection method for superstructures, unaffected under severe meteorological environments

5. To establish the maintenance and control system to preserve high degree of safety and invulnerable serviceability of completed Honshu-Shikoku bridge routes

Development for above mentioned technological problems is illustrated in the succeeding paragraphs.

FRONTIERS OF STRUCTURAL DESIGN OF HONSHU-SHIKOKU BRIDGES

In connection with the Honshu-Shikoku Bridge Project, fundamental structural problems are to establish the feasible design specifications for any sort of bridge structures to be employed for this project. Great achievement in this aspect has been made by number of theoretical and experimental investigations. However the most important problem among them is the **safety and reliability evaluation** of dynamic behavior of long span suspended structures such as suspension bridge and cable stayed bridge. In this paper a particular stress is placed on development of wind and earthquake resistant design of super-long span suspension bridges.

Wind Resistant Design of Honshu-Shikoku Bridges: It is well known that the Tacoma Narrows Bridge in Washington collapsed in 1940 by wind blowing of wind speed of 19 m/sec. The main cause of this failure is considered due to so-called torsional flutter of stiffening girder of H-lettered shaped cross section. Since then tremendous efforts have been paid to establish the feasible wind-resistant design method of suspension bridge. The problem mainly consists of two parts, namely (1) how to estimate the design wind speed and (2) how to design aerodynamically stable cross section of stiffening girder.

The large portion of the Seto Inland Sea where the Honshu-Shikoku bridges ought to be constructed is typhoon prone area, so that a special task committee was organized to investigate meteorological characteristics of natural strong winds statistically as well as aerodynamic behaviors of long span suspension bridge. One of the main difficulties was how to estimate the design basic wind speed for so complicate topographical typhoon prone area as the Seto Inland Sea, where a number of the most severe typhoons attacked in past sixty years.

The second problem was how to design the aerodynamically stable cross section of stiffening girder and pylons of the longest suspension bridge in the world. In order to achieve this task, a series of theoretical and experimental research works have been performed in collaboration of researchers from academia, public research institutes and industrial firms. Stress was placed on systematic collection of wind tunnel test results for various types of bluff cross sections and on analysis of aerodynamic behaviors of each cross sections by use of the most advanced and sophisticated experimental techniques. It should be noted that large scale wind flume was installed to investigate aerodynamic responses of the one tenth scaled bridge model subjected to natural turbulent winds and also the large wind tunnel of 41 m x 17.5 m (width x height) was installed in Tsukuba in 1991 in order to examine the three dimensional aerodynamic behaviors of the Akashi Strait Bridge by use of one hundredth model in both uniform and turbulent flows. The investigation for aerodynamic problems in connection with the Akashi Strait Bridge, the longest suspension

bridge in the world, is now ongoing as the most important technical subject. Almost 30 year studies on aerodynamic design of bridge structures including suspension bridge, cable-stayed bridge, truss bridge and ordinary tubular box girder bridge have clarified number of fundamental aerodynamic characteristics of bluff bodies, namely,

(1) Vortex-induced vibrations can be classified into the following three categories, (a) the flow-induced vibration associate with perfectly separated flow (termed as *perfectly separation type*), (b) the flow-induced vibration associate with separated bubble at windward edge (termed as *separation at windward edge type* or *movement excitation type*) and (c) the flow induced vibration associate with reattached bubble (termed as *reattachment type*). Vortex-induced vibrations of majority of bridge structures are the separation at windward edge type or the movement excitation type, the onset velocity of which can be so easily estimated by the recommended onset velocity and the slenderness ratio (the ratio of width to height of cross section) relation.

(2) The longer the span length of suspension bridge, the higher the tower of suspension bridge becomes. The height of tower of the Akashi Strait Bridge is approximately 300 m. The natural frequency of free vibrations of the lowest mode may vary from 0.08 Hz to 0.445 Hz depending on construction steps. And there is another particular point as for aerodynamic characteristics of this highly elevated structure, namely aerodynamic responses of bluff section for large angle of attack to correspond to action of all wind directions.

(3) In connection with construction of cable-stayed bridges a new type of aerodynamic problems has become so controversial, namely so called *rain vibrations of inclined cables*. It has been clarified that this vibration is generated and enhanced by both formation of rivulets on both upper and lower surfaces of cable and also formation of secondary axial flow at leeward stagnation region to be considered as galloping type oscillation.

(4) For aerodynamic design of various types of so flexible structures as suspension bridges and cable-stayed bridges, new method has been developed to evaluate fatigue damages due to such wind induced vibrations as vortex-induced oscillations and buffeting by turbulent flows.

Earthquake Resistant Design of Honshu-Shikoku Bridges: To design any sort of civil engineering structures, earthquake takes an important role to determine *structural system and its safety*. A task committee was organized to investigate fundamental characteristics of expected earthquake in Seto Inland Sea Area and to make the earthquake resistant design specifications for the Honshu-Shikoku Bridges. Based on previous earthquake activity records, rather remote earthquake (namely earthquake to occur at 100 to 150 km remote from the structure of concern) is thought to take a dominant role, which corresponds to *earthquake of magnitude or Richter scale 8 at the return period of 100 years.* The response spectrum method is employed for aseismic design of the Honshu Shikoku Bridges, based on the response spectrum obtained from large number of previous earthquake records and corrected to be feasible enough for design purpose to be the design acceleration response spectrum. This spectrum is used to estimate the *maximum seismic response* of structure for application of so called design earthquake force. The structural model for analysis of suspension bridge includes girder, cables, towers as well as foundations by introducing equivalent mass and spring. In order to assure feasibility of used model, the time history analysis was also made by using number of time history records of actual earthquakes.

Design of Railway Long Span Suspension Bridge: In the central route, namely the Kojima-Sakaide route, the Main Island and the Shikoku Island is now connected by Seto Bridge including three suspension bridges which are Shimotsui Seto Bridge, North Bisan Seto Bridge and South Bisan Seto Bridge. The longest is **South Bisan Seto Bridge** of center span of 1,100 m,

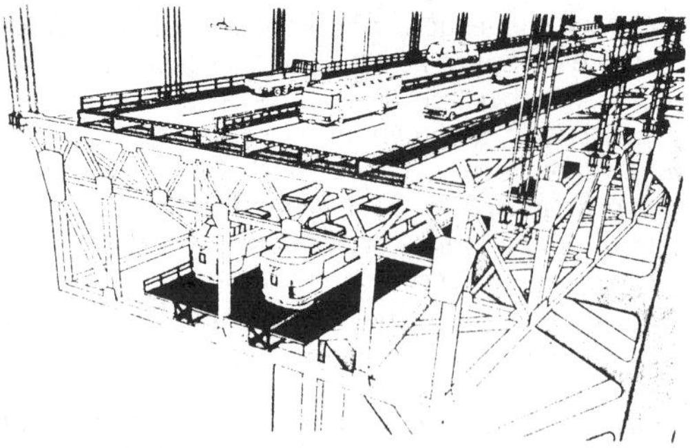

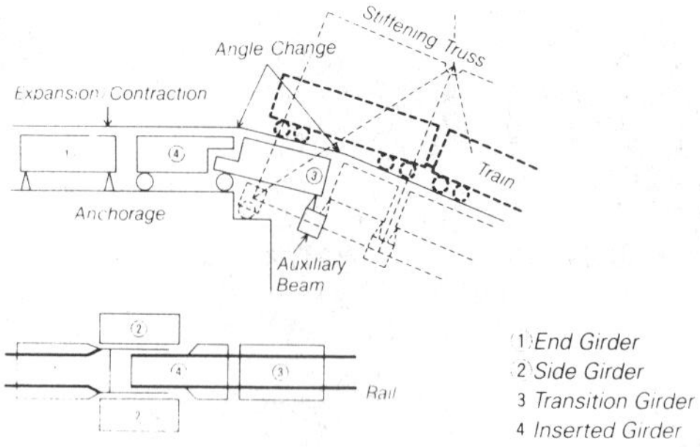

which is the *largest in the world as a highway and railroad combined bridge*. Stiffening girders of suspension bridges were so designed to provide the upper deck for four lane highway and the lower deck for railroad for the conventional line of double track. A particular attention was paid for the structure of railroad so as to be able to add the Shinkansen line of double track in future.

In connection with design of a railway long suspension bridge, two major problems have been raised at the beginning of investigations. The first problem is *travelling safety of high speed trains* at ends of girders where considerably large amount of relative angular displacement and relative longitudinal displacement are expected, since the ordinary type of suspension bridge has a stiffening girder simply supported at both ends. And the second problem is *fatigue damage of local structural parts* by large number of repetition of heavy train loadings. In order to have sufficient degree of safety of travelling train, an attempt was made to reduce the amount of relative angular and longitudinal displacements structurally by

(1) employing *continuous girder* over main and side spans,
(2) *reducing the span ratio of side to main span*, namely much shorter side span than normal length of side span required by ordinary design of suspension bridge, and
(3) introducing newly developed *transition girder expansion joints* at both ends of continuous stiffening girders.

As the consequence, the girders were connected to the towers so as to decrease the eccentricity between the tower and the cable and a protruding bracket was thus installed for fixing of hanger rope and girder to the outside of stiffening girder. A series of travelling and durability experiments for the transition girder expansion joint were performed by using high speed travelling train of 180 km/h of the Shinkansen Standards to assure its function and travelling safety. The tests indicated that the transition girder expansion joint was quite effective up for $\pm$ 75 cm of relative longitudinal displacement and 10 $^0/_{00}$ of relative angular displacement at the maximum. More precisely, three kinds of transition girder railroad expansion joints were developed as

* 1500-type of contractility capacity of $\pm$ 750 mm

* special 800 type of contractility capacity of $\pm$ 400 mm in regular condition and $\pm$ 550 mm in earthquakes

* 600-type of contractility capacity of $\pm$ 300 mm

The 1500-type was installed on both bridge ends of each of the three suspension bridges in the Seto Bridges, the special 800-type mounted on both ends of each of cable-stayed bridge and Ban-no-su viaduct trusses, and the 600-type mounted on both ends of the Yo-shima Bridge. As for railway construction, a dynamic problem was raised in connection with the Bannosu Viaduct, which was located at the most northern part of Shikoku Island. This viaduct of 2.9 km is directly connected to the South Bisan Seto Bridge. Steel box girders are used for highway and PC box girders continuous over three to five spans are used for railroad. The problem came from ground condition of soft and thicker stratum so that the supporting bedrock was only found at the Motoyo strata or granite of 70 m depth. Since 74 m deep foundation piles were required and the viaduct itself was highly elevated, train running characteristics were investigated when an earthquake occurred. Based on the dynamic analysis of whole viaduct structures including ground condition, the span length was so determined not to exceed the derailment limit of railroad or to differentiate the natural frequencies of viaduct from natural frequencies of running train.

Fatigue Proof Design of Welded Parts: Main strategy for design of Kojima-Sakaide route of Honshu-Shikoku Bridge was to alleviate the dead load or own weight of structures by employing high strength steels since the bridge was required to carry both highway load and railway load. It is well known that heat-treated high strength carbon steels used for Honshu-Shikoku Bridges are mechanically so sensitive for various types of notch, temperature and also welding that an investigation was carried out to clarify fatigue limit at various types of welded parts and to lead to the specification for *fatigue proof design of welded members*. Because of number of new type of welded structural parts to be used for this national project, a large scale fatigue testing machine of maximum loading amplitude of 400 t was used for this purpose. The theme of this research project were following;

(1) experimental evaluation of fatigue limit of various welded joints
(2) allowance of initial defects
(3) effective detective method of initial defects
(4) development of welding method to increase fatigue limit
(5) development of welding materials for high strength carbon steels

A particular stress was placed to develop a new *non-destructive automatic detector by use of supersonic waves*. It is interesting to note development of supersonic wave detector. In order assure detective capability of detectors for various types of initial defects, it was necessary to produce large number of different initial defects for welded parts and this leads to clarify various causes of initial defects at welding. Consequently development of new non-destructive detector was so useful to develop fatigue proof welding method by eliminating clarified causes for initial defects.

FRONTIERS OF CONSTRUCTION TECHNOLOGY OF HONSHU-SHIKOKU BRIDGE

For construction of long span suspension bridge and cable-stayed bridge, safe and feasible construction methods for such substructures as foundation or piers for towers and anchorage block takes also as important role for overall safety of structures as those for superstructures. In connection with the Honshu-Shikoku Bridge construction project, a huge mass of undersea concrete constructions having to be completed in a considerably short limited period, a series of land experiments were repeatedly performed as well as undersea experiments, placing a stress on prepacked concrete construction methods which were used in actual works.
The geological structure, for example, in the strait for the Seto Bridges comprises basically granite, particularly weathered granite exposed over a region from Mt. Washizu of Honshu side to the offshore in the south of Mitsugo Island. A thick layer of sedimentary layer covers a range in the northern part of Shikoku side. The sea depth for the route varies from 10 m to 50 m at maximum and tidal current reaches 5 knots at maximum. And furthermore the site is located at the major international navigation route with approximately 1,000 ships passing per day and crowded with fishery. Thus the so called rigid foundation was employed except the pile foundation in Bannosu area. There are a total of 23 foundations for major 6 bridges in the central route, namely the Seto Bridges, including 5 foundations on land. Out of 18 undersea foundations, 7 foundations were constructed by means of the cofferdam method and 11 by means of the laying-down caisson method.

Excavation for Foundations: As the preliminary works for foundations, surface layer was ought to be removed. And for this purpose the detonating fuse system was used for submerged blasting. The OD drill, mounted on the self-elevating platform (SEP), was used to drill holes in which PVC pipes were installed for explosives. The deck of SEP " Banjaku " is 47 m x 35 m at the center of which there is an opening of 21 m x 10 m, to be used for offshore work up to a depth of 50 m. The *OD drilling machine* is a pneumatic rotary impact type machine with dual tubes. The outer tube (drill tube) normally stops in drill hole in the sedimentary layer at rock base and then the impact type inner tube (drill rod) is used to drill hole. Also the *Wellman drilling machine*, a hydraulically rotary drilling machine, was used for sedimentary and thick weathered rock layer.
Large scale submerged blasting was prohibited at some of construction sites because of an oil refinery plant located within 400 m to 1 km. However, note that the foundation was the greatest in scale, a water depth of 50 m, an excavation dimensions on the bottom of 85 m x 74 m with volume of about 600,000 m^3 in undersea excavation. A method for this purpose was developed to the successful result, namely *one second delayed firing method*. At a founda-

tion construction for sedimentary soil and rock deposited on the seabed as thick as about 30 m, the overburden undersea blasting was used without removal of surface layer.
After underwater blasting, seabed was excavated by grab dredger. The used **grab dredger** is one of the largest and most powerful Japanese sand grabbing bucket(13 m wide when opened and weight of 125 t), excavating 25 m^3. By introduction of this powerful grab dredger, tremendous amount of seabed soil was securely excavated even in fast tidal current in shorter period. Besides, even hard bedrock was found to be possibly excavated by this large powerful bucket claws, which greatly effected on blasting technology proceeding to excavation. In other words, it was only necessary to create through-cracks in bedrock to be excavated by large pitches of explosives charged in holes.

Concrete Works:
In order to proceed major concrete work of Honshu Shikoku Bridge construction at sea, a *mortar plant barge*, 90 m length, 32 m width, 7.5 m depth, 4.5 m draft and 11,500 t displacement, was built, installed with all machinery and supplies. The mortar plant consists of 3 lines of the plants of an output of 2,000 lit/min each, the 2 lines of which are normally used while leaving the other line as backup. Twenty four mortar pumps are incorporated, among which 20 pumps are used at 200 lit/min x 20 units = 4,000 lit/min, leaving the other 4 pumps for backup and cleaning, etc.
The *prepacked concrete method* was used for the concrete work of undersea foundations for major concrete works of Honshu-Shikoku Bridge construction for secure casting of large amount of high quality concrete. In concrete works, particular attentions were placed on (1) *quality of aggregates* as well as mortar, (2) *workability or liquidity* of mortar so that no spacing between coarse aggregates should be left, (3) *robustness*, and (4) *thermal stress control*. There are some trade-offs between above required characteristics so that the optimum proportion of various mixtures were sought through experiments.
A fleet of mortar injection barges comprise a main barge of mortar plant together with material supply barge loaded with cement, sand, water, etc. and work barge loaded with overflow water treating facilities. And as much as 12,000 tons of mortar material was so used per day that cement and sand were transported using cement tanker of 2,000 to 3,000 tons and barge of 3,000 m^3. Since mortar injection work continued to a maximum of 3 days and nights, 150 engineers and 300 workers a day were required. Note that there is no more important than education and training of personnel for operation of machinery to prevent failure.

Large Block Erection of Superstructures:
In a last few decades, large block erection method has been frequently used in Japan for construction of bridges of various types near sea coast or wherever floating crane was able to access to the construction site. A majority of bridges in Honshu-Shikoku Connecting Route being constructed in Seto Inland Sea, the large block erection method is particularly effective as erection method for superstructures. *Advantages to use large erection method* are for example summarized as follows;

(a) alleviation and simplification of construction works at the sites
(b) reduction of period required for construction work at the sites
(c) reduction of the number of fastening or joints at the sites
(d) simplification of construction facilities and machinery at the sites
(e) *assurance of safety* due to progression of quality and *accuracy of constructed structures*

This large block erection method is so effective if construction plan be made carefully and all preceding works be made without any delay.
For construction of the Seto Bridges, the large block erection method was frequently employed for construction of various types of structures including laying-down caissons, parts of pylons of suspension bridges and those of cable-stayed bridges, girders of cable-stayed bridges and those of truss bridges. The maximum weight of one block at construction of Seto Bridges reached to 6,100 tons, carried by two floating crane boats. However it should be noted that *disad-*

vantage of this large block erection method was to be largely influenced by *weather conditions*.

Vibration Attenuator:

For a certain period of construction of suspension bridges, towers stand freely and it is so flexible like a cantilever that it vibrates laterally in longitudinal direction of bridge axis. As the results of wind tunnel tests, amplitude of vortex induced vibration in deflectional mode is expected to reach to several meters at the top of tower and amplitude in torsional mode to reach to more than 5^{0}. For Seto Bridges, two kinds of attenuators were used, namely (1) a combination of an oil damper and a weight and (2) a sliding block system. Both the oil damper system and the sliding block system were used in both South Bisan Bridge and North Bisan Bridge. For construction of the **Akashi Strait Bridge**, another counter-measure system for vortex-induced vibrations is now going to be used, namely *Tuned Mass Damper System*.

Cables and Cable Erection:

There are two different kinds of bridges constructed for Honshu-Shikoku Bridges of which cables are used as main structural members, namely suspension bridge and cable-stayed bridge.

Cable is the most important structural and mechanical part of suspension bridge to carry both its own weight and all sort of live load to transmit them to foundations via pylons. The S.B.B. (South Bisan Bridge) cable has the largest diameter of 1,070 mm among Seto Bridges and receives the effective axial tensile force of 44,700 tons. The wires of 4 mm in diameter were wrapped around the circumference of the cable to be painted to be corrosion proof for long period. Cable bands were installed at the interval of about 13 m to hang the stiffening girder by hanger ropes with saddles. For both S.B.B. (South Bisan Bridge) and N.B.B. (North Bisan Bridge), main cables consist of prefabricated parallel wire strands (termed as the *PS method*), while for S.S.B. (Shimotsui Seto Bridge) the aerial spinning method (termed as the *AS method*) was used. In the PS method, 127 wires of 5 mm diameter were preliminary bundled to form a strand in the factory and the strand was rolled around a reel to convey to the site. 271 strands for S.B.B. and 234 strands for N.B.B., respectively, were bundled to make a main cable. Thus the number of wires is 127/strand x 271 strands = 34,417 for S.B.B. And at the site, four wires were drawn out at the same time and 552 wires were bundled as one strand. For S.S.B.(Shimotsui Seto Bridge), 44 strands were bundled to be a main cable, thus the number of wires per cable is 552/strand x 44 = 24,288. Note a small mechanical difference between the PS method and the AS method but significant difference between two methods for total number of strands and it resulted in difference for strand fixing area at the anchorage. This was one of the reasons that the aerial spinning method was employed for the Shimotsui Seto Bridge, since cable anchoring area was so restricted at the Honshu side.

As already known, cables of the cable-stayed bridges are very different from those for suspension bridges. For cable-stayed bridge, each cable was manufactured in the factory according to the design requirement to transport to the construction site. A cable for cable-stayed bridge consists of the required number of galvanized steel wires of 7 mm in diameter, sockets (*Hi-Am anchor*) and PE pipes (*polyethylene covering tube*). Galvanized steel wires were made of heat treated extended wires. Each wire was so cut to keep the designed length precisely and a bundle of predetermined number of wires was formed in approximately parallel and wrapped with a spacer strand in a spiral form, covered with a polyethylene pipe. Both ends of wires, heat treated, were fixed together with molding epoxy resin, zinc powder and steel ball in the sockets. After molding the cable, it was heated, filled with tar-epoxy in the PE pipe. The outer surface of the socket was coated with inorganic zinc rich primer.

For construction of Honshu-Shikoku suspension bridges, a particular attention was placed on the cable erection method since bridge structures were so large and on the other side the period for construction was so limited. Cables were successfully erected by the PS method for the first time in the world for suspension bridges of more than 1,000 m of center span length by means of well organized quality control system for construction. For cables of cable-stayed bridges, one should note that variation of induced tensile forces for cables was expected to be much larger to compare with the variation of induced tensile forces in cables of suspension bridges and much attention was therefore required on fatigue resistance of each cable at fixing point of socket. In addition to the above mentioned problems, the Seto Bridge was required to function as both highway and railway bridge, so the fatigue problem of cables was of the primary importance. This was the main reason that Hi-Am anchor socket was extensively employed for cable-stayed bridges of Honshu-Shikoku Bridges.

TRAFFIC CONTROL & MAINTENANCE SYSTEM

Tests for Actual Bridges:
In design stage, number of mechanical parameters have been assumed to be certain values based on large amount of documented data for previously constructed bridges in Japan as well as those in many foreign countries. For example, *wind resistant safety* of Honshu-Shikoku bridges is largely dependent on damping capacity of both deflectional and torsional modes of vibrations. Thus, the vibration test was performed by use of newly developed vibrators to examine the vibrational characteristics of actual bridges. The first test was performed for the **Ohnaruto Bridge** (suspension bridge between Shikoku Island and Awaji Island in the eastern route) and in the central route the tests were made for the **South Bisan Seto Bridge** of suspension bridge and for **Hitsuishi Island Bridge** of cable-stayed bridge. Displacements, natural frequencies and damping capacity were measured by means of number of sensors such as accelerometers installed at bridges and data were processed on line. The displacement of girder was measured at point L/2 (central point) and L/4 (quarter point)of center span by use of an optical displacement meter.
The Seto Bridge was designed and constructed not only as express highway bridge but also high-speed railway bridge by introduction of *transition girder expansion joint*. In order to examine the function of transition girder expansion joint and to clarify operational characteristics of train running on railroad on bridges, the *train running tests* were undertaken to investigate particularly the *comfort and safety against derailment*. Four trains were used and the heaviest train was 1,000 tons. The test rains run on railroad at the maximum speed of 120 km/hour. The results of this test confirmed the *safety for train travelling* on this large size bridges.

Traffic Control & Maintenance System:
For traffic control and maintenance for Seto Bridges, the following systems were developed;
Information Collection System
Information Processing & Display System
Information Supply System
Bridge Behavior Observation System
Inspection Vehicle
Security System

The information collection facility consists of (1) a traffic counter to measure traffic capacity, car speed and exclusive ratio, (2) weather observation system to measure rain and snowfall, wind direction as well as wind speed, the field of vision and freezing on the road, (3) camera to monitor vehicle flows and traffic accidents and emergency telephones to obtain information of accidents or troubles and (4) seismometer. The information from such sensors is collected at the Central Station of Information Processing and Display System, processed and recorded on the graphic panel, weather monitoring panel or printer to be used for traffic control.
Any kind of highway information that have been collected and processed is supplied to drivers on the highway by use of highway information display and communication system, for example, to introduce the car velocity regulation depending on the weather conditions.

The Seto Bridge consists of three suspension bridges, two cable-stayed bridges and some other types of bridges to span approximately 10 km between Honshu and Shikoku Island. The bridge behavior observation system was established to observe mainly dynamic behavior of bridges under action of earthquakes and typhoons. The quantitative records of various sensors such as seismometers, acceleration meters, anemometers and displacement meters are collected and transmitted to the Sakaide Operation Office.

CONCLUDING REMARKS

In this paper the frontiers of Honshu-Shikoku Bridge constructions are briefly reviewed and some fundamental problems investigated in connection with this project are presented.

Stresses were placed on safety and reliability at planning and design stages as well as those at construction stage. During last decades there were significant progress in fabrication of structural steels including cable fabrication and welding technologies. Computer-aided construction technologies have been also progressed and it is interesting to note that introduction of new powerful construction machinery changes construction method itself and development of defect detector revealed number of causes to produce defects. However there are still unsolved problems for design and construction of so highly elevated and so long structures as suspended structures. A typical example is wind-induced vibration problems of the *Akashi Strait Bridge*. In order to ascertain aerodynamic safety of the Akashi Strait Bridge, an improved wind tunnel test is performed by use of the super large size wind tunnel constructed in March, 1991 at Tsukuba, Japan.

The Honshu-Shikoku Bridge Authority devotes its major efforts to construction of the Akashi Strait Bridge of 3,910 m (960 + 1990 + 960), the longest suspension bridge in the world, and construction of the western route or the Imabari route. By the end of the 20 th century, three routes will have been connected by a number of newly constructed bridges. In connection with the Honshu-Shikoku Bridge Construction, a number of new technologies have been developed based on almost 30 year investigations in various fields which resulted in successful construction of more safe and more reliable bridge structures. New technologies developed for this Honshu-Shikoku Bridge Construction Project are valuable not only for bridge engineering but also for other fields of engineerings.

ACKNOWLEDGEMENT: The author would like to express his sincere thanks for cooperation of the Honshu-Shikoku Bridge Authority for preparation of this paper. Particularly the author would like to appreciate Mr. M. Yasuda, the former chief of Design Section of the Authority (currently the Director, Tarumi Construction Office) for his invaluable comments on this manuscript.

REFERENCES:

1) Ishiyama,S.(1988); *Technical Development and Construction of Honshu-Shikoku Bridges*, Japan Construction Information Center, Vol. 3, No. 1, 1988, pp 30-36 (in Japanese)

2) **The Bridge and Foundation Engineering**, Special Issue on Honshu-Shikoku Bridges, Vol. 18, No. 8, 1984 (in Japanese)

3) **The Seto Ohashi Bridge** (1988), published by the Honshu-Shikoku Bridge Authority (in Japanese and in English)

4) **The Seto Ohashi Bridge and Foundation** (1988), published by the Honshu-Shikoku Bridge Authority (in Japanese and in English)

Plenary panel discussion: How safe is safe enough?

Structural Safety & Reliability, Schuëller, Shinozuka & Yao (eds) © 1994 Balkema, Rotterdam, ISBN 90 5410 357 4

Plenary panel discussion on 'How safe is safe enough?'

H.G. Payer
HSVA, Hamburg, Germany

H. Huppmann
Munich Reinsurance, Germany

Chr. Jochum
Hoechst, Frankfurt, Germany

H.O. Madsen
Det Norske Veritas, Copenhagen, Denmark

K. Nittinger
Lufthansa, Hamburg, Germany

H. Shibata
Science Council of Japan, Tokyo, Japan

W. Wild
DARA, Bonn, Germany

H.-J. Wingender
NUKEM, Altenau, Germany

H. G. PAYER, HSVA, Hamburg

Ladies and Gentlemen: The topic of this panel is "How safe is safe enough?". Its objective is to make a connection between the possibilities of reliability and safety analysis today and the application in practice. When we face our future, here on our little planet, we have to face two facts: one that the population continues to increase and second - and that is a very positive one - that the average living standard will probably rise - it will have to rise. I have my personal doubts that the living standard of the developed world will rise any further but on an average it will have to increase, it will have to be improved to stop endless suffering and to stop poverty, one of the main causes for instability in the world. These two facts lead to the requirement to carry out big projects - big projects in the field of energy generation, of transportation, of housing, just to mention a few. To master the future we have to take certain risks. Today we are quite aware that there is a trend in the public of the developed world to oppose any project involving any risk. This is an attitude which opposes development and which can lead to very unreasonable decisions. There is a saying "celibacy is not hereditary", I hope you understand what I mean. If we are not prepared for risk we will probably die. We have to distinguish between the personal risk a person is prepared to accept and risk he is exposed to as part of society. Here in Tyrol this can be very well illustrated. The rock climber today exposes himself to very high risks. In fact he does not even rely on the strength of the rope anymore, the free climber just relies on himself. Our willingness to step into a car is often cited as an example how easily we do accept personal risks. The reason for the disparity in the willingness to accept personal risks and the opposition to risk on a broad basis, I think, lies in the fact that man likes to control his own fate. With big projects like nuclear plants or a large tanker, man has the feeling that he is exposed to an unknown risk. The press very often strengthens this feeling. Man is not aware of the very careful planning, of the detailed analysis, of the efforts in reliability and safety, which generally form the background of big projects today. Therefore it is very important to make the public aware of all these efforts that lie behind our big projects. We have the risks under control.

This panel consits of experts from different fields- decision makers and not necessarily experts in reliability analysis. This I think will be the interesting part that you, the audience, are the experts and up here are the decision makers which will use your numbers to make important decisions, which form the link between theory and practical application.

H. HUPPMANN, Munich Reinsurance

The insurance industry is a network - a network of direct insurers and reinsurers, and this network is the professional risk carrier which enables an individual and an enterprise to transfer risks to another subject, to the insurance industry. The risks are financially transferred. Your life if you have a life insurance, your property if you have a property insurance, the risk of erecting a plant, a hydro power station, a nuclear power station, the risk of running an enterprise, your liability risk with respect to the environment or to individuals, these risks are always assumed as financial burdens, and only this financial burden is carried by the insurance industry. This is important because only in this sense is the insurance industry a partner for you, the individuals who are insured or who may insure themselves, or for an enterprise which wants to avoid unlimited or limited risks.

In this risk transfer, which is one of the main objectives of risk management, the insurance industry - direct insurers and reinsurers - have similar problems as the ones you are aware of. The know-how is limited, the data are insufficient. We can only rely on past experience, and we must find ways to look into the future But we are engineers or economists, not prophets. This means that the insurance industry too is dependent on limited know-how, for instance when a bridge or a railway system is built. Yesterday in a session, the way my colleague, Mr. Shibata, presented probabilistic or deterministic assessments for the risk of high-speed trains to us was very impressive. Afterwards his colleague from Japan described the risk of the innovative magnetically levitated train, which presently is being tested. Both assessments have to rely on past experience like we do in the insurance industry. We know that this experience from the past is insufficient.
In our assessment of future risks we have to rely on limited experience, and this is traditionally done by applying a safety margin. The engineers do know the safety factors. However, in applying them we have the same difficulties you have. We live in a world of economic competition, and due to this competition it is very difficult to get the technically required safety margins agreed in the process of risk transfer that means in the premium rates, deductibles and other conditions in the insurance of industrial risks.

Life insurers have excellent statistical data available. Health insurers can adjust the premium rates for the risk of illness of an individual. An engineering insurer, on the other hand, who insures a large hydropower station for ten years through erection and test operation carries this risk from the very beginning and has no chance of adjusting the terms and rates during this ten-year period. That is the reason why we apply all the methods you are discussing, knowing that these methods such as probabilistic and deterministic risk assessment, safety analysis, reliability analysis are limited, because they are inherently connected with failure rates based on our limited experience from the past or on our lack of data. But we have to insure the new technologies at present for the future.

And as you know, many risks are even then hazardous to mankind, to you, the insured, and to the insurance industry. Let us have a look at one actual example: The risks of natural hazards or "acts of god". The fear that man-made threats to our habitat, the globe, increase the risk of natural hazards. Typhoons, windstorms - as far as I know not yet earthquakes (they are mainly acts of god, not acts of human beings) - are an increasing risk and consequently losses a tremendous financial burden. They are an enormous hazard for the population and the enterprises living and working in the related areas. It is very difficult to set up models to quantify the risks of natural hazards for the insurance industry in a changing and developing world. Growing population, increasing standards of living, the density of population in urban areas like the megapolis of Los Angeles or Tokyo, the expansion of the insured values of assets all contribute to the high exposure. It is possibly known to you that the years 1991 and 1992 were the most severe years for the insurance industry worldwide, threatened and burdened by losses due to natural hazards.

The insurance industry has only a limited capacity, hence limited possibilities to carry the risks of the natural hazards. These capacities are limited because in the public the understanding of the mechanism of risk transfer to the insurance industry is so limited, and consequently the technically required premium rating does not find acceptance. The public is only aware of splendid insurance premises and therefrom simply assumes that profit margins must be high. When, however, you read the Financial Times about the London market, you learn that Lloyd's has suffered severe losses through three consecutive years. As a prediction nobody would have considered such negative results realistic. This shows you that the risk of the insurance industry being the professional risk carrier in the world, is a very real and serve risk which can only be very insufficiently calculated. Therefore I can only call for sufficient safety margins in our assumptions about risks and exposures, determined in the way engineers are educated to act in their daily design calculations and construction practice in order to be able to overcome any uncertainties and grey zones.

I, on behalf of the insurance industry, cannot only rely on the different quantitative analyses. They can only serve for the purpose of orientations. A leading steam turbine manufacturer has calculated the probability of a total loss of a modern steam turbine in a nuclear power station to be 10^{-9} per year on total loss in one billion years of operation. In the insurance industry we evaluate the actual data of all major losses world-wide. Total losses of steam turbines are published in the newspapers, we evaluate this information. The actual probability of a total loss in a steam turbine varies between 0.4 to 0.5 x 10^{-3} per year. This is an example for the difference between theoretically evaluated probabilistic risk assessment and the statistically true values. The insurance industry must rely on facts. If we are not guided by statistics, but misled by individual, therefore limited, experience, we may accept risks in a careless way with insufficient terms and conditions. Underwriting losses to the insurance industry directly influences the capacity of the world market for risk transfer. We finally would loose the ability to operate successfully, to work professionally at present and for the future.

For us a risk is a financial threat. For any insured risk we must be able to pay for the maximum loss possible, and nobody cares if our original assessment of the probable maximum loss was too low. Neither does the low probability for a catastrophic loss help the inurance industry, when it occurs it has to be indemnified in cash. No conference statements about Poisson, lognormal or Bayesian distributions nor any explanations with hindsight, why these distributions or parameter values applied did not prove to be faultless tools in our daily work of risk assessment, reduce our responsibility towards the insured client and for the stability and continuity of the insurance industry. But we must use the methods you develop for our comparison of different but similar risks. We can make use of these sohisticated methods to reduce our lack of data, to make the gap between well-known technological risks and innovative new ones smaller. Therefore we are very interested in the progress made by your efforts.

CHR. JOCHUM, Hoechst, Frankfurt

The Chemical Industry is one of the safest industries. The frequency rate of accidents is declining since decades and is far below industry average. This is the result of a risk management, which may be summarized by the few phrases on the chart (Fig. 1):

First of all we try to minimize hazards by using chemicals with the lowest possible hazards in the lowest possible amounts and by choosing chemical reactions which are fail safe, that means which are tolerant to anomalous conditions. In many cases we do not come close enough to that aim of intrinsic safe processes. However, the techniques are available to control hazardous chemicals and reactions.

Due to the big variety of chemicals and reaction conditions the design of our equipment and of our processes has to be tailor made. Safety cannot be added up at the end, it has to be integrated into research, design and day to day operations.

However, a low probability of failure does

RISK MANAGEMENT

- Minimize Hazards
- Minimize Probability of Failures

  by integrating safety into
  - *research*
  - *design*
  - *operations*

Dr. Christian Jochum
Hoechst AG, Germany

Fig. 1

SAFETY IS ON OPEN SYSTEM - BUT ECONOMY IS NOT

- We will never reach Zero Incidents
- Our State of the Art will never meet public expectations
- We need a dialogue to close that gap

Dr. Christian Jochum
Hoechst AG, Germany

Fig. 2

not mean that nothing goes wrong. Minor emissions due e.g. to leakages, spills, scrubber malfunctions or operator errors are the problems which are typical for the chemical industry. Therefore we provide the techniques and the organization to mitigate the effects.

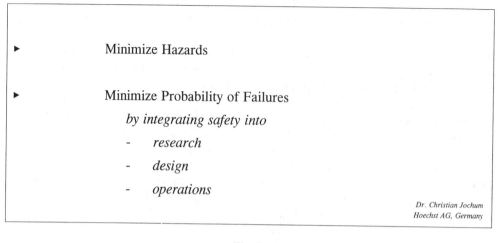

Major incidents are seldom, but my company and the neighbours had to learn earlier this year that they may happen and that the laws of probability do not exclude even a cluster of those events. We all experienced again that public perception does not discriminate between small controllable incidents and major incidents. Quite contrary to people's attitude towards lifestyle risks or even the risk e.g. mining, heavy industry, etc., in chemical industry for many of them only zero incidents means safe enough. This is our ultimate goal too. We know, that we will come closer to it, but we will never reach it. Therefore, especially in metropolitan areas with more service businesses than producing industries, the level of safety which we accept as state-of-the-art will not be safe enough for our neighbours.

This conflict becomes even worse, because we cannot exclude economic considerations from our safety concepts. A complete secondary containment around the chemical unit certainly reduces the probability of emissions. However, incidents may still happen and our expenses may become so high that we cannot survive in a global market. Shutting down a plant for economic reasons will result in "zero incidents", but this is not the solution.
Safety is an open system, but economy is not (Fig. 2). As technology proceeds we will have safer plants. We will concentrate our resources on reducing the probability of major incidents, but they still will happen. We will also reduce the number of minor incidents, but here we have to consider the economic feasibility, too. We will inevitably stay behind the expectations of the environmentalists and other opinion leaders. Therefore we have to find a compromise - not on our own, but in an open dialogue with all members of society. This is one of the main components of "Responsibility Care", a worldwide initiative of chemical industry. Originally coming from North America and the United Kingdom , it had a difficult start in the different social and political environment especially in Germany. I believe that the incidents which we had will bring my company into a leadership for that dialogue in Germany. The start has been made. We installed a neighbourhood council a few weeks ago. We invited even our strongest opponents among the green groups, so this is not a coffee-"klatsch". But I believe this is the only way to come closer to a consensus on this panel's questions: "How safe is safe enough?".

H.O. MADSEN, DNV, Copenhagen

I will address the topic - how safe is safe enough - from the point of view of the offshore industry, that is oil and gas exploration and exploitation. The background of our organisation - Det Norske Veritas - is mainly anchored in the North Sea. What we have experienced in the North Sea, we however experience in other areas around in the world: in Canada, in the Adriatic Sea, to some extent in the Mexican Gulf and also in offshore Africa and in the Far East Asia.

For operations in the North Sea there is now a significant experience on qualitative and quantitative risk assessment available. This work began following an accident - as we very often see in other sectors that an accident triggers new developments and requirements. The accident was the blowout of a platform with significant environmental damage. At that time there were studies on structured risk analyses by oil companies and authorities, but nothing was mandatory. Following a second serious accident, the capsizing of the Alexander Kielland platform in March 1980 with 123 lives lost, such risk analyses became mandatory. Concepts like safety evaluation, progressive limit states, etc. are direct results of these accidents.

In the Norwegian Sector these concepts became standard practice. The companies and the authorities including the certifying authorities learned from the first analyses and assisted in further developing the method. The development also took place - maybe at a little slower pace - in the UK sector of the North Sea and in North America. There were no requirements for quantification of risk nor a formal regulatory process related to this in these areas.

In the UK sector of the North Sea a third accident - the Piper Alpha accident in July 1988 with 163 lives lost - resulted in the Cullen report and subsequently in requirements for a Safety Case. Today in the North Sea there are requirements for qualitative risk analyses and the certifying authorities are involved in evaluating these.

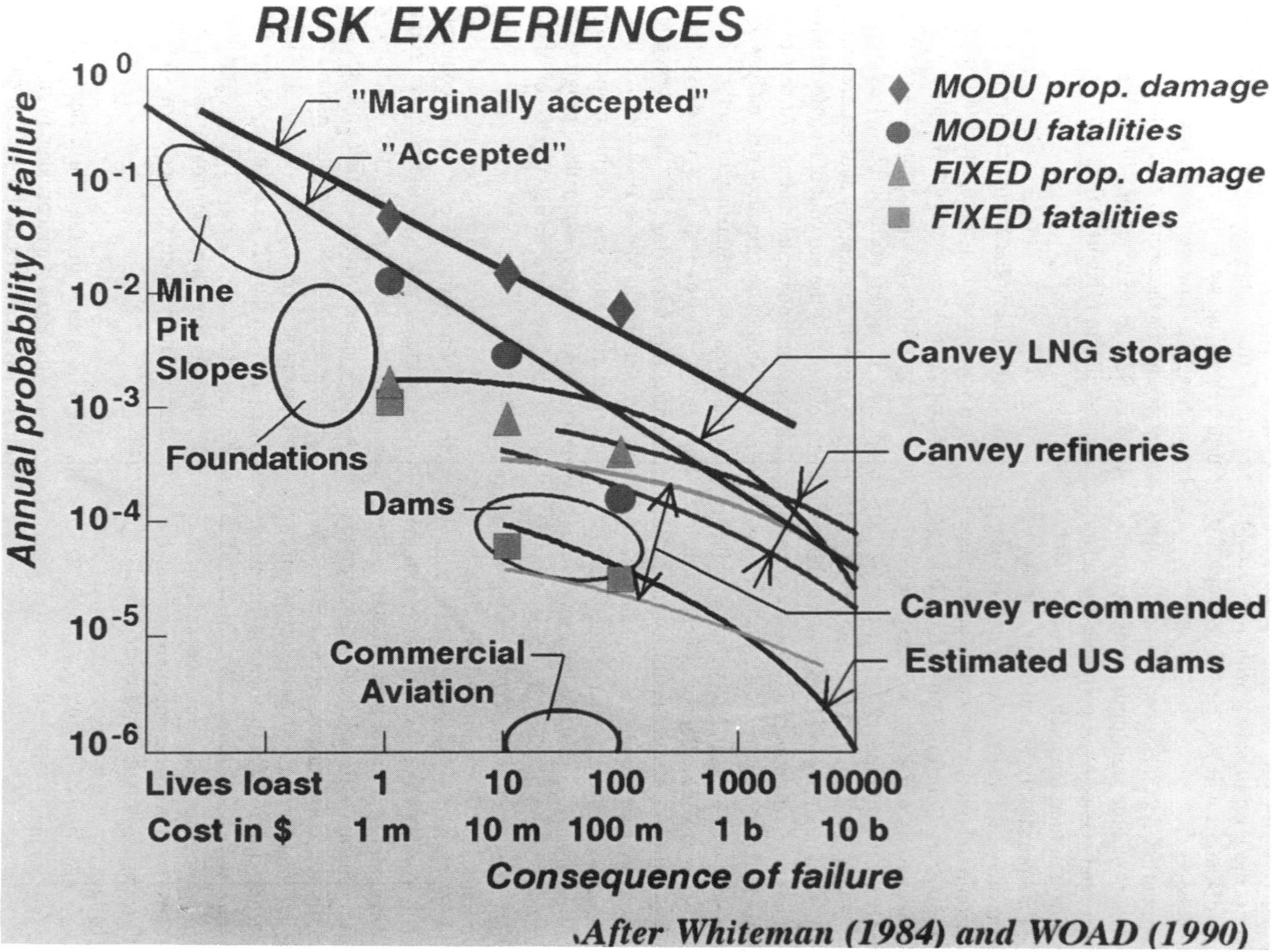
RISK EXPERIENCES
Annual probability of failure
10 0
10-1
10-2
10-3
10-4
10-5
10-6
"Marginally accepted"
"Accepted"
Mine
Pit
Slopes
Foundations
Dams
Commercial
Aviation
MODU prop. damage
MODU fatalities
FIXED prop. damage
FIXED fatalities
Canvey LNG storage
Canvey refineries
Canvey recommended
Estimated US dams
Lives loast 1 10 100 1000 10000
Cost in $ 1 m 10 m 100 m 1 b 10 b
Consequence of failure
After Whiteman (1984) and WOAD (1990)

Fig. 3

The risk analyses are prepared by the oil companies or their consultants. It is very important that a fair and consistent evaluation of these analyses takes place. A standard must therefore exist for risk analysis, and the whole industry is working together to achieve this. This also applies to common data bases for accident and equipment failure rate data.

One tool is the OHRAT - offshore hazard and risk analysis tool - consisting of a number of computer programs. This development is supported by many oil companies, consultants, the authorities, etc., and allows an easy access to risk analyses and consequence analysis, so that these can be used actively as management tool. Another important tool - developed over the last 10-15 years - is the OREDA data base, which gives statistics on failure rates for important types of equipment like generators, heat exchangers, pumps, valves, subsea equipment, etc.. A lot of data from many years of service experience is now included, and the data base is recently also being used by companies operating onshore plants.

A third data base is the world offshore accident data base - WOAD. Fig. 3 shows the accident rate experience for various types of industries including the offshore industry. Both fixed and mobile offshore drilling units - MODUs - are included. The x-axis has a scale for lives lost as well as a scale for damage cost. There is no intentional relative scaling of the two scales. All accidents from 1990 and 20 years back are included, so the data base is quite large. It is observed that the risk for floating systems is larger than for fixed systems. For floating systems a number of accidents are related to temporary phases during transit and installation.

The figure shows two curves marked "mariginally acceptable" and "accepted" respectively. They show what is an acceptable - or rather tolerable - risk level. The philosophy is that a single curve is not given but rather an acceptable range is provided. Some years ago the authoroties in Norway - the Norwegian Pretoleum Directorate - gave only one curve. The experience with this was not very good because all efforts in risk analyses were spent on reaching the numbers given on this curve and the general quality of the risk analyses was not good enough. By having an acceptable range a different attitude to risk analyses has emerged. The main purpose of a risk analysis is to *analyse* risk. This should further take place at an early stage in a project where the results of the risk analysis can have effect on outline, design, etc., and not just be a number calculated when the design is finalised. Within the acceptable area it is also important that risk reduction is sudied in the form of cost benefit analyses. With this new philosophy risk analysis has been an important and useful tool in the design of offshore structures. The operators submit their safety cases - this is the name used in the UK sector - to the authorities and this is followed by a discussion of the risk level and risk reduction possibilities.

The offshore oil and gas industry is somewhat different from other industries in that an accident is not in "anybody's backyard". The activity is far away from populated areas and the consequences of an oil spill is generally not so severe. In this sense the industry may not have the same pressure from society about risk, but the industry has to show some self discipline. The industry has, however, for a long time realised that it pays to have a high safety level. Workers accident rates are thus much lower for offshore installations than for similar installations onshore. When exploration moves to more Northern areas towards the Arctic, the environmental damage from an accident becomes much larger and maybe irreversible. Environmentalist groups and others will then put stronger requirements to an even higher safety level, but this will also be in the interest of the oil companies.

One last topic which is very important, for offshore structures as well as for other structures, is what to do with existing

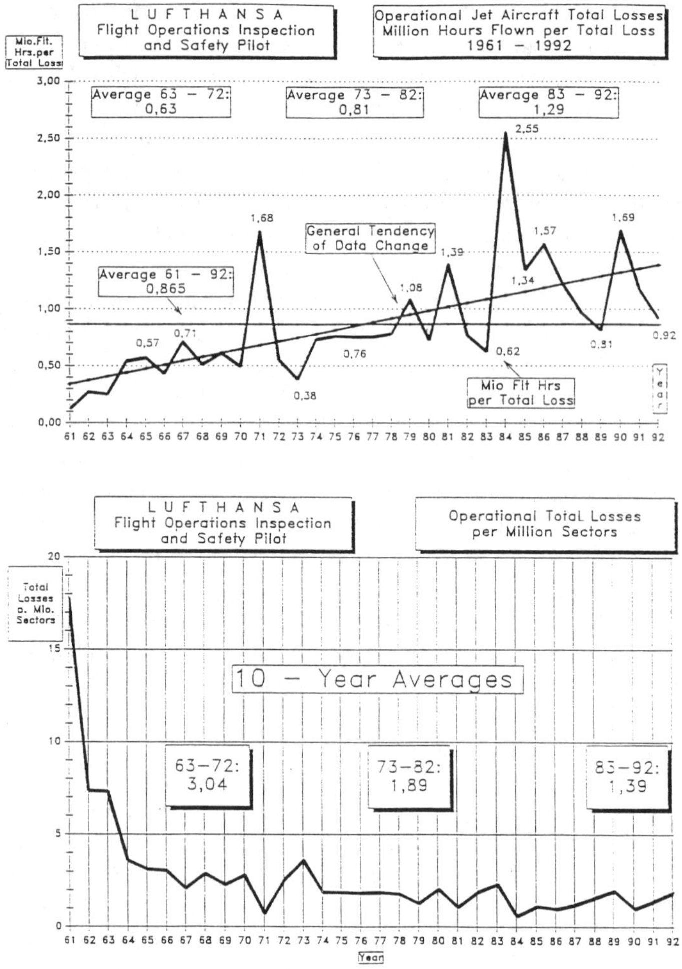

Fig. 4

installations when new knowledge becomes available or new regulations are introduced. For ships one has the "Grandfather Clause", which means that when the safety of a ship is evaluated it is against the rules that were in force at the time of construction of the ship. For ship structures this clause is now under serious discussion and important changes will likely be seen. For offshore structures it has always been the case that new regulations must be considered and evaluated and in general this causes some modifications. It is, however, very difficult to state to which degree new requirements should be fulfilled by an existing installation. This has to do with the risk reduction and the associated cost. No general consensus exists on analysis methods nor decision philosophy and this is thus an important area for future research.

K. NITTINGER, Lufthansa, Hamburg

I would like to start my very brief remarks with a quotation from *Lufthansa's* cooperate philosophy. The quotation is as follows: "Meeting the requirements of our customers is of paramount importance. They are the measure of our activities. We offer punctual, reliable and safe air services. Our aim is to provide the best possible quality". And I think that this remark shows that quality and safe operation is not only a necessary constituent of safe address quotation, but it also has developed into a strong sales argument. So far, so good. What have we achieved in the past? This is a very brief diagram (Fig. 4) where you can see that over the past 30 years the amount of losses per million flight hours - or you can fly more flight hours until you have an accident - and the rate has doubled from 0.63 to 1.29 over the last 30 years. So you can fly double the hours until you have the risk or the chance to have an accident. Or in flight sectors, this is the lower graph you can also see that the total losses per million sectors have decreased from around 3 to 1.39. These are the achievements of the industry in general. Safety aspects in aircraft design and production are a very important supporter of this development. Regulations for aircraft design requires that the occurrence of any failure condition preventing safe flight continuation is extremely improbable. This means that the probability of such failure is less than 10^{-9} per flight hour (Fig. 5). As you can see here there is a very clear definition, and this definition is not only based on national rules, this is also based on international rules, where frequent failures, reasonably probable remote or extreme remote failures are allowed, what effects they have and where the multiple task failure may occur, that is the 10^{-9} case, I think this was also mentioned by Mr. Huppmann. What I have learned from Mr. Madsen, it is also very interesting to me. I think all systems should fly in the future, because being them on ground or floating is more dangerous. So why this high level of safety was achieved. Compliance with the requirements when you are developing or producing an aircraft must be shown by analysis and necessary tests. Design principles that have to achieve such high safety standards are mainly for systems redundancy, and not only double redundancy, in many cases triple or quadruple redundancy, and for structures it is a failure tolerant design. All of these principles are not applicable at the safe life design. Furthermore, flight tests have to demonstrate certain failure conditions (Fig. 6). So these failure conditions are simulated, and, on top of that, fatigue tests have to demonstrate the structural integrity during the aircraft life. And the manufacturers have to assure that design conditions and production quality assurance and safe operation criteria are met and permanently monitored. So a very comprehensive information system is established worldwide for all aviation related activities. All parties involved, being authorities or manufacturers or airlines, are informed about real or potential safety problems. There are international reporting systems established where the persons involved can make their claims anonymous, so that the problem of being penalized personally is avoided and that is to create more openness in the system. Potential safety problems mostly become obvious, when so-called incidents occur. Incidents are events in flight operation, which could in combination with other technical or non technical factors lead to safety problems, such as interrupted take-off's, flight diversions for technical reasons, in-flight shut downs, etc.. And therefore a well functioning international reporting systems covering incidents and major technical findings on aircrafts is absolutely necessary and is requested by law. Furthermore, there are very comprehensive recording systems in-house, just to mention one example, there are hundreds of thousands of individual parts, which are at time and cycle controlled. Every single turbine blade is time and cycle controlled. But I do not want to forget the human factors, performing this technical works. Within my organization there are several thousand employees, which are working on safety related items. And the qualification system accepted by the authority must assure that any work on aircraft systems, which is relevant for safe aircraft operation is performed by persons with a specified high qualification. And this is just an example how we do it in our company: it takes 8 years to be trained to a mechanic grade 3 (Fig. 7). And that is a rather long time - others do it different.

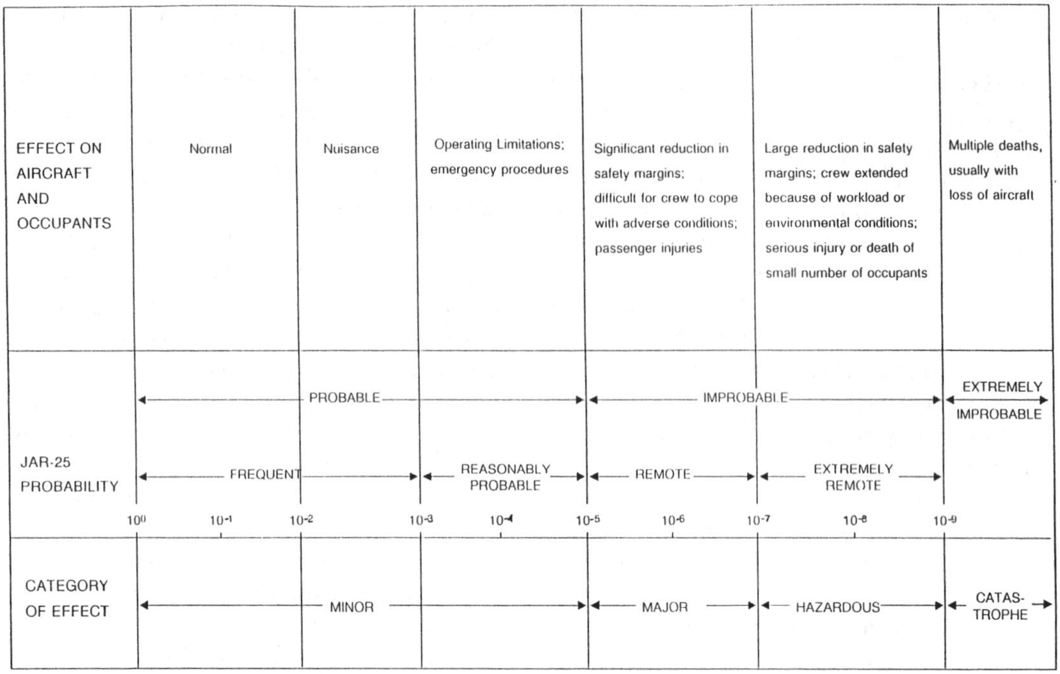

RELATIONSHIP BETWEEN PROBABILITY AND SEVERITY OF EFFECTS

Fig. 5

And I also want to make my very last statement that is: an airline is not only an airline, but it is also an industrial enterprise. And your rules on organization and responsibility should not only reflect flight safety - I am personally not only responsible for that - but also responsible for environmental safety, for radiation protection, for dangerous goods and last but not least for export control. So probably I am the most endangered species in this room to go into jail.

H. SHIBATA, Science Council of Japan, Tokyo

The Science council of Japan is a governmental organization attached to the Prime Minister's Office, and acts as the representative of Japanese Scientists. Under the 5th Division , Engineering, the Liaison Committee on Safety Engineering has been functioning to establish the safety in Japanese industries as engineering and science. Also, as one of ad hoc committees, "Megasystems and Humans" has been organized for the 15th term of the Council (July 1991 to July 1994). I am a member of both committees and have been working with other members, who belong to all seven divisions, consisting of cultural, social and natural scientists in the ad hoc committee, and with other scientists and engineers from research organizations, universities and industries in the Liaison Committee.

I plan to organize an International Liaison Committee for Safety Engineering and Science to establish the fundamental concept of "what is safe" for all engineering fields. I have been working for the safety of nuclear power plants since 1960, however, I feel that there are various practices to achieve safety in various industries. The first practice and concept was established in railway engineering in 1850's, and the fundamental concept in that field is the same up to now, and also the same as those in most industries.

MSG Logic for Aircraft Systems

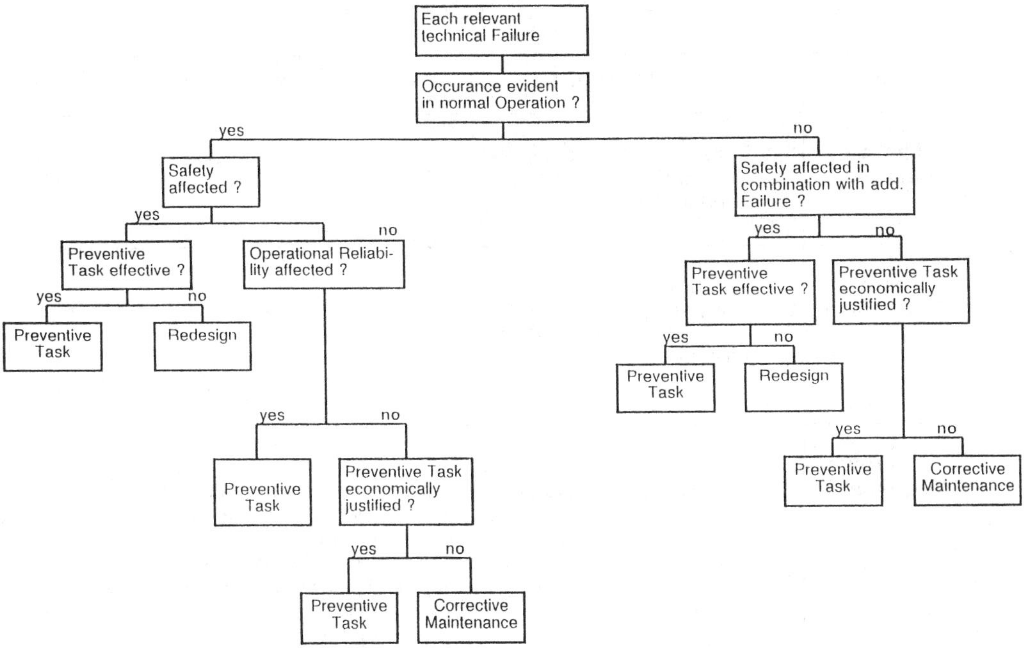

Fig. 6

These two areas, nuclear engineering and railway engineering, are typical ones in safety engineering, and we can recognize similar concepts as well as quite different ones in various fields. However, I believe that there must be a common concept through all kinds of fields and that we shall find it. There are many international organizations of engineering and science, which are related into the safety. I plan to organize an international "Liaison Committee for Safety Engineering and Science" in the future and as the first step to discuss how to organize it, I am preparing a forum to discuss the possibility of this committee in next April or May in Tokyo as an event of the Liaison Committee on safety engineering.

In general, two fundamental ways to establish the safety in various engineering fields. As I mentioned in the previous paragraph, one is the deterministic safety logic approach, and the other one is a probabilistic approach as we observed in railway engineering and in nuclear engineering respectively. For the public, the term "safe" still seems to be an absolute one, and the deterministic logic has been used in various engineering fields not only for railway engineering. As we have been doing, the probabilistic approach is gradually to be common, but for the public it is still difficult to understand it. The severe accident has been discussed for nuclear business in relation to recent accidents. Usually some scenarios are assumed for the design of a containment, equipment and others to mitigate the results of the accident. A typical one is a vent of the containment vessel. These scenarios are described as deterministic, and they do not have any sense except for the design condition of related items. But they are supported by probabilistic sense as an extreme case, and quite different to that in railway engineering. If some unexpected event will occur in a railway operation, they examine it and try to find a new back-up device to prevent such an event

Qualifications of Aircraft Mechanics and Engineers on Aircraft Maintenance

Title	Qualification	Competences
Aircraft Mechanic	3 1/2 years apprenticeship plus special knowledge on aircraft maintenance/overhaul	1. Routine work (code: none)
Aircraft Mechanic I	1. As above 2. Special type training 3. 1 1/2 years of experience after apprenticeship	1. As above 2. Routine work, coded: A 3. Minor repairs under supervision of aircraft mechanic II
Aircraft Mechanic II	1. As above 2. Additional 6 months on line stations 3. Additional years as mechanic I 4. At least 21 years old	1. As above 2. Routine work, coded: B 3. First approval of work, coded: C 4. Repair work with the exception of vital systems like: - engines - controls - landing gear, etc. 5. Special checks 6. Runup (if authorized)
Aircraft Mechanic III	1. As above 2. Additional 1 1/2 years as aircraft mechanic II	1. As above 2. Change, repair and test of vital systems like: - engines - controls - landing gear, etc. 3. Taxying of aircraft (if authorized)

Fig. 7

anymore. In the nuclear case, they evaluate the probability of occurrence of a new event, and how to reduce the value. As the safety goal, the railway engineers intend to no accident, even it is not realistic in a probabilistic sense, and it is more understandable for the public.

We, the member of the sub-committee which will be organized very soon, are intending to solve such subjects and to establish an international communication/Liaison Committee for the safety engineering and science as an attached material.

Most of the subjects raised by Dr. H.G. Payer are included in the discussion between two extreme engineering fields above. I believe that the fusion of these two concepts is necessary, and I try to understand this subject through the study on PSA of the Japanese High-speed Transportation System as presented in this ICOSSAR'93.

W. WILD, DARA, Bonn

I think that the question which we want to discuss this morning, "how safe is safe enough?", can be answered from my point of view quite easily for the case of unmanned space missions. This is the fact because a failure of an unmanned space mission has almost no impact on the well-being of people. Therefore we only lose money if an unmanned space mission fails. The money which we lose is of the order of some hundred million dollars on the average it is true. But because we have only an economical problem we can find a purely economical answer and this answer is that you have to work at the minimum of total costs. These total costs include the development costs of the satellite, the production costs, the launching costs, the operational costs and last not least the insurance costs. The insurance costs for a commercial satellite mission are nowadays on a level of 17% and I do not think that we can achieve a better value very soon. On the other hand we have a slow but constant improvement; the insurance costs have been above 20 % a few years ago and now they are - as I have just mentioned - at approximately 17%.

The reason for this improvement is of course the fact that we have a learning curve. One of the disadvantages in space is obviously that we have only very few objects of the same kind. There exist some dozens or even some hundreds of identical launchers but only very few identical satellites. In the case of launchers it is certainly a disadvantage for the Europeans and the Japanese compared to the Americans and especially to the Russians that the total number of European and of Japanese rockets is much smaller than that of American and Russian rockets because there are a lot of military and governmental missions in the United States and in Russia. The Russians had 2.370 successful launchers during the period from 1957 to 1991 and in spite of all the difficulties in Russia they had still 54 launches in 1992, whereas the numbers for the United States are 972 and 19 and for Europe 61 and 7 respectively.

Nevertheless the European Ariane rocket, has done a very good job; it has today achieved a reliability of 95 to 96%. Therefore the Ariane launcher has won more than 60% of the commercial market.

Of course there are some exceptions to my general statement that for unmanned missions the economic issue is the measure how safe space systems should be. One of these exceptions is for instance the encounter of a satellite like Giotto with a comet like Halley. Halley approaches the sun only every 70 years and if there is a failure of the mission one cannot repeat the experiment before another 70 years have passed. Another example are missions to planets very distant from the sun. In this case one cannot use your solar cells for power supply but has to use the energy which comes from the decay of radioactive elements. If such a mission fails there could result some pollution of the earth and its atmosphere. Despite of the fact that this possible pollution is very small one has difficulties with the protest of the public. We have experienced this a few years ago when the Gallileo satellite was launched which is planned to go to the planet Jupiter.

Let me now shortly attack another problem. The issue of space debris becomes more and more important. In the future we certainly have to include into our cost calculations the costs of avoiding space debris. Especially in orbits which are frequently used namely the geostationary orbit 36.000 km above earth and the so-called Low Earth Orbit (LEO) some 400 km above earth it is necessary to avoid space debris. This can be done by positioning the satellite when its mission has ended in a so-called graveyard orbit where it is sufficiently far away from active satellites or by bringing it down to the lower atmosphere where it is burnt up.

Finally I want to discuss the case of manned space flight. Of course purely economical considerations are inadequate in this case. One has to take care of the lives of the astronauts. Despite of the fact that these people know that they run a risk - like the test pilots of airplane

Table 1

Disasters - world wide
1984 through 1992

HOW SAFE IS SAFE ENOUGH ?

DOMAIN	FATALITIES ANNUAL AVERAGE	MAXIMUM PER YEAR	MINIMUM PER YEAR	MAXIMUM PER EVENT
AVIATION	1429	2336 (85)	483 (84)	520 (85)
FIRE AND EXPLOSION	701	1615 (84)	373 (85)	700 (89)
MARINE	1822	4184 (87)	1051 (84)	3000 (87)
MINING	325	589 (89)	143 (87)	300 (91)
MISCELL.	4285 (2271)	20394(89)	407 (91)	20000 (89)
NATURAL	36910 (22971)	148400(91)	3262 (84)	131000(91)
RAIL	553	1127 (89)	83 (92)	462 (89)
TRAFFIC	525	696 (92)	350 (85)	100 (92)

Reference: Encyclopaedia Britannica, Book of the Year
Disaster : more than 10 fatalities
not included: war, famine

prototypes - we must try to reduce this risk as much as possible. Therefore we must achieve a man-rated reliability which is much higher than the reliability needed for unmanned space missions. This is the main reason, why shuttle missions are delayed so often. Only very recently the launch of a German satellite called Astrospas was stopped three seconds before take off because of some irregularities on the shuttle system. The same situation happened with the German manned D2 mission a few month ago. These events show on one hand that the shuttle has become a very safe system; even three seconds before take off when the main engines are already in action the launch can be stopped, On the other hand the safety requirements make the shuttle a very vulnerable system; the number of unsuccessful launching attempts is steadily increasing. The hope that a reusable system like the shuttle would reduce the costs of space missions has not been fulfilled. On the contrary it is more economical to launch satellites by not reusable rockets than by shuttle.

In the case where we use manned laboratories for experiments in space - especially in order to investigate the effects of microgravity on men - it is of course impossible to replace the manned space flights by unmanned once. But the general tendency is to reduce the activities in manned space flights and to do everything which can be done unmanned by means of unmanned space activities. Therefore nowadays all the applications of space technology in telecommunications and most of the

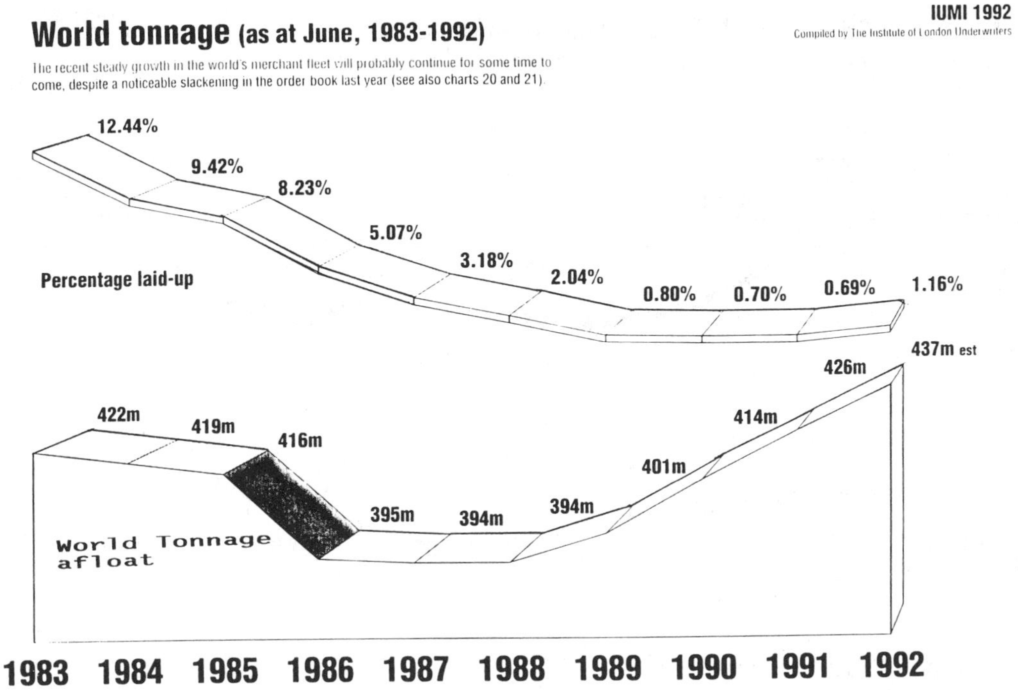

Fig. 8

investigations in earth observation and astronomy but more and more also the microgravity experiments on material objects are performed in an unmanned way.

H.J. WINGENDER, NUKEM, Altenau

"A person has a horizon of zero radius; that is what he claims to be his point of view." (George Christoph Lichtenberg, 1742-1799).

At first, I should apologize because I come to the nuclear power plant safety rather at the end of my short contribution. Let me start first with a few words about safety and risk in general. They are phenomena existing in such a way that you can neither generate nor eliminate any one of them. However, local and temporary alterations of some aspects of these phenomena are resulting from our efforts to avoid unwanted events. Such actions can be: move away from endangered areas, reduce or stop activities whose consequences are classified as at least partly controllable, be prepared to mitigate consequences of events which cannot be prevented from occurring. Now, any such action will eventually lead to a situation in which new patterns of risk and safety are to be faced. Coming back to the four types of action; two of them, move from the area and stop hazardous activities, are of basically behavioural nature; the two others, prevent accidents and mitigate their consequences are technical concepts. The disadvantage of a technical means is that non-technical approaches are by far more easily evidenced as successful. That is one of the problems which we, as engineers in the nuclear field, are facing; and I suppose, other engineers have similar problem.

Now, how successful have we been in risk control? Table 1 shows a compilation of experience over a nine year period. Fatalities from disasters are listed according to eight domains. In the first column the annual averages over the period are listed (in brackets: singular worst event is extreme and thus excluded from

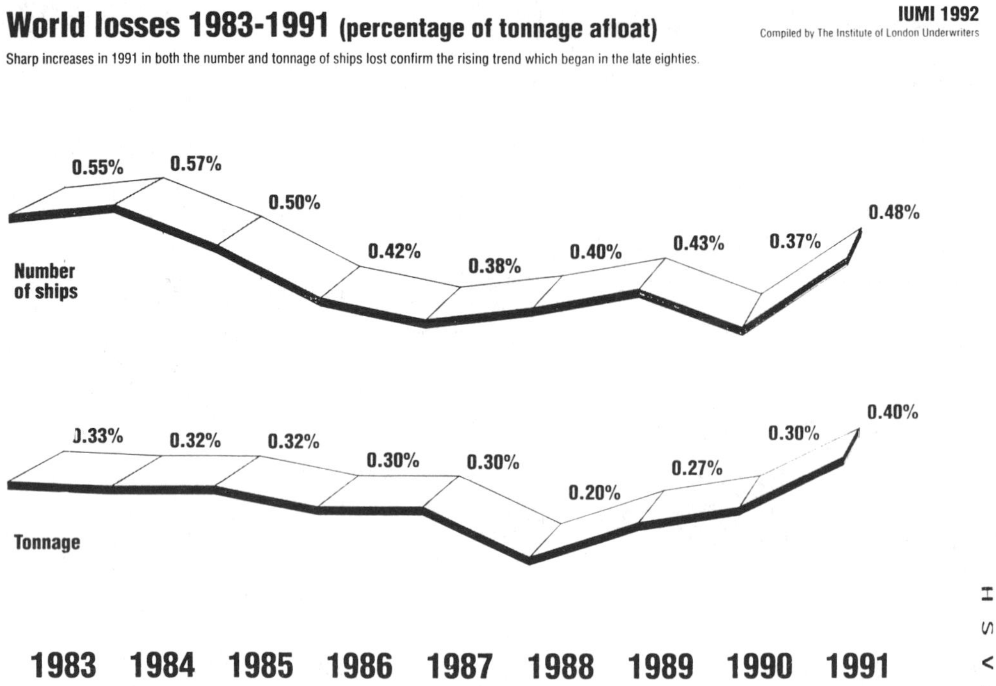

HSVA

Fig. 9

average), second and third column show the maximum and minimum annual values (year in brackets) for the period (year of event in brackets). For seven domains the respective consequences of the worst single events exceed the annual minimum, and they even exceed or come close to the annual average for six of the eight domains. In consequence, the observed frequency of single events whose consequences are close to the annual mean of a domain is rather high.

Of course, these data are incomplete in a few ways, in particular because of the limit of ten fatalities defining the minimum disaster. With respect to road traffic, for instance, in a single country like Germany, the actual annual death toll is about ten times higher than the world annual average as listed from disasters.

A closer view at a more detailed set of these data unveils that disasters in industry, such as the Piper Alpha fire; the Bhopal accident; the Chernobyl catastrophy; or the Herald of Free Enterprise do not occur amongst the worst single events in this list.

So far one side of the hard facts; now to a second aspect: I do not hesitate to state that the general attitude - I emphasize the term "general" - towards most of the disasters listed is one of nonchalance, tolerance and of being untouched. Many people have become used to living with these risks, partly because they do live in the endangered areas and have familiarized with the facts, partly because they live outside the respective region and ban the strange facts from their minds. On the other hand, the attitude of the public towards disasters or even not disastrous events will dramatically change if events affecting the environment or industrial accidents are concerned. They are qualified as incomparable with normally accepted risks and are to be banned at any risk. It is this attitude which is scaring me because it is inconsiderate and hazardous.

I am of the opinion that safety is a matter

requiring a fairly broad and considered approach to achieve and to maintain it. For instance, it may well be that a temporary continuation of operating Chernobyl type reactors can be safer than their immediate shut down, if safety/risk is considered in the light of the actual, overall situation in the country concerned. Thus, it is not surprising that these reactors will be kept in operation for a number of years; and that this number will depend not at the least on the support that will be available to this country. A second point: It is my opinion that nuclear plants are fairly safe in the West. Actually, the resources of nuclear safety enhancement have become small; nuclear safety is close to its ceiling and it is highly demanded. In consequence, any further increment of safety becomes increasingly more expensive; espenses which will be dearly missed when we need them at our disposal for what will scare us tomorrow and had better scared us today. The third and last point: Having a possibly beneficial means at hand, such as nuclear energy, even if it is looked at or, as some say, to be looked at with suspicion, it is an option which should not be discarded lightheartedly. Such a well considered attitude has recently been taken by quite a few nuclear opponents who are concerned about global warming and who have come to the conclusion that nuclear could be one of the options needed for the remidiation of this problem.

I cannot possibly conclude without mentioning my concern about a too narrow and much too focussed consideration of safety and risk matters - even on my part.

H.G. PAYER

Not all of you may be aware that the ships are and will remain the main carrier of

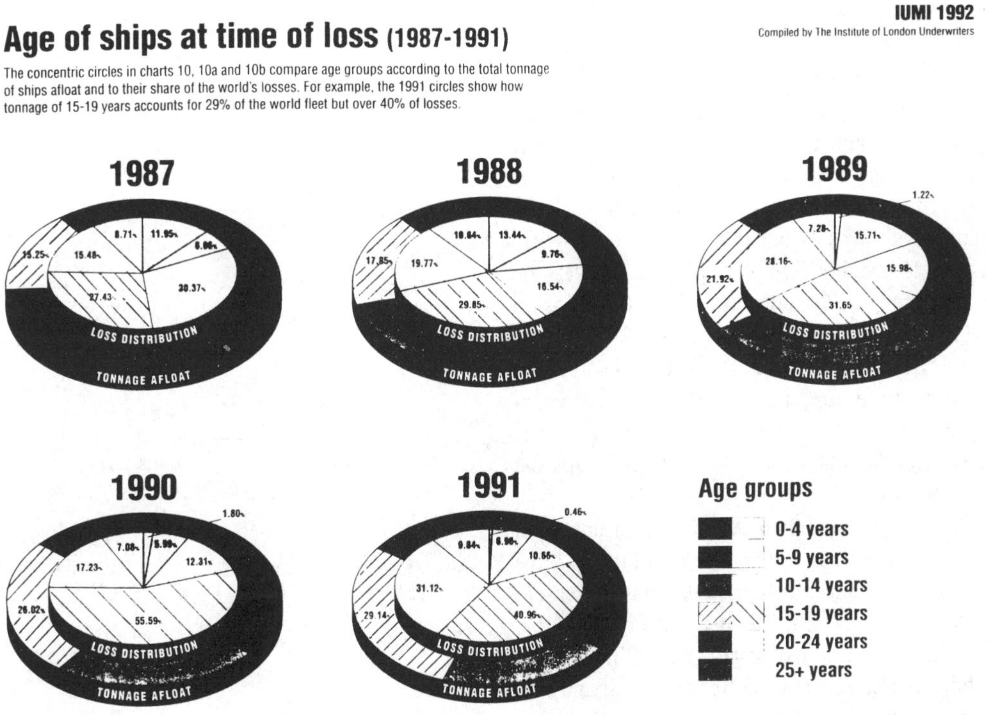
IUMI 1992
Compiled by The Institute of London Underwriters
Age of ships at time of loss (1987-1991)
The concentric circles in charts 10, 10a and 10b compare age groups according to the total tonnage of ships afloat and to their share of the world's losses. For example, the 1991 circles show how tonnage of 15-19 years accounts for 29% of the world fleet but over 40% of losses.
1987
1988
1989
1990
1991
LOSS DISTRIBUTION
TONNAGE AFLOAT
Age groups
0-4 years
5-9 years
10-14 years
15-19 years
20-24 years
25+ years

Fig. 10

transportation in the world. Today more than 95% of world trade is being carried by ships. Ships are not outdated, ships are very sophisticated conveyances. A modern ship carries more electronics than the big airplanes. And shipping is by far - I tend to say the environmentally - best compatible means of transport, if you consider fuel consumption per ton mile. The problem with ships, again, is the size. Any accident can have catastrophic consequences. Let us look at some statistics. The world fleet today is steadily growing (see Fig. 8) the world tonnage afloat in the lower curve and the percentage of laid-up ships in the upper curve - laid-up ships are ships which are not in service and are waiting for a cargo - and in the past years the statistics show the tendency of increased tonnage and less laid-up ships. How about their safety performance: Fig. 9 gives an overview from 1983 to 1991 of losses of ships. There are all types of ships, where the extent of the damage depends very much on the type of the ship: tankers and gas carriers are with the most consequence if they are lost. In the tonnage lost there is a rising trend since the late eighties to about 0.4%.

And finally, there is a correlation between the age of ships and their losses. In the concentric circles of Fig. 10 for each year are compared the age groups of the ships according to total tonnage of ships afloat to their share of world losses. If we look for instance at 1991, the shaded group in the outer circle are the ships 15 - 19 years of age. They make up 29% of the world fleet. In the inner circle you see the losses, where the age group 15 - 19 years accounts for 40% of the world fleet losses. This shows clearly that with the age of the ships the risks increase.

Dr. Madsen has touched on one important point in this context that is that the regulations have become better and our knowledge has increased, but this has not been forced for the older ships. From the facts that the world fleet is increasing, that the average age of the fleet is rising and that accidents are more likely with older ships we may conclude that the overall safety in shipping is declining. It is to be expected that in direct consequence of recent dramatic accidents a more coordinated attempt will be made worldwide to improve ship safety. This is very well possible with reliability aspects being increasingly incorporated into the rules of ship classification societies. With this very brief overview I would like to conclude our presentation here.

Random vibration

Structural Safety & Reliability, Schuëller, Shinozuka & Yao (eds) © 1994 Balkema, Rotterdam, ISBN 90 5410 357 4

Localization of stress wave propagation in disordered multi-wave structures

S.T. Ariaratnam & Wei-Chau Xie
University of Waterloo, Ont., Canada

ABSTRACT: The localization of stress wave propagation in disordered multi-wave structures is investigated in this paper. The smallest localization factor is of particular interest, and is related to the smallest positive Lyapunov exponent. The localization of a large beam-like lattice truss modelled as an equivalent continuous Timoshenko beam is studied.

1 INTRODUCTION

Many structures, especially for application in space, are designed to be composed of identically constructed units which are assembled end-to-end to form a large spatially periodic structure, e.g. long space antennae or periodic truss structures. These periodic structures behave like band-pass filters when propagating stress-waves. For waves in certain frequency bands, known as pass-bands, the amplitudes do not undergo any attenuation as the wave propagates, where as for certain other frequency bands, known as stop-bands, the amplitudes are attenuated.

However, due to defects in manufacture and assembly, no structure designed to be periodic can be perfectly periodic, but will be disordered. In disordered, nearly periodic structures, amplitudes of waves with frequencies in the pass-bands will also be attenuated; the steady-state response of the structure will decay exponentially away from the source. This is known as the localization phenomenon, first pointed out by Phillip Anderson in Solid State Physics; the exponential rate of decay of amplitude is called the localization factor. There are d waves in a d-wave disordered periodic structure; each wave attenuates at a certain exponential rate or corresponds to a certain localization factor, which implies that each wave will extend to a certain localization length. The smallest localization factor or the largest localization length is of particular interest for multi-wave structures, since it corresponds to the wave which has potentially the least amount of decay or which carries energy along the multi-wave structure farther than any other wave.

Hodges (1982) was the first to recognize the relevance of localization theory to dynamical behaviour of engineering periodic structures. There have been many research papers on the localization of vibration of structures. In a series of publications, Pierre and Dowell (1987), Pierre (1990), Cha and Pierre (1991) studied the localization of vibration modes in mono-coupled disordered structural systems. Cai and Lin (1990) developed a perturbation scheme to calculate the localization factor for wave propagation based on a generic periodic structure, while Kissel (1988), Ariaratnam and Xie (1991, 1993) used a travelling wave approach to investigate localization effects in one dimensional periodic engineering structures. A transfer matrix formulation employing wave transfer matrices was used to model disordered periodic structures. Furstenberg's theorem (1963) on the asymptotic properties of products of random matrices was applied to calculate the localization effect as a function of frequency. In these one-dimensional examples, whose transfer matrices are of dimension 2, the localization factor is related to the largest Lyapunov exponent. Little work has been done on the localization behaviour of multi-wave periodic structures. Kissel (1991) derived the localization factor as a function of the transmission matrix for multi-wave disordered systems using the *Multiplicative Ergodic Theorem* of Oseledec (1968). The difficulty in the study of localization of multi-wave periodic structures is that, for such structures, the localization factor is related to the smallest positive Lyapunov exponent rather than the largest. Therefore, the numerical algorithm usually employed for evaluating the largest Lyapunov exponent for single wave periodic structures cannot be used for multi-wave structures.

In this paper, an algorithm due to Wolf *et al.* (1985) is modified to determine all the Lyapunov exponents of randomly disordered multi-wave periodic structures. As an application the localization factor of a large beam-like lattice truss modelled as an equivalent continuous Timoskenko beam is evaluated.

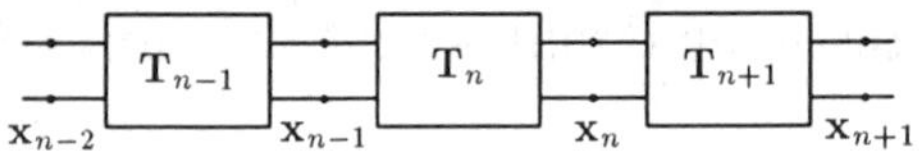

Figure 1 Transfer matrix

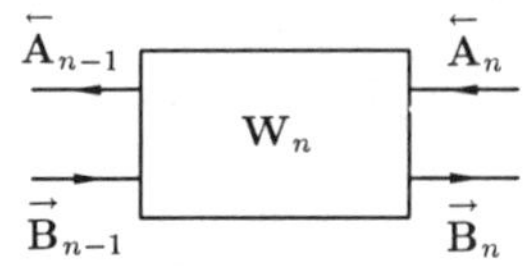

Figure 2 Wave transfer matrix

2 TRANSFER MATRIX AND LYAPUNOV EXPONENTS FOR MULTI-WAVE STRUCTURES

Consider an element numbered n in a multi-wave structure (Figure 1). The element is modelled by the transfer matrix, $\mathbf{T}$, which relates a state vector $\mathbf{x}_{n-1}$ on the left-side of element n to that on the right-side, $\mathbf{x}_n$, by the linear transformation

$$\mathbf{x}_n = \mathbf{T}\mathbf{x}_{n-1}. \tag{2.1}$$

For a d-wave structure, the state vector $\mathbf{x}_n$ usually involves d generalized displacements and d generalized forces; the transfer matrix is a square matrix of dimension $2d \times 2d$. The transfer matrix can be derived from the dynamic equations of motion of the structure by, possibly, the finite difference or finite element method. The derivation of transfer matrices of various engineering stuctures may be found in Pestel and Leckie (1963). If the periodic structure is symmetric, the transition from the section "n" to the section "$n+1$" is the same as that from "$n+1$" to "n". It follows therefore that $\det(\mathbf{T}) = \det(\mathbf{T}^{-1})$, so that $\det(\mathbf{T}) = 1$, where $\det(\mathbf{T})$ denotes the determinant of $\mathbf{T}$.

For a perfectly periodic structure, each element of the structure is identical; therefore the transfer matrix for each element is the same. The state vector after n elements is related to that at the beginning by

$$\mathbf{x}_n = \mathbf{T}^n\mathbf{x}_0. \tag{2.2}$$

When the periodic structure is disordered randomly due to variability in geometry, material and manufacturing conditions, the transfer matrix for each element is not the same, but is a function of the parameter of disorder. In this case, the state vector after n elements is related to the initial state vector by

$$\mathbf{x}_n = \mathbf{T}_n\mathbf{T}_{n-1}\cdots\mathbf{T}_1\mathbf{x}_0, \tag{2.3}$$

where $\mathbf{T}_1, \mathbf{T}_2, \cdots, \mathbf{T}_n$ are now *random matrices.*

Considering the system (2.3) as describing a discrete random dynamical system, the Lypunov exponent of its orbit is defined by

$$\lambda(\mathbf{x}_0) = \lim_{n\to\infty} \frac{1}{n} \log \|\mathbf{x}_n(\mathbf{x}_0)\|, \tag{2.4}$$

where $\|\cdot\|$ is the Eucledian vector norm. The Lyapunov exponent describes the average rate of exponential decay or growth of state vector $\mathbf{x}_0$. There will in general be several Lyapunov exponents $\lambda_1 \geq \lambda_2 \geq \cdots \geq \lambda_{2d}$, depending on the initial random vector $\mathbf{x}_0$, where $2d$ is the dimension of the matrices $\mathbf{T}_1, \mathbf{T}_2, \cdots$.

In order to obtain the properties of the Lyapunov exponents, it is easier to use the wave transfer matrix formulation. The wave transfer matrix $\mathbf{W}_n$ is a linear transformation relating the left (or backward) and the right (or forward) travelling wave amplitudes, $\overleftarrow{\mathbf{A}}$ and $\overrightarrow{\mathbf{B}}$, at two neighbouring sections of the element n of the structure (Figure 2)

$$\left\{ \begin{array}{c} \overleftarrow{\mathbf{A}}_n \\ \overrightarrow{\mathbf{B}}_n \end{array} \right\} = \mathbf{W}_n \left\{ \begin{array}{c} \overleftarrow{\mathbf{A}}_{n-1} \\ \overrightarrow{\mathbf{B}}_{n-1} \end{array} \right\}, \tag{2.5}$$

where

$$\overleftarrow{\mathbf{A}}_n = \{\overleftarrow{A}_n^{(1)}, \overleftarrow{A}_n^{(2)}, \cdots, \overleftarrow{A}_n^{(d)}\}^T,$$
$$\overrightarrow{\mathbf{B}}_n = \{\overrightarrow{B}_n^{(1)}, \overrightarrow{B}_n^{(2)}, \cdots, \overrightarrow{B}_n^{(d)}\}^T.$$

The wave transfer matrix is a simplectic matrix and is derivable from the transfer matrix by a linear tranformation.

When a periodic structure is disordered, the wave amplitudes after n elements are related to those at the beginning through a matrix in the form of a product of random matrices:

$$\begin{aligned} \left\{ \begin{array}{c} \overleftarrow{\mathbf{A}}_n \\ \overrightarrow{\mathbf{B}}_n \end{array} \right\} &= \mathbf{V}_n \left\{ \begin{array}{c} \overleftarrow{\mathbf{A}}_0 \\ \overrightarrow{\mathbf{B}}_0 \end{array} \right\} \\ &= \mathbf{W}_n\mathbf{W}_{n-1}\cdots\mathbf{W}_1 \left\{ \begin{array}{c} \overleftarrow{\mathbf{A}}_0 \\ \overrightarrow{\mathbf{B}}_0 \end{array} \right\}. \end{aligned} \tag{2.6}$$

where $\mathbf{W}_1, \mathbf{W}_2, \cdots, \mathbf{W}_n$ are the wave transfer matrices of the individual elements.

It can be shown that the Lyapunov exponents are given by (Bougerol and Lacroix, 1985)

$$\lambda_i = \lim_{n\to\infty} \frac{1}{n} \log \sigma_i(\mathbf{V}_n), \qquad i = 1, 2, \cdots, 2d, \tag{2.7}$$

where $\sigma_1 \geq \sigma_2 \geq \cdots \geq \sigma_{2d}$ are the singular values of the matrix $\mathbf{V}_n$, i.e. $\sigma_i(\mathbf{V}_n)$ is equal to the positive square root of the ith eigenvalue of the matrix $\mathbf{V}_n^*\mathbf{V}_n$, * denoting the operation of transposition and taking complex conjugate.

Since the wave transfer matrices $\mathbf{W}_1, \mathbf{W}_2, \cdots, \mathbf{W}_n$ are simplectic matrices, $\mathbf{V}_n = \mathbf{W}_n\mathbf{W}_{n-1}\cdots\mathbf{W}_1$ is also simplectic. It can be easily shown that the singular values of a simplectic matrix occur in reciprocal pairs, i.e

$$\sigma_1, \ \sigma_2, \ \cdots, \ \sigma_d, \ \sigma_d^{-1}, \ \cdots, \ \sigma_2^{-1}, \ \sigma_1^{-1},$$

where $\sigma_1 \geq \sigma_2 \geq \cdots \geq \sigma_d \geq 1$. Therefore, the Lyapunov exponents given by equation (2.7) have the following property

$$\begin{aligned} \lambda_1 \geq \lambda_2 \geq \cdots \geq \lambda_d \geq (\lambda_{d+1} = -\lambda_d) \\ \geq (\lambda_{d+2} = -\lambda_{d-1}) \geq \cdots \geq (\lambda_{2d} = -\lambda_1). \end{aligned} \tag{2.8}$$

The positive Lyapunov exponents correspond to right-travelling waves, while the negative Lyapunov exponents correspond to left-travelling waves. It can be seen that the localization factors are by definition related to the Lyapunov exponents, since both quantities characterize the average exponential decaying rates of the amplitudes of propagating waves. The smallest positive Lyapunov exponent λ_d is of particular interest for multi-wave structures, since λ_d represents the wave which has potentially the least amount of decay or which carries energy along the multi-wave structure farther than any other wave.

From the *Multiplicative Ergodic Theorem* of Oseledec (1968), it is known that a randomly chosen initial state vector, $\mathbf{x}_0$, will evolve in general at the rate of the largest Lyapunov exponent λ_1. It is therefore relatively easy to determine the largest Lyapunov exponent λ_1. Since the smallest positive Lyapunov exponent λ_d is needed for a multi-wave structure, this gives rise to the principal difficulty in the study of the localization behaviour of multi-wave structures.

3 ALGORITHM FOR CALCULATING ALL LYAPUNOV EXPONENTS

The algorithm for determining all the Lyapunov exponents for continuous dynamical systems by Wolf *et al.* (1985) will be modified to determine all the Lyapunov exponents for the random discrete dynamical system:

$$\mathbf{x}_n = \mathbf{T}_n \mathbf{x}_{n-1}, \qquad n = 1, 2, \cdots. \tag{3.1}$$

where $\mathbf{T}_n$ is the transfer matrix of the nth element.

To evaluate the largest Lyapunov exponent λ_1, a unit initial vector $\mathbf{v}_0$ is chosen, i.e. $\|\mathbf{v}_0\| = 1$. The solution of (3.1) is evaluated iteratively or element by element. At the kth element, let

$$\mathbf{u}_k = \mathbf{T}_k \mathbf{v}_{k-1}. \tag{3.2}$$

The new vector $\mathbf{u}_k$ is now normalized to give

$$\mathbf{v}_k = \frac{\mathbf{u}_k}{\|\mathbf{u}_k\|}, \qquad \text{or} \qquad \mathbf{u}_k = \|\mathbf{u}_k\| \mathbf{v}_k. \tag{3.3}$$

Therefore, at the Nth element, the magnitude of $\mathbf{v}_0$ evolves to

$$\|\mathbf{T}_N \mathbf{T}_{N-1} \cdots \mathbf{T}_1 \mathbf{v}_0\| = \|\mathbf{u}_N\| \cdot \|\mathbf{u}_{N-1}\| \cdots \|\mathbf{u}_1\|. \tag{3.4}$$

Since it is known that the length of a unit vector evolves on the average as $e^{\lambda_1 N}$, the largest Lyapunov exponent is then given by

$$\lambda_1 = \lim_{N \to \infty} \frac{1}{N} \log(\|\mathbf{u}_N\| \cdot \|\mathbf{u}_{N-1}\| \cdots \|\mathbf{u}_1\|),$$

or

$$\lambda_1 = \lim_{N \to \infty} \frac{1}{N} \sum_{k=1}^{N} \log(\|\mathbf{u}_k\|). \tag{3.5}$$

In general, the m-dimensional volume defined by the first m principal axes evolves in the system (3.1) on the average as $e^{(\lambda_1+\lambda_2+\cdots+\lambda_m)N}$. Thus $\lambda_1 + \lambda_2 + \cdots + \lambda_m$ can be found by following the evolution of any m linearly independent vectors. However, almost all vectors tend to point towards the local direction of most rapid growth as they evolve, i.e. at the rate of the largest Lyapunov exponent. This problem can be overcome by applying the Gram-Schmidt orthonormalization procedure. To evaluate the mth Lyapunov exponent, m orthogonal unit vectors $\mathbf{v}_0^{(1)}, \mathbf{v}_0^{(2)}, \cdots, \mathbf{v}_0^{(m)}$ are chosen as the initial conditions. The solution of (3.1) is evaluated iteratively for each initial vector. At the kth element:

$$\mathbf{u}_k^{(i)} = \mathbf{T}_k \mathbf{v}_{k-1}^{(i)}, \qquad i = 1, 2, \cdots, m. \tag{3.6}$$

The vectors $\mathbf{u}_k^{(i)}, i = 1, 2, \cdots, m$ are usually not orthogonal. The Gram-Schmidt orhtonormalization procedure is now applied:

$$\begin{aligned}
\hat{\mathbf{u}}_k^{(1)} &= \mathbf{u}_k^{(1)}, \qquad \mathbf{v}_k^{(1)} = \frac{\hat{\mathbf{u}}_k^{(1)}}{\|\hat{\mathbf{u}}_k^{(1)}\|}, \\
\hat{\mathbf{u}}_k^{(2)} &= \mathbf{u}_k^{(2)} - (\mathbf{u}_k^{(2)}, \mathbf{v}_k^{(1)})\mathbf{v}_k^{(1)}, \qquad \mathbf{v}_k^{(2)} = \frac{\hat{\mathbf{u}}_k^{(2)}}{\|\hat{\mathbf{u}}_k^{(2)}\|}, \\
&\cdots \quad \cdots \\
\hat{\mathbf{u}}_k^{(m)} &= \mathbf{u}_k^{(m)} - (\mathbf{u}_k^{(m)}, \mathbf{v}_k^{(m-1)})\mathbf{v}_k^{(m-1)} \\
&\quad - (\mathbf{u}_k^{(m)}, \mathbf{v}_k^{(m-2)})\mathbf{v}_k^{(m-2)} - \cdots - (\mathbf{u}_k^{(m)}, \mathbf{v}_k^{(1)})\mathbf{v}_k^{(1)}, \\
\mathbf{v}_k^{(m)} &= \frac{\hat{\mathbf{u}}_k^{(m)}}{\|\hat{\mathbf{u}}_k^{(m)}\|},
\end{aligned} \tag{3.7}$$

where $(\cdot, \cdot)$ denotes the vector dot-product.

After multiplying by a product of transfer matrices $\prod_{k=1}^{N} \mathbf{T}_k$ or after N elements, the volume of an m-dimensional unit hypersphere is

$$\prod_{k=1}^{N} \Big(\prod_{i=1}^{m} \|\hat{\mathbf{u}}_k^{(i)}\| \Big).$$

From the definition of the Lyapunov exponents, one has

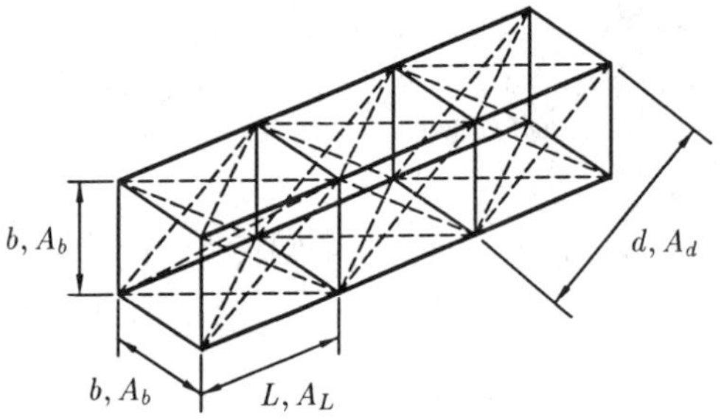

	Cross Sectional Area	Length	Mass Density	Young's Modulus
Long. Bars	A_L	L	ρ_L	E_L
Diag. Bars	A_d	d	ρ_d	E_d
Battens	A_b	b	ρ_b	E_b

Figure 3 Beam-like lattice truss

$$e^{(\lambda_1+\lambda_2+\cdots+\lambda_m)N} = \prod_{k=1}^{N}\Big(\prod_{i=1}^{m}\|\hat{\mathbf{u}}_k^{(i)}\|\Big), \quad \text{for } N \to \infty,$$

since, as mentioned before, the m-dimensional volume defined by the first m principal axes evolves on the average as $e^{(\lambda_1+\lambda_2+\cdots+\lambda_m)N}$. Therefore

$$\begin{aligned}\lambda_1+\lambda_2+\cdots+\lambda_m &= \lim_{N\to\infty}\frac{1}{N}\sum_{i=1}^{m}\Big(\sum_{k=1}^{N}\log\|\hat{\mathbf{u}}_k^{(i)}\|\Big)\\ &= \lambda_1+\lambda_2+\cdots+\lambda_{m-1} + \lim_{N\to\infty}\frac{1}{N}\sum_{k=1}^{N}\log\|\hat{\mathbf{u}}_k^{(m)}\|,\end{aligned}$$

so that

$$\lambda_m = \lim_{N\to\infty}\frac{1}{N}\sum_{k=1}^{N}\log\|\hat{\mathbf{u}}_k^{(m)}\|. \tag{3.8}$$

By following the evolution of a $2d$-dimensional hypersphere, each of the $2d$ Lyapunov exponents can be calculated.

In the implementation of this algorithm, certain properties of each element of the structure such as span length L, bending stiffness EI, etc. are assumed to be randomly disordered and are therefore random variables. For each element, random numbers are generated for these quantities and the corresponding random transfer matrix is evaluated.

4 STRESS WAVE LOCALIZATION OF LARGE BEAM-LIKE LATTICE TRUSSES

As an example of multi-wave structures, a large beam-like lattice truss used in large-area space structures such as a space solar power station (Figure 3) is considered. The lattice structure is replaced by an equivalent continuous Timoshenko beam in which the effects of shear deformation and rotary inertia are taken into consideration. In terms of the dimensions and the properties of the original lattice truss, the properties of the equivalent Timoshenko beam are determined as follows (Noor and Anderson, 1979):

mass per unit length μ:

$$\mu = \rho A = 4\Big(\rho_L A_L + \frac{b}{L}\rho_b A_b + \frac{2d}{L}\rho_d A_d\Big),$$
$$d = \sqrt{b^2+L^2},$$

flexural rigidity EI:

$$EI = b^2\Big(E_L A_L + \frac{L^3}{\mu_1 d^3}E_d A_d\Big), \quad \mu_1 = 1 + 2\frac{b^3}{d^3}\frac{E_d A_d}{E_b A_b},$$

shear stiffness GA_s

$$GA_s = \frac{4b^2L}{d^3}E_d A_d,$$

radius of gyration r:

$$r^2 = \frac{\rho I}{\rho A}, \qquad \rho I = b^2\Big(\rho_L A_L + \frac{2b}{3L}\rho_b A_b + \frac{4d}{3L}\rho_d A_d\Big).$$

It is assumed that the length L of the logitudinal bars is a uniformly distributed random number with mean L_0 and standard deviation σ_L; the length of the diagonal bars is assumed to vary accordingly and satisfy $d = \sqrt{b^2+L^2}$. The large beam-like lattice truss may be considered as a randomly disordered periodic structure with each element being an equivalent Timoshenko beam of the corresponding lattice truss element. By employing the notations:

$$\sigma = \frac{\mu\omega^2L^2}{GA_s}, \ \tau = \frac{\mu r^2\omega^2L^2}{EI}, \ \beta^4 = \frac{\mu\omega^2L^4}{EI}, \ k = \frac{EI}{L},$$

the transfer matrix of a Timoshenko beam element is as follows:

Case 1. $-\frac{1}{2}(\sigma+\tau) + \sqrt{\beta^4 + \frac{1}{4}(\sigma-\tau)^2} < 0$

$$\mathbf{T} = \begin{bmatrix} c_0+\sigma c_2 & Lc_1 & -\frac{L}{k}c_2 & -\frac{L^2}{\beta^4 k}c_5 \\ -\frac{\beta^4}{L}c_3 & c_0+\tau c_2 & \frac{1}{k}c_6 & -\frac{L}{k}c_2 \\ -\frac{\beta^4 k}{L}c_2 & kc_7 & c_0+\tau c_2 & Lc_1 \\ \frac{\beta^4 k}{L^2}c_8 & -\frac{\beta^4 k}{L}c_2 & -\frac{\beta^4}{L^2}c_3 & c_0+\sigma c_2 \end{bmatrix} \tag{4.1}$$

where

$$\lambda_1 = \sqrt{\tfrac{1}{2}(\sigma+\tau) - \sqrt{\beta^4 + \tfrac{1}{4}(\sigma-\tau)^2}},$$

$$\lambda_2 = \sqrt{\tfrac{1}{2}(\sigma+\tau) + \sqrt{\beta^4 + \tfrac{1}{4}(\sigma-\tau)^2}},$$

$$\Lambda = \frac{1}{\lambda_1^2 - \lambda_2^2} = -\frac{1}{2\sqrt{\beta^4 + \tfrac{1}{4}(\sigma-\tau)^2}},$$

$$c_0 = \Lambda(\lambda_1^2 \cos\lambda_2 - \lambda_2^2 \cos\lambda_1),$$

$$c_1 = \Lambda(\lambda_1 \sin\lambda_1 - \lambda_2 \sin\lambda_2),$$

$$c_2 = \Lambda(\cos\lambda_1 - \cos\lambda_2), \quad c_3 = \Lambda\Big(\frac{\sin\lambda_1}{\lambda_1} - \frac{\sin\lambda_2}{\lambda_2}\Big),$$

$$c_4 = \Lambda(\lambda_1^3 \sin\lambda_1 - \lambda_2^3 \sin\lambda_2),$$

$$c_5 = \sigma c_1 - (\sigma\tau - \beta^4)c_3, \quad c_6 = c_1 - \sigma c_3,$$

$$c_7 = \sigma c_1 - c_4, \quad c_8 = c_1 - \tau c_3.$$

Case 2. $-\frac{1}{2}(\sigma+\tau) + \sqrt{\beta^4 + \frac{1}{4}(\sigma-\tau)^2} > 0$

$$\mathbf{T} = \begin{bmatrix} c_0 - \sigma c_2 & Lc_1 & \frac{L}{k}c_2 & -\frac{L^2}{\beta^4 k}c_5 \\ \frac{\beta^4}{L}c_3 & c_0 - \tau c_2 & \frac{1}{k}c_6 & \frac{L}{k}c_2 \\ \frac{\beta^4 k}{L}c_2 & kc_7 & c_0 - \tau c_2 & Lc_1 \\ \frac{\beta^4 k}{L^2}c_8 & \frac{\beta^4 k}{L}c_2 & \frac{\beta^4}{L^2}c_3 & c_0 - \sigma c_2 \end{bmatrix}, \tag{4.2}$$

where

$$\lambda_1 = \sqrt{-\tfrac{1}{2}(\sigma+\tau) + \sqrt{\beta^4 + \tfrac{1}{4}(\sigma-\tau)^2}},$$

$$\lambda_2 = \sqrt{\tfrac{1}{2}(\sigma+\tau) + \sqrt{\beta^4 + \tfrac{1}{4}(\sigma-\tau)^2}},$$

$$\Lambda = \frac{1}{\lambda_1^2 + \lambda_2^2} = \frac{1}{2\sqrt{\beta^4 + \tfrac{1}{4}(\sigma-\tau)^2}},$$

$$c_0 = \Lambda(\lambda_1^2 \cosh\lambda_2 + \lambda_2^2 \cos\lambda_1),$$

$$c_1 = \Lambda(\lambda_1 \sinh\lambda_1 + \lambda_2 \sin\lambda_2),$$

$$c_2 = \Lambda(\cosh\lambda_1 - \cos\lambda_2),$$

$$c_3 = \Lambda\Big(\frac{\sinh\lambda_1}{\lambda_1} - \frac{\sin\lambda_2}{\lambda_2}\Big),$$

$$c_4 = \Lambda(\lambda_1^3 \sinh\lambda_1 - \lambda_2^3 \sin\lambda_2),$$

$$c_5 = \sigma c_1 + (\sigma\tau - \beta^4), \quad c_6 = c_1 + \sigma c_3,$$

$$c_7 = \sigma c_1 + c_4, \quad c_8 = c_1 + \tau c_3.$$

For the periodic structure, i.e. $\sigma_L = 0$, the eigenvalues are solutions of the equation $\det(m\mathbf{I} - \mathbf{T}) = 0$. It is known that the eigenvalues of the transfer matrix $\mathbf{T}$ are in reciprocal pairs, i.e. the eigenvalues are of the form m_1, $1/m_1$, m_2, $1/m_2$. Hence, after expansion of the determinant, the eigenvalues are the roots of

$$m^4 - \mathrm{tr}(\mathbf{T})m^3 + \alpha m^2 - \mathrm{tr}(\mathbf{T})m + 1 = 0, \tag{4.3}$$

where

$$\alpha = t_{11}t_{22} + t_{11}t_{33} + t_{11}t_{44} + t_{22}t_{33} + t_{22}t_{44} + t_{33}t_{44} - (t_{12}t_{21} + t_{13}t_{31} + t_{14}t_{41} + t_{23}t_{32} + t_{24}t_{42} + t_{34}t_{43}).$$

Dividing both sides of the eigenvalue equation by m^2 gives

$$M^2 - \mathrm{tr}(\mathbf{T})M + (\alpha - 2) = 0, \tag{4.4}$$

where $M = m + 1/m$.

The solutions of equation (4.4) are given by

$$M_1 = \frac{\mathrm{tr}(\mathbf{T}) + \sqrt{[\mathrm{tr}(\mathbf{T})]^2 - 4(\alpha-2)}}{2}, \quad M_2 = \frac{\mathrm{tr}(\mathbf{T}) - \sqrt{[\mathrm{tr}(\mathbf{T})]^2 - 4(\alpha-2)}}{2}. \tag{4.4}$$

Case 1. If $[\mathrm{tr}(\mathbf{T})]^2 \geq 4(\alpha - 2)$, M_1 and M_2 are real

In terms of M_1 and M_2, the eigenvalues of equation (4.3) are of the form:

$$m_1 = \frac{M_1}{2} + \sqrt{\Big(\frac{M_1}{2}\Big)^2 - 1}, \quad \frac{1}{m_1} = \frac{M_1}{2} - \sqrt{\Big(\frac{M_1}{2}\Big)^2 - 1}, \quad m_2 = \frac{M_2}{2} + \sqrt{\Big(\frac{M_2}{2}\Big)^2 - 1}, \quad \frac{1}{m_2} = \frac{M_2}{2} - \sqrt{\Big(\frac{M_2}{2}\Big)^2 - 1}. \tag{4.5}$$

(1) $|M_1| \geq 2, |M_2| \geq 2$.

The eigenvalues of the transfer matrix are real and of the form $e^{\pm a}$ or $e^{\pm a + i\pi}$ ($a \in \mathbf{R}$); the correponding frequencies are in the stop-band and the wave amplitudes of all waves after traversing n elements are attenuated by the factor $e^{\pm an}$, in which the real exponent a implies non-travelling or attenuating waves.

(2) $|M_1| < 2, |M_2| < 2$.

The eigenvalues of the transfer matrix are complex and of the form $e^{\pm ik}$; the corresponding frequenices are in the pass-band and all waves travel in the form of $e^{\pm ikn}$, where k is the real wave number, the positive sign indicating left-travelling waves and the negative sign right-travelling waves.

(3) $|M_1| \geq 2, |M_2| < 2$.

$m_1, 1/m_1$ are real and of the form $e^{\pm a}$ or $e^{\pm a + i\pi}$ ($a \in \mathbf{R}$); the correponding frequencies are in the stop-band and the wave is non-travelling. $m_2, 1/m_2$ are complex and of the form $e^{\pm ik}$; the corresponding frequenices are in the pass-band and the wave is a travelling wave.

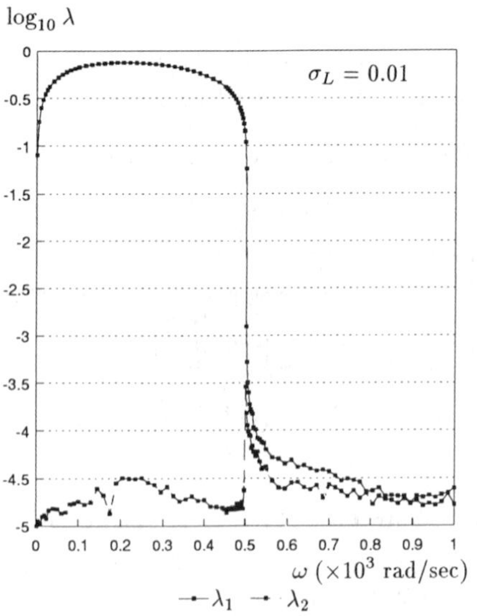

Figure 4 Lyapunov exponents for Timoshenko beam

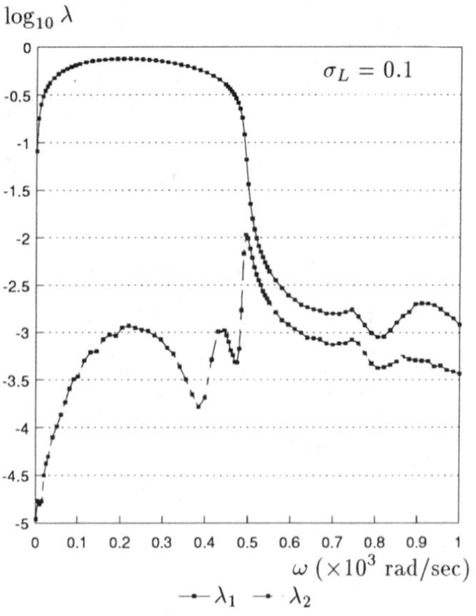

Figure 5 Lyapunov exponents for Timoshenko beam

(4) $|M_1| < 2, |M_2| \geq 2$.

$m_1, 1/m_1$ are complex and of the form $e^{\pm ik}$; the correponding frequencies are in the pass-band and the wave is a travelling wave. $m_2, 1/m_2$ are real and of the form $e^{\pm a}$ or $e^{\pm a + i\pi}$ ($a \in \mathbf{R}$); the correponding frequencies are in the stop-band and the wave is non-travelling.

Case 2. If $[\mathrm{tr}(\mathbf{T})]^2 < 4(\alpha - 2)$, M_1 and M_2 are complex

The eigenvalues of the transfer matrix are complex and of the form $e^{\pm ik}$; the corresponding frequencies are in the pass-band and all waves are travelling waves.

As a numerical example, the following values are chosen:

$$L_0 = 7.5 \text{ m}, \quad b = 5 \text{ m}, \quad A_L = 80 \times 10^{-6} \text{ m}^2,$$
$$A_b = 60 \times 10^{-6} \text{ m}^2, \quad A_d = 40 \times 10^{-6} \text{ m}^2,$$
$$E_L = E_b = E_d = 71.7 \times 10^9 \text{ N/m}^2,$$
$$\rho_L = \rho_b = \rho_d = 2768 \text{ kg/m}^3.$$

It is found that for all values of ω, $[\mathrm{tr}(\mathbf{T})]^2 > 4(\alpha - 2)$ in equations (4.5). When $\omega < 498.54$, $|M_1| > 2$, $|M_2| < 2$, which leads to Case 1(3), i.e. one wave is travelling and the other is non-travelling. When $\omega > 498.54$, $|M_1| < 2$, $|M_2| < 2$, which leads to Case 1(2), i.e. both waves are travelling.

Using the algorithm described in Section 3, all four Lyapunov exponents of the randomly disordered Timoshenko beam are determined for $0 < \omega \leq 1000$, $\sigma_L = 0.01$, and $\sigma_L = 0.1$, respectively, with the transfer matrix given by equations (4.1) or (4.2). From Section 2, it is known that the Lyaponov exponents of the randomly disordered Timoshenko beam are of the form λ_1, λ_2, $\lambda_3 = -\lambda_2$, $\lambda_4 = -\lambda_1$, ($\lambda_1 > \lambda_2$). This property may be used to check the correctness of the algorithm, and the numerical results obtained for the Lyapunov exponents show that this condition is satisfied very well. The results for the two positive Lyapunov exponents are plotted in Figures 4 and 5. For all frequencies, all the waves are localized due to the randomness in the length of the horizontal bars. The larger the disorder parameter σ_L, the larger the localization factors. The smallest positive Lyapunov exponent λ_2 is of particular interest, since it gives the smallest localization factor, which corresponds to the wave that carries energy along the structure farther than the wave corresponding to λ_1.

5 CONCLUSION

The localization of multi-wave structures has been studied in this paper. For a perfectly periodic multi-wave structure, in a given frequency band, some of the waves may be travelling waves and the others non-travelling waves. When the structure is randomly disordered due to imperfection in geometries or material properties, all the waves are attenuated. The exponential rates of decay of the amplitudes of waves propagating in the structure are characterized by the Lyapunov exponents. An algorithm was introduced to determine all the Lyapunov exponents. The smallest positive Lyapunov exponent is of particular interest; it is related to the smallest localization factor or the largest localization length, which corresponds to the wave that has potentially the least amount of decay or that carries energy along the multi-wave structure farther than any other wave. As an application, the localization of wave propagation in a large beam-like lattice truss used in large-area space structures was studied. The skeleton lattice truss was modelled as an equivalent continuous Timoshenko beam.

ACKNOWLEDGEMENT

The research for this paper was supported, in part, by the Natural Sciences and Engineering Research Council of Canada through Grants No. A-1815 (STA) and OGPO121355 (WCX).

REFERENCES

Ariaratnam, S.T. and Xie, W.-C., 1991. Lyapunov exponents in stochastic structural dynamics. *Lyapunov Exponents*, Oberwolfach, Germany, 1990, L. Arnold, H. Crauel and J.-P. Eckmann (eds.), *Lecture Notes in Mathematics*, Springer-Verlag, Berlin: 271-291.

Ariaratnam, S.T. and Xie, W.-C., 1993. Wave localization in randomly disordered nearly periodic long continuous beams. *Journal of Sound and Vibration*, (submitted).

Bougerol, P. and Lacroix, J., 1985. *Products of random matrices with applications to Schrödinger operators.* Birkhäuser.

Cai, G.Q. and Lin, Y.K., 1991. Localization of wave propagation in disordered periodic structures. *AIAA Journal* **29**(3): 450-456.

Cha, P.D. and Pierre, C., 1991. Vibration localization by disorder in assemblies of monocoupled, multimode component systems. *ASME Journal of Applied Mechanics* **58**: 1072-1081.

Furstenberg, H., 1963. Noncommuting random products. *Transactions of the American Mathematical Society* **108**(3): 377-428.

Hodges, C.H. 1982. Confinement of vibration by structural irregularity. *Journal of Sound and Vibration* **82**(3): 411-424.

Kissel, G.J., 1988. *Localization in disordered periodic structures.* PhD thesis, Department of Aeronautics and Astronatics, Massachusetts Institute of Technology, USA.

Kissel, G.J., 1991. Localization factor for multichannel disordered systems. *Physical Review A* **44**(2): 1008-1014.

Noor, A.K. and Andersen, C.M., 1979. Analysis of beam-like lattice trusses. *Computer Methods in Applied Mechanics and Engineering* **20**: 53-70.

Oseledec, Y.I., 1968. A multiplicative ergodic theorem. Lyapunov characteristic number for dynamical systems. *Transactions of the Moscow Mathematical Society* **19**: 197–231, (English translation).

Pestel, E.C. and Leckie, F.A., 1963. *Matrix methods in elastomechanics.* McGraw-Hill, New York.

Pierre, C., 1990. Weak and strong vibration localization in disordered structures: a statistical investigation. *Journal of Sound and Vibration* **139**: 111-132.

Pierre, C. and Dowell, E.H., 1987. Localization of vibrations by structural irregularity. *Journal of Sound and Vibration* **114**(3): 549-564.

Wolf, A., Swift, J., Swinney, H. and Vastano, A., 1985. Determining Lyapunov exponents from a time series. *Physica* **D16**: 285-317.

Structural Safety & Reliability, Schuëller, Shinozuka & Yao (eds) © 1994 Balkema, Rotterdam, ISBN 90 5410 357 4

Broad-band vibrations of driven components of complex structures

A.K. Belyaev
State Technical University, St. Petersburg, Russia

H.J. Pradlwarter
Innsbruck University, Austria

ABSTRACT: The concept allows us to consider any driven component separately. The presence of other structural members and of coupling is to be taken into account by means of a frequency drift and of an effective damping factor. The three methods: Bolotin´s method of integral estimates, Statistical Energy Analysis and theory of high-frequency vibration are combined in order to forecast the vibrations of structural members of complex systems

1 INTRODUCTION

A detailed description of vibration of a complex structure is made difficult, first, by the complexity of the structure´s shape, then, by its being assembled of separate substructures and, finally, by the presence of various secondary systems fixed to the primary structure. Even if one could obtain an "exact" solution the very interpretation of this result is a problem too. The field of vibration of up-to-date complex structures under broad-band excitation is a complicated function of time and spatial coordinates, since so many modes are excited in the structure.

Under these circumstances it seems reasonable to use integral descriptions. First, the Bolotin method of integral estimates (Bolotin 1984) is worth mentioning. The central idea of the method implies the consideration of a driven substructure as an absolutely isolated one. This method was applied to estimate a certain mean value of vibration of thin-walled elements under broad-band excitation. Its application results in obtaining rather uncomplicated formulae in closed form. The main shortcoming of the Bolotin approach is the absolute isolation of substructures considered. On the contrary, Statistical Energy Analysis (SEA) being a version of simple, diffusive, transport theory is aiming mainly at studying the mechanical properties of substructures´ coupling (Lyon 1975). A complex structure is viewed as a set of coupled substructures and a dynamicist only deals with the spatially averaged vibration level within each substructure. The coupling itself is under the closest investigation in SEA.

Certain integral methods are offered by Belyaev and Palmov 1986 to forecast the field of high-frequency vibration of complex structures. The main distinctions between low- and high-frequency vibrations as well as some essential features of high-frequency vibration are given in paper by Belyaev 1991. As shown in these papers one succeeds in obtaining a relatively uncomplicated boundary-value problem for high-frequency vibration. The properties of an actual structure are reflected in this integral theory in the form of certain averaged rigidity and averaged mass characteristics and generalized spectra. This enables us to avoid dipping into too many details of the structure. The application of the theory results in obtaining some generalized characteristics of the vibration field. This level of description can be considered sufficient for many cases. Nonetheless it is often necessary to know the vibration of a particular member of the structure. It cannot be managed with the integral methods alone, because this member itself is not represented in the dynamical model. The reasonable concept lies in precise consideration of this member alone, with the rest of the structure being described integrally. The combination of integral and precise description, namely, locality principle in structural dynamics was proposed by Belyaev and Palmov 1984, see also Belyaev 1990. This paper is intended to combine the three aforementioned approaches.

2 THE BOUNDARY PROBLEM

The field of high-frequncy vibration of actual complex structures is no doubt a three-dimensional one. Nevertheless as shown by Pervozvanskii 1986 the high-frequency vibrations propagate along the wavegiudes as uniaxial waves. It means that for extended complex structures one can apply one-dimensional theory of high-frequency vibration,

Belyaev 1992. In Fig. 1 a complex structure is depicted as a rod $0 \le x \le L$ described integrally. The substructure of interest is represented as a rod $l \le x \le r$ to be described with conventional methods of the vibration theory.

The coupling is commonly considered to be conservative (Lyon 1975, Fahy 1974, Hodges and Woodhouse 1986). Gaskets and contacts in engineering mechanics as two of the most typical cases of coupling will be under our consideration. Contacting surfaces are undoubtedly rough that results in a very low local contact rigidity and in massless coupling. The same conclusions are valid for gaskets. This allows us to represent the coupling as a spring having a stiffness K.

The dynamics of the subsystem $l \le x \le r$ is governed by the equation

$$(C_2 u_2')' - \mu_2 \ddot{u}_2 + f = 0 \; ; \quad \dot{} = \frac{\partial}{\partial t}, \; ' = \frac{\partial}{\partial x} \tag{2.1}$$

where $C_2(x)$ and $\mu_2(x)$ are longitudinal rigidity and mass density per unit length, $u_2(x,t)$ is the absolute displacement , $f(x,t)$ is external distributed force. Let $v_n(x)$, $n = 1,2,..\infty$ be the normal modes of free rod $l \le x \le r$, which are supposed to be normalized. In order to obtain the equation for generalized coordinate $q_n(t)$ one multiplies (2.1) by $v_n(x)$ and integrates within this rod. After the integration by parts we have

$$\ddot{q}_n + \Omega_n^2 q_n = \int_l^r f(x,t) v_n(x) dx - [C_2 u_2'(l,t)] v_n(l)$$

$$q_n(t) = \int_l^r \mu_2(x) v_n(x) u_2(x,t) dx \tag{2.2}$$

where the properties of the normal modes $v_n'(l) = v_n'(r) = 0$ as well as the boundary condition $u_2'(r) = 0$ have been taken into account, and Ω_n is an eigenfrequency. Let us assume that external load, generalized coordinates etc. can be represented by their spectral representations

$$f(x,t) = \int_{-\infty}^{\infty} f(x,\omega) e^{i\omega t} d\omega \quad etc \tag{2.3}$$

where from now on the same designation for the spectra will be retained. Substitution of (2.3) into (2.2) gives us the expression for the generalized coordinates´ spectrum

$$q_n(\omega) = \frac{\int_l^r f(x,\omega) v_n(x) dx - [C_2 u_2'(l,\omega)] v_n(l)}{-\omega^2 + \Omega_n^2} \tag{2.4}$$

The differential equation for integral description of the part $0 \le x \le L$ is given as

$$0 < x < L \quad C_1 \frac{d^2 u_1}{d^2 x} + M_1 \omega^2 (1 - i\kappa)^2 u_1 = 0 \tag{2.5}$$

where M_1 and C_1 are an averaged mass of the length unit and an averaged longitudinal rigidity of the structure, $\kappa(\omega)$ is a dimensionless frequency-dependent parameter responsible for absorption of high-frequency vibration by the structure, u_1 is absolute displacement of framework of complex structure, Belyaev 1991, 1992. The balance of forces in coupling is

$$C_1 u_1'(L) = K[u_2(l) - u_1(L)] = C_2 u_2'(l) \tag{2.6}$$

As shown by Belyaev, 1991 the field of high-frequency vibration is not actually sensitive to boundary conditions at unloaded remote boundaries of an extended structure. This allows us to put for instance the simplest boundary condition

$$x = 0 \quad u_1 = 0 \tag{2.7}$$

3. EFFECTIVE CHARACTERISTICS OF DRIVEN COMPONENTS

The solution of (2.5), (2.7) is

$$u_1(x,\omega) = A \sin \omega(1 - i\kappa) x / a_1 \tag{3.1}$$

where $a_1 = \sqrt{C_1 / M_1}$ is the velocity of the energetic centre of the propagating disturbance. Substituting (2.6) into (2.4) we can easily obtain

$$u_2(l,\omega) = \sum_{n=1}^{\infty} v_n(l) q_n(\omega) = \sum_{n=1}^{\infty} \frac{v_n(l)}{-\omega^2 + \Omega_n^2} \tag{3.2}$$

$$\left\{ \int_l^r f(x,\omega) v_n(x) dx - K[u_2(l,\omega) - u_1(L,\omega)] v_n(l) \right\}$$

The condition (2.6) helps us to express (3.1) in such a way

$$u_1(L,\omega) = C_1 u_1'(L) \frac{\tan \omega(1 - i\kappa) L / a_1}{C_1 \omega (1 - i\kappa) / a_1} =$$

$$= K[u_2(l,\omega) - u_1(L,\omega)] \frac{\tan \omega(1 - i\kappa) L / a_1}{C_1 \omega (1 - i\kappa) / a_1} \tag{3.3}$$

Subtracting (3.3) from (3.2) allows us to obtain expression for $u_2(l,\omega) - u_1(L,\omega)$, the latter being subsituted into (2.4) gives the following expression for generalized coordinates

See Appendix for Eq. (3.4)

In order to find the response of a structural member let us make use of the method of integral estimates, Bolotin 1984. A certain mean value of vibration of a driven component could be evaluated using asymptotic expressions for eigenfrequencies and normal modes. A generalization of this method done by Palmov 1979 does not require any information about normal modes at all. Following Palmov 1979 let us agree to consider the case $\mu(x) = const$. In this case an averaged value of the square of displacement looks like this

$$\langle u_2^2 \rangle = \frac{1}{r - l} \int_l^r u_2^2(x,t) dx = \tag{3.5}$$

$$= \frac{1}{\mu(r-l)} \sum_{n=1}^{\infty} \sum_{j=1}^{\infty} q_j q_n \int_l^r \mu v_n v_j dx = \frac{1}{\mu(r-l)} \sum_{n=1}^{\infty} q_n^2(t)$$

Let us suppose that the external force f is a stationary delta-correlated spatial white noise, i.e.

$$\langle f(x,\omega) f^*(x',\omega') \rangle = S_f(\omega)\delta(\omega-\omega')\delta(x-x') \quad (3.6)$$

where $S_f(\omega)$ is the spectral density of external force and * denotes the complex conjugate of a quantity. Provided that $\mu(x) = const$ and the normal modes are normalized, the substitution of (3.4) into (3.5) gives

See Appendix for Eq. (3.7)

where internal damping is introduced with viscous damping coefficient ξ_n.

In order to simplify this formula one should observe that $\langle u_2^2 \rangle$ is formed by an infinite number of resonant curves corresponding to the vibration of a single-degree-of-freedom system. If the modal overlap is high the theory of high-frequency vibration (Belyaev and Palmov 1986, Belyaev 1991) can be applied. Our interest lies in the case when the adjacent resonant curves are rather distant from each orther (low modal overlap). In this case considerable simplification is possible. In the vicinity of an eigenfrequency Ω_n the resonant term prevails and this allows us to neglect the nonresonant terms, to obtain

$$\langle u_2^2 \rangle = \frac{1}{\mu^2(r-l)} \sum_{n=1}^{\infty} \quad (3.8)$$

$$\int_{-\infty}^{\infty} \frac{S_f(\omega) d\omega}{\left| -\omega^2 + i\xi_n\omega + \Omega_n^2 + \frac{K C_1 \omega v_n^2(l)(1-i\kappa)}{C_1\omega(1-i\kappa) - i a_1 K} \right|^2}$$

which is asymptotically valid for extended complex structures at high frequencies (i.e. $\omega\kappa L a_1^{-1} >> 1$). The latter term in the denominator represents the effect of backward influence of the whole structure on the substructure vibrations. The result (3.8) can be rewritten in such a way

$$\langle u_2^2 \rangle = \frac{1}{\mu^2(r-l)} \sum_{n=1}^{\infty} \int_{-\infty}^{\infty} \frac{S_f(\omega) d\omega}{\left| -\omega^2 + i\hat{\xi}_n\omega + \hat{\Omega}_n^2 \right|^2} \quad (3.9)$$

where an effective eigenfrequency $\hat{\Omega}_n$ and an effective damping $\hat{\xi}_n$ are introduced

$$\hat{\Omega}_n^2 = \Omega_n^2 + \frac{K v_n^2(l) \left[1 + \frac{\kappa a_1 K}{C_1 \Omega_n (1+\kappa^2)} \right]}{\left[1 + \frac{\kappa a_1 K}{C_1 \Omega_n (1+\kappa^2)} \right]^2 + \left[\frac{a_1 K}{C_1 \Omega_n (1+\kappa^2)} \right]^2} \quad (3.10)$$

$$\hat{\xi}_n = \xi_n + \frac{\frac{a_1 K^2 v_n^2(l)}{C_1 \Omega_n^2 (1+\kappa^2)}}{\left[1 + \frac{\kappa a_1 K}{C_1 \Omega_n (1+\kappa^2)} \right]^2 + \left[\frac{a_1 K}{C_1 \Omega_n (1+\kappa^2)} \right]^2} \quad (3.11)$$

The formula (3.9) is a standard one for broad-band vibrations of a rod. Hence, instead of thinking of the whole structure as two coupled mechanical systems with distinctive characteristics and properties, it becomes convenient to consider only a substructure of interest. The influence of the discarded structure and of coupling is to be taken into consideration by means of effective eigenfrequencies $\hat{\Omega}_n$ and effective damping $\hat{\xi}_n$.

4 APPLICATION OF THE CONCEPT

1. Vibrating combustion of a solid propellant rocket engine.

Such a combustion may be considered as a wide-band spatially uncorrelated random process. For the propellant´s compartment the condition of constant mass per length unit is satisfied. Hence, the result (3.9) is applicable. For a numerical example the following parameters were taken: $\mu = M_1 = 100$ kgm^{-1}, $C_1 = 8 \cdot 10^7$ N, $C_2 = 4 \cdot 10^7$N, $K = 8 \cdot 10^7$Nm^{-1}, $\kappa = 0.2$, $\xi_n = 198.5$s^{-1}, $r - l = 1$m, $L = 5$m. The theoretical results obtained by means of Eqs. (3.9)-(3.11) for the averaged square of the absolute value of the transfer function $|H|^2 = S_{\langle u_2^2 \rangle} S_f^{-1}$ are shown in Fig.2. The upper curve corresponds to the absolutely isolated engine compartment (*K=0 Theoretical*), while the curve $K \neq 0$ *Theoretical* was calculated with the help of (3.9). The theoretical results have been checked by a numerical Finite Element Approach. The first body $(0 \le x \le L)$ was discretized by 30 and the second one $(l \le x \le r)$ by 20 degrees of freedom. The damping parameter κ of the first body is represented by means of a complex stiffness matrix $C_1^* = C_1 / (1-i\kappa)^2 = \left[(1-\kappa^2)/(1+\kappa^2)^2 + i2\kappa/(1+\kappa^2)^2 \right] C_1$ where C_1 is the real classical stiffness matrix. The damping matrix of the second body is given as $D_2 = M_2 \cdot \xi_n$ where M_2 is the mass matrix. The averaged transfer function $|H(\omega)|^2$ has been determined by solving the complex matrix equation $\tilde{H}(\omega) = \left[K + i\omega D + \omega^2 M \right]^{-1} \cdot \tilde{f}(\omega)$ and averaging $|\tilde{H}(\omega)|^2$ over the extend of the second body. The result of numerical work is represented in Fig.2 by curve referred to as $K \neq 0$*Numerical*. The discordance between these two theoretical curves is considerable for the first three resonances (notice that the vertical scale is logarithmic). The discrepancy between the analytical result (3.9) and the numerical approach is observable at high frequencies and it is displayed by means of a frequency shift. This phenomenon is a well-known one and the shift can be easily eliminated by increasing of a number of degrees of freedom. Fig.3 illustrates the effect of the entire structure parameters´ variations on the vibrations of the structural member. The various sets of parameters of the first body and of the coupling were taken: κ=0.1 and 0.2 as well as 0.4, $K=8\cdot10^7 Nm^{-1}$ and $4\cdot10^7 Nm^{-1}$, L=5m and 10m, the end $x=0$ is free and fixed, while the parameters of the substructure were not changed. Despite the wide variations of the structure´s parameters and of the coupling the substructure remains unaffected in the high-frequency domain. That means that at the high frequencies the vibrations localize within structural members.

2. Cover of an engine´s head

The vibrational design of engines´ covers is important for the prediction of engine noise. The head and its cover form an "acoustical chamber". Since there is a little attenuation due to damping and the reflection of propagating waves is strong, the wave is reflected many times before it is absorbed. It results in a highly reverberant field, which is more or less uniformly distributed throughout the cover, Hodges and Woodhouse 1986. Such a "rain-on-the-roof" loading (after Maidanik) is a typical spatial white noise.
Let the cover be a box all five walls of which being thin plates h thick. In this case $\mu = \rho h = const$ is a mass per unit area, where ρ is density. After the averaging of the field of velocities we have the following result instead of (3.9)

$$\left\langle \dot{u}_2^2 \right\rangle = \frac{1}{\mu^2 A} \sum_{n=1}^{\infty} \int_{-\infty}^{\infty} \frac{\omega^2 S_p(\omega) d\omega}{\left| -\omega^2 + i\hat{\xi}_n \omega + \hat{\Omega}_n^2 \right|^2} \tag{4.1}$$

where $A = a_1 a_2 + 2(a_1 a_3 + a_2 a_3)$ is the area of the box faces and $S_p(\omega)$ is the spectral density of the external pressure.
For lightly damped material of the cover and a broad-band external driving, the integral (4.1) can be evaluated, Bolotin 1984, to give

$$\left\langle \dot{u}_2^2 \right\rangle = \frac{\pi}{\mu^2 A} \sum_{n=1}^{\infty} S_p\left(\hat{\Omega}_n\right) / \hat{\xi}_n(\hat{\Omega}_n) =$$

$$= \frac{\pi}{\mu^2 A} \int_0^{+\infty} S_p(\Omega) / \hat{\xi}(\Omega) \; dN / d\Omega \, d\Omega \tag{4.2}$$

where the density of eigenfrequencies $dN/d\Omega$ is introduced. Non-degeneration of the dynamic fringe effect for plates allows an approximate matching of solutions for adjacent subregions. Consequently, the eigenfrequencies´ density of the cover is equal to the sum of the densities of these plates, Bolotin 1984, i.e.

$$\frac{dN}{d\Omega} = \frac{A}{4\pi} \sqrt{\frac{\rho h}{D}} \tag{4.3}$$

where D is the flexural rigidity of the plate. The final result becomes

$$\left\langle \dot{u}_2^2 \right\rangle = \frac{1}{4\rho h \sqrt{\rho h D}} \int_0^{+\infty} S_p(\Omega) / \hat{\xi}(\Omega) \, d\Omega$$

$$S_{\left\langle \dot{u}_2^2 \right\rangle}(\Omega) = \frac{\pi}{4\rho h \sqrt{\rho h D}} \frac{S_p(\Omega)}{\hat{\xi}(\Omega)} \tag{4.4}$$

where the spectral density of spatially averaged velocities of the cover $S_{\left\langle \dot{u}_2^2 \right\rangle}$ is introduced.

For numerical calculations the following parameters were taken: a_1=0.15 m, a_2=0.4 m, a_3=0.08 m, $h=0.8\cdot10^{-3}$ m, $E=2.1\cdot10^{11}$ Nm-2, $\rho=7.8\cdot10^3$ kgm^{-3}, ν=0.3, $K=10^8$ Nm^{-2}, $C_1=1.35\cdot10^9$ N, M_1=150 kgm^{-1}, L=1 m. The critical ratio of damping is chosen to be $5\cdot10^{-3}$ at all frequencies $5\cdot10^2$ cps $\leq f \leq 4\cdot10^3$ cps. For these parameters of the cover the eigenfrequency separation is equal to 17 cps, that allows us to make use of the formula (4.4). We can neglect the frequency drift (3.10), since the asymptotic density of eigenfrequencies is used. In Fig. 4 the curves of the square of the absolute value of transfer function $|H|^2 = S_{\left\langle \dot{u}_2^2 \right\rangle} S_p^{-1}$ are shown. The upper curve corresponds to the absolutely isolated cover (Bolotin´s method), the lower one being calculated using the offered concept.

5 CONCLUDING REMARKS

Three methods: Bolotin´s method of integral estimates, Statistical Energy Analysis and the theory of high-frequency vibration were combined in order to forecast the vibrations of structural members of complex systems. This allows to overcome shortcomings of these methods. So, the application of the theory of high-frequency vibration results in obtaining certain integral fields of vibration, but a structural element itself is not represented in this theory. Quite the contrary, in the method of integral estimates a structural element is considered, but it is assumed to be absolutely isolated. The main result of the proposed concept is: structural members may be considered separately. The presence of other structural members and of coupling is to be taken into account by means of a frequency drift and of an effective damping factor. Out of these two parameters the effective damping factor is more important. Under a broad-band driving a great many modes are excited, so that we can use an asymptotic density of eigenfrequencies which is not sensitive to a frequency drift.

Two practical problems were solved using this concept. The analysis of Figs 2-4 allows us to make some conclusions about the localization of vibrations, see e.g. Hodges and Woodhouse 1986. Analysing these figures one can visualize the frequency domain where the results of the conventional methods and of the present approach can be considered as coinciding with desired accuracy. It is also evident that the localization of vibrations actually takes place at high frequencies and the tendency to localization becomes stronger with the growth of frequency. This conclusion fully agrees with the central result of the papers by Pierre 1990 and Cha and Pierre 1991. The present study allows us to indicate a frequency domain wherein a strong localization of vibration within structural members occurs.

ACKNOWLEDGEMENT

The authors are greatly indebted to Professor G.I.Schuëller, University of Innsbruck, Austria for his encouragement and constant attention. The research was supported by Austrian Industrial Research Promotion Fund under the contract No 6/546. Deep gratitude is expressed by the first author to the Alexander von Humboldt Foundation, Germany for the financial support.

REFERENCES

Belyaev, A.K. 1990. On the application of the locality principle in structural dynamics. Acta Mechanica 83: 213-222.

Belyaev, A.K. 1991. Vibrational state of complex mechanical structures under broad-band excitation. Int. J. of Solids and Structures 27: 811-823.

Belyaev, A.K. 1992 Dynamic simulation of high-frequency vibration of extended complex structures. Int. J. Mechanics Structures and Machines 20: 155-168.

Belyaev, A.K. & Palmov, V.A. 1984. Locality principle in structural dynamics. In M. Petyt & H.F.Wolfe (eds.), Proceedings of the 2nd Int. Conf. on Recent Advances in Structural Dynamics, 1, p. 229-238, University of Southampton, England.

Belyaev, A.K. & Palmov, V.A. 1986. Integral theories of random vibration of complex structures. In I. Elishakoff & R. Lyon (eds.), Random Vibration - Status and Recent Developments, p. 19-38. Amsterdam, Elsevier.

Bolotin, V.V. 1984. Random vibrations of elastic systems. The Hague, Nijhoff.

Cha, P.D. & Pierre, C. 1991. Vibration localization by disorder in assemblies of monocoupled, multimode component systems. ASME J. Applied Mechanics 58: 1072-1081.

Fahy, F.J. 1974. Statistical energy analysis - a critical review. Shock and Vibration Digest 6: 14-33.

Hodges, C.H. & Woodhouse, J.1986. Theories of noise and vibration transmission in complex structures. Reports in Progress in Physics. 49: 107-170.

Lyon, R.H. 1975. Statistical energy analysis of dynamical systems: Theory and applications. Cambridge, MIT Press.

Palmov, V.A. 1979. Integral methods for the analysis of vibration of dynamic structures (in Russian). Advances in Mechanics 2: 3-24.

Pervozvanskii, A.A. 1986. High-frequency vibrations in case of considerable dissipation (in Russian). Reports of the USSR Academy of Science 288: 1068-1070.

Pierre, C. 1990. Weak and strong vibration localization in disordered structures: a statistical investigation. J. Sound Vibration 139: 111-132.

APPENDIX: EQUATIONS (3.4) and (3.7)

$$q_n(\omega) = \frac{\left\{1 + K\left[\sum_{\substack{j=1 \\ j\neq n}}^{\infty} \frac{v_j^2(l)}{-\omega^2 + \Omega_j^2} + \frac{\tan\omega(1-i\kappa)L/a_1}{C_1\omega(1-i\kappa)/a_1}\right]\right\}\int_l^r f v_n dx - K v_n(l) \sum_{\substack{j=1 \\ j\neq n}}^{\infty} \frac{v_j(l)}{-\omega^2 + \Omega_j^2}\int_l^r f v_j dx}{(-\omega^2 + \Omega_n^2)\left\{1 + K\left[\sum_{j=1}^{\infty} \frac{v_j^2(l)}{-\omega^2 + \Omega_j^2} + \frac{\tan\omega(1-i\kappa)L/a_1}{C_1\omega(1-i\kappa)/a_1}\right]\right\}} \tag{3.4}$$

$$\left\langle u_2^2 \right\rangle = \frac{1}{\mu(r-l)} \sum_{n=1}^{\infty} q_n^2(t) = \tag{3.7}$$

$$= \frac{1}{\mu^2(r-l)} \sum_{n=1}^{\infty} \int_{-\infty}^{\infty} \frac{S_f(\omega)d\omega}{\left|-\omega^2 + i\xi_n\omega + \Omega_n^2\right|^2 \left|1 + K\left[\sum_{j=1}^{\infty} \frac{v_j^2(l)}{-\omega^2 + i\xi_j\omega + \Omega_j^2} + \frac{\tan\omega(1-i\kappa)L/a_1}{C_1\omega(1-i\kappa)/a_1}\right]\right|^2}$$

$$\left\{\left|1 + K\left[\sum_{\substack{j=1 \\ j\neq n}}^{\infty} \frac{v_j^2(l)}{-\omega^2 + i\xi_j\omega + \Omega_j^2} + \frac{\tan\omega(1-i\kappa)L/a_1}{C_1\omega(1-i\kappa)/a_1}\right]\right|^2 + K^2 v_n^2(l) \sum_{\substack{j=1 \\ j\neq n}}^{\infty} \frac{v_j^2(l)}{\left|-\omega^2 + i\xi_j\omega + \Omega_j^2\right|^2}\right\}$$

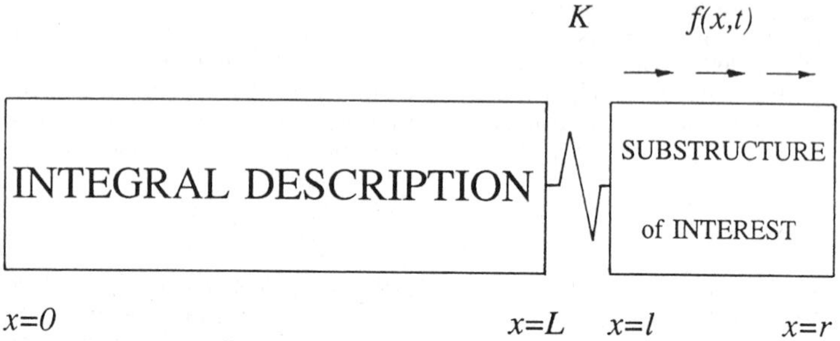

Fig. 1. Dynamic model

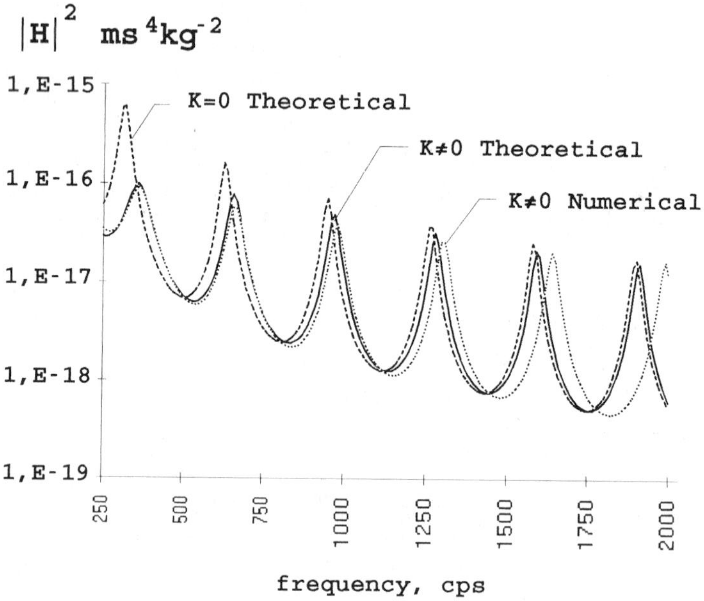

Fig. 2. Transfer function versus frequency for the methods under discussion.

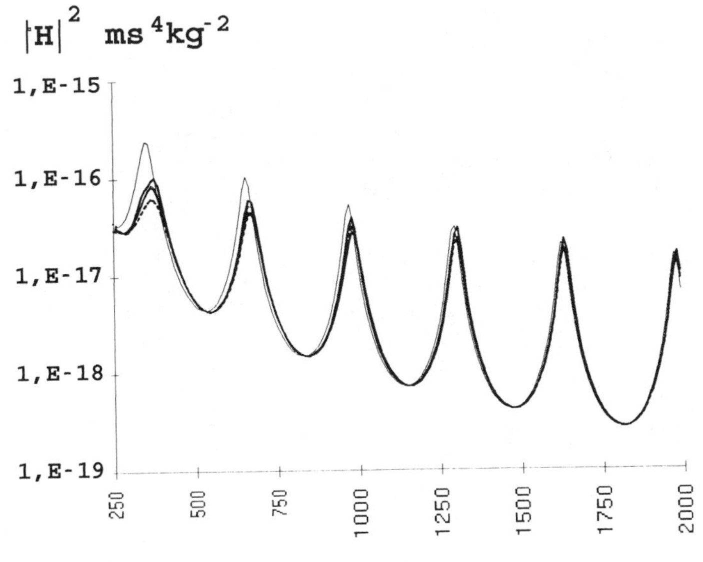

Fig. 3. Transfer function versus frequency for various sets of parameters of the structure.

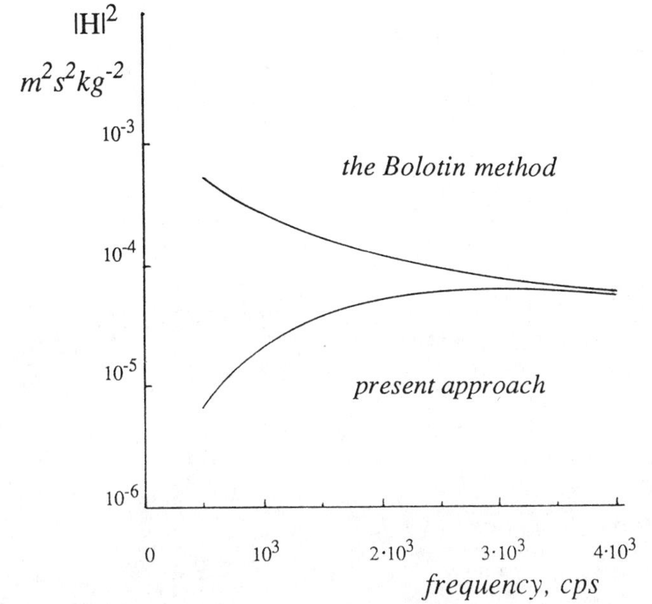

Fig. 4. Comparison of the Bolotin method and of the present approach

Structural Safety & Reliability, Schuëller, Shinozuka & Yao (eds) © 1994 Balkema, Rotterdam, ISBN 90 5410 357 4

Stochastic stability of uncertain two-dimensional structures

Christian G. Bucher
Institute of Engineering Mechanics, University of Innsbruck, Austria

1 Introduction

There is an increasing demand on the cost-effectiveness of engineering structures which leads to design problems related to stability. While in traditional design the possible loss of stability is accounted for primarily in static and deterministic cases, the application of new technologies as developed e.g. for aerospace structures require the consideration of situations in which both stochasticity and dynamics play an important role.
In this paper, stochasticity refers to both system properties such as mass, stiffness or boundary conditions as well as time-varying quantities such as external loads or parametric excitations. Clearly, overall stability may be (adversely) affected by both classes of uncertainties.
Dynamic analysis has traditionally been confined to the investigation of the effect of external loads on the structural response. There is a considerable number of computational procedures available for the solution of such problems. Since external loads do not change the basic properties of the structure (the structure is time-invariant) numerical solutions based on the Finite Element Method are most suitable (e.g. Deodatis and Shinozuka, 1991, Ghanem and Spanos, 1991; Bucher and Brenner, 1992). However, as soon as the structure undergoes parametric excitations the methods are far less well developed.

This paper deals with the analyis of two-dimensional structures (plates) with stochastic properties under the influence of random parametric excitations. The stability is investigated on one hand in terms of static instability (divergence) and on the other hand in terms of the top Lyapunov exponent i.e. by the criterion of almost sure (sample) stability. The random system properties are taken into account by applying a Monte Carlo simulation technique.

2 Statement of problem

The analysis of a continuous type system under parametric excitation can usually be described by a set of partial differential equations. In the following, a plate with in-plane random loading and stochastic boundary conditions (elastic supports with random field spring constants) is investigated. Within the framework of Kirchhoff plate theory, the differential equation governing the transversal vibration of the plate $w(x,y)$ is given by

$$K\Delta\Delta w + \mu\frac{\partial^2 w}{\partial t^2} + n_{xx}\frac{\partial^2 w}{\partial x^2} + n_{xy}\frac{\partial^2 w}{\partial x \partial y} + n_{yy}\frac{\partial^2 w}{\partial y^2} = 0 \tag{1}$$

In this equation, K denotes the plate bending stiffness, μ is the mass per unit area, n_{xx}, n_{xy}, and

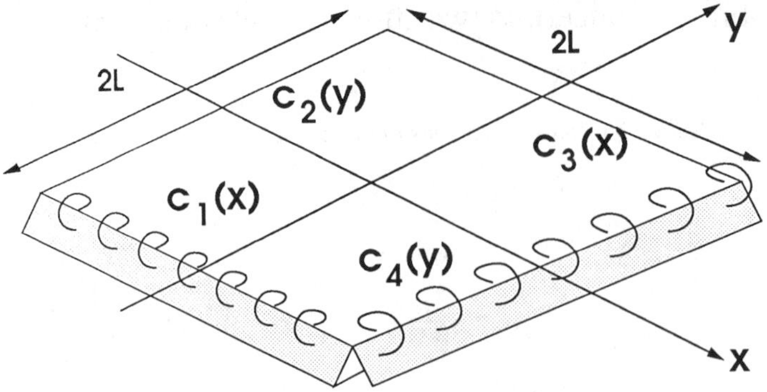

Fig. 1: Square plate with elastic boundary springs

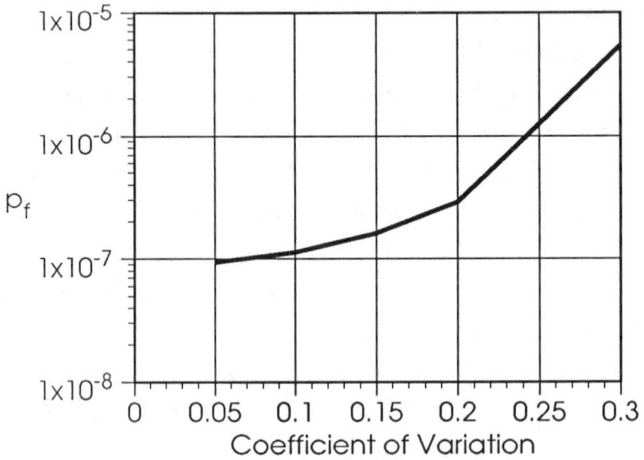

Fig. 2: Probability of static instability vs. c.o.v. of boundary springs (Case 1)

n_{yy} denote the in-plane forces, possibly depending on time. This equation can be solved by specifying a set of appropriate boundary conditions. In the current investigation, the boundary conditions (as mentioned above) are assumed to be stochastic. Clearly, the above equation, together with homogeneous boundary conditions, admits the trivial solution $w(x,y) = 0$. This equation can be utilized to decide on the stability of the solution in terms of the properties of the in-plane loads and of the boundary constants.

3 Method of Analysis

The solution for $w(x,y)$ is expanded in a power series

$$w(x,y,t) = \sum_{i=1}^{n} \sum_{k=1}^{m} A_{ik}(t)\, x^i\, y^k \qquad (2)$$

which, after applying a Galerkin procedure (i.e. multiplying the differential equations as well as the set of boundary conditions by powers of x and y and integrating over the domain), leads to a set of ordinary differential equations for the coefficients $A_{ik}(t)$. Rearranging these equations

Table 1: Statistical Data for Numerical Example

		F_{xx}	F_{yy}	ζ	c_1, c_2, c_3, c_4
Case 1 static	Mean	10.5	10.5	-	1.0
	c.o.v.	0.0571	0.0571	-	0.05 - 0.3
Case 2 static	Mean	9.0	9.0	-	1.0
	c.o.v.	0.20	0.20	-	0.05 - 0.3
Case 3 dyn.	Mean	7.0	7.0	0.01	1.0
	c.o.v.	0.0571	0.0571	0.2	0.05 - 0.3
Case 4 dyn.	Mean	6.0	6.0	0.01	1.0
	c.o.v.	0.20	0.20	0.30	0.05 - 0.3

and introducing modal analysis finally leads to the form

$$\ddot{X}_r + 2\zeta_r\,\omega_r\,\dot{X}_r + \omega_r^2\,X_r + (d_{xx,rs}\,n_{xx}(t) + d_{xy,rs}\,n_{xy}(t) + d_{xx,rs}\,n_{yy}(t))X_s;\; r = 1, n\cdot m \qquad (3)$$

(summation over repeated indices is implied). In the above equation, ζ_r is the modal damping ratio of the r-th mode.

If, for ease of presentation, it its assumed that the temporal variations of the in-plane forces are caused by the same random process $\xi(t)$, then the equations of motion (3) can be reduced to one equation

$$\ddot{X}_1 + 2\zeta_1\,\omega_1\,\dot{X}_1 + \omega_{eff}^2\,X_1 + d_{11}\,X_1\,\xi(t) \qquad (4)$$

Note that in this equation, static instability (divergence) is indicated by the effective circular frequency ω_{eff} becoming zero.

Applying the method of stochastic averaging, the equations (4) can be transformed into an equivalent Itô-equation for amplitude and phase. Details are omitted here, since they are well documented in the literature (e.g.Sri Namachchivaya and Lin, 1988). Stability of the trivial solution of the original PDE (1) is then assessed by calculating the top Lyapunov exponent for the system (4). Since the system in general becomes unstable in only one mode (usually the lowest mode of vibration) the formulation for the top Lyapunov λ exponent considering only this mode becomes very simple. It should be mentioned, that of course the consideration of higher modes introduces coupling through the random parametric excitation and consequently changes the Lyapunov exponents (Sri Namachchivaya et al. 1991, Bucher, 1991). However, for the purpose of the current investigation this may be omitted since the relative effect of random boundary conditions it is expected to remain largely unaltered by modal coupling. In that case, the expression for the top Lyapunov exponent becomes (compare e.g. Sri Namachchivaya and Lin, 1988)

$$\lambda = -\zeta_1\omega_1 + \frac{1}{4}d_{11}\,S_{\xi\xi}(2\omega_{eff}) \qquad (5)$$

In the above equations, the coefficients contain random quantities arising from the uncertainties of system parameters (in this case the damping and the boundary conditions). Hence the Lypunov exponents as calculated are conditional on the particular realization of the random data.

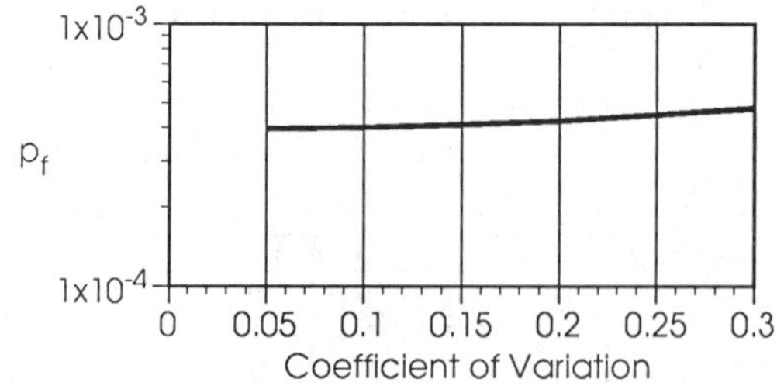

Fig. 3: Probability of static instability vs. c.o.v. of boundary springs (Case 2)

4 Numerical Example

In the following, it is assumed that the plate as shown in Fig. 1 is supported along the edges, so that one set of boundary conditions becomes

$$w(-1,y) = w(1,y) = w(x,-1) = w(x,1) = 0 \qquad (6)$$

Due to the elastic springs $c_1(x)$, $c_3(x)$, $c_2(y)$, $c_4(y)$, the rotations and bending moments are randomly coupled along the boundaries so that (neglecting the effect of Poisson's ratio)

$$K\frac{\partial^2 w}{\partial y^2} - c_1(x)\frac{\partial w}{\partial y} = 0 \; at \; y = -L;$$

$$K\frac{\partial^2 w}{\partial y^2} - c_3(x)\frac{\partial w}{\partial y} = 0 \; at \; y = L;$$

$$K\frac{\partial^2 w}{\partial x^2} - c_2(y)\frac{\partial w}{\partial x} = 0 \; at \; x = -L;$$

$$K\frac{\partial^2 w}{\partial x^2} + c_4(y)\frac{\partial w}{\partial x} = 0 \; at \; x = L; \qquad (7)$$

In the numerical calculations the following values of the parameters are assumed: K = 1, $n_{xx} = F_{xx}\,[1 + \xi(t)]$, $n_{xy} = 0$, $n_{yy} = F_{yy}\,[1 + \xi(t)]$. The parameters F_{xx} and F_{yy} are assumed to be random variables with a log-normal distribution. The random process $\xi(t)$ is assumed to be a white noise with a constant spectral density $S_{\xi\xi} = 0.04$. The modal damping ζ_1 of the critical mode is assumed to be log-normally distributed. The elastic springs are assumed to be Gaussian random fields with a mean value of 1. Their spatial autocorrelation function is assumed in the form (e.g. Shinozuka and Deodatis, 1988)

$$R_{cc}(x_1, x_2) = \sigma_c^2 \exp(-\frac{|x_1 - x_2|}{L_c}) \qquad (8)$$

in which σ_c is the variance of the random field and L_c is the correlation length. In a first step, the four random fields c_1, c_2, c_3, c_4 are assumed to be mutually independent. For the case of statically acting forces $n_{xx} = n_{yy}$, $n_{xy} = 0$ and deterministic spring constants the buckling load is $n_{crit} = 12.646$. Two sets of statistical parameters for the loads, the modal damping and the boundary springs are considered in the following. Reliability analyses for the cases of static instability (divergence) as well as dynamic instability(top Lyapunov exponent becoming positive) are performed using an advanced Monte Carlo simulation procedure as available in the software package ISPUD (Schuëller and Stix, 1987; Bourgund and Bucher, 1986).

The resulting probabilites of failure are shown in the above two figures relating the c.o.v. of the random support springs to the probability of the system becoming statically unstable. Figs. 4 and 5 show similar results for the dynamic stability

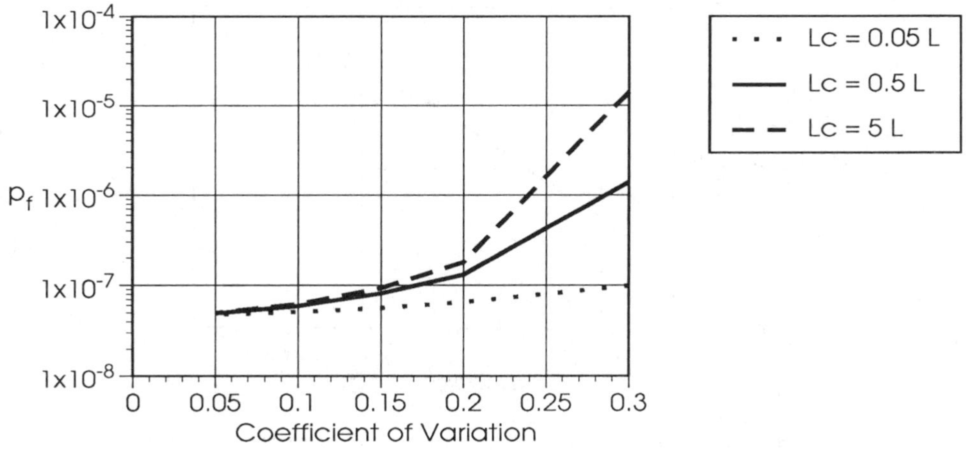

Fig. 4: Probability of dynamic instability vs. c.o.v. of boundary springs (Case 3)

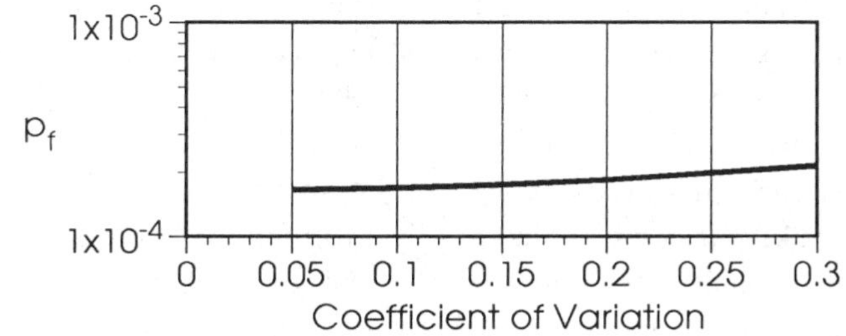

Fig. 5: Probability of dynamic instability vs. c.o.v. of boundary springs (Case 4)

problem. In addition, Fig. 4 shows the effect of varying correlation length L_c on the probability of instability.

The last Fig.6 explicitly compares the results for the dynamic case 4 with a c.o.v = 0.20 of the boundary spring constants for different values of the correlation length L_c

5 Conclusions

A Monte-Carlo based semi-analytical technique for the stability analysis of plates has been presented. Expressions for the top Lyapunov exponent governing the long-term behavior of the plate were found to be conditional on the realizations of system parameters. The inherent stochasticity of these parameters was taken into account by including them into a probabilistic model suitable for reliability analysis. The failure criterion "instability" as described by the sign of the top Lyapunov exponent formed the basis for calculation of the probaility of failure.

The numerical example seems to indicate that the effect of random boundary conditions on the stability of plates is pronounced only in situations where both loads and system damping are very accurately known, i.e. they are almost deterministic. Otherwise it appears that these uncertainties have significantly more influence on the reliability (in terms of both static and dynamic stability) of such structural systems.

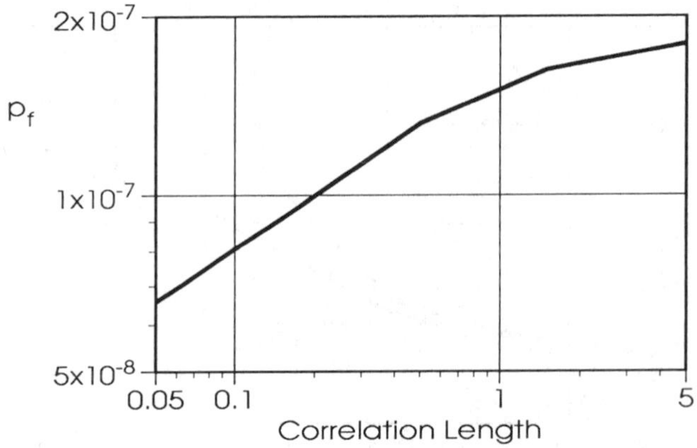

Fig. 6: Probability of dynamic instability vs. correlation length of boundary springs

This becomes even more pronounced when imperfect correlation (short correlation length) of the system uncertainties is taken into account, since such a correlation structure means that the spatial uncertainties virtual average out. In view of realistic representation of stochasticity it appears to be very unlikely that system uncertainties can play a role comparable to the loading uncertainties. It may be anticipated that these results play an important role in the future developments in computational stochastic mechanics as outlined e.g. by Schuëller et al., 1992.

Acknowledgements

This research has been supported by the Austrian Industrial Reserach Promotion Fund (FFF) under Project No. 6/636 which is gratefully acknowledged. The author also would like to thank *N. Sri Namachchivaya* for most fruitful discussions.

References

Bourgund, U., Bucher, C.G.: "Importance Sampling Procedure Using Design Points - ISPUD - A User´s Manual", Report 8-86, Inst. Eng. Mech., University of Innsbruck, 1986.

Bucher, C.G.: "Sample stability of multi-degree-of-freedom systems", in L.Arnold et al. (eds): *Lyapunov Exponents* (Lecture Notes in Mathematics), Springer, Berlin-Heidelberg, 1991, pp 322 - 330.

Bucher, C.G. and Brenner, C.E.: "Stochastic Response of Uncertain Systems", *Archive of Applied Mechanics*, 62 (1992), pp 507 - 516.

Deodatis, G. and Shinozuka, M.: "The weighted integral method. II: Response variability and reliability", *J. Eng. Mech.*, Proc ASCE, Vol. 117, No.8, August 1991, pp 1865 - 1877.

Ghanem, R., Spanos, P.D.: *Stochastic Finite Elements*, Springer New York, 1991.

Schuëller, G.I. and Stix, R.: "A critical appraisal of methods to determine failure probabilities", *Structural Safety*, 4, 1987, pp 293 - 309.

Schuëller, G.I., Pradlwarter, H.J. and Bucher, C.G.: "Efficient Computational Procedures for Reliability Estimates of MDOF-Systems", *Int. J. Non-Linear Mechanics*, Vol.26, No.6, 1991, pp 961 - 974.

Shinozuka, M., Deodatis, G.: "Response Variability of Stochstic Finite Element Systems",

J. Eng. Mech., Proc. ASCE, Vol 114, No.3, March, pp 499 - 519.

Sri Namachchivaya, N., Pai. M.A, Doyle, M.: "Stochastic approach to small disturbance stability in power systems.", in L.Arnold et al. (eds): *Lyapunov Exponents* (Lecture Notes in Mathematics), Springer, Berlin-Heidelberg, 1991, pp. 292 - 308.

Sri Namachchivaya, N. and Lin, Y.K.: "Application of stochastic averaging for nonlinear dynamical systems with high damping", *Prob. Eng. Mech.*, 3, 1988, pp 159 - 167.

Structural Safety & Reliability, Schuëller, Shinozuka & Yao (eds) © 1994 Balkema, Rotterdam, ISBN 90 5410 357 4

Nonlinear forced vibration of rectangular plates with random geometric imperfections

Siu-Tong Choi & Y.L. Liao
National Cheng Kung University, Tainan, Taiwan

ABSTRACT: Effects of random initial imperfections and in-plane stress on the nonlinear dynamic response of rectangular plates subjected to external random pressure are investigated in this paper using the Monte Carlo Method. The imperfection being studied is only initial out-of-plane deflection and is assumed to be Gaussian random field. Lindstedt-Poincare's perturbation method is employed to solve the nonlinear differential equation derived from vibration analysis of imperfect rectangular plates. The method of equivalent linearization is applied to obtain an approximate solution for nonlinear random vibration of imperfect plates with postbuckled condition. Snap-through vibration of the imperfect plate is discussed.

1 INTRODUCTION

Thin-walled structures, such as plates and shells, are often used as critical structural components in many engineering applications. Extensive research work has been carried out for linear and nonlinear analyses of flat plates and perfect shells of various types of shapes, loadings, and boundary conditions over the past decades (Timoshenko and Woinowsky-Krieger 1959; Leissa 1969; Chia 1980). Most analytical work was done based on ideal models, i.e., structures do not have any imperfections. However, initial imperfections exist in most structures and have significant effect on buckling and vibrational behaviors of thin plates and shells.

Effects of geometric imperfections on large amplitude vibration of plates and shells and on buckling behavior of axially loaded cylindrical shells were studied by various researchers (Rosen and Singer 1974; Hansen 1975; Watawala and Nash 1983; Hui and Leissa 1983; Hui 1984). Hui (1984) found that the presence of geometric imperfection in a plate may significantly raise its free vibration frequencies and may cause the plate to exhibit a soft-spring behavior. This is contrary to the commonly accepted theory that large-amplitude vibrations of plates are of the hardening type. In the studies mentioned above, geometric imperfections of structures were assumed to be of some deterministic forms.

Since geometric imperfections in structures are of random nature, it appears that a proper solution could be obtained only in statistical terms. Buckling behavior was studied for cylindrical shells with random imperfections (Amazigo 1969; Roorda and Hansen 1972; Hansen 1977; Elishakoff and Arbocz 1982). Reliability functions of such axially compressed shells were obtained through the Monte Carlo method (Elishakoff and Arbocz 1982). Wang (1990) used a similar approach to study effects of random geometric imperfections on vibrational behavior of rectangular plates, with the assumption that the imperfections are Gaussian random fields. However, effects of random geometric imperfection on nonlinear forced vibration of plates has received relatively little attention. Large amplitude vibration of buckled or initially curved plates could have snap-through motion. Ng (1988) examined the effects of snap-through motion on dynamic response of postbuckled plates.

This paper presents a numerical investigation on effects of initial random geometric imperfections on the nonlinear response of rectangular plates subjected to external random pressure. The imperfection being studied is only initial out-of-plane deflection. Governing equations are

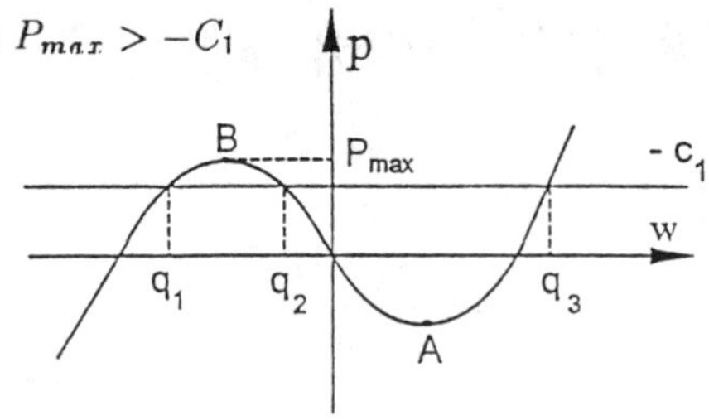

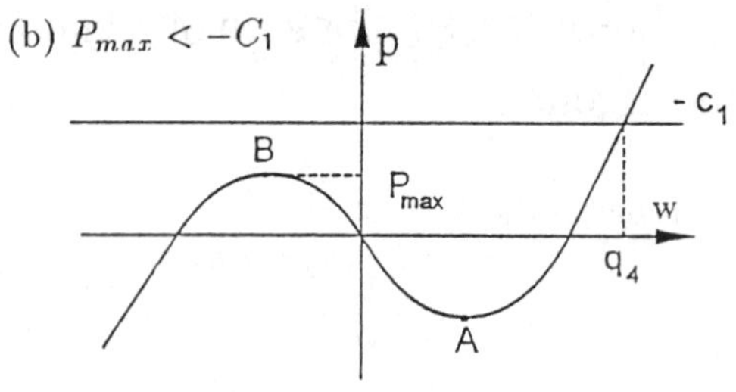

Fig. 1 Static load-deflection curve.

those of von Karman's moderately large deflection equations of motion for thin plates, modified to account for an initial geometric imperfection. The nonlinear compatibility equation is satisfied exactly by a suitable choice of the stress function with appropriate in-plane boundary conditions, and the nonlinear equilibrium equation is satisfied approximately using the Galerkin approach. Imperfections of plates are assumed to be Gaussian random fields and are simulated numerically. Perturbation method and method of equivalent linearization are used for free and forced vibration analyses, respectively. Snap-through vibration of the imperfect plate is discussed. Numerical results are obtained through Monte Carlo Method.

2 THEORETICAL MODEL

Consider a rectangular plate with initial imperfection $\overline{w}_0(\overline{x},\overline{y})$. The equations of motion based on von Karman's large deflection theory of thin plates, modified to account for an initial imperfection, are adopted and written in non-dimensional form as (Hui and Leissa 1983):

$$\nabla^4 W + 4\pi^4 \frac{\partial^2 W}{\partial t^2} + 4\pi^4 c^* \frac{\partial W}{\partial t} = 2c_0 \Big[f_{,xx}(W+w_0)_{,xx} + f_{,yy}(W+w_0)_{,yy} - 2f_{,xy}(W+w_0)_{,xy} \Big] + 4\pi^4 P \quad (1)$$

$$\nabla^4 f = 2c_0 \Big[(W_{,xy})^2 - W_{,xx}W_{,yy} - w_{0,xx}W_{yy} - w_{0,yy}W_{,xy} + 2w_{0,xy}W_{,xy} \Big] \quad (2)$$

where $\nabla^2 = \frac{\partial^2}{\partial x^2} + \frac{\partial^2}{\partial y^2}$, $c^* = \frac{\overline{\eta}}{2\pi^2\sqrt{\rho D}}$, $c_0 = \sqrt{3(1-\nu^2)}$, ν is the Poisson ratio, ρ is the mass of the plate per unit area, $\overline{\eta}$ is the damping coefficient, and D is the flexural rigidity of the plate, The non-dimensional quantities W, w_0, f, P, x, y, and t are defined as:

$$(W, w_0) = (\overline{W}, \overline{w}_0)/h, \qquad f = 2c_0F/Eh^3$$
$$(x,\ y) = (\overline{x},\ \overline{y})/b, \qquad t = \omega_0\overline{t},$$
$$P(x,y,t) = \overline{P}(\overline{x},\overline{y},\overline{t})/q_0$$
$$\omega_0^2 = \frac{4\pi^4 D}{\rho b^4} \qquad q_0 = \frac{4\pi^4 Dh}{b^4}$$

where $\overline{W}$ is the normal displacement, F is stress function, h is the plate thickness, E is Young's modulus, b is the length of the plate in the y-direction, and $\overline{P}$ is external random pressure.

The imperfect plate is assumed to be simply supported along all four edges. It is further assumed that there is no in-plane shear stress along all the edges and the in-plane displacements normal to the edges are constant. Therefore, boundary conditions are taken to be:

$$x=0 \text{ or } \alpha:\ W = W_{,xx} = f_{,xy} = 0,\ f_{,yy} = -N_x/h$$
$$y=0 \text{ or } 1:\ W = W_{,yy} = f_{,xy} = 0,\ f_{,xx} = -N_y/h$$

where $\alpha = a/b$, a is the length of the imperfect plate in the x-direction. N_x and N_y are the applied in-plane forces at the boundaries in the x- and y-directions, respectively. N_x and N_y are positive when they are compressive.

3 LARGE AMPLITUDE VIBRATION

In this study, the geometric imperfection of the plate is assumed to be a Gaussian random field. A method of digital simulation of Gaussian random fields has been developed by Elishakoff (1979). In this method, the original random field problem is reduced to one of the simulation of normal vectors.

The imperfect plate has been assumed to be simply supported. Therefore, the initial random

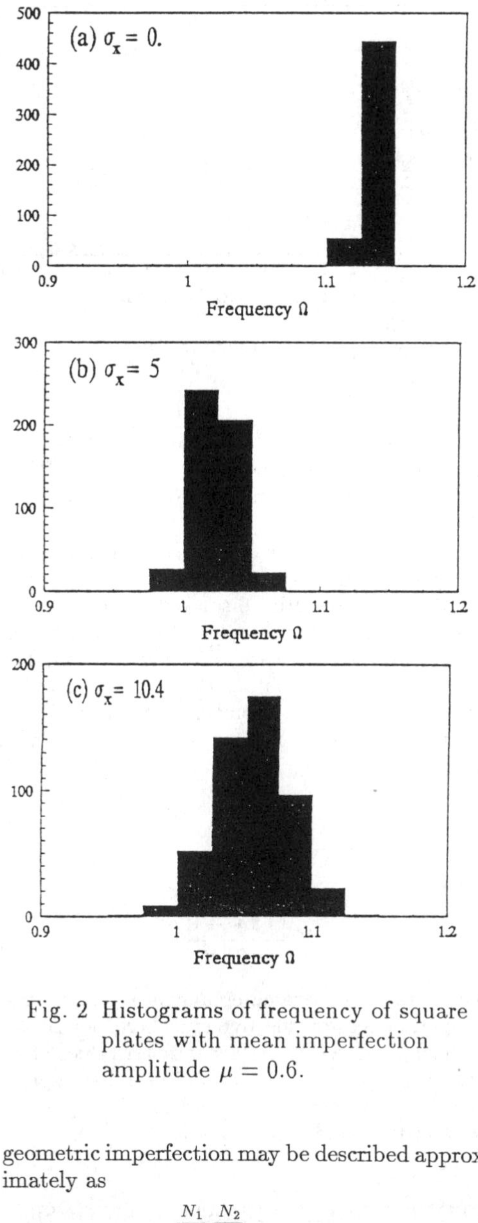

Fig. 2 Histograms of frequency of square plates with mean imperfection amplitude $\mu = 0.6$.

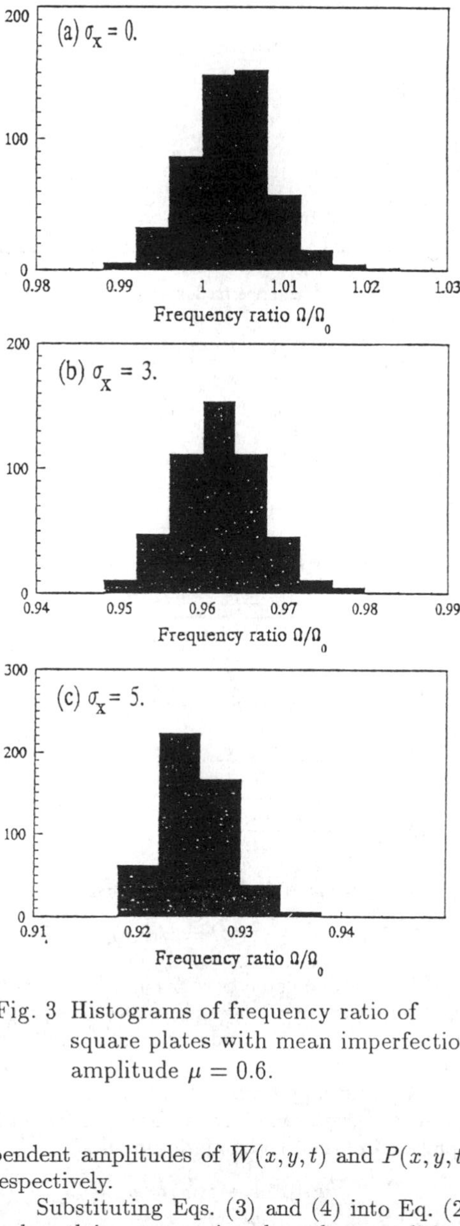

Fig. 3 Histograms of frequency ratio of square plates with mean imperfection amplitude $\mu = 0.6$.

geometric imperfection may be described approximately as

$$w_0(x,y) = \sum_{i=1}^{N_1}\sum_{j=1}^{N_2} A_{ij} \sin I\pi x \sin j\pi y \qquad (3)$$

where $I = i/\alpha$; A_{ij} are assumed to be Gaussian random variables. It is further assumed that the vibration mode and the external pressure field are given as

$$W(x,y,t) = w(t) \sin M\pi x \sin n\pi y \qquad (4a)$$

$$P(x,y,t) = p(t) \sin M\pi x \sin n\pi y \qquad (4b)$$

where $M = m/\alpha$; $w(t)$ and $p(t)$ are the time- dependent amplitudes of $W(x,y,t)$ and $P(x,y,t)$, respectively.

Substituting Eqs. (3) and (4) into Eq. (2), and applying appropriate boundary conditions, we obtain the stress function, $f(x,y,t)$, in terms of $w(t)$ and A_{ij} as

$$f = \frac{w^2(t)}{32}\left[\left(\frac{n}{M}\right)^2 \cos 2M\pi x + \left(\frac{M}{n}\right)^2 \cos 2n\pi y\right]$$

$$+ \frac{w(t)}{4}\left\{\sum_{i=1}^{N_1}\sum_{j=1}^{N_2} A_{ij}\left[H^1_{ijmn} C_{(i+m,j+n)}\right.\right.$$

$$+ H^2_{ijmn} C_{(i-m,j-n)} + H^3_{ijmn} C_{(i+m,j-n)}$$

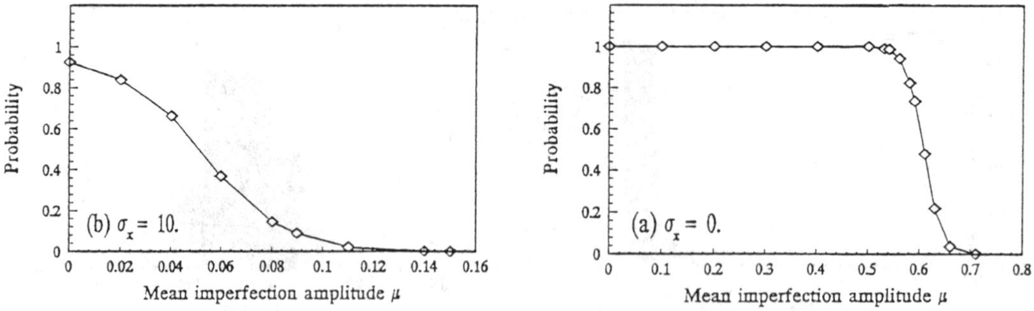

Fig. 4 Probability of hard-spring behavior of imperfect square plates.

$$+ H^4_{ijmn}C_{(i-m,j+n)}\Big]\Big\}$$

where

$$C_{(i,j)} = \cos I\pi x \sin j\pi y$$

$$H^1_{ijmn} = -\frac{\Delta^1_{ijmn}}{\Delta(i+m,j+n)}$$

$$H^2_{ijmn} = -\frac{\Delta^1_{ijmn}}{\Delta(i-m,j-n)}$$

$$H^3_{ijmn} = \frac{\Delta^2_{ijmn}}{\Delta(i+m,j-n)}$$

$$H^4_{ijmn} = \frac{\Delta^2_{ijmn}}{\Delta(i-m,j-n)}$$

$$\Delta^k_{ijmn} = [In + (-1)^k jM]^2, \; k = 1,2$$

$$\Delta(i,j) = (I^2 + j^2)^2, I = i/\alpha, M = m/\alpha$$

With the complete solution of the stress function, we may now substitute $W(x,y,t)$, $w_0(x,y)$, $P(x,y,t)$ and $f(x,y,t)$ into the nonlinear equilibrium equation (1), and use the Galerkin approach to obtain the following second-order nonlinear ordinary differential equation in time for the amplitude $w(t)$:

$$\ddot{w} + c^*\dot{w} + f(w) = p \tag{5}$$

where $(\cdot)$ denotes d/dt, and

$$f(w) = C_1 + C_2w + C_3w^2 + C_4w^3 \tag{6}$$

is a function of w with random coefficients C_1, C_2, C_3 and C_4 which are given in the Appendix.

Neglecting the damping and forcing terms in Eq. (5), and assuming the center of oscillation as W_S, with $w = Z + W_S$, we have

$$\ddot{Z} + \sum_{n=1}^{3} \alpha_n Z^n = 0 \tag{7}$$

where $\alpha_n = \frac{1}{n!}f^{(n)}(W_S)$, and $f^{(n)}$ is the nth derivative of f with respect to w. Since W_S is the center of oscillation, we have $f(W_S) = 0$, and $f^{(1)}(W_S) > 0$.

For free vibration analysis of the system due to initial disturbance, initial velocity is taken to be $V_0 = 0$ and initial displacement is S_0. The system nonlinearity is assumed to be weak, such that Lindstedt-Poincare's perturbation method is applicable. We obtain the ratio of the nonlinear to linear free vibration frequency, Ω/Ω_0, related to the vibration amplitude A as (Nayfeh and Mook 1979)

$$\frac{\Omega}{\Omega_0} = 1 + \eta A^2 + O(\varepsilon^3) \tag{8}$$

where

$$\eta = \frac{9\alpha_1\alpha_3 - 10\alpha_2^2}{24\alpha_1^2} \tag{9}$$

η is a random parameter and is a characteristic parameter for the dynamic behavior of the imperfect plate. For sufficient small values of vibration amplitude A, the nonlinear hard-spring or soft-spring character is indicated by positive or negative value of η, respectively.

4 POSTBUCKLED VIBRATION ANALYSIS

Buckled imperfect plates exhibit a strong nonlinear character. The static portion of Eq. (5) is

$$C_4w^3 + C_3w^2 + C_2w + C_1 = p \tag{10}$$

The relation between p and w as given in Eq. (10) is shown in Fig. 1, where P_{max} is the static snap-through load. The abscissas of points A and B obtained from $\partial p/\partial w = 0$ in eq. (10) are:

$$(q_A, q_B) = \frac{-C_3 \pm \sqrt{C_3^2 - 3C_2C_4}}{3C_4} \quad (11)$$

The system has negative stiffness between points A and B in Fig. 1.

4.1 *Undamped free vibration*

In this paper, we are interested in studying effects of the geometric imperfection and in-plane stress on snap-through vibration of the imperfect plate. Therefore, we mainly consider the case for which there are three equilibrium points, q_1, q_2 and q_3, in Fig. 1, i.e., when $P_{max} > -C_1$.

Letting $w = q + q_2$, and substituting into Eq. (5), we have, for undamped free vibration:

$$\ddot{q} + \beta_1 q + \beta_2 q^2 + \beta_3 q^3 = 0 \quad (12)$$

where $\beta_n = \frac{1}{n!} f^{(n)}(q_2)$. The three equilibrium positions are now: $\overline{q}_1 = q_1 - q_2$, $\overline{q}_2 = 0$, $\overline{q}_3 = q_3 - q_2$.

Initial conditions are assumed to be $\dot{q}(0) = 0$, $q(0) = S_0 - q_2 = q_0$, and $S_0 > q_3$. With the substitution of $\ddot{q} = \dot{q}\frac{d\dot{q}}{dq}$, Eq. (12) can be integrated. Letting $\dot{q} = 0$, we obtain the extreme points of vibration as

$$q = q_0 \quad \text{and} \quad q^3 + b_1 q^2 + b_2 q + b_3 = 0 \quad (13)$$

where

$$b_1 = \frac{\beta_3 q_0 + 4/3\beta_2}{\beta_3}$$

$$b_2 = \frac{\beta_3 q_0^2 + 4/3\beta_2 q_0 + 2\beta_1}{\beta_3}$$

$$b_3 = \frac{\beta_3 q_0^3 + 4/3\beta_2 q_0^2 + 2\beta_1 q_0}{\beta_3}$$

For undamped free vibration, there exists a critical value of initial displacement such that when the initial displacement is less than the critical value, the system vibrates about q_3. When the initial displacement is greater than the critical value, snap-through motion of the system occurs, and oscillation of the system passes through both equilibrium positions q_1 and q_3. The mean square displacement associated with this critical value of initial displacement is denoted by d_{min}^2. This is the minimum value of mean square displacement for all persistent snap-through motion.

4.2 *Forced vibration*

The external pressure is assumed to be a Gaussian white noise with constant spectral density S_{pp}. Depending on the magnitudes of the external excitation, the postbuckled plate may have the following types of response:

(1) Vibration without snap-through

Let $w = q_3 + \Delta q$, where Δq is the dynamic displacement with respect to q_3. Using the method of equivalent linearization, we obtain the mean square value of Δq as

$$<\Delta q^2> = \frac{-a_1 + \sqrt{a_1^2 + 12a_3\overline{\alpha}}}{6a_3} \quad (14)$$

where $a_n = \frac{1}{n!} f^{(n)}(q_3)$, and $\overline{\alpha} = \frac{\pi S_{pp}}{c^*}$ is a non-dimensional parameter for external pressure.

When the system vibrates about q_3, the following inequality has to be satisfied:

$$<\Delta q^2> < (q_3 - q_A)^2 \quad (15)$$

From Eqs. (14) and (15), we obtain

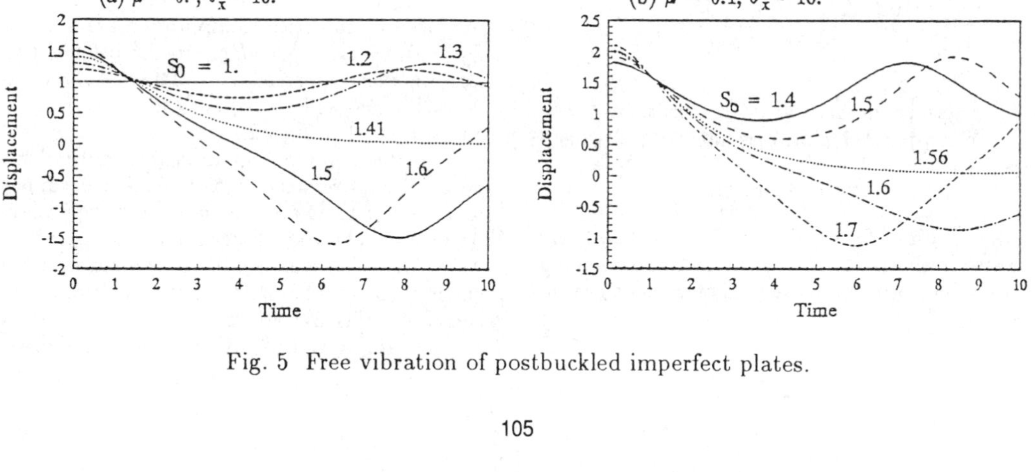

Fig. 5 Free vibration of postbuckled imperfect plates.

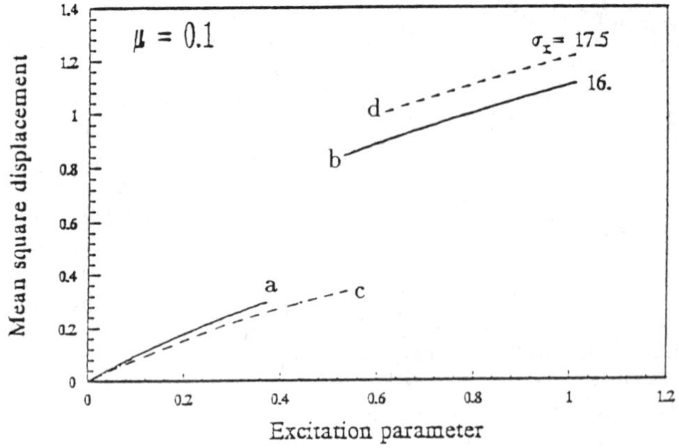

Fig. 6 Mean square displacement versus excitation parameter $\overline{\alpha}$ for $\mu = 0.1$.

$$\overline{\alpha} < \frac{\left[6a_3(q_3 - q_A)^2 + a_1\right]^2 - a_1^2}{12a_3} = L_1 \quad (16)$$

Eq. (14) is valid if $\overline{\alpha}$ satisfies Eq. (16).

(2) Persistent snap-through vibration

When snap-through motion occurs, the system does not vibrates about q_3. Instead, its oscillation passes through both equilibrium positions, q_1 and q_3. Letting $w = q + q_2$ and using the method of equivalent linearization, we obtain the mean square values of q as

$$< q^2 > = \frac{-\beta_1 + \sqrt{\beta_1^2 + 12\beta_3\overline{\alpha}}}{6\beta_3} \quad (17)$$

For the system to have persistent snap-through vibration, $< q^2 >$ in Eq. (17) must satisfy

$$< q^2 > \geq d_{min}^2 \quad (18)$$

Solving Eqs. (17) and (18), we obtain

$$\overline{\alpha} \geq \frac{(6\beta_3 d_{min}^2 + \beta_1)^2 - \beta_1^2}{12\beta_3} = L_2 \quad (19)$$

(3) Intermittent snap-through vibration

Snap-through motion is intermittent when $\overline{\alpha}$ is such that

$$L_1 \leq \overline{\alpha} < L_2 \quad (20)$$

where L_1 and L_2 are given in Eqs. (16) and (19), respectively. In this case, the mean position and the values of mean square displacement are unsteady.

5 MONTE CARLO SIMULATION AND NUMERICAL RESULTS

For a numerical example, we assume the random initial imperfection is fully separable, and its autocorviance function is of exponential-cosine type in both x- and y-directions, i.e., the non-dimensional autocovariance function is

$$\kappa(x_1, x_2; y_1, y_2) = \Delta^2 e^{-B_1|x_1 - x_2| - B_2|y_1 - y_2|} \cdot \cos B_3(x_1 - x_2)\cos B_4(y_1 - y_2) \quad (21)$$

and the mean function is

$$E[w_0] = \mu \sin K\pi x \sin l\pi y, \ K = k/\alpha \quad (22)$$

where μ is the non-dimensional mean imperfection amplitude. This form of mean function emphasizes the effect of the (k, l) imperfection mode.

In this study, Monte Carlo method is used to create 500 plate samples. Technique for simulating the random geometric imperfection is given in Elishakoff (1979) and Liao (1992). Parameters used in numerical computation are fixed at: $B_1 = B_2 = \pi/2$, $B_3 = B_4 = \pi$, $\Delta^2 = 0.005$ (unless otherwise specified), $\nu = 0.3$, $\alpha = 1$, $\sigma_y = 0$, $N_1 = N_2 = 8$ in Eq. (3), $m = n = 1$ in Eq. (4), and $k = l = 1$ in Eq. (22).

For fixed values of parameters, natural frequencies and frequency ratios of the 500 samples of imperfect plate are calculated and their distributions are shown in Figs. 2 and 3, respectively. As σ_x increases, the degree of divergence of frequency increases (Fig. 2) and the frequency ratio decreases (Fig. 3). For fixed values of σ_x, there exists a critical value μ_c of μ; the degree of divergence of frequency ratio is high when μ is near μ_c, and is low when μ is far away from μ_c. Numerical result has shown that the random parameter η in Eq. (9) is approximately zero when $\mu \approx \mu_c$, which means that the nonlinear behavior of the imperfect plate is either of hard-spring type or soft-spring type.

Fig. 4 shows the probability of hard-spring behavior of imperfect plates at $\sigma_x = 0.$ and $\sigma_x = 10$. The probability decreases as the mean imperfection amplitude increases. For example, in Fig. 4(b), when $\mu = 0.$, the probability is only 0.92, which is to say the probability of hard-spring behavior of imperfect plate is less than 1 when $\sigma_x = 10$.

Numerical results for postbuckled vibration analysis are shown in Figs. 5 and 6. For simplicity, it is assumed that the geometric imperfection is deterministic, i.e., $\Delta^2 = 0$. Fig. 5 shows the time-histories of free vibration response. In Fig. 5(a), when initial displacement $S_0 < 1.41$, the system vibrates about $q_3 = 1$, which is vibration without snap-through, and the period of vibration increases as S_0 increases. When $S_0 = 1.41$, the period of vibration tends to be infinite theoretically. When $S_0 > 1.41$, increase of S_0 results in decrease of vibration period, and the system vibrations passes through two equilibrium positions, $q_1 = 1$ and $q_3 = -1$. The system now vibrates about $q_2 = 0$, and this is snap-through motion. Similarly, in Fig. 5(b), the system vibrates with or without snap-through motion when S_0 is greater than or less than 1.56, respectively. It should be noted that imperfect plate exhibits soft-spring type behavior for free vibration without snap-through motion. When snap-through motion occurs, center of vibration is changed and the vibration behavior becomes hard-spring type.

Mean-square displacements versus excitation parameter $\overline{\alpha}$ are shown in Fig. 6. The lower two curves correspond to vibration without snap-through, whereas the upper two curves correspond to persistent snap-through vibration. Intermittent snap-through vibration occurs when $\overline{\alpha}$ is between points a and b for $\sigma_x = 16.$, and between points c and d for $\sigma_x = 17.5$. The mean position and mean-square displacement are unsteady for intermittent snap-through vibration. It is found that as σ_x increases, the maximum excitation parameter for vibration without snap-through increases (from point a to point c), and the minimum excitation parameter for persistent snap-through vibration increases (from point b to point d).

6 Conclusion

A statistical method using the Monte Carlo technique for the analysis of effects of random initial imperfections and in-plane stress on the nonlinear dynamic response of simply-supported rectangular plates is presented. The method used in this study is adequate in dealing with the uncertainties of geometric imperfection in structures. Probabilities of hard-spring behavior of imperfect plate are obtained. Snap-through vibration of imperfect plate subjected to external white noise excitation are quantified.

ACKNOWLEDGEMENT

The authors are grateful to the support of this research by the National Research Council of the Republic of China under contract No. NSC81-0401-E-006-576.

REFERENCES

Amazigo, J. C. 1969. Buckling under axially compression of long cylindrical shells with random axisymmetric imperfections. *Quarterly of Applied Mathematics* 26: 537–566.

Chia, C.-Y. 1980. *Nonlinear analysis of plates.* McGraw-Hill, New York.

Elishakoff, I. 1979. Simulation of space-random fields for solution of stochastic boundary-value problems. *Journal of Acoustical Society of America* 65:399–403

Elishakoff, I. & J. Arbocz 1982. Reliability of axially compressed cylindrical shells with random axisymmetric imperfections. *International Journal of Solids and Structures* 18: 563–585.

Hansen, J. S. 1975. Influence of general imperfections in axially loaded cylindrical shells. *International Journal of Solids and Structures* 11: 1223–1233.

Hansen, J. S. 1977. General random imperfections in the buckling of axially loaded cylindrical shells. *AIAA Journal* 15: 1250–1256.

Hui, D. 1984. Influence of geometric imperfections and in-plane constraints on nonlinear vibrations of simply supported cylindrical panels. *ASME Journal of Applied Mechanics* 51: 383–390.

Hui, D. & A. W. Leissa 1983. Effects of geometric imperfections on vibrations of biaxially compressed rectangular flat plates. *ASME Journal of Applied Mechanics* 50: 750–756.

Leissa, A. W. 1969. *Vibrations of plates.* NASA SP-160.

Liao, Y. L. 1992. *Nonlinear vibration of rectangular plates with random geometric imperfections*, M.S. Thesis, Institute of Aeronautics & Astronautics, National Cheng Kung University, Tainan, Taiwan, R.O.C.

Nayfeh, A. H. & D. T. Mook 1979. *Nonlinear Oscillations.* John Wiley, New York.

Ng, C. F. 1988. The influence of snap-through motion on the random response of curved panels to intense acoustic excitation. *Proceedings of the Third International Conference on Recent Advances in Structural Dynamics*, Southampton, England.

Roorda, J. & J. S. Hansen 1972. Random buckling behavior in axially loaded cylindrical shells with axisymmetric imperfections. *Journal of Spacecraft and Rockets* 9: 88–91.

Rosen, A. & J. Singer 1974. Effects of axisymmetric imperfections on the vibrations of cylindrical shells under axial compression. *AIAA Journal* 12: 995–997.

Timoshenko, S. & S. Woinowsky-Krieger 1959. *Theory of plates and shells.* McGraw-Hill, New York.

Wang, F.-Y. 1990. Monte Carlo analysis of nonlinear vibration of rectangular plates with random geometric imperfections. *International Journal of Solids and Structures* 26: 99–109.

Watawala, L. & W. Nash 1983. Influence of initial geometric imperfections on vibrations of thin circular cylindrical shells. *Computers and Structures* 16: 125–130.

APPENDIX

Coefficients C_1, C_2, C_3 and C_4 in Eq. (6) are

$$C_1 = -\frac{2c_0}{4\pi^2}A_{mn}(M^2\sigma_x + n^2\sigma_y)$$

$$C_2 = \frac{1}{4}\left[(M^2+n^2)^2 - \frac{c_0^2}{4}\cdot \sum_{i=1}^{N_1}\sum_{j=1}^{N_2} A_{ij}F_{ijmn} - \frac{2c_0}{\pi^2}(M^2\sigma_x + n^2\sigma_y)\right]$$

$$C_3 = \frac{3c_0^2}{16}\left[M^4(A_{mn} - A_{m3n}) + n^4(A_{mn} - A_{3mn})\right]$$

$$C_4 = \frac{c_0^2}{16}(M^4 + n^4)$$

$$F_{ijmn} = \sum_{k=1}^{4} H_{ijmn}^k F_{ijmn}^k$$

$$\begin{aligned}
F_{ijmn}^1 &= \Delta_{ijmn}^1[A_{ij} + A_{(i+2m)(j+2n)}] \\
&\quad - \Delta_{ijmn}^5[A_{i(j+2n)} + A_{(i+2m)j}] \\
F_{ijmn}^2 &= \Delta_{ijmn}^1[A_{ij} + A_{(2m-i)(2n-j)} \\
&\quad + A_{(i-2m)(j-2n)} - A_{(2m-i)(j-2n)} \\
&\quad - A_{(i-2m)(2n-j)}] + \Delta_{ijmn}^6[A_{i(2n-j)} \\
&\quad + A_{(2m-i)j} - A_{i(j-2n)} - A_{(i-2m)j}] \\
F_{ijmn}^3 &= \Delta_{ijmn}^2[-A_{ij} + A_{(i+2m)(2n-j)} \\
&\quad - A_{(i+2m)(j-2n)}] + \Delta_{ijmn}^7[A_{(i+2m)j} \\
&\quad + A_{i(j-2n)} - A_{i(2n-j)}] \\
F_{ijmn}^4 &= \Delta_{ijmn}^2[-A_{ij} + A_{(2m-i)(2n+j)} \\
&\quad - A_{(i-2m)(j+2n)}] + \Delta_{ijmn}^8[A_{i(j+2n)} \\
&\quad + A_{(i-2m)j} - A_{(2m-i)j}]
\end{aligned}$$

$$\Delta_{ijmn}^{(4+k)} = \left[In - (-1)^k 2Mn + jM\right]^2, k = 1,2$$

$$\Delta_{ijmn}^{(6+k)} = \left[In - (-1)^k 2Mn - jM\right]^2, k = 1,2$$

$A_{ij} = 0$, if $i \le 0$ or $i > N_1$, or $j \le 0$, or $j > N_2$

where
$$(\sigma_x, \sigma_y) = \frac{-2c_0 b^3}{Eh^3}(N_x, N_y)$$

Structural Safety & Reliability, Schuëller, Shinozuka & Yao (eds) © 1994 Balkema, Rotterdam, ISBN 90 5410 357 4

Ideas of probability and convexity combined for analyzing parameter uncertainty

Isaac Elishakoff
Florida Atlantic University, Boca Raton, Fla., USA
Pierluigi Colombi
University di Pavia, Italy

ABSTRACT: This paper is devoted to a fundamental problem of accounting for parameter uncertainties in random vibration of structures. In contrast to overwhelming majority of random vibration studies where perfect knowledge is assumed for the parameters of the excitation, this crucial conjecture is dispensed with. The probabilistic characteristics of the excitation are assumed to be given as depending on some parameters which are not known in advance. We postulate that some imprecise knowledge is available; namely, that these parameters belong to a bounded, convex set. In the case when this convex set is represented by an ellipse, closed form solutions are derived for the upper and lower bounds of the mean-square displacement of the structure.

In this study, apparently, for the first time in the literature the system uncertainty in the random vibrations is formulated stated as an "anti-optimization" problem of finding the least favorable values of the mean-square response. Hopefully, the approach developed herein opens a novel avenue for tackling parameter uncertainty often encountered in various branches of engineering.

1. SINGLE-DEGREE-OF-FREEDOM SYSTEM

Motion of the system is governed by the equation

$$m\ddot{X} + c\dot{X} + kX = F(t) \qquad (1)$$

where m is the mass, k = spring constant, d = damping coefficient. The excitation F(t) is a weakly stationary random process with zero mean and an autocorrelation function:

$$R_F(\tau) = E(F^2)e^{-\alpha|\tau|}[\cos(\beta\tau) + (\alpha/\beta)\sin(\beta|\tau|)] \qquad (2)$$

where $E(F^2)$ is a mean-square excitation, parameters α and β are parameters of excitation, and τ denotes the time delay. Two parameters, α and β model the presence of number of parameters in the fitted or hypothesized probabilistic characteristics of the random acoustic excitation. The spectral density of the excitation $\Phi_{FF}(\omega) = (1/2\pi)\int_{-\infty}^{\infty} R_{FF}(\tau)e^{i\omega\tau}d\tau$ is

$$\Phi_{FF}(\omega) = \frac{E(F^2)}{\pi}\frac{2\alpha(\alpha^2+\beta^2)}{(\alpha^2+\beta^2)^2+2(\alpha^2-\beta^2)\omega^2+\omega^4} \qquad (3)$$

The mean square response of the system is written as

$$E(X^2) = \int_{-\infty}^{\infty}\Phi_{FF}(\omega)\ |H(\omega)|^2\ d\omega \qquad (4)$$

where H(ω) is the frequency response function

$$H(\omega) = \left[m(i\omega)^2 + c(i\omega) + k\right]^{-1} = \left\{m[(i\omega)^2 + 2\zeta_1(i\omega)\omega_n + \omega_n^2]\right\}^{-1} \qquad (5)$$

In Eq. (5) ζ_1 is the viscous damping factor, $\zeta_1 = c/2m\omega_n$, $\omega_n = \sqrt{k/m}$ is the natural frequency.
Eq. (4) becomes

$$E(X^2) = \frac{2\alpha(\alpha^2+\beta^2)E(F^2)}{m^2\pi}\int_{-\infty}^{\infty}\frac{1}{(\alpha^2+\beta^2)^2+2(\alpha^2-\beta^2)\omega^2+\omega^4}\times \qquad (6)$$

$$\times \frac{d\omega}{|(i\omega)^2 + 2\zeta_1 i\omega\, \omega_n + \omega_n^2|^2}$$

The result obtained through the use of the residue theorem reads:

$$E(X^2) = f(\alpha,\beta)\ E(F^2) \tag{7}$$

$$f(\alpha,\beta) = \frac{2\alpha}{m} \frac{(\alpha^2+\beta^2)\,(a_3 - a_1 a_2)}{a_4(a_3^2 + a_1^2 a_4 - a_1\, a_2\, a_3)} \tag{8}$$

with notations

$$\begin{aligned} a_1 &= 2(\alpha + \zeta_1\, \omega_n) \\ a_2 &= \alpha^2 + \beta^2 + 4\zeta_1\omega_n\alpha + \zeta_1^2\omega_n^2 + \omega_d^2 \\ a_3 &= 2(\zeta_1\omega_n\, \alpha^2 + \zeta_1\omega_n\beta^2 + \alpha\zeta_1^2\omega_n^2 + \alpha\omega_d^2) \\ a_4 &= (\alpha^2 + \beta^2)(\omega_\alpha^2 + \zeta_1^2\omega_n^2), \omega_d = \omega_n(1-\zeta_1^2)^{1/2} \end{aligned} \tag{9}$$

Note that when the maxima of the frequency response function and the spectral density of excitation are well separated, and the damping coefficient ζ_1 is much smaller than unity, the method of approximate integration [1] can be employed to yield the following approximation:

$$E(X^2) \approx \Phi_{FF}(\omega_n) \int_{-\infty}^{\infty} |H(\omega)|^2 d\omega + |H(\kappa_1)|^2 \int_{-\infty}^{\infty} \Phi_{FF}(\omega) d\omega \tag{10}$$

Here ω_n is the frequency at which the frequency response function has its maximum, and κ_1 is the frequency which maximizes $\Phi_{FF}(\omega)$. In the vicinity of ω_n, the spectral density of excitation is assumed to be a slowly varying function, whereas in the vicinity of β, the frequency response function varies slowly. This version of approximate integration generalizes the one discussed in Ref. 1. Indeed, for moderate values of α only the first term will suffice, where as for small values of α inclusion of the second term is necessary. Evaluation of Eq. (10) yields

$$\begin{aligned} E(X^2) &\approx \Phi_{FF}(\omega_n) \int_{-\infty}^{\infty} |H(\omega)|^2 d\omega + E(F^2)\,|H(\gamma)|^2 \\ &= E(F^2) f(\alpha,\beta) \end{aligned} \tag{11}$$

$$f(\alpha,\beta) = \frac{2\alpha(\alpha^2+\beta^2)}{(\alpha^2+\beta^2)^2 + 2(\alpha^2-\beta^2)\omega_n^2 + \omega_n^4} \cdot \frac{1}{2\zeta_1\omega_n^3 m} + \frac{1}{(\omega^2-\kappa_1)^2 + (2\zeta_1\omega_n\kappa_1)^2} \tag{12}$$

The error due to approximation attains its maximum in the vicinity $\kappa_1 \approx \omega_n \approx \beta$. In this region the exact expression given in Eqs. (8) and (9), although cumbersome, should be used. When ω_n and β are far apart, approximation given by Eq.(11) holds.

Assume now that α and β are uncertain parameters. We basically have three ways of describing the situation:

a) The parameters α and β are possible values taken by the respective random variables A and B. This situation occurs when we have sufficient experimental data on the random variables A and B so that the hypotheses about their joint density function $p_{AB}(\alpha,\beta)$ can be checked.

Consider the case when A and B are random variables. Their joint probability density $p_{AB}(\alpha,\beta)$ is assumed to be given. Then, according to the total probability formula

$$E(X^2) = \int_{-\infty}^{\infty}\int_{-\infty}^{\infty} E(X^2|A=\alpha, B=\beta)\, p_{AB}(\alpha,\beta) d\alpha d\beta \tag{13}$$

and the integration extends over the region where α and β are varying. In Eq.(13) the value $E(X^2|A = \alpha, B = \beta)$ is a conditional mean-square displacement calculated on a condition that the random variable A takes on value α, and random variable B takes on value β.

b) There is no information on the loading which is treated deterministically. Instead a fragmentary knowledge is present; namely, that some governing parameters belong to a some bounded set. In a sufficiently large class of situations this set is convex. In these circumstances an alternative analysis, namely, convex modelling, recently developed by Ben-Haim and Elishakoff in monograph [2], should be applied (reader may also consult with references 3-5).

c) There is a probabilistic information on the loading process, and the form of the probabilistic characteristics can be estimated. However, there is a limited knowledge available on some governing parameters. In these circumstances probabilistic method can be successfully combined with the convex modelling. This is done in the following section.

2. UNCERTAIN PARAMETERS BELONGING TO A CONVEX SET

Consider now a more realistic situation when α are uncertain but not "enough" information is available to obtain their probabilistic characteristics. Under these circumstances, the new, convex modelling of uncertainty should be applied [2]. Hereinafter, we will in part follow Elishakoff and Ben-Haim [3]. Ref.3 dealt with a different problem; namely, a problem of a impact buckling of shell due to uncertain geometrical imperfections. We assume that we possess only scarce information on α and β; namely, that their parameters belong to some bounded convex sets.

$$Z(\Omega,\theta) = \{(\alpha,\beta): \quad G(H) \leq \theta^2\} \tag{14}$$

where vector H has its elements α and β, and G is a quadratic form, θ^2 is a positive constant. The proper choice of the form G(H) and of constant θ is performed based on experimental information on loading, to determine, for example, the smallest ellipse enclosing all available, although scant data.

We assume that H_0 is a nominal vector, with nominal values of α_0 and β_0; for example, H_0 may correspond to average values of α and β. Structures which operate under similar conditions will experience common patterns of forces, belonging to the ellipsoidal set.

The mean square response for the force parameters $\alpha_0 + \delta_1$, $\beta_0 + \delta_2$ to the first order in δ_1 and δ_2 is

$$E[X^2(\alpha_0+\delta_1,\beta_0+\delta_2)] = E(F^2)\, f(\alpha_0+\delta_1,\beta_0+\delta_2) \tag{15}$$

where

$$f(\alpha_0+\delta_1,\beta_0+\delta_2) = f(\alpha_0,\beta_0) + \frac{\partial f}{\partial \alpha}\Big|_{H=H_0}\delta_1 + \frac{\partial f}{\partial \beta}\Big|_{H=H_0}\delta_2 + \dots \tag{16}$$

where $H_0^T = (\alpha_0,\beta_0)$.

We want to evaluate lower and upper limits of the mean square value as δ_1 and δ_2 vary within an ellipse. For convenience we define the gradient vector:

$$(grad\ f)_0^T = \left(\partial f/\partial\alpha\ ,\ \partial f/\partial\beta\right)_0 \tag{17}$$

where T denotes transposition and subscript means that the derivatives are evaluated at the nominal values, α_0 and β_0. Consider a particular case of Eq.(14), when the deviations δ_1 and δ_2 from the nominal parameters forming a vector δ vary in the ellipse

$$Z(\Omega,\theta^2) = \{\delta: \quad \delta^T\Omega\delta \leq \theta^2\} \tag{18}$$

where Ω is a positive-definite matrix

$$\Omega = \begin{bmatrix} \omega_{11} & \omega_{12} \\ \omega_{21} & \omega_{22} \end{bmatrix} \tag{19}$$

We are interested in the maximum possible mean-square response the system may possess:

$$E(X^2;\Omega,\theta) = \max_{\delta\in Z(\Omega,\theta^2)} E(F^2) \times [f(\alpha_0,\beta_0) + (grad\ f_0,\delta)] \tag{20}$$

where (f_1, f_2) is a inner product so that two last terms in Eq.(20) represent $(\text{grad } f)_0^T\delta$.

Eq.(20) demands finding the maximum of the linear functional $(\text{grad } f)_0^T\delta$ on the convex set $Z(\Omega, \theta^2)$. According to the Kelly-Weiss theorem [2] the extreme value will occur on the set of extreme points of Z which is a collection of vectors $\eta = (\eta_1,\eta_2)$ in the set

$$C(\Omega,\theta^2) = \{\eta: \ \eta^T\Omega\eta = \theta^2\} \tag{21}$$

Eq.(20) becomes

$$E(X^2;\Omega,\theta) = \max_{\eta\in C(\Omega,\theta^2)} E(F^2) [f(\alpha_0,\beta_0) + (grad f)_0^T\eta] \tag{22}$$

We wish to maximize

$$\Psi \equiv f(\alpha_0,\beta_0) + (grad\ f)_0^T\eta \tag{23}$$

subject to the requirement that vector η satisfies the equality $\eta^T\Omega\eta = \theta^2$. We employ the Lagrange multiplier method [2-3]. We construct Lagrangean as

$$H = f(\alpha_0,\beta_0) + (grad\ f)_0^T\eta + \lambda(\eta^T\Omega\eta - \theta^2) \tag{24}$$

A necessary condition for extremum reads:

$$\partial H/\partial\eta = (grad\ f)_0 + 2\lambda\Omega\eta = 0 \tag{25}$$

We premultiply this equation by Ω^{-1} from the left to arrive at

$$\eta = -\frac{1}{2\lambda}\Omega^{-1}(grad\, f)_0 \qquad (26)$$

Substituting Eq.(26) into the constraint itself results in

$$(grad f)_0^T(\Omega^{-1})^T\Omega\Omega^{-1}(grad f)_0/4\lambda^2 = \theta^2 \qquad (27)$$

or, since Ω is a symmetric matrix, we obtain the following expression for the Lagrange multiplier squared

$$\lambda^2 = \frac{1}{4\theta^2}(grad\, f)_0^T\Omega^{-1}(grad\, f)_0 \qquad (28)$$

In view of Eq.(25) we obtain the vector η which makes $E(X^2;\, \Omega,\theta)$extremal:

$$\eta = \pm\frac{\theta}{\sqrt{(grad\, f)_0^T\Omega^{-1}(grad f)_0}}\Omega^{-1}(grad f)_0 \qquad (29)$$

Maximum mean-square displacement becomes

$$\max_{\alpha,\beta} E[X^2(\alpha,\beta)] = E[X^2(\alpha_0,\beta_0)] \times \qquad (30)$$

$$\left[\left(1 + \theta\sqrt{(grad\, \bar{f}\,)_0^T\Omega^{-1}(grad\, \bar{f}\,)_0}\right)\right] ,\bar{f}=f/f_0$$

whereas the minimum mean-square displacement reads

$$\min_{\alpha,\beta} E[X^2(\alpha,\beta)] = E[X^2(\alpha_0,\beta_0)] \times \qquad (31)$$

$$\left[\left(1 - \theta\sqrt{(grad\bar{f})_0^T\Omega^{-1}(grad\bar{f})_0}\right)\right]$$

Equation (30) yields the maximum possible response the system may possess. This maximum response should be directly incorporated into the design procedures of structures, instead of the nominal response.

CONCLUSION

In this study a new philosophy, namely that of rigorously combining probabilistic and convex approaches to uncertainty, is presented, for accounting the uncertainty in random excitation parameters. Instead of nominal responses, the least favorable responses, derived in this study, should be employed for design purposes. The combination of the present method and the finite element method appears to be imperative, to consider complex structures for which analytical evaluation of the modal characteristics is not feasible. Once such a combination is materialized, the present method is expected to become an important tool for accounting the system uncertainties when sufficient information on probabilistic characteristics of the excitation is absent, to justify the totally probabilistic approach. The present hybrid probabilistic-convex approach appears to be an attractive alternative to fully probabilistic treatment.

ACKNOWLEDGEMENT

This study has been supported by the NASA Kennedy Space Center, through Cooperative Agreement No. NCC10-0005, S-1. Technical monitor Mr.R.Caimi. We thank Professor Y.K.Lin and Mr.L.Zhu for reading the manuscript and making constructive suggestions. The extension of this study, as well as the application to acoustically excited multi-span structures is underway and will be published elsewhere.

REFERENCES

I. Elishakoff, Probabilistic Methods in the Theory of Structures, (Wiley-Interscience, New York, 1983).

Y. Ben-Haim and I. Elishakoff, Convex Models of Uncertainty in Applied Mechanics, (Elsevier Science Publishers, Amsterdam, 1990).

I. Elishakoff and Y. Ben-Haim, Dynamics of a thin cylindrical shell under impact with limited deterministic information on its initial imperfections, Journal of Structural Safety, 8 (1990) 103-112.

I. Elishakoff, Convex versus probabilistic modeling of uncertainty in structural dynamics, Keynote Lecture, in "Structural Dynamics: Recent Advances", M. Petyt, M.F. Wolfe and C. Mei,

editors (Elsevier Applied Science Publishers, London) (1991), 3-21.
I. Elishakoff, On the uncertainty triangle, The Shock and Vibration Digest, 22 (10), No. 1 (1990) 1 (editorial).
I. Elishakoff and P. Colombi, Combination of probabilistic and convex models of uncertainty when scarce knowledge is present on acoustic excitation parameters, Computer Methods in Applied Mechanics and Engineering, 104(1993), 187-209.

Structural Safety & Reliability, Schuëller, Shinozuka & Yao (eds) © 1994 Balkema, Rotterdam, ISBN 90 5410 357 4

A new computational approach for nonlinear random vibration

R.Ghanem
Civil Engineering Department, State University of New York at Buffalo, N.Y., USA
P.D.Spanos
Rice University, Houston, Tex., USA

ABSTRACT: An approach is developed for the numerical solution of random vibration problems. It is based on treating random variables as functions in a certain Hilbert space. Stochastic processes are then described as curves defined in these spaces, and concepts from deterministic approximation theory are applied to represent the solution as a series involving a known basis of stochastic processes, and a set of unknown coefficients which are deterministic functions of time. Then, a Galerkin projection procedure is utilized to derive a set of ordinary differential equations which can be solved numerically to determine the coefficients in the series . The versatility of the proposed approach is demonstrated by its application to a nonlinear vibration problem involving the probability density of a non-Markovian oscillator response.

1 Introduction

With the recent wide-spread diffusion of a new generation of computational resources, it seems warranted to take a fresh look at the various alternatives for solving nonlinear random vibration problems. A number of these alternatives that used to be classified as unrealistic can be readily pursued using modern computer technology. While providing a reasonably rigorous formulation of a general computational solution scheme for nonlinear random vibration problems, this paper sets the stage for a perspective on computational stochastic mechanics as a sub-branch of approximation theory, whereby desirable levels of accuracy can be reached. This is achieved by considering the random aspect of the problem to be the manifestation of a new dimension, along which probability measure, as opposed to Lebesgue measure, is defined. This perspective allows for viewing random variables as functions along this dimension, and random processes as functions defined over the cross product of this dimension with an index space. In the context of random vibration, this index space can be identified with the time interval over which the problem is defined. Since random variables are viewed as functions, techniques from deterministic functional analysis can then be used to cast the problem as a problem of approximation theory. In particular, projection properties of Hilbert spaces are used to obtain an approximation of the solution on a finite dimensional subspace. The convergence of these projections is facilitated by the completeness property of Hilbert spaces. A number of attempts at viewing the problem from the perspective of functional analysis have already been reported. Those approaches, however, attempted to obtain analytical solutions for the resulting systems of deterministic equations (Jahedi and Ahmadi, 1983). This has impeded the application of the formalism to complicated problems since the analytical forms they yielded in that case were unmanageable. These complicated mathematical forms were associated with the analytical evaluation of kernels of multiple integrals. In this paper, this problem is circumvented by performing a discretization along the random dimension prior to invoking an approximation through a sequence of projection operators. This discretization is performed using spectral measures, leading to a generalized Fourier expansion with coordinate functions defined globally over the whole random dimension. Note that other expansions, involving local discretizations using an appropriate probability measure can also be considered.

2 Representation of Stochastic Processes

Essential to the computational nature of the approach presented in this paper, is the representation of stochastic processes in a format that lends itself to manipulations on a digital computer. Since its early days the theory of stochastic processes was intimately linked to generalized harmonic analysis, and to the spectral theory of generalized functions. A most important result in this regard has been the spectral representation of random processes given by the equation

$$x(t,\theta) = \int_{-\infty}^{\infty} g(t,\omega) dZ(\omega,\theta) , \quad (1)$$

where $dZ(\omega,\theta)$ is a stochastic measure, the symbol θ refers to elements in the sample space and the function $g(t,\omega)$ is dependent on the correlation function of the process. The usefulness of this representation is restricted to Gaussian processes and is of limited scope in phenomena featuring nonlinear transformations of Gaussian processes. A generalization of equation (1) can be obtained using a theorem by Cameron and Martin (1947) for the Fourier-like expansion of square integrable functions using the Wiener measure. Wiener (1958) used that expansion for the case of random processes. The result is an expansion, in the form of multiple integrals, featuring only the Gaussian measure, independently of the process being expanded. That expansion is generally known as the Wiener-Hermite expansion and is given by the equation

$$\begin{aligned} x(t,\theta) &= G_0 + \sum_{i=0}^{\infty} G_i \Gamma_i(\beta(t,\theta)) \\ &+ \sum_{m_1=0}^{\infty} \sum_{m_2=0}^{\infty} G_{m_1 m_2} \Gamma_{m_1 m_2}(\beta(t,\theta)) + \ldots (2) \end{aligned}$$

where $\beta(t,\theta)$ denotes the Brownian motion. The Hermite coefficients $G_{m_1 \ldots m_n}$ featured in the above equation uniquely specify the process being expanded. The kernel $\Gamma_{m_1 \ldots m_n}(\beta(t,\theta))$, can be obtained as the product of n random Hermite polynomials $\Psi_{m,p}(\beta,\theta)$ according to the equation

$$\Gamma_{m_1 \ldots m_p}(\beta(t,\theta)) = \Psi_{m_1,1}(\beta(t),\theta) \times \ldots \times \Psi_{m_p,p}(\beta(t),\theta) \quad (3)$$

where $\Psi_{m,p}(\beta(t),\theta)$ are given by the following equation,

$$\Psi_{m,p}(\beta(t),\theta) = H_m \left[\int_0^1 \alpha_p(t) d\beta(t,\theta) \right] . \quad (4)$$

In the above equation, $\{\alpha_p(t)\}$ constitutes any basis in the space L_2. A significant drawback of these representations is the appearance in them of integrals that cannot be calculated numerically. Only statistical averages of these integrals can be evaluated. Furthermore, it has proven extremely difficult in the nonlinear case to include more than two terms in the Wiener-Hermite expansion, thus severely restricting its range of applicability (Jahedi and Ahmadi, 1983.) An approach to circumvent these difficulties is to substitute countable representations of stochastic processes in lieu of the above continuous ones. This can be achieved in a straightforward manner using the Karhunen-Loeve expansion of a stochastic process (Spanos and Ghanem, 1989; Ghanem and Spanos, 1990). Accordingly, a stochastic process $x(t)$ can be expanded in the form

$$x(t,\theta) = \sum_{i=0}^{\infty} \sqrt{\lambda_i} \xi_i(\theta) \phi_i(t) , \quad (5)$$

where $\{\xi_i(\theta)\}$ is a set of orthogonal random variables, λ_i and $\phi_i(t)$ are obtained as the solution to an integral eigenvalue problem (Ghanem and Spanos, 1991),

$$\lambda_i \phi_i(t) = \int_{\mathcal{D}} R_{xx}(t,s) \phi_i(s) ds , \quad (6)$$

where $\mathcal{D}$ denotes the domain of the problem ($t \in \mathcal{D}$.) In this equation $R_{xx}(t,s)$ denotes the covariance of the process $x(t)$ which may be non-stationary. The Karhunen-Loeve expansion is valid for arbitrary stochastic processes. For non-Gaussian processes, the random variables $\{\xi_i(\theta)\}$ are non-Gaussian, and they can be calculated from realizations of the process using the following equation,

$$\xi_i(\theta) = \frac{1}{\sqrt{\lambda_i}} \int_{\mathcal{D}} x(t,\theta) \phi_i(t) dt . \quad (7)$$

In the case of a Gaussian process, however, these random variables form a vector with orthogonal Gaussian components; this property expedites the numerical simulation of the process $x(t)$ using equation (4). In the case of a stationary process, it can be shown that the Karhunen-Loeve expansion reduces to a spectral representation extensively utilized in stochastic mechanics (Shinozuka and Jan, 1972). Specifically,

$$x(t) = \sum_{i=1}^{\infty} \xi_i e^{j\omega_i t}, \quad (8)$$

where $j = \sqrt{-1}$. Furthermore, the general form of the Karhunen-Loeve expansion, given in equation (4) applies for non-stationary processes, and in particular for processes defined over finite domains. For non-Gaussian processes, as often encountered in nonlinear random vibrations, the Karhunen-Loeve involves non-Gaussian variables and leads to substantial complications; this is specially true if a large number of realizations is not available to permit the use of equation (6). In that case, a generalization of the Wiener-Hermite expansion can be used to obtain an expansion in terms of polynomials of Gaussian random variables. In this context, the polynomial chaos expansion has been suggested (Ghanem and Spanos, 1991) as a viable representation for non Gaussian stochastic processes. The expansion is given by the equation

$$
\begin{aligned}
x(t,\theta) &= \sum_{i=0}^{\infty} x_i(t)\Gamma_i(\xi_i) + \sum_{i=0}^{\infty}\sum_{j=i}^{\infty} x_{ij}(t)\Gamma_{ij}(\xi_i,\xi_j) \\
&+ \sum_{i=0}^{\infty}\sum_{j=i}^{\infty}\sum_{k=j}^{\infty} x_{ijk}(t)\Gamma_{ijk}(\xi_i,\xi_j,\xi_k) + \ldots \quad (9)
\end{aligned}
$$

where the general stochastic process $x(t,\theta)$ is characterized by its projections $\{x_i, x_{ij}, \ldots\}$ on the denumerable set of functions given by $\{\Gamma_i(\theta), \Gamma_{ij}(\theta), \ldots\}$

3 Nonlinear Random Vibration

The implementation of the above representations into the equations of motion of randomly excited nonlinear dynamical systems is investigated in this section. The general approach is first presented, followed by a specific application to the Duffing oscillator.

Many engineering systems can be modeled using the following equation of motion

$$
\begin{aligned}
\ddot{x} + 2\zeta\omega_0\dot{x} + \omega_0^2 x + g(x,\dot{x}) &= f(t,\theta), \\
x(0) = x_0,\ \dot{x}(0) &= v_0, \quad (10)
\end{aligned}
$$

where $f(t,\theta)$ is a random excitation, $g(x,\dot{x})$ is a nonlinear function of its arguments, and the initial conditions x_0 and v_0 are specified with probability one. Once a suitable basis in the space of random variables has been identified, the excitation can be expanded along a basis spanning a finite-dimensional subspace. This leads to the following representation

$$
f(t,\theta) = \sum_{i=0}^{N} \Psi_i f_i(t) \, . \quad (11)
$$

In particular, the set consisting of the Polynomial Chaoses of successive orders, has been shown to provide a complete basis in the space of second order random variables (Ghanem and Spanos, 1991). This basis will be used in the following theoretical development. Alternatively, since in general, the probabilistic characteristics of the process $f(t,\theta)$ are known, the above expansion can be obtained using its Karhunen-Loeve expansion. In particular, if the process $f(t,\theta)$ is Gaussian, the Polynomial Chaos expansion and the Karhunen-Loeve expansion will be identical, and only the Gaussian terms in equation (10) will not vanish. Denoting by M the number of nonvanishing terms, the corresponding terms in equation (10) can be grouped into the 1st M positions in the summation. Preserving the generality of the theoretical development, however, equation (10) will be used without identifying a particular set $\{\Psi_i\}$. The response process can also be expanded along the same basis, leading to the equation

$$
x(t) = \sum_{i=0}^{N} \Psi_i q_i(t). \quad (12)
$$

Note that in general, irrespective of the particular set $\{\Psi_i\}$ used, the expansion for $x(t)$ will involve non-zero contributions from all the terms, and not just the Gaussian components. In the above equations, $\{f_i(t)\}$ and $\{q_i(t)\}$ are two sets of deterministic functions. The problem can then be restated as one of seeking to determine the functions $\{q_i(t)\}$. Substituting the above expansions into the equation of motion yields

$$
\sum_{i=0}^{N} \Psi_i \left(\ddot{q}_i + 2\zeta\omega_0\dot{q}_i + \omega_0^2 q_i\right) + g(\{q_n\},\{\dot{q}_n\}) = \sum_{i=0}^{N} \Psi_i f_i(t) \, . \quad (13)
$$

Next, the error in this equation, resulting from truncating the infinite series after a finite number of terms, is minimized by requiring it to be orthogonal to the finite dimensional space spanned by the basis used in the expansion. Orthogonality in this case is construed to be with respect to the inner product defined by the covariance operator. This orthogonality requirement leads to a set of N nonlinear deterministic differential equations of the form

$$
\begin{aligned}
&<\Psi_i\Psi_m> \left(\ddot{q}_i + 2\zeta\omega_0\dot{q}_i + \omega_0^2 q_i\right) \\
&+<\Psi_m\ g(\{q_n\},\{\dot{q}_n\})> = <\Psi_i\Psi_m> f_i(t) \\
&m = 0, ..., N \, , \quad (14)
\end{aligned}
$$

with initial conditions given by

$$
\begin{aligned}
q_0(0) = q_0, &\quad \dot{q}_0 = v_0 \\
q_i(0) = \dot{q}_i(0) = 0, &\quad i > 1 \, . \quad (15)
\end{aligned}
$$

These conditions on the components $q_i(t)$ ensure that the initial conditions on the solution process $x(t)$ are satisfied with probability one. Note that if the Karhunen-Loeve expansion is used to represent the excitation process, the following equation is valid

$$
f_i(t) = 0, \quad i > M \, . \quad (16)
$$

The solution of the problem can be obtained by solving the set of coupled deterministic nonlinear equations. It is worth noting that the level of coupling between these equations depends on the exact set $\{\Psi_i\}$ used to expand the random functions. Clearly, using the Karhunen-Loeve expansion of the response process would minimize the number of equations to be solved, since that expansion is optimal. This is not possible, however, since the covariance function of the response is not known apriori. The Polynomial Chaos basis will be used in the present development to expand the various random quantities. This basis is chosen because of the availabe pertinent mathematical background, such as the orthogonality of its elements. This property is expected to induce desirable stability and convergence behavior of the derived solution. The application of the proposed technique to the case of the Duffing oscillator is next demonstrated.

4 Duffing Oscillator

The equation governing the motion of a Duffing oscillator is

$$\ddot{x}+2\zeta\omega_0\dot{x}+\omega_0^2 x\left(1+\mu x^2\right) = f(t,\theta), \quad x(0)=\dot{x}(0)=0 \,. \tag{17}$$

The excitation $f(t,\theta)$ is assumed to be a random process applied over a time interval $[0,T]$. It is specified over this interval by its correlation function $R_{ff}(\Delta t)$ which is given by the equation,

$$R_{ff}(\Delta t) = \sigma_f^2 e^{-c\Delta t} \,. \tag{18}$$

In the above equation, σ_f denotes the standard deviation of the process, and c denotes the inverse of its correlation length. The spectral density function corresponding to this process is

$$S_{ff}(\omega) = \frac{c}{c^2+\omega^2} \,. \tag{19}$$

Strictly speaking, the above process is non-stationary since it is applied over a finite time interval. Therefore, the spectral density function given by the above equation must be interpreted to represent the process as this time interval becomes infinite. Corresponding to this infinite duration process, the response of the associated linear system, $\mu = 0$, can be obtained, and its variance can be evaluated using the following equation,

$$\sigma_x^2 = \frac{1}{2\pi}\int_{-\infty}^{\infty} |H(\omega)|^2 S_{ff}(\omega) d\omega \,. \tag{20}$$

In the above equation, $H(\omega)$ represents the frequency response of the linear equation of motion and its modulus is given by the equation

$$|H(\omega)|^2 = \frac{1}{((\omega^2-\omega_0^2)^2+(2\zeta\omega_0)^2)} \,. \tag{21}$$

For the purpose of performing a parametric study of the proposed technique, it will prove expedient to normalise the equation of motion using nondimensional time $\tau = \omega_0 t$, and nondimensional displacement $y = x/\sigma_x$ (Iwan and Spanos, 1978.) This procedure leads to the following equation of motion

$$\ddot{y}+2\zeta\dot{y}+y\left(1+\epsilon y^2\right) = \frac{f(t,\theta)}{\sigma_x\omega_0^2}\,, \quad y(0)=\dot{y}(0)=0 \,, \tag{22}$$

where $\epsilon = \mu\sigma_x^2$. Using nondimensional time, the autocorrelation function can be rewritten as

$$R_{ff}(\Delta\tau) = \sigma_f^2 e^{-c_0\Delta\tau} \,, \tag{23}$$

where now, c_0 is a nondimensional quantity and is equal to c/ω_0. Using the above nondimensional quantities, an expression for the variance of the linear system, corresponding to $\epsilon = 0$, can be derived by relying on formulas for random vibration integrals given by Spanos (1987). Specifically, it is found that

$$\sigma_x^2 = \pi\frac{-32c_0\zeta^4+16\zeta^3+\left(8c_0^3+40c_0\right)\zeta^2-16\zeta-8c_0^3-8c_0}{c_0\left(-4c_0^2\zeta^2+c_0^4+2c_0^2+1\right)\left(16\zeta^3-16\zeta\right)}\sigma_f^2 \,. \tag{24}$$

Expanding $f(t,\theta)$ in its Karhunen-Loeve series, and expanding the response process $y(t)$ in its Polynomial Chaos series involving the deterministic coefficients $q_i(t)$, the equation of motion becomes,

$$\sum_{i=0}^{N}\Psi_i\left(\ddot{q}_i(t)+2\zeta\dot{q}_i(t)+q_i(t)\right) \tag{25}$$
$$+\epsilon\sum_{i=0}^{N}\sum_{j=0}^{N}\sum_{k=0}^{N}\Psi_i\Psi_j\Psi_k q_i(t)q_j(t)q_k(t) = \sum_{i=0}^{M}\Psi_i f_i(t).$$

Multiplying the above equation by Ψ_m and averaging yields a system of nonlinear deterministic differential equations

$$\ddot{q}_m(t)+2\zeta\dot{q}_m(t)+q_m(t) =$$
$$-\epsilon\sum_{i=0}^{N}\sum_{j=0}^{N}\sum_{k=0}^{N}\frac{c_{ijkm}}{<\Psi_m^2>}q_i(t)q_j(t)q_k(t)+f_m(t)\,,$$
$$y_m(0)=\dot{y}_m(0)=0, \qquad m=1,\ldots,N \,. \tag{26}$$

In the above equation, $c_{ijkm} = <\Psi_i\Psi_j\Psi_k\Psi_m>$, and $f_m(t)$ vanishes for $m > M$. The coefficients c_{ijkm} can be evaluated very efficiently using certain properties of the Hermite polynomials (Ghanem and

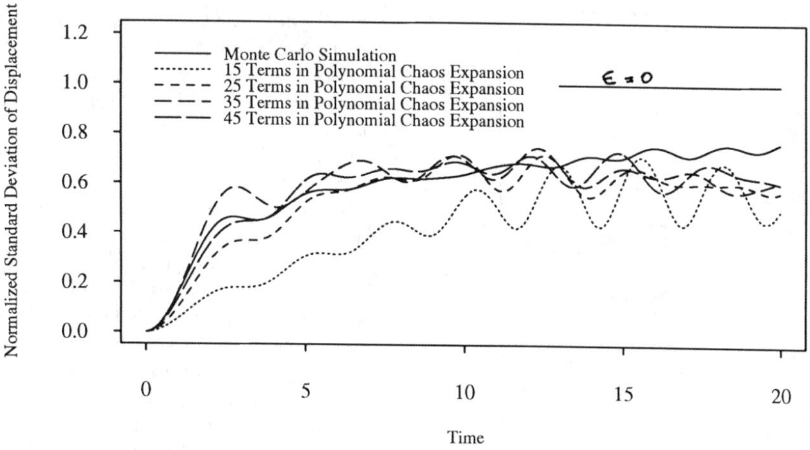

Figure 1: Standard Deviation of the Response Using 1000 Samples in the Monte Carlo Simulation; Two Terms in the Karhunen-Loeve Expansion; ϵ = 0.5; Stationary Excitation.

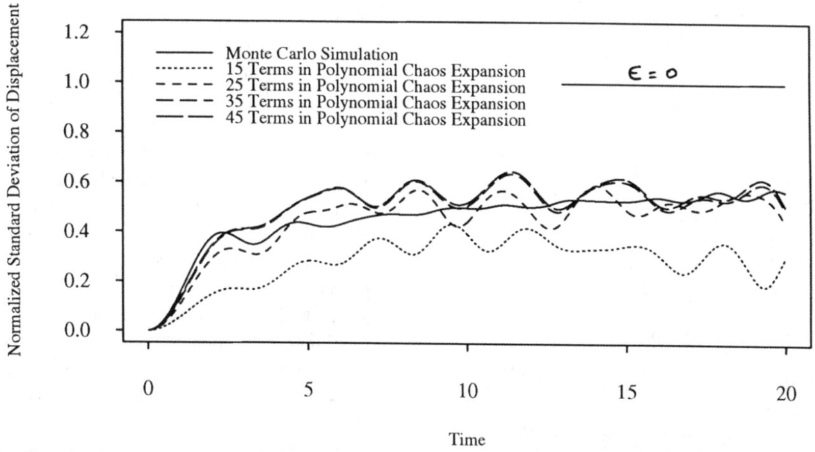

Figure 2: Standard Deviation of the Response Using 1000 Samples in the Monte Carlo Simulation; Two Terms in the Karhunen-Loeve Expansion; ϵ = 2; Stationary Excitation.

Spanos, 1991). They have been evaluated and tabulated using the symbolic computation software *MACSYMA*. Once this has been done, they can be used for various problems without having to be recalculated. The above system of equations can be readily solved using standard techniques for deterministic differential equations. Recall that in the limit as its duration becomes infinite, the process $f(t, \theta)$ becomes a stationary process. Therefore, its Karhunen-Loeve expansion tends to the well known spectral expansion given by equation (7). Thus, it can be expected that the individual components $f_i(t)$ in the Karhunen-Loeve expansion for $f(t, \theta)$ should represent the spectral content of the process around a particular frequency. Care should be taken while truncating the Karhunen-Loeve expansion, to keep those terms that are amplified by the dynamical system. These terms are associated with frequencies in the neighborhood of some effective natural frequency of the system.

5 Numerical Example

The proposed approach was numerically implemented for a value of the normalised damping ratio ζ equal to 0.02, and for a range of the nonlinearity parameter ϵ ranging from 0.1 to 10. Thirty terms were used in the linear-Gaussian component of the Polynomial Chaos expansion of the response process. The nonlinear component consisted of quadratic and

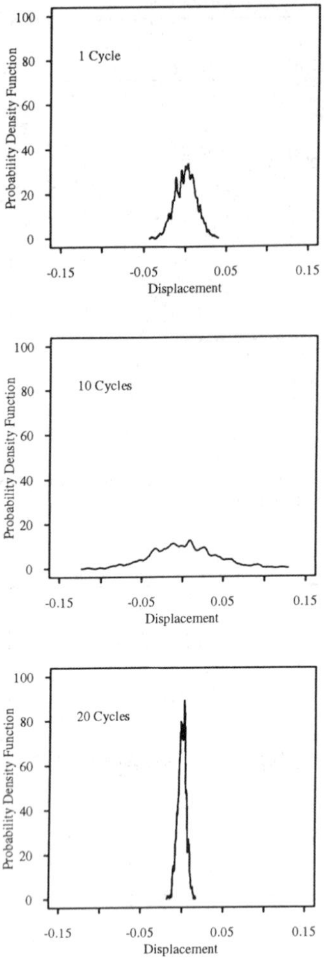

Figure 3: Probability Density Function of the Response Using 1000 Samples in the Monte Carlo Simulation; Two Terms in the Karhunen-Loeve Expansion; $\epsilon = 2$; $\zeta = 0.2$; Non-Stationary Excitation.

cubic terms in the two dominant linear terms. These were determined using an iterative scheme. Initially, the linear terms were ordered according to their proximity to the effective natural frequency of the system, equal to 1 for the nondimensional problem. A first solution to the problem was obtained, involving only the Gaussian terms, representing the linearized solution, and which was therefore obtained very efficiently. Based on this solution, the linear terms were reordered according to the magnitude of the computed Polynomial Chaos coefficients for the solution. The nonlinear component of the response was then obtained by projecting the solution on the higher order polynomial chaoses generated by the four leading Gaussian components. If following the solution of the problem, the ordering of the leading terms had changed, computations were repeated with the new ordering. This step was not needed for the calculations presented herein. It can be shown that equations (25) can be separated into two groups. The equations associated with an odd power in the Gaussian variables are uncoupled from those associated with an even power. This property was noted by Jahedi and Ahmadi (1983). However, since they used an integral expansion, it was not possible to include more than two terms in the expansion. Since the excitation process had a mean of zero, however, the components of even order, mean and quadratic order terms, had no contribution to the response process. Figure (1) shows the time dependent standard deviation of the response process for the case $\epsilon = 0.5$. On the same plot, the analytical results are compared with results obtained through Monte Carlo simulation. The agreement between the two solutions is remarkable. Figure (2) shows similar results corresponding to a value of $\epsilon = 2$.

Once the coefficients in the Polynomial Chaos expansion have been calculated, they can be used very efficiently to obtain realizations of the response process. This is achieved by simulating the Polynomial Chaoses, and evaluating the summations in equation (11). Once a sample population has been obtained for the response, nonparametric density estimation techniques can be used to calculate the probability density of functions of quantities of interest. The proposed approach is also applicable to nonlinear dynamical systems excited by a modulated white noise process. Corresponding results are shown in Figure (3). This figure corresponds to the evolution in time of the probability density function of the response corresponding to a modulating function $e(t)$ given by the equation

$$e(t) = e^{-0.7t} - e^{-0.9t} \, . \tag{27}$$

The probability density function is shown at values of the non-dimensional time equal to 1, 10, and 20. Note the expected behavior of the probability density function, changing from a delta function, to a wide curve, and then back to a delta function when the response has been damped out enough.

Acknowledgement

The partial support of this work from a grant MSM-902 from the NSF is gratefully acknowledged.

Bibliography

[1] Cameron, R.H. and Martin, W.T., "The orthogonal development of nonlinear functionals in series of Fourier-Hermite functionals", *Ann. Math*, Vol. 48, pp. 385-392, 1947.

[2] Ghanem, R. and Spanos, P., *Stochastic Finite Elements: A Spectral Approach,* Springer-Verlag, 1991.

[3] Ghanem, R. and Spanos, P., "Polynomial chaos in stochastic finite elements," *ASME, Journal of Applied Mechanics,* pp. 197-202, Vol. 57, No. 1, 1990.

[4] Ghanem, R. and Spanos, P., "Spectral stochastic finite element formulation for reliability analysis," *ASCE, Journal of Engineering Mechanics,* pp. 2351-2372, Vol. 117, No. 10, 1990.

[5] Iwan, W., and Spanos, P., "Response envelope statistics for nonlinear oscillators with random excitation," pp. 170-174, *Journal of Applied Mechanics,* Vol. 45, March 1978.

[6] Jahedi, A. and Ahmadi, G., "Application of Wiener-Hermite expansion to nonstationary random vibration of a Duffing oscillator," *Journal of Applied Mechanics,* Vol. 50, pp. 436-442, June 1983.

[7] *MACSYMA, Reference Manual, Version 12,* Symbolics Inc., 1986.

[8] Shinozuka, M. and Jan, C.M., "Digital simulation of random Processes and its applications", *Journal of Sound and Vibration*, Vol. 25, No. 1, 1972.

[9] Spanos, P.D., "An approach to calculating random vibration integrals," *ASME, Journal of Applied Mechanics,* pp. 409-413, Vol. 54, 1987.

[10] Spanos, P. and Ghanem, R., "Stochastic finite element expansion for random media," *ASCE, Journal of Engineering Mechanics,* p. 1035-1053, Vol. 115, No. 5, May 1989.

[11] Wiener, N., *Nonlinear Problems in Random Theory*, Technology Press of the Massachussets Institute of Technology and John Wiley and Sons Inc., New York, 1958.

Structural Safety & Reliability, Schuëller, Shinozuka & Yao (eds) © 1994 Balkema, Rotterdam, ISBN 90 5410 357 4

Response of a cable system to vector valued random excitation

D.M.Ghiocel
Stevenson & Associates, Ohio, USA

D.A.Gasparini
Case Western Reserve University, Ohio, USA

ABSTRACT: The paper deals with random vibration of a simple planar model that is believed to capture the essential dynamic behavior of a geometrically non-linear cable dome. Excitation is modeled as a correlated, two-dimensional white process. For perfect negative correlation; i.e. $\rho = -1$, the excitation is purely additive while for perfect positive correlation; i.e. $\rho = +1$, the excitation is purely parametric. Otherwise, the excitation has both additive and parametric components. Mean square responses and their stability are determined for a linear model and for a non-linear model using Gaussian closure to truncate the set of moment equations. For purely additive excitation, i.e. $\rho = -1.0$, both the linear and non-linear systems have stable responses for any finite excitation strength. For purely parametric excitation, i. e. $\rho = +1.0$, if the excitation strength is below a certain value, both the linear and non-linear systems have trivial (zero) stationary responses. Above that value, the linear system has an unstable mean-square response.

1 INTRODUCTION

The Florida Suncoast Dome is a 209 m diameter prestressed cable dome recently completed in the United States. The structural system of the dome is extremely efficient, weighing only 287 N/ m^2. It is a geometrically nonlinear structure with a radial symmetry. Its dynamic behavior depends strongly on the magnitude of the prestressing. To increase understanding of such systems an experimental and analytical study of the dynamic behavior of the Florida Suncoast Dome was recently completed.

The analytical studies reported herein were performed using the simple planar model shown in figure 1. It is believed that the planar model captures the essential dynamic behavior of this radially symmetric type of cable dome, whose (linearized) fundamental mode is shown qualitatively in figure 3. An analytical formulation for the planar cable system was defined by Mesarovic and Gasparini (1992). It is shown therein that the simplest analytical model for the cable system has one dominant centrally symmetric generalized coordinate, q_1, statically coupled with two other coordinates, q_2 and q_3 (see figure 2).

A companion paper by the same authors explores the nature and stability of the dynamic response of the planar cable structure to harmonic excitation. That work is extended herein to excitation modelled as a correlated two-dimensional white process including

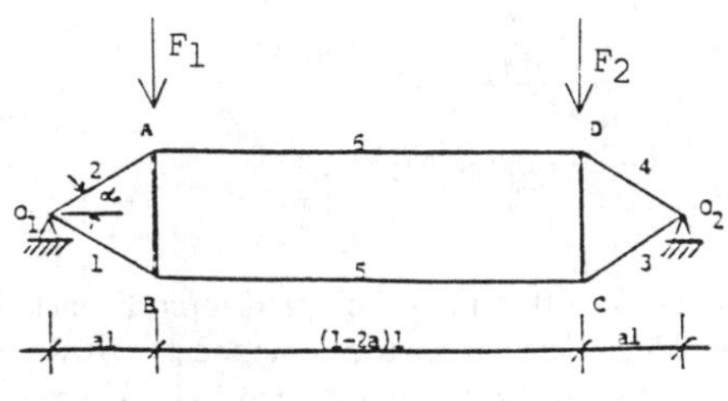

Figure 1. Cable system 6 dof model

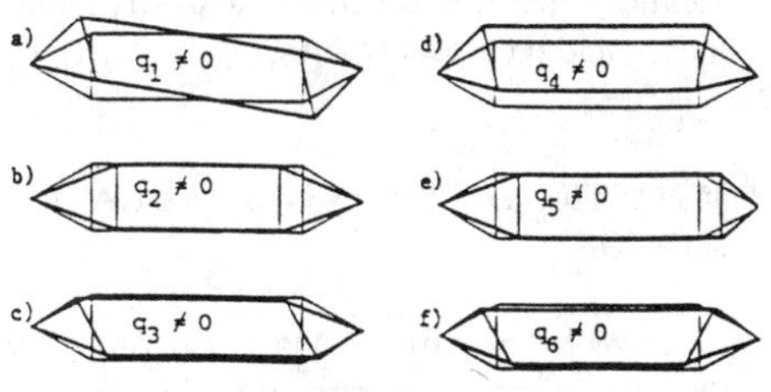

Figure 2. Generalized coordinates.

Table 1 Mean-square response of the linear system

Correlation Coefficient ρ	Excitation strength Q	Mean-square stability	Mean-square responses $E[q_1^2]$ and $E[p_1^2]$
$\rho = -1.00$	Any finite value	Stable	$E[q_1^2] = AQC^2/\xi$ $E[p_1^2] = QC^2/\xi$
$-1.00 < \rho < +1.00$	$Q < \frac{2\xi}{AD^2(1+\rho)}$	Stable	$E[q_1^2] = \frac{AQC^2(1-\rho)}{2\xi - AQD^2(1+\rho)}$ $E[p_1^2] = \frac{QC^2(1-\rho)}{2\xi - AQD^2(1+\rho)}$
	$Q \geq \frac{2\xi}{AD^2(1+\rho)}$	Unstable	
$\rho = +1.00$	$Q < \frac{\xi}{AD^2}$	Stable	$E[q_1^2] = 0.00$ $E[p_1^2] = 0.00$
	$Q \geq \frac{\xi}{AD^2}$	Unstable	

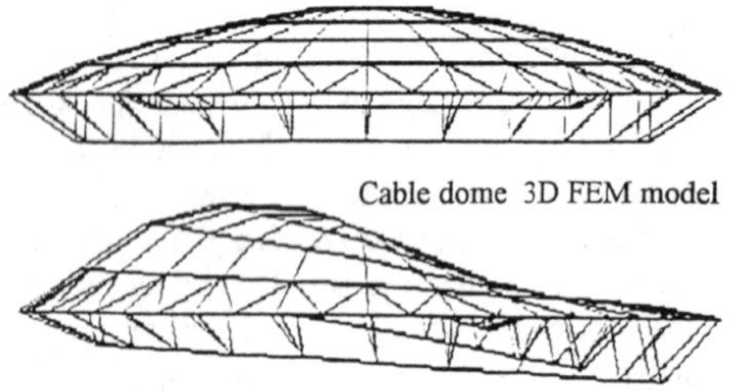

Figure 3. Dominant antisymmetric modal shape

both additive and parametric components. Specifically, second order moments of response are determined for excitation components having a large range of strengths and correlations, and for systems with various geometries and prestressing levels. Digital simulation is used to assess the accuracy of the analytical moment solutions and to estimate distributions and their peak values.

2 STOCHASTIC DIFFERENTIAL EQUATIONS OF MOTION

It was shown by Mesarovic and Gasparini (1992) that the governing equations for the dominant centrally symmmetric generalized coordinate, q_1 (see figure 2a) and generalized momentum, p_1, can be expressed as:

$$q_1 = \frac{3\cos^2\alpha}{1+2\cos^2\alpha} p_1$$
$$p_1 = -q_1 - \frac{\mu(1+\mu)}{2}\left(1 - \frac{\mu}{\mu+\cos^3\alpha}\right)q_1^3 - 2\xi p_1 + F^* \quad (1)$$

in which $\mu = 2a/(1-2a)$ and F^* is the normalized generalized force for the q_1 coordinate. In the present investigation, using the virtual work principle, the generalized force F^* for two concentrated vertical forces (see figure 1) is determined to be:

$$F^* = \frac{1}{4(1+\mu)}\left[\frac{F_1}{EA\,\varepsilon_0}\left(\frac{1}{\sqrt{\varepsilon_o}} - q_1\tan\alpha\right) - \frac{F_2}{EA\,\varepsilon_0}\left(\frac{1}{\sqrt{\varepsilon_o}} + q_1\tan\alpha\right)\right] \quad (2)$$

It can be seen that the generalized force F^* includes both additive and parametric components. Now the normalized force vector is assumed to be a vector of correlated white noises:

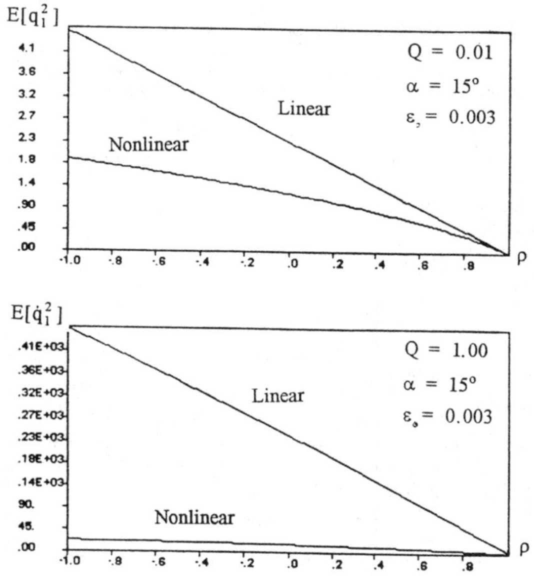

Figure 4. Nonlinear vs. linear response as a function of ρ

$$\begin{Bmatrix} F_1 / EA\varepsilon_o \\ F_2 / EA\varepsilon_o \end{Bmatrix} = \begin{Bmatrix} W_1 \\ W_2 \end{Bmatrix} = \mathbf{W} \tag{3}$$

Regrouping , eq. (2) becomes:

$$F^* = \frac{1}{4(1+\mu)}[(W_1 - W_2)\frac{1}{\sqrt{\varepsilon_o}} - (W_1 + W_2)q_1 \tan\alpha] \tag{4}$$

Letting

$$A = \frac{3\cos^2\alpha}{1+2\cos^2\alpha};\ B = \frac{\mu(1+\mu)}{2}\frac{\cos^3\alpha}{\mu+\cos^3\alpha};$$
$$C = \frac{1}{4(1+\mu)}\frac{1}{\sqrt{\varepsilon_o}};\ D = \frac{1}{4(1+\mu)}\tan\alpha. \tag{5}$$

Then the governing equations can be written in the Ito stochastic differential form (Arnold,1974):

$$d\mathbf{X}(t) = \mathbf{f}(t,\mathbf{X}(t))dt + \mathbf{G}(t,\mathbf{X}(t))d\mathbf{W}(t) \tag{6}$$

in which $\mathbf{X} = \{q_1 \quad p_1\}^T$

$$\mathbf{f}(t,\mathbf{X}) = \begin{Bmatrix} Ap_1 \\ -q_1 - Bq_1^3 - 2\xi p_1 \end{Bmatrix} \text{ and}$$
$$\mathbf{G}(t,\mathbf{X}) = \begin{bmatrix} 0 & 0 \\ C - Dq_1 & -(C + Dq_1) \end{bmatrix} \tag{7}$$

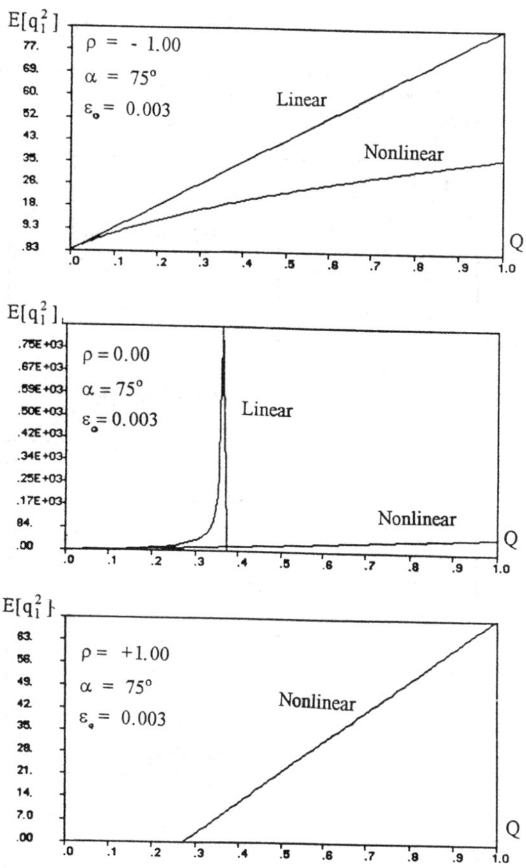

Figure 5. Nonlinear vs. linear response as a function of Q

If both of the white noise components have the same strength, Q, and a zero time-lag correlation coefficient, ρ, then the strength matrix, **Q,** of the vector process may be written as:

$$\mathbf{Q} = Q\begin{bmatrix} 1 & \rho \\ \rho & 1 \end{bmatrix} \tag{8}$$

Following the conventional stochastic calculus (Cumming, 1967), the first and second order moments for eq.(6) are:

$$\begin{aligned}
\dot{E}[q_1] &= E[f_1] \\
\dot{E}[p_1] &= E[f_2] \\
\dot{E}[q_1^2] &= 2E[q_1 f_1] + E[(\mathbf{GQG}^T)_{11}] \\
\dot{E}[q_1 p_1] &= E[q_1 f_2 + p_1 f_1] + E[(\mathbf{GQG}^T)_{12}] \\
\dot{E}[p_1^2] &= 2E[p_1 f_2] + E[(\mathbf{GQG}^T)_{22}]
\end{aligned} \tag{9}$$

Table 2. Mean-square response of the nonlinear system

Correlation coefficient ρ	Excitation strength Q	Mean-square stability	Mean-square responses $E[q_1^2]$ and $E[p_1^2]$
$\rho = -1.00$	Any finite value	Unique solution Stable	$E[q_1^2] = \frac{1}{6B\xi}[-\xi + (\xi + 12AB\xi QC^2)^{1/2}]$ $E[p_1^2] = QC^2/\xi$
$-1.00 < \rho < +1.00$	Any finite value	Unique solution Stable	$E[q_1^2] = \frac{1}{12B\xi}\{AQD^2(1+\rho) - 2\xi + \{[AQD^2(1+\rho) - 2\xi]^2 + 24AB\xi QC^2(1-\rho)\}^{1/2}\}$ $E[p_1^2] = \frac{1}{2\xi}Q\{C^2(1-\rho) + D^2(1+\rho)E[q_1^2]\}$
$\rho = +1.00$	$Q < \xi/AD^2$	Unique solution Stable	$E[q_1^2] = 0.00$ $E[p_1^2] = 0.00$
	$Q = \xi/AD^2$	Critical stability (Bifurcation point)	
	$Q > \xi/AD^2$	Multiple solutions One Stable (nontrivial), One Unstable (trivial)	$E[q_1^2] = \frac{1}{3B\xi}(AQD^2 - \xi)$ $E[p_1^2] = \frac{1}{3B\xi^2}QD^2(AQD^2 - \xi)$

Table 3. Moment equation solution vs. Monte-Carlo simulation of nonlinear mean square response ($\varepsilon_0 = 0.003$)

Correlation coefficient ρ	Angle α	Excitation strength Q	Moment eqs. $E[q_1^2]$	Simulation $E[q_1^2]$	Moment eqs. $E[p_1^2]$	Simulation $E[p_1^2]$
-1.00	15°	0.10	7.24	8.62	46.30	46.62
-1.00	15°	1.00	24.32	27.26	463.0	462.2
-1.00	75°	0.10	6.57	7.76	46.30	47.26
-1.00	75°	1.00	35.23	38.55	463.0	468.2
0.00	15°	0.10	4.94	6.25	23.17	26.56
0.00	15°	1.00	17.03	19.52	232.3	256.0
0.00	75°	0.10	4.16	4.24	27.17	27.23
0.00	75°	1.00	43.79	44.25	655.0	619.2
+0.80	15°	0.10	2.92	2.19	9.28	4.92
+0.80	15°	1.00	10.56	9.15	93.43	53.80
+0.80	75°	0.10	2.04	1.21	12.42	6.92
+0.80	75°	1.00	54.24	45.31	932.1	723.3

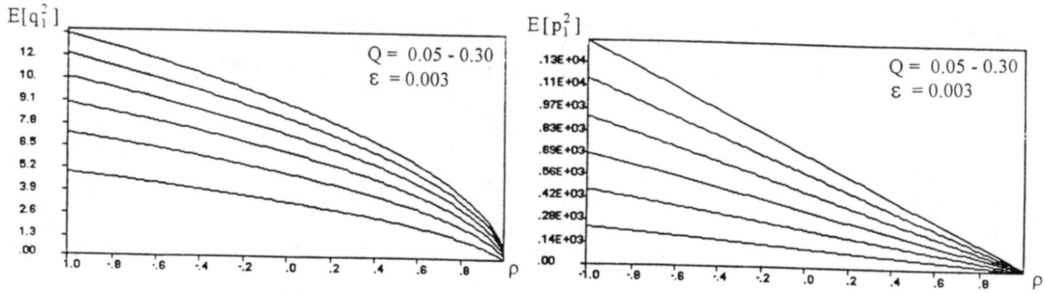

Figure 6. Nonlinear mean-square response q_1 and p_1 for $\alpha = 15°$ and $\varepsilon = 0.003$

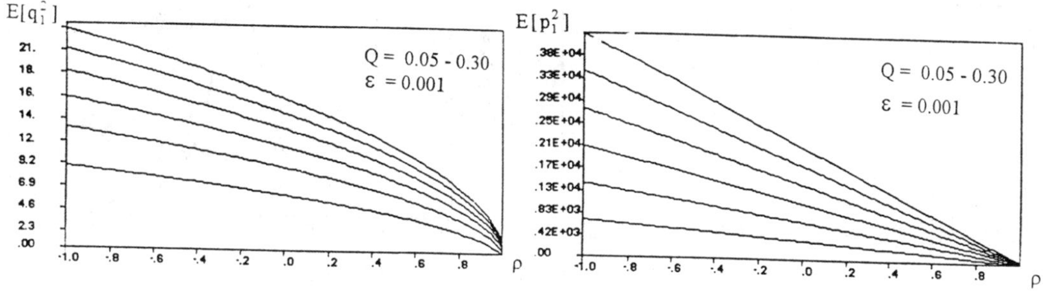

Figure 7. Nonlinear mean-square response q_1 and p_1 for $\alpha = 15°$ and $\varepsilon = 0.001$

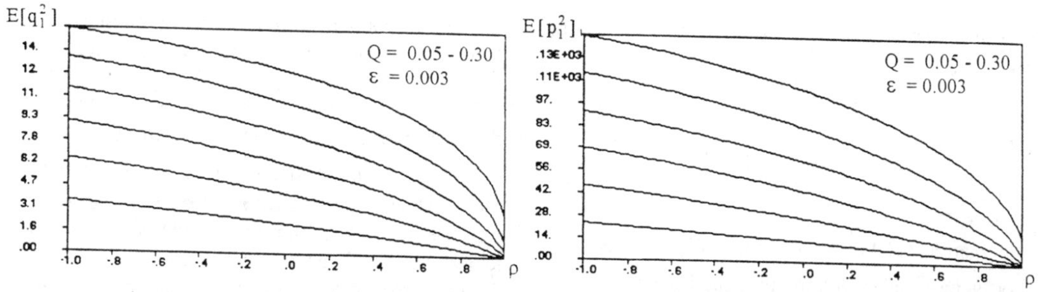

Figure 8 Nonlinear mean-square response q_1 and p_1 for $\alpha = 75°$ and $Q = 0.05 - 0.30$

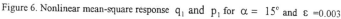

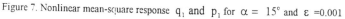

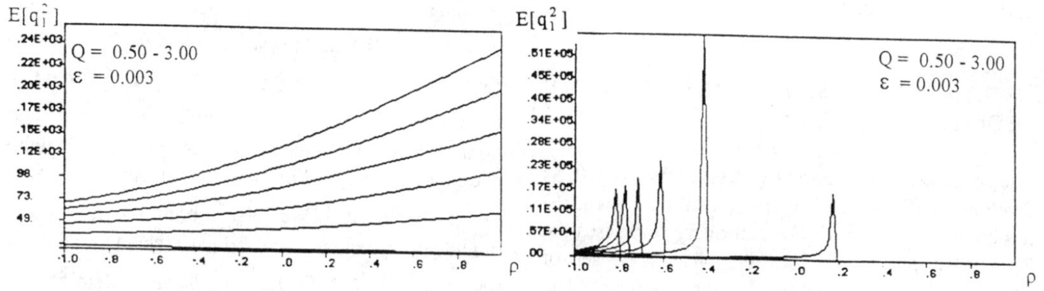

a) Nonlinear system

b) Linear system

Figure 9 Nonlinear vs. linear mean-square response q_1 for $\alpha = 75°$ and $Q = 0.50 - 3.00$

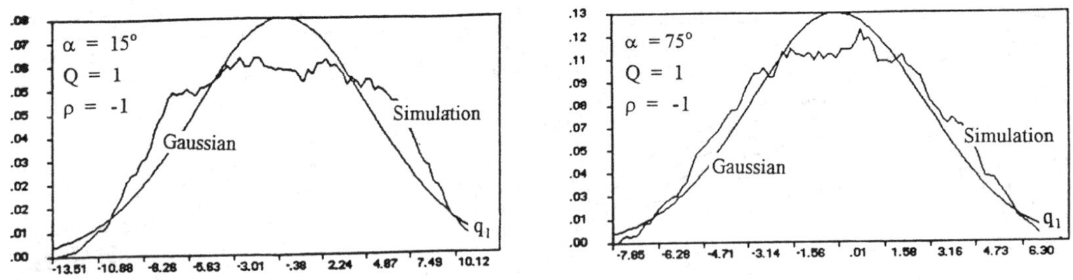

a) Correlation coefficient ρ = -1.00

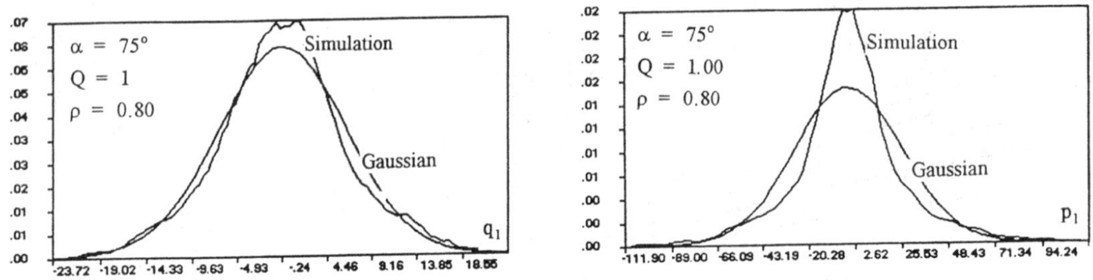

b) Correlation coeficient ρ = 0.80

Figure 10. Simulated PDF of response q_1

in which, for the cable system model, the matrix $\mathbf{GQG}^T$ is

$$\mathbf{GQG}^T = \begin{bmatrix} 0 & 0 \\ 0 & 2Q[C^2(1-\rho)+D^2(1+\rho)q_1^2] \end{bmatrix} \quad (10)$$

The matrix $\mathbf{GQG}^T$ shows that if ρ= -1.00, the coordinate q_1 is excited in purely additive way, while if $\rho = +1.00$ the coordinate q_1 is excited in a purely parametric way. For any other ρ, the excitation has both additive and parametric components.

3 STATIONARY RESPONSE AND STABILITY BOUNDS

The components of vector **f** (t,**X**) and matrix **G**(t,**X**) given in eq. (7) can be directly substituted in eq. (9) to obtain the moments. For stationary responses the first order derivatives are zero, leaving a set of nonlinear algebraic equations for the moments. The stability of the solutions of those equations may be checked using the Routh-Hurwitz criterion. Because of the nonlinear term in the equation of coordinate q_1, the equations for the first and second order moments contain higher order moments and therefore a truncation procedure must be used.

Tables 1 and 2 contain mean square responses and stability bounds for a linear system obtained by dropping the nonlinear term q_1^3 from eq. (1) and for the nonlinear system using Gaussian closure to truncate the set of moment equations. The two tables clearly indicate that responses and their stability are functions of the excitation parameters, the strength, Q, and importantly, the zero time-lag correlation coefficient, ρ, as well as of the system parameters, damping, ξ, prestressing level, ε_o and geometry of the cable system, α and μ; C and D are functions of ε_o, α and μ. For purely additive excitation, i.e. $\rho = -1.00$, both the linear and nonlinear systems have stable response for any finite excitation strength. For purely parametric excitation, i.e. ρ=+1.00, if $Q < \xi / AD^2$, both the linear and nonlinear systems have trivial (zero) stationary responses. If $Q \geq \xi / AD^2$ the linear system has an unstable response, while the nonlinear system has a positive stable solution. By simulation it was shown that the responses are nonstationary unbounded for the linear system and nonstationary bounded for the

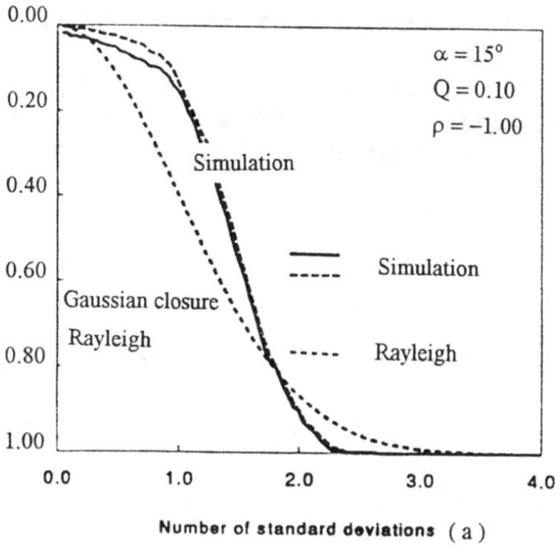

Figure 11. CDF of q_1 peak values

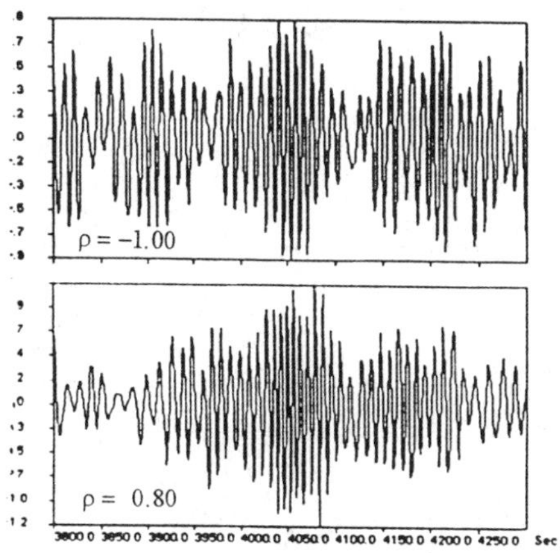

Figure 12. Time history of generalized displacement q_1

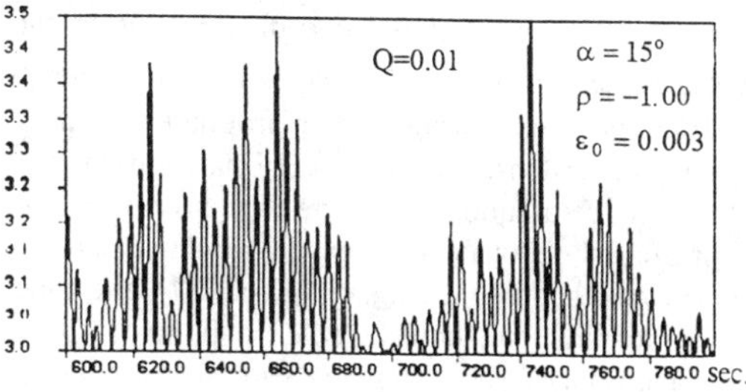

Figure 13. Time history of strain ε x 1000

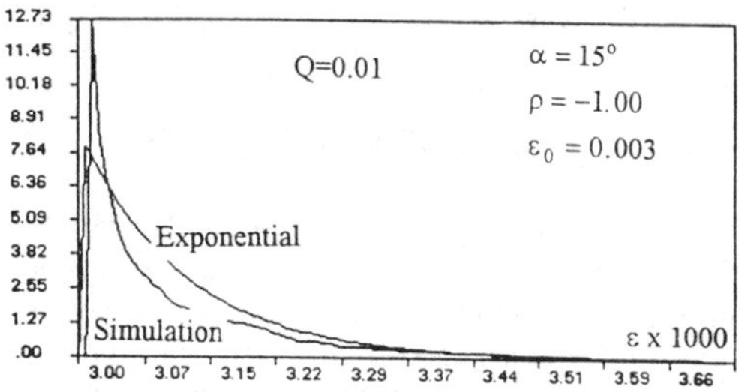

Figure 14. Simulated PDF of strain ε x 1000

nonlinear one. If the excitation components are not perfectly positively correlated, both systems have nontrivial stable responses for low strengths Q. The linear system becomes unstable if $Q \geq 2\xi / AD^2(1+\rho)$.

To reveal more clearly the effects of geometric nonlinearity, the q_1 mean square responses have been plotted for the linear and nonlinear systems, as shown in figures 4 and 5. Figure 4 indicates that for $\alpha = 15^\circ$ the nonlinear hardening system has a considerably smaller response than the linear one. For the linear system the response is a linear function of ρ. Figure 5 shows that for large angles, $\alpha = 75^\circ$, the linear system may be unstable if ρ is greater than -1.00, while the nonlinear system remains stable for all correlation and strength values. The effect of prestrain is identified by comparing figures 6 and 7. Increasing prestrain clearly decreases the response.

The effect of angle α is illustrated by comparing figures 6 and 8. Figure 9 shows that for very high strengths the response actually increases with ρ becoming unstable if the system is linear. For all analyses damping was $\xi = 0.02$.

To assess the accuracy of the analytical predictions obtained using Gaussian closure, digital simulations were performed. Table 3 compares analytical and simulated mean square responses for various parameters. In general there is a satifactory agreement, although the analytical predictions are conservative for $\rho = +0.80$, especially for coordinate p_1. Figure 10 shows the probability density for the simulated response in comparison with a Gaussian density having the same mean and variance. As excitation becomes more parametric the nature of the response changes toward a greater probability of times with small or large amplitudes and a smaller probability for intermediate amplitudes. This is shown by the 500 sec window of unnormalized response q_1 in figure 12.

Figure 11 illustrates cumulative distributions of the peak values of q_1. Also shown is a Rayleigh distribution which would be appropriate for truly Gaussian response. Figure 11 shows that the Rayleigh distribution may be inappropriate for estimating the probability of exceeding thresholds for nonlinear response. In figures 13 and 14 the total strain in the longitudinal cable is plotted. The probability density is compared with an exponential density having the same variance. Even for small strength Q the probability density is not bell-shaped. This is because the strain is a quadratic function of the coordinate q_1.

4 CONCLUSIONS AND FURTHER WORK

The objective of the work was to study the behavior of geometrically nonlinear cable systems to random correlated excitations. The principal parameters were the strength and correlation of the excitation components and the geometry and prestrain of the cable system. The nature of the excitation depends on component correlation. The presence of parametric components influences qualitatively and quantitatively the nonlinear response. Fluctuating strains in cables have a probability distribution close to an exponential one. Analytical predictions of mean square response obtained using Gaussian closure are in satisfactory agreement with simulation results. They are unconservative or conservative depending on the correlation between excitation components.

The study is currently being extended to nonwhite correlated excitation. Continuation techniques are being used to explore parameter spaces that have multiple, stable and unstable, response moments.

ACKOWLEDGEMENTS

The support of the U.S. National Science Foundation, Grant MSM-8817461, is gratefully acknowledged.

REFERENCES

Arnold, L. (1974). *Stochastic differential equations:theory and applications.* John Wiley and Sons:New York.

Cumming, I.G. (1967). Derivation of the moments of a continuous stochastic system. Int. J. Control, 5(1).

Mesarovic, S. and Gasparini, D.A. (1992). Dynamic behavior of nonlinear cable system (part I and II). *J. Engrg. Mech. Div.*, ASCE, Vol. 118, No. 5, May.

Structural Safety & Reliability, Schuëller, Shinozuka & Yao (eds) © 1994 Balkema, Rotterdam, ISBN 90 5410 357 4

Transient response of linear systems based on properties of conditional Gaussian processes

Mircea Grigoriu
Cornell University, Ithaca, N.Y., USA

ABSTRACT:

The determination of the second-moment characteristics of the response of a linear system to random noise is simple during the period of stationary vibrations. However, the computational effort for finding these characteristics increases significantly when the response is not stationary. An efficient method is developed for finding the second-moment characteristics of the transient response of a time-invariant stable linear system that is subject to a stationary Gaussian input. The initial state of the system can be deterministic or random and the input can be white or colored. The method is based on properties of the conditional Gaussian random variables and the Slepian model and second-moment characteristics of the stationary response. Several examples including white and colored stationary Gaussian inputs are used to demonstrate the proposed method.

1 INTRODUCTION

Consider a linear system with a n-dimensional state vector $\mathbf{X}(t)$ satisfying the differential equation

$$\begin{cases}\dot{\mathbf{X}}(t)=\mathbf{a}\mathbf{X}(t)+\mathbf{b}\mathbf{W}(t)\\ \mathbf{X}(0)=\mathbf{x}\end{cases} \quad (1)$$

where $\mathbf{a}$ is a (n, n) matrix with eigenvalues of negative real part, $\mathbf{b}$ denotes a (n, m) matrix, and $\mathbf{W}(t)$ is a m-dimensional stationary white noise with mean $\boldsymbol{\mu}_W = E\mathbf{W}(t)$ and covariance function

$\mathbf{c}_W(\tau)=E(\mathbf{W}(t+\tau)-\boldsymbol{\mu}_W)(\mathbf{W}(t)-\boldsymbol{\mu}_W)' = \mathbf{q}\delta(\tau)$.

General methods are available for finding the first two moments of the transient response of $\mathbf{X}(t)$ and of the stationary response $\mathbf{X}_s(t)$ by analysis in either the frequency or the time domain (Soong & Grigoriu 1993). For example, the mean $\boldsymbol{\mu}(t)=E\mathbf{X}(t)$ and covariance $\boldsymbol{\gamma}(t)= E(\mathbf{X}(t)-\boldsymbol{\mu}(t))(\mathbf{X}(t))-\boldsymbol{\mu}(t))'$ of $\mathbf{X}(t)$ in Eq. 1 satisfy the differential equations

$$\begin{cases}\dot{\boldsymbol{\mu}}(t)=\mathbf{a}\boldsymbol{\mu}(t)+\mathbf{b}\boldsymbol{\mu}_W\\ \boldsymbol{\mu}(0)=\mathbf{x}\end{cases} \quad (2)$$

and

$$\begin{cases}\dot{\boldsymbol{\gamma}}(t)=\mathbf{a}\boldsymbol{\gamma}(t)+\boldsymbol{\gamma}(t)\mathbf{a}'+\mathbf{b}\mathbf{q}\mathbf{b}'\\ \boldsymbol{\gamma}(0)=\mathbf{0}\end{cases} \quad (3)$$

The mean and covariance, μ and γ, of $\mathbf{X}_s(t)$ are time invariant. They satisfy algebraic equations that are independent of the initial conditions and can be derived from Eqs. 2 and 3 by taking $\dot{\boldsymbol{\mu}}(t)=\mathbf{0}$, $\dot{\boldsymbol{\gamma}}(t)=\mathbf{0}$, $\boldsymbol{\mu}(t)=\boldsymbol{\mu}$, and $\boldsymbol{\gamma}(t)=\boldsymbol{\gamma}$.

A method is presented in this paper for calculating the second-moment characteristics of $\mathbf{X}(t)$ from the corresponding descriptors of $\mathbf{X}_s(t)$ for the case in which the input white noise $\mathbf{W}(t)$ is Gaussian. The method is general. It can be applied for deterministic or random initial conditions and can be extended to colored Gaussian inputs (Grigoriu 1992). The method is based on properties of conditional Gaussian processes. An alternative method for deriving probabilistic characteristics of $\mathbf{X}(t)$ from $\mathbf{X}_s(t)$ is in Ref. 3. However, this method can be applied to linear systems with homogeneous initial conditions, i. e., $\mathbf{X}(0)=\mathbf{0}$ in Eq. 1.

2 CONDITIONAL GAUSSIAN PROCESSES

From Eqs. 2 and 3, the mean and covariance matrices, μ and γ, of the stationary response $\mathbf{X}_s(t)$ satisfy the equations

$$\mathbf{a}\boldsymbol{\mu}+\mathbf{b}\boldsymbol{\mu}_W=0 \quad (4)$$

and

$$\mathbf{a}\boldsymbol{\gamma}+\boldsymbol{\gamma}\mathbf{a}'+\mathbf{b}\mathbf{q}\mathbf{b}'=0 \quad (5)$$

The covariance function of $\mathbf{X}_s(t)$ is

$$\mathbf{c}(\tau) = E(\mathbf{X}_s(t+\tau) - \boldsymbol{\mu})(\mathbf{X}_s(t) - \boldsymbol{\mu})' = \boldsymbol{\theta}(\tau)\gamma \qquad (6)$$

in which transition matrix $\boldsymbol{\theta}(t)$ is the solution of

$$\begin{cases} \dot{\boldsymbol{\theta}}(t) = \mathbf{a}\,\boldsymbol{\theta}(t) \\ \boldsymbol{\theta}(t) = \mathbf{I} \end{cases} \qquad (7)$$

where $\mathbf{I}$ is the identity matrix. The second-moment characteristics $\mathbf{X}_s(t)$ in Eqs. 4-6 are independent of the initial conditions and specify $\mathbf{X}_s(t)$ fully because the noise is Gaussian and so is the response.

Consider the conditional process

$$\hat{\mathbf{X}}(t) = \mathbf{X}_s(t) | \mathbf{X}_s(0) = \mathbf{x} \qquad (8)$$

The probability of this process can be obtained from second-moment characteristics of the vector $\{\mathbf{X}_s(t), \mathbf{X}_s(s), \mathbf{X}_s(0)\}$ by conditioning because $\mathbf{X}_s(t)$ is a Gaussian process. The components of this vector have mean μ. The covariance of values of $\mathbf{X}_s(t)$ at any two time instances, t and s, is $\mathbf{c}$(t-s). Using properties of conditional Gaussian variables, the mean and covariance functions of $\hat{\mathbf{X}}(t)$ are (Leadbetter, Lindgren & Rootzen 1980)

$$\hat{\boldsymbol{\mu}}(t) = \boldsymbol{\mu} + \boldsymbol{\theta}(t)\,(\mathbf{x} - \boldsymbol{\mu}) \qquad (9)$$

$$\hat{\boldsymbol{\gamma}}(t) = \boldsymbol{\gamma} - \boldsymbol{\theta}(t)\,\boldsymbol{\gamma}\,\boldsymbol{\theta}(t)' \qquad (10)$$

The conditional moments $\hat{\boldsymbol{\mu}}(t)$ and $\hat{\boldsymbol{\gamma}}(t)$ approach $\mathbf{x}$ and $\mathbf{0}$, respectively, as $t \to 0$. The covariance function of $\hat{\mathbf{X}}(t)$ is

$$\begin{aligned} \hat{\mathbf{c}}(t,s) &= E(\hat{\mathbf{X}}(t) - \hat{\boldsymbol{\mu}}(t))(\hat{\mathbf{X}}(s) - \hat{\boldsymbol{\mu}}(s))' \\ &= \boldsymbol{\theta}(t-s)\,\boldsymbol{\gamma} - \boldsymbol{\theta}(t)\,\boldsymbol{\gamma}\,\boldsymbol{\theta}(s)', \qquad t > s \end{aligned} \qquad (11)$$

From Eqs. 9-11, $\hat{\mathbf{X}}(t)$ is a nonstationary Gaussian process. It can be given in the form

$$\hat{\mathbf{X}}(t) = \hat{\boldsymbol{\mu}}(t) + \boldsymbol{\xi}(t) \qquad (12)$$

in which $\xi(t)$ is a zero-mean nonstationary Gaussian process with covariance function

$$E\boldsymbol{\xi}(t)\boldsymbol{\xi}(s)' = \hat{\mathbf{c}}(t,s), \qquad t > s \qquad (13)$$

The process $\xi(t)$ does not depend on the initial conditions $\mathbf{X}(0) = \mathbf{x}$. The dependence on $\mathbf{x}$ is incorporated in the conditional expectation $\hat{\boldsymbol{\mu}}(t)$.

Processes $\mathbf{X}(t)$ in Eq. 1 and $\hat{\mathbf{X}}(t)$ in Eq. 8 are equal in distribution for $t > 0$. The proof of this statement follows from the observation that $\hat{\mathbf{X}}(t)$ and $\hat{\mathbf{W}}(t) = \mathbf{W}(t) | \mathbf{X}_s(0) = \mathbf{x}$ satisfy the differential equation

$$\begin{cases} \dot{\hat{\mathbf{X}}}(t) = \mathbf{a}\hat{\mathbf{X}}(t) + \mathbf{b}\hat{\mathbf{W}}(t) \\ \hat{\mathbf{X}}(0) = \mathbf{x} \end{cases} \qquad (14)$$

that can be obtained from Eq. 1 by conditioning. An alternative form of this equation is

$$\begin{cases} \dot{\hat{\mathbf{X}}}(t) = \mathbf{a}\hat{\mathbf{X}}(t) + \mathbf{b}\mathbf{W}(t) \\ \hat{\mathbf{X}}(0) = \mathbf{x} \end{cases} \qquad (15)$$

because $\mathbf{W}(t)$ is a Gaussian white noise so that the condition $\mathbf{X}_s(0) = \mathbf{x}$ does not affect the input and the noises $\hat{\mathbf{W}}(t)$ and $\mathbf{W}(t)$ coincide. Therefore, $\hat{\mathbf{X}}(t)$ and $\mathbf{X}(t)$ are equal in distribution as Gaussian processes with the same mean and covariance functions.

The results in this section can be extended to the case in which the input is colored provided it can be represented as one or more components of the output of a linear filter to white noise. Moreover, the analysis can also account for the randomness in the initial conditions (Grigoriu 1992).

3 NUMERICAL EXAMPLES

Consider first a half linear oscillator with state X(t) satisfying the differential equation

$$\begin{cases} \dot{X}(t) = -\alpha X(t) + W(t) \\ X(0) = x, \quad \alpha > 0 \end{cases} \qquad (16)$$

in which W(t) is a stationary white noise Gaussian process with mean μ_w and covariance function $E(W(t+\tau) - \mu_w)(W(t) - \mu_w) = \pi G_0 \delta(\tau)$. The stationary response $X_s(t)$ has mean $\mu = \mu_w / \alpha$, variance $\gamma = \pi G_0 / (2\alpha)$, and covariance function $c(\tau) = E(X(t+\tau) - \mu)(X(t) - \mu) = \gamma e^{-\alpha\tau}$, $\tau > 0$. The transition matrix of the system is $\theta(t) = e^{-\alpha t}$, $\tau > 0$, so that the mean and covariance functions of the transient response are

$$\hat{\mu}(t) = \mu + (x - \mu)e^{-\alpha t} \qquad (17)$$

and

$$\hat{c}(t+\tau, t) = \gamma(e^{-\alpha\tau} - e^{-\alpha(2t+\tau)}), \quad \tau > 0 \qquad (18)$$

These moments approach the stationary values μ and $\gamma e^{-\alpha\tau}$ as $t \to \infty$.

Consider now a simple oscillator with damping ratio ζ and natural frequency ω that is subject to a white noise process W(t) as in Eq. 16. The displacement X(t) of the oscillator satisfies the differential equation

$$\begin{cases} \ddot{X}(t) + 2\zeta\omega\dot{X}(t) + \omega^2 X(t) = W(t) \\ X(0) = x, \quad \dot{X}(0) = z \end{cases} \tag{19}$$

The equation of motion of the oscillator can also be given by Eq. 1 with

$$\mathbf{a} = \begin{bmatrix} 0 & 1 \\ -\omega^2 & -2\zeta\omega \end{bmatrix}, \quad , \mathbf{b} = \begin{bmatrix} 0 \\ 1 \end{bmatrix} \tag{20}$$

and $\mathbf{X}(t) = \{X_1(t) = X(t),\ X_2(t) = \dot{X}(t)\}'$. The stationary response $\mathbf{X}_s(t)$ has mean $\mu = \{\mu_1 = \mu_w / \omega^2, \mu_2 = 0\}'$ and covariance γ with components $\gamma_{11} = \pi G_0 / (4\zeta\omega^3)$, $\gamma_{12} = \gamma_{21} = 0$ and $\gamma_{22} = \pi G_0 / (4\zeta\omega)$. The covariance function of the process is $\mathbf{c}(\tau) = \boldsymbol{\theta}(\tau)\boldsymbol{\gamma} =$ and has the components

$$c_{11}(\tau) = \frac{\pi G_0}{4\zeta\omega^3} e^{-\zeta\omega\tau} \left[\cos(\omega_d \tau) + \frac{\zeta}{\sqrt{1-\zeta^2}} \sin(\omega_d \tau) \right]$$

$$c_{12}(\tau) = \frac{\pi G_0}{4\zeta\omega^3} e^{-\zeta\omega\tau} \frac{\omega}{\sqrt{1-\zeta^2}} \sin(\omega_d \tau)$$

$$c_{21}(\tau) = \frac{\pi G_0}{4\zeta\omega^3} e^{-\zeta\omega\tau} \frac{\omega^3}{\sqrt{1-\zeta^2}} \sin(\omega_d \tau)$$

$$c_{22}(\tau) = \frac{\pi G_0}{4\zeta\omega^3} e^{-\zeta\omega\tau} \omega^2 \left[-\frac{\zeta}{\sqrt{1-\zeta^2}} \sin(\omega_d \tau) + \cos(\omega_d \tau) \right] \tag{21}$$

where $\omega_d = \omega\sqrt{1-\zeta^2}$ and $\boldsymbol{\theta}(\tau) = \{\theta_{ij}(\tau)\}$ with components

$$\theta_{11}(\tau) = e^{-\zeta\omega\tau} \left[\cos(\omega_d \tau) + \frac{\zeta}{\sqrt{1-\zeta^2}} \sin(\omega_d \tau) \right]$$

$$\theta_{12}(\tau) = e^{-\zeta\omega\tau} \frac{1}{\omega_d} \sin(\omega_d \tau)$$

$$\theta_{21}(\tau) = e^{-\zeta\omega\tau} \frac{\omega}{\sqrt{1-\zeta^2}} \sin(\omega_d \tau)$$

$$\theta_{22}(\tau) = e^{-\zeta\omega\tau} \left[-\frac{\zeta}{\sqrt{1-\zeta^2}} \sin(\omega_d \tau) + \cos(\omega_d \tau) \right] \tag{22}$$

is the transition matrix of the system.

The mean and covariance function of the transient response $\mathbf{X}(t)$ can be obtained from Eqs. 9, 11, 21, and 22. For example, the mean of the transient displacement is

$$\begin{aligned} \hat{\mu}_1(t) &= \mu_1 + \theta_{11}(t)(x - \mu_1) + \theta_{12}(t)\ z \\ &= \frac{\mu_w}{\omega^2} + e^{-\zeta\omega t}\{[\cos(\omega_d t) \\ &+ \frac{\zeta}{\sqrt{1-\zeta^2}} \sin(\omega_d t)](x - \mu_1) + \frac{1}{\omega_d} \sin(\omega_d t)\ z\} \end{aligned} \tag{23}$$

It approaches the stationary values μ_w / ω^2 as $t \to \infty$. The covariance function of the transient displacement has the expression

$$\begin{aligned} \hat{c}_{11}(t+\tau, t) &= \theta_{11}(\tau)\gamma_{11} - \theta_{11}(t+\tau)\gamma_{11}\theta_{11}(t) \\ &\quad - \theta_{12}(t+\tau)\gamma_{22}\theta_{12}(t) \\ &= \frac{\pi G_0}{4\zeta\omega^2}\{e^{-\zeta\omega\tau}(\cos\bar{\tau} + \frac{\zeta}{\sqrt{1-\zeta^2}}\sin\bar{\tau}) \\ &\quad - e^{-\zeta\omega(t+\tau)}[\frac{1}{1-\zeta^2}\cos\bar{\tau} - \frac{\zeta^2}{1-\zeta^2}\cos(2\bar{t} + \bar{\tau}) \\ &\quad + \frac{\zeta}{\sqrt{1-\zeta^2}}\sin(2\bar{t} + \bar{\tau})]\} \end{aligned} \tag{24}$$

where $\bar{t} = \omega_d t$, $\bar{\tau} = \omega_d \tau$, and $\tau > 0$. The variance of the transient displacement at a time t is

$$\begin{aligned} \hat{\gamma}_{11}(t) &= \hat{c}_{11}(t, t) \\ &= \frac{\pi G_0}{4\zeta\omega_0^3}\left\{1 - \frac{e^{-\zeta\omega t}}{1-\zeta^2}\left[1 - \zeta^2\cos 2\bar{t} + \zeta\sqrt{1-\zeta^2}\sin 2\bar{t}\right]\right\} \end{aligned} \tag{25}$$

The covariances in Eqs. 24-25 converge to their stationary expressions $(\pi G_0 / 4\zeta\omega_0^3)e^{-\zeta\omega\tau}(\cos\bar{\tau} + (\zeta/\sqrt{1-\zeta^2})\sin\bar{\tau})$ and $(\pi G_0 / 4\zeta\omega_0^3)$, respectively, as t increases indefinitely.

4 CONCLUSION

A method was developed for calculating the probability law of the transient response of linear systems subject to stationary Gaussian white noise. The method involves two phases. First, the second-moment characteristics of the stationary response are determined. This phase requires the solution of algebraic equations. Second, properties of conditional Gaussian processes are used to find the mean and covariance functions of the transient response. Two examples are presented to illustrate the proposed method of analysis.

REFERENCES

Grigoriu, M. 1992. Transient Response of Linear Systems to Stationary Gaussian Inputs. *Probabilistic Engineering Mechanics* 7: 159-164.

Leadbetter, M.R., G. Lindgren & H. Rootzen 1980. *Extremes and Related Properties of Random Sequences and Processes*, New York: Springer-Verlag.

Madsen, H.O., S. Krenk & N.C. Lind 1986. *Methods of Structural Safety. New Jersey:* Prentice-Hall.

Soong, T.T. & M. Grigoriu 1993. *Random Vibrations of Mechanical and Structural Systems*, New Jersey: Prentice Hall.

Structural Safety & Reliability, Schuëller, Shinozuka & Yao (eds) © 1994 Balkema, Rotterdam, ISBN 90 5410 357 4

Response of linear and nonlinear systems to non-Gaussian white noise

Mircea Grigoriu
Cornell University, Ithaca, N.Y., USA

ABSTRACT:

Methods are examined for the analysis of the response of linear and nonlinear systems to Poisson white noise. The response of a linear system to Poisson white noise is a filtered Poisson process so that the general theory of these processes can be applied for solution. However, there is no practical approach for the response analysis of nonlinear systems subject to Poisson white noise. The generalized Itô calculus and Kolmogorov equations can be used in analysis but these formulations are usually complex. An alternative method is presented for the response analysis of nonlinear systems subject to Poisson white noise. The method is based on a mapping with random time step relating responses immediately following consecutive pulses. An integral equation and a Markov model are developed for finding characteristics of the system responses. Examples of simple non-linear conservative systems with potentials having single and double wells are used to illustrate numerically the Markov model.

1 INTRODUCTION

Consider a linear or nonlinear system with state vector $X(t), t \geq 0$, that is subject to a Poisson white noise. The noise can be viewed as the formal derivative of a compound Poisson process

$$Y(t) = \begin{cases} 0 & , \quad N(t) = 0 \\ \sum_{k=1}^{N(t)} Y_k \, u(t - \tau_k) & , \quad N(t) > 0 \end{cases} \tag{1}$$

in which $\{N(t), t \geq 0\}$ is a homogeneous Poisson counting process of intensity $\lambda > 0$, $\{Y_k\}$ denote independent identically distributed variables following a distribution F, $\{\tau_k\}$ are the arrival times of the Poisson process, and u(s) = 1 or 0 for s > 0 or s < 0. The noise represents a sequence of random pulses of magnitude $\{Y_k\}$ arriving at random instances $\{\tau_k\}$.

The response of linear system to Poisson white noise is a filtered Poisson process so that the general theory of these processes can be applied for solution. However, there is no practical method for calculating the distribution of the state process X(t) for nonlinear systems. A generalized version of the Fokker-Planck equation can be used for finding the distribution of X(t) (Gilman & Skorohod 1972; Tylikowski & Marowski 1986). However, the solution of this equation is complex. Theoretical results are available for the distribution of X(t) when the state process is scalar and the random variables $\{Y_k\}$ are nonnegative (Harrison & Resnick 1976).

This paper presents two methods for characterizing the response X(t). The methods are based on an integral equation for the distribution of the state process and a Markov model that can be applied to find this distribution and the probability of failure of a system.

2 SYSTEM RESPONSE

Let h(x, s) be the free vibration response of a dynamic system to initial condition x at a time $s \geq 0$. This function can be obtained in closed form and is linear in x for linear systems. The free vibration response of nonlinear systems is not generally available in closed form but can be determined numerically.

Consider a system subject to a sequence of random pulses defined formally as the derivative of Y(t) in Eq. 1. Denote by $T_k = \tau_k - \tau_{k-1}$, k = 1, 2, ... the random interarrival times between successive Poisson events, where $\tau_0 = 0$. Let $X_k^- = X(\tau_k^-)$ and $X_k^+ = X(\tau_k^+)$ be the values of the state vector immediately prior and immediately after the arrival of pulse k. Then,

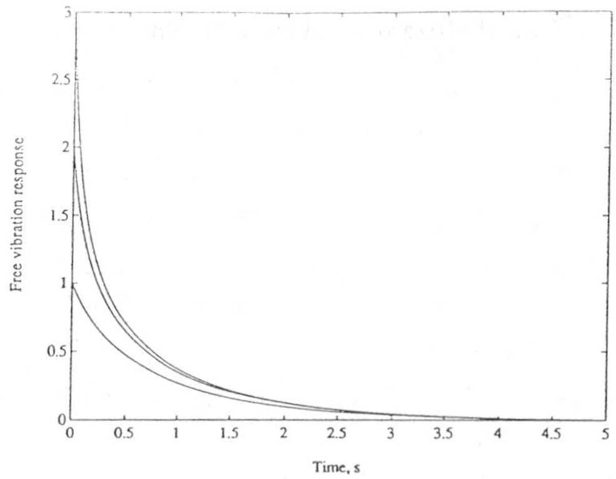

Figure 1. Free vibration response of the (-, +) oscillator in Eq. 5 with ρ=1, ε=10, and x = 1; 2; 3.

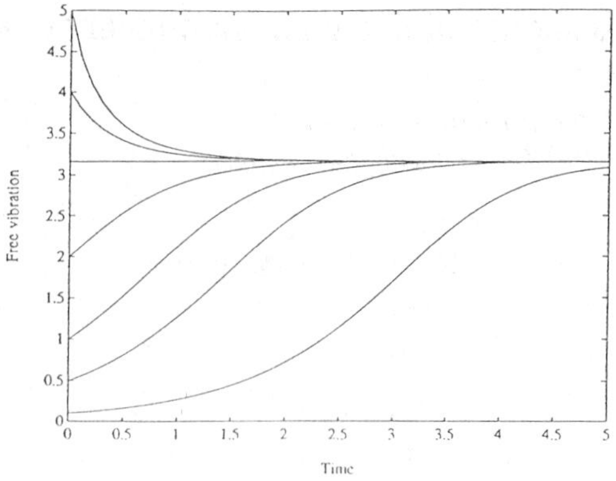

Figure 2. Free vibration response of the (+, -) oscillator in Eq. 5 with ρ=1, ε=0.1, and x = 0.1; 0.5; 1.0; 2.0; 3.162; 4.0; 5.0.

$$X_k^- = h(X_{k-1}^+, T_k) \tag{2}$$

because the input is zero in the time interval (τ_{k-1}, τ_k) so that the system vibrates freely during this interval. Responses X_k^- into X_k^+ can be related by

$$X_k^+ = X_k^- + g(Y_k) \tag{3}$$

in which function g depends on the system dynamic characteristics. The mapping

$$X_k^+ = h(X_{k-1}^+, T_k) + g(Y_k) \tag{4}$$

can be obtained from Eqs. 2-3 and provides a relationship between response values immediately following successive random pulses. A similar equation can be developed for the random series $\{X_k^-\}$.

Consider, for example, the half oscillators with state X(t) satisfying the stochastic differential equations

$$dX(t) = \mp\rho X(t)(1 \pm \varepsilon X(t)^2)dt + dY(t) \quad , \quad \rho > 0,\ \varepsilon > 0,\ t > 0 \tag{5}$$

where Y(t) is the process in Eq. 1. The homogeneous solutions of Eq. 5 can be obtained in closed form for an arbitrary initial condition X(0) = x and time $s \geq 0$, so that the free vibration functions of the oscillators are

$$h(x,s) = \frac{x}{\sqrt{\mp\varepsilon x^2 + (1 \pm \varepsilon x^2)e^{\pm 2\rho s}}} \tag{6}$$

Figures 1 and 2 show the variation of these functions in time for several values of x, ρ, and ε. The free vibration solutions of the oscillators with signs (-, +) and (+, -) in front of parameters (ρ, ε) in Eq. 5 differ significantly because the potentials of these oscillators are characterized by a single and a double well, respectively. The single well is located at x = 0 while the double wells are located at $x = \pm 1/\sqrt{\varepsilon}$. The asymptotic values of h(x, s) as $s \to \infty$ are 0 and $\pm 1/\sqrt{\varepsilon}$ for the (-, +) and (+, -) oscillators. The mapping in Eq. 4 becomes

$$X_k^+ = \frac{X_{k-1}^+}{\sqrt{\mp\varepsilon(X_{k-1}^+)^2 + [1 \pm \varepsilon(X_{k-1}^+)^2]e^{\pm 2\rho T_k}}} + Y_k \tag{7}$$

because g(y) = y is the identity function in this case. The free vibration response and the mapping of the (-, +) oscillator simplify to

$$h(x,s) = xe^{-\rho s} \tag{8}$$

and

$$X_k^+ = X_{k-1}^- e^{-\rho T_k} + Y_k \tag{9}$$

when $\varepsilon = 0$, in which case the system is linear. From Figs. 1 and 2, extreme responses can occur immediately following a random pulse or at any instant for the (-, +) and (+, -) oscillators, respectively.

3 RESPONSE DISTRIBUTION

An integral equation and a Markov model are developed to characterize the response of the nonlinear system in Eq. 5.

3.1 Integral Equation

The distribution of X_k^+ conditional on $X_{k-1}^+ = x$ and $T_k = s$ is

$$F_{X_k^+|X_{k-1}^+,T_k}(y|x,s) = F(y - h(x,s)) \tag{10}$$

in which the free vibration response h(x, s) has the expression Eq. 6 for the system in Eq. 5 and F is the common distribution of pulses $\{Y_k\}$. By differentiating Eq. 10 relative to y one finds the density of the conditional variable $X_k^+|\{X_{k-1}^+, T_k\}$ which is

$$f_{X_k^+|X_{k-1}^+,T_k}(y|x,s) = f(y - h(x,s)) \tag{11}$$

with the notation f = F'. The unconditional densities X_k^+ and X_{k-1}^+ can be related by

$$f_{X_k^+}(y) = \int ds f_{T_k}(s) \int dx f_{X_{k-1}^+}(x) f(y - h(x,s)) \tag{12}$$

This integral equation becomes

$$f_{X^+}(y) = \int ds f_{T_k}(s) \int dx f_{X^+}(x) f(y - h(x,s)) \tag{13}$$

if time series $\{X_k^+\}$ approaches stationarity as $k \to \infty$ and $f_{X^+}(y) = \lim_{k\to\infty} f_{X_k^+}(y)$ denotes its stationary density.

The integral equations in Eqs. 12 and 13 can be applied to determine the transient and the stationary densities of state process X(t) immediately following a pulse. Similar equations can be developed for X(t) immediately prior to a pulse.

3.2 Markov Model

Consider a partition in cells C_i of the state space associated with a dynamic system. Let

$$p_{ij}^{(k)}(s) = P(X_k^+ \in C_j | X_{k-1}^+ \in C_i, T_k = s) \tag{14}$$

be the transition probability of state vector X^+ from cell C_i to cell C_j during an interarrival time T_k assumed to be equal to s. This transition probability can be calculated simply because the only source of uncertainty is the pulse Y_k. For example,

$$p_{ij}^{(k)}(s) = P(h(x_i,s) + Y_k \in C_j) \tag{15}$$

for the half oscillators in Eq. 5 if it is assumed that X_{k-1}^+ coincides with, for example, the central point x_i of C_i. The assumption is satisfactory if the cells are small.

Let $\mathbf{P}^{(0)}$ be the initial probability vector characterizing the state vector X(0). This vector becomes

$$\mathbf{P}^{(n)}(s_1,\dots,s_n) = \left[\prod_{k=2}^{n} \mathbf{p}^{(k)}(s_k)\right]\mathbf{P}^{(0)} \tag{16}$$

following n pulses with interarrival times $\{T_k = s_k\}$, k = 1, ..., n, where $\mathbf{p}^{(k)}(s) = \{p_{ij}^{(k)}(s)\}$ is the transition probability matrix when the interarrival time T_k is equal to s.

The conditional probability $\mathbf{P}^{(n)}(s_1, \dots, s_n)$ of the response $X_{N(t)}^+$ immediately following the last pulse in (0, t) depends on the number of Poisson events N(t) = n and the values of interarrival times $\{T_1 = s_1, \dots, T_n = s_n\}$. The unconditional probability of $X_{N(t)}^+$ is

$$\mathbf{P} = E\ \mathbf{P}^{(N(t))}(T_1,\dots,T_{N(t)}) \tag{17}$$

This average can be calculated by simulation approximately. Let $\mathbf{P}_i^{(n_i)}(s_1^{(i)},\dots,s_{n_i}^{(i)})$ be the probability of $X_{N(t)}^+$ corresponding to a realization $N(t) = n_i$ and $\{T_1 = s_1^{(i)}, \dots T_n = s_{n_i}^{(i)}\}$, i = 1, 2, ..., n_s, where n_s denotes the number of realizations. Then **P** can be approximated by

$$\hat{\mathbf{P}} = \frac{1}{n_s}\sum_{i=1}^{n_s} \mathbf{P}^{(n_i)}(s_1^{(i)},\dots,s_{n_i}^{(i)}) \tag{18}$$

An alternative approximation of **P** as $\lambda t \to \infty$ can be obtained from the equality

$$\mathbf{P} = E\mathbf{p}(T)\ \mathbf{P} \tag{19}$$

provided that the oscillator response is stationary.

The Markov model can also be applied to determine the first passage time distribution of $\{X_k^+\}$ by adding an absorbing cell and by accounting for the number of transitions that can occur during a specified period. Moreover, the probability of state vector X(t) can also be determined at an arbitrary instant t.

4 NUMERICAL EXAMPLE

Suppose that the pulses $\{Y_k\}$ follow a Gaussian dis-

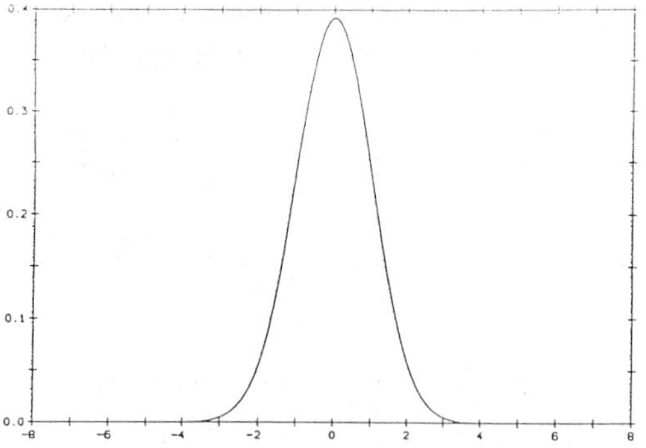

Figure 3. The stationary density of X_k^+ for the (-, +) oscillator in Eq. 5 with ρ=1, ε=10, λ=0.5, μ=0.0, and σ=1.0.

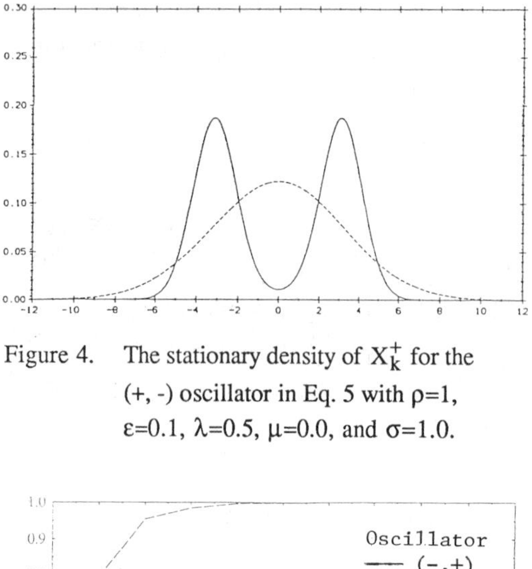

Figure 4. The stationary density of X_k^+ for the (+, -) oscillator in Eq. 5 with ρ=1, ε=0.1, λ=0.5, μ=0.0, and σ=1.0.

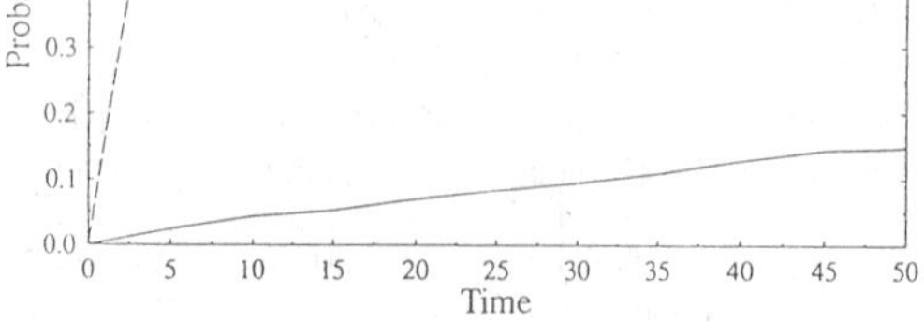

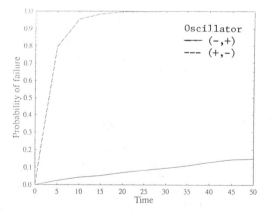

Figure 5. Probability of failure in (0, t) relative to safe set (-3, 3) for the oscillators (-,+) and (+,-) in Eq. 5 with ρ=1; ε=10; λ=0.5 and ρ=1; ε=0.1; λ=0.5, respectively.

tribution with mean μ and variance σ^2. Let $C_1=(-\infty,a_2)$, $C_{2N}=(a_{2N-1},+\infty)$, and $C_i=(a_i,a_{i+1})$, for $a_i=(-N+i-1)\Delta x$ and i=2, ..., 2N-1, be a partition of the real line in 2(N-1) cells of equal size Δx and two cells of much larger measure, the cells C_1 and C_{2N}.

The conditional transition probability from cell C_i to cell C_j in Eq. 15 can be calculated from

$$p_{ij}(s)=\Phi\left(\frac{a_{j+1}-h(x_i,s)-\mu}{\sigma}\right)-\Phi\left(\frac{a_j-h(x_i,s)-\mu}{\sigma}\right) \tag{20}$$

in which Φ is the distribution of the standard Gaussian variable. The corresponding unconditional transition probability has the expression

$$p_{ij}=\int_0^\infty ds\lambda e^{-\lambda s}[\Phi\left(\frac{a_{j+1}-h(x_i,s)-\mu}{\sigma}\right) -\Phi\left(\frac{a_j-h(x_i,s)-\mu}{\sigma}\right)] \tag{21}$$

Figures 3 and 4 show stationary distributions of X_k^+ of the (-, +) and (+, -) oscillators in Eq. 5 for λ=0.5, μ=0.0, and σ=1.0. The analysis is based on the Markov model in Eqs. 14-19. The resultant distributions are nearly Gaussian for the (-, +) oscillator but differ significantly from the Gaussian model for the (+, -) oscillator. These findings are consistent with the free vibration response in Figs. 1 and 2. Indeed, $X_k^+=h(X_{k-1}^+,T_k)+Y_k$ is approximately equal to Y_k with a large probability for the oscillator in Fig. 1 because its free vibration response takes values close to zero for $s=ET_k=1/\lambda=2.0$ when α=1 and ε=10. On the other hand, the relationship between X_k^+ and the train of pulses Y_k is much more complex for the oscillator in Fig. 2. In this case the stationary state X_k^+ follows a bimodal density.

Figure 5 shows probabilities of failure relative to a safe set (-a, a), a = 3.0, for the (-, +) and (+,-) oscillators with ρ = 1; ε = 10; λ = 0.5 and ρ = 1; ε =0.1; λ = 0.5, respectively. An absorbing cell was used to represent all cells outside the safe set.

5 CONCLUSIONS

Two methods were presented for calculating a stationary probability distribution of the response of nonlinear systems subject to non-Gaussian white noise. The methods are based on an integral equation and a Markov model. Numerical results based on the Markov model are presented for two nonlinear half oscillators. They indicate that the stationary distribution can depart from the Gaussian distribution significantly.

REFERENCES

Gihman, I.I. & Skorohod, A.V. 1972. *Stochastic Differential Equations*, New York: Springer-Verlag.

Harrison, J.M. & Resnick, S.I. 1975. The stationary distribution and first exit probabilities of a storage process with general release rule," *Mathematics of Operations Research* 1,4:347-358.

Tylikowski, A. & Marowski, W. 1986. Vibration of a non-linear single degree of freedom system due to Poissonian impulse excitation. *International Journal of Non-Linear Mechanics* 21, 3: 229-238.

Structural Safety & Reliability, Schuëller, Shinozuka & Yao (eds) © 1994 Balkema, Rotterdam, ISBN 90 5410 357 4

Stochastic fluid-structure interaction of large floating islands

T. Hamamoto
Department of Architecture, Musashi Institute of Technology, Tokyo, Japan

H. Kamura
Applied Technology Research Center, NKK Corporation, Kawasaki, Japan

ABSTRACT: The dynamic responses of an artificial large floating island subjected to wind-induced waves and seaquakes are evaluated using both long-term and short-term descriptions of loading and stochastic response analysis of the linearized dynamic model. The short-term description is concerned with the details of the load time history during each load intensity, while the long-term description is associated with the recurrence pattern of all possible load intensities. The short-term response which is conditional on each load intensity is evaluated by a stationary random vibration analysis, taking into account structural flexibility and fluid-structure interaction. Then the long-term response is evaluated by integrating the contribution of each short-term response weighted with the occurrence rate with respect to all possible load intensities. Numerical examples are presented to discuss the short-term and long-term responses of the floating island against both wind-induced waves and seaquakes at different sites.

1 INTRODUCTION

With the recent trend toward the utilization of ocean space, a number of projects related to artificial large floating islands which serve as floating cities or airports have been proposed. The designs of these large floating structures demand a rational prediction of structural response in an ocean environment. As the areal expanse of floating island increases, structural deformation becomes more significant than rigid body motion. As a result, the spatial distribution of structural response tends to be rather complicated than relatively small-sized floating structures (Tanaka *et al.*, 1986). Thus, the evaluation of the spatial response distribution of large floating structures is a key issue to assure structural safety and keep human comfort on the island.

In the wave response analysis, the dynamic interaction between elastic deformation of a rectangular floating structure and sea water has been investigated (Wen, 1974). So far, however, only rigid body motion has been taken into account for circular floating structures (*e.x.*, Garret, 1971). In the seaquake response analysis, on the other hand, Liou, Penzien and Yeung (1988) studied the rigid body motion of a circular floating structure with tension-legs.

The short-term response of tension-leg platforms subjected to wind-induced waves has been evaluated using a probabilistic approach (Soong and Prucz, 1984). However, the spatial response distribution of large floating structures easily varies according to load intensity. The long-term response may be obtained by integrating the short-term response weighted with the occurrence rate with respect to all possible load intensities. The long-term response is independent of load intensity and may be regarded as the averaged, spatial response distribution at a specific site. In this study, both short-term and long-term responses are evaluated with the aids of short-term and long-term descriptions of wind-waves and seaquakes. The long-term description of external loadings is represented by the occurrence rate of each load intensity. The short-term description of external loadings is represented by the spectral density functions in terms of each load intensity.

Numerical examples are presented to discuss the spatial distribution of response statistics of a large floating island with or without anchor system against both wind-induced waves and seaquakes at different sites.

2 DESCRIPTION OF WAVES AND SEAQUAKES

2.1 WAVE LOADING

2.1.1 SHORT-TERM DESCRIPTION

It is clear that the sea-surface is not a stationary process because of changing meteorological conditions. For relatively short time period, however, the assumption of stationarity is reasonable. The spectral density function of wave height in a fully developed sea is given by the Pierson-Moskowitz (1964) spectrum,

$$S_{\eta\eta}(\sigma) = \frac{\alpha g^2}{\sigma^5} \exp\left[- \beta \left(\frac{g}{\sigma \cdot V}\right)^4 \right] \quad 0 \le \sigma \le \infty , \tag{1}$$

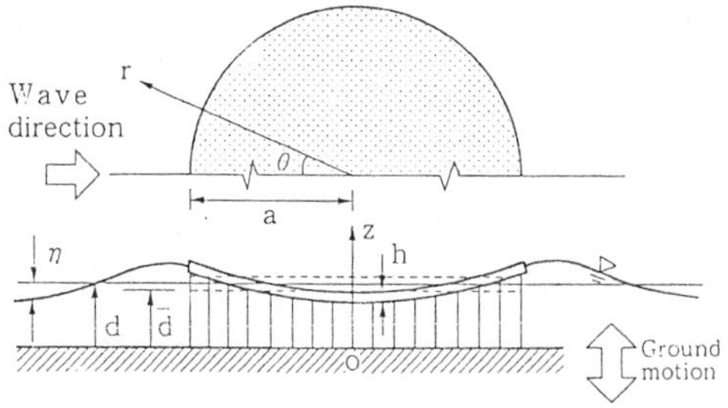

Fig.1 Analytical model.

in which V is the mean wind velocity at 19.5m above the sea surface, g is the acceleration due to gravity, and α and β are the spectral parameters. Common values α=0.0081 and β =0.74 are adopted in this study.

2.1.2 LONG-TERM DESCRIPTION

Corresponding to the short-term description of wind-induced wave, the long-term description of wind-induced wave is described by the frequencies of occurrence of the maximum mean wind velocity above the sea-surface at a site. It is often assumed that the frequency of occurrence of mean wind velocity follows the Weibull distribution (Davenport, 1967). In this case, the maximum mean wind velocity, Vp , of independent samples taken from the parent population approaches a type I (Gumbel) asymptotic distribution of the largest extreme given by

$$F_{V_p} = \exp\left\{-\exp\left[-\frac{(v-\mu)}{\sigma}\right]\right\}, \tag{2}$$

in which μ and σ are the location and dispersion parameters of the distribution, respectively.

2.2. SEAQUAKE LOADING

2.2.1 SHORT-TERM DESCRIPTION

Seaquake loading is a hydrodynamic pressure due to shock waves consisting solely of compressional waves traveling through the water by earthquake, since water cannot transmit shear wave. Earthquake ground motion is generally nonstationary with respect to frequency content and intensity. In this study, however, motions during each loading event are modeled as limited duration segments of stationary random process.

The short-term description of seaquake loading is described by the spectral density function of vertical ground acceleration at sea bottom as follows(Clough and Penzien, 1975):

$$S_{\ddot{U}_g\ddot{U}_g}(\sigma) = G_0 \cdot H_1(\sigma) \cdot H_2(\sigma) \qquad 0 \le \sigma \le \infty, \tag{3}$$

in which

$$H_1(\sigma) = \frac{(\sigma/\omega_k)^4}{\left\{1-(\sigma/\omega_k)^2\right\}^2 + 4\xi_k^2(\sigma/\omega_k)^2}, \qquad H_2(\sigma) = \frac{1+4\xi_g^2(\sigma/\omega_g)^2}{\left\{1-(\sigma/\omega_g)^2\right\}^2 + 4\xi_g^2(\sigma/\omega_g)^2},$$

Go is the spectral intensity, ω_k and ξ_k are the high-pass filter parameters, and ω_g and ξ_g are the low-pass filter parameters and may be interpreted as the predominant ground circular frequency and ground damping ratio, respectively. In this study, we assume ω_k = 1.0 rad/sec, ξ_k = 0.6, ω_g = 10.0 rad/sec, ξ_g = 0.6.

2.2.2 LONG-TERM DESCRIPTION

The long-term description of seaquake is represented by the seismic hazard curve at a site. The seismic hazard curve is a plot of annual exceedance probability versus the peak ground acceleration or velocity. Ozaki *et al.* (1978) showed that the peak ground velocity at bed-rock, Vo , follows the type II extreme value distribution,

$$P[V_0 \ge v] = 1 - F_{V_0}(v) = 1 - \exp\left[-\left(\frac{\mu}{v}\right)^k\right], \tag{4}$$

in which μ and k are the size and shape parameters of the distribution, respectively. Moreover, the relationship between the peak ground velocity at ground surface, Vp, and that at bed-rock, Vo, is assumed as

$$V_p = 5\sqrt{T_g}\, V_0 = 5\sqrt{\frac{2\pi}{\omega_g}}\, V_0, \tag{5}$$

in which T_g is the predominant ground period at sea bottom.

From Eqs. (3) and (5), the relation between the spectral intensity, Go, and the peak velocity, Vp, may be obtained as

$$G_0 = \frac{4}{\pi}\left(\frac{V_P}{Z}\right)^2 \cdot \frac{f_1}{f_2}, \tag{6}$$

in which

$f_1 = \omega_k \left\{\xi_k\xi_g(\omega_k^2 - \omega_g^2)^2 + 4\xi_k^2\xi_g^2\omega_k\omega_g(\omega_k^2 + \omega_g^2) + 4\xi_k\xi_g\omega_k^2\omega_g^2(\xi_k^2 + \xi_g^2)\right\}$,

$f_2 = \left\{2\xi_k\omega_k\omega_g^3 + \xi_g\omega_g^2(\omega_k^2 + \omega_g^2)\right\}$, and Z is the peak factor which is assumed to be 3.0 in this study.

3 FLUID-STRUCTURE INTERACTION

The analytical model of a floating island under wind-induced waves and seaquakes is shown in Fig. 1. The floating island is modeled as an elastic circular plate with or without anchor system. The anchor system is assumed to be composed of a series of tension-legs uniformly distributed on the bottom surface of the island. To evaluate the short-term response, the stochastic response analysis of the floating island is carried out, taking into account structural flexibility and fluid-structure interaction.

3.1 HYDRODYNAMIC PRESSURE

Due to wind-induced waves or seaquakes, the hydrodynamic pressure is generated on the wetted surface of the floating island. The hydrodynamic pressure may be evaluated as a linear combination of the pressure component, p_f, acting on the motionless island subjected to wind-induced waves and seaquakes and the pressure component, p_m, due to the island motion which contains rigid body motion and elastic deformation. Moreover, p_f may be divided into two components, p_i and p_s, due to incident waves and scattered waves, respectively. Because of the geometric simplicity, each pressure component can be obtained in closed form based on a linear potential flow theory (Tanaka *et al.*, 1990, Hamamoto *et al.*, 1991A).

3.1.1 PRESSURE COMPONENT p_f

(1) FREE-FIELD PROBLEM FOR WIND-WAVES

The hydrodynamic pressure acting on the bottom surface of the motionless island subjected to wind-induced waves is evaluated in the free-field. The linearlized governing equation for an incompressible and inviscid fluid is

$$\nabla^2 \phi_i = 0\,, \tag{7}$$

in which ϕ_i is the velocity potential function of incident wave and ∇^2 is the Laplace operator.

The boundary conditions are as follows:
Free water surface condition,

$$\frac{\partial \phi_i}{\partial z} + \frac{1}{g}\frac{\partial^2 \phi_i}{\partial t^2} = 0 \qquad \text{at } z = d, \tag{8a}$$

$$\eta = -\frac{1}{g}\frac{\partial \phi_i}{\partial t} \qquad \text{at } z = d, \tag{8b}$$

Sea-bed condition,

$$\frac{\partial \phi_i}{\partial z} = 0 \qquad \text{at } z = 0, \tag{9}$$

in which t is time and η is the height of sea surface measured from the still water level and may be expressed as the superposition of harmonic wave components as follows:

$$\eta = \sum_{j=1}^{\infty} \sqrt{2S_{\eta\eta}(\sigma_j)\Delta\sigma}\cdot\exp\left[i(\sigma_j t - k_j r\cos\theta + \phi_j\right], \tag{10}$$

in which σ_j, k_j and ϕ_j are the circular frequency, wave number and phase of the *j*-th wave component, respectively, and $S_{\eta\eta}(\sigma_j)$ is the wave height power spectral density function.

(2) FREE-FIELD PROBLEM FOR SEAQUAKES

The hydrodynamic pressure during a seaquake is associated with the water-transmitted seismic vibration from sea-bed to the floating island. The incident wave consists of vertical pulses which induce a series of compression and tension waves in the fluid medium. The amplification of the seaquake force through the water must be evaluated by taking into account the compressibility of water. The linearlized governing equation for an incompressible and inviscid fluid is

$$\nabla^2 \phi_i = \frac{1}{c^2}\frac{\partial^2 \phi_i}{\partial t^2}\,, \tag{11}$$

in which c is the compressional wave velocity.
The boundary conditions are as follows:
Free water surface condition,

$$\frac{\partial \phi_i}{\partial z} + \frac{1}{g}\frac{\partial^2 \phi_i}{\partial t^2} = 0 \qquad \text{at } z = d, \tag{12}$$

Sea-bed condition,

$$\frac{\partial \phi_i}{\partial z} = \frac{\partial U_g}{\partial t} \qquad \text{at } z = 0, \tag{13}$$

in which U_g is the vertical ground displacement at sea-bed and may be expressed in terms of the vertical acceleration power spectral density function.

(3) SCATTERING PROBLEM

Due to the existence of the floating island, the incident wave is reflected and scattered at the interface between the island and sea water. For the frequency of interest, the sea water may be assumed to be incompressible for both wind-induced waves and seaquakes. The governing equation for an incompressible and inviscid fluid is the Laplace's equation:

Table 1. Natural frequencies of wet mode (rad/sec).

	Without Anchor		With Anchor	
n	m=1	m=2	m=1	m=2
0	0.013	0.250	0.155	1.521
1	0.038	0.605	1.374	2.434
2	0.080	1.180	1.801	2.958
3	0.171	1.948	2.058	3.323
4	0.462	2.817	2.785	5.124

Table 2. Parameter values of the extreme value distribution parameters for wind-induced waves and seaquakes at each site.

	Wind- wave		Sea-quake	
	σ	μ (m/sec)	k	μ(cm/sec)
SiteA	1.60	25.0	2.10	0.56
SiteB	1.80	30.0	1.80	0.23

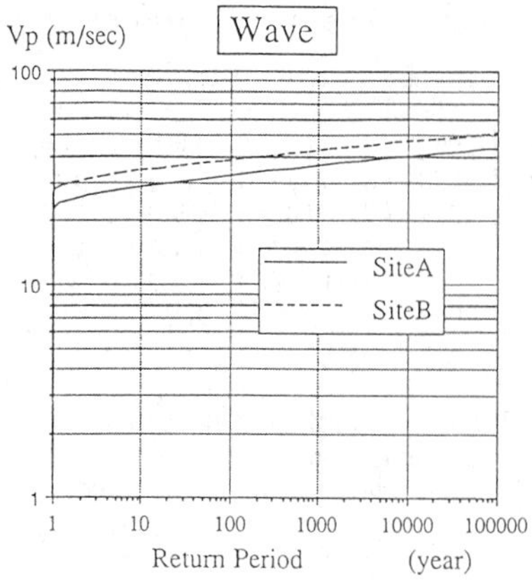

Fig.2a Return period and its expectation for wind-waves.

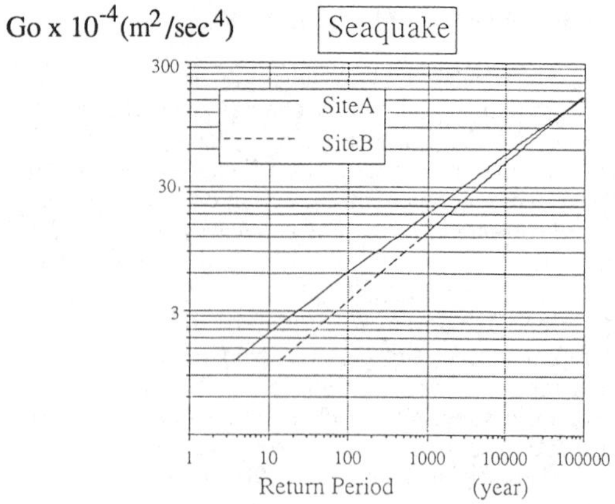

Fig.2b Return period and its expectation for seaquakes.

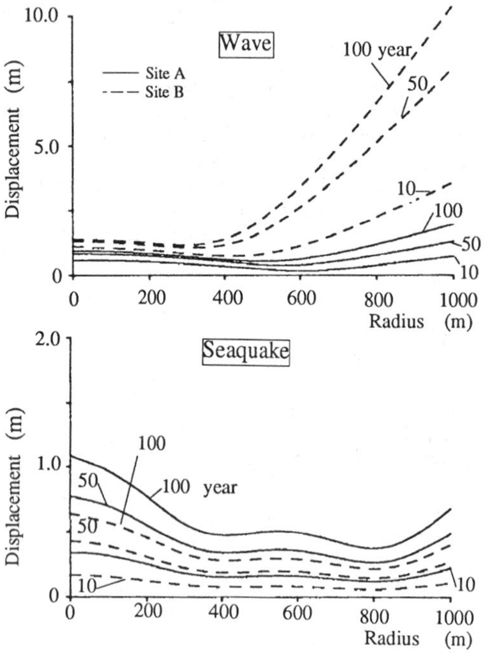

Fig.3a Standard deviations of short-term displacement response distribution for the case without anchor.

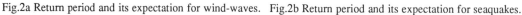

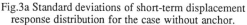

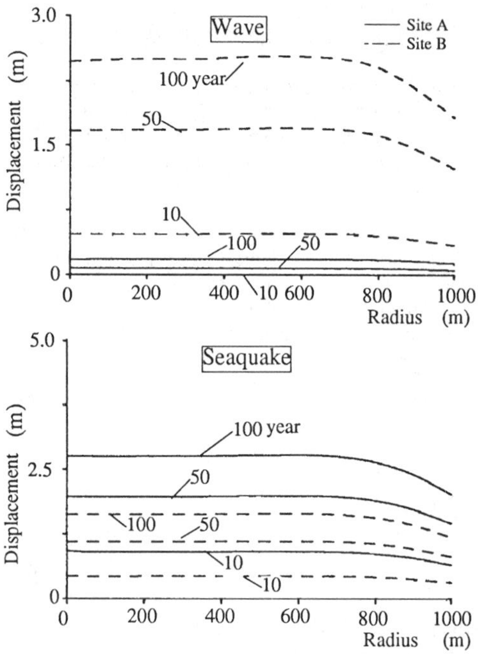

Fig.3b Standard deviations of short-term displacement response distribution for the case with anchor.

$$\nabla^2 \phi_s = 0 , \tag{14}$$

in which ϕ_s is the velocity potential function of scattered wave.

The boundary conditions are as follows:
Free water surface condition,

$$\frac{\partial \phi_s}{\partial z} + \frac{1}{g}\frac{\partial^2 \phi_s}{\partial t^2} = 0 \qquad \text{at } z = d, \tag{15}$$

Sea-bed condition,

$$\frac{\partial \phi_s}{\partial z} = 0 \qquad \text{at } z = 0, \tag{16}$$

Island-water interface condition,

$$\frac{\partial \phi_s}{\partial z} = -\frac{\partial \phi_i}{\partial z} \qquad \text{at } z = \bar{d}, \tag{17}$$

Radiation condition,

$$\lim_{r \to \infty} \sqrt{r}\left(\frac{\partial \phi_s}{\partial r} - ik\phi_s\right) = 0 \quad \text{at } r \to \infty. \tag{18}$$

The pressure component p_f may be obtained by the Bernoulli's equation,

$$p_f = -\rho_w \left(\frac{\partial \phi_i}{\partial t} + \frac{\partial \phi_s}{\partial t}\right)\Big|_{z=\bar{d}} , \qquad (19)$$

in which ρ_w is the mass density of sea water.

3.1.2 PRESSURE COMPONENT p_m

When subjected to wind-induced waves or vertical ground motion, the floating island responds with its rigid body motion and elastic deformation. Using the mode superposition approach, the total displacement of the floating island can be expressed as

$$\zeta = \sum_{n=0}^{\infty} \sum_{m=1}^{\infty} \zeta_{nm}(r) \cos(n\theta)\, q_{nm}(t), \qquad (20)$$

in which n is the circumferential Fourier wave number, m is the radial mode number, $\zeta_{nm}(r)$ is the nm-th wet-mode shape in the radial direction along $\theta = 0$, and $q_{nm}(t)$ is the nm-th generalized coordinate. The wet-mode shape of the floating island is obtained in the free vibration analysis of an island-anchor-water coupled system.

Due to the low frequency nature of the island response, the effect of the compressibility of water may be disregarded. When a floating island is moving vertically with harmonic motion, the velocity potential function ϕ_m for inviscid flow can be obtained by solving the Laplace's equation,

$$\nabla^2 \phi_m = 0 . \qquad (21)$$

The boundary conditions are as follows:
Free water surface condition,

$$\frac{\partial \phi_m}{\partial z} + \frac{1}{g}\frac{\partial^2 \phi_m}{\partial t^2} = 0 \qquad \text{at } z = d, \qquad (22)$$

Sea-bed condition,

$$\frac{\partial \phi_m}{\partial z} = 0 \qquad \text{at } z = 0, \qquad (23)$$

Island-water interface condition,

$$\frac{\partial \phi_m}{\partial z} = \frac{\partial \zeta}{\partial t} \qquad \text{at } z = \bar{d}, \qquad (24)$$

Radiation condition,

$$\lim_{r \to \infty} \sqrt{r}\left(\frac{\partial \phi_m}{\partial r} - ik\phi_m\right) = 0 \qquad \text{at } r \to \infty. \qquad (25)$$

The pressure component p_m may be obtained by the Bernoulli's equation,

$$p_m = -\rho_w \frac{\partial \phi_m}{\partial t}\Big|_{z=\bar{d}} - \rho_w g \zeta. \qquad (26)$$

3.2 TENDON FORCE

If tension-legs are attached to the island, the tendon force acts on the island in addition to hydrodynamic pressure. In case a series of anchors are uniformly distributed on the bottom surface of the island, the dynamic tendon force is given by

$$p_d(r,\theta,t) = -k_d(r,\theta)\left\{\zeta(r,\theta,t) - U_g(t)\right\}, \qquad (27)$$

in which $k_d(r,\theta)$ is the stiffness per unit area.

4 MODAL EQUATIONS OF MOTION

Having obtained the hydrodynamic pressure induced by wind-induced waves or seaquakes, the equation of motion of the floating island may be obtained in the frequency domain. The dynamic behavior of the floating island is governed by the Lagrange's equation:

$$\frac{d}{dt}\left(\frac{\partial T}{\partial \dot{q}_{nm}(t)}\right) - \frac{\partial T}{\partial q_{nm}(t)} + \frac{\partial S}{\partial q_{nm}(t)} = Q^L_{nm}(t) + Q^D_{nm}(t), \qquad (28)$$

in which $\cdot$ denotes time derivative, T and S are the kinematic and strain energy of the island, respectively, $Q^L_{nm}(t)$ is the nm-th generalized loading force and $Q^D_{nm}(t)$ is the nm-th generalized damping force of the island.

Making use of the orthogonal properties of wet-mode shapes of the floating island, the nm-th uncoupled modal equations of motion can be obtained as follows:

$$(M^P_{nm} + M^W_{nm})\ddot{q}_{nm}(t) + (C^P_{nm} + C^W_{nm})\dot{q}_{nm}(t) + (K^P_{nm} + K^W_{nm} + K^T_{nm})q_{nm}(t) = Q^F_{nm}(t) + Q^T_{nm}(t), \qquad (29)$$

in which M^P_{nm}, C^P_{nm} and K^P_{nm} are the nm-th generalized mass, damping and stiffness of the island, M^W_{nm}, C^W_{nm} and K^W_{nm} are the nm-th generalized added mass, added damping and added stiffness, respectively, due to dynamic interaction between the island and sea water, K^T_{nm} is the nm-th generalized added stiffness due to the attachment of anchor system, $Q^F_{nm}(t)$ is the nm-th generalized force associated with the pressure component p_f, $Q^T_{nm}(t)$ is the nm-th generalized added excitation due to direct propagation of seismic waves through anchor system. It is noted that the generalized added mass and damping are frequency-dependent.

Dividing both sides of Eq.(29) by $(M^P_{nm} + M^W_{nm})$ and making use of the relation,

$$K^P_{nm} + K^W_{nm} + K^T_{nm} = \overline{\omega}^2_{nm}(M^P_{nm} + M^W_{nm}), \qquad (30)$$

the following equation may be obtained,

$$\ddot{q}_{nm}(t) + 2(\xi_{nm} + \vec{\xi}_{nm})\overline{\omega}_{nm}\dot{q}_{nm}(t) + \overline{\omega}^2_{nm}q_{nm}(t) = \frac{Q^F_{nm}(t) + Q^T_{nm}(t)}{M^P_{nm} + M^W_{nm}} , \qquad (31)$$

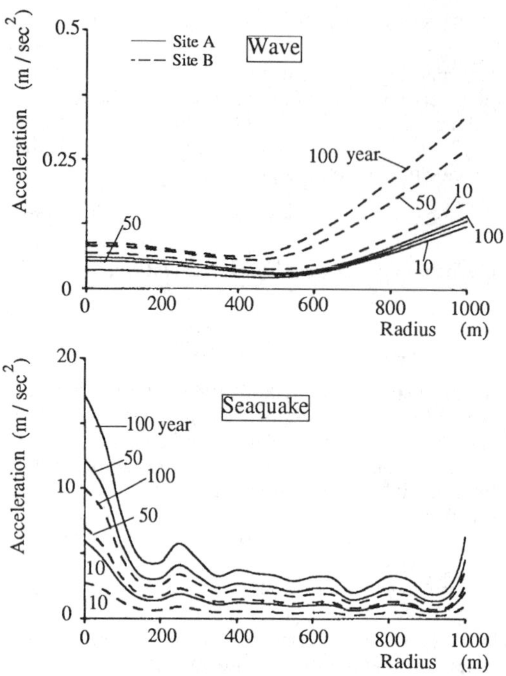

Fig.4a Standard deviations of short-term vertical acceleration response distribution for the case without anchor.

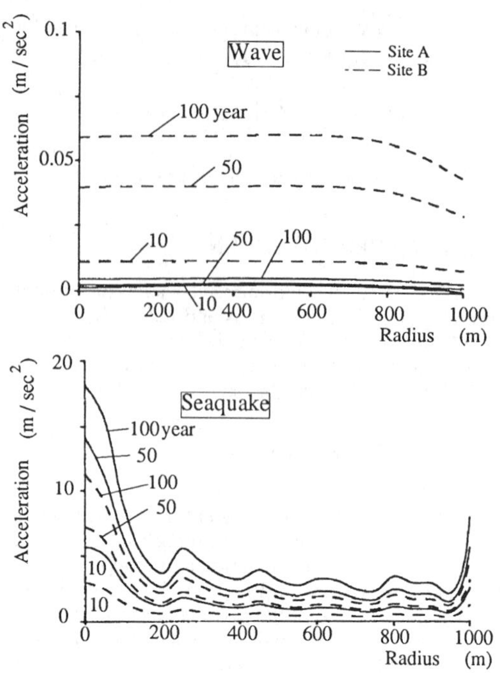

Fig.4b Standard deviations of short-term vertical acceleration response distribution for the case with anchor.

in which $\overline{\omega}_{nm}$ is the nm-th wetted natural circular frequency, and $\bar{\xi}_{nm}$ and $\vec{\xi}_{nm}$ are the nm-th material damping ratio of the island and hydrodynamic radiation damping ratio due to fluid-structure interaction, respectively.

5 SHORT-TERM RESPONSE

The short-term description of wind-induced waves or seaquakes serves as the input to the linearized dynamic model. On the basis of a linear random vibration theory, the response power spectrum of displacement ζ may be obtained as

$$S_{\zeta\zeta}(r,\theta;\sigma) = \sum_{n=0}^{\infty}\sum_{m=1}^{\infty} \zeta_{nm}^2(r)\cos^2(n\theta)\,|H_{nm}(\sigma)|^2\, S_{Q_{nm}Q_{nm}}(\sigma), \tag{32}$$

in which $S_{Q_{nm}Q_{nm}}(\sigma)$ is the power spectral density function of the sum of $Q_{nm}^F(t)$ and $Q_{nm}^T(t)$, and $|H_{nm}(\sigma)|^2$ is the nm-th transfer function given by

$$|H_{nm}(\sigma)|^2 = \frac{1}{(M_{nm}^P + M_{nm}^W)^2\left[(\overline{\omega}_{nm}^2 - \sigma^2)^2 + 4\,(\bar{\xi}_{nm} + \vec{\xi}_{nm})^2\,\overline{\omega}_{nm}^2\,\sigma^2\right]}, \tag{33}$$

In Eq.(32), the cross spectra related to coupling between different modes of vibration are disregarded because of small damping of the system.

The variance of the short-term displacement response is obtained by white noise approximation,

$$\overline{\zeta}^2(r,\theta) = \int_0^{\infty} S_{\zeta\zeta}(r,\theta;\sigma)\,d\sigma \approx$$

$$\sum_{n=0}^{\infty}\sum_{m=1}^{\infty} \zeta_{nm}^2(r)\cos^2(n\theta)\,\frac{\pi}{4(M_{nm}^P + M_{nm}^W)^2\,(\bar{\xi}_{nm} + \vec{\xi}_{nm})\,\overline{\omega}_{nm}^3}\,S_{Q_{nm}Q_{nm}}(\overline{\omega}_{nm}). \tag{34}$$

Other response statistics such as hydrodynamic pressure, acceleration and internal force can be obtained in the same manner.

6 LONG-TERM RESPONSE

In the case of large floating islands, the natural frequencies related to structural deformation are closely-spaced and enter within the central region of the wave height power spectrum. Therefor, the predominant modes of vibration excited by each load intensity easily varies. As a results, the short-term response becomes very sensitive to load intensity.

The long-term response is represented as a composite of short-term response for all possible load intensities. To obtain the long-term response, information about the relative likelihood of different load intensity is combined with the short-term response for each load intensity. The long-term response is obtained by summing the contribution of short-term response due to each load intensity as follows:

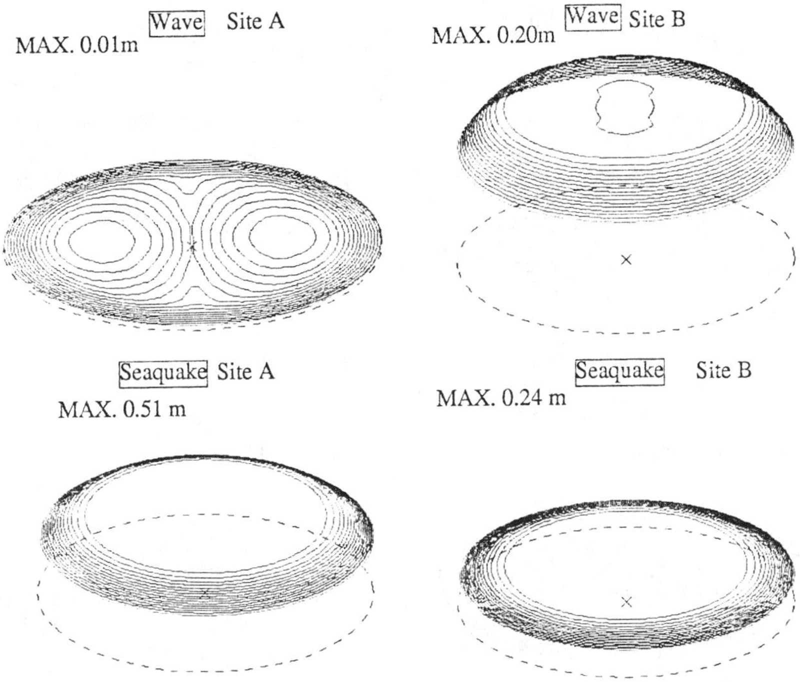

Fig.5 Spatial distribution of long-term displacement response distribution for the case with anchor.

$$\bar{\zeta}(r,\theta) = \int_0^{\infty} \bar{\zeta}(r,\theta | v) f_{V_p}(v) dv \qquad \text{for waves,} \tag{35}$$

$$\bar{\zeta}(r) = \int_0^{\infty} \bar{\zeta}(r | g) f_{G_0}(g) dg \qquad \text{for seaquakes ,} \tag{36}$$

in which $\bar{\zeta}(r,\theta | v)$ and $\bar{\zeta}(r | g)$ are displacement responses conditional on the maximum mean wind velocity, V_p, and the vertical earthquake intensity, Go, respectively, and $f_{V_p}(v)$ and $f_{G_0}(g)$ are probability density functions of V_p and Go, respectively. Other response quantities such as hydrodynamic pressure, acceleration and internal force can be obtained in the same manner.

7 NUMERICAL RESULTS AND DISCUSSION

The short-term and long-term responses of an example floating island are evaluated. For the computations, dimensions and material constants of an island-anchor-sea water system are assumed as follows: radius of island = 1000 m, thickness of island = 50 m, Young's modulus of island = 2.0×10^5 kg / cm^2, Poison's ratio of island = 0.15, mass density of island = 0.4×10^{-6} kg sec^2 / cm^4, material damping ratio in air = 0.05 for all modes of vibration, spring stiffness of anchor system = 1.0×10^{-1} kg / cm^3, water depth = 200 m, mass density of sea water = 1.046×10^{-6} kg sec^2 / cm^4 and compressional wave velocity in water = 1500 m / sec. Natural frequencies of the wet mode of floating island are shown in Table 1.

The long-term spatial response distribution is evaluated at two different sites. SiteA is more seismically active than siteB. More severe wind-induced wave is likely to occur at siteB compared to siteA. Figs.2a and 2b show the relationships between return period and its expected load intensity for both waves and seaquakes at both sites. The distribution parameters are shown in Table 2. For the calculation of long-term response, the integration interval of Vp and Go is taken from 0 to 1000 year with respect to the return-period.

Before going into the long-term response, the short-term responses are calculated for three different return periods, *i.e.*, 10, 50 and 100 year-return-periods at both sites. Figs.3a and 3b show the standard deviations of short-term displacement response ($\theta = 0$) for both cases with and without anchor system. The elastic deformation is significant for the case without anchor system, while the rigid body motion is predominant for the case with anchor system against both wave and seaquake loadings. The attachment of anchor system effectively reduces the wave response, while adversely amplifies the seaquake response. This is mainly due to the direct

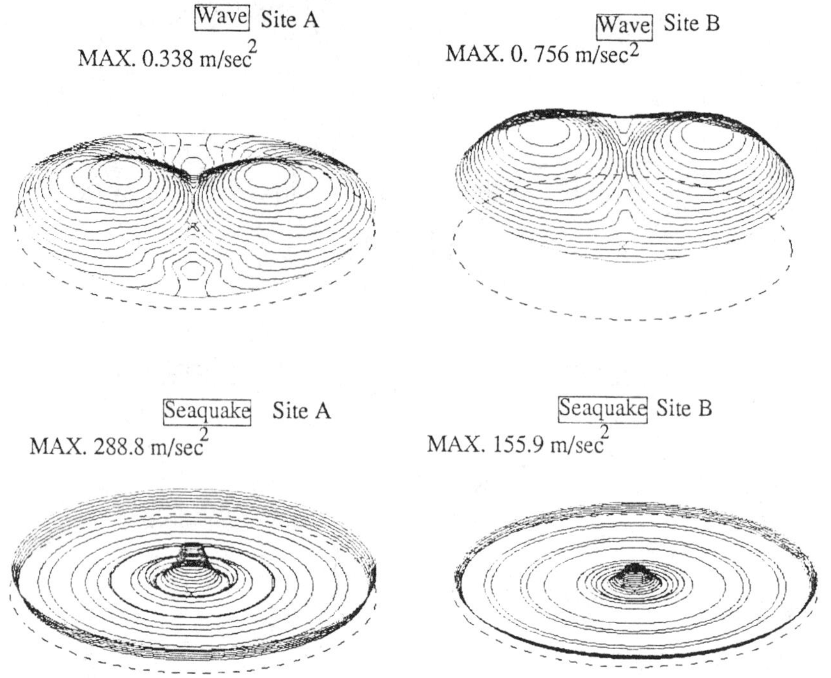

Fig.6 Spatial distribution of long-term vertical acceleration response distribution for the case with anchor.

propagation of seismic waves through anchor system during seaquakes. In the seaquakes response, the distribution pattern of displacement response is the same for all return periods, although the longer the return period, the larger the response. In the wave response, on the other hand, the distribution pattern is very sensitive to each load intensity and significantly varies depending on the return period. This is mainly due to the change in the contribution of excited modes of vibration. Figs.4a and 4b show the standard deviations of short-term vertical acceleration response ($\theta = 0$) for both cases with and without anchor system. As in the case with the displacement response, the same trends are observed both in the wave and seaquake responses. The effect of higher modes of vibration can be clearly seen in the seaquake response for both cases with and without anchor system. On the other hand, only the effect of lower modes of vibration is observed in the wave response for both cases with and without anchor system.

Fig.5 shows the spatial distributions of the standard deviation of long-term displacement response for the case with anchor system at both sites. The distribution pattern varies depending on the site. The seaquake response becomes much larger than the wave response at siteA, although the wave and seaquake responses are almost same at siteB. Fig.6 shows the spatial distributions of the standard deviation of long-term acceleration response for the case with anchor system. The seaquake response is much larger than the wave response irrespective of an attachment of anchor system.

8 CONCLUSIONS

A random vibration based approach for evaluating the short-term and long-term responses of a large floating island subjected to wind-induced waves and seaquakes has been developed. The short-term response is conditional on each load intensity and the spatial distribution of response statistics easily varies according to load intensity. On the other hand, the long-term response is an averaged, spatial distribution of response statistics during a specific time period at a site and is useful at the stage of preliminary design, such as the selection of structural type, size and material, because we can rationally compare different load effects at a given site. The long-term response is also useful for site selection of the floating island by comparing the wave and seaquake responses at different sites.

REFERENCES

Clough, R.W., Penzien, J., 1975, "Dynamics of Structures", McGraw-Hills.

Davenport, A.G., 1967, "The Dependence of Wind Loads on Meteorological Parameters", in 'Wind

Effects on Buildings and Structures Vol.1', pp 19-82.
Garret, C.R.J., 1971, "Wave Forces on a Circular Dock", J. of Fluid Mechanics, Vol.46, pp.129-139.
Hamamoto, T., Takahasi,H., Tanaka,Y., 1991A, "Stochastic Response of Anchored Flexible Floating Islands Subject to Wind-Waves and Seaquakes", Proc. of 1st ISOPE, Vol.3, pp.261-268.
Hamamoto,T., Tanaka,Y., 1991B, "Dynamic Behaviors of Circular Flexible Floating Island Subjected to Wind-Induced Waves", Proc. of IASS, Vol.3, pp.261-268.
Liou, G-S., Penzien, J., Yeung, R.W., 1988, "Response of Tension-Leg Platforms to Vertical Seismic Excitation", Earthquake Engineering and Structural Dynamics, Vol.16, pp.157-182.
Ozaki, M., Kitagawa, Y., Hattori, S., "Study on Regional Characteristics of Earthquake Motions in Japan (Part 1)", Trans. of Architectural Inst. of Japan, No.266, pp.31-40, (in Japanese).
Pierson, W.J., Moskowitz, Z., 1964, "A Proposed Spectral Form for Fully Developed Wind Seas Based on the Similarity Theory of S.A. Kitaigorodoskii", J. of Geophysical Research, Vol.69, pp. 5158-5190.
Soong, T.T., Prucz, Z., 1984, "Reliability and Safety of Tension Leg Platforms", Engineering Structures, Vol.6, pp.142-149.
Tanaka, Y., Hamamoto, T., Kamura, H., 1986, "Stochastic Wave Response Analysis of Floating Offshore Elastic Plates", Proc. of 5th Int. Offshore Mechanics and Arctic Symposium, pp.433-440.
Tanaka, Y., Hamamoto, T., Kamura, H., 1990, "Dynamic Behaviors of Circular Flexible Floating Island Subjected to Seaquakes", Proc. of 1st Pacific/Asia Offshore Mechanics Symposium, Vol.3, pp.311-318.
Wen, Y-K., 1974, "Interaction of Ocean Waves with Floating Plate", ASCE, Vol.100(EM2), pp.375-395.

Structural Safety & Reliability, Schuëller, Shinozuka & Yao (eds) © 1994 Balkema, Rotterdam, ISBN 90 5410 357 4

Modeling chaotic processes with linear filters

G. D. Jeong
Purdue University, West Lafayette, Ind., USA

ABSTRACT: Time series of interest in civil engineering are studied to distinguish between chaos and randomness. A chaotic moving average model that represents the time series as a convolution of a linear filter and an uncorrelated chaotic process is used. Seismic, hydrologic, and climatic series are analyzed. The results are compared to those given by the Grassberger-Procaccia algorithm (GPA). GPA can also identify the presence of low-dimensional chaotic dynamics.

1 INTRODUCTION

The reliability of structures is often based on time series models of natural phenomena. General articles and books on chaos theory (Jensen, 1987, Crutchfield et al., 1986, Gliek, 1987) suggest that random-appearing natural time series may be chaotic processes. These time series are responses of complex systems. The difference between random response and chaotic response is that random response is the result of random external effects whereas chaotic response is intrinsic to the dynamics of the system itself. Chaos theory has shown that relatively simple nonlinear equations can generate complex behavior (May, 1976; Shaw, 1984). A number of investigations have been undertaken to determine the possibility of modeling the irregular behavior exhibited by many natural time series by such simple nonlinear models in the fields of biology, meteorology and hydrology.

The Grassberger-Procaccia (1983) algorithm (GPA) is designed to investigate the dynamics of a process by using the time series of the process. The dynamics of the process is represented by the number of degrees of freedom which is estimated by using the GPA. If the number of degrees of freedom is low, then the process is considered as chaotic; otherwise, it is classified as stochastic. There are no quantitative rules to decide how high the dimension of a process should be in order to consider it to be stochastic. The number of degrees of freedom determined by the GPA is an indication of the number of parameters in a nonlinear model which would generate a time series similar to the one which is being investigated. Chaotic behavior in climatic records has been investigated by using GPA by Nicolis and Nicolis (1986), Fraedrich (1986), Essex et al. (1988), Tsonis and Elsner (1988) and a rainfall record by Rodriguez-Iturbe et al. (1989).

Although the GPA is a popular algorithm, its data requirement can be considerable. This aspect has been discussed extensively in the literature and has led to the development of other methods of estimation of the chaotic behavior of time series. One of the more promising of these methods was developed by Scargle (1989, 1990). Scargle's method is based on extension of Wold's theorem (Wold, 1938) that separates the deterministic and stochastic components of a time series. Experiments based on Scargle's method (1990) have been successful in identifying chaotic time series. Data requirements of Scargle's method are not as stringent as that of GPA.

The objective of the present study is to investigate the chaos behavior of seismic, hydrologic, and climatic time series using GPA and Scargle's method. A secondary result from this investigation is a comparison of GPA and Scargle's method. This paper is organized as follows. The GPA and Scargle's method are discussed in the second section. The data used in the study are described in the third section. Finally, the results are presented and discussed, and a set of conclusions are drawn.

2 THEORY

The literature related to chaos theory is vast and is growing rapidly. Only the theory relevant to the present investigation is discussed herein.

2.1 Grassberger-Procaccia algorithm

Let $x(t)$ be the process of interest. The process is sampled at N equally-spaced time intervals to form a time series $x_i=x(t_i)$ where $t_i=t_0+i\Delta t$ for $i=0,\ldots,N-1$. Theoretically, the value of Δt selected should not affect the computational re-

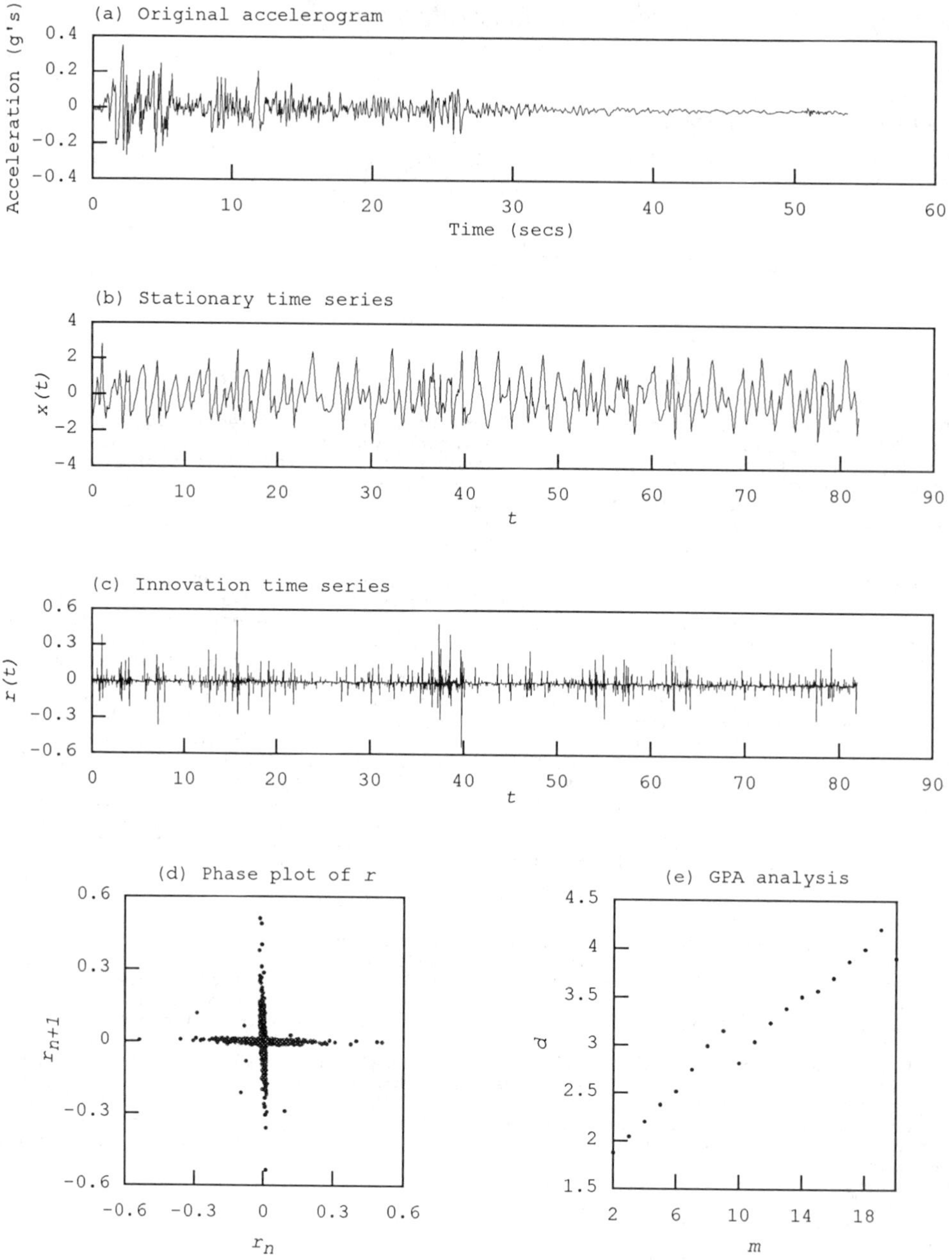

Figure 1. North-south component of 1940 El Centro earthquake

sults. A series of vectors ξ_i are constructed from m consecutive values of the time series x_i, x_{i+1}, ..., x_{i+m-1}. The dimension of the vectors m is known as the embedding dimension. Then the correlation integral $C(r,m)$ is computed using

$$C(r,m) = \lim_{N\to\infty} \frac{1}{N^2} \sum_{i=1}^{N} \sum_{j=1}^{N} H(r - |\xi_i - \xi_j|) \quad (1)$$

where $H(\cdot)$ is the Heavyside function with $H(s)=1$ if $s>0$ and $H(s)=0$ if $s<0$. The double sum counts the number of pairs (i,j) whose distance $|\xi_i - \xi_j|$ is less than r. For small values of r, the correlation integral exhibits a power law dependence on r of

$$\lim_{r\to 0} C(r,m) = ar^d \quad (2)$$

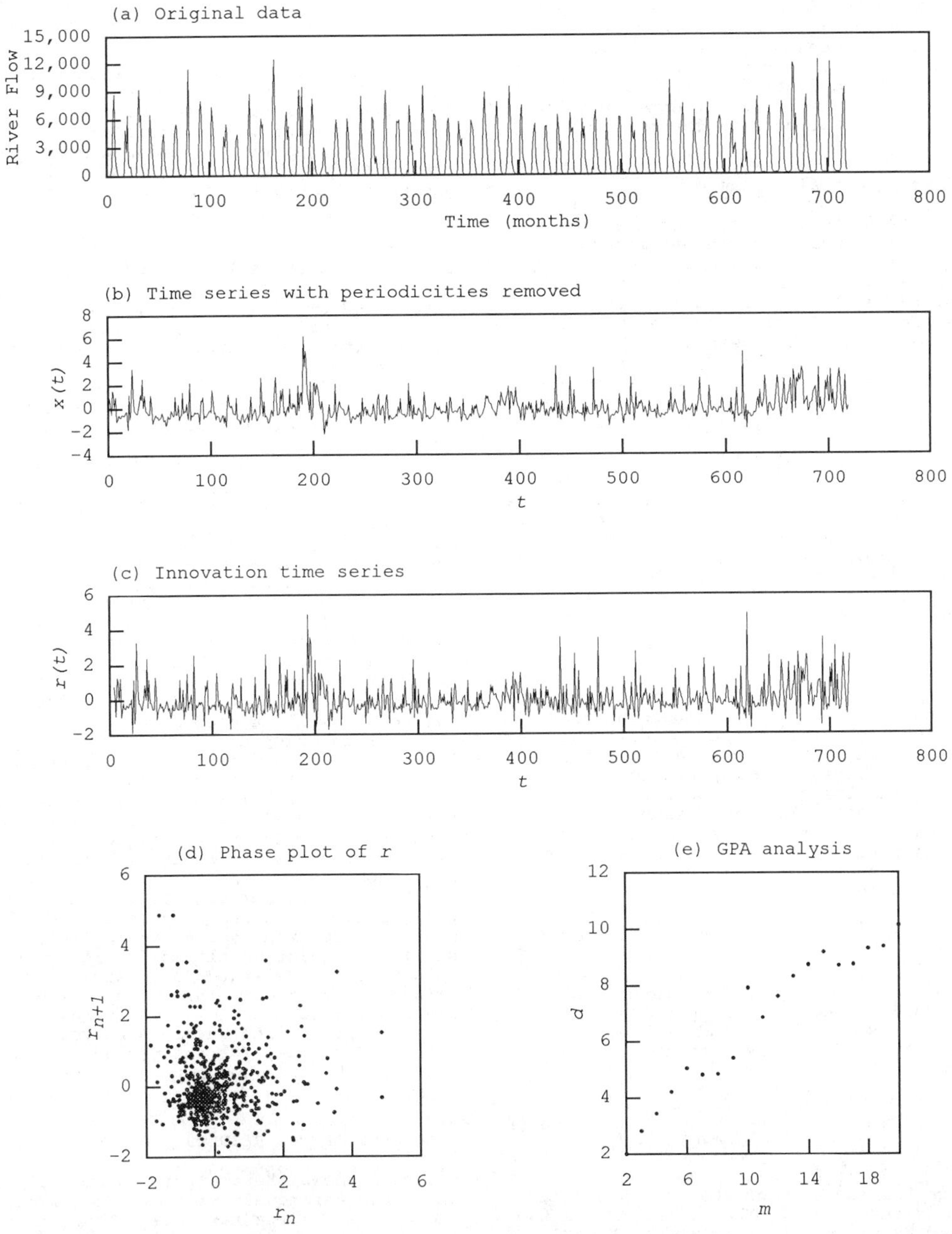

Figure 2. Monthly flow data for the Krishna river

The correlation dimension is defined by using the slope *d* of the log *C*(*r*,*m*) versus log *r* curve (Moon, 1987). For chaotic data, *d* will reach a certain value as *m* is increased and will then become independent of *m*. Fraedrich (1986) has shown that for random data, the exponent *d* will increase indefinitely as embedding dimension *m* is increased.

The GPA is implemented in the following fashion. Starting with an embedding dimension of *m*=2, the correlation integral *C*(*r*,*m*) is computed for different values of *r*. The results are plotted in a log-log plot. The slope *d* of the linear portion of the log *C*(*r*,*m*) versus log *r* where *r* is small is estimated. The process is repeated for increasing values of the embedding dimension m. If d reaches an approximately constant value as the embedding dimension increases, the process is chaotic and that constant value is the correlation dimension. If d continues to increase as the embedding dimension increases, the process is random.

2.2 Scargle's algorithm

The Wold decomposition theorem states that a stationary process can be decomposed into a deterministic and a random process and the random part has a moving average form (Wold, 1938). Scargle (1989, 1990) has generalized Wold's theorem to also include chaotic processes.

Let D be a deterministic process, R an uncorrelated chaotic process and W a white-noise process. Then a stationary process X is represented by Scargle as

$$X = W * B + R * C + D \qquad (3)$$

where B and C are constant filters and the asterisk represents the convolution operation. Scargle calls the term $R * C$ a *chaotic moving average* and R the *chaotic innovation*. It is assumed that the chaotic process r_n satisfies a recurrence equation of the form

$$r_{n+1} = f(r_n, r_{n-1}, \ldots, r_{n-m+1}) \qquad (4)$$

where $f(.)$ is a nonlinear function and m is the dimension of the embedding space. The function $f(.)$ is the optimal predictor of the future values of the process x_n from past data. The representations in Eq. (4) is the basis of Scargle's method.

The deterministic component D can be estimated in time series and eliminated. If the random component can be hypothesized to be only chaotic then the model in Eq. (3) reduces to

$$X = R * C \qquad \text{or} \qquad x_n = \sum_k c_k \, r_{n-k} \qquad (5)$$

In other words, x_n is a moving average of the uncorrelated chaotic process r_n.

In Scargle's method, an autoregressive (AR) filter A is estimated of the form

$$A = (a_{-q}, a_{-q+1}, \ldots, a_{-1}, \; 1, a_1, \ldots a_p). \qquad (6)$$

where the location of the origin of time is indicated by 1. The filter A is a two-sided AR filter of order (p,q) which is the inverse of the filter C. The order of the filter $p+q$ is systematically increased to select the parameters of the filter that maximize the degree of chaos in r_n and yield the smallest prediction error according to the Akaike Final Prediction error criterion (Akaike, 1974). The details of the method are found in Scargle (1990).

Once the autoregressive parameters are estimated the chaotic innovations r_n are estimated easily by using

$$R = A * X \qquad \text{or} \qquad r_n = \sum_k a_k x_{n-k} \qquad (7)$$

The characteristics of the chaotic innovations are then examined by investigating the phase plots of r_{n+1} versus r_n. If r_n is a chaotic process, the nonlinear function $f(.)$ in Eq. (4) will be evident from the phase plot; however, if r_n is a random process, the phase plot will be a scatter diagram.

3 THE DATA USED IN THIS STUDY

Several seismic, hydrologic, and climatic data were analyzed in this study. A representative record from each type of data will be presented here.

The AR model is applicable only to stationary time series. Strong motion earthquake accelerograms are not stationary processes. In order to apply the above theory, the baseline, amplitude and frequency variation of a time series must be eliminated. The value of each point in the process is first shifted by the mean and scaled by the standard deviation of the values within a time window about that point. Then, a number points are added to each time interval in proportion to the number of zero crossings in a time window about the interval. The resulting stationary series is then analyzed by the techniques described above. The North-South component of the 1940 El Centro earthquake will be used here.

As an example of hydrologic data, the monthly flow data of the Krishna river (720 points) will be analyzed. The Krishna monthly river flow series shows strong periodic effects of monsoons and a twelve month period. The periodicities were removed prior to performing the analysis for chaos.

Tree ring series have been used as proxy climatic series. Therefore any chaos in tree ring series would imply similar behavior in climate. Tree ring data from the Salt and Verde Basins in Arizona were used in this study. The tree ring series do not show any obvious periodicities. The Nutrioso Pinyon Pine data (394 points) for the period of 1587-1980 will be analyzed here.

4 SUMMARY OF THE RESULTS

Figure 1 shows the analysis for the 1940 El Centro earthquake record. The first time series is the raw data. The time series x was obtained after stabilizing the baseline, amplitude, and frequency. Scargle's method was applied to this series to give the filter A=(-0.508, 1, -0.498). The innovation series r was calculated by convolution of the filter with the x series. For a chaotic process, the phase plot would approximate the nonlinear function of Eq. (4). For a random time-series, the phase plot is a scatter diagram. The phase plot of the innovation series for this time series indicates a random process with long tails. The final

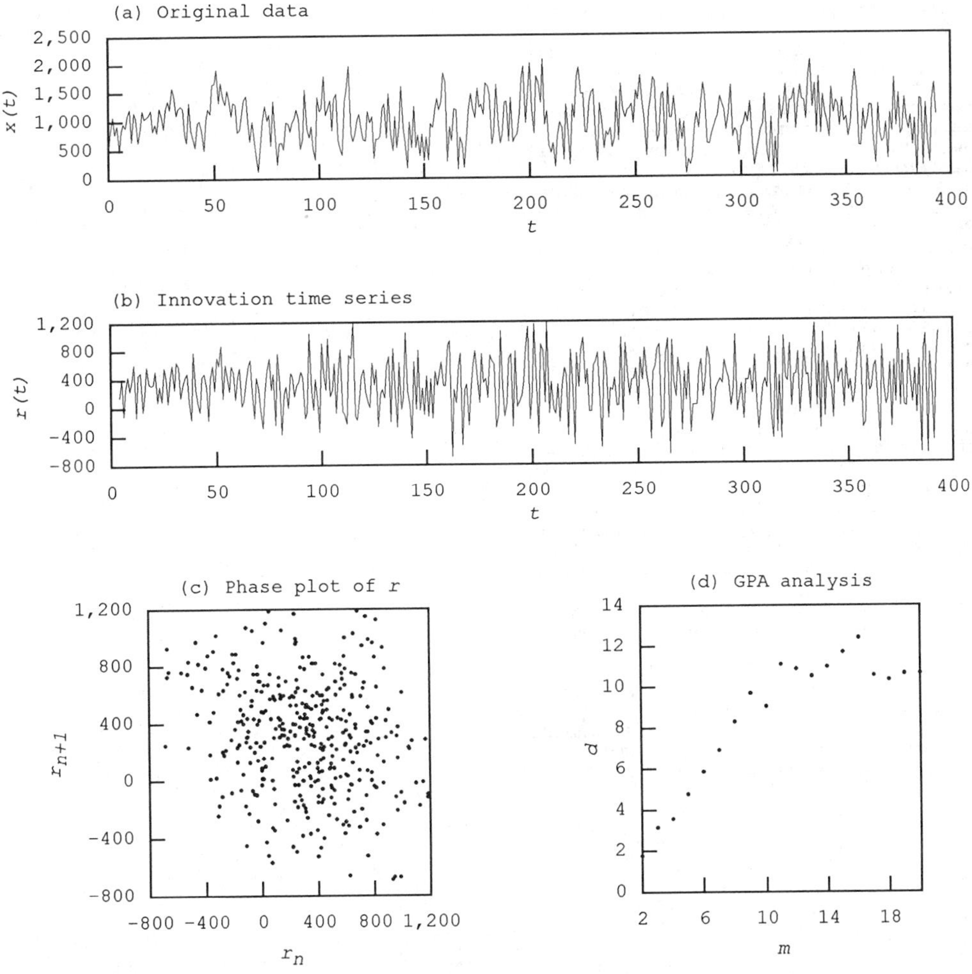

Figure 3. Tree ring data for Nutrioso Pinyon Pine

plot, which was derived from GPA, shows that the correlation exponent *d* increases with increasing embedding dimension *m*. This also indicates a random time series.

Figure 2 shows the same analysis for the Krishna river flow data. Here, the *x* time series was obtained by removing the twelve month periodicity from the raw data. The filter parameters for this series were *A*=(-0.112, -0.048, -0.164, 1, -0.146). The analysis is repeated in Figure 3 for the Nutrioso Pinyon Pine data. Since the raw data do not show any obvious periodicities, the raw data was analyzed directly. The filter parameters were A=(-0.257, 1, -0.182, -0.214, -0.037). The phase plots and correlation exponent plots for both these series indicate randomness. The correlation exponent for the tree ring data appears to have reached a constant value; however it is of high dimension and may be due to a limited number of data points.

Other time series data like the ones here have also been analyzed. None has shown to be a chaotic process as describe above. These results substantiate the use of stochastic models for describing the natural phenomena studied here. Furthermore, GPA and Scargle's method give comparable results.

REFERENCES

Akaike, H. 1974. A new look at statistical model identification. *IEEE Trans. Auto. Contr.* AC-19:716-722.

Crutchfield, R.P., J.D. Farmer, N.H. Packard, R.S. Shaw 1986. Chaos. *Scientific American* 225:46-57.

Essex, C., T Lookman, and M.A.H. Nerenberg 1988. The climate attractor over short time scales. *Nature* 333:545-547.

Fraedrich, K. 1986. Estimating the dimensions of weather and climate

attractors. *Jour. Atmos. Sci.* 43:419-432.
Gliek, J. 1987. *Chaos: Making a New Science*. New York: Viking.
Grassberger, P. and I. Procaccia 1983. Measuring the strangeness of strange attractors. *Physica* 9D:189-208.
Jensen, R.V. 1987. Classical chaos. *American Scientist* 75:168-181.
May, R.M. 1976. Simple mathematical models with very complicated dynamics. *Nature*, 261:459-467.
Moon, F.C. 1987. *Chaotic vibrations*. New York: John Wiley and Sons.
Nicolis, C. and G. Nicolis 1986. Is there a climatic attractor? *Jour. Atmos. Sci.* 43:419-432.
Rodriguez-Iturbe, I., B. Febres de Power, M.B. Sharifi and K.P. Georgakakos 1985. Chaos in rainfall. *Water Resources Research* 25:1667-1675.
Scargle, J.D. 1989. An introduction to chaotic and random time series analysis. *Int. Jour. of Imaging Systems and Technology* 1:243-259.
Scargle, J.D. 1990. Studies in astronomical time series analysis. IV. Modeling chaotic and random processes with linear filters. *The Astrophysical Journal* 359:469-482.
Shaw, R. 1984. *The dripping faucet as a model chaotic system.* Santa Cruz:Aerial Press.
Tsonis, A.A. and J.B. Elsner 1988. The weather attractor over very short times scales. *Nature* 333:545-547.
Wold, H. 1938. *A study in the analysis of stationary time series*. Uppsala: Almquist and Wiksell.

Structural Safety & Reliability, Schuëller, Shinozuka & Yao (eds) © 1994 Balkema, Rotterdam, ISBN 90 5410 357 4

Determining Lyapunov exponents by solving the corresponding Fokker-Planck equation

G. Karch & W. Wedig
Institut für Technische Mechanik, Universität Karlsruhe, Germany

ABSTRACT: Determining stability of sochastic dynamical systems the first Lyapunov exponent has to be calculated. A transformation onto a hypersphere leads to a representation of the Lyapunov exponents in terms of expectation values of the stationary part of the process. These expectation values are coefficients of an orthogonal expansion of the corresponding Fokker-Planck equation. Using symbolic computation, the transformation and expansion can be managed automatically. The remaning numerical computation is to solve a set of sparse linear algebraical equations.

1 INTRODUCTION

In 1967 a concept developed by Hasminskii opened a new chapter in stochastic stability. He derived a characteristic number to determine almost sure stability. As later proved (see e.g. Arnold 1984) this number is equal to the top Lyapunov exponent witch became a topic of many authors in the past decade.

The concept of Hasminskii was to introduce polar coordinates to separate the process on a unit hypersphere from the instationary amplitude process. The top Lyapunov exponent can then be expressed in terms of expectation values of the process on the hypersphere if this process is ergodic.

The idea was catched on by Kozin, Nishioka, Wedig et al. In the case of a two-dimensional system they evaluated the probability density function of the scalar phase process. A classification of the singularities appearing in the scalar diffusion equation of the phase process was first discussed in (Kozin 1971) and (Kozin 1974). The probability density is there expressed (see also Nishioka 1976) in terms of the scale and speed measure. In (Wedig 1991) the corresponding Fokker-Planck equation is solved to determine the probability density. Therefore a Fourier expansion of the density function is introduced. Applying Galerkins method a system of algebraic equations for the coefficients of the expansion is performed. The top Lyapunov exponent can be derived as a function of these coefficients.

The aim of this contribution is to extend the described ideas to systems with more than a single degree of freedom. Therefore angular coordinates are introduced to describe the process on the unit hypersphere. The stochastic differential equations are derived. To solve the corresponding Fokker-Planck equation the singularities are to be discussed. A Fourier expansion leads to a representation of the Lyapunov exponent in terms of the coefficients of this expansion.

This manuscript is based on a dissertation submitted to the Fakultät für Maschinenbau, Universität Karlsruhe.

2 REMARKS TO SCALAR DIFFUSION PROCESSES

The basis of Hasminskiis concept is the existence of a ergodic phase process. Because of the singularities this assumption is not easy to prove. In (Kozin 1974) it is proved that in the case of only unidirectional shunts there is one ergodic component over the entire unit circle. In (Nishioka 1976) the most general form of a linear second order system of the form

$$d\vec{X}_t = \mathbf{A}\vec{X}_t dt + \mathbf{B}\vec{X}_t dW_t^1 + \mathbf{C}\vec{X}_t dW_t^2,$$
$$\dim(\vec{X}_t) = 2 \qquad (1)$$

is discussed.

Concerning mechanical systems with a single degree of freedom and the usual introduction of the state variables displacement and velocity the matrices have the form

$$\mathbf{A} = \begin{pmatrix} 0 & 1 \\ a_{21} & a_{22} \end{pmatrix}, \quad \mathbf{B} = \begin{pmatrix} 0 & 0 \\ b_{21} & b_{22} \end{pmatrix}$$

$$\text{and} \quad \mathbf{C} = \begin{pmatrix} 0 & 0 \\ c_{21} & c_{22} \end{pmatrix}. \tag{2}$$

A transformation with $\vec{y} = \mathbf{U}\vec{x}$ ($\mathbf{U}$ a regular and constant matrix) always leads to the form

$$d\vec{Y}_t = \mathbf{A}^*\vec{Y}_t dt + \mathbf{B}^*\vec{Y}_t dW_t^1 + \mathbf{C}^*\vec{Y}_t dW_t^2 \tag{3}$$

with the two cases

$$(a): \mathbf{B}^* = \begin{pmatrix} b_1 & 0 \\ b_2 & b_1 \end{pmatrix}, b_2 \neq 0 \quad \text{or}$$

$$(b): \mathbf{B}^* = \begin{pmatrix} b_1 & 0 \\ 0 & b_2 \end{pmatrix}, b_1 \neq b_2,$$

If $b_{22} = 0$ and $\mathbf{U}$=$\mathbf{I}$ (Identity matrix) the system (1) is of the type (3) case (a). If $b_{22} \neq 0$ and

$$\mathbf{U} = \begin{pmatrix} 1 & 0 \\ b & 1 \end{pmatrix}, \quad b = \frac{b_{21}}{b_{22}}$$

the matrices of (3) case (b) are

$$\mathbf{B}^* = \begin{pmatrix} 0 & 0 \\ 0 & b_{22} \end{pmatrix}, \quad \mathbf{A}^* = \begin{pmatrix} 0 & 1 \\ a_{21} & b(1 - a_{21}) - a_{22} \end{pmatrix}$$

$$\text{and} \quad \mathbf{C}^* = \begin{pmatrix} 0 & 0 \\ c_{21} & c_{22} - bc_{21} \end{pmatrix}.$$

In both cases there are only left shunts and following the statement in (Kozin 1974) there is one ergodic component over the full domain of definition.

3 SOLUTION OF THE MULTIDIMENSIONAL FOKKER-PLANCK EQUATION

Consider a d-dimensional linear system of stochastic differential equations in the sense of Itô,

$$d\vec{X}_t^k = \sum_{i=1}^{d} f^{ik} X_t^i \, dt + \sum_{\ell=1}^{m} \sum_{i=1}^{d} g^{k\ell i} X_t^i \, dW_t^\ell \tag{4}$$

where f^{ik} and $g^{k\ell i}$ are constants and W_t^ℓ are independent Wiener processes. Following the concept of Hasminskii Itô's formula is applied to

$$\varphi^1 = \begin{cases} \arctan \frac{x^2}{x^1} & , \quad x^1 \geq 0, \\ \arctan \frac{x^2}{x^1} + \pi & , \quad x^1 < 0, \end{cases}$$

$$\varphi^j = \arctan \frac{x^{j+1}}{\sqrt{\sum_{i=1}^{j} (x^i)^2}}, \quad j \in [2, d-1], \tag{5}$$

$$a = \|\vec{x}\|.$$

Hence the (Itô) stochastic differential equations for the angular process $\vec{\Phi}_t$ are

$$d\Phi^k = \left(\sum_{i=1}^{d} u_{,i}^k f^i + \frac{1}{2} \sum_{i=1}^{d} \sum_{j=1}^{d} \sum_{\ell=1}^{m} u_{,ij}^k g^{i\ell} g^{j\ell} \right) dt + \\ + \sum_{i=1}^{d} \sum_{\ell=1}^{m} u_{,i}^k g^{i\ell} dW^\ell, \quad k \in [1, d-1]. \tag{6}$$

The equation for the logarithmic amplitude is

$$d(\ln A) = \left(\sum_{i=1}^{d} v_{,i} f^i + \frac{1}{2} \sum_{i=1}^{d} \sum_{j=1}^{d} \sum_{\ell=1}^{m} v_{,ij} g^{i\ell} g^{j\ell} \right) dt + \\ + \sum_{i=1}^{d} \sum_{\ell=1}^{m} v_{,i} g^{i\ell} dW^\ell. \tag{7}$$

In these equations the following terms are given:

$$v_{,i} = \frac{1}{A} \sin \Phi^{i-1} \prod_{\ell=i}^{d-1} \cos \Phi^\ell,$$

$$v_{,ij} = \frac{1}{A^2} (\delta_{ij} - 2\, z_{d,i,j}),$$

$$u_{,i}^k = \begin{cases} \frac{\cos \Phi^k}{A \prod_{\ell=k+1}^{d-1} \cos \Phi^\ell} & , i=k+1, \\ -\frac{\sin \Phi^k \sin \Phi^{i-1} \prod_{\ell=i}^{k-1} \cos \Phi^\ell}{A \prod_{\ell=k+1}^{d-1} \cos \Phi^\ell} & , i<k+1, \\ 0 & , \text{otherwise}, \end{cases}$$

$$u_{,ij}^k = \begin{cases} -2 \frac{\sin \Phi^k \cos \Phi^k}{A^2 \prod_{\ell=k+1}^{d-1} \cos^2 \Phi^\ell} & , i=j=k+1, \\ \frac{\sin \Phi^{j-1} (1 - 2\cos^2 \Phi^k) \prod_{\ell=j}^{k-1} \cos \Phi^\ell}{A^2 \prod_{\ell=k+1}^{d-1} \cos \Phi^\ell} & , j<k+1, i=k+1, \\ \frac{\sin \Phi^{i-1} (1 - 2\cos^2 \Phi^k) \prod_{\ell=i}^{k-1} \cos \Phi^\ell}{A^2 \prod_{\ell=k+1}^{d-1} \cos \Phi^\ell} & , i<k+1, j=k+1, \\ \frac{\sin \Phi^k z_{k,i,j} (1 + 2\cos^2 \Phi^k) - \sin \Phi^k}{A^2 \prod_{\ell=k+1}^{d-1} \cos^2 \Phi^\ell \cos \Phi^k} & , i=j<k+1, \\ \frac{\sin \Phi^k z_{k,i,j} (1 + 2\cos^2 \Phi^k)}{A^2 \prod_{\ell=k+1}^{d-1} \cos^2 \Phi^\ell \cos \Phi^k} & , i \neq j<k+1, \\ 0 & , \text{otherwise}, \end{cases}$$

$$z_{k,i,j} = \sin \Phi^{i-1} \sin \Phi^{j-1} \prod_{\ell=i}^{k-1} \cos \Phi^\ell \prod_{\ell=j}^{k-1} \cos \Phi^\ell,$$

$$f^k = \sum_{i=1}^{d} f^{ki} A \sin \Phi^{i-1} \prod_{j=i}^{d-1} \cos \Phi^j,$$

$$g^{k\ell} = \sum_{i=1}^{d} g^{k\ell i} A \sin \Phi^{i-1} \prod_{j=i}^{d-1} \cos \Phi^j,$$

$$\sin \Phi^0 = 1.$$

To evaluate the products it is considered that the conventions

$$\prod_{\ell=k}^{k-1} \cos \Phi^\ell = 1$$

or generally

$$\prod_{\ell=k}^{k-n} c^\ell = \frac{c^k}{\prod_{i=0}^{n-1} c^{k-i}}$$

are valid. The process $\vec{\Phi}_t$ lives on

$$\begin{aligned} \Pi^{d-1} &= \{\vec{\varphi} \in \mathcal{R}^{d-1} | -\frac{\pi}{2} \leq \varphi^1 < \frac{3\pi}{2}, \\ & -\frac{\pi}{2} \leq \varphi^k \leq \frac{\pi}{2}, \; k \in [2, d-1]\}. \end{aligned}$$

There are two problems arising for the following evaluations.

First the space Π^{d-1} is bounded. Following the statements of Feller (1952) it depends on the type of the boundaries whether the Fokker-Planck equation is valid in the classical form given from Kolmogorov.

Second within the space Π^{d-1} there arise singularities. As discussed above the ergodicity of the angular process $\vec{\Phi}_t$ depends on the type of these singularities.

If the underlying problem is a mechanical system with $\frac{d}{2}$ degrees of freedom and the usual coordinates displacement and velocity are introduced the coordinate φ^1 describe the motion of the "first" degree of freedom. It is suggested that the behavior of the angular process in this direction in analogous to that one described above for the oscillator.

The boundaries in the direction of the coordinates φ^k, $k \in [2, d-1]$ are the subspaces $M^{d-k} = \left\{ \vec{x}_t \in R^d \,|\, x^1 = \ldots = x^k = 0 \right\}$. They are all enclosed in M^{d-2} which describe the state of the process $\vec{X}_t$ where the displacement and the velocity of the "first" degree of freedom are both vanishing at the same time. If this never take place the boundaries in the direction of the coordinates φ^k, $k \in [2, d-1]$ are all natural boundaries in the sense of Feller. In that case the Fokker-Planck equation is valid in the classical form if the singularities within $-\frac{\pi}{2} < \varphi^k < \frac{\pi}{2}$, $k \in [2, d-1]$. are of "good nature".

This assumption is a lack from the mathematical point of view which is not removed jet. These singularities are to be characterized and the consequences are to be discussed in further work. In spite of this lack the results obtained from the procedure described in the following can be verified by a numerical integration of the stochastic differential equation (4).

To solve the corresponding Fokker-Planck equation to (6) the multi dimensional Fourier expansion

$$p(\vec{\varphi}) = \sum_{k_1, \ldots, k_{d-1} = -\infty}^{\infty} c_{k_1 \ldots k_{d-1}} e^{\jmath (k_1 \varphi^1 + 2 \sum_{\ell=2}^{d-1} k_\ell \varphi^\ell)},$$

is introduced, with $\jmath = \sqrt{-1}$.

Applying Galerkins method a system of algebraic equations can be deduced. Therefore the terms appearing in the Fokker-Planck equation which are not π-periodic harmonic functions of the coordinates φ^k, $k \in [2, d-1]$ are to be expanded into π-periodic Fourier series. Taking into account that the distribution function is normalized the coefficient

$$c_{0 \ldots 0} = \frac{1}{2\pi^{d-1}} \int_{\Pi^{d-1}} p(\vec{\varphi}) d\vec{\varphi} = \frac{1}{2\pi^{d-1}}.$$

is determined. Introducing this condition into the algebraic equations the coefficients of the expansion can be calculated.

The top Lyapunov exponent

$$\lambda = E\left\{ \sum_{i=1}^{d} v_{,i} f^i + \frac{1}{2} \sum_{i=1}^{d} \sum_{j=1}^{d} \sum_{\ell=1}^{m} v_{,ij} g^{i\ell} g^{j\ell} \right\}$$

is a function of the coefficients of the Fourier expansion

$$c_{k_1 \ldots k_{d-1}} = \frac{1}{2\pi^{d-1}} \int_{\Pi^{d-1}} p(\vec{\varphi}) e^{-\jmath (k_1 \varphi^1 + 2 \sum_{\ell=2}^{d-1} k_\ell \varphi^\ell)} d\vec{\varphi}.$$

4 APPLICATION TO A SPECIAL PROBLEM

To study the described procedure it is applied to a system of third order. The stochastic differential equations of a mechanical system with two degrees of freedom

$$\begin{aligned} & m_1 \ddot{x} + k_1 \dot{x} + k_2 (\dot{x} - \dot{y}) + c(1 + \sigma z(t)) x = 0, \\ & m_2 \ddot{y} + k_2 (\dot{y} - \dot{x}) = 0 \end{aligned} \tag{8}$$

is considered. There are two stiff bodies with the masses m_1 and m_2 two damping coefficients k_1

and k_2 and a restoring coefficient $c(1+\sigma z(t))$, where $z(t)$ is assumed to be a white noise process. Introducing state variables the stochastic differential equations

$$dX_t^1 = X_t^2 dt,$$
$$dX_t^2 = (-(\delta_1+\delta_2)X_t^2 + \delta_2 X_t^3 - X_t^1)dt - \sigma X_t^1 dW_t,$$
$$dX_t^3 = (-\delta_3 X_t^3 + \delta_3 X_t^2)dt$$

are performed. The dimensionless coefficients are $\delta_1 = \frac{k_1}{\sqrt{cm_1}}$, $\delta_2 = \frac{k_2}{\sqrt{cm_1}}$ and $\delta_3 = \frac{k_2}{m_2}\sqrt{\frac{m_1}{c}}$. The time variable t is normed with $\sqrt{\frac{c}{m_1}}$

The introduction of angular coordinates with the transformation (5) leads to the nonlinear stochastic differential equations

$$d\vec{\Phi}_t = \vec{f}\left(\vec{\Phi}_t\right) dt + \vec{g}\left(\vec{\Phi}_t\right) dW_t, \quad \dim(\vec{\Phi}_t) = 2$$

with the elements of the drift vector

$$f^1 = \frac{-(2\delta_1+2\delta_2+\sigma^2)\sin 2\Phi_t^1}{4} - \frac{\sigma^2 \sin 4\Phi_t^1}{8} + \delta_2 \cos\Phi_t^1 \tan\Phi_t^2 - 1,$$

$$f^2 = \frac{(\delta_3-\delta_2)\sin\Phi_t^1}{2} + \frac{(\delta_2+\delta_3)\cos 2\Phi_t^2 \sin\Phi_t^1}{2} + \frac{(4\delta_1+4\delta_2-8\delta_3-\sigma^2)\sin 2\Phi_t^2}{16} - \frac{(2\delta_1+2\delta_2+\sigma^2)\cos 2\Phi_t^1 \sin 2\Phi_t^2}{8} - \frac{\sigma^2\cos 4\Phi_t^1 \sin 2\Phi_t^2}{16} + \frac{\sigma^2(1-\cos 4\Phi_t^1)\sin 4\Phi_t^2}{64}$$

and the elements of the diffusion matrix

$$B^{11} = \sigma^2 \frac{3+4\cos 2\Phi_t^1 + \cos 4\Phi_t^1}{8}$$
$$B^{12} = \sigma^2 \frac{-(2\sin 2\Phi_t^1 + \sin 4\Phi_t^1)\sin 2\Phi_t^2}{16}$$
$$B^{21} = \sigma^2 \frac{-(2\sin 2\Phi_t^1 + \sin 4\Phi_t^1)\sin 2\Phi_t^2}{16}$$
$$B^{22} = \sigma^2 \frac{(-1+\cos 4\Phi_t^1)(-1+\cos 4\Phi_t^2)}{64}$$

Hence the corresponding Fokker-Planck equation follows as

$$\frac{\sigma^2(1-\cos 4\varphi^1)(-1+\cos 4\varphi^2)}{128} p_{,\varphi^2\varphi^2} + \frac{\sigma^2(2\sin 2\varphi^1 + \sin 4\varphi^1)\sin 2\varphi^2}{16} p_{,\varphi^1\varphi^2} - \frac{\sigma^2(3+4\cos 2\varphi^1+\cos 4\varphi^1)}{16} p_{,\varphi^1\varphi^1} +$$
$$+ \left[\frac{\delta_3-\delta_2}{2}\sin\varphi^1 + \frac{\delta_2+\delta_3}{2}\cos 2\varphi^2 \sin\varphi^1 + \left(\frac{\delta_1+\delta_2}{4} - \frac{\delta_3}{2} - \frac{\sigma^2}{16}\right)\sin 2\varphi^2 + \left(\frac{\delta_2-\delta_1}{4} + \frac{\sigma^2}{8}\right)\cos 2\varphi^1 \sin 2\varphi^2 + \frac{\sigma^2\cos 4\varphi^1 \sin 2\varphi^2}{16} - \frac{3\sigma^2 \sin 4\varphi^2}{64} + \frac{3\sigma^2\cos 4\varphi^1 \sin 4\varphi^2}{64}\right] p_{,\varphi^2} +$$
$$+ \left[-1 - \left(\frac{\delta_1+\delta_2}{2} - \frac{3\sigma^2}{4}\right)\sin 2\varphi^1 + \frac{\sigma^2\cos 2\varphi^2 \sin 2\varphi^1}{4} + \frac{3\sigma^2 \sin 4\varphi^1}{8} + \frac{\sigma^2 \cos 2\varphi^2 \sin 4\varphi^1}{8} + \delta_2 \cos\varphi^1 \tan\varphi^2\right] p_{,\varphi^1} +$$
$$+ \left[\left(-\delta_1 - \delta_2 + \frac{\sigma^2}{2}\right)\cos 2\varphi^1 + \frac{\sigma^2\cos 4\varphi^1}{2} + \left(\frac{\delta_1+\delta_2}{2} - \delta_3 - \frac{\sigma^2}{8}\right)\cos 2\varphi^2 + \left(-\frac{\delta_1+\delta_2}{2} + \frac{\sigma^2}{4}\right)\cos 2\varphi^1 \cos 2\varphi^2 + \frac{3\sigma^2\cos 4\varphi^1 \cos 2\varphi^2}{8} - \frac{\sigma^2\cos 4\varphi^2}{16} + \frac{\sigma^2 \cos 4\varphi^1 \cos 4\varphi^2}{16} - \delta_2 \sin\varphi^1 \tan\varphi^2 + (\delta_3-\delta_2)\sin\varphi^1 \sin 2\varphi^2\right] p = 0.$$

The appearing function $\tan\varphi^2$ has to be expressed in terms of harmonic functions. Because of the unboundedness a Taylor expansion is introduced. The resulting polynomial is than expanded into a Fourier series. The expansion is valid because the natural boundaries $\varphi^2 = \pm\frac{\pi}{2}$ can be excluded. Applying Galerkins method a linear system of algebraic equations for the coefficients c_{ij} is deduced. It is solved with a standard numerical algorithm.

In Fig. 1 (left side) the distribution density function is given. The function $\tan\varphi^2$ is expanded into a Taylor series of 5-th. order and a Fourier series of order four respectively. The distribution density $p(\varphi^1, \varphi^2)$ is expanded into a Fourier series of order 25 in each direction. For a discussion of the ratio of convergence refer to (Karch 1992). To compare with this analytical result, in Fig. 1 (right side) the density function is also shown evaluated by a numerical integration of the stochastic differential equations (which is descibed in (Karch 1992).

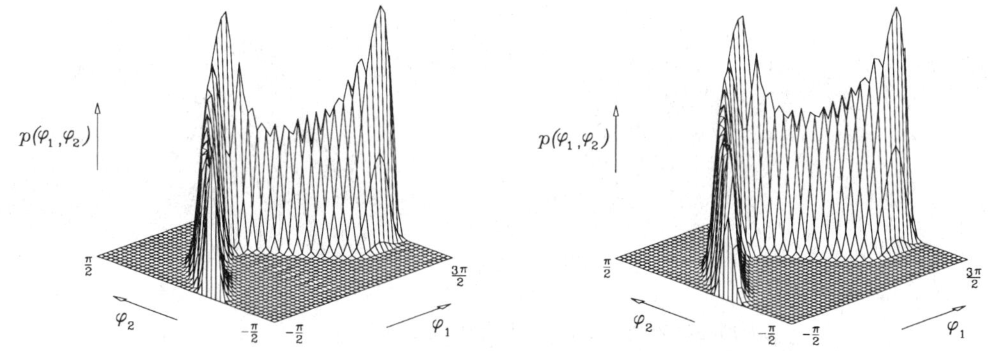

Figure 1: distribution density $p(\varphi^1, \varphi^2)$

The top Lyapunov exponent can be expressed in terms of the coefficients c_{ij} as follows:

$$\begin{aligned}
\lambda &= \frac{-\delta_1}{4} - \frac{\delta_2}{4} - \frac{\delta_3}{2} + \frac{5\sigma^2}{64} + 2\pi^2\Bigg[\frac{\sigma^2 c_{(-4,-2)}}{256} + \\
&+ \frac{\sigma^2 c_{(-4,-1)}}{64} + \frac{3\sigma^2 c_{(-4,0)}}{128} + \frac{\sigma^2 c_{(-4,1)}}{64} + \\
&+ \frac{\sigma^2 c_{(-4,2)}}{256} + \frac{k_1}{2} c_{(-2,-1)} + k_1 c_{(-2,0)} + \frac{k_1}{2} c_{(-2,1)} - \\
&- k_2 c_{(-1,-1)} + k_2 c_{(-1,1)} - \frac{\sigma^2 c_{(0,-2)}}{128} + k_3 c_{(0,-1)} + \\
&+ k_3 c_{(0,1)} - \frac{\sigma^2 c_{(0,2)}}{128} + k_2 c_{(1,-1)} - k_2 c_{(1,1)} + \\
&+ \frac{k_1}{2} c_{(2,-1)} + k_1 c_{(2,0)} + \frac{k_1}{2} c_{(2,1)} + \frac{\sigma^2 c_{(4,-2)}}{256} + \\
&+ \frac{\sigma^2 c_{(4,-1)}}{64} + \frac{3\sigma^2 c_{(4,0)}}{128} + \frac{\sigma^2 c_{(4,1)}}{64} + \frac{\sigma^2 c_{(4,2)}}{256}\Bigg],
\end{aligned}$$

with

$$\begin{aligned}
k_1 &= \frac{\delta_1 + \delta_2}{8} + \frac{\sigma_2}{16}, \\
k_2 &= \frac{\delta_2 + \delta_3}{8}, \\
k_3 &= \left(-\frac{\delta_1 + \delta_2}{8} + \frac{\delta_3}{4} + \frac{\sigma_2}{32}\right).
\end{aligned}$$

To verify the assumptions about the singularities and boundaries the procedure is performed to the 4-th. order system

$$\begin{aligned}
dX_t^1 &= X_t^2 dt, \\
dX_t^2 &= (-\delta_3 X_t^2 + \delta_3 X_t^4 - \eta_2 X_t^1 + \eta_2 X_t^3)dt, \\
dX_t^3 &= X_t^4 dt, \\
dX_t^4 &= (-(\delta_1 + \delta_2)X_t^4 + \delta_2 X_t^2 - (1 + \eta_1)X_t^3 + \\
&+ \eta_1 X_t^1)dt - \sigma X_t^3 dW_t.
\end{aligned}$$

These equations are yielded by introducing a additional spring into the described mechanical system (8). The marginal density functions $p(\varphi^i, \varphi^j)$ of the joint density function $p(\varphi^1, \varphi^2, \varphi^3)$ are plotted at the left side in Fig. 2, 3 and 4 evaluated with an expansion of 8-th. order in each direction. The corresponding numerical results are given at the right side in Fig. 2, 3 and 4 respectively. Obviously the boundaries in the direction of the coordinates φ^2 and φ^3 are natural ones in the sense of Feller. The singularities in these directions do not affect the validity of the Fokker-Planck equation and are in that sense of "good nature".

5 CONCLUSIONS

To calculate the top Lyapunov exponent of stochastic differential equations a method is presented to solve analytically the Fokker-Planck equation corresponding to the stationary part of the state process. Therefore angular coordinates are introduced yielding nonlinear stochastic differential equations on a bounded space. Within this space singularities arise. The boundaries and singularities are discussed with the resulting assumption that the Fokker-Planck equation is valid in its classical form.

The evaluation of the distribution density function for a mechanical system of two degrees of freedom is performed. Not only for the system of third order but also for that one of fourth order the assumptions are verified comparing the analytical results with results of a numerical integration.

From the mathematical point of view in further work a classification of singularities has to be performed for multi dimensional systems similar to the work of Kozin and Nishioka. This would

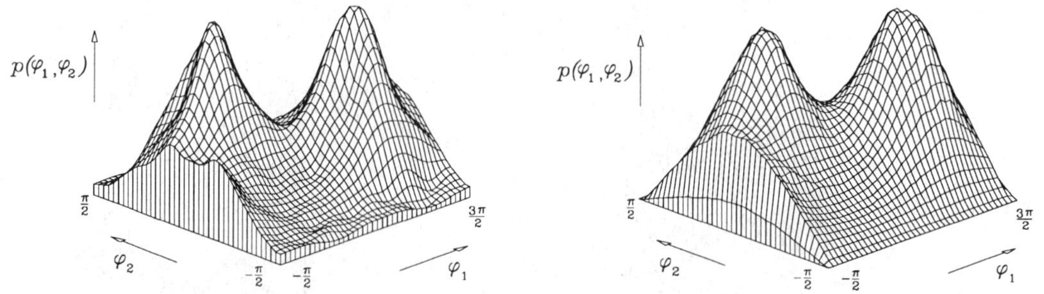

Figure 2: marginal distribution density $p(\varphi^1, \varphi^2)$

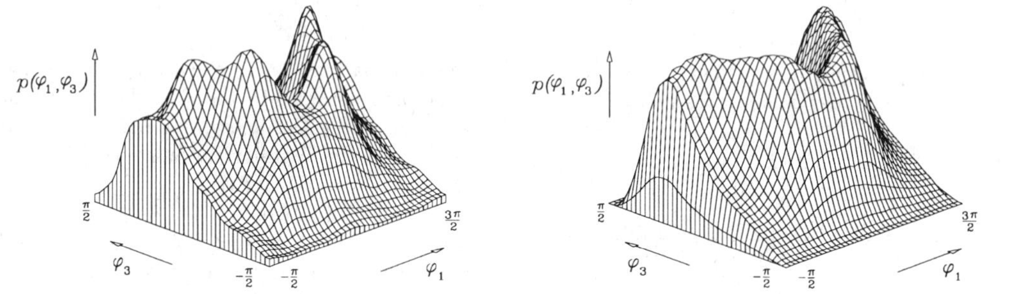

Figure 3: marginal distribution density $p(\varphi^1, \varphi^3)$

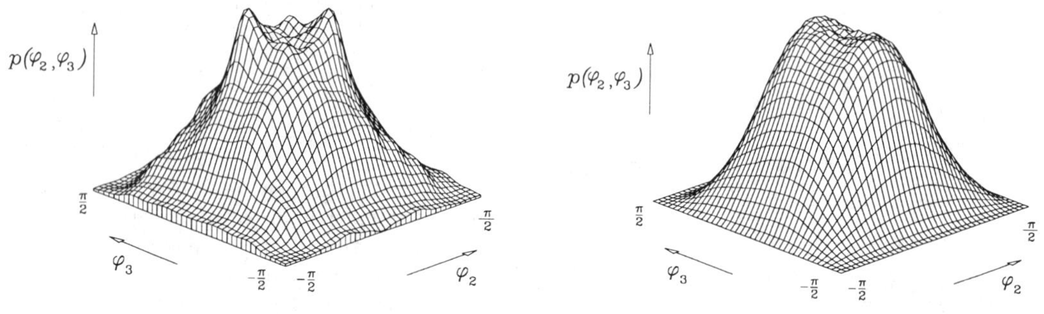

Figure 4: marginal distribution density $p(\varphi^2, \varphi^3)$

avoid the necessity to verify the analytical results with a numerical evaluation.

The whole evaluations described in this contribution can be performed with the help of symbolic computation. The application of commercial software is useful but there are problems with storage and computational time. It seems to be much more efficient to develop special adapted symbolic programs for the computations.

REFERENCES

Arnold, L., Oeljeklaus, E., & Pardoux, E. 1984. Almost sure and moment stability for linear Itô equations. *Lecture Notes in Mathematics* 1186:129 – 159. Springer.

Feller, W. 1952. The parabolic differential equations and the associated semigroups of transformations. *Annals of Mathematics* 55(3):468 – 519.

Karch, G. 1992. Zur Sabilität linearer, stochastisch parametererregter Schwingungssyteme. *Dissertation*, Universität Karlsruhe.

Kozin, F. & Mitchell, R. 1974. Sampel stability of second order linear differential equations with wide band noise coefficients. *SIAM J. Appl. Math.* 27:571– 605.
Kozin, F. & Prodromou, S. 1971. Necessary and sufficient conditions for almost sure sample stability of linear Itô equations. *SIAM J. Appl. Math.* 21:413 - 425.
Nishioka, K. 1976. On the stability of two-dimensional linear stochastic systems. *Kodai Mathematics Seminar Reports* 27:211 - 230.
Wedig, W. 1991. Pitchfork and hopf bifurcations in stochastic systems - effective methods to calculate Lyapunov exponents. In P. Kree and W. Wedig, editors, *Effective Stochastic Analysis.* Springer, to appear.

Structural Safety & Reliability, Schuëller, Shinozuka & Yao (eds) © 1994 Balkema, Rotterdam, ISBN 90 5410 357 4

Stability in probability of some nonlinear stochastic systems

Y.K.Lin & G.Q.Cai
Center for Applied Stochastics Research, Florida Atlantic University, Boca Raton, Fla., USA

ABSTRACT: Motion stabilities of some nonlinear stochastic systems are investigated, using the theory of diffusive Markov processes. It is shown that the stochastic differential equation for the motion amplitude or energy envelope of a single-degree-of-freedom nonlinear system is generally associated with singular boundaries, where either the diffusion coefficient vanishes, or the drift coefficient becomes unbounded. Criteria are established to characterize the behaviors of sample functions near each type of singular boundary, and the asymptotic stability conditions are obtained in terms of convergence in probability. Examples are given for illustration.

1 INTRODUCTION

Stability of a randomly excited dynamical system is an important topic in many fields of engineering and science. For example, motion of a long-span flexible bridge can become unstable in turbulent winds. Since motion stability is defined in terms of boundedness and convergence, and since convergence of a sequence of random variables can be interpreted in different ways, different definitions for stochastic stability are possible (Kozin 1969). The following definition will be adopted for the purpose of the present discussion:

Stability in Probability. Let $X(t)$ be a stochastic vector process, describing the motion of a system, and let $\|X(t;x_0,t_0)\|$ be a suitable norm of $X(t)$, given $X(t_0)=x_0$ and $t\geq t_0$. The trivial solution is said to be stable in probability if, for every pair of $\varepsilon_1, \varepsilon_2 > 0$, there exists a $\delta(\varepsilon_1,\varepsilon_2,x_0,t_0)>0$ such that

$$\text{Prob}[\,\|X(t;x_0,t_0)\|\geq\varepsilon_1\,]\leq\varepsilon_2, \quad t\geq t_0 \tag{1}$$

provided $\|x_0\|\leq\delta$, where $x_0 = X(t_0)$ is deterministic. The trivial solution is said to be asymptotically stable in probability, if, in addition, the right hand side of (1) tends to zero as t approaches infinity.

Stochastic stability has been investigated extensively for linear systems (e.g., Kozin 1963, Khasminskii 1967, Infante 1968, Mitchell and Kozin 1974, Wedig 1988, Ariaratnam and Ly 1989, Ariaratnam and Xie 1990, 1992), but very few results are known for nonlinear systems. However, if a system is one-dimensional, or reducible to one-dimensional, then the problem becomes simpler. The reduction of dimensionality can often be accomplished by using the stochastic averaging procedure (Stratonovich 1963, Khasminskii 1966), or the quasi-conservative averaging procedure (Landa and Stratonovich 1962, Khasminskii 1964). The case of a nonlinearly damped oscillator was investigated in detail by Brouwers (1986), by means of the transient nonstationary probability distribution, which is generally difficult to obtain in most other cases. Kozin and Sunahara (1987) and Sri Namachchivaya (1989) made use of an alternative procedure by examining the behaviors of the sample paths at and near the boundaries. Since the diffusion coefficient usually vanishes at the trivial solution (referred to hereafter as the singular boundary of the first kind), tables are provided by Zhang and Kozin (1990) to facilitate the identification of the boundary behaviors.

It will be shown in the present paper that a second kind of singular boundary exists where the drift coefficient becomes unbounded, and that the sample behaviors at both boundaries should be examined for stability investigations. Examples are given for illustration.

2 ONE-DIMENSIONAL DIFFUSION AND SINGULAR BOUNDARIES

Consider a one-dimensional diffusive Markov process, governed by an Itô stochastic differential equation

$$dX(t) = m(X)dt + \sigma(X)dB(t) \tag{2}$$

where $m(X)$ and $\sigma(X)$ are known as the drift and diffusion coefficients, respectively, and $B(t)$ is a unit Wiener process. According to Feller (1952), four types of boundaries exist:

1. **Regular boundary**: The process can either reach the boundary from an interior point, or reach an interior point from the boundary.
2. **Exit boundary**: The process can reach the boundary from an interior point, but cannot reach an interior point from the boundary.
3. **Entrance boundary**: The process can reach an interior point from the boundary, but cannot reach the boundary from an interior point.
4. **Natural boundary**: The process cannot reach the boundary from an interior point, nor can it reach an interior point from the boundary.

These boundaries can be identified through two time measures, the time measure to reach a boundary, starting from an interior point, and the time measure to reach an interior point, starting from a boundary (Itô and McKean 1965). A refinement has been proposed by Karlin and Taylor (1981) by splitting the class of natural boundaries into the class of attractive natural boundaries and the class of unattractive natural boundaries. However, for the purpose of stability investigation, a further refinement is necessary (Lin and Cai 1993) in which the unattractive natural boundaries are divided into the repulsively natural boundaries and the strictly natural boundaries.

Some conclusions can be made readily from the above boundary classifications. A trivial solution is asymptotically stable in probability, if the trivial solution is an exit or attractively natural, and if the other boundary is an entrance or repulsively natural. A stationary solution exists within the interval on which the Markov process is defined, if each of the two boundaries is either an entrance or repulsively natural. The stationary probability density is a unit delta function at an exit boundary, if the other boundary is either an entrance or repulsively natural. A stationary solution does not exist, if each of the two boundaries is either an exit, or attractively natural, or strictly natural. The presence of a regular boundary renders the solution non-unique without further stipulation on the behaviors of sample functions after reaching the boundary (Feller 1952); however, this permits certain assumptions to be incorporated at a regular boundary according to the physical problem at hand.

The calculation of the two time measures is generally complicated, especially at a singular boundary. However, only their boundedness, which can be determined based on the limiting behaviors of the drift and diffusion coefficients near the boundary, is needed for the boundary classification.

Consider a singular boundary $X = x_s$. The subscript s can be either l or r, denoting either the left or the right boundary. Three important parameters associated with a singular boundary: diffusion exponent α_s, drift exponent β_s and character value c_s are defined as follows for those cases of interest in the present investigation:

1. Singular Boundary of the First Kind — $\sigma(x_s) = 0$

$$\sigma^2(x) = 0\,|x - x_s|^{\alpha_s}, \quad as\ x \to x_s \tag{3}$$

$$m(x) = 0\,|x - x_s|^{\beta_s}, \quad as\ x \to x_s \tag{4}$$

$$c_l = \lim_{x \to x_l^+} \frac{2m(x)(x - x_l)^{\alpha_l - \beta_l}}{\sigma^2(x)} \tag{5}$$

$$c_r = -\lim_{x \to x_r^-} \frac{2m(x)(x_r - x)^{\alpha_r - \beta_r}}{\sigma^2(x)} \tag{6}$$

Table 1 Classification of Singular Boundary of the First Kind*

STATE	CONDITIONS				CLASS
$\sigma(x_s) = 0$ $(\alpha_s > 0)$ $m(x_s) \neq 0$ $(\beta_s = 0)$ (shunt)	$\alpha_s < 1$				regular
	$\alpha_s = 1$	$m(x_l) < 0$ or $m(x_r) > 0$			exit
		$m(x_l) > 0$ or $m(x_r) < 0$	$0 < c_s < 1$		regular
			$c_s \geq 1$		entrance
	$\alpha_s > 1$	$m(x_l) < 0$ or $m(x_r) > 0$			exit
		$m(x_l) > 0$ or $m(x_r) < 0$			entrance
$\sigma(x_s) = 0$ $(\alpha_s > 0)$ $m(x_s) = 0$ $(\beta_s > 0)$ (trap)	$\alpha_s < 1 + \beta_s$	$\alpha_s < 1$			regular
		$1 \leq \alpha_s < 2$			exit
		$\alpha_s \geq 2$			attractive natural
	$\alpha_s > 1 + \beta_s$	$\beta_s < 1$	$m(x_l^+) < 0$ or $m(x_r^-) > 0$		exit
			$m(x_l^+) > 0$ or $m(x_r^-) < 0$		entrance
		$\beta_s \geq 1$	$m(x_l^+) < 0$ or $m(x_r^-) > 0$		attractive natural
			$m(x_l^+) > 0$ or $m(x_r^-) < 0$		repulsively natural
	$\alpha_s = 1 + \beta_s$	$\beta_s < 1$	$c_s > \beta_s$	$c_s \geq 1$	entrance
				$c_s < 1$	regular
			$c_s \leq \beta_s$		exit
		$\beta_s \geq 1$	$c_s > \beta_s$		repulsively natural
			$c_s \leq \beta_s$	$c_s \geq 1$	strictly natural
				$c_s < 1$	attractively natural

* modified from the original tables of Zhang and Kozin (1990)

2. Singular Boundary of the Second Kind at Infinity — $m(x_s) = \infty$, $|x_s| = \infty$

$$\sigma^2(x) = 0|x|^{\alpha_s}, \quad as \ |x| \to \infty \tag{7}$$

$$m(x) = 0|x|^{\beta_s}, \quad as \ |x| \to \infty \tag{8}$$

$$c_l = \lim_{x \to -\infty} \frac{2m(x)|x|^{\alpha_l - \beta_l}}{\sigma^2(x)} \tag{9}$$

$$c_r = -\lim_{x \to \infty} \frac{2m(x)|x|^{\alpha_r - \beta_r}}{\sigma^2(x)} \tag{10}$$

In (3), (4), (7) and (8), $0|\cdot|$ denotes the order of magnitude of $|\cdot|$. The classification of a singular boundary x_s can then be identified from the values of α_s, β_s and c_s with the aid of Tables 1 and 2. It may be noted that α_s and β_s are also called the shunt index and the trap index (Zhang and Kozin 1990), respectively. For more details, see Lin and Cai (1993).

3 A NONLINEARLY DAMPED OSCILLATOR

Consider the following nonlinearly damped oscillator

$$\ddot{X}+\eta\,|\dot{X}|^{\delta}\operatorname{sgn}\dot{X}+\omega_0^2 X = XW_1(t)+\dot{X}W_2(t) \qquad \eta>0,\ \delta\geq 0 \tag{11}$$

where $W_1(t)$ and $W_2(t)$ are independent Gaussian white noises with spectral densities K_1 and K_2 respectively. We assume that both the damping coefficient η and the spectral densities are small, so that the stochastic averaging procedure is applicable. Letting

$$\begin{aligned} X(t) &= A(t)\cos\theta, \quad \theta=\omega_0 t+\phi(t) \\ \dot{X}(t) &= -\omega_0 A(t)\sin\theta \end{aligned} \tag{12}$$

equation (11) can be replaced by an Itô stochastic differential equation for $A(t)$ in the same form as (2), namely,

$$dA = m(A)dt+\sigma(A)dB(t) \tag{13}$$

where

$$m(A)=\frac{\pi}{8}(3K_1+5K_2)A-\frac{\eta\omega_0^{\delta-1}}{\sqrt{\pi}}\frac{\Gamma(\frac{\delta}{2}+1)}{\Gamma(\frac{\delta+3}{2})}A^{\delta} \tag{14}$$

$$\sigma(A) = \left[\frac{\pi}{4}(K_1+3K_2)\right]^{1/2}A \tag{15}$$

It is seen that the left boundary $A=0$ is singular of the first kind, and the right boundary $A=\infty$ is singular of the second kind. The stability analysis, however, must proceed separately for three cases, $\delta=1$, $0\leq\delta<1$ and $\delta>1$.

The case of $\delta=1$ corresponds to a linear system. The diffusion exponents, drift exponents, and character values at the two boundaries $A=0$ and $A=\infty$ are obtained from equations (3) through (10) as follows

$$\alpha_l=2,\ \beta_l=1,\ c_l=\frac{-4\eta\omega_0^2+\pi(3K_1+5\omega_0^2K_2)}{\pi(K_1+3K_2\omega_0^2)} \tag{16}$$

$$\alpha_r=2,\quad \beta_r=1,\quad c_r=-c_l \tag{17}$$

Table 2 Classification of Singular Boundary of the Second Kind at Infinity

STATE	CONDITION				CLASS
$\lvert m(\infty)\rvert=\infty$ $(\beta_s>0)$ $\sigma(\infty)<\infty$ $(\alpha_s=0)$	$m(-\infty)<0$ or $m(+\infty)>0$				exit
	$m(-\infty)>0$ or $m(+\infty)<0$				entrance
$\lvert m(\infty)\rvert=\infty$ $(\beta_s>0)$ $\sigma(\infty)=\infty$ $(\alpha_s>0)$	$\beta_s>\alpha_s-1$	$m(-\infty)<0$ or $m(+\infty)>0$			exit
		$m(-\infty)>0$ or $m(+\infty)<0$			entrance
	$\beta_s<\alpha_s-1$				regular
	$\beta_s=\alpha_s-1$	$\beta_s\leq 1$	$c_s>-\beta_s$		repulsively natural
			$c_s\leq-\beta_s$	$c_s\geq -1$	strictly natural
				$c_s<-1$	attractively natural
		$\beta_s>1$	$c_s>-\beta_s$	$c_s\geq -1$	entrance
				$c_s<-1$	regular
			$c_s\leq-\beta_s$		exit

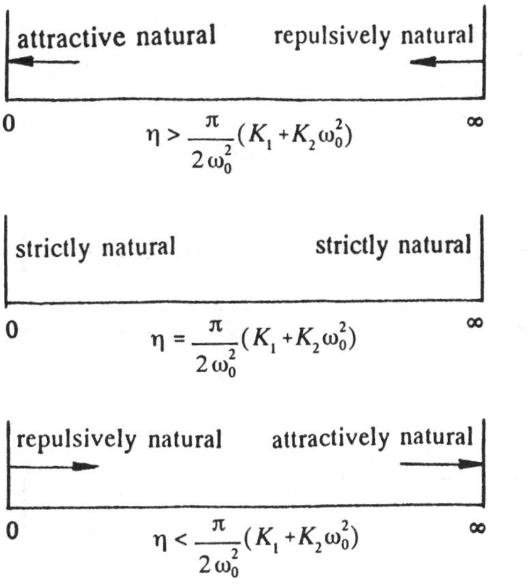

Fig. 1 Boundary behaviors of sample functions of a linear system, $\delta = 1$

The two singular boundaries can be identified using Tables 1 and 2 and represented in Figure 1 schematically. If $\eta > \pi(K_1 + \omega_0^2 K_2)/(2\omega_0^2)$, the trivial solution is asymptotically stable, although it is only approachable, not accessible from an interior point. A stationary probability density does not exist otherwise.

For the case $0 \leq \delta < 1$, the diffusion exponents, drift exponents and character value required to identify the two boundaries are

$$\alpha_l = 2, \quad \beta_l = \delta < 1 \tag{18}$$

$$\alpha_r = 2, \quad \beta_r = 1, \quad c_r = -3 \tag{19}$$

According to Tables 1 and 2, the left boundary $A = 0$ is an exit, and the right boundary $A = \infty$ is attractively natural. A sample function can either reach the left boundary and stay there forever, or tend to infinity asymptotically. Therefore, the asymptotic stability of the trivial solution is not guaranteed in probability. Moreover, a stationary probability density does not exist, regardless of the η value.

For the case $\delta > 1$, it can be shown that

$$\alpha_l = 2, \quad \beta_l = 1, \quad c_l = 3 \tag{20}$$

$$\alpha_r = 2, \quad \beta_r = \delta > 1 \tag{21}$$

Therefore, the left boundary is repulsively natural and the right boundary is an entrance. The trivial solution is unstable in probability, and an stationary probability density exists.

It is interesting to note that the trivial solution of system (11) is asymptotically unstable in probability for either $\delta > 1$ or $\delta < 1$, and that stability is possible only when a linear damping term, say $\gamma \dot{X}$, is added on the left hand side of equation (11). For details, see Lin and Cai (1993) in which the question of stability in statistical moments is also addressed.

4 AN OSCILLATOR WITH NONLINEAR RESTORING FORCE

Systems with nonlinear restoring force will now be investigated. Consider the following equation of motion

$$\ddot{X} + \eta\dot{X} + k|X|^{\delta}\operatorname{sgn}X = XW_1(t) + \dot{X}W_2(t) \tag{22}$$
$$\eta > 0,\ k > 0,\ \delta \geq 0$$

We assume that both damping and the spectral density of the excitation are sufficiently small, so that the quasi-conservative averaging procedure is applicable. The total energy of the oscillator is given by

$$\Lambda = \frac{1}{2}\dot{X}^2 + U(X) = \frac{1}{2}\dot{X}^2 + \frac{\eta}{\delta+1}|X|^{\delta+1} \tag{23}$$

which can be approximated as a diffusive Markov process, governed by the following Itô stochastic differential equation

$$d\Lambda = m(\Lambda)dt + \sigma(\Lambda)\,dB(t) \tag{24}$$

where the drift and diffusion coefficients obtained from quasi-conservative averaging are given by

$$m(\Lambda) = 2(-\eta + 2\pi K_2)\frac{\delta+1}{\delta+3}\Lambda + \pi K_1\left(\frac{\delta+1}{k}\right)^{\frac{2}{\delta+1}} \frac{B\left(\frac{3}{\delta+1}, \frac{1}{2}\right)}{B\left(\frac{1}{\delta+1}, \frac{1}{2}\right)}\Lambda^{\frac{2}{\delta+1}} \tag{25}$$

$$\sigma^2(\Lambda) = 24\pi K_2 \frac{(\delta+1)^2}{(3\delta+5)(\delta+3)}\Lambda^2$$

$$+ 2\pi K_1 \left(\frac{\delta+1}{k}\right)^{\frac{2}{\delta+1}} \frac{B\left(\frac{3}{\delta+1}, \frac{3}{2}\right)}{B\left(\frac{1}{\delta+1}, \frac{1}{2}\right)} \Lambda^{\frac{\delta+3}{\delta+1}} \quad (26)$$

where $B(\cdot,\cdot)$ denotes the Beta function. Obviously, $\Lambda = 0$ and $\Lambda = \infty$ are singular boundaries of the first and second kinds, respectively.

First we consider the case of $K_1 = 0$ and $K_2 \neq 0$. In this case, equations (25) and (26) reduce to

$$m(\Lambda) = 2(-\eta + 2\pi K_2)\frac{\delta+1}{\delta+3}\Lambda \quad (27)$$

$$\sigma^2(\Lambda) = 24\pi K_2 \frac{(\delta+1)^2}{(3\delta+5)(\delta+3)}\Lambda^2 \quad (28)$$

Therefore,

$$\alpha_l = 2, \ \beta_l = 1, \ c_l = \frac{(-\eta+2\pi K_2)(3\delta+5)}{6\pi K_2(\delta+1)} \quad (29)$$

$$\alpha_r = 2, \ \beta_r = 1, \ c_r = -c_l \quad (30)$$

The boundary behaviors are shown schematically in Figure 2. The trivial solution is asymptotically stable if $\eta > 4\pi K_2/(3\delta+5)$. A stationary probability density does not exist otherwise. The bifurcation point depends on the power order δ of the nonlinear restoring force.

If $K_1 \neq 0$, then three cases, $\delta = 1$, $0 \le \delta < 1$ and $\delta > 1$, must be investigated separately. The analysis for the linear case $\delta = 1$ is similar to that in section 3, with ω_0^2 replaced by k. The trivial solution is asymptotically stable in probability if $\eta > \pi(K_1 + K_2 k)/(2k)$.

For the case $0 \le \delta < 1$, it is found that

$$\alpha_l = 2, \ \beta_l = 1, \ c_l = \frac{(-\eta+2\pi K_2)(3\delta+5)}{6\pi K_2(\delta+1)}, \quad \text{if } K_2 \neq 0 \quad (31)$$

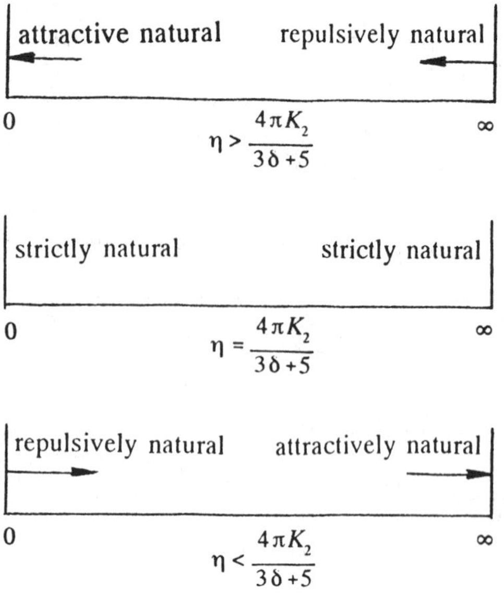

Fig. 2 Boundary behaviors of sample functions of a system with nonlinear restoring force, $K_1 = 0$ and $K_2 \neq 0$

$$\alpha_l = \frac{\delta+3}{\delta+1} > 2, \ \beta_l = 1, \ m(0^+) < 0, \quad \text{if } K_2 = 0 \quad (32)$$

whereas

$$\alpha_r = \frac{\delta+3}{\delta+1}, \ \beta_r = \frac{2}{\delta+1}, \ c_r = -\frac{\delta+7}{2(\delta+1)} \quad (33)$$

independent of K_2. Then, the right boundary $\lambda = \infty$ is always an exit, while the left boundary can be attractively natural, strictly natural or repulsively natural depending on the K_2, η and δ values. Thus, the system is unstable in probability regardless of the η value.

For the case $\delta > 1$,

$$\alpha_l = \frac{\delta+3}{\delta+1}, \ \beta_l = \frac{2}{\delta+1}, \ c_l = \frac{\delta+7}{2(\delta+1)} \quad (34)$$

independent of K_2,

$$\alpha_r = 2, \ \beta_r = 1, \ c_r = \frac{(-\eta+2\pi K_2)(3\delta+5)}{6\pi K_2(\delta+1)}, \quad \text{if } K_2 \neq 0 \quad (35)$$

$$\alpha_r = \frac{\delta+3}{\delta+1} > 2, \quad \beta_r = 1, \quad m(\infty) < 0,$$

$$\text{if } K_2 = 0 \qquad (36)$$

indicating that the left boundary is regular or an entrance. The trivial solution is unstable in probability.

It is of interest to note that the stability analysis for system (22) must be carried out separately for cases $K_1 = 0$ and $K_1 \neq 0$. If $K_1 = 0$ and $K_2 \neq 0$, the asymptotic stability condition depends on the power order of the nonlinear stiffness term. If $K_1 \neq 0$, then the trivial solution of system (22) is asymptotically unstable in probability for either $\delta > 1$ or $\delta < 1$. In this case, stability is possible only if a linear stiffness term is added to equation (22).

5 CONCLUSIONS

The theory of one-dimensional diffusive Markov process is applied to investigate the asymptotic stability of nonlinear systems under Gaussian white-noise excitations. The one-dimensional variable selected in the analysis can be either the amplitude or the energy envelope of a critical mode. Boundaries of such a diffusive process can be either regular, an exit, an entrance, attractively natural, repulsively natural, or strictly natural. A smooth stationary probability density exists if and only if each of the two boundaries is either an entrance or repulsively natural. For a singular boundary, where either the diffusion coefficient vanishes or the drift coefficient is unbounded, the classification can be identified in terms of the drift exponent, diffusion exponent and character value.

Conditions for the asymptotic stability in probability are obtained. When excitations are purely multiplicative, a trivial solution exists, and is asymptotically stable in probability if it is either an exit or attractively natural, while the other boundary is either an entrance or repulsively natural. Nonlinear oscillators with nonlinear damping and nonlinear stiffness are investigated, respectively, in detail. For an oscillator with nonlinear damping, the trivial solution is unstable in probability if the damping term does not contain a linear component. For a oscillator with nonlinear restoring force, the condition for the asymptotic stability in probability depends on whether or not a multiplicative excitation appears in the displacement term. If the multiplicative excitation appears only in the velocity term, the condition for the asymptotic stability in probability is dependent on the power order of the restoring force. If a multiplicative excitation exists in the displacement term, then the trivial solution is unstable in probability unless the restoring force has a linear component.

ACKNOWLEDGEMENT

This research has been supported by the National Science Foundation (NSF) under Grant BCS-9015272. Opinions, findings, and conclusions or recommendations expressed are those of the writers, and do not reflect the views of NSF.

REFERENCES

Ariaratnam, S.T. & Ly, B.L. 1989. Almost-sure stability of some linear stochastic systems. Journal of Applied Mechanics 56: 175-178.

Ariaratnam, S.T. & Xie, W.C. 1990. Lyapunov exponent and rotation number of a two-dimensional nilpotent stochastic system. Dynamics and Stability of Systems 5: 1-9.

Ariaratnam, S.T. & Xie, W.C. 1992. Lyapunov exponents and stochastic stability of coupled linear systems under real noise excitation. Journal of Applied Mechanics 59: 664-673.

Brouwers, J.J.H. 1986. Stability of a non-linearly damped second-order system with randomly fluctuating restoring coefficient. International Journal of Non-Linear Mechanics 21: 1-13.

Feller, W. 1952. The parabolic differential equation and the associated semigroups of transformations. Annals of Mathematics 55: 468-519.

Infante, F.F. 1968. On the stability of some linear nonautonomous systems. Journal of Applied Mechanics 35: 7-12.

Itô, K. & McKean, Jr. H.P, 1965. Diffusion

Processes and Their Sample Paths. Academic Press, New York.
Karlin, S. & Taylor, H.M. 1981. A Second Course in Stochastic Processes. Academic Press, New York.
Khasminskii, R.Z. 1964. On the behavior of a conservative system with small friction and small random noise. (in Russian), Prikladnaya Matematika i Mechanica (Appl. Math and Mech.) 28: 1126-1130.
Khasminskii, R.Z. 1966. A limit theorem for the solution of differential equations with random right hand sides. Theory of Probability and Application 11: 390-405.
Khasminskii, R.Z. 1967. Sufficient and necessary conditions of almost sure asymptotic stability of a linear stochastic system. Theory of Probability and Application 12: 144-147.
Kozin, F. 1963. On almost sure stability of linear systems with random coefficients. Journal of Mathematics and Physics 42: 59-67.
Kozin, F. 1969. A survey of stability of stochastic systems. Automatica 5: 95-112.
Kozin, F. & Sunahara, Y. 1987. An application of the averaging method to noise stabilization of nonlinear systems. Proceedings of 20th Midwestern Mechanics Conference, Vol.14(a): 291-298. West Lafayette, Indiana.
Landa, P.S. & Stratonovich, R.L. 1962. Theory of stochastic transitions of various systems between different states. (in Russia), Vestnik MGU (Proc. of Moscow University), Series III (1): 33-45.
Lin, Y.K. & Cai, G.Q. 1993. Stochastic stability of nonlinear systems. Submitted to International Journal of Non-Linear Mechanics.
Mitchell, R.R. & Kozin, F. 1974. Sample stability of second-order linear differential equations with wide-band noise coefficients. SIAM Journal of Applied Mathematics 27: 571-604.
Sri Namachchivaya, N. 1989. Instability theorem based on the nature of the boundary behavior for one-dimensional diffusion. Solid Mechanics Archives 14: 131-142.
Stratonovich, R.L. 1963. Topics in the Theory of Random Noise, Vol. 1. Gordon and Breach, New York.
Wedig, W.V. 1988. Lyapunov exponents of stochastic systems and related bifurcation problems. In S.T.Ariaratnam, G.I.Schuëller & I.Elishakoff (eds.), Stochastic Structural Dynamics: 315-327. Elsevier Science, London-New York.
Zhang, Z.Y. & Kozin, F. 1990. On almost sure sample stability of nonlinear stochastic dynamic systems. Proceedings of IEEE International Conference on System Engineering, Pittsburg.

Structural Safety & Reliability, Schuëller, Shinozuka & Yao (eds) © 1994 Balkema, Rotterdam, ISBN 90 5410 357 4

Nonlinear analysis of thin-walled structures subjected to stationary random loads

P. Melnik-Melnikov
Institute of Engineering Mechanics, University of Innsbruck, Austria

V. Bazhenov & E. Dekhtyaruk
Institute of Structural Mechanics, Kiev Civil Engineering University, Ukraine

ABSTRACT: Numerical techniques for random vibration analysis of thin-walled structures with large displacements (geometric nonlinearity) are proposed. In a first step the finite element technique and modal superposition is utilized in order to create a coupled system of nonlinear differential equations, which describes the behaviour of the above mentioned structures. In this case special functions, which allow the consideration of membrane stresses, are included in the basis of modal superposition besides linear natural modes. Second step in the solution is based on Gaussian equivalent linearization technique in random analysis. The numerical procedure was verified by analysing an non-symmetric SDOF nonlinear system and nonlinear two modal vibration of a beam. Results for a plate supported by two ribs and subjected to stationary random loading are presented and discussed.

1 INTRODUCTION

Quite often the loads, which cause vibrations of thin-walled structures, are random in time and space. Examples of such loads can be found in flight vehicle, earthquake, transport and offshore engineering. The increasing intensity of loads on one hand and design tendency to diminish the mass of structures on the other, leads to displacements in thin-walled structures, which may be compared with its average thickness. These comparatively large displacements should be treated in terms of geometrically nonlinear analysis.

Stochastic response of nonlinear systems can be estimated using various methods, which might be grouped in three categories: analytical methods limited to small idealised systems (e.g. solution of the Fokker-Planck equation), linearization methods (including closure techniques) and Monte-Carlo simulation methods of a very general nature. According to review paper by Schuëller at al. (1993) [1] at present only Monte-Carlo simulation and statistical linearization are available for practical use in nonlinear analysis of MDOF systems. Significant computational cost of straightforward Monte-Carlo simulation has limited analysis by this method.

Method of equivalent linearization up to now proved to be very feasible and useful approximate approach , especially when dealing with weakly and moderately nonlinear systems (Roberts and Spanos, 1990 [9]). Recently, the non-Gaussian equivalent linearization was successfully applied to strongly nonlinear MDOF systems with hysteretic behaviour (Schuëller at al. (1991) [10]).

Method especially well suited being applied to the finite element type systems (Spanos, (1981) [2]). One of the first steps in this direction has been made by Busby and Weingarten, (1973) [3] for geometrically nonlinear vibration of a beam under stationary random loading.

2 EQUATION OF MOTION

The equation of motion for a discretized thin-walled structure with large displacements can be written as:

$$M\ddot{u} + C\dot{u} + Ku + K_n(u) = P(t), \qquad (1)$$

where M, C, K are the mass, damping and stiffness matrices, respectively; K_n is a vector function of the structure displacements u(t); dot above a variable denotes differentiation with respect to the time t.

Reduction of this system by the modal superposition method is made under following assumptions. Deformations of bending and membrane groups are small, the bending components of displacements have the same order with the thickness of structure, effects of elastic waves in longitudinal directions are neglected, material is isotropic, homogeneous, governed by the Hooke's law.

In this case, if the basis of modal superposition includes only the linear natural modes, the obtained dynamic model is inadequate. This is connected with the fact that linear combination of bending natural modes cannot properly describe the membrane displacements and stresses under the large displacements of structures. Therefore it is necessary to add the special functions [4] in the basis of modal superposition. Applying the theory developed in [4] one can yield the couple set of nonlinear differential equations which describes the reduced dynamic model in generalized coordinates:

$$\ddot{x}_i + 2\varepsilon_i\omega_i\dot{x}_i + \omega_i^2 x_i + \sum_{jk} K_2^{ijk} x_j x_k + \sum_{jkl} K_3^{ijkl} x_j x_k x_l = \mu Q_i(t) \qquad (2)$$

$$i, j, k, l = 1, \ldots, N$$

where ε_i is a viscous damping coefficient; ω_i is an i-th natural frequency; K_2^{ijk} and K_3^{ijkl} are components of $(N \times N \times N)$ and $(N \times N \times N \times N)$ matrices of quadratic and cubic nonlinearities; μ - parameter, which indicates the intensity of loads ; x_i - i-th generalized displacement; $Q_i(t)$ - i-th component of stationary Gaussian vector process, which is described by matrix of cross-spectral densities $S_{ij}(\omega)$; N - number of generalized degrees of freedom.

3 EQUIVALENT LINEARIZATION PROCEDURE

Due to the asymmetry of the equation (2) the solution may not have a zero mean value. Therefore in general:

$$Q_i(t) = Q_i^m + \hat{Q}_i(t), \quad x_i(t) = x_i^m + \hat{x}_i(t) \qquad (3)$$

where Q_i^m, x_i^m are the time independent deterministic mean values, $\hat{Q}_i(t), \hat{x}_i(t)$ are zero-mean random processes of i-th generalized force and displacement, respectively.

Substituting equation (3) into equation (2) and applying ensemble averaging one can obtain:

$$\omega_i^2 x_i^m + \sum_{jk} K_2^{ijk}(x_j^m x_k^m + \theta_{jk}) +$$

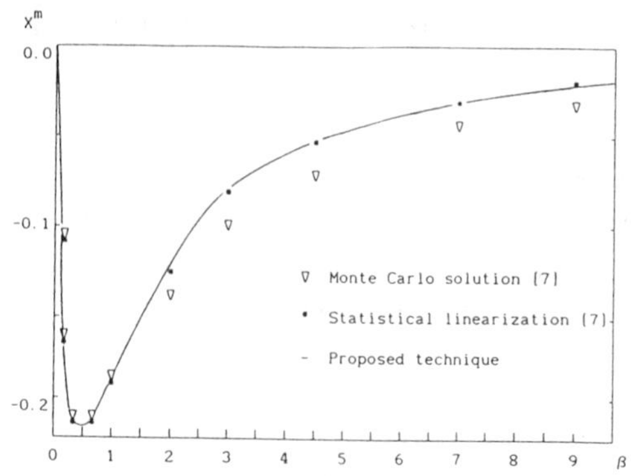

Fig.1 Mean value of the system displacement versus the nonlinearity parameter β .

$$\sum_{jkl} K_3^{ijkl}(x_j^m x_k^m x_l^m + x_j^m \theta_{kl} + x_k^m \theta_{jl} + x_l^m \theta_{jk}) = \mu Q_i^m \qquad (4)$$

where $\theta_{jk} = \langle \hat{x}_j(t)\hat{x}_k(t) \rangle$ are components of covariance matrix; $\langle\ \rangle$ denotes the operation of ensemble averaging. Equations (4) are nonlinear algebraic equations and can be solved numerically for various values of parameter μ . Unfortunately, they contain the unknown components θ_{jk} . The last can be found from solving the equivalent linear system

$$\ddot{\hat{x}}_i + 2\varepsilon_i\omega_i\dot{\hat{x}}_i + \omega_i^2\hat{x}_i + \sum_n K_e^{in}\hat{x}_n = \mu\hat{Q}_i(t), \qquad (5)$$

$$i, n = 1, \ldots, N$$

where K_e^{in} are the components of the equivalent stiffness matrix K_e. It has been shown that if the loads $Q_i(t)$ are Gaussian, the minimization criteria for mean square error between equations (2) and (5) gives the following expression for K_e^{in} [2]:

$$K_e^{in} = \left\langle \frac{\partial F_i}{\partial \hat{x}_n} \right\rangle, \qquad (6)$$

where $F_i = \omega_i^2 x_i^m + \sum_{jk} K_2^{ijk} x_j x_k + \sum_{jkl} K_3^{ijkl} x_j x_k x_l$ (7)

Substitution of the equation (7) into (6) leads to the following relations:

$$K_e^{in} = \sum_j (K_2^{ijn} + K_2^{inj}) x_j^m + \qquad (8)$$

$$+\sum_{jk}(K_3^{ijkn}+K_3^{ijnk}+K_3^{injk})(x_j^m x_k^m+\theta_{jk})$$

The spectral method [5] for solving a coupled system of linear differential equations gives the following expression for θ_{jk},

$$\theta_{jk}=\mu^2\int_{-\infty}^{\infty}\sum_n\sum_l H_{jn}(\omega)H_{kl}^*(\omega)S_{nl}(\omega)d\omega, \qquad (9)$$

where $H_{jn}(\omega)$ are components of the $(N\times N)$ complex admittance matrix :
$H(\omega)=\left[\mathrm{diag}(-\omega^2+2i\varepsilon_i\omega_i\omega+\omega_i^2)+K_e\right]^{-1}$; symbol * denotes complex conjugate value.

Due to the dependence of the K_e on the x_i^m and θ_{jk} the equation (9) is coupled with equations (8) and (4). When the matrix K_e is known, equation (9) can be solved numerically after subdivision of the frequency domain. Then, the cross-spectral density matrix of the response is calculated for each discrete frequency and after is integrated to obtain θ_{jk}. It should be mentioned that all correspondent frequencies, which fall within the frequency domain of the external loads, should be included in the modal superposition.

Summarizing the preceding analysis, the problem of solving equation (2) is replaced by a solution of the coupled system of equations (4), (8) and (9).

4 STEP-BY-STEP TECHNIQUE

In case, when $K_2^{ijk}=0$ (this valid for plates and beams) and $Q_i^m=0$, i=1,..,N equation (4) has the trivial solution $x_i^m=0$, and only equations (8) and (9) need to be solved. This can be done using iteration approach. In the general case, when $K_2^{ijk}\neq 0$, and $Q_i^m\neq 0$, the combination of parameter incrementation method [6] (for solving equation (4)) and iteration approach (for solving equations (8) and (9)) is proposed.

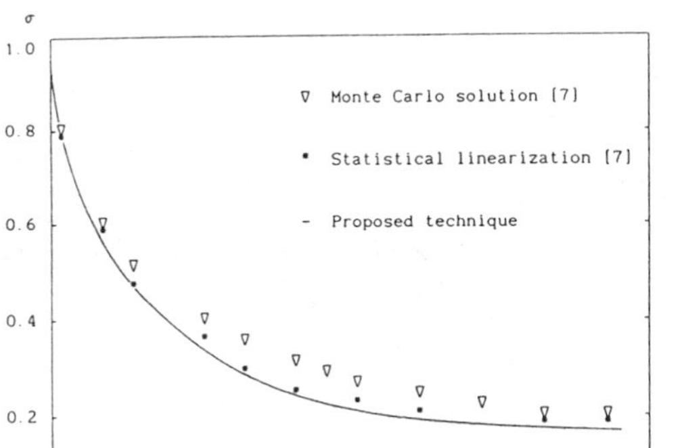

Fig.2 Standard deviation of the system displacement versus the nonlinearity parameter β

5 NUMERICAL EXAMPLES

In order to estimate the accuracy and efficiency of the proposed numerical technique the non-symmetric nonlinear oscillator was considered:

$$\ddot{x}+\varepsilon\dot{x}+x(1+3\beta+3\beta x+\beta x^2)=\mu w(t), \qquad (10)$$

where ε=0.2 denotes the damping coefficient; w(t) is stationary zero mean random process with spectral density $S(\omega)=1/2\pi$; β serves to indicate parameter of nonlinearity; $\mu^2=0.4$.

This equation was previously investigated by Spanos in paper [7]. Mean value x^m and standard deviation $\sigma=\sqrt{\theta}$ for various values of nonlinearity parameter β are shown on fig.1-2. The maximum discrepancy with results, obtained by statistical linearization method in [7], for the mean value is 3% (for β =2), and for the standard deviation is 2% (for β =5). A good agreement of results in a wide range of varying parameter β can be noticed.

Another comparison was made with regard to the Busby and Weingarten results [3] for the nonlinear vibrations of clumped beam, subjected to stationary random loading. The beam has the following properties: Young's modulus = 21.08×10^{10} Pa; Poisson's ratio =0.3; density = 8.304×10^3 kg/m^3; length = 1.524 m; height = 3.175×10^{-3} m; width = 3.175×10^{-3} m. Damping coefficient ε_i was assumed to be 5% for all modes of vibration.

The beam was divided into 30 equal finite elements. The set of nonlinear differential equations produced by applying finite element technique and modal superposition with two lowest symmetrical modes (1st and 3d modes respectively) was written in [3] as

$$\ddot{x}_1+\varepsilon_1\dot{x}_1+\omega_1^2x_1+F_{111}x_1^3+ \\ F_{113}x_1^2x_3+F_{131}x_3^2x_1+F_{133}x_3^3=Q_1(t) \qquad (11)$$

$$\ddot{x}_3+\varepsilon_3\dot{x}_3+\omega_3^2x_3+F_{333}x_3^3+ \\ F_{331}x_3^2x_1+F_{313}x_1^2x_3+F_{311}x_1^3=Q_3(t), \qquad (12)$$

where in the terms of equation (2)

$$F_{111}=K_3^{1111},\quad F_{133}=K_3^{1222}, \\ F_{113}=K_3^{1112}+K_3^{1211}+K_3^{1121}, \qquad (13) \\ F_{131}=K_3^{1122}+K_3^{1221}+K_3^{1212},$$

$$F_{333}=K_3^{2222},\quad F_{311}=K_3^{2111}, \\ F_{331}=K_3^{2221}+K_3^{2122}+K_3^{2212}, \qquad (14) \\ F_{313}=K_3^{2112}+K_3^{2211}+K_3^{2121}.$$

The values of this coefficients are listed in Table 1.

Table 1. Coefficients for equation of motion of clamped beam.

Coef.	ω_1^2	F_{111}	F_{113}	F_{131}	F_{133}
Prop. techn.	1951	88915	-222129	910117	-655106
Busby, Weingarten [3]	1942	89366	-223322	921096	-663954
Discrepancy in %	0.5	0.5	0.5	1.2	1.3

Coef.	ω_3^2	F_{333}	F_{331}	F_{313}	F_{311}
Prop. techn.	57698	6273848	-1770759	820069	-66708
Busby, Weingarten [3]	56690	6408522	-1795506	830296	-67102
Discrepancy in %	1.8	2.1	1.4	1.2	0.6

Mean square deflections at mid span of clamped beam, using equivalent linearization were obtained. The maximum discrepancy between results is less than 2%.

Numerical results for a square plate supported by two ribs and subjected to uniformly distributed over surface broad band stationary excitation have been calculated. The cut-off frequency ω_c is equal 1000 rad/s. The plate, simply supported on the sides without ribs and free on the others two sides, has the following properties: Young's modulus = $7.\times10^5$ MPa; Poisson's ratio= 0.3, density = 2.75×10^3 kg/m^3; dimensions a=b=0.4m, thickness h =$1.\times10^{-3}$ m; height of ribs = 0.5×10^{-2} m; width of ribs = 0.2×10^{-2} m. In this study, a 10×12 mesh of 24 DOF shell elements was used. Eight lowest natural frequencies and modes were obtained. They are listed in Table 2.

Table 2. Eight lowest natural frequencies and modes for plate, supported by two ribs.

Modes	1	2	3	4	5	6	7	8
Frequencies rad/s	174.5	335.4	481.5	527.7	757.5	803.6	974.5	1181.3

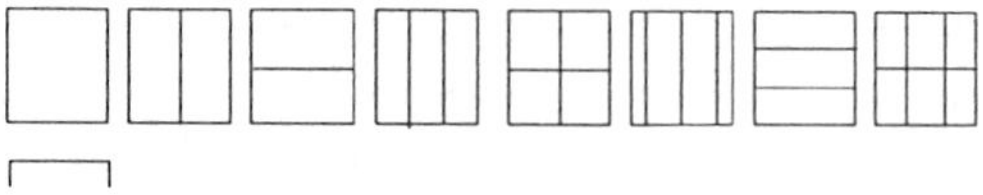

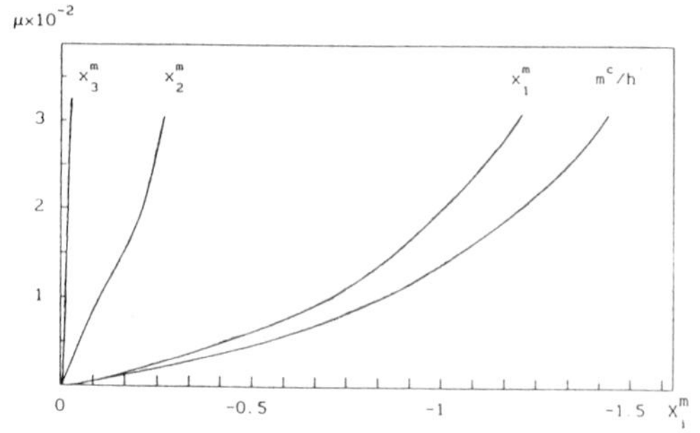

Fig.3 Mean values of generalized coordinates and mean value of deflection at the centre of the plate versus parameter μ.

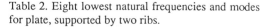

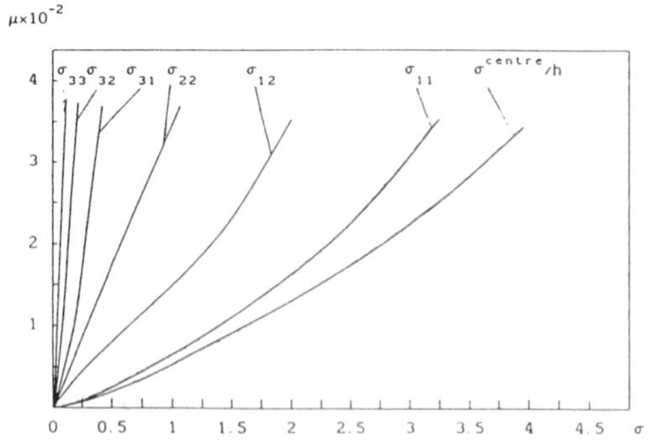

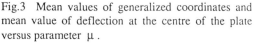

Fig.4 Standard deviation of deflection at the centre of the plate and coefficients σ_{ij} versus parameter μ.

The seven lowest natural frequencies are less than cut-off frequency of excitation. Therefore, and due to the symmetry of space configuration of the load only symmetrical natural modes (the 1 st, 4th and 7th, correspondingly) were taken into consideration in the process of creating the reduced dynamic model in the form of equation (2).

The components θ_{ij} were calculated by integrating the cross-spectral densities over frequency domain (up to ω_c) using Simpson's rule.

Number of points in the frequency domain varied from 800 to 2000 in order to obtain results with given accuracy. In figures 3 and 4 the mean values x_i^m and coefficients $\sigma_{ij} = \sqrt{|\theta_{ij}|}$ of generalized coordinates as well as mean value m^c and standard deviation σ^c of deflections at the centre of plate are shown as the functions of the parameter μ .

6 CONCLUSIONS

A modal superposition approach and statistical linearization technique have been presented in order to evaluate the mean values and second order statistics of a nonlinear finite element type system, subjected to stationary random loading. Based on results of the two first examples, the accuracy and efficiency of the proposed numerical procedure were verified. The proposed numerical technique was developed as the part of STADYS software [8].

ACKNOWLEDGEMENT

The first author are greatly indebted to Prof. Schuëller , University of Innsbruck, Austria for his encouragement and constant attention.

REFERENCES

[1] Schuëller ,G.I., Pradlwarter H.J. and Pandey M.D. Methods for Reliability Assessment of Nonlinear Systems under Stochastic Loading - a review, Proc. EURODYN' 93, Trondheim, Norway, June 1993, pp.751-759.

[2] Spanos, P-T.D. Stochastic Linearization in Structural Dynamics, Applied Mechanics Reviews, Vol.34, N1, pp.1-8, 1981.

[3] Busby, H.R.Jr and Weingarten, V.I. Response of Nonlinear Beam to Random Excitation, Journal of the Engineering Mechanics Division , Proc. ASCE, Vol.99, pp.55-68, 1973.

[4] Dekhtyaruk, E.S. and Lumelski, E.D. Numerical Forming of Nonlinear Dynamic Models of Plates and Panels, Strength of Materials and Theory of Structures, Kiev, Vol.45, pp.5-9, 1984 (in Russian).

[5] Bolotin, V.V. Random Vibrations of Elastic Systems, Martinus Nijhof Publishers, Hague , p. 468, 1984.

[6] Guliaev, V.I., Bazhenov, V.A. and Dekhtyaruk, E.S. Stability of Periodical Processes in Nonlinear Mechanical Systems, High school, Lvov, 1983 (in Russian).

[7] Spanos, P-T.D. Monte Carlo Simulations of Responses of Non-Symmetric Dynamic System to Random Excitations, Computers & Structures, Vol. 13, pp.371-376, 1981.

[8] Isakhanov, G.V., Dekhtyaruk, E.S., Melnik-Melnikov, P.G. and Sinyavski, A.L. Software for Random Vibration Analysis of Structures, p.115-119, Proc. Int. Symp. Strength of Materials and Structural Components at Sonic and Ultrasonic Loading Frequencies, Kiev, 1984, Naukova Dumka, 1986, (in Russian).

[9] Roberts, J.B. and Spanos, P.D. Random Vibration and Statistical Linearization. John Wiley & Sons, p.407, 1990.

[10] Schuëller, G.I., Pradlwarter, H.J. and Bucher C.G. Efficient Computational Procedures for Reliability Estimates of MDOF-systems. Int. J. Non-Linear Mechanics, Vol.26, N° 6, pp.961-974, 1991.

Structural Safety & Reliability, Schuëller, Shinozuka & Yao (eds) © 1994 Balkema, Rotterdam, ISBN 90 5410 357 4

Statistical linearization methods for prediction of extreme response of nonlinear structures

A. Naess
Faculty of Civil Engineering, The Norwegian Institute of Technology, Trondheim, Norway

ABSTRACT: The paper describes statistical linearization methods that are better suited for the purpose of predicting extreme responses than the standard mean square linearization method. Standard statistical linearization applied to a nonlinear dynamic system is based on a mean square deviation measure to derive the 'equivalent' linear system. Experience indicates that the obtained linear system may have approximately the same mean square response as the original nonlinear system. For predicting extreme response this procedure is not, however, equally suitable. In the case of strong nonlinearities in the dynamic model, application of the standard method of linearization may lead to substantial overestimation of extreme responses. It is shown by example studies that the proposed methods may offer significant improvement in the ability to predict extreme responses.

1 INTRODUCTION

The representation of an engineering structure by a linear and time invariant model for the purpose of dynamic analyses, is a widely adopted practice. Since the case of linear dynamic systems subjected to random loading in many respects can be analysed rather easily and satisfactorily, it is a fairly obvious and attractive idea to try to replace a given nonlinear system by an 'equivalent' linear one. However, it is, or should be, clearly understood that such a practice has its limitations. A substantial amount of research work has carried out on the response statistics of nonlinear dynamic systems over the past decades, and even if we are still just in the beginning of getting to grips with this vast field, a considerable body of knowledge is available. One of the easy conclusions that can be drawn from this knowledge is that linearization as such, cannot be expected to serve as a panacea for simplifying nonlinear systems. It is also quickly realized that the parameters of an 'equivalent' linear system must necessarily depend on the purpose of the analyses.

The criteria proposed for constructing an 'equivalent' linar system are invariably based on the minimization of some specific deviation measure. This technique was first developed for deterministic nonlinear problems (Bogoliubov and Mitropolski, 1963). Booton (1954), Kazakov (1954) and Caughey (1963) adapted this technique to apply to stochastic systems. Their procedure is now variously known as 'stochastic linearization', 'equivalent linearization' or 'statistical linearization'. The standard way of implementing this technique is to minimize a mean square measure of the difference between the nonlinear and the equivalent linear equations (Roberts and Spanos, 1990). This method will therefore be denoted by the MSSL method (MSSL - mean square stochastic linearization). General experience indicates that this way of carrying out a stochastic linearization leads to estimates of the response variance that is often fairly accurate. However, for the purpose of predicting the large response levels, expressed for example by exceedance probabilities of high levels, or extreme values, this method leads in many cases to overly conservative estimates. Hence for the purpose of design, the MSSL method is not adequate. Since the feature of replacing nonlinear equations of motion by linear equations is very attractive, it is appropriate to ask the question whether the MSSL method can be extended to provide a linearization procedure that yields better predictions of extreme responses.

Considering the fact that the MSSL method is based on minimizing a mean square deviation measure, it is not surprising that good agreement is often obtained for second order response moments, e. g. variance. In order to shift the emphasis to higher response levels, an obvious modification would be to look at mean n'th order deviation measures. It is, however, quickly realized that simply raising the power in the deviation measure used in the MSSL method is not a practical approach. For a large class of dynamic models, typical for offshore applications, it will be shown that an alternative deviation measure can be established that is equivalent to the one used in the

MSSL method. And this alternative measure does in fact provide for a practical extension to higher order deviation measures.

In previous papers, Naess et al. (1990, 1992) gave the results of some initial efforts to develop a stochastic linearization procedure specifically designed for making predictions of large responses. The purpose of the present paper is to complement and extend these results to gain further insight into the potential of the proposed linearization method.

2 NONLINEAR EQUATIONS OF MOTION

The following multi-degree-of-freedom (MDOF) equation of motion describing the (generalized) displacement response of a nonlinear structure with r degrees of freedom, will be assumed

$$\boldsymbol{M}\,\ddot{\boldsymbol{X}} + \boldsymbol{f}(\boldsymbol{X},\dot{\boldsymbol{X}}) = \boldsymbol{F}(t) \tag{1}$$

Here $\boldsymbol{M}$ denotes an $r \times r$ generalized mass matrix. $\boldsymbol{f}(\boldsymbol{X},\dot{\boldsymbol{X}})$ is an r-dimensional nonlinear vector function of (generalized) displacement and velocity. $\boldsymbol{F}(t)$ denotes the (generalized) excitation force vector, and $\boldsymbol{X}(t)$ the corresponding (generalized) displacement response vector.

In engineering applications, it is very often the case that the nonlinear function $\boldsymbol{f}(\boldsymbol{X},\dot{\boldsymbol{X}})$ can be written as a sum of a damping term, $\boldsymbol{g}(\dot{\boldsymbol{X}})$ say, and a restoring force term, $\boldsymbol{h}(\boldsymbol{X})$ say. Accordingly, in this paper the attention is limited to the particular case

$$\boldsymbol{f}(\boldsymbol{X},\dot{\boldsymbol{X}}) = \boldsymbol{g}(\dot{\boldsymbol{X}}) + \boldsymbol{h}(\boldsymbol{X}) \;. \tag{2}$$

It will be recognized from the subsequent developments that the methods described are particularly well adapted to deal with the case when the damping and restoring force terms can be represented by polynomial type expansions of the following kind, using the summation convention that repeated indices are summed from 1 to r.

$$g_i(\dot{X}) = g_{0i}^{j}\, sign(\dot{X}_j) + g_{1i}^{j}\dot{X}_j + g_{2i}^{jk}\dot{X}_j|\dot{X}_k| + \\ + g_{3i}^{jkl}\dot{X}_j\dot{X}_k\dot{X}_l + \ldots \tag{3}$$

and

$$h_i(X) = h_{s0}^{j} + h_{a0i}^{j}\, sign(X_j) + h_{a1i}^{j}X_j + h_{s1i}^{j}|X_j| + \\ + h_{s2i}^{jk}X_jX_k + h_{a2i}^{jk}X_j|X_k| + h_{a3i}^{jkl}X_jX_kX_l + \ldots \tag{4}$$

Here $sign(\cdot)$ denotes the sign function, $g_{..}^{...}$, $h_{s..}^{...}$ and $h_{a..}^{...}$ are suitable constants. For practical applications, these models should cover a large range of actual damping and restoring force behaviour. It is seen that $\boldsymbol{g}(\cdot)$ has been assumed to be an antisymmetric function, which is usually appropriate, i.e. $\boldsymbol{g}(-\dot{\boldsymbol{X}}) = -\boldsymbol{g}(\dot{\boldsymbol{X}})$. This property will be assumed throughout without specific mention.

As stated in the Introduction, one of the aims of the present work has been to develop rational methods for replacing equation (1) by the following linear, time-invariant dynamic model

$$\boldsymbol{M}\ddot{\boldsymbol{X}} + \boldsymbol{C}\dot{\boldsymbol{X}} + \boldsymbol{K}\boldsymbol{X} + \boldsymbol{d} = \boldsymbol{F}(t) \tag{5}$$

where $\boldsymbol{C}$ and $\boldsymbol{K}$ are real, constant $r \times r$ matrices and $\boldsymbol{d}$ is a constant r-dimensional vector. More specifically, the aim has been to determine the quantities $\boldsymbol{C}$, $\boldsymbol{K}$ and $\boldsymbol{d}$ so that the extreme responses predicted by using equation (5) are reasonably accurate estimates of the extreme responses obtained from equation (1).

3 STOCHASTIC LINEARIZATION

Let $\boldsymbol{F}(t)$ denote a stationary, random excitation force vector. If $E[\boldsymbol{F}(t)]$ is not assumed equal to zero, then the standard stochastic linearization procedure (Roberts and Spanos, 1990; Atalik and Utku, 1976) can be formulated as follows for the nonlinear equation of motion (1): 'Equivalent' quantities $\boldsymbol{C}_{(2)}$, $\boldsymbol{K}_{(2)}$ and $\boldsymbol{d}_{(2)}$ are chosen so as to minimize the mean square deviation measure

$$\varepsilon_{(2)} = E[\,\|\boldsymbol{f}(\boldsymbol{X},\dot{\boldsymbol{X}}) - \boldsymbol{C}_{(2)}\dot{\boldsymbol{X}} - \boldsymbol{K}_{(2)}\boldsymbol{X} - \boldsymbol{d}_{(2)}\|_2^2\,] \tag{6}$$

so that

$$E[\,\boldsymbol{f}(\boldsymbol{X},\dot{\boldsymbol{X}}) - \boldsymbol{C}_{(2)}\dot{\boldsymbol{X}} - \boldsymbol{K}_{(2)}\boldsymbol{X} - \boldsymbol{d}_{(2)}\,] = \boldsymbol{0} \;, \tag{7}$$

that is,

$$\boldsymbol{d}_{(2)} = \boldsymbol{d}(\boldsymbol{K}_{(2)}) = E[\,\boldsymbol{f}(\boldsymbol{X},\dot{\boldsymbol{X}})] - \boldsymbol{K}_{(2)}\,E[\boldsymbol{X}] \tag{8}$$

and under the assumption that $\boldsymbol{X}(t)$ is a stationary Gaussian vector process. Here the norm $\|\boldsymbol{x}\|_n = (\Sigma_{i=1}^{r}|x_i|^n)^{1/n}$, $\boldsymbol{x} = (x_1,\ldots,x_r)^T$ (n = positive number, T - transposition) has been used. It is obvious that by choosing the matrices $\boldsymbol{C}_{(2)}$ and $\boldsymbol{K}_{(2)}$ according to this criterion, the main emphasis is on minimizing the mean square response deviation, which is also shown by experience. Therefore, one cannot in general expect the standard stochastic linearization procedure to work when the goal is to obtain agreement on the level of large or extreme responses. To emphasize the use of a mean square deviation measure, this type of stochastic linearization will be denoted MSSL (mean square stochastic linearization).

The failure of the MSSL as a general tool for providing accurate estimates of the larger response levels, is clearly due to the fact that the minimization of the deviation measure $\varepsilon_{(2)}$ puts the emphasis on mean square deviation. To shift the emphasis to higher response levels with the same type of deviation measure as $\varepsilon_{(2)}$, it might seem natural to extend the MSSL and look at the deviation measures

$$\varepsilon_{(n)} = E[\|f(X,\dot{X}) - C_{(n)}\dot{X} - K_{(n)}X - d(K_{(n)})\|_n^n], \quad (9)$$

$$n = 3,4,...$$

However, this does not appear to be a practical approach since the procedure becomes quite involved for $n > 2$. Apart from that, there is also another shortcoming of this deviation measure in the sense that the same n-value applies to all components of the nonlinear function $f(\cdot,\cdot)$, which would clearly not be appropriate in general. Having assumed that $f(X,\dot{X}) = g(\dot{X}) + h(X)$, it turns out that a basis for more tractable extensions of the deviation measure $\varepsilon_{(2)}$ are obtained by observing that $\varepsilon_{(2)}$ is equivalent to the following deviation measure

$$\tilde{\varepsilon}_{(2)} = \varepsilon_1 + \varepsilon_2 \quad (10)$$

where

$$\varepsilon_1 = E[\|g(\dot{X}) - C_{(2)}\dot{X}\|_2^2], \quad (11)$$

and

$$\varepsilon_2 = E[\|h(X) - K_{(2)}X - d(K_{(2)})\|_2^2] \quad (12)$$

These comments point to an alternative and far more tractable approach than that offered by equation (9). By adopting the notation $|x|^n = \sum_{i=1}^{r} |x_i|^{n_i}$, where $n = (n_1,\ldots,n_r)$, $n_i = 2,3,4,\ldots$, the following deviation measures are now introduced

$$e_{(m,n)} = E[|g(\dot{X}) - C_{(m)}\dot{X}|^m] +$$

$$+ E[|h(X) - K_{(n)}X - d(K_{(n)})|^n], \quad m_i, n_i = 2,3,4 \quad (13)$$

where advantage has been taken of the possibility to use different powers in the two terms of the deviation measure, which would be appropriate in general.

The stochastic linearization procedure obtained by minimizing $e_{(m,n)}$ with respect to the matrices $C_{(m)} = (c_{ij}^{(m_i)})$ and $K_{(n)} = (k_{ij}^{(n_i)})$, will be denoted by ESL(m,n), (ESL - extended stochastic linearization). To simplify notation, $e_{(n)} = e_{(n,n)}$ and ESL(n) = ESL(n,n). It may be noted that, in general, the deviation measures $\varepsilon_{(n)}$ and $e_{(n,\ldots,n)}$ are not equivalent for $n > 2$. This, however, should not in itself cause undue concern since, a priori, there is no compelling reason to prefer $\varepsilon_{(n)}$ over $e_{(n,\ldots,n)}$, disregarding the fact that $\varepsilon_{(n)}$ seems to be too complicated for practical use.

In principle, the components of the exponent vectors m and n could be any real positive numbers, but for practical analysis it is convenient to restrict them to integer values. And, in fact, even odd integers are difficult to handle directly. Hence, from a practical, computational point of view, the candidates for extension are the even integers $m_i, n_i = 4, 6, \ldots$.

It is easily anticipated that the computational complexity of applying a specific deviation measure $e_{(m,n)}$ grows with increasing components of m and n. In this paper attention is limited to $m_i, n_i = 4$, 6 and 8, which seems to represent a range of values that would cover many applications. Admittedly, this statement is based on a qualified assessment of rather limited factual experience with the ESL procedures. Subsequent examples will serve to (partly) substantiate the statement.

Ideally, it would be desirable if the choice of the parameter vectors m,n in $e_{(m,n)}$, in each specific case could be determined on the basis of pertinent characteristics of the nonlinear problem at hand. It has not been within the reach of the present rather limited study to try to attain such a goal.

As for the case of the MSSL method, it is usually impractical to solve the problem of minimizing the right hand side of equation (13) directly. The reason for this is that there is a coupling between the optimum 'equivalent' parameters $c_{ij}^{(m_i)}$ and $k_{ij}^{(n_i)}$ and statistical moments of $X(t)$ and $\dot{X}(t)$. This difficulty can be avoided by using the following iteration procedure. Initial estimates of the necessary statistical moments of $X(t)$ and $\dot{X}(t)$ are provided by applying the standard MSSL method. These estimates are then used in the deviation measure $e_{(m,n)}$, which can be minimized to produce numerical estimates of the parameters $X(t)$ and $\dot{X}(t)$. With the resulting $C_{(m)}$ and $K_{(n)}$ substituted into equation (5), new estimates of the statistical moments of $X(t)$ and $\dot{X}(t)$ can be provided. Thus an iteration loop is established that provides numerical estimates of $C_{(m)}$ and $K_{(n)}$ by minimizing $e_{(m,n)}$ and then uses equation (5) to produce new estimates of the necessary statistical moments of $X(t)$ and $\dot{X}(t)$. When convergence is obtained, the desired numerical estimates of $C_{(m)}$ and $K_{(n)}$ are established. So far, the convergence properties of this procedure has not been fully tested. A discussion of the existence and uniqueness of an equivalent linear system would also be appropriate, cf. Roberts and Spanos (1990) or Spanos and Iwan (1978).

So far it has been tacitly assumed that each excitation force has been of a broad-band type. In the general situation, assuming that the excitation forces are Gaussian and that a standard MSSL can be performed in a mean-

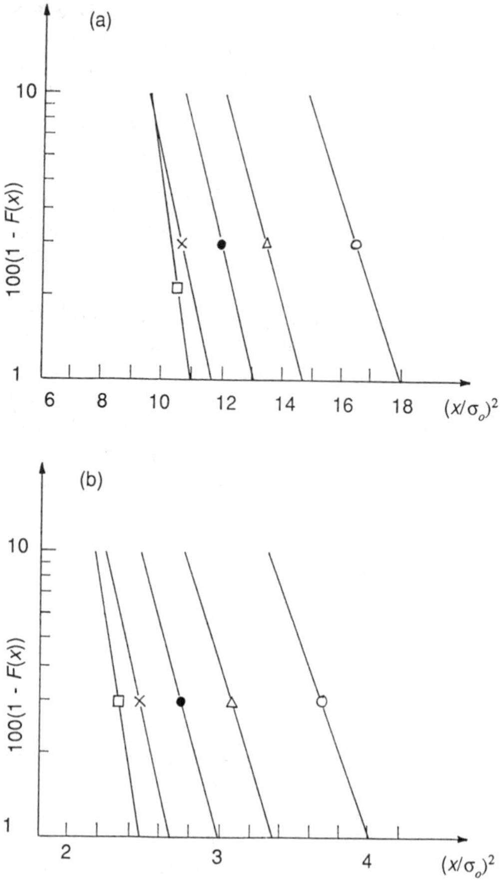

Figure 1(a)-(b) Exceedance probability of displacement response $X(t)$ during a three hour period for Example 1. (a): $\sigma_o = 1$. (b): $\sigma_o = 10$. Symbol key: Exact: □, MSSL: ○, ESL(4): Δ, ESL(6): ○, ESL(8): ×

ingful way, it is proposed to associate to each specific ESL$(\boldsymbol{m},\boldsymbol{n})$-method a set of shifted excitation force processes. Each $F_i(t)$ is replaced by a Gaussian process $F_i^{(n_i)}(t)$ with the same mean value as $F_i(t)$, but with a shifted (one-sided) spectral density defined as follows, $S_{F_i^{(n_i)}}(\omega) = = S_{F_i}(\omega - \omega_{(n_i)} + \omega_{(2)})$. This will ensure a proper balance between the different linearized systems.

4 NUMERICAL EXAMPLES

To illustrate the proposed linearization method, we shall choose a class of simple, yet relevant, single-degree-of-freedom nonlinear oscillators subjected to Gaussian white noise excitation for which the response statistics can be calculated in a straight-forward manner. The displacement response $X(t)$ of the oscillator is determined by the normalized equation of motion

$$\ddot{X} + \beta\dot{X} + h(X) = F(t) \qquad (14)$$

where $h(X)$ denotes a nonlinear restoring force term, and $F(t)$ denotes a zero-mean Gaussian white noise normalized load process with a one-sided spectral density $S_F(\omega) = S_o$. The corresponding linearized equation of motion is written as ($\omega_{(n)}^2 = k_{(n)}$)

$$\ddot{X} + \beta\dot{X} + \omega_{(n)}^2 X = F(t) \qquad (15)$$

where $n = 2,4,6$ or 8 according to whether MSSL, ESL(4), ESL(6) or ESL(8) is used to determine the equivalent parameters.

To study the performance of the three linearization procedures for prediction of large responses, a comparison is made between the extreme value distributions and exceedance probabilities obtained by these methods and the corresponding quantities given by the exact solution. Since the primary purpose of the subsequent examples is comparison of results, the extreme value distribution of a stationary process $X(t)$ is assumed, for simplicity, to be given as (Lin, 1967)

$$F(x) = Prob\{M(T) \le x\} = \exp\{-\nu^+(x)T\} \qquad (16)$$

Here $M(T) = \max\{X(t);\ 0 \le t \le T\}$ is the largest value of $X(t)$ during a time interval of length T. $\nu^+(x)$ denotes the mean x-upcrossing rate of $X(t)$. A level x_p with an exceedance probability p during time T is defined by

$$F(x_p) = 1 - p \qquad (17)$$

In the subsequent examples, the function $h(\cdot)$ will be given specific, simple shapes for which the upcrossing rate of $X(t)$ can be easily calculated (Lin, 1967) . The notation

$$\sigma_o^2 = \frac{\pi S_o}{2\beta\omega_o^2} \qquad (18)$$

is also introduced. σ_o can be construed as the standard deviation of a reference (normalized) displacement response. ω_o is a specified frequency. The time duration T is chosen to be 3 hours, and the interval of exceedance probabilities p adopted here is 0.01 - 0.1, which is a rather typical range.

In the first three examples $\omega_o = 1$, and the displacement can be normalized so that $h(1) = 1$. Invariably, the extreme value distributions only depend on x and σ_o, specifically $F(x) = F(x/\sigma_o;\sigma_o)$.

Example 1 In this example, $h(x) = x|x|$. It is obtained that $\omega_{(n)}^2 = a_{(n)}\sigma_{(n)}^2$, where (Lin, 1967) $\sigma_{(n)}^2 =$

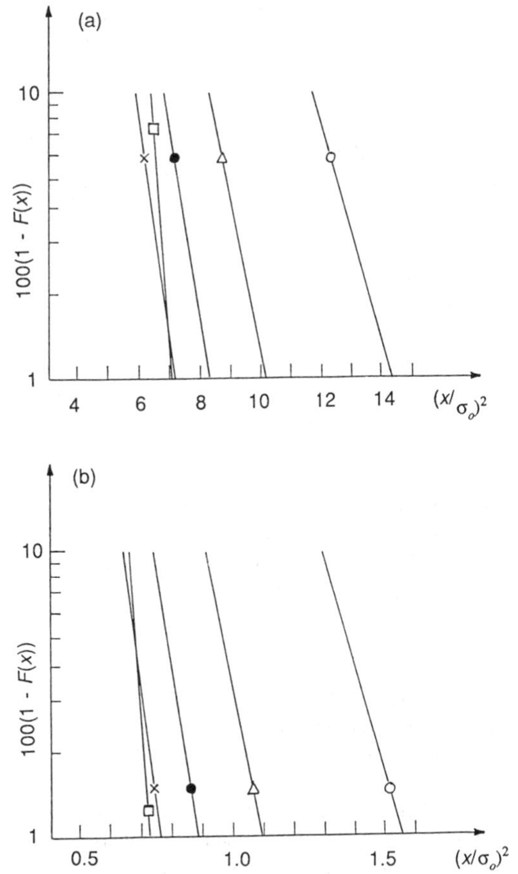

Figure 2(a)-(b) Exceedance probability of displacement response $X(t)$ during a three hour period for Example 2. (a): σ_o = 1. (b): σ_o = 10. Key as in Fig. 1.

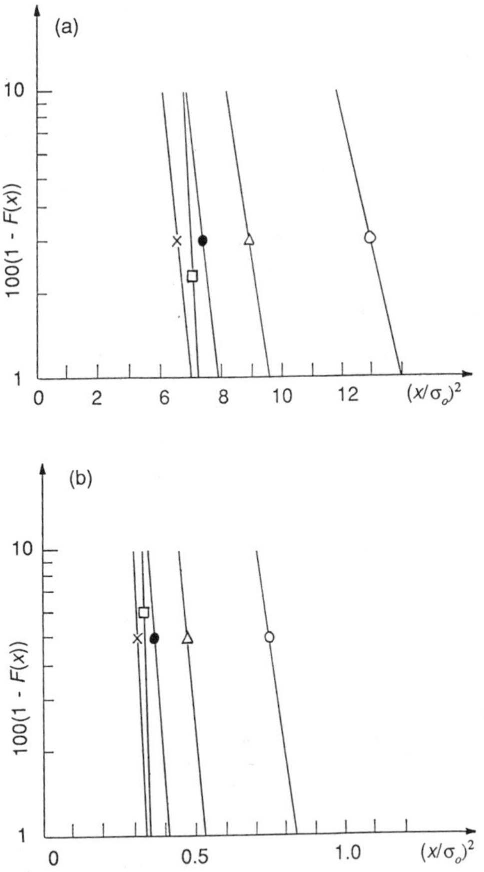

Figure 3(a)-(b) Exceedance probability of displacement response $X(t)$ during a three hour period for Example 3. (a): σ_o = 1. (b): σ_o = 10. Key as in Fig. 1.

$(\pi S_o)/(2\beta\omega_{(n)}^2)$ and $a_{(2)} = 1.60$, $a_{(4)} = 2.21$, $a_{(6)} = 2.68$ and $a_{(8)} = 3.10$. This gives $\omega_{(n)} = b_{(n)}\sigma_o^{1/3}$, where $b_{(2)} = 1.17$, $b_{(4)} = 1.30$, $b_{(6)} = 1.39$ and $b_{(8)} = 1.46$. For the purpose of illustration, the obtained exceedance probabilities for the two cases $\sigma_o = 1$ and $\sigma_o = 10$ have been given in Fig. 1.

Example 2 Here, $h(x) = x^3$, and it is found that $\omega_{(n)} = b_{(n)}\sigma_o^{1/2}$, where $b_{(2)} = 1.32$, $b_{(4)} = 1.58$, $b_{(6)} = 1.76$ and $b_{(8)} = 1.89$.. The obtained exceedance probabilities for the two cases $\sigma_o = 1$ and $\sigma_o = 10$ have been given in Fig. 2.

Example 3 Here, $h(x) = x^3|x|$, and it is obtained that $\omega_{(n)} = b_{(n)}\sigma_o^{3/5}$, where $b_{(2)} = 1.45$, $b_{(4)} = 1.84$, $b_{(6)} = 2.10$ and $b_{(8)} = 2.30$. The obtained exceedance probabilities for the two cases $\sigma_o = 1$ and $\sigma_o = 10$ have been given in Fig. 3.

The last example deals with the classical case of a white noise excited Duffing oscillator.

Example 4 Let $h(x) = \omega_o^2 x(1 + \alpha x^2)$, where α is a positive constant. The linearization process then leads to the results $\omega_{(n)}^2 = \omega_o^2(1 + a_{(n)}\alpha\sigma_{(n)}^2)$, where $a_{(2)} = 3$, $a_{(4)} = 6.27$, $a_{(6)} = 9.54$ and $a_{(8)} = 12.82$. Note the almost completely linear dependence of $a_{(n)}$ as a function of n. In this example $\omega_o = 0.6$. The extreme value distributions then depend on only x, σ_o and α, specifically $F(x) = F(x/\sigma_o; \alpha\sigma_o^2)$. It is therefore convenient to introduce the parameter $\alpha^* = \alpha\sigma_o^2$. Results for α^* = 0.05 and 0.5 are given in Fig. 4, cf. Naess et al. (1992).

5 COMMENTS AND CONCLUSIONS

The paper describes a proposed method for stochastic

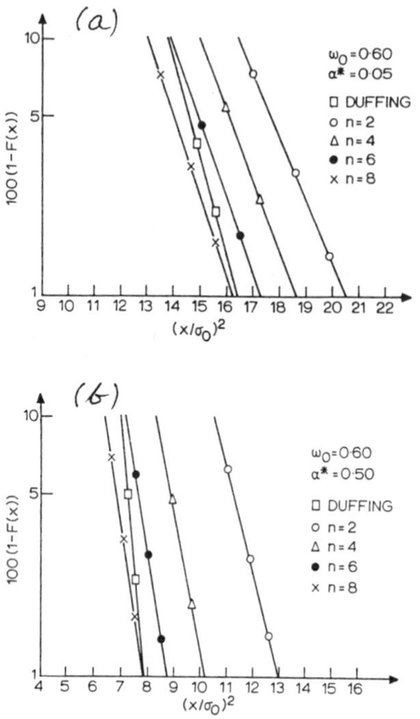

Figure 4(a)-(b) Exceedance probability of displacement response $X(t)$ of a Duffing oscillator during a three hour period, cf. Example 4. (a): $\omega_o = 0.6$, $\alpha^* = \alpha\sigma_o^2 = 0.05$. (b): $\omega_o = 0.6$, $\alpha^* = \alpha\sigma_o^2 = 0.5$. Key as in Fig. 1.

linearization that aims at prediction of extreme responses. The goal has been to establish a useful tool for design purposes. The standard method of stochastic linearization (MSSL), which is based on a mean square deviation measure, very often is not appropriate for the prediction of large responses. However, due to the appealing properties of a stochastic linearization procedure, it would clearly be of considerable practical value if such a procedure could be found that would allow reasonably accurate prediction of extreme responses.

Although the conclusions that can be drawn from the investigations presented here, are not definitive, the example studies presented show that the proposed procedures of extended stochastic linearization (ESL(***m,n***), m_i, n_i = 2,4,6,...) can lead to a substantial improvement in extreme response predictions compared to the standard MSSL method. It should also be noted that over the range of nonlinear models studied, the performance of the ESL is surprisingly stable in the sense that in all cases the same ESL method would give fairly accurate estimates at a given exceedance probability level. If this is true over a sufficiently broad range of nonlinear models, the ESL methods would provide a very useful practical tool. Further studies to map out the properties of the ESL methods will be pursued. Work is also in progress to apply the ESL methods to fatigue damage estimation problems.

ACKNOWLEDGEMENTS: The author gratefully acknowledges the assistance of Mr. Nils O. Esborg in carrying out some of the numerical calculations presented in this paper.

REFERENCES

Atalik, T.S. and Utku, S. (1976), Stochastic Linearization of Multi-Degree-of-Freedom Non-Linear Systems. *Earthquake Engineering and Structural Dynamics,* 4, 411-420.

Bogoliubov, N. and Mitropolsky, A. (1963), *Asymptotic Methods in the Theory of Nonlinear Oscillations,* 2. edition, Gordon and Breach, New York.

Booton, R.C. (1954), Non-Linear Control Systems with Random Inputs, *IRE Trans Circuit Theory,* CT-1, pp. 9-18.

Caughey, T.K. (1963), Equivalent Linearization Techni ques, *J Acoust Soc Am,* Vol. 35, pp. 1706-1711.

Caughey, T.K. (1964), On the Response of a Class Non linear Oscillators to Stochastic Excitation, *Proc Coll Int du Centre Nat de la Recherche Scient,* No 148, Marseille, France, pp. 393-402.

Caughey, T.K. (1986), On the Response of Non-Linear Oscillators to Stochastic Excitation, *Probabilistic Engineering Mechanics,* Vol. 1, No.1, pp. 2-4.

Kazakov. I.E. (1954), Approximate method for the stati stical analysis of nonlinear systems. *Trudy VVIA,* No. 394.

Lin, Y.K. (1967), *Probabilistic Theory of Structural Dynamics.* McGraw-Hill, New York.

Naess, A., Galeazzi, F. and Dogliani, M. (1990), Stochastic Linearization Method for Prediction of Extreme Response of Offshore Structures. *Proc. 1st European Offshore Mechanics Symposium (EUROMS '90),* Trondheim, Norway.

Naess, A., Galeazzi, F. and Dogliani, M. (1992), Extreme response predictions of nonlinear compliant offshore structures by stochastic linearization. *Applied Ocean Research,* Vol.14, No. 2, pp. 71-81.

Roberts, J.B. and Spanos, P.D. (1990), *Random Vibration and Statistical Linearization.* John Wiley & Sons.

Spanos, P.D. and Iwan, W. D. (1978), On the Existence and Uniqueness of Solutions Generated by Equivalent Linearization. *Int. Journal of Non-Linear Mechanics,* Vol. 13, pp. 71-78.

Structural Safety & Reliability, Schuëller, Shinozuka & Yao (eds) © 1994 Balkema, Rotterdam, ISBN 90 5410 357 4

First-passage study of a hysteretic system via quasi-conservative stochastic averaging method

M. Noori, M. Dimentberg, Z. Hou, R. Christodoulidou & A. Alexandrou
Mechanical Engineering Department, Worcester Polytechnic Institute, Mass., USA

ABSTRACT: The quasi-conservative stochastic averaging (QCSA) method is applied to a Bouc-Wen-Baber hysteretic system (BWB) under Gaussian white noise excitations. The stationary probability density of the system's response amplitude is obtained for different excitation levels and different damping ratios and are compared with the studies presented by Cai and Lin (1985). The first-passage time problem for the hysteretic system is then studied using the method of QCSA. The results for the expected time to failure as a function of the initial total energy are shown, for different excitation levels and for different threshold values of amplitude. The relationship between the expected time to failure when initial total energy is zero and excitation level is presented for a wide range of hysteresis shape parameters.

1 INTRODUCTION

Random vibration of nonlinear hysteretic systems has been the topic of extensive study (see Baber and Noori (1986), Baber and Wen (1981), Lutes (1970), Noori et al (1986), Roberts and Spanos (1990), and Wen (1989)). However, to the best of the authors' knowledge only a limited work has been reported on the prediction of higher order moments for the nonstationary response analysis of hysteretic systems (see Davoodi and Noori (1990), Lin and Wu (1984), Liu (1990), Wu and Lin (1989)). These work do not seem to be promising for reliability analysis since they result in negative probability (Davoodi and Noori (1990)). The most recent work has been the extension of the method of energy dissipation balancing to compute the probability density of the response of a BWB hysteretic system by Cai and Lin (1988) and Lin and Cai (1988). Another important problem in the random vibration analysis of structural systems in general is the first passage problem as addressed by Ariaratnam (1973), Crandall (1966) and Grossmayer (1977). However, only a few studies have been presented for first passage study of hysteretic systems (Spencer and Bergman (1985)). Herein, the QCSA by Dimentberg (1988) is extended to study the stationary response analysis of a BWB hysteretic system. Results are compared with those by Cai and Lin (1988). Furthermore, the first passage problem of the same hysteresis is examined using QCSA for prescribed values of the threshold levels of amplitude and the expected time to reach the threshold value are obtained.

2 FORMULATION OF THE PROBLEM

The nonlinear system to be studied herein is shown in Figure 1. The governing equation of motion is given as

$$\ddot{X} + 2\varsigma\omega_0\dot{X} + F(t, X, \dot{X}) = \Xi(t) \qquad (1)$$

where $\Xi(t)$ is a stationary Gaussian white noise, ς and ω_0 are the linear damping ratio and natural frequency, X, $\dot{X}$ and $\ddot{X}$ are displacement, velocity and acceleration responses respectively, and F(.) is the restoring

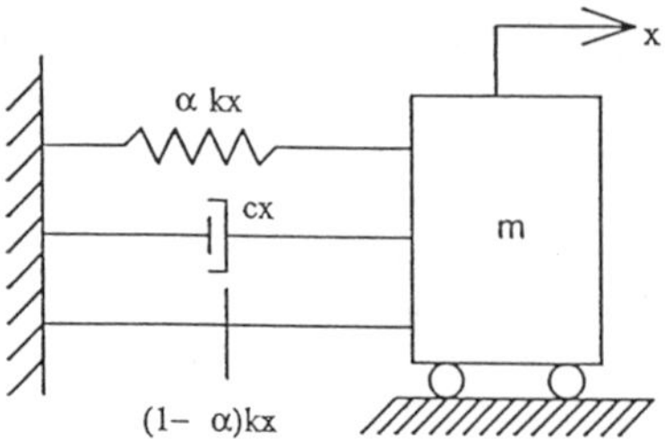

Figure 1 - BWB SDOF System under study.

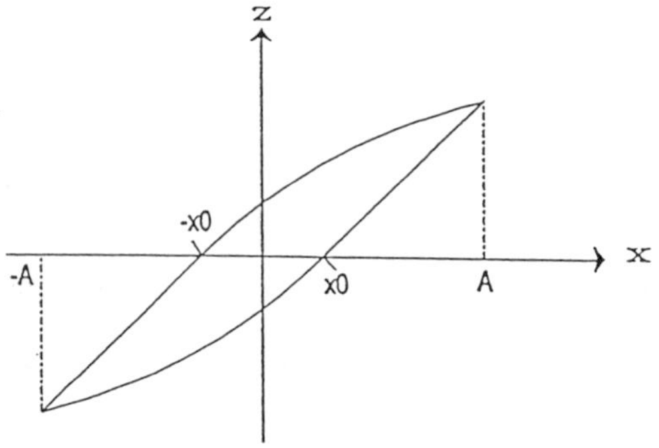

Figure 2 - Hysteretic component of restoring force vs. deformation. Special case of the BWB model with n=1, b=1, $\gamma=\beta=0.5$, $\alpha=1/21$.

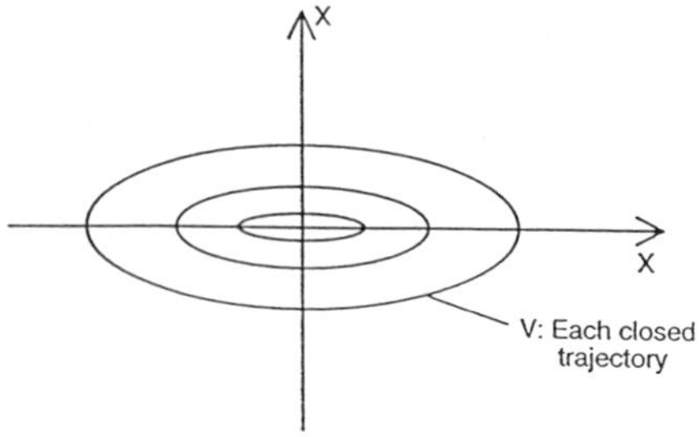

Figure 3 - Phase trajectories.

force as given by

$$F(t,X,\dot{X}) = \alpha\omega_0^2 X + (1-\alpha)\,\omega_0^2 Z(X,\dot{X}) \quad (2)$$

which is a combination of a linear and a hysteretic restoring force components, and $0<\alpha<1$ is the ratio of post-yield to pre-yield stiffness in the BWB hysteresis part. The non-dimensionalized version of the equation of motion can be written as

$$\ddot{x} + 2\varsigma\,\dot{x} + f(\tau,x,\dot{x}) = \xi(\tau) \quad (3)$$

$$f(\tau,x,x) = \alpha x + (1-\alpha)\, z(x,\dot{x}) \quad (4)$$

where $x = X/\Delta$ and $\dot{x} = dx/d\tau$ are the non-dimensional displacement and velocity respectively, $z = Z/\Delta$ is the non-dimensional hysteretic restoring force as shown in Figure 2, $\tau = \omega_0 t$, $\xi(\tau)=\Xi(\tau/\omega_0)/(\omega_0^2\,\Delta)$ and $f(\tau) = F(\tau/\omega_0)/(\omega_0^2\,\Delta)$ where Δ is a characteristic displacement. The hysteretic component of the restoring force for stationary analysis is given by the form described in Cai and Lin (1988). It should be noted that the hysteretic force is assumed to be small as compared with the inertia and mean stiffness of the system.

The method of QCSA is introduced in the following manner (see also Zhu and Lin (1974)). Let's consider a system with general form of nonlinearity for the restoring force and with the governing equation of motion

$$\ddot{x} + \varepsilon(x,\dot{x}) + g(x) = \omega_0^2\, x\xi(t) + \zeta(t) \quad (5)$$

where $\xi(t)$ is a parametric and $\zeta(t)$ is an external stationary zero-mean Gaussian White Noise, with intensity factors D_ξ and D_ζ respectively. These two excitations are uncorrelated as discussed by Dimentberg (1988). The parameter ω_0 is the natural frequency of small-amplitude free oscillation $\omega_0^2 = g'(0)$ when $g(x)$ is a linear non-hysteretic part of the restoring force. The method of QCSA by Dimentberg (1988) is an approximation scheme used to derive stationary probability density and to study the first passage problem. To do so, let's consider a new variable $V(t)$, the total energy of the system, for the system described by equation (5), using the following description

$$V(t) = \frac{\dot{x}^2}{2} + G(x), \quad G(x) = \oint g(x')dx' \quad (6)$$

By applying Ito stochastic differential rule we obtain the following equation:

$$\dot{V}(t) = \dot{x}(\ddot{x} + g(x))$$
$$= \dot{x}[-\varepsilon(x,\dot{x}) + \omega_0 2x\xi(t) + \zeta(t)] \quad (7)$$

The total energy, given by equations (6) & (7), with a small parameter in the right hand side, is averaged over the natural period $T_0(V)$, defined by equation of the corresponding conservative system, obtained from equation (5) with $\varepsilon = 0$, and $D_\xi = D_\zeta = 0$. Assuming that this equation has a set of solutions with closed loop trajectories in the phase plane $x\,\dot{x}$ as shown in Figure 3. The integration with respect to time, implied by the averaging method, is replaced by integration with respect to state variable along these closed paths. While V is constant within each cycle. After averaging the "drift" and "diffusion" coefficients of V(t) the following shortened equation is obtained. Details of derivation are discussed by Christodoulidou (1992), and Dimentberg (1988).

$$\dot{V} = \varepsilon \frac{[g(V)+G(V)]}{T_0(V)} + \sqrt{\varepsilon}\,\frac{\sigma_*(V)}{\sqrt{T_0(V)}}\xi(t) \qquad (8)$$

where $T_0(V)$, $\sigma(V)$, G(V), and g(V) are described by Christodoulidou (1992). The stationary probability can be obtained from the corresponding FPK equation as follows (Christodoulidou (1992))

$$P(V) = CT_0(V)\exp[\int_0^V \frac{g(V')-\Phi(V')}{\sigma_*^2(V)}dV'] \quad (9)$$

where C is a normalization constant and remaining terms are as explained in Christodoulidou (1992). Equation (9) is the stationary solution which holds for any general form of nonlinear restoring force. In this case $\Phi(V) = 0$ (Christodoulidou (1992)). Also, g(0) = dG(0)/dx, where g(x)>0 for every x>0, g(x) = restoring force <0 for every x<0 and $g(x) \rightarrow \pm\infty$ when $x \rightarrow \pm\infty$; g(x) is the hysteretic restoring force and G(x) = is the potential energy. Derivations for the total energy V of the quasi-conservative system, for the area within the hysteresis loop, $\bar{H}(V)$, which is equal to the dissipated hysteretic energy, and the procedure for calculating x_0 for particular values of A are presented by Christodoulidou (1992). It should be noted that in this work there is no parametric excitation and thus, parameter D (with no subscript) is used to represent the intensity of

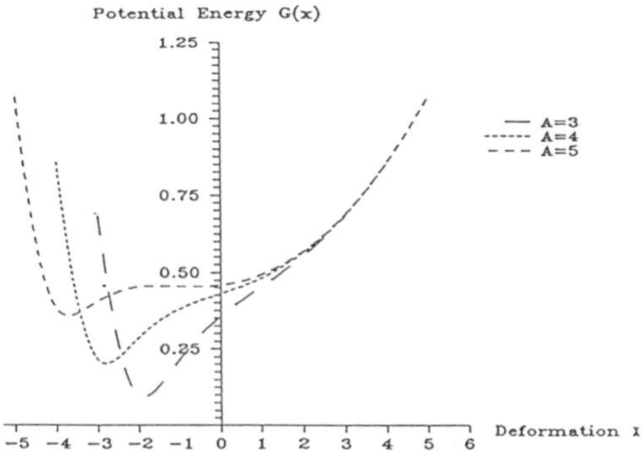

Figure 4 - Potential Energy vs. Deformation for amplitudes 3 to 5. Note that these curves agree with those obtained by Cai and Lin.

external excitation. The value of $\bar{H}(V)$ is as follows

$$\bar{H}(V) = -(A - x_0)^2 + (2\frac{x_0}{\gamma}) \qquad (10)$$

and the potential energy is given by

$$G(x) = \begin{cases} G_-(x) = (1/2)\alpha\, x^2 + [\,(1-\alpha)/2](x + x_0)^2, & -A \le x \le -x_0, \quad \gamma = \beta \\ G_+(x) = (1/2)\alpha x^2 + [(1-\alpha)/(8\gamma^2). \; [1 - e^{-2\gamma(x+x_0)}]^2, & -x_0 \le x \le A, \quad \gamma = \beta \end{cases} \qquad (11)$$

Plots of potential energy are presented in Figure 4 for $\gamma = \beta = 0.5$, $\alpha = 1/21$ and different values of amplitude for increasing x. The method of QCSA is only valid if the area within the hysteresis loop is much smaller than the total energy V, where $V = G_+(A) = G_-(-A)$ (Christodoulidou (1992)). The above equation (11) can be used to estimate the stationary probability densities for the amplitude A for different excitation levels and for different system parameters along with the following equations:

$$p(V) = p_s(x,\dot{x}) = \exp\{-(\frac{2\varsigma}{\pi K})V - \int_0^V \frac{\bar{H}(V')}{2\pi K\int_{-a'}^{a'}\sqrt{2[V'-G(x)]}dx}dV'\} \qquad (12)$$

and

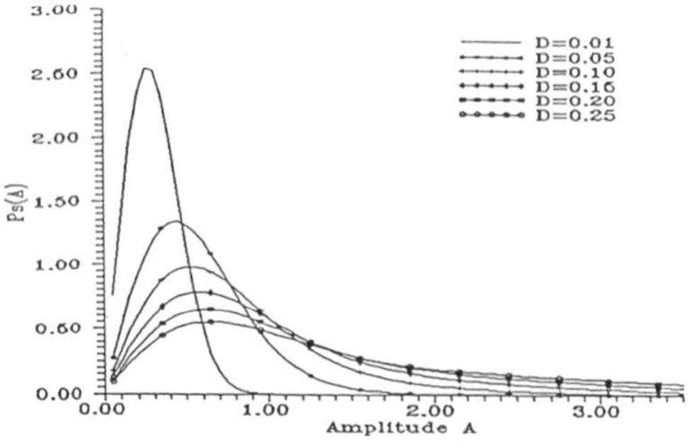

Figure 5a - Stationary probability density of response amplitude of a smooth hysteretic system with α=1/21, γ= β=0.5, and damping ratio ζ=0; weak to intermediate excitations D=0.01 to 0.25.

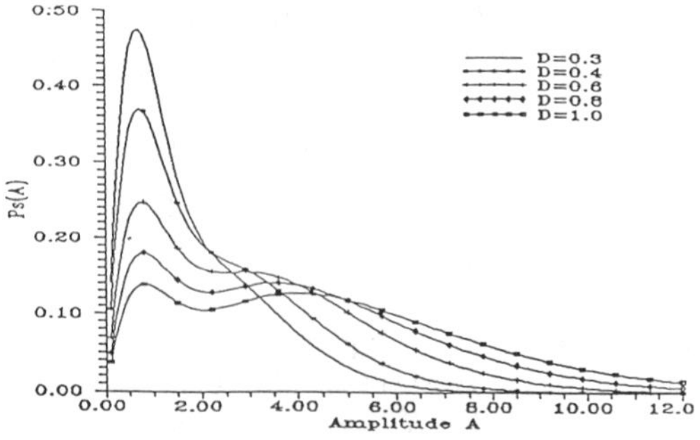

Figure 5b - Stationary probability density of response amplitude of a smooth hysteretic system with α=1/21, γ= β=0.5, and damping ratio ζ=0; intermediate to strong excitations D=0.3 to 0.1.

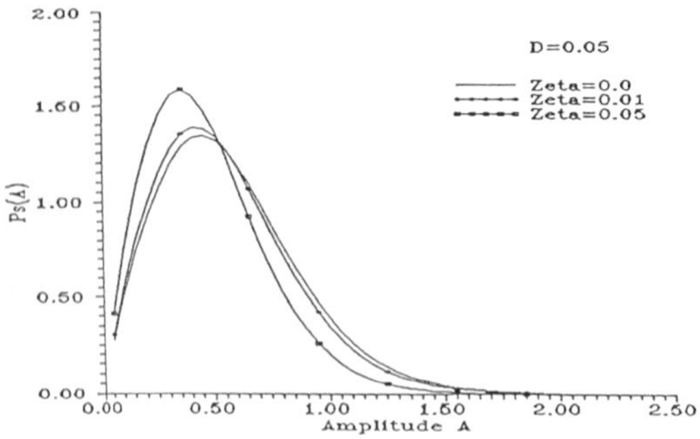

Figure 6a - Stationary probability density of response amplitude of a smooth hysteretic system with α=1/21, γ= β=0.5, and different values of damping ratios; excitation level D=0.05.

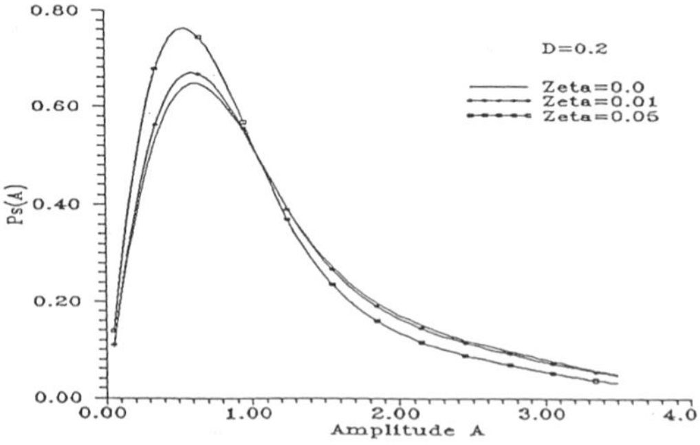

Figure 6b - Stationary probability density of response amplitude of a smooth hysteretic system with α=1/21, γ= β=0.5, and different values of damping ratios; excitation level D=0.2

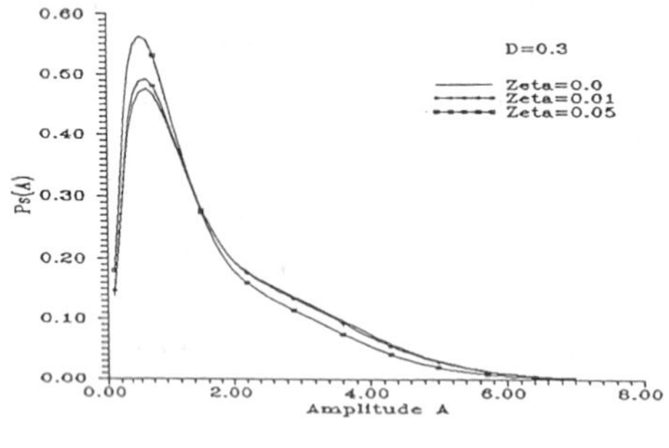

Figure 7a - Stationary probability density of response amplitude of a smooth hysteretic system with α=1/21, γ= β=0.5, and different values of damping ratios; excitation level D=0.3.

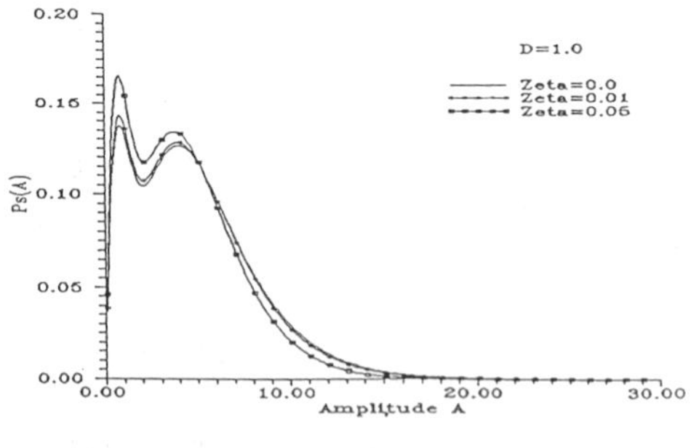

Figure 7b - Stationary probability density of response amplitude of a smooth hysteretic system with α=1/21, γ= β=0.5, and different values of damping ratios; excitation level D=0.1.

$$p_S(A)= C\,|\,g(A)\,|\,\rho[V(A)]\int_{-A}^{A}\frac{dx}{\sqrt{2[V-G(x)]}} \quad (13)$$

where $p_S(x,\dot{x})$ and $P_S(A)$ denote an approximate stationary probability density for the hysteretic system and the stationary probability density of the amplitude A. K is the spectral constant with $2\pi K = D$ being the external excitation strength. Note that a' and $-a'$ are the roots of the equation G(x) = V, g(A) is the restoring force in the system, linear and hysteretic combined, for the value of x = A (Christodoulidou (1992)). C is a normalization constant evaluated elsewhere (Christodoulidou (1992)). Equations (12) and (13) are solved numerically and the results are presented in the Numerical Studies section.

Numerous approximation procedures have been suggested for first passage analysis of linear and memoryless nonlinear systems by Grossmayer (1977) and Spencer and Bergman (1985). The problem for moments of a first-passage time is generally governed by a partial differential equation with the number of arguments being equal to the number of state variables. This may result in numerical difficulties as in the work by Spencer and Bergman (1985), or in the necessity to use MCS (Ariaratnam (1973)), whereas by using averaging-over-the-period the number of state variables is reduced to only one (amplitude for the linear system and total energy for the non-linear one). Other work on the estimation of first passage time have been reported by Ariaratnam (1973), Crandall (1966) and Vanmarke (1974). Spencer and Bergman (1985) developed a method to determine statistical moments of time to first passage incorporating the modified Bouc hysteresis model using a Petrov-Galerkin finite element method. The results obtained from the finite element method were compared to results obtained from Monte Carlo Simulation. In the present study, the QCSA introduced above will be used for the first-passage problem, in order to obtain probabilistic predictions.

Let V(t) be the total energy at any time t. Let the system failure be defined as the event that V(t) reaches the value V*, which is a critical state. Let T(V) be the expected time that V(t) will reach the critical state for the first time and let $V(t=0)=V_0$. Therefore, the probability that the system will not fail within the interval $(0,\tau]$ is a function of τ, V_0 and V* as follows:

$$P(\tau,V_0,V^*)= P[V(t)<V^*,0<t<\tau \,|\, V(0)=V_0]` \quad (14)$$

In order to have a better understanding of the subject being studied in this section, consider first the scalar diffusion Markov process x(t), with drift and diffusion coefficients a(x) and b(x), respectively. The transition probability density p(x,t;y,s) of x(t), considered as a function of the initial values y,s, satisfies the backward Kolmogorov equation and depends only on the difference between t and s. Let the interval (x_2,x_1) be the admissible states domain $0<V<V^*$. Consider a probability P(x,t) where x(t) does not reach either x_1 or x_2 during time interval (0,t] provided that initially $x_2<x<x_1$. Therefore, the probability is

$$P(x,t) = \int_{x2}^{x1} p'(x',x,t)dx' \quad (15)$$

where p'(x',x,t) is a conditional probability density of the value of x(t), at time instant t, denoted by x', provided that this process never reaches either x_1 or x_2, and x is the value of x(t) at the initial time s, which is set to be zero. Note that this probability density is not the same as p(x',x,t), which is the transition probability density, due to the condition for no arrivals of x(t) at either of the boundary points x_1 and x_2. Thus, p'(x',x,t) should satisfy both the FPK and the backward Kolmogorov equations, the same way as the transition probability density does for a system without boundaries, for time instants before the first of such arrivals. The difference between the two probability densities manifests itself only after the first arrival at the boundary (x_1 or x_2) and is reflected in different boundary conditions. The probability density p'(x',x,t) is equal to zero at $x=x_1$ and $x=x_2$, as the process x(t) reaches these boundaries. Using the backward Kolmogorov equation we can obtain the following differential equation (Christodoulidou (1992):

$$\frac{\partial P}{\partial t} = a(x)\frac{\partial P}{\partial x} + \frac{1}{2}b(x)\frac{\partial^2 P}{\partial x^2} \quad (16)$$

The above equation must satisfy initial

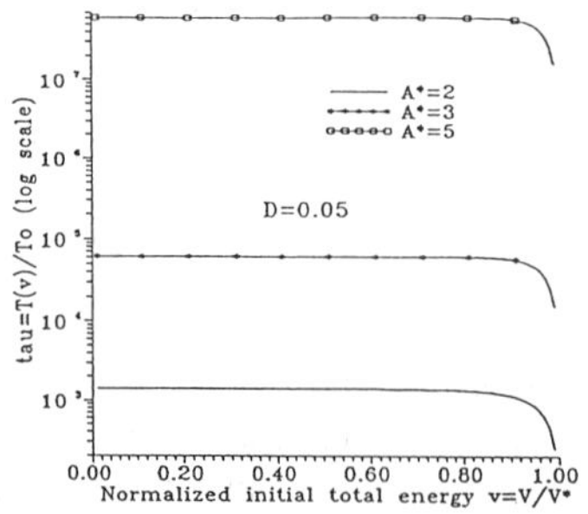

Figure 8a - First-passage time problem: comparison between different critical values of amplitude·

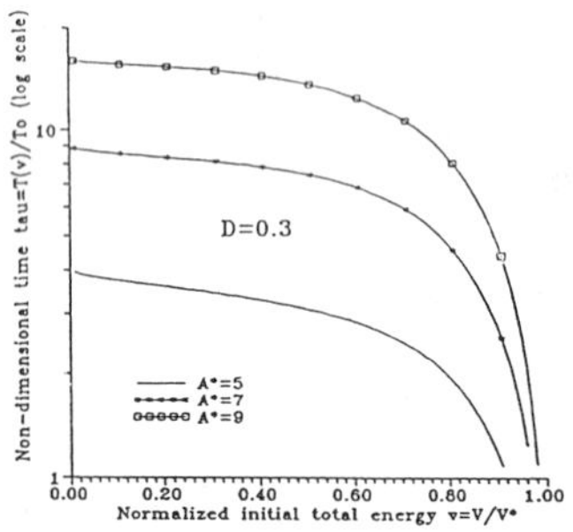

Figure 8b - First-passage time problem: comparison between different critical values of amplitude·

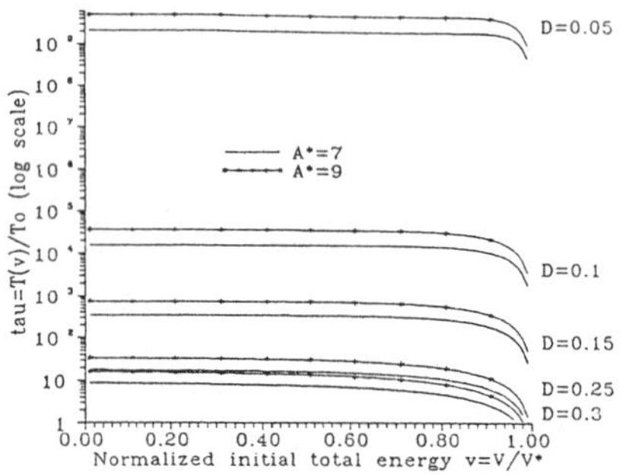

Figure 9 - First-passage time problem: comparison between different excitation levels for different critical values of amplitudes.

condition P(0,x)=1, since $x_2<x<x_1$ and the boundary conditions $P(t,x_2)=P(t,x_1)=0$. Note that if the system has reached the boundary, the probability of the event is equal to 1. Equation (16) must also satisfy the probability P'(t,x)=1-P(t,x) for at least one arrival of x(t) at the boundary of the interval (x_2, x_1), assuming that for t=0, $x_2<x<x_1$. The probability P'(t,x) has the following initial and boundary conditions; P'(0,x) =0, and $P'(t,x_2)=P'(t,x_1)=1$. Consider then the random time for the system to reach the boundary, which may be denoted by Θ. The cumulative probability function of Θ is by definition P'(Θ,x). Furthermore, a differential equation for the first-order moment T=<Θ> of Θ, the expected time to reach a given boundary, is derived. Hence,

$$T(x) = \int_0^\infty t \frac{\partial P'(t,x)}{\partial t} dt \tag{17}$$

Differentiating equation (17) with respect to t for P'(t,x), multiplying by t and integrating over t from zero to infinity, and using the initial condition P'(0,x)=0 along with the condition P'(∞,x)=1, and equation (17), we arrive (Christodoulidou 1992)) to the following differential equation for T(x), the expected time for the response to reach the boundary:

$$\frac{b(x)}{2}\frac{d^2 T}{d x^2} + a(x)\frac{dT}{dx} = -1 \tag{18}$$

Equations (17) and (18) may be applied to the specific case of the BWB hysteresis model used for analysis in this report. All the necessary steps associated with this case of first-passage problem, a detailed description of the steps to evaluate the boundary conditions used to solve the differential equation (18), and coefficients a(x) and b(x) are reported by Christodoulidou (1992) . In order to make the analysis simpler, V is normalized by dividing V with V* and introducing v=V/V* as the new variable of integration. T, the time it takes the amplitude to reach the threshold (critical) value when the system would fail, is simplified by non-dimensionalization. The expected T is divided by the natural period $\omega_0=2\pi$ and $\tau=T/2\pi$ is introduced as the non-dimensional time.

3 NUMERICAL STUDIES

Numerical studies presented herein were conducted to: 1) obtain probability distributions of certain response variables for the hysteretic system under study using the method of QCSA; 2) study the first passage problem of the BWB model and demonstrate the versatility and application of the QCSA method for such application. Figure 5 (a) illustrates the probability densities. For lower values of D, the probability distribution is closer to the Rayleigh distribution, which indicates that the linear component of the restoring force is dominant. Figure 5 (b) illustrates the probability densities for stronger excitation. Contribution of both linear and hysteretic components of the restoring force is evident, giving rise to two peaks in the amplitude probability density. Figures 6 (a)-(b) and 7 (a)-(b) demonstrate as the linear damping ratio ζ increases, the peak of the probability distribution is higher and shifted to the left, and the tail of the distribution is shortened. The general shape of the distribution is not changed as the damping ratio changes. The distribution is closer to Rayleigh for lower values of D and becomes bi-modal as the excitation level increases. These results agree qualitatively with those in(Cai and Lin (1988). Figures 8-9 present the first-passage time as a function of normalized initial total energy for two levels of excitation and a wide range of system critical amplitudes (a measure of critical yielding) for ζ=0. For a low level of excitation, D=0.05 , as the critical amplitude increases, the first-passage time increases, see Figure 8(a). This figure illustrates a type of "boundary-layer effect" which is explained mathematically by a small factor D multiplied by the highest-order derivative and can be physically explained as follows: in the case of a small excitation level and a large corresponding T, the stationary probability density will be established before the system reaches the barrier almost everywhere except at the immediate vicinity of the barrier. As Figure 8(b) indicates, under a higher level of excitation, where T is smaller, the above effect is less pronounced. Moreover, in contrast to Figure 8 (a), as the energy is dissipated, the time required to reach first passage decreases. This is because the system is deteriorating faster under a higher level of excitation. Figure 9 shows the first-passage time as a function of total energy for a wider range of excitation levels and for a higher value of critical amplitude threshold. Figure 10 shows the

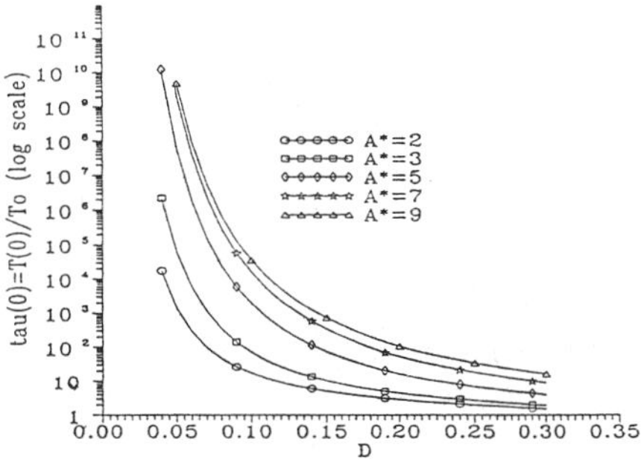

Figure 10 - First-passage time problem: expected time at v=0 vs. excitation levels for different critical values of amplitude.

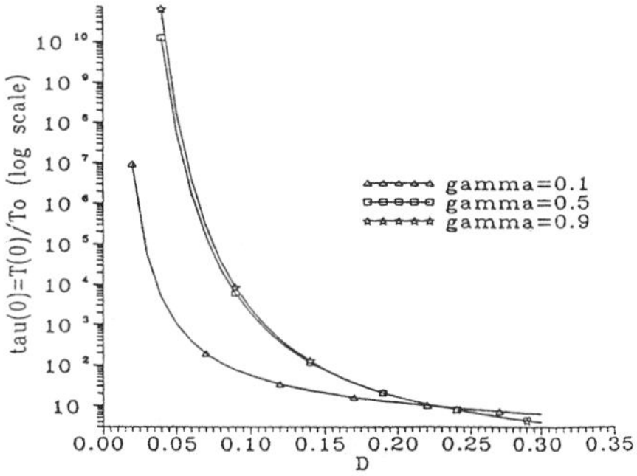

Figure 11 - First-passage time problem: expected time at v=0 vs. excitation levels for different values of γ.

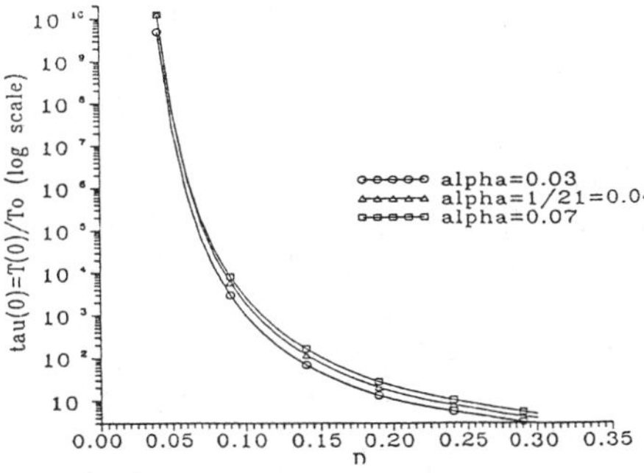

Figure 12 - First-passage time problem: expected time at v=0 vs. excitation levels for different values of α.

expected time to first crossing when initial total energy is zero as a function of excitation level and for a wide range of yield levels A* (evaluated in (Christodoulidou 1992)), α=1/21 and γ=0.5. As these plots indicate, the initial time to failure increases with an increase in the level of critical yielding. Figure 11 shows the effect of the shape parameter γ on the]relationship between the initial time to failure and the excitation strength, with A*=5 and α =1/21. Similarly, Figure 12 illustrates the relationship between the initial time and the excitation level for varying post-yield/pre-yield ratios α, γ=0.5 and A*=5. It is seen that as α increases, the expected time to reach failure when initial time is zero, increases. It should be noted that, as demonstrated in this work, by using the QCSA method the formulation of the problem is reduced to a single stochastic differential equation which makes it easier for first passage problem application.

REFERENCES

Baber, T.T., and Noori, M.N. 1986. Modeling general hysteresis behavior and random vibration application. *ASME, J. of Vibration, Acoustics, Stress and Reliability in Design*, 108: 411-420.

Baber, T.T. and Wen, Y.K. 1981. Random vibration of hysteretic degrading systems. *ASCE, J. of the Eng. Mech.* 107: 1069-1087.

Lutes, L.D. 1970. Approximate technique for treating random vibration of hysteretic systems. *J. of the Acous. Soc. of America*. 48:1:299-306.

Noori, M.N., Choi, J.D. and Davoodi, H. 1986. Zero and nonzero mean random vibration analysis of a new general hysteresis model. *Probabilistic Engineering Mechanics*. 1: 4: 192 201.

Roberts, J.B. and Spanos, P.D.1990. *Random vibration and statistical linearization*. England: John Wiley & Sons Ltd.

Wen, Y.K. 1989. Methods for random vibration for inelastic structures. *Applied Mechanics Review*. 42: 2: 39-52.

Davoodi, H. and Noori, M. 1990. Extension of an ito-based general approximation technique for random vibration of a BWB general hysteresis model, part II: Non-Gaussian analysis. *Journal of Sound and Vibration*. 140: 2: 319-339.

Ibrahim, R.A. 1985. *Parametric Random Vibration*, John Wiley & Sons, New York.

Lin, Y.K. and Wu, W.F. 1984. Applications of cumulant closure to random vibration problems. T.C.Huang and P.D. Spanos (eds), *Proceedings of the Symposium of Random Vibrations, ASME Winter Annual Meeting, New Orleans, AMD. 65.*

Liu, Q. 1990. *Response analysis of a hysteretic system under random excitations*. Mechanical Engineering Department, University of New Brunswick, Canada.

Wu, W.F. and Lin, Y.K. 1984. Cumulant-neglect closure for nonlinear oscillators under random parametric and external excitations. *International Journal of Nonlinear Mechanics*. 19: 349-362.

Cai, G.Q. and Lin, Y.K. 1988. On exact stationary solutions of equivalent non-linear stochastic systems. *International Journal of Nonlinear Mechanics*. 23: 315-325.

Cai, G.Q. and Lin, Y.K. 1988 . A new solution technique for randomly excited hysteretic structures. *Tech. Report on Research Grant No. NCEER-88-0012*, State University of New York, Buffalo, NY.

Lin, Y.K. and Cai, G.Q. 1988. Exact stationary response solution for second order nonlinear systems under parametric and external white noise excitations: part II. *ASME, Journal of Applied Mechanics*. 55: 702-705.

Ariaratnam, S.T. and Pi, H.N. 1973. On the first-passage time for envelope crossing for a linear oscillator. *International Journal of Control*. 18: 1: 89-96.

Crandall, S.H., Chandiramani, K.L. and Cook, R.G.1966. Some first-passage problems in random vibration. *ASME, Journal of Applied Mechanics*. 33: E3:532-538.

Grossmayer, R. 1977. A seismic reliability and first passage failure. *CISM Courses and Lectures No. 225: Random Excitation of Structures by Earthquakes and Atmospheric Turbulence*. Springer-Verlag, New York. 110-200.

Spencer, B.F., Jr. and Bergman, L.A. 1985. On the reliability of a simple hysteretic system. *ASCE, Journal of the Engineering Mechanics Division*. 111: 12: 1502-1514.

Dimentberg, M.F. 1988. *Statistical Dynamics of Nonlinear and Time-Varying Systems*, Research Studies Press, Taunton, U.K.

Christodoulidou, R. 1992. *Stationary response and first passage study of BWB hysteresis model using quasi-conservative stochastic averaging method*. M.S. Thesis, Mechanical Engineering Department, Worcester Polytechnic Institute.

Vanmarke, E.H., and Veneziano, D. 1974. Probabilistic seismic response of simple inelastic systems. *Proceedings of the Fifth World Conference on Earthquake Engineering*, Rome, Italy. 2851-2863.

Zhu, W.Q. and Lin, Y.K. 1991. Stochastic averaging of energy envelope. *ASCE, Journal of the Engineering Mechanics Division.* 117: 8

Structural Safety & Reliability, Schuëller, Shinozuka & Yao (eds) © 1994 Balkema, Rotterdam, ISBN 90 5410 357 4

Spectral analysis and synthesis of non-Gaussian processes

Roland H. Peinelt
D. Swarovski, Wattens, Austria (Formerly: Institute of Engineering Mechanics, University of Innsbruck, Austria)
Christian G. Bucher
Institute of Engineering Mechanics, University of Innsbruck, Austria

ABSTRACT: A concept for analysis and synthesis of Non-Gaussian processes based on the estimation of higher order spectra is presented. The approach utilizes extensions of well established frequeny-domain methods such as power spectral analysis. In this paper, such a frequency-domain based approach first analyzes higher order spectra of random processes and subsequently utilizes this information for the digital generation of non-Gaussian processes.
Especially for the step-by-step analysis of nonlinear systems it is necessary to obtain sample functions of the loading processes. Such digitally sampled realizations of non-Gaussian processes may then be utilized advantageously for reliability analyses.
Several numerical examples show the application of a respective software package. Results compare Gaussian and non-Gaussian models with respect to spectral contents and probability densities.

1 INTRODUCTION

Loading processes encountered in engineering analyses are frequently assumed to be Gaussian. While this assumption is certainly justifiable in many cases, it is only an approximation for loads caused by e.g. wave forces, earthquake loads, or wind effects etc. For these loads a non-Gaussian model appears to reflect reality in an appropriate way.
The computational methods for analyzing and synthesizing non-Gaussian processes are by far less developed than those for their Gaussian counterparts. There is an apparent gap between state of knowledge and state of computational capability. This gap is bridged by recent approaches which extend spectral analysis beyond the power spectrum to higher order spectra, i.e. the bispectrum and the trispectrum.
The paper presents such a frequency-domain based approach which first analyzes higher order spectra of random processes and subsequently utilizes this information for the digital generation of non-Gaussian processes by matching bi- and/or trispectra.

2 MOMENT FUNCTIONS AND HIGHER ORDER SPECTRA

2.1 Theoretical Concepts

The stochastic process $x(t)$ under consideration is regarded to be stationary and ergodic at least up to the fourth order. Under these assumptions the moment functions can be defined as follows. The autocorrelation function $R_{XX}(\tau)$ is given by:

$$R_{XX}(\tau) = E[X(t)X(t+\tau)] \quad (1)$$

The so-called bicorrelation $R_{XXX}(\tau_1, \tau_2)$ (the third order moment function) is

$$R_{XXX}(\tau_1, \tau_2) = E[X(t)X(t+\tau_1)X(t+\tau_2)] \quad (2)$$

and the tricorrelation $R_{XXXX}(\tau_1,\tau_2,\tau_3)$ is defined as

$$R_{XXXX}(\tau_1,\tau_2,\tau_3) = \mathrm{E}\left[X(t)X(t+\tau_1)X(t+\tau_2)X(t+\tau_3)\right] \tag{3}$$

Higher order spectra, i.e. the bi-spectrum S_{XXX} and the tri-spectrum S_{XXXX} can be defined as multidimensional Fourier transforms of the higher order moment functions (analogous to the Wiener-Khintchine-relations) or, alternatively, based on convolution in the frequency domain. This yields the following representation, shown here for S_{XXX}

$$S_{XXX}(\omega_1,\omega_2) = \lim_{T\to\infty} \frac{2\pi}{T} E\left[X(\omega_1,T)X(\omega_2,T)X^*(\omega_1+\omega_2,T)\right] \tag{4}$$

$$X(\omega,T) = \frac{1}{2\pi}\int_{-T/2}^{T/2} x(t)e^{-i\omega t}dt \tag{5}$$

In the above equations (.)* denotes complex conjugates.
From these definitions it is obvious that higher order spectra are a measure for the statistical dependence between the harmonic components (see e.g. Hasselmann et al., 1963).

2.2 Numerical Evaluation

For the numerical estimation of higher order spectra, experience shows it is advantageous from the computational point of view to use the direct method as described below.
Given n samples of the stochastic process each consisting of N points $x_k^{(j)}$, $k=0,...N-1$, $j=1,n$ the discrete Fourier transform $\tilde{X}(\omega_m,T)$ is calculated for each sample. This yields n higher order periodograms, again shown here for the bi-spectrum:

$$\tilde{S}^{(j)}_{XXX}(\omega_{1,m},\omega_{2,n}) = \frac{2\pi}{T} \left(\tilde{X}^{(j)}(\omega_{1,m},T)\tilde{X}^{(j)}(\omega_{2,n},T)\tilde{X}^{(j)*}(\omega_{1,m}+\omega_{2,n},T)\right) \tag{6}$$

$$\tilde{X}^{(j)}(\omega_m,T) = \frac{T}{2\pi}\frac{1}{N}\sum_{i=0}^{N-1} x_k{}^{(j)}e^{-i\omega_m t_k} \tag{7}$$

where T denotes the length of the samples, the x_k are the values of the sample function at time $t_k = k\Delta t$ with $\Delta t = T/N$ and ω_m, $m=1, N/2$ are the Fourier frequencies. To reduce the variance of this estimates one may average over the n samples. An overview about bispectral estimation and corresponding literature is given e.g. by Subba Rao and Gabr, 1984; Nikias and Raghuveer, 1987. Some observations on tri-spectra are given e.g. by Schuëller and Bucher, 1988.

3 SYNTHESIS OF NON-GAUSSIAN PROCESSES

3.1 Linear Systems under non-Gaussian white noise

Herein non-Gaussian white noise $\zeta(t)$ is defined as a stationary process with

$$E[\zeta(t)] = 0; \quad E[\zeta(t)\zeta(t+\tau)] = R_{\zeta\zeta}(\tau) = 2\pi S_0\delta(t);$$

$$E[\zeta(t)\zeta(t+\tau_1)\zeta(t+\tau_2)] = R_{\zeta\zeta\zeta}(\tau_1,\tau_2) = (2\pi)^2 D_0\delta(\tau_1)\delta(\tau_2) \tag{8}$$

where $\delta(.)$ denotes the Dirac-Delta function. Its power spectrum and bispectrum are both flat, i.e.:

$$S_{\zeta\zeta}(\omega) = S_0; \qquad S_{\zeta\zeta\zeta}(\omega_1,\omega_2) = D_0 \tag{9}$$

If a linear filter, defined by its frequency response function $H(\omega)$ is excited by a non-Gaussian process $\zeta(t)$, the higher order spectra of its output process $y(t)$ are defined by

$$S_{YY}(\omega) = |H(\omega)|^2 S_0;$$

$$S_{YYY}(\omega_1,\omega_2) = D_0 H(\omega_1)H(\omega_2)H^*(\omega_1+\omega_2) \tag{10}$$

Note that by analyzing the power spectrum (and fitting a linear filter to the data) only the magnitude $|H(\omega)|$ can be recovered by assuming that the observed process is the output of a linear system under (not necessarily white) noise. The complete transfer function of a linear system is given by

$$H(\omega) = |H(\omega)| e^{i\Theta(\omega)} \quad (11)$$

The phase estimation can be performed by means of the bispectrum (Matsuoka and Ulrych, 1984).

$$S_{YYY}(\omega_1,\omega_2) = D_0 |H(\omega_1)| \, |H(\omega_2)| \, |H^*(\omega_1+\omega_2)| \, e^{i(\Theta(\omega_1)+\Theta(\omega_2)-\Theta(\omega_1+\omega_2))} \quad (12)$$

Using such systems processes of given power spectrum and given third order statistical moment can be generated. Of course it cannot be expected that many physical processes can be described by such simple systems because the possibility of fitting higher order spectra is very limited once the power spectrum is fixed.

3.2 Nonlinear Static Transformation of a Gaussian Process

A relatively simple method of approximating the probability density function of a stochastic process consists in the static (memoryless) transformation of a Gaussian process (see e.g. Grigoriu, 1984; Winterstein 1985). In a first step the PDF of the process $y(t)$ is estimated. Then a static transformation $x = g^{-1}(y)$ has to be found which transforms the non-Gaussian process $y(t)$ into a Gaussian process $x(t)$. Applying this transformation to the realizations of $y(t)$, corresponding realizations of $x(t)$ can be obtained. Since $x(t)$ is assumed to be a Gaussian linear process, well-known methods for analysis and simulation can be applied. One disadvantage of this method is the fact that the power spectrum of the simulated process does not necessarily need to coincide with the measured one. Another, perhaps minor, disadvantage is that the power spectrum cannot be calculated analytically in all cases. This is the reason why frequently a transformation in terms of an algebraic polynomial $g(x)$ is used:

$$g(x) = \sum_{k=1}^{M} c_k x^{k-1} \quad (13)$$

In this case, the power spectrum S_{YY} can be expressed analytically (Ammon 1990). Details can be found in this reference as well as (Peinelt, 1992).
This method aims at two goals, namely the approximation of the PDF and the power spectrum of the process. The shape of the bispectrum is completely neglected.

3.3 Volterra Functional Models

A considerable number of processes encountered in mechanical systems can be represented as Volterra series. These series have been described as "power series with memory" relating the output of a nonlinear system to the "powers" of the input (Bedrosian and Rice 1971, Bendat 1990). The output $y(t)$ of such systems can be written as

$$y(t) = \sum_{n=1}^{\infty} \int_{-\infty}^{\infty} \dots \int_{-\infty}^{\infty} h_n(\tau_1,\dots,\tau_n) \cdot \cdot (t-\tau_1) \dots \cdot (t-\tau_n) d\tau_1 \dots d\tau_n \quad (14)$$

$x(t)$ is a Gaussian process and the convolution kernels $h(\cdot)$ describe the system. An extensive discussion of this topic can be found in Schetzen, 1980. The first kernel $h_1(\tau)$ is the simple impulse response function of a linear system. The n-fold Fourier transforms of these kernels

$$H_n(\omega_1,\dots\omega_n) = \underbrace{\int_{-\infty}^{\infty} \int_{-\infty}^{\infty}}_{n-fold} h_n(\tau_1,\dots\tau_n) \, e^{-i(\omega_1\tau_1+\dots\omega_n\tau_n)} d\tau_1 \dots d\tau_n \quad (15)$$

are called n-th order transfer functions. These systems can be viewed as direct extension of linear systems theory. An important property of such systems is that the higher order spectra

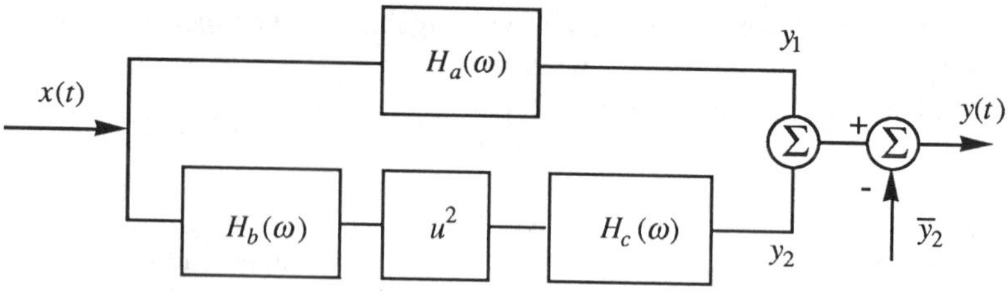

Fig. 1: System with Quadratic Nonlinearity

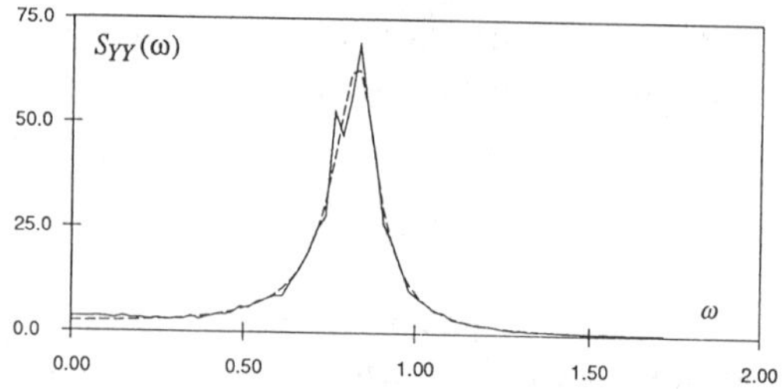

Fig. 2: Power Spectrum
— Simulation ------ Linear Model

can be expressed analytically for Gaussian input. For the case of simple, weak nonlinearities it is even possible to choose the parameters of the nonlinear system such that both power spectral density and higher order spectra, respectively, fit to the measured process. For a specific nonlinear system as shown in Fig. 1 the second order transfer function $H_2(\omega_1, \omega_2)$ becomes

$$H_2(\omega_1,\omega_2) = H_b(\omega_1)H_b(\omega_2)H_c(\omega_1+\omega_2) \qquad (16)$$

The power spectrum $S_{YY}(\omega)$ and the bispectrum $S_{YYY}(\omega_1, \omega_2)$ become

$$S_{YY}(\omega) = |H_a(\omega)|^2 S_{XX}(\omega) + 2|H_c(\omega)|^4 \int_{-\infty}^{\infty} |H_b(\Omega)|^2$$

$$S_{XX}(\Omega)|H_b(\omega-\Omega)|^2 S_{XX}(\omega-\Omega)d\Omega \qquad (17)$$

$$S_{YYY}(\omega_1,\omega_2) = 2\{S_{XX}(\omega_1)S_{XX}(\omega_2)H_a(\omega_1)H_a(\omega_2)$$

$$H_b^*(\omega_2)H_c(\omega_1+\omega_2)$$

$$+S_{XX}(\omega_1+\omega_2)S_{XX}(\omega_2)H_a^*(\omega_1+\omega_2)$$

$$H_a(\omega_2)H_b(\omega_1+\omega_2)H_b^*(\omega_2)H_c(\omega_1)$$

$$+S_{XX}(\omega_1+\omega_2)S_{XX}(\omega_1)H_a^*(\omega_1+\omega_2)$$

$$H_a(\omega_1)H_b(\omega_1+\omega_2)H_b^*(\omega_1)H_c(\omega_2)\} \qquad (18)$$

Here $x(t)$ is a weak Gaussian excitation so that terms of higher order could be neglected. Details are shown in (Peinelt, 1992).

4. NUMERICAL EXAMPLE

A nonlinear system described by the following equations is considered

$$\ddot{y} + 2\beta\dot{y} + g(y) = x(t) \qquad (19)$$

$$g(y) = \begin{cases} y & y \geq 0 \\ c \cdot y & y \leq 0; c = 0.5 \end{cases} \qquad (20)$$

which corresponds to a SDOF oscillator with nonlinear spring. $x(t)$ is assumed to be Gaussian white noise. Based on 100 digitally simulated realizations ($\Delta t = 0.5$, $N = 1024$) of $y(t)$ the power spectrum $S_{YY}(\omega)$ and the bispectrum $S_{YYY}(\omega_1,\omega_2)$ are estimated. The results are shown below. Fig. 2 also contains the power spectrum of a fitted Gaussian process as dashed line.

The simulated time series were modeled by a simple Volterra functional model as sketched in Fig. 1. Additionally, the individual linear systems were chosen to be proportional, i.e. $H_a(\omega) = H_b(\omega) = -1/c_2 H_c(\omega)$. The transfer function $H_a(\omega)$ is chosen as

Table 1: Parameter values for optimized model

	a_1	b_3	b_2	b_1	c_2
Linear Model	1.553	1.445	0.2347	1.	-
Bilinear Model	1.069	1.	0.1649	0.6908	0.0134

$$H(\omega) = \frac{a_1}{-b_3\omega^2 + b_2 i\omega + b_1} \quad (21)$$

The free parameters a_1, b_1, b_2, b_3 and c_2 were chosen to minimize the error functional

$$J = F\sum_{i=1}^{NP}\left[S_{XX}(\omega_i) - \tilde{S}_{XX}(\omega_i)\right]^2 + \sum_{i=1}^{NP}\left[\left|S_{XXX}(\omega_i,\omega_i) - \tilde{S}_{XXX}(\omega_i,\omega_i)\right|^2\right] + \sum_{i=1}^{NP}\left[\mathrm{Re}\{S_{XXX}(\omega_i,0)\} - \mathrm{Re}\{\tilde{S}_{XXX}(\omega_i,0)\}\right]^2 \equiv \min \quad (22)$$

in which F is chosen to put equal weights on both power spectrum and bispectrum. Note that for reasons of computational efficiency the bispectrum only along the lines $\omega_1 = \omega_2$ and $\omega_2 = 0$ is utilized. The resulting parameter values are given in Table 1 .

Fig. 4 shows the bispectrum of the bilinear system. It can be seen that the bilinear model quite closely reproduces the simulated data (Fig. 3).

In addition, the probability density function of simulation, linear model (Gaussian) and bilinear model are compared in Fig. 5

The bilinear model represents the asymmetry of the PDF very accurately.

5 CONCLUSIONS

The example shows in quantitative terms that a non-Gaussian random process clearly is not sufficiently well described by its power spectrum. Consequently several methods for simulation of non-Gaussian processes were discussed. Their respective merits and de-merits can be summarized as follows.

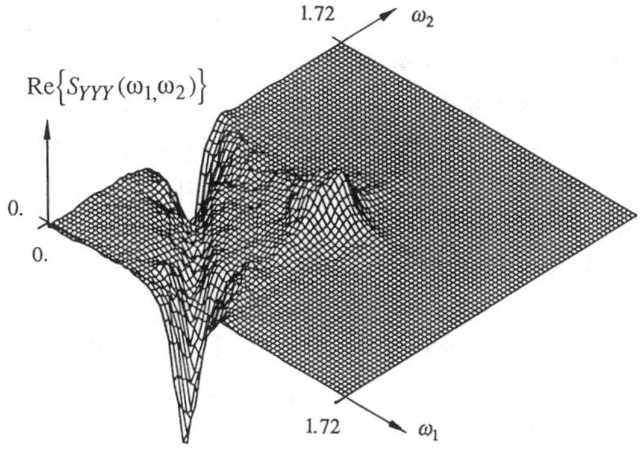

Fig. 3: Estimate for Bispectrum from 100 samples (Real Part)

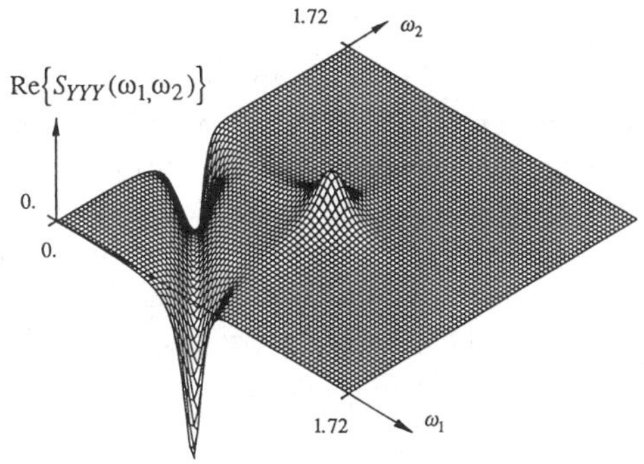

Fig. 4: Bispectrum of bilinear model (Real Part)

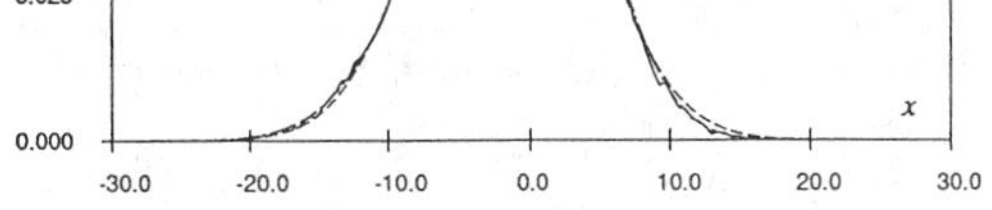

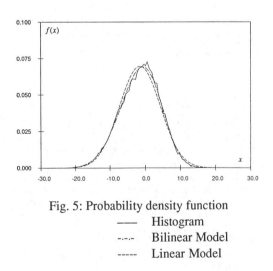

Fig. 5: Probability density function
—— Histogram
-.-.- Bilinear Model
----- Linear Model

Linear systems under non-Gaussian white noise can reproduce non-zero third moments as well as the power spectrum. However, there is almost no possibility of reproducing a prescribed bispectrum. The method can be used advantageously particularly if the phase information on the linear system is of importance.

Nonlinear static transformation allows for representing both the one-time probability density function and the power spectrum. For polynomial transformations there are even analytical expressions available for the power spectrum. However, the bispectrum is completely disregarded within this approach.

Finally, Volterra functional models take into account both power spectrum and bispectrum simultaneously. In this way, interactions of frequency components are represented correctly. The PDF is represented only in terms of statistical moments. Compared to the first two models, Volterra functional models are more flexible but require considerably higher computational effort. It appears that this model is capable of reproducing higher order spectra and probability densities with good accuracy.

ACKNOWLEDGEMENT

This research has been partially supported by the Austrian Industrial Research Promotion Fund under contract no. 6/636 which is gratefully acknowledged by the authors.

REFERENCES

AMMON, D. (1990), "Approximation and Generation of Gaussian and Non-Gaussian Stationary Processes", *Structural Safety*, 8, pp. 153-160.

BEDROSIAN, E., RICE, S.O. (1971), "The Output Properties of Volterra Systems (Nonlinear Systems with Memory) Driven by Harmonic and Gaussian Inputs", *Proc. of the IEEE*, Vol.59, No. 12, Dec., pp. 1688-1706.

BENDAT, J.S. (1990), *Nonlinear System Analysis and Identification from Random Data*, John Wiley & Sons, New York.

GRIGORIU, M. (1984), "Crossings of Non-Gaussian Translation Processes", *J. Eng. Mech.*, Proc. ASCE, Vol.110, No.4, April, pp 610 - 620.

HASSELMANN, K., MUNK, W., MacDONALD, G. (1963), "Bispectra of ocean waves", in: Time Series Analysis, M. Rosenblatt (Ed.), Wiley, New York, pp. 125-130.

MATSUOKA, T., ULRYCH, T.J. (1984), "Phase Estimation Using the Bispectrum", *Proc. IEEE*, Vol. 72, No.10, Oct., pp. 1403-1411.

NIKIAS, C.L., RAGHUVEER, M.R. (1987), "Bispectrum Estimation: A Digital Signal Processing Framework", *Proc. of the IEEE*, Vol. 75, No. 7, July, pp. 869-891.

PEINELT, R.H. (1992), "Analyse und Simulation Nicht-Gauss'scher Prozesse", Report No. 30, Institut für Mechanik, Universität Innsbruck.

SCHETZEN, M, (1980), *The Volterra and Wiener Theories of Nonlinear Systems*, Wiley-Interscience, New York.

SCHUËLLER, G.I., BUCHER, C.G. (1988), "Non-Gaussian Response of Systems Under Dynamic Excitation", in: Stochastic Structural Dynamics - Progress in Theory and Application, S.T. Ariaratnam et al. (Eds.), Elsevier Appl. Science Publ., Barking, Essex, pp. 219-239.

SUBBA RAO, T., GABR, M.M. (1984), *An Introduction to Bispectral Analysis and Bilinear Time Series Models*, Springer, New York.

WINTERSTEIN, S.R. (1985), "Non-normal Response and Fatigue Damage", *J. Eng. Mech.*, Proc. ASCE, Vol 111, No.10, October, pp 1291 - 1295.

Structural Safety & Reliability, Schuëller, Shinozuka & Yao (eds) © 1994 Balkema, Rotterdam, ISBN 90 5410 357 4

Estimation of continuous-time nonlinear system parameters from measured random response histories

J.B.Roberts, J.F.Dunne & A.Debonos
School of Engineering, University of Sussex, Falmer, Brighton, East Sussex, UK

ABSTRACT: The problem of estimating the parameters in a randomly excited single degree of freedom non-linear system, from measurements of the response alone, is addressed. It is shown that, in the case where the bandwidth of the excitation is significantly greater than that of the response, it is possible to estimate the overall damping, as measured by the equivalent linear damping factor, and the linear and nonlinear stiffness parameters, using a spectral technique. To separate the contributions from linear and nonlinear damping terms a further estimation stage is required, based on an analysis of the probability distribution of the energy envelope. To validate the proposed estimation method it is applied to some simulated data, corresponding to both white noise and correlated noise inputs.

1 INTRODUCTION

The estimation of the parameters in continuous-time models of mechanical and structural systems is an important aspect of model verification and is often needed when the parameters cannot be calculated accurately by theoretical means. Good estimates of the model parameters may be needed for a number of reasons - for example in reliability assessment or for control purposes. The usual approach to continuous-time parameter estimation (e.g. see Gawthrop et al 1988, Kountzeris et al 1991, Rice and Fitzpatrick 1991 a & b) assumes that measurements of both the response and the excitation are readily available. There are circumstances, however, where it is either impractical or impossible to measure the excitation process; estimation must then be based on response measurements only, together with a stochastic model of the excitation. Examples include the dynamic response of structures to wind loading or seismic excitation, the monitoring of land vehicles traversing random surfaces, and the motion of ships and offshore structures due to wave excitation.

Rigorous testing of previously proposed time-domain methods for parametric estimation, applied to a nonlinear oscillator, has shown that they usually give poor results in the situation where the excitation process is unknown (Debonos 1993). For example, the Invariant Imbedding method returns a very inaccurate estimate of damping, even when the system is linear. A similarly disappointing result is obtained when using the Maximum Likelihood method. Moreover, discrete time methods, which seems to offer some promise, do not allow accurate recovery of continuous time parameters. These difficulties have led the authors to develop a new approach for the case of an oscillator with linear and nonlinear damping and stiffness (Roberts et al 1992,1993). It is based on a use of a theoretical result for the energy envelope, derived by stochastic averaging theory (Roberts and Spanos 1986), in combination with results obtained by using the statistical linearisation method (Roberts and Spanos 1990). The technique requires the equivalent linear damping to be obtained from the response data. Of the several methods proposed by the authors, the one based on the response spectrum is most satisfactory (Debonos 1993); however it only works well in the case of linear stiffness.

In this paper a more generalised spectral estimation method is employed. A similar approach has been used previously for the case of known excitation (Rice and Fitzpatrick 1991 a & b). The difference in the present work is that an equivalent white noise is used to model the unmeasured excitation and the spectral method is used to estimate the intensity of this noise, in addition to the damping and stiffness parameters. A considerable advantage of the method is that it takes full account of nonlinear stiffness effects. The difficulty in separating linear and nonlinear damping contributions using spectral methods, observed by Rice and Fitzpatrick (1991 b), is overcome here in a new way by combining the method with an analysis of the probability distribution of the energy envelope.

The method is validated here through an analysis of noise-free simulated data, where the true parameter values are known a-priori. It has also been validated by applying it to ship model data, obtained from tests in a wave-tank (Debonos 1993). The latter shows that the method is not sensitive to the effect of measurement noise provided that this noise is in a frequency range which is well separated from the response frequency range.

2 THEORETICAL BASIS OF THE METHOD

A general form of the equations of motion of an n degree of freedom system is

$$\underset{\sim}{g}(\underset{\sim}{q},\dot{\underset{\sim}{q}},\ddot{\underset{\sim}{q}},\underset{\sim}{\lambda}) = \underset{\sim}{Q} \tag{1}$$

where $\underset{\sim}{q}$ is an n-vector of response functions, $\underset{\sim}{Q}$ is an n-vector of excitation functions, $\underset{\sim}{g}$ is an n-vector of nonlinear functions and $\underset{\sim}{\lambda}$ is an m-vector of the parameters which need to be estimated. On multiplying equation (1) by itself, delayed through a time lag, taking expectations throughout to form a correlation relationship and then Fourier transforming all terms, spectral relationships can be established. These will involve the joint (cross) spectra of $\underset{\sim}{Q}$ and the joint (cross) spectra of $\underset{\sim}{q}$ and functions of $\underset{\sim}{q}$ and its derivatives. If parametric forms for the spectra of the excitation $\underset{\sim}{Q}$ are assumed, then the relevant excitation parameters, and all the parameters in $\underset{\sim}{\lambda}$, can be estimated by minimising the difference between the two sides of equation (1), in a least square sense. This requires only a knowledge of sample functions of the responses, since the spectral terms can be estimated from such functions.

2.1 *Equation of motion*

As an illustration a nonlinear oscillator with the following equation of motion will be considered:

$$\ddot{x} + a_1\dot{x} + n_1\dot{x}|\dot{x}| + a_2 x + n_2 x^3 = f(t) \tag{2}$$

It will be assumed here that $f(t)$ can be modelled as a stationary stochastic process. If the bandwidth of this input process is significantly greater than that of the response, $f(t)$ may be approximated as an ideal white noise. Then

$$E\{f(t)f(t+\tau)\} = D^2\delta(\tau) \tag{3}$$

where D is the "strength" of the excitation, E is the expectation operator and $\delta(\tau)$ is Dirac's delta function. $D^2 = 2\pi S_0$, where S_0 is the constant spectral level of the white noise process which approximates $f(t)$.

Later it is convenient to consider linearised versions of equation (2). This involves the replacements

$$a_1\dot{x} + n_1\dot{x}|\dot{x}| \rightarrow a_{1eq}\dot{x},\; a_2 x + n_2 x^3 \rightarrow a_{2eq}x \tag{4}$$

where a_{1eq} and a_{2eq} are the equivalent linear damping and stiffness parameters, respectively.

Standard statistical linearisation (Roberts and Spanos 1990) gives the results

$$a_{1eq} = a_1 + n_1\left(\frac{8}{\pi}\right)\sigma_{\dot{x}} \ , \ a_{2eq} = a_2 + 3n_2\sigma_x^{\ 2} \quad (5)$$

where $\sigma_{\dot{x}}$ and σ_x are, respectively, the standard deviations of $\dot{x}$ and x. Moreover, applying standard linear theory to the totally linearised version of equation (2) gives (using the white noise approximation for $f(t)$)

$$\sigma_{\dot{x}}^{\ 2} = \frac{D^2}{2a_{1eq}} \qquad \sigma_x^{\ 2} = \frac{D^2}{2a_{1eq}a_{2eq}} \quad (6)$$

2.2 *Spectral identification method*

The specific identification problem which will be addressed here is as follows: how can a sample function of the response, $x(t)$, over a period $0 \leq t \leq T$, be processed to yield estimates of the system parameters a_1, n_1, a_2, n_2 and D?

A spectral relationship can be derived fairly readily from equation (2) by following the procedure outlined earlier. Using standard spectral relationships for the derivatives of processes, the following equation is eventually obtained:

$$F(\omega) = F_\ell(\omega) + F_n(\omega) = S_o \quad (7)$$

where

$$F_\ell(\omega) = (\omega^4 + a_1^{\ 2}\omega^2 - 2a_2\omega^2 + a_2^{\ 2})S_{xx} \quad (8)$$

$$F_n(\omega) = n_1^{\ 2}S_{yy} + n_2^{\ 2}S_{zz} + 2n_1n_2S_{yzr} + 2(a_2 - \omega^2)(n_1S_{xyr} + n_2S_{xzr}) + 2a_1\omega(n_1S_{xyi} + n_2S_{xzi}) \quad (9)$$

Here $y = x|\dot{x}|$, $z = x^3$ and $S_{uv}(\omega)$ are (cross) spectra defined by the Fourier relationship

$$S_{uv}(\omega) = \frac{1}{2\pi}\int_{-\infty}^{\infty} R_{uv}(\tau)e^{-i\omega\tau}d\tau,$$

$$\text{where } R_{uv}(\tau) = E\{u(t)v(t+\tau)\} \quad (10)$$

R_{uv} is a (cross) correlation function and $S_{uv}(\omega)$ is a complex function of frequency; the additional subscripts, r and i, appearing in equation (9) denote, respectively, the real and imaginary parts of the cross-spectrum. $F_\ell(\omega)$ and $F_n(\omega)$ represent, respectively, the linear and nonlinear contributions to equation (7).

All the spectral quantities in equations (7) to (9) can be estimated by processing a response sample function, using standard spectral estimation techniques based on the FFT algorithm. Numerical differentiation of the response sample function, to obtain $\dot{x}$, is necessary in some cases. To obtain estimates of the parameters one can then minimise the cost function, J, where

$$J = \sum_{i=1}^{N}\left[S_o - \hat{F}(\omega_i)\right]^2 \quad (11)$$

using a standard nonlinear optimisation algorithm. The hat symbol denotes that $\hat{F}(\omega)$ is computed according to equations (7) to (9), but using estimated spectra, $\hat{S}_{uv}(\omega)$, rather than theoretical spectra. Clearly the number, N, of discrete frequencies, ω_i, at which $\hat{F}(\omega_i)$ is evaluated, must be greater than the number of parameters to be estimated (here 5).

2.3 *Separation of damping components*

The estimation scheme outlined above enables one to successfully estimate the stiffness parameters, a_2 and n_2, the overall damping level, as expressed by a_{1eq}, and the excitation strength, D. However, it is not capable of separating out the individual contributions from the linear and nonlinear damping components, as will be demonstrated shortly.

This problem can be overcome by introducing a further stage to the estimation procedure, based on the probability distribution of the response. For nonlinear damping an analytical expression for the distribution of x is unknown, even in the case of white noise excitation. However, an

TABLE 1

Summary of the four cases (A, B, C and D)

Case	Input	a_1	n_1	a_2	n_2	D	a_{1eq}	a_{2eq}
A	white	0.012	0.35	0.504	0	0.075	0.104	0.504
B	white	0.012	0.35	0.504	3.0	0.132	0.148	1.023
C	correlated	0.012	0.35	0.504	0	0.081	0.104	0.504
D	correlated	0.012	0.35	0.504	3.0	0.140	0.148	1.023

approximate theoretical expression for the stationary probability density function, $w(E)$ of the energy envelope process, $E(t)$, can be found; this is accurate when the bandwidth of the excitation process is significantly greater than that of the response (Roberts and Spanos 1986). Here $E(t)$ is defined as the sum of the kinetic and potential energies of the oscillator. The result is (Roberts et al 1992)

$$\log_e\left[\frac{w(E)}{A(E)}\right] = C - \frac{2a_1}{D^2}E - \frac{2n_1}{D^2}g(E) \qquad (12)$$

where $A(E)$ and $g(E)$ depend on the stiffness parameters a_2 and n_2 and can be evaluated from expressions given in (Roberts et al 1992). C is a constant.

Fitting equation (12) to estimates of $\log_e[w(E)/A(E)]$, derived from the histogram of the estimated energy levels, and using the estimates of a_2 and n_2 obtained by the spectral method described earlier, enables estimates of a_1/D^2 and n_1/D^2 to be generated.

In applying the spectral estimation method it is natural to replace the damping terms with an equivalent linear damping term, according to equation (4). Two possibilities then exist for finding the values of a_1 and n_1. Firstly, and most obviously, one can use the estimated value of D directly. Alternatively, one can combine the estimate of a_{1eq} obtained from the spectral estimation method with the estimates of a_1/D^2 and n_1/D^2 from the energy histogram, and employ equations (5) and (6). This yields a nonlinear algebraic equation for D, which is readily solved; hence a_1 and n_1 can be found. This latter method was found to be more accurate, in general, and is used to obtain the results given in section 3. In this context it is worth noting that the expression for $\sigma_{\dot{x}}^2$ given by equation (6) is actually valid in cases where the damping is replaced by an equivalent linear term, but the stiffness is not linearised.

3 VALIDATION THROUGH SIMULATION

3.1 *Simulation method*

To test the proposed method, sample functions of $x(t)$ were generated by numerically integrating equation (2), using a Runge-Kutta algorithm. Two types of Gaussian excitation processes were used in the simulation studies - white noise and correlated excitation. The method of generating sample functions of these processes is described by Roberts et al (1993). Four particular cases are studied here, as summarised in Table 1. In all cases the equivalent damping level, as computed from the statistical linearisation method, is 7.4% of critical.

Fig. 1(a) shows, for comparison purposes, a typical response spectrum (Case B), as estimated from a sample response function, together with the corresponding spectrum of the simulated white noise excitation. Fig. 1(b) shows the estimated power spectrum of the response for Case D, together with the spectrum of the simulation correlated excitation (same as for Case B).

It is evident that in both cases the bandwidth of the correlated excitation is significantly greater than that of the response, as required by the energy envelope method for separating

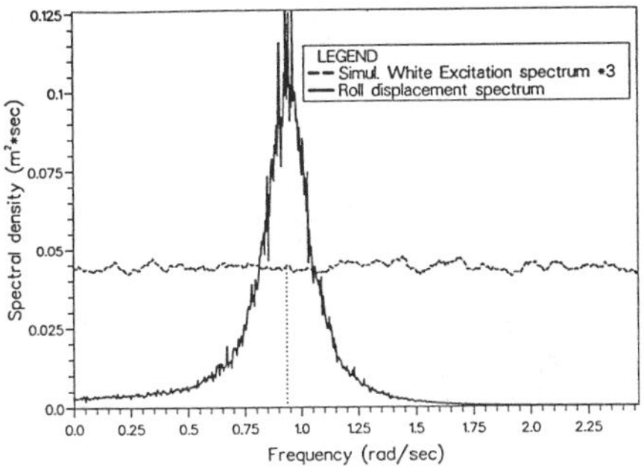

Fig. 1(a) Excitation and response spectra: white noise excitation

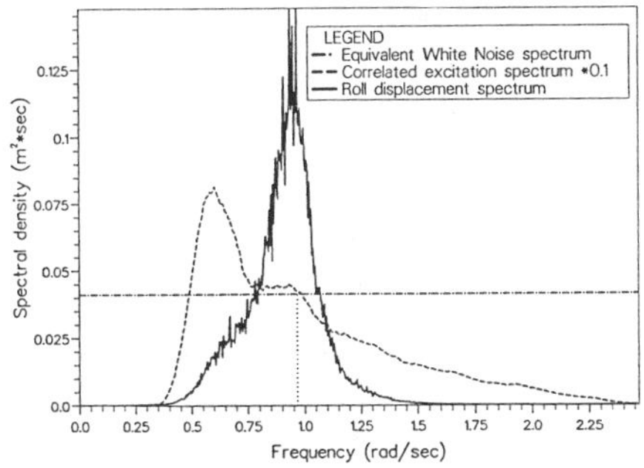

Fig. 1(b) Excitation and response spectra: corelation noise excitation

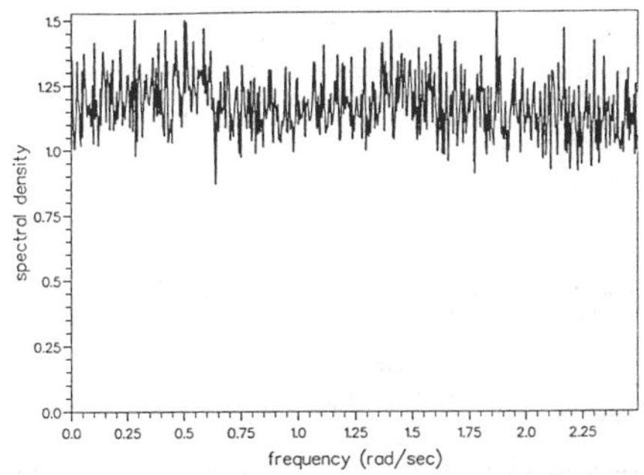

Fig. 2 Variation of $\hat{F}(\omega)$: linear stiffness

the damping contributions. In the case of correlated excitation the D values in Table 1 correspond to the level of the excitation spectrum at the frequency at which the response spectrum peaks. Thus they give the strengths of the equivalent white noises (see Fig. 1(b)) which can be compared with the D values returned by the estimation technique. To compute D the theoretical target excitation spectrum is used and the peak frequency is estimated using a_{2eq} (see equation (5)).

3.2 *Spectral Based Estimation*

To illustrate the strengths and weaknesses of the spectral estimation method, outlined in Section 2, some typical plots relating to the left hand side of equation (7) are presented, for the case of white noise excitation. Here the estimate $\hat{F}(\omega)$ is obtained by using the estimates of the (cross) spectral terms derived from a simulated sample function of the response.

Initially a case where the stiffness is linear will be considered ($a_1 = 0.02$, $n_1 = 0.458$, $a_2 = 0.504$, $n_2 = 0$, and $D = 0.087$). The pre-selected values of the parameters were used to compute $\hat{F}(\omega)$ and, as expected, $\hat{F}(\omega)$ was found to be effectively constant with respect to frequency (when allowance is made for statistical variability) and in fact was virtually identical to the input spectrum, as derived from the corresponding sample function of the excitation. However if one computes $\hat{F}(\omega)$ using the same sample function of the response, but a linear damping model with a_{1eq} calculated according to statistical linearisation (see equation (5)), then a very similar result is obtained, (as shown in Fig. 2). This shows clearly that it is not possible to distinguish between linear and nonlinear damping through an analysis based on spectra. The spectral estimation method does, however, offer a powerful method of estimating the equivalent linear damping parameter, a_{1eq} , as will be shown later.

When the stiffness is nonlinear, a similar result to Fig. 2 is obtained if linear damping is assumed and a_{1eq}, a_2, n_2 and D are used to compute $\hat{F}(\omega)$, according to equations (7) to (9). In contrast Fig. 3 shows a typical variation of $\hat{F}(\omega)$ with frequency, when the stiffness is actually nonlinear ($a_1 = 0.14$, n_1 =

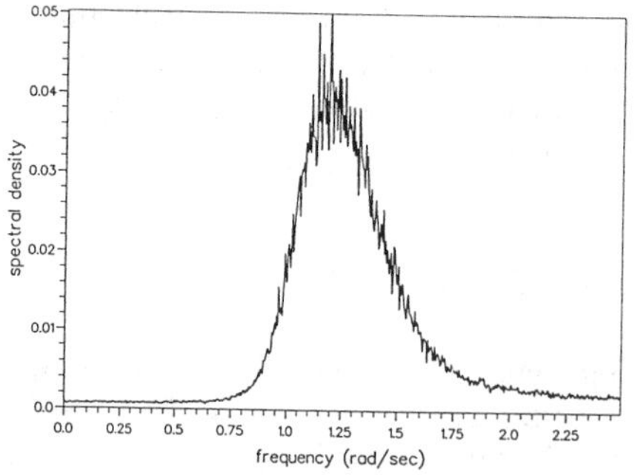

Fig. 3 Variation of $\hat{F}(\omega)$: nonlinear stiffness

TABLE 2

Stiffness and equivalent linear damping estimation-frequency domain method

Case	$\hat{a}_2$	$\hat{n}_2$	$\hat{D}$	a_{1eq}	$\hat{a}_{2eq}$
A	0.510	-0.148	0.075	0.099	0.504
	(0.032)	(0.217)	(0.002)	(0.011)	(0.014)
B	0.468	2.95	0.143	0.155	0.942
	(0.054)	(0.338)	(0.005)	(0.016)	(0.032)
C	0.497	-0.102	0.080	0.097	0.499
	(0.031)	(0.214)	(0.005)	(0.013)	(0.013)
D	0.568	2.20	0.119	0.129	0.935
	(0.076)	(0.500)	(0.014)	(0.047)	(0.030)

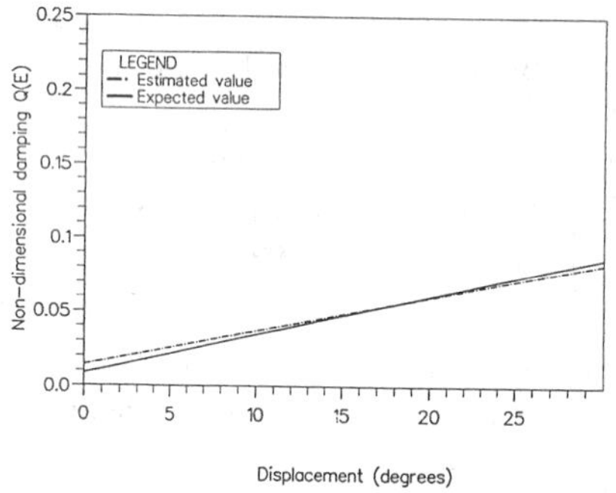

Fig. 4a Nonlinear damping function: Case A

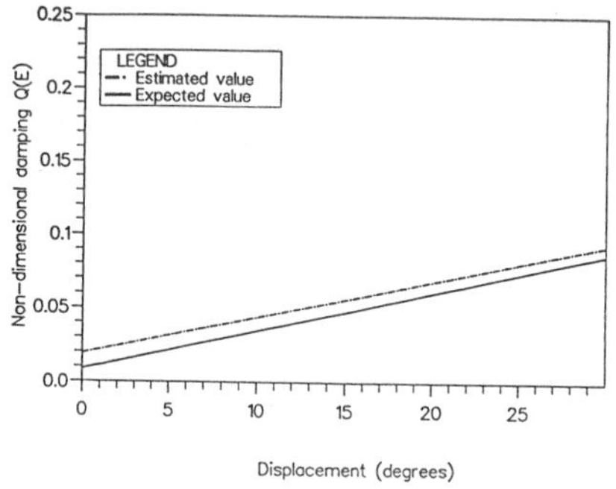

Fig. 4b Nonlinear damping function: Case B

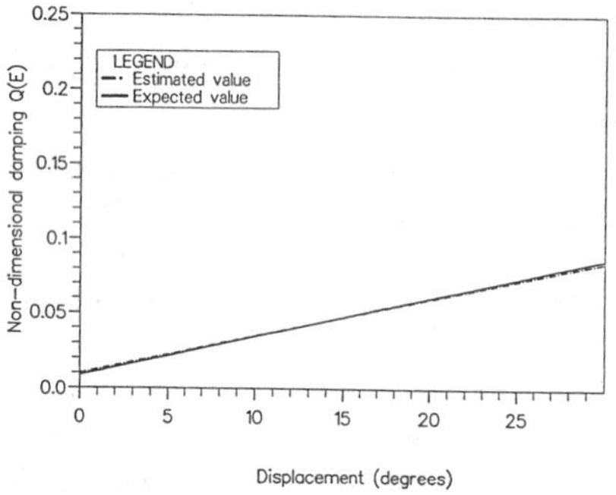

Fig. 4c Nonlinear damping function: Case C

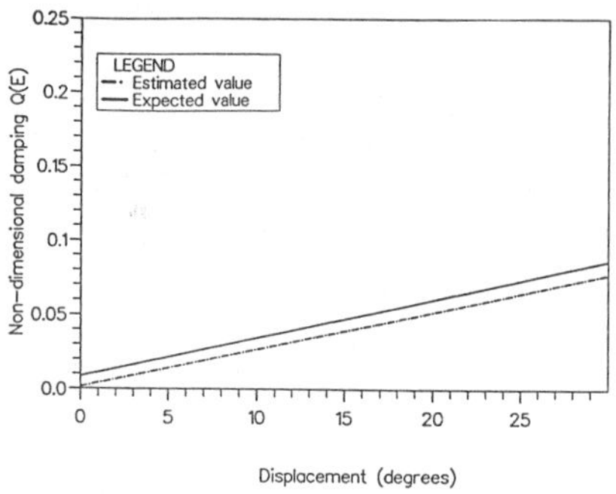

Fig. 4d Nonlinear damping function: Case D

0, $a_2 = 0.504$, $n_2 = 3$, $D = 0.087$), but a totally linear model is used to compute $\hat{F}(\omega)$, with a_{1eq} and a_{2eq} calculated from equation (5). Here there is a very significant variation with frequency, showing that the method is capable of distinguishing between linear and nonlinear stiffness.

3.3 *Parameter estimates*

For all four cases the mean and statistical variability of the parameter estimates generated by the spectral method were examined by processing 100 independent response histories, each of duration 1800 seconds, with a time step of 0.09 seconds. An equivalent linear damping representation was used throughout. Spectra were computed at 70 frequencies, in the vicinity of the peak in the response spectrum.

Table 2 shows the mean and standard deviation (the latter in brackets) of the estimates of a_2, n_2, D, a_{1eq}.
For comparison purposes estimates of a_{2eq} are also shown, as derived from a completely linear model fit to the data. In all cases the estimated parameters agree reasonably well with the true values, shown in Table 1, when the level of statistical variability is taken into account. This level is particularly high for a_2 and n_2, due to the significant degree of correlation between the estimates, as discussed by Roberts et al (1993).

Estimates of a_1 and n_1 were derived by using the energy probability distribution, as described earlier. Figs. 4(a) to (d) show, for cases A to D respectively, the variation of the nonlinear damping function, $Q(E)$ (see Roberts et al 1992), plotted against response amplitude, as computed from the estimated values of a_1 and n_1. For comparison the solid lines show the true variation of $Q(E)$, obtained using the true values of a_1 and n_1. In all four cases a good agreement is obtained.

4 CONCLUSIONS

By combining a spectral estimation method with an analysis of the energy probability density function a technique for estimating all the parameters in a nonlinear oscillator equation can be developed, requiring only a stochastic model of the excitation. Good estimates are obtained from simulated data, even when the excitation is correlated.

ACKNOWLEDGMENTS

The work described in this paper was jointly funded by the Marine Technology Directorate and by British Maritime Technology (BMT). The authors also gratefully acknowledge the assistance received from Dr. R. Standing and Dr. A. Morrall of BMT.

REFERENCES

Debonos, A.1993. Estimation of Non-linear Ship Roll Parameters using Stochastic Identification Techniques. D.Phil. Thesis, University of Sussex, .

Gawthrop, P. J., A. Kountzeris & J. B. Roberts 1988. Parametric Identification of Non-linear Ship Roll Motion from Forced Roll Data. *Journal of Ship Research, SNAME* 32: 101-111.

Kountzeris, A., J. B. Roberts & P. J. Gawthrop 1991. Estimation of Ship Roll Parameters from Motion in Irregular Seas. *Transactions of the Royal Institution of Naval Architects* 132: 253-266.

Rice, H. J. & J. A. Fitzpatrick 1991 a. The Measurement of Nonlinear Damping in Single-Degree-of-Freedom Systems. *Journal of Vibration and Acoustics, ASME* 113: 132-140.

Rice, H. J. & J. A. Fitzpatrick 1991 b. A Procedure for the Identification of Linear and Nonlinear Multi-degree-of-freedom Systems. *Journal of Sound and Vibration* 149(3): 397-411.

Roberts, J. B., J. F. Dunne & A. Debonos May 1992. Estimation of Ship Roll Parameters in Random Waves. *Journal of Offshore Mechanics and Arctic Engineering, ASME* 114:114-121.

Roberts, J.B., , J. F. Dunne & A. Debonos 1993. Stochastic Estimation Methods for Non-linear Ship Roll Motion. *Probabilistic Engineering Mechanics* (to be published).

Roberts, J. B. & P. D. Spanos 1986. Stochastic Averaging: An Approximate Method for Solving Random Vibration Problems. *International Journal of Non-linear Mechanics* 21: 111-134.

Roberts, J. B. & P. D. Spanos 1990. Random Vibration and Statistical Linearization. J. Wiley & Sons, Chichester, UK, .

Structural Safety & Reliability, Schuëller, Shinozuka & Yao (eds) © 1994 Balkema, Rotterdam, ISBN 90 5410 357 4

Some asymptotic properties of nonstationary processes in dynamical systems

Jerzy Skrzypczyk
Institute of Theoretical Mechanics, Silesian Technical University, Poland

ABSTRACT: This paper reports on some results concerning the properties of multidimensional stochastic systems described by random integral Volterra equations of the II-kind. In particular, for dynamic systems containing linear time-varying part and an arbitrary positive-slope nonlinearity, it is proved, under reasonable conditions that the response to a stationary stochastic excitation applied at t=0 is ultimately stationary with the same stationary character as the excitation, regardless of the initial state of the system.

1 INTRODUCTION

Equations of the form

$$x(t,\omega) = z(t,\omega) + \int_0^t k(t-\tau,\omega)f(\tau,x(\tau,\omega),\omega)d\tau \quad , \quad t \in R^1_+ \qquad (1)$$

in which $z(t,\omega)$ is a 2-nd order vector stochastic process, are often encountered in the study of dynamical systems containing linear time-invariant elements and an arbitrary finite number of time-varying nonlinear elements. Typically, x represents the system response and z takes into account both the independent energy sources and the initial conditions at t=0 (Bharucha-Reid 1972, Holtzman 1968, Przybylak & Skrzypczyk 1989, Skrzypczyk 1974a, b, 1986, Tsokos & Padgett 1971, 1974).

The related equation

$$y(t,\omega) = z(t,\omega) + \int_{-\infty}^t k(t-\tau,\omega)\tilde{f}(\tau,y(\tau,\omega),\omega)d\tau \quad , \quad t \in R^1 \qquad (2)$$

is also often encountered. It arises when it is convenient for mathematical reasons to formulate a model of the system such that the response and excitation are defined for all $t \in R^1$ (Sandberg 1964, Sandberg & Benes 1964).

In connection with the actual determination of the steady-state response, the principal ideas evidently being that if the physical system is stable in some suitable sense, then the effect of the initial conditions at t=0 should eventually "die out", and, moreover, that the steady-state response of the system should be obtained "at once" if the stationary excitation is applied at " $t=-\infty$ ".

The purpose of this paper is to report on some mathematical results concerning the properties of (1) and (2) that are pertinent, to a considerable extent, to engineering questions of the type discussed.

2 PRELIMINARIES

In this report we deal with random Volterra integral equations of the 2-nd kind of the form (1) and (2). We assume further that the following assumptions are fulfilled :

(A1) ω is an element of some abstract probability measure space $(\underline{\omega},\beta,P)$;

(A2) $z(t,\omega)$ is a vector-valued, measurable stochastic process defined on $R^1 x\underline{\omega}$, with values in R^n;

(A3) the random Volterra kernel $k(t,\omega)$ is a measurable, random, matrix-valued function defined for $t\in R^1_+=[0,\infty[$ and equal zero for $t\in R^1_-=]-\infty,0[$ for almost all $\omega \in \underline{\omega}$, with values in $L(R^n,R^n)$;

(A4) the nonlinear random function $f(t,x,\omega)$ is the measurable mapping

$R^1_+ \times R^n \times \underline{\omega} \longrightarrow R^n$ and similarly $\tilde{f}(t,y,\omega)$: $R^1 \times R^n \times \underline{\omega} \longrightarrow R^n$ and $f(t,.,.)=\tilde{f}(t,.,.)$ for all $t \in R^1_+$ mod(P);

(A5) $x(t,\omega), y(t,\omega)$ are unknown stochastic processes.

We denote further $E(\cdot) := \int_{\underline{\omega}} (\cdot) P(d\omega)$,

$\|x(t,.)\|_2 := \left(E\left\{ |x(t,.)|^2 \right\} \right)^{1/2}$, $|\cdot|$ denotes a norm in Euclidean space. The usual notations $L^q(\underline{\omega},\beta,P;R^n)$ and $L^q(\underline{\omega},\beta,P;L(R^n,R^n))$, $1 \le q \le \infty$, are used for the space of q-order random variables with values in R^n and $L(R^n,R^n)$ respectively (Bensoussan 1971, Skrzypczyk 1989, Tsokos 1971).

We assume that stochastic processes $x(t,\omega), y(t,\omega), z(t,\omega)$ are functions of $t \in R^1$ with values in $L^2(\underline{\omega},\beta,P;R^n)$. The random functions $f(t,x(t,\omega),\omega), \tilde{f}(t,y(t,\omega),\omega)$ are (under some additional conditions) random operator variables with values in $L^2(\underline{\omega},\beta,P;R^n)$.

Suppose further that the random kernel $k(t,\omega)$ is P-essentially bounded for all $t \in R^1_+$. This is sufficient to assure that the multiplication of $k(\cdot,\omega)$ and $f(\cdot,\cdot,\omega)$ is an element of the space $L^2(\underline{\omega},\beta,P;R^n)$.

We use further the simplified notation $L^2(\underline{\omega},\beta,P) = L^2(\underline{\omega},\beta,P;R^n)$.

DEFINITION 1. Let $C(R^1_+;L^2(\underline{\omega},\beta,P))$ denotes the Banach space of stochastic measurable 2-nd order processes, continuous in mean square with metric

$$\|x(t,\omega)\|_{C(R^1_+;L^2(\underline{\omega},\beta,P))} = \sup_{t \in R^1_+} \|x(t,\omega)\|_2$$

Suppose that we define the class N of stochastic processes $x_0(t,\omega)$ with the property $\underset{t \to \infty}{\text{l.i.m.}}\, x_0(t,\omega) = 0$. The class $N(R^1_+;L^2(\underline{\omega},\beta,P))$ of all random functions, which are continuous in mean square for all $t \in R^1_+$ and tend to zero in mean square is the linear, complete subspace. Divide the space of continuous functions $C(R^1_+;L^2(\underline{\omega},\beta,P))$ into equivalence classes.

So two random functions $x_1(t,\omega)$ and $x_2(t,\omega)$ belong to the same class of equivalence iff their difference $x_1 - x_2$ belongs to the space $N(R^1_+;L^2(\underline{\omega},\beta,P))$.

DEFINITION 2. The set of equivalent classes we denote as difference space $C/N = C/N(R^1_+;L^2(\underline{\omega},\beta,P))$. In the space C/N we introduce the norm

$$\|x(t,\omega)\|_{C/N(R^1_+;L^2(\underline{\omega},\beta,P))} = \overline{\underset{t \to \infty}{\text{l.i.m.}}}\, |x(t,\infty)| = \lim_{T \to \infty} \sup_{t \ge T} \|x(t,\omega)\|_2$$

So, two signals belong to the same class iff they have the same mean "steady state" i.e. the elements of the space C/N are stochastic processes bounded and continuous in mean square sense for all $t \in R^1_+$ and defined with accuracy to processes which tend to zero in mean square as $t \longrightarrow \infty$.

In further considerations another type of stochastic signals appears. It belongs to the so called space with a mixed norm (Bensoussan 1971, Kantorovitsch & Akilov 1977, Dunford & Schwartz 1958).

DEFINITION 3. Denote by $L^{p,q}(R^1_+) = L^p(R^1_+;L^q(\underline{\omega},\beta,P))$, $1 \le p < \infty$, $1 \le q < \infty$ the set of all measurable stochastic processes $x(t,\omega)$, for which

$$\int_0^\infty \left(\|x(t,\omega)\|_q \right)^p dt < \infty$$

The norm in the space $L^{p,q}$ will be introduced as follows:

$$\|x(t,\omega)\|_{p,q} = \left(\int_0^\infty \left(\| x(t,\omega) \|_q \right)^p dt \right)^{1/p}$$

All introduced to considerations functional spaces are of Banach type and all definitions remain true if we replace the subspace R^1_+ with R^1 one.

3 EXISTENCE AND UNIQUENESS OF SOLUTIONS

THEOREM 1. Let :

(B1) $k(t,\omega)$, $tk(t,\omega)$, $t^2k(t,\omega)$ belong to $L^1 \cap L^2(R^1_+; L^\infty(\underline{\omega},\beta,P;L(R^n,R^n)))$;

(B2) for each $x \in R^n$ a function $f(t,x,\omega)$ is in mean-square continuous , $f(t,0,\omega)=0$ mod(P), and there exist a matrix $\Lambda \in L(R^n,R^n)$ and a nonnegative real number $r>0$ with the property that:

$$\left\| f(t,x_1,\omega) - f(t,x_2,\omega) - \Lambda(x_1 - x_2) \right\| \le r\left\| x_1 - x_2 \right\| \quad \text{mod(P)}$$

for all $x_1, x_2 \in R^n$, $t \in R$;

(B3) $$\text{P-ess} \inf_{\omega \in \omega_0} \inf_{\text{Re } s \ge 0} \left| \det\left[I - \hat{K}(s,\omega)\Lambda \right] \right| > 0$$

$$\text{P-ess} \sup_{\omega \in \omega_0} \max_i \sup_{\text{Re } s \ge 0} r^2 \mu_i(s,\omega) < 1$$

where $\mu_i(s,\omega)$, $i=1,\ldots,n$ are the eigenvalues of the matrix GG^*, and

$$G(s,\omega) = [I-\hat{K}(s,\omega)\Lambda]^{-1}\hat{K}(s,\omega) ,$$

$$\hat{K}(s,\omega) = \int_0^\infty \exp(-st)k(t,\omega)dt .$$

Then for every $z(t,\omega) \in C/N(R^1_+; L^2(\underline{\omega},\beta,P;R^n))$ there exist a unique solution of the equation (1) $x \in C/N$ and a constant α such that

$$\overline{\lim_{t\to\infty}} \|x_1(t,\omega)-x_2(t,\omega)\|_2 \le 2\alpha \overline{\lim_{t\to\infty}} \|z_1(t,\omega)-z_2(t,\omega)\|_2$$

where x_1, x_2 are solutions of eq. (1) related to excitations z_1 and z_2 respectively.

Proof. It's based on the following two lemmas and its idea is similar to that of (Przybylak, Skrzypczyk 1989).

LEMMA 1. Consider the random Volterra integral equation of the type :

$$x(t,\omega) = z(t,\omega) + \int_0^t k(t-\tau,\omega)q(\tau)x(\tau,\omega)d\tau , \quad t \in R^1_+ \tag{3}$$

where $q: R^1_+ \longrightarrow L(R^n,R^n)$. Let assumptions (B1) and (B3) are satisfied and additionally:

(C1) $m \le |q(t)| \le M$ for all $t \in R^1_+$.

If $z_n(t,\omega)$ is the stochastic process equal zero mod(P), beyond the limits of the interval $[n,n+1[$, and $x_n(t,\omega)$ is the corresponding solution of (3), then there exists a constant α (it depends only of $m, M, k(t,\omega)$) such that

$$\|x_n(t,\omega)\|_2 \le \frac{\alpha}{(1+t-n)^2} \sup_{n\le t\le n+1} \|z_n(t,\omega)\|_2 , \quad t \ge n \tag{4}$$

LEMMA 2. Let the assumptions of lemma 1 be satisfied. Then for every equation of the form (3) and for every stochastic process $z(t,\omega)$ which belongs to the space $C/N(R^1_+; L^2(\underline{\omega},\beta,P))$ the solution $x(t,\omega)$ of the equation (3) satisfies inequalities :

$$\| x(t,\omega)\|_2 \le 2\alpha \sup_{\tau\le t} \| z(\tau,\omega)\|_2$$

$$\overline{\lim_{t\to\infty}} \| x(t,\omega)\|_2 \le 2\alpha \overline{\lim_{t\to\infty}} \| z(t,\omega)\|_2$$

where α is the same constant introduced in lemma 1.

The final result is a simple conclusion based on these lemmas.■

REMARK 1. For the stochastic system (2) the same result remains true for the space $C/N(R^1; L^2(\underline{\omega},\beta,P;R^n))$.

4 ASYMPTOTIC PROPERTIES

Theorem 2, below focuses attention on a relation between the solutions of equations (1) and (2). This theorem is later used in order to obtain conditions under which the solution of (1), belonging to $C/N(R^1_+; L^2(\underline{\omega},\beta,P;R^n))$, approaches a stationary steady state as $t \longrightarrow \infty$.

THEOREM 2. Let

$$x_1(t,\omega) = z_1(t,\omega) + \int_0^t k(t-\tau,\omega)f(\tau,x_1(\tau,\omega),\omega)d\tau , \quad t \in R^1_+ \tag{5}$$

$$x_2(t,\omega) = z_2(t,\omega) +$$

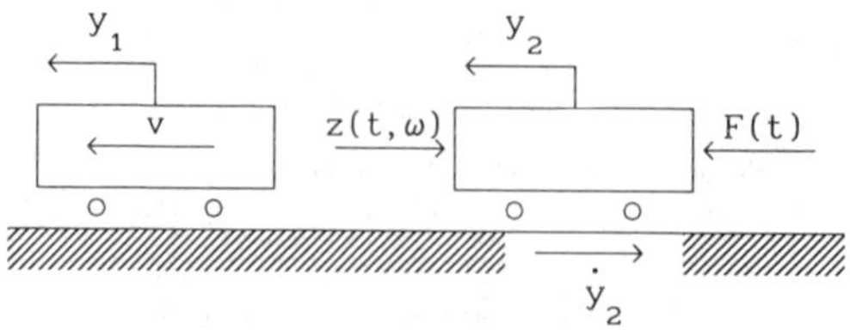

Fig. 1. Modelling the two car behaviour

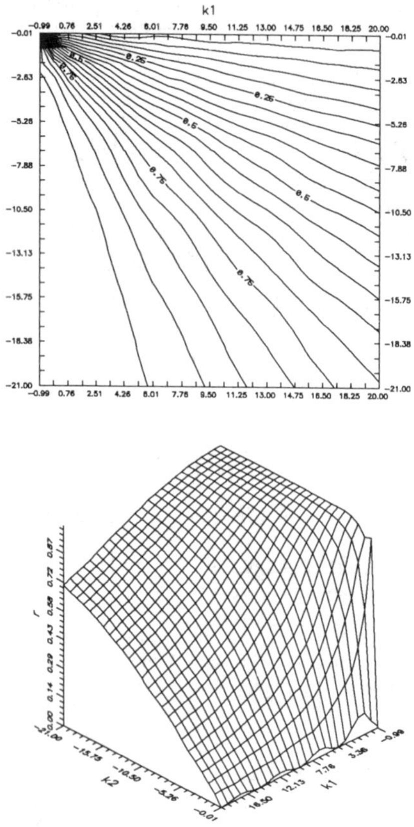

Fig. 2. Addmissible boundary-value parameters r, k_1, k_2 for stable systems

$$\int_{-\infty}^{t} k(t-\tau,\omega)\tilde{f}(\tau, x_2(\tau,\omega),\omega)d\tau \quad , t\in R^1 \qquad (6)$$

Suppose that the assumptions of the theorem 1 are satisfied, and additionally

(D1) $z_1(t,\omega)-z_2(t,\omega)\in C/N(R^1_+;L^2(\underline{\omega},\beta,P;R^n))$ and $\lim_{t\to\infty} \|z_1(t,\omega)-z_2(t,\omega)\|_2=0$;

(D2) for arbitrary

$x_2(t,\omega)\in C/N(R^1_+;L^2(\underline{\omega},\beta,P;R^n))$ and

$$\int_{-\infty}^{0} k(t-\tau,\omega)\tilde{f}(\tau, x_2(\tau,\omega),\omega)d\tau \in$$

$C/N(R^1_+;L^2(\underline{\omega},\beta,P;R^n))$ and

$$\lim_{t\to\infty} \left\| \int_{-\infty}^{0} k(t-\tau,\omega)\tilde{f}(\tau, x_2(\tau,\omega),\omega)d\tau \right\|_2=0.$$

Then

$$\lim_{t\to\infty} \| x_1(t,\omega)-x_2(t,\omega)\|_2 = 0$$

Proof. Subtracting eqs. (5) and (6) and applying results of theorem 1 we get the following inequality

$$\overline{\lim_{t\to\infty}} \|x_1(t,\omega)-x_2(t,\omega)\|_2 \le 2\alpha \overline{\lim_{t\to\infty}} \|h(t,\omega)\|_2$$

where

$$h(t,\omega) = z_1(t,\omega)-z_2(t,\omega) - \int_{-\infty}^{0} k(t-\tau,\omega)f(\tau, x_2(\tau,\omega),\omega)d\tau$$

and the statement of the theorem follows from assumptions (D1-D2) and inequality (7).■

REMARK 2. The integral

$$\int_{-\infty}^{0} k(t-\tau,\omega)\tilde{f}(\tau, x_2(\tau,\omega),\omega)d\tau \longrightarrow 0 \text{ as } t\longrightarrow\infty$$

(in mean square) if :

a) $k\in L^2(R^1_-;L^2(\underline{\omega},\beta,P;L(R^n,R^n)))$, and $f(x_2)\in L^{2,2}(R^1_-)$ for arbitrary stochastic process $x_2\in L^{2,2}(R_-)$; eventually

b) $k\in L^1(R^1_-;L^2(\underline{\omega},\beta,P;L(R^n,R^n)))$, $E\left\{|f(x_2)|^2\right\}$ is bounded for all $t\in R^1_-$.■

In many cases input signal $z(t,\omega)$ in dynamical systems described by the equation (1) is decomposed into a sum of two signals $z_1(t,\omega)$ and $z_2(t,\omega)$ such that $z_1(t,\omega)$ is a second-order stationary and

$z_2(t,\omega)$ has a finite mean energy i.e. $z_2(t,\omega) \in L^{2,2}(R^1_+)$, $z_2(t,\omega)=0$ for $t<0$ and $\|z_2(t,\omega)\|_2 \longrightarrow 0$ as $t \longrightarrow \infty$. In such cases it is often of considerable importance to determine whether $x(t,\omega)$ approaches a steady-state response $\tilde{x}(t,\omega)$ that is stationary as $t \longrightarrow \infty$. As a specific application the following result is proved.

THEOREM 3. We consider the dynamical system described by the equation

$$x(t,\omega) = z_1(t,\omega) + z_2(t,\omega) + \int_0^t k(t-\tau) f(x(\tau,\omega))d\tau \quad , t \in R^1_+ \qquad (8)$$

where $z_1(t,\omega)$ is a 2-nd order stationary process, defined for $t \in R^1$, $z_2(t,\omega) \in L^{2,2}(R^1_+)$, $z_2(t,\omega)=0$ for $t<0$, and $\| z_2(t,\omega)\|_2 \longrightarrow 0$ if $t \to \infty$.

If, in addition to the hypothesis stated above, the assumptions of the theorem 1 are satisfied, then a solution $\tilde{x}(t,\omega)$ of the equation

$$\tilde{x}(t,\omega)=z_1(t,\omega)+\int_{-\infty}^t k(t-\tau,\omega) f(\tilde{x}(\tau,\omega))d\tau \quad , t \in R^1 \qquad (9)$$

is a 2-nd order stationary stochastic process. Moreover the difference $x(t,\omega)-\tilde{x}(t,\omega)$ is an element of $L^{2,2}(R^1_+)$ and

$$\lim_{t \to \infty} \| x(t,\omega)-\tilde{x}(t,\omega)\|_2 = 0$$

Proof. The solution of (9) exists and is a stochastic stationary process, continuous and bounded in mean square (Skrzypczyk 1986). It is sufficient to prove that

$$\int_{-\infty}^0 k(t-\tau,\omega) f(\tilde{x}(\tau,\omega))d\tau \in L^{2,2}(R^1_+).$$

Notice, that

$$\int_0^\infty \left\| \int_{-\infty}^0 k(t-\tau,\omega) f(\tilde{x}(\tau,\omega))d\tau \right\|_2^2 dt \le$$

$$\int_0^\infty \int_t^\infty \left|\left|\left| k(\tau,\omega) \right|\right|\right|^2 (1+\tau)^4 d\tau \int_t^\infty \frac{\| f(\tilde{x})\|_2^2}{(1+\tau)^4} d\tau dt$$

Following condition (B2) we have

$$\sup_{t \in R^1} \| f(\tilde{x}(t,\omega))\|_2 \le \gamma \sup_{t \in R^1} \| \tilde{x}(t,\omega)\|_2 < \infty$$

and additionally from assumption (B1) it follows, that $(1+\tau)^2 k(\tau,\omega)$ belongs to $L^1(R^1_+; L^\infty(\underline{\omega},\beta,P;R^n)) \cap L^2(R^1_+; L^\infty(\underline{\omega},\beta,P;R^n))$, and in consequence

$$\int_0^\infty \left| \int_{-\infty}^0 k(t-\tau,\omega) f(\tilde{x}(\tau,\omega))d\tau \right|^2 dt \le$$

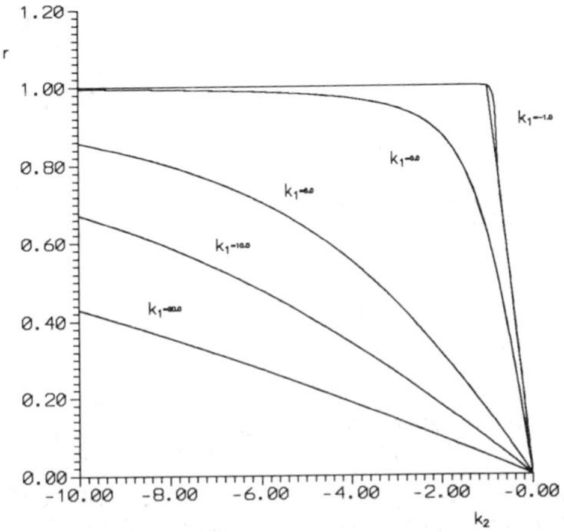

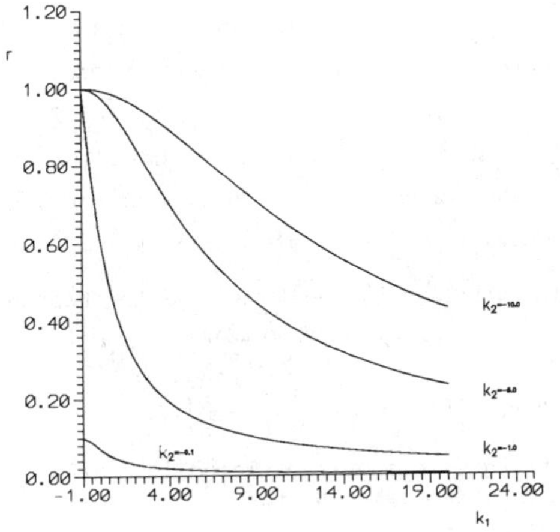

Fig. 3. Tolerance bands for the sector parameter r for different values of k_1, k_2

$$\alpha_1 \int_0^\infty \int_t^\infty \frac{1}{(1+\tau)^4} d\tau dt = \frac{1}{6} \alpha_1$$

where α_1 is a constant. Since the assumptions of remark 2 are fulfilled the thesis of the theorem is a simple conclusion. ■

Additionally some illustrative considerations are presented.

EXAMPLE . Consider two vehicles moving one behind the other along a straight line as shown in fig. 1. Then we have the following equations

$$\dot{y}_1 = v$$

$$m\ddot{y}_2 = -\dot{y}_2 + F(t) - z(t,\omega)$$

where m denotes the mass of the second vehicle, $F(\cdot)$ the accelerating force as control input, v the velocity of the first vehicle and $z(\cdot,\omega)$ the stochastic stationary disturbance. With $m=1$, $x_1 = \dot{y}_2$, $x_2 = y_1 - y_2$ we have

$$\frac{d}{dt}\begin{bmatrix} x_1 \\ x_2 \end{bmatrix} = \begin{bmatrix} -1 & 0 \\ -1 & 0 \end{bmatrix}\begin{bmatrix} x_1 \\ x_2 \end{bmatrix} + \begin{bmatrix} 1 \\ 0 \end{bmatrix} F(t) + \begin{bmatrix} 0 \\ 1 \end{bmatrix} v + \begin{bmatrix} -1 \\ 0 \end{bmatrix} z(t,\omega)$$

Consider the design of a nonlinear state-feedback controller

$$F(t) = f(x_1, x_2) = -f_1(x_1) - f_2(x_2)$$

We assume further that the following "sector inequalities" are satisfied

$$m_i \le \frac{f_i(x) - f_i(y)}{x - y} \le M_i \quad i=1,2 \quad \forall\ x,y \in R^1$$

where m_i, M_i are some constants. Let $k_i = (M_i + m_i)/2$, $f_i(x) = k_i x - g_i(x)$, $i=1,2$. The introduced functions g_i have the same sector property as f_i with changed constants i.e. $\tilde{m}_i = -(M_i - m_i)/2$, $\tilde{M}_i = (M_i - m_i)/2$, or in the form used in theorem 1

$$|g_i(x) - g_i(y) - \lambda_i (x-y)| \le r|x-y| ,$$

where $\lambda_i = 0$, $r = \max_i \frac{1}{2}\left(M_i - m_i\right)$ for $i=1,2$.

The equations of the closed-loop transformed system take the form

$$\frac{d}{dt}\begin{bmatrix} x_1 \\ x_2 \end{bmatrix} = \begin{bmatrix} -1-k_1, & -k_2 \\ -1 & , 0 \end{bmatrix}\begin{bmatrix} x_1 \\ x_2 \end{bmatrix} + \begin{bmatrix} g_1(x_1) + g_2(x_2) \\ 0 \end{bmatrix} + \begin{bmatrix} 0 \\ 1 \end{bmatrix} v + \begin{bmatrix} -1 \\ 0 \end{bmatrix} z(t,\omega)$$

The stability requirements are imposed to avoid nonstationary behaviours between the neighbouring vehicles within a long vehicle string. In the notation of theorem 1 we get

$$\hat{K}(s) = \begin{bmatrix} s/\delta & -k_2/\delta \\ -1/\delta & (1+k_1+s)/\delta \end{bmatrix}$$

To assure the assumption (B1) it's needed that the real parts of eigenvalues of the system matrix $\begin{bmatrix} -1-k_1, & -k_2 \\ -1 & , 0 \end{bmatrix}$ are negative i.e. that $k_1 > -1$ and $k_2 < 0$.

It is sufficient to assure that the conditions of theorem 3 are satisfied too. Since the first part of condition (B3) is always satisfied the analysis was concentrated about the second part i.e. about bounds of sector inequality introduced in (B2). Since $r \ge 0$ the upper bound of r was calculated numerically. Results are presented in fig. 2, 3.

Notice that the initial deviation $x_2(0)$ has the same effect as a disturbance

$$d(t) = x_2(0)\delta(t).$$

The discussed values of parameters r, k_1, k_2 assure us that the effects of the initial conditions at t=0 should "die out" and the response of the system "asymptotically" stationary.

REFERENCES

Bensoussan, A. 1971. Filtrage optimal des systemes lineaires. Dunod Paris.

Bharucha-Reid, A.T. 1972. Random integral equations. Academic Press. New York and London.

Dunford, I. & Schwartz, J.T. 1958. Linear operators. Part I. Interscience Publ. New York, London.

Gihman, I.I. & Skorohod, A.V. 1980. The theory of stochastic processes I. Springer-Verlag. Berlin-Heidelberg-New York

Harris, C.J. & Valenca, J.M.E. 1983. The stability of input-output dynamical systems. Academic Press London, New York

Holtzman, J.M. 1968. Analysis of statistical linearization of nonlinear control systems. SIAM J. Control, Vol. 6, N. 2.

Kantorovitsch, L.W. & Akilov, G.P. 1977. Functional analysis. Izd. Nauka. Moskwa

Kudrewicz, J. 1970. Frequency methods in the theory of nonlinear dynamical systems. WNT. Warsaw (in polish)

Przybylak, S. & Skrzypczyk, J. 1989. On some asymptotic properties of nonstationary systems. Sci. Fasc. Silesian Tech. Univ. ser. Math.-Physics, n. 28, Gliwice. 335-352 (in polish)

Sandberg, I.W. 1964. On the L^2-boundedness of solutions of nonlinear functional equations. BSTJ. Vol. 43. 1581-1599.

Sandberg, I.W. & Benes, V.E. 1964. On the properties of nonlinear integral equations that arise in the theory of dynamical systems. BSTJ. Vol. 43. 2839 -2853.

Skrzypczyk, J. 1975. Stability of dynamical systems described by random integral equations. Doct. Dissert. Silesian Techn. University. Gliwice. (in polish).

Skrzypczyk, J. 1974a. L^p_m-stability in the mean of nonlinear stochastic integral equations. Sci. Fasc. Silesian Tech. Univ., ser. Automatic. n. 28. Gliwice. 5-24 (in polish).

Skrzypczyk, J. 1974b. Stability of solutions of nonlinear stochastic integral equations in Banach spaces. Sci. Fasc. Silesian Tech. Univ. ser. Automatic. n. 28. Gliwice. 25-36 (in polish).

Skrzypczyk, J. 1986. Statistical linearization of nonlinear dynamic systems described by integral equations over locally compact Abelian groups. Conf. Nonlinear and Random Vibrations. Oberwolfach.

Skrzypczyk, J. 1989. A note on nonlinear Volterra integral equations over locally compact groups. Sci. Fasc. Silesian Tech. Univ. ser. Math.-Physics. n. 28. Gliwice. 353-363.

Tsokos, Ch. P. & Padgett, W. J. 1971. Random integral equations with applications to stochastic systems. Springer-Verlag. Berlin.

Tsokos, CH. P. & Padgett, W. J. 1974. Random integral equations with applications to life sciences and engineering. Academic Press. New York, London.

Structural Safety & Reliability, Schuëller, Shinozuka & Yao (eds) © *1994 Balkema, Rotterdam, ISBN 90 5410 357 4*

Stochastic linearization method with random parameters and power spectral density calculation

C. Soize
ONERA, Chatillon, France

ABSTRACT: This paper describes the calculation of the Power Spectral Density Function (PSDF) for stationary responses of non-linear second order dynamical systems excited by a white or a broadband gaussian noise. We present an approach based on a stochastic linearization method with random parameters which is an efficient way of approximating the PSDF. We show the gain obtained on Duffing's equation by this method, in comparing it with the statistical linearization method.

1 INTRODUCTION

Our purpose here is to calculate the Power Spectral Density Function (PSDF) for the stationary response of second-order non-linear random oscillators. Statistical linearization method (Roberts and Spanos 1990) is very efficient because the PSDF can be calculated for multi-dimensional non-linear dynamical systems. Unfortunately, when strong non-linear effects exist, this method still yields correct approximations of the root mean-square of responses but overestimates the peak values of the PSDF and underestimates the spectral bandwidths (Miles 1989, Soize 1991, Bouc 1991). To improve the linearization technique, recently, P.D. Spanos and M.G. Donley (1991) proposed an equivalent statistical quadratization for a class of non-linear single-degree-of-freedom systems.

Within the context of identification of dynamical systems a general methodology and practical techniques have been studied by J.S. Bendat(1990) for analyzing a wide class of non-linear systems consisting of parallel linear, bilinear and trilinear systems. R.N. Miles (1989) developed a heuristic approach based on the concept of an equivalent linear system with random coefficients. More recently, R. Bouc (1991) proposed another method for constructing an approximation of the PSDF of weakly damped, strongly non-linear random oscillators. This method is based on an equivalent linear system with coefficients depending on the amplitude process of the non-linear oscillations, and the amplitude process is derived from the averaging stochastic method.

For many years, numerous publications have appeared on methods for solving this type of problem, but it is clear that the greatest difficulty is trying to extend them to multi-dimensional cases. Except for the Monte Carlo numerical simulation method, which can solve practically any non-linear stochastic problem, though with a high cost for large multi-dimensional systems, only the statistical linearization method is really efficient at the present time for use on non-linear stochastic system with many degrees of freedom. Thus the incentive to extend the classical statistical linearization method to improve the prediction of the PSDF for those cases where this method fails.

In this paper we present a stochastic linearization method with random parameters for calculating the PSDF. We have limited the developments to one-degree-of-freedom non-linear dynamical systems, but we believe that extension to multi-dimensional cases is conceivable.

2 NON-LINEAR STOCHASTIC DIFFERENTIAL EQUATION OF THE PROBLEM

We consider the one-dimensional non-linear Stochastic Differential Equation (SDE) :

$$M\ddot{Q}(t) + 2\xi M\omega_0\dot{Q}(t) + M\omega_0^2 Q(t) + \varepsilon f(Q(t), \dot{Q}(t)) = g_0\dot{W}(t) \quad , \qquad (1)$$

in which M, ξ, ω_0 and g_0 are greater than zero, $\varepsilon \geq 0$, $q, \dot{q} \mapsto f(q, \dot{q})$ is a real-valued function on $\mathbb{R} \times \mathbb{R}$ and $\dot{W}$ a normalized Gaussian white noise with power spectral measure $(1/2\pi)d\omega$. We assume that the function f verifies the necessary conditions for Eq. (1) to have a unique stationary solution $Q(t)$, which is a mean-square continuous second-order process, centered, and once mean-square differentiable on $\mathbb{R}$. We will denote its PSDF by $\omega \mapsto S_Q(\omega)$. Let $\mathbb{M}_{\mu,\nu}$ be the moments defined for all μ and ν in $\mathbb{N}$ by

$$\mathbb{M}_{\mu,\nu} = E(Q(t)^\mu \dot{Q}(t)^\nu) = \int_{\mathbb{R}}\int_{\mathbb{R}} q^\mu \dot{q}^\nu p_{Q(t),\dot{Q}(t)}(q, \dot{q})\, dq\, d\dot{q} \quad , \qquad (2)$$

where $p_{Q(t),\dot{Q}(t)}(q,\dot{q})\, dq\, d\dot{q}$ is the probability law of the random variable $(Q(t), \dot{Q}(t))$ on $\mathbb{R}^2$ for all fixed t. We will assume that there are two integers, $m_{max} \geq 1$ and $n_{max} \geq 1$, such that for all $m \in \{0, 1, \ldots, m_{max}\}$ and $n \in \{0, 1, \ldots, n_{max}\}$ we have $\mathbb{M}_{2m,2n} < +\infty$, and that all the odd-order moments are zero, i.e. $\mathbb{M}_{2m+1,n} = \mathbb{M}_{m,2n+1} = 0$ for all m and n in $\mathbb{N}$.

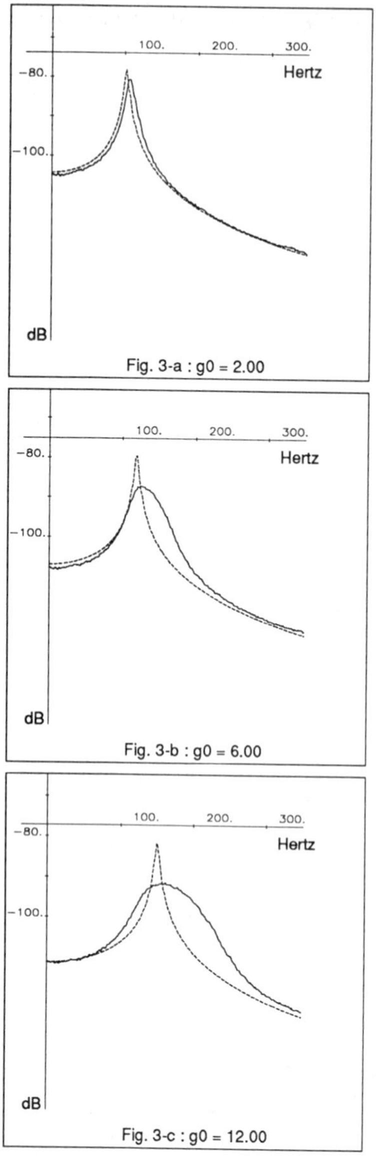

Fig.1 : CASE 1

———— numerical simulation

- - - - statistical linearization

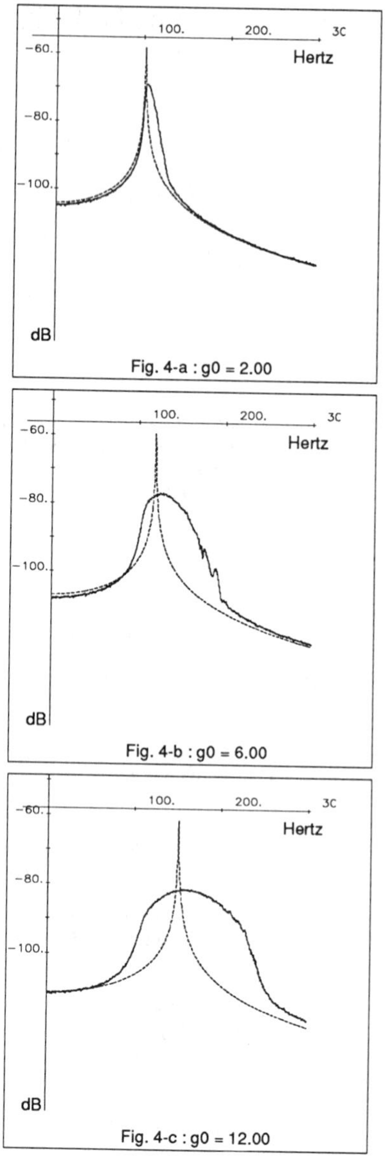

Fig.2 : CASE 2

———— numerical simulation

- - - - statistical linearization

3 LINEAR STOCHASTIC DIFFERENTIAL EQUATION WITH RANDOM PARAMETERS FOR CONSTRUCTING AN EQUIVALENT SYSTEM

The linear SDE with random parameters that will be used to approximate Eq. (1) is written

$$M\ddot{Y}(t) + 2\xi M\omega_0(1+\varepsilon\Theta)\dot{Y}(t) + M\omega_0^2(1+\varepsilon\Lambda)Y(t) = g_0\dot{W}(t) \quad , \qquad (3)$$

where M, ξ, ω_0, g_0, ε and $\dot{W}$ are the previously defined quantities. The $\mathbb{R}^2$-vector (Θ, Λ) is a random variable, independent of the process $\dot{W}$, and its probability law $P_{\Theta,\Lambda}(d\theta, d\lambda)$ on $\mathbb{R}^2$ verifies the following property concerning its support :

$$Supp\, P_{\Theta,\Lambda} \subset [\theta_0, +\infty[\times[\lambda_0, +\infty[\quad , \qquad (4)$$

where $1 + \varepsilon\theta_0 > 0$ with $\theta_0 \leq 0$ and $1 + \varepsilon\lambda_0 > 0$ with $\lambda_0 \leq 0$, and

$$\int_{\mathbb{R}}\int_{\mathbb{R}} \frac{P_{\Theta,\Lambda}(d\theta, d\lambda)}{(1+\varepsilon\theta)^{m+n}(1+\varepsilon\lambda)^m} < +\infty \quad , \qquad (5)$$

for all $m \in \{0, 1, \ldots, m_{max}\}$ and $n \in \{0, 1, \ldots, n_{max}\}$. Let us note that this model allows us to write $P_{\Theta,\Lambda}(d\theta, d\lambda) = \delta_0 \otimes P_\Lambda(d\lambda)$ or $P_{\Theta,\Lambda}(d\theta, d\lambda) = P_\Theta(d\theta) \otimes \delta_0$, where δ_0 is the Dirac measure at the origin of $\mathbb{R}$. Consequently, we can use a model with only one random parameter (stiffness or damping), or with two random parameters (stiffness and damping). With the hypotheses defined by Eqs. (4) and (5), the SDE (3) has a stationary solution $Y(t)$ which is a mean-square continuous second-order centered process, once mean-square differentiable on $\mathbb{R}$. For all t fixed in $\mathbb{R}$, the $\mathbb{R}^2$-valued random variable $(Y(t), \dot{Y}(t))$ has a non gaussian probability density function $p_{Y(t),\dot{Y}(t)}(y, \dot{y}))$ which is explicitly known (Soize 1991), and the PSDF $S_Y(\omega)$ of the $Y(t)$ process is written

$$S_Y(\omega) = \frac{2\xi C_0}{\pi\omega_0} \times \int_{\mathbb{R}}\int_{\mathbb{R}} \frac{P_{\Theta,\Lambda}(d\theta, d\lambda)}{\{(1+\varepsilon\lambda - (\frac{\omega}{\omega_0})^2)^2 + 4\xi^2(\frac{\omega}{\omega_0})^2(1+\varepsilon\theta)^2\}} \qquad (6)$$

with $C_0 = g_0^2/(4\xi M^2\omega_0^3)$. Let $\mathcal{M}_{\mu,\nu}$ be the moments defined for μ and $\nu \geq 0$ by

$$\begin{aligned}\mathcal{M}_{\mu,\nu} &= E(Y(t)^\mu \dot{Y}(t)^\nu) \\ &= \int_{\mathbb{R}}\int_{\mathbb{R}} y^\mu \dot{y}^\nu p_{Y(t),\dot{Y}(t)}(y, \dot{y})\, dy\, d\dot{y} \quad . \qquad (7)\end{aligned}$$

Then, for all $m \in \{0, \ldots, m_{max}\}$ and $n \in \{0, \ldots, n_{max}\}$, we have

$$\mathcal{M}_{2m,2n} = \frac{(2m)!}{m!\,2^m}\frac{(2n)!}{n!\,2^n}\omega_0^{2n} C_0^{m+n} \times \int_{\mathbb{R}}\int_{\mathbb{R}} \frac{P_{\Theta,\Lambda}(d\theta, d\lambda)}{(1+\varepsilon\theta)^{m+n}(1+\varepsilon\lambda)^m} < +\infty, \qquad (8)$$

$$\mathcal{M}_{2m+1,n} = \mathcal{M}_{m,2n+1} = 0 \ , \ \forall m \in \mathbb{N} \ , \ \forall n \in \mathbb{N} . \qquad (9)$$

In particular we deduce from Eqs. (8) and (9) that $C_{Y\dot{Y}} = E(Y(t)\dot{Y}(t)) = 0$, and that $C_Y = E(Y(t)^2)$ and $C_{\dot{Y}} = E(\dot{Y}(t)^2)$ can be written as

$$C_Y = C_0 \int_{\mathbb{R}}\int_{\mathbb{R}} \frac{P_{\Theta,\Lambda}(d\theta, d\lambda)}{(1+\varepsilon\theta)(1+\varepsilon\lambda)} \quad , \qquad (10)$$

$$C_{\dot{Y}} = \omega_0^2 C_0 \int_{\mathbb{R}} \frac{P_\Theta(d\theta)}{1+\varepsilon\theta} \quad .$$

4 STOCHASTIC LINEARIZATION METHOD WITH RANDOM PARAMETERS AND CALCULATION OF THE POWER SPECTRAL DENSITY FUNCTION

The proposed method of stochastic linearization with random parameters consists in constructing the probability measure $P_{\Theta,\Lambda}(d\theta, d\lambda)$ in such a way as to minimize the distance between the two stationary stochastic processes $Q(t)$ and $Y(t)$ verifying Eq. (1) and Eq. (3) respectively. We have studied in detail (Soize 1991) several possible choices for this distance, and we summarize the main results below. Firstly, it is very easy to see that the processes Q and Y cannot be stochastically equivalent in the wide sense, i.e. no probability measure $P_{\Theta,\Lambda}$ exists such that the two processes Q and Y have identical systems of marginal probability laws. Secondly, we have proved that the following problem is correctly formulated. Find the probability measure $P_{\Theta,\Lambda}$, solution of the integral equation on $\mathbb{R}^2$:

$$\int_{\mathbb{R}}\int_{\mathbb{R}} p_{Y_{\theta,\lambda}(t),\dot{Y}_{\theta,\lambda}(t)}(q, \dot{q}|\theta, \lambda) P_{\Theta,\Lambda}(d\theta, d\lambda) = p_{Q(t),\dot{Q}(t)}(q, \dot{q}) \quad , \qquad (11)$$

where the conditional probability density function $p_{Y_\theta,\lambda(t),\dot{Y}_\theta,\lambda(t)}(q, \dot{q}|\theta, \lambda)$ is an explicitly known Gaussian measure. This criterion, defined by Eq. (11), corresponds to the equality $P_{Y(t),\dot{Y}(t)} = P_{Q(t),\dot{Q}(t)}$ of the invariant measures. Existence of a probability measure verifying Eqs. (4), (5) and (11) is a problem which is opened at the present time for the general case. Thirdly one can write the equality of a finite number of moments. As all the odd-order moments are zero, one can find the probability measure $P_{\Theta,\Lambda}$ verifying Eqs. (4) and (5), and solve a finite number of integral equations defined by the equalities $\mathbb{M}_{2m,2n} = \mathcal{M}_{2m,2n}$. This problem is close to the previous one, and is related to the representation of $P_{\Theta,\Lambda}$ as it is developed in (Soize 1991). Finally, as the aim of this developments is to construct the PSDF, it is quite natural to determine $P_{\Theta,\Lambda}$ by writing the equality of the second-order moments $\mathbb{M}_{2,0} = \mathcal{M}_{2,0}$, $\mathbb{M}_{0,2} = \mathcal{M}_{0,2}$, and obviously, the normalization condition. So, we obtain the following formulation.

Problem $(\mathcal{P})$. Find the probability measure $P_{\Theta,\Lambda}(d\theta, d\lambda)$ on $\mathbb{R}^2$ having to verify condition (4) and such that

$$\int_{\mathbb{R}}\int_{\mathbb{R}} \frac{P_{\Theta,\Lambda}(d\theta, d\lambda)}{(1+\varepsilon\theta)(1+\varepsilon\lambda)} = \mathcal{S}_{2,0} \quad , \qquad (12)$$

$$\int_{\mathbb{R}}\int_{\mathbb{R}} \frac{P_{\Theta,\Lambda}(d\theta, d\lambda)}{(1+\varepsilon\theta)} = \mathcal{S}_{0,2}$$

where $\mathcal{S}_{2,0}$ and $\mathcal{S}_{0,2}$ are the positive constants such that $\mathbb{M}_{2,0} = C_0\mathcal{S}_{2,0}$ and $\mathbb{M}_{0,2} = \omega_0^2 C_0 \mathcal{S}_{0,2}$.

A priori, the solution of problem $(\mathcal{P})$ is not unique. Therefore, we are going to define a class of representation of the probability measure $P_{\Theta,\Lambda}(d\theta, d\lambda)$.

Definition 1. One defines the family of probabilities $P_X(dx) = p_X(x)\,dx$, with three parameters $x_1 \in \mathbb{R}$, $\alpha > 0$ and $\beta > 0$, such that for all $x \in \mathbb{R}$

$$p_X(x) = \alpha(1+\varepsilon x)W_X(x) \quad , \qquad (13)$$

$$W_X(x) = \mathbf{1}_{[x_1,+\infty[}(x)(x - x_1)e^{-\beta(x-x_1)^2} \quad .$$

This one-order finite expansion (13) in orthogonal polynomials with respect to the weight W_X implies that the density $p_X(x)$ is a continuous function defined on $\mathbb{R}$ with

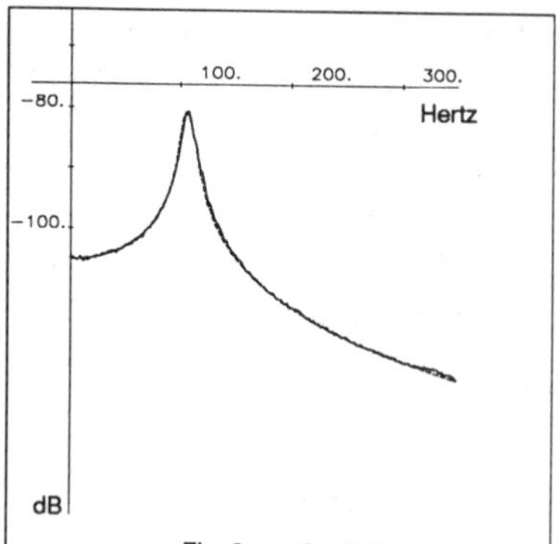

Fig. 3-a : g0 = 2.00

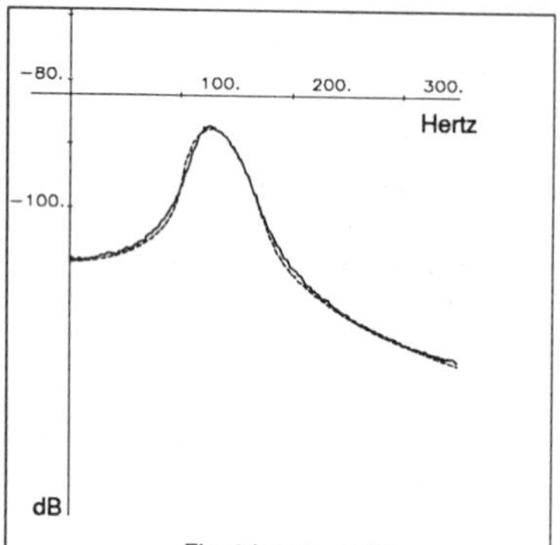

Fig. 3-b : g0 = 6.00

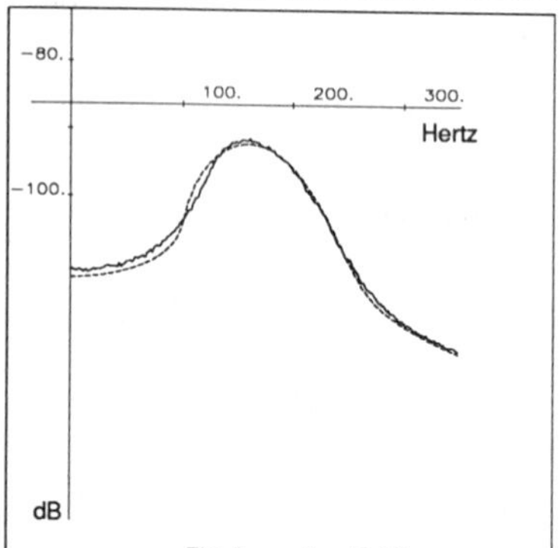

Fig. 3-c : g0 = 12.00

Fig.3 : CASE 1

——— numerical simulation

- - - - SLM with random parameters

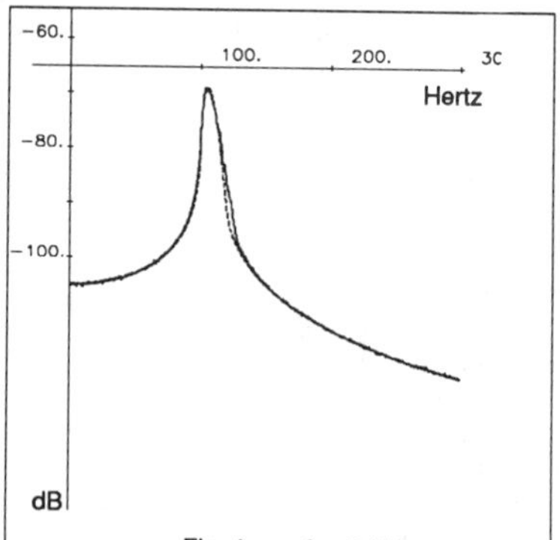

Fig. 4-a : g0 = 2.00

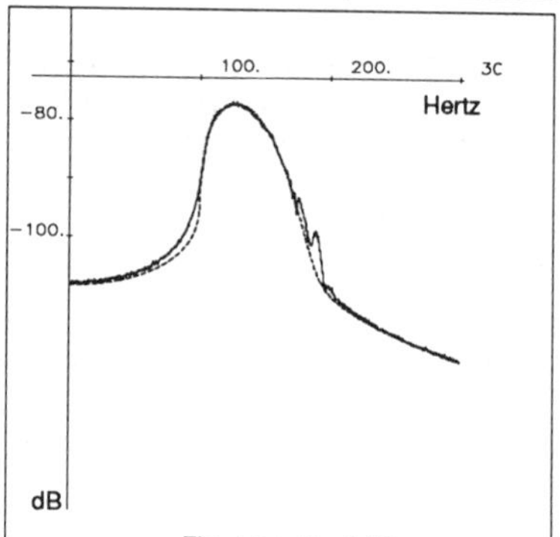

Fig. 4-b : g0 = 6.00

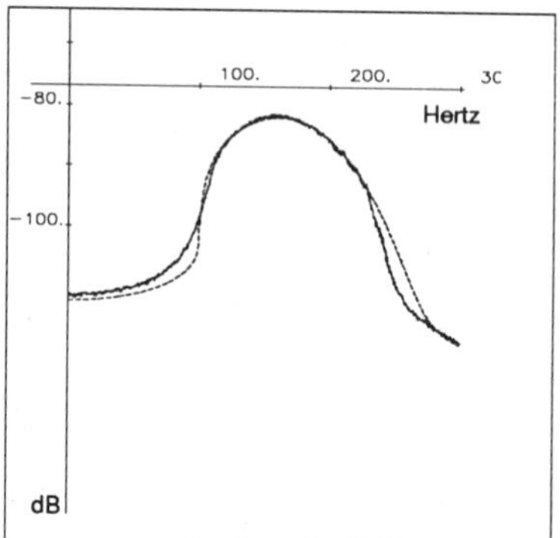

Fig. 4-c : g0 = 12.00

Fig.4 : CASE 2

——— numerical simulation

- - - - SLM with random parameters

support $[x_1, +\infty[$ and with all the m-order moments finite. We have proved (Soize 1991) the three following results for the three types of modeling.

Proposition 1. If $\mathcal{S}_{0,2} = 1$, i.e. $\mathbb{M}_{0,2} = \omega_0^2 C_0$, we can write $P_{\Theta,\Lambda}(d\theta, d\lambda) = \delta_0 \otimes P_\Lambda(d\lambda)$ and we look for $P_\Lambda(d\lambda)$ in the family of Definition 1, using the three parameters $\lambda_1 > -1/\varepsilon$, $\alpha_\Lambda > 0$, $\beta_\Lambda > 0$:

$$P_\Lambda(d\lambda) = \alpha_\Lambda (1+\varepsilon\lambda) \mathbf{1}_{[\lambda_1,+\infty[}(\lambda)\, (\lambda - \lambda_1)\, e^{-\beta_\Lambda(\lambda-\lambda_1)^2}\, d\lambda \, . \tag{14}$$

For all $\mathcal{S}_{2,0} > 0$, if λ_1 is chosen such that $0 < 1 + \varepsilon\lambda_1 < \mathcal{S}_{2,0}^{-1}$, then the problem $(\mathcal{P})$ has a unique solution which is given by Eq. (14) with $\beta_\Lambda = \pi\varepsilon^2 \mathcal{S}_{2,0}^2 (1 - \mathcal{S}_{2,0}(1+\varepsilon\lambda_1))^{-2}/4$ and $\alpha_\Lambda = 2\beta_\Lambda \mathcal{S}_{2,0}$. The approximation $S_Y(\omega)$ given by Eq. (6) of the PSDF $S_Q(\omega)$ is written

$$S_Y(\omega) = \frac{4\xi C_0}{\pi\omega_0} \mathcal{S}_{2,0} \times$$

$$\int_0^{+\infty} \frac{x\,(1+\varepsilon\lambda_1+Ax)\,e^{-x^2}\,dx}{\{(1+\varepsilon\lambda_1+Ax-(\frac{\omega}{\omega_0})^2)^2+4\xi^2(\frac{\omega}{\omega_0})^2\}} \quad (15)$$

with $A = 2\pi^{-1/2}(\mathcal{S}_{2,0}^{-1} - (1+\varepsilon\lambda_1)) > 0$.

Proposition 2. If $\mathcal{S}_{0,2} = \mathcal{S}_{2,0}$, i.e. $\mathbb{M}_{0,2} = \omega_0^2 \mathbb{M}_{2,0}$, we can write $P_{\Theta,\Lambda}(d\theta, d\lambda) = P_\Theta(d\theta) \otimes \delta_0$ and we look for $P_\Theta(d\theta)$ in the family of Definition 1, using the three parameters $\theta_1 > -1/\varepsilon$, $\alpha_\Theta > 0$, $\beta_\Theta > 0$:

$$P_\Theta(d\theta) = \alpha_\Theta(1+\varepsilon\theta)\mathbf{1}_{[\theta_1,+\infty[}(\theta)\,(\theta-\theta_1)\,e^{-\beta_\Theta(\theta-\theta_1)^2}\,d\theta\,. \quad (16)$$

For all $\mathcal{S}_{0,2} > 0$, if θ_1 is chosen such that $0 < 1+\varepsilon\theta_1 < \mathcal{S}_{0,2}^{-1}$, then the problem $(\mathcal{P})$ has a unique solution which is given by Eq. (16), with $\beta_\Theta = \pi\varepsilon^2\mathcal{S}_{0,2}^2(1-\mathcal{S}_{0,2}(1+\varepsilon\theta_1))^{-2}/4$ and $\alpha_\Theta = 2\beta_\Theta\mathcal{S}_{0,2}$. The approximation $S_Y(\omega)$ given by Eq. (6) of the PSDF $S_Q(\omega)$ is written

$$S_Y(\omega) = \frac{4\xi C_0}{\pi\omega_0} \mathcal{S}_{0,2} \times$$

$$\int_0^{+\infty} \frac{y\,(1+\varepsilon\theta_1+By)\,e^{-y^2}\,dy}{\{(1-(\frac{\omega}{\omega_0})^2)^2+4\xi^2(\frac{\omega}{\omega_0})^2(1+\varepsilon\theta_1+By)^2\}} \quad (17)$$

with $B = 2\pi^{-1/2}(\mathcal{S}_{0,2}^{-1} - (1+\varepsilon\theta_1)) > 0$.

Proposition 3. If the hypotheses of Propositions 1 and 2 do not hold, we consider Θ and Λ as independent random variables. We can write $P_{\Theta,\Lambda}(d\theta, d\lambda) = P_\Theta(d\theta) \otimes P_\Lambda(d\lambda)$, with $P_\Theta(d\theta)$ given by Eq. (16) and $P_\Lambda(d\lambda)$ given by Eq. (14). For all $\mathcal{S}_{0,2} > 0$ and $\mathcal{S}_{2,0} > 0$ such that $\mathcal{S}_{0,2} \neq \mathcal{S}_{2,0}$ and $\mathcal{S}_{0,2} \neq 1$, if λ_1 is chosen such that $0 < 1+\varepsilon\lambda_1 < \mathcal{S}_{0,2}\mathcal{S}_{2,0}^{-1}$ and θ_1 as in Proposition 2, then the problem $(\mathcal{P})$ has a unique solution which is given by Eqs. (16) and (14) with α_Θ and β_Θ as in Proposition 2, and with $\beta_\Lambda = \pi\varepsilon^2\mathcal{S}_{2,0}^2(\mathcal{S}_{0,2}-\mathcal{S}_{2,0}(1+\varepsilon\lambda_1))^{-2}/4$ and $\alpha_\Lambda = 2\beta_\Lambda\mathcal{S}_{2,0}\mathcal{S}_{0,2}^{-1}$. The approximation $S_Y(\omega)$ given by Eq. (6) of the PSDF $S_Q(\omega)$ is written

$$S_Y(\omega) = \frac{8\xi C_0 \mathcal{S}_{2,0}}{\pi\omega_0} \int_0^{+\infty}\int_0^{+\infty}$$

$$\frac{(1+\varepsilon\lambda_1+Ax)(1+\varepsilon\theta_1+By)xye^{-x^2-y^2}\,dx\,dy}{\{(1+\varepsilon\lambda_1+Ax-(\frac{\omega}{\omega_0})^2)^2+4\xi^2(\frac{\omega}{\omega_0})^2(1+\varepsilon\theta_1+By)^2\}} \quad (18)$$

with B as in Proposition 2 and $A = 2\pi^{-1/2}(\mathcal{S}_{0,2}\mathcal{S}_{2,0}^{-1} - (1+\varepsilon\lambda_1))$.

5 APPLICATION TO THE DUFFING EQUATION

We consider the Duffing equation defined by Eq. (1) with $f(q,\dot{q}) = M\omega_0^2 q^3$, and with $M = 1$, $\omega_0 = 2\pi\times 100$, $g_0 = 2$, $= 6$ or $= 12$. Two cases are studied, **Case 1** : $\xi = 0.025$, $\varepsilon = 344\,514.18$ and **Case 2** : $\xi = 0.0025$, $\varepsilon = 34\,451.418$. We know that $\mathcal{S}_{2,0} = \gamma_\varepsilon U(1,\gamma_\varepsilon)/U(0,\gamma_\varepsilon) < 1$ and $\mathcal{S}_{0,2} = 1$, with $\gamma_\varepsilon = (2\varepsilon C_0)^{-1/2}$ and $U(a,x)$ the parabolic cylinder function. These formulas yield, for the two cases, $\mathcal{S}_{0,2} = 1$ and $\mathcal{S}_{2,0} = 0.87942$, $= 0.57920$ and $= 0.36596$ for $g_0 = 2$, $= 6$ and $= 12$ respectively. Figs. 1 to 4 give the normalized levels in dB : $10\log_{10}(S(\omega)/S_0)$ with $S_0 = g_0^2/(2\pi)$, of PSDF $S(\omega)$ as a function of the frequency ω expressed in hertz in the band 0 Hz - 350 Hz. Figs. 1 and 2 compare, for the two cases, the PSDF $S(\omega) = S_Q(\omega)$ calculated by Monte Carlo numerical simulation of Eq. (1), with the PSDF $S(\omega) = S_{Q,lin}(\omega)$ calculated by the statistical linearization method (Roberts and Spanos 1990). We see in Fig.1-a that the latter method yields correct predictions for small-amplitude responses (small amplitude of the excitation $g_0 = 2$ and a not-too-small damping $\xi = 0.025$), but wild correct predictions when the amplitude of the responses increase, as shown in Figs.1-c and 2-c. Figs. 3 and 4 show, for the same two cases, the PSDF $S(\omega) = S_Q(\omega)$ previously calculated by Monte Carlo numerical simulation of Eq. (1), with the PSDF $S(\omega) = S_Y(\omega)$ calculated by the stochastic linearization method with random parameters, i.e. by Eq. (15). We see that correct predictions are obtained for small and large amplitude responses.

6 CONCLUSIONS

We have presented a Stochastic Linearization method with Random Parameters (RPSL method) which allows us to calculate an approximation of the Power Spectral Density Functions (PSDF) of stationary responses of stochastic non-linear second-order dynamical systems.

We have seen that this RPSL method yields correct predictions of the PSDF for the stationary reponse of the stochastic Duffing equation, whereas the PSDF is not correctly predicted in this case by the usual statistical linearization method.

Presently, the theory has been developed only for one-dimensional second-order systems with a white noise external excitation, but this can immediately be extended to construct the PSDF of the stationary response of the same type of system with a colored stochastic external excitation. In this case we can use the statistical linearization method first, for calculating an approximation of the second-order moments required by the RPSL method,and then use a lightly modified RPSL method to predict the PSDF.

In the context of one-dimensional second-order dynamical systems, we think that this RPSL method should be advantageous for approximating the PSDF of these systems with additional parametric stationary stochastic excitations.

Finally, the extension of this RPSL method to the multi-dimensional case must be studied, considering that the underlying basis of this method is the linear theory, which is very efficient in constructing the PSDF in multi-dimensional cases.

REFERENCES

Bendat, J.S. 1990. *Nonlinear system analysis and identification from random data.* John Wiley and Sons, New York.

Spanos, P.D. and Donley, M.G. 1991. Equivalent statistical quadratization for non-linear systems. *J. of Eng. Mech.*, ASMC, 117(6): 1289–1310.

Roberts, J.B. and Spanos, P.D. 1990. *Random vibration and statistical linearization.* John Wiley and Sons, New York.

Miles, R.N. 1989. An approximate solution for the spectral response of Duffing's oscillator with random input. *J. of Sound and Vibration*, 132(1): 43–49.

Soize, C. 1991. On the calculation of spectral density functions of stationary responses for non-linear stochastic dynamical systems. In Contrôle actif vibro-acoustique et dynamique stochastique, *Publications du LMA (CNRS, Marseille, France)*, ISSN 0750-7356, 127: 297–344.

Bouc, R. 1991. The power spectral density of a weakly damped strongly non-linear random oscillation and stochastic averaging. In Contrôle actif vibro-acoustique et dynamique stochastique, *Publications du LMA (CNRS, Marseille, France)*, ISSN 0750-7356, 127: 373–395.

Structural Safety & Reliability, Schuëller, Shinozuka & Yao (eds) © 1994 Balkema, Rotterdam, ISBN 90 5410 357 4

Hilbert transform approach for linear system spectral moment determination

P.D. Spanos & S.M. Miller
Rice University, Houston, Tex., USA

ABSTRACT: The stationary response of a linear system excited by a random process is considered. Attention is focused on the determination of the spectral moments of the response by considering one-sided power spectral density functions. Both even order and odd order spectral moments are determined. This is achieved by decomposing the one-sided spectral density function of the response into its even and odd parts. Using linear system theory, differential equations which govern the various auto- and cross-correlation functions and their Hilbert transforms are derived. Successive differentiation of these equations leads to linear equations for the even and odd order spectral moments for the spectral density of the response. It is also shown that higher order spectral moments of both even and odd order can be found by using a recursive relation. Several examples are provided involving, among others, the clipped white spectrum and the Kanai-Tajimi seismic spectrum.

1 INTRODUCTION

A class of integrals often encountered in random vibrations and many other scientific fields is considered. Within the field of random vibrations, these integrals may represent the spectral moments of the response of a linear, time-invariant system to a stationary random excitation. Using the properties of correlation functions and the Hilbert transform, linear algebraic equations involving the response spectral moments are derived. The present approach is a generalization of that presented by Spanos (1983, 1987) and includes the well-known James et al. (1965) integral of control engineering as a special case.

The paper provides general formulae for use in calculating the response spectral moments if the system excitation is white-noise, band-limited white noise, or has a "Gaussian" shaped spectrum. Furthermore, the applicability of the method for other spectra is discussed. Finally, the response of a standard linear oscillator to an excitation possessing the well-known and often-used Kanai-Tajimi earthquake spectrum is considered.

2 DIFFERENTIAL EQUATIONS FOR CORRELATION FUNCTIONS

Consider an mth order linear system with random excitation $f(t)$. The system response is governed by the differential equation

$$\lambda_m x^{(m)}(t) + \lambda_{m-1} x^{(m-1)}(t) + \cdots + \lambda_0 x(t) = f(t). \quad (1)$$

Here, $x^{(k)}(t) = \frac{d^k}{dt^k} x(t)$, and λ_k $(k = 0, 1, 2, \ldots, m)$ are time-invariant constants. Assume that $f(t)$ is a stationary random process with autocorrelation function $R_{ff}(\tau)$. Recall that the autocorrelation function and the double-sided power spectral density, $S_{ff}(\omega)$, form a Fourier transform pair. That is, they satisfy the equations

$$S(\omega) = \frac{1}{2\pi} \int_{-\infty}^{\infty} R(\tau) e^{-i\omega\tau} d\tau \,, \quad (2)$$

$$R(\tau) = \int_{-\infty}^{\infty} S(\omega) e^{i\omega\tau} d\omega. \quad (3)$$

In addition, the one-sided PSD, $G_{ff}(\omega)$, is related to its double-sided counterpart by the equation $G(\omega) = 2U(\omega)S(\omega) = S(\omega) + S(\omega)\text{sgn}(\omega)$, where $U(\omega)$ is the unit-step function and $\text{sgn}(\omega)$ is the signum function. Furthermore, the response spectral density of the linear system is related to the input spectral density by the equation

$$S_{xx}(\omega) = \frac{S_{ff}(\omega)}{|\Lambda(i\omega)|^2}, \tag{4}$$

where $\Lambda(s) = \lambda_m s^m + \lambda_{m-1}s^{m-1} + \cdots + \lambda_0$ is the characteristic polynomial of the linear system.

To derive differential equations which govern $R_{xx}(\tau)$ and $R_{xf}(\tau)$, one must first note the relations $E[x(t)x^{(k)}(t-\tau)] = (-1)^k R_{xx}^{(k)}(\tau)$ and $E[x^{(k)}(t)f(t-\tau)] = R_{xf}^{(k)}(\tau)$, where $(\cdot)^{(k)} = \frac{\partial^k}{\partial\tau^k}(\cdot)$ and $E[\,\cdot\,]$ denotes the expectation operator. Thus, evaluating the system differential equation at time $t-\tau$, multiplying the expression by $x(t)$, and applying the expectation operator yield

$$\tilde{\lambda}_m R_{xx}^{(m)}(\tau) + \tilde{\lambda}_{m-1}R_{xx}^{(m-1)}(\tau) + \cdots + \tilde{\lambda}_0 R_{xx}(\tau) = R_{xf}(\tau), \tag{5}$$

where $\tilde{\lambda}_k = (-1)^k\lambda_k$ $(k = 0,1,2,\ldots,m)$. Furthermore, multiplying the system differential equation by $f(t-\tau)$ and applying the expectation operator yield

$$\lambda_m R_{xf}^{(m)}(\tau) + \lambda_{m-1}R_{xf}^{(m-1)}(\tau) + \cdots + \lambda_0 R_{xf}(\tau) = R_{ff}(\tau). \tag{6}$$

Finally, differentiating equation (5) j times with respect to τ, and differentiating equation (6) n times with respect to τ give

$$\tilde{\lambda}_m R_{xx}^{(j+m)}(\tau) + \tilde{\lambda}_{m-1}R_{xx}^{(j+m-1)}(\tau) + \cdots + \tilde{\lambda}_0 R_{xx}^{(j)}(\tau) = R_{xf}^{(j)}(\tau), \tag{7}$$

and

$$\lambda_m R_{xf}^{(n+m)}(\tau) + \lambda_{m-1}R_{xf}^{(n+m-1)}(\tau) + \cdots + \lambda_0 R_{xf}^{(n)}(\tau) = R_{ff}^{(n)}(\tau), \tag{8}$$

respectively. Clearly, these equations are valid only if the required differentiation is meaningful.

Note that equation (6) is a linear differential equation with excitation $R_{ff}(\tau)$ and response $R_{xf}(\tau)$. Thus, one may express $R_{xf}(\tau)$ as a convolution of the system impulse response function, $h(\tau)$, and $R_{ff}(\tau)$. Using Leibniz' rule to differentiate this convolution with respect to τ, one finds

$$R_{xf}^{(k)}(\tau) = \int_{-\infty}^{\infty} h^{(k)}(\tau-u)R_{ff}(u)du\,. \tag{9}$$

Next, consider the Hilbert transform $\hat{x}(t)$ of $x(t)$ (Papoulis, 1984). This is defined by the equation

$$\hat{x}(t) = \frac{1}{\pi}\int_{-\infty}^{\infty}\frac{x(\tau)}{t-\tau}\,d\tau = x(t) * \frac{1}{\pi t}. \tag{10}$$

Note that the Hilbert transform operation commutes with differentiation and with convolution. Now, taking the Hilbert transform of equations (7), (8), and (9) gives

$$\tilde{\lambda}_m \hat{R}_{xx}^{(j+m)}(\tau) + \tilde{\lambda}_{m-1}\hat{R}_{xx}^{(j+m-1)}(\tau) + \cdots + \tilde{\lambda}_0 \hat{R}_{xx}^{(j)}(\tau) = \hat{R}_{xf}^{(j)}(\tau), \tag{11}$$

$$\lambda_m \hat{R}_{xf}^{(n+m)}(\tau) + \lambda_{m-1}\hat{R}_{xf}^{(n+m-1)}(\tau) + \cdots + \lambda_0 \hat{R}_{xf}^{(n)}(\tau) = \hat{R}_{ff}^{(n)}(\tau), \tag{12}$$

and

$$\hat{R}_{xf}^{(k)}(\tau) = \int_{-\infty}^{\infty} h^{(k)}(\tau-u)\hat{R}_{ff}(u)du\,, \tag{13}$$

respectively, for $k = 0,1,2,\ldots,m-1$. Again, these equations are only valid if the differentiation is meaningful.

3 EQUATIONS FOR THE SPECTRAL MOMENTS

Reserving the notation $S(\omega)$ for spectral densities which are possibly nonzero over the whole range of positive and negative frequencies, an equivalent double-sided spectral density is introduced such that $S_{zz}(\omega) = G_{xx}(\omega) = S_{xx}(\omega) + S_{xx}(\omega)\mathrm{sgn}(\omega)$. In essence, the random process $z(t)$ is introduced such that $S_{zz}(\omega) = G_{xx}(\omega)$. Taking the inverse Fourier transform of $S_{zz}(\omega)$ yields

$$\begin{aligned} R_{zz}(\tau) &= R_{xx}(\tau) + iR_{xx}(\tau) * \left(\frac{1}{\pi t}\right) \\ &= R_{xx}(\tau) + i\hat{R}_{xx}(\tau), \end{aligned} \tag{14}$$

where $\hat{R}_{xx}(\tau)$ denotes the Hilbert transform of $R_{xx}(\tau)$.

Now, consider the kth spectral moment of the system response. This can be expressed as

$$\begin{aligned} M_k &= \int_0^{\infty}\omega^k G_{xx}(\omega)\,d\omega = \int_{-\infty}^{\infty}\omega^k S_{zz}(\omega)\,d\omega \\ &= (-i)^k R_{zz}^{(k)}(0), \end{aligned} \tag{15}$$

where $k = 0,1,2,\ldots$, and the inverse Fourier transform of $S_{zz}(\omega)$ has been differentiated and evaluated at $\tau = 0$. Note that for real-valued random processes, the autocorrelation function is an even function. Furthermore, it can be shown that the Hilbert transform of an even function is an odd function. Thus, it is clear that the odd order derivatives of $R_{xx}(\tau)$ and the even order derivatives of $\hat{R}_{xx}(\tau)$ must be zero when evaluated at

$\tau = 0$. Then, substituting equation (14) into equation (15), the expressions for the even and the odd spectral moments become uncoupled. Specifically, it is found that

$$M_{2k} = (-1)^k R_{xx}^{(2k)}(0)\,, \quad R_{xx}^{(2k+1)}(0) = 0\,, \tag{16}$$

and

$$M_{2k+1} = (-1)^k \hat{R}_{xx}^{(2k+1)}(0)\,, \quad \hat{R}_{xx}^{(2k)}(0) = 0\,. \tag{17}$$

Next consider equation (7) when $\tau = 0$. This becomes merely a linear algebraic equation involving the even order response spectral moments. Thus, it is possible to obtain m linear algebraic equations by setting j equal to $m-1, m-2, m-3, \ldots, 0$. A similar procedure can be applied to equation (11) to derive a set of linear algebraic equations involving the response spectral moments of odd order. Specifically, the following systems of equations are obtained

$$\begin{bmatrix} \lambda_{m-1} & -\lambda_{m-3} & \lambda_{m-5} & \cdot & \cdot & \cdot & \cdot \\ -\lambda_m & \lambda_{m-2} & -\lambda_{m-4} & \cdot & \cdot & \cdot & \cdot \\ 0 & -\lambda_{m-1} & \lambda_{m-3} & \cdot & \cdot & \cdot & \cdot \\ \cdot & \cdot & \cdot & \cdot & \cdot & \cdot & \cdot \\ \cdot & \cdot & \cdot & \cdot & \lambda_2 & -\lambda_0 & 0 \\ \cdot & \cdot & \cdot & \cdot & -\lambda_3 & \lambda_1 & 0 \\ \cdot & \cdot & \cdot & \cdot & \lambda_4 & -\lambda_2 & \lambda_0 \end{bmatrix} \begin{Bmatrix} M_{2m-2} \\ M_{2m-4} \\ M_{2m-6} \\ \cdot \\ M_4 \\ M_2 \\ M_0 \end{Bmatrix} = \begin{Bmatrix} r_{m-1} \\ r_{m-2} \\ r_{m-3} \\ \cdot \\ r_2 \\ r_1 \\ r_0 \end{Bmatrix} \tag{18}$$

and

$$\begin{bmatrix} -\lambda_m & \lambda_{m-2} & -\lambda_{m-4} & \cdot & \cdot & \cdot & \cdot \\ 0 & -\lambda_{m-1} & \lambda_{m-3} & \cdot & \cdot & \cdot & \cdot \\ 0 & \lambda_m & -\lambda_{m-2} & \cdot & \cdot & \cdot & \cdot \\ \cdot & \cdot & \cdot & \cdot & \cdot & \cdot & \cdot \\ \cdot & \cdot & \cdot & \cdot & -\lambda_3 & \lambda_1 & 0 \\ \cdot & \cdot & \cdot & \cdot & \lambda_4 & -\lambda_2 & \lambda_0 \\ \cdot & \cdot & \cdot & \cdot & -\lambda_5 & \lambda_3 & -\lambda_1 \end{bmatrix} \begin{Bmatrix} M_{2m-1} \\ M_{2m-3} \\ M_{2m-5} \\ \cdot \\ M_5 \\ M_3 \\ M_1 \end{Bmatrix} = \begin{Bmatrix} \hat{r}_{m-1} \\ \hat{r}_{m-2} \\ \hat{r}_{m-3} \\ \cdot \\ \hat{r}_2 \\ \hat{r}_1 \\ \hat{r}_0 \end{Bmatrix} \tag{19}$$

where $r_k = R_{xf}^{(k)}(0)$, $\hat{r}_k = \hat{R}_{xf}^{(k)}(0)$, and $k = 0, 1, 2, \ldots, m-1$.

Therefore, the response spectral moments M_0, $M_1, \ldots, M_{2m-1}$ can be determined by solving these two sets of linear equations. Furthermore, note that the matrices which appear in equations (18) and (19) depend solely on the properties of the system. Thus, they do not change for different input processes. The vectors which appear on the right-hand side of these equations, however, do depend upon the input process. Specifically, using equations (9) and (13), one finds

$$r_k = \int_0^\infty h^{(k)}(u) R_{ff}(u)\, du\,, \tag{20}$$

$$\hat{r}_k = -\int_0^\infty h^{(k)}(u) \hat{R}_{ff}(u)\, du\,. \tag{21}$$

Note that these must be calculated for each input process being considered. Nevertheless, general analytical expressions for r_k and $\hat{r}_k$ can be developed for commonly encountered random processes.

Furthermore, if higher order response spectral moments are desired, it is possible to use equations (7), (8), (11), and (12) to obtain recurrence relations. Specifically,

$$\tilde{\lambda}_m R_{xx}^{(j+m)}(0) + \tilde{\lambda}_{m-1} R_{xx}^{(j+m-1)}(0) = r_j - \sum_{k=0}^{m-2} R_{xx}^{(j+k)}(0)\,, \tag{22}$$

and

$$\tilde{\lambda}_m \hat{R}_{xx}^{(j+m)}(0) + \tilde{\lambda}_{m-1} \hat{R}_{xx}^{(j+m-1)}(0) = \hat{r}_j - \sum_{k=0}^{m-2} \hat{R}_{xx}^{(j+k)}(0)\,, \tag{23}$$

where r_j and $\hat{r}_j$ for $j > m-1$ can be recursively calculated from the expressions

$$r_{n+m} = \left[R_{ff}^{(n)}(0) - \sum_{k=0}^{m-1} \lambda_k r_{n+k} \right] / \lambda_m\,, \tag{24}$$

and

$$\hat{r}_{n+m} = \left[\hat{R}_{ff}^{(n)}(0) - \sum_{k=0}^{m-1} \lambda_k \hat{r}_{n+k} \right] / \lambda_m\,. \tag{25}$$

To relate the correlation functions in these equations with the response spectral moments, equations (16) and (17) can be used. In dealing with equations (22) and (23), it is important to recall that the odd order derivatives of $R_{xx}(\tau)$ and the even order derivatives of $\hat{R}_{xx}(\tau)$ equal zero at $\tau = 0$.

4 IMPULSE RESPONSE DETERMINATION

Consider the characteristic polynomial of a linear, time-invariant system. Note that it may be written in the form $\Lambda(s) = \lambda_m \prod_{j=1}^{n}(s+s_j)^{m_j}$, where $-s_j$ is the jth distinct root of the characteristic equation and has multiplicity m_j. Note also that the individual multiplicities must sum to the order of the system. That is, $\sum_{j=1}^{n} m_j = m$. For simplicity, consider only the case where the characteristic equation has only simple roots. That is, $m_j = 1$ for all $j = 1, 2, 3, \ldots, m$. For this case, the impulse response function can be written in the form (Korn and Korn, 1969)

$$h(t) = \sum_{j=1}^{m} h_j e^{-s_j t}, \quad h_j = \left[\frac{s+s_j}{\Lambda(s)}\right]_{s=-s_j} . \tag{26}$$

Differentiating this equation p times with respect to t yields

$$h^{(p)}(t) = \sum_{j=1}^{n} h_j(-s_j)^p e^{-s_j t} . \tag{27}$$

Given this expression for $h^{(p)}(t)$, it is now feasible to find r_p and $\hat{r}_p$. In this regard, note that the determination of $\hat{r}_p$ requires the Hilbert transform of $R_{ff}(\tau)$. This can be found either by calculating the Hilbert transform of $R_{ff}(\tau)$ directly or by performing the inverse Fourier transform of $-i\omega^p S_{ff}(\omega)\text{sgn}(\omega)$.

5 APPLICATIONS FOR SOME KNOWN RANDOM SPECTRA

In this section, expressions are developed for excitations described by the white-noise, the band-limited white-noise, and the "Gaussian" shaped spectra.

5.1 *White-Noise Spectrum*

First, consider the input $f(t)$ to be a white-noise process. Recall that this process has the double-sided spectral density

$$S_{ff}(\omega) = 1 \quad , -\infty < \omega < \infty , \tag{28}$$

and autocorrelation function $R_{ff}(\tau) = 2\pi\delta(\tau)$, where $\delta(\tau)$ refers to the Dirac delta function. Using equation (10), one finds that

$$\hat{R}_{ff}(\tau) = 2/\tau . \tag{29}$$

Inspecting the definition of r_p and using the properties of the delta function, it becomes clear that r_p is simply a multiple of the pth derivative of the impulse response function evaluated at $t = 0$. Thus,

$$r_p = \begin{cases} 0 & , 0 \le p \le m-2 \\ \frac{\pi}{\lambda_m} & , p = m-1 \end{cases} . \tag{30}$$

Furthermore, it can be shown that $\hat{r}_p$ is given by the equation

$$\hat{r}_p = \begin{cases} 2\sum_{j=1}^{m} h_j(-s_j)^p \ln s_j & , 0 \le p \le m-2 \\ \text{not finite} & , p = m-1 \end{cases} . \tag{31}$$

Note that $\hat{r}_{m-1}$ does not exist. This result is expected due to the mathematical peculiarity of the spectrum of the white-noise process. Indeed, it is known that the spectral moments of the response of a linear system subjected to white-noise excitation of order higher than $2m-2$ are infinite. Obviously, since in this case M_{2m-1} is not finite, one may simply disregard the first row and first column of the matrix in equation (19) and solve for the lower order spectral moments, which are finite.

5.2 *Band-Limited White-Noise Spectrum*

Next consider a band-limited white-noise input $f(t)$. Recall that this random process has a double-sided spectral density given by the equation

$$S_{ff}(\omega) = \begin{cases} 1 & , 0 \le \omega_a \le |\omega| \le \omega_b \\ 0 & , \text{elsewhere} \end{cases} . \tag{32}$$

Thus, its autocorrelation function is given by the equation $R_{ff}(\tau) = 2(\sin\omega_b\tau - \sin\omega_a\tau)/\tau$. Here, ω_b and ω_a are the upper and lower cutoff frequencies of the spectral density, respectively. Furthermore, $\hat{R}_{ff}(\tau)$ can be expressed in the form

$$\hat{R}_{ff}(\tau) = 2[\cos(\omega_a \tau) - \cos(\omega_b \tau)]/\tau \,. \tag{33}$$

Then, relying on Gradshteyn and Ryzhik (1980), one finds

$$r_p = 2\sum_{j=1}^{m} h_j(-s_j)^p \tan^{-1}\left(\frac{(\omega_b - \omega_a)s_j}{s_j^2 + \omega_a\omega_b}\right), \tag{34}$$

$$\hat{r}_p = \sum_{j=1}^{p} h_j(-s_j)^p \ln\left(\frac{s_j^2 + \omega_a^2}{s_j^2 + \omega_b^2}\right). \tag{35}$$

5.3 *"Gaussian" Spectrum*

Finally, consider a random process with double-sided spectral density given as

$$S_{ff}(\omega) = e^{-\omega^2} \,. \tag{36}$$

The corresponding autocorrelation function is expressed as $R_{ff}(\tau) = \sqrt{\pi}\, e^{-\frac{\tau^2}{4}}$. Taking the inverse Fourier transform of $-i\mathrm{sgn}(\omega)S_{ff}(\omega)$ gives

$$\hat{R}_{ff}(\tau) = -i\sqrt{\pi}\, e^{-\frac{\tau^2}{4}}\Phi(\frac{i\tau}{2}), \tag{37}$$

where $\Phi(x)$ is the error function (Gradshteyn and Ryzhik, 1980). Although $\hat{R}_{ff}(\tau)$ appears to be a complex-valued function, it can be shown to be purely real-valued, as expected.

Next, these expressions may be substituted into equation (21) to obtain

$$r_p = \sum_{j=1}^{p} h_j(-s_j)^p \pi\, e^{s_j^2}\,[1 - \Phi(s_j)] \,, \tag{38}$$

$$\hat{r}_p = -\sum_{j=1}^{p} h_j(-s_j)^p e^{s_j^2} \mathrm{Ei}(s_j^2) , \tag{39}$$

where $\mathrm{Ei}(x)$ refers to the exponential integral function (Gradshteyn and Ryzhik, 1980).

6 EXAMPLE: KANAI-TAJIMI EARTHQUAKE SPECTRUM

Consider a standard linear oscillator governed by the differential equation

$$\ddot{x}(t) + 2\zeta\omega_0\dot{x}(t) + \omega_0^2 x(t) = f(t) \,. \tag{40}$$

It is readily seen that the characteristic polynomial in this case is $\Lambda_2(s) = s^2 + 2\zeta\omega_0 s + \omega_0^2$. Furthermore, assume that the excitation $f(t)$ possesses the Kanai-Tajimi earthquake spectrum (Roberts and Spanos, 1990). That is, its spectrum has the form

$$\begin{aligned} S_{ff}(\omega) &= \frac{G_0\left[1 + 4\zeta_g^2(\omega/\omega_g)^2\right]}{\left\{[1 - (\omega/\omega_g)^2]^2 + 4\zeta_g^2(\omega/\omega_g)^2\right\}} \\ &= \frac{G_0\left[1 + 4\zeta_g^2(\omega/\omega_g)^2\right]}{|\Lambda_1(i\omega)|^2}, \end{aligned} \tag{41}$$

where G_0 is a scaling parameter, ζ_g is the ground critical damping factor, and ω_g is the predominant ground frequency. It is then obvious that the system response spectral moments can be expressed as

$$\begin{aligned} M_k = G_0 &\int_0^\infty \frac{2\,\omega^k}{|\Lambda(i\omega)|^2}\, d\omega \\ &+ 4G_0\left(\frac{\zeta_g}{\omega_g}\right)^2 \int_0^\infty \frac{2\,\omega^{k+2}}{|\Lambda(i\omega)|^2}\, d\omega \,, \end{aligned} \tag{42}$$

where $\Lambda(s) = \lambda_4 s^4 + \lambda_3 s^3 + \lambda_2 s^2 + \lambda_1 s + \lambda_0$ with

$$\lambda_4 = \frac{1}{\omega_g^2}, \quad \lambda_3 = \frac{2}{\omega_g^2}(\zeta_0\omega_0 + \zeta_g\omega_g) \,, \tag{43}$$

$$\lambda_2 = 1 + \left(\frac{\omega_0}{\omega_g}\right)^2 + 4\zeta_0\zeta_g\left(\frac{\omega_0}{\omega_g}\right), \tag{44}$$

$$\lambda_1 = 2\omega_0\left[\zeta_0 + \zeta_g\left(\frac{\omega_0}{\omega_g}\right)\right], \tag{45}$$

and

$$\lambda_0 = \omega_0^2 \,. \tag{46}$$

Since this polynomial is found by multiplying $\Lambda_1(s)$ and $\Lambda_2(s)$, its roots are merely the roots of $\Lambda_1(s)$ and $\Lambda_2(s)$. That is,

$$s_{1,2} = \zeta_g\omega_g \pm i\omega_g\sqrt{1 - \zeta_g^2} \tag{47}$$

and

$$s_{3,4} = \zeta_0\omega_0 \pm i\omega_0\sqrt{1 - \zeta_0^2} \,. \tag{48}$$

Furthermore, recall that the formulae already developed for a white-noise system excitation can be used to evaluate the required integrals. Thus, the response spectral moments of a linear oscillator excited by a stationary random process with Kanai-Tajimi spectrum are given by the equation

$$M_k = G_0 M_k' + 4\,G_0\left(\frac{\zeta_g}{\omega_g}\right)^2 M_{k+2}' , \tag{49}$$

where M_k' and M_{k+2}' for $k = 0, 1, 2, 3, 4$ can be found from the equations

$$\begin{bmatrix} \lambda_3 & -\lambda_1 & 0 & 0 \\ -\lambda_4 & \lambda_2 & -\lambda_0 & 0 \\ 0 & -\lambda_3 & \lambda_1 & 0 \\ 0 & \lambda_4 & -\lambda_2 & \lambda_0 \end{bmatrix} \begin{Bmatrix} M_6' \\ M_4' \\ M_2' \\ M_0' \end{Bmatrix} = \begin{Bmatrix} \pi/\lambda_4 \\ 0 \\ 0 \\ 0 \end{Bmatrix} \tag{50}$$

and

$$\begin{bmatrix} -\lambda_3 & \lambda_1 & 0 \\ \lambda_4 & -\lambda_2 & \lambda_0 \\ 0 & \lambda_3 & -\lambda_1 \end{bmatrix} \begin{Bmatrix} M_5' \\ M_3' \\ M_1' \end{Bmatrix} = \begin{Bmatrix} \hat{r}_2 \\ \hat{r}_1 \\ \hat{r}_0 \end{Bmatrix}. \tag{51}$$

Note that in this case, M_p' for $p > 6$ will not be finite. Consequently, the response spectral moments M_k for $k > 4$ are not finite.

7 CONCLUDING REMARKS

Two sets of linear equations involving the response spectral moments of a randomly excited linear system have been derived. Equation (18) can be solved to determine any or all of the even order response spectral moments, while equation (19) can be used to determine any or all of the odd order response spectral moments. Both sets of linear equations are quite versatile. Note that equation (18) can be used to derive the James et al. (1965) formula used extensively in control engineering calculations. Since equation (18) is applicable whenever the r_k terms can be found, regardless of the excitation spectra, it provides an important generalization of this formula. Furthermore, a similar procedure is available through equation (19) for the determination of the odd order response spectral moments. Analytical formulae of this kind were previously unavailable.

It should also be noted that equations (18) and (19) can be readily incorporated into problems where the input spectra are modeled as the response of a linear filter to a random excitation. In most cases, a white-noise is used as input to such a filter, but the given equations can be used with more general input spectra. Note that these remarks are particularly applicable to problems involving a cascade of linear structures.

Finally, the formulae can be used to evaluate rather complex integrals, as illustrated by the formulae for the Gaussian spectrum. Thus, this approach can be quite useful in a variety of other technical fields irrespective of the presence of a random vibrations context.

REFERENCES

Gradshteyn, I. S. and I. M. Ryzhik, 1980, *Tables of Integrals, Series and Products*, Academic Press, New York.

James, H. M., N. B. Nichols, and R. S. Philips, eds., 1965, *Theory of Servomechanisms*, Dover, New York.

Korn, A. G., and T. M. Korn, 1969, *Mathematical Handbook for Scientists and Engineers*, McGraw-Hill, New York.

Papoulis, A., 1984, *Probability, Random Variables, and Stochastic Processes*, McGraw-Hill, New York.

Roberts, J. B., and P. D. Spanos, 1990, *Random Vibration and Statistical Linearization*, John Wiley and Sons, New York.

Spanos, P-T. D., 1983, Spectral Moments Calculation of Linear System Output, *ASME Journal of Applied Mechanics*, Vol. 50, pp. 901-903.

Spanos, P-T. D., 1987, An Approach to Calculating Random Vibration Integrals, *ASME Journal of Applied Mechanics*, Vol. 54, pp. 409-413.

Spanos, P. D., and S. M. Miller, 1992, Linear System Spectral Moments Determination, *Probabilistic Mechanics, and Structural and Geotechnical Reliability*, Proceedings of the Sixth Specialty Conference, Denver, Colorado, Y. K. Lin, ed., ASCE, New York, pp. 192-195.

Structural Safety & Reliability, Schuëller, Shinozuka & Yao (eds) © 1994 Balkema, Rotterdam, ISBN 90 5410 357 4

Stochastic seismic ground motion modeling with imperfectly stratified earth medium

Ruichong Zhang
Princeton University, N.J., USA

ABSTRACT: Earthquake wave propagation in an imperfectly stratified linear elastic earth medium, generated by a buried seismic dislocation source, is investigated. Using a first order perturbation approach, it is shown that the total wave field may be obtained as superposition of a mean wave field and a scattered wave field. The mean wave field corresponds to the mean interface, which is assumed to be a perfect horizontal plane. The scattered wave field is obtained from a set of fictitious discontinuity sources distributed on the fictitious mean interface, which has the same effect as the interface irregularities. A numerical example is carried out for illustration.

1 INTRODUCTION

In modeling earthquake ground motion, the earth medium is often idealized as being horizontally stratified with layerwise uniform physical properties (e.g. Luco and Apsel 1983, Zhang et al. 1991a,b). However, the actual interface between neighboring layers cannot be a perfect horizontal plane. Whether or not the perfect-plane approximation is adequate depends upon magnitudes of the irregularities in terms of deviations, slopes, etc., and the wavelength of the propagating waves in a medium (Ishimaru, 1978).

If the variations of the rough interface are small compared with the dominant wave length of the seismic waves, the perturbation technique is perhaps the most efficient in solving the wave scattering problem. Kennett (1972) used a first-order perturbation to solve the problem of in-plane and out-of-plane scattered wave. The propagation of Rayleigh, Scholte and Stoneley waves on a random interface was analyzed by Ostoja-Starzewski (1987). However, both Kennett and Ostoja-Stazewski restricted their analyses to a two-dimensional space, namely, the rough interface to be two-dimensional, in which case the SH wave may be separated from P and/or SV waves. Specifically, an incident SH wave will generate only the scattered SH waves, and an incident P and/or SV wave will likewise generate only scattered P-SV waves. The two-dimensional wave scattering problem have also been analyzed by Aki and Larner (1970), Varadan et al (1987a,b), Geli et al. (1988) and Kawase (1990), although the early work may date back to Rayleigh (1907, 1945).

Since a realistic rough interface is three-dimensional, the possibility of an incident SH wave generating scattered P and SV waves, and vice versa, must be considered. Recently, Liu (1991a,b,c,d) has developed a method for obtaining the exact solution for wave scattering problem. However, the three-dimensional scattering problem with vector wave field has not been studied yet.

In this paper, the first order perturbation method will be applied to solve the three dimensional wave scattering problem in a layered medium with a rough interface and with a seismic source located in one layer (Zhang, 1992, 1993). The

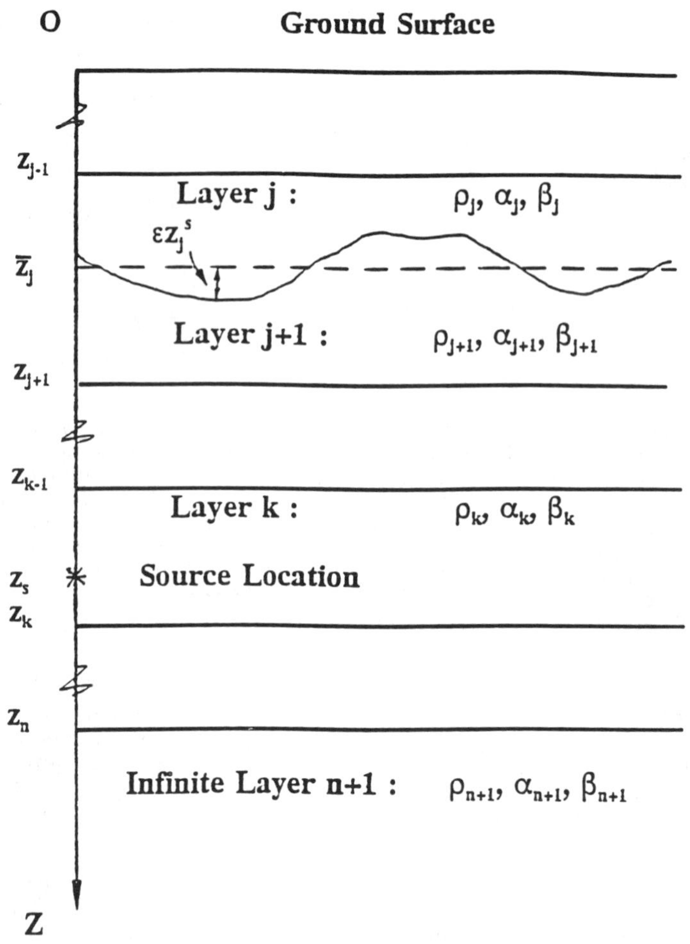

Fig. 1 A layered earth medium with a rough interface

total wave field is separated into two parts; one is the mean wave field, i.e. the solution obtained for the flat mean interface, the other is the scattered wave field, accounting for the effects of interface irregularities. The response in the scattered wave field is assumed to be small compared with that in the mean wave field. The solution for the mean wave field is obtained first using finite integral approach. The solution for the scattered wave field is then found similarly in terms of the mean wave field. The statistic characteristics of wave field can finally be obtained.

2 DECOMPOSITION OF WAVE FIELD AND FIRST-ORDER PERTURBATION APPROACH

As shown in Fig. 1, the earth is modeled as a horizontally stratified medium except for a rough interface between layers j and j+1, with a seismic source located in layer k. Then the earthquake wave motion may be governed by the following general form

$$L\begin{Bmatrix}\mathbf{u}(x,y,z,t)\\ \boldsymbol{\tau}(x,y,z,t)\end{Bmatrix} = \begin{Bmatrix}\Delta\mathbf{u}(x,y,z_s,t)\\ \mathbf{0}\end{Bmatrix} \quad (1)$$

where $\mathbf{u}(x,y,z,t)$ and $\boldsymbol{\tau}(x,y,z,t)$ denote respectively the displacement and stress fields, $\Delta\mathbf{u}(x,y,z_s,t)$ represents the seismic discontinuity source occurred at location (x,y,z_s)

and at time t, and L is linear operator, characterizing the imperfectly stratified linear elastic earth medium and being function of the physical parameters for each uniform earth medium such as P and S wave speeds, etc. and the structure of earth medium such as depth of the interfaces.

The profile of the rough interface z_j can be expressed as

$$z_j(x,y) = \bar{z}_j + \epsilon z_j^s(x,y) \tag{2}$$

where $\bar{z}_j$ is the average value of $z_j(x,y)$, taken to be independent of x and y, and ϵz_j^s is the deviation from the average value, assumed to be of order of ϵ. Correspondingly, the displacements and stresses may also be decomposed into two parts, i.e.

$$\mathbf{u}(x,y,z,t) = \bar{\mathbf{u}}(x,y,z,t) + \epsilon \mathbf{u}^s(x,y,z,t) \tag{3}$$

$$\boldsymbol{\tau}(x,y,z,t) = \bar{\boldsymbol{\tau}}(x,y,z,t) + \epsilon \boldsymbol{\tau}^s(x,y,z,t) \tag{4}$$

In (3) and (4) each overbar denotes the corresponding average value (called mean wave field), obtainable as the solution corresponding to a perfectly flat interface $\bar{z}_j$, and each superscript s denotes the corresponding perturbation due to rough interface z_j (called scattered wave field). Since the linear operator L is function of depth of the rough interface z_j, it may also be expanded as a Taylor series with respect to $\bar{z}_j$, i.e.

$$L = \bar{L} + \epsilon L^s (1 + \epsilon \eta_2 + \epsilon^2 \eta_3 \cdots) \tag{5}$$

where $\bar{L}$ denotes the mean linear operator, characterizing the perfectly stratified earth medium, ϵL^s stands for the first order scattered linear operator due to the rough interface, and $\epsilon^i \eta_i L^s$ (i=2,3,⋯) corresponds to the i-th order scattered linear operators. In this paper, the first-order perturbation approach is applied, which is considered adequate to reveal the fundamental characteristics of the ground motion with the rough interface.

Substituting (3)-(5) into (1) and neglecting the second and higher order terms of smallness, one may obtain two sets of equations for the original problem, i.e.

$$\bar{L} \begin{Bmatrix} \bar{\mathbf{u}}(x,y,z,t) \\ \bar{\boldsymbol{\tau}}(x,y,z,t) \end{Bmatrix} = \begin{Bmatrix} \Delta\mathbf{u}(x,y,z_s,t) \\ \mathbf{0} \end{Bmatrix} \tag{6}$$

$$\bar{L} \begin{Bmatrix} \mathbf{u}^s(x,y,z,t) \\ \boldsymbol{\tau}^s(x,y,z,t) \end{Bmatrix} = -L^s \begin{Bmatrix} \bar{\mathbf{u}}(x,y,z,t) \\ \bar{\boldsymbol{\tau}}(x,y,z,t) \end{Bmatrix} \equiv \begin{Bmatrix} \Delta\mathbf{u}(x,y,\bar{z}_j,t) \\ \Delta\boldsymbol{\tau}(x,y,\bar{z}_j,t) \end{Bmatrix} \tag{7}$$

where sign ≡ denotes the definition and

$$\Delta u_\zeta = -[\bar{u}_{\zeta,z}]_-^+ z_j^s \ , \ \zeta = x \ , \ y \text{ or } z \tag{8}$$

$$\Delta\tau_{xz} = -[\bar{\tau}_{xz,z}]_-^+ z_j^s + [\bar{\tau}_{xx}]_-^+ z_{j,x}^s + [\bar{\tau}_{xy}]_-^+ z_{j,y}^s \tag{9}$$

$$\Delta\tau_{yz} = -[\bar{\tau}_{yz,z}]_-^+ z_j^s + [\bar{\tau}_{yy}]_-^+ z_{j,y}^s + [\bar{\tau}_{xy}]_-^+ z_{j,x}^s \tag{10}$$

$$\Delta\tau_{zz} = -[\bar{\tau}_{zz,z}]_-^+ z_j^s \tag{11}$$

in which the subscript ",ζ" denotes $\partial/\partial\zeta$, ζ=x, y and z, and for each function f,

$$[\,f(x,y,z,t)\,]_-^+ = f(x,y,z=\bar{z}_j^+,t) - f(x,y,z=\bar{z}_j^-,t) \tag{12}$$

The physical meanings of (6) and (7) are quite apparent and can be explained as follows. Both mean and scattered wave fields share the same linear operator $\bar{L}$, which indicates that they exist in the same perfectly stratified medium without presence of a rough interface. The mean wave field is obtained first from (6) with a seismic discontinuity source $\Delta\mathbf{u}$ located at depth z_s. The scattered wave field can then be obtained from (7) with fictitious discontinuity source $(\Delta\mathbf{u},\Delta\boldsymbol{\tau})$ acting at mean depth of rough interface $\bar{z}_j$. The fictitious discontinuity source $(\Delta\mathbf{u},\Delta\boldsymbol{\tau})$ has equivalent effects of the rough interface and can be expressed in

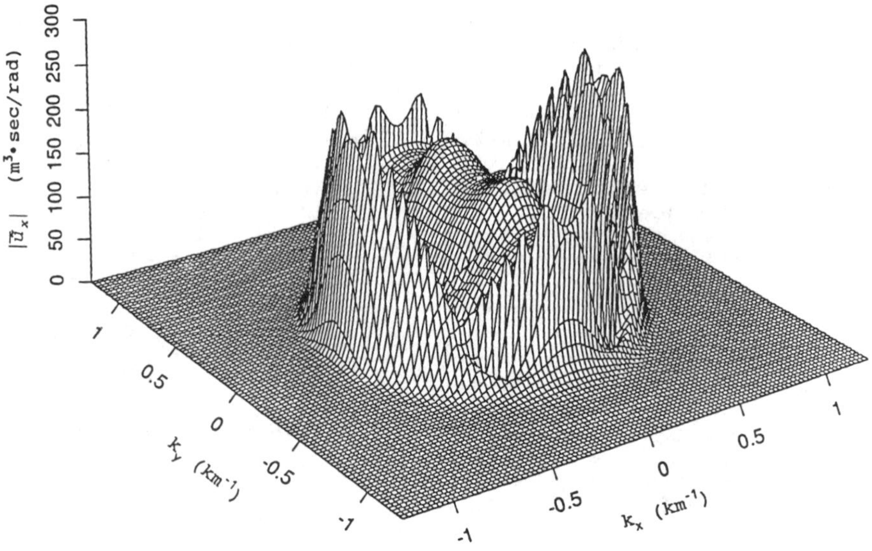

Fig. 2 Ground displacement response of mean wave field in the x direction in the wave number domain with ω=2 rad/sec

terms of the mean wave field. The total wave field ($\mathbf{u},\boldsymbol{\tau}$) in a layered medium with a rough interface and with a buried seismic discontinuity source may finally be obtained as superposition of these two wave fields.

The response in a stratified earth medium with discontinuity source, i.e. solution of (6) and (7), may be obtained using finite Fourier transform approach and wave propagation approach. It is noted that the coupling between P-SV wave and SH wave may be found from the fictitious discontinuity source of (8)-(11) when they are transformed to frequency-wave number domain. The detailed derivation of response solution and its analysis of wave coupling may be found in Zhang (1992).

3 STOCHASTIC CHARACTERISTICS OF SEISMIC GROUND MOTION

We assumed that the profile of the rough interface $z_j(x,y)$ is a homogeneous random process in both x and y directions. Then $z_j^s(x,y)$ in (2) may be represented by a double Fourier series

$$z_j^s(x,y) = \frac{\pi}{x_t'}\frac{\pi}{y_t'}\sum_{m'=-\infty}^{\infty}\sum_{n'=-\infty}^{\infty} V_{m'n'}(k_x',k_y')\exp[i(k_x'x+k_y'y)] \quad (13)$$

where $k_x'=m'\pi/x_t'$ and $k_y'=n'\pi/y_t'$, x_t' and y_t' are the selected truncation distances in the x and y directions, k_x' and k_y' are the respective wave numbers, and $V_{m'n'}(k_x',k_y')$ are orthogonal random variables with the spectral density in the wave number domain $\Phi_{m'n'}(k_x',k_y')$. Correspondingly, the response in the scattered wave field may be constructed by

$$Y^s(x,y,z,t) = \frac{\pi}{x_t'}\frac{\pi}{y_t'}\sum_{m'=-\infty}^{\infty}\sum_{n'=-\infty}^{\infty} V_{m'n'}(k_x',k_y')\,H_Y^s(x,y,z,t;k_x',k_y') \quad (14)$$

where Y^s may be either displacement or stress response associated with the scattered wave field, and H_Y^s is the response for the rough interface being of the form of $z_j^s=\exp[i(k_x'x+k_y'y)]$. The counterpart for (14) in the frequency-wave number domain is

$$\tilde{Y}^s(k_x,k_y,z,\omega) = \frac{\pi}{x_t'}\frac{\pi}{y_t'}\sum_{m'=-\infty}^{\infty}\sum_{n'=-\infty}^{\infty} V_{m'n'}(k_x',k_y')\,\tilde{H}_Y^s(k_x,k_y,z,\omega;k_x',k_y') \quad (15)$$

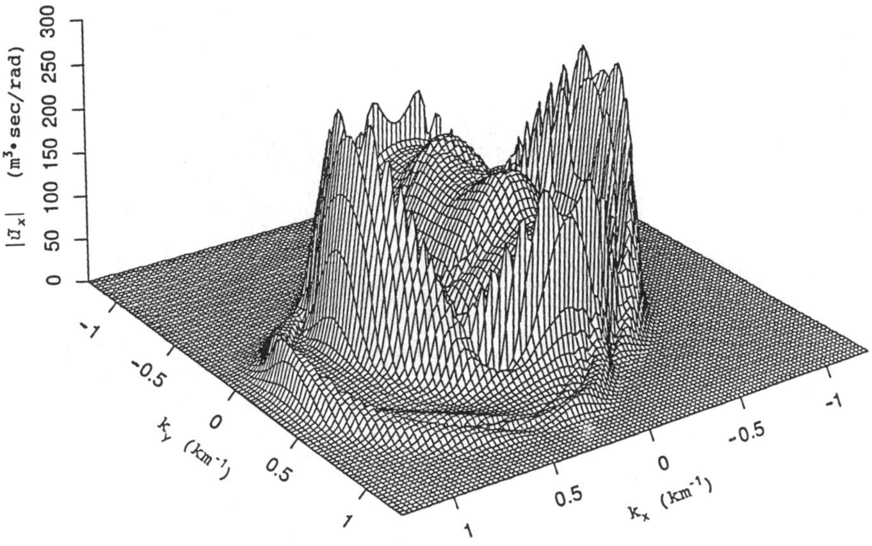

Fig. 3 Total ground displacement response in the x direction in the wave number domain with ω=2 rad/sec

where $\tilde{H}_Y^s$ is the frequency-wave number response function, corresponding to H_Y^s. The frequency-wave number spectra of the ground motion in a layered medium with a random rough interface can then be found as

$$\begin{aligned} S(k_x,k_y,\omega) &= E\{|\tilde{Y}(k_x,k_y,z=0,\omega)|^2\} \\ &= \bar{S}(k_x,k_y,\omega) + S^s(k_x,k_y,\omega) \end{aligned} \tag{16}$$

where the mean and scattered spectra of ground motion are, respectively,

$$\bar{S}(k_x,k_y,\omega) = |\bar{Y}(k_x,k_y,z=0,\omega)|^2 \tag{17}$$

$$\begin{aligned} S^s(k_x,k_y,\omega) &= \\ &= \epsilon^2 E\{|\tilde{Y}^s(k_x,k_y,z=0,\omega)|^2\} \\ &= \epsilon^2 \frac{\pi}{x_t'} \frac{\pi}{y_t'} \sum_{m'=-\infty}^{\infty} \sum_{n'=-\infty}^{\infty} \Phi_{m'n'}(k_x',k_y') \\ &\quad |\tilde{H}_Y^s(k_x,k_y,z=0,\omega;k_x,k_y)|^2 \end{aligned} \tag{18}$$

It should be noted that the mean wave field is deterministic, and that the scattered wave field itself has a zero mean. The correlation function of the scattered wave field alone is given by

$$\begin{aligned} C(\Delta x,\Delta y,\Delta t) &= \frac{\pi}{t_t} \frac{\pi}{x_t} \frac{\pi}{y_t} \sum_{l=-\infty}^{\infty} \sum_{m=-\infty}^{\infty} \sum_{n=-\infty}^{\infty} \\ &\quad S^s(k_x,k_y,\omega) \exp[ik_x\Delta x+k_y\Delta y+\omega\Delta t)] \end{aligned} \tag{19}$$

where t_t, x_t and y_t are the truncated time and truncated distances in the x and y directions, respectively, and $\omega=l\pi/t_t$, $k_x=m\pi/x_t$ and $k_y=n\pi/y_t$. The mean-square ground motion attributable to the scattered wave field is obtained by setting Δx=0, Δy=0, and Δt=0 in (19). The special case of (19) is the frequency-dependent correlation function, represented by

$$\begin{aligned} \tilde{C}(\Delta x,\Delta y,\omega) &= \frac{\pi}{x_t} \frac{\pi}{y_t} \sum_{m=-\infty}^{\infty} \sum_{n=-\infty}^{\infty} \\ &\quad S^s(k_x,k_y,\omega) \exp[i(k_x\Delta x+k_y\Delta y)] \end{aligned} \tag{20}$$

4 NUMERICAL EXAMPLE

As a numerical example, let the earth be modeled as a three-layer half space with a rough interface between the first and second layers and with a buried point seismic source in the second layer. The physical properties of the three layers are given in Table 1. The seismic source is modeled as point source of double couple, characterizing a shear dislocation in the x direction acting on the seismic fault, which is parallel to the ground surface. The associated slip function is assumed to be a ramp function with rise time equal

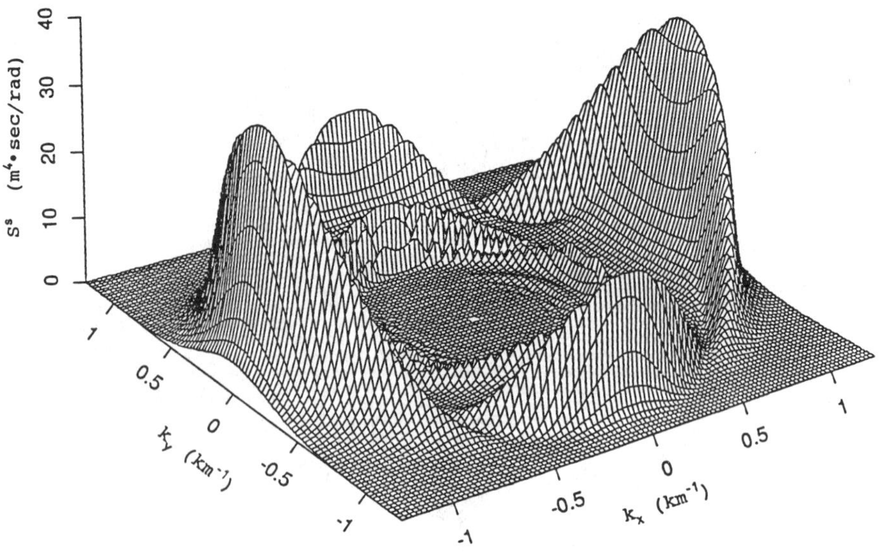

Fig. 4 Scattered wave number spectrum of ground displacement response in the x direction with ω=2 rad/sec

Table 1 Physical properties of a three-layer medium

	ρ (kg/m^3)	α (m/s)	β (m/s)	γ_α	γ_β	z (km)
Layer 1	1670	3100	1790	0.08	0.04	0 - 3
Layer 2	2280	5000	2887	0.04	0.02	3 - 10
Layer 3	2580	6000	3464	0.02	0.01	10 - ∞

to 0.628 s. The seismic moment is 1.9×10^{15} Nm.

We first look at effects of the seismic ground motion due to a deterministic rough interface, which is assumed as $\bar{z}_1=3$ km and $\epsilon z_1^s(x,y)=0.2 \exp(ik_x'x+ik_y'y)$ km, where $k_x'=0.4$ km^{-1} and $k_y'=0.3$ km^{-1}. Fig. 2 shows the amplitude of ground displacement response of the mean wave field in the x direction in the wave-number domain with ω=2 rad/sec while Fig. 3 presents the corresponding total displacement response with the rough interface described as above. The effects of wave scattering on the ground motion due to the rough interface can be seen clearly from these two figures, which is directly dependent upon wave numbers (k_x' and k_y') of the rough interface. More detailed study on the issue of wave scattering due to a deterministic rough interface can be found in Zhang (1992 and 1993).

Assume the spectral density of the rough interface in the wave number domain have the following form,

$$\Phi_{m'n'}(k_x', k_y') = \frac{\sigma^2}{16\pi} C^4 (k_x'^2 + k_y'^2) \exp[-\frac{C^2}{4}(k_x'^2 + k_y'^2)] \tag{21}$$

where σ^2 denotes the variance of the rough interface and C is a parameter. In the numerical example, $\sigma^2=212.69^2$ m^2 and C=3000 m. We can then obtain spectral density of the scattered ground displacements using (18), whose component in the x direction is shown in Fig. 4. The amplitude of

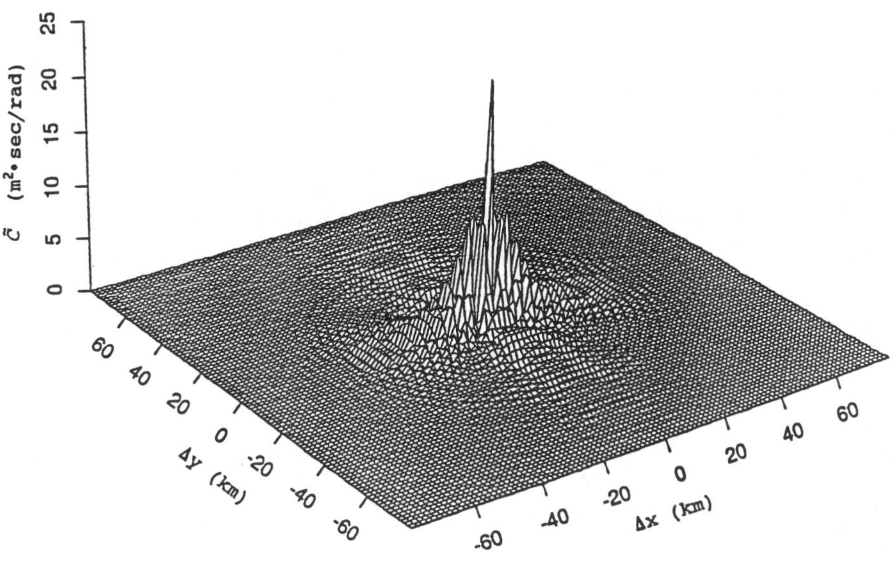

Fig. 5 Amplitude of frequency-dependent correlation function of scattered ground displacement response in the x direction with ω=2 rad/sec

corresponding frequency-dependent correlation function of scattered response is plotted in Fig. 5. Their quadrant symmetric property in both wave number domain and space domain is attributable to the selected symmetric point seismic source as well as the assumed symmetric spectral density of the rough interface in the wave number domain.

5 CONCLUDING REMARKS

Stochastic ground motion in a layered medium with a rough interface subjected to a point seismic dislocation source is investigated in this paper. With the use of the first-order perturbation approach as well as wave propagation approach, the statistical characteristics of the ground motion with the imperfectly stratified medium can be evaluated. Extensive numerical results for the effects of the rough interface on the ground motion will be compiled in other papers.

ACKNOWLEDGEMENTS

The research proposal reported in this paper was supported by the National Center for Earthquake Engineering Research under Grants NCEER 90-2003, 91-3313A and 92-3302A. It is a part of the author's Ph.D. dissertation, prepared under his advisor, Professor Y.K. Lin, Schmidt Chair in Engineering of Florida Atlantic University, whose valuable guidance and encouragement are gratefully acknowledged. Opinions, findings, and conclusions expressed herein are those of the writer and do not necessarily reflect the views of sponsors.

REFERENCES

Aki, K. and Larner, K.L. 1970. Surface motion of a layered medium having an irregular interface due to incident plane SH wave, J. Geophysical Res. 75, 5, 933-954.

Geli, L., Bard, P and Jullien, B. 1988. The effects of topography on earthquakes ground motion: A review and new results, Bull. Seis. So. Am. 78, 1, 42-63.

Ishimaru, A. 1978. Wave Propagation and Scattering in Random Media, Vol. 1 and 2. Academic Press, New York, San Francisco, London.

Kawase, H. 1990. Effects of

topography and subsurface irregularities on strong ground motion, ORI Report 90-02, Ohsaki Research Institute, Inc.
Kennett, B.L.N. 1972. Seismic wave scattering by obstacles on interfaces, Geophys. J.R. astr. Soc. 28, 249-266.
Liu, K.C. 1991a. Wave scattering in discrete random media by the discontinuous stochastic field method, I: basic method and general theory, J. Sound and Vibration, 147(2), 301-311.
Liu, K.C. 1991b. Wave scattering in discrete random media by the discontinuous stochastic field method, II: contribution of the second order moment of the β-field, J. Sound and Vibration, 147(2), 313-321.
Liu, K.C. 1991c. Wave scattering in discrete random media by the discontinuous stochastic field method, III: contribution of the third order moment of the β-field, J. Sound and Vibration, 147(2), 323-334.
Liu, K.C. 1991d. Wave scattering from random rough surfaces by method of discontinuous stochastic fields, J. Sound and Vibration, 148(1), 171-175.
Luco, J.E. and Apsel, R.J. 1983. On the Green's function for a layered half-space. Part I, Bull. Seis. Soc. Am. 73, 4, 909-929.
Ostoja-Starzewski, M. 1987. Propagation of Rayleigh, Scholte and Stoneley waves along random boundaries, Prob. Eng. Mech. 2, 2, 64-73.
Rayleigh, L. 1907. On the dynamical theory of gratings, Proc. Roy. Soc. London, A, 79, 399.
Rayleigh, L. 1945. The theory of Sound, (2 vols) New York, Dover Publications.
Varadan, V.K., Lakhtakia, A., Varadan, V.V., and Langston, C.A., 1987a. Radiation characteristics of elastodynamic line sources buried in layered media with periodic interfaces. II. SH-wave analysis, Bull. Seis. Soc. Am., 77, 6, 2181-2191.
Varadan, V.K., Lakhtakia, A., Varadan, V.V., and Langston, C.A., 1987b. Radiation characteristics of elastodynamic line sources buried in layered media with periodic interfaces. II. P- and SV-wave analysis", Bull. Seis. Soc. Am., 77, 6, 2192-2211.
Zhang, R. 1992. Stochastic Earthquake Ground Motion Modeling, Ph.D. dissertation, Florida Atlantic University, Boca Raton, Florida.
Zhang, R. 1993. Stochastic seismic ground motion modeling with imperfectly stratified earth medium, (to appear in J. Sound and Vibration).
Zhang, R., Yong, Y. and Lin, Y.K. 1991a. Earthquake ground motion modeling I: Deterministic point source, J. Eng. Mech., 117, 9, 2114-2132.
Zhang, R., Yong, Y. and Lin, Y.K. 1991b. Earthquake ground motion modeling II: Stochastic line source", J. Eng. Mech., 117, 9, 2133-2148.

Random vibration (ongoing research)

Structural Safety & Reliability, Schuëller, Shinozuka & Yao (eds) © 1994 Balkema, Rotterdam, ISBN 90 5410 357 4

Prediction of noise emission of engines using random vibration analysis of non-classically damped structures

P. Fischer & H. J. Pradlwarter
Institute of Engineering Mechanics, University of Innsbruck, Austria

H. H. Priebsch
AVL LIST GmbH, Graz, Austria

ABSTRACT: An efficient method is presented to predict the structure borne noise emission of engines by using random vibration analysis. The noise emission is quantified by the mean square of the velocities normal to the surface within defined frequency bands. The developed procedure is capable to analyze classically and non-classically damped systems. The numerical example shows the mean square velocities of a cover of the cylinder head of a diesel engine where complex modal analysis is utilized to describe the local damping characteristics of the system.

1 INTRODUCTION

In general, the acceptance level of environmental noise is constantly decreasing. Therefore, the reduction of noise becomes an important issue for the design of engines. Traditional efforts for noise reduction are usually based on experimental studies and normal mode analysis. The aim of this work is to provide effective computational procedures to predict structure borne noise emission of engines.

Measurements at the engine show response characteristics typical to realizations of a stochastic process. Hence random vibration theory is applied for calculating the response of the system. The engine is modeled by Finite Elements (FE) with a large number of degrees of freedom describing all the relevant details. For a realistic description of the dynamic response of the engine, the assessment of the damping plays an important role. The engine consists of substructures connected by joints. The dissipation of vibrational energy by damping is concentrated locally in the joints, whereas the damping in the substructures has only minor effects. Therefore, the damping in the joints is described by damping elements. This requires the application of non classical complex modal analysis for assessing the modal damping properties and the prediction of the dynamic behavior of such systems. Additionally, the vibration modes are often strongly overlapping, where the damping is not represented well by the assumption of classical damping (see e.g. Park et al. 1992, Xu & Igusa 1991 and Igusa et al. 1984).

The noise sensitivity of the human ear is frequency dependent. Therefore, an integral description of the noise emission within arbitrary defined frequency bands is considered. This description can be realized by using the mean square of the velocities normal to the surface of the structure. The mean squares can be obtained by integrating the spectral density of the velocities in the frequency domain. In the literature, e.g. Igusa et al. 1984, Maldano & Singh 1991 and Der Kiureghian & Neuenhofer 1992, there are several methods available showing the use of complex modal analysis for assessing response properties of non classical damped systems. However, these methods are not directly applicable to noise emission problems.

2 ANALYSIS

The noise emission of vibrating surfaces is quantified by the mean square of the velocities normal to the surface. These quantities can be computed by integrating a product of the matrix of transfer functions $\boldsymbol{V}(\omega)$ and the spectral matrix of excitation $\boldsymbol{S}_{ff}(\omega)$ in the frequency domain, eq. (1). A star denotes the complex conjungate.

$$E[\dot{\boldsymbol{x}}(t)\dot{\boldsymbol{x}}^{*T}(t)] = \int_{\omega=-\infty}^{\infty} \boldsymbol{V}(\omega)\boldsymbol{S}_{ff}(\omega)\boldsymbol{V}^{*T}(\omega)d\omega \tag{1}$$

Similar equations are given e.g. in Lin 1976 or Robson 1980. This formulation is restricted to linear systems under stationary excitation. The system is described by its transfer functions, which can be based either on classical or non classical damping. The numerical integration of eq.(1) requires the calculation of the matrix of transfer functions for each frequency step (Sharan et. al. 1982).

The nature of the problem involves non classical dynamic analysis. An effective algorithm for calculating non classical transfer functions is realized by assuming symmetric structural matrices $\boldsymbol{M}$, $\boldsymbol{C}$ and $\boldsymbol{K}$, which lead to a state space formulation with symmetric state matrices $\boldsymbol{D}$ and $\boldsymbol{E}$, eq. (2) and (3).

$$\begin{bmatrix} C & M \\ M & O \end{bmatrix} \begin{Bmatrix} \dot{\boldsymbol{x}}(t) \\ \ddot{\boldsymbol{x}}(t) \end{Bmatrix} - \begin{bmatrix} -K & O \\ O & M \end{bmatrix} \begin{Bmatrix} \boldsymbol{x}(t) \\ \dot{\boldsymbol{x}}(t) \end{Bmatrix} = \begin{Bmatrix} \boldsymbol{f}(t) \\ O \end{Bmatrix} \tag{2}$$

$$[\boldsymbol{D}]\{\dot{\boldsymbol{y}}(t)\} - [\boldsymbol{E}]\{\boldsymbol{y}(t)\} = \{\boldsymbol{g}(t)\} \tag{3}$$

The corresponding matrix of complex eigenvectors can be subdivided into an upper displacement part $\boldsymbol{\Phi}^D$ and a lower velocity part $\boldsymbol{\Phi}^D\boldsymbol{\Lambda}$, where $\boldsymbol{\Lambda}$ is the diagonal matrix of complex eigenvalues, e.g. Traill-Nash 1981, Lang 1989.

$$\boldsymbol{\Phi} = \begin{bmatrix} \boldsymbol{\Phi}^D \\ \boldsymbol{\Phi}^D\boldsymbol{\Lambda} \end{bmatrix} \tag{4}$$

Utilizing this property and taking into account that the matrix $\boldsymbol{\Phi}^D$ consists of conjungate complex pairs of eigenvectors, the solution for the components $v_{r,s}(\omega)$ of the non classical matrix of transfer functions can be found by eq. (5).

$$\begin{aligned} v_{r,s}(\omega) &= \sum_{j=1}^{nmod} (p_j(\omega) + \overline{p}_j(\omega))\eta^{re}_{r,s,j} \\ &\quad + i(p_j(\omega) - \overline{p}_j(\omega))\eta^{im}_{r,s,j} \end{aligned} \tag{5}$$

where

$$\eta_{r,s,j} = \Phi_{r,j}\Phi_{s,j} \tag{6}$$

$$h_j(\omega) = \frac{1}{i\omega - \lambda_j} \tag{7}$$

$$p_j(\omega) = \lambda_j h_j(\omega) \tag{8}$$

$$\overline{p}_j(\omega) = p_j(\omega|\lambda^{im} = -\lambda^{im}_j) \tag{9}$$

Φ are the components of the complex modal matrix of displacements, where the eigenvectors with even mode numbers have been removed. These are conjungate complex to the eigenvectors with odd mode numbers, therefore no information is lost. Similary, λ_j are the components of the reduced eigenvalue matrix. The superscripts re and im denote the real and imaginary part of a complex number, respectively. Equation (5) requires the same number of summations as the corresponding approximation of classical damping. The summation is performed over a truncated set of modes.

For some applications it may not be required to use the exact solution by nonclassical modes. In this case, the complex eigenproperties can be approximated by the normal modes of the system. The relationship depends on the normalisation of the modal matrices defined by eq. (10).

$$\boldsymbol{\Phi}^T\boldsymbol{D}\boldsymbol{\Phi} = \boldsymbol{I} \qquad \boldsymbol{\Psi}^T\boldsymbol{M}\boldsymbol{\Psi} = \boldsymbol{I} \tag{10}$$

The matrix $\boldsymbol{\Psi}$ is the $(n * n)$ modal matrix of normal modes of the classical eigenvalue problem, where n is the number of degrees of freedom of the system. Following the relation leads to equations (11) and (12),

$$\Phi_{k,j} = \Psi_{k,j}\sqrt{\frac{1}{2\Omega_j\sqrt{1-\xi_j^2}}}\, e^{i\frac{\pi}{4}} \tag{11}$$

$$\lambda_j = -\xi_j\Omega_j + i\Omega_j\sqrt{1-\xi_j^2} \tag{12}$$

where ξ_j and Ω_j are the modal damping coefficients and the undamped eigenfrequencies of the classical damped problem.

The numerical effort can be further reduced by considering only a subset of response DOFs. Focusing on noise emission, the mean square of the velocities are only required for the degrees of freedom belonging to the surface of the structure. Therefore the size of the matrices of eq. (1) reduces considerably. The reduction is given by eq. (13). The variables $nresp$ and $nexc$ denote the

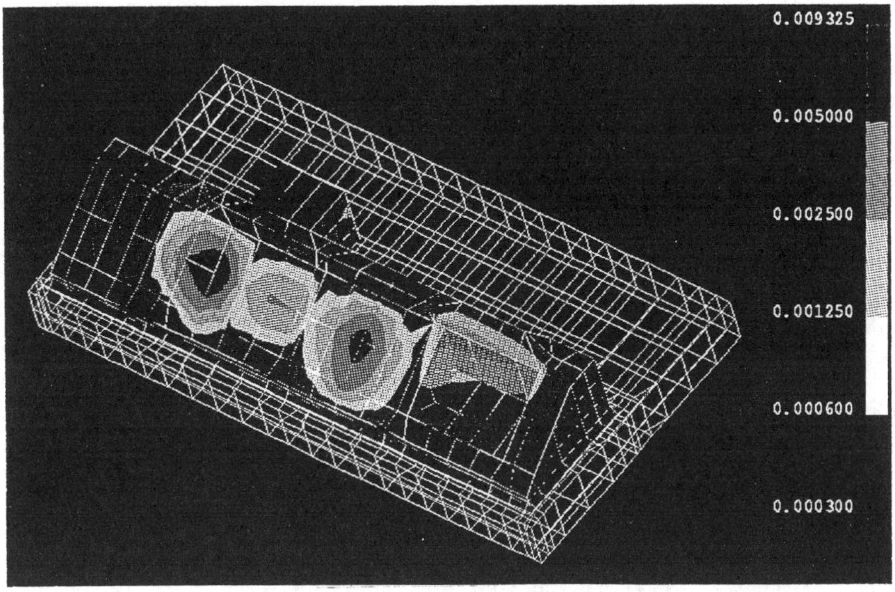

Figure 1: Noise Emission: 1600 - 1750 Hz

number of response and excitation DOFs, respectively.

$$
\begin{array}{lccc}
E[\dot{\boldsymbol{x}}_{(t)}\dot{\boldsymbol{x}}^{*T}_{(t)}] & (n*n) & \Rightarrow & (nresp*nresp) \\
\boldsymbol{V}_{(\omega)} & (n*n) & \Rightarrow & (nresp*nexc) \\
\boldsymbol{S}_{ff(\omega)} & (n*n) & \Rightarrow & (nexc*nexc)
\end{array}
\qquad (13)
$$

From (1) and (5) follows, that only those components of the eigenvectors are required, which describe excitation or response locations. This is used to build a reduced set of effective eigenvectors, which have displacement directions normal to the noise emitting surface and in direction of the exciting forces. This results in the calculation of the noise emission normal to the surface of the structure.

3 NUMERICAL EXAMPLE

The example shows a device, which was designed to test the computational procedures and to verify the calculation results. It consists of the cover of the cylinder head of a diesel engine that is bolted to a steel plate with a gasket between the cover and the plate. The structure is excited by a single force at the bottom of the plate, approximately central under the cylinder head cover. In case of the presented example, white noise excitation with the spectral density of $1N^2sec$ has been applied. In principle any spectral density, e.g. also that of a measured force, could be used. Figure1 shows the calculation results of the variance of the velocities normal to the surface of the cylinder head cover. The considered frequency interval is between 1600 and $1750Hz$, where 6 modes of the reduced complex modal matrix are included in the computation. The step size of the numerical integration was $4Hz$. The required computation time demonstrates the effectiveness of the presented method: the CPU time for calculating the variances was about 20 seconds on a VAX Station 3100.

4 CONCLUSIONS

The presented method is an efficient tool for calculating the noise emission of random excited, large non classical damped structures. The noise emission is described by the expectation of the square of the velocities, which are calculated within an arbitrary frequency domain. Depending on the required quality of the result, either complex or normal modes can be used. The method requires only a slightly increased computation time compared to classical damped systems.

5 ACKNOWLEDGEMENT

This research is supported by the Austrian Industrial Research Promotion Fund (FFF) under contract no. 6/636, which is gratefully acknowledged by the authors.

REFERENCES

Park, S. ,Park, I. & Ma, F. 1992. Decoupling Approximation of Nonclassically Damped Structures. AIAA Journal. 30(9): 2348-2351.

Xu, K. & Igusa, T. 1991. Dynamic Characteristics of Non-Classically Damped Structures. Earthquake Engineering and Structural Dynamics. 20: 1127-1144.

Igusa, T. , Der Kiureghian, A. & Sackman, J. L. 1984. Modal Decomposition Method for Stationary Response of Non-Classically Damped Systems. Earthquake Engineering and Structural Dynamics. 12: 121-136.

Maldano, G. O. & Singh, M. P. 1991. An Improved Response Spectrum Method for Calculating Seismic Design Response. Part 2: Non-Classically Damped Structures. Earthquake Engineering and Structural Dynamics. 20: 637-649.

Der Kiureghian, A. & Neuenhofer, A. 1992. Response Spectrum Method for Multi-Support Seismic Excitations. Earthquake Engineering and Structural Dynamics. 21: 713-740.

Lin, W. K. 1976 Probalistic Theory of Structural Dynamics. New York: Robert E. Krieger Publ. Co., Huntington.

Robson, J. D. 1980 The Response Relationships of Random Vibration Analysis. Journal of Sound and Vibration. 73(2): 312-315.

Sharan, A. M., Sankar, S. & Sankar, T. S. 1982. A New Approach for the Calculation of Response Spectral Density of a Linear Stationary Random Multidegree of Freedom System. Journal of Sound and Vibration. 83(4): 513-519.

Traill-Nash, R. W. 1981. Modal Methods in the Dynamics of Systems with Non-Classical Damping. Earthquake Engineering and Structural Dynamics. 9: 153-169.

Lang, G. F. 1989. Demystifying Complex Modes. Sound and Vibration. January: 36-40

Structural Safety & Reliability, Schuëller, Shinozuka & Yao (eds) © 1994 Balkema, Rotterdam, ISBN 90 5410 357 4

Noise perturbations of a nonlinear dynamical system with multiple steady states

R.Valéry Roy
Department of Mechanical Engineering, University of Delaware, Newark, Del., USA

Abstract: We examine the white noise effects in a harmonically forced nonlinear system with two oscillatory steady states. We predict the asymptotic form of the probability density in each domain of attraction, and of the mean exit-time from each domain in the limit of weak noise intensity.

One the dominant characteristics of nonlinear systems of engineering interest is the non-uniqueness of their response for a given set of system and excitation parameters. Small noise perturbations may have large effects on such a system and may bring about transitions between its coexisting states.[4-6] The system may experience a fluctuation sufficiently large so that the trajectory will leave the domain of attraction of a stable state with probability one. Kramers (1940) was the first to predict the noise-induced escape rate of trajectories out of a potential system of the form $\ddot{x}+\beta\dot{x}+U(x)=\xi(t)$ where the noise process $\xi(t)$ has the autocorrelation function $\langle\xi(t)\xi(t+\tau)\rangle = D\delta(\tau)$:

$$\kappa_K = \frac{\omega_A}{4\omega_S}\left(-\beta+\sqrt{\beta^2+4\omega_S^2}\right)\, e^{-\frac{\beta}{D}E_c} \quad (1)$$

where $\omega_A^2 = U''(x_A)$ at the stable state $A(x_A, 0)$ and $\omega_S^2 = -U''(x_S)$ at the unstable state $S(x_S, 0)$. To illustrate such noise-induced phenomenon in a non-potential system, we consider the response of a harmonically forced Duffing oscillator in the presence of weak Gaussian white noise. The equation of motion is taken in the form

$$\ddot{x}+2\eta\omega_o\dot{x}+\omega_o^2 x+\gamma\omega_o^2 x^3 = \Gamma\cos\Omega t+\xi(t) \quad (2)$$

where the noise process $\xi(t)$ has the autocorrelation function $\langle\xi(t)\xi(t+\tau)\rangle = D\delta(\tau)$. As is well-known,[3] for some values of the parameters η, γ, Γ and ω_o/Ω, this system may have two stable oscillatory states with widely differing amplitudes and phases. In the absence of noise perturbations, the lifetime of each of these states is infinite. The addition of random fluctuations in the system will cause transitions between these two states, and the lifetime of each state will decrease as the noise level increases. These transitions can be studied by first transforming the variables $(x(t),\ \dot{x}(t))$ into the van der Pol variables $(a(t),\ b(t))$ as

$$x(t) = a(t)\cos\Omega t + b(t)\sin\Omega t, \quad (3)$$

$$y(t) = \dot{x}(t) = \Omega\left(-a(t)\sin\Omega t + b(t)\cos\Omega t\right). \quad (4)$$

If the dominant response of interest is at the frequency of the forcing function, and if it is not perturbed too far from the steady state, then higher harmonics can be neglected and the new variables $(a(t),\ b(t))$ can be expected to be slowly-varying random processes. Under these conditions, the corresponding governing equations of motion can be averaged to yield , after non-dimensionalizing,

$$\frac{du}{d\tilde{t}} = -pu - v + q\,v(u^2+v^2) + \epsilon^{1/2}\,\xi_1(\tilde{t})$$

$$\frac{dv}{d\tilde{t}} = u - pv - q\,u(u^2+v^2) + 1 + \epsilon^{1/2}\,\xi_2(\tilde{t}) \quad (5)$$

with $u = \tilde{\Omega}a/\tilde{\Gamma}$, $v = \tilde{\Omega}b/\tilde{\Gamma}$, $\tilde{t} = \tilde{\Omega}t$, $p = \eta\omega_o/\tilde{\Omega}$, $q = 3\gamma\Gamma^2\omega_o^2/(4((\Omega^2-\omega_o^2)^3)$, $\tilde{\Gamma} = \Gamma/(2\Omega)$, $\tilde{\Omega} = (\Omega^2-\omega_o^2)/(2\Omega)$, and $\epsilon = D\tilde{\Omega}/(2\Omega^2\tilde{\Gamma}^2)$. $\xi_1(t)$ and $\xi_2(t)$ are two independent white Gaussian noise processes such that $\langle\xi_i(t)\xi_i(t+\tau)\rangle = \delta(\tau)$ $(i = 1, 2)$. It can readily be shown that the unperturbed system governed by Eq.(5) admits three fixed points if the parameters $(p,\ q)$ satisfy the condition $(27/2q-1-9p^2)^2 < (1-3p^2)^3$. Two of these fixed points (cor-

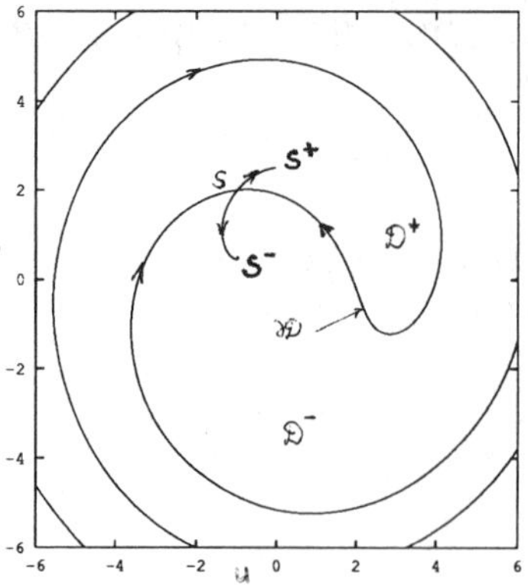

Fig.(1): Phase portrait of (3) for $p = 0.4$ and $q = 0.16$. Only shown are the four trajectories asymptotic to the saddle point S.

responding to the smallest and largest amplitude in the van der Pol plane $(u - v)$) can be shown to be stable, and they correspond to the two stable coexisting periodic solution of (2). The third solution is unstable, and corresponds to a saddle point of the trajectories in the $(u - v)$ plane. An example of a $(u - v)$ phase portrait is shown in Fig.(1): S^+ and S^- denote respectively the largest (resonant) and smallest (non-resonant) fixed points, and S denote the saddle point. Also shown are the stable and unstable invariant manifolds, that is, the four trajectories asymptotic to the saddle point. Critical to this phase portrait are the stable invariant manifolds $\partial\mathcal{D}$ which separates the two domains of attraction $\mathcal{D}^+$ and $\mathcal{D}^-$.

We are interested in the noise-induced transitions from S^+ to S^- and vice-versa, in the limit $\epsilon \rightarrow 0$, and, for that purpose, we approximate the stationary probability density function $p_\epsilon(u, v)$, solution of a Fokker-Planck-Kolmogorov equation, in the WKB approximation $w(u, v) \exp(-\Psi(u, v)/\epsilon)$.[2,4] Then the pseudo-potential Ψ can be shown to be solution of a zero-energy Hamilton-Jacobi equation associated with the Hamiltonian

$$H(u, v, p_u, p_v) = \Big(-pu - v + qv(u^2 + v^2)\Big)p_u + \quad (6)$$

$$\Big(u - pv - qu(u^2 + v^2) + 1\Big)p_v + \frac{1}{2}(p_u^2 + p_v^2). \quad (7)$$

The trajectories (called rays[2]) associated with this Hamiltonian system are governed by the equations

$$\dot{u} = \Big(-pu - v + qv(u^2 + v^2)\Big) + p_u$$

$$\dot{v} = \Big(u - pv - qu(u^2 + v^2) + 1\Big) + p_v$$

$$\dot{p_u} = -(-p + 2quv)p_u - (1 - q(3u^2 + v^2))p_v$$

$$\dot{p_u} = -(-1 + q(u^2 + 3v^2))p_u - (-p - 2quv)p_v. \quad (8)$$

Along these rays the functions Ψ and w are governed by the equations

$$\dot{\Psi} = \frac{1}{2}(p_u^2 + p_v^2)$$

$$\dot{w} = -\Big(-2p + \frac{1}{2}\Psi_{uu} + \frac{1}{2}\Psi_{vv}\Big)\, w \quad (9)$$

where the partial derivatives $\Psi_{uu} = \partial^2\Psi/\partial u^2$ and $\Psi_{vv} = \partial^2\Psi/\partial v^2$ still remain to be found. Each ray is initialized in a neighborhood of a chosen stable fixed point $\mathbf{x}_e = (u_e, v_e)$ (S^+ or S^-), for example, on a circle of radius $\delta \ll 1$:

$$u = u_e + \delta\cos\theta, \; v = v_e + \delta\sin\theta. \quad (10)$$

Each value of θ specifies a ray $(u(t,\theta), v(t,\theta))$ in the $(u - v)$ plane obtained by integration of (8). Initial conditions for Ψ and w are then set according to the asymptotic Gaussian shape of $p_\epsilon(u, v)$ near $\mathbf{x}_e$, that is, $\Psi = \frac{1}{2}\delta^2(k_{11}\cos^2\theta + 2k_{12}\cos\theta\sin\theta + k_{22}\sin^2\theta)$, and $w = 1$. Then, accordingly

$$p_u = \frac{\partial\Psi}{\partial u} = \delta(k_{11}\cos\theta + k_{12}\sin\theta), \quad (11)$$

$$p_v = \frac{\partial\Psi}{\partial v} = \delta(k_{12}\cos\theta + k_{22}\sin\theta). \quad (12)$$

The elements k_{11}, k_{12} and k_{22} of symmetric matrix $\mathbf{K}$ are found by solving the matrix equation $\mathbf{B}\,\mathbf{K}^{-1} + \mathbf{K}^{-1}\mathbf{B}^T = -\mathbf{I}$, where the matrix $\mathbf{B}$ is defined by linearizing the flow field $\mathbf{b}(u, v) = (b_1, b_2)^T \approx \mathbf{B}(\mathbf{x} - \mathbf{x}_e)$ of the noise-free system about the fixed point $\mathbf{x}_e$. In order to obtain the second partial derivatives of Ψ, the ray equations (8) must be differentiated with respect to parameter θ.

Therefore, the solution of the stationary FPK equation has been reduced to the solution of a system of ten ordinary differential equations with the appropriate initial conditions at each stable fixed point, yielding the values of $\Psi(u(t,\theta), v(t,\theta))$ and $w(u(t,\theta), v(t,\theta))$ along the rays $(u(t,\theta), v(t,\theta))$ specified by varying the parameter θ in the interval $0 \leq \theta < 2\pi$. Several contours for the pseudo-potential Ψ and ray trajectories are shown in Fig.(2a) and Fig.(2b) for rays initialized at fixed point S^- and fixed point S^+ respectively. Critical in this procedure is the determination of the value of θ for which the corresponding ray reaches the

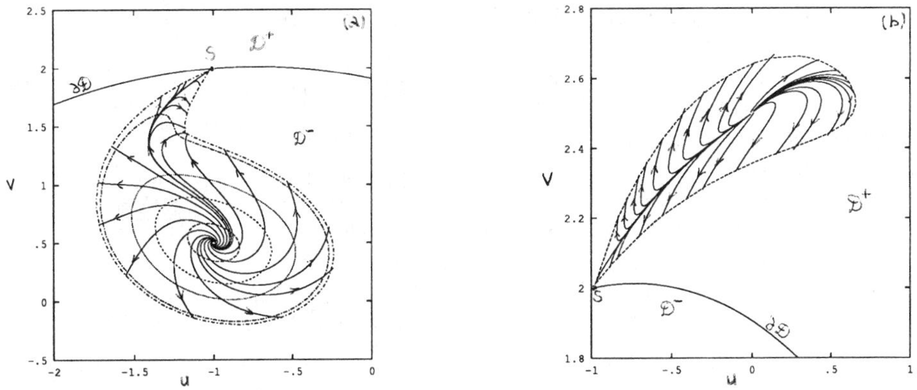

Fig.(2): Contours of Ψ and rays emanating from fixed point S^- (case (a)) and fixed point S^+ (case (b)) for $p = 0.40$ and $q = 0.16$. The most probable path of exit from S^- to S is given by $\theta = 2.718350106904$, and that from S^+ to S is given by $\theta = 3.4816259457613937$ (for a circle of radius $\delta = 0.01$).

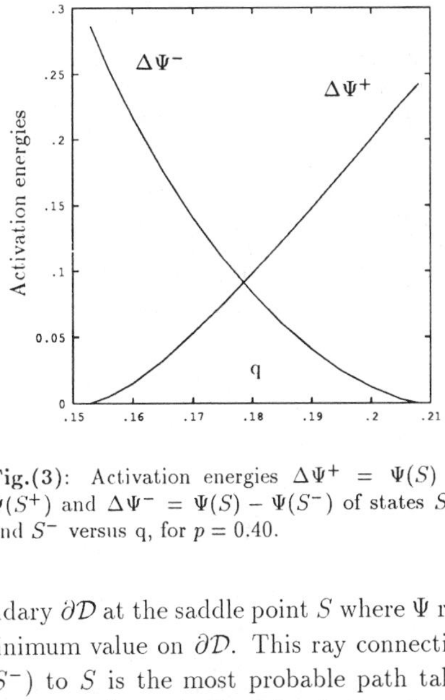

Fig.(3): Activation energies $\Delta\Psi^+ = \Psi(S) - \Psi(S^+)$ and $\Delta\Psi^- = \Psi(S) - \Psi(S^-)$ of states S^+ and S^- versus q, for $p = 0.40$.

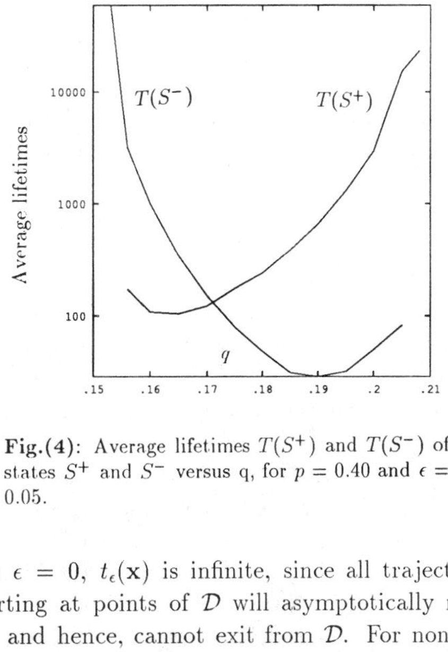

Fig.(4): Average lifetimes $T(S^+)$ and $T(S^-)$ of states S^+ and S^- versus q, for $p = 0.40$ and $\epsilon = 0.05$.

boundary $\partial\mathcal{D}$ at the saddle point S where Ψ reaches it minimum value on $\partial\mathcal{D}$. This ray connecting S^+ (or S^-) to S is the most probable path taken by the system to leave the domain of attraction of the chosen fixed point, and it gives the value reached by Ψ, w, Ψ_{uu}, Ψ_{uv} and Ψ_{vv} at the saddle point.

In order to have a measure of the relative stability between S^+ and S^-, we determine the asymptotic value of mean exit-time in the limit $\epsilon \to 0$, that is, the average time of first exit , $t_\epsilon(\mathbf{x})$, from a given domain of attraction $\mathcal{D}$ of the trajectories starting at points $\mathbf{x}$ of $\mathcal{D}$. $t_\epsilon(\mathbf{x})$ is known to be solution of $\mathcal{L}^* t_\epsilon(\mathbf{x}) = -1$ with the boundary condition $t_\epsilon(\mathbf{x}) = 0$ on $\partial\mathcal{D}$, where $\mathcal{L}^* = (-pu - v + qv(u^2+v^2))\partial/\partial u + (u - pv - qu(u^2+v^2))\partial/\partial v + \epsilon/2(\partial^2/\partial u^2 + \partial^2/\partial v^2)$.

For $\epsilon = 0$, $t_\epsilon(\mathbf{x})$ is infinite, since all trajectories starting at points of $\mathcal{D}$ will asymptotically reach $\mathbf{x}^*$, and hence, cannot exit from $\mathcal{D}$. For non-zero but small ϵ ($0 < \epsilon \ll 1$), we expect that $t_\epsilon(\mathbf{x})$ will reach a large value uniformly for all $\mathbf{x} \in \mathcal{D}$, except within a thin layer along the boundary $\partial\mathcal{D}$. Therefore, we write[5] $t_\epsilon(\mathbf{x}) = T(\mathbf{x}_e)\,\tau_\epsilon(\mathbf{x})$ where the boundary-layer function $\tau_\epsilon(\mathbf{x})$ is assumed to take the value one in the inner part of $\mathcal{D}$, and the value zero on $\partial\mathcal{D}$. The constant $T(\mathbf{x}_e)$ is assumed to be of the order $\exp(K/\epsilon)$ for $\epsilon \to 0$ ($K > 0$), and can be shown to be given by

$$T(\mathbf{x}_e) = -\frac{1}{2}\epsilon \frac{\int_{\mathcal{D}} p_\epsilon(\mathbf{x})\, d\mathbf{x}}{\int_{\partial\mathcal{D}} p_\epsilon(\mathbf{x})\, \nabla\tau_\epsilon \cdot \boldsymbol{\nu}\, ds}. \qquad (13)$$

We evaluate these two integrals asymptotically by the method of Laplace and by using the WKB approximation of p_ϵ in $\mathcal{D}$: the main contribution of the numerator's integral in (13) is at the fixed point $\mathbf{x}_e$, and that of the denominator's is at the saddle point S. Hence, in the limit $\epsilon \to 0$

$$\int_\Omega p_\epsilon(\mathbf{x})\, d\mathbf{x} \sim 2\pi\epsilon\, H^{-1/2}(\mathbf{x}_e), \qquad (14)$$

where $H(\mathbf{x}_e) = \det(\partial^2\Psi(\mathbf{x}_e)/\partial x_i \partial x_j) = k_{11}k_{22} - k_{12}^2$, and

$$\int_{\partial\Omega} p_\epsilon(\mathbf{x})\, \nabla\tau_\epsilon \cdot \boldsymbol{\nu}\, ds \sim -\left(\frac{\partial\tau}{\partial\rho}\right)_S (2\pi\epsilon)^{1/2}|\Psi_{ss}|^{-1/2} w(S) e^{-\Psi(S)/\epsilon}, \qquad (15)$$

In Eq.(13) and (15), $\nabla = (\partial/\partial u, \partial/\partial v)^T$, $\boldsymbol{\nu} = (\nu_1, \nu_2)^T$ is the unit outer normal to $\partial\mathcal{D}$, $(\rho = dist(\partial\mathcal{D}, \mathbf{x}), s)$ is a local coordinate system near the boundary. The functions $\partial\tau/\partial\rho$, $\Psi_{ss} = \partial^2\Psi/\partial s^2 = \nu_1^2\Psi_{uu} + 2\nu_1\nu_2\Psi_{uv} + \nu_2^2\Psi_{vv}$, Ψ and w must all be evaluated at the saddle point S as described above by the method of rays. Finally, it can be shown that

$$\left(\frac{\partial\tau}{\partial\rho}\right)_S = \left(\frac{4\lambda_p}{\epsilon\pi}\right)^{1/2} \qquad (16)$$

where λ_p is the unique positive eigenvalue of the noise-free system linearized about S. For more details of these theoretical developments see Roy (1993).

We illustrate the application of Eq.(13) for the mean-exit times $T(S^+)$ and $T(S^-)$ from the domains of attraction $\mathcal{D}^+$ and $\mathcal{D}^-$ of S^+ and S^- respectively. The value of the parameter p is fixed to 0.4 and the parameter q is varied in the interval $q_0(p) \leq q \leq q_1(p)$ according to the bistability condition. Fig.(3) shows the values of $\Delta\Psi^+ = \Psi(S) - \Psi(S^+)$ and $\Delta\Psi^- = \Psi(S) - \Psi(S^-)$ which might be interpreted as the minimum "activation energies" or potential heights for the transitions $S^+ \to S^-$ and $S^- \to S^+$ respectively. The variation of the "lifetimes" $T(S^+)$ and $T(S^-)$ of states S^+ and S^- upon q is shown on Fig.(4) for $\epsilon = 0.05$.

Acknowledgements

The support of the Engineering Foundation and AFOSR under Grant No. RI-B-91-09 is greatfully acknowledged.

References

[1] Kramers, H. A., "Brownian Motion in a Field of Force and the Diffusion Model of Chemical Reactions", *Physica*, **Vol. 7**, pp. 284-304, (1940).

[2] Cohen, J. K., and Lewis, R. M., "A Ray Method for the Asymptotic Solution of the Diffusion Equation", *J. Inst. Math. Appl.*, **Vol. 3**, pp. 266-290, (1967).

[3] Hayashi, C., *Nonlinear Oscillations in Physical Systems*, McGraw-Hill, New York, (1964).

[4] Ludwig, D., "Persistence of Dynamical Systems Under Random Perturbations", *Siam Review*, **Vol. 17**, No. 4, pp. 605-640, (1975).

[5] Matkowsky, B. J., and Schuss, Z., "The Exit Problem for Randomly Perturbed Dynamical Systems", *SIAM J. Appl. Math.*, **Vol. 33**, pp. 365-382, (1977).

[6] Wentzell, A. D., and Freidlin, M. I., "On small Random Perturbations of Dynamical Systems", *Russian Math. Surveys*, **25**, pp. 1-55, (1970).

[7] Roy, R. V., "Noise Perturbations of Nonlinear Dynamical Systems", *Computational Stochastic Mechanics*, (A. H.-D. Cheng, and C. Y. Yang, Eds.), Computational Mechanics Publications, (Southampton, UK), Elsevier, (1993).

Statistical uncertainty

Structural Safety & Reliability, Schuëller, Shinozuka & Yao (eds) © 1994 Balkema, Rotterdam, ISBN 90 5410 357 4

Effect of controller uncertainty on the stability of a distributed parameter system

Lawrence A. Bergman & W. Brent Hall
University of Illinois, Urbana, Ill., USA

ABSTRACT: It is well known that parameter uncertainty can degrade the performance of a well-designed control system, sometimes leading to system instability. In the context of structural control, performance degradation and instability imply excessive vibration and even structural failure. The ability of a controller to maintain the stability of a system in spite of parameter uncertainty is measured by its robustness. The application of reliability concepts to the robustness of distributed parameter systems is the subject of this work. Herein, Boyce's problem of the taut string controlled by a single actuator will be re-examined. However, the string will be both viscously and viscoelastically damped, and the control law will contain both proportional and rate feedback terms. In addition, the effect of time delay will be included explicitly.

1 INTRODUCTION

It is well known that parameter uncertainty can degrade the performance of a well-designed control system, sometimes leading to system instability. In the context of structural control, performance degradation and instability imply excessive vibration and even structural failure. The ability of a controller to maintain the stability of a system in spite of parameter uncertainty is measured by its robustness. In a recent series of papers, (Spencer, *et al.*, 1992, 1993) have presented a systematic and probabilistically consistent approach for determining the likelihood that instability will arise from parameter uncertainty in a discrete linear system of arbitrary dimension. Their formulation employs Hurwitz determinants and second order reliability methods to estimate the probability of instability of the closed loop system, which is represented by a finite number of subsystems with highly dependent failure modes in series.

The extension of these concepts to distributed parameter systems is, unfortunately, not immediate. The mere fact that these systems are infinite dimensional precludes the use of much of the machinery available for discrete systems, unless the distributed system is first discretized, which itself introduces error into the analysis, or is represented by an eigenfunction expansion, which requires truncation after some finite number of modes, also a potential source of error. In fact, the system will behave as one with an infinite number of subsystems with highly dependent failure modes in series.

To determine the reliability of such systems, root locus analysis will be employed, requiring the repetitive solution of a transcendental characteristic equation over a range of the parameter under investigation. The loci then provide a mapping from the probability distribution of the random parameter to the probability distribution of the system eigenvalues. This approach was utilized nearly 30 years ago (Boyce, 1962). Boyce examined eigenvalue distributions for undamped taut strings and Euler-Bernoulli beams, each subjected to the action of a single point actuator. He demonstrated that, for the case of uncertainty in the actuator gain alone, a simple, closed form mapping leading to the distributions of the eigenvalues of the system could be determined directly from the distribution of the actuator gain, and for uncertainty in the remaining parameters, approximate distributions could be obtained through the application of perturbation methods (Boyce, 1968). In related work, the distributions of the open loop eigenvalues of taut strings having random parameters have been studied (Iyengar and Athreya, 1975; Iyengar and Manohar, 1989).

Herein, Boyce's treatment of the taut string controlled by a single actuator will be re-examined. However, the string will be both viscously and viscoelastically damped, and the control law will contain both proportional and rate feedback terms. In addition, the effect of time delay will be included explicitly. The characteristic equation of the system will be derived in a fashion similar to that described in recent papers (Pang, Tsao and Bergman, 1992a,b), and root locus analysis will be performed. The real parts of the loci will be found in terms of the uncertain parameters, in this case a random gain or time delay with a prescribed distribution. The zero exceedences of the real parts will be mapped back into the parameter space, thereby avoiding the need to obtain explicitly the probability distributions of the eigenvalues. The probability of a zero exceedence will then be obtained directly from the probability distribution of the uncertain parameter, giving the probability of failure for an individual mode.

Certain conjectures will be made regarding the usefulness of such analyses, and the impact of these results upon the analysis of discretized systems will be discussed.

2.0 ANALYSIS

Following the analysis of Boyce (Boyce, 1968), consider a taut string fixed at one end and free at the other, the latter connected to ground through a single actuator, shown in Fig. 1. The (nondimensionalized) boundary value problem is given by

$$(1+a_1\frac{\partial}{\partial t})\, w''(x,t) = (a_0+\frac{\partial}{\partial t})\, \dot{w}(x,t),\ x \in (0,1) \quad (1)$$

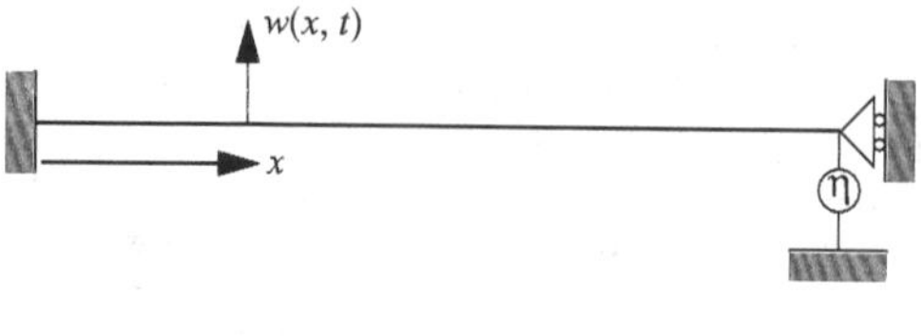

Figure 1. Taut string with single actuator.

$$w(0,t) = 0 \quad (2)$$

$$w'(1,t) + k_p w(1,t) + k_v \dot{w}(1,t) = 0 \quad (3)$$

$$w(x,0) = \dot{w}(x,0) = 0 \quad (4)$$

Here, a_0 and a_1 are viscous and Voigt damping coefficients, and k_p and k_v are proportional and rate gains, respectively.

Taking the Laplace transform gives

$$(1+a_1 s)\, W''(x,s) = (a_0+s)\, sW(x,s) \quad (5)$$

$$W(0,s) = 0 \quad (6)$$

$$W'(1,s) + k_p (1+\alpha s)\, W(1,s) = 0 \quad (7)$$

where

$$\alpha = \frac{k_v}{k_p} \quad (8)$$

Equation (5) can be rewritten as

$$W''(x,s) + \lambda^2 W(x,s) = 0 \quad (9)$$

where

$$\lambda^2 = -\left(\frac{a_0 s + s^2}{1+a_1 s}\right) \quad (10)$$

The solution is given by

$$W(x,s) = A_1 \sin\lambda x + A_2 \cos\lambda x \quad (11)$$

which, when substituted into the boundary conditions, gives

$$h(\lambda) = -\lambda\cot\lambda = \eta \quad (12)$$

where

$$\eta = k_p (1+\alpha s) \quad (13)$$

Note that this is precisely Boyce's result, except that in his analysis damping is neglected and the rate gain is presumed to be zero. Thus, all of his eigenvalues are purely imaginary, while herein

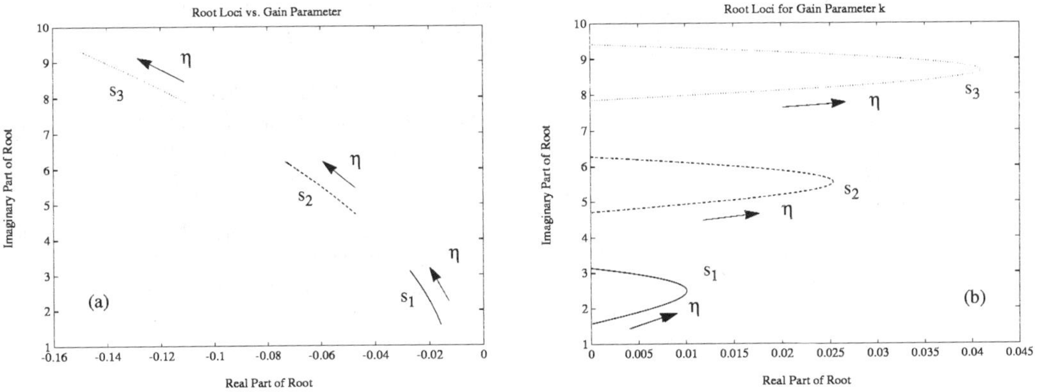

Figure 2. System 1-- a. damping, no time delay, and b. time delay, no damping

the eigensolution will generally be complex. For his case Boyce obtained a compact result for the probability distribution of eigenvalues, namely

$$F_{\lambda}(t) = G_{\eta}[h(t)] \qquad (14)$$

in which the cumulative distribution functions F_{λ} and G_{η} are associated with λ and η, respectively.

Our analysis is further complicated by the addition of time delay τ which appears in the characteristic equation of the closed loop system as an exponential term such that

$$-\lambda \cot\lambda = e^{-\tau s}\eta \qquad (15)$$

Thus, this simple mapping procedure (14) will not be possible in general, and we seek an alternate approach.

From eq. (15) combined with eq. (10) and (13), the root loci of the system can be determined. In the analyses to follow, the damping coefficients a_0 and a_1 are deterministic, and the gain ratio α is set equal to zero leading to purely proportional control. The proportional gain k_p and the time delay τ are alternately allowed to become random variables with prescribed distributions.

The probability of failure is the probability of the real part of any one eigenvalue being positive (Spencer, *et al.*, 1993). This can be determined from the marginal distributions of the loci, but this is not done here. Instead, the real parts of the loci are mapped into the random parameter space, and the probability is determined there. For each mode the mapping produces a set of failure boundaries that subdivide the parameter space into safe and unsafe regions. Each failure boundary corresponds to a zero crossing of the real part of the root locus. The probability of failure for an individual mode is then equal to the probability content of all the partitions of the parameter space corresponding to a zero exceedence.

The system probability of failure is obtained from the set of failure regions for all modes in the parameter space. In particular, the system failure region is the union of modal failure regions. Because the number of eigenvalues is infinite for distributed parameter systems like this one, the probability of system success or failure may not be obtainable from an analysis of a finite number of modes. Reliability analysis of the full system will be possible, in principle, if a finite number of failure modes can found that contain all other failure modes as subsets in the parameter space. This occurrence will depend upon the values of deterministic parameters, such as the damping parameters a_0 and a_1, and on the bounds, if any, of the probability distributions of random parameters. The wide variety of possible scenarios is illustrated in the following examples.

3.0 RESULTS

In order to demonstrate the behavioral extremes that can arise and the complex interplay between damping, time delay, and gain, several typical systems will be discussed.

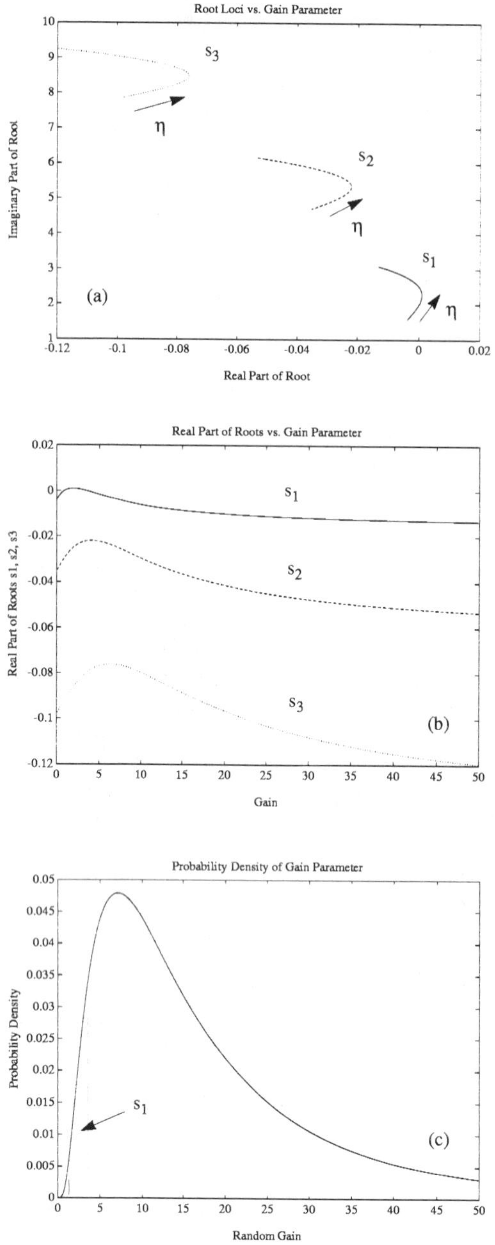

Figure 3. System 2

3.1 System 1:

We include this rather arcane case to show that, in the presence of any level of damping and in the absence of time delay, the real part of the system eigenvalues is confined to the left-half plane, and the system is safe (robust) for any positive value of gain η. Conversely, in the presence of time delay and the absence of damping, the system is unsafe for all values of gain. The root loci for the first three modes of the damped system are shown in Fig. 2a. For this system the damping parameters are $a_0 = 0.0236$ and $a_1 = 0.00318$, the time delay $\tau = 0$, and the gain η varies between 0.01 and 100. The root loci for the system with time delay $\tau = 0.01$, no damping, and the same range of gains are shown in Fig. 2b.

3.2 System 2:

Here, we demonstrate that a damped system can be destabilized by a small time delay. For this system the damping parameters are $a_0 = 0$ and $a_1 = 0.00318$, the time delay is $\tau = 0.01$, and the gain η varies between 0.01 and 100. The damping parameters are such that only the root locus of the first system mode crosses into the right-half plane, as shown in Fig. 3a, since it is known that the Voigt damping term dominates with increasing frequency, eventually causing overdamped behavior (Pang, Tsao, and Bergman, 1992b). The real parts of the loci are plotted versus gain in Fig. 3b, and the zero crossings of the first mode are mapped onto the probability distribution of gain in Fig. 3c. The gain parameter is assumed to be lognormally distributed with mean 20 and standard deviation 20. In this figure and in subsequent ones a high standard deviation has been chosen for the sake of a clear graphical presentation. The probability of failure of the first mode is represented by the shaded band in the figure, for values of η between 0.921 and 3.465, and is equal to 0.045 for the first three modes of the system.

3.3 System 3:

Here, we show that time delay can be a predominant factor in control system design. For this system the damping parameters are $a_0 = 0.0236$ and $a_1 = 0.00318$, the gain is $\eta = 1$, and the time delay τ varies between 0.01 and 10. Despite significant damping, increasing the time delay will eventually cause each mode to destabilize, as shown in Fig. 4a. Interestingly, the loci become fairly intricate and seem to exhibit almost periodic behavior. The real parts of the eigenvalues versus time delay are shown in Fig. 4b, and the zero exceedences are mapped onto the distribution of time delay in Fig. 4c, where the time delay is assumed to be lognormal with mean 4 and standard deviation 2. In this example, the parameter space (time delay) is subdivided into many overlapping and intersecting failure regions, except for very small time delays. It is clear that not only the first mode but also the higher modes are sensitive to the time delay and that, given latitude, time delay will almost surely cause the system to destabilize.

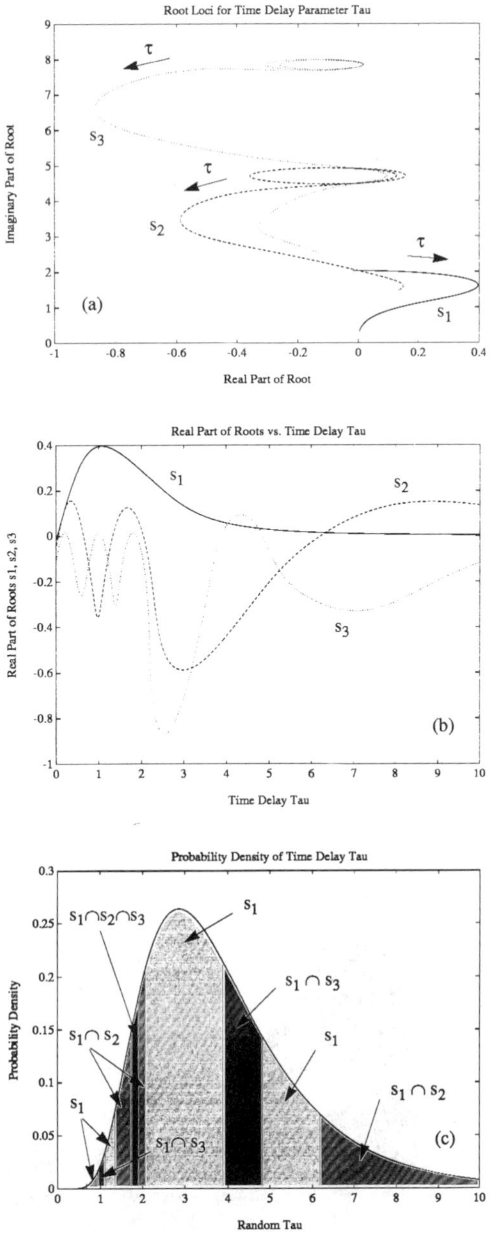

Figure 4. System 3

3.4 System 4:

We include this case to show that viscous damping alone is generally not sufficient to stabilize a system with time delay. For this system the damping parameters are $a_0 = 0.0236$ and $a_1 = 0$, the time delay is $\tau = 0.01$, and the gain η varies between 0.01 and 100. The root loci for the first three modes of the system are shown in Fig. 5a, and the real parts are plotted versus gain in Fig. 5b. Although damping has stabilized the system in the first mode, the second and third modes still have real parts with zero exceedences, and therefore probabilities of failure, as illustrated in Fig. 5c for a lognormally distributed gain with mean 20 and standard deviation 20. It appears that the root loci of the higher modes have increasingly larger excursions into the right half plane and that the lower mode failure regions are subsets of the higher mode ones. Therefore, system reliability for all modes will be low and perhaps impossible to evaluate. For the first three modes of the system, failure occurs between the first zero upcrossing at a gain of 1.232 and the last down crossing at a gain of 70.91, both for mode 3. The probability of failure of the full system is greater than the three mode probability of failure of 0.936.

3.5 System 5:

This system is the same as System 1a but with a time delay of $\tau = 0.1$. As the gain η is increased, each of the loci for the first three modes crosses the imaginary axis and then returns to the left hand side, as shown in Fig. 6a. The real parts are plotted versus gain in Fig. 6b and the zero crossings and failure regions are mapped onto the lognormal gain in Fig. 6c, which has a mean of 20 and a standard deviation of 20. Here, the failure regions overlap for the first three modes, and failure occurs between the first zero upcrossing of mode 1 at a gain of 0.177 and the last downcrossing of mode 3 at a gain of 48.989. The probability of failure is 0.932,

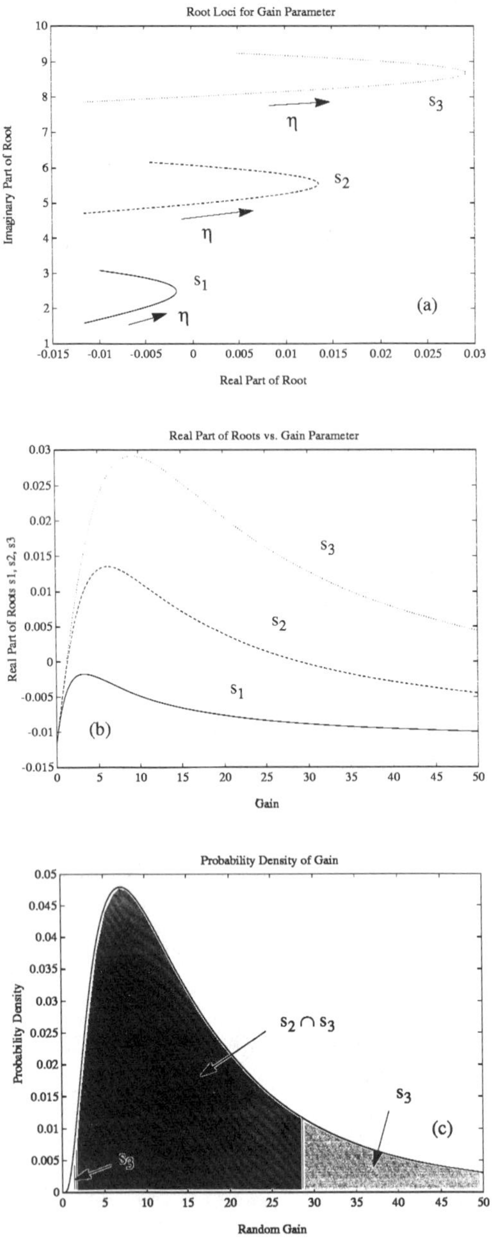

Figure 5. System 4

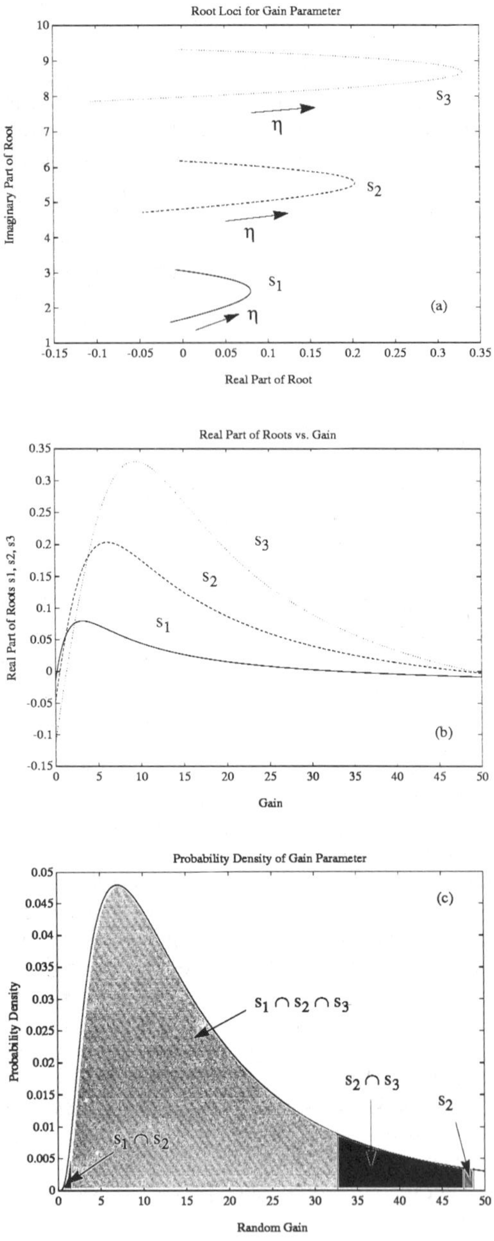

Figure 6. System 5

which will increase as additional modes are included.

4.0 CONCLUSIONS

We have demonstrated a probabilistically consistent method by which to examine the robustness of controlled distributed parameter systems, reflecting the effect of uncertainty in the system parameters upon closed loop system stability. The method utilizes the changes in the real part of the eigenvalues of the system as a function of the uncertain parameter, given by the root loci, to characterize regions of potential instability. These regions are projected onto the probability space of the uncertain parameter, giving the modal probabilities of failure, or instability, in the uncertain parameter space.

We have applied the method to the taut string problem first studied by Boyce and have demonstrated that, for certain combinations of parameters, the probability of failure of the system can be readily determined. However, for other parameter sets, it is difficult to deduce how the infinity of modes will behave by examining the first few. Thus, the global stability of the system cannot be immediately inferred. One might also expect a similar result for a discretized distributed parameter system, where an erroneous conclusion regarding robustness is drawn based upon an analysis of a finite dimensional representation of the system. Here, though, as previously shown (Spencer, *et al.*, 1992b), an estimate of the system probability of failure can be readily obtained.

REFERENCES

Boyce, W.E. 1962. Random vibration of strings and bars. Proc. 4th U.S. National Congress of Applied Mechanics: 77-85.

Boyce, W.E. 1968. Random eigenvalue problems. In A.T. Bharucha-Reid (ed.), Probabilistic Methods in Applied Mathematics 1: 1-73.

Iyengar, R.N. and K.B. Athreya 1975. A diffusion process approach to a random eigenvalue problem. Journal of the Indian Institute of Science 57: 185-191.

Iyengar, R.N. and C.S. Manohar 1989. Probability distribution of the eigenvalues of the random string equation. Journal of Applied Mechanics (ASME) 56: 202-207.

Pang, S.T., T-C. Tsao, and L.A. Bergman 1992. Active and passive damping of Euler-Bernoulli beams and their interactions. Proc. American Control Conference, Chicago: 2144-2149. Accepted for publication in the Journal of Dynamics, Measurement, and Control ASME).

Pang, S.T., T-C. Tsao, and L.A. Bergman 1992. Interaction effects in the hybrid control of Euler-Bernoulli beams. Proc. Ninth ASCE Engineering Mechanics Conference, College Station, Texas: 820-823.

Spencer, B.F., Jr., C. Montemagno, M.K. Sain, and P.M. Sain 1992. Reliability of controlled structures subject to real parameter uncertainties. Proc. ASCE Specialty Conference on Probabilistic Mechanics and Structural and Geotechnical Reliability, Denver: 369-372.

Spencer, B.F., Jr., M.K. Sain, J.C. Kantor, and C. Montemagno 1992. Probabilistic stability measures for controlled structures subject to real parameter uncertainties. Smart Materials and Structures 1: 294-305.

Structural Safety & Reliability, Schuëller, Shinozuka & Yao (eds) © *1994 Balkema, Rotterdam, ISBN 90 5410 357 4*

Analysis of random response of structures with uncertain parameters: Combination of perturbation method and substructure synthesis method

Takuzo Iwatsubo & Shozo Kawamura
Kobe University, Japan

Shin-ichiro Hata
Yamaha Co., Ltd, Kobe University, Japan

ABSTRACT: In this paper, the method to obtain the random responses of the structure with uncertain parameters is proposed. The proposed method is the combination method of the perturbation method and the substructure synthesis method, that can reduce the calculating time and the memory space of the computer. As a numerical example, a simple piping system is considered as a structure and divided into two substructures. The damping coefficient and the spring stiffness of the support are considered as the uncertain parameters. Then the variations of the eigenvalues and the eigenvectors, the auto-covariance function and the power spectral density function of the responses are calculated by using the proposed method. As a result, the proposed method is useful to analyze the sensitivities of eigenvalues and eigenvectors. And it is also useful to analyze the random response in terms of the accuracy and the calculation time.

1 Introduction

In the structures, mass, stiffness and damping coefficients have uncertainty in the design process or manufacturing process. And some structures, such as the piping system, are also exposed to the danger of the earthquake excitation.

The authors have already reported about the reliability design of the structures with uncertain parameters to earthquake excitation(1),(2). In that papers, the random responses of the structure are obtained by using the hierarchy method(3), which is first applied to the reliability problem by one of the authors(1). In another paper(4), in order to reduce the calculation time and the memory space of the computer, the random responses are obtained by using the combination method of the hierarchy method and the substructure synthesis method(5).

In this paper, the method to obtain the random responses of the structure with uncertain parameters is presented. The present method is the combination method of the perturbation method(6) and the substructure synthesis method, that can reduce the calculating time and the memory space of the computer. The basic procedures are as follows. (1) The structure is divided into some substructures. (2) The perturbation method is applied to obtain the eigenvalues, the eigenvectors and their sensitivities of each substructure(7). (3) The Guyan's static reduction ma-

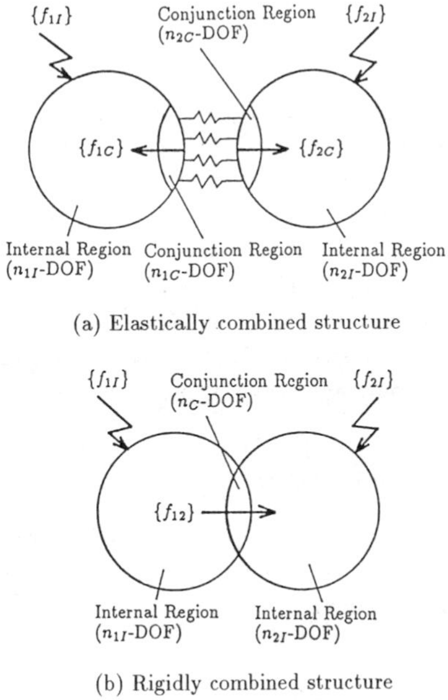

Fig.1 Structure separated into two substructures

trix of each substructure are calculated. (4) The equation of motion of the overall system and its sensitivity matrix are obtained by using the substructure synthesis method. (5) The eigenvalues, the eigenvectors and their variations of the overall system are calculated from the equation of motion. (6) The random response is calculated by using the impulsive response functions.

A simple piping system is considered as a structure and divided into two substructures. The damping coefficient and the spring stiffness of the support are considered as the uncertain parameters. Then the variations of the eigenvalues and the eigenvectors, the auto-covariance function and the power spectral density function of the responses are calculated by using the present method. And the present method is examined for its validity in terms of the accuracy and the calculation time.

2 Analysis by using substructure synthesis method

2.1 Equation of motion

The equations of motion of n degrees of freedom structure with uncetrain parameters are

$$[M]\{\ddot{v}(t)\} + [C]\{\dot{v}(t)\} + [K]\{v(t)\} = \{f(t)\} \quad (1)$$

where $[M]$, $[C]$ and $[K]$ are the symmetric mass, damping and stiffness matrices which have the uncertain paremeters $\alpha_i(i = 1, \cdots, N_\alpha)$, respectively. The avarage value $\overline{\alpha_i}$ and the standard deviation σ_{α_i} of the parameter α_i are assumed to be known and the uncertain parameters are assumed to be statistically independent. $[C]$ is not necessarily the proportional damping matrix. $\{v(t)\}$ is the relative displacement vector from the reference plane, $\{f(t)\}$ is the random excitation vector. When the random excitation $\{f(t)\}$ is regarded as a seismic excitation which is assumed to be identical at any excitating points, it is expressed as follows:

$$\{f(t)\} = -[M]\{I\}g(t) \quad (2)$$

where $\{I\}$ is the vector expressed the location where the seismic wave is applied, and $g(t)$ is the acceleration of the seismic wave.

2.2 Representation of perturbation equation of each substructure

For the simplicity of the explanation, the structure which is composed of two substructures shown in Fig.1 is considered. Fig.1(a) and (b) are the cases that two substructures are combined elastically and rigidly, respectively. The small terms over the second order sensitivities assumed to be ignored. And there is no interfernce between two substructures since the uncertain parameters are assumed to be statistically independent.

First, the case that two substructures are combined elastically shown in Fig.1(a) is considered. The equation of motion of the substructure 1 which is $n_1(= n_{1I} + n_{1C})$ degrees of freedom system (the internal region is n_{1I} degrees of freedom and the conjunction region is n_{1C} degrees of freedom) is expressed as follows:

$$[M_1]\{\ddot{v_1}\} + [C_1]\{\dot{v_1}\} + [K_1]\{v_1\} = \{f_1\} \quad (3)$$

where $[M_1]$, $[C_1]$ and $[K_1]$ are the mass, damping and stiffness matrices which have the uncertain paremeters $\alpha_i(i = 1, \cdots, N_\alpha)$, respectively and they are expressed as follows:

$$\begin{aligned} [M_1] &= \begin{bmatrix} [M_{1II}] & [M_{1IC}] \\ [M_{1CI}] & [M_{1CC}] \end{bmatrix}, \\ [C_1] &= \begin{bmatrix} [C_{1II}] & [C_{1IC}] \\ [C_{1CI}] & [C_{1CC}] \end{bmatrix}, \\ [K_1] &= \begin{bmatrix} [K_{1II}] & [K_{1IC}] \\ [K_{1CI}] & [K_{1CC}] \end{bmatrix} \end{aligned} \quad (4)$$

And they can be expanded in the form of the power series in $\varepsilon_i(i = 1, \cdots, N_\alpha)$, which is the small variation of the uncertain parameter $\alpha_i(i = 1, \cdots, N_\alpha)$. $\{v_1\}$ and $\{f_1\}$ are the response vector and the force vector expressed as follows:

$$\{v_1\} = \begin{Bmatrix} v_{1I} \\ v_{1C} \end{Bmatrix}, \ \{f_1\} = \begin{Bmatrix} f_{1I} \\ f_{1C} \end{Bmatrix} \quad (5)$$

where $\{v_{1I}\}$ and $\{v_{1C}\}$ are the responses of the internal region and the conjunction region, $\{f_{1I}\}$ is the external force acting to the internal region and $\{f_{1C}\}$ is the internal force acting to the internal region from the conjunction region.

When the substructure synthesis method[5] is used, the response of the internal region is approximated by the linear combination of the response of the rigidly connected internal region and the elastic deformations.

The response of the rigidly connected internal region $\{v_{1I}\}_{con.}$ are obtained as follows. The equation of motion of the free vibration of the internal region is obtained from Eqs.(3) and (4) as follows:

$$[M_{1II}]\{\ddot{v}_{1I}\}_{con.} + [K_{1II}]\{v_{1I}\}_{con.} = \{0\} \quad (6)$$

The sensitivity of the eigenvalues and the eigenvectors can be obtained as follows[7]:

$$\begin{aligned} \lambda_{1i} &= \lambda_{1i}^{(0)} + \sum_{j=1}^{N_\alpha} \varepsilon_j \lambda_{1i}^{(j)}, \\ \{\phi_{1i}\} &= \{\phi_{1i}^{(0)}\} + \sum_{j=1}^{N_\alpha} \varepsilon_j \{\phi_{1i}^{(j)}\} \quad (i = 1, \cdots, n_{1I}) \end{aligned} \quad (7)$$

Then the response of the rigidly connected internal region $\{v_{1I}\}_{con.}$ can be obtained as follows :

$$\{v_{1I}\}_{con.} = [\Phi_1]\{\xi_1\} \quad (8)$$

where $\{\xi_1\}$ is the modal response vector of the rigidly connected internal region and $[\Phi_1]$ is the modal matrix which can be expanded in the form of the power series in $\varepsilon_i (i = 1, \cdots, N_\alpha)$.

Next the elastic deformation $\{v_{1I}\}_{ela.}$ are obtained from Eqs.(3) and (4) as follows:

$$\begin{aligned} \{v_{1I}\}_{ela.} &= -[K_{1II}]^{-1}[K_{1IC}]\{x_{1C}\} \\ &\equiv [T_1]\{x_{1C}\} \end{aligned} \quad (9)$$

where $[T_1]$ is the Guyan's static reduction matrix which can be expanded in the form of the power series in $\varepsilon_i (i = 1, \cdots, N_\alpha)$.

Hence the response of the substructure 1 $\{v_1\}$ is approximated as follows :

$$\begin{aligned} \{v_1\} &= \begin{Bmatrix} v_{1I} \\ v_{1C} \end{Bmatrix} = \begin{bmatrix} [\Phi_1] & [T_1] \\ 0 & [I] \end{bmatrix} \begin{Bmatrix} \xi_1 \\ x_{1C} \end{Bmatrix} \\ &\equiv [T_{p1}]\{u_1\} \end{aligned} \quad (10)$$

Using Eq.(10), Eq.(3) can be represented in the form of the perturbation equation as follows:

$$[\tilde{M}_1]\{\ddot{u}_1\} + [\tilde{C}_1]\{\dot{u}_1\} + [\tilde{K}_1]\{u_1\} = \{\tilde{f}_1\} \quad (11)$$

where

$$\begin{aligned} [\tilde{M}_1] &= [T_{p1}]^T[M_1][T_{p1}] = \begin{bmatrix} [I] & [\tilde{M}_{1IC}] \\ [\tilde{M}_{1CI}] & [\tilde{M}_{1CC}] \end{bmatrix}, \\ [\tilde{C}_1] &= [T_{p1}]^T[C_1][T_{p1}] = \begin{bmatrix} [\tilde{C}_{1II}] & [\tilde{C}_{1IC}] \\ [\tilde{C}_{1CI}] & [\tilde{C}_{1CC}] \end{bmatrix}, \\ [\tilde{K}_1] &= [T_{p1}]^T[K_1][T_{p1}] = \begin{bmatrix} [\Lambda_1] & 0 \\ 0 & [\tilde{K}_{1CC}] \end{bmatrix}, \\ \{\tilde{f}_1\} &= [T_{p1}]^T\{f_1\} = \begin{Bmatrix} \tilde{f}_{1I} \\ \tilde{f}_{1C} \end{Bmatrix} \end{aligned} \quad (12)$$

where $[\Lambda_1]$ is the diagonal matrix of the eigenvalues λ_{1i}. These matrices and the vector in Eq.(12) can be expanded in the form of the power series in $\varepsilon_i (i = 1, \cdots, N_\alpha)$.

The equation of motion of the substructure 2 which is $n_2 (= n_{2I} + n_C)$ degrees of freedom system is expressed as follows:

$$[M_2]\{\ddot{v}_2\} + [C_2]\{\dot{v}_2\} + [K_2]\{v_2\} = \{f_2\} \quad (13)$$

where

$$\{f_2\} = \begin{Bmatrix} f_{2I} \\ f_{2C} \end{Bmatrix} \quad (14)$$

In the same procedure as the case of the substructure 1, the modal matrix $[\Phi_2]$ and the Guyan's static reduction matrix $[T_2]$ can be obtained. Then the response $\{v_2\}$ can be approximated as follows :

$$\begin{aligned} \{v_2\} &= \begin{Bmatrix} v_{2I} \\ v_{2C} \end{Bmatrix} = \begin{bmatrix} [\Phi_2] & [T_2] \\ 0 & [I] \end{bmatrix} \begin{Bmatrix} \xi_2 \\ x_{2C} \end{Bmatrix} \\ &\equiv [T_{p2}]\{u_2\} \end{aligned} \quad (15)$$

Using Eq.(15), Eq.(13) can be represented in the form of the perturbation equation as follows:

$$[\tilde{M}_2]\{\ddot{u}_2\} + [\tilde{C}_2]\{\dot{u}_2\} + [\tilde{K}_2]\{u_2\} = \{\tilde{f}_2\} \quad (16)$$

where $[\tilde{M}_2]$, $[\tilde{C}_2]$, $[\tilde{K}_2]$ and $\{\tilde{f}_2\}$ are obtained by the same procedure as Eq.(12).

2.3 Synthesis of perturbation equation of each substructure

The method for synthesizing two substructures is explained. When the substructures are combined elastically shown in Fig.1(a), the equation which governing the conjunction region is as follows:

$$\begin{bmatrix} 0 & 0 & 0 & 0 \\ 0 & [K'_{1C}] & 0 & [K'_{12C}] \\ 0 & 0 & 0 & 0 \\ 0 & [K'_{21C}] & 0 & [K'_{2C}] \end{bmatrix} \begin{Bmatrix} v_{1I} \\ v_{1C} \\ v_{2I} \\ v_{2C} \end{Bmatrix} = \begin{Bmatrix} 0 \\ -f_{1C} \\ 0 \\ -f_{2C} \end{Bmatrix}$$

$$\text{or} \quad [K'_C]\begin{Bmatrix} v_1 \\ v_2 \end{Bmatrix} = \begin{Bmatrix} f'_1 \\ f'_2 \end{Bmatrix} \quad (17)$$

where $[K'_{1C}]$, $[K'_{12C}]$, $[K'_{21C}]$ and $[K'_{2C}]$ are the stiffness matrices of the conjunction region.

Synthesizing Eqs.(11), (16) and (17), the next equation can be obtained.

$$[\tilde{M}]\{\ddot{y}\} + [\tilde{C}]\{\dot{y}\} + [\tilde{K}]\{y\} = \{\tilde{f}\} \quad (18)$$

where

$$
\begin{aligned}
[\tilde{M}] &= \begin{bmatrix} [\tilde{M}_1] & 0 \\ 0 & [\tilde{M}_2] \end{bmatrix}, \ [\tilde{C}] = \begin{bmatrix} [\tilde{C}_1] & 0 \\ 0 & [\tilde{C}_2] \end{bmatrix}, \\
[\tilde{K}] &= \begin{bmatrix} [\tilde{K}_1] & 0 \\ 0 & [\tilde{K}_2] \end{bmatrix} + [K_C'], \\
\{y\} &= \left\{ \{\xi_1\}^T \ \{v_{1C}\}^T \ \{\xi_2\}^T \ \{v_{2C}\}^T \right\}^T, \\
\{\tilde{f}\} &= \begin{Bmatrix} [T_{p1}]^T \{f_1 + f_1'\} \\ [T_{p2}]^T \{f_2 + f_2'\} \end{Bmatrix}
\end{aligned} \quad (19)
$$

This synthesising procedure is equivalent to the transformation of Eqs.(3), (13) and (17) by using the next equation.

$$
\begin{Bmatrix} v_1 \\ v_2 \end{Bmatrix} = \begin{bmatrix} [\Phi_1] & [T_1] & 0 & 0 \\ 0 & [I] & 0 & 0 \\ 0 & 0 & [\Phi_2] & [T_2] \\ 0 & 0 & 0 & [I] \end{bmatrix} \begin{Bmatrix} \xi_1 \\ v_{1C} \\ \xi_2 \\ v_{2C} \end{Bmatrix} \equiv [T_p]\{y\} \quad (20)
$$

where $[T_p]$ can be expanded in the form of the power series in $\varepsilon_i (i = 1, \cdots, N_\alpha)$.

Next, the case that two substructures are combined rigidly shown in Fig.1(b) is considered. The force vectors $\{f_1\}$ and $\{f_2\}$ in Eqs.(5) and (14) are as follows:

$$
\{f_1\} = \begin{Bmatrix} f_{1I} \\ -f_{12} \end{Bmatrix}, \ \{f_2\} = \begin{Bmatrix} f_{2I} \\ f_{12} \end{Bmatrix} \quad (21)
$$

where $\{f_{12}\}$ is the internal force acting to the substructure 2 from the substructure 1 through the conjunction region. When the substructures are combined rigidly, the relation of the response in the conjunction region is as follows :

$$
\{v_{1C}\} = \{v_{2C}\} \equiv \{v_C\} \quad (22)
$$

The synthesising procedure is equivalent to the transformation of Eqs.(3), (13) and (22) by using the next equation.

$$
\begin{Bmatrix} v_1 \\ v_2 \end{Bmatrix} = \begin{bmatrix} [\Phi_1] & [T_1] & 0 \\ 0 & [I] & 0 \\ 0 & [T_2] & [\Phi_2] \\ 0 & 0 & [I] \end{bmatrix} \begin{Bmatrix} \xi_1 \\ v_C \\ \xi_2 \end{Bmatrix} \equiv [T_p]\{y\} \quad (23)
$$

When the number of the mode in the substructure 1 is reduced to $\tilde{n}_{1I}$ from n_{1I} and the one in the substructure 2 is reduced to $\tilde{n}_{2I}$ from n_{2I}, Eq.(18) becomes $\tilde{n}(= \tilde{n}_{1I} + \tilde{n}_{2I} + n_C)$ degrees of freedom equation. Therefore the sensitivity analysis and the random response analysis is carried out for the reduced order equation.

2.4 Sensitivity analysis of eigenvalue and eigenvector

Since the matrices $[\tilde{M}]$, $[\tilde{C}]$ and $[\tilde{K}]$ in Eq.(18) can be expanded in the form of the power series in $\varepsilon_i (i = 1, \cdots, N_\alpha)$, the sensitivities of the eigenvalue and eigenvector can be obtained easily[7]. The eigenvalue λ_i and the eigenvectors $\{\phi_i\}$ are expressed as follows:

$$
\begin{aligned}
\lambda_i &= \lambda_i^{(0)} + \sum_{j=1}^{N_\alpha} \varepsilon_j \lambda_i^{(j)}, \\
\{\phi_i\} &= \{\phi_i^{(0)}\} + \sum_{j=1}^{N_\alpha} \varepsilon_j \{\phi_i^{(j)}\} \ (i = 1, \cdots, 2\tilde{n})
\end{aligned} \quad (24)
$$

The eigenvectors can be transformed into the physical space by using Eq.(20) or (23).

2.5 Random response

To calculate the responses considering the uncertainty, the impulse response function of Eq.(1) is obtained. First the vector $\{v(t)\}$ in Eq.(1) is assumed as follows:

$$
\{v(t)\} = \{v^{(0)}(t)\} + \sum_{j=1}^{N_\alpha} \varepsilon_j \{v^{(j)}(t)\} \quad (25)
$$

Substituting Eq.(25) into Eq.(1), the following equations are obtained.

$$
\begin{aligned}
&[M^{(0)}]\{\ddot{v}^{(0)}(t)\} + [C^{(0)}]\{\dot{v}^{(0)}(t)\} + [K^{(0)}]\{v^{(0)}(t)\} \\
&= \{f(t)\}
\end{aligned} \quad (26)
$$

$$
\begin{aligned}
&[M^{(0)}]\{\ddot{v}^{(j)}(t)\} + [C^{(0)}]\{\dot{v}^{(j)}(t)\} + [K^{(0)}]\{v^{(j)}(t)\} = \\
&- [M^{(j)}]\{\ddot{v}^{(0)}(t)\} - [C^{(j)}]\{\dot{v}^{(0)}(t)\} - [K^{(j)}]\{v^{(0)}(t)\} \\
&\equiv \{f^{(j)}(t)\}
\end{aligned} \quad (27)
$$

From Eq.(26), the average value of the impulse response function matrix $[H^{(0)}(t)]$ can be obtained. Then the average value of the resonse $\{v^{(0)}(t)\}$ and the first order sensitivity $\{v^{(j)}(t)\}$ are obtained as follows:

$$
\begin{aligned}
\{v^{(0)}(t)\} &= \int_0^t [H^{(0)}(t-\tau)]\{f(\tau)\}d\tau, \\
\{v^{(j)}(t)\} &= \int_0^t [H^{(0)}(t-\tau)]\{f^{(j)}(\tau)\}d\tau
\end{aligned} \quad (28)
$$

Here the impulse response function matrix $[\tilde{H}^{(0)}(t)]$ of Eq.(18) is related to the impulse response function matrix in the physical space $[H^{(0)}(t)]$ by the following equation:

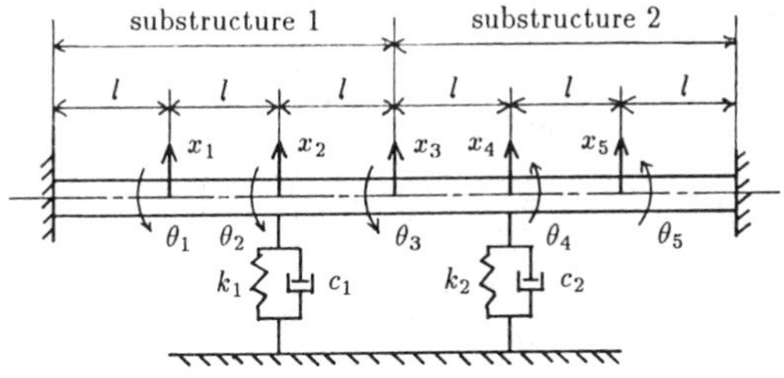

Fig.2 Piping system (c_1, c_2, k_1 and k_2 are uncertain)

Table 1 Dimensions of piping system

Length of each element	10.0 [m]
Density	7.7×10^3 [kg/m^3]
Modulus of Elasticity	206.0×10^9 [N/m^2]
Outside Diameter	1.0 [m]
Inside Diameter	0.9 [m]
Spring Constant of Support $k_1, k_2(k_{s0})$	1.0×10^7 [N/m]
Damping Constant of Support $c_1, c_2(c_{s0})$	5.0×10^4 [N· sec/m]
Spring Constant of Beam Part k_b	1.3×10^6 [N/m]
Damping Constant of Beam Part c_b	6.7×10^2 [N· sec/m]

Table 2 Eigenvalues (λ), natural frequencies (ω[rad/sec.]) and damping ratios (ζ) for deterministic system

mode	λ	ω	ζ
1	$(-0.104\times10^1)+j(0.232\times10^2)$	0.233×10^2	0.045
2	$(-0.161\times10^1)+j(0.383\times10^2)$	0.383×10^2	0.042
3	$(-0.100\times10^1)+j(0.592\times10^2)$	0.592×10^2	0.017
4	$(-0.328\times10^1)+j(0.998\times10^2)$	0.999×10^2	0.033
5	$(-0.699\times10^1)+j(0.149\times10^3)$	0.149×10^3	0.047
6	$(-0.132\times10^2)+j(0.229\times10^3)$	0.229×10^3	0.058
7	$(-0.259\times10^2)+j(0.317\times10^3)$	0.319×10^3	0.081
8	$(-0.498\times10^2)+j(0.438\times10^3)$	0.441×10^3	0.113
9	$(-0.903\times10^2)+j(0.594\times10^3)$	0.601×10^3	0.150
10	$(-0.152\times10^3)+j(0.765\times10^3)$	0.780×10^3	0.195

$$[H^{(0)}(t)] = [T_p^{(0)}][\tilde{H}^{(0)}(t)][T_p^{(0)}]^T \tag{29}$$

Then $\{v^{(0)}(t)\}$ and $\{v^{(j)}(t)\}$ can be expressed as follows:

$$\begin{aligned} \{v^{(0)}(t)\} &= [T_p^{(0)}]\int_0^t \{\tilde{h}^{(0)}(t-\tau)\}g(\tau)d\tau, \\ \{v^{(j)}(t)\} &= [T_p^{(0)}]\int_0^t [\tilde{H}^{(0)}(t-\tau)][T_p^{(0)}]^T \\ & \quad \{f^{(j)}(\tau)\}d\tau \end{aligned} \tag{30}$$

where

$$\{\tilde{h}^{(0)}(t)\} = -[\tilde{H}^{(0)}(t)][T_p^{(0)}]^T[M^{(0)}]\{I\} \tag{31}$$

The correlation function of the response $[R_x(t_1,t_2)]$ is described as follows :

$$\begin{aligned} [R_x(t_1,t_2)] &= \overline{\{v(t_1)\}\{v(t_2)\}^T} \\ &= \{v^{(0)}(t_1)\}\{v^{(0)}(t_2)\}^T \\ &+ \sum_{j=1}^{N_\alpha}\sigma_{\alpha_j}^2\{v^{(j)}(t_1)\}\{v^{(j)}(t_2)\}^T \end{aligned} \tag{32}$$

where the terms $\{v^{(0)}(t_1)\}$ $\{v^{(0)}(t_2)\}^T$ and $\{v^{(j)}(t_1)\}$ $\{v^{(j)}(t_2)\}^T$ can be written as follows:

$$\begin{aligned} &\{v^{(0)}(t_1)\}\{v^{(0)}(t_2)\}^T \\ &= [T_p^{(0)}]\int_0^{t_1}\int_0^{t_2}\{\tilde{h}^{(0)}(t_1-\tau_1)\}\{\tilde{h}^{(0)}(t_1-\tau_1)\}^T \\ &\quad R_g(\tau_1,\tau_2)d\tau_1 d\tau_2[T_p^{(0)}]^T \end{aligned} \tag{33}$$

$$\begin{aligned} &\{v^{(j)}(t_1)\}\{v^{(j)}(t_2)\}^T \\ &= [T_p^{(0)}]\int_0^{t_1}\int_0^{t_2}[\tilde{H}^{(0)}(t_1-\tau_1)][T_p^{(0)}]^T\{f^{(j)}(\tau_1)\} \\ &\quad \{f^{(j)}(\tau_2)\}^T[T_p^{(0)}][\tilde{H}^{(0)}(t_2-\tau_2)]^T \\ &\quad d\tau_1 d\tau_2[T_p^{(0)}]^T \end{aligned} \tag{34}$$

where $R_g(t_1,t_2)$ is the autocorrelation function of $g(t)$.

Moreover, the power and cross spectral density function of the response can be calculated by using the power spectral density function of $g(t)$.

3. Numerical example

3.1 Piping system

In order to check the accuracy of the proposed method, it is applied to the piping system shown in Fig.2. The piping system is fixed at both ends, and these supports are modeled as a stiffness and damping system. And it is separated into two substructures at the center and two substructures are combined rigidly. The original system is 10 degrees of freedom system. The dimensions of the piping system are shown in Table 1. The damping matrix of the pipe $[C_p]$ is the proportional damping matrix expressed as follows :

$$[C_p] = \beta[K_p] \tag{35}$$

where β is the constant value set up as 0.0005 and $[K_p]$ is the stiffness matrix of the pipe. It is assumed that the damping constant and spring constant of the support $c_i(i=1,2)$ and $k_i(i=1,2)$ are uncertain, i.e. its mean value and the standard deviation are given as $\overline{c_i}$, $\sigma_{c_i}(i=1,2)$ and $\overline{k_i}$, $\sigma_{k_i}(i=1,2)$, and the other parameters are deterministic. The equations of motion of the substructure 1 and 2 are shown as Eqs.(3) and (13).

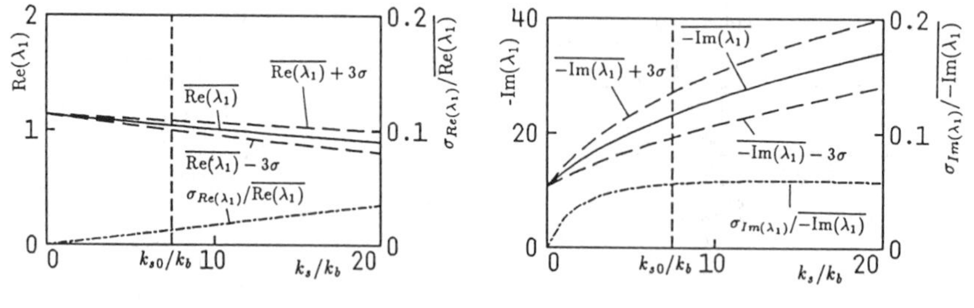

(a) Variation of the real part (b) Variation of the imaginary part

Fig.3 Variation and coefficient of variation of the first eigenvalue to the spring constant of support k_s
($\sigma_{k_1}/\overline{k_1}=\sigma_{k_2}/\overline{k_2}=0.3$, $\sigma_{c_1}/\overline{c_1}=\sigma_{c_2}/\overline{c_2}=0.0$)

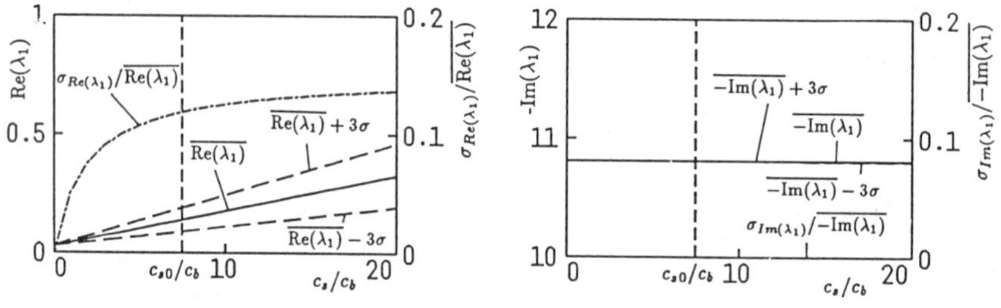

(a) Variation of the real part (b) Variation of the imaginary part

Fig.4 Variation and coefficient of variation of the first eigenvalue to the damping constant of support c_s
($\sigma_{c_1}/\overline{c_1}=\sigma_{c_2}/\overline{c_2}=0.3$, $\sigma_{k_1}/\overline{k_1}=\sigma_{k_2}/\overline{k_2}=0.0$)

The vectors $\{v_1\}$ and $\{v_2\}$ are expressed as follows:

$$\begin{aligned}\{v_1\} &= \{x_1,\ l\theta_1,\ x_2,\ l\theta_2,\ x_3,\ l\theta_3\}^T,\\ \{v_2\} &= \{x_4,\ l\theta_4,\ x_5,\ l\theta_5,\ x_3,\ l\theta_3\}^T \end{aligned} \tag{36}$$

The elements x_3, $l\theta_3$ are the conjunction region so that $n_{1I}=4$, $n_{2I}=4$, $n_C=2$.

First, in the case that the piping system is assumed to be deterministic, i.e. $\sigma_{c_i}=0(i{=}1,2)$ and $\sigma_{k_i}=0(i{=}1,2)$, the eigenvalues $\lambda_i(i=1,\cdots,20)$ and eigenvectors $\{\phi_i\}(i=1,\cdots,20)$ are calculated and the natural frequencies $\omega_i(i=1,\cdots,20)$ and damping ratios $\zeta_i(i=1,\cdots,20)$ are obtained by using the following equation :

$$\omega_i=\sqrt{\mathrm{Re}(\lambda_i)^2+\mathrm{Im}(\lambda_i)^2},\ \zeta_i=-\mathrm{Re}(\lambda_i)/\omega_i \tag{37}$$

Table 2 shows the eigenvalues, natural frequencies and damping ratios.

The numerical analysis is carried out for the following six cases.

Case1: the deterministic case.
The system is 10 degrees of freedom system.

Case2: the stochastic case by using the hierarchy method(2). The system is 30 degrees of freedom system.

Case3: the stochastic case by using the perturbation method(6). The system is 10 degrees of freedom system.

Case4: the stochastic case by using the present method. The system is 10 degrees of freedom system, i.e. the all four modes in $\{\xi_1\}$ and $\{\xi_2\}$ are used.

Case5: the stochastic case by using the present method. The system is 8 degrees of freedom system, i.e. the first, second and third modes in $\{\xi_1\}$ and $\{\xi_2\}$ are used.

Case6: the stochastic case by using the present method. The system is 6 degrees of freedom system, i.e. the first and second modes in $\{\xi_1\}$ and $\{\xi_2\}$ are used.

Table 3 Coefficients of variation of real part and imaginary part of eigenvalues for stochastic system ($\sigma_{k_1}/\overline{k_1}= \sigma_{c_1}/\overline{c_1}=0.3,\ \sigma_{k_2}/\overline{k_2}= \sigma_{c_2}/\overline{c_2}=0.0$)

(a) Case 3

mode	$\sigma_{Re(\lambda)}$	$\sigma_{Re(\lambda)}/\overline{Re(\lambda)}$	$\sigma_{Im(\lambda)}$	$\sigma_{Im(\lambda)}/\overline{Im(\lambda)}$
1	0.1521	0.1458	1.304	0.0561
2	0.2084	0.1293	1.084	0.0283
3	0.0213	0.0212	0.071	0.0012
4	0.1311	0.0400	0.262	0.0026
5	0.2347	0.0336	0.315	0.0021

(b) Case 4

mode	$\sigma_{Re(\lambda)}$	$\sigma_{Re(\lambda)}/\overline{Re(\lambda)}$	$\sigma_{Im(\lambda)}$	$\sigma_{Im(\lambda)}/\overline{Im(\lambda)}$
1	0.1521	0.1458	1.304	0.0561
2	0.2084	0.1293	1.084	0.0283
3	0.0213	0.0212	0.071	0.0012
4	0.1311	0.0400	0.262	0.0026
5	0.2347	0.0336	0.315	0.0021

(c) Case 5

mode	$\sigma_{Re(\lambda)}$	$\sigma_{Re(\lambda)}/\overline{Re(\lambda)}$	$\sigma_{Im(\lambda)}$	$\sigma_{Im(\lambda)}/\overline{Im(\lambda)}$
1	0.1521	0.1458	1.305	0.0562
2	0.2084	0.1293	1.084	0.0283
3	0.0211	0.0211	0.072	0.0012
4	0.1319	0.0402	0.265	0.0027
5	0.2319	0.0332	0.319	0.0021

(d) Case 6

mode	$\sigma_{Re(\lambda)}$	$\sigma_{Re(\lambda)}/\overline{Re(\lambda)}$	$\sigma_{Im(\lambda)}$	$\sigma_{Im(\lambda)}/\overline{Im(\lambda)}$
1	0.1520	0.1457	1.302	0.0561
2	0.2082	0.1292	1.083	0.0283
3	0.0212	0.0211	0.065	0.0011
4	0.1311	0.0399	0.258	0.0026
5	0.2548	0.0351	0.306	0.0020

3.2 Sensitivitiy analysis of eigenvalues and eigenvectors

Figure 3 shows the average value and the coefficient of variation of the first eigenvalue when the spring constant of support $k_s(=\overline{k_1}=\overline{k_2})$ is changed. Here k_b and c_b are the spring constant and damping constant of the beam part of piping system calculated from the first eigenvalue. k_{s0} and c_{s0} are the values used in this paper. As in Fig.3, the spring constant of support k_s has much influence on the variation of the imaginary part of eigenvalue. Figure 4 shows the ones when the damping constant of support $c_s(=\overline{c_1}=\overline{c_2})$ is changed. As in Fig.4, the damping constant of support c_s has much influence on the variation of the real part of eigenvalue. As in Figs.3 and 4, it is shown the coefficient of variation is relatively small.

Next, when the coefficients of variation are set as $\sigma_{c_1}/\overline{c_1}= \sigma_{k_1}/\overline{k_1}=0.3$ and $\sigma_{c_2}/\overline{c_2}= \sigma_{k_2}/\overline{k_2}=0.0$, the coefficients of variation of eigenvalue are calculated for Case3 ~ 6. The results are shown in Table.3. As in Table 3, the results in Case 4 agrees with the ones in Case 3 and the results of the higher modes have little error in Case 5 and Case 6. And the coefficients of variation of the real part of eigenvalue are bigger than the ones of the imaginary part of eigenvalue since the average values of the imaginary part are bigger than the ones of the real part. Moreover the coefficients of variation of the real part and imaginary part of eigenvalue are smaller than the ones of the spring constant and damping constant.

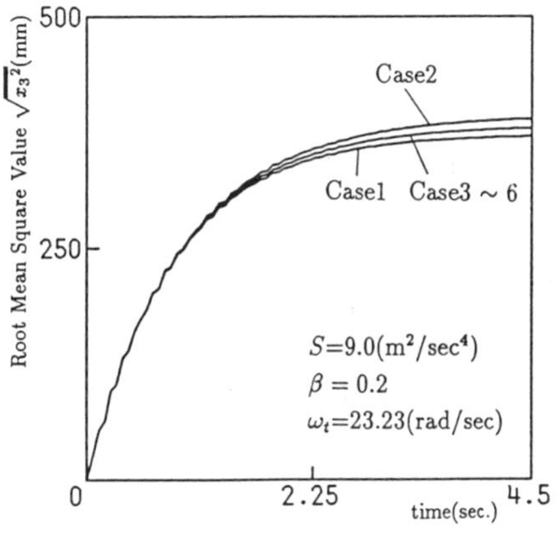

Fig.5 Random response to stationary narrow band process

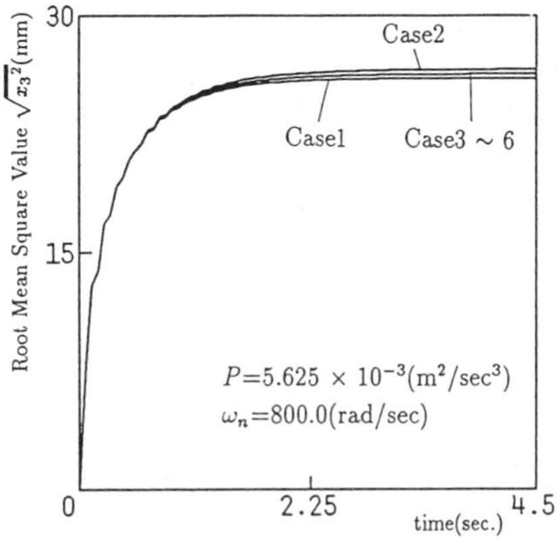

Fig.6 Random response to stationary wide band process

3.3 Random response

The mean square values of the responses to the stationary narrow band process and the stationary wide band process are calculated by using Eqs.(32) ~ (34).

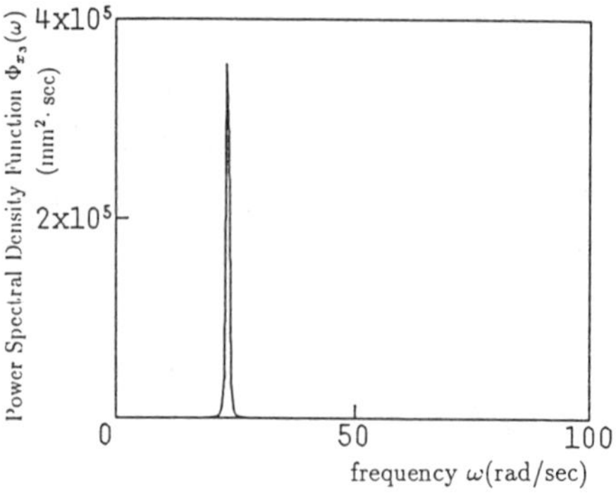

Fig.7 Power spectral density function to stationary narrow band process

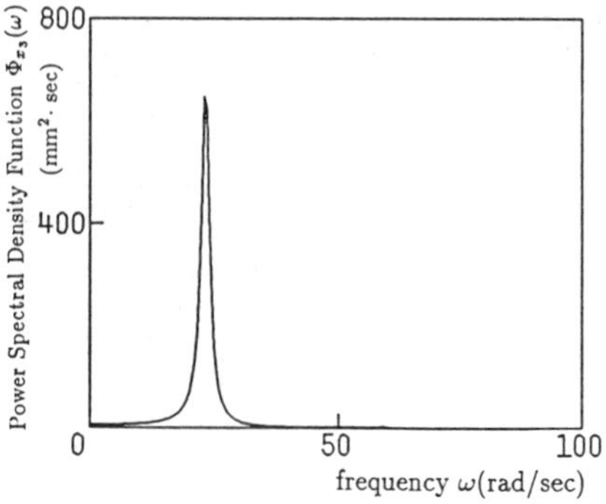

Fig.8 Power spectral density function to stationary wide band process

Since the variation of damping has much influence on the magnitude of random response, the coefficients of variation are set as $\sigma_{c_1}/\overline{c_1}=\sigma_{c_2}/\overline{c_2}=0.3$ and $\sigma_{k_1}/\overline{k_1}=\sigma_{k_2}/\overline{k_2}=0.0$. The stationary narrow band process is expressed as follows:

$$\begin{aligned} R_g(\tau) &= S\exp(-\beta\tau)\cos(\omega_t\tau),\ (\tau=\mid t_1-t_2\mid) \\ \Phi_g(\omega) &= \frac{2S\beta(\omega_t^2+\beta^2+\omega^2)}{(\omega_t^2+\beta^2+\omega^2)^2-4\omega_t^2\omega^2} \end{aligned} \quad (39)$$

And the stationary wide band process is expressed as follows:

$$\begin{aligned} R_g(\tau) &= 2P\sin(\omega_n\tau)/\tau,\ (\tau=\mid t_1-t_2\mid) \\ \Phi_g(\omega) &= \begin{cases} 2P\pi & (\mid\omega\mid<\omega_n) \\ P\pi & (\mid\omega\mid=\omega_n) \\ 0 & (\mid\omega\mid>\omega_n) \end{cases} \end{aligned} \quad (40)$$

In order to compare the random response to the narrow band process with that to wide band process, the mean square values of two processes are set to be equal, i.e., the following relation is used.

$$S = 2P\omega_n \quad (41)$$

Fig.5 shows the root mean square values at the nodal point 3 to the stationary narrow band process for Case 1 ~ 6. As in Fig.5, the result for Case 1 is smaller than the others since the uncertainty is ignored. The results for Case 2 is a little bigger than the ones for Case 3 ~ 6 since the second order variations of the unceterain parameters are approximated in the hierarchy method. The results for Case 3 ~ 6 agree well since the dominant frequency ω_t is near the first eigenvalue. Fig.6 shows the root mean square values to the stationary wide band process. As in Fig.6, the similar results are obtained as the case for the stationary narrow band process. In this case, also, there is no significant difference between Case 3 and Case 4 ~ 6 because the damping ratios for the higher order modes are relatively high shown in Table 2 and the variations of eigenvalues are relatively small to the values of c_{s0} and k_{s0} used in this paper.

Figs.7, 8 are the power spectral density function $\Phi_{x3}(\omega)$ to the stationary narrow and wide band process in Case 4. As similar in the results of the root mean sqaure values, since the damping ratios for the higher order modes are relatively high, the power spectrl density function can be seen near the first eigen frequency.

3.4 Comparison of calculating time and memory space

The memory space for calculating the eigenvalues is compared. When the memory space in Case3 is assumed to be unit, the one in Case4 is almost 1.0 times of the unit, the one in Case5 is about 0.7 and the one in Case6 is about 0.4.

The calculating time for obtaining the correlation function is compared. When the calculating time in Case3 is assumed to be unit, the one in Case4 is almost 1.0 times of the unit, the one in Case5 is about 0.5 and the one in Case6 is about 0.15.

4. Conclusion

In this paper, the method to obtain the random responses of the structure with uncertain parameters is proposed. The proposed method is the combination method of the perturbation method and the substructure synthesis method, that can reduce the calculating time and the memory space of the computer. As a numerical example, a simple piping system is considered as a structure and divided into two substructures. The damping coefficient and the spring stiffness of the support are considered as the uncertain parameters. Then the variations of the eigenvalues and the eigenvectors, the auto-covariance function and the power spectral density function of the responses are calculated by using the proposed method. As a result, the proposed method is useful to analyze the sensitivities of eigenvalues and eigenvectors. And it is also useful to analyze the random response in terms of the accuracy and the calculation time.

Reference

(1) T.Iwatsubo, I.Kawahara, N.Nakagawa and R.Kawai, (1979). Reliability Design of Rotating Machine against Earthquake Excitation, Bulletin of JSME, Vol.22, No.173, pp.1632-1639.
(2) T.Iwatsubo, S.Kawamura and H.Mori, (1990). Reliability Design of Piping System against Earthquake Excitation, JSME International Journal (Series 3), Vol.33, No.3, pp.366-370.
(3) A.T.Bharuca-Reid, et al.,(1968). Probabilistic Methods in Applied Mathematics, Vol.1, p.48, Academic press.
(4) T.Iwatsubo, S.Kawamura and H.Mori,(1991). Analysis of Random Response of Structures with Uncertain Parameters (Combination of Hierarchy Method and Substructure Synthesis Method), SMiRT11 Transaction, Vol.K, pp.423-428.
(5) R.R.Craig Jr. and M.C.C.Bampton, (1968). Coupling of Substructures for Dynamic Analyses, AIAA Journal, Vol.6, No.7, pp.1313-1319.
(6) Bellman.R,(1964). Perturbation Technique in Mathematics, Physics and Engineering, Holt, Rinehart and Winston.
(7) R.L.Fox and M.P.Kapoor,(1968). Rate of Change of Eigenvalues and Eigenvectors, AIAA Journal, Vol.6, No.12, p.2426.

Structural Safety & Reliability, Schuëller, Shinozuka & Yao (eds) © 1994 Balkema, Rotterdam, ISBN 90 5410 357 4

Response of stochastically loaded Bernoulli-Euler beams with randomly varying bending stiffness

H.U. Köylüoğlu & A.Ş.Çakmak
Department of Civil Engineering and Operations Research, Princeton University, N.J., USA

S.R.K.Nielsen
Department of Building Technology and Structural Engineering, University of Aalborg, Denmark

ABSTRACT: First and second order statistical characteristics of the response of a stochastic linear system, with stochastically varying system parameters, under stochastic loadings are studied. The variabilities of the deflection and the moment fields of statically determinate Bernoulli-Euler beams with randomly varying bending stiffness field, subject to statical random external loading field are determined. The exact solution shows that the variance of the deflection field will be the integral of the sum of three terms, which are the product of the power spectral density function of the loading with a variability response function, the product of the power spectral density function of the bending stiffness with another variability response function and the product of both power spectral densities with a third variability response function. This expression makes it possible to establish spectral-distribution free upper bounds for the response variability of this stochastic system which is very important since detailed information concerning the probabilistic nature of the bending stiffness and the loading fields are usually not available. The derived formulas can also be applied to the cases for which only the loading or the bending stiffness is modeled as a random field. The analytical results derived for the variability response functions are compared to the ones obtained from numerical simulations.

1. INTRODUCTION

In the past decade, there has been significant interest in the analysis of response variability of stochastic structural systems with system parameters modelled as random fields (i.e. [1,2,3,4,6,7,8,9,12] and many others). The analysis of deterministic systems subject to stochastic input is not a new issue, however, there has been very few studies in which both the input and the system parameters are modelled as random fields. The purpose of this study is to determine the statistical characteristics of the output field of a stochastic linear system with randomly varying system parameters subject to a random input field. The system considered is the statically determinate Bernoulli-Euler beam, with randomly varying bending stiffness field, subject to statical random external loading field. The deflection and the moment fields are considered as the output of the problem. The Green function formulation is used for the exact solution. It should be noted that there has been studies on statically determinate beams with only randomly varying bending stiffness by Vanmarcke and Grigoriu (1983), Shinozuka (1987), and, Deodatis and Shinozuka (1989).

The exact solution shows that the variance of the deflection field will be the integral of the sum of three terms, which are the product of the power spectral density function of the loading field with a variability response function, the product of the power spectral density function of the bending stiffness with another variability response function and the product of both power spectral density functions with a third variability response function. The variability response functions can be interpreted like the frequency response function in random vibration analysis, Deodatis and Shinozuka (1989). This expression makes it possible to establish spectral-distribution free upper bounds of the response variability of stochastic systems. These bounds are very important since detailed information concerning the probabilistic nature of material properties and loading are usually not available. The derived formulas can also be applied to the cases for which only the loading

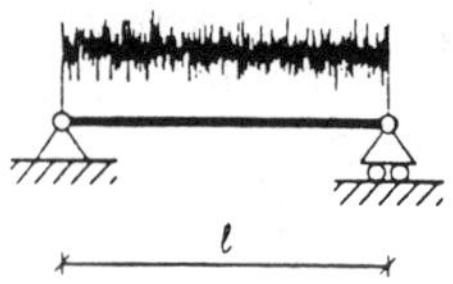

$$g_m(x,\xi) = g_v(x,\xi) = \begin{cases} \left(1-\frac{\xi}{l}\right)x & , \quad 0 < x \leq \xi \\ \left(1-\frac{x}{l}\right)\xi & , \quad \xi \leq x < l \end{cases}$$

Figure 1. Simply supported beam and the corresponding Green function.

$$l = 1., \quad \mu_P(x) = 1., \quad \frac{1}{\mu_B(x)} = 1. \implies P(x) = 1. + p(x) \quad \frac{1}{B(x)} = 1. + b(x)$$

$$S_{pp}(\kappa) = \frac{\sigma_{pp}^2}{2(\kappa_{u,p} - \kappa_{l,p})} \qquad \kappa_{l,p} \leq |\kappa| \leq \kappa_{u,p}$$

$$S_{bb}(\kappa) = \frac{\sigma_{bb}^2}{2(\kappa_{u,b} - \kappa_{l,b})} \qquad \kappa_{l,b} \leq |\kappa| \leq \kappa_{u,b}$$

$$\sigma_{pp} = \sigma_{bb} = 0.2, \quad \kappa_{u,p} = \kappa_{u,b} = 2., \quad \kappa_{l,p} = \kappa_{l,b} = 1.$$

$$\implies S_{pp}(\kappa) = S_{bb}(\kappa) = 0.02 \qquad 1. \leq |\kappa| \leq 2.$$

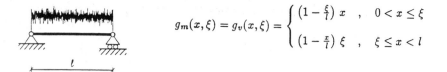

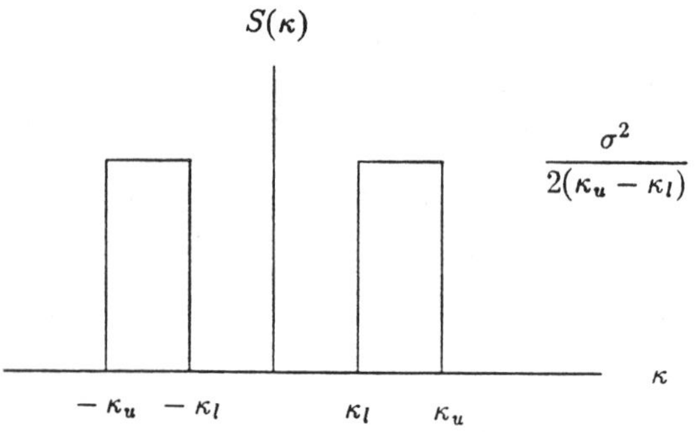

Figure 2. Stochastic model for the bending stiffness and the loading random fields and system parameters.

or bending stiffness is modelled as a random field. The analytical results derived for the variability response functions are verified by comparing them with the ones obtained from numerical simulations. The fast Monte Carlo simulation technique of Deodatis and Shinozuka (1989) is used in the numerical simulations and convergence is obtained for extremely small sample sizes, e.g. 5 simulations.

2. DIFFERENTIAL EQUATIONS FOR BERNOULLI-EULER BEAMS

The deflection field, $v(x)$, and the moment field, $m(x)$, of a statically loaded Bernoulli-Euler beam of length l fulfil the following relations along the beam (for all $x \in (0, l)$)

$$\frac{d^2}{dx^2} m(x) = -P(x) \tag{1}$$

$$\frac{d^2}{dx^2} v(x) = -\frac{m(x)}{B(x)} \tag{2}$$

$$m(x) = \int_0^l g_m(x,\xi)[-P(\xi)]d\xi \tag{3}$$

$$v(x) = \int_0^l g_v(x,\xi)[-\frac{m(\xi)}{B(\xi)}]d\xi \tag{4}$$

where $P(x)$ and $B(x)$ denote the loading field and the bending field respectively.

Equations (1) and (2) are the fundamental equations of Bernoulli-Euler beam theory. Equations

(3) and (4) are the integral forms of the first and second equations where $g_m(x,\xi)$ and $g_v(x,\xi)$ denote the one-sided Green's functions of the second order differential operator of equations (1) and (2) satisfying equations (5.a) and (5.b), respectively.

$$\frac{d^2}{dx^2}g_m(x,\xi) = \delta(x-\xi) \qquad \frac{d^2}{dx^2}g_v(x,\xi) = \delta(x-\xi) \tag{5.a,b}$$

$g_m(x,\xi)$ and $g_v(x,\xi)$ also satisfy the mechanical and geometrical boundary conditions respectively.

The deflection field, $v(x)$, and the moment field $m(x)$ are treated as random fields due to the random loading field, $P(x)$, and due to the random bending stiffness field $B(x)$. It is assumed that the differential equations of the Bernoulli-Euler beam theory are valid when the loading and the bending stiffness fields are modelled as random fields. For simplicity, the loading and the bending stiffness fields are assumed to be mutually independent and the analytical formulas are derived accordingly. If they are stochastically dependent, formulas can be improved easily. The mutually independent random fields $B(x)$ and $P(x)$ can be written in the form

$$P(x) = \mu_P(x) + p(x) \tag{6}$$

$$\frac{1}{B(x)} = \frac{1}{\mu_B(x)} + b(x) \tag{7}$$

$\mu_P(x)$ and $\frac{1}{\mu_B(x)}$ signify the mean value functions of the loading and the reciprocal of the bending stiffness fields. Then, $\{p(x), x \in (0,l)\}$ and $\{b(x), x \in (0,l)\}$ become one-dimensional, zero-mean, homogeneous random fields. Since the bending stiffness cannot take on negative values, $b(x)$ is assumed to be bounded from below as $b(x) > -\frac{1}{\mu_B(x)}$ with probability 1.

In this study, expressions for the first and the second order moment properties of the bending moment and the deflection fields are developed. The method requires knowledge of the corresponding first and second order moment properties of the bending stiffness and the loading fields, i.e. of the mean values $\frac{1}{\mu_B(x)}$, $\mu_P(x)$, of the standard deviations $\sigma_{\frac{1}{B}} = \sigma_b, \sigma_P = \sigma_p$ and of the autocorrelation functions $\rho_{\frac{1}{B}}(x)$ and $\rho_P(x)$.

3. STATICALLY DETERMINATE BEAMS

For statically determinate beams, the Green functions $g_m(x,\xi)$ and $g_v(x,\xi)$, as determined by equations (5.a) and (5.b), are deterministic and independent of the bending stiffness field. When equation (6) is substituted in equation (3), the moment field becomes

$$m(x) = \int_0^l g_m(x,\xi)[-\mu_P(\xi)]d\xi + \int_0^l g_m(x,\xi)[-p(\xi)]d\xi \tag{8}$$

where the first term is the mean value function and the second term is the zero-mean random part of the moment field. Here, only the forcing function of the differential equation is random.

Due to the deterministic nature of the Green function in (3), statistical moments of the bending moment of arbitrary order can be immediately obtained.

$$\mu_m(x) = \int_0^l g_m(x,\xi)[-\mu_P(\xi)]d\xi \tag{9}$$

$$\begin{aligned} cov[m(x_1), m(x_2)] &= \\ \int_0^l\int_0^l & g_m(x_1,\xi_1)g_m(x_2,\xi_2)E[p(\xi_1)p(\xi_2)]d\xi_1 d\xi_2 = \\ \int_0^l\int_0^l & g_m(x_1,\xi_1)g_m(x_2,\xi_2)\int_{-\infty}^{\infty} S_{pp}(\kappa)e^{i\kappa(\xi_1-\xi_2)}d\kappa d\xi_1 d\xi_2 \\ &= \int_{-\infty}^{\infty} S_{pp}(\kappa)VRF_I(x_1,\kappa)VRF_I^*(x_2,\kappa)d\kappa \end{aligned} \tag{10}$$

where $S_{pp}(\kappa)$ is the power spectral density function of $p(x)$ and $VRF_I(x,\kappa)$ is the variability response function for the moment field of statically determinate beams, defined as

$$VRF_I(x,\kappa) = \int_0^l g_m(x,\xi)e^{i\kappa\xi}d\xi \tag{11}$$

For the variance, equation (10) can be simplified to

$$var[m(x)] = 2\int_0^{\infty} S_{pp}(\kappa)|VRF_I(x,\kappa)|^2 d\kappa \tag{12}$$

The last statement of equation (12) takes into ac-

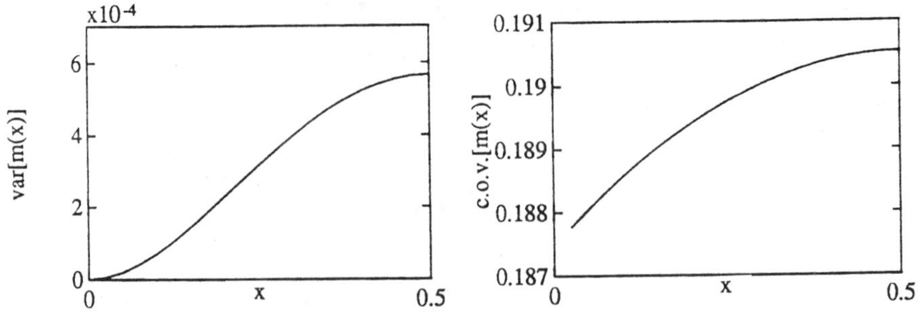

Figure 3. a) Variance of the moment field, b) Coefficient of variation of the moment field.

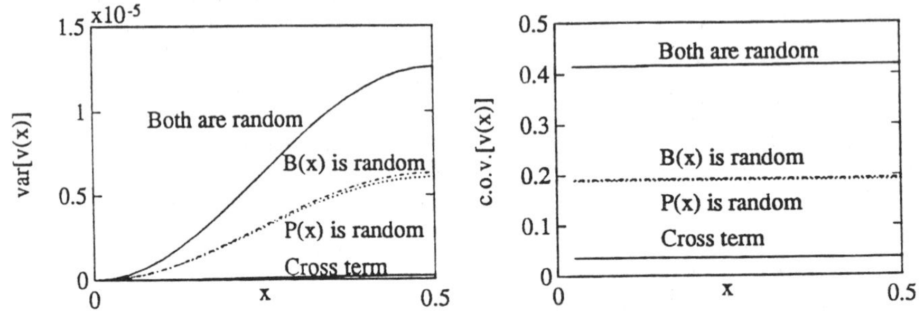

Figure 4. a) Variance, b) Coefficient of the variation of the deflection field.

count the fact that both $S_{pp}(\kappa)$ and $|VRF_I(x,\kappa)|^2$ are even functions of κ. In equation (12), the power spectral density function for the moment field of statically determinate beams is also given in terms of power spectral density function of the loading and response variability function for the moment field of statically determinate beams, that is

$$S_{m(x)m(x)}(\kappa) = S_{pp}(\kappa)|VRF_I(x,\kappa)|^2 \tag{13}$$

Hence, the variance of the moment field of statically determinate beams is simply determined from the distribution of the average power of $p(x)$ over the wave number spectrum and the variability response function which depends on the differential equation and the boundary conditions.

The deflection of statically determinate beams can be evaluated from equation (4) similarly. Again, only the forcing function of the differential equation is random. Introducing equation (7) into equation (2) provides

$$v(x) = -\int_0^l g_v(x,\xi)[-\frac{m(\xi)}{\mu_B(\xi)}]d\xi - \int_0^l g_v(x,\xi)b(\xi)m(\xi)d\xi \tag{14}$$

where both terms are random. Since $m(x)$ and $b(x)$ are stochastically independent, the mean value and the covariances can be evaluated as follows.

$$\mu_v(x) = \int_0^l g_v(x,\xi)[-\frac{\mu_m(\xi)}{\mu_B(\xi)}]d\xi \tag{15}$$

$$cov[v(x_1), v(x_2)] =$$

$$\int_0^l\int_0^l g_v(x_1,\xi_1)g_v(x_2,\xi_2)\Big(\frac{cov[m(\xi_1),m(\xi_2)]}{\mu_B(\xi_1)\mu_B(\xi_2)} \tag{16}$$

$$+ E[b(\xi_1)b(\xi_2)]E[m(\xi_1)m(\xi_2)]\Big)d\xi_1 d\xi_2$$

Next, the variance of the deflection can be written in terms of the the power spectral density functions, $S_{bb}(\kappa)$ of $b(\xi)$ and $S_{pp}(\kappa)$ of $p(\xi)$.

$$var[v(x)] = \int_{-\infty}^{\infty}\int_{-\infty}^{\infty} S_{bb}(\kappa_1)S_{pp}(\kappa_2)$$

$$\int_0^l g_v(x,\xi_1)e^{i\xi_1\kappa_1}VRF_I(\xi_1,\kappa_2)d\xi_1$$

$$\int_0^l g_v(x,\xi_2)e^{-i\xi_2\kappa_1} VRF_I^*(\xi_2,\kappa_2)d\xi_2 d\kappa_1 d\kappa_2$$
$$+\int_{-\infty}^{\infty} S_{bb}(\kappa_1)\int_0^l g_v(x,\xi_1)\mu_m(\xi_1)e^{i\xi_1\kappa_1}d\xi_1$$
$$\int_0^l g_v(x,\xi_2)\mu_m(\xi_2)e^{-i\xi_2\kappa_1}d\xi_2 \tag{17}$$
$$+\int_{-\infty}^{\infty} S_{pp}(\kappa_2)\int_0^l \frac{g_v(x,\xi_1)}{\mu_B(\xi_1)} VRF_I(\xi_1,\kappa_2)d\xi_1$$
$$\int_0^l \frac{g_v(x,\xi_2)}{\mu_B(\xi_2)} VRF_I^*(\xi_2,\kappa_2)d\xi_2 d\kappa_2$$
$$= 4\int_0^{\infty}\int_0^{\infty} S_{bb}(\kappa_1)S_{pp}(\kappa_2)|VRF_{II}(x,\kappa_1,\kappa_2)|^2 d\kappa_1 d\kappa_2$$
$$+2\int_0^{\infty} S_{bb}(\kappa_1)|VRF_{III}(x,\kappa_1)|^2 d\kappa_1$$
$$+2\int_0^{\infty} S_{pp}(\kappa_2)|VRF_{IV}(x,\kappa_2)|^2 d\kappa_2$$

where $VRF_{II}(x,\kappa_1,\kappa_2)$, $VRF_{III}(x,\kappa_2)$ and $VRF_{IV}(x,\kappa_2)$ are (with $VRF_I(x,\kappa_2)$) the variability response functions for the deflection of statically determinate beams.

$$VRF_{II}(x,\kappa_1,\kappa_2) = \int_0^l g_v(x,\xi)e^{i\xi\kappa_1} VRF_I(\xi,\kappa_2)d\xi \tag{18.a}$$
$$VRF_{III}(x,\kappa_1) = \int_0^l g_v(x,\xi)\mu_m(\xi)e^{i\xi\kappa_1}d\xi \tag{18.b}$$
$$VRF_{IV}(x,\kappa_2) = \int_0^l \frac{g_v(x,\xi)}{\mu_B(\xi)} VRF_I(\xi,\kappa_2)d\xi \tag{18.c}$$

It should be noted that $S_{bb}(\kappa_1)$, $S_{pp}(\kappa_2)$ and $|VRF_i(x,\kappa_1,\kappa_2)|^2, i = II, III, IV$ are even functions of κ_1 and/or κ_2.

Within equation (17), the power spectral density function for the deflection of statically determinate beams, $S_{v(x)v(x)}$, is also given in terms of the power spectral density function of the loading and the bending stiffness, as well as the variability response function of the moment and the deflection. It is a two-dimensional power spectral density function, one-dimension along the random loading spectrum and the other along the bending stiffness spectrum.

$$S_{v(x)v(x)}(\kappa_1,\kappa_2) = S_{bb}(\kappa_1)S_{pp}(\kappa_2)|VRF_{II}(x,\kappa_1,\kappa_2)|^2 + S_{bb}(\kappa_1)|VRF_{III}(x,\kappa_1)|^2 + S_{pp}(\kappa_2)|VRF_{IV}(x,\kappa_2)|^2 \tag{19}$$

Hence, the variance function of the deflection field of statically determinate beams depends on the distribution of the average power of $b(\xi)$ and $p(\xi)$ over the wave number spectrum and the variance response functions I, II, III and IV which carry the information about the differential relationship and the boundary conditions. When only the bending stiffness is modelled as a random field, expressions derived for the variance of the deflection field reduce to the ones given by Deodatis and Shinozuka (1989).

4. SPECTRAL-DISTRIBUTION FREE UPPER BOUNDS

Formulas (12) and (17) make it possible to establish spectral-distribution free upper bounds for the response variability of the stochastic system. This is very important since detailed information concerning probabilistic nature of the material properties and loading is usually not available. Finding spectral-distribution free upper bounds for response variability is mathematically a maximization problem with some constraints on the integrand. The integrals of equation (12) or (17) are to be maximized with the constraint that the area under the power spectral density function is the variance of the random field. To evaluate spectral distribution free upper bounds for response variability at any point on the beam, it is necessary to find the maximum of the variability response function in the frequency domain and then, assuming that all power is concentrated at this frequency, apply power as a Dirac delta function at this frequency. This way of establishing a conservative upper bound for the response variability was first suggested by Shinozuka (1987).

Therefore, for any specific point $x = x_0$ along the beam, the variance of moment assumes its maximum value for the following form of the power spectral density of the loading:

$$S_{pp}(\kappa) = 0.5\,\sigma_{pp}^2\,(\delta(\kappa-\kappa^*) + \delta(\kappa+\kappa^*)) \tag{20}$$

where $\delta(.)$ is the Dirac delta function and κ^* is the wave number at which $|VRF_I(x_0,\kappa)|^2$ assumes its maximum value:

$$|VRF_I(x_0,\kappa^*)|^2 \geq |VRF_I(x_0,\kappa)|^2 \quad 0 \leq \kappa < \infty \tag{21}$$

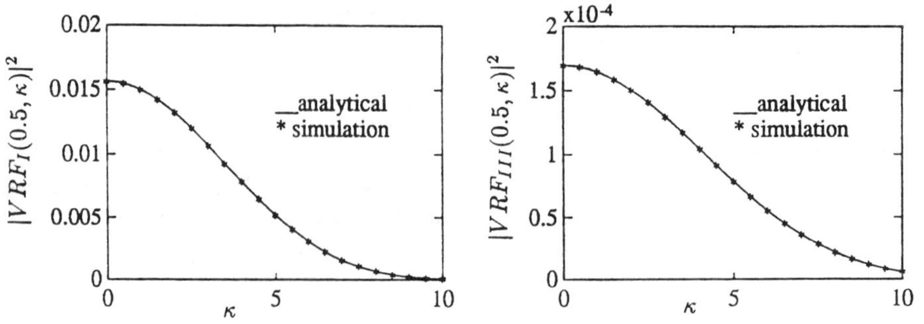

Figure 5. a) Variability response function I at the center of the beam, b) Variability response function III at the center of the beam.

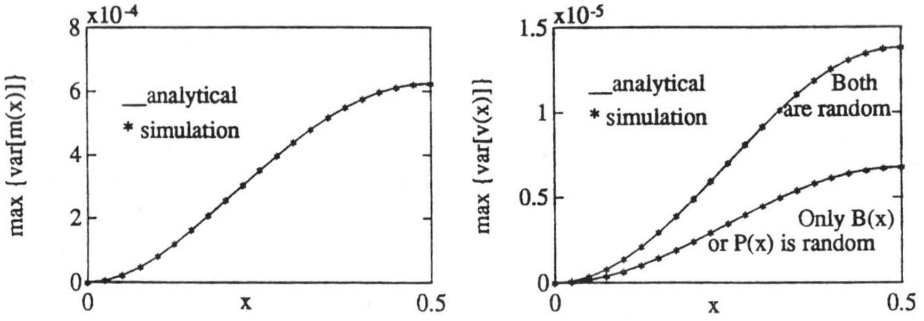

Figure 6. Spectral-distribution free upper bounds a) for the moment field, b) for the deflection field.

An upper bound for $var[m(x = x_0)]$ is then obtained as

$$\max\left(var[m(x=x_0)]\right) = \sigma_{pp}^2 \; |VRF_I(x_0,\kappa^*)|^2 \qquad (22)$$

With this upper bound for the variance, it is straightforward to obtain upper bounds for the standard deviation and coefficient of variation.

Similarly, at the same specific point $x = x_0$, an upper bound for the variance of the deflection can be evaluated using the Dirac delta function form of the power spectral density for the bending stiffness and the loading so that all three terms of equation (17) assume their maximum value before they are summed up.

$$S_{pp}(\kappa) = 0.5\sigma_{pp}^2\left(\delta(\kappa-\kappa_{2,i}^*)+\delta(\kappa+\kappa_{2,i}^*)\right) \quad i=2,4$$
$$S_{bb}(\kappa) = 0.5\sigma_{bb}^2\left(\delta(\kappa-\kappa_{1,j}^*)+\delta(\kappa+\kappa_{1,j}^*)\right) \quad j=2,3$$
$$(23.a,b)$$

Here, $\kappa_{2,i}^*$ $\; i = 2,4$ is the wave number at which $|VRF_{II}(x_0,\kappa_1,\kappa_2)|^2$ and $|VRF_{IV}(x_0,\kappa_2)|^2$ assume their maximum value on the loading frequency axis, κ_2. $\kappa_{1,j}^*$ $\; j = 2,3$ is defined similarly for $|VRF_{II}(x_0,\kappa_1,\kappa_2)|^2$ and $|VRF_{III}(x_0,\kappa_1)|^2$ along the bending stiffness frequency axis, κ_1.

Thus, an upper bound for $var[v(x = x_0)]$ is obtained as

$$\max\left(var[v(x=x_0)]\right)=\sigma_{pp}^2\sigma_{bb}^2|VRF_{II}(x_0,\kappa_{1,2}^*,\kappa_{2,2}^*)|^2$$
$$+\sigma_{bb}^2|VRF_{III}(x,\kappa_{1,3}^*)|^2 + \sigma_{pp}^2|VRF_{IV}(x,\kappa_{2,4}^*)|^2 \qquad (24)$$

With this upper bound for the variance, it is straightforward to obtain upper bounds for the standard deviation and coefficient of variation.

5. NUMERICAL EXAMPLES

A simply supported beam is studied for which the forms of the power spectral density functions for the loading and bending stiffness are assumed to be band-limited white noise processes. The sample beam and its Green function are shown in Figure 1, and the system parameters are chosen as in Figure 2. Mathematica [13] is used for most of the analytical and numerical computations.

Since the mean solution for this linear stochastic problem is equivalent to a deterministic problem with mean inputs, the results for the mean solution will not be addressed. The results are symmetric with respect to the centre of the beam due to the symmetry of the forcing function and the boundary conditions of the differential equations. Hence, only one half of the beam is considered in the plots. The variance and the coefficient of variation of the moment field are shown on Figures 3.a and 3.b, respectively. The contributions of the random loading as well as bending stiffness fields and the cross term on the variance and the coefficient of variation of the deflection field are shown on Figures 4.a and 4.b. It is concluded from these figures that neither the moment nor the deflection field is a homogeneous random field.

For the same sample problem, spectral-distribution free upper bounds of the response variability are also evaluated. The complex square of the variability response functions I and III are plotted on Figures 5.a and 5.b. The analytical evaluation of these functions might be considered lengthy, therefore in order to check the results and propose an alternative way of finding the square of the variability response functions, thus the upper bounds, of this problem, the fast Monte Carlo simulation technique of Deodatis and Shinozuka (1989) is applied. The stochastic field having its power spectral density as a Dirac delta function has been digitally simulated as

$$p(x) = \sqrt{2}\sigma_{pp}\cos(\kappa^* x + \phi) \tag{25}$$

where ϕ is the random phase angle uniformly distributed in $(0, 2\pi)$. The random phase angle is not generated using a random number generator, instead N equally spaced phase angles in $(0, 2\pi)$ domain are used. This is the gist of the fast Monte Carlo simulation technique which is proved to be very effective as shown in the Figures 5.a, 5.b, 6.a and 6.b. The convergence to the exact value is obtained after only 5 simulations. Next, given the complex squares of the variability response functions I,II,III and IV as well as the variance of the loading and the bending stiffness fields, the conservative upper bounds for response variability are evaluated, analytically and using the fast Monte Carlo simulation technique, as shown in Figures 6.a and 6.b. It should be noted that much lower upper bounds can be obtained and second order moment computations can be verified using the variability response functions if at least the peak frequencies of the spectral density functions are known. Similar analyses can be introduced for statically indeterminate beams with some approximations.

6. CONCLUSION

The first and second order statistical characteristics of the deflection and the moment fields of statically determinate Bernoulli-Euler beams with randomly varying bending stiffness subject to statical random external loadings are analytically determined. Since detailed information concerning the probabilistic nature of the bending stiffness and the loading fields is usually not available, spectral-distribution free upper bounds for the response variability are analytically obtained and compared to simulation results.

7. REFERENCES

Adomian, G. 1983. Stochastic Systems. Academic Press, Inc. New York, N.Y., USA.

Deodatis, G. and Shinozuka, M. 1989. Bounds on Response Variability of Stochastic Systems. J. Engrg. Mech., ASCE, 115 (11), 2543-2563.

Deodatis, G. 1990. Bounds on Response Variability of Stochastic Finite Element Systems. J. Engng. Mech., ASCE, 116 (3), 565-585.

Deodatis, G. 1990. Bounds on Response Variability of Stochastic Finite Element Systems: Effect of Statistical Dependence. Prob. Engng. Mech., 5 (2), 88-98.

Eringen, A. C. 1957. Response of Beams and Plates to Random Loads. J. Appl. Mech., ASME, 24 (1957), 46-52.

Ghanem, R. G., Spanos, P. D. 1991. Stochastic Finite Elements: A Spectral Approach. Springer-Verlag New York Inc., N.Y.,USA.

Liu, W. K., Belytschko, T., Mani, A. 1986. Random Field Finite Elements. Int. J. Num. Meth. in Engrg. Vol 23, 1831-1845.

Shinozuka, M. 1987. Structural Response Variability. J. Engrg. Mech., ASCE, 113 (6), 825-842.

Shinozuka, M. and Deodatis, G. (1987) Simulation of Stochastic Processes by Spectral Repre-

sentation, Appl. Mech. Rev., ASME, 40 (9).
Vanmarcke, E. et al. 1986. Random Fields and Stochastic Finite-elements. Structural Safety 3, 143-166, Amsterdam, The Netherlands.
Vanmarcke, E. 1983. Random Fields. MIT Press, Cambridge, Massachusetts, USA.
Vanmarcke, E., Grigoriu, M. 1983. Stochastic Finite Element Analysis of Simple Beams. J. Engrg. Mech., ASCE, 109 (5), 1203-1214.
Wolfram, S. 1988. Mathematica, A System for Doing Mathematics by Computer. Addison-Wesley Publishing Company.

Structural Safety & Reliability, Schuëller, Shinozuka & Yao (eds) © 1994 Balkema, Rotterdam, ISBN 90 5410 357 4

Effects of load uncertainty on performance of a jack-up foundation

Farrokh Nadim
Norwegian Geotechnical Institute, Oslo, Norway

Sverre Haver
Statoil a.s., Trondheim, Norway

Olav Mo
DNV Industri Norge AS, Høvik, Norway

ABSTRACT: The effects of the uncertainties in storm loads and soil parameters on the reliability of a jack-up foundation were studied. The study involved the development of a stochastic wave climate model, time domain simulations of jack-up response, and a probabilistic description of soil properties. The results of the study showed that the uncertainty in soil properties and the uncertainty in storm loads contribute equally to the total uncertainty in the jack-up foundation performance.

1 INTRODUCTION

The uncertainties in loads used as input to the geotechnical analyses of offshore structures can be very important. Probabilistic analyses run at NGI and Veritec (Nadim and Lacasse, 1992; Løseth and Bjerager, 1989) have shown that the loads and storm characteristics often represent the most significant uncertain parameters that determine the reliability of a foundation design, and its probability of non-performance. This is true for gravity base, jack-up and piled offshore structures.

This paper presents a detailed quantitative study of the effects of the uncertainties in storm loads on the reliability of a jack-up foundation. The jack-up foundation was studied because jack-up rigs have the highest accident rate among the mobile offshore platform types, averaging 2.6% of the fleet annually over the 1955-1980 period (McClelland et al., 1982). In a risk analysis of jack-up rigs, Sharples et al. (1989) concluded that about 22% of all jack-up rig mishaps are associated to "soils".

The storm-induced forces on a jack-up rig in deep water are influenced greatly by the (non-linear) dynamic response of the structure. Therefore, to perform a probabilistic bearing capacity analysis of the jack-up foundation, the following steps were followed: a) development of a stochastic wave climate model for the jack-up site, b) time domain simulations of the jack-up response for selected sea states, c) development of a stochastic model for the forces on the jack-up foundation based on the unconditioning of the results of time domain simulations (Step b) with respect to the wave climate model (Step a), and d) probabilistic bearing capacity analysis using the load model from Step c and a probabilistic description of the soil strength parameters.

2 WAVE CLIMATE MODEL

The largest wave induced response is likely to occur during the largest storm events. It was therefore assumed that the annual largest foundation forces can be approximated by the largest foundation forces during the annual largest storm events. The year to year variability of this event for the Ekofisk area in the North Sea was considered by Haver (1991). The investigation was based on a peak-over-threshold technique including all storms exceeding a significant wave height of 6.5 m. The data included all storms in the 10-year period from July 1980 to June 1990.

The annual largest storm peak is characterized by the significant wave height,

H_{mo}, and the corresponding spectral peak period, T_p. The inherent randomness of these quantities are modelled by (Haver, 1991):

$$F_{H_{mo}}(h) = \exp\left\{-\phi \exp\left\{-\left(\frac{h}{\theta}\right)^{\gamma} + \left(\frac{h_o}{\theta}\right)^{\gamma}\right\}\right\} \qquad (1)$$

where $h_o = 6.5$ m is the storm threshold, $\phi = 5.8$ is the expected number of storms exceeding the level h_o per year, and the model parameters $\theta = 2.03$ m (scale parameter) and $\gamma = 1.25$ (shape parameter) are estimated from the Ekofisk data.

$$f_{T_p|H_{mo}}(t|h) = \frac{1}{\sqrt{2\pi}\,\sigma_{\ln T_p}\,t} \exp\left\{-\frac{1}{2}\left(\frac{\ln t - \mu_{\ln T_p}}{\sigma_{\ln T_p}}\right)^2\right\} \qquad (2)$$

where $\mu_{\ln T_p}$ and $\sigma_{\ln T_p}$ are related to the mean μ_{T_p}, and standard deviation, σ_{T_p}, of T_p as follows:

$$\mu_{\ln T_p} = \ln \mu_{T_p} - \frac{1}{2} \ln\left[1 + \left(\frac{\sigma_{T_p}}{\mu_{T_p}}\right)^2\right] \qquad (3)$$

$$\sigma_{\ln T_p} = \left\{\ln\left[1 + \left(\frac{\sigma_{T_p}}{\mu_{T_p}}\right)^2\right]\right\}^{1/2} \qquad (4)$$

Based on the Ekofisk data the following models were adopted for μ_{T_p} and σ_{T_p}:

$$\mu_{T_p} = 6.59\, h^{0.382} - 2.60 \qquad (5)$$

$$\sigma_{T_p} = 0.09\, \mu_{T_p} \qquad (6)$$

where h is the significant wave height in metres and μ_{T_p} is the mean period in seconds.

For failure modes involving an accumulation mechanism, the time histories for the sea state characteristics may be of importance. Subsequently, the following mean histories were adopted:

$$\frac{\bar{h}(\tau)}{\bar{h}(o)} = \begin{cases} 1.0 - 0.0509\,|\tau| + 0.0000144\,|\tau|^2 - 0.000014\,|\tau|^3 & \text{for } \tau \le 0 \\ 1.0 - 0.0372\,\tau + 0.000905\,\tau^2 - 0.000008\,\tau^3 & \text{for } \tau > 0 \end{cases} \qquad (7)$$

$$\frac{\bar{t}_p(\tau)}{\bar{t}_p(o)} = \begin{cases} 1.0 - 0.252\,|\tau| + 0.00084\,|\tau|^2 - 0.000009\,|\tau|^3 & \text{for } \tau \le 0 \\ 1.0 & \text{for } 0 < \tau \le 3 \\ 1.0 - 0.0104(\tau - 3) + 0.00022(\tau - 3)^2 - 0.0000024(\tau - 3)^3 & \text{for } \tau > 3 \end{cases} \qquad (8)$$

where τ is time in hours relative to the storm peak, h(o) is the significant wave height at storm peak, $t_p(o)$ is the associated spectral peak period, and a bar over a variables denotes the mean value.

In a reliability analysis, the uncertianties related to the environmental description should be included. The most important sources of uncertainties are related to the adequacy of the probabilistic model for the annual largest storm peak. For the Ekofisk area, these uncertainties can be modelled by introducing the shape parameter, γ, as a log-normal variable with a mean value equal to the base case value (=1.25) and a standard deviation of 0.71. The corresponding values for the scale parameter, θ, in metres can be calculated by:

$$\theta = (7.69^{\gamma} - 6.5^{\gamma})^{1/\gamma} \qquad (9)$$

3 TIME DOMAIN SIMULATIONS OF JACK-UP RESPONSE

The jack-up platform used in the study is shown in Fig. 1. The jack-up is supported by three legs on spud cans (conical footings). The supporting soil consists of dense North Sea sand. The mean water depth at the site is 82 m and the wave climate is similar to the wave climate for the Ekofisk area. The jack-up has a submerged weight of 184.5 MN and a total mass of 21230 tonnes.

Non-linear time domain analyses were performed to evaluate the response of the jack-up under different sea states. The purpose of the time domain simulations is to model both the non-linearities (P - Δ effects and non-linear loading) and the dynamics. The analyses were carried out by Veritec (1991) using the computer program FENRIS (Veritec, 1983). A 3-dimensional global beam element model of the jack-up, supported by soil springs at the seabed, was used in the simulations. The first eigenperiod of the jack-up was calculated to be 6.58 s (bending mode). The calculated eigen-

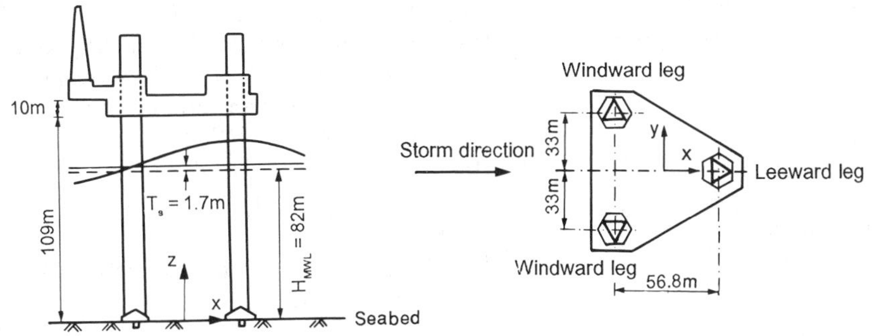

Fig. 1 Geometry of the jack-up studied

values are quite sensitive to the boundary conditions assumed at the seabed. The first eigenperiod is 4.23 s for a fixed footing boundary condition, and 7.90 s for a pinned footing (free rotation about horizontal axes) boundary condition.

The hydrodynamic damping was included through the relative velocity formulation of Morison's equation. To represent the additional structural and soil damping, the Rayleigh damping formulation (damping matrix proportional to a linear combination of stiffness and mass matrices) was applied. The combination factors for the damping matrix were chosen to give 1% modal damping for the first two bending modes in the global x-direction.

Three sea states, characterized by ($h = 10$ m, $t_p = 11.5$ s), ($h = 10$ m, $t_p = 13.3$ s), and ($h = 15$ m, $t_p = 15.5$ s), were simulated. The analyses were performed for long crested seas propagating in the positive x-direction (Fig. 1). In addition to the storm loading, a tidal current of 0.5 m/s (constant with depth) and a wind driven current of 0.5 m/s at the sea surface, decreasing linearly to zero at 33.7 m above the seabed, were applied. The marine growth thickness was considered to vary linearly from zero at the sea surface to 50 mm at 23 m above the seabed, reducing to 25 mm at the seabed.

Morison's equation with the relative fluid-structure velocity formulation was used to calculate the wave forces (Sarpkaya and Isaacson, 1981). The wave kinematics were established by the Airy wave theory. The integration of the force intensity was performed up to the instantaneous wave surface, considering the fluid velocity and acceleration above mean water level equal to the values at the mean water level in the same horizontal position. The hydrodynamic load coefficients used in Morison's equation were: $C_D = 0.63$ / $C_M = 2.0$ for smooth circular members, $C_D = 1.20$ / $C_M = 1.5$ for rough circular members, and $C_D = 2.00$ / $C_M = 1.5$ for chords (including racks).

A total of 4096 time steps of 0.44 s duration were applied in each simulation. This time step length was assumed to be sufficient for a proper estimation of the statistics of the foundation force process. The time integration of the dynamic equilibrium equation was performed by the unconditionally stable Newmark method (trapezoidal rule). To eliminate the start transients from the time histories, the first 200 time steps of each simulated time history were not included in the statistical post-processing. This gave a net sample length of approximately 29 minutes for each simulation.

The following statistics of the foundation reaction forces for the three sea states were obtained (Veritec, 1991): mean level, μ, standard deviation, σ, coefficient of skewness, γ_1, coefficient of kurtosis, γ_2, and average zero up-crossing period, t_z. In a reliability study, these characteristics will be needed for all possible combinations of the significant wave height and spectral peak period which correspond to an annual largest storm event. For this purpose, smoothed functions were fitted to the data obtained for the reaction forces on the leeward leg. The following functional forms were assumed for the statistical parameters of horizontal force, vertical force, and overturning moment acting on the leeward spud can foundation:

Mean value:

$$\mu = \mu_o + a\cdot(h - 10) \tag{10}$$

Standard deviation:

$$\sigma = (h/10)^n\cdot(a_o - a_1 t_p) \tag{11}$$

Zero up-crossing period (sec.):

$$t_z = 7.60 - 0.418\, t_p + 0.0253\, t_p^2 \tag{12}$$

Coefficient of skewness:

$$\gamma_1 = (h/10)^m\cdot(b_o + b_1 t_p) \tag{13}$$

Coefficient of kurtosis:

$$\gamma_2 = (h/10)^{0.89}\cdot(1.5 + 0.16\cdot t_p),\ \gamma_2 \geq 3 \tag{14}$$

In the above equations, h is the significant wave height in metres and t_p is the spectral peak period in seconds.

The zero up-crossing period and the coefficient of kurtosis were assumed to be more or less the same for all three response quantities. For the remaining characteristics, the various coefficients of Eqs. 10 through 14 are given in Table 1.

For a Gaussian process, the distribution of the largest value reads:

$$F_{Y_{max}|H_{mo},T_p}(y|h,t_p) \approx \exp\left\{-\frac{\Delta\tau}{t_z}\exp\left\{-\frac{1}{2}\left(\frac{y}{\sigma}\right)^2\right\}\right\} \tag{15}$$

where $\Delta\tau$ is the duration of the stationary event. The response processes under consideration are clearly non-Gaussian. It was assumed that these non-Gaussian processes can be approximated by a series of Hermite polynomials of a standard Gaussian process, Winterstein (1987). The distribution function for the largest (or smallest) maximum of the non-Gaussian process is then obtained by transforming the fractiles of Eq. (15) as follows, Winterstein (1987):

$$\tilde{y}_{Non-Gaussian} = \alpha\left\{\tilde{y} + k_3(\tilde{y}^2 - 1) + k_4(\tilde{y}^3 - 3\tilde{y})\right\} \tag{16}$$

where $\tilde{y} = y/\sigma$ and the coefficients α, k_3 and k_4 are given by:

$$\alpha = \left(1 + 2k_3^2 + 6k_4^2\right)^{-1/2} \tag{17}$$

$$k_3 = \gamma_1/[6(1 + 6k_4)] \tag{18}$$

$$k_4 = \frac{1}{18}\left[\sqrt{\left(1 + \frac{3}{2}(\gamma_2 - 3)\right)} - 1\right] \tag{19}$$

These equations describe the dynamic extreme value. The resulting load level is obtained by adding the mean values. The distribution functions obtained by this procedure are conditional with respect to the sea state parameters. Marginal distributions for the annual largest forces are obtained by combining the results above with the probabilistic models describing the environmental conditions, as described in the following section.

4 PROBABILISTIC BEARING CAPACITY ANALYSIS

The bearing capacity of a jack-up spud can foundation is strongly dependent on the installation procedure for the unit. The typical installation procedure for spud cans is describedby Schotman and Efthymiou (1989).

Table 1 Coefficients describing the foundation forces on the leeward spud can

Force component	Mean, μ		Standard dev., σ			Skewness, γ_1		
	μ_o	a	a_0	a_1	n	b_0	b_1	m
Horizontal force (kN)	-433	-37.8	1414.3	32.2	1.29	1.327	-0.128	0.87
Vertical force (kN)	61474	75.4	6347.3	202.2	1.13	-0.5326	0.0517	0.99
Overturning moment (kNm)	-7613	-720.6	41381	1203.3	1.21	0.844	-0.082	1.00

The main feature of the installation is vertical preloading of the jack-up. The foundation reaction during preloading on any one leg is required to be greater than the maximum vertical reactions arising from gravity loads and 100% of the design environmental loads. It may be noted that satisfying this criterion is no guarantee for satisfactory foundation performance during the design storm because a horizontal force will also act on the foundation at the time of the maximum vertical reactions. A vertical preload equal to twice the submerged operational weight of the jack-up was assumed in the analysis. This corresponds to a preload of 123 MN on each spud can.

With the assumed direction of the storm, the bearing capacity of the leeward leg (see Fig. 1) is far more critical than those of the windward legs. Therefore, the geotechnical stability analysis was performed for the annual maximum force on the leeward spud can (1-year event). To obtain the marginal distribution of the largest force on the spud can, the simplified response functions fitted to the time domain simulations (Eqs. 10 - 14), and the wave climate description for Ekofisk were used. The marginal distributions are obtained by integrating the conditional distributions (Eqs. 15 and 16) over the domains of the random variables. In the evaluation of the marginal distributions, 11 time segments of 3-hour duration each, with the middle (i.e. sixth) segment corresponding to the storm peak, were considered. The integrations were carried out numerically.

Figures 2, 3, and 4 show respectively the computed marginal distributions for the storm-induced component of the annual maximum vertical force, horizontal force, and overturning moment on the spud can supporting the leeward leg. It can be seen the Gumbel distribution provides an excellent fit for all three force components.

Nadim and Lacasse (1992) developed a procedure for performing reliability analysis and estimating the nominal probability of failure for a jack-up spud can foundation with no moment fixity (zero overturning moment) on clay. The jack-up considered in this study is located at a dense sand site and the spud can has some moment fixity. Therefore, a modified version of the procedure was used. The procedure includes three main steps: 1) a-priori calculation of bearing capacity the spud can under different horizontal and vertical load combinations (this is commonly referred to as the interaction diagram), 2) updating of the interaction diagram from vertical preloading, and 3) bearing capacity analysis under environmental loads.

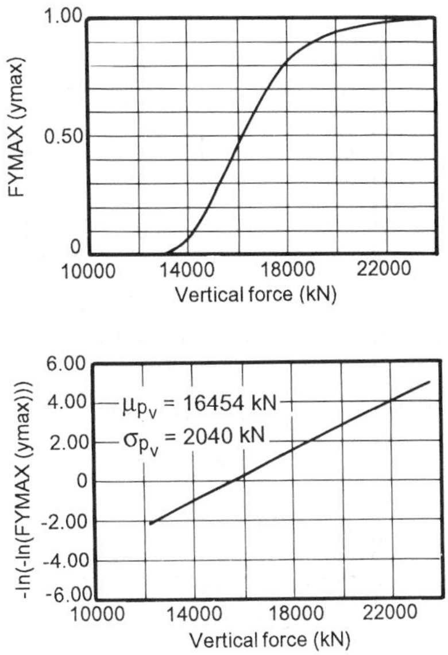

Fig. 2 Computed distribution for annual maximum vertical force on leeward spud can (storm component).

The vertical preload defines the static foundation capacity under pure vertical loading immediately after installation. The foundation capacity under other load combinations (i.e., the interaction diagram) was evaluated by the bearing capacity formula of Brinch Hansen (1970). For the purpose of this study, the Brinch Hansen bearing capacity formula was written as:

Bearing Capacity:
$$f(\phi, M, \text{other parameters}) \qquad (20)$$

where ϕ is the friction angle of soil, M is the overturning moment, and other parameters are soil unit weight, foundation embedment and shape, etc. In the reliability analysis, it was assumed that the uncertainty in the spud can bearing capacity is primarily due to the uncertainties in the soil friction angle and in the

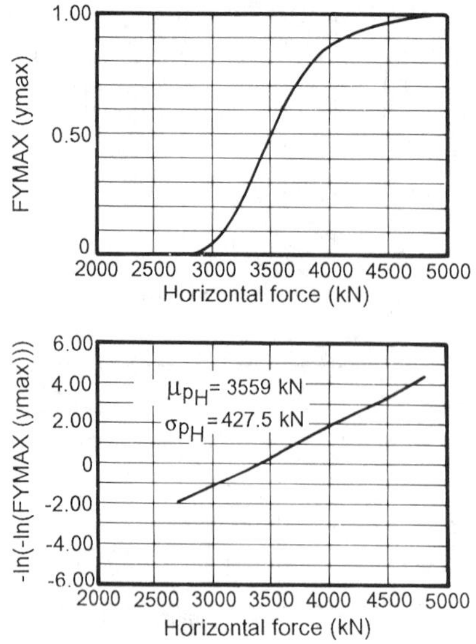

Fig. 3 Computed distribution for annual maximum horizontal force on leeward spud can.

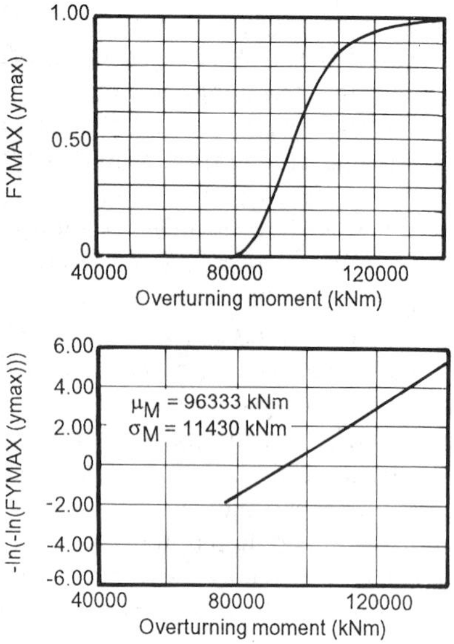

Fig 4 Computed distribution for annual maximum overturning moment on leeward spud can.

Table 2 Computed sensitivity factors for spud can supporting the leeward leg

	Parameter	Sensitivity factor, α	Contribution to total uncertainty, α^2
Load	Maximum vertical and horizontal foundation forces[(1)]	- 0.734	54%
Soil	Random variable describing the uncertainty in interaction diagram[(2)]	0.679	46%

(1) Assumed perfectly correlated
(2) Includes effects of uncertainty in overturning moment due to moment fixity

overturning moment. A mean value of μ_ϕ = 40° and a standard deviation of σ_ϕ = 1° were used for the friction angle under storm loading. Fully drained conditions were assumed. A drained friction angle of 40°overturning moment is discussed later.

The mean value and variance of any point on the bearing capacity diagram were estimated by is on the lower side of what is typical for dense North Sea sands. The uncertainty in the the first-order, second-moment (FOSM) approximation (Ang and Tang, 1975). The FOSM approach does not provide any information regarding the distribution function for the bearing capacity diagram. Based on other case studies performed for similar problems (Nadim and Lacasse, 1992), a log-normal distribution function for any given point on the interaction diagram was assumed. This means that the spud can bearing capacity for a given ratio of horizontal to vertical load is a random variable with the mean value shown in Fig. 6 and log-normal distribution.

The spud cans for the jack-up under consideration have a special geometry shown in Fig. 5. The spud cans penetrate very little into the dense sand under the assumed preload. Bearing capacity calculations showed that an effective radius of about 4 m will provide the necessary contact area to carry the preload of

123 MN on each spud can. A large overturning moment on the spud can will cause more contact between the spud can and the soil foundation. Therefore, in the analysis it was assumed that only 25% of the calculated overturning moment will affect the bearing capacity diagram; i.e., the mean and standard deviation of the overturning moment used in the derivation of the interaction diagram were only 25% of those shown in Fig. 4. This reduction was based on judgement and no detailed analysis for its verification was performed.

Figure 6 shows the updated (on the basis of preloading) bearing capacity interaction diagram obtained using the modified Nadim and Lacasse (1992) procedure. The updating reduces the uncertainty in the bearing capacity under pure vertical loading, but has little effect on the uncertainty in the bearing capacity under other failure modes. This is mainly due to the assumption of independence of model errors, which results in a low correlation coefficient between the different failure modes.The evaluation of foundation reliability is performed by the first-order reliability method (FORM). Details of this method can be found in many references (e.g. Madsen et al., 1986) and will not be repeated here. The limit state function was defined as

$$g = R - S \tag{21}$$

where R and S are the lengths of vectors shown on Fig. 6.

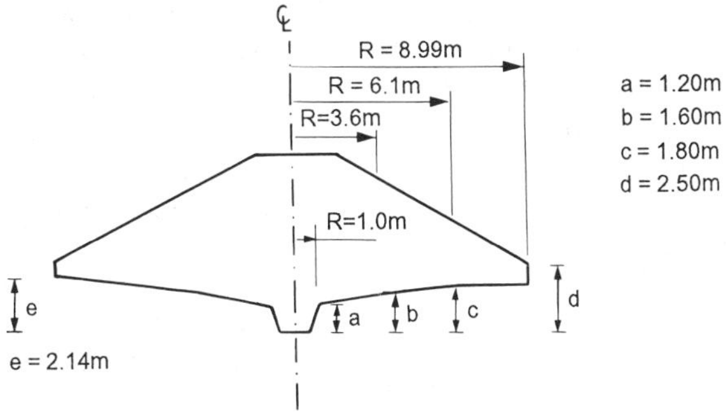

Fig 5 Geometry of the spud can

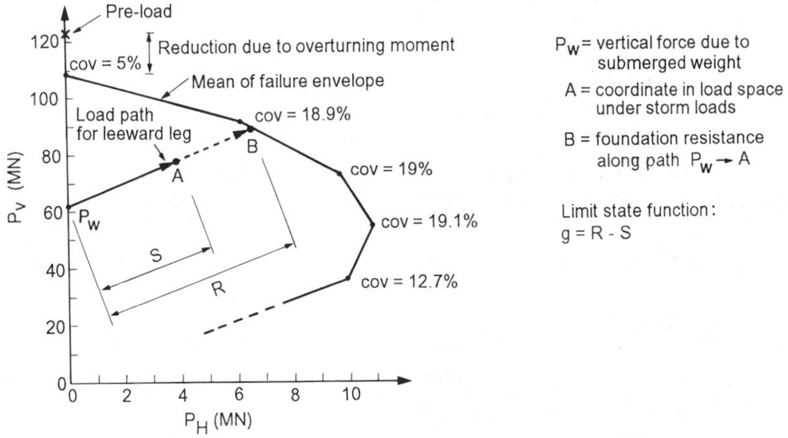

Fig. 6 Probabilistic interaction diagram used in reliability analysis. Effects of an overturning moment of 24 MNm and updating under pre-loading are included in the diagram.

5 RESULTS OF THE ANALYSIS

Assuming perfect correlation between the maximum horizontal and vertical forces (Veritec, 1991), a reliability index of $\beta = 1.98$ and a nominal annual probability of failure of $P_f = 2.40 \cdot 10^{-2}$ were computed for the spud can supporting the leeward leg.

The computed sensitivity factors are shown in Table 2. It can be seen that there is essentially no difference between the contributions of the uncertainty in loads and of the uncertainty in soil resistance to the total uncertainty in the jack-up foundation performance.

ACKNOWLEDGEMENT

The work described in this paper was supported by Statoil, Elf Petroleum Norge A/S, A/S Norske Shell, Mobil Research and Development Corporation, U.K. Health and Safety Executive, the Norwegian Petroleum Directorate, and NGI. Their support and permission to publish this paper are gratefully acknowledged. The opinions expressed in the paper are those of the authors and do not necessarily reflect the views or policy of the supporting organisations.

REFERENCES

Ang, A.H-S., and W.H. Tang 1975. *Probability concepts in engineering planning and design.* John Wiley & Sons, Inc.

Brinch Hansen, J. 1970. *A revised and extended formula for bearing capacity.* The Danish Geotechnical Institute, Copenhagen, Bull. No. 28, pp. 5-11.

Haver, S. 1991. *Wave climate description for foundation design.* Statoil report No. 91047

Løseth, R. and P. Bjerager 1989. Reliability of offshore structures with uncertain properties under multiple load processes. *Proc., OTC'89,* Houston, Texas.

Madsen, H.O., S. Krenk and N.C. Lind 1986. *Methods of Structural Safety.* Prentice Hall Inc., Englewood Cliffs, New Jersey, USA.

McClelland, B., A.G. Young, and B.D.Remmes 1982. Avoiding jack-up rig foundation failures. *Geotech. Eng.*, Vol. 13, No. 2, pp. 151-188.

Nadim F. and S. Lacasse 1992. Probabilistic bearing capacity analysis of jack-up structures. *Canadian Geotechnical Journal,* Vol 29, No. 4, pp. 580-588.

Schotman, G.J.M., and M. Efthymiou 1989. Aspects of the stability of jack-up spud-can foundations. *Proc., 2nd Int. Conf. on Jack-Up Drilling Platform, Design, Construction and Operation.* pp. 245-269.

Sarpkaya, T. and M. Isaacson 1981. *Mechanics of wave forces on offshore structures.* Van Nostrand Reinhold Company.

Sharples, B.P.M., J.C. Trickey, and W.T. Bennett 1989. Risk analysis of jack-up rigs. *Proc., 2nd Int. Conf. on Jack-Up Drilling Platform, Design, Construction and Operation.* pp. 101-123.

Veritec 1983. *FENRIS: Finite Element Nonlinear Integrated System.* Technical Report No. 83-6064 System Manual.

Veritec 1991. *Geotechnical reliability analysis for offshore structures - Time domain analysis of a jack-up platform.* Veritec report No. 91-3461.

Winterstein, S.R. 1987. *Moment-based Hermite models of random vibrations.* Tech. Univ. of Denmark, Series R, No. 219, Copenhagen.

Structural Safety & Reliability, Schuëller, Shinozuka & Yao (eds) © 1994 Balkema, Rotterdam, ISBN 90 5410 357 4

Approximate analysis of response statistics of uncertain MDOF linear systems

Costas Papadimitriou & Bing Wang
Texas A&M University, Tex., USA

Lambros S. Katafygiotis
The University of Akron, USA

ABSTRACT: A new probabilistic methodology is presented for obtaining the variability and statistics of the dynamic response of non-classically damped MDOF linear systems with uncertain parameters. Random variables are used to quantify the uncertainties associated with the system parameters. Complex modal analysis is employed and quadratic polynomial approximations are first used to express modal quantities with respect to the system random variables. Each modal response is then expanded in a series of orthogonal polynomials of these random variables. A system of linear ordinary differential equations for the coefficients of each series expansion is derived using a weighted residual method. A procedure is then presented to calculate the statistics of the uncertain response. Compared to existing techniques, the proposed method drastically reduces the computational effort and computer storage required to solve for the response statistics. Furthermore, the methodology can treat a variety of probability distributions assumed for the system random variables. The performance and accuracy of the method are illustrated by examples.

1 INTRODUCTION

The analysis of the dynamic behavior of a structure involves the selection of a mathematical model, such as a finite element model, which accurately describes the input-output behavior of the system. There is uncertainty in the parameters of this model due to the numerous assumptions made when modeling the material properties, constitutive laws, geometry, and boundary conditions. In the case where the structure is subjected to a deterministic excitation, the uncertainty in the model parameters will be propagated and produce an uncertain response. When the excitation is modeled as a stochastic process, as with seismic or wind excitations, the resulting response is also a stochastic process. Its statistics (e.g., the covariance of the response) become uncertain because of the uncertainty in the model parameters. This model uncertainty can adversely affect the structural response and reliability both in the case of deterministic and stochastic input. Therefore, it must be accounted for in the dynamic analysis, and the resulting uncertainty in the structural response must be quantified.

Probabilistic finite element methods have recently been proposed (Deodatis 1991, Spanos and Ghanem 1989 , Yamazaki *et al.* 1986) to treat the uncertainties in the structural characteristics and/or the applied loads. These studies have modeled the spatial variations of the structural properties in terms of random variables. Solution techniques were also developed to evaluate the response variability and its statistics due to the variability of the system parameters. Most studies have dealt with the solution of static problems and only a few methods have been proposed to determine the dynamic response variability of systems with uncertain properties. Among them, simulation techniques (Shinozuka 1972) are very powerful but they are very costly and time consuming. Perturbation techniques (Liu *et al.* 1987), although they work well for static or eigenvalue problems, suffer from inaccuracy when applied to dynamic problems.

Recently, in order to overcome these limitations, approaches with better accuracy, based on orthogonal series expansions of the response have been proposed (Ghanem and Spanos 1990, Jensen and Iwan 1991, 1992). The latter methodologies, requiring much less computational time than the simulation approach, have been used for both static and dynamic problems. However, there are certain limitations on the applicability of these methods on dynamic systems. Specifically, they require the solution of a very large system of coupled ordinary differential equations. The dimension of this system increases almost exponentially with the number of the uncertain parameters involved, and therefore, the computational time required to solve this system increases significantly. For the case where several uncertain parameters are involved, the implementation of the method becomes even more complicated because of the excessive computer storage requirements. Furthermore, there are still certain limitations associated with the type of probability distributions that these methods can handle.

In this paper, we propose an alternative method, based on complex modal analysis, that is computationally more efficient by taking advantage of the fact that only a few modes contribute to the structural response. The methodology is general, applicable to both classically and non-classically damped systems, and computationally feasible even for systems with a very large number of degrees of freedom. The methodology can handle nonlinear dependencies between the system matrices and the system random variables, as well as it can treat a variety of probability distributions assumed for the random variables.

2 MODELING OF UNCERTAINTIES

The equation of motion of an n degree-of-freedom linear model of a structure subjected to external forces can be cast in the form

$$M\ddot{\underline{x}} + C\dot{\underline{x}} + K\underline{x} = \underline{f}(t) \qquad (1)$$

in which M, C and K are the nxn mass, damping and stiffness matrices, $\underline{x}$ is the displacement response vector, and $\underline{f}(t)$ is the forcing vector of the structure. The mass and stiffness matrix of the structure is assembled from the mass and stiffness matrix of individual elements or substructures comprising the structure. A nondimensional set $\underline{\theta} = \{\theta_1, \theta_2, \cdots, \theta_{N_\theta}\}$ of independent random variables with zero mean and unit variance is employed to quantify the uncertainties in the structural properties. In past studies, the mass and stiffness matrices of the structure have been parameterized in the form (Katafygiotis 1991)

$$D(\underline{\theta}) = D_0 + \sum_{i=1}^{N_\theta} D_i \theta_i \qquad (2)$$

where the matrix $D(\underline{\theta})$ represents either the mass matrix $M(\underline{\theta})$ or the stiffness matrix $K(\underline{\theta})$ of the system. The matrix D_0 represents the contribution to $D(\underline{\theta})$ of the deterministic part of the system properties. Each dimensionless parameter θ_i scales the contribution of the uncertain part of a structural property to the matrix $D(\underline{\theta})$ of the system. In a stochastic finite element formulation, equation (2) also arises in the modeling of the spatial variation of the randomness of various structural properties such as mass or stiffness (Deodatis 1991, Jensen and Iwan 1991, Spanos and Ghanem 1989).

In contrast to the mass and stiffness matrices, the damping matrix $C(\underline{\theta})$ of the system is not assembled from element damping values because of the lack of sufficient information. Instead, it is computed by empirical or system identification methods, in which values are assigned to each modal damping coefficient of the structure or the substructures synthesizing the superstructure. The damping matrix is then constructed using the modal properties of the substructures. Unless Rayleigh damping is assumed for each of the substructures, the damping matrix is nonlinearly dependent on the random parameters θ_i, $i = 1, \cdots, N_\theta$ defined in (2). Existing solution techniques (Jensen and Iwan 1991, 1992, Katafygiotis and Beck 1991) do not handle nonlinear dependence of the system matrices M, K, C on the random parameter set $\underline{\theta}$.

3 FORMULATION FOR DETERMINISTIC EXCITATION

In what follows, the structure is considered to be non-classically damped and the mass matrix $M(\underline{\theta})$, stiffness matrix $K(\underline{\theta})$ and damping matrix $C(\underline{\theta})$ are assumed to be nonlinear functions of the random parameter set $\underline{\theta}$. The mode superposition technique is used to express the response $\underline{y}(t)$ in terms of the modal responses. An advantage of the mode superposition method is that it is often sufficient to fully analyze only the first few modes that contribute to the response.

Defining the $2n$ vector $\underline{y}(t) = (\underline{x}, \dot{\underline{x}})^T$, the equation of motion of the system (1) is written in the $2n$ state-space form

$$\dot{\underline{y}}(t) + A(\underline{\theta})\, \underline{y}(t) = \underline{G}(t) \qquad (3)$$

where $A(\underline{\theta}) \equiv A$ is a matrix of order $2nx2n$ which depends on the $M(\underline{\theta})$, $K(\underline{\theta})$ and $C(\underline{\theta})$ matrices. In structural dynamics applications, the matrix A is generally non-symmetric and, therefore, the associated eigenvalues and eigenvectors are generally complex. Physically, these complex eigenvalues are related to the frequencies and dampings of vibrations of the system. Let $\Phi = [\underline{\phi}_1, \cdots, \underline{\phi}_{2n}]$ be the eigenvector matrix of A and Λ be the diagonal matrix of the eigenvalues $\lambda_1, \lambda_2, \cdots, \lambda_{2n}$ of A. The associated $2nx2n$ eigenvalue problem is

$$A\Phi = \Phi\Lambda \qquad (4)$$

The complex eigenvalues and eigenvectors occur in conjugate pairs. Associated with the eigenvalue problem (4), one can write the adjoint eigenvalue problem

$$A^T\Psi = \Psi\Lambda \qquad (5)$$

where $\Psi = [\underline{\psi}_1, \cdots, \underline{\psi}_{2n}]$ is the eigenvector matrix of A^T. Assuming that $\lambda_r \neq \lambda_s$ for $r \neq s$, the following orthogonality conditions between the two eigenvalue problems hold:

$$\Psi^T\Phi = S \qquad \text{and} \qquad \Psi^T A\Phi = \Lambda\, S \qquad (6)$$

where S is a diagonal matrix with diagonal elements s_r.

Let $\underline{\xi}(t) = (\xi_1, \cdots, \xi_{2n})^T$ be the modal coordinate vector obtained from the transformation

$$\underline{y}(t) = \Phi\underline{\xi}(t) = \sum_{r=1}^{2n} \underline{\phi}^{(r)}\xi_r = 2\sum_{r=1}^{n} Re\{\underline{\phi}^{(r)}\xi_r\} \quad (7)$$

where each term in the second sum represents the contribution of a pair of complex conjugate modes to the response of system (3). The orthogonality conditions (6) are used to transform the first-order system of equations (3) into an uncoupled system of equations for the modal coordinates. This is accomplished by substituting (7) into (3), and in turn premultiplying the resulting equation by Ψ^T. The equation for each complex modal coordinate ξ_r is finally obtained in the form

$$\dot{\xi}_r + \lambda_r\xi_r = s_r^{-1}\underline{\psi}^{(r)T}\underline{G} \quad (8)$$

in which $s_r = \underline{\psi}^{(r)T}\underline{\phi}^{(r)}$. The vectors

$$\underline{\phi}^{(r)} = (\phi_1^{(r)}, \phi_2^{(r)}, ..., \phi_{2n}^{(r)})^T \quad (9)$$

and

$$\underline{\psi}^{(r)} = (\psi_1^{(r)}, \psi_2^{(r)}, ..., \psi_{2n}^{(r)})^T \quad (10)$$

are the r-th eigenvectors of the matrices A and A^T, respectively.

Introducing for convenience the new variables

$$\eta_m^{(r)} = 2Re\{\xi_r\phi_m^{(r)}\} \text{ and } \vartheta_m^{(r)} = 2Im\{\xi_r\phi_m^{(r)}\} \quad (11)$$

the response y_m of the structure at the mth DOF is finally obtained from (7) as a superposition of the modal responses:

$$y_m = \sum_{r=1}^{N_m} \eta_m^{(r)}, \qquad m = 1, 2, \cdots, N_d, \quad (12)$$

Making use of (8) and (11), $\eta_m^{(r)}$ and $\vartheta_m^{(r)}$ satisfy the following system of two coupled and real mode equations

$$\left\{ \begin{array}{c} \dot{\eta}_m^{(r)} \\ \dot{\vartheta}_m^{(r)} \end{array} \right\} + \left[\begin{array}{cc} \lambda_R^{(r)} & -\lambda_I^{(r)} \\ \lambda_I^{(r)} & \lambda_R^{(r)} \end{array} \right] \left\{ \begin{array}{c} \eta_m^{(r)} \\ \vartheta_m^{(r)} \end{array} \right\} = \left\{ \begin{array}{c} Re\left\{\underline{\beta}_m^{(r)T}\right\} \\ Im\left\{\underline{\beta}_m^{(r)T}\right\} \end{array} \right\} \underline{G} \quad (13)$$

in which $\lambda_R^{(r)}$ and $\lambda_I^{(r)}$ are the real and imaginary parts of the r-th eigenvalue $\lambda^{(r)}$, and $\underline{\beta}_m^{(r)T} = \underline{\psi}^{(r)T}\phi_m^{(r)}/s_r$. The modal variable $\eta_m^{(r)}$ represents the contribution from the rth mode to the response y_m at the mth degree of freedom of the structure. The contribution of the response from only the first N_m ($N_m \leq n$) modes is considered in (12). Generally, $N_m \ll n$. Furthermore, in practical applications only a very small number of DOF are excited externally, which simplifies the right-hand side of equation (13). Therefore, only a very small number of the elements in $\underline{\beta}_m^{(r)}$ are needed to completely define the rth modal response. To illustrate this, consider the response of a building to a base acceleration $(-\ \ddot{a})$. The forcing term in the modal equation (13) takes the simple form $(Re\{\gamma_m^{(r)}\},\ Im\{\gamma_m^{(r)}\})^T\ \ddot{a}$, where the scalar $\gamma_m^{(r)} = \underline{\psi}^{(r)T}(\underline{0}^T, \underline{1}^T)^T\phi_m^{(r)}/s_r$.

3.1 Variability of Modal Properties

Each modal response in (13) can be completely described by the modal quantities $\lambda_R^{(r)}$ and $\lambda_I^{(r)}$, which are related to the frequency and damping coefficient of the mode, and the effective participation factors $\underline{\beta}_m^{(r)}$. Next, the dependence of the above modal quantities on the parameter set $\underline{\theta}$ is addressed. Past studies on random eigenvalue problems indicate that the modal frequencies and effective participation factors of classically damped systems are smooth functions of the system parameter set $\underline{\theta}$ (Katafygiotis 1991, Shinozuka and Astill 1972). Similar variations are also expected for non-classically damped systems. The quadratic polynomial approximation

$$f_r(\underline{\theta}) \approx c_{0.f_r} + \sum_{i=1}^{N_\theta} c_{i.f_r}\theta_i + \sum_{i=1}^{N_\theta}\sum_{j=1}^{N_\theta} c_{ij.f_r}\theta_i\theta_j,$$

$$r = 1, 2, \cdots, N_m, \quad (14)$$

is used to express a modal quantity f_r in terms of the system random variables $\theta_1, \theta_2, \cdots, \theta_{N_\theta}$. In equation (14), f_r represents $\lambda_R^{(r)}$ or $\lambda_I^{(r)}$ or one of the elements in $\underline{\beta}_m^{(r)}$. The coefficients $c_{0.f_r}$, $c_{i.f_r}$, and $c_{ij.f_r}$ $i, j = 1, 2, \cdots, N_\theta$ are determined so that a good fit between the multivariate quadratic expansion and the exact surface $f_r(\underline{\theta})$ is achieved. In this study the coefficients are chosen so that the quadratic approximation takes exact values at N_{c_f} points appropriately selected on the N_θ-dimensional space of the parameters $\theta_1, \theta_2, \cdots, \theta_{N_\theta}$, where $N_{c_f} = \frac{N_\theta(N_\theta+1)}{2} + N_\theta + 1$ is the total number of the above coefficients. In the present analysis, the following points are chosen (Katafygiotis 1991): $\underline{\theta} = \underline{0}$, $\underline{\theta} = \pm\underline{\alpha}_i$, i=1,2,$\cdots$,$N_\theta$, and $\underline{\theta} = \underline{\alpha}_i + \underline{\alpha}_j$, i=1,2,$\cdots$ N_θ, j=i+1,$\cdots$,N_θ, where $\underline{\alpha}_i = \sqrt{3}(\delta_{1i}, \delta_{2i}, \cdots, \delta_{N_\theta i})$ and δ_{ij} is the Kronecker delta function.

3.2 Modal Response Variability

Next, the modal response variability is analytically modeled in terms of the random variables describing the system uncertainties. Each modal response is expanded in a series of orthogonal polynomial functions of the random variable set $\underline{\theta}$. Let

$[P_{l_p}(\theta_p)]$, $l_p = 0, 1, 2, \cdots, \infty$ denote a set of orthogonal polynomials satisfying the orthogonality relationship

$$\int_{\theta_p} w(\theta_p) P_{l_p}(\theta_p) P_{l'_p}(\theta_p) d\theta_p = a_{l_p} \delta_{l_p l'_p},$$

$$p = 1, 2, \cdots, N_\theta, \qquad (15)$$

where l_p is the degree of $P_{l_p}(\theta_p)$, $w(\theta_p)$ is the weighting function, and a_{l_p} is a constant dependent upon the given set of orthogonal polynomials (Korn 1968). The real modal response quantities defined by (11) are expanded in the form

$$\left\{ \begin{array}{c} \eta_m^{(r)} \\ \vartheta_m^{(r)} \end{array} \right\} = \sum_{0 \le L \le N_p} \left\{ \begin{array}{c} \chi_{l_1 l_2 \cdots l_{N_\theta}}^{(r,m)} \\ v_{l_1 l_2 \cdots l_{N_\theta}}^{(r,m)} \end{array} \right\} \prod_{p=1}^{N_\theta} P_{l_p}(\theta_p),$$

$$r = 1, 2, \cdots, N_m, \; m = 1, 2, \cdots n, \qquad (16)$$

where $L = \sum_{p=1}^{N_\theta} l_p$, N_p is the order of approximation in the random space of the variables $\underline{\theta}$, and the unknown coefficients $[\chi_{l_1 l_2 \cdots l_{N_\theta}}^{(r,m)}]_{0 \le L \le N_p}$ and $[v_{l_1 l_2 \cdots l_{N_\theta}}^{(r,m)}]_{0 \le L \le N_p}$ are subsets of the N_θ-dimensional tensors $[\chi_{l_1 l_2 \cdots l_{N_\theta}}^{(r,m)}]$ and $[v_{l_1 l_2 \cdots l_{N_\theta}}^{(r,m)}]$ respectively, that satisfy $0 \le L \le N_p$. It can be shown that the number of the unknown coefficients in the expansion is

$$N_d = 2(N_p + N_\theta)!/(N_p!)/(N_\theta!) \qquad (17)$$

The weighted residuals method is used to compute the unknown coefficients in the expansion (16). For this, the orthogonal expansion (16) and the quadratic expansion (14) are substituted into the system of equations (13). The orthogonality condition (15) along with appropriate recursive formulas (Korn 1968) for the polynomials involved in (16) are then used to derive a system of first-order ordinary differential equations for the unknown coefficients. This procedure is described in reference (Wang 1992), where it was applied to a second-order modal equation corresponding to the special case of a classically-damped system. Rearranging the elements $[\chi_{l_1 l_2 \cdots l_{N_\theta}}^{(r,m)}]_{0 \le L \le N_p}$ into an N_d-dimensional vector $\underline{z}_1^{(r,m)}$ and the elements $[v_{l_1 l_2 \cdots l_{N_\theta}}^{(r,m)}]_{0 \le L \le N_p}$ into an N_d-dimensional vector $\underline{z}_2^{(r,m)}$, the evolution of $\underline{z}_1^{(r,m)}$ and $\underline{z}_2^{(r,m)}$ is governed by the set of equations

$$\left\{ \begin{array}{c} \dot{\underline{z}}_1 \\ \dot{\underline{z}}_2 \end{array} \right\} + \left[\begin{array}{cc} \Gamma(\underline{c}_{\lambda_R}) & -\Gamma(\underline{c}_{\lambda_I}) \\ \Gamma(\underline{c}_{\lambda_I}) & \Gamma(\underline{c}_{\lambda_R}) \end{array} \right] \left\{ \begin{array}{c} \underline{z}_1 \\ \underline{z}_2 \end{array} \right\}$$

$$= \left\{ \begin{array}{c} \underline{\Upsilon}(\underline{c}_{\gamma_R}) \\ \underline{\Upsilon}(\underline{c}_{\gamma_I}) \end{array} \right\} \ddot{a} \qquad (18)$$

in which the matrix $\Gamma(\underline{c}_f)$ and the vector $\underline{\Upsilon}(\underline{c}_f)$ are functions of the coefficients $\underline{c}_f = (c_{0.f}, c_{i.f}\, c_{ij.f})$ involved in the quadratic expansion (14). The dependence on (r, m) has been omitted in the above equations for clarity. For simplicity, in deriving the forcing term in the system (18), it has been assumed that equation (1) models the response of a building to a base acceleration $(-\ddot{a})$. A computational advantage of the complex mode analysis is that the dimension N_d of the resulting system (18) is independent of the dimension of the original system (3). Also, it should be noted that the matrix $\Gamma(\underline{c}_f)$ is sparse, which results in further reduction of the computational effort and coputer storage requirements for solving the system (18). Thus, very large systems are treated efficiently by this approach, especially when only a few modes significantly contribute to the response.

3.3 System Response Variability and Statistics

Substituting equation (16) for the modal response variability into the mode superposition formula (12), and interchanging the order of summation, the variability of the response $y_m(t)$ due to the variability of the system parameters is obtained in the form

$$y_m(t) = \sum_{1 \le L \le N_p} \left(\sum_{r=1}^{N_m} x_{l_1 l_2 \cdots l_{N_\theta}}^{(r,m)}(t) \right) \prod_{p=1}^{N_\theta} P_{l_p}(\theta_p) \qquad (19)$$

The response variability of the system is fully characterized by the coefficients $[x_{l_1 l_2 \cdots l_{N_\theta}}^{(r,m)}]_{0 \le L \le N_p}$, $r = 1, \cdots, N_m$ of the orthogonal polynomial expansion (16). After the coefficients $[\chi_{l_1 l_2 \cdots l_{N_\theta}}^{(r,m)}]_{0 \le L \le N_p}$ of the orthogonal expansion (16) have been obtained, formula (19) can be used to obtain fast Monte Carlo simulations of the response for a given $\underline{\theta}$.

The statistics of the response can be computed by utilizing the orthogonality property (15) of the polynomials $P_{l_p}(\theta_p)$. This analysis can be carried out efficiently by choosing the weighting function $w(\theta_p)$ to be equal to the probability distribution of the parameter θ_p. Assuming that the probability distribution for each θ_p is normal or uniform, the Hermite or Legendre polynomials can be chosen respectively for $P_{l_p}(\theta_p)$ to satisfy the orthogonality conditions. In such cases, the mean and variance of the response take the simple forms

$$E[y_m(t)] = a_0^{N_\theta} \sum_{r=1}^{N_m} x_{00\cdots 0}^{(r,m)} \qquad (20)$$

$$Var[y_m(t)] = \sum_{1 \le L \le N_p} \bar{x}_{l_1 l_2 \cdots l_{N_\theta}}^{(m)}(t) \prod_{p=1}^{N_\theta} a_{l_p} \qquad (21)$$

where

$$\bar{x}_{l_1 l_2 \cdots l_{N_\theta}}^{(m)} = \sum_{r=1}^{N_m} \sum_{s=1}^{N_m} x_{l_1 l_2 \cdots l_{N_\theta}}^{(r,m)} x_{l_1 l_2 \cdots l_{N_\theta}}^{(s,m)} \qquad (22)$$

For other common types of probability distributions, it is not possible to define a set of polynomials that will satisfy the above conditions. In such cases, appropriate probability transformations can be used to transform the system's random variables into a new set $\underline{\theta}$ of statistically independent, standard normal variables.

The order of the polynomials included in the orthogonal expansion (16) to obtain results with acceptable accuracy depends on the nonlinear relationship between the modal response and the parameter set $\underline{\theta}$, as well as the range of variation of the parameter set $\underline{\theta}$. In addition to the displacement or velocity, other response variables such as acceleration and stress can also be characterized in a similar manner. Furthermore, higher statistical moments of system response can also be computed in a way similar to that of first and second moments.

4 EXAMPLES

To illustrate the results and verify the accuracy of the proposed methodology, an example system consisting of an equipment attached to the top floor of an N-story building is considered. The building is subjected to a base acceleration. The variables describing the motion of the system are the displacements x of each floor mass m relative to the base, and the displacement u of the equipment mass m_s relative to mass m of the top floor of the N-story structure. The building is assumed to be classically damped with the jth modal frequency and damping ratio denoted by ω_j and ζ_j, respectively. The equipment is modeled by a SDOF oscillator with mass $m_s = \eta m_1$, frequency $\omega_s = \alpha\omega_1$ and damping ratio $\zeta_s = c_s/(2m_s\omega_s)$, where m_1 and ω_1 are the modal mass and modal natural frequency of the first mode of the building, respectively. In general, the combined building-equipment system is expected to be a non-classically damped system.

Representative values of the system parameters are used to get insight into the accuracy of the approximate formulation. The north-south component of the 1940 El Centro earthquake record is taken as the base acceleration. Letting N=1, the simple but basic system of two oscillators connected in series is obtained. The mass ratio of the nominal system takes the values $\eta = 0.05$ so that the system is representative of a 2-DOF primary-secondary structure. The interstory stiffness of the first oscillator is taken to be $k = 100m$. The uncertain parameters are chosen to be the stiffness k_s of the secondary system and the damping ratios ζ_1 and ζ_s of each oscillator. The following nominal values are assumed for these uncertain parameters: $\zeta_1 = \zeta_s = 5\%$ and $\alpha = 0.85$, where $\alpha = \sqrt{k_s/(m_s\omega_1)}$.

Figure 1 compares the mean and the standard deviation obtained from the proposed method and the "exact" numerical integration in the random parameter space. Each random variable θ is assumed to be Gaussian distributed. A representative case is considered in which the coefficient of variation of the random variables is chosen to be 0.1, and the order of the polynomials is chosen to be $N_p = 6$. Very good agreement is observed for the mean response. The low frequency variation of the standard deviation matches the results obtained from the numerical integration very well. The accuracy for the much higher but less important frequency variation of the standard deviation can be improved by using larger values of N_p. Specifically, very good agreement was obtained for the choice of $N_p = 8$. Assuming that the θ_is are uniformly distributed, an excellent agreement was found between the results of the proposed methodology with $N_p = 6$ and the exact one.

The variability of the response at times t= 5.8 and 20 sec with respect to the variability of the stiffness k_s of the secondary oscillator is shown in Figures 2 and 3, respectively. Figure 2 shows the effect of the order N_p of the approximation on the variability for uniform and Gaussian distributed θs. Observing Figures 2 and 3, it is evident that higher values of N_p are needed to accurately predict the variability of the response at later times. Uniformly distributed variables provide a better fit to the response variability than the Gaussian distributed variables. This is because of the wider range of variation of Gaussian distributed variables for the same coefficient of variation. Reducing the coefficient of variation of the Gaussian random variables to 0.067, a much better fit is obtained (curve 4 in Figure 3) which is comparable to that obtained for the uniformly distributed random variables. Comparing Figure 3 with Figure 1 at time t=20, it can be seen that low values for N_p provide a very good fit of the first two moments. However, the variability of the response in not accurately predicted. The fit for the variability can be improved to any degree of accuracy by using sufficiently large values for N_p.

A second example of an N=10 story primary structure is also considered. The interstory stiffnesses of the primary structure are taken as $k = 2000m$. The uncertain variables and coefficients of variations are the same as before. The nominal values of the mass ratio and the frequency ratio are chosen to be $\eta = 0.05$ and $\alpha = 1$ which correspond to a secondary system tuned to the first mode of the building. The nominal values of the modal damping ratios ζ_j of the building and the damping ratio ζ_s of the equipment are chosen to be 5%. A lognormal distribution is assumed for the random variables. In order to treat this distribution by the proposed approach, probability transformations are used to transform the system's basic random variables into a new set $\underline{\theta}$ of statistically independent, standard normal variables.

The adequacy of the quadratic approximation (14) is examined for this example. It was found that the first two modes of the combined system are sufficient to provide accurate estimates of the response. The variability of the participation factors $Re\{\gamma_m^{(r)}\}$ and $Im\{\gamma_m^{(r)}\}$ is shown in Figure 4 for the first two contributing modes. Comparison between the exact variability and that obtained from the quadratic approximation (14) indicates that a quadratic approximation (14) may not always be adequate in approximating the modal variability. Higher order polynomial approximations should be used to replace (14). Figure 5 compares the mean and the standard deviation obtained from the proposed method and the "exact" numerical integration. Again a very good agreement is observed. Although the fit for $Re\{\gamma_m^{(r)}\}$ in Figure 4 is not satisfactory, the values of $Re\{\gamma_m^{(r)}\}$ compare to $Im\{\gamma_m^{(r)}\}$ are too small to significantly contribute to the response in this case. Finally, it is worth mentioning that the computational effort involved in the proposed method is significantly less than that involved in the numerical integration, especially for small values of N_p.

5 FORMULATION FOR STOCHASTIC EXCITATION

For a zero-mean Gaussian stochastic input process, the response of linear systems with deterministic parameters is also a Gaussian process, completely defined by the covariance matrix. The covariance matrix of the response is often used as a simplified description of the response process to provide useful information about the level of the system response. Furthermore, it has been extensively used in analytical random vibration formulations to provide approximate estimates of exceedance probabilities. For systems with uncertain properties, the covariance matrix will also be uncertain. The variability of the response moments is important because it can be combined with existing random vibration formulations of the first passage problem to compute system reliability in the presence of system uncertainties.

A computationally efficient method is next developed to determine the variability of the covariance matrix due to the variability of the system parameter set $\underline{\theta}$. The excitation is considered to be a nonstationary white noise process. General excitation processes, modeled as the output of a linear filter with a white noise input, can also be treated by analyzing an augmented system which consists of the original system (3) and the linear filter and excited by a nonstationary white noise process.

The covariance matrix of the response satisfies the well-known Liapunov differential matrix equation (Lin 1967). Rearranging the elements of the covariance matrix into an $n(2n+1)$-dimensional vector $\underline{B}$, which includes only the unique elements of the covariance matrix, the evolution of the elements of the covariance of the response is governed by the set of equations (Lutes 1986)

$$\dot{\underline{B}}(t) + Z(\underline{\theta})\ \underline{B}(t) = \underline{F}(t) \qquad (23)$$

The matrix $Z(\underline{\theta})$ depends on the matrices $K(\underline{\theta})$, $C(\underline{\theta})$, and $M(\underline{\theta})$ in equation (1). The system (23) is of the same form as system (3) with the coefficients being functions of the set $\underline{\theta}$ of random variables. Therefore, the previous formulation, developed for the deterministic excitation case, is directly applicable to the stochastic excitation case.

As in the case of system (3), the eigensolutions corresponding to the matrices Z and Z^T in (23) have to be computed N_{c_f} different times. Very simple algebraic expressions exist (Lutes 1986, Papadimitriou and Lutes 1993), which relate the eigensolutions of the matrices Z and Z^T in the system (23) with the eigensolutions of the matrices A and A^T in system (3). These relationships are very important since the eigensolutions of the large eigenvalue problems associated with Z and Z^T are avoided. Instead, only the eigensolutions of A and A^T are required, while the corresponding eigensolutions of Z and Z^T can easily be obtained without much computational effort and essentially no additional work space in a computer. After the eigensolutions of the system (23) have been obtained, the modal analysis described for the deterministic excitation is used to obtain the variability and the statistics of each element in the covariance matrix. It should be noted that a small fraction of the eigenvalues of Z would be real (Papadimitriou and Lutes 1993). This simplifies the modal equation (13), and subsequently simplifies the analysis for obtaining the modal response variability. Furthermore, because of the very low frequency content of the excitation in equation (23), accurate results for the covariance response are obtained by dynamic analysis of only the first few lower frequency modes of the system (23) (Papadimitriou and Lutes 1993).

For the example structures considered before and for modulated white-noise base excitation, it was found that lower values of N_p are sufficient to obtain very good results. This is attributed to the low frequency content of the excitation in (23) and the much smoother variation of the covariance response versus the system parameters over the range of interest. Therefore, despite the significantly larger dimension of system (23) than that of system (3), the computational effort involved in the stochastic excitation case is substantially less than that of the deterministic excitation.

6 CONCLUSIONS

A computationally efficient methodology based on complex modal analysis has been developed to calculate the response variability and statistics of

MDOF linear systems with uncertain parameters. The methodology treats both deterministic and random excitations. The computational effort and accuracy compares favorably with existing numerical and analytical methods. The methodology can be used to study the effect of the modeling uncertainties on the dynamic response and reliability of a wide range of complex engineering systems.

REFERENCES

Deodatis, G. 1991. "Weighted Integral Method. I: Stochastic Stiffness Matrix," *Journal of Engineering Mechanics*, 117(8), 1851-1864.

Ghanem, R. and Spanos, P. 1990. "Polynomial Chaos in Stochastic Finite Element," *J. Appl. Mech., ASME,* 57, 197-202.

Jensen, H. and Iwan, W.D. 1991. "Response Variability in Structural Dynamics," *Earthquake Engineering and Structural Dynamics,* 20, 949-959.

Jensen, H. A. and Iwan, W. D. 1992. "Response of Systems with Uncertain Parameters to Stochastic Excitation," *Journal of Engineering Mechanics*, 118(5), 1012-1025.

Katafygiotis, L. S. 1991. "Treatment of Model Uncertainties in Structural Dynamics," Dissertation, California Institute of Technology, *EERL Report No. 91-01.*

Katafygiotis, S.L. and Beck, J.L. 1991. "An Efficient Treatment of Model Uncertainties for the Dynamic Response of Structures, " *Computational Stochastic Mechanics*, P.D. Spanos and C.A. Prebbia (Ed), Elsevier Science, New York, 661-672.

Korn, G. A. 1968. "Mathematical Handbook for Scientists and Engineers." 2nd Edition, McGraw-Hill Book Co., New York.

Lin, Y. K. 1967. *Probabilistic Theory of Structural Dynamics*, McGraw-Hill Book Co., New York.

Liu, W.K., Mani, A. and Belytsckho, T. 1987. "Finite Elements Methods in Probabilistic Mechanics," *Probabilistic Engineering Mechanics,* 2(4), 201-213.

Lutes, L.D. 1986. "State Space Analysis of Stochastic Response Cumulants," *Probabilistic Engineering Mechanics*, Vol. 1, No. 2, 94-98.

Papadimitriou, C. and Lutes, L.D. 1993. "Approximate Analysis of Higher Cumulants for MDF Random Vibration," *Probabilistic Engineering Mechanics,* to appear.

Shinozuka, M. 1972. "Monte Carlo Solution of Structural Dynamics," *International Journal of Computers and Structures*, 2, 855-874.

Shinozuka, M., and Astill, J. 1972. "Random Eigenvalue Problems in Structural Mechanics," *AIAA J.,* 10(4), 456-462.

Spanos, P.D. and Ghanem, R. 1989. "Stochastic Finite Element Expansion for Random Media," *Journal of Engineering Mechanics,* 115(5), 1035-1053.

Yamazaki, F., Shinozuka, M., and Dasgupta, G. 1986. "Neumann Expansion for Stochastic Finite Element Analysis," M2-Rep., Dept. of Civ. Engrg. and Engrg. Mech., Columbia, Univ., New York, N.Y.

Wang, B. 1992. "An Efficient Method for Treating Uncertainties in Structural Dynamics," Thesis Report, Dept. of Civil Engrg., Texas A&M University.

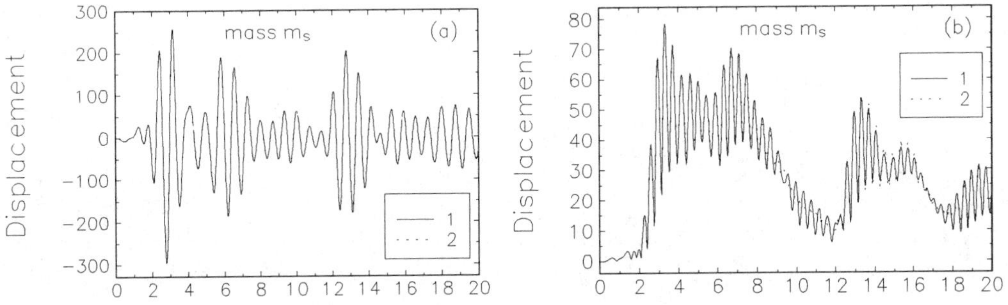

Figure 1. Comparison between the proposed method (1) and the "exact" numerical integration method (2); (a) mean response, (b) standard deviation; Gaussian probability density function; $N = 1$, $\eta = 5\%$, $\alpha = 0.85$, $N_p = 6$.

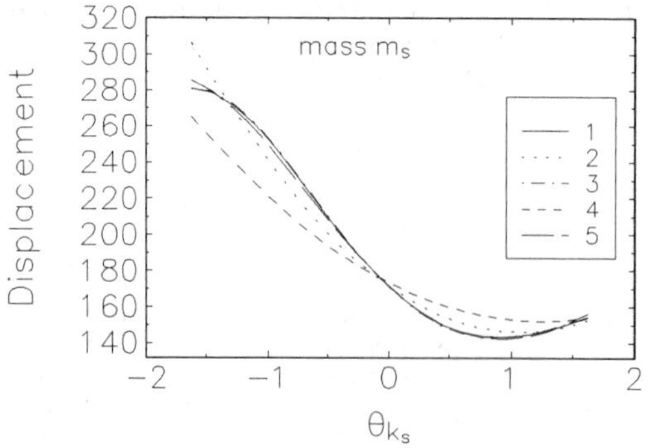

Figure 2. Variability of the response due to the variability of the stiffness k_s. All other $\theta_i = 0$; $t =$ 5.8 sec; exact (1); proposed method with uniform (2, 3) and Gaussian (4, 5) random variables; $N_p =$ 2 (2, 4); $N_p = 4$ (3, 5);

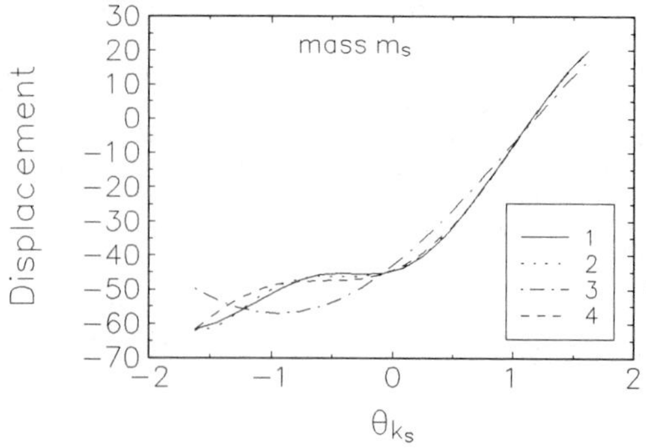

Figure 3. Variability of the response due to the variability of the stiffness k_s. All other $\theta_i = 0$; $t =$ 20 sec, $N_p = 6$; exact (1); proposed method with uniform (2) and Gaussian (3) random variables (c.o.v.=0.1), and Gaussian (4) (c.o.v.=0.067).

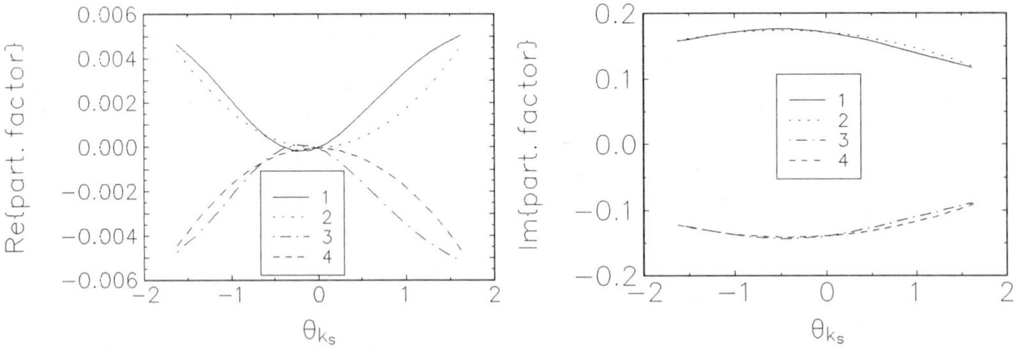

Figure 4. Variability of the effective participation factor $\gamma_m^{(r)}$ due to the variability of the stiffness k_s. All other $\theta_i = 0$; First mode (1, 2); second mode (3, 4); exact (1, 3); quadratic approximation (2, 4).

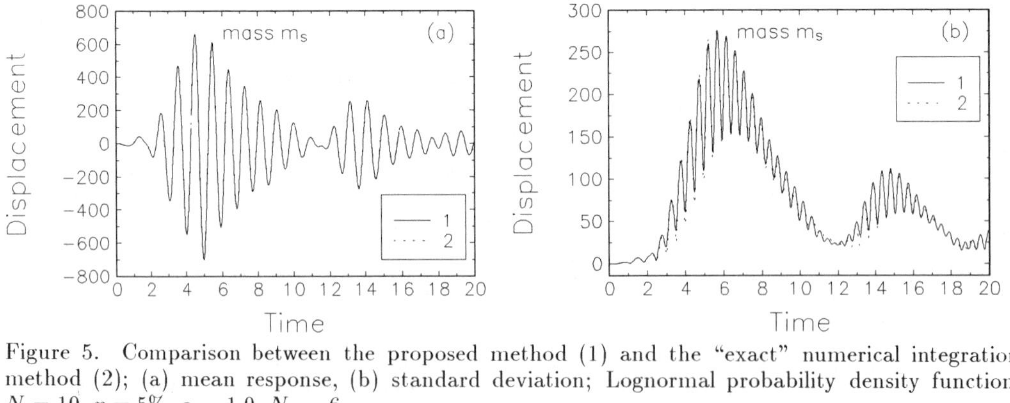

Figure 5. Comparison between the proposed method (1) and the "exact" numerical integration method (2); (a) mean response, (b) standard deviation; Lognormal probability density function; $N = 10$, $\eta = 5\%$, $\alpha = 1.0$, $N_p = 6$.

Statistical uncertainty (ongoing research)

Structural Safety & Reliability, Schuëller, Shinozuka & Yao (eds) © 1994 Balkema, Rotterdam, ISBN 90 5410 357 4

Mean square response of random beam under random load

I. Czmoch & K. Handa
Chalmers University of Technology, Göteborg, Sweden

ABSTRACT: The paper deals with the evaluation of statistics for the deflection function of a simply supported, linearly elastic beam, whose flexibility is considered as a stationary random process. The bending moment function is assumed to be dependent on two random variables. The mean value and the variance function of deflections are obtained analytically by applying the Green's function method. The theoretical solution is compared with the mean square statistics of deflections measured for 59 timber beams.

1. INTRODUCTION

The subject of the analysis is simply supported beam, whose flexibility is considered as a stationary random process. The randomness of flexibility is described by mean value, standard deviation, correlation function and spectral density function. The bending moment function is assumed to be dependent on two random variables, e.g. concentrated forces, concentrated bending moments, intensity of uniform load. The method of Green's function has been applied to evaluate the mean square statistics for the deflection function of a random beam under random load. The theoretical solution is compared with the corresponding statistics of deflections measured for 59 timber beams.

2. GENERAL SOLUTION

Simply supported beam, made of linear elastic material is considered. The flexibility of beam f(x) is assumed to be the stationary random process with the mean value m_f, standard deviation σ_f, covariance function $C_f(\xi)$ and spectral density function $S_f(\kappa)$. It is assumed that the bending moment M(x) depends on two generalised random forces A and B

$$M(x) = A\,a(x) + B\,b(x) \tag{1}$$

where a(x) and b(x) are deterministic normalized bending moment functions produced by the forces A and B, e.g. concentrated forces, concentrated bending moments, intensity of uniform load. It is assumed that the second moment representation of the random variables A and B is known

$$E[A] = m_A \qquad E[B] = m_B \tag{2}$$

$$Var[A] = \sigma_A^2 \quad Var[B] = \sigma_B^2 \quad Cov[A,B] = \sigma_{AB}$$

By applying the Green's function method to the differential equation governing the bending of the beam

$$\frac{d^2w}{dx^2} = M(x)\,f(x) \tag{3}$$

the deflection function w(x) can be obtained from the integral

$$w(x) = \int_0^L h(x,\xi)\,M(\xi)\,f(\xi)\,d\xi \tag{4}$$

where L is the beam span and $h(x,\xi)$ is the Green's function. For a simply supported beam the Green's function is equal to

$$h(x,\xi) = x\left(1-\frac{\xi}{L}\right) \qquad \text{if} \quad x \le \xi$$
$$h(x,\xi) = \xi\left(1-\frac{x}{L}\right) \qquad \text{if} \quad x \ge \xi \tag{5}$$

Taking into account equations (1), (4), (5) the expected value of the deflection function w(x) is found to be equal to

$$E[w(x)] = \int_0^L h(x,\xi)\,(m_A\,a(\xi) + m_B\,b(\xi))\,m_f\,d\xi \tag{6}$$

The variance of the deflection function can be represented as the sum of two integrals

$$Var[w(x)] = I_1 + I_2 \tag{7}$$

$$I_1 = \int_0^L\int_0^L h(x,\xi)\,h(x,\zeta)\,(E[M(\xi)M(\zeta)] - E[M(\xi)]\;E[M(\zeta)])m_f^2\;d\xi\,d\zeta \tag{8}$$

$$I_2 = \int_0^L\int_0^L h(x,\xi)\,h(x,\zeta)\,E[M(\xi)M(\zeta)]\,C_f(\xi-\zeta)\,d\xi\,d\zeta$$

The first integral I_1 expressed the influence of variability of loads on the variance of deflections. The integral I_1 can be easily determined for the bending moment function M(x) defined by equation (1).

The second integral I_2 expressed the influence of variability of load and flexibility on the variance of deflections. The integral I_2 can be evaluated by different methods, e.g. numerical integration, probabilistic FEM approximation. Another approach was suggested by Shinozuka (1987). The covariance function $C_f(x)$ is substituted by the spectral density function $S_f(\kappa)$ in accordance with the Wiener-Khintchine theorem. Analysing the variance of random beam under deterministic load Shinozuka introduced the variability response function defined by the integral

$$V(x,\kappa) = \int_0^L h(x,\xi)\,M(\xi)\,e^{-i\kappa\xi}\,d\xi \tag{9}$$

where $h(x,\xi)$ is the Green's function and $M(\xi)$ is the bending moment function. The variability response function $V(x,\kappa)$ depends on the support conditions of the beam and on the load configuration. This function can be compared with the frequency response function $H(i\omega)$ of the linear dynamic system.

Hence, taking into account the bending moment function (1) the integral I_2 can be expressed in terms of two variability response functions

$$V_A(x,\kappa) = \int_0^L h(x,\xi)\,a(\xi)\,e^{-i\kappa\xi}\,d\xi \tag{10}$$

$$V_B(x,\kappa) = \int_0^L h(x,\xi)\,b(\xi)\,e^{-i\kappa\xi}\,d\xi \tag{11}$$

Finally, the variance of the deflection function for random beam under two random forces A and B is equal to

$$\begin{aligned} Var[w(x)] = {} & m_f^2\left(\sigma_A^2 C_a^2(x) + \sigma_B^2 C_b^2(x) + 2\sigma_{AB} C_a(x) C_b(x)\right) \\ & + \int_{-\infty}^{\infty} S_f(\kappa)\left\{\left(\sigma_A^2 + m_A^2\right)\left|V_A(x,\kappa)\right|^2\right. \\ & + \left(\sigma_B^2 + m_B^2\right)\left|V_B(x,\kappa)\right|^2 \\ & \left. + 2\left(\sigma_{AB} + m_A m_B\right) V_A^*(x,\kappa)\,V_B(x,\kappa)\right\} d\kappa \end{aligned} \tag{12}$$

where $V_A^*(x,\kappa)$ denotes the conjugate function for the variability response function $V_A(x,\kappa)$ and the functions $C_a(x)$, $C_b(x)$ have the form

$$C_a(x) = \int_0^L h(x,\xi)\,a(\xi)\,d\xi \qquad C_b(x) = \int_0^L h(x,\xi)\,b(\xi)\,d\xi$$

3. COMPARISON OF THE ANALYTICALLY DETERMINED VARIANCE OF DEFLECTIONS WITH THE VARIANCE OF THE MEASURED DEFLECTIONS

A set of 59 timber beams of cross-section 45x120 mm was subjected to experimental test, cf. Czmoch (1991). The deflection curve was measured at equally spaced points (every 5 mm) along each beam

$$\{x_i, u_{ik}\} \quad i = 1,...,519 \quad k = 1,...,59 \tag{13}$$

where $x_i = i * 0.005$ [m]

u_{ik} - deflection of the k-th beam at the i-th point

The mean value and the variance of the measured deflections at the i-th point were calculated over the set of 59 beams

$$E[u_i] = \frac{1}{59}\sum_{k=1}^{59} u_{ik} \tag{14}$$

$$Var[u_i] = \frac{1}{58}\sum_{k=1}^{59}\left(u_{ik} - E[u_i]\right)^2 \tag{15}$$

Each beam was subjected to concentrated bending moments at supports. The applied loads were measured for each beam and the mean square statistics of two bending moments A and B were determined

$$m_A = 620.7 \text{ [Nm]} \qquad m_B = 560.8 \text{ [Nm]} \tag{16}$$
$$\sigma_A = 59.3 \text{ [Nm]} \qquad \sigma_B = 72.4 \text{ [Nm]}$$
$$\sigma_{AB} = 3772.2 \text{ [N}^2\text{m}^2\text{]} \qquad \rho_{AB} = 0.88$$

Next, the flexibility functions $f_k(x)$ were evaluated for the measured deflection curves (21) on the basis of the formula

$$f_k(x) = \frac{\kappa_k(x)}{M_k(x)} \tag{17}$$

where $\kappa_k(x)$ is the curvature function and $M_k(x)$ is the bending moment function for the k-th beam. The functions $f_k(x)$ were filtered from the high frequencies caused by error in the measured deflections (cf. Czmoch 1991).

The flexibility functions $f_k(x)$ can be considered as realizations of the stationary random process. The global mean value m_f and the variance with respect to m_f were calculated for the set of 59 beams

$$m_f = 1.28*10^{-5} \qquad \sigma_f = 0.262*10^{-5} \text{ [1/Nm}^2\text{]} \tag{18}$$

The spectral density function $S_f(\kappa)$ was estimated for the fluctuations of the flexibility with respect to the global mean value m_f.

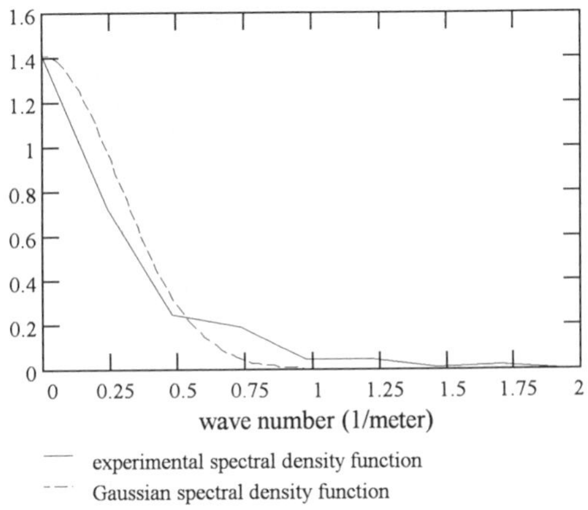

Fig. 1 Normalized spectral density functions

The normalized spectral density function of the flexibility

$$s_f(\kappa) = \frac{S_f(\kappa)}{\sigma_f^2} \tag{19}$$

is presented in Fig. 1 together with the normalized Gaussian spectral density function

$$s_G(\kappa) = \frac{S_G(\kappa)}{\sigma_f^2} = \theta \exp\left(-\pi\, \theta^2\, \kappa^2\right) \tag{20}$$

where θ is the scale of fluctuations (Vanmarcke 1983), defined as the ordinate of the normalized spectral function for the wave number equal to zero, i.e. $\theta = S_f(0)$. The scale of fluctuations θ was found for the set of 59 timber beams to be equal to 1.4 m.

Finally, the variance of the deflection function of the random beam under two random bending moments at supports have been evaluated according to equation (12) for:

a) the experimental spectral density function $S_f(\kappa)$
b) the Gaussian spectral density function $S_G(\kappa)$ with the scale of fluctuations θ=1.4m

Fig. 2 presents the standard deviation of the measured deflection (equation 15) versus the standard deviation of deflection calculated for the random beam (equation 12). Both theoretical

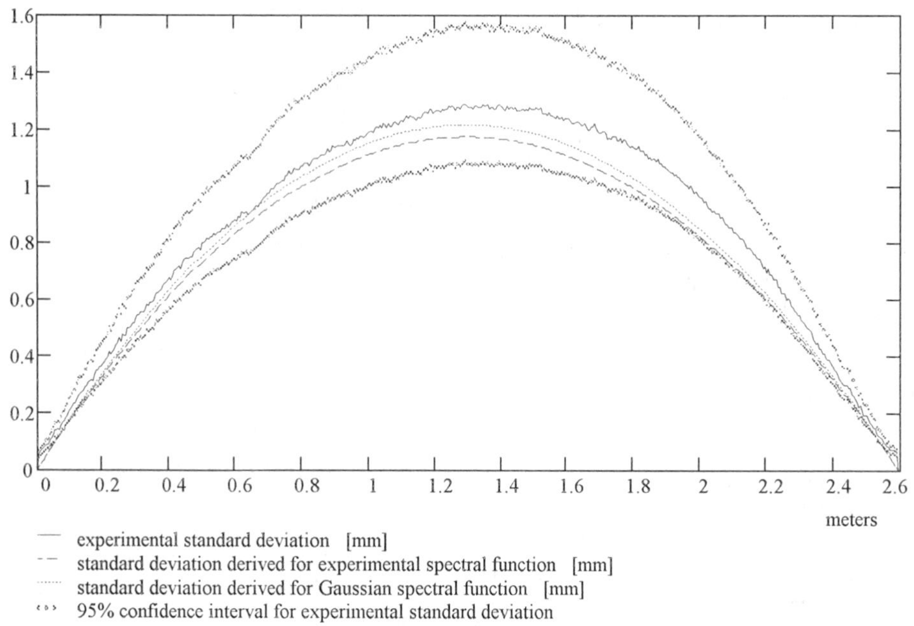

Fig. 2 Comparison between the theoretical standard deviation and experimental standard deviation of the deflection function for the set of 59 timber beams

standard deviation curves are within the 95% confidence interval for the standard deviation calculated for the experimental data.

4. CONCLUSIONS

The good correspondence between the theoretical variance and the experimental variance of deflection function means that both the randomness of the flexibility of timber beams and the randomness of loads have been estimated with appropriate accuracy.

The paper shows that the scale of fluctuations θ is an important measure of variability of flexibility. By assuming the scale of fluctuations θ=1.4 m estimated from experimental data, the Gaussian spectral density function $S_G(\kappa)$ has been fitted to the experimental spectral density function of flexibility $S_f(\kappa)$. It can be noticed a good correspondence between the variance of the deflection function calculated for the Gaussian spectral density function and the variance of experimental deflection curves.

5. ACKNOWLEDGEMENTS

The financial support for this project from the Swedish Council for Building Research (BFR) is gratefully acknowledged.

6. REFERENCES

Czmoch, I. 1991. Lengthwise Variability of Bending Stiffness of Timber Beams. *1991 International Timber Engineering Conference*, London, United Kingdom

Shinozuka, M. 1987. Structural response variability. *Journal of Engineering Mechanics. ASCE*. 1987. 113(6):825-842

Vanmarcke, E. 1983. *Random Fields: Analysis and Synthesis*, The MIT Press Cambridge Massachusetts London. England

Structural Safety & Reliability, Schuëller, Shinozuka & Yao (eds) © 1994 Balkema, Rotterdam, ISBN 90 5410 357 4

Definition and measure of uncertainties in structural reliability analysis

Wei-Liang Jin
Department of Civil Engineering, Dalian University of Technology, People's Republic of China

Eberhard Luz
Institute of Mechanics for Civil Engineering, Stuttgart University, Germany

ABSTRACT: According to the arrangement of the process of structural reliability analysis(SRA), two concepts, parameter (local) uncertainty and system (global) uncertainty, are redefined in this paper. The former is only concerned basic variables of limit state function in SRA with fundamental features of objective uncertainty and subjective uncertainty, while the latter will be dealt the internal connection and external influence of limit state function with obvious subjective uncertainty. In order to set up a new method of generalized structural reliability assessment, the relativistic information theory is introduced into this paper to measure parameter and system uncertainties of SRA. An example will be shown the application of this method.

1 INTRODUCTION

The treatment of uncertainty has become an important problem in SRA. According to the characteristics of uncertainty, uncertainty in structural engineering can be divided into objective and subjective parts. Objectivity is from all countable information of events where the uncertainty is of randomness and is often measured by probability reliability theory. Subjectivity is from all uncountable information of events due to human factors where it is of fuzziness and ignorance or incomplete of knowledge and it is possible to be described by use of the fuzzy theory too(Hu & Chen, 1992). SRA with objective uncertainty and subjective uncertainty is usually called the generalized SRA. But it should be noted that objective uncertainty is only concerned with basic variables of events, while subjective uncertainty is dealt with not only system or environment of events, but also basic variables of events. Brown(1980a) presented a margin method of combining subjective uncertainty with objective uncertainty. But this method is not shown each uncertainty's effect in the process of reliability analysis. Therefore, it is neccessary to redefine new concepts of uncertainties to reflect each uncertainty effect in the process of SRA, which is one of this paper's objective.

Because SRA is a process of information, the fundamental method of information theory can be applied into SRA. Brown(1980b) considered with subjective and objective uncertainties in SRA by use of the concept of entropy, and calculated the failure probability with subjective uncertainty. This method gives a new analysis path for generalized SRA. But considering that each uncertainty, especially subjective uncertainty, has a relativity, it is necessary to introduce the relativistic information theory into generalized SRA. The relativistic information theory presented by Jumarie(1979) is based on the following considerations: (1) information is by itself an observation process; (2) this observation process involves interference between the information source S and the observer R; (3) the observer has his own semantic framework. Then, Jumarie has set up a relativistic information theory by means of Lorentz transmitting theory. It is obvious that the basic considerations by Jumarie are also suitable for generalized SRA. Another objective of this paper is to measure parameter uncertainty and subjective uncertainty proposed by authors by means of the relativistic information theory.

2 DEFINITION OF UNCERTAINTIES

We know that uncertainties embody in every part of basic variables, limit state function and failure probability in SRA. The effect of uncertainty will result in the difference of structural safety assessment. Though the definition of subjective and objective uncertainties have also been reflected the fundamental features of uncertainty, it doesn't show the effects of uncertainty in the process of SRA. Therefore, from the viewpoint of arrangement of SRA, we have the following definitions:

Parameter Uncertainty — is that due to inexact knowledge of basic variables or parameters of SRA assuming a precisely defined system. It comes from randomness of generating basic variable, and fuzziness or incompletion of definition of basic variable. Subjectivity of variable definition is mainly shown in the definition of probability distribution function for basic variable and the values of mean and variance of distribution function etc. Hence, parameter uncertainty has an outstanding fuzzy randomness. Kam and Brown(1983) took into account the influence of original distribution variable or parameter due to the distribution range and generating possibility of variable by use of the fuzzy entropy. Obvious, the change of variable distribution due to parameter uncertainty will certainly result in the change of original structural reliability assessment.

System Uncertainty — is that due to the inadequacy of the theoretical model(limit state function) or structural failure probability of SRA assuming precisely defined basic variables or parameters describing structural reliability. It is the rest of parameter uncertainty in SRA. This uncertainty is mainly due to human active influences, for instance, fuzziness of theoretical calculation (mathematical modeling, numerical calculation, and design experience), human factors (construction experience, external environments), and imperfection (information, knowledge, and latent factors) etc.(Blockley, 1980). So, it is of outstanding characteristics of subjective uncertainty and will directly affect to assess structural safety correctly.

It is obvious that parameter and system uncertainties mentioned above will directly influence to the process of SRA. The treatment of parameter uncertainty is the base of the whole SRA, while the treatment of system uncertainty is the extension of generalized SRA. There are not only connections but also differences between both uncertainties in the process of SRA. Meanwhile, these definitions are of benefit to make the generalized SRA by means of the relativistic information theory.

3 RELATIVISTIC INFORMATION THEORY

For a system S observed by R, which is called relativistic information observing process S/R, it can consist of four fundamental states $\{H_i(S/R), H_o(S/R), V(S/R), W(S/R)\}$. Here, $H_i(S/R)$ is internal entropy of S/R which is the amount of knowledge that R has about the internal structure of S/R, while $H_o(S/R)$ is external entropy of S/R which is the amount of knowledge that S has about the internal structure of its environment $\bar{S}$. $V(S/R)$ and $W(S/R)$ are objective and transformation potential of S/R, respectively. Assume there is another observing process S/R', which has an internal relation $G = \{u(R/R'), 0, 0\}$ with R and $V(S/R)$, $W(S/R)$ are constant under the action of Lorentz transmitting process, the internal and external entropies of S/R' will then be as follows

$$H_i(S/R') = \rho(R/R')[H_i(S/R) + u(R/R') * H_o(S/R)] \quad (1)$$

$$H_o(S/R') = \rho(R/R')[H_o(S/R) + \frac{1}{C^2}u(R/R') * H_i(S/R)] \quad (2)$$

in which $u(R/R')$ is called subjectivity of R with respect to R'; $\rho(R/R') = 1/\sqrt{1 - u^2(R/R')/C^2}$; C is an universal constant.

When $R = S$, Eqs. (1) and (2) will become

$$H_i(S/R') = \rho(S/R')[H_i(S/S) + u(S/R') * H_o(S/S)] \quad (3)$$

$$H_o(S/R') = \rho(S/R')[H_o(S/S) + \frac{1}{C^2}u(S/R') * H_i(S/S)]. \quad (4)$$

Because the system S is by itself a conservation system in SRA with $C = 1$, equations mentioned above can be written as

$$\begin{aligned} H(S/R') &= \rho(S/R')[H(S) + u(S/R') * H(S)] \\ &= \sqrt{\frac{1 + u(S/R')}{1 - u(S/R')}} H(S) \\ &= \pi(S/R') * H(S); \end{aligned} \quad (5)$$

in which $H(S) = H_i(S/S) = H_o(S/S)$ is Shannon entropy (objective entropy) in information theory; $u(S/R')$ is called relativistic fuzzy function or subjective function if S/R' is a relativistic fuzzy set; $\pi(S/R')$ is a possibility function about S/R'. Eq. (5) is also called subjective entropy of S/R', which is an important concept of relativistic information theory, and will become an important and fundamental equation in generalized SRA.

If R is independent of R', Eq. (5) can be presented as follows

$$\begin{aligned} H(S/R, R') &= \sqrt{\frac{1 + u(S/R, R')}{1 - u(S/R, R')}} H(S) \\ &= \pi(S/R, R') * H(S) \end{aligned} \quad (6)$$

$$u(S/R, R') = \frac{u(S/R) + u(S/R')}{1 + u(S/R) * u(S/R')}. \quad (7)$$

Eq.(7) is often called composition law of relativistic information theory too.

For the relationship of relativistic fuzzy function $u(S/R)$ with fuzzy membership function $\mu(S/R)$, it can be written as

$$u(S/R) = \frac{\dot{\mu}(S/R)}{\max \dot{\mu}(S/R)}, \quad -1 \leq u(S/R) \leq 1 \quad (8)$$

in which $\dot{\mu}(S/R)$ is derivative of $\mu(S/R)$ in the Zadeh sense. When $0 \leq u(S/R) \leq 1$, positive subjective, $u(S/R)$, will increase prior information shown in Eq.(5). Otherwise, negative subjective will decrease prior information. Particularly, $u(S/R) = 0$ represents that there is not subjective feature in S/R. In other words, S can be replaced completely by S/R and $H(S/R) = H(S)$.

4 MEASURE OF UNCERTAINTIES

4.1 Parameter uncertainty

According to definition of parameter uncertainty, objective uncertainty of basic variable x can be represented by Shannon entropy

$$H(x) = -\sum_{i=1}^{n} p(x_i)\ln p(x_i) \tag{9}$$

in which $p(x_i)$ is i-th dispersed distribution probability of x. Assume that subjectivity of parameter uncertainty comes mainly from subjectivity of probability distribution for x and subjectivity of variable's distribution range due to R with relativistic fuzzy function $u_X(x/R)$, then subjective entropy of x/R will become

$$H(x/R) = -\sum_{i=1}^{n} \pi(x_i/R)p(x_i)\ln p(x_i). \tag{10}$$

Here subjectivity due to R is considered that it deals with the process of variable x generation. The formular of $\pi(x_i)$ will be shown in Eq.(5). So, it is said that the influent degree of parameter uncertainty for basic variable x has been shown in Eq. (10).

Let the equivalent Shannon entropy denote the influence of subjectivity of parameter uncertainty for x, one has

$$\begin{aligned} \bar{H}(x) &= H(x/R) \\ &= -\sum_{i=1}^{n} \bar{p}(x_i)\ln \bar{p}(x_i) \end{aligned} \tag{11}$$

in which $\bar{p}(x_i)$ is equivalent distribution probability with parameter uncertainty associating with $p(x_i)$.

4.2 System uncertainty

In SRA, the limit state function $Z = g(X)$ of structure is a function of basic variables $X(=\{x\})$. Assume that x is a relativistic fuzzy variable with parameter uncertainty, let $g: R \to \Re, X \to g(X)$ denote a function of the relativistic fuzzy variable $(X, u_X(x/R))$ and let $g^{-1}(X)$ denote the inverse of $g(X)$. Then, $Z = g(X)$ is also a relativistic fuzzy variable and one has

$$u_Z^P[g(X) = Z/R] = g(u_X[X = g^{-1}(Z)/R]). \tag{12}$$

Hence, subjective entropy of structure due to parameter uncertainty can be written as follows:

$$\begin{aligned} H^P(Z/R) &= \sqrt{\frac{1+u_Z^P(Z/R)}{1-u_Z^P(Z/R)}}H(Z) \\ &= \pi^P(Z/R) * H(Z) \end{aligned} \tag{13}$$

in which $H(Z)$ is Shannon entropy of probability reliability analysis, $H(Z) = -p_f \ln p_f - (1-p_f)\ln(1-p_f)$; p_f is failure probability of structure obtained by probability reliability analysis.

Considering the influence of system uncertainty, let $u_Z^S(Z/R')$ denote the relativistic fuzzy function due to system uncertainty. Then, the whole relativistic fuzzy function of structure with parameter uncertainty and system uncertainty can be obtained by use of composition law (Eq.(7))

$$u_Z(Z/R, R') = \frac{u_Z^P(Z/R) + u_Z^S(Z/R')}{1 + u_Z^P(Z/R) * u_Z^S(Z/R')}. \tag{14}$$

Similarly to Eqs. (13) and (6), subjective entropy of structure with parameter and system uncertainties has

$$H(Z/R, R') = \pi(Z/R, R')H(Z) \tag{15}$$

Introducing Eq.(11) into Eq.(15), one has the equivalent failure probability of structure, which will be used as an assessment of structural safety with subjective uncertainty on the base of probability reliability analysis.

5 EXAMPLE

For the limit state function of structure, $Z = g(X) = x - y$, let x/R and y/R denote two subjective variables of structural resistance and load with the relevant subjective function $u_X(x/R)$ and $u_Y(y/R)$, respectively. Then we have to measure $(z/R) = (x/R) - (y/R)$ and determine the subjectivity of the event $x - y \le z_0$ where $z_0 = 0$ in SRA. Therefore, the conditional relativistic fuzzy function for z/R is

$$\begin{aligned} u_Z^P(z/y/R) &= u_Z^P(Z \le z, Y = y/R) \\ &= \frac{v_Y(y/R) + u_X(z+y/R)}{1 + v_Y(y/R) * u_X(z+y/R)} \end{aligned} \tag{16}$$

and the relativistic fuzzy function for z/R has

$$u_Z^P(z/R) = th[\int_{-\infty}^{+\infty} u_Z^P(z/y/R)dy] \tag{17}$$

in which $v_Y(y/R) = 2u_Y(y/R)/[1 + u_Y^2(y/R)]$ is called the second relativistic fuzzy function for y/R.

Now let us calculate the failure probability of structure. Assume $X \sim N(200, 30)$ and $Y \sim N(100, 20)$, the failure probability obtained by probability reliability analysis is $p_f = 2.7408 \times 10^{-3}$ and the relevant Shannon entropy is $H(Z) = 1.8908 \times 10^{-2}$. Assume that the probability of variable x or y being greater than $\theta(= \mu - 1.96\sigma)$ is about 0.975, $p(x, y \ge \theta) = 0.975$ and $1-p = 0.025$, the corresponding relativistic fuzzy functions for variables x or y are defined as follows

$$u_X(x_1/R) = u_Y(y_1/R) = 0.2 \quad x_1, y_1 \ge \theta$$
$$u_X(x_2/R) = u_Y(y_2/R) = 0.1 \quad x_2, y_2 < \theta$$

Meanwhile, according to definition of Eq.(17), one has

$$\int_{-\infty}^{+\infty} u_Z^P(z/y/R)dy = p(y_1)u_Z^P(z/y_1/R) + p(y_2)u_Z^P(z/y_2/R) \quad (18)$$

Thus, from Eqs.(16), (17) and (18), the relativistic fuzzy functions for z/R have $u_{Z\le 0}^P(z/R) = 0.4417$ and $u_{Z>0}^P(z/R) = 0.4952$. Therefore, the subjective entropy of the event considered the effect of parameter uncertainty is $H^P(z/R) = 3.0695 \times 10^{-2}$, and the relevant equivalent failure probability is $\bar{p}_f^P = 4.8523 \times 10^{-3}$. The ratio of failure probability is $\bar{p}_f^P/p_f = 1.7704$.

Now let's take into account the effect of system uncertainty. Assume $u_Z^S(z/R')$ is the relativistic fuzzy function of z/R' about system uncertainty, the relativistic fuzzy function of the event can then be obtained by Eq.(14). If $u_{Z\le 0}^S(z/R') = 0.1$ and $u_{Z>0}^S(z/R') = 0.05$, one has $u_{Z\le 0}(z/R, R') = 0.5188$ and $u_{Z>0}(z/R, R') = 0.5320$. Meanwhile, from Eqs.(5) and (15), the subjective entropy of the event considered the effect of parameter uncertainty and system uncertainty $H(z/R, R') = 3.3679 \times 10^{-2}$, and the relevant equivalent failure probability $\bar{p}_f = 5.4188 \times 10^{-3}$. The ratio of failure probability $\bar{p}_f/p_f = 1.9770$. Therefore, it can be seen that the possibility of structural failure due to positive subjective uncertainty would be greater than the failure probability of structure obtained by probability reliability analysis.

6 REMARKS

According to the arrangement of the process of SRA, the concepts presented in this paper, parameter and system uncertainties, can be shown the effect of each uncertainty in SRA. Considering that subjectivity always exists in the process of SRA, these definitions of parameter uncertainty and system uncertainty are different with those defined by Blockley(1980). Both uncertainties can be measured by the relativistic information theory which will make up some lack of information or knowledge imperfection in SRA.

As Brown(1980b) said, SRA is also a process of information treatment. However, the process of obtaining information in SRA is a relativistic process. So, it is necessary to make a generalized SRA by use of the relativistic information theory. For the method proposed in this paper, subjective uncertainty is thought to exist in parameter and system uncertainties. Objectivity of parameter uncertainty can be measured by Shannon entropy, while subjectivity of both uncertainties can be represented by relativistic fuzzy function or subjective function. Then, according to the limit state function, whole relativistic fuzzy function of structure can be constructed by composition law. The subjective entropy, which is a product of objective entropy and relativistic fuzzy function, will be used to assess the safety of structure.

The most specific feature of this method is that each subjective function about uncertainties of SRA can be composed into a whole relativistic fuzzy function by the composition law, and the treatments of objective and subjective uncertainty can be done independently. The difference of this method with Brown's method is that the whole relativistic fuzzy function is combined with every subjective function of parameter and system uncertainties, rather than only single subjective function like to Brown's method. Therefore, this method can be shown the effect of every uncertainty in SRA.

7 CONCLUSIONS

Uncertainty in SRA can be divided into parameter uncertainty and system uncertainty. The former is only concerned basic variables or parameters in SRA with random fuzziness, while the latter is dealt limit state function and failure probability etc. in SRA with evident fuzziness or ignorance. For objectivity of parameter uncertainty, it can be measured by Shannon entropy. While subjectivity of both uncertainties will be represented by relativistic fuzzy function or subjective function, and it can be combined objective entropy into subjective entropy so that measuring the influent degree of subjective uncertainty in SRA. It is the method proposed in this paper that can be used to measure and assess generalized structural reliability by means of the relativistic information theory. An example has been shown this method's feasibility and application.

8 ACKNOWLEDGEMENT

This research is supported by Alexander von Humboldt Foundation of Germany and Institute of Mechanics(Bauwesen) of University Stuttgart which are gratefully acknowledged by authors.

REFERENCES

Blockley, D.I. 1980. *The Nature of Structural Design and Safety*. John Wiley & Sons.

Brown, C.B. 1980a. The merging of fuzzy and crisp information. *J. Engrg. Mech.*, ASCE, 106(1), 123-133.

Brown, C.B. 1980b. Entropy constructed probabilities. *J. Engrg. Mech.*, ASCE, 106(4), 633-640.

Hu, Yuxian & Chen Han-Yao. 1992. Probabilistic treatment of uncertainties from incomplete knowledge in SHA. *Proc. of 10th world conference on earthquake engineering*, Madrid, Spain. 351-355.

Jumarie, Cuy M. 1979. *Subjectivity, information, systems*. Gordon and Breach Science Publishers. Montreux.

Kam, Tai-Yan & Brown, Colin B. 1983. Updating parameters with fuzzy entropies. *J. Engrg. Mech.*, ASCE, 109(6), 1334-1343.

Structural Safety & Reliability, Schuëller, Shinozuka & Yao (eds) © 1994 Balkema, Rotterdam, ISBN 90 5410 357 4

Modelling of statistical uncertainty due to small sample size

W. Manners
University of Leicester, UK

ABSTRACT: This paper discusses a simple problem of statistical uncertainty: if a small sample of data exists which is known to be from a single normally distributed population, how should the uncertainty about the values of the mean and standard deviation of that distribution be modelled, and how should a predictive distribution for a future sample be obtained? Classical, Bayesian and likelihood approaches to this question are discussed. A new alternative is also proposed, and its extension to non-normal cases is briefly discussed. For the normal case, the results of the various approaches are generally similar, and none is free of conceptual difficulties.

1 INTRODUCTION

Calculations for the reliability of engineering structures and systems often have to use data of limited quantity or doubtful quality. A probabilistic model derived from such data is therefore subject to uncertainty about how accurately it models the variability inherent in the variable in question. It is often important that this 'statistical uncertainty' be included in reliability calculations so that further research, leading to better data, will tend to reduce the uncertainty used in the calculations, and hence, for example, give more economic designs for the same target reliability. If statistical uncertainty is not included, then additional research is as likely to make future designs more conservative as it is to make them more economic, and there is no incentive to collect better data.

This paper discusses the simplest statistical uncertainty problem, that is the effect of the quantity of data. A particular case is considered in which the data, $\mathbf{x}=(x_1,x_2,...,x_n)$, are known, or assumed, to be random samples of a normally distributed variable. The mean, μ, and standard deviation, σ, of this 'parent' distribution are however unknown, and it is therefore the uncertainty about their values that needs to be modelled.

2 THE CLASSICAL APPROACH

The best known approach to this problem is to calculate the statistics

$$m = (1/n) \sum_{i=1}^{n} x_i \qquad v = \sum_{i=1}^{n} (x_i-m)^2 \tag{1}$$

Now, m and v/(n-1) are unbiassed estimators for μ and σ^2 respectively, and statistical uncertainty is measured by 'confidence limits'. The philosophical background to this is the use of a relative frequency view of probability by classical statisticians. This does not allow the 'degree of belief' that the analyst has in the correctness of possible values of μ and σ to be modelled by probabilities, and so the concept of confidence has to be introduced. This then makes it impossible to combine the statistical uncertainty with that inherent in the parent distribution.

3 THE BAYESIAN APPROACH

Statisticians of a Bayesian persuasion have no such inhibitions. A probability distribution for $\Theta=(\mu,\sigma^2)$, given the data $\mathbf{x}$, can be obtained (Raiffa and Schlaifer, 1968) by using Bayes' Theorem, thus:

$$f_{\Theta|X}(\theta,\mathbf{x}) \propto f_{X|\Theta}(\mathbf{x},\theta)\ f_\Theta(\theta) \qquad (2)$$

where $f_{X|\Theta}(\mathbf{x},\theta)$ is the likelihood of the data given Θ, and $f_\Theta(\theta)$ is a prior distribution for Θ.

Furthermore, a 'predictive distribution' for y, a future sample from the parent distribution, is given by

$$f_{Y|X}(y,\mathbf{x}) = \int f_{Y|\Theta}(y,\theta)\ f_{\Theta|X}(\theta,\mathbf{x})\ d\theta \qquad (3)$$

where $f_{Y|\Theta}(y,\theta)$ is, of course, the parent distribution, and the integration is over all Θ.

One problem with this approach is the need for a prior distribution for Θ, i.e. a distribution for Θ chosen before the data is available. Using a 'non-informative' conjugate prior, Aitchison and Dunsmore (1975) show that the distribution of $\Theta = (\mu,\sigma^2)$ is a combination of normal and χ^2-squared distributions, and the predictive distribution for y is a t-distribution. Writing the density of the latter as being proportional to $(1+a^2)^{-(f+1)/2}$ where f is the number of degrees of freedom, the present author, following Aitchison and Dunsmore's method obtains f=n, and

$$a = (y-m)/[(1+1/n)v]^{1/2} \qquad (4)$$

An example given by Aitchison and Dunsmore, however, implies that they believe f=n-1 to be the correct answer in this case. This is discussed in Manners (1993).

Butler (1986) discusses a number of other approaches to the same problem and obtains the same final result, but with the number of degrees of freedom varying from (n-1) to (n-3). He also comments that (n-1) can be obtained by Bayesian methods using Jeffries' priors.

Assuming that (n-1) is correct, one way of expressing this result is to say that moving from the use of the parent distribution with unbiassed estimators for its parameters to the use of the predictive distribution which includes statistical uncertainty means changing from a normal distribution to a t-distribution, and increasing the variance from v/(n-1) to v(n+1)/[n(n-3)], a factor of $\sqrt{\{(n+1)(n-1)/[n(n-3)]\}}$ on standard deviation.

4 AN ALTERNATIVE APPROACH

In the author's opinion, the need for a prior distribution is a significant drawback with the Bayesian approach. Information does not often come in two stages, and even if it does, it seems more natural to put all the available information together into a single analysis. The frequent use of vague or non-informative priors seems like an ingenious solution to a problem that should not exist. The approach outlined below avoids this problem completely, while continuing to utilise a 'degree-of-belief' view of probability.

The first step is to separate the act of sampling, which, although random, is precisely defined, from the unknown features of the distribution, i.e. the values of its mean and standard deviation. This is done by initially creating a sample $\mathbf{u}=(u_1,u_2,...,u_n)$ drawn from an N(0,1) distribution. For any μ,σ a data set $\mathbf{x}$ can be created by

$$x_i = \mu + \sigma u_i \qquad (5)$$

The statistics m,v are now given by

$$m = (1/n)\ \Sigma x_i = \mu + (\sigma/n)\ \Sigma u_i \qquad (6)$$

$$v = \Sigma(x_i-m)^2 = \sigma^2\ \Sigma u_i\ -\ (\sigma^2/n)(\Sigma u_i)^2 \qquad (7)$$

where the summations are from i=1 to n. These equations can be rearranged to give

$$(v/\sigma^2) = \Sigma u_i^2\ -\ (1/n)(\Sigma u_i)^2 \qquad (8)$$

$$\mu = m\ -\ (\sigma/n)\ \Sigma u_i \qquad (9)$$

Thus, for given m and v, μ and σ are defined in the 'sampling space' (u-space). If (v/σ^2) is constant, equation (8) describes a hypercylinder in u-space, straight in one dimension and curved with radius $\sqrt{v}/\sigma$ in the other (n-1) dimensions. Since the u_i are N(0,1) variables, equation (8) therefore means that (v/σ^2) is χ^2-distributed with (n-1) degrees of freedom. Equation (9) means that $\mu|\sigma$ is Normal with mean m and variance σ^2/n. Hence a joint distribution is obtained for μ and $1/\sigma^2$, which, ignoring normalising constants takes the following form:

$$f_{\mu,1/\sigma^2|x}(\mu,\sigma,m,v) \propto \sigma^{-(n-2)}\exp\{-[v+n(\mu-m)^2]/2\sigma^2\} \qquad (10)$$

From this a predictive distribution is obtained using equation (3), giving, in fact, the same result as that claimed by Aitchison and Dunsmore (i.e. equation (4) with f=n-1).

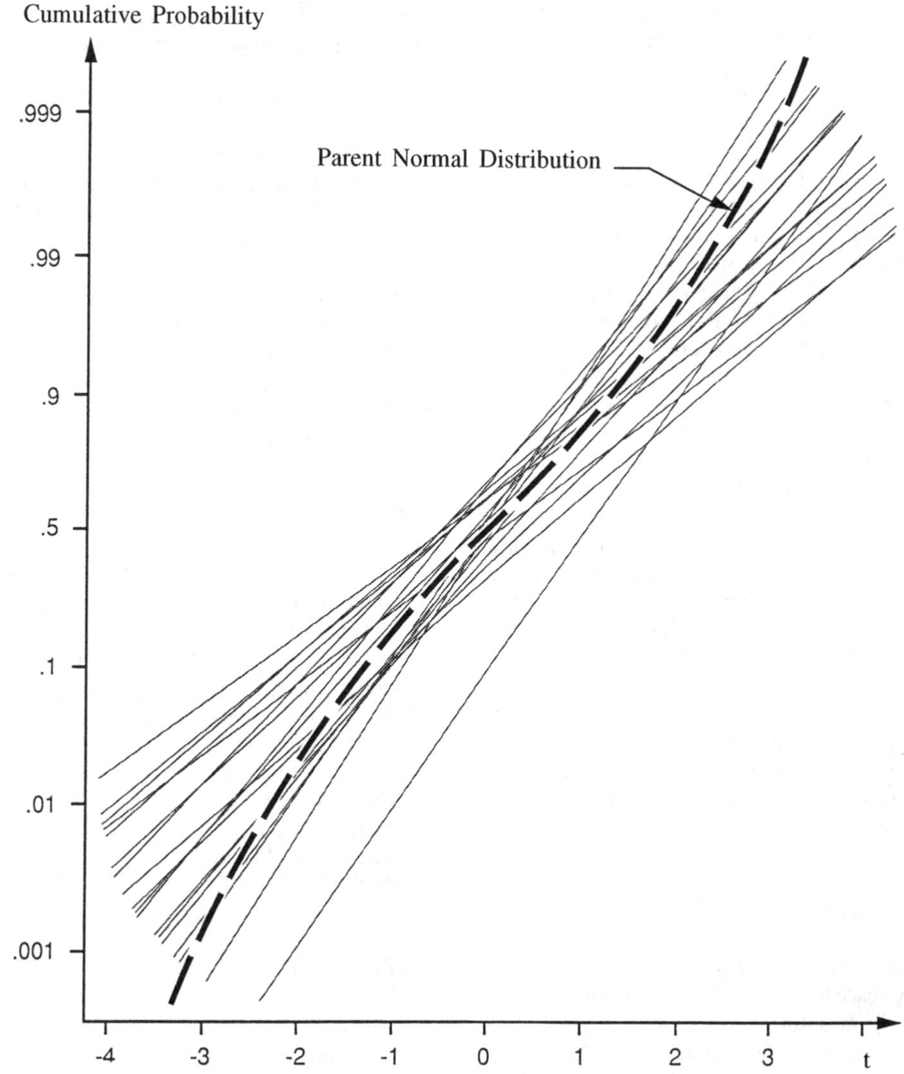

Fig. 1. Twenty Predictive Distributions derived from the same Normal Distribution

It is also noteworthy that, although justified differently, the calculations performed to find the distribution of μ and σ are not unlike those used in classical statistics in the derivation of confidence limits.

The most difficult step to justify is the reduction of the data, **x**, to the statistics m and v. A reduction to two statistics can be argued for on the grounds that this is necessary to preserve the 'degrees of freedom' in the problem, i.e. to allow all points in the sampling space to lead to values for μ and σ. A possible justification for the use of m and v is that they are 'sufficient statistics' for **x**, but this argument is weakened by the observation that the concept of sufficient statistics arises most naturally from the use of the likelihood function, which is not used in this derivation. It is also possible reduce the data to two statistics in other ways. For example, if R is the range between the largest and smallest of the data samples, **x**, then

$$R = \sigma(u_{(n)} - u_{(1)}) \tag{11}$$

where $u_{(n)}$ and $u_{(1)}$ are the greatest and least of n samples from an N(0,1) distribution. Equation (11) can be rewritten as

$$1/\sigma = (u_{(n)} - u_{(1)})/R \qquad (12)$$

and hence, if the range between the greatest and least of n samples from an N(0,1) distribution had a tractable algebraic form, it would be possible to use equation (12) to give a distribution for $1/\sigma$ using the range, R, as the statistic derived from the data. Although this approach is not sensible on a practical level, it is difficult to say that it is invalid on a theoretical level, but it will, in general, give a different distribution for σ.

The effect of using a t-distribution with (n-1) degrees of freedom for equation (4) is demonstrated in fig.1. Twenty different simulations each consisting of ten data values were taken from a normal distribution, and the figure shows the cumulative distributions for the twenty resulting t-distributions, plotted on axes such that t-distributions with nine degrees of freedom plot as straight lines. The parent normal distribution is also shown, and it can be seen that as the tails of the distributions are approached the t-distributions are increasingly likely to be conservative in comparison with the parent distribution.

5 NON-NORMAL CASES

This new approach can be extended to non-normal variables. In general it is unlikely that closed form solutions will exist in such cases, and numerical methods will have to be used. In a similar way to the use in the First-Order Reliability Method (FORM) of a standard normal variable space, it may be advantageous to continue to use a standard normal sampling space, because of its rotational symmetry.

As an example, if the variable is known to have a two-parameter Weibull distribution with parameters η and β, then equation (5) becomes

$$\ln(x_i) = \ln(\eta) + (1/\beta)\ln(-\ln(1-\Phi(u_i))) \qquad (13)$$

The use of the first and second moments of $\ln x_i$ as statistics would result in calculations similar to equations (6) to (9), leading to equations defining β and $\ln(\eta)$ in u-space. As mentioned above, numerical integration or some appropriate approximate method would then be necessary to obtain the required distribution for the parameters β and η, and further calculations would give a predictive distribution for a future sample.

As with the normal case above, there would be other possible ways in which a predictive distribution could be found. Indeed the situation seems to be very similar to that of obtaining point estimates of the parameters of a distribution from a sample of data; each of the various methods for doing the latter (method of moments, likelihood, plotting on appropriate graph paper, etc.) could be used as a way of defining the parameters in a sampling space and obtaining a distribution for them.

6 CONCLUSIONS

It has been shown that similar conclusions are obtained from several different starting points concerning the question of how the statistical uncertainty due to small sample size should be modelled, at least for a variable known to be normally distributed. None of the approaches discussed are without their apparently arbitrary elements, which makes the similarity of the results all the more surprising. It is also perhaps disappointing that such a well defined problem should not have a well-accepted method of solution.

REFERENCES

Aitchison,J. & I.R.Dunsmore 1975. *Statistical Prediction Analysis*. Cambridge:CUP.

Butler,R.W. 1986. Predictive Likelihood Inference with Applications.*J.R.Statist.Soc.B* 48:1-38.

Manners,W. 1993. *Predictive Distributions for Normally Distributed Variables*. Leicester University Engineering Department Report.

Raiffa,H. & R.Schlaifer 1968.*Applied Statistical Decision Theory*. Cambridge, Mass.:MIT Press.

Structural Safety & Reliability, Schuëller, Shinozuka & Yao (eds) © 1994 Balkema, Rotterdam, ISBN 90 5410 357 4

On mechanical modeling uncertainties in view of real failure data

Ruy C. Ramos de Menezes
Companhia Estadual de Energia Elétrica, Porto Alegre, Brazil (Formerly: Institute of Engineering Mechanics, University of Innsbruck, Austria)

C. E. Brenner
Institute of Engineering Mechanics, University of Innsbruck, Austria

ABSTRACT: This work presents an analytical/computational treatment of mechanical modeling uncertainties. The methodology is illustrated by an example of a transmission line steel tower for which failure data from full-scale prototype tests are available. The tower is modeled considering uncertainties inherent in material properties described in a data-based probabilistic manner. The discrepancies between real observed behavior and computational results are taken into account by a mechanical model uncertainty vector.

1 INTRODUCTION

The differences between observed structural behavior and results obtained by the mathematical treatment of an idealized mechanical model may be defined as mechanical model uncertainties. They are due to both, the lack of understanding of the phenomenon to be modeled, known as ignorance, and the use of simplified models, denoted as error of simplification. Mechanical model uncertainties are fundamentally different from physical uncertainties. Whereas the physical variability is intrinsic to nature, and hence beyond control, the uncertainties due to modeling are extrinsic and may be reduced to some extent. This can be done, e.g., by collecting additional information, refinements concerning the model to reduce errors of simplification and experiments to improve the understanding of the physical phenomena. At present, the stochastic finite element method (SFEM) (see e.g. Brenner (1991)) is considered to be one of the most sophisticated mechanical modeling procedures, as structural uncertainties and correlation effects can be included in the analysis.

2 BASIC FORMULATION

The structural reliability problem is usually formulated in terms of a limit state function $g(\boldsymbol{X})$ of a random vector $\boldsymbol{X} = (X_1, X_2, \ldots, X_n)$ which describes the physical uncertainties such as the randomness in material parameters, geometry and loads. By convention, the limit state function is formulated such that $g(\boldsymbol{X}) \leq 0$ denotes failure of the structure and $g(\boldsymbol{X}) > 0$ denotes its survival. The boundary between failure and safe set is known as the limit surface. By definition, reliability is the complement of the probability of failure p_f, which is given by the multidimensional integral

$$p_f = \int_{g(\boldsymbol{X}) \leq 0} f_{\boldsymbol{X}}(X) d\boldsymbol{X} \tag{1}$$

where $f_{\boldsymbol{X}}(X)$ is the joint probability function of $\boldsymbol{X}$. The function $g(\boldsymbol{X}) = 0$ represents the *idealized* model of the limit state of the structure, which is described by the analyst according to the current knowledge about the laws governing the underlying physical processes. For a meaningful evaluation of the probability of failure, it is essential to incorporate the model uncertainties in a reliability analysis. The degree of reliability of a structure is intrinsic to a structure and its environment. Therefore, the measure of reliability should remain uninfluenced, if neither the structure nor the environment is changed. This is not the case when the knowledge improves. Hence, if the knowledge is not perfect, the probability of failure itself is uncertain and can be only assessed in a probabilistic sense. In this case, the dispersion of its value is a measure of the quality of knowledge. This fact suggests that the uncertainties arising from mechanical modeling can be represented by a vector of uncertain, i.e., random parameters $\boldsymbol{A} = (\alpha_0, \alpha_1, \alpha_2, \ldots)$. These parameters may or may not have physical meaning and are usually not observable. They may represent, e.g., the uncertainty in modeling the load effect and the strength of components. For the hypothetical state of perfect knowledge, the vector $\boldsymbol{A}$ would be deterministically known. Otherwise, the state of knowledge is imperfect and its degree of imperfection is characterized by the distribution of $\boldsymbol{A}$, denoted $f_{\boldsymbol{A}}(\boldsymbol{\alpha})$. In this approach, the limit state function is written as $g(\boldsymbol{X}, \boldsymbol{A})$. The vector $\boldsymbol{X} = (X_1, X_2, \ldots, X_n)$ represents only the physical uncertainties, i.e., the inherent variability of the structure, and $\boldsymbol{A} = (\alpha_0, \alpha_1, \alpha_2, \ldots)$ represents the uncertainties due to mechanical modeling. The uncertain parameters α_i could also be described within the vector of basic random variables $\boldsymbol{X}$, however, due to their fundamental differences discussed earlier, it is convenient to keep them in a separate set. Then, given a particular state of mechanical modeling by $\boldsymbol{\alpha}$, the conditional probability of failure may be expressed by

$$p_f(\boldsymbol{\alpha}) = \int_{g(\boldsymbol{X}, \boldsymbol{\alpha}) \leq 0} f_{\boldsymbol{X}}(X) d\boldsymbol{X} \quad (2)$$

It is obvious, that for an uncertain $\boldsymbol{\alpha}$, $p_f(\boldsymbol{\alpha})$ is also uncertain. The expected value of the conditional probability of failure is expressed by

$$p_f = \int p_f(\boldsymbol{\alpha}) f_{\boldsymbol{A}}(\boldsymbol{\alpha}) d\boldsymbol{\alpha} \quad (3)$$

The probability function $f_{\boldsymbol{A}}(\boldsymbol{\alpha})$ can be obtained, at least theoretically, by conventional techniques from a function of random variables using available data. The statistics of the parameters $f_{\boldsymbol{A}}(\boldsymbol{\alpha})$ may be obtained by comparing predictions of a number of models, representing the current state of knowledge for a specific engineering problem, with results from carefully controlled experiments.

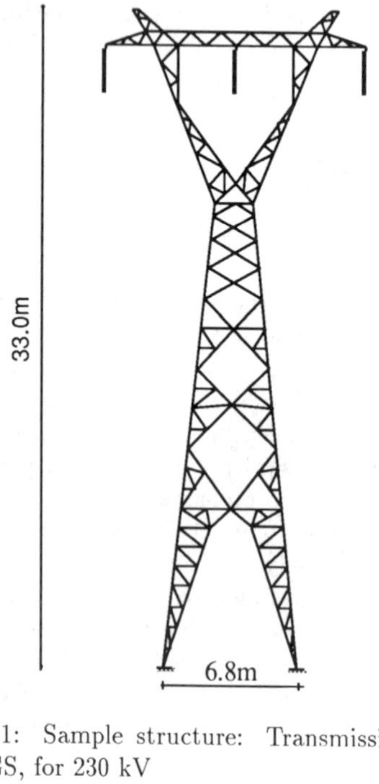

Figure 1: Sample structure: Transmission line tower GS, for 230 kV

3 EXAMPLE

A transmission line tower (figure 1) is used as an example to show the effects of mechanical model uncertainties. As Young's modulus and yield stresses clearly show variability, their effects are considered in the analysis. To establish a model which is capable of reproducing closely the statistics of the failure load as observed in prototype tests, the correlation model of the basic random variables, i.e., the Young's modulus and the yield stress, respectively, should be based on the manufacturing process. In view of these aspects, a simple correlation model is herein adopted for the random variables describing Young's modulus and yield stress (Ramos de Menezes (1992)). At this step, the application of a SFE method would allow the use of a more sophisticated correlation model (see e.g. Brenner (1993)). Introducing the random variable Y_1 as the tower strength in terms of the ratio "modeled failure load / design load", it can be expressed as

$$Y_1 = R_{\boldsymbol{X}}(X_{E1}, \ldots, X_{E16}, X_{Y1}, \ldots, X_{Y16}) \quad (4)$$

where X_{Ei} and $X_{Yi}, i = 1, \ldots, 16$ are the random variables describing the Young's modulus and the yield stress of the structural elements (bars) of different type and steel, respectively. Since just the global response of the tower from the prototype

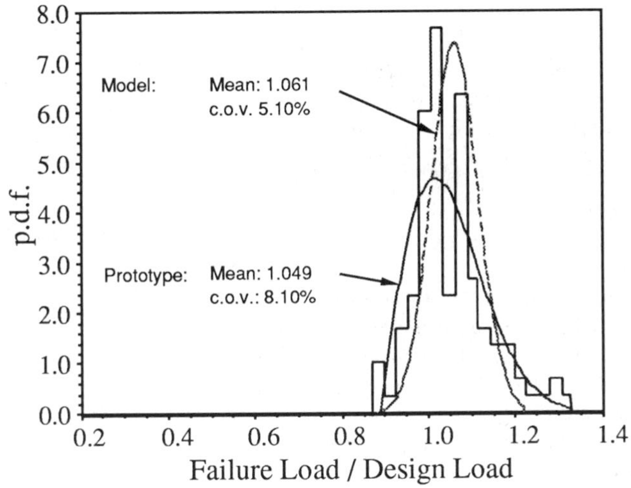

Figure 2: Modeled failure load. Prototype tests after Riera et al. (1990)

tests is available, only the uncertainty in modeling Y_1 may be evaluated. The probabilistic description for the yield stress is taken from data of tests in specimens published in Ramos de Menezes (1990). In addition the Young's modulus is considered to be normally distributed with a coefficient of variation of 6% as given by Galambos and Ravindra (1978). Considering the discussion in section 2, the mechanical model uncertainties are taken into account as follows: The adopted mechanical model, assuming a realistic description of the random variables involved, gives a response Y_1 as previously described. Comparing the distribution of Y_1 with the distribution of Y_t, as obtained from the results of prototype tests, some discrepancies - which indicate that the model is not perfect - are observed (see figure 2).

Thus, in this case, the mechanical model uncertainties may be expressed by a single variable α_1, which accounts for the discrepancies by

$$Y_t = \alpha_1 Y_1 \tag{5}$$

From this, the probabilistic description of α_1 in terms of its moments results in a mean value of 0.989 and c.o.v. of 6.28%.

4 FINAL REMARKS

An analytical/computational treatment of mechanical modeling uncertainties is presented. Considering these uncertainties, the uncertainties of the probability of failure can be assessed, where the dispersion of p_f may serve as a qualitative measure of the mechanical model. The variation of the estimate of the probability of failure becomes smaller when the mechanical model improves, since the model uncertainties vector $\boldsymbol{A}$ tends to become negligibly small. A transmission line steel tower, modeled by taking into account the statistical uncertainties of the material properties, is used as an example. The discrepancies between observed failure data from prototype testing and the computational results are described by mechanical model uncertainties. Possible improvements of the mechanical model include the application of the stochastic finite element method to account for correlation effects (see also Brenner (1993)).

ACKNOWLEDGEMENT

This work was partially supported by the Austrian Ministry of Foreign Affairs (North-South Dialogue Program, ÖAD project EH-894), by the "Companhia Estadual de Energia Elétrica do Rio Grande do Sul (CEEE)", by the Brazilian Ministry of Education (CAPES) and by the Austrian Industrial Research Promotion Fund (FFF) under contract No. 6/636, which is gratefully acknowledged. Thanks are due to G.I. Schuëller, director of the Institute of Engineering Mechanics at the University of Innsbruck for the scientific supervision of this work.

REFERENCES

Brenner, C.E. 1991. *Stochastic finite element methods (literature review)*. Internal working report 35-91, Inst. Engrg. Mech., Univ. of Innsbruck, Innsbruck, Austria

Brenner, C.E. 1993. On methods for nonlinear problems including system uncertainties. Submitted for publication, *Proc. ICOSSAR '93*, Innsbruck, Austria.

Galambos, T.V.; Ravindra, M.K. 1978. Properties of steel for use in LRFD. *Journal of the Structural Division* 104, ASCE, No.ST9: 1459-1468

Ramos de Menezes, R.C. 1990. Statistical data of tests in Brazilian steel angles for transmission line towers. In G.I.Schuëller (ed.), *Statistical analysis for observed material and structural data*. Vol. III, Inst. Engrg. Mech., Univ. of Innsbruck, Innsbruck, Austria.

Ramos de Menezes, R.C. 1992. *Failure-data-based reliability assessment considering mechanical model uncertainties*. Thesis, Inst. Engrg. Mech., Univ. of Innsbruck, Innsbruck, Austria.

Riera, J.D.; Ramos de Menezes, R.C.; Silva, V.R.; Ferreira da Silva, J.B. 1990. *Evaluation of the probability distribution of the strength of transmission line steel towers based on tower test results*. Report 22-13, CIGRE - Int. Conf. on High Voltage Electric Systems, Paris.

Stochastic finite element methods

Structural Safety & Reliability, Schuëller, Shinozuka & Yao (eds)© 1994 Balkema, Rotterdam, ISBN 90 5410 357 4

On methods for nonlinear problems including system uncertainties

C.E. Brenner
Institute of Engineering Mechanics, University of Innsbruck, Austria

ABSTRACT: In this paper, a selective review with respect to efficiency, accuracy and capabilities is given on the topic of stochastic finite element formulations for nonlinear problems. The requirements of nonlinear dynamic structural analysis of complex and large structures are found to be met by a stochastic finite element procedure using a point discretization scheme together with a Monte Carlo simulation and estimation techniques. The results may be utilized to assess the reliability of a structure. A transmission line tower is used as a numerical example.

1 INTRODUCTION

Within engineering analysis, idealized mechanical models serve to describe the physical behavior of a structure. To keep the degree of approximation small, nonlinear behavior as well as structural uncertainties must be taken into account by appropriate mechanical models. In particular, this is important for a credible reliability analysis.

An advanced method of analysis, capable of including structural uncertainties, is the stochastic finite element method (SFEM). Several SFE methods have been presented so far, e.g. the weighted integral method (see Deodatis (1991), Deodatis and Shinozuka (1991), Takada (1991) and Bucher and Brenner (1992)) which has some desirable features with respect to the accuracy of representation of the underlying random field, and to efficiency. To the present, however, only very few proposed SFE methods – usually at the cost of either accuracy or efficiency (see e.g. Liu et al. (1986), Hisada and Noguchi (1989), Deodatis and Shinozuka (1988)) – take into account nonlinearities.

An overview on the extension of some selected stochastic finite element formulations to treat nonlinearities is presented in this paper. Existing computational methods are assessed with respect to efficiency and accuracy as well as their capabilities when dealing with nonlinear dynamic behavior. A stochastic finite element procedure using a point discretization scheme together with a MC simulation is chosen to meet the requirements of complex nonlinear dynamic structural analysis. By applying estimation techniques, the response variability is determined which eventually may be used for a reliability assessment. A transmission line tower is used as numerical example to show the method's computational capabilities.

2 EXISTING SFE METHODS FOR NONLINEAR PROBLEMS

At present, three fundamentally different approaches – utilizing stochastic finite elements – for the assessment of the response variability of nonlinear structures taking into account statistical variation of structural properties may be distinguished. Their capabilities with respect to the treatment of complex nonlinear dynamic structures are reviewed in the sequel.

For nonlinear dynamic problems, the probabilistic finite element method (PFEM) was introduced by Liu et al. (1986). Further contributions were made e.g. by Hisada and Noguchi (1989) including path dependent and elasto-plastic effects and by Teigen et al. (1991) who adopted the method for material and geometrical nonlinearities of concrete structures. The governing structural equation terms are expanded by means of a usually first- or second-order Taylor series expansion around the mean values of the basic random variables. Solving the system of equations by equating terms of the same order, the mean, covariances and gradients of the response can be obtained as functions of the means and covariances of the random variables. Advantages of this method are that semi-analytical solutions for the covariances are available and that it is applicable to geometrical and material nonlinearities. The selection of only few important eigenvalues, the utilization of the adjoint methodology and the superposition of random fields improve its computational efficiency (Liu et al. (1987)). Unfortunately, the PFEM becomes computationally inefficient when a large number of random variables is involved. Besides, it is feasible only for coefficients of variation which are not "too large" (Liu et al. (1987)). Furthermore, nonlinear methods usually require the problem dependent, time consuming computation of the partial derivatives of the response by finite differences. According to Liu et al. (1986), the accuracy of the solution deteriorates with increasing number of the actual time step.

A first- and second-order reliability method including finite element analysis of geometrically nonlinear stochastic structures was presented by Liu and Der Kiureghian (1989). For each different set of random variables the method requires the FE computation of the response and its Jacobian with respect to the random variables. An iterative scheme is applied to determine the limit state function and the design point for a first- or second-order reliability computation. Although, for small structures, the method appears to be quite useful, its disadvantages are the required formulation of a limit-state function and the repeated nonlinear FE computations for different failure modes as well as for each new set of random variables. This is a severe restriction which prevents the method to be applicable to realistic, i.e. larger type practical problems.

The SFEM in combination with a simulation procedure and an estimation technique has been used for the analysis of nonlinear dynamic problems and to solve for the response variability e.g. by Deodatis and Shinozuka (1988) and later by Brenner (1993). Advantages of a stochastic finite element formulation along with a Monte Carlo technique are its applicability to different types of distributions and a large class of nonlinear stochastic structural problems, e.g. buckling and elastic/plastic problems. The disadvantage of direct Monte Carlo simulation is the extraordinary high computational effort, as the nonlinear FE equation system must be solved for each realization of the random variables involved. Although the application of the Neumann series expansion and/or selective MC procedures may reduce the computational effort significantly, so far, they were not applied to nonlinear dynamic problems as yet.

3 ON THE EXTENSION OF EXISTING SFE METHODS TO NONLINEAR PROBLEMS

Within the last decade, some stochastic finite element methods, in addition to those mentioned above, were developed for the treatment of linear static or dynamic problems, partially applicable to one- or two-dimensional systems, involving few random variables and a few degrees of freedom, only. However, most of the structures are of a more complex nature i.e. involving many degrees of freedom and many random variables (Ramos de Menezes and Brenner (1993)). Especially in the region where failure is most likely to occur, the structure often exhibits nonlinear, in fact highly nonlinear behavior which must not be disregarded within a credible reliability analysis. When planning the extension of existing linear SFE methods for the analysis of nonlinear problems, it must be kept in mind, that all the disadvantages of the linear case (Brenner (1991)) apply to the nonlinear case as well.

A SFE method, most suitable for linear problems, is the weighted integral method. However, unfortunately, this method cannot be extended to nonlinear dynamic problems without difficulties because, if the total Lagrangian formulation is examined, the 2nd Piola-Kirchhoff stresses are not independent of the input random variables as they affect the response of the system. Hence, a weighted integral decomposition of the nonlinear part of the stiffness matrix is not possible.

Another quite elaborated procedure for static and dynamic problems is the homogeneous chaos expansion based SFEM. The expansion of the nonlinear stiffness matrix in terms of a Karhunen-Loeve expansion is not possible for the same reasons as stated in context with the weighted integral method, described earlier.

4 NONLINEAR DYNAMICS OF LARGE STRUCTURES

Complex structures usually consist of several different types of structural elements. Considering the manufacturing process of mass products, one element type with certain dimensions is produced in large numbers, e.g., hundreds or thousands. This implies possibly the existence of large correlation between elements of same type and dimensions when manufactured, e.g., utilizing the same tools. During the subsequent manufacturing and assembly process of a structure, an element of certain type and dimension will be selected more less randomly from a number of equal elements, thus correlation effects may be changed. This should be accounted for in the analysis

Point methods as discretization tools for large structures appear to be feasible as they retain the distribution function and the autocorrelation function of the underlying random field and moreover, they are time effective. However, it has to be kept in mind, that their applicability is restricted by the quality of representation of the stochastic field, which for the case of accuracy of the autocorrelation function, is mainly determined by the distance between two points.

Assuming a particular quality of the manufacturing process, the autocorrelation function R_{ff} of the underlying random field $f(\boldsymbol{x})$ is known. Now, given a vector of random variables $\boldsymbol{H}(\boldsymbol{x})$, obtained by any point discretization method, e.g. a midpoint method, and assuming for simplicity $f(\boldsymbol{x})$ to be a zero mean homogeneous isotropic Gaussian random field, the covariance matrix $\boldsymbol{C}_{HH}$ can be established. One element of the covariance matrix is given by

$$C_{HH}(i,j) = R_{ff}(||\boldsymbol{x}_i - \boldsymbol{x}_j||) \tag{1}$$

Making use of an eigenvalue orthogonalization, that is, transforming the problem into the uncorrelated space yields

$$\boldsymbol{C}_{HH} = \boldsymbol{\Phi}\boldsymbol{\Lambda}\boldsymbol{\Phi}^T \tag{2}$$

where the $\boldsymbol{C}_{YY} = \boldsymbol{\Lambda}$ is the diagonal covariance matrix in the uncorrelated space. Usually only very few random variables whose variances are the largest eigenvalues suffice to describe the fluctuations of the underlying random field to e.g. 90 percent (Bucher et al. (1989)). Thus, the computational effort can be reduced by retaining only the corresponding k random variables in the analysis. The transformation into the correlated space yields the realization of the vector $\boldsymbol{H}$ as a function of the vector of uncorrelated random variables $\boldsymbol{Y}_r$ and the reduced eigenvector matrix $\boldsymbol{\Phi}_k$.

$$\boldsymbol{H}_r = \boldsymbol{\Phi}_k \boldsymbol{Y}_r \tag{3}$$

At this stage, the change of correlation introduced by the assembly process can be taken into account by assuming the relation between random variables and a structural element to be random. If a midpoint method is used, this simplifies to randomly mixing the indices of the elements of vector $\boldsymbol{H}_r$, what can be accomplished by generating a suitable random integer vector $\boldsymbol{I}$ by means of which the elements of vector $\boldsymbol{H}_r$ are rearranged. This reads for one vector element as

$$H_m(i) = H_r(I(i)) \tag{4}$$

The eigenvalue orthogonalization has to be carried out only once for each structural element type resulting from different manufacturing processes and for each stochastic field contributing to the structural uncertainty.

Within the simulation loop, the realizations of all random variables are generated using the computed eigenvalues and eigenvectors of the orthogonalization. Now, a sample of the nonlinear finite element equation system, which, in the most general case may be written as

$$\boldsymbol{M}\ddot{\boldsymbol{u}} + \boldsymbol{r}(\boldsymbol{u},\dot{\boldsymbol{u}}) = \boldsymbol{F}(t) \tag{5}$$

can be established using the generated sample values. Employing the total Lagrangian formulation, an implicit time integration scheme and a Newton iteration, the FE equation system for any time $t+\Delta t$ and the iteration step i yields

$$\begin{aligned} \boldsymbol{M}\,{}^{t+\Delta t}\ddot{\boldsymbol{u}}^{(i)} + {}^{t}_{0}\boldsymbol{C}\,{}^{t+\Delta t}\dot{\boldsymbol{u}}^{(i)} + ({}^{t}_{0}\boldsymbol{K}_L + {}^{t}_{0}\boldsymbol{K}_{NL})\Delta\boldsymbol{u}^{(i)} \\ = {}^{t+\Delta t}\boldsymbol{F} - {}^{t+\Delta t}_{0}\boldsymbol{R}^{(i-1)} \end{aligned} \tag{6}$$

The left superscript describes the time at which a variable is measured. The time of reference with respect to which the measurement was performed is given by the left subscript. ${}^{t+\Delta t}\boldsymbol{F}$ is the external force acting at the time $t + \Delta t$. The vector ${}^{t+\Delta t}_{\ 0}\boldsymbol{R}^{(i-1)}$ contains the computed internal nonlinear nodal forces – which are equivalent to the element stresses – for time $t + \Delta t$ and the iteration step $(i - 1)$ of the Newton Iteration.

Solving the above equations and utilizing estimation techniques, the second moment statistics of the response can be computed for the j-th element of the response vector and the m-th time step by

$$m_u(j,m) = \frac{1}{n_{sim}} \sum_{l=1}^{n_{sim}} u_l(j,m)$$

$$1 \le j \le n_{DOF} \; ; \; 1 \le m \le t_{max} \tag{7}$$

$$\sigma_u^2(j,m) =$$

$$= \frac{1}{n_{sim} - 1} \left(\sum_{l=1}^{n_{sim}} u_l^2(j,m) - n_{sim} m_u^2(j,m) \right) \tag{8}$$

where n_{DOF} and n_{sim} denote the number of degrees of freedom and simulations, respectively. The two summations are executed within the simulation loop.

5 NUMERICAL EXAMPLE

A transmission line tower may serve as an example for a large structure as shown in figure 1. Further structural details are given by Ramos de Menezes (1992). The FE model describing the tower in terms of a two node nonlinear 3D-beam element with 12 degrees of freedom per element consists of 457 nodes, 2718 degrees of freedom and 1073 beam elements with 16 different cross-sections and types of steels. It takes into account nonlinear boundary conditions, geometric nonlinearities due to large deflections and rotations but small strains as well as elastic, elastic-plastic and plastic buckling behavior. At this stage, hysteretic behavior is still neglected.

In the following, Youngs modulus, the cross-sectional area, the moment of inertia, and the yield stress are considered to be random quantities described by homogeneous isotropic Gaussian random fields $f(x)$ with autocorrelation functions R_{ff} of the type (Vanmarcke et al. (1986))

$$R_{ff}(x_1, x_2) = \sigma_{ff}^2 \exp \frac{-|x_1 - x_2|}{l_c} \tag{9}$$

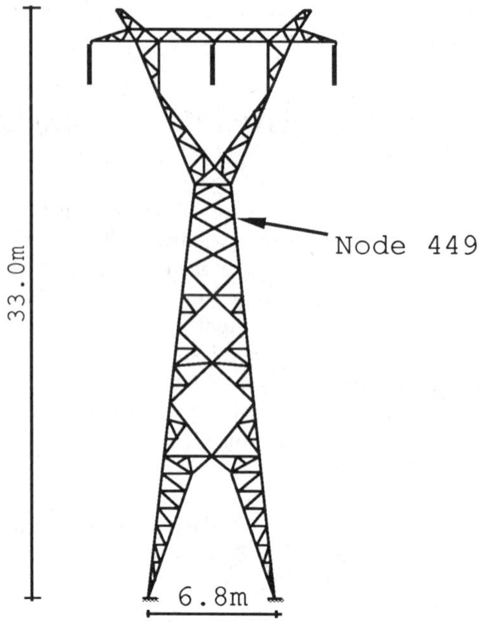

Figure 1: Schematic sketch of the transmission line tower (Ramos de Menezes (1992))

The bars of the tower are manufactured as very long parts which are cut for transportation purposes into pieces of 20 meters length. Therefore, a correlation length of $l_c = 20$ meters is assumed for all random quantities. Bars of different cross-section and material are treated as statistically independent as well as different random quantities with the exception of the cross-sectional area and the moments of inertia which are fully correlated as the latter depends on the first. The midpoint method is utilized to discretize the random fields. This leads to about 5000 random variables. One-dimensional structures, representing the manufacturing process, consisting of all bars of the tower of same cross-section and material are used to establish the correlation matrices. The assembly process is taken into account by randomly relating the structural elements to the generated random variables for every 1D-structure. The coefficients of variation (c.o.v) for the different random parameters are given in table 1.

The damping matrix is assumed to be constant with respect to time, since no statistical data is available. It is modeled by the Rayleigh damping where the first two modes are damped by 1 percent. For the computation of the damping matrix, the initial tangent stiffness matrix (time $t = 0$, no displacement except resulting from the dead weight) is used.

Quite frequently, design criteria for vertical loads on the structure for galloping conditions are incorporated (Committee on Electrical Transmission Structures of the Committee on Analysis and Design of Structures of the Structural Division (1982)). Galloping is observed in the case of strong winds when transmission lines are covered with ice. The wind generates vertical oscillations of the iced conductors with a frequency which is relatively high in comparison to the first natural frequency of the conductor. As a simple model to represent the forces produced by galloping, the load was assumed to be of sinusoidal shape having a frequency of $\omega = 2.5\frac{1}{s}$.

$$F(t) = F_{des}\sin(\omega t) \qquad (10)$$

The amplitude F_{des} reflects the design load of the tower in the static case.

To solve the time-dependent equation system, a Newmark integration scheme is applied. A modified Newton-Raphson iteration is utilized to obtain the solution of the nonlinear equations for each time step. By solving the finite element equation system repeatedly for all time steps and applying estimation techniques, the mean and variances of all time steps and for all nodes are computed. All computations have been carried out by means of COSSAN, a structural reliability analysis code (COSSAN (1994)).

32 Monte Carlo simulations have been performed. In one case, the Newton-Raphson iteration diverged which is interpreted as failure of the structure due to the structural uncertainties. The conditional (with respect to structures which did not fail at a certain time step) first and second moments of the displacements are obtained for 50 time steps, which are shown in figure 2 as function of the time step number for node 449. A closer look at the six plots of figure 2 reveals that the structural uncertainties in combination with the nonlinear structural behavior show no, or at most minor influence on the mean values of the response. However, the standard deviation is strongly affected by the nonlinearities, resulting from buckling or the slip of hinges. This may be clearly seen in the plot for the displacement in z-direction (fig. 2). Therefore, it can be concluded that the structural uncertainties of the tower play an important role, since strongly nonlinear behavior or even failure, observed in some simulations, can be attributed to them. A reliability assessment based on a linear analysis including uncertainties could lead to erroneous results. Hence, in this context, the nonlinear behavior of a structure and its uncertainties must be taken into account.

6 CONCLUDING REMARKS

The capabilities of some existing SFE methods for reliability analysis concerning the treatment of large complex nonlinear dynamic structures are reviewed. A short overview is given on the extension of some existing stochastic finite element formulations to treat nonlinearities. A stochastic finite element procedure using a point discretization scheme and a MC simulation is found to fit best the requirements of complex nonlinear dynamic structural analysis. By applying estimation techniques, the response variability in terms of second moments is determined. A transmission line tower is used as numerical example to demonstrate the method's computational capabilities. The results show that nonlinearities strongly affect the standard deviation, hence, the nonlinear behavior of a structure must not be neglected in a reliability analysis.

Further work directs to the reduction of the computational effort by possibly employing Neumann expansion, selective MC simulation schemes as well as the response surface methodology. This will allow a credible reliability analysis.

ACKNOWLEDGEMENT

This work was partially supported by the Austrian Industrial Research Promotion Fund (FFF) under contract No. 6/636. Thanks are due to G.I. Schuëller, director of the Institute of Engineering Mechanics at the University of Innsbruck, and C.G. Bucher for the scientific supervision of this work.

REFERENCES

Brenner, C.E. 1991. *Stochastic finite element methods (literature review)*. Internal working report 35-91, Inst. Engrg. Mech., Univ. of Innsbruck, Innsbruck, Austria.

Table 1: Coefficients of variation for the random parameters

Parameter	C.O.V [%]	Source
Young's Modulus	6	(Galambos and Ravindra (1978))
Yield Stress	≈ 10	(Ramos de Menezes (1990))
Cross-Sectional Area	0.74	based on (DIN (1970))
Moments of Inertia	1	based on (DIN (1970))

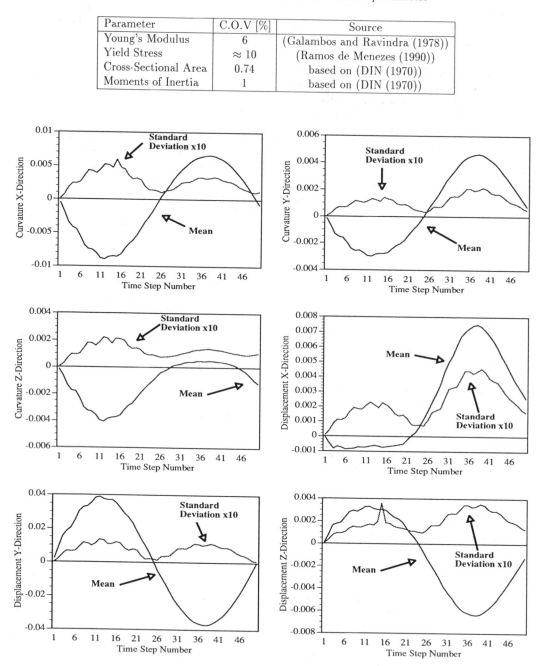

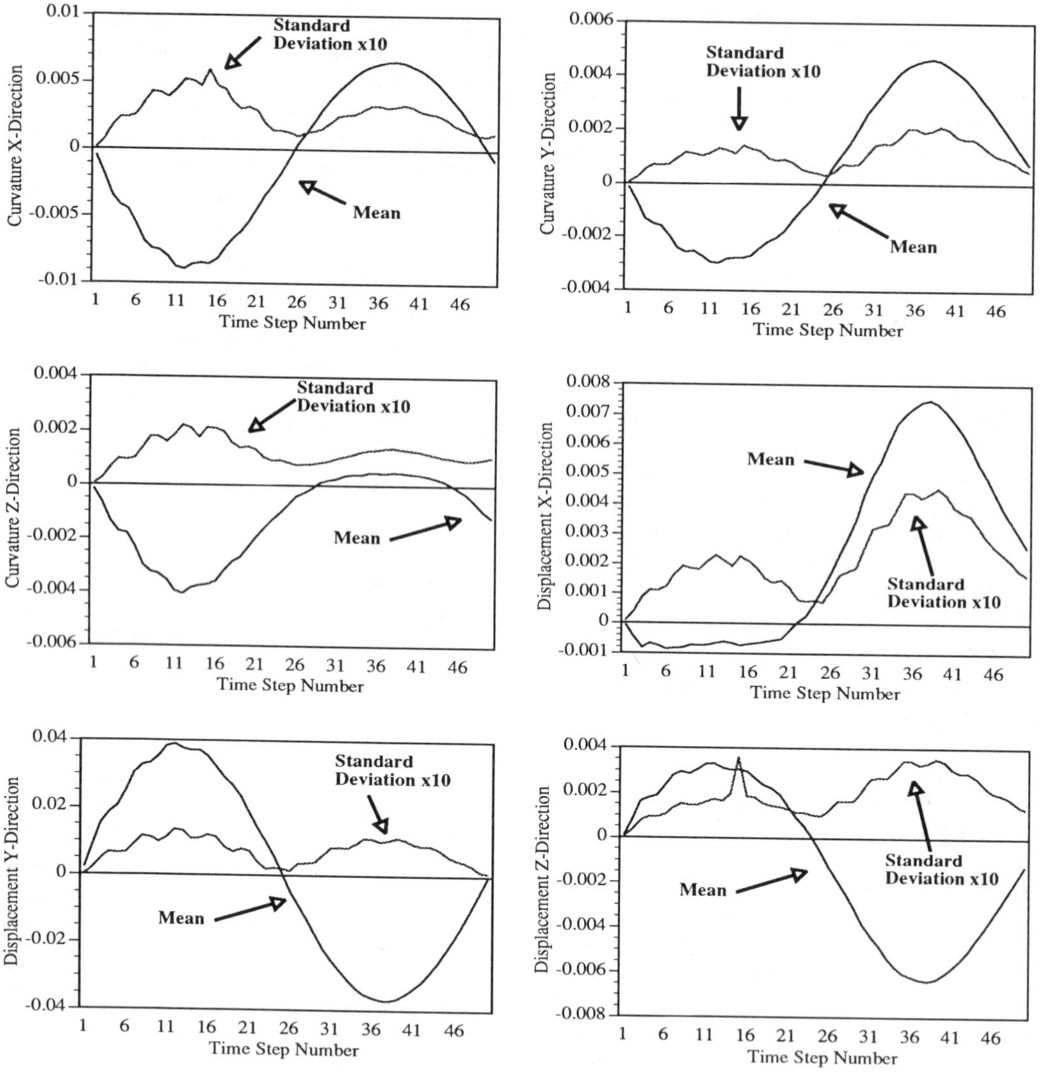

Figure 2: Response of node 449: Displacements and curvatures in x- ,y- and z-direction

Brenner, C.E. 1993. A stochastic method for nonlinear dynamic problems. To appear, *ZAMM 73,*, issue 6.

Bucher, C.G.; Brenner, C.E. 1992. Stochastic response of uncertain systems. *Arch. Appl. Mech.* 62: 507–516.

Bucher, C.G.; Pradlwarter, H.J.; Schuëller, G.I. 1989. COSSAN - Ein Beitrag zur Software-Entwicklung für die Zuverlässigkeitsbeurteilung von Strukturen. *VDI, Bericht zur Zuverlässigkeit von Komponenten Techn. Systeme*: 271–281.

Committee on Electrical Transmission Structures of the Committee on Analysis and Design of Structures of the Structural Division 1982. Loadings for electrical transmission structures. *J. Struct. Division* 108, No. ST5: 1088–1105.

COSSAN 1994. -Computational stochastic structural analysis. To appear, *Handbook*, Draft, Inst. Engrg. Mech., Univ. of Innsbruck, Innsbruck, Austria.

Deodatis, G. 1991. The Weighted Integral Method. I: Stochastic stiffness matrix. *J. Engrg Mech.* 117: 1851–1864.

Deodatis, G.; Shinozuka, M. 1988. Stochastic FEM analysis of nonlinear dynamic problems. In M.Shinozuka (ed.). *Stochastic mechanics.* Vol.III, Dep. of Civil Engrg. & Operations Research, Princeton Univ., Princeton, USA: 27–54.

Deodatis, G.; Shinozuka, M. 1991. The Weighted Integral Method. II: Response variability and reliability. *J. Engrg. Mech.* 117: 1865–1877.

DIN 1970. Deutsche Industrie Norm: DIN 7168, Freimaßtoleranzen.

Galambos, T.V.; Ravindra, M.K. 1978. Properties of steel for use in LRFD. *Journal of the Structural Division* 104, ASCE, No.ST9: 1459-1468.

Hisada, T.; Noguchi, H. 1989. Development of a nonlinear stochastic FEM and its application. In Ang, A.H-S.; Shinozuka, M.; Schuëller, G.I.(eds.), *Structural safety & reliability.* Proc. 5th ICOSSAR, San Francisco, USA: 1097–1104.

Liu, P.-L.; Der Kiureghian, A. 1989. *Finite-element reliability methods for geometrically nonlinear stochastic structures.* Report UCB/SEMM-89/05, Struct. Engrg. Mech. and Materials Dep. of Civ. Engrg., Univ. of California, Berkeley, Ca., USA.

Liu, W.K.; Belytschko, T.; Mani, A. 1986. Probabilistic finite elements for nonlinear structural dynamics. *Comp. Meth. Appl. Mech. Engrg.* 56: 61-81.

Liu, W.K.; Mani, A.; Belytschko, T. 1987. Finite element methods in probabilistic mechanics. *Prob. Engrg. Mech.* 2: 201–213.

Ramos de Menezes, R.C. 1990. Statistical data of tests in Brazilian steel angles for transmission line towers. in G.I.Schuëller (ed.), *Statistical analysis for observed material and structural data.* Vol. III, Inst. Engrg. Mech., Univ. of Innsbruck, Innsbruck, Austria.

Ramos de Menezes, R.C. 1992. *Failure-data-based reliability assessment considering mechanical model uncertainties.* Thesis, Inst. Engrg. Mech., Univ. of Innsbruck, Innsbruck, Austria.

Ramos de Menezes, R.C.; Brenner, C.E. 1993. On mechanical modeling uncertainties in view of real failure data. Paper for presentation at *ICOSSAR'93*, Innsbruck, Austria.

Takada, T. 1991. *Fundamental study on application of stochastic field theory to engineering.* ORI Report 91-02, Ohsaki Research Institute, Shimizu Corporation.

Teigen, J.G.; Frangopol, D.M.; Sture, S.; Felippa, C.A. 1991. Probabilistic FEM for nonlinear concrete structures. I: Theory. *J.Struct.Engrg* 117: 2674–2689.

Vanmarcke, E.; Shinozuka, M.; Nakagiri, S.; Schuëller, G.I.; Grigoriu, M. 1986. Random fields and stochastic finite elements. *Structural Safety* 3: 143–166.

Structural Safety & Reliability, Schuëller, Shinozuka & Yao (eds) © 1994 Balkema, Rotterdam, ISBN 90 5410 357 4

Approximate stochastic approach for structures with random media and loading

M. El-Meligy & P. Śniady
Institute of Civil Engineering, Technical University of Wrocław, Poland

ABSTRACT: A methodology has been proposed for the stochastic analysis of structures with random media and subjected to random loading. The variability of both media and loading are treated as stochastic fields. These fields are approximated as discretized stochastic fields and through expansion to orthogonal random variables. The stochastic finite element solution is formulated in such a way that the fluctuating components of both media and loading, which are statistically independent, are treated individually. The response component related to the random media is obtained through a Neumann expansion for the inverse operator. Expressions for second-order statistics of displacements, strains and internal forces are developed. numerical example is introduced to illustrate the proposed method.

1 INTRODUCTION

The uncertainty in the response of any structural system is mainly resulting from the stochastic nature of both media and loads. Many situations exist in which the stochasticity ought to be represented by random vector field. Each field can be discretized using one of the various methods of random field discretization; the midpoint method implemented by Der Kiureghian and Ke (1988) and Yamazaki, Shinozuka and Dasgupta (1988) , the spatial averaging method formulated by Vanmarcke and Grigoriu (1983), the interpolation method developed by Liu, Belytschko and Mani (1986), the expansion to basis random variables method formulated by Lawrence (1987), the Karhunen-Loeve expansion method developed by Traina et. al. (1986) and Spanos and Ghanem (1989, 1991), the weighted integral method developed by Deodatis (1990) , Deodatis and Shinozuka (1990) and Takada (1990) , and the optimal linear estimation method recently formulated by Li and Der Kiureghian (1992). Then, the response statistics are obtained using one of the stochastic finite element methods; the perturbation methods (see Handa and Anderson (1981), Hisada and Nakagiri (1981) and Liu, Belytschko and Mani (1986)), the Neumann expansion methods (see Shinozuka and Deodatis (1988), Yamazaki, Shinozuka, and Dasgupta (1988) and Spanos and Ghanem (1989,1991)), and the weighted integral method. All these methods, but the Neumann expansion methods, have been used to solve structures with random characteristics and subjected to random loads.

In the present paper, the random characteristics and the random loads are modelled as random vector fields. Each random field is expanded to a set of uncorrelated random variables using Karhunen-Loeve expansion. The stochastic finite element is formulated in conjunction with the Neumann matrix expansion for the inverse operator in order to get the response statistics up to the second order; i. e., the expectation, variance, and covariance of displacements, strains and forces for structures with random characteristics and loads. This work is an extension to the work of Spanos and Ghanem (1989) in which the second-order displacements statistics were obtained for structures with one random field representing the modulus of elasticity and subjected to static deterministic loads.

2 SFEM FORMULATION

In the following the complete formulation of the proposed approach will be presented.

2.1 Assumptions

The basic assumption regarding the SFEM formulation herein are as follows:

1. The uncertainty in the response is due to both the random characteristics and random loads.
2. The random fields representing the system characteristics and those representing loads are statistically independent.
3. All the random fields affecting the stochastic response of the system are stationary Gaussian random fields; each random field is sufficiently characterized by its mean, coefficient of variation, and autocorrelation function.
4. The random material property that is involved in this study is the modulus of elasticity which may be represented by different correlated random fields over the subdomains of the structure under consideration.

2.2 Random fields vector representation

Consider the vector of zero mean random fields X(x) which is a function of the spatial co-ordinate x, with the cross-covariance function matrix C. Each random field $X_i(x)$ is expanded as

$$X_i(x) = \sum_n V_{in} \sqrt{\lambda_{in}} \varphi_{in}(x) \tag{1}$$

where the λ_{in}'s and the φ_{in}'s are the eigenvalues and eigenfunctions, respectively, of the covariance function for the random field $X_i(x)$ (see Traina, Miller and Masri (1986)) and the V_{in}'s are random variables with the following second-order statistics:

$$\varepsilon[V_{in}] = 0 \text{ , and}$$
$$\varepsilon[V_{in} V_{im}] = \delta_{nm} \tag{2}$$

where ε is the expectation operator and δ is the kronecker delta.

In practical situations, the series in Eq. (1) is truncated after few terms and good results can be obtained by arranging the eigenvalues in a mono-decreasing manner. However, according to Der Kiureghian and Liu (1991), reducing the number of random variables by truncating the series after few terms may lead to erroneous results. Since each random variable V_{in} in Eq. (1) covers the whole subdomain of the random field $X_i(x)$ through the eigenfunctions $\varphi_{in}(x)$, the correlation between any two random variables representing the two different random fields $X_i(x)$ and $X_j(x)$ can be approximated in the form

$$\varepsilon[V_{in} V_{jm}] = R_{ij}(\tau) \tag{3}$$

regardless of n and m, where $R_{ij}(\tau)$ is the cross-correlation function between the random fields $X_i(x)$ and $X_j(x)$ and τ is the separation distance between the midpoints of these random fields.

2.3 Formulation for beam element stiffness

Using the standard finite element analysis method for linear static systems, the local random element stiffness matrix k can be formulated as follows:

$$k = \int_{\Omega_e} B^T C B dx \tag{4}$$

where B is the deterministic element strain-displacement matrix, C is the random elasticity matrix, Ω_e is the domain of element e, and T is indication for the transpose of a matrix. The C matrix can be written as

$$C = E_i(x) D \tag{5}$$

where $E_i(x)$ is a random field representing the modulus of elasticity over the domain Ω_i and

$\Omega_e \subset \Omega_i$. Using Eq. (1) to express $E_i(x)$, then substituting in Eq. (4), the random element stiffness matrix is obtained as

$$k = \bar{k} + \sum_{n=1}^{t} V_{in}^{m} k_{in} \tag{6}$$

where

$$\bar{k} = \bar{E}_i \int_{\Omega_e} B^T DB dx \tag{7}$$

is the deterministic local element stiffness matrix, $\bar{E}_i(x)$ is the deterministic value of $E_i(x)$, t is the number of terms after which the series is truncated, and

$$k_{in} = \sqrt{\lambda_{in}^{m}} \int_{\Omega_e} \varphi_{in}^{m}(x) B^T DB dx \tag{8}$$

is the in-th random component of the local element stiffness matrix, the superscript m refers to material property, and the subscript i refers to the random field under consideration. The integrations in equations (7) and (8) are performed in this study using the mathematical manipulation program "DERIVE".

2.4 *Formulation for element load vector*

Consider an element subjected to random distributed load $W_j(x)$, the equivalent element load vector f can be expressed as

$$f = \int_{\Omega_e} W_j(x) N(x) dx \tag{9}$$

where N(x) is the vector of shape functions. Substituting from Eq. (1) into Eq. (9), one can obtain the local random element load vector as

$$f = \bar{f} + \sum_{p=1}^{t} V_{jp}^{l} f_{jp} \tag{10}$$

where

$$\bar{f} = \bar{W}_j \int_{\Omega_e} N(x) dx \tag{11}$$

is the deterministic local equivalent element load vector, $\bar{W}_j$ is the deterministic value of $W_j(x)$, and

$$f_{jp} = \sqrt{\lambda_{jp}^{l}} \int_{\Omega_e} \varphi_{jp}^{l}(x) N(x) dx \tag{12}$$

is the random jp-th component of the local element load vector, the superscript l refers to loads, and the subscript j refers to the loading random field under consideration.

2.5 *Overall formulation*

Transforming the local element stiffness matrices and the equivalent element load vectors to the global co-ordinate system and summing them up, one can obtain the structure finite element equations

$$[\bar{K} + \sum_{i=1}^{M} \sum_{n=1}^{t} V_{in}^{m} K_{in}] q = \bar{F} + \sum_{j=1}^{L} \sum_{p=1}^{t} V_{jp}^{l} F_{jp} \tag{13}$$

where q is the structure displacement vector, $\bar{K}$ is the deterministic structure stiffness matrix, K_{in} is the in-th random component of the structure stiffness matrix, $\bar{F}$ is the deterministic load vector, F_{jp} is the jp-th random component of the load vector, M is the number of random fields representing the modulus of elasticity, and L is the number of random fields representing loads.

The displacement vector q can be obtained as

$$q = [I + \sum_{i=1}^{M} \sum_{n=1}^{t} V_{in}^{m} Z_{in}]^{-1} \bar{K}^{-1} [\bar{F} + \sum_{j=1}^{L} \sum_{p=1}^{t} V_{jp}^{l} F_{jp}] \tag{14}$$

where I is the identity matrix, and

$$Z_{in} = \bar{K}^{-1} K_{in} \tag{15}$$

The matrix Neumann expansion to the inverse operator takes the following form

$$[I + \sum_{i=1}^{M} \sum_{n=1}^{t} V_{in}^{m} Z_{in}]^{-1} = \sum_{N=0}^{\infty} (-1)^N [\sum_{i=1}^{M} \sum_{n=1}^{t} V_{in}^{m} Z_{in}]^N \tag{16}$$

hence

$$q=\sum_{N=0}^{\infty}(-1)^N[\sum_{i=1}^{M}\sum_{n=1}^{t}V_{in}^{m}Z_{in}]^N.$$
$$\overline{K}^{-1}[\overline{F}+\sum_{j=1}^{L}\sum_{p=1}^{t}V_{jp}^{l}F_{jp}] \qquad (17)$$

Applying the expectation operator to both sides of Eq. (17) and keeping in mind the assumption of statistical independence between random fields representing media and those representing loads, the expected displacement vector E_q can be expressed as

$$E_q=\overline{q}+E_{qm}+E_{ql} \qquad (18)$$

where

$$\overline{q}=\overline{K}^{-1}\overline{F} \qquad (19)$$

is the deterministic component of E_q,

$$E_{ql}=\varepsilon[\sum_{j=1}^{L}\sum_{p=1}^{t}V_{jp}^{l}P_{jp}]=0 \qquad (20)$$

is the component of the expected displacement vector due to the stochasticity of loads,

$$P_{jp}=\overline{K}^{-1}F_{jp} \qquad (21)$$

and

$$E_{qm}=\varepsilon[\sum_{N=1}^{\infty}(-1)^N[\sum_{i=1}^{M}\sum_{n=1}^{t}V_{in}^{m}Z_{in}]^N\overline{q}] \qquad (22)$$

Expanding Eq. (22), hence

$$E_{qm}=[\sum_{i=1}^{M}\sum_{c=1}^{M}\sum_{n=1}^{t}\sum_{d=1}^{t}\varepsilon[V_{in}^{m}V_{cd}^{m}]Z_{in}Z_{cd}+\ldots]\overline{q} \qquad (23)$$

The correlation matrix of displacement R_q can be expressed as

$$R_q=R_{qm}+R_{ql} \qquad (24)$$

where

$$R_{qm}=Q+\sum_{i=1}^{M}\sum_{c=1}^{M}\sum_{n=1}^{t}\sum_{d=1}^{t}\varepsilon[V_{in}^{m}V_{cd}^{m}].$$
$$[Z_{in}Z_{cd}Q+Z_{in}QZ_{cd}^{T}+QZ_{in}^{T}Z_{cd}^{T}]+\ldots \qquad (25)$$

is the displacement correlation matrix due to the stochasticity in media,

$$R_{ql}=\sum_{j=1}^{L}\sum_{g=1}^{L}\sum_{p=1}^{t}\sum_{h=1}^{t}\varepsilon[V_{jp}^{l}V_{gh}^{l}]P_{jp}P_{gh}^{T} \qquad (26)$$

is the component of the displacement correlation matrix due to loading stochasticity, and

$$Q=\overline{q}\overline{q}^{T} \qquad (27)$$

The displacement covariance matrix C_q can be written as

$$C_q=R_q+E_qE_q^{T} \qquad (28)$$

One should note that the expansions in equations (23) and (25) are expressed up to the first term only. However, including higher-order terms is straightforward and will increase the accuracy especially for higher values of the coefficient of variation. On the other hand, including higher-order terms increases the required storage capacity and CPU time enormously. It can be seen also that the response statistics due to the randomness in both of structure characteristics and loads can be obtained individually after performing the deterministic analysis.

2.6 Expressions for strains and forces

The displacement-strain relationship can be written as

$$s=Bq^{e} \qquad (29)$$

where q^e and s are the element vectors of displacements and strains, respectively. Applying the expectation operator to Eq. (29), hence

$$E_s=BE_q^{e} \qquad (30)$$

where E_s is the vector of expected element strains and E_q^e is the vector of the expected element displacements which can be obtained from the global vector of expected displacements E_q. The correlation matrix of element strains R_s can be expressed as

$$R_s = BR_q^e B^T \tag{31}$$

where R_q^e is the correlation matrix of element displacements which can be obtained from the corresponding position in R_q in the global co-ordinate system then transforming to the local co-ordinate system.

The expected element force vector E_f and the correlation matrix of element forces R_f can be approximated as

$$E_f = \bar{k}E_q^e \tag{32}$$

and

$$R_f = \bar{k}R_q^e\bar{k}^T + \sum_{n=1}^{t} k_{in}E_q^e E_q^{eT} k_{in} \tag{33}$$

3 NUMERICAL EXAMPLE

A turbo Pascal code for the analysis of plane frame structures with random media and loading has been written following the formulation described in the preceding sections. The example application is carried out using this code. The example is the 1-bay, 1-story plane frame shown in Fig. (1). The frame is subjected to the uniformly distributed loads W_1 and W_2 acting on the left column and the beam, with expected values 5 KN/m and 50 KN/m, and coefficients of variation 0.15 and 0.1, respectively.

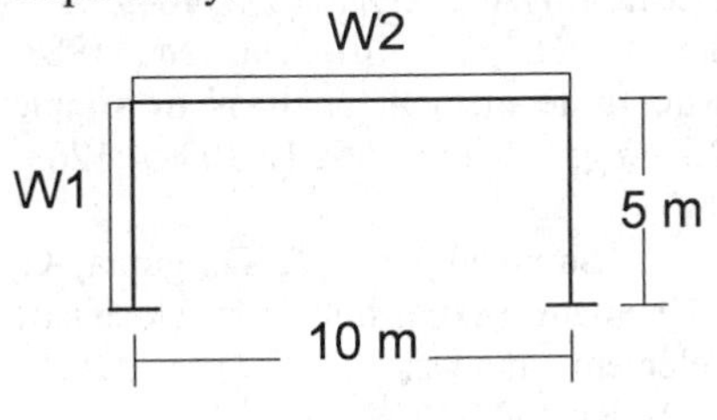

Fig. 1 One-story one-bay plane frame

The moduli of elasticity of the columns and the beam E_1 and E_2, respectively, have the same expected value 2×10^6 KN/m^2 and the same coefficient of variation 0.1. The correlation structure of the Gaussian random fields W_1, W_2, E_1, and E_2 are described by the same autocorrlation function

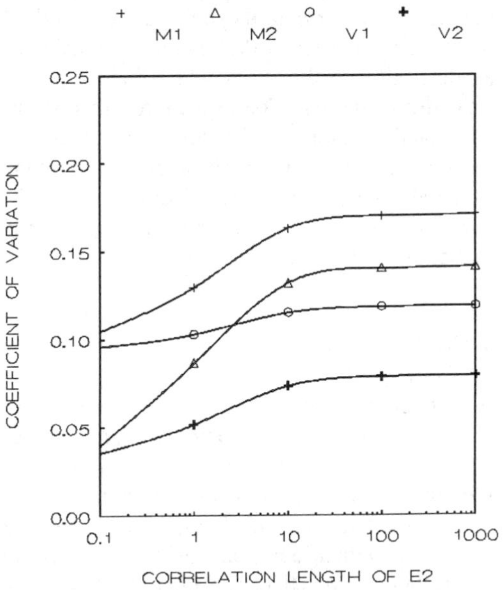

Fig. 2 Effect of correlation length of E_2 on the coefficient of variation of the response

$$R = \mathrm{Exp}[-\frac{|\tau|}{b}] \tag{34}$$

where b is the correlation length which is assumed equal to the length of the corresponding structural component for W_1, W_2, and E_1 and varied from 0.1 m to 1000 m for E_2. The cross-correlation coefficient between W_1 and W_2 is assumed to be 0.2 while between E_1 and E_2 is taken as 0.5. The applied loads are considered once as random and another as deterministic. In Fig. (2) shown is the effect of varying the correlation length of E_2 on the resulting coefficients of variation of the moment M and the vertical deflection V at the midpoint of the beam. Curves indicated by M1 and V1 represent the relation between the correlation length of E_2 and the coefficient of variation of M and V, respectively, when the loads are considered as random, whereas those indicated by M2 and V2 are the corresponding curves when the structure is subjected to deterministic loads. It is seen that the effect is significant for the values of the correlation length less than the beam length(i. e; less than 10 m). Furthermore, the effect of varying the correlation length is more significant when the loading is considered as deterministic than that when the loading is treated as random. The reason for this is the additional factors affecting the stochastic response when loads are considered as random. It can also be seen that the effect of varying the correlation length is greater on the moment than that on the deflection.

4 CONCLUSIONS

A stochastic finite element approach for analysing structures with random characteristics and subjected to random loads within the framework of Neumann matrix expansion is presented. The random fields are discretized using Karhunen-loeve expansion which is, to the authors knowledge, the most efficient way for discretizing random fields. The use of Neumann matrix expansion with Karhunen-loeve expansion for analysing structures with random media and loading increases the accuracy especially on using more higher-order terms in Neumann expansion. Although the formulation is presented for framed structures, the extension to two dimensional structures is straightforward.

REFERENCES

Deodattis, G. 1990. Weighted integral method. I: Stochastic stiffness matrix. *J. Engrg. Mech.*, ASCE, 117(8): 1851-1864.

Deodatis, G. and Shinozuka, M. 1990. Weighted integral method. II: Response variability and reliability. *J. Engrg. Mech.* , ASCE, 117(8): 1865-1877.

Der Kiureghian, A. and Ke, J.-B. 1988. The stochastic finite element method in structural reliability. *Probabilistic Engrg. Mech.*, 3(2): 83-91.

Der Kiureghian, A. and Liu, P.-L. 1991. First and second-order finite element reliability methods. in: W. K. Liu and T. Belytschko (Eds.), *Computational Mechanics of Probabilistic and reliability Analysis*: 282-298. Elmepress International, Switzerland.

Handa, K. and Anderson, K. 1981. Application of finite element method in the statistical analysis of structures. *Proc. 3rd ICOSSAR*: 409-417. Trondheim: Norway.

Hisada, T. and Nakagiri, S. 1981. Stochastic finite element method developed for structural safety and reliability. *Proc. 3rd ICOSSAR*: 395-408. Trondheim: Norway.

Lawrence, M. A. 1987. Basis random variables in finite element analysis. *Int. J. Numer. Methods in Engrg.*, 24: 1849-1863.

Li, C.-C. and Der Kiureghian, A. 1992.Optimal discrerization of random fields for SFEM. *Proc. 6th Conf. on Probabilistic Mechanics and Structural and Geotechnical Reliability*:29-32. Denver:USA.

Liu, W. K., Belytschko, T. & Mani, A. 1986. Random field finite elements. *Int. J. Numer. Methods in Engrg.*, 23(10): 1831-1845.

Shinozuka, M. and Deodatis, G. 1988. Response variability of stochastic finite element systems. *J. Engrg. Mech.*, ASCE, 114(3): 499-519.

Spanos, P. D. and Ghanem, R. 1989. Stochastic finite element expansion for random media. *J. Engrg. Mech.*, ASCE, 115(5): 1035-1053.

Spanos, P. D. and Ghanem, R. 1991. *Stochastic finite elements: A spectral approach,* Springer-Verlag: New York.

Takada, T. 1990. Weighted integral method in stochastic finite element analysis. *Probabilistic Engrg. Mech.*, 5(3): 146-156.

Vanmarcke, E. H. and Grigoriu, M. 1983. Stochastic finite element analysis of simple beams. *J. Engrg. Mech.*, ASCE, 109(5):1203-1214.

Yamazaki, F., Shinozuka, M. & Dasgupta, G. 1988. Neumann expansion for stochastic finite element analysis. *J. Engrg. Mech.*, ASCE, 114(8): 1335-1354.

Traina, M. -I, Miller, R. K. & Masri, S. F. 1986. Orthogonal decomposition and transmission of nonstationary random process. *Probabilistic Engrg. Mech.* 1(3):136-149.

Structural Safety & Reliability, Schuëller, Shinozuka & Yao (eds) © 1994 Balkema, Rotterdam, ISBN 90 5410 357 4

Nonlinear SFEM-based reliability for space structures

Liwei Gao & Achintya Haldar
Department of Civil Engineering & Engineering Mechanics, University of Arizona, Ariz., USA

ABSTRACT: An efficient stochastic finite element-based procedure is proposed for the reliability analysis of three-dimensional structures with geometrical nonlinearity. The method is based on the Advanced First Order Second Moment reliability analysis procedure. The assumed stress field approach is used in the finite element formulation to compute nonlinear structural response and the corresponding response gradients. The material properties, geometry and external loads are considered to be random variables. The failure criteria of the structure are expressed in terms of the ultimate and serviceability limit state functions. The proposed method avoids dealing with the complicated failure mechanisms and stable configuration approaches, and directly provides the information on the structural reliability.

1. INTRODUCTION

The use of the finite element method to evaluate the reliability of complicated structures where the limit state functions are not available in close form has become very common. This led to the development of the stochastic finite element method (SFEM). However, most of these studies were limited to two-dimensional linear structures, and thus are limited to some extent for practical applications. Since the SFEM is based on the deterministic finite element method (FEM), the efficiency of the SFEM will be greatly improved when state of the art FEM is used. Since a nonlinear three-dimensional (3-D) structure needs to be analyzed repeatedly in a stochastic finite element format, the efficiency of the deterministic FEM is particularly important for the stochastic model. It has been established in the literature that the assumed stress-based FEM is very efficient in solving nonlinear problems, since the tangent stiffness can be expressed in explicit form and needs fewer elements to model a structure (Kondoh and Atluri 1987; Haldar and Nee 1988; Shi and Atluri 1988). These desirable features of the assumed stress method have been utilized in the proposed SFEM to estimate the reliability of nonlinear structures.

2. FEM OF 3-D STRUCTURES

Details of the assumed stress-based nonlinear FEM formulation can not be described here due to lack of space, but can be found in the literature (Haldar and Nee 1988; Kondoh and Atluri 1987; Shi and Atluri 1988). Only the essential steps required for the SFEM formulation are discussed very briefly below.

The commonly used linear iterative strategy for solving nonlinear structural problems can be expressed as:

$$\boldsymbol{K}^{(n)}\Delta\boldsymbol{D}^{(n)} = \boldsymbol{F}^{(n)} - \boldsymbol{R}^{(n-1)} \tag{1}$$

where $\boldsymbol{K}^{(n)}, \Delta\boldsymbol{D}^{(n)}$, and $\boldsymbol{F}^{(n)}$ are the tangent stiffness matrix, displacement vector, and external load vector at the nth iteration, respectively; and $\boldsymbol{R}^{(n-1)}$ is the internal force vector at the (n-1)th iteration. The tangent stiffness matrix of a beam-column element using the assumed stress method can be expressed as:

$$\boldsymbol{K} = \boldsymbol{A}^{t}_{\sigma do}\boldsymbol{A}^{-1}_{\sigma\sigma}\boldsymbol{A}_{\sigma do} + \boldsymbol{A}_{ddo} \tag{2}$$

where $A_{\sigma\sigma}^{-1}$ is the elastic property matrix; $A_{\sigma do}^{t}$ is the transformation matrix; and A_{ddo} is the geometric stiffness matrix. The internal force vector can be expressed as:

$$R = -A_{\sigma do}^{t} A_{\sigma\sigma}^{-1} R_{\sigma} + R_{do} \qquad (3)$$

Since K and R can be expressed in explicit form, geometric nonlinear problems can be solved very efficiently using Eq. 1.

3. EVALUATION OF GRADIENTS

The Advanced First-Order Second-Moment method (AFOSM) is used in this paper to estimate the reliability of structures. In this context, G(**x**,**u**,**s**) is defined as the limit state function, where vector **x** denotes the set of basic random variables pertaining to a structure (e.g. loads, material properties and structural geometry), the vector **u** denotes the set of displacements involved in the limit state function and the vector **s** denotes the set of load effects (except the displacement) involved in the limit state function (e.g. stresses, internal forces). The displacement **u** can be expressed as $u = QD$, where **D** is the global displacement vector and **Q** is a transformation matrix. In general, **x**, **u** and **s** are related through a transformation which is available in an algorithmic sense, e.g., a finite element code. For a specified failure criterion, the limit state function G(**x**,**u**,**s**)>0 defines the safe state, G(**x**,**u**,**s**) < 0 defines the failure state, and G(**x**,**u**,**s**) = 0 denotes the limit state surface.

For reliability computation, it is convenient to transform the variables **x** into the standard normal space **Y**= *Y*(**x**) such that the elements of **Y** are statistically independent and have a standard normal distribution. The limit state surface in the standard normal space is replaced with the tangent plane at the point with minimum distance from the origin, denoted the design point, and the reliability index $\beta = \sqrt{y^t y}$, is the minimum distance from the origin. To find the minimum distance point(s), **y** can be formulated as a constrained optimization problem. For nonlinear problems, this constrained optimization problem does not have a closed-form solution, and the following iteration scheme can be followed to obtain the design point:

$$y_{i+1} = \left(y_i^t a_i + \frac{G(y_i)}{|\nabla G(y_i)|} \right) a_i \qquad (4)$$

where

$$\nabla G(y) = \left[\frac{\partial G(y)}{\partial y_1}, \ldots, \frac{\partial G(y)}{\partial y_n} \right]^t$$

and a_i is the unit vector normal to the performance surface having direction away from the origin. It can be estimated as:

$$a_i = -\frac{\nabla G(y_i)}{|\nabla G(y_i)|} \qquad (5)$$

The implementation of the AFOSM in the reliability estimation requires the evaluation of the gradient of the limit state function $\nabla G(y_i)$. By the chain rule of differentiation, this gradient can be shown to be:

$$\nabla G = \left[\frac{\partial G}{\partial s} J_{s,x} + \left(Q \frac{\partial G}{\partial u} + \frac{\partial G}{\partial s} J_{s,D} \right) J_{D,x} + \frac{\partial G}{\partial x} \right] J_{y,x}^{-1} \qquad (6)$$

where $J_{i,j}$ are the Jacobians of transformation and all other variables are described earlier. The three gradients on the right-hand side of Eq. 6 are easy to compute since G(**x**,**u**,**s**) is an explicit function of **x**, **u** and **s**. Because of the triangular nature of the transformation, $J_{y,x}$ and its inverse are also easy to compute. If **s** is not an explicit function of the basic random variables **x**, then $J_{s.x} = 0$. The Jacobian of the transformation $J_{s,D}$ and $J_{D,x}$, however, is not easy to compute since **s**, **D** and **x** are implicit functions of each other. The adjoin method (Ryu et al 1985) is used here to compute the product of the second term in Eq. 6 directly rather than evaluating its constituent parts. This method is shown to be very accurate and efficient.

With the adjoin method, an auxiliary vector λ can be introduced such that

$$\lambda^t K = Q \frac{\partial G}{\partial u} + \frac{\partial G}{\partial s} J_{s,D} \qquad (7)$$

It can be shown that:

$$\lambda^t K J_{D,x} = \left(Q\frac{\partial G}{\partial u} + \frac{\partial G}{\partial s} J_{s,D} \right) J_{D,x} = \lambda^t \left(\frac{\partial F}{\partial x} - \frac{\partial R}{\partial x} \right) \tag{8}$$

where $\partial F/\partial x$ is easy to obtain if the external load is not affected by the structural response, and $\partial \boldsymbol{R}/\partial x$ can be derived from Eq. 3. Substituting Eq. 8 into Eq. 6, ∇G can be obtained.

The practical implementation of the SFEM-based reliability analysis needs to be done in a computer environment. The amount of memory required in the computer grows rapidly with the number of basic random variables. Sensitivity analysis (Der Kiureghian and Ke 1988; Mahadevan and Haldar 1989) can be used to reduce the number of basic random variables. Among the structural parameters, only the Young's modulus E, area A, Poisson ratio μ, yield stress F_y and the moments of inertia of a cross section J, I_2 and I_3, along with the external load **F**, are considered as basic random variables in this study. Then, one can rewrite Eq. 3 as

$$\boldsymbol{R} = -\boldsymbol{A}^t_{\sigma do} \boldsymbol{R}_{A\sigma} + \boldsymbol{R}_{do} \tag{9}$$

where

$$\boldsymbol{R}_{A\sigma} = \begin{Bmatrix} n - \dfrac{EAH}{l} \\ m_1 - \dfrac{GJ}{l}\left({}^2\theta_1^* - {}^1\theta_1^*\right) \\ {}^1m_2 + \dfrac{2EI_2}{l}\left(2\,{}^1\theta_2^* + {}^2\theta_2^*\right) \\ {}^2m_2 - \dfrac{2EI_2}{l}\left({}^1\theta_2^* + 2\,{}^2\theta_2^*\right) \\ {}^1m_3 + \dfrac{2EI_3}{l}\left(2\,{}^1\theta_3^* + {}^2\theta_3^*\right) \\ {}^2m_3 - \dfrac{2EI_3}{l}\left({}^1\theta_3^* + 2\,{}^2\theta_3^*\right) \end{Bmatrix} \tag{10}$$

Since $\boldsymbol{R}_{do}$ and $\boldsymbol{A}^t_{\sigma do}$ are not functions of basic variables, the derivative of $\boldsymbol{R}$ with respect to $\boldsymbol{x}$ can be expressed as

$$\frac{\partial \boldsymbol{R}}{\partial \boldsymbol{x}}\bigg|_{D,\sigma} = -\boldsymbol{A}^t_{\sigma do} \frac{\partial \boldsymbol{R}_{A\sigma}}{\partial \boldsymbol{x}} \tag{11}$$

where

$$\frac{\partial \boldsymbol{R}_{A\sigma}}{\partial \boldsymbol{x}} = \begin{bmatrix} \dfrac{\partial \boldsymbol{R}_{A\sigma}}{\partial E} & \dfrac{\partial \boldsymbol{R}_{A\sigma}}{\partial A} & \dfrac{\partial \boldsymbol{R}_{A\sigma}}{\partial \mu} & \dfrac{\partial \boldsymbol{R}_{A\sigma}}{\partial J} & \dfrac{\partial \boldsymbol{R}_{A\sigma}}{\partial I_2} & \dfrac{\partial \boldsymbol{R}_{A\sigma}}{\partial I_3} & \boldsymbol{0} \end{bmatrix} \tag{12}$$

and

$$\frac{\partial \boldsymbol{R}_{A\sigma}}{\partial E} = \begin{Bmatrix} -\dfrac{AH}{l} \\ -\dfrac{J}{2(1+\mu)l}\left({}^2\theta_1^* - {}^1\theta_1^*\right) \\ \dfrac{2I_2}{l}\left(2\,{}^1\theta_2^* + {}^2\theta_2^*\right) \\ -\dfrac{2I_2}{l}\left({}^1\theta_2^* + 2\,{}^2\theta_2^*\right) \\ \dfrac{2I_3}{l}\left(2\,{}^1\theta_3^* + {}^2\theta_3^*\right) \\ -\dfrac{2I_3}{l}\left({}^1\theta_3^* + 2\,{}^2\theta_3^*\right) \end{Bmatrix} \tag{13}$$

$$\frac{\partial \boldsymbol{R}_{A\sigma}}{\partial A} = \begin{Bmatrix} -\dfrac{EH}{l} \\ 0 \\ 0 \\ 0 \\ 0 \\ 0 \end{Bmatrix} \tag{14}$$

$$\frac{\partial \boldsymbol{R}_{A\sigma}}{\partial \mu} = \begin{Bmatrix} 0 \\ \dfrac{-EJ}{2(1+\mu)^2 l}\left({}^2\theta_1^* - {}^1\theta_1^*\right) \\ 0 \\ 0 \\ 0 \\ 0 \end{Bmatrix} \tag{15}$$

$$\frac{\partial R_{A\sigma}}{\partial J} = \begin{Bmatrix} 0 \\ -\frac{G}{l}\left({}^2\dot\theta_1 - {}^1\dot\theta_1\right) \\ 0 \\ 0 \\ 0 \\ 0 \end{Bmatrix} \quad (16)$$

$$\frac{\partial R_{A\sigma}}{\partial I_2} = \begin{Bmatrix} 0 \\ 0 \\ \frac{2E}{l}\left(2\,{}^1\dot\theta_2 + {}^2\dot\theta_2\right) \\ -\frac{2E}{l}\left({}^1\dot\theta_2 + 2\,{}^2\dot\theta_2\right) \\ 0 \\ 0 \end{Bmatrix} \quad (17)$$

$$\frac{\partial R_{A\sigma}}{\partial I_3} = \begin{Bmatrix} 0 \\ 0 \\ 0 \\ 0 \\ \frac{2E}{l}\left(2\,{}^1\dot\theta_3 + {}^2\dot\theta_3\right) \\ -\frac{2E}{l}\left({}^1\dot\theta_3 + 2\,{}^2\dot\theta_3\right) \end{Bmatrix} \quad (18)$$

Normally, when the strength limit state functions are under consideration, only the internal force vector is involved in s as s = $A\sigma$, where **A** is a constant matrix; and σ is the internal force vector and can be represented as:

$$\sigma = \left[\, n \quad m_1 \quad {}^1m_2 \quad {}^2m_2 \quad {}^1m_3 \quad {}^2m_3 \,\right]^t \quad (19)$$

Thus, one can obtain

$$J_{s,D} = \frac{\partial s}{\partial D} = A\frac{\partial \sigma}{\partial D} = A\left[\frac{\partial \sigma}{\partial d} \quad 0\right] \quad (20)$$

where **d** is the nodal displacement vector in the global coordinate for the element and can be defined as:

$$d = \left[{}^1u_1\ {}^1u_1\ {}^1u_3\ {}^1\theta_1\ {}^1\theta_2\ {}^1\theta_3\ {}^2u_1\ {}^2u_2\ {}^2u_3\ {}^2\theta_1\ {}^2\theta_2\ {}^2\theta_3\right] \quad (21)$$

The relationship between the increment of internal nodal force $\Delta\sigma$ and the increment of global displacement ΔD can be expressed as

$$\Delta\sigma = A_{\sigma\sigma}^{-1}\left(A_{\sigma do}\Delta D - R_\sigma\right) \quad (22)$$

Eq. 22 needs to be solved iteratively. Once the algorithm converges, **D** and $\Delta\sigma$ will be zero. Eq. 22 becomes $R_\sigma = 0$. Then the relationship between σ and **d** can be shown to be:

$$\sigma = \begin{Bmatrix} \frac{EAH}{l} \\ \frac{GJ}{l}\left({}^2\dot\theta_1 - {}^1\dot\theta_1\right) \\ -\frac{2EI_2}{l}\left(2\,{}^1\dot\theta_2 + {}^2\dot\theta_2\right) \\ \frac{2EI_2}{l}\left({}^1\dot\theta_2 + 2\,{}^2\dot\theta_2\right) \\ -\frac{2EI_3}{l}\left(2\,{}^1\dot\theta_3 + {}^2\dot\theta_3\right) \\ \frac{2EI_3}{l}\left({}^1\dot\theta_3 + 2\,{}^2\dot\theta_3\right) \end{Bmatrix} \quad (23)$$

The derivative of the internal forces with respect to the displacements becomes:

$$\left[\frac{\partial \sigma}{\partial d}\right]^t = \left[\frac{\partial n}{\partial d} \quad \frac{\partial m_1}{\partial d} \quad \frac{\partial\, {}^1m_2}{\partial d} \quad \frac{\partial\, {}^2m_2}{\partial d} \quad \frac{\partial\, {}^1m_3}{\partial d} \quad \frac{\partial\, {}^2m_3}{\partial d}\right] \quad (24)$$

where

$$\frac{\partial n}{\partial d} = \frac{AE}{rl}\begin{Bmatrix} -(DX_1 + \tilde u_1) \\ -(DX_2 + \tilde u_2) \\ -(DX_3 + \tilde u_3) \\ 0 \\ 0 \\ 0 \\ (DX_1 + \tilde u_1) \\ (DX_2 + \tilde u_2) \\ (DX_3 + \tilde u_3) \\ 0 \\ 0 \\ 0 \end{Bmatrix} \quad (25)$$

$$\frac{\partial m_1}{\partial d} = \frac{GJ}{l} \left\{ \begin{matrix} 0 \\ 0 \\ 0 \\ -(T_0)_{11} \\ -(T_0)_{12} \\ -(T_0)_{13} \\ 0 \\ 0 \\ 0 \\ (T_0)_{11} \\ (T_0)_{12} \\ (T_0)_{13} \end{matrix} \right\} \tag{26}$$

$$\frac{\partial^1 m_2}{\partial d} = \frac{-2\,EI_2}{l} \left\{ \begin{matrix} 3\dfrac{(T_0)_{11}\lambda_1 - (T_0)_{31}\lambda_2}{\lambda_1^2 + \lambda_2^2} \\ 3\dfrac{(T_0)_{12}\lambda_1 - (T_0)_{32}\lambda_2}{\lambda_1^2 + \lambda_2^2} \\ 3\dfrac{(T_0)_{13}\lambda_1 - (T_0)_{33}\lambda_2}{\lambda_1^2 + \lambda_2^2} \\ 2(T_0)_{21} \\ 2(T_0)_{22} \\ 2(T_0)_{23} \\ -3\dfrac{(T_0)_{11}\lambda_1 - (T_0)_{31}\lambda_2}{\lambda_1^2 + \lambda_2^2} \\ -3\dfrac{(T_0)_{12}\lambda_1 - (T_0)_{32}\lambda_2}{\lambda_1^2 + \lambda_2^2} \\ -3\dfrac{(T_0)_{13}\lambda_1 - (T_0)_{33}\lambda_2}{\lambda_1^2 + \lambda_2^2} \\ (T_0)_{21} \\ (T_0)_{22} \\ (T_0)_{23} \end{matrix} \right\} \tag{27}$$

$$\frac{\partial^2 m_2}{\partial d} = \frac{2\,EI_2}{l} \left\{ \begin{matrix} 3\dfrac{(T_0)_{11}\lambda_1 - (T_0)_{31}\lambda_2}{\lambda_1^2 + \lambda_2^2} \\ 3\dfrac{(T_0)_{12}\lambda_1 - (T_0)_{32}\lambda_2}{\lambda_1^2 + \lambda_2^2} \\ 3\dfrac{(T_0)_{13}\lambda_1 - (T_0)_{33}\lambda_2}{\lambda_1^2 + \lambda_2^2} \\ (T_0)_{21} \\ (T_0)_{22} \\ (T_0)_{23} \\ -3\dfrac{(T_0)_{11}\lambda_1 - (T_0)_{31}\lambda_2}{\lambda_1^2 + \lambda_2^2} \\ -3\dfrac{(T_0)_{12}\lambda_1 - (T_0)_{32}\lambda_2}{\lambda_1^2 + \lambda_2^2} \\ -3\dfrac{(T_0)_{13}\lambda_1 - (T_0)_{33}\lambda_2}{\lambda_1^2 + \lambda_2^2} \\ 2(T_0)_{21} \\ 2(T_0)_{22} \\ 2(T_0)_{23} \end{matrix} \right\} \tag{28}$$

$$\frac{\partial^1 m_3}{\partial d} = \frac{-2\,EI_3}{l} \left\{ \begin{matrix} 3\dfrac{(T_0)_{21}\lambda_2 - (T_0)_{11}\lambda_3}{\lambda_3^2 + \lambda_2^2} \\ 3\dfrac{(T_0)_{22}\lambda_2 - (T_0)_{12}\lambda_3}{\lambda_3^2 + \lambda_2^2} \\ 3\dfrac{(T_0)_{23}\lambda_2 - (T_0)_{13}\lambda_3}{\lambda_3^2 + \lambda_2^2} \\ 2(T_0)_{31} \\ 2(T_0)_{32} \\ 2(T_0)_{33} \\ -3\dfrac{(T_0)_{21}\lambda_2 - (T_0)_{11}\lambda_3}{\lambda_3^2 + \lambda_2^2} \\ -3\dfrac{(T_0)_{22}\lambda_2 - (T_0)_{12}\lambda_3}{\lambda_3^2 + \lambda_2^2} \\ -3\dfrac{(T_0)_{23}\lambda_2 - (T_0)_{13}\lambda_3}{\lambda_3^2 + \lambda_2^2} \\ (T_0)_{31} \\ (T_0)_{32} \\ (T_0)_{33} \end{matrix} \right\} \tag{29}$$

$$\frac{\partial^2 m_3}{\partial d} = \frac{2EI_3}{l}\begin{Bmatrix} 3\frac{(T_0)_{21}\lambda_2 - (T_0)_{11}\lambda_3}{\lambda_3^2+\lambda_2^2} \\ 3\frac{(T_0)_{22}\lambda_2 - (T_0)_{12}\lambda_3}{\lambda_3^2+\lambda_2^2} \\ 3\frac{(T_0)_{23}\lambda_2 - (T_0)_{13}\lambda_3}{\lambda_3^2+\lambda_2^2} \\ (T_0)_{31} \\ (T_0)_{32} \\ (T_0)_{33} \\ -3\frac{(T_0)_{21}\lambda_2 - (T_0)_{11}\lambda_3}{\lambda_3^2+\lambda_2^2} \\ -3\frac{(T_0)_{22}\lambda_2 - (T_0)_{12}\lambda_3}{\lambda_3^2+\lambda_2^2} \\ -3\frac{(T_0)_{23}\lambda_2 - (T_0)_{13}\lambda_3}{\lambda_3^2+\lambda_2^2} \\ 2(T_0)_{31} \\ 2(T_0)_{32} \\ 2(T_0)_{33} \end{Bmatrix} \qquad (30)$$

where,

$$l = \sqrt{(DX_1)^2 + (DX_2)^2 + (DX_3)^2}$$

$$r = \sqrt{(DX_1+\tilde{u}_1)^2 + (DX_2+\tilde{u}_2)^2 + (DX_3+\tilde{u}_3)^2}$$

$$H = r - l$$

$$\lambda_1 = (T_0)_{31}\tilde{u}_1 + (T_0)_{32}\tilde{u}_2 + (T_0)_{33}\tilde{u}_3$$
$$\lambda_2 = (T_0)_{11}\tilde{u}_1 + (T_0)_{12}\tilde{u}_2 + (T_0)_{13}\tilde{u}_3 + l$$
$$\lambda_3 = (T_0)_{21}\tilde{u}_1 + (T_0)_{22}\tilde{u}_2$$

$$\tilde{u}_i = {}^2u_i - {}^2u_i \qquad (i = 1,2,3)$$

$${}^\alpha\theta_1^* = (T_0)_{1i}\,{}^\alpha\theta_i$$
$${}^\alpha\theta_2^* = (T_0)_{2i}\,{}^\alpha\theta_i + \tan^{-1}\frac{\tilde{v}_3}{l+\tilde{v}_1}$$
$${}^\alpha\theta_3^* = (T_0)_{3i}\,{}^\alpha\theta_i - \tan^{-1}\frac{\tilde{v}_2}{l+\tilde{v}_1} \qquad (\alpha = 1,2)$$
$$\tilde{v}_i = (T_0)_{ij}\,\tilde{u}_j$$

By now, all quantities required for the computation of $\nabla G(\mathbf{y})$ are available in very simple explicit forms. Eq. 4 can be used to compute the design point, and the corresponding reliability index can be obtained. To improve the stability of the nonlinear finite element analysis, the Newton-Raphson method with arc-length procedure is used in this study.

4. NUMERICAL EXAMPLE

To illustrate the applicability of the proposed method, a three-dimensional frame is considered. The probabilistic descriptions of all basic variables are listed in Table 1. The geometric dimensions of the frame are shown in the figure. Both the strength (the combination of axial compression and biaxial bending) and serviceability (vertical deflection under load V) limit state functions are considered in this example. For the strength limit state, the LRFD (*Load and Resistance Factor Design*) has to satisfy the following two interaction equation:

$$\frac{P_u}{\phi_c P_n} \geq 0.2$$
$$\frac{P_u}{\phi_c P_n} + \frac{8}{9}\left(\frac{M_{ux}}{\phi_b M_{nx}} + \frac{M_{uy}}{\phi_b M_{ny}}\right) \leq 1.0 \qquad (31)$$

$$\frac{P_u}{\phi_c P_n} < 0.2$$
$$\frac{P_u}{2\phi_c P_n} + \left(\frac{M_{ux}}{\phi_b M_{nx}} + \frac{M_{uy}}{\phi_b M_{ny}}\right) \leq 1.0 \qquad (32)$$

The serviceability limit state can be defined as:

$$v/v_{limit} \leq 1.0 \qquad (33)$$

where v_{limit} is assumed to be 1 inch in this study. Adequate lateral bracing is assumed to ensure that the flexural strength of the member is equal to its plastic moment capacity. The results show that the safety of member AB under combined axial compression and bi-axial bending controls the failure of the structure. The results for linear and nonlinear analyses of the space frame are given in Table 2 and 3, respectively. Monte Carlo simulation is also used in this study to verify the proposed algorithm. By 4000 simulations, the reliability indexes for the linear and nonlinear cases are found to be 2.12 and 2.09, respectively.

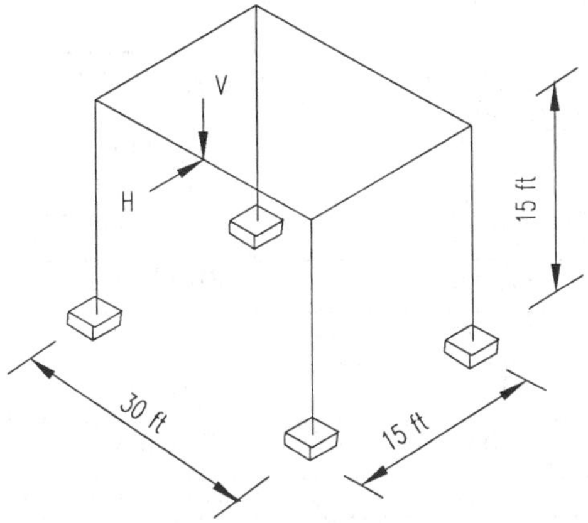

3-D Frame Structure in Example

5. CONCLUSIONS

An efficient SFEM is proposed for the reliability analysis of 3-D structures with geometric and material nonlinearities using the AFOSM method. The assumed stress field approach is used in computing nonlinear structural responses and the corresponding response gradients. The comparison of the reliability indexes among the linear structure and nonlinear structure are stressed in the numerical example. For stability-related structures, the nonlinearity has a considerable influence on the structural reliability. The proposed method is very efficient and robust in evaluating risk of nonlinear structures.

ACKNOWLEDGMENTS

This paper is based upon work partly supported by the National Science Foundation under Grants No. MSM-8896267. Financial support received from the American Institute of Steel Construction (AISC), Chicago, is appreciated. Any opinions, findings, and conclusions, or recommendations expressed in this publication are those of the writers and do not necessarily reflect the views of the sponsors.

REFERENCES

Der Kiureghian, A., and Ke, J-B.(1988) "The stochastic finite element method in structural reliability," *Probabilistic Engrg. Mech.* **3**(2), 83--91.

Haldar, A. and Nee, K.M.(1988) "Elasto-plastic large deformation analysis of PR steel frames for LRFD," *Comput. Struct.*, **34**(5), 811--823.

Kondoh, K. and Atluri, S.N. (1987) "Large deformation, elasto-plastic analysis of frames under non-conservative loading, using explicitly derived tangent stiffness based on assumed stress," *Comput. Mech.*, **2**(1), 1--25.

Mahadevan, S. and Haldar, A. (1989) "Efficient algorithm for stochastic structural optimization," *J. Struct. Engrg.*, ASCE, **115**(7), 1579--1598.

Ryu, Y.S., Harian, M., Wu, C.C. and Arora, J.S.(1985) "Structural design sensitivity analysis of nonlinear response," *Comput. Struct.*, **21**(1/2), 245--255.

Shi, G. and Atluri, S.N. (1988) "Elasto-plastic large deformation analysis of space--frames," *Int. J. for Numerical Methods in Engineering*, **26**, 589--615.

Table 1. Description of Basic Random Variables In Example

Random Variables	Mean	C.O.V.	Distribution
E (ksi)	29000	0.06	Lognormal
A (in^2)	19.7	0.05	Lognormal
J (in^4)	2.39	0.05	Lognormal
I_2 (in^4)	954	0.05	Lognormal
I_3 (in^4)	119	0.05	Lognormal
Z_x (in^3)	130	0.05	Lognormal
Z_y (in^3)	35.5	0.05	Lognormal
F_y (ksi)	37.8	0.10	Lognormal
V(kips)	35.0	0.15	Lognormal
H(kips)	7.0	0.15	Lognormal

Table 2. Reliability Analysis Results For Example (Linear) (β=2.284)

Random Variables	Initial Point	Final Point	Sensitivity Index
E (ksi)	29000	28946.67	-0.00032
A (in^2)	19.7	19.67	-0.00439
J (in^4)	2.39	2.387	-0.00008
I_2 (in^4)	954	952.73	-0.00071
I_3 (in^4)	119	118.85	-0.00024
Z_x (in^3)	130	126.39	-0.23573
Z_y (in^3)	35.5	34.80	-0.16332
F_y (ksi)	37.8	36.051	-0.40404
V(kips)	35.0	44.20	0.71780
H(kips)	7.0	8.18	0.48914

Table 3. Reliability Analysis Results For Example (Nonlinear) (β=2.225)

Random Variables	Initial Point	Final Point	Sensitivity Index
E (ksi)	29000	28947.67	-0.00008
A (in^2)	19.7	19.67	-0.00236
J (in^4)	2.39	2.39	-0.00000
I_2 (in^4)	954	952.80	-0.00007
I_3 (in^4)	119	118.85	-0.00000
Z_x (in^3)	130	128.23	-0.11220
Z_y (in^3)	35.5	35.15	-0.07796
F_y (ksi)	37.8	36.95	-0.19252
V(kips)	35.0	45.49	0.82313
H(kips)	7.0	8.22	0.51645

Structural Safety & Reliability, Schuëller, Shinozuka & Yao (eds) © *1994 Balkema, Rotterdam, ISBN 90 5410 357 4*

Stochastic Finite Element Analysis of soil layers with random interface

R.Ghanem
Civil Engineering Department, State University of New York at Buffalo, N.Y., USA

V. Brzakala
Institute of Geotechnics, Technical University, Wrocław, Poland

ABSTRACT: This paper addresses the problem of a medium with two layers separated by an interface randomly fluctuating in space. The medium is subjected to in-plane stresses simulating the effect of a surface foundation. The second moments characteristics of the interface are used to formulate the problem. The Karhunen-Loeve and the Polynomial Chaos expansions are utilized to transform the problem into a computationally tractable form, therefore resulting in a system of linear algebraic equations to solve. Due to the geometric nature of the randomness, however, the stiffness matrix is nonlinear in the randomness. This leads to a nonlinear problem. It is demonstrated how this nonlinearity can be accommodated using the proposed approach. Numerical examples are also presented.

1 Introduction

Recent developments in the Stochastic Finite Element Method have prompted a fresh look at various engineering problems featuring uncertainty in the properties of the system under consideration. These systems are common in all branches of engineering. In many situations of interest, some average value of these properties is determined through parameter estimation or system identification. These techniques, however, do not permit the full utilization of field data, as they usually involve only the first two statistical moments of the data. Furthermore, in most such approaches, the uncertain material property is assumed not to vary in space, and therefore implicitly assumed to be a random variable with no spatial fluctuations.

Clearly, incorporating the spatial fluctuations of measured data into an identification algorithm would greatly enhance the reliability of the resulting model. System identification of large scale engineering systems is by itself is daunting task. It involves choosing a model for the system that captures the important aspects of the behavior of the system. These aspects, of course, depend on the task the system is expected to perform. With the addition of a spatially varying stochastic component to the physical model, it becomes very important to clearly identify those variables of the stochastic model that can be expected to affect the behavior of the system. This in turn will decide on the class of models to which system identification is to be applied.

Stochastic Finite Elements provides a computational tool to combine the complexity of large engineering systems with uncertainty models for the material properties. Thus an environment is created whereby the sensitivity of the response of complex systems to various parameters of the uncertainty model can be assessed. This in turn can play a meaningful role in identifying those parameters the estimation of which can enhance the reliability of the mathematical model.

2 Geometrical Randomness

In most cases, available information about the medium in which a physical process takes place is incomplete. The importance of this uncertainty is emphasized in applications to geophysics and geotechnical engineering. In those cases, the margin of uncertainty is particularly wide. It refers to in-situ measurements which are often based on a small number of bore holes. The uncertainty is magnified by measurement errors, human errors coupled with fuzzy decision making and processing of the information.

The physical model to be treated in the paper involves a soil medium consisting of two layers, layer (1) at the bottom, and layer (2) at the top. The medium is subjected to a an applied load on a portion of the surface, resulting in a plane strain situation. Each of the two layers has well defined properties such as a modulus of elasticity $E^{(i)}$ and a Poisson Ratio $\nu^{(i)}$, where index i refers to a particular layer. The interface between the two layers, however, is assumed to be the realization of a stochastic process $H(\mathbf{x};\omega)$. Here ω refers to an element in the space of random events, and is used to index a random dimension; $\mathbf{x}$ denotes the position of the medium. This random process is assumed to have a mean value $\bar{H}(\mathbf{x})$ and covariance function $C^H(\mathbf{x},\mathbf{y})$ given by the following two equations

$$\bar{H}(\mathbf{x}) = E\{H(\mathbf{x};\omega)\} \tag{1}$$

and

$$C^H(\mathbf{x},\mathbf{y}) = E\{(H(\mathbf{x};\omega) - \bar{H}(\mathbf{x}))(H(\mathbf{y};\omega) - \bar{H}(\mathbf{y}))\} . \tag{2}$$

Clearly, in the two-dimensional case, $H(\mathbf{x};\omega)$ is a one-dimensional process with $\mathbf{x}$ spanning the horizontal extent of the domain. Since the

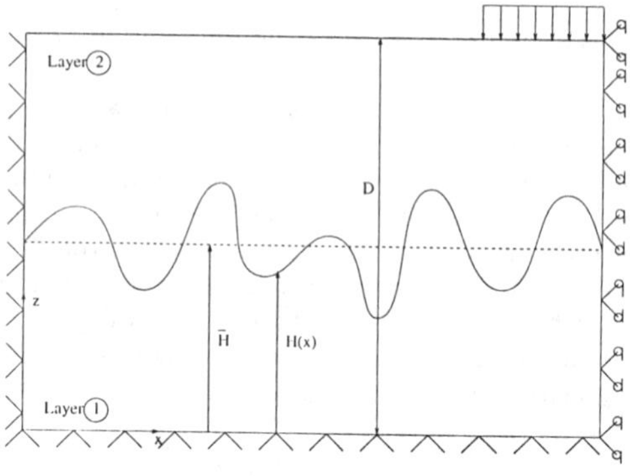

Figure 1: Two-Layer Soil with a Random Interface.

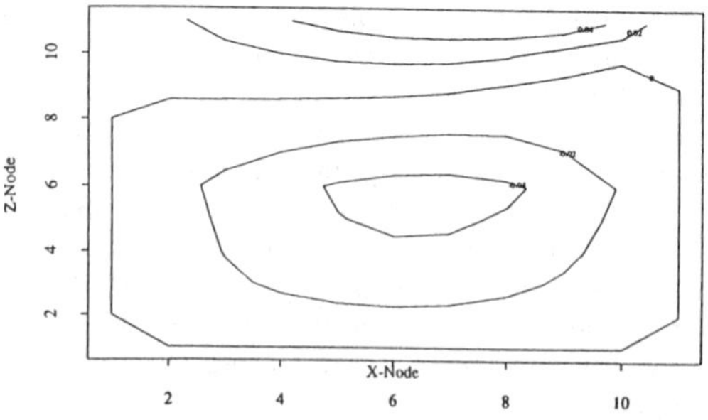

Figure 2: Contour Lines for the Mean of the Horizontal Displacement. Order of the Polynomial Chaos Expansion = 2; σ_H = 0.02; b = 0.1.

covariance function is a positive definite function, and based on a theorem by Mercer (1909), it can be represented using the following expansion

$$C^H(\mathbf{x},\mathbf{y}) = \sum_{n=1}^{\infty} \lambda_n^H \phi_n^H(\mathbf{x})\phi_n^H(\mathbf{y}) , \tag{3}$$

where $\phi_n^H(\mathbf{x})$ and λ_n^H denote the eigenfunctions and eigenvalues of the integral equation,

$$\int_{\mathcal{D}_1} C^H(\mathbf{x},\mathbf{y})\phi_n^H(\mathbf{x})d\mathbf{x} = \lambda_n^H \phi_n^H(\mathbf{y}) . \tag{4}$$

It can be shown that the above eigenfunctions are orthogonal and complete in the space L^2. They can therefore be used to expand the process $H(\mathbf{x};\omega)$. The resulting expansion is known as the Karhunen-Loeve expansion (Loeve, 1948; Spanos and Ghanem, 1989) and is given by,

$$\begin{aligned} H(\mathbf{x};\omega) &= \bar{H}(\mathbf{x}) + \Delta H(\mathbf{x};\omega) \\ &= \bar{H}(\mathbf{x}) + \sum_{n=1}^{\infty} \sqrt{\lambda_n^H}\xi_n^H(\omega)\phi_n^H(\mathbf{x}) . \end{aligned} \tag{5}$$

It can be shown that the above expansion is optimal in the mean-square sense, and that the random variables ξ_n^H form an orthogonal Gaussian vector of mean zero and covariance equal to the identity matrix.

3 Representation of Response Process

Since the covariance of the response process is not known apriori, the Karhunen-Loeve expansion cannot be used for its representation. Following Ghanem and Spanos (1990, 1991), the Polynomial Chaos expansion is used to represent the response process. This expansion is based on a theorem by Cameron and Martin (1947), and provides a means for expanding second order random variables in terms of orthogonal polynomials in Gaussian variables. Thus, for a general second order random variable $X(\omega)$, viewed as a function of ω, it can be represented in the form,

$$\begin{aligned} X(\omega) &= a_0\Gamma_0 \\ &+ \sum_{i_1=1}^{\infty} a_{i_1}\Gamma_1(\xi_{i_1}(\omega)) \\ &+ \sum_{i_1=1}^{\infty}\sum_{i_2=1}^{i_1} a_{i_1 i_2}\Gamma_2(\xi_{i_1}(\omega),\xi_{i_2}(\omega)) \\ &+ \sum_{i_1=1}^{\infty}\sum_{i_2=1}^{i_1}\sum_{i_3=1}^{i_2} a_{i_1 i_2 i_3}\Gamma_3(\xi_{i_1}(\omega),\xi_{i_2}(\omega),\xi_{i_3}(\omega)) + . \end{aligned} \tag{6}$$

In this equation, the symbol $\Gamma_n(\xi_{i_1},\ldots,\xi_{i_n})$ denotes the polynomial chaos of order n in the variables $(\xi_{i_1},\ldots,\xi_{i_n})$. The general expression for generating these polynomials is

$$\Gamma_n(\xi_{i_1},\ldots,\xi_{i_n}) = e^{\frac{1}{2}\boldsymbol{\xi}^T\boldsymbol{\xi}}(-1)^n \frac{\partial^n}{\partial\xi_{i_1}\ldots\xi_{i_n}} e^{-\frac{1}{2}\boldsymbol{\xi}^T\boldsymbol{\xi}}. \tag{7}$$

Equation (6) can be rewritten for notational convenience in the following form

$$X(\omega) = \sum_{j=0}^{\infty} \hat{a}_j \Psi_j(\{\xi_k(\omega)\}_{k=1}^{\infty}), \tag{8}$$

where there is a one-to-one correspondence between the functions $\Gamma_n(\xi_{i_1},\ldots,\xi_{i_n})$ and $\Psi(\{\xi_k(\omega)\}_{k=1}^{\infty})$, as well as between the coefficients in both expansions. Noting that the polynomial chaoses form an orthogonal basis in the L_2 space of random variables, the coefficients $\hat{a}_j$ in this last expansion can be explicitly obtained in the form,

$$\hat{a}_j = \frac{<X,\Psi_j>}{<\Psi_j,\Psi_j>}, \qquad j = 0,1,\ldots, \tag{9}$$

where $<.,.>$ denotes inner product in the Hilbert space of random variables.

4 Finite Element Formulation

Following Akin (1982), the global stiffness matrix for this problem can be assembled from elemental stiffness matrices, each one of them defined over a finite element. The general form of these elemental matrices is given by the following equation,

$$\begin{aligned}\mathbf{K}^e &\equiv \mathbf{K}^e[E,\nu] \\ &= \int_{D^e} E(\mathbf{x};\omega)\mathbf{B}^{eT}(\mathbf{x})\mathbf{P}^e(\mathbf{x};\omega)\mathbf{B}^e(\mathbf{x})d\mathbf{x}, \end{aligned} \tag{10}$$

where

$$\mathbf{P}^e = \frac{1}{(1+\nu)(1-2\nu)}\begin{bmatrix} 1-\nu & \nu & 0 \\ \nu & 1-\nu & 0 \\ 0 & 0 & \dfrac{1-2\nu}{2}\end{bmatrix}, \tag{11}$$

and

$$\mathbf{B}^e(\mathbf{x}) = \begin{bmatrix} \dfrac{\partial}{\partial x} & 0 \\ 0 & \dfrac{\partial}{\partial z} \\ \dfrac{\partial}{\partial z} & \dfrac{\partial}{\partial x}\end{bmatrix} \mathbf{H}^e(\mathbf{x}). \tag{12}$$

Here $\mathbf{H}^e(\mathbf{x})$ denotes a matrix of the finite element interpolation functions; bilinear interpolation is used in the following examples. Furthermore, the argument ω indicates that since the interface between the two layers is a spatially random process, the modulus of elasticity and Poisson's ration will themselves be random processes. Furthermore, these two random processes are binary since they can occupy one of two possible values. Denoting by $\mathcal{I}[.]$ the indicator function that equals one if its argument is positive, and zero otherwise, equation (10) can be rewritten as

$$\begin{aligned}\mathbf{K}^e &= \mathbf{K}^e\left[E^{(1)}(\mathbf{x})\mathcal{I}[H(\mathbf{x};\omega)-z];\nu^{(1)}\right] \\ &+ \mathbf{K}^e\left[E^{(2)}(\mathbf{x})\mathcal{I}[z-H(\mathbf{x};\omega)];\nu^{(2)}\right]. \end{aligned} \tag{13}$$

where superscripts (1) and (2) over E and ν refer to the particular layer of interest, and the variable z denotes the vertical position of a point in element e. Obviously, the elemental stiffness matrix is a random matrix and it depends in a nonlinear manner on the Gaussian process describing the interface

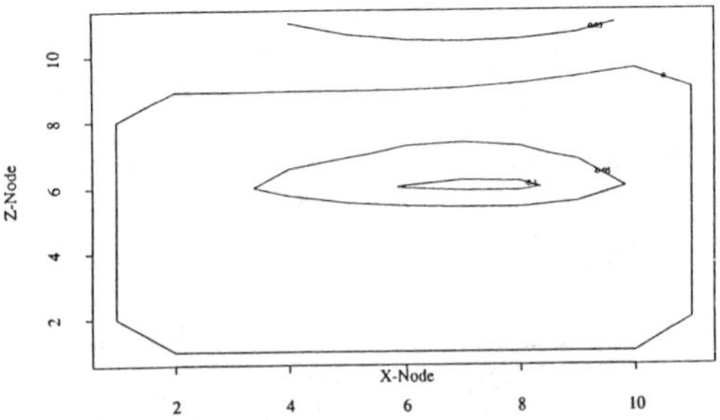

Figure 3: Contour Lines for the Mean of the Horizontal Displacement. Order of the Polynomial Chaos Expansion = 2; σ_H = 0.16; b = 0.1.

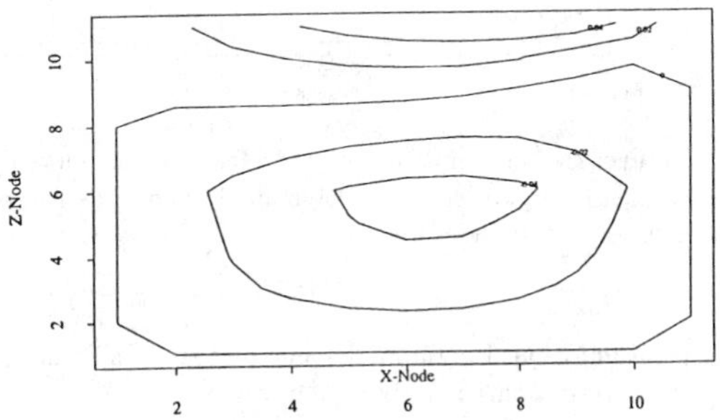

Figure 4: Contour Lines for the Mean of the Horizontal Displacement. Order of the Polynomial Chaos Expansion = 2; σ_H = 0.02; b = 1.

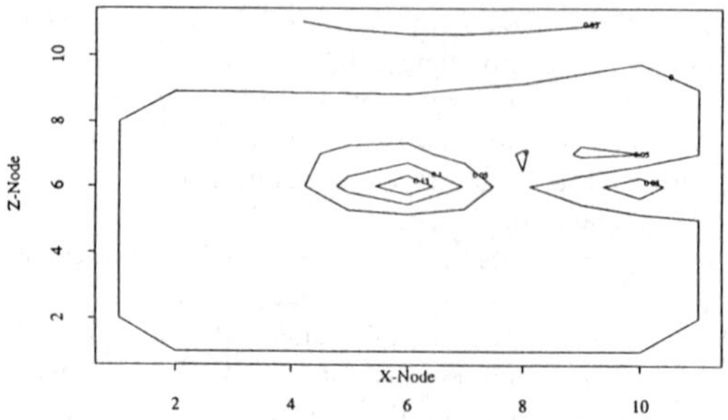

Figure 5: Contour Lines for the Mean of the Horizontal Displacement. Order of the Polynomial Chaos Expansion = 2; σ_H = 0.16; b = 1.

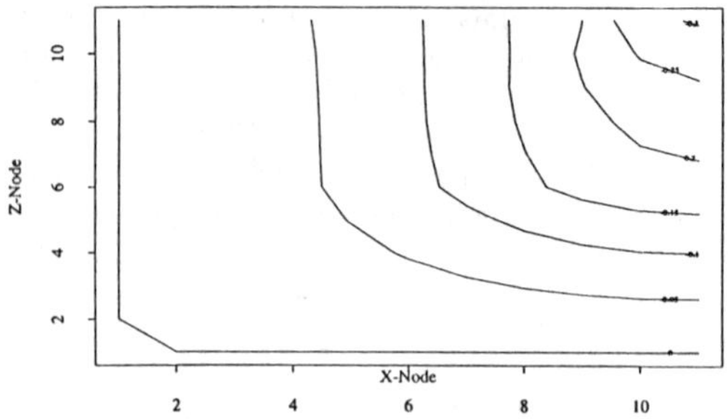

Figure 6: Contour Lines for the Mean of the Vertical Displacement. Order of the Polynomial Chaos Expansion = 2; σ_H = 0.02; b = 0.1.

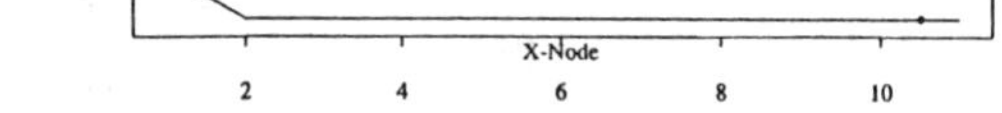

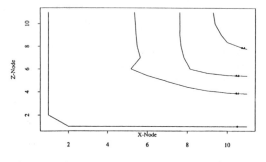

Figure 7: Contour Lines for the Mean of the Vertical Displacement. Order of the Polynomial Chaos Expansion = 2; σ_H = 0.16; b = 0.1.

position. The Polynomial Chaos expansion is used to represent this matrix in the form

$$\mathbf{K}^e = \sum_{i=0}^{\infty} \Psi_i(\omega)\mathbf{K}_i^e. \tag{14}$$

Using equation (9), the coefficient matrices $\mathbf{K}_i^e$ are given by the equation,

$$\begin{aligned}\mathbf{K}_i^e &= \frac{<\mathbf{K}^e, \Psi_i>}{<\Psi_i, \Psi_i>} \\ &= \frac{1}{<\Psi_i, \Psi_i>} \times \\ \sum_{k=1}^{2} \mathbf{K^e} &\left[<\Psi_i E^{(k)}(\mathbf{x})\mathcal{I}\left[(-1)^k(z - H(\mathbf{x};\omega))\right]>; \nu^{(k)}\right] \\ &= \frac{1}{<\Psi_i, \Psi_i>} \sum_{k=1}^{2} \mathbf{K^e}\left[I_{i0}^{(k)} E^{(k)}; \nu^{(k)}\right].\end{aligned} \tag{15}$$

In this equation, the symbol $I_{i0}^{(k)}$ denotes the quantity,

$$I_{i0}^{(k)} = E\left\{\Psi_i \mathcal{I}\left[(-1)^k\left(\beta - \sum_{m=1}^{r^H} \xi_m^H \kappa_m(\mathbf{x})\right)\right]\right\}. \tag{16}$$

where the following notation has been used,

$$\kappa_m(\mathbf{x}) \equiv \kappa_m(x) = \frac{\sqrt{\lambda_m^H}\phi_m^H(x)}{\sigma_H}, \quad m = 1, \ldots, r^H, \tag{17}$$

$$\beta = \frac{z - \bar{H}(x)}{\sigma_H}, \tag{18}$$

and

$$\sigma_H^2 = \sum_{m=1}^{r^H} \lambda_m^H \left(\phi_m^H(x)\right)^2. \tag{19}$$

Furthermore, the Karhunen-Loeve expansion for the interface process, equation (5) has been truncated at the r^H term. The summation in equation (16) above can be simplified by using the following coordinate transformation,

$$\boldsymbol{\eta} = \mathbf{R}\boldsymbol{\xi}^H, \tag{20}$$

where $\mathbf{R}$ is an orthonormal rotation matrix (r_{ij}) satisfying the following property,

$$r_{1m} = \kappa_m(\mathbf{x}), \tag{21}$$

so that the following relationship holds,

$$\eta_1 = \sum_{m=1}^{r^H} \xi_m^H \kappa_m(\mathbf{x}). \tag{22}$$

In this new coordinate system, the integrals $I_{i0}^{(k)}$ become

$$I_{i0}^{(2)} = \frac{1}{2\pi}\int_{-\infty}^{\beta} \Psi_i(\boldsymbol{\xi}(\boldsymbol{\eta}))e^{-\eta_1^2/2}d\eta_1, \tag{23}$$

and

$$I_{i0}^{(1)} = \frac{1}{2\pi}\int_{\beta}^{\beta_u} \Psi_i(\boldsymbol{\xi}(\boldsymbol{\eta}))e^{-\eta_1^2/2}d\eta_1, \qquad (24)$$

where

$$\beta_u = \frac{D - \bar{H}(x)}{\sigma_H}, \qquad (25)$$

and D denotes the elevation of the surface measured from the bottom of the second layer. The upper limit β_u in the expression for $I_{i0}^{(1)}$, implies a zero-stiffness model for the air overlaying the topmost soil layer. In the form given by equations (23) and (24), the above integrals can be evaluated in closed form for a given value of r^H, in terms of the coefficients $\kappa_m(\mathbf{x})$ Once the components $\mathbf{K}_i^{(e)}$ have been computed, the matrix $\mathbf{K}^{(e)}$ can be readily evaluated. Formally, the elemental stiffness matrices can be assembled into the global stiffness matrix, resulting in the following system of linear equations,

$$\mathbf{K}\mathbf{u} = \mathbf{f}, \qquad (26)$$

where $\mathbf{f}$ denotes the applied load on the surface the medium. The dimension of the vectors in equation (27) is equal to the number of degrees of freedom in the discrete system, denoted by n. However, and in view of the Polynomial Chaos expansion of the elemental stiffness matrices, the global stiffness matrix can be represented as follows,

$$\mathbf{K}(\omega) \approx \mathbf{K}^L(\omega) = \sum_{i=0}^{L} \mathbf{K}_i \Psi_i(\omega), \qquad (27)$$

where the superscript L denotes that the expansion has been truncated after the Lth term. A truncation at the Lth term in the expansion can be accomplished in a number of ways. Depending on the number of terms that are carried in the Karhunen-Loeve expansion of the random process representing the interface position. Thus, keeping r^H terms in that expansion, $\left\{\xi_1^H, \ldots, \xi_{r^H}^H\right\}$, and carrying out the Polynomial Chaos expansion through order P, the number L of terms that are featured in the expansion for the stiffness matrix is given by the equation

$$L = \sum_{s=1}^{P} \frac{1}{s!} \prod_{n=0}^{s-1} (r^H + n). \qquad (28)$$

Similarly to the stiffness matrix, the response is a random process and can be represented using the Polynomial Chaos expansion in the form,

$$\mathbf{u}(\omega) \approx \mathbf{u}^L(\omega) = \sum_{i=0} \mathbf{u}_i^L \Psi_i(\omega). \qquad (29)$$

Based on equation (9), the coefficients in the expansions for the stiffness and response quantity

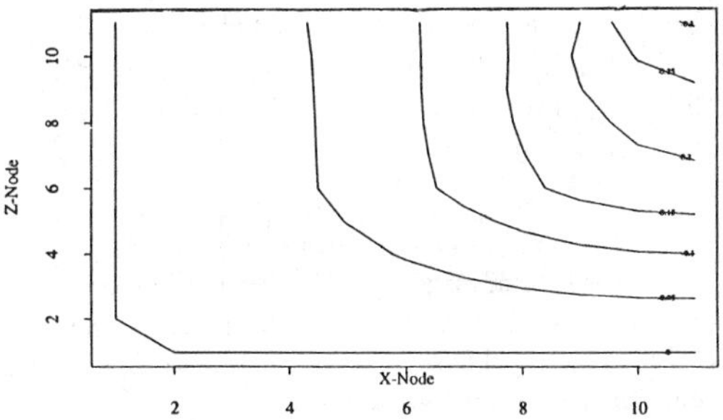

Figure 8: Contour Lines for the Mean of the Vertical Displacement. Order of the Polynomial Chaos Expansion = 2; σ_H = 0.02; b = 1.

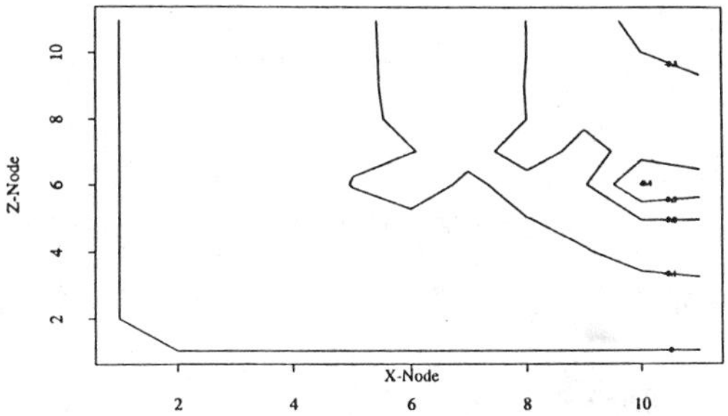

Figure 9: Contour Lines for the Mean of the Vertical Displacement. Order of the Polynomial Chaos Expansion = 2; σ_H = 0.16; b = 1.

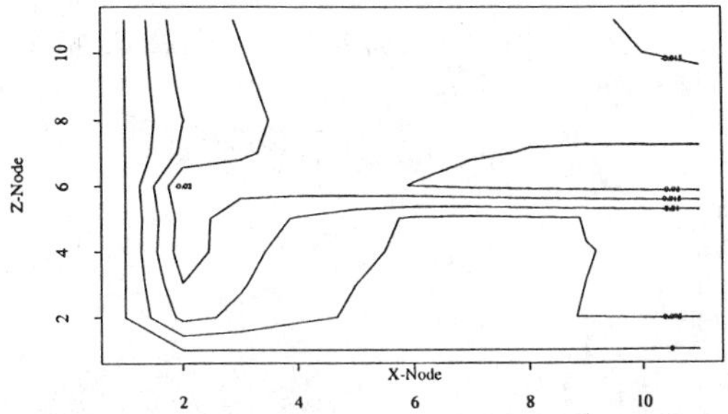

Figure 10: Contour Lines for the Coefficient of Variation of the Vertical Displacement. Order of the Polynomial Chaos Expansion = 2; σ_H = 0.02; b = 0.1.

are given by the following expressions

$$\mathbf{K}_i = \frac{<\mathbf{K}, \Psi_i>}{<\Psi_i, \Psi_i>}, \qquad (30)$$

and

$$\mathbf{u}_i = \frac{<\mathbf{u}, \Psi_i>}{<\Psi_i, \Psi_i>}. \qquad (31)$$

Clearly, matrices $\mathbf{K}_i$ can be obtained by assembling matrices $\mathbf{K}_i^{(e)}$. The truncation of the Polynomial Chaos expansion at the Lth term results in an error given by the equation

$$\epsilon(\omega) = \mathbf{K}^L(\omega)\mathbf{u}^L(\omega) - \mathbf{f}. \qquad (32)$$

This error is minimized by requiring it to be orthogonal to the space spanned by the approximating set of Polynomial Chaoses. This results in L sets of equations, each having the form

$$<\mathbf{K}^L\mathbf{u}^L - \mathbf{f}, \Psi_k> = \mathbf{0} \ , \quad k = 0, \ldots, L. \qquad (33)$$

Substituting the expansions for $\mathbf{K}^L$ and $\mathbf{u}^L$, results in the following equation,

$$\sum_{i=0}^{L}\sum_{j=0}^{L} d_{ijk}\mathbf{K}_j\mathbf{u}_i = \mathbf{f}_k \ , \qquad (34)$$

where

$$d_{ijk} = <\Psi_i(\omega)\Psi_j(\omega)\Psi_k(\omega)> \ , \qquad (35)$$

and

$$\mathbf{f}_k = <\mathbf{f}, \Psi_k(\omega)> = \begin{cases} \mathbf{f} & \text{for } k = 0 \\ \mathbf{0} & \text{for } k > 0 \end{cases} . \qquad (36)$$

The above procedure results in the following system of deterministic linear algebraic equations to be solved for the coefficient vectors $\mathbf{u}_i$

$$\mathbf{K}^*\mathbf{u}^* = \mathbf{f}^*, \qquad (37)$$

where $\mathbf{K}^*$, $\mathbf{u}^*$, and $\mathbf{f}^*$ consist of L blocks, each of dimension n,

$$\mathbf{u}^* = [\mathbf{u}_0, \ldots, \mathbf{u}_L]^T \ , \qquad (38)$$

$$\mathbf{f}^* = [\mathbf{f}_0, \ldots, \mathbf{f}_L]^T \ , \qquad (39)$$

and

$$\mathbf{K}^* = [\mathbf{K}_{kj}] \ . \qquad (40)$$

In particular, the kj block of $\mathbf{K}$ is given by the equation,

$$\mathbf{K}_{kj} = \sum_{i=0}^{L} d_{ijk}\mathbf{K}_i \ , \quad j,k = 0, \ldots, L. \qquad (41)$$

5 Numerical Example

The numerical example to be treated in this paper involves a soil medium consisting of two layers.

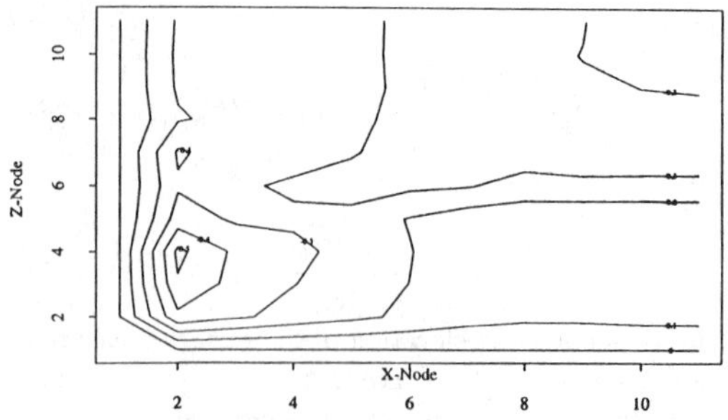

Figure 11: Contour Lines for the Coefficient of Variation of the Vertical Displacement. Order of the Polynomial Chaos Expansion = 2; σ_H = 0.16; b = 0.1.

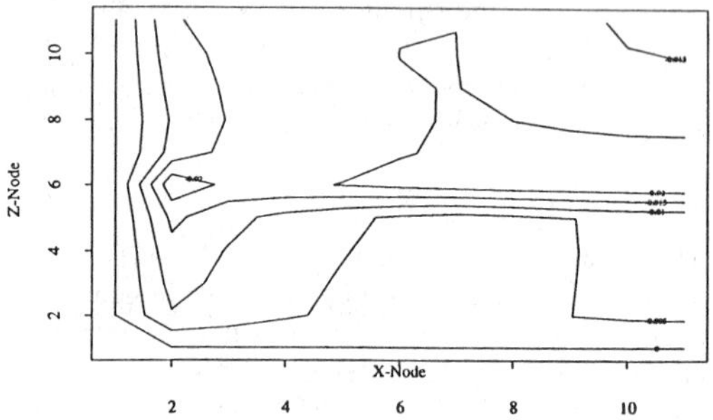

Figure 12: Contour Lines for the Coefficient of Variation of the Vertical Displacement. Order of the Polynomial Chaos Expansion = 2; σ_H = 0.02; b = 1.

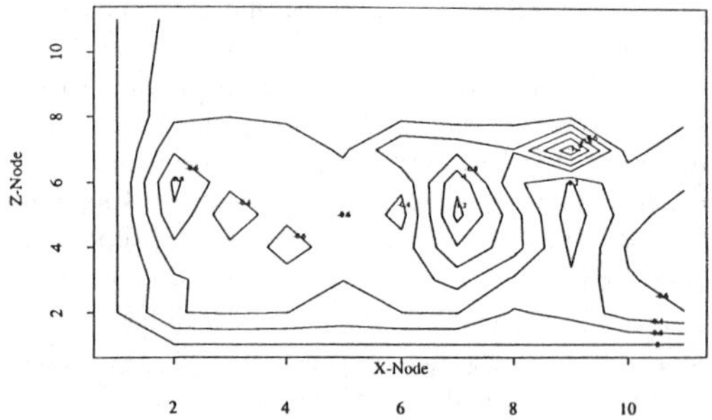

Figure 13: Contour Lines for the Coefficient of Variation of the Vertical Displacement. Order of the Polynomial Chaos Expansion = 2; σ_H = 0.16; b = 1.

The lower layer has a modulus of elasticity equal to 100 MPa and a Poisson ratio of 0.2, while the upper layer has a modulus of elasticity equal to 20 MPa and a Poisson Ratio equal to 0.3. A load of magnitude 100 MPa is applied at the edge of the upper boundary of the domain, on one quarter of its total length. The mean value of the interface is assumed to be halfway through the depth of the medium, with random fluctuations that follow the Gaussian distribution, and having a covariance function given by the equation

$$C^H(\mathbf{xy}) = \sigma_H^2 e^{|\mathbf{x}-\mathbf{y}|/b} , \qquad (42)$$

where b denotes the correlation length. A parametric study is conducted to assess the effect of the standard deviation and the correlation length on the variability of the displacement within the medium. Both vertical and horizontal displacements are investigated. A mesh of 36 elements is used in the finite element discretization. Figure (1) shows the physical set-up. The finite element discretization involves a uniform rectangular mesh with ten elements on each side, resulting in a total of one hundred elements. Figures (2)-(9) show contour lines for the mean horizontal and vertical displacement within the domain, for selected values of the standard deviation of the interface location, and selected values of its correlation length. Both of these values are shown on the figures. The substantial effect of the coefficient of variation of the material properties on the mean of the response quantities indicates a highly non-nonlinear relationship between the response surface and the basic random variables describing the interface process. Furthermore, note the increasing effect of the correlation length with increasing magnitude of the coefficient of variation of the interface process. Figures (10)-(13) show related results for the coefficient of variation of both horizontal and vertical displacements.

6 Conclusion

Based on the results of this paper, it is concluded that a meaningful reliability assessment of the settlement of foundations on randomly layered formations requires knowledge of both the correlation length and the standard deviation of the random interface process. This paper has presented a new algorithm that facilitates the analytical and computational tasks involved in such reliability calculations.

7 Acknowledgements

The work of the first author was sponsored by a grant No. 9325-27792 from the Center for Infrastructure Research and Development. The work of the second author was sponsored by a grant No. 7-7239-92-03 from the Polish State Committee for Scientific Research.

Bibliography

[1] Akin, J.E. *Application and Implementation of Finite Element Methods,* Academic Press, 1982.

[2] Cameron, R.H. and Martin, W.T., "The orthogonal development of nonlinear functionals in series of Fourier-Hermite functionals," *Ann. Math,* Vol. 48, pp. 385-392, 1947.

[3] Ghanem, R. and Spanos, P. "A new computational approach for nonlinear random vibration," accepted for publication in *Probabilistic Engineering Mechanics.*

[4] Ghanem, R. and Spanos, P. *Stochastic Finite Elements: A Spectral Approach,* Springer-Verlag, New York, 1991.

[5] Loeve, M., "Fonctions aleatoires du second ordre," supplement to P. Levy, *Processus Stochastic et Mouvement Brownien,* Paris, Gauthier Villars, 1948.

[6] Mercer, J., "Functions of positive and negative type and their connection with the theory of integral equations," *Phil. Tran. Roy. Soc. London, Series A,* Vol.209, pp. 415-446, 1909.

[7] Spanos, P. and Ghanem, R., "Stochastic finite element expansion for random media," *ASCE, Journal of Engineering Mechanics,* p. 1035-1053, Vol. 115, No. 5, May 1989.

Structural Safety & Reliability, Schuëller, Shinozuka & Yao (eds) © *1994 Balkema, Rotterdam, ISBN 90 5410 357 4*

An approach to stochastic linear soil-structure-interaction under dynamic loading

H. Grundmann & H. Waubke
Lehrstuhl für Baumechanik, Technische Universität München, Germany

ABSTRACT: A Stochastic Finite Element approach is presented for the evaluation of the stochastic stiffness coefficients representing a layered elastic half space below a foundation under periodic loading. It's characteristics are an application of Fourier-Hankel transform technique and a representation of the stochastic quantities by means of a Karhunen Loewe expansion and a series of polynomial chaoses in the space of the Hankel transforms. The conditions of compatibility of the displacements at the interface of the structure and the half space are fulfilled as usual by introduction of partial loadings, which leads to a set of stochastic linear equations.

1 INTRODUCTION

Whereas in general the randomness of the system parameters is less important in comparison with the stochastic nature of the loading and the resistance of the material this does not hold in general in the case of periodic or nearly periodic loadings. In this case, caused by an extreme sensitivity of the system in the neigbourhood of the resonance points, the behaviour may be altered significantly, when the system parameters and depending on them, the natural frequencies of systems change in a random manner. One case of specific interest is that of a foundation of a machine resting on the soil. Here the behaviour is dependent on the mass of the foundation and on the properties of the half space below. To take account of the soil-structure-interaction in an efficient manner, one can describe the participation of the soil by frequency dependent stiffness and damping coefficients. The corresponding deterministic work has been done already for different situations, especially for circular foundation on an elastic half space or on a single layer with a rigid base. [1], [2], [3]. An inspection of the corresponding results shows that in several cases of a layered half space a rather small change of the system parameters may cause large variations of the evaluated parameters. Taking regard of the fact that the basic coefficients of the soil, such as the shear modulus, may have large stochastic variations, it becomes clear, that this randomness may be significant for the reliability of such a system.

In order to evaluate the stiffness and damping coefficients describing the interaction between a rigid foundation and the soil one has to solve a mixed boundary value problem: Below the foundation the displacement is prescribed whereas on the soil surface outside of the foundation the stresses have to

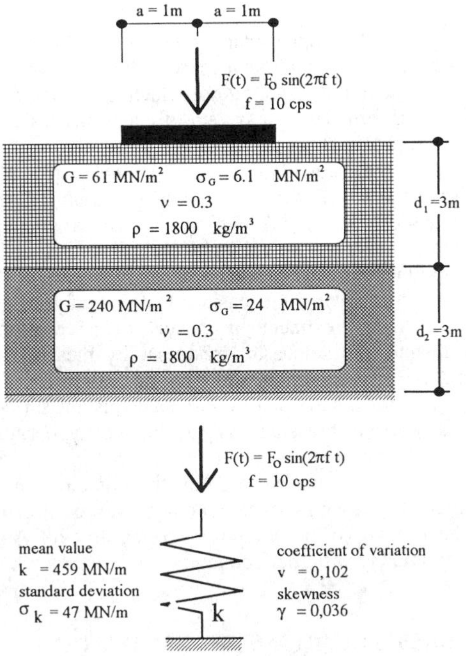

Fig. 1: Parameters of the layered soil and the substituting spring

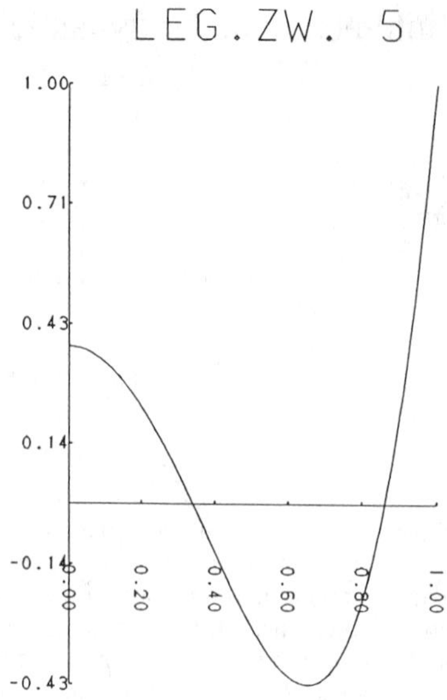

Fig. 2: The 4th Legendre polynomial

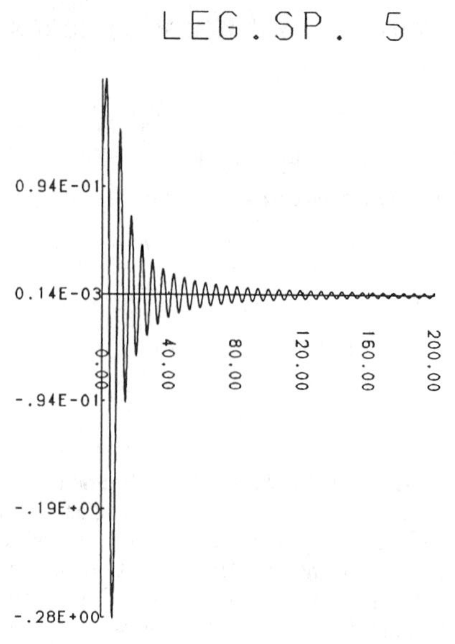

Fig. 3: Hankel transform of the 4th Legendre polynomial

disappear. An approximate solution can be found using a set of unknown partial loadings at the contact area, the magnitude of which is evaluated by means of formulating a system of equations assuring in an approximate sense that the surface deformation below the foundation corresponds to the displacement of the rigid contact area of the foundation. As partial loadings circular or annular stresses or stresses distributed according to specific form functions can be used.

In the following the basic equations of the half space, their treatment by means of a transform technique, the stochastic treatment by means of a spectral FE approach and the derivation of the solution for a deterministic circular constant surface loading will be presented only rather shortly, because these questions are shown in more detail in a former paper [4]. In some more detail the combination of different circular surface loadings will be treated which have to be chosen such as to fulfil the compatibility with the foundation.

2 BASIC EQUATIONS, DETERMINISTIC TREATMENT

The basic equations of the dynamic of the three dimensional linear elastic continuum are the well known Lamé equations

$$\mu\, u^i|^j_j + (\lambda+\mu) u^j|^i_j - \rho u^{i\bullet\bullet} = 0 \tag{1}$$

in which u^i means the components of the displacement vector and λ and μ are the Lamé constants representing the elastic behaviour of the material. If these constants are introduced as complex variables it is possible to take account of damping in the case of harmonic excitation.

It is advantageous to introduce cylindrical coordinates if the response to some circular surface loading shall be developed. After the corresponding transformation of the basic equations three potentials φ, ψ and χ are used [5], by which the three displacements can be expressed. Introducing this substitution into equation (1) three wave equations

$$\nabla^2\varphi - \frac{\omega^2}{c_p^2}\varphi = 0 \qquad \nabla^2\psi - \frac{\omega^2}{c_s^2}\psi = 0 \tag{2a,b}$$

$$\nabla^2\chi - \frac{\omega^2}{c_s^2}\chi = 0 \tag{2c}$$

can be derived.

Describing the variation of the potential amplitudes with respect to the circumferential coordinate by a Fourier series and writing the n-th member of the series in form of a product, φ_n for example, as

$$\varphi_n = \varphi_{nr}(r)\,\varphi_{nz}(z), \tag{3}$$

leads to the final form of the equations, for instance for φ_n

$$r^2\, \varphi_{nr,rr} + r\, \varphi_{nr,r} + \left(k^2 r^2 - n^2\right) \varphi_{nr} = 0, \tag{4}$$

and

$$\varphi_{nz,zz} + \left(\frac{\omega^2}{c_p^{\;2}} - k^2 \right) \varphi_{nz} = 0, \tag{5}$$

where

$$(\;)_{,r} = \frac{\partial (\;)}{\partial r} \quad and \quad (\;)_{,z} = \frac{\partial (\;)}{\partial z}$$

is written for abbreviation, and k is an arbitrary parameter at the moment. Similar equations are found for ψ_{nr}, ψ_{nz}, χ_{nr} *and* χ_{nz}.

Equation (4) shows, that $\varphi_{nr}(r)$ (and χ_{nr}, ψ_{nr} as well) can be expressed by Bessel functions, whereas for the functions φ_{nz}, $(\varphi_{nz}, \varphi_{nz})$, depending on whether

$\left(\omega^2 / c_p^2 - k^2\right)$ (*or* $\left(\omega^2 / c_s^2 - k^2\right)$ respectively)

is positive or negative, according to equation (5) a sinusoidal or an exponential behaviour arises. It is remarkable in particular, that with respect to the vertical coordinate equations are found which correspond completely to those which arise at the corresponding place in a description in cartesian coordinates, employing a Fourier transform with respect to the horizontal coordinates. This has far reaching consequences: It allows to use the solutions found for the cartesian case as essential elements for the composition of the final solution, details are found in [5]. As representation of the final form of the solution only the expressions for the radial displacement u_n and of the vertical stress σ_{nz} are shown here,

$$u_n = \frac{1}{k} I_n(kr)_{,r}\, u(z) + \frac{n}{kr} I_n(kr)\, v(z)$$

$$\sigma_{nz} = I_n(kr)\, i\, \sigma_z(z) \qquad \left(i^2 = -1\right). \tag{6 a,b}$$

In these expressions the I_n(kr) are Bessel functions whereas $u(z)$, $v(z)$ and $\sigma_z(z)$ are displacements and stresses respectively describing the variation of the corresponding quantities with respect to the vertical direction for each value of the wave number k in completely the same manner as this is done by the same quantities when cartesian coordinates are used and a Fourier transformation is applied with respect to x and y.

As a consequence for the description of the variation of the solution with respect to the vertical direction in the case of a layered half space completely the same one-dimensional FE-approach can be employed here as is used in the plane case. To take advantage of this possibility it is necessary to operate in a transformed space, which means to

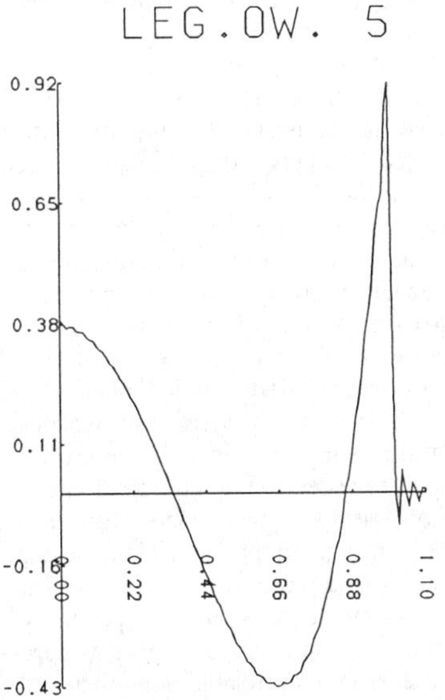

Fig. 4: Backward transformation of Hankel transform of the 4th Legendre polynomial

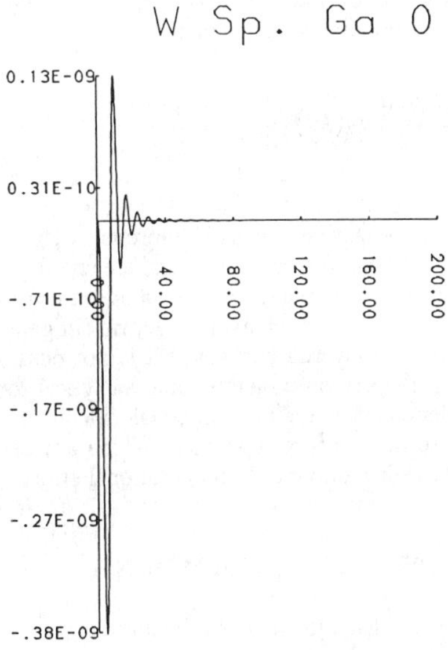

Fig. 5: Hankel transform of the mean value displacement w corresponding to the 4th Legendre polynomial

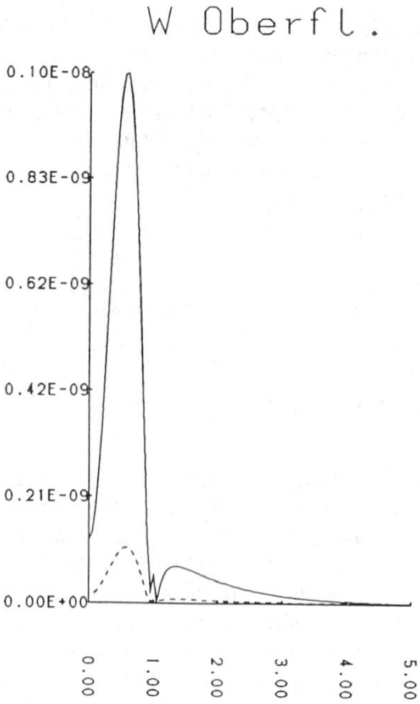

Fig. 6: Displacement w for the 4th Legendre polynomial mean value __ and standard deviation ....

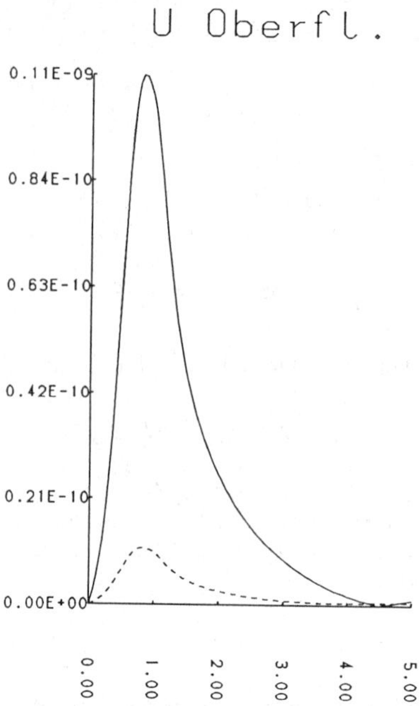

Fig. 7: Displacement u for the 4th Legendre polynomial mean value __ and standard deviation ....

apply a Fourier series with respect to the circumferential direction and a Hankel transform to the applied loading with respect to the radial direction.

In the case of a constant vertical disc loading e.g. the corresponding loading in the transformed space is described as

$$p(k) = \frac{p_o a}{k} I_1(ka)$$

a being the radius of the disc. Having evaluated the interesting quantities by a FE approach with respect to the vertical direction in dependency on k the desired response quantities depending on r are to be found by the inverse Hankel transform. In general in the case of a layered half space it is not possible to describe the response quantities in analytical form in their dependency on k. Numerical calculations are therefore necessary at this point which are decisive for the needed amount of computational effort.

3 STOCHASTIC MODELISATION

In the case that the stochastic quantities such as the Lamé constants would be variable in all three directions x, y and z a solution would only be possible by means of a three dimensional FE calculation, which however seems to be an unrealistic approach caused by a too expensive amount of computer time.

Therefore it seems to be reasonable at the moment to assume that stochastic variations of the random properties of the soil are to be taken into account only in the vertical direction, whereas in the horizontal directions the behaviour of the soil will be considered as constant. In this case it becomes possible to employ the transform techniques described a head with respect to the horizontal coordinates, while at the same time in the vertical direction a variation of the soil parameters according to a stochastic process may be admitted. In the simplest case, which will be considered here in the presented example, the parameters of the different layers can be considered as independent stochastic variables. In the more general case, which was dealt with for the plane case in [6], the parameters are stochastic processes in the interior of each layer, the values however being independent from one layer to the next, an assumption which seems to be reasonable for a layered soil. In the last case a Karhunen-Loewe expansion as proposed by Ghanem and Spanos [7] can be used to give a representation of the process in uncorrelated quantities.

The governing equations for a description of the variation of the solution with respect to the coordinate z can be brought into the form [6].

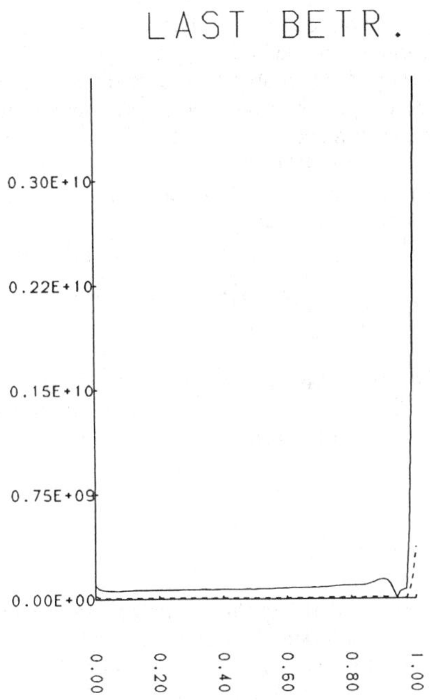

Fig. 8: Evaluated stress distribution at the surface mean value __ and standard deviaton ....

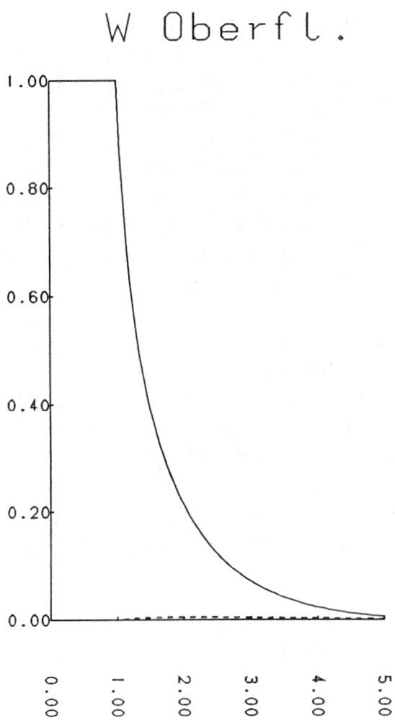

Fig. 9: Vertical displacement w mean value __ and standard deviation ....

$$\boldsymbol{K}(k)\boldsymbol{u}(k) = \boldsymbol{p}(k)$$

in which $\boldsymbol{K}(k)$ is the stochastic stiffness matrix, $\boldsymbol{u}(k)$ is the vector of the stochastic nodal displacements and $\boldsymbol{p}(k)$ are the deterministic or stochastic nodal loadings in the transformed space. As one can see immediately, after subdividing the stiffness matrix into a mean value part and a stochastic part with vanishing mean, the response depends on the stochastic parameters in a nonlinear manner. As it was proposed in [7], one can take account of this in an elegant manner by introducing polynomial chaoses $\psi_j(\varsigma_r)$ as building blocks for a description of the stochasticity of the response in the form

$$u_i = \sum_j c_{ij} \; \psi_j[\xi_r(\theta)]. \tag{9}$$

The polynominal chaoses are built up from orthonormal Gaussian variables $\varsigma(\vartheta)$. They obey certain orthonormality conditions. By introducing (9) into the governing FE equation this is reduced to a set of linear equations for the unknown coefficients. These are found by forming inner products with respect to the probability space. After having solved these equations for a lot of different values of the wave number k the inverse Hankel transformation can be performed. As a result of these operations the displacement $w(r,\vartheta)$ at all different points of the surface is known in it´s dependency on the given deterministic surface loading and on the stochastic soil parameters [4].

4 DETERMINING THE STIFFNESS AND DAMPING COEFFICIENTS OF A FOUNDATION

The stiffness and the damping coefficient of the soil below a rigid foundation as frequency dependent quantities can be evaluated as the real and the imaginary part respectively of the total force or moment, which are necessary at a specific frequency to produce a plain displacement or rotation of the magnitude 1 in that region of the soil which is lying below the rigid foundation. To find in an approximate manner these quantities, the contact area may be subdivided approximately into a great number of subdiscs, on each of which taking account of symmetry an unknown loading is applied. The magnitude of these loadings will be calculated by a set of equations by which is assured, that the superposition of all loads will produce in some approximate sense the named displacement or rotation with magnitude unity. The resulting force or moment of the partial loadings lead to the stiffness and damping parameters of interest. For this and for more general deterministic interaction problems shape functions for the loading have been developed in [8], which lead to an efficient approach.

In the case, that only the coefficients for vertical movements of a circular foundation are to be calculated, as will be done in the example, concentric loadings can be used, in the form of concentric disc loadings of different diameters, annular loadings or loadings of different shape on the whole contact area.

If the soil parameters are stochastic quantities the response of the soil is stochastic too, even for a deterministic loading. In order to solve the actual task, the intensity of a partial loading in the first step will be considered as deterministic, fixed with some nomalized magnitude "1". The response quantities found by a procedure as described ahead for a layered soil with random properties are brought into the form

$$w_i(r,\theta)=\sum_j w_{ij}(r)\,\psi_j[\xi_r(\theta)], \qquad (10)$$

in which θ is used to indicate the stochastic nature of the displacements. The quantities $w_{ij}(r)$ are the deterministic response coefficients found by means of the stochastic FE method and the Hankel transform technique sketched ahead. The chaos functions $\psi_j[\varsigma_r(\upsilon)]$ contain in a certain polynomial dependency the random character of the response. The index i is used to indicate that the response belongs to the i-th partial loading of magnitude 1.

Next the compatibility with the prescribed unit displacement of the foundation has to be assured leading to a system of linear equations of the type

$$\boldsymbol{W}\,\boldsymbol{P}=\boldsymbol{1}. \qquad (11)$$

The matrix $\boldsymbol{W}$ contains for the different partial loadings the diplacements $w_i(r,\upsilon)$ at different selected points r_j for which the compatibility condition shall be fulfilled. The unknown stochastic loading coefficients are assembled in the matrix $\boldsymbol{P}$. On the r.h.s. of the equation the prescribed deterministic displacement or rotation is introduced. Since the coefficients of the matrix $\boldsymbol{W}$ are stochastic, being expressed by a series of polynomial chaoses, it seems to be advantageous, to build up the unknown stochastic load intensities also by means of the same polynomial chaoses

$$p_i=\sum_j p_{ij}\;\psi_j[\xi_r(\theta)]. \qquad (12)$$

With this approach, after forming suitable inner products with respect to the probability space, the solution can be found by a system of linear equations for the unknown coefficients p_{ij}.

Knowing these quantities the resultant R (or M respectively) can be evaluated by an integration over the whole contact area taking account of the each partial loading according to it's respective shape function $f_j(r)$. The operation leads to the stiffness and damping coefficients:

$$R=K_w=\sum_i\sum_j\int p_{ij}\,f_j(r)\;\psi_j[\xi_r(\theta)]dA. \qquad (13)$$

After the stiffness is known in dependency on the polynomial chaoses, the corresponding stochastic parameters can be calculated in the way described in [7].

5 AN EXAMPLE

As an example the stiffness coefficient for vertical movements of a circular foundation, resting on a soil built up by two layers on a rigid base were considered. In each layer constant random values of the elastic coefficients were assumed. The elastic parameters are G = 61 MN/m^2, σ_G = 6 MN/m^2 for the upper layer and G = 240 MN/m^2, σ_G 24 = MN/m^2 for the lower. The thickness d = 3 m,

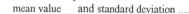

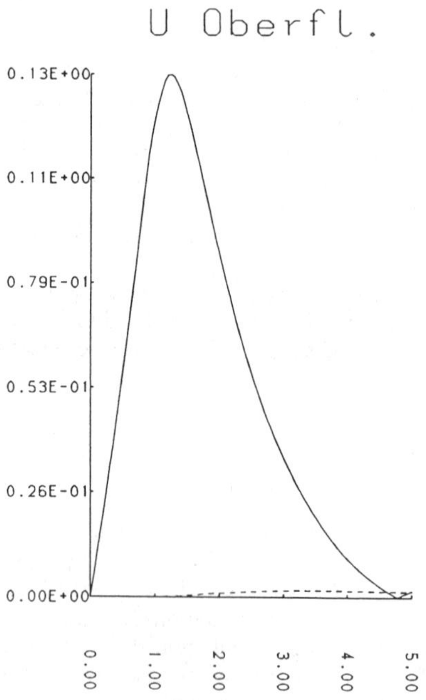

Fig. 10: Horizontal displacement u mean value __ and standard deviation ....

Poissons ration $\nu = 0,3$ and the mass density $\rho = 1,8$ t/m^3 are considered to be deterministic with the same values in both layers. The loading is variable in time with a frequency f = 10 cps. The radius of the foundation is chosen a = 1 m, (fig. 1).

The requirement of constant displacement below the foundation, which can be fulfilled only in an approximate manner, can be taken into consideration in different ways, for example firstly by fulfilling the compatibility in the sense of a collocation at several points or secondly by demanding that the deviation of the prescribed value be minimized in the mean square for a lot of points at the interface. In general the second procedure would have been preferable. However this was not so clear for the problem at hand: As the calculations have shown, in the error minimizing operations the contributions of the mean value part and of the random part, were mixed up in that way, that the stochastic parts were mobilized, to remove errors mainly related to the mean value parts. Since no criteria could be formulated till now, to avoid this, the collation procedure, where the named difficulty doesn't arise, was used in the example.

To facilitate the calulations the partial loadings in a first attempt were introduced in form of concentric constant disc loadings with different diameters. This however lead to a rather high numerical sensitivity, caused by the fact that differences between the displacements at a typical collocation point caused by one disc loading and the next were too small. This sensitivity could not be removed by the aid of annular loadings. The corresponding problems were solved by introducing an other type of partial loadings, chosen as distributed along the radius of the foundation according to Legendre-polynomials.

The calculations with respect to the depth z were performed with 6 elements in the upper and 3 elements in the lower layer. For the stochastic part of the solution 5 polynomial chaoses were used for each unknown quantity, therefore 6 unknown factors were needed. Into the equations concerning the compatibility between soil and foundation also 5 polynomial chaoses were introduced.

The simple example was chosen because it allowed several controls. It was possible of course in a first step to control the mean value part of the program by comparison with known results for this problem as published for instance in [5]. After the stochastic part was applied in it´s basic form, which was already tested earlier in comparison with the example of a beam presented in [7], the last control could be performed by using once again the results published in [5]. This could be done in an approximate sense by taking the results for the stiffness parameters corresponding to two neighbouring values of the shear modulus in order to form with them an approximate value of the first derivative. This could be used to find by the aid of a Taylor series an estimative value for the variance, which could be compared with the variance found by applying the procedure of this paper.

The result of the calculations, the mean value of the stiffness coefficient substituting the soil, its standard deviation and the corresponding skewness are given on Figure 1. Figure 2 shows the Hankel transform of the 4th Legendre-polynomial and figure 3 the corresponding Hankel transform. The inverse transformation, which should equal fig. 2, can be seen on figure 4. The Hankel transform of the mean value of the corresponding vertical displacement w(r) is shown on fig. 5, and the corresponding displacements w(r) and u(r) at the surface of the half space on fig. 6 and 7. The stresses σ_z fulfilling the compatibility (mean value and corresponding standard deviation) and the displacements w(r) and u(r) are drawn on the figure 8 und figure 9 respectively .

6 CONCLUSIONS

Random soil parameters may influence the reliability of a foundation subjected to a periodic loading, since they may alter the resonant behavior. In order that a computation should be manageable with respect to expenditure of computer time it must be assumed that the stochastic quantities do not vary with respect to the horizontal directions. With respect to the depth they may obey a stochastic process. In that case a transition to uncorrelated variables is possible by applying the Karhunen-Loewe expansion. In the case that one can assume that stochastic quantity to be constant with respect of the thickness of each layer and these quantities to be independent from one layer to the next (what seems reasonable) the treatment of the problem will simplified in a considerable manner.

In a favourable approach one can work in a transformed space using in the case of cylindrical coordinates a Fourier series expansion and a Hankel transform. In the transformed space only a one dimensional treatment of the stochastic quantities is necessary namely with respect to the vertical direction by means of a FE application. Here for the description of the stochastic nature of the response a polynomial chaos expansion is introduced. After the inverse Bessel transform the surface displacements as stochastic quantities are known depending on the loading. These results can be used to combine the loadings applied at all parts of the later interface between the foundation and the soil in such a way that a prescribed displacement or rotation of this interface is acchieved in an approximate sense. The resultants of the applied partial loadings correspond to the desired stiffness and damping coefficients. With respect to their stochasticity they are described by an other polynomial chaos expansion. The results could be controlled. The amount of computational time was high, it is caused particularly by the numerical inverse Hankel transform. Taking regard of typical properties of the spectral quantity to be transformed, it must be possible to reduce this amount considerably.

LITERATURE

[1] Gazetas, G.: "Analysis of machine foundations: state of the art", Soil Dynamics and Eearthquake Engineering, 1983, Voll. 2, No. 1.

[2] Studer, J.A.; Ziegler, A.: "Bodendynamik, Grundlagen, Kennziffern, Probleme". Springer-Verlag, 1986.

[3] Gaul, L.: "Zur Berechnung der Vertikal- und Kippschwingungen eines starren Fundamentes auf viskoelastischem Halbraum". Dissertation, TU Hannover 1976.

[4] Grundmann, H.; Waubke, H.: "Approximative Solution for the Response of a Layered Stochastic Linear Soil to a Dynamic Excitation at the Surcface", to be presented at EURODYN'93, Trondheim, Norway. June 1993.

[5] Wolf, J. P.: "Dynamic Soil Structure Interaction". Prentice Hall, Inc., Eaglewood Cliffs, N.J. 07632, 1985.

[6] Grundmann, H.; Waubke, H.: "Stochastische Finite Elemente zur Ermittlung der Bewegungen eines mehrschichtigen ebenen Kontinuums unter periodischer Belastung". Magnus Festschrift 1992.

[7] Ghanem R.G.; Spanos, P. D.: "Stochastic Finite Elements: A Spectral Approach" Springer-Verlag, 1991.

[8] Zirwas, G.: "Impedance matrix of viscoelastic (layered) halfspace". Proc. of the Int. Conf. on Comp. Eng. Science (1991), pp 1293 - 1296, Atlanta: 1991.

Structural Safety & Reliability, Schuëller, Shinozuka & Yao (eds) © 1994 Balkema, Rotterdam, ISBN 90 5410 357 4

Conditional simulation of a stochastic field

M. Hoshiya
Musashi Institute of Technology, Tokyo, Japan

ABSTRACT: A Kriging method has been utilized to interpolate a sample field $\underline{Z}^*(Xr)$ on the basis of observation data $\underline{Z}(Xi)$ at specific spatial points $Xi, i=1\sim N$ and a simple method of the conditional simulation for the same purpose is proposed based on a step by step expansion procedure.

1 INTRODUCTION

Kriging methodology has been developed by many researchers mainly in geostatistics (Krige(1966), Matheron(1973), Journel(1974) and etc.) that provides the best linear unbiased estimate built on data of a stochastic field. Substantial contents as to the theory and applications of Kriging are well documented in texts (Davis and McCullagh(1975), David(1977), Journel and Huijbregts(1978), Ripley(1981), Matheron (1989), and Cressie(1991)). An estimator by Kriging is a smoothed one since it is a weighted linear interpolating procedure and therefore it does not completely restore the stochastic variability. In order to visualize the real aspect, a conditional simulation method has been devised that allows one to realize a sample field at other locations which satisfies with the properties of a stochastic field and is compatible with measured values at sampling locations.

This conditional simulation method is based on the following equation.

$$W^*(X) = \hat{W}(X)+[S(X)-\hat{S}(X)] \qquad (1)$$

where $W^*(X)$ is a simulated process at an unrecorded location X, and $\hat{W}(X)$ is a Kriging estimate at X.

Equation(1) indicates that since the sample value $\underline{W}(X)$ of $W(X)$ is not available, the error value of $\hat{\underline{W}}(X)$, that is, $\underline{W}(X)-\hat{\underline{W}}(X)$ remains unknown. Because of this reason, the error value is estimated based on unconditionally simulated data $\underline{S}(X)$ and the Kriging estimate $\hat{\underline{S}}(X)$ from $\underline{S}(X)$. However, if the stochastic properties of the error $W(X)-\hat{W}(X)$ are clarified, the error may be simulated. In fact, this paper finds its basis on the following equation, instead of equation(1).

$$W^*(X) = \hat{W}(X)+[W(X)-\hat{W}(X)] \qquad (2)$$

This paper, utilizing the orthogonality property between the first and second terms of equation(2) derives a very simple method of conditional simulation and the verification is made through a numerical example.

2 PROBLEM

The problem is how logically and simply we simulate a set of sample realizations $\underline{Z}^*(Xr)$ at unrecorded spatial locations $Xr, r=1\sim M$ of a stochastic gaussian field, conditional on recorded data $\underline{Z}(Xi)$ at spatial locations $Xi, i=1\sim N$, where the field is homogeneous with known mean $m(X)$ and covariances $C(Z(Xk), Z(Xm))$.

Since the stochastic field is expressed as $Z(X)=m(X)+W(X)$ where $E[W(X)]=0$, and since $C(Z(Xk),Z(Xm))=C(W(Xk),W(Xm))$, hereafter the simulation of $W(Xr), r=1\sim M$, instead of $Z(Xr)$ is considered on the basis of equation(2).

Rewriting equation(2), we have

$$W^*(Xr) = \hat{W}(Xr)+\varepsilon(Xr), \qquad r=1\sim M \qquad (3)$$

where

$$\hat{W}(Xr) = \sum_{i=1}^{N}\lambda i(Xr)W(Xi) \qquad (4)$$

and

$$\varepsilon(Xr) = W(Xr)-\hat{W}(Xr) \qquad (5)$$

It is indicated that equation(4) is a linear interpolation of $W(Xr)$ in terms of $W(Xi)$ which are recorded as $\underline{W}(Xi)$ at N spatial locations. The unknown coefficients

$\lambda_i(X_r)$ are to be determined to minimize the mean square of the error function $\varepsilon(X_r)$ of equation(5).

It is important that if the conditional simulation is properly carried out, then simulated one $\underline{W}^*(X_r)$ should coincide precisely with recorded one $\underline{W}(X_i)$ when r=i, and also a set of a sample field mixing up with recorded $\underline{W}(X_i)$, $i=1\sim N$ and simulated $\underline{W}^*(X_r)$, $r=1\sim M$ should be consistent with the given field property of $C(W(X_k),W(X_m))$.

3 INTERPOLATION

Kriging technique is used to interpolate $\hat{W}(X_r)$ based on unbiased and least error variance estimation. From equation(4), we have $E[\hat{W}(X_r)]=E[W(X_r)]=0$. Thus, equation(4) is unbiased.

The variance of $\varepsilon(X_r)$ is from equations (4) and (5),

$$\sigma^2[\varepsilon(X_r)]$$
$$= E[W^2(X_r)]-2\sum_{i=1}^{N}\lambda_i(X_r)C(W(X_r),W(X_i))$$
$$+\sum_{i=1}^{N}\sum_{j=1}^{N}\lambda_i(X_r)\lambda_j(X_r)C(W(X_i),W(X_j)) \quad (6)$$

minimizing equation(6), we have

$$C(W(X_r),W(X_m)) = \sum_{i=1}^{N}\lambda_i(X_r)C(W(X_i),W(X_m)),\quad m=1\sim N \quad (7)$$

and in this case, equation(6) becomes

$$\sigma^2[\varepsilon(X_r)]$$
$$= E[W^2(X_r)]-\sum_{i=1}^{N}\lambda_i(X_r)C(W(X_r),W(X_i)) \quad (8)$$

It is noted that equation(7) is a N-dimensional simultaneous algebraic equation and the coefficients $\lambda_i(X_r)$ may be solved unless the field is perfectly correlated. Since we find from equation(7) that $\lambda_i(X_r)=1$ if $i=r$ and $\lambda_i(X_r)=0$ if $i\neq r$, equation(2) becomes when the unrecorded location X_r coincides with one of the recorded locations $X_r=X_j$,

$$\hat{W}(X_j) = \lambda_j(X_j)W(X_j) = W(X_j) \quad (9)$$

And in this case, equation(8) becomes

$$\sigma^2[\varepsilon(X_j)]$$
$$= E[W^2(X_j)]-\lambda_j(X_j)C(W(X_j),W(X_j))$$
$$= 0 \quad (10)$$

With equation(9) and (10), equation(3) becomes when the sample field is substituted,

$$\underline{W}^*(X_j) = \underline{W}(X_j) \quad (11)$$

This indicates that the interpolation coincides precisely with the observed one at X_j.

4 CONDITIONAL SIMULATION I

Simulation is based on equation(3). If equation(7) holds, the properties of the error function $\varepsilon(X_r)$, $r=1\sim M$ are the following two orthogonality conditions.

$$E[\hat{W}(X_r)\varepsilon(X_r)] = 0 \quad (12)$$
$$E[W(X_j)\varepsilon(X_r)] = 0$$
$$(X_j \text{ is a recorded location}) \quad (13)$$

and

$$E[\varepsilon(X_r)\varepsilon(X_s)]$$
$$= C(W(X_r),W(X_s))-\sum_{i=1}^{N}\lambda_i(X_s)C(W(X_r)W(X_i)) \quad (14)$$

where X_r and X_s are unrecorded locations.

Equation(14) is not zero in general. Equations(12) and (13) mean that since $E[\varepsilon(X_r)]=0$, $\varepsilon(X_r)$ is not correlated with the interpolation term $\hat{W}(X_r)$ and with $W(X_j)$ at recorded locations. Since a gaussian field is considered, this means that $\varepsilon(X_r)$ is independent of $\hat{W}(X_r)$ and $W(X_j)$. However, $\varepsilon(X_r)$, $r=1\sim M$ are mutually correlated because of equation(14).

With these properties, the simulation on $W(X_r)$, $r=1\sim M$ may be carried out in the following way.

First, evaluate $\lambda_i(X_r)$, $i=1\sim N$, $r=1\sim M$ based on equation(7). Then, obtain the sample field $\underline{\hat{W}}(X_r)$, $r=1\sim M$ by substituting recorded data $\underline{W}(X_i)$ to equation(4) as follows.

$$\underline{\hat{W}}(X_r) = \sum_{i=1}^{N}\lambda_i(X_r)\underline{W}(X_i) \quad (15)$$

Next, simulate a sample field $\underline{\varepsilon}(X_r)$, $r=1\sim M$ with zero mean and the covariance of equation(14). Finally, a set of simulated ones is obtained as

$$\underline{W}^*(X_r) = \underline{\hat{W}}(X_r)+\underline{\varepsilon}(X_r),\quad r=1\sim M \quad (16)$$

5 CONDITIONAL SIMULATION II

This method is an artistic excellence based on the orthogonality properties of equations(12) and (13). First obtain $\underline{W}^*(X_r)$ at a single location by simulating $\underline{\varepsilon}(X_r)$ independently from $\hat{W}(X_r)$ at the same location based on $E[\varepsilon(X_r)]=0$ and $\sigma^2[\varepsilon(X_r)]$. Then, this simulated one $\underline{W}^*(X_r)$ is included into the recorded sample field $\underline{W}(X_i)$, $i=1\sim N$. Thus, N+1 recorded data are now used to interpolate $\hat{W}(X_s)$ at another unrecorded location. Similarly, $\varepsilon(X_s)$ is independently simulated, and $\underline{W}^*(\overline{X_s})$ is obtained as $\underline{W}^*(X_s)=\underline{\hat{W}}(X_s)+\underline{\varepsilon}(X_s)$.

In this manner of a step by step expansion, a multiply correlated sample field may be realized.

6 INTERPOLATION

(1) If X_r is remote from recorded loca-

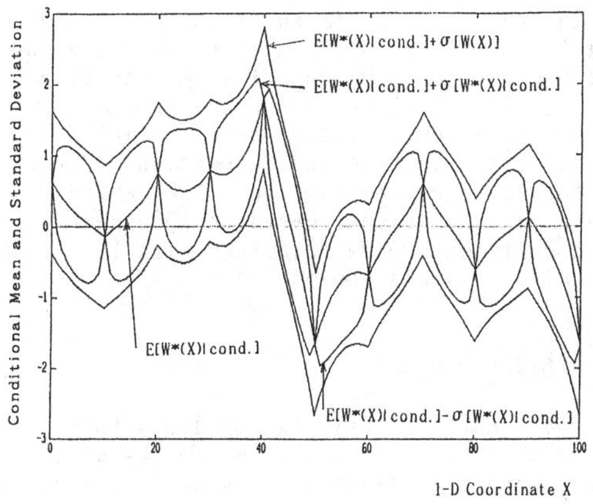

Fig.1 Analytical solutions

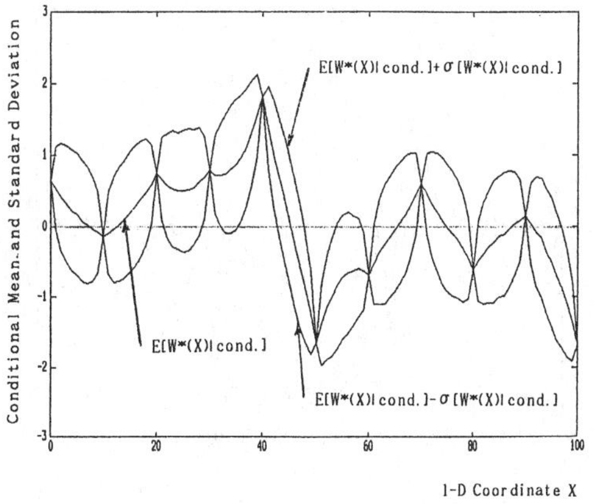

Fig.2 Solutions by simulation by a step by step expansion method (1000 sample fields)

tions Xi, i=1~N, correlation between W(Xr) and W(Xi) may be generally very weak. If there should be no correlation, then C(W(Xr),W(Xi))=0 and from equation(7), $\lambda i(Xr)$, i=1~N becomes zero. Thus, $\hat{W}(Xr)=0$. On the other hand, equation(14) becomes $E[\varepsilon(Xr)\varepsilon(Xs)]=C(W(Xr),W(Xs))$ indicating that the covariance of the error functions becomes the covariance of the stochastic field.

In this special case, equation(3) becomes $W^*(Xr)=\varepsilon(Xr)$. This means that the solution becomes identical to that of unconditional simulation.

(2) The method of conditional simulation I requires Cholesky decomposition or modal decomposition of an error covariance matrix consisting of equation(14) in order to simulate the multiply correlated $\varepsilon(Xr)$, whereas the method of conditional simulation II avoids this step and becomes a simple procedure.

(3) The conditional mean of $W^*(Xr)$ is given by

$$E[W^*(Xr)|\underline{W}(Xi), i=1\sim N] = \sum_{i=1}^{N}\lambda i(Xr)\underline{W}(Xi) \tag{17}$$

and the conditional variance of $W*(Xr)$ is given by

$$\sigma^2[W^*(Xr)|\underline{W}(Xi), i=1\sim N] = \sigma^2[\varepsilon(Xr)] \tag{18}$$

These two equations are simply obtained from the nature of linearity of equation(3). It should be pointed out that the conditional variance is independent of recorded data.

(4) The expansion of equation(3) to a spatial and temporal stochastic field may be straightforward based on the following equation.

$$W^*(Xr,k) = \sum_{i=1}^{N}\sum_{j=-M}^{M}\lambda ij(Xr)W(Xi,k+j)+\varepsilon(Xr,k) \tag{19}$$

where k stands for a discrete time with $t=k\cdot\Delta$, in which Δ is a discrete time interval.

It should be noted that equation(19) seems to be an autoregressive model in the form. However, it is quite different in the fact that $\varepsilon(Xr,k)$ is not a white noise as in equation(14).

7 NUMERICAL VERIFICATION

The following covariances are postulated for a one dimensional homogeneous field.

$$C(W(Xk),W(Xm)) = \exp\{-|Xk-Xm|/5.0\} \tag{20}$$

Then, eleven locations with equal interval are chosen along a line, and $\underline{W}(Xi)$, i=1~11 are simulated on equation(20) by the Cholesky decomposition method. Thus, they are employed as recorded data at eleven locations. Next, $\underline{W}^*(Xr)$ at inbetween locations with unit interval are conditionally simulated by the step by step expansion. Figure 1 shows analytical solutions of the conditional mean of equation(17) and the conditional variance of equation(18) together with the theoretical bounds whereas Figure 2 shows the corresponding solutions based on 1000 simulated samples at each locations. The curves in the both figures are almost identical. Figure 3 shows that the conditional mean and variance approach zero and unity respectively if the location is taken remote from recorded locations. In other words, extrapolation at remote locations is same to solutions of unconditional simulation.

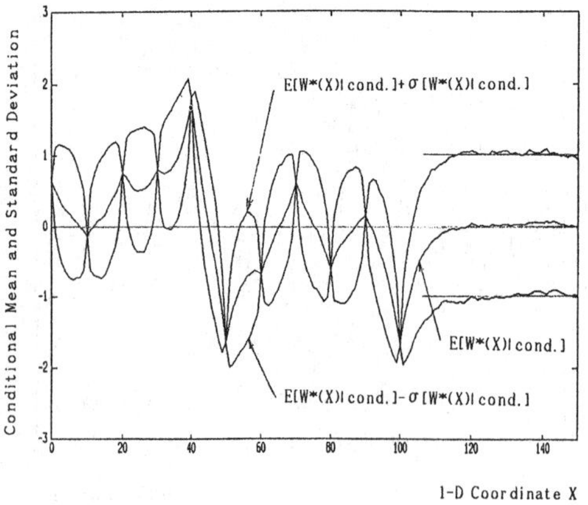

Fig.3 Inter and extra-polation

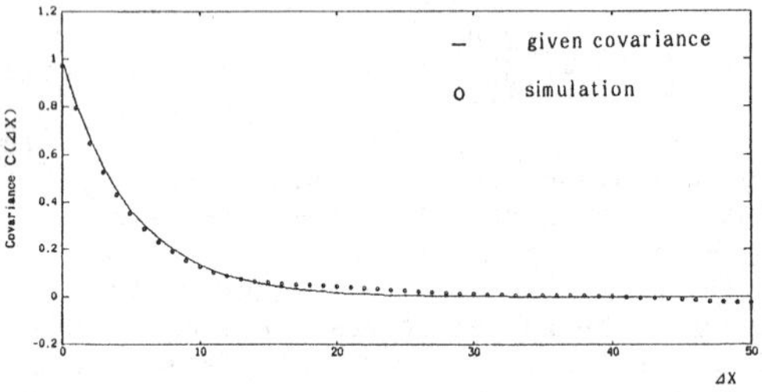

(a) Left to right expansion

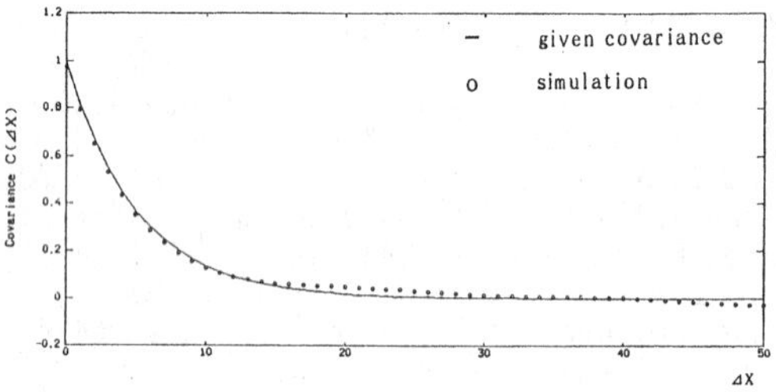

(b) Right to left expansion

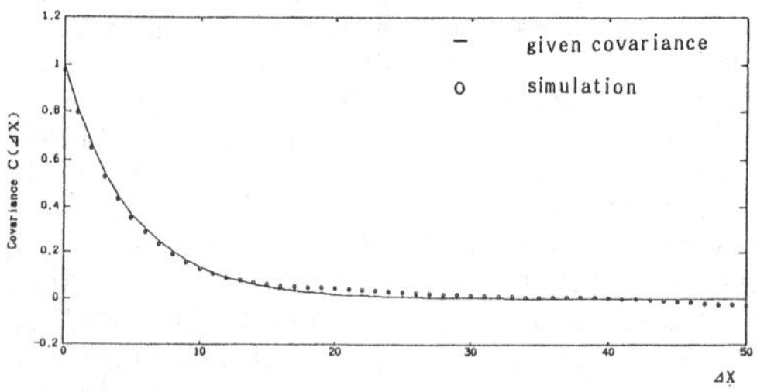

(c) Alternative expansion

Fig.4 Covariance of stochastic field

In order to justify solution invariance of the sampling location order of a step by step expansion procedure in the conditional simulation II, three different expansions are carried out, namely left to right expansion, right to left expansion and alternative expansion toward the central lacation. Covariance based on 1000 sample fields in each case coincides with the specified covariance.

8 CLOSING REMARKS

(1) Conditional simulation II discussed herein is original and efficient as well in engineering applications compared to the common practice based on equation(1) where the Cholesky decomposition of a covariance matrix of dimension (N+M)*(N+M) is required to simulate S(X).

(2) It shoud be noted that the problem of conditional simulation discussed herein is, in the basic notion and theory, identical to finding the solution of Wiener-Hopf integral equation in least mean square estimation (Papoulis(1965)) and also to the solution of Kalman-Filter identification and prediction (Sakai and Hoshiya(1987)).

APPRECIATIONS

The author appreciates Mr.Kuwana for his carrying out the numerical calculation, and Dr.Maruyama for his invaluable discussion.

REFERENCES

Cressie, N.A.C. 1991. Statistics for Spatial Data, John Wiley.

David, M. 1977. Geostatistical Ore Reserve Estimation, Elsevier Scientific Publishing Co.

Davis, J.C. and McCullagh, M.J. 1975. Display and Analysis of Spatial Data, NATO Advanced Study Institute, John Wiley.

Journel, A.G. 1974. Geostatistics for Conditional Simulation of Ore Bodies, Economic Geology, Vol.69: 673-687.

Journel, A.G. and Huijbregts, Ch.J. 1979. Mining Geostatistics, Academic Press Inc.

Krige, D.G. 1966. Two-dimensional Weighted Moving Averaging Trend Surfaces for Ore Evaluation, Proc. of Sym. on Math., Stat. and Computer Appl. for Ore Evaluation, Johannesburg, South Africa: 13-38.

Matheron, G. 1973. The Intrinsic Random Functions and Their Applications, Adv. Appl. Prob.5: 439-468.

Matheron, G. 1989. Estimating and Choosing, Springer-Veriag.

Papoulis, A. 1965. Probability, Random

Variables and Stochastic Processes, McGraw-Hill, New York.
Ripley, R.D. 1981. Spatial Statistics, John Wiley.
Sakai, K. and Hoshiya, M. 1987. Prediction and Control of Behaviors on Driving Shields Using Kalman-Filter Theory, VI-7, J. of JSCE, Sept.

Structural Safety & Reliability, Schuëller, Shinozuka & Yao (eds) © 1994 Balkema, Rotterdam, ISBN 90 5410 357 4

Application of stochastic finite element method to system reliability analysis of offshore structures

Joo-Sung Lee
University of Ulsan, Republic of Korea

ABSTRACT: This paper is concerned with the structural system reliability analysis of offshore platforms. The stochastic finite element method (SFEM) has been applied for a more explicitly account for the uncertainty in the post-ultimate behaviour of failed members in assessing structural system reliability. Several case studies have been carried out to investigate the effect of the residual strength and its uncertainty on the system reliability of a fixed type offshore structure. Comparison of system reliability is made between the cases when the post-ultimate behaviour is treated as probabilistic and when it is done as deterministic. This show the significance of considering the uncertainty of the post-ultimate behaviour in the system reliability analysis.

1. INTRODUCTION

In system reliability analysis it has been well recognised that the post-ultimate behaviour of a member after failure much affect the residual strength of structural system and consequently on the level of system reliability (e.g, see Moses & Rashedi 1983, Melchers & Tang 1985, Lee 1989). The post-ultimate behaviour is usually characterised by the post-ultimate slope, θ' and the residual strength parameter, η as shown in Fig.1 which are commonly treated as deterministic variables at present. In system reliability analysis the post-ultimate behaviour has been in usual treated as deterministic. The principal structural members found in offshore platforms are mainly cylindrical members. If one examine the previous experimental works for such structural members it would be found that the post-ultimate behaviour after a member has failed basically shows non-ductile behaviour and also there is enough uncertainty in the post-ultimate behaviour. It seems to be, therefore, natural to take into account the uncertainty of the post-ultimate behaviour in the system reliability analysis of such structural system.

At present commonly used is the reliability analysis procedure that structural analysis is carried out just once, and then the load effects, say stress and displacement, are directly input into the reliability analysis procedure. And so the effect of variation of material and geometric variables on the variation of load effects are disregarded in the structural analysis and the level of uncertainties in load effects are assumed to be the same as those of applied loads. In this paper, this method is termed as the "ordinary reliability method". To account for their effects on the variation of load effects, the stochastic finite element methods have been proposed and applied to reliability assessment of structures (Der Kiureghian & Ke 1985, Handa & Karrholm 1975, Hisada & Nakagiri 1981, Shinozuka 1987). One of major shortcomings of the stochastic finite element method is that it is computationally much more expensive than the ordinary reliability method since the structural analysis should be repeated so many times to get the gradient of limit state equation to random variables (at every iteration steps when iterative method is used). In spite of this the stochastic finite element method seems to be an adequate method to account for the uncertainties in the post-ultimate behaviour of a structural member.

In system analysis of this study the post-ultimate behaviour of a member after failure is modelled into the two-state model as in Fig.1(a), that is, the residual strength parameter, η only is treated as another random variable, which is defined as the ratio of residual strength after ultimate state to the ultimate strength. For the system reliability analysis, the extended incremental load method is used to get the limit state equation of failure mode and the identifying procedure in Lee (1989) is used. Several case studies have been carried out with varying the mean and COV of the residual strength parameter to investigate their quantitative effects on the system reliability.

2 FORMULATION OF THE PRESENT METHOD FOR STOCHASTIC FINITE ELEMENT ANALYSIS

An essential point in applying the stochastic finite element method is to find the partial derivatives of limit state equation to random variables. The formulation of the present stochastic finite element method is as follow (Lee, 1991). Let divide the

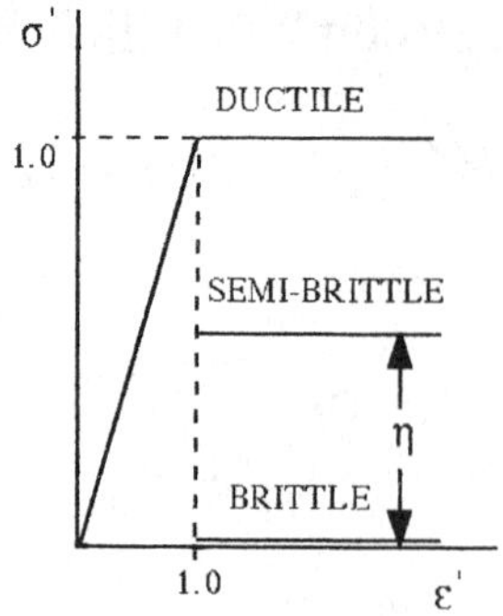

(a) two-state model

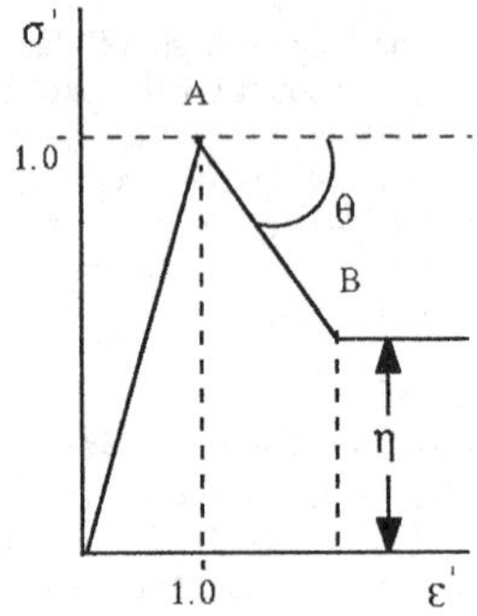

(b) three-state model

Fig. 1 Typical model of post-ultimate behaviour

random variables into two groups: resistance variable vector $\{r\}$ and load variable vector $\{q\}$, that is, random variable vector $\{x\}$ is:

$$\{x\} = (\{r\}, \{q\}) \tag{1}$$

and the limit state equation is expressed in terms of random variables as:

$$g(\{x\}) = g(\{r\}, \{q\}) \tag{2}$$

Using the displacement method of structural analysis for a linear system the stiffness equation is given as:

$$[K]\{U\} = \{F\} \tag{3}$$

where $[K]$ is the stiffness matrix of a total structural system. $\{U\}$ and $\{F\}$ are nodal displacement and nodal force vectors, respectively. For a linear structural system the load effects $\{q\}$ would be stress at elements or nodal displacement. For a component "e" considered now the load effect is obtained from:

$$\{q^{(e)}\} = [B]\{u^{(e)}\} \tag{4}$$

in which superscript (e) is the element which contains the component considered now and matrix $[B]$ is the load effect-nodal displacement relation matrix. At the current design point $\{x^*\}$ the limit state equation is $g(\{x^*\}) = g(\{r^*\}, \{q^*\})$ where $\{r^*\}$ is explicitly known in terms of $\{x^*\}$, and $\{q^*\}$ is given using Eqs(3) and (4) as:

$$\{q^*\} = [B]^T\left([K]^{-1}\{F\}\right)^{(e)}_{x = x^*} \tag{5}$$

in which the curled bracket of vectors $\{x\}$ and $\{x^*\}$ are omitted. Superscript (e) is added to denote that the nodal displacement vector, referred to element (e), is sorted out using the element definition data. The partial derivatives of limit state equation (2) to random variables is given by:

$$\left\{\frac{dg}{dx_i}\right\}_{x = x^*} = \left[\left\{\frac{dg}{dr_k}\right\}^T\left\{\frac{dr_k}{dx_i}\right\} + \left\{\frac{dg}{dq_k}\right\}^T\left\{\frac{dq_k}{dx_i}\right\}\right]_{\substack{x = x^* \\ r = r^* \\ q = q^*}} \tag{6}$$

Using Eq.(5) the derivative $\{dq_k/dx_i\}$ is given as:

$$\begin{aligned}\left\{\frac{dq}{dx_i}\right\}_{x = x^*} &= \left[\left\{\frac{dB}{dx_i}\right\}^T\{u^{(e)}\} + [B]^T\left\{\frac{dU}{dx_i}\right\}^{(e)}\right]_{x=x^*} \\ &= \left[\frac{d[B]^T}{dx_i}\{u^{(e)}\} + [B]^T\left(\frac{d[K]^{-1}}{dx_i}\{F\} + [K]^{-1}\frac{d\{F\}}{dx_i}\right)^{(e)}\right]_{x=x^*}\end{aligned} \tag{7}$$

It is easily shown that

$$\frac{d[K]^{-1}}{dx_i} = -[K]^{-1}\frac{d[K]}{dx_i}[K]^{-1} \tag{8}$$

Then Eq.(7) becomes

$$\left\{\frac{dq}{dx_i}\right\}_{x = x^*} = \left[\left\{\frac{dB}{dx_i}\right\}^T\{u^{(e)}\} + [B]^T\left[[K]^{-1}\left(-\frac{d[K]}{dx_i}\{U\} + \frac{d\{F\}}{dx_i}\right)^{(e)}\right]\right]_{x = x^*} \tag{9}$$

where $\{U\}$ is obtained from Eq.(3) at the current design point of random variables. Once after calculating the partial derivatives of limit state equation the iterative procedure (Rackwitz & Fiessler, 1978) does work. The above formulation of the stochastic finite element method has a merit that the available computer code for the ordinary reliability analysis can be linked to this procedure without much modification.

3 NUMERICAL EXAMPLE

Fig.2 shows a three-story offshore structure model from Thoft-Christensen & Murotsu (1986), and its geometric properties and the applied load are listed in Table 1.

The behaviour of a structural member found in offshore structures should be differently modelled according to the structural member type and the applied load type. The offshore structure model shown in Fig.2 is composed of unstiffened cylindrical members. The post-ultimate behaviour of such type of structural member can be modelled into as follows.

1. In the case that a member is under axial tension, the behaviour is ductile and deterministic.
2. In the case that a member is under axial compression, since the post-ultimate slope is nearly negatively infinite, the effect can be disregarded, and so the behaviour can be modelled into the two-state model as shown in Fig.1(a).
3. In the case that a member is under bending, the member shows close to ductile behaviour with some unloading as shown in Fig.3 (thin solid line) and so it can be modelled into the three-state behaviour as the thick solid line in Fig.3. With referring to the test results the post-ultimate behaviour has small uncertainty, such that its COV is less than 10%. COV of η and θ' is assumed to be 10% in this study.

Case study with varying mean and COV of the residual strength parameter, η, when a member is under compression, has been carried out with in the following range:

mean of η : μ_η = 0.4 to 0.9

COV of η : V_η = 0 to 40%

Thoft-Christensen & Murotsu (1986) assumed the joints of brace members to be freely moved and so, the structure showed unlikely low system reliability in such a way that the minimum path reliability index was around 1.3. In this paper the joints are assumed to be clamped (this would be more reasonable). COV of wave load is assumed to be 40% and that of deck load 10%. All variables are assumed to be normally distributed. When a component is under combined loading condition of axial tension or compression and bending, following interaction equation is employed as a strength model (Warwick & Faulkner 1988).

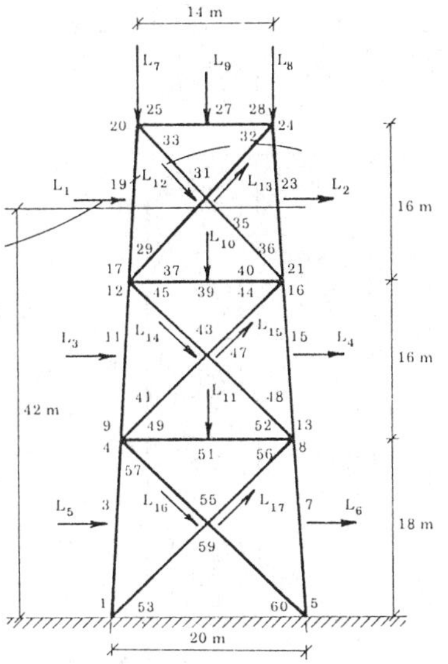

Fig.2 Example offshore structure
(after Thoft-Christensen & Murotsu 1986)

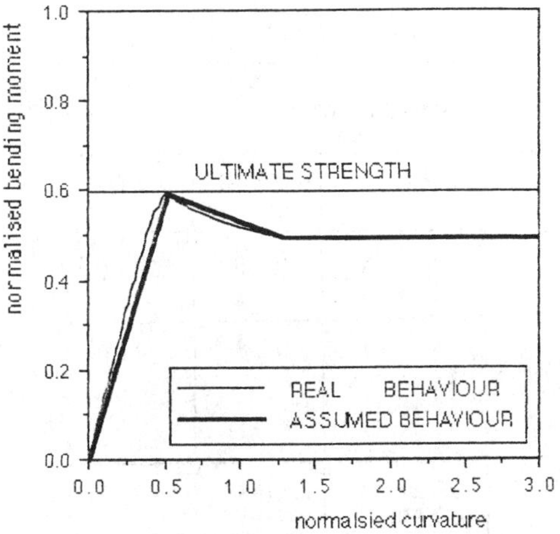

Fig.3 Assumed post-ultimate behaviour of a member under bending

(1) when axial stress is tensile

$$\frac{\sigma_x}{\sigma_Y} + \frac{M}{M_Y} \leq 1.0 \qquad (10)$$

where σ_x : applied axial stress

σ_Y : yield stress

M : applied bending moment

M_Y : first yield bending moment

Table 1 Data of offshore structure in Fig.2 (unit: m, KN)

o Load data

type of load no.	mean value
wave load	
L_1	368.1
L_2	334.1
L_3	181.5
L_4	169.6
L_5	142.7
L_6	137.6
L_9	4.5
L_{10}	5.5
L_{11}	3.7
L_{12}	139.0
L_{13}	205.0
L_{14}	72.3
L_{15}	88.9
L_{16}	56.9
L_{17}	59.9
deck load	
L_7	2490.0
L_8	2490.0

o Strength data

component no.	sectional area	strength
1, 3, 4 5, 7, 8	0.0810	5286.0
9,11,12 13,15,16 17,19,20 21,23,24	0.0638	3842.0
25,27,28 37,39,40	0.0154	476.9
29,31,32 33,35,36 41,43,44 45,47,48	0.0200	798.5
49,51,52	0.0167	517.8
53,55,56 57,59,60	0.0247	980.4

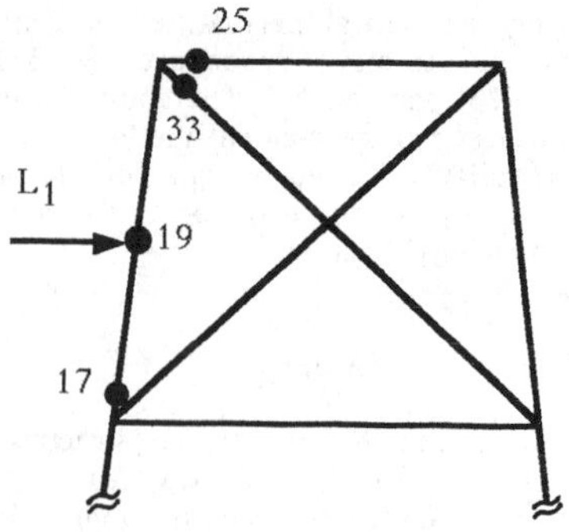

Fig.4 Failure path 19-33-17-25 (figures are failed component numbers)

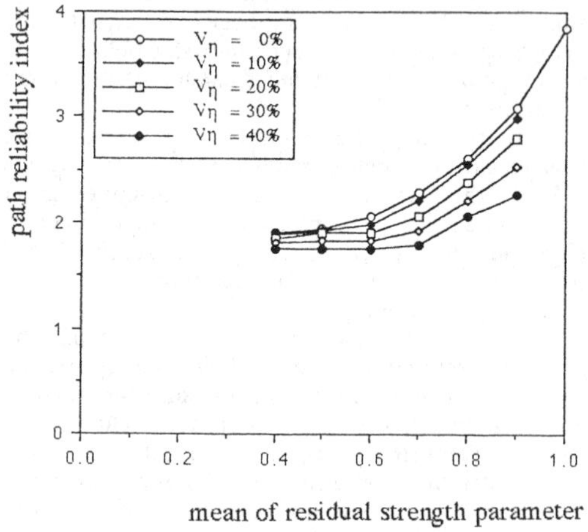

(a) β_{path} vs μ_η

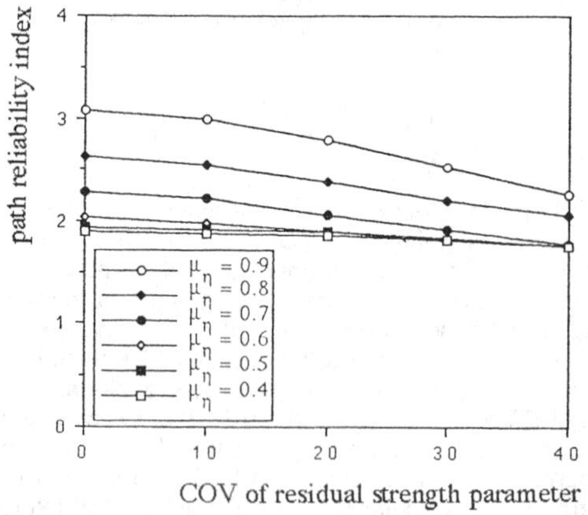

(b) β_{path} vs V_η

Fig.5 β_{path} vs μ_η and V_η for offshore structure in Fig.3

(2) when axial stress is compressive

$$\frac{\sigma_x}{\sigma_u} + k\frac{C_m M}{M_P} \leq 1.0 \qquad (11)$$

where σ_u : ultimate compressive stress
M_P : plastic moment capacity of section
$C_m = 0.6 - M_B/M_A \geq 0.4$
k : elastic magnification factor

$$k = \frac{1}{1 - \frac{\sigma_x}{\sigma_{cr}}}$$

σ_{cr} : elastic buckling stress

Sturctural failure is assumed to occur if following inequality is satisfied:

$$\frac{\text{determinant of } [K_d]}{\text{determinant of } [K_o]} \leq \varepsilon \qquad (12)$$

where $[K_o]$ and $[K_d]$ are the stiffness matrix of intact and damaged structures, respectively. ε is a prescribed small number. Here, ε is taken as 10^{-3}.

For all cases of μ_η and V_η the most important failure mode has been obtained as the same one in Fig.4, path 19-33-17-25 (figures are component numbers). This is because there is a concentrated load at the mid span of the column and the brace member on the third floor, the applied loads are greater than others and especially the column has the weaker strength than others. And so it can be easily expected that the components on the member are likely to fail before others do.

The case study results are shown in Fig.5 as the relation between path reliability index (β_{path}), μ_η, and V_η. It can be seen that there is a significant influence of μ_η on β_{path} as its value is close to 1.0, and when μ_η is less than about 0.6, there is not much change in β_{path} to V_η. For examples, in the case of the fixed value of V_η , as 10% as μ_η is reduced by 0.1 from 0.9 to 0.4, the decreasing rate of path reliability index is about 15%, 12%, 10%, 4% and 2%, respectively at each decreasing step. In the case that the fixed V_η is great,say 30%, the decreasing rate is about 13%, 11%, 6%, 0%, 0%, respectively at each decreasing step. As it can be seen in Fig.5(b) the path reliability index is more sensitive to change in V_η as μ_η is closer to 1.0. For the fixed value of μ_η the maximum difference of path reliability index within the present case study range, i.e., the difference between the cases when V_η is 0% and 40%, is about 27% when μ_η is 0.9, and only about 7% when μ_η is 0.4. From Fig.5(a) it can be also seen that the path reliability index converges to a certain value as μ_η decreases to zero. This may indicate the minimum system strength with regardless to μ_η value of a component behaviour.

Since the failre path shown in Fig.4 is the most important or dominant one, it can be said that the change of system reliability index to changes in μ_η and V_η for the present offshore structure model shows the similar tendency as the path reliability index.

Similar results can be also found for the floating type offshore structure in Lee (1990). Fig.6 is the system reliability analysis results for a TLP model, called TLP-B which is an variant of the Hutton TLP in the North Sea. The results were obtained when η was deterministic and by using the three-state model as well as two state model for the component behaviour (see Fig.1).

4 CONCLUSION

This paper is an attempt to apply the stochastic finite element method to the structural system reliability analysis to show the effect of residual strength of a failed structural component. It has been found that the residual strength does much affect the system reliability level of offshore structure, and its uncertainty, therefore, should likely be considered in the reliability analysis for a more realistic estimation of the structural safety.

This paper is basically aimed to investigate the effect of uncertainty in the residual strength of failed components on the path reliability of failure mode

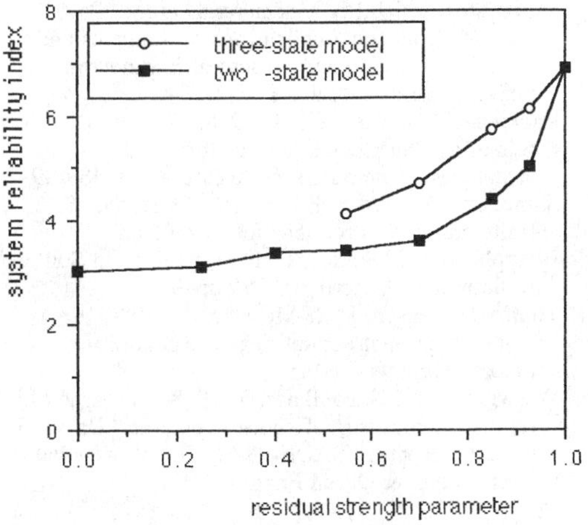

Fig.6 System reliability index to change in residual strength Parameter for TLP-B (after Lee 1990)

with the two-state model. More work would be needed in applying the method to reliability analysis of structural system before applying to real structures, and the uncertainties in the post-ultimate slope as well as in the residual strength should be quantified for various types of structure members under possible loading conditions through experiments and non-linear structural analysis.

REFERENCES

Der Kiureghian, A. & Ke, J.B. 1985. Finite element based reliability analysis of frame structures. Proc. 4th Intl. Conf. on Applications of Statistics and Probability in Soil and Structural Engineering, Univ. of Firenze, Italy. 1: 395-404

Handa, K. & Karrholm, G. 1975. Application of finite element method in the statistical analysis of structures. : Chalmer Univ. of Tech.

Hisada, T. & Nakagiri, S. 1981. Stochastic finite element method developed for structural safety and reliability. Proc. 3rd Intl. Conf. on Structural Safety and Reliability, Trondheim: 395-408

Lee, J.S. 1989. Reliability analysis of continuous structural systems. Ph.D. Thesis, Dept. of Naval Architecture & Ocean Engineering: Univ. of Glasgow, U.K.

Lee, J.S. 1990. System reliability of TLP structures to resistance variables. Proc. 9th OMAE Conf., Houston , Texas, U.S.A. 2: 115-123

Lee, J.S 1991. Reliability analysis of structural systems by using the stochastic finite element method. Proc. 4th IFIP WG7.5 Conf. on Reliability and Optimization of Structural Systems, Munich, Germany: 267-284

Melchers, R.E. & Tang, L.K. 1985. Failure modes in complex stochastic systems. Proc. 4 th Intl. Conf. on Structural Safety and Reliability, Kobe, Japan. 1:97-106

Moses, F. & Rashedi, M.R. 1983. The application of system reliability to structural safety. Proc. 4th Intl. Conf. on Applications of Statistics and Probability in Soil and Structural Engineering, Univ. of Firenze, Italy. 1: 573-584

Rackwitz, R. & Fiessler, B. 1978. Structural reliability under combined random load sequences. Computers & Structures. 9: 489-494

Shinozuka, M. 1987. Basic issues in stochastic finite analysis. Proc. 5th Intl. Conf. on Application of Statistics and Probability in Soil and Structural Engineering. 1: 506-519

Thoft-Christensen, P. & Murotsu, Y. 1986. Application of structural systems reliability theory: Springer-Verlag

Warwick, D.M. & Faulkner,D. 1988. Strength of tubular members in offshore structures. Univ. of Glasgow Report, NAOE-88-36, Dept. of Naval Architecture & Ocean Engg.

Structural Safety & Reliability, Schuëller, Shinozuka & Yao (eds) © 1994 Balkema, Rotterdam, ISBN 90 5410 357 4

Probabilistic analysis of micro-fabricated multicrystalline beams

Dariush Mirfendereski & Armen Der Kiureghian
Department of Civil Engineering, University of California at Berkeley, Calif., USA

ABSTRACT: This paper reports on detailed approaches to the probabilistic modeling and analysis of the mechanical response of multicrystalline structural beams in the context of micro-electro-mechanical systems (MEMS) using simulation, random fields and stochastic finite element techniques. MEMS are used as micro sensors and actuators with structural dimensions of the order of a micron. Due to the microscopic sizes of these structures, the random orientation and shape of the material crystal grains have significant influences on their mechanical response.

Random, multicrystalline beam structures are simulated and analyzed by the finite element method, employing randomly oriented orthotropic material properties to model each individual crystal grain. The simulation studies are used to assess the influence of size effects on the beam response. Accurate estimation of these characteristics is essential for functioning of the beam as a reliable micro-sensor. Results indicate that, for the typical sizes of MEMS, the spatial randomness in the micro-structure has a significant influence on the response characteristics of the beam, thus demonstrating the necessity for a probabilistic approach to their design and analysis.

For a more efficient probabilistic approach than simulation, a continuous-parameter random field characterization of the crystalline micro-structure is employed. Auto-correlation functions for equivalent elastic properties of the micro-structure field are generated by a moving window approach in the field of simulated or observed crystals. The equivalent random field properties are then used in conjunction with stochastic finite element methods to compute the reliability of the beam. Results from the simulation and random field approaches are in close agreement, thus validating the more efficient continuous-parameter equivalent random field approach.

INTRODUCTION

Micro-fabricated multicrystalline beams are used in Micro-Electro-Mechanical Systems, or MEMS (O'Connor 1992), where their mechanical response to excitations is of interest. In MEMS, electronic and mechanical components are combined to make miniature structures such as sensors that deploy automobile air bags in collisions or which monitor blood pressure inside the human heart during operations, as well as valves and actuators, all with structural dimensions as small as a micron. Based on processing techniques borrowed from the semiconductor industry, these tiny structures can be inexpensively mass-produced, adding the benefit of low cost to their size advantages. Following the developments of the transistor, the integrated circuit, and the microprocessor, the development of MEMS promises to be the next major milestone in electronics.

Structural beams used in MEMS are typically made from polycrystalline materials, e.g. polysilicon. It has been shown that for some classes of these beams, called multicrystalline beams, the effects of random crystal size, shape, and orientation result in the existence of significant random spatial variability of material parameters even though the elastic properties of each individual crystal are well known (Mirfendereski and Der Kiureghian,

1992). This is in contrast to most traditional probabilistic structural analysis problems, e.g. as applied to prestressed concrete segmental box-girder bridges (Bazant and Kim, 1991), where the effect of random spatial variability can be neglected for all practical purposes since it is considerably smaller than the effect of random statistical variability of material parameters.

In their previous study, the authors used simulations of the multicrystalline structure to model the random spatial variability of material parameters. This simulation approach resulted in phenomenological insights into the problem, showing the significance of the random spatial variabilities. However, as is explained below, this approach is computationally inefficient. For a more efficient probabilistic analysis of MEMS, random field models of the materials are developed based on simulations of the random multicrystalline structure. A continuous-parameter equivalent random field approach is developed here, where the auto-correlation function for the equivalent flexural rigidity of the beam is generated by a moving-window approach in the field of simulated or observed crystals. After checking the validity of this random field model, it is used in conjunction with finite element reliability methods to compute the reliabilities of the multicrystalline beam structures of interest.

MULTICRYSTALLINE SIMULATIONS

Multicrystalline structures in MEMS can be such that the typical linear dimension of a crystal within the structural material is of the same order of magnitude as the smallest dimension of the structure—hence the term "multicrystalline" as opposed to "polycrystalline." Fig. 1 shows such a structure in which each crystal has a distinct crystallographic orientation with respect to the global coordinate system. The crystal material properties are typically known but anisotropy, thus resulting in a discontinuous, non-homogeneous field of material properties for the beam.

The multicrystalline structure in Fig. 1 is based on a simple but versatile method of randomly dividing a two dimensional region into "crystals" called the Voronoï tessellation

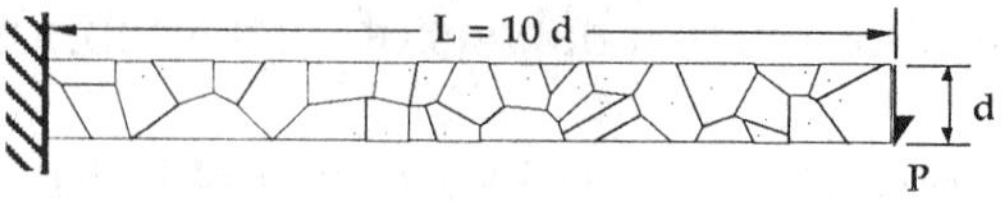

Fig. 1 Example "Multicrystalline" Structure and Loading

(Aurenhammer 1991). Given some number of generated points in the plane—here based on a Poisson point process (Stoyan et al., 1987)—their Voronoï tessellation, or diagram, divides the plane according to the *nearest-neighbor rule,* where each generated point is associated with the region of the plane closest to it. The arrangement shown in Fig. 1 with d = 1 has $\lambda = 3$ for the density of the stationary Poisson point process, i.e. 3 points per unit area. Alternatively, an empirical model similar to the tessellation shown in Fig. 1, but derived using image analysis techniques with actual MEMS, could be developed and used in subsequent analyses. Current areas of research in image analysis of the microstructure of materials include studies of techniques that interpret micrographs of multicrystalline microstructures as a tessellation similar to that shown in Fig. 1 (Williams et al., 1991).

The solution of the boundary value problem shown in Fig. 1 would require a model appropriately describing the material behavior. Homogenization is often used in obtaining bounds on macroscopic properties for polycrystalline materials (Hirsekorn 1990). In all homogenization schemes, however, the existence of a representative volume element (RVE) of the medium—"a cube sufficiently large at the micro-scale and sufficiently small at the macro-scale" (Hashin 1983)—is assumed. For example, a cubic millimeter of steel containing tens of thousands of crystals is an appropriate RVE in conventional structural uses of steel. Conversely, for the multicrystalline structure in Fig. 1, the requirement for the existence of a RVE is clearly not met and therefore homogenization schemes cannot be applied. Mirfendereski et al. (1992) have demonstrated this by showing how the "rigorous" Voigt and Reuss bounds are violated.

Since homogenization is not valid, new approaches to solving the problem need to be developed that take into account the uncertain discontinuities in and inhomogeneity of the material properties. Discrete simulations of the material microstructure of the type shown in Fig. 1 can be used initially for analyzing the problem. Based on a random multicrystalline realization, the crystals are individually modeled with oriented anisotropy material properties (orthotropic for silicon which is widely used in MEMS), reflecting the desired material texture, usually given as the orientation distribution function or ODF (Bunge 1982). Given the types of MEMS structures that are being fabricated (Wise 1991) and the associated range of imposed force and displacement boundary conditions, a linear elastic assumption for the material behavior is appropriate with the following stress-strain relations applying in the principal directions for materials with cubic crystal symmetry such as silicon:

$$\begin{bmatrix} \sigma_{11} \\ \sigma_{22} \\ \tau_{12} \end{bmatrix} = \begin{bmatrix} C_{11} & C_{12} & 0 \\ C_{12} & C_{11} & 0 \\ 0 & 0 & C_{44} \end{bmatrix} \begin{bmatrix} \varepsilon_{11} \\ \varepsilon_{22} \\ \gamma_{12} \end{bmatrix} \tag{1}$$

Given a rotation ψ in the plane, the stress-strain relations are transformed (Lekhnitskii, 1963) with the transformed matrix coefficients, denoted by superscript *, defined as follows:

$$C^*_{11} = \frac{1}{4}[(1-\alpha)(C_{11}-C_{12})\cos 4\psi + (3\alpha+1)C_{11} + (1-\alpha)C_{12}] \tag{2a}$$

$$C^*_{12} = \frac{1}{4}[(1-\alpha)(C_{12}-C_{11})\cos 4\psi + (3\alpha+1)C_{12} + (1-\alpha)C_{11}] \tag{2b}$$

$$C^*_{13} = \frac{1}{4}\left[(1-\alpha)(C_{11}-C_{12})\sin 4\psi\right] \tag{2c}$$

$$C^*_{22} = C^*_{11} \tag{2d}$$

$$C^*_{23} = -C^*_{13} \tag{2e}$$

$$C^*_{33} = \frac{1}{4}[(1-\alpha)(C_{12}-C_{11})\cos 4\psi + (1+\alpha)(C_{11}-C_{12})] \tag{2f}$$

where α, the degree of anisotropy, is defined as $\alpha = 2C_{44}/(C_{11} - C_{12})$.

The finite element model for the multicrystalline structure defined above employs an easy to generate, efficient, and convergent finite-element discretization mesh that also maintains the integrity of the crystal structure such that different orientation angles, ψ, can be assigned for each.

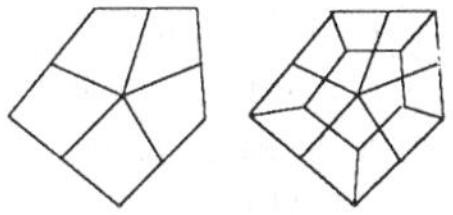

(a) (b)

Fig. 2 Subdivisions of a Typical Crystal into Quadrilateral Elements

The per crystal subdivisions into elements is shown in Figs. 2(a) and 2(b) in which the latter represents a mesh refinement of the subdivision shown in the former. Each n-sided polygon that defines the crystal boundaries is thus divided into n or 3n quadrilaterals, respectively. The random geometry of the crystalline polygons generally results in linearly distorted quadrilateral elements for which 9-noded isoparametric elements, used in this study, represent better cartesian polynomials and are generally preferable to 8-noded elements in minimizing the finite-element discretization errors. Material homogeneity is assumed within each crystal, allowing the material coefficients to be taken out of the integrations involving the shape functions, thus simplifying the finite element analysis. In this study, the general purpose, research-oriented finite element code FEAP is used for this purpose (Zienkiewicz and Taylor, 1989).

Results are simulated for an example structure with a uniformly random distribution of orientation angles, i.e. with no texture, associated with each realization of the Voronoï tessellation forming the overall beam geometry shown in Fig. 1. The coefficient of variation (C.O.V.) for the tip displacement was found to be approximately 3% for α = 1.562, i.e. representing silicon, and 6% for α = 5.0. Although these variabilities may appear small for conventional structures, they are significant for MEMS, where precision in predicting stiffness properties and response is critical due to their functional requirements.

The above simulation study represents a direct form of analyzing the problem and provides insights into the phenomenological characteristics of the problem. However, since the finite element discretization requires the generation of a new mesh for each simulation, this approach is computationally intensive and has limited applicability.

Furthermore, the simulation approach can be prohibitively costly for computing probabilities associated with different levels of performance. In the following sections, an alternative approach employing continuous-parameter random fields and stochastic finite elements is presented.

RANDOM FIELD CHARACTERIZATION

For a continuous-parameter random field characterization of the crystalline microstructure, the auto-correlation function for equivalent elastic properties of the field are generated by a moving window approach in fields of simulated crystals (Fig. 3). These equivalent random field properties are then used in conjunction with finite element reliability methods (Der Kiureghian & Ke, 1988) to compute the reliability of beam structures.

For beam structures, a one dimensional characterization of the flexural rigidity (EI) appears to be appropriate. To characterize the point-to-point random field, infinitesimally small window widths need to be taken, i.e. with W approaching zero. Each window has an equivalent average EI associated with its position x. Taking I to be constant throughout the beam, the equivalent Young's Modulus in the direction of the beam axis needs to be measured.

This can be achieved using two alternative finite element-based methods: subjecting the window to (a) uniaxial tension, and (b) pure flexure and deducing the equivalent Young's Modulus from the loading and resulting deformations. The results for the effective Young's modulus obtained by these loading conditions are found to be approximately equal.

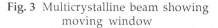

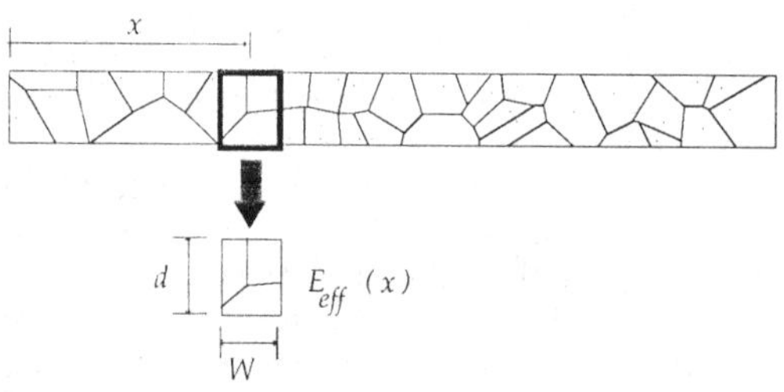

Fig. 3 Multicrystalline beam showing moving window

Alternatively, an analytical approach could be used. The results for the orientation-dependent axial Young's modulus, $E(\psi)$, for a homogeneous window, i.e. composed of only one crystal, would equal the axial Young's modulus obtained by setting $\sigma_{11} = \tau_{12} = 0$ in the transformed stress-strain relations. $E(\psi)$ is found to obey a simple analytic relation with respect to ψ, given by:

$$E(\psi) = \frac{1}{a \cos 4\psi + b} \tag{3}$$

where ψ is the orientation angle of the single-crystal window and a and b are constants related to C_{11}, C_{12}, and α (or C_{44}) as follows:

$$a = \frac{(\alpha - 1)}{4(C_{11} - C_{12})\alpha} \tag{4}$$

$$b = \frac{(3\alpha + 1)C_{11} + (1 - \alpha)C_{12}}{4(C_{11}^2 - C_{12}^2)\alpha} \tag{5}$$

Now, considering a multicrystalline window, e.g. as shown in Fig. 3, with n crystals, an approximation for the effective Young's modulus, E_{eff}, of the window is:

$$E_{eff} = \frac{1}{A_{Total}} \sum_{i=1}^{n} A_i E(\psi_i) \tag{6}$$

where $E(\theta_i)$ is defined in Eq. 3, A_i are the individual crystal areas within the window, and A_{Total} is the sum of all A_i, equal to the area of the widow. This approximation takes into account certain aspects of the geometry of the multicrystalline structure within each window in addition to crystallographic orientation information and the results compare favorably (to within 0.5%) with exact finite element analyses. Eq. 6 is subsequently used in the generation of realizations of the effective beam axial Young's modulus random field. Fig. 4 shows one realization of the effective axial Young's modulus along the beam, associated with a particular random crystal geometry and ODF.

The random field formulation requires a model describing the correlation function. The effective Young's modulus, or any other material property, obtained as realizations of a random field using the windowing method outlined previously is a *moving average* (MA) field. From a continuous parameter stationary random field $E(x)$ with mean μ and standard deviation σ, a family of moving average fields $E_W(x)$ can be derived by (Vanmarcke, 1983):

$$E_W(x) = \frac{1}{W}\int_{x-W/2}^{x+W/2} E(u)du \quad (7)$$

where W denotes the window width or averaging length.

Properties of moving average fields are based on the parent, or point-to-point random field. The mean is not affected by the averaging operation, however, the variance of $E_W(x)$ may be expressed as follows:

$$Var[E_W] = \sigma_W^2 = \gamma(W)\sigma^2 \quad (8)$$

where $\gamma(W)$, denoted the *variance function* of $E(x)$, is a dimensionless function measuring the reduction of the point variance under local averaging. In the limit as W tends to infinity, the expression $W\gamma(W)$ equals a constant, θ, known as the *scale of fluctuation*. The covariance function, $B_W(\chi)$, of local average fields is given by:

$$B_W(\chi) = \frac{\sigma^2}{2W^2}\left[\Delta(W+\chi) - 2\Delta(\chi) + \Delta(W-\chi)\right] \quad (9)$$

where $\Delta(W) = W^2\gamma(W)$. Thus knowing $\gamma(W)$ and σ of the parent random field, the autocorrelation coefficient function of a moving average field such as the one produced by the windowing method outlined previously can be derived for different values of W using Eq. 9.

Use is made of a family of models for the variance function, $\gamma(W)$, described by Vanmarcke (1983) as:

$$\gamma_m(W) = \left[1 + \left(\frac{W}{\theta}\right)^m\right]^{-1/m} \quad (10)$$

Substituting for $\gamma(W)$ in Eq. 9, we obtain an analytical expression for the moving average correlation function for given window width W as shown below:

$$B_W(\chi) = \frac{\sigma^2}{2W^2}\left\{(W+\chi)^2\left[1+\left(\frac{|W+\chi|}{\theta}\right)^m\right]^{-1/m} - 2\chi^2\left[1+\left(\frac{|\chi|}{\theta}\right)^m\right]^{-1/m} + (W-\chi)^2\left[1+\left(\frac{|W-\chi|}{\theta}\right)^m\right]^{-1/m}\right\} \quad (11)$$

Thus, given $\{\theta,m,\sigma\}$ the correlation function for a specified W is fully defined. The set of

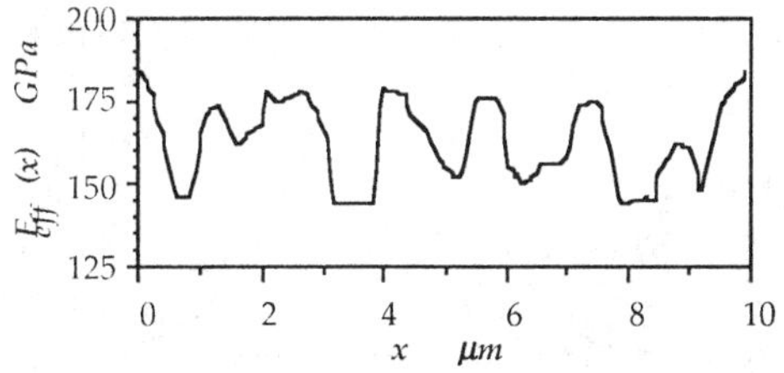

Fig. 4 One Realization of the Random Field $E_{eff}(x)$

Table 1 Effects of Texture and Average Grain Size on $\{\theta,m,\sigma\}$

λ	ODF	m	σ	θ
1	U(0,π)	2.5	13.016 x 10^9	0.9718
3	U(0,π)	2.5	11.944 x 10^9	0.5626
9	U(0,π)	2.5	10.740 x 10^9	0.2663
3	N(0,π/8)	2.5	10.888 x 10^9	0.5195
3	N(0,π/40)	2.5	1.008 x 10^9	0.5929

values $\{\theta,m,\sigma\}$, as well as the mean μ of the random field, are empirically derived from the generated data for various values of λ. Some of the results are summarized in Table 1, thus defining the covariance function (Eq. 11) for each case. Increased values of λ indicate higher densities of crystals, i.e. smaller crystals within the beam. For the uniformly distributed ODF's, the mean, μ = 164.256 x 10^9.

The structural model needs to employ appropriate beam elements to model the irregular variation in $E(x)$ shown in Fig. 4. Nonuniform beam elements are used here which allow a linear variation of $E(x)$ within each element, with constant second moment of area, I. The beam element degrees of freedom 1,2,3, and 4 are the vertical displacements and end rotations for the left and right nodes respectively. To solve for the elements of the stiffness matrix, the following three basic integrals need to be evaluated:

$$J_i = \frac{1}{I}\int_0^L \frac{x^i}{E(x)}dx \quad , i = 0,1,2 \quad (12)$$

Eq. 12, for i = 0, 1, 2, is solved analytically for linear variations in E(x). The element stiffness matrix is given by:

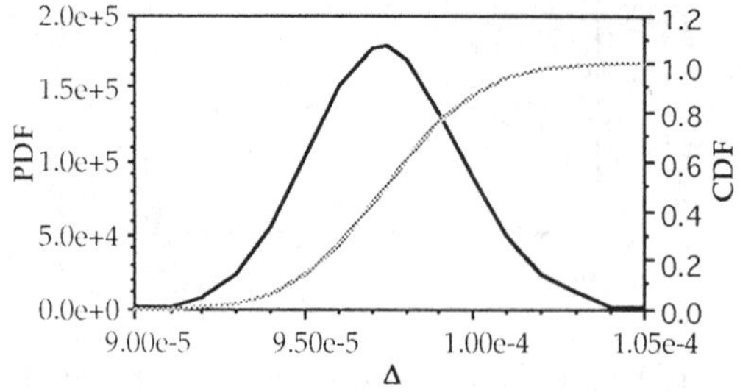

Fig. 5 PDF & CDF of Cantilever Tip Displacement, Δ

$$\mathbf{K} = \frac{1}{D}\begin{bmatrix} J_0 & J_1 & -J_0 & J_0 L - J_1 \\ J_1 & J_2 & -J_1 & J_1 L - J_2 \\ -J_0 & -J_1 & J_0 & -J_0 L + J_1 \\ J_0 L - J_1 & J_1 L - J_2 & -J_0 L + J_1 & J_0 L^2 - 2J_1 L + J_2 \end{bmatrix} \quad (13)$$

where L is the length of the beam element and D is defined by:

$$D = J_0 J_2 - J_1^2 \quad (14)$$

The above formulation enables a continuous modeling of any irregular variation in $E(x)$ and, with sufficiently fine discretizations, produces close representations of the multicrystalline beam behavior.

The random field formulation for the variation of $E(x)$ (Eq. 11), is now discretized using the "EOLE" method, recently developed by Li and Der Kiureghian (1993), which is based on the eigenvalues and eigenvectors of the covariance matrix of the nodal random variables (Eq. 15), and is represented by independent random variables ζ_i (Eq. 16).

$$\Sigma_{EE}\, \phi_i = \theta_i\, \phi_i \quad (15)$$

$$E^*(x) = \mu_E + \sum_{i=1}^{r} \frac{\zeta_i}{\sqrt{\theta_i}} \phi_i^T \Sigma_{E(x)E} \quad (16)$$

Reliability Analysis

This discretized form is used in conjunction with the beam element formulation presented above, to perform first order reliability and sensitivity calculations to evaluate the PDF and CDF of structural response. For the cantilever problem shown in Fig. 1, (with $\lambda = 3$, $\alpha = 1.56$, i.e. silicon, U(0,π) ODF, d = P = 1, and a beam thickness of 0.25) 20 random variables were used in discretizing the random field, together with 100 beam elements, consistent with the correlation length requirements of the random field. The PDF and CDF of the cantilever tip displacement, Δ, are shown in Fig. 5, where the coefficient of variation (C.O.V.) of response is approximately equal to 3.14%.

These results compare well with the simulations performed previously, thus validating the more efficient continuous-parameter equivalent random field approach. Having established and verified the equivalent random field characteristics of the class of beams in question (i.e. with a particular crystal grain size distribution relative to the beam depth and a particular ODF), in contrast with simulation methods, reliability analyses can be easily performed on beam structures with various force and displacement boundary conditions.

Summary & Conclusions

Methods for probabilistic analysis of micro-fabricated multicrystalline beams by the use of simulation and first-order reliability methods are developed. The simulation method employs two-dimensional finite element analysis on random realizations of the crystalline field characterized by random shape and orientation of crystals. In order to employ a more efficient approach based on FORM, the random crystalline field is identified by a continuous-parameter equivalent random field through a moving-averaging approach. The continuous field is then discretized by eigenvalue expansion to be represented by a finite number of random variables (20 in the case of the example study). A one-dimensional finite element formulation together with conventional FORM analysis is employed to compute the CDF and PDF of micro-beam response. The approach provides an efficient means for reliability study of micro-electro-mechanical systems consisting of beam elements with arbitrary geometry and boundary conditions.

Acknowledgment

The authors gratefully acknowledge support for this study from the National Science Foundation under Grant No. ECS-9023714. G.

Johnson and M. Ferrari have provided continued support and valuable input throughout the course of this study.

REFERENCES

Aurenhammer, F. 1991. Voronoï diagrams—a survey of a fundamental geometric data structure. *ACM Computing Surveys* 23: 345–405.

Bunge, H. J. 1989. *Texture Analysis in Materials Science*. Berlin: Butterworth.

Der Kiureghian, A. & B. -J. Ke 1988. The stochastic finite element method in structural reliability. *Probabilistic Engineering Mechanics* 3: 83.

Friedman, Z. & J. B. Kosmatka 1992. Exact stiffness matrix of a nonuniform beam—I. Extension, torsion, and bending of a Bernoulli-Euler beam. *Computers and Structures* 42: 671–682.

Hashin, Z. 1983. Analysis of composite materials, a survey. *Journal of Applied Mechanics* 50: 481–505.

Hirsekorn, S. 1990. Elastic properties of polycrystals—a review.*Textures and Microstructures* 12: 1–14.

Li, C.-C. & A. Der Kiureghian 1993. An optimal discretization of random fields. *Journal of Engineering Mechanics, ASCE,* to appear.

Mirfendereski, D. & A. Der Kiureghian 1992. Random response of multi-crystalline structures. *Proc. 9th Engineering Mechanics Conference, ASCE,* College Station, Texas.

Mirfendereski, D., A. Der Kiureghian & M. Ferrari 1992. Analysis of microfabricated textured multicrystalline beams: II. Probabilistic approach. *MRS Spring Meeting,* San Francisco, California.

O'Connor, L. 1992. MEMS: Microelectromechanical systems. *Mechanical Engineer,* February 1992.

Stoyan, D., W. S. Kendall & J. Mecke 1987. *Stochastic Geometry and Its Applications*. Berlin, GDR: John Wiley & Sons.

Vanmarcke, E. 1983. *Random Fields: Analysis and Synthesis,* Cambridge, Massachusetts: The MIT Press.

Williams, D.B., A. R. Pelton & R. Gronsky [Eds.] 1991. *Images of Materials,* New York: Oxford University Press (Ch. 12 by J. C. Russ).

Wise, K. D. 1991. Integrated microelectromechanical systems: a perspective on MEMS in the 90s. *Proc. IEEE Microelectromechanical Systems,* Nara Japan, pp. 33–38.

Zienkiewicz, O. C. & R. L. Taylor 1989. *The Finite Element Method, Vol. 1,* London: McGraw-Hill.

Structural Safety & Reliability, Schuëller, Shinozuka & Yao (eds) © 1994 Balkema, Rotterdam, ISBN 90 5410 357 4

Experimental determination of system stochasticity

N. Stubbs, J.T. Kim & K.G. Topole
Texas A&M University, Tex., USA

ABSTRACT: The feasibility of estimating the spatial stochasticity of a structural element is examined. The structural element is discretized and expressions are developed to estimate the stiffness parameter of each discretized unit as a function of the modal parameters of the structural element and the modal parameters of a reference structure. A numerical experiment is performed to validate the approach. A free-free stochastic beam is formed by discretizing the beam into 24 elements and the elements are randomly assigned stiffnesses from a log normal distribution. Modal properties of the stochastic beam are obtained via a finite element analysis. The reference beam is defined to be a uniform free-free beam with the average stiffness of the stochastic beam. Knowing only the modal responses of the stochastic beam and the uniform beam, the theory is used to recapture the stiffness distribution of the stochastic beam. Results indicate excellent agreement between the prediction and the simulation.

1 INTRODUCTION

This paper deals with the problem of estimating, in situ, the spatial variation in the physical properties of solids and structures. Such variations, which either exist naturally (e.g., in soils) or are the results of the inherent variation in the fabrication of structural elements (e.g., cold-formed members), partially account for the uncertainty in the response of a structure (See e.g., Vanmarcke, et al, 1986). Whereas, significant data has been collected to estimate the first and second moments of random material and geometric parameters describing structural elements (See, e.g., Ellingwood and Galambos, 1982), comparatively little effort has been directed at assessing the spatial stochasticity of structural parameters. This paper makes a step in that direction by examining the feasibility of experimentally determining the spatial stochasticity of a structural system. The approach followed here consists of three major steps: (a) providing a precise definition of the problem, (b) developing the appropriate theory for determining the stochasticity of the system, and (c) presenting a numerical experiment to demonstrate the feasibility of the approach. We feel that the originality and contribution of this work lies in, respectively, the use of modal information to estimate spatial stochasticity and the in situ determination of spatial stochasticity, proper.

2 PROBLEM STATEMENT

The problem addressed here may be defined as follows: Given (a) a structural element that exhibits spatial stochasticity in some physical parameter, e.g., stiffness; and (b) a set of measurements representing the modal response of that structural element; estimate the spatial stochasticity of the parameter. Here the parameter of interest will be some form of stiffness, e.g., bending or axial. In this paper, examples will be limited to wide sense stationary discrete processes {X(t)} and the spatial stochasticity represented by the covariance $\sigma(k)$ and mean μ functions estimated by

$$\hat{\sigma}(k) = \frac{1}{n} \sum_{t=1}^{n-k} (X_t - \hat{\mu})(X_{t+k} - \hat{\mu}) \tag{1}$$

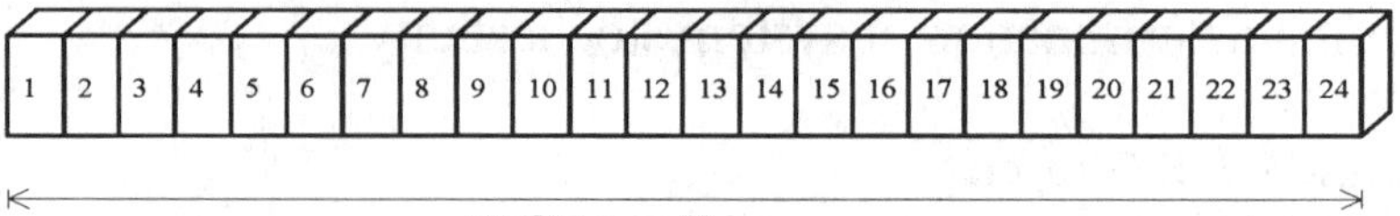

$E = 206.0$ Gpa $\quad$ Area $= 5.12 \times 10^{-4} m^2$

$\nu = 0.3$ $\quad$ $I_y = 4.37 \times 10^{-8} m^4$

$\rho = 7.65 \times 10^3$ Kg/m^3 $\quad$ $I_z = 1.09 \times 10^{-8}$ m^4

Figure 1. Example Free-Free Beam

Table 1: Simulated Moduli

Element No.	Simulated Modulus (Pa) $\nu = 0$	$\nu = .01$	$\nu = .1$	$\nu = .5$
1	0.20600E+12	0.21026E+12	0.21865E+12	0.21351E+12
2	0.20600E+12	0.20823E+12	0.20376E+12	0.38983E+12
3	0.20600E+12	0.20616E+12	0.22989E+12	0.30940E+12
4	0.20600E+12	0.20864E+12	0.19347E+12	0.27584E+12
5	0.20600E+12	0.20348E+12	0.24136E+12	0.49610E+12
6	0.20600E+12	0.20669E+12	0.24567E+12	0.48599E+12
7	0.20600E+12	0.20329E+12	0.24842E+12	0.13543E+12
8	0.20600E+12	0.20287E+12	0.20800E+12	0.21283E+12
9	0.20600E+12	0.20505E+12	0.20442E+12	0.25105E+12
10	0.20600E+12	0.20475E+12	0.20216E+12	0.24174E+12
11	0.20600E+12	0.20662E+12	0.20219E+12	0.11178E+12
12	0.20600E+12	0.20283E+12	0.16857E+12	0.22962E+12
13	0.20600E+12	0.20910E+12	0.19848E+12	0.35336E+12
14	0.20600E+12	0.20564E+12	0.18127E+12	0.14145E+12
15	0.20600E+12	0.20702E+12	0.21348E+12	0.21605E+12
16	0.20600E+12	0.20862E+12	0.27309E+12	0.73022E+11
17	0.20600E+12	0.20752E+12	0.19109E+12	0.33812E+12
18	0.20600E+12	0.20560E+12	0.17502E+12	0.11776E+12
19	0.20600E+12	0.20683E+12	0.20772E+12	0.17508E+12
20	0.20600E+12	0.20819E+12	0.20224E+12	0.10903E+12
21	0.20600E+12	0.20417E+12	0.17838E+12	0.22409E+12
22	0.20600E+12	0.20681E+12	0.18957E+12	0.33027E+12
23	0.20600E+12	0.20650E+12	0.18480E+12	0.28858E+12
24	0.20600E+12	0.20708E+12	0.22190E+12	0.28261E+12

and

$$\hat{\mu} = \frac{1}{n} \sum_{t=1}^{n} X_t \tag{2}$$

where n is the number of observation points.

3 THEORY OF APPROACH

Suppose that we are given an arbitrary structure with NB elements and N nodes. Assume that the structure behaves linearly. The ith modal stiffness, K_i, of the structure is given by

$$K_i = \Phi_i^T C \Phi_i \tag{3}$$

where Φ_i is the ith modal vector and C is the system stiffness matrix. The contribution of the jth member of the ith modal stiffness, K_{ij}, is given by

$$K_{ij} = \Phi_i^T C_j \Phi_i \qquad (4)$$

where C_j is the contribution of the jth member to the system stiffness matrix. The fraction of energy for the ith mode that is concentrated in member j is given by

$$F_{ij} = K_{ij} / K_i \qquad (5)$$

Let Equation (3) represent the reference structure. Then the structure to be identified can be characterized by the following quantities with asterisks

$$F_{ij}^* = K_{ij}^* / K_i^*;$$

$$[K_{ij}^* = \Phi_i^{*T} C_j^* \Phi_i^*, K_i^* = \Phi_i^{*T} C^*] \qquad (6)$$

On dividing Equation (4) by Equation (3), we obtain

$$\frac{F_{ij}^*}{F_{ij}} = \frac{K_{ij}^*}{K_{ij}} \frac{K_i}{K_i^*} \qquad (7)$$

The matrices C_j and C_j^* in Equation (3) may be written in the form

$$C_j = E_j C_{j0};\ C_j^* = E_j^* C_{j0} \qquad (8)$$

where the scalar E_j is a parameter representing the material stiffness properties and the matrix C_{j0} involves only geometric quantities (and possibly terms containing Poisson's ratio).

It has been numerically demonstrated that the modal strain energy for an unmodified element of a structure does not change when the structure is modified elsewhere (See Stubbs et al, 1992). It has also been numerically demonstrated that modal deformations at a given location increase with decreasing stiffness and decrease with increasing stiffness (Stubbs et al, 1992). Both of these observations are consistent with the hypothesis that the modal strain energy in an element remains the same before and after damage.

If we assume that the fraction of modal strain energy in a given member is the same for both referenced and identified structures (subjected to the same excitation), on substituting Equation (6) into Equation (5) and rearranging, we obtain

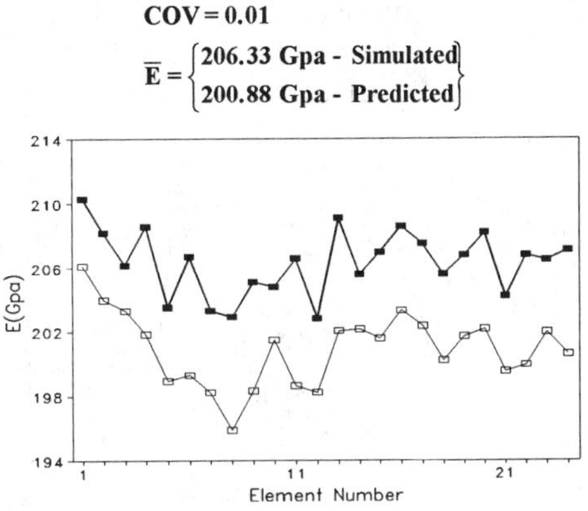

(a)

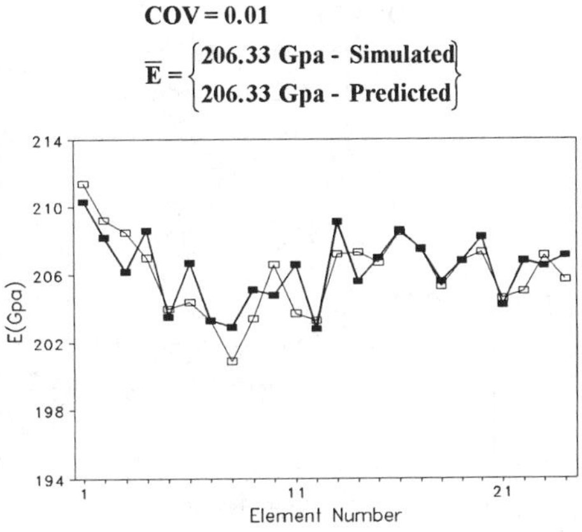

(b)

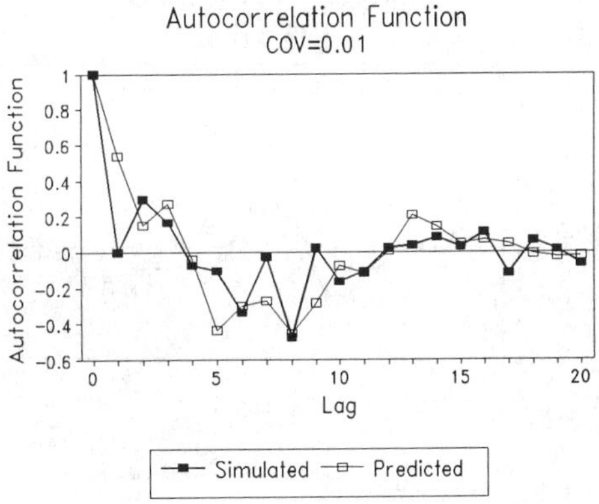

(c)

Figure 2. Comparison of simulated and predicted properties for ν=0.01.

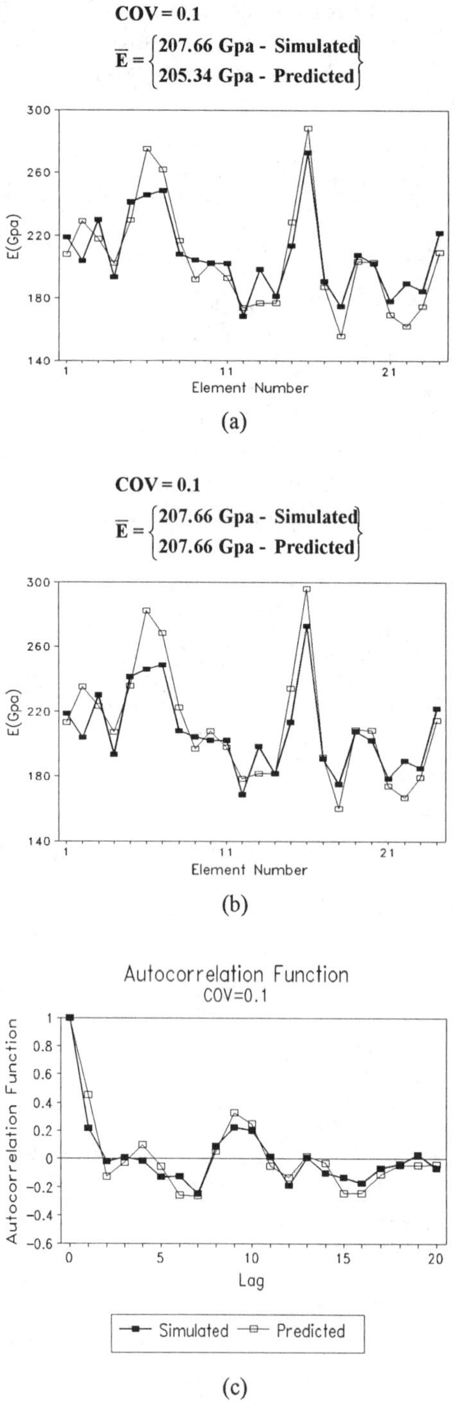

Figure 3. Comparison of simulated and predicted properties for ν=0.1.

$$E_j^* = \frac{E_j[\Phi_i^T C_{jo} \Phi_i] K_i^*}{[\Phi_i^{*T} C_{jo} \Phi_i^*] K_i} \tag{9}$$

$$= E_j(1 + \alpha_j)$$

where α_j is the fractional change in stiffness at location j. Thus, if we model the system to be identified using some reference structure with properties E_j, C_{jo}, and K_{ij} and we measure the modal properties Φ^*_i and K^*_i of the system to be identified; the difference in stiffnesses of the jth element is α_j. Note that

$$-1 \le \alpha_j \le \infty \tag{10}$$

and for stiffness increase $\alpha_j < 0$ and visa versa.

4 A NUMERICAL EXPERIMENT

The free-free beam shown in Figure 1 will be used to validate the proposed approach. The beam is subdivided into 24 elements of equal size. While the stiffness of each element is constant, the distribution of the element stiffnesses follows a log-normal distribution. Each element location is assigned a number, beginning with No. 1 on the left end of the beam and ending with No. 24 on the right end of the beam. Three effective beams were studied here corresponding to distributions with a mean modulus E of 206 GPa and with coefficient of variations equal to $\nu = 0.01$, $\nu = 0.1$, and $\nu = 0.5$. The corresponding moduli, obtained from a random number generator, are listed in the second, third and fourth columns of Table 1. The first column in Table 1 lists the mean modulus or the seed distribution. Dynamic finite element analyses were performed on the beams and mode shapes, modal stiffness, and eigenvalues were computed. In this problem the mode shape is defined by a 25x1 matrix whose elements represent the vertical deflection of the beam at the end of each element.

Assuming only a knowledge of the modal parameters of the stochastic beam, the problem here is to measure the distribution of stiffnesses. The following approach is used:

1. Identify the modal parameters of the referenced uniform beam.
2. Use the theory in the last section to

generate a realization of the stochastic field, and
3. Estimate the first and second moments of the stochastic field.

The following assumptions were made in selecting the reference uniform beam. First, the geometric and mass properties of the stochastic beam were assumed known. Second, the modulus of the uniform beam was estimated from the value of the first eigenvalue of the stochastic beam using the equation

$$\hat{E} = \frac{\lambda_1 m}{c^2 I} \tag{11}$$

where m is the mass density per unit length, I is the second moment of area, and constant C = 9.274. With the latter value of Ê, a dynamic analysis was performed on the uniform beam to yield Φ_i, λ_i, and K_i.

Equation (9) for the free-free beam reduces to

$$(EI)_k^* = (EI)_k \left[\frac{\lambda_i^* M_i^*}{\lambda_i M_i} \; \frac{\int_k (\Phi_i^{\prime\prime})^2 dx}{\int_k (\Phi_i^{*\prime\prime})^2 dx} \right]$$

$$= (EI)_k (1 + \alpha_k) \tag{12}$$

Note that the subscript k refers to the kth element and the integral of the curvature is over the element. The values of $(EI)_k$ developed here represent the average obtained from using the first five bending modes of the beam.

Results for the three simulated experiments are shown in Figures 2-4 for the samples with $\nu = 0.01$, $\nu = 0.1$, and $\nu = 0.5$, respectively. In Figure 2a, for example, the dark squares represent the simulated value of the modulus while the open squares represent the predicted values using Equations 11 and 12. Figure 2b represents the same information but we now assume that the mean initial modulus is known. Figure 2c compares the autocorrelation function of the simulated stiffness, normalized, by the variance, with the correlation function of the predicted stochasticity.

5 DISCUSSION OF RESULTS

The accuracy of the approach can be assessed

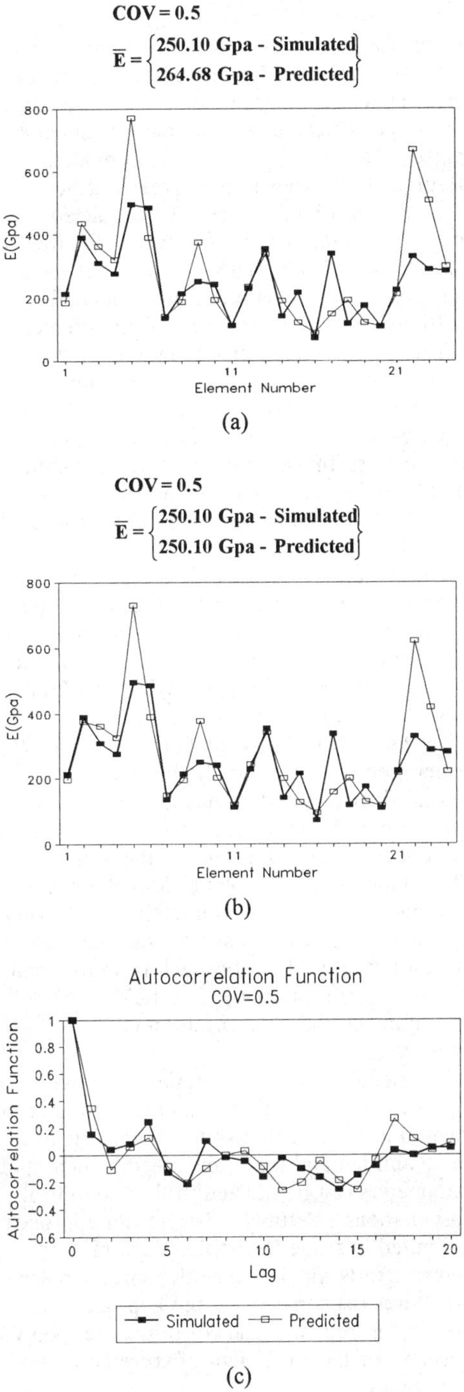

Figure 4. Comparison of simulated and predicted properties for ν=0.5.

from a review of Figures 2a, 3a, and 4a. In Figure 2a, there is a bias of approximately 6 GPa or 3 percent. In Figures 3a and 4a, the error appears totally random. In all cases the predicted pattern closely follows the simulated pattern. If the value of the mean modulus is assumed to be known, the agreement between the simulated modulus and the predicted modulus is quite good, as shown in Figures 2b, 3b, and 4b. Finally the agreement between the autocorrelation function of the simulated stiffness, see Figure 2c - 4c, and the predicted stiffness is very good especially for the cases of $\nu = 0.1$ and $\nu = 0.5$. To give a further indication of the accuracy of the results, a least squares regression was performed on the predicted stiffnesses and the simulated stiffness. The resulting coefficient of correlation for the $\nu = 0.01$, $\nu = 0.1$, and $\nu = 0.5$, were respectively, 0.80, 0.92, and 0.78.

The standard error of estimate of the regression line provides an indication of the precision of the predictions. The values obtained for the cases $\nu = 0.01$, $\nu = 0.1$, and $\nu = 0.5$, were respectively, 0.8 GPa, 13.1 GPa, and 95.3 GPa.

To the authors' knowledge, the problem of experimentally determining the spatial stochasticity of a system has not received much attention in the literature. Most approaches to data have concentrated on the theoretical formulation of the problem. Moreover, the approach used here to estimate a realization of system stochasticity is used for the first time for such a purpose. The approach is limited only by the ability to measure, in a field environment, the appropriate modal information.

The significance and potential contribution of the methodology and results presented here are related to the fact that system stochasticity can be quantified and more meaningful stochastic parameters can be utilized in the theoretical formulations. Certainly, the present approach is limited in scope to simple structures and future efforts should (a) study more complex structures (b) corroborate the approach with field measurements, and (c) refine the theory to account for the coefficient of correlation less than one.

REFERENCES

Ellingwood, B. & T.V. Galambos, 1982. Probability-Based Criteria for Structural Design. Structural Safety 1:15-26.

Stubbs, N., J.T. Kim, & K.G. Topole 1992. An Efficient and Robust Algorithm for Damage Localization in Offshore Platforms. Tenth Structural Congress '92. ASCE. pp. 543-546. San Antonio, Texas, April 13-15.

Vanmarcke, E., M. Shinozuka, S. Nakagiri, G.I. Schueller, & M. Grigoriu 1986. Random Fields and Stochastic Finite Elements. Structural Safety. 3:143-166.

Structural Safety & Reliability, Schuëller, Shinozuka & Yao (eds) © 1994 Balkema, Rotterdam, ISBN 90 5410 357 4

Stochastic finite element analysis of structures

R.Y.Tan & I.Y.Lu
Department of Civil Engineering, National Taiwan University, Taipei, Taiwan

ABSTRACT: This paper derives the stochastic stiffness matrix for a truss and a beam elements, so that the conventional finite element method can be utilized to evaluate the response variability of a structure whose material property exhibits spatial random variation. Upon the application of static condensation and the first–order perturbation method, we obtain the nodal displacements which consist of not only the deterministic component but also the random component.

1. INTRODUCTION

Recently, several studies have investigated the response variability of the relatively simple structures with continuous random variation of material property (Handa & Andresson, 1981; Hisada & Nakagiri, 1985; Shinozuka, 1987). Although those concepts are extended even further to more complicated structures, the associated analytical formulas become rather cumbersome. Therefore, the primary focus of this study is to modify the procedure of the conventional finite element method which has been recognized as a powerful method for structural analysis, so that the response variability of a structure whose material property exhibits spatial random variation can be obtained by a more systematic manner.

In this paper, Young's modulus is assumed to constitute a homogeneous one-dimensional and uni–variate random process. A stochastic element stiffness matrix is derived. In this connection, an element is divided into several sub-elements whose number depends on the scale of fluctuation of the process. In any sub-element, the elastic characteristics is represented by the "local spatial average" of the process. Consequently, the stiffness matrix of the sub-element is obtained. Using the technique of static condensation, we eliminate the additional degrees-of-freedom stemmed from the division of sub-elements to produce a new element stiffness matrix whose size is equal to the conventional one. The global stiffness matrix is then established by the well-known direct stiffness method. The stiffness and the displacement matrices in the system equation are expanded with respect to the probabilistic variables which reflect the material spatial variations, i.e. local averages of all sub-elements. Upon the application of the first–order perturbation method, the nodal displacements are derived. It is shown that the covariance of nodal displacements is a function of the covariance of local spatial averages. When an appropriate variation function, which depicts the dependence of covariance of spatial averages on the size of interval of average, is employed , all the random terms involved in the response variability are obtained.

2. TRUSS ELEMENT

Consider a truss element in Fig.1 with a deterministic axial load, and having a modulus of elasticity varying randomly along its length, which is given by

$$E(x) = \bar{E}\,(1 + g(x)) \tag{1}$$

where $E(x)$ and $g(x)$ are one-dimensional homogeneous random processes, and $\bar{E}$ is $\varepsilon[E(x)]$. Note that $\varepsilon[.]$= the expectation. $\bar{E}g(x)$ represents the deviation of the modulus of elasticity around its mean value. The function $g(x)$ is a zero-mean stochastic field with autocorrelation function $R_{gg}(\xi)$.

The conventional derivation of stiffness matrix starts from the assumption of a displacement function, i.e.

$$u(x) = N_e\, u_e + N_{e+1}\, u_{e+1} \tag{2}$$

in which the shape functions N_e and N_{e+1} are given by

$$N_e = (x_{e+1} - x) / L \quad N_{e+1} = (x - x_e) / L \tag{3}$$

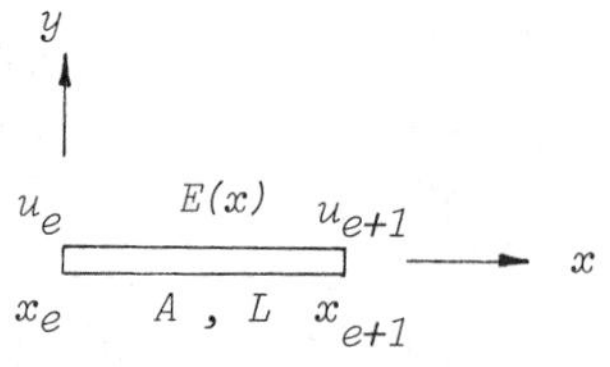

Fig.1 A truss element

However, in the present case, Eqs.(2) and (3) are not appropriate because of the variation of elasticity modulus.

To tackle this, we introduce the concept of local average (Vanmarcke & Grigoriu, 1983). The i-th local average of g(x) is defined as

$$\bar{g}_i(x_i) = \frac{1}{l_i}\int_{x_i-\frac{l_i}{2}}^{x_i+\frac{l_i}{2}} g(x)\,dx \qquad (4)$$

where l_i = length of the local average, and x_i = centroid of the length. It can be shown that

$$\varepsilon[\bar{g}_i(x_i)] = 0 \quad \mathrm{Var}[\bar{g}_i(x_i)] = \sigma_g^2 \cdot \gamma(l_i) \qquad (5)$$

where σ_g^2 = variance of g(x), and $\gamma(l_i)$= variance function.

$$\gamma(l_i) = \frac{2}{L_i}\int_0^{l_i} \left(1-\frac{\xi}{L_i}\right) R_{gg}(\xi)/\sigma_g^2\, d\xi \qquad (6)$$

Dependent upon the scale of fluctuation of the process, this element is further divided into N sub-elements. The elasticity modulus in each sub-element may be represented by a local average of E(x), $\bar{E}_i$, which is a specific value of a random variable. By this measure, the shape functions similar to Eq.(3) can be used. Note that L is replaced by l (l = L/N) under this condition.

The stiffness matrix of the i-th sub-element in the e–th element is

$$[\bar{K}_{ei}] = \int_0^l \begin{bmatrix} -1/l \\ 1/l \end{bmatrix} \bar{E}_i \begin{bmatrix} -1/l \\ 1/l \end{bmatrix}^t A\,dx = \frac{\bar{E}A}{l}\begin{bmatrix} 1 & -1 \\ -1 & 1 \end{bmatrix} + \frac{\bar{E}A}{l}\bar{g}_i\begin{bmatrix} 1 & -1 \\ -1 & 1 \end{bmatrix} \qquad (7)$$

The first and the second terms in Eq.(7) are the deterministic (conventional) and random components, respectively. If N=2, the assembled element stiffness matrix is

$$\frac{\bar{E}A}{l}\begin{bmatrix} C_1 & -C_1 & 0 \\ -C_1 & C_1+(1+\bar{g}_2) & -(1+\bar{g}_2) \\ 0 & -(1+\bar{g}_2) & (1+\bar{g}_2) \end{bmatrix}$$

in which

$$C_1 = 1+\bar{g}_1. \qquad (8)$$

Using the technique of static condensation, we eliminate the internal degree of freedom to yield a new element stiffness matrix, having the same size of the conventional one. The outcome is

$$[\bar{K}_e] = \frac{\bar{E}A}{l}C_2\begin{bmatrix} 1 & -1 \\ -1 & 1 \end{bmatrix} \qquad C_2 = \frac{C_1(1+\bar{g}_2)}{C_1+(1+\bar{g}_2)} \qquad (9)$$

It is not difficult to show that the general form of the above equation is written as

$$[\bar{K}_e] = \frac{\bar{E}A}{l}C_N\begin{bmatrix} 1 & -1 \\ -1 & 1 \end{bmatrix} \quad C_N = \frac{C_{N-1}(1+\bar{g}_N)}{C_{N-1}+(1+\bar{g}_N)} \qquad (10)$$

3. PERTURBATION METHOD

When all element stiffness matrices are determined, we combine them to find the global stiffness matrix. Then, applying the boundary condition, the equilibrium equation is

$$[K]\,[U] = [P]_0 \qquad (11)$$

in which [K] is the stiffness matrix, [U] is the free or unknown displacement vector, and $[P]_0$ is the external force vector. it should be noted that [K] involves a set of random components, i.e. all local averages in each element.

Eq.(11) can be solved by the first-order perturbation method. The stiffness matrix is expanded in a Taylor series about the mean value of local average $\bar{g}_{ei}$. The symbol $\bar{g}_{ei}$ is used in place of $\bar{g}_i$ in order to indicate that this local average is associated with the e-th element. When the first-order approximation is adopted, we have

$$[K] = [K]|_E + \sum_{e=1}^{Ne}\sum_{i=1}^{N}\frac{\partial[K]}{\partial\bar{g}_{ei}}\Big|_E \cdot \bar{g}_{ei} \qquad (12)$$

The displacement vector is also expanded. Substituting those expansions into Eq.(11) and simplifying them, we have (Lu, 1991)

$$[U] = [U]_0 - \sum_{e=1}^{Ne}\sum_{i=1}^{N}[K]_0^{-1}[K]_0^{(e)}[U]_0 \cdot \bar{g}_{ei}/N \qquad (13)$$

in which $[U]_0$ = conventional displacement vector,

i.e. without the consideration of random property, $[K]_0$ = conventional stiffness matrix, and $[K]_0^{(e)}$ is a sub–matrix of $[K]_0$ and consists of the components contributed by the e–th element.

The covariance of displacements can be obtained with the aid of Eq.(13). It is clear that this covariance is a function of the covariance of local averages. It is shown that the latter is expressed as follows (Vanmarcke & Grigoriu, 1983):

$$\mathrm{Cov}(\bar{g}_{e_1 i_1}, \bar{g}_{e_2 i_2}) = \frac{1}{2}\sigma_g^2 \frac{N^2}{l_1 l_2}[L_0^2\gamma(L_0) - L_1^2\gamma(L_1) + L_2^2\gamma(L_2) - L_3^2\gamma(L_3)] \quad (14)$$

where L_i are distances defined in Fig. 2. Therefore, with a specific autocorrelation function $R_{gg}(\xi)$ of the stochastic field, we can calculate the covariance of displacements upon the application of Eqs. (6) and (14).

4. BEAM ELEMENT

Similarly, a beam element is divided into N sub–elements, shown in Fig. 3. The stiffness matrix of the i–th sub–element in the e–th element is

$$[\bar{K}_{ei}] = \frac{EI}{l^3}(1+\bar{g}_i)\begin{bmatrix} 12 & 6l & -12 & 6l \\ 6l & 4l^2 & -6l & 2l^2 \\ -12 & -6l & 12 & -6l \\ 6l & 2l^2 & -6l & 4l^2 \end{bmatrix} \quad (15)$$

in which I is the moment of inertia. On the basis of the condensation technique, the stochastic stiffness matrix for a beam element can be derived. Because of the algebraic complexity, the derivation is accomplished by an application of MACSYMA TM (1988). The element stiffness matrix turns out to be the following form (lu, 1991):

$$[\bar{K}_e] = \sum_{p=1}^{N} (R_1^2 \cdots R_{p-1}^2\, R_p\, R_{p+1}^2 \cdots R_n^2)\, [\bar{K}_e(p)] / (D_1 + D_2) \quad (16)$$

in which

$$[\bar{K}_e(p)] = \frac{N^3 EI}{L^3}\begin{bmatrix} 12 & & \text{SYM} & \cdot \\ 6(2p-1)l & [4+12p(p-1)]l^2 & \cdot & \cdot \\ -12 & -6(2p-1)l & 12 & \cdot \\ [12N-6(2p-1)]l & [(6N-4)+12(N-P)(p-1)]l^2 & -[12N-6(2p-1)]l & [4+12(N-P)(N-P+1)]l^2 \end{bmatrix} \quad (17)$$

$$R_p = 1 + \bar{g}_p \quad (18)$$

$$D_1 = \sum_{i=1}^{N} \prod_{j=1,\, j\neq i}^{N} R_j^2 \qquad N = 2, 3, \cdots \quad (19)$$

$$D_2 = \sum_{k=1}^{N-1} \sum_{q=1}^{N-k} (12k^2+2)(R_1^2 \cdots R_{q-1}^2\, R_q R_{q+1}^2 \cdots R_{q+k-1}^2\, R_{q+k}\, R_{q+k+1}^2 \cdots R_N^2) \quad (20)$$

Eqs. (16)–(20) are further illustrated. For instance, let N=3, we have

$$D_1 = R_2^2 R_3^2 + R_1^2 R_3^2 + R_1^2 R_2^2$$

In Eq.(20), k is an integer to indicate the 'seperation distance' between two linear terms, i.e. R_q and R_{q+k} . When k=1 , we have

$$\sum_{q=1}^{N-k} (12k^2 + 2)(R_1^2 \cdots R_N^2) = 14\, R_1\, R_2\, R_3^2 + 14 R_1^2\, R_2\, R_3$$

When k=2 , we have

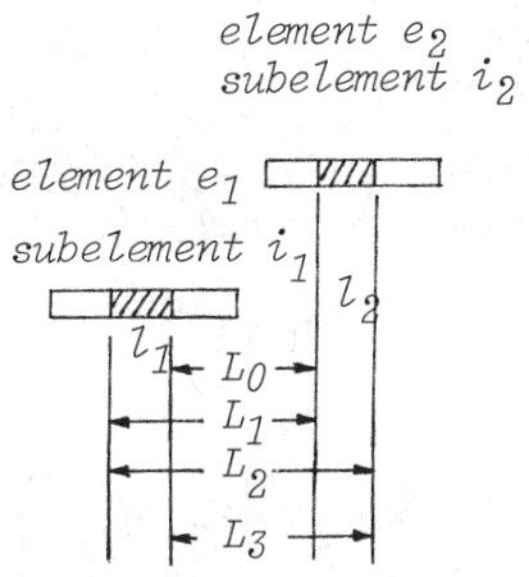

Fig.2 Definition of the interval L_i

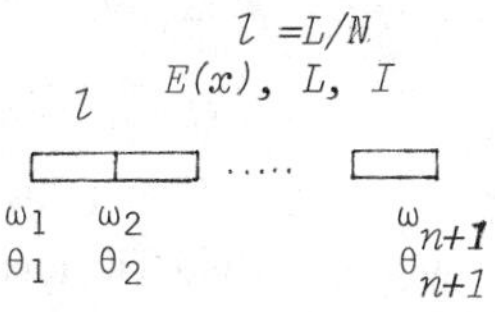

Fig.3 A beam element

$$\sum_{q=1}^{N-k} (12k^2 + 2)(R_1^2 \cdots R_N^2) = 50\, R_1R_2^2R_3$$

As a resut,

$$D_2 = 14\, R_1R_2R_3^2 + 50\, R_1R_2^2R_3 + 14R_1^2R_2R_3$$

When the element in Eq. (16) is denoted as s(i,j), we have

$$s(1,1) = \frac{N^3\overline{E}I}{L^3} \cdot \frac{12R_1R_2^2R_3^2 + 12R_1^2R_2R_3^2 + 12R_1^2R_2^2R_3}{D_1 + D_2}$$

Similarly,

$$s(2,1) = \frac{N^2\overline{E}I}{L^2} \cdot \frac{6R_1R_2^2R_3^2 + 18R_1^2R_2R_3^2 + 30R_1^2R_2^2R_3}{D_1 + D_2}$$

When all R_p are equal to 1, i.e. the deterministic case, Eq.(16) is nothing but the conventional stiffness matrix of a beam element.

5. NODAL DISPLACEMENT

The system stiffness matrix are associated with all pertinent local averages in elements. Expansion of this matrix leads to

$$[K] = [K]\,|_E + \sum_{e=1}^{Ne} \sum_{i=1}^{N} \frac{\partial [K]}{\partial \overline{g}_{ei}}\Big|_E \cdot \overline{g}_{ei}$$

$$= [K]_0 + \sum_{e=1}^{Ne} \sum_{i=1}^{N} [K_{ei}^*] \cdot \overline{g}_{ei} \qquad (21)$$

To explain $[K_{ei}^*]$ in Eq. (21), we define a matrix related with the e–th element

$$[\overline{K}_{ei}^*] = \frac{2N^4 - 2N + 2 - F(i)}{N^4} \frac{\overline{E}I}{L^3} \cdot \begin{bmatrix} 12 & 6L & -12 & 6L \\ 6L & 4L^2 & -6L & 2L^2 \\ -12 & -6L & 12 & -6L \\ 6L & 2L^2 & -6L & 4L^2 \end{bmatrix} - \frac{1}{N^4}[\overline{K}_e(i)] \qquad (22)$$

where $[\overline{K}_e(i)]$ is from Eq.17 and F(i) is a function of N. $[\overline{K}_{ei}^*]$ is transformed to the global coordinate. $[K_{ei}^*]$ is the matrix whose components are chosen from the transformed matrix. Note that only the components corresponding to the free or unrestrained degrees of freedom are adopted.

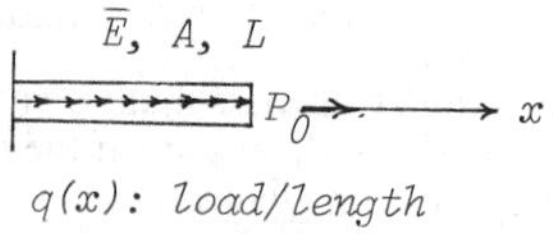

Fig.4 A truss analyzed

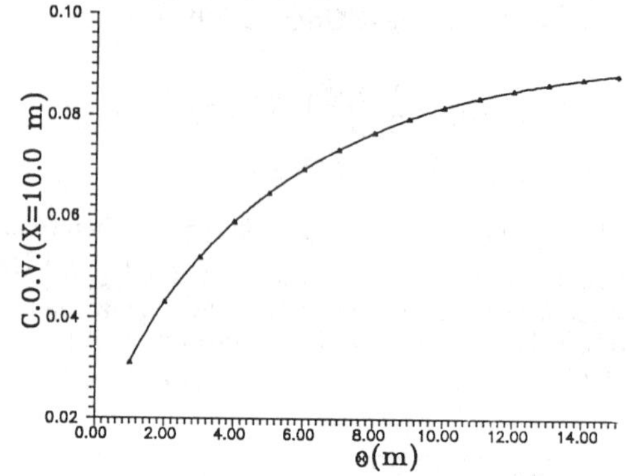

Fig.5 Coeffiecint of variation of the truss displacement

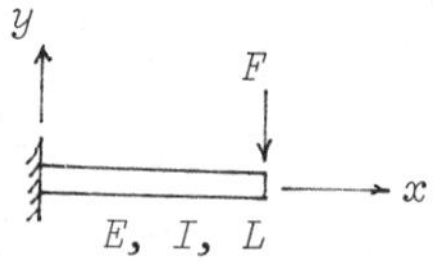

Fig.6 A beam analyzed

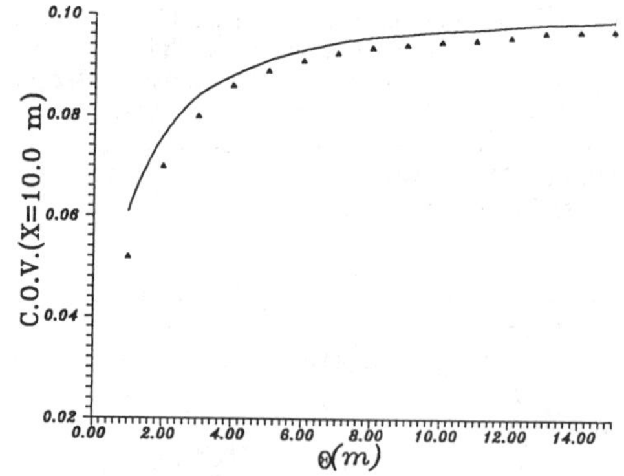

Fig.7 Coefficient of variation of the beam displacement

The displacement vector is also expanded. Substituting those expansions into the equilibrium equation and simplifying them, we obtain

$$[U] = [U]_0 - \sum_{e=1}^{Ne} \sum_{i=1}^{N} [K]_0^{-1}[K_{ei}^*][U]_0 \cdot \overline{g}_{ei} \qquad (23)$$

6. NUMERICAL EXAMPLES

A truss shown in Fig.4 is analyzed. Let $\bar{E} = 2 \times 10^{11}$ N/m^2, A= 10×10^3 mm^2, L= 10m, $\sigma_g = 0.1$, q(x)=0, and P_0 = 10KN. The reason to choose this example is that its exact solution is available (Shinozuka, 1987). The number of sub-elements is an integer no less than L/θ, in which θ is the scale of fluctuation. A form of autocorrelation function given as follows is used.

$$R_{gg}(\xi) = \sigma_g^2 \left(1 - \frac{|\xi|}{\theta}\right) \qquad |\xi| \le \theta \tag{24}$$

The corresponding variance function is

$$\begin{aligned} \gamma(L) &= 1 - \frac{|L|}{3\theta} \qquad |L| \le \theta \\ &= \frac{\theta}{|L|}\left(1 - \frac{\theta}{3|L|}\right) \quad |L| \ge \theta \end{aligned} \tag{25}$$

Fig.5 plots the coefficient of variation of the displacement at x=10 m as a function of θ. The curve is the exact solution for comparison. The observation confirms the applicability of the derived stochastic element stiffness matrix for a turss.

A beam shown in Fig. 6 is considered. Let $\bar{E} = 2 \times 10^{11}$ N / m^2, I = 50×10^{-6} m^4, L = 10m , F = 10 KN. Fig 7 delineates the C.O.V. of the vertical displacement at x=10m as a function of θ.

7. CONCLUSIONS

This study proposes a stochastic element stiffness matrix with consideration of the material spatial variation. The derivation is made for a truss element and a beam element. The essence of the present approach is that it reserves all the advantageous properties inherent from the conventional finite element method. Indeed, it provides a more tractable procedure to determine the response variability of a highly complex structure with spatial variation in material properties.

REFERENCES

Handa, K. and K. Andresson 1981. Application of finite element methods in statistical analysis of structures,Proc. of the 3rd ICOSSAR,409-417.

Hisada, T. and S. Nakagiri 1985. Role of the stochastic finite element method in structural safety and reliability, Proc. of the 4th ICOSSAR, 1, 385-394.

Shinozuka, M. 1987. Structural response variability, J. Eng. Mech., ASCE, 113.

Vanmarcke, E. and M. Grigoriu 1983. Stochastic finite element analysis of simple beam, J. Eng. Mech., ASCE,109,1203-1214.

Lu, I. Y. 1991. A study on stochastic finite element method, M.S. Thesis, Department of Civil Engineering, National Taiwan University.

MACSYMA TM 1988. User's Guide, Symbolics, Inc.

Structural Safety & Reliability, Schuëller, Shinozuka & Yao (eds) © 1994 Balkema, Rotterdam, ISBN 90 5410 357 4

Variability response functions and upper bounds of response variability of 2D stochastic systems

Friedrich J.Wall
Hilti AG, Schaan, Principality of Liechtenstein (Formerly: Princeton University, N.J., USA)

George Deodatis & Masanobu Shinozuka
Princeton University, N.J., USA

ABSTRACT: The concept of variability response function based on the weighted integral method is extended to two-dimensional plane stress/plane strain stochastic problems, in order to calculate their response variability (in terms of second moments of response quantities) with great accuracy even when using relatively coarse finite element meshes. The concept of variability response function is used to establish spectral-distribution-free upper bounds of the response variability. In addition, the variability response function based on the local averaging method is introduced to reduce the computational effort associated with the weighted integral method. The two methods are compared to estimate the relative accuracy of the more approximate local averaging method. The response variability is calculated using a first-order Taylor expansion approximation of the response quantities. One of the most important findings is the fact that the coefficient of variation of certain response quantities can be much larger than the coefficient of variation of the elastic modulus (the input quantity).

1 INTRODUCTION

The challenge of realistically accounting for uncertainties that are inherent in all structural systems has attracted a lot of attention in recent years. Considerable material and/or geometric random inhomogeneities are either introduced during the manufacturing process or can simply be found as an inherent characteristic of the material itself. The latter case is especially true for concrete and soil. A rigorous description of these random properties can only be achieved using stochastic fields. Because of the system stochasticity outlined above, the response of the system is going to be stochastic as well. The analysis of the response variability of stochastic structural systems consists of evaluating the first and second moment of their response.

As analytical solutions are restricted to simple linearly elastic structures, numerical techniques have to be used to deal with more complex stochastic structural systems. In most cases, the only available numerical approach is a class of methods known as "Stochastic Finite Element Methods" (SFEM). Four papers containing literature reviews of SFEM are mentioned here: Vanmarcke et al. (1986), Benaroya and Rehak (1988), Brenner (1991) and Der Kiureghian et al. (1991).

The present paper extends the weighted integral method (Deodatis, 1990; Takada, 1990a,b; Deodatis, 1991; Deodatis and Shinozuka, 1991) and the concept of variability response function (Deodatis and Shinozuka, 1989; Deodatis, 1990) to two-dimensional plane stress/plane strain stochastic problems.

2 ANALYSIS OF RESPONSE VARIABILITY

The well-known bilinear rectangular element is used in the analysis of plane stress/strain problems considered in this paper. This element has four nodes and eight degrees of freedom (two displacement degrees of freedom per node). The elastic modulus is considered to vary over the area of element (e) as:

$$E^{(e)}(x,y) = E_0^{(e)}[1 + f^{(e)}(x,y)] \quad (1)$$

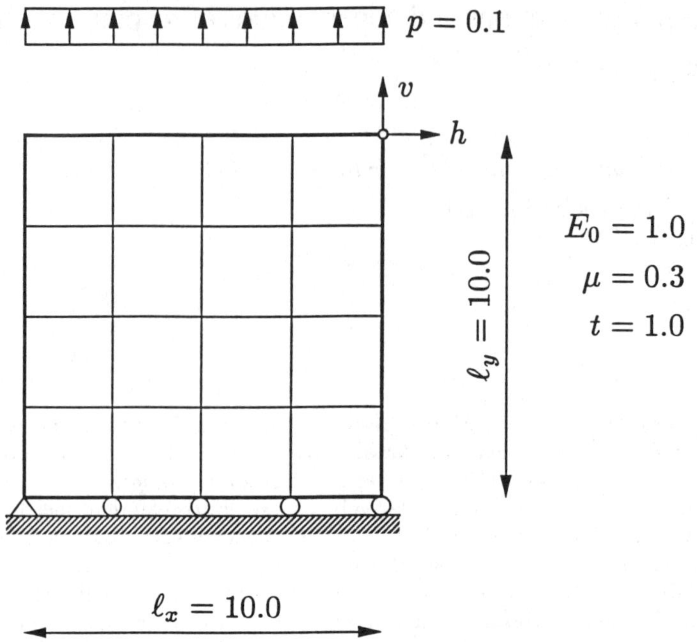

Fig. 1 Square Plate Subjected to In-Plane Loading

where $E_0^{(e)}$ = mean value of elastic modulus and $f^{(e)}(x,y)$ = 2D, zero-mean, homogeneous stochastic field. In order to prevent the occurrence of non-positive values of the elastic modulus, bounds are imposed on $f^{(e)}(x,y)$ (Wall and Deodatis, 1993). Using the principle of stationary potential energy, the stochastic element stiffness matrix is computed as (Deodatis, 1991 and Wall and Deodatis, 1993):

$$\mathbf{K}^{(e)} = \mathbf{K}_0^{(e)} + X_1^{(e)}\Delta\mathbf{K}_1^{(e)} + X_2^{(e)}\Delta\mathbf{K}_2^{(e)} + \ldots + X_6^{(e)}\Delta\mathbf{K}_6^{(e)} \quad (2)$$

where all matrices in the right-hand-side of Eq. 2 are deterministic and the weighted integrals $X_k^{(e)}$ are random variables defined as:

$$X_k^{(e)} = \int_{A^{(e)}} x^i y^j f^{(e)}(x,y)dA^{(e)} \quad (3)$$

with $A^{(e)}$ denoting the area of the finite element and subscript k being related to the powers i and j as: $(k \to (i,j))$: 1→(0,0), 2→(1,0), 3→(0,1), 4→(1,1), 5→(2,0), 6→(0,2). Due to the zero-mean property of the weighted integrals $X_k^{(e)}$, $\mathbf{K}_0^{(e)}$ represents the mean value of the stochastic element stiffness matrix. Using standard finite element analysis methodology, the equations of equilibrium can be written as:

$$\mathbf{KU} = \mathbf{P} \quad (4)$$

where $\mathbf{K}$ is the stochastic global stiffness matrix assembled from the stochastic element stiffness matrices, $\mathbf{U}$ is the stochastic global displacement vector and $\mathbf{P}$ is the deterministic global force vector.

To analyze the response variability, the displacement vector $\mathbf{U}$ is approximated by a first-order Taylor expansion around the mean values of the weighted integrals:

$$\mathbf{U} \approx \mathbf{U}_0 + \sum_{e=1}^{N}\sum_{k=1}^{6}\left(X_k^{(e)} - \bar{X}_k^{(e)}\right)\left[\frac{\partial \mathbf{U}}{\partial X_k^{(e)}}\right]_{\mathcal{E}} \quad (5)$$

where both subscripts $_0$ and $_{\mathcal{E}}$ denote evaluation at the mean values of the weighted integrals, N is the total number of finite elements and $\bar{X}_k^{(e)}$ denotes the mean value of $X_k^{(e)}$.

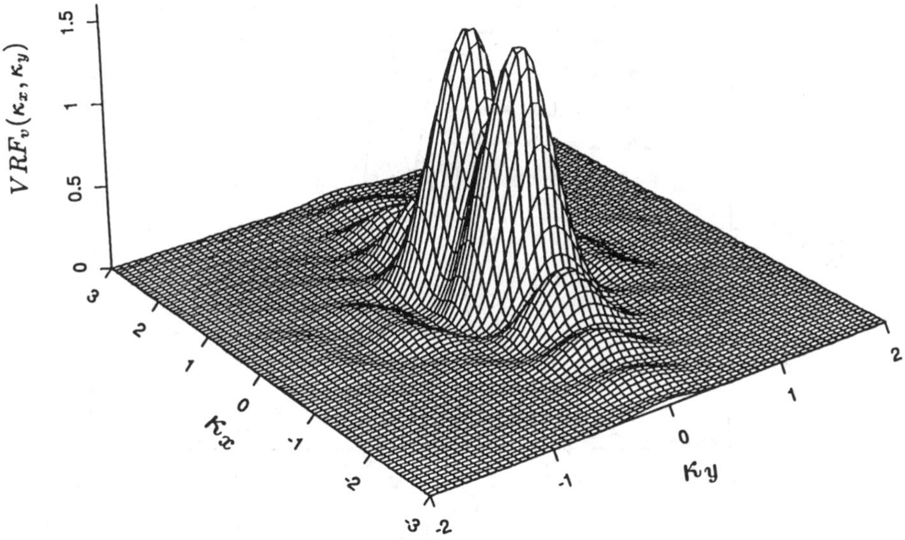

Fig. 2 Variability Response Function of Vertical Displacement v as a Function of κ_x and κ_y

2.1 Variability response function

After some algebra (Wall and Deodatis, 1993), the variance vector of $\mathbf{U}$ can be expressed in the following form:

$$\mathrm{Var}\,\{\mathbf{U}\} = \int_{-\infty}^{\infty}\int_{-\infty}^{\infty} S_{ff}(\kappa_x,\kappa_y)\ \mathbf{VRF}(\kappa_x,\kappa_y)\ d\kappa_x\, d\kappa_y \qquad (6)$$

Equation 6 is derived under the assumption that the elastic modulus of all finite elements is characterized by the same stochastic field, having power spectral density function $S_{ff}(\kappa_x,\kappa_y)$ with κ_x and κ_y denoting the wave numbers. $\mathbf{VRF}(\kappa_x,\kappa_y)$ is a vector whose components are the first-order approximations of the variability response function (Deodatis and Shinozuka 1989) at the corresponding degree of freedom of the structure. In the following, $\mathbf{VRF}(\kappa_x,\kappa_y)$ will be simply called variability response function. After some algebra, $\mathbf{VRF}(\kappa_x,\kappa_y)$ can be expressed in closed form as:

$$\mathbf{VRF}(\kappa_x,\kappa_y) = \sum_{e=1}^{N}\sum_{f=1}^{N}\sum_{k=1}^{6}\sum_{n=1}^{6} \mathrm{diag}\left(\mathbf{K}_0^{-1}\Delta\mathbf{K}_k^{(e)}\mathbf{U}_0\right)\mathbf{K}_0^{-1}\Delta\mathbf{K}_n^{(f)}\mathbf{U}_0 \{[Q_{ek}Q_{fn} + W_{ek}W_{fn}] \cos(\Delta x_{fe}\kappa_x + \Delta y_{fe}\kappa_y) - [Q_{fn}W_{ek} - Q_{ek}W_{fn}] \sin(\Delta x_{fe}\kappa_x + \Delta y_{fe}\kappa_y)\} \qquad (7)$$

where diag( ) represents a diagonal matrix whose diagonal components consist of the vector within the parentheses and $\Delta\mathbf{K}_k^{(e)}$ denotes now the extended version of the matrices shown in Eq. 2. In Eq. 7, Δx_{fe} and Δy_{fe} are measures of the relative position of elements (e) and (f) in the x- and y-directions, respectively, and the Q's and W's are expressions involving trigonometric functions of products of wave numbers multiplied by dimensions of finite elements (Wall and Deodatis, 1993).

At this point it should be mentioned that the variability response function is non-negative at any point of the (κ_x,κ_y) plane and symmetric with respect to the origin of the (κ_x,κ_y) coordinate system.

2.2 Spectral-distribution-free bounds

Equation 6 indicates that the variance vector of $\mathbf{U}$ will be a function of the particular form of the power spectral density function

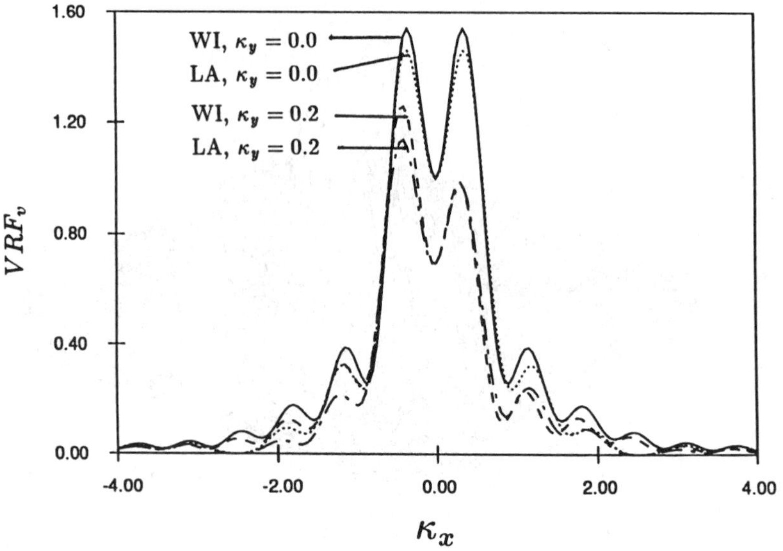

Fig. 3 Variability Response Function of Vertical Displacement v as a Function of κ_x, for $\kappa_y = 0.0$ and $\kappa_y = 0.2$

$S_{ff}(\kappa_x, \kappa_y)$ (note the similarity of the variability response function to the frequency response function in random vibration analysis). Unfortunately, beyond a reasonable estimation of its mean value (which is assumed to be zero without loss of generality) and coefficient of variation, very little information is usually available about the probabilistic characteristics of stochastic field $f(x, y)$. It is this unavailability of detailed information that makes it engineering-wise very significant to establish "spectral-distribution-free" bounds of the response variability of the system. The paramount importance of the variability response function lies in the fact that it can be used to establish such bounds for cases where the power spectral density (or autocorrelation) function of stochastic field $f(x, y)$ is not known. The calculation of these "spectral-distribution-free" bounds is achieved following a procedure similar to the one described in Deodatis and Shinozuka (1989).

The following relation holds for any form of the power spectral density function $S_{ff}(\kappa_x, \kappa_y)$:

$$\int_{-\infty}^{\infty}\int_{-\infty}^{\infty} S_{ff}(\kappa_x, \kappa_y)\, d\kappa_x\, d\kappa_y = \sigma_{ff}^2 \qquad (8)$$

where σ_{ff} is the standard deviation of stochastic field $f(x, y)$ or equivalently the coefficient of variation of the elastic modulus.

Considering now Eq. 6, it is easy to evaluate the maximum value of the variance of the i-th component of the nodal displacement vector $\mathbf{U}$ as:

$$\max\left\{\mathrm{Var}[U_i]\right\} = \sigma_{ff}^2 \cdot VRF_i(\kappa_{x,i}^*, \kappa_{y,i}^*) \quad \text{for} \quad i = 1, 2, \ldots, M \qquad (9)$$

where M is the total number of degrees of freedom and $(\kappa_{x,i}^*, \kappa_{y,i}^*)$ is the set of wave numbers at which the variability response function corresponding to the i-th degree of freedom U_i takes its maximum value:

$$VRF_i(\kappa_{x,i}^*, \kappa_{y,i}^*) \geq VRF_i(\kappa_x, \kappa_y) \quad -\infty < \kappa_x < \infty\,; \quad -\infty < \kappa_y < \infty \qquad (10)$$

Since the expression shown in Eq. 9 does not depend on the power spectral density function $S_{ff}(\kappa_x, \kappa_y)$, it constitutes a spectral-distribution-free upper bound for the variance of the nodal displacements and the following expression can be written:

$$\mathrm{Var}[U_i] \leq \sigma_{ff}^2 \cdot VRF_i(\kappa_{x,i}^*, \kappa_{y,i}^*) \quad \text{for} \quad i = 1, 2, \ldots, M \qquad (11)$$

Furthermore, for non-zero mean values of U_i, the upper bound of the coefficient of variation of the i-th nodal displacement is established as:

$$\mathrm{COV}\,[U_i] \leq \sigma_{ff} \frac{\sqrt{VRF_i(\kappa^*_{x,i}, \kappa^*_{y,i})}}{|\mathcal{E}\{U_i\}|} \quad \text{for} \quad i = 1, 2, \ldots, M \qquad (12)$$

2.3 Variability response function based on local averages

Using the weighted integral method, the variability response function was established analytically in Eq. 7. Because of the accurate and consistent representation of stochastic field $f(x, y)$ in terms of random variables called weighted integrals, a relatively large number of terms is involved in the expression for $\mathbf{VRF}(\kappa_x, \kappa_y)$ displayed in Eq. 7. In many cases, however, an approximate representation of stochastic field $f(x, y)$ in terms of a smaller number of random variables may be sufficiently accurate.

Such an approximate representation of $f(x, y)$ can be achieved using the local averaging method (Vanmarcke, 1983). Although the concept of local averaging has been used in the context of stochastic finite element analysis for some time, it has never been utilized to evaluate the variability response function and to establish upper bounds of the response variability. These two tasks will be accomplished in the following.

According to the local averaging method, the stochastic field $f(x, y)$ is integrated over the (rectangular) finite element area $A^{(e)}$ (Vanmarcke, 1983) and the result is then divided by $A^{(e)}$. Therefore, the stochastic field is reduced to only one random variable $\alpha^{(e)}$, per finite element:

$$\alpha^{(e)} = \frac{1}{A^{(e)}} \int_{A^{(e)}} f^{(e)}(x, y) dA^{(e)} \qquad (13)$$

where $\alpha^{(e)}$ is called local average. The elastic modulus of finite element (e) is then expressed as:

$$E^{(e)} = E_0^{(e)} (1 + \alpha^{(e)}) \qquad (14)$$

and the stochastic element stiffness matrix is written in the form:

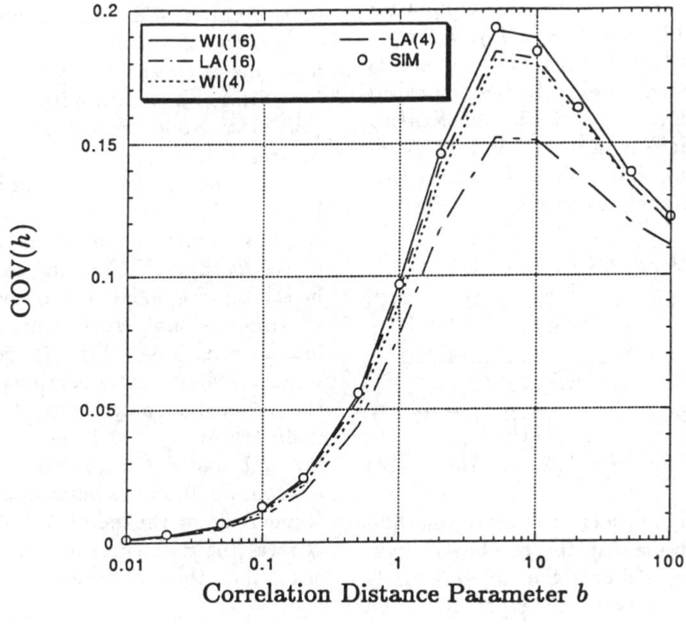

Fig. 4 Coefficient of Variation of Horizontal Displacement h as a Function of Correlation Distance Parameter b

$$\mathbf{K}^{(e)} = \mathbf{K}_0^{(e)}(1 + \alpha^{(e)}) \quad (15)$$

Following the same procedure as in the weighted integral approach, the nodal displacement vector $\mathbf{U}$ is approximated by a first-order Taylor expansion around the mean values of the local averages. Eventually, after some algebra (Wall and Deodatis, 1993), the variability response function based on local averages is established as:

$$\mathbf{VRF}_{LA}(\kappa_x, \kappa_y) = \sum_{e=1}^{N}\sum_{f=1}^{N}$$
$$\text{diag}\left(\mathbf{K}_0^{-1}\mathbf{K}_0^{(e)}\mathbf{U}_0\right)\mathbf{K}_0^{-1}\mathbf{K}_0^{(f)}\mathbf{U}_0$$
$$\frac{1}{A^{(e)}A^{(f)}}\{[Q_eQ_f + W_eW_f]$$
$$\cos(\Delta x_{fe}\kappa_x + \Delta y_{fe}\kappa_y)$$
$$- [Q_fW_e - Q_eW_f]$$
$$\sin(\Delta x_{fe}\kappa_x + \Delta y_{fe}\kappa_y)\} \quad (16)$$

where $\mathbf{K}_0^{(e)}$ and $\mathbf{K}_0^{(f)}$ are the extended versions of the (deterministic) element stiffness matrices of elements (e) and (f), respectively. Due to the fact that the first weighted integral and the local average differ only by a constant factor of $1/A^{(e)}$ (compare Eq. 3 to Eq. 13), Q_i and W_i, $(i = e, f)$ in the above equation are identical to Q_{i1} and W_{i1} in Eq. 7.

Upper bounds of the response variability are now established in exactly the same way as was done in Eqs. 11 and 12. Substituting $VRF_{LA,i}$ for VRF_i in Eqs. 11 and 12 yields the corresponding expressions for the upper bounds based on local averages:

$$\text{Var}_{LA}[U_i] \leq \sigma_{ff}^2 \cdot VRF_{LA,i}(\kappa_{x,i}^*, \kappa_{y,i}^*) \quad \text{for} \quad i = 1, 2, \ldots, M \quad (17)$$

$$\text{COV}_{LA}[U_i] \leq \sigma_{ff}\frac{\sqrt{VRF_{LA,i}(\kappa_{x,i}^*, \kappa_{y,i}^*)}}{|\mathcal{E}\{U_i\}|} \quad \text{for} \quad i = 1, 2, \ldots, M \quad (18)$$

Since, in general, the local averaging method tends to under-represent the stochastic variability within an element, the upper bounds based on local averages are expected to be smaller than the more accurate ones based on weighted integrals.

2.4 Error estimation of first-order Taylor expansion approximation

In order to quantify the error introduced by the first-order Taylor expansion approximation indicated in Eq. 5, Monte Carlo simulations are performed. These simulations are based on the covariance matrix of the vector containing the $6N$ weighted integrals involved in the problem. Specifically, after generating a sample function of this vector using the modal decomposition method (e.g. Yamazaki 1987), all element stiffness matrices are computed using Eq. 2, and a sample global nodal displacement vector is obtained by solving Eq. 4. This procedure is repeated a large number of times in order to accurately estimate the statistics (e.g. coefficient of variation) of the response. Such Monte Carlo simulation results are presented in the Numerical Examples section.

3 NUMERICAL EXAMPLES

The structure analyzed is a square plate under vertical in-plane loading. The discretized structure together with its support conditions and loading are shown in Fig. 1. The FE-mesh consists of 4×4=16 elements and the total number of degrees of freedom is 50. The coefficient of variation (COV) of the elastic modulus is chosen to be $\sigma_{ff} = 0.10$.

First, the variability response function (VRF) of the vertical displacement of the upper right node (denoted by v) is plotted as a function of the wave numbers κ_x and κ_y in Fig. 2. In order to compare the variability response function based on the weighted integral method (WI) to the corresponding one based on the local averaging method (LA), two-dimensional cross-sections of the three-dimensional VRF (Fig. 2) are most suitable. Figure 3 shows cross-sections of the VRF for the vertical displacement of the upper right node (v) at $\kappa_y = 0.0$ and $\kappa_y = 0.2$, for both the WI and LA methods. Although there are some differences between the two methods (especially in the range $1.0 \leq |\kappa_x| \leq 3.0$), in general their agreement is quite good considering that the LA method is an approximate one.

In order to calculate the spectral-distribution-free upper bound of the response variability, the maximum value of the VRF has to be evaluated. This is done numerically, since no analytical result is available. The following values are established for the numerical example considered here:

Weighted integral method :

$\mathrm{COV}(v) \leq 0.125; \quad \mathrm{COV}(h) \leq 0.412$

Local averaging method :

$\mathrm{COV}(v) \leq 0.121; \quad \mathrm{COV}(h) \leq 0.396$

Note that h is the horizontal displacement of the upper right node. The difference in the results between the two methods is less than five percent. For this specific example, therefore, it can be claimed that both methods yield sufficiently accurate results for the spectral-distribution-free upper bounds. At this juncture, it is pointed out that a FE-mesh consisting of 4×4=16 elements is considered relatively fine for the plate and loading examined here. Reducing the number of finite elements to 2×2=4, the difference in the results between the two methods increases. Considering the vertical displacement v only, the upper bounds based on the 2×2=4 element mesh are $\mathrm{COV}(v) \leq 0.121$ for the weighted integral method and $\mathrm{COV}(v) \leq 0.106$ for the local averaging method. Here, the difference is more than ten percent.

To evaluate the response variability, the power spectral density function characterizing the stochastic field $f(x, y)$ is chosen to be:

$$S_{ff}(\kappa_x, \kappa_y) = \frac{\sigma_{ff}^2}{\pi^2} \cdot \frac{b^2}{(1 + b^2\kappa_x^2) \cdot (1 + b^2\kappa_y^2)} \tag{19}$$

Equation 6 is now used to estimate the coefficient of variation (COV) of the horizontal displacement of the upper right node (h), for different values of the correlation distance parameter b (the integral appearing in Eq. 6 is calculated numerically). The results of five cases are compared in Fig. 4: weighted integral method using the 4×4=16 mesh [WI(16)] and the 2×2=4 mesh [WI(4)], local averaging method using the 4×4=16 mesh [LA(16)] and the 2×2=4 mesh [LA(4)] and Monte Carlo simulation using the 4×4=16 mesh [SIM].

Although the curves for WI(16), LA(16) and WI(4) agree very well, there is a considerable deviation for curve LA(4), especially for $b > 2$. This fact indicates that a sufficiently large number of finite elements is required for the local averaging method to yield accurate results. Another finding drawn from Fig. 4 is that the coefficient of variation of the response (in this case h) can be much larger (almost by a factor of 2) than the coefficient of variation of the elastic modulus. Finally, the comparison with Monte Carlo simulation shows that the error introduced by the first-order Taylor expansion is negligible for the system under consideration.

4 CONCLUSIONS

The concept of variability response function (VRF) was introduced for plates under plane stress/strain based on both the weighted integral and the local averaging methods. The VRF allows the calculation of spectral-distribution-free upper bounds of the response variability and provides insight of the underlying mechanisms controlling the response variability of stochastic systems. It was found that if the finite element mesh is relatively fine (e.g. due to the expected stress gradient), the local averaging method can produce sufficiently accurate results, with substantial computational savings, for both upper bound and response variability estimates. It was also established that the coefficient of variation of the response quantities can be considerably higher that the coefficient of variation of the elastic modulus.

ACKNOWLEDGMENTS

This work was supported by the National Science Foundation under Grant # BCS-9257900. The Erwin-Schrödinger Fellowship awarded to F.J. Wall under the Austrian Science Foundation (FWF) Grant # J0617-TEC is gratefully acknowledged by the first author.

REFERENCES

Benaroya, H. and Rehak, M. (1988). "Finite element methods in probabilistic structural analysis: A selective review," *Applied Mechanics Reviews*, 41(5), 201–213.

Brenner, C.E. (1991). "Stochastic finite element methods," *Internal Working Report*

No. 35-91, Institute of Engineering Mechanics, University of Innsbruck, Austria.

Deodatis, G. (1990). "Bounds on response variability of stochastic finite element systems," *Journal of Engineering Mechanics*, ASCE, 116(3), 565–585.

Deodatis, G. (1991). "Weighted integral method. I: Stochastic stiffness matrix," *Journal of Engineering Mechanics*, ASCE, 117(8), 1851–1864.

Deodatis, G. and Shinozuka, M. (1989). "Bounds on response variability of stochastic systems," *Journal of Engineering Mechanics*, ASCE, 115(11), 2543–2563.

Deodatis, G. and Shinozuka, M. (1991). "Weighted integral method. II: Response variability and reliability," *Journal of Engineering Mechanics*, ASCE, 117(8), 1865–1877.

Der Kiureghian, A., Li, C.C. and Zhang, Y. (1991). "Recent developments in stochastic finite elements," *Lecture Notes in Engineering IFIP 76, Proceedings of Fourth IFIG WG 7.5 Conference*, (Eds. Rackwitz and Thoft-Christensen), Germany, Springer-Verlag.

Takada, T. (1990a). "Weighted integral method in stochastic finite element analysis," *Probabilistic Engineering Mechanics*, 5(3), 146–156.

Takada, T. (1990b). "Weighted integral method in multi-dimensional stochastic finite element analysis," *Probabilistic Engineering Mechanics*, 5(4), 158–166.

Vanmarcke, E. (1983). "Random Fields: Analysis and Synthesis," The MIT Press, Cambridge, MA.

Vanmarcke, E., Shinozuka, M., Nakagiri, S., Schuëller, G.I. and Grigoriu, M. (1986). "Random fields and stochastic finite elements," *Structural Safety*, 3(3+4), 143–166.

Wall, F.J. and Deodatis, G. (1993). "Variability response functions and upper bounds of response of 2D stochastic systems," *submitted for publication to the ASCE Journal of Engineering Mechanics.*

Yamazaki, F. (1987). "Simulation of stochastic fields and its application to finite element analysis," *ORI Report 87-04*, Ohsaki Research Institute, Japan.

Stochastic finite element methods (ongoing research)

Structural Safety & Reliability, Schuëller, Shinozuka & Yao (eds) © 1994 Balkema, Rotterdam, ISBN 90 5410 357 4

Probabilistic failure predictions of plain concrete structures on the basis of the extended Leon model

Yong-Hak Lee, Dan M. Frangopol & Kaspar Willam
Department of Civil Engineering, University of Colorado, Boulder, Colo., USA

ABSTRACT: In this study stochastic nonlinear finite elements are applied to plane stress analysis of plain concrete panels. The concrete nonlinearity is included in the stochastic finite element approach. The essential material properties of concrete are assumed to be random variables and are defined in terms of random fields. Two types of limit state criteria, a displacement-based criterion and a hardening-based criterion, are considered.

1 INTRODUCTION

In structural reliability analysis the main purpose is to calculate the probability of failure of a structure relative to a limit state criterion (Ang and Tang 1984). During the past decade there has been a rapid growth for incorporating probabilistic analysis in a finite element framework (Vanmarcke 1983, Der Kiureghian 1985, Liu and Der Kiureghian 1989, Teigen et al. 1991, among others). In this context, this study uses FORM (first order reliability method) for the reliability analysis of a plain concrete panel under plane stress. Therefore, it is essential to get the gradients of the limit state performance function with respect to the basic random variables. Those gradients can be analytically calculated from the nonlinear finite element analysis considering the material nonlinearity in concrete provided by the constitutive model. The plane stress version of the extended Leon model (Pramono and Willam 1989, Etse and Willam 1993) is adopted, which describes the deterministic nonlinear and inelastic response behavior of plain concrete in the pre- and post-peak regime.

In order to describe the uncertainty of the material properties, the modulus of elasticity E, Poisson's ratio ν, the uniaxial compressive concrete strength f'_c, and the uniaxial tensile concrete strength f'_t, are assumed to be Gaussian random variables. Additionally, the applied loads p are also assumed to be Gaussian. Furthermore, to include the spatial variability of each material random variable, the spatial averaging method (Vanmarcke 1983) is adopted.

2 LIMIT STATE AND GRADIENTS

Two types of limit state criteria are considered in this study. The first one is related to the nodal displacement of the finite element solutions which induces the most critical deformation. The second one is related to the hardening parameter of the extended Leon model which has the most critical value at a specific Gauss point.

The performance functions associated with these two criteria can be expressed as follows:

- The displacement-based performance function

$$g_1(u(\mathbf{V})) = u_0 - u(\mathbf{V}) \tag{1}$$

- The hardening-based performance function

$$g_2(k(\epsilon_p(\mathbf{V}))) = k^* - k(\epsilon_p(\mathbf{V}), x_p(\sigma_m(\boldsymbol{\sigma}(\boldsymbol{\epsilon}(\mathbf{u}(\mathbf{V})), \mathbf{V})), \mathbf{V})) \tag{2}$$

where the vector of random variables is $\mathbf{V} = (E, \nu, f'_c, f'_t, p)^T$, u_0 designates the prescribed deformation specified at a nodal point, and k^* designates the prescribed hardening parameter at a Gauss point in a finite element.

The hardening parameter k in eqn.(2) is described empirically as a monotonically increasing elliptic function of the variables ϵ_p and x_p (Pramono and Willam 1989)

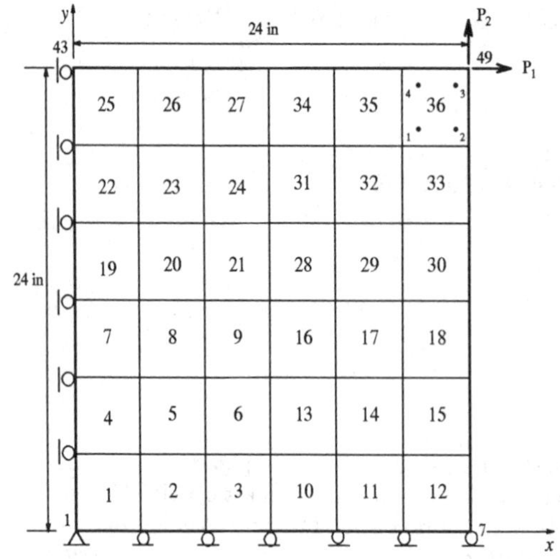

Figure 1. Finite Element Mesh of a Plain Concrete Panel Subjected to Corner Loads

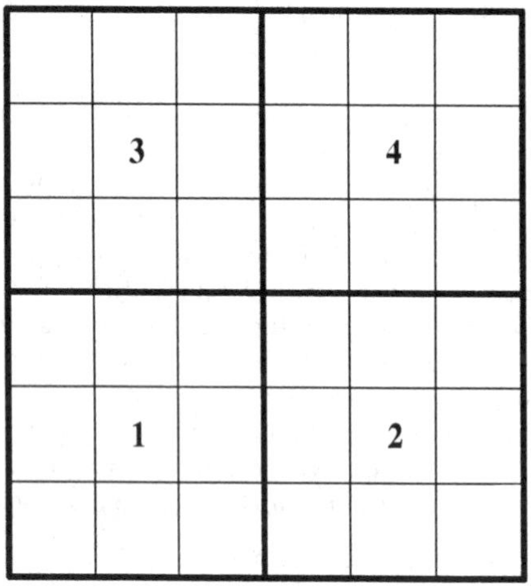

Figure 2. Four Random Meshes of the Plain Concrete Panel

$$k = k_0 + \frac{1-k_0}{x_p}\sqrt{2\epsilon_p x_p - \epsilon_p^2} \tag{3}$$

where $0 \leq k_0 \leq 1$ denotes the initial value of the hardening parameter which defines the elastic limit, ϵ_p denotes the equivalent plastic strain as the Euclidean norm of the plastic strain, and x_p is a ductility measure which defines the accumulated plastic strain at peak (*i.e.*, k=1) in terms of the lateral confinement.

Since in this study the algorithm for the reliability analysis is based on FORM, which uses a first order Taylor series expansion, the gradients of the performance functions with respect to each random variable are needed. The gradients of the displacement-based and the hardening-based performance functions with respect to each random variable can be calculated through the chain rule of differentiation applied to eqns.(1) and (2), respectively, as follows:

$$\frac{\partial g_1(u(\mathbf{V}))}{\partial \mathbf{V}} = \frac{\partial g_1(u)}{\partial u}\frac{\partial u}{\partial \mathbf{V}} \tag{4}$$

$$\frac{\partial g_2(k(\epsilon_p(\mathbf{V})))}{\partial \mathbf{V}} = \frac{\partial g_2(k)}{\partial k}[\frac{\partial k}{\partial \epsilon_p}\frac{\partial \epsilon_p}{\partial \boldsymbol{\epsilon}_p}\frac{\partial \boldsymbol{\epsilon}_p}{\partial \mathbf{V}}|_{\epsilon_p} + \frac{\partial k}{\partial x_p}\frac{\partial x_p}{\partial \mathbf{V}}|_{x_p} + \frac{\partial k}{\partial x_p}\frac{\partial x_p}{\partial \sigma_m}\frac{\partial \sigma_m}{\partial \boldsymbol{\sigma}}\frac{\partial \boldsymbol{\sigma}}{\partial \mathbf{V}}|_{x_p}] \tag{5}$$

In the above equations the partial derivatives of the displacement, strain, stress, and plastic strain vectors with respect to the vector $\mathbf{V}$, $\frac{\partial u}{\partial \mathbf{V}}$, $\frac{\partial \boldsymbol{\epsilon}}{\partial \mathbf{V}}$, $\frac{\partial \boldsymbol{\sigma}}{\partial \mathbf{V}}$ and $\frac{\partial \boldsymbol{\epsilon}_p}{\partial \mathbf{V}}$, respectively, can be analytically calculated from the global equilibrium equations of the nonlinear finite element analysis and the stress-strain relations in the inelastic range.

3 NUMERICAL EXAMPLE

The reliability of a plain concrete panel subjected at the upper right-hand corner to two loads P_1 and P_2 along x and y directions, respectively, is analyzed under plane stress conditions (Figure 1). Four material random variables including the modulus of elasticity ($\bar{E} = 3500ksi, V_E = 0.08$), Poisson's ratio ($\bar{\nu} = 0.23, V_\nu = 0.08$), the uniaxial compressive concrete strength ($\bar{f}_c' = 4.3ksi, V_{f_c'} = 0.1$), and the uniaxial tensile concrete strength ($\bar{f}_t' = 0.421ksi, V_{f_c'} = 0.12$) are considered. Each material random variable is assumed to be normally distributed and defined in terms of random fields. Also, the applied loads are assumed to be Gaussian random variables. The panel is idealized by 36 finite elements (Figure 1) and 4 random meshes (Figure 2). The mean values and coefficients of variation of P_1 and P_2 are as follows: $\bar{P}_1 = \bar{P}_2 = -12.0kips$ and $V_{P_1} = V_{P_2} = 0.2$. Four-node quadrilateral finite elements are used for the finite element analysis. The autocorrelation function is assumed to be Gaussian (Vanmarcke 1983)

$$\rho(X,Y) = exp[-\frac{X^2+Y^2}{b^2}] \tag{6}$$

Table 1. Reliability Analysis Results for a Plain Concrete Panel

	Displacement Criterion		Hardening Criterion		Hardening Criterion (without load variability)
Reliability Index	2.06		3.85		4.56
Number of Iterations	12		17		15
Random Variables	Mean	Design Point	Mean	Design Point	Design Point
E_1	3500 ksi	3486.32	3500 ksi	3498.87	3497.52
E_2	3500 ksi	3479.56	3500 ksi	3495.47	3493.10
E_3	3500 ksi	3467.34	3500 ksi	3495.47	3493.10
E_4	3500 ksi	3443.17	3500 ksi	3487.63	3481.16
ν_1	0.23	0.23	0.23	0.23	0.23
ν_2	0.23	0.23	0.23	0.23	0.23
ν_3	0.23	0.23	0.23	0.23	0.23
ν_4	0.23	0.23	0.23	0.23	0.23
$(f'_c)_1$	4.3 ksi	4.23	4.3 ksi	4.125	4.089
$(f'_c)_2$	4.3 ksi	4.10	4.3 ksi	3.821	3.724
$(f'_c)_3$	4.3 ksi	4.10	4.3 ksi	3.821	3.724
$(f'_c)_4$	4.3 ksi	3.76	4.3 ksi	2.992	2.726
$(f'_t)_1$	0.421 ksi	0.4212	0.421 ksi	0.4216	0.4217
$(f'_t)_2$	0.421 ksi	0.4216	0.421 ksi	0.4226	0.4228
$(f'_t)_3$	0.421 ksi	0.4216	0.421 ksi	0.4226	0.4228
$(f'_t)_4$	0.421 ksi	0.4228	0.421 ksi	0.4255	0.4260
P_1	-12.0 kips	-14.85	-12.0 kips	-13.12	-12.0
P_2	-12.0 kips	-10.70	-12.0 kips	-13.12	-12.0

Note : The subscripts ($_{1}$,$_{2}$,$_{3}$,$_{4}$) correspond to the random meshes shown in Fig.2.

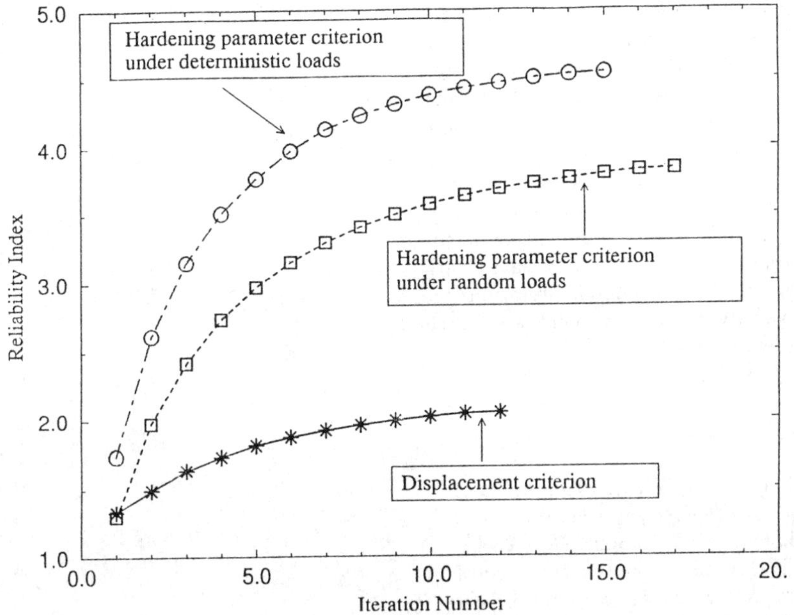

Figure 3. Convergence of the Computational Process for Reliability Analysis

where X and Y are the distances between two points of a random mesh along x- and y-direction, respectively, and $b = 0.4 \times 24in = 9.6in$ is a measure of correlation length.

The problem is analyzed using FORM with two performance functions, the displacement-based performance function (1) and the hardening-based performance function (2). In order to examine the relative importance of material random variables, the hardening-based performance function was used by considering two cases separately: random loads and deterministic (without coefficients of variation) loads.

The displacement-based performance function is expressed based on eqn.(1)

$$g(\mathbf{X}) = u_0 - u_{49}(\mathbf{V}) \tag{7}$$

where $u_0 = -4.278 \times 10^{-3} in$ is the prescribed displacement and u_{49} is the actual displacement along x-direction at node 49 (see Fig.1).

The hardening-based performance function is expressed based on eqn.(2)

$$g(k) = k^* - k \tag{8}$$

where $k^* = 0.6$ is the prescribed hardening parameter and k is the actual hardening parameter defined at the Gauss point 3 of element 36 (see Fig.1).

The value of the prescribed displacement u_0 corresponds to the hardening criterion under the proportional loading condition which induces $k^* = 0.6$ at the Gauss point 3 of element 36.

The computed design points and the associated reliability indices are shown in Table 1. The convergence of the computational process for reliability analysis in each of the three cases considered is shown in Figure 3. The starting iteration point corresponds to the mean values of basic random variables. As expected, the reliability index corresponding to the displacement-based criterion is lower than that corresponding to the hardening-based criterion. Also, due to the high nonlinearity of the performance function associated with the hardening-based criterion, more iterations are needed to evaluate the design point associated with the hardening case.

REFERENCES

Ang, A. H-S, and Tang, W. H. 1984. *Probabilistic Concepts in Engineering Planning and Design* Vol.II, John Wiley & Sons, New York.

Der Kiureghian, A. 1985. Finite element method in structural safety studies. *Structural Safety Studies.* eds. J.T.P. Yao et al. ASCE. New York. 40-52.

Etse, G. and Willam, K. 1993. A fracture energy-based constitutive theory for inelastic bahavior of plain concrete. *Journal of Engineering Mechanics.* ASCE. submitted for publication.

Liu, P-L. and Kiureghian, A. 1989. Finite-Element Reliability Methods for Geometrically Nonlinear Stochastic Structures, *Report No. UCB /SEMM-89/05.* Dept. of Civil Engrg., Univ. of California, Berkeley, Calif.

Pramono, E. and Willam, K. 1989. Fracture energy-based plasticity formulation of plain concrete, *Journal of Engineering Mechanics*, ASCE. 115(6) : 1183-1204.

Teigen, J.G., Frangopol, D.M., Sture, S. and Felippa, C. 1991. Probabilistic FEM for nonlinear concrete structures. I : Theory. *Journal of Structural Engineering.* ASCE. 117(9) : 2674-2689.

Vanmarcke, E. H. 1983. *Random Fields: Analysis and Synthesis*, MIT Press, Cambridge, Mass.

Structural Safety & Reliability, Schuëller, Shinozuka & Yao (eds) © 1994 Balkema, Rotterdam, ISBN 90 5410 357 4

Probabilistic FEM analysis of reinforced concrete beams

J. Piliszek & I. Czmoch
Chalmers University of Technology, Göteborg, Sweden

ABSTRACT: The paper deals with the application of the probabilistic finite element method to the reliability analysis of reinforced concrete beams. The formulation takes into account the stochastic variability along beam of material properties, geometrical parameters and loads. Non-linear behaviour of reinforced concrete beam is described by a trilinear moment-curvature relationship. The probability of unreliable performance is calculated by means of the Hasofer-Lind β–index method. The developed probabilistic FEM analysis has been implemented in the computer program BETA.

1. INTRODUCTION

The probabilistic finite element method (PFEM) has been intensively developed during last years. A review of papers was done by Vanmarcke et al. (1986) and Der Kiureghian et al. (1991). Fundamental papers about non-linear PFEM were published by Liu, W. K. et al. (1986) and Liu, P.-L. et al. (1989). An extension of general probabilistic finite element method to the analysis of non-linear concrete structures was presented by Teigen et al. (1991), who considered the problem of geometric and material non-linearity in RC frame. Three random fields were taken into account: material, geometry and load. Randomness of basic variables were characterised by mean values and covariance matrices. The perturbation method and the first order Taylor series expansion about the arbitrary reference values of basic variables were applied to determine the mean value and standard deviation of structural response.

2. DESCRIPTION OF PROBABILISTIC FEM ANALYSIS

The basic assumptions of the presented model are as follows:

(i) Randomness of structure is described by means of a set of primary random processes which are considered mutually stochastically independent. In general, each random process may be assumed as a non-Gaussian process along the beam. Each random process is defined in terms of the marginal distribution (which can vary along a beam) and the covariance function.
(ii) Each continuous random process is discretized to a set of random variables associated with finite elements. Any method of discretization may be applied.
(iii) Discrete process associated with degrees of freedom at nodes of finite elements can be introduced in the analysis.
(iv) The random process with the marginal non-Gaussian CDF is replaced by the equivalent random process with the marginal Gaussian CDF, which is obtained by the Nataf transformation (Nataf 1962), (Der Kiureghian 1986) and the normal-tail approximation (Ditlevsen 1981).
(v) The reliability of beam is defined by means of the FORM Hasofer-Lind β-index (1974).
(vi) The performance function, which separates reliable and unreliable states of structure, is defined in terms of discretized primary random variables and random load effects (i.e. displacements, strains and internal forces).
(vii) The finite element formulation is based on

a 2-D beam element with two degrees of freedom at each node (i.e. transversal displacement and rotation).

(viii) The reliability analysis is based on the iterative Rackwitz-Fiessler (1978) algorithm. The non-linear FEM equations are solved by means of the modified Newton-Raphson method for each iteration step. Than the gradient of performance function is calculated. To minimize the required storage of computer and cost of calculation the gradient vector of performance function is independently evaluated for particular random processes.

The presented approach is more complicated and more computer time consuming than the procedure applied by Teigen et al. (1991). Furthermore, it should be emphasise that methods based on the arbitrary choice of coordinates of expansion of the performance function do not guarantee the minimum β value and are not invariant.

3. APPLICATION OF PROBABILISTIC FEM ANALYSIS

The general model described above was applied to analyse the short-time deflections of reinforced concrete beams. The non-linear behaviour of reinforced concrete beam is described by a trilinear moment-curvature relationship (Alwis 1990), which takes account of three states: the linear-elastic, the cracked and the collapse state. The two characteristic points of the relationship are defined as follows (Fig. 1)

a) cracking point (κ_{cr}, M_{cr})

$$\kappa_{cr} = \frac{f_{ct}}{y_t} \qquad M_{cr} = E_c I_g \kappa_{cr} \qquad (1)$$

where f_{ct} and E_c are respectively, the tensile strength and elastic modulus of concrete; Ig is moment of inertia of the uncracked section, including the contribution of steel; y_t is the distance from the neutral axis to the extreme tensile edge of the section,

b) yielding point (κ_y, M_y)

κ_y, M_y correspond to the yield point of

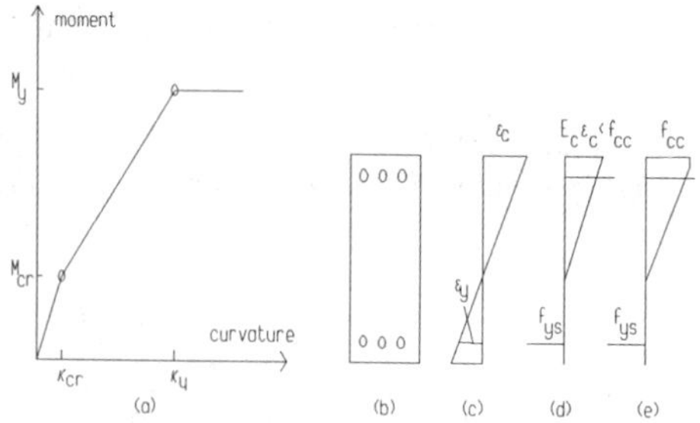

Fig. 1 Trilinear relationship; (a) curve; (b) cross-section; strain distribution at yielding point; (c) stress distribution for elastic concrete (d) stress distribution for yielding concrete

tensile steel for fully cracked cross-section; two possibilities are considered, the compressive stress at the extreme compressive fibre is equal to the compressive strength of concrete in structure or is less than this value (Fig. 1)

The compressive strength of concrete in structure, f_{cc}, the tensile strength of concrete in structure, f_{ct}, and the modulus of elasticity of concrete in structure , E_c, are assumed as functions of the compressive strength of concrete measured on the cylinder specimens 150x300 mm, f_{cs},

$$\begin{aligned} f_{cc} &= C_{f_{cc}} f_{cs} && [\text{MPa}] \\ f_{ct} &= C_{f_{ct}} 0.3 \sqrt[3]{(f_{cc}-8)^2} && [\text{MPa}] \\ E_c &= C_{E_c} 9500 \sqrt[3]{f_{cc}} && [\text{MPa}] \end{aligned} \qquad (2)$$

where $C_{f_{cc}}$, $C_{f_{ct}}$, C_{E_c} are the conversion factors introduced in order to cover the model uncertainties.

In the probabilistic model of RC beams the following mutually independent random processes are considered:

- compressive strength of concrete measured on specimens,
- yield strength of bottom reinforcement,
- yield strength of top reinforcement,
- cross-sectional area of bottom reinforcement,
- cross-sectional area of top reinforcement,
- distance between bottom surface of element and centre of gravity of bottom reinforcement,
- distance between top surface of element and

centre of gravity of top reinforcement,
- depth of cross-section,
- width of cross-section,
- Young modulus of bottom reinforcement,
- Young modulus of top reinforcement,
- conversion factor between the compressive strength measured on specimens and the compressive strength of concrete in structure,
- conversion factor between the compressive strength measured on specimens and the tensile strength of concrete in structure,
- conversion factor between the compressive strength measured on specimens and Young modulus of concrete in structure,
- self-weight of material,
- distributed load,
- point load at nodes,
- geometric positions of nodes,
- initial deflections of supports.

The developed probabilistic FEM analysis has been implemented in the computer program BETA. The probability of unreliable performance can be determined for any performance function (i.e. the ultimate limit state or serviceability limit state) defined in terms of primary variables, internal forces, displacements and generalized strains.

4. EXAMPLE

The reliability of mid span displacements for the beam shown in Fig. 2 is analysed.
The compressive strength of concrete in structure f_{cc} is considered as a random process. The beam is subjected to random point load F. The statistical parameters of the random field f_{cc} and random variable F are assumed as follows:
f_{cc} - Normal or log-Normal distribution;

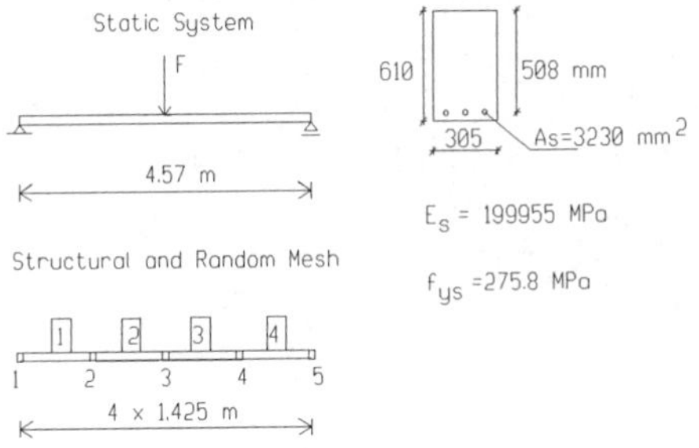

Fig. 2 Definition of problem

mean value=24.82 MPa; C.O.V.=0.15,
F - Normal or Extreme I type distribution;
mean value varies; C.O.V.=0.20.

It has been assumed that the compressive strength f_{cc} is not correlated along the beam. The continuous random process f_{cc} has been discretized according to the mid-point method (Der Kiureghian et al. 1991).
In order to compare the mean value FORM with the Hasofer-Lind method, the performance function for each load level is a function of mean value of point load F, m_F , and has the form

$$g(...) = m_a + 1.645\,\sigma_a - a \tag{3}$$

where m_a is the mean value of the mid span displacement for the load level equal to the mean value of load m_F; σ_a is the standard deviation of the mid span displacements for the load level equal to the mean value of load m_F . Hence, the mean value FORM β index is constant and equal to 1.645 for each load level.

The calculations have been carried out for two cases:
- strength of concrete f_{cc} is deterministic; load is random,
- strength of concrete f_{cc} is random; load is deterministic.

The Hasofer-Lind β–index for the first case is shown in Fig. 3. If the mean value m_a for certain load level corresponds to the uncracked state and the checking point value of random variable, a, corresponds to the cracked state, the Hasofer-Lind β–index differs considerably from the mean value FORM β–index. Such situation occurred for the finite elements 2 and 3 at the mean value of load $m_F \approx 0.05$ MN and for the finite elements 1 and 4 at the mean value of load $m_F \approx 0.17$ MN. Since the displacement-load relationships are linear within the uncracked state or the cracked state (Fig. 1), the Hasofer-Lind β–index is constant for other mean values of load. Moreover, it is equal to the mean value FORM β–index, if load is modelled by the Normal distribution.

The Hasofer-Lind β–index for the random strength of concrete and deterministic load is shown in Fig. 4. Since the relationship displacement-strength of concrete is non-linear for

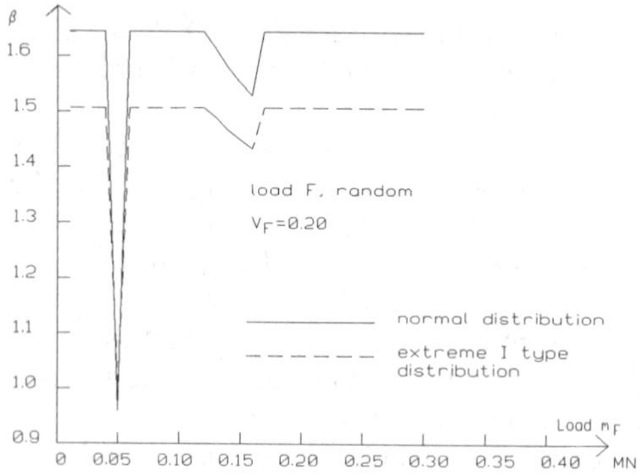

Fig. 3 β index versus load

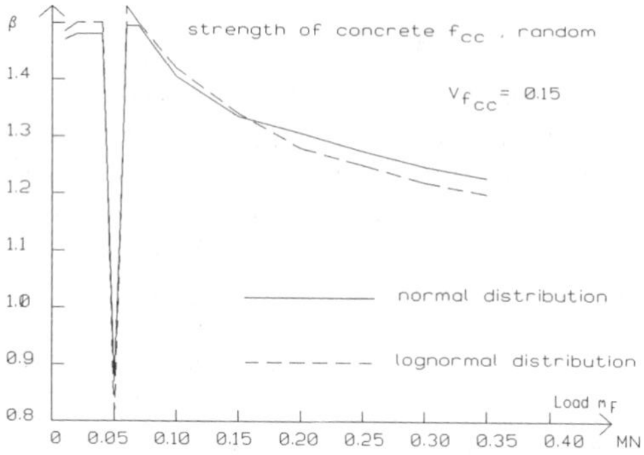

Fig. 4 β index versus load

whole range of applied load, the Hasofer-Lind β–index differs from the mean value FORM β–index for any load level. A sudden drop of the Hasofer-Lind β–index occurs for the mean value of load m_F associated with the transition between the uncracked and cracked state for the finite elements 2 and 3.

5. ACKNOWLEDGEMENTS

This paper has been written in association with the project from the Swedish Council for Building Research (BFR). The financial support is gratefully acknowledged.

6. REFERENCES

Alwis, W.A.M. 1990. Trilinear Moment-Curvature Relationship for Reinforced Concrete Beams, *ACI Journal*, May-June 1990, 276-283

Ditlevsen, O. 1981. Principle of Normal Tail Approximation, *J. Eng. Mech., ASCE*, 107(12), 1981, 1191-1208

Hasofer, A..M., Lind, L. 1974. Exact and Invariant Second Moment Code Format, *J. Eng. Mech., ASCE*, 100(2), 1974, 111-121

Der Kiureghian, Liu, P.-L. 1986. Structural Reliability under Incomplete Probability Information, *J. of Struct. Mech., ASCE*, 112(1), 1986, 85-104

Der Kiureghian, A., Li, C.-C., Zhang, Y. 1991. Recent Developments in Stochastic Finite Elements, in Reliability and Optimization of Structural Systems 91, *Proc. of 4th IFIP WG 7.5 Conference*, Münich, Germany, 19-38

Liu, W.K., Belytschko, T., Mani, A. 1986. Probabilistic Finite Element Methods for Non-linear Structural Dynamics, *Comp. Methods Appl. Mech. Eng.*, 56(1), 1986, 61-81

Liu, P.-L., Der Kiureghian, A. 1989. Finite Element Reliability of Geometrically Nonlinear Uncertain Structures, *J. Eng. Mech., ASCE*, 117(8), 1989, 1806-1825

Nataf, A.. 1962. Détermination des Distribution dont Les Marges sont Donnés; *Comptes Rendues de L''Academie des Sciences*, Vol. 225, Paris, 1962, 42-43

Teigen, J.G., Frangopol, D. M., Sture, S., Fellipa, C.A. 1991. Probabilistic FEM for Nonlinear Concrete Structures I: Theory & II: Applications, *J. Struct. Eng., ASCE*, 117(9), 1991; 2674-2707

Rackwitz, R., Fiessler, B. 1978. Structural Reliability under Combined Random Load Sequences; *Comp. & Struct.*, 9, 1978, 489-494

Vanmarcke, E., Shinozuka, M., Nakagiri, S., Schuëller, G.I. and Grigoriu; M. 1986. Random Fields and Stochastic Finite Elements, *Structural Safety*, 3(3+4), 1986, 143-166

Structural Safety & Reliability, Schuëller, Shinozuka & Yao (eds) © *1994 Balkema, Rotterdam, ISBN 90 5410 357 4*

Reliability analysis of complex structures with random material properties and random geometry

S. Reh
Kernforschungszentrum Karlsruhe GmbH, Institut für Materialforschung II, Germany

A. Brückner-Foit
Universität Karlsruhe, Institut für Zuverlässigkeit und Schadenskunde, Germany

F. Böhm
Hilti AG, Konzern-Forschung, Schaan, Principality of Liechtenstein

ABSTRACT: Stochastic finite element methods are a combination of the mechanical analysis of complex components using finite element (FE) methods and the probabilistic assessment of these components, e.g. the calculation of the failure probability. In general, random influences arise from material, loading and geometry parameters which are subjected to scatter. If the computation time of one FE-analysis becomes too large then there are two major probabilistic methods, which are still suitable, namely the reliability method of first or second order and the response surface method. Both methods are briefly described in the present contribution. For the reliability methods the gradient of the structural response with respect to the random parameters of the problem has to be determined. This gradient can be evaluated using the "method of adjoint variables". As an illustration, both probabilistic methods are applied to a model turbine blade.

1 INTRODUCTION

FE-methods are required for the stress analysis of complex components under static loading. The primary problem is the computation of the displacement field $\underline{u}$ by solving the equilibrium equation

$$\underline{\underline{K}}(\underline{b})\cdot\underline{u} = \underline{f} \ , \qquad (1)$$

where $\underline{\underline{K}}$ is the stiffness matrix, $\underline{b}$ is the vector of all random parameters and $\underline{f}$ is the vector of external forces. In this contribution the loading is considered to be deterministic. The entries of $\underline{b}$ may arise from random variables or from random fields (Kiureghian, 1989). The random fields are discretized and described in terms of correlated random variables (Kiureghian, 1989). Hence, the number of entries of $\underline{b}$ may become very large.

Since very fine finite element meshes resulting in large CPU time are required for complex components computationally efficient methods for the probabilistic analysis are needed. Two suitable methods are available in literature, namely the response surface method (RSM) and the first and second order reliability methods. Here, the first order reliability method (FORM) and the reponse surface methods are used.

RSM can be applied to various kinds of problems without spending much time on the implemention, but the number of random variables that can be taken into account is limited, whereas the programming effort for FORM is large, but it can be applied to problems with very large numbers of random parameters.

2 PROBABILITY METHODS

The definition of failure of a component is given by the performance function $g(\underline{b}, \underline{u})$ which follows the convention:

$$\begin{aligned} g(\underline{b}, \underline{u}) &\leq 0 : \text{component has failed} \\ g(\underline{b}, \underline{u}) &> 0 : \text{component is still operable.} \end{aligned} \qquad (2)$$

The boundary $g(\underline{b}, \underline{u}) = 0$ is called the limit state surface. The failure probability is then expressed by the multidimensional integral of the joint probability density function $f_{\underline{b}}(\underline{b})$ over the failure domain $g(\underline{b}, \underline{u}) \leq 0$:

$$P_f = \int_{g(\underline{b},\,\underline{u}) \leq 0} f_{\underline{b}}(\underline{b}) \cdot db_1 \cdot \ldots \cdot db_n \quad . \tag{3}$$

This integral cannot be solved analytically especially if the failure domain is given only in an algorithmic form by a FE calculation.

2.1 Response surface methods

Response surface methods solve eq. (3) by an approximation of the performance function g. According to the approach of Faravelli (Böhm et al., 1992; Faravelli, 1989) this approximation is given by

$$g \simeq \tilde{g} = \theta_0 + \underline{b}'^{\mathrm{T}} \cdot \underline{\theta}_1 + \underline{b}'^{\mathrm{T}} \cdot \underline{\underline{\theta}}_2 \cdot \underline{b}' + \varepsilon \tag{4}$$

where the θ's are the coefficients of the polynomial and ε is the experimental error. Only the spacial average of random fields is included in the vector of input variables $\underline{b}'$ and the effect of the local variability is taken into account by the error term ε. Hence, the dimension of $\underline{b}'$ is much smaller compared with the original vector of random variables $\underline{b}$.

Response surface methods are based on an experimental design which is used to obtain suitable data for determining the coefficients θ_0, $\underline{\theta}_1$, $\underline{\underline{\theta}}_2$ and for calculating the error term. Here, a central composite design (Böhm et al., 1992) was used.

2.2 First order reliability method

The first order reliability method solves eq. (3) by approximating the limit state surface in the space of the independent standard normal variables $\underline{r}$ by a hyper-plane in the so-called design point $\underline{r}^*$. This is the point on the limit state surface with minimal distance to the origin of the $\underline{r}$-space. Then eq. (3) reduces to

$$P_f \simeq \Phi(-\beta) \quad , \tag{5}$$

where Φ is the cumulative distribution function of the standard normal distribution and β is the length of the vector $\underline{r}^*$. This implies that the correlated and arbitrarily distributed variables $\underline{b}$ are transformed into the standard normal variables $\underline{r}$ (Kiureghian, 1989).

The design point, i.e. the point on the limit state surface with minimal distance to the origin of the $\underline{r}$-space, can be found if the corresponding constraint optimization problem is solved. For this purpose the gradient of the performance function with respect to $\underline{b}$ has to be calculated.

3 METHOD OF ADJOINT VARIABLES

As shown in (Reh et al., 1992) the method of adjoint variables is a very efficient way to compute the gradient of the performance function, because in addition to the solution of the displacement field $\underline{u}$ in (1) only one FE calculation is required. The additional FE problem is given by

$$\underline{\underline{K}}(\underline{b}) \cdot \underline{\lambda} = \left(\frac{\partial g(\underline{b}, \underline{u})}{\partial \underline{u}} \bigg|_{\underline{b}} \right)^T \quad , \tag{6}$$

where $\underline{\underline{K}}$ is the same stiffness matrix as in (1). This equation can be derived from the Lagrange function solving the constrained optimization problem for the design point, where $\underline{\lambda}$ is introduced as a Lagrange multiplier.

Once equation (6) is solved the gradient of the performance function with respect to the random variables follows from:

$$\frac{\mathrm{d}g(\underline{b}, \underline{u})}{\mathrm{d}\underline{b}} = \frac{\partial g(\underline{b}, \underline{u})}{\partial \underline{b}} \bigg|_{\underline{u}} + \sum_{i=1}^{n_{el}} \left(\underline{\lambda}_{(i)}^{\ T} \frac{\partial \underline{f}_{(i)}(\underline{b})}{\partial \underline{b}} - \underline{\lambda}_{(i)}^{\ T} \frac{\partial \underline{\underline{K}}_{(i)}(\underline{b})}{\partial \underline{b}} \underline{u}_{(i)} \right) , \tag{7}$$

where the sum is taken with respect to all n_{el} elements of the FE mesh and the corresponding vectors and matrices are elemental quantities.

The derivative of the element stiffness matrix with respect to random material parameters is easy, because it depends explicitly on material parameters (Reh et al., 1992). If the random variable b_j is a geometrical parameter such as the thickness of the structure, which may be a random field, then this is more involved, because $\underline{\underline{K}}_{el}$ depends on the coordinates of the element nodes. Additional information has to be supplied by the user on how these nodal coordinates vary with the specific geometrical pa-

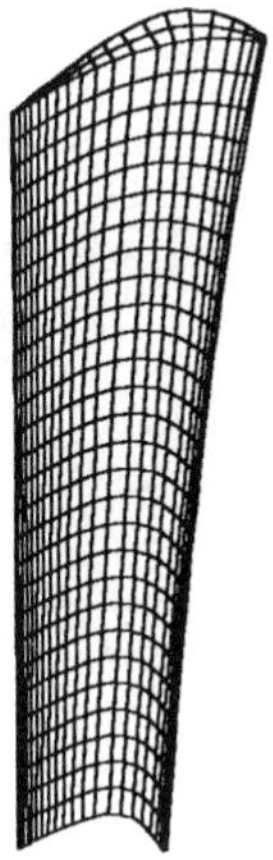

Figure 1. FE mesh of blade

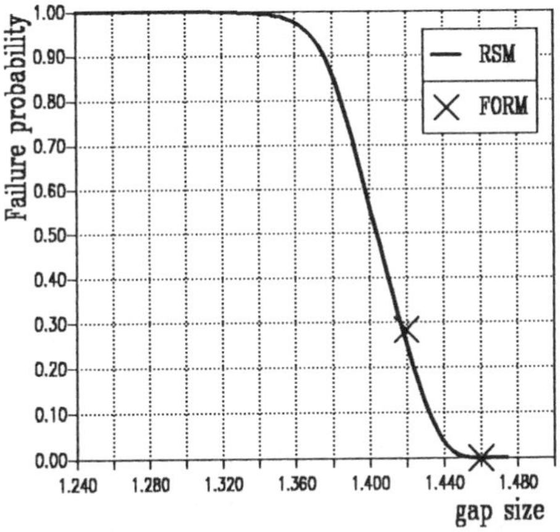

Figure 2. Failure probability versus gap size

rameter. In any case, all derivatives in (6) and (7) can be calculated in closed form.

4 EXAMPLE

4.1 Description of the example

A guide vane of a turbine as illustrated in Figure 1 is considered as application example. The material is a single crystal nickel-based superalloy. The pressure difference on the blade surfaces due to the gas flow around the blade was modelled with equivalent nodal forces. The orientation of the material axes can be described by the three Euler angles ϕ, θ and κ. All these angles are treated as random variables. A lognormal distribution was fitted for θ based on the data published by J.H. Laflen (Laflen, 1983). Due to symmetry ϕ and κ are uniformly distributed in the range from 0° to 90°. In addition, the length of the blade l is a random quantity with uniform distribution within the range of $\pm 0.3\%$ around the nominal value.

Since there is a contact problem of the stator this blade belongs to with the rotating parts of the turbine the performance function g indicates failure if the maximal displacement $u_{\max}$ on the suction side of the blade exceeds the value of the gap Δ_{gap} between rotor and stator, i.e. $g = \Delta_{gap} - u_{\max}$.

In Figure 2 the failure probability is plotted versus the gap size Δ_{gap} for stochastic analyses using RSM and FORM, respectively. The FORM analyses have been performed only for lower failure probabilities, because this is the most important region. Calculations with random fields of geometric parameters are under way.

4.2 Discussion of the results

The results illustrate that RSM and FORM agree with each other quite well. The relative difference was below 10%. As expected the accuracy of FORM was much better at very low failure probability. The required number of FE calculations to obtain these results was 25 for RSM and 32 for FORM. The slight advantage of RSM is caused by the limited number of random variables (only four) in this example.

5 SUMMARY

In the present contribution the reliability analysis of components with scattering material and geometrical parameters was outlined. The basics of the response surface method and the first order reliability methods have been explained briefly and the method of adjoint variables for calculating the gradient of the performance function was discussed.

The application example demonstrates that both RSM and FORM are suitable for cost effective reliability analyses of complex structures described by finite element models.

REFERENCES

Kiureghian A. Der, Liu, P.-L. 1989. *Finite Element Reliability Methods for Geometrical Nonlinear Stochastic Structures*, Report No. UCB/SEMM-89/05, Department of Civil Engineering, University of California, Berkeley

Böhm, F., Brückner-Foit, A. 1992. On a criterion for accepting a response surface model, *Probabilistic Engineering Mechanics* 7: 183-190

Faravelli, L. 1989. Response Surface Approach for Reliability Analysis, *Journal of Engineering Mechanics* 115, 12: 2763-2781

Reh, S., Böhm, F., Brückner-Foit, A., Riesch-Oppermann, H. 1991. First Order Reliability Analysis Using Stochastic Finite Element Methods, *Computational Stochastic Mechanics* Edited by P. D. Spanos and C. A. Brebbia, Computational Mechanics Publications, Southampton

Laflen, J. H. 1983. Analysis of an Idealized Directionally Solidified FCC Material, *Journal of Engineering Materials and Technology*, 105, 10: 307-312.

Structural Safety & Reliability, Schuëller, Shinozuka & Yao (eds) © 1994 Balkema, Rotterdam, ISBN 90 5410 357 4

A practical reliability evaluation method of structures with spatial uncertainties modeled by random functions

W. Shiraki
Tottori University, Japan

ABSTRACT: In this study, a practical reliability evaluation method of structural systems with spatial uncertaintues modeled by two-variate random fields is proposed. Using the results of stochastic finite element method, the response in each finite element of a structural system is modeled as a two-dimensional random pulse space-process. The level crossing theory of two-dimensional pulse space-processes is applied to evaluate the mean upcrossing rate of the response by extending the level crossing theory of one-dimensional pulse processes. As a numerical example, a bending problem of a steel plate resting on the elastic foundation with statistical non-uniformity such as the bearing coefficient is treated, and the reliability analysis of the plate is performed for the lateral deflection and the bending moment.

1. INTRODUCTION

In structural reliability problems, recently, the uncertainties for loads or material properties of structures have been treated as random fields or random processes of time and space coordinates, Kiureghian et al. (1988), Yamazaki et al. (1988) and Vanmarke et al. (1983). Such processes that develop simultaneously in space and time are called spatiotemporal processes. In many enginerring problems, two special cases are usually considered such as random function of only time (random process) and random function of only space coordinates (random field).

In this paper, a practical reliability evaluation method of structural systems with spatial uncertainties modeled by two-variate random fields is proposed.

First, the probabilistic characteristics such as the expectation functions and the auto-covariance functions of responses of a stochastic structural system are evaluated by using the stochastic finite element method. The set of response in each finite element of the system is modeled as a two-dimensional random pulse space-process.

Next, extending the level crossing theory of one-dimensional random pulse processes, the mean upcrossing rate of the two-dimensional random pulse space-process is evaluated. Using this result the estimation of reliability of structural system becomes relatively simple.

Finally, this proposed method is applied to the reliability analysis of a steel plate resting on an elastic foundation with statistical non-uniformity such as the bearing coefficient which is assumed to be a two-variate random field. In numerical examples, the probability of exceedance for the deflection and bending moment of the plate are respectively evaluated, and the effectiveness of this proposed methods is demonstrated.

2. DISCRETIZATION OF RANDOM FIELDS USING LOCAL AVERAGES

For the finite element anlysis, it is necessary to discrete the random field of interest into random variables in each finite element. In this study, the random field is discretized by means of the local average which is evaluated by taking spatial average of the field over each finite element, as suggested by Vanmarcke (1983).

A two-dimensional, statistically homogeneous, continuous random field of the spatial coordinates, t_1 and t_2, is considered and denoted by $X(t_1, t_2)$ here. The local average of a rectangular finite element with the centroid located at (t_1, t_2) is defined as

$$X_A(t_1, t_2) = \frac{1}{A}\int_{t1-T1/2}^{t1+T1/2}\int_{t2-T2/2}^{t2+T2/2} X(t_1, t_2)\, dt_1 dt_2 \quad (1)$$

where T_1 and T_2 are the side length along the respective coodinate axis and $A=T_1T_2$ is the area of the finite element. When the original random field $X(t_1, t_2)$ has the expectation m and standard deviation σ, the expectation and variance of the local average $X_A(t_1, t_2)$ are represented by Eqs. (2) and (3), respectively.

$$E[X_A] = m \quad (2)$$

$$Var[X_A] = \sigma^2 \gamma(T_1, T_2) \quad (3)$$

In Eq. (3), $\gamma(T_1, T_2)$ is called as the variance function of $X(t_1, t_2)$ and given by

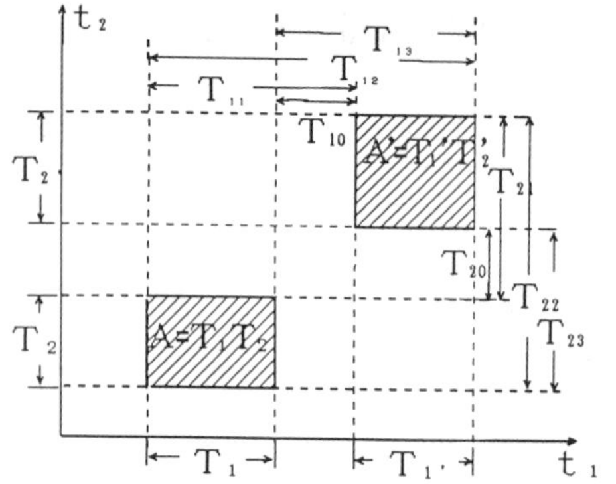

Fig. 1 Distances characterizing relative location of rectangular areas A and A'

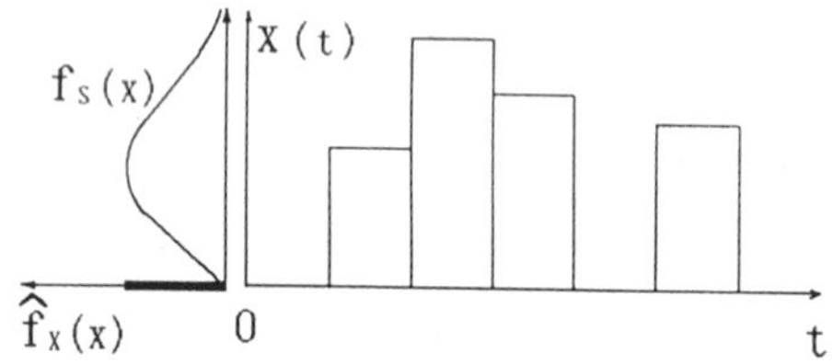

Fig. 2 One-dimensional rectangular pulse process

$$\gamma(T_1,T_2)=\frac{1}{T_1T_2}\int_{-T1}^{+T1}\int_{-T2}^{+T2}\left(1-\frac{|\tau_1|}{T_1}\right)\left(1-\frac{|\tau_2|}{T_2}\right)\rho(\tau_1,\tau_2)d\tau_1 d\tau_2 \quad (4)$$

where τ_1 and τ_2 are the distances between two points of interest along the respective coordinate axis, and $\rho(\tau_1,\tau_2)$ is the auto-correlation function of $X(t_1, t_2)$.

Finally, the covariance of the two averaged random fields, X_A and X_A', which correspond to the two rectangular finite elements located arbitrarily, can be expressed as Eq. (5) using the significations shown in Fig. 1 as follows:

$$Cov[X_{Ai},X_{Aj}]=\frac{1}{A_iA_j}\frac{\sigma_c^2}{4}\sum_{k=0}^{3}\sum_{l=0}^{3}(-1)^k(-1)^l T_{1k}^2T_{2l}^2\gamma(T_{1k},T_{2l}) \quad (5)$$

Taking advantage of the finite element characteristics assessed by the local averages of spatial uncertainties, the stochastic finite element analysis using the first-order perturbation technique is performed. And the probabilistic characteristics and correlation properties of the responses of the stochastic structural systems.

3. RELIABILITY ANALYSIS USING THEORY OF RANDOM PULSE PROCESSES

Using the results of the first-order pertubation-based stochastic finite element analysis, the response in each finite element of a stochastic structural system is modeled as the random pulse space-process. Then, the failure probability of the system is evaluated on the basis of the mean upcrossing rate of the pulse space-process by extending the level crossing theory of one-dimensional pulse processes.

In the followings, the level crossing theory of one-dimensional pulse processes is shown in short at first. Then, the extension of the one-dimensional level crossing problems to the two-dimensional ones is demonstrated. It is very difficult to solve accurately in practice the level crossing problems of random fields or responses of stochastic structual systems. However, it is clarified that the application of the level crossing theory to the stochastic responses modeled as pulse processes makes it easy.

The mean upcrossing rate $\nu_x^+(r)$ of a random rectangular pulse process X(t) of time coordinate, depicted in Fig. 2, to a given threshold r is expressed as

$$\nu_X^+(r) = \lim_{\Delta t\to 0}\frac{1}{\Delta t}P[\,X(t) < r \text{ and } X(t+\Delta t) > r\,] \quad (6)$$

For a smooth continuous process, the Rice's formula

$$\nu_X^+(r) = \int_0^{\infty}\dot{x}f_{X\dot{X}}(r,\dot{x})d\dot{x} \quad (7)$$

is obtained from Eq. (6). In the above equation $\dot{X}$ is the time derivative process, and $f_{X\dot{X}}$ is the join probability density function of the value of the original and its derivative process. For a renewal rectangular pulse process, Eq. (6) is rewritten as follows, Larrabee et al. (1981)

$$\nu_X^+(r) = \lim_{\Delta t\to 0}\frac{1}{\Delta t}\int_{-\infty}^{r}\hat{f}_X(x)\lambda_f\Delta t\hat{G}_S(r)dx \quad (8)$$

In Eq. (8), $\hat{f}_X(x)$ is the arbitary-point-in-time probability density function, and λ_f is the mean arrival rate of X(t). And then, $\lambda_f\Delta t$ means the probability that the new pulse arrives between t and $t+\Delta t$. And $\hat{G}_S(r)$ is the probability that the peak of new pulse exceeds r. If the new pulse always arrives at the time interval Δt, the integration in Eq. (8) turns out the product of the probability that the prior pulse remains below r and the probabilty that the new pulse exceeds r.

Analogously to the concepts mentioned above, the one-dimensional level crossing problems are extended to the level crossing problems of two-dimensional nonhomogeneous pulse space-processes. For a two-dimen sional space-process, illustrated in Fig. 3, the event that a rectangular space-process with the centroid at coordinated $(t_1+\Delta t_1/2, t_2+\Delta t_2/2)$ exceeds a threshold r along the respective coordinate axis in the area $\Delta t_1\Delta t_2$ is defined by the condition that $X(t_1-\Delta t_1/2, t_2-\Delta t_2/2)<r$ and $X(t_1+\Delta t_1/2, t_2-\Delta t_2/2)<r$ and $X(t_1-\Delta t_1/2, t_2+\Delta t_2/2)<r$ and $X(t_1+\Delta t_1/2, t_2+\Delta t_2/2)>r$. The mean upcrossing rate in $\Delta t_1\Delta t_2$ is defined by

$$\nu_X^+(r|t_1,t_2)=\frac{1}{\Delta t_1\Delta t_2}P[\,X_{11}<r \text{ and } X_{12}<r \text{ and } X_{21}<r \text{ and } X_{22}>r\,] \quad (9)$$

where $X_{11}=X(t_1-\Delta t_1/2, t_2-\Delta t_2/2)$, $X_{12}=X(t_1+\Delta t_1/2, t_2-\Delta t_2/2)$, $X_{21}=X(t_1-\Delta t_1/2, t_2+\Delta t_2/2)$ and $X_{22}=X(t_1+\Delta t_1/2, t_2+\Delta t_2/2)$. By introducing the joint probability density function of the four adjoint pulses, denoted by $f_{x_{11}x_{12}x_{21}x_{22}}$, Eq. (9) is rewritten by

$$\nu_X^+(r \mid t_1, t_2)$$

$$= \frac{1}{\Delta t_1 \Delta t_2} \int_{-\infty}^{r} dx_{11} \int_{-\infty}^{r} dx_{12} \int_{-\infty}^{r} dx_{21} \int_{r}^{\infty} f_{X_{11}X_{12}X_{21}X_{22}}(x_{11}, x_{12}, x_{21}, x_{22})\, dx_{22} \quad (10)$$

If the event of upcrossing of the pulse space-process may be approximated by the Poisson process, the failure probability indicated by $P_f^*(r)$ of the two-dimensional stochastic structural system is evaluated by

$$P_f^*(r) = 1 - \exp\left(-\sum_{i=1}^{n} \nu_{X_i}^+(r \mid t_1, t_2) \cdot \Delta t_{1i} \Delta t_{2i} \right) \quad (11)$$

where n is the number of rectangular pulse, i.e. the number of finite elements in the analysis.

In general, it is very difficult to solve accurately level crossing problems utilizing the classic formula such as the Rice's formula expressed as Eq. (7). This comes from the difficulty of analytical estimations of the probabilistic properties of derivative processes, included in the classic formula. Taking advantage of the concept of pulse processes, however, there are no terms of the derivatives. Consequently, the reliability analysis of stochastic structural systems can be easily performed.

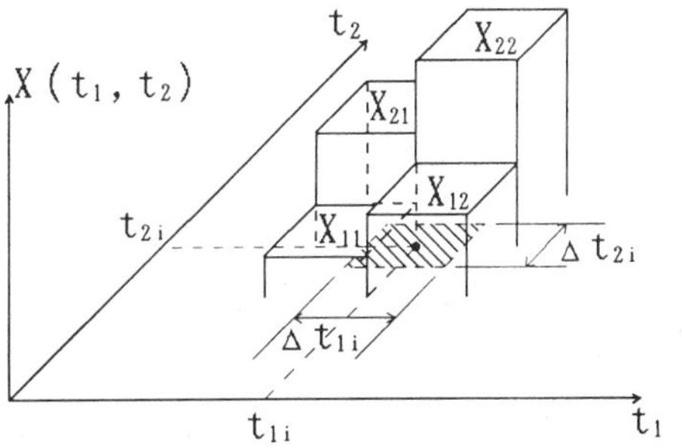

Fig. 3 Level crossing of two-dimensional rectangular pulse-process space process X_{22}

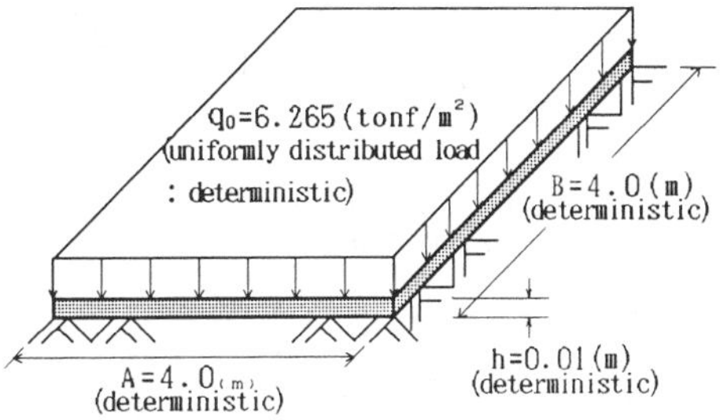

Fig. 4 Analytical model

4. NUMERICAL EXAMPLES

As numerical examples, the bending problems of a steel plate resting on the elastic foundation of the Winkler's model is considered, as shown in Fig. 4. In the problems, the bearing coefficient of the foundation is modeled as the two-dimensional, statistically homogeneous, continuous Gaussian random field, represented by $g(x_1, x_2)$, with the mean m_g=626.5tonf/m², the standard deviation σ_g and the auto-correlation function of the form

$$\rho_g(\tau_1, \tau_2) = \exp\left[-\left\{ \left(\frac{\tau_1}{\delta_1}\right)^2 + \left(\frac{\tau_2}{\delta_2}\right)^2 \right\} \right] \quad (12)$$

In Eq. (12), δ_1 and δ_2 are the parameters relating to the correlation distances for x_1 and x_2-direction, respectively, and $\delta_1=\delta_2$=1.127m is used. And τ_1 and τ_2 are the distances between two points of interest along respective coordinate axis. The steel plate with deterministic flexual rigidity is simply supported and subjected to a deterministic uniformly distributed load whose intensity is q_0=6.265 tonf/m². The parameters associated with the steel plate such as Young's modulus, Poisson's ratio and thickness are taken as 2.1×10^7tonf/m², 0.3 and 0.01m, respectively, and all these parameters are treated as deterministic quantities.

In the stochastic finite element analysis for the evaluation of the failure probability for the lateral deflection, the system, i.e. the deterministic steel plate and the random elastic foundation, was divided into 64 finite elements with 81 nodal points (8×8 meshes). Then the side length of each square finite element becomes 0.50(m), one-half of the correlation distance of the random elastic foundation.

In the evaluation of the failure probability for deflection, three parameters such as the coefficient of variation (c.o.v.) of the bearing coefficient of the foundation, defined by σ_g/m_g, the type of probability density function (p.d.f.) of the deflection in each finite element and the threshold are considered. Three variations of the c.o.v., i.e. 0.10, 0.15, 0.20, are treated. And the normal distribution type is assumed since the type of p.d.f. of responses can not be obtained by the stochastic finite element method, by which only the first and second order statistical moment of response are obtained. Three thresholds, i.e. level 7, 8 and 9, are determined as follows: the threshold of seven times the value of the maximum standard deviation of all finite elements for the case of c.o.v. of 0.20 is referred to level 7. Levels 8 and 9 are also determined in the same way of level 7. As a result, level 7=0.01354m, level 8=0.01547m and level 9=0.01741m are used as the three thresholds in the calculation.

In the reliability analysis for moment, three same parameters as for deflection are taken into consideration, that is the c.o.v. of the bearing coefficient of the foundation, the type of p.d.f. of the moment response and thresholds. The failure probability for the moment perpendicular to x_1-axis is calculated for the case with the threshold 10(=0.1691(tof·m)), the normal p.d.f. and c.o.v. of 0.20. The threshold is determined with respect to the maximum standard deviation of the moment of all finite elements in the same way for deflection. The system is divided into 16×16 meshes.

The integration calculus in Eq. (10) is performed by means of the IFM(Iterative Fast Monte Carlo) procedure, Shiraki(1989). The integration calculus with small error can be carried by the use of the IFM software package. The numerical error is about 2%.

Table 1 Failure probability for lateral deflection for level 7 (Cases 1, 2 and 3)

C.O.V. \ Pf	Pf^1	Pf*	Pf^0
0.10	0.6363E-02	0.1252E-01	0.5885E-01
0.15	0.2750E-05	0.5472E-05	0.1931E-04
0.20	0.2042E-10	0.4151E-10	0.1297E-09

Table 2 Failure probability for lateral deflection for level 8 (Cases 4, 5 and 6)

C.O.V. \ Pf	Pf^1	Pf*	Pf^0
0.10	0.4812E-01	0.1145	0.5364
0.15	0.1209E-02	0.3621E-02	0.1471E-01
0.20	0.5407E-05	0.1708E-04	0.5611E-04

Table 3 Failure probability for lateral deflection for level 9 (Cases 7, 8 and 9)

C.O.V. \ Pf	Pf^1	Pf*	Pf^0
0.10	0.1056	0.2446	0.8949
0.15	0.1147E-01	0.3840E-01	0.1774
0.20	0.4806E-03	0.1906E-02	0.7029E-02

Table 4 Failure probability for bending moment perpendicular to X_1-axis

	$Pf^{(1)}$	Pf^*	$Pf^{(0)}$
level 10	0.396178E-04	1.364229E-04	4.953146E-04

First, the system calculation results of system failure probability for the lateral deflection are listed on Tables through 1 to 3. In these Tables, P_f^0 and P_f^1 are the failure probabilities which are calculated for the assumption that each pluse might be mutually independent and completely correlated, respectively. P_f^0 and P_f^1 are given by Eqs. (13) and (14), respectively.

$$P_f^0 = 1 - \prod_{i=1}^{n} (1 - P_{fi}) \tag{13}$$

$$P_f^1 = \max\{ P_{fi} \} \tag{14}$$

where P_{fi} is the probability that each pulse independently exceeds a given threshold, and n is the number of pulses, i.e. the number of finite elements.

Similarly, Table 4 lists the failure probabilities for the bending moment perpendicular x_1-axis.

From these results, it is clarified that the proposed method in this study can perform the reliability analysis of the structure in the two-variate random field.

5. SUMMARY AND CONCLUSIONS

In this study a practical reliability evaluation method of structural system with spatial uncertainties modeled by two-variate random fields is proposed.

In numerical examples, the reliability analysis of a steel plate resting on the random elastic foundation of the well-known Winkler's model is performed by the proposed method. In the analysis, the bearing coefficient is treated as two-dimensional random field. And the failure probability of the plate is evaluated for deflection as well as bending moment using the mean upcrossing rate to the given thresholds. The failure probability, obtained by this proposed method was compared as the failure probability obtained on the assumption of the mutually independent deflection (or bending moment) in each finite element and the failure probability obtained on the assumption of the completely correlated deflection (or moment).

From these results, it is clarified that the proposed method in this study can perform the reliability analysis of the structural systems with spatial uncertainties represented by random field, without considering the derivative random process required in the formulation of the conventional level crossing problems such as Rice's formula.

PEFERENCES

Kiureghian, A.D. & Ke, J.-B. 1988. Stochastic finite element method in structural reliability. J. Prob. Eng.Mech. 3(2):88-91.

Larrabee, R.D. & Cornell, A. 1981. Combination of various load processes. J. Struc. Eng. ,ASCE, 107(1) :223 -239.

Shiraki, W. 1989. An extension of Iterative Fast Monte -Carlo(IFM) procedure and its applications to time-variant structural reliability analysis. Proc. 5th ICOSSAR:1015-1018.

Vanmarke, E. & Grigoriu, M. 1983. Stochastic finite element analysis of simple beams. J. Eng. Mech. , ASCE, 109(5):1203-1214.

Vanmarke, E. 1983. Random fields : analysis and synthesis. MIT Press.

Yamazaki, F. & Shinozuka, M. 1988. Safety analysis of stochastic finite element systems by Monte Carlo simulation. J. Structural. Eng./Earthquake Eng. 5(2) :313-323.

Structural Safety & Reliability, Schuëller, Shinozuka & Yao (eds) © *1994 Balkema, Rotterdam, ISBN 90 5410 357 4*

Variability response functions of stochastic dynamic systems

T.Takada
Ohsaki Research Institute, Shimizu Corporation, Chiyoda-ku, Tokyo, Japan

ABSTRACT: Stochastic wave propagation of a simple system with finite boundaries is investigated in the present paper. A perturbation and Green function technique is utilized to analytically derive a variability response function associated with the mean wave field. Finally, possible applications of the function is discussed in a filed of structural dynamics.

1. INTRODUCTION

Wave propagation problems of media with spatially varying constitutive parameters has been great concern for last three decades in optics, seismology, electro-magnetics, and etc. For this problem, stochastic approach is recognized as a very useful method that requires only lower-order moments of the parameters to identify stochastic behavior of wave motion since deterministic approaches require complete information of the parameters, which are difficult to obtain in reality, and lead to enormous amount of computational efforts.

The problem using only variability statistics of the parameters is known as stochastic wave propagation problems (Sobczyk 1985). The parameters are often idealized as multi-dimensional stochastic field to solve the problems. The governing equation, however, becomes a stochastic differential equation involving the stochastic field and many approximate techniques have been proposed due to its mathematical difficulty so far.

Analytical methods directly solving the equation are based on a perturbation method and a closure method (Askar and Cakmak 1988, Aki and Richards 1980). The perturbation method has often been used since it has close relationship with physical phenomena called multiple wave scattering. Among them, the Born approximation (single scattering theory) which takes only the first-order perturbed term is famous. The stochastic media analyzed by this method, however, are limited to semi-infinite or infinite media with making use of radiation condition so far.

A stochastic finite element method or a finite difference method, on the other hand, is proposed in structural engineering as discrete methods (Hisada and Nakagiri 1981, Vanmarcke and Grigoriu 1983). These methods are based on the spatial discretization of stochastic fields involved and take advantage of the perturbation and/or Monte Carlo simulation technique to evaluate wave motion statistics. Another type of discretization based on the Galerkin method has been proposed for different problems by the author recently (Takada 1991).

The stochastic methods mentioned above, however, require statistical property of the involved stochastic fields. This might be a big obstacle for the methods to be carried out. Considering the variability of soil properties, for instance, it is necessary to require a large amount of field measurements to determine the statistical nature of the soil. If one considers artificial materials or substances such as man-made buildings, it might be, however, possible to artificially control the nature of variability of constitutive materials or mechanical properties. In other words, the stochastic methods can be useful tools to identify the most appropriate feature of variability in a sense that the structural maximum response, for instance, is reduced when one positively takes account of the variaiblity into his analysis. This means the other way of usage of the stochastic methods which used to be tools for the purpose of investigation of actual phenomena.

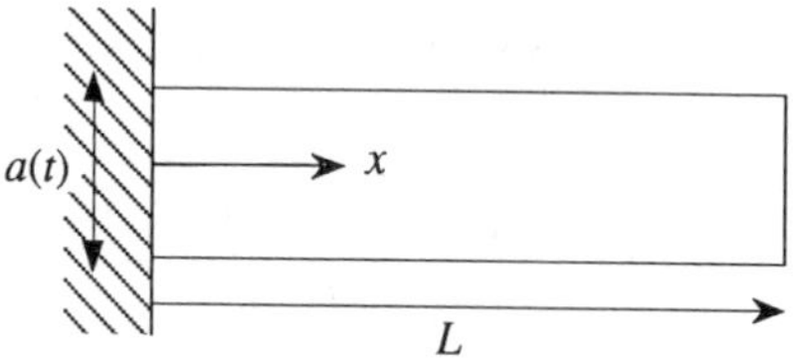

Figure 1 A stochastic bar subjected to external disturbance

These are the background of this study and the present paper shows an analytical method based on the perturbation technique and a Green function to evaluate mean and covariance wave fields, and derives variability response functions which are other representation of variability of wave motions and importantly have nothing to do with the stochastic fields involved, and finally discusses on the engineering applications of them in structural dynamics.

2. FUNDAMENTAL EQUATION

The stochastic system to be analyzed is a one-dimensional bar with fixed boundary condition at $x = 0$, as is shown in Fig. 1. The wave velocity along the bar varies spatially and is assumed to be modeled as a one-dimensional stochastic field. An acceleration disturbance $a(t)$ is applied to the fixed end and induces wave motion within the bar. The equation of motion $w(x,t)$ now can be written as follows:

$$\frac{\partial^2 w}{\partial t^2} - \frac{\partial}{\partial t}\left[c^2 \frac{\partial w}{\partial x}\right] = -a(t) \qquad (1)$$

where c is a wave velocity and is assumed to be the following form.

$$c^2(x) = c_0^2 \{1 + f(x)\} \qquad (2)$$

where c_0 is a mean wave velocity and $f(x)$ a random component due to heterogeneity of the bar, which is idealized as zero mean, 1D stochastic field and is assumed to be statistically homogeneous and differentiable at least once. Either its auto-correlation function $R_{ff}(\xi)$ or the corresponding power spectral density function $S_{ff}(\kappa)$ is given.

Assuming the temporal stationary response, namely letting $a(t) = Ae^{i\omega t}$ and $w(x,t) = W(x)\,e^{i\omega t}$, (1) becomes the following equation which is easier to treat.

$$\frac{dW}{dx}\left\{(1+f)\frac{dW}{dx}\right\} + k_0^2 W = k_0^2 A \qquad (3)$$

where k_0 is a deterministic quantity and is called a refractive index defined as

$$k_0 = \frac{\omega}{c_0} \qquad (4)$$

3. USE OF PERTURBATION AND GREEN FUNCTIONS

To solve (3), a perturbation and Green function approach is utilized. Assuming $\epsilon << 1$, (2) is rewritten into

$$c^2(x) = c_0^2\{1 + \epsilon f(x)\} \qquad (5)$$

Similarly, the wave motion $W(x)$ is perturbed with respect to ϵ as follows:

$$W(x) = W_0(x) + \epsilon W_1(x) + \epsilon^2 W_2(x) + \cdots \qquad (6)$$

Substituting the above two equations into (3) we obtain the following recurrence equations.

$$W_0'' + k_0^2 W_0 = k_0^2 A \qquad (7a)$$

$$W_1'' + k_0^2 W_1 = -\left(fW_0'\right)' \qquad (7b)$$

$$W_2'' + k_0^2 W_2 = -\left(fW_1'\right)' \qquad (7c)$$

$$\vdots$$

where the prime stands for the derivative. It is observed from the above equations that (7a) is an ordinary deterministic wave equation. Therefore, the Green function $G(x,\xi)$ satisfying the following wave equation can facilitate the problem solving.

$$W_0'' + k_0^2 W_0 = -\delta(x - \xi) \qquad (8)$$

where $\delta(\,.\,)$ is a Dirac's delta function.

For the bar shown in Fig. 1, the Green function can be easily obtained as

$$G(x,\xi) = \begin{cases} \frac{1}{k_0 \cos k_0 L} \sin k_0 x \cos(L-\xi) & : \ 0 \le x \le \xi \\ \frac{1}{k_0 \cos k_0 L} \sin k_0 \xi \cos(L-x) & : \ \xi \le x \le L \end{cases} \qquad (9)$$

Once W_0 is obtained, W_1 and W_2 are consecutively evaluated as follows:

$$W_1(x) = \int_0^L G(x,\xi)\{f(\xi)\,W_0'(\xi)\}'\,d\xi$$
$$= \left[G(x,\xi)\,f(\xi)\,W_0'(\xi)\right]_0^L$$

$$- \int_0^L G'_\xi(x,\xi)\, f(\xi)\, W'_0(\xi)\, d\xi$$

$$= - \int_0^L G'_\xi(x,\xi)\, f(\xi)\, W'_0(\xi)\, d\xi \quad (10)$$

$$W_2(x) = \int_0^L G(x,\xi)\, \{f(\xi)\, W'_1(\xi)\}'\, d\xi$$

$$= \left[G(x,\xi)\, f(\xi)\, W'_1(\xi)\right]_0^L$$

$$- \int_0^L G'_\xi(x,\xi)\, f(\xi)\, W'_1(\xi)\, d\xi$$

$$= - \int_0^L G'_\xi(x,\xi)\, f(\xi)\, W'_1(\xi)\, d\xi \quad (11)$$

In the above, $G'_\xi(x,\xi) = \partial G(x,\xi)/\partial \xi$, the boundary conditions: $W(0) = 0, W'(L) = 0$ are taken into account in the derivation.

4. MEAN WAVE FIELD

Taking up to the second-order term in (6), the mean wave field of $W(x)$ is evaluated below.

$$\langle W(x)\rangle = W_0(x) + \epsilon^2 \langle W_2(x)\rangle \quad (12)$$

where
$\langle W_2(x)\rangle$

$$= \int_0^L \int_0^L G'_\xi(x,\xi)\, G''_{\eta\xi}(\xi,\eta)\, W'_0(\eta)\, R_{ff}(\xi,\eta)\, d\eta d\xi \quad (13)$$

with $G''_{\eta\xi}(\xi,\eta) = \partial^2 G(\xi,\eta)/\partial\xi\partial\eta$.

(13) obviously shows that the mean wave field can be expressed in terms of the auto-correlation function of $f(x)$ and deterministic terms associated with the Green function. Using the power spectral representation instead of R_{ff}, (13) becomes,

$$\langle W_2(x)\rangle = \int_{-\infty}^{+\infty} S_{ff}(\kappa)\, V_{MF}(x,\kappa)\, d\kappa \quad (14)$$

where
$V_{MF}(x,\kappa)$

$$= \int_0^L \int_0^L G'_\xi(x,\xi)\, G''_{\eta\xi}(\xi,\eta)\, W'_0(\eta)\, e^{i\kappa(\xi-\eta)}\, d\eta d\xi \quad (15)$$

The function V_{MF} has nothing to do with the probabilistic nature of $f(x)$ and is called "a variability response function associated with mean wave field (Deodatis and Shinozuka 1989)". V_{MF} is determined only from the dimension of the bar,

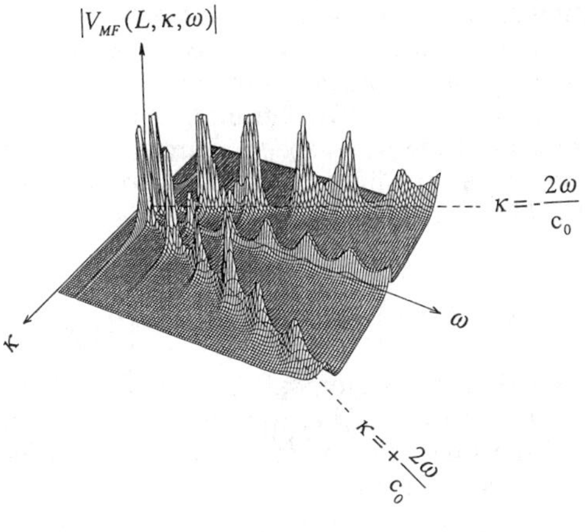

Figure 2 A variability response function associated with mean wave field

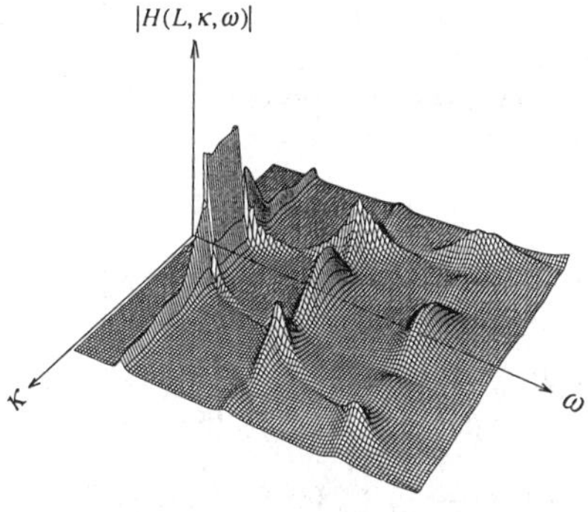

Figure 3 A variability response function associated with covariance wave field

the refractive index defined in (4) and boundary conditions.

For the simple case of Fig.1, (15) can be obtained in an closed form. Figure 2 plots the absolute value of V_{MF} at the free end in the plane of the input frequency ω and wave number κ. From this figure, $|V_{MF}|$ takes large value at the following range.

$$\kappa = \pm\frac{2\omega}{c_0} = \pm 2k_0 \quad (16)$$

In other words, there exists a certain relationship between the input frequency of the external disturbance and periodicity of the heterogeneity of the bar, which reduces the mean wave field. This does imply that if the disturbance is a narrow-band process, there exists a correlation property of the fluctuation of the bar that minimizes the magnitude of the wave motion. Regarding the bar analyzed herein as a high-rise building, it might be possible to reduce the structural response by controlling the correlation property of the wave speed along the building. That is the possibility of increase of apparent structural damping effect.

5. COVARIANCE WAVE FIELD

Similarly to the secion 4, it is possible to derive the covariance function from (10). The covariance function regarding the bar displacement can be written as follows.

$$\begin{aligned} C_{WW}(x,y) &= \langle W_1(x)\, W_1(y)\rangle \\ &= \int_0^L \int_0^L G'_\xi(x,\xi)\, G'_\eta(y,\eta) \\ &\times W'_0(\xi)\, W'_0(\eta)\, R_{ff}(\xi,\eta)\, d\eta d\xi \end{aligned} \quad (17)$$

When one uses the power spectrum density function instead of $R_{ff}(\xi,\eta)$, the equation becomes

$$C_{WW}(x,y) = \int_{-\infty}^{+\infty} H_{CF}(x,\kappa)\, H^*_{CF}(y,\kappa)\, S_{ff}(\kappa)\, d\kappa \quad (18)$$

where $|H_{CF}(x,\kappa)|$ is called a first-order variability function defined in the following.

$$H_{CF}(x,\kappa) = \int_0^L G'_\xi(x,\xi)\, W'_0(\xi)\, e^{i\kappa\xi} d\xi \quad (19)$$

Figure 3 is a plot of $|H_{CF}(x,\kappa)|$ in $\kappa-\omega$ plane.

6. CONCLUSIONS

A finite and stochastic simple bar system is analyzed based on the perturbation and Green function technique. Deriving the solution for the mean wave field and transforming it into a spatial wave number domain, it is presented that there exists a certain relationship between the input frequency and the wave number of fluctuation of the wave speed, which drastically reduces the wave motion of the bar.

For future study, this method will be able to be extended to 2D systems to investigate the behavior of wave propagation in heterogeneous media. And also, this method will have to be compared with other methods such as SFEM.

REFERENCES

Aki and Richards, 1980 *Quantitative Seismology*, Vol.II, Freeman

Askar, A. and Cakmak, A.S., 1988 Seismic waves in random media, J. of Probabilistic Engineering Mechanics, Vol.3, No.3

Deodatis, G. and Shinozuka, M., 1989 "Bounds on Response Variability of Stochastic Systems," J. of EM, ASCE, Vol. 115, No. 11, pp.2543-2563

Hisada, T. and Nakagiri, S., 1981 "Stochastic Finite Element Methods Developed for Structural Safety and Reliability," Proc. of the 3rd ICOSSAR, pp.395-408

Sobczyk, K., 1985 *Stochastic Wave Propagation*, Elesevier Science Publisher

Takada, T., 1991 Galerkin method to analyze systems with stochastic flexural rigidity, Proc. of 1st International Conference on Computational Stochastic Mechanics, Corfu Greece, pp.511-522

Vanmarcke, E. and Grigoriu, M., 1983 "Stochastic Finite Element Analysis of Simple Beams," J. of EM, ASCE, Vol. 109, No. 5, pp.1203-1214

Structural Safety & Reliability, Schuëller, Shinozuka & Yao (eds) © 1994 Balkema, Rotterdam, ISBN 90 5410 357 4

Reliability estimates of plates with uncertain material properties

F.J.Wall
Hilti AG, Corporate Research, Schaan, Principality of Liechtenstein (Formerly: Department of Civil Engineering & Operations Research, Princeton University, N.J., USA)

G. Deodatis
Department of Civil Engineering & Operations Research, Princeton University, N.J., USA

C.G. Bucher
Institute of Engineering Mechanics, University of Innsbruck, Austria

ABSTRACT: The response variability and reliability of two-dimensional stochastic systems is analyzed. The proposed method utilizes a stochastic finite element approach which is based on the concept of weighted integrals. Introducing a first-order Taylor expansion, the response variability is evaluated in terms of first and second moments of the displacement response. A first reliability estimate of the system is given by its safety index, which is computed using an advanced first-order second-moment approach. Secondly, the response surface methodology in conjunction with Monte Carlo simulation is applied to determine the reliability of the structure under investigation.

1 INTRODUCTION

Over the past few years, stochastic structural analysis has become an important subject in academic as well as applied research activities. The necessity of accounting for material and/or geometric inhomogeneities - inherent in almost any structural system - in realistic structural analysis has stimulated the development of the stochastic finite element method (SFEM). This approach extends the classical finite element method, which has become common practice for many engineering applications, by combining it with modern probabilistic methods. The resulting methodology promises to be an efficient and reasonable tool to deal with system stochasticity.

Recently, a stochastic finite element method based on the concept of weighted integrals has been introduced to calculate response variability and reliability of stochastic trusses and frames (Deodatis (1991), Deodatis and Shinozuka (1991)). This underlying concept allows for the accurate and consistent reduction of the stochastic field, describing the randomness of the system properties, to a number of random variables called weighted integrals.

In the present study the weighted integral method is extended to two-dimensional stochastic systems. The states of plane stress and plane strain of plates under deterministic in-plane loading are analyzed using rectangular finite elements. In this context the variability of the displacement response is evaluated. The main objective, however, is to investigate the capabilities of the proposed methodology with respect to the reliability assessment of 2D stochastic systems.

2 METHOD OF ANALYSIS

Within the present investigation of plates focus is placed on the states of plane stress and plane strain. The in-plane loading of the rectangular plates under consideration is assumed to be known deterministically. Using finite element procedures, the governing equation for linear systems subjected to static loading reads

$$\mathbf{KU} = \mathbf{F} \tag{1}$$

where $\mathbf{K}$= global stiffness matrix, $\mathbf{U}$= global nodal displacement vector, and $\mathbf{F}$= global load vector. Taking into account the random variation of the system properties yields a stiffness matrix which is stochastic in nature. From eq.(1) it is therefore obvious that the displacement vector $\mathbf{U}$ will be random, too. This is true even in the case of deterministic loading. In the following analysis only one system property, namely the modulus of elasticity E, will be considered random. The remaining quantities, e.g. Poisson's ratio, are kept constant at a given (mean) value. It is assumed that the random variation of the elastic modulus over the area of a finite element can adequately be described by

$$E^{(e)}(x,y) = E_0^{(e)}[1 + f^{(e)}(x,y)] \tag{2}$$

In above equation $E_0^{(e)}$= mean value of the elastic modulus, $f^{(e)}(x, y)$= two-dimensional, zero mean, homogeneous stochastic field, and x,y= local coordinates. To prevent the occurrence of negative values of the modulus of elasticity the stochastic field $f^{(e)}(x, y)$ is assumed to be bounded as follows:

$$-1+\eta \leq f^{(e)}(x, y) \leq 1-\eta \quad ; \quad 0<\eta<1 \tag{3}$$

where the upper bound on $f^{(e)}(x, y)$ is introduced for reasons of symmetry.

Applying the principle of minimum potential energy, the stochastic element stiffness matrix for a four-node rectangular element is computed as (Wall (1993)):

$$\mathbf{K}^{(e)} = \mathbf{K}_0^{(e)} + X_1^{(e)}\Delta\mathbf{K}_1^{(e)} + X_2^{(e)}\Delta\mathbf{K}_2^{(e)} + \ldots$$
$$\ldots + X_6^{(e)}\Delta\mathbf{K}_6^{(e)} \tag{4}$$

where $\mathbf{K}_0^{(e)}$, $\Delta\mathbf{K}_k^{(e)}$= deterministic matrices, and $X_k^{(e)}$= weighted integrals. Assuming a constant element thickness, the weighted integrals $X_k^{(e)}$ are defined as:

$$X_k^{(e)} = \int_{A^{(e)}} x^i y^j f^{(e)}(x,y)\, dA^{(e)} \tag{5}$$

where $A^{(e)}$ denotes the area of the finite element *(e)*. In eq.(5) subscript k is related to the powers i and j of the local coordinates as: (k→(i,j)): 1→(0,0), 2→(1,0), 3→(0,1), 4→(1,1), 5→(2,0), 6→(0,2). Given the stochastic stiffness matrix, the covariances of the nodal displacements are found to be (Wall (1993))

$$\mathrm{Cov}\{\ \} = \sum_{e=1}^{N}\sum_{f=1}^{N}\sum_{k=1}^{6}\sum_{n=1}^{6} \mathbf{K}_0^{-1}\Delta\mathbf{K}_k^{(e)}\mathbf{U}_0$$
$$\cdot \mathbf{U}_0^T(\Delta\mathbf{K}_n^{(f)})^T(\mathbf{K}_0^{-1})^T \mathcal{E}\{X_k^{(e)}X_n^{(f)}\} \tag{6}$$

where the displacement vector has been approximated by a first-order Taylor expansion around the (zero) mean values of the weighted integrals. In eq.(6) the subscript 0 denotes quantities calculated at the mean values of the weighted integrals and $\mathcal{E}\{\cdot\}$ means mathematical expectation. The covariances of the weighted integrals may be evaluated either in the space domain or in the wave number domain. A major advantage of wave number domain calculations is the availability of the variability response function (Wall et al (1993)).

In order to show the suitability of the proposed stochastic finite element concept for reliability assessment, two different approaches are pursued in the following. First, a reliability estimate is determined in terms of the safety index. The second approach shows the application of the response surface technique together with Monte Carlo simulation.

Assuming that the serviceability limit, i.e. a possible critical state of the system, is reached whenever the displacement at a specific node exceeds a certain threshold, the limit state function may be established as:

$$g_{U_j}(\mathbf{X}) = U_{jc} - U_j(\mathbf{X}) = 0 \tag{7}$$

where $\mathbf{X}$= vector of weighted integrals, and U_{jc}= threshold value of nodal displacement. Within the first order reliability method the failure probability is determined by $p_f \approx \Phi(-\beta)$, where β= safety index, and Φ= standard cumulative distribution. The safety (or reliability) index β represents the minimum distance from the origin in the standard normal space to a point (design point) on the limit state function. In literature several methods are available to calculate the design point. According to the iterative algorithm proposed in Madsen et al (1986), the vector $\mathbf{X}$ - at iteration m - is upgraded as:

$$\mathbf{X}^{(m)} = \mathbf{C_X}\nabla g_{U_j}[\mathbf{X}^{(m-1)}]$$
$$\cdot \frac{\mathbf{X}^{(m-1)T}\nabla g_{U_j}[\mathbf{X}^{(m-1)}] - g_{U_j}[\mathbf{X}^{(m-1)}]}{\nabla g_{U_j}[\mathbf{X}^{(m-1)}]^T \mathbf{C_X}\nabla g_{U_j}[\mathbf{X}^{(m-1)}]} \tag{8}$$

In above equations $\mathbf{C_X}$= covariance matrix of weighted integrals, and $\nabla g_{U_j}[\mathbf{X}^{(m-1)}]$= gradient vector evaluated at $\mathbf{X}^{(m-1)}$. Upon reaching convergence the corresponding safety index can be evaluated as:

$$\beta[\mathbf{X}^{(m)}] = [\mathbf{X}^{(m)T}\mathbf{C}_{\mathbf{X}}^{-1}\mathbf{X}^{(m)}]^{1/2} \tag{9}$$

Alternatively, reliability estimates may be obtained by applying the response surface (RS) methodology (see e.g. Bucher and Bourgund (1990)) in conjunction with Monte Carlo simulations. Herein, the actual limit state function $g(\mathbf{X})$ is replaced by an equivalent approximate function $\bar{g}(\mathbf{X})$. The reasoning for this step is that an analytic expression for the limit state function, required by many reliability methods, cannot always be provided. This is especially true when using finite element procedures, where only points on the limit state function can be calculated. These points, however, may be used to formulate an approximate analytical function, i.e. the response function (surface). Based on this function advanced Monte Carlo simulation techniques can efficiently be utilized to obtain reliability estimates. Within this investigation the response surface is conveniently defined as a second order polynomial

$$\bar{g}(\mathbf{X}) = a + \mathbf{bX} + \mathbf{X}^T\mathbf{cX}$$
$$= a + \sum_{i=1}^{M} b_i X_i + \sum_{i=1}^{M}\sum_{j=1}^{M} c_{ij} X_i X_j \tag{10}$$

where M is the number of random variables, and a, b_i and c_{ij} are constants. Points on the failure surface are found by varying at most two random variables at a time, while keeping all the others constant at their (zero) mean values. First, however, only one variable is varied at a time. The search for the point on the response surface is performed in the positive as well as negative direction. Using standard normal variates, which are uniquely related to the weighted integrals, the distances of the resulting points on the response surface from the origin provide a measure of relative importance of the corresponding variables. Therefore, the closest points identify the important basic random variables. Only these random variables are finally taken into account. As the number of important random variables is much smaller than the total number of weighted integrals involved in the problem, the algorithm remains efficient. Given the response surface the failure probability is evaluated using the program package ISPUD (Bourgund and Bucher (1986); Bucher et al (1989)).

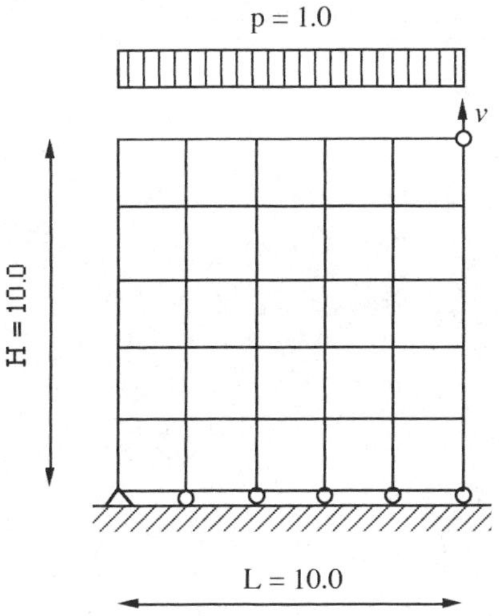

Fig. 1: Structural system

3 NUMERICAL EXAMPLE

The system considered herein is a square plate under uniform vertical in-plane loading (see Fig.1). The load p is assumed to be known deterministically.

The spectral density characterizing the stochastic field $f(x,y)$ is chosen to be:

$$S_{ff}(\kappa_x, \kappa_y) = \frac{\sigma_{ff}^2 b^2}{\pi^2(1 + b^2\kappa_x^2)(1 + b^2\kappa_y^2)} \tag{11}$$

with corresponding correlation function:

$$R_{ff}(\xi, \eta) = \sigma_{ff}^2 \exp\left(-\frac{|\xi| + |\eta|}{b}\right) \tag{12}$$

where σ_{ff} is the standard deviation of $f(x,y)$, κ_x and κ_y are wave numbers, b is a constant related to the correlation distance of $f(x,y)$, and ξ and η denote separation distances along the x and y axes, respectively. The coefficient of variation of the elastic modulus is σ_{ff}=0.10 for all cases examined. Setting b=1.0, the failure probabilities corresponding to different critical threshold values of the vertical displacement v of the upper right node (see Fig.1) are displayed in Table 1. The threshold values U_{jc} are given in terms of mean value (m=1.0) and standard deviation (σ=0.033) of v.

Table 1. Probability of failure

U_{jc}	p_f (FOSM)	p_f (RS)
m+1σ	1.6 x 10^{-1}	1.7 x 10^{-1}
m+2σ	2.7 x 10^{-2}	2.7 x 10^{-2}
m+3σ	2.4 x 10^{-3}	2.2 x 10^{-3}
m+4σ	1.1 x 10^{-4}	8.7 x 10^{-5}
m+5σ	2.9 x 10^{-6}	1.6 x 10^{-6}

4 CONCLUDING REMARKS

The weighted integral method is applied to a two-dimensional stochastic system (plane stress and plane strain) using bilinear rectangular finite elements. Based on a first-order Taylor expansion of the displacement vector, the response variability of the system is analyzed. Furthermore, the suitability of the suggested stochastic finite element (SFE) concept for reliability assessment is demonstrated. For the system under consideration only small differences between the results of the first-order second-moment approach and the response surface method are observed.

ACKNOWLEDGMENT

This work was supported by the National Science Foundation under Grant No. BCS-9257900. The Erwin-Schrödinger Fellowship awarded F.J. Wall under Austrian Science Foundation (FWF) Grant J0617-TEC, enabling him to do research at Princeton University, is gratefully acknowledged.

REFERENCES

Bourgund, U., C.G. Bucher 1986. Importance sampling procedure using design points (ISPUD) - A user's manual. *Report No. 8-86*. Institute of Engineering Mechanics, University of Innsbruck, Austria.

Bucher, C.G., U. Bourgund 1990. A fast and efficient response surface approach for structural reliability problems. *Structural Safety* 7: 57-66.

Bucher, C.G., J. Nienstedt & W. Ouypornprasert, 1989. Adaptive strategies in ispud V3.0 - A user's manual. *Report No. 25-89*. Institute of Engineering Mechanics, University of Innsbruck, Austria.

Deodatis, G. 1991. Weighted integral method. I: Stochastic stiffness matrix. *Journal of Engineering Mechanics* 117(8): 1851-1864.

Deodatis, G., M. Shinozuka 1991. Weighted integral method. II: Response variability and reliability. *Journal of Engineering Mechanics* 117(8): 1865-1877.

Madsen, H.O., S. Krenk & N.C. Lind 1986. *Methods of structural safety*. Prentice-Hall: Englewood Cliffs, NJ.

Wall, F.J. 1993. Probabilistic response of stochastic 2D systems under static loading. Department of Civil Engineering & Operations Research, Princeton University, Princeton, NJ.

Wall, F.J., G. Deodatis & M. Shinozuka 1993. Variability response functions and upper bounds of response variability of 2D stochastic systems. *Proc. ICOSSAR'93*. Aug. 9-13, 1993, Innsbruck, Austria.

Human factors and vulnerability

Structural Safety & Reliability, Schuëller, Shinozuka & Yao (eds) © 1994 Balkema, Rotterdam, ISBN 90 5410 357 4

Quality assurance in structural design

S. Engelund & R. Rackwitz
Technical University of Munich, Germany

ABSTRACT The fact that a large number of structural failures is caused by human errors committed during the design emphasizes the necessity of an efficient and documented quality assurance system. This paper is concerned with checks of the design, only. General models which can be used to determine the occurrence and detection probabilities of human errors are proposed. On this basis various known quality assurance systems implemented in different countries and for different projects are compared with respect to efficiency optimal total cost and operational failure probabilities.

1. INTRODUCTION

Due to the increasing demands for high "quality" of structural engineering and the fact, that a large number of structural failures are caused by human errors committed during the design (see e.g. Matousek and Schneider [3]), much effort is made to develop and to execute schemes for the assurance of "quality" of structural design. A number of different schemes for the quality assurance of engineering structures are implemented in different countries and for projects of different type.

Given a complete description of the purpose, performance and requirements, the design process can be modeled by three phases: Concept, Design and Check (see figure 1)

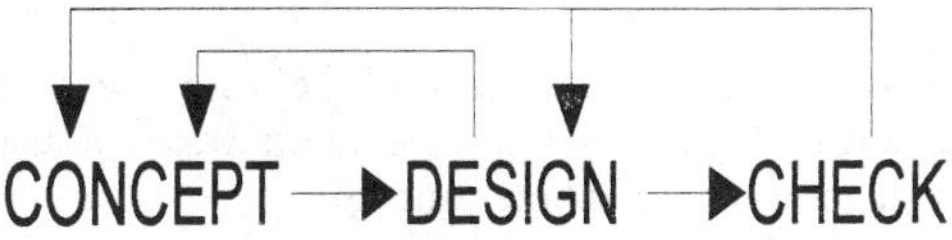

Figure 1: The design process.

In the concept phase all necessary decisions and assumptions, which form the basis of the specific design task, are made. In the second phase the actual design involving the calculations and the appropriate detailing together with the necessary documentation is performed by a person or a group of persons, who also perform self-checking to a certain degree. In the third phase the results obtained in the first and second phase are checked by a person of group of persons. The various activities in the three phases can be performed by the same person or group of persons or different ones. They necessarily are performed sequentially but there are feed backs to earlier phases. The way in which the phases and their interactions are organized together with the assignment of persons or groups of persons to various activities, the specification of the professional qualifications and the distribution of responsibilities defines a quality assurance system. Any such system of course requires a suitable agreement about financial organization, payments, reimbursements etc. Obviously a variety of quality assurance systems is possible and are effectively in use

2. CHECKING SCHEMES

It is not suitable to discuss all checking schemes presently in use. The following three systems are considered as typical and might cover by and large the majority of systems.

SCHEME 1: In this system all tasks are performed by the same engineer. Usually certain qualifications are required depending on the type and complexity of the structure to be designed. The advantage of this system is that the engineer does not have to use time in order to understand the problem. The efficiency of checking, therefore, is high throughout the whole checking process. However, the basic assumptions and decisions are only seldom questioned once they are made.

SCHEME 2: No special qualifications are demanded of the engineer who performs the initial design and the calculations, but the check has to be made by and independent and experienced engineer. This engineer usually has to fulfill some pre specified qualification requirements. In this case the checker will use some time in the beginning of the check as he is trying to understand the problem. Because of the independence, it is expected that the errors in the initial assumptions and decisions are detected.

SCHEME 3: In this system the checking is performed by a person or group of persons within the same company, where the design was performed. The checking group, however, is organizationally independent from the design group. In this case there is often a kind of negative correlation between the designer and checker, so that when the designer is inexperienced the checker is experienced and vice versa. As by the second system the checker will need some time to understand the problem before the check becomes efficient. By this system, however, the designer and checker are likely to share the same experience and tradition of doing things. This reduces the probability, that the checker detects the errors in the assumptions and decisions.

3. VALUATION CRITERIA

A judgment and ranking of the three methods can be made on various grounds but a fair judgement appears possibly only by comparing the optimal expected cost to society, $E[C_T]$ for a given structural element, which has been dimensioned according to a specific set of rules.

$$E[C_T] = C_D(t_D;A_D) + C_{QA}(t_{QA};A_C) + C_C + P_f(t_D;A_D;t_{QA};A_C)C_F \quad (1)$$

where $C_D(t_D;A_D)$ are the costs of the design as a function of the time spent on the design, t_D and the qualification of the designer A_D, $C_{QA}(t_{QA};A_C)$ are the cost of quality assurance as a function of the time spent on quality assurance, t_{QA} and the qualifications of the checker, A_C. C_C is the cost of construction, C_F is the cost of failure and P_f is the probability of failure. The cost of design of quality assurance are proportional to the time spent on the tasks

4. HUMAN ERROR

Human error can be defined as violations of the design rules, which will change the reliability, serviceability or economy of the structure. A more exact and widely accepted definition of human error is, that it is a noncompliance with code, or more generally, a significant deviation from acceptable practice. This definition excludes unforeseen events, "acts of god", and it is in compliance with legal practice. It further excludes gross negligence and criminal actions. But it does not account for errors in the accepted practice, and it leaves the problem of defining, what is acceptable practice.

The load on a structural element, *S*, will by assumption not be affected by human errors. The coefficient of variation of the resistance, *R*, is a constant, which depends on the quality control of the material. The purpose of the design process is to determine the mean value of the resistance *R*, so that the probability of failure becomes "small". The effect of the human errors can, therefore, be introduced by the variable ε, which describes the change of the mean value of the resistance.

$$\Psi_R = \mu_R \varepsilon \quad (2)$$

μ_R is the mean value of the resistance in case no errors have occurred, and ε describes the effect of the errors in the design. ε is a function of the number of errors and the magnitude of each of these errors. The magnitude of an error, *e*, is defined as

$$e = \frac{\mu_R^{\,e}}{\mu_R} \quad (3)$$

where μ_R^e is the mean value of the resistance with the error included and μ_R the mean value of the resistance of the error-free structure.

It is convenient to introduce two classes of errors: concept and design errors. The concept and design errors are errors, which are made in the concept and design phase of the design process, respectively. The following limited number of concept errors, which are assumed to be the most important are included in the model.

- An error in the geometry of the structure
- An error in the boundary conditions
- An error in loads or load cases
- An error in the selection of the relevant failure mode
- An error in the selection of the structural analysis model

It is evident, that all the tasks in the concept phase will either be performed correctly or wrongly. Each of these types of errors occur only once. For the analysis of a design task it is in most cases sufficient to include the following types of design errors

- Error in calculator calculations
- Error in table look-up
- Error in ranking

The calculations are performed in a number of steps, each of which correspond to a mathematical operation (multiplication, addition, ...). The probability of making an error in each of these steps is assumed to be roughly the same and all errors are assumed to be independent. Then the probability that j errors remain in the design can then be modeled by a Poisson process

$$P_E(j) = \frac{(\lambda p_E)^j}{j!}\exp(-\lambda p_E) \tag{4}$$

where λ is the intensity, which depends on the complexity of the design task and p_E is the probability that a calculation error remains. This model has also been suggested by Nessim [5]. The logarithm of the magnitude of the error is normally distributed with mean 0 and standard deviation 2.03 (Melchers and Harrington [4]). The decision about which table look-up errors and errors by ranking to include in the analysis, has to be based on information about the occurrence probability and the effect of each of the potential errors.

5. CHECKING MODEL

The checking consists of two parts: a self checking taking place during the design process and a detailed checking after the design. For self checking the model suggested by Stewart and Melchers [7] is used. For detailed checking it is assumed that its success depends on the complexity of the design task, the magnitude of the error, the qualification of the checker and the time spent in checking. Let $q_E(e,t)de$ be the probability of an error in the (small) interval de, when the time spent checking is t.

$$q_E(e,t) = f_{E_o}(e)\,P(\text{nondetection of } E \text{ at } t = t | E = e \text{ at } t = 0) \tag{5}$$

where f_{E_o} is the density of the error before checking. It should be noted that by assumption the error is certain to occur. The density function of the error after the time t spent checking is

$$f_E(e,t) = \frac{q_E(e,t)}{\int q_E(e,t)de} \tag{6}$$

In the following the notation will be used

$$P(\text{nondetection of } E \text{ at } t = t | E = e \text{ at } t = 0) = Q(e,t) \tag{7}$$

Following Lind's model [1] the rate of decrease of this function is proportional to the probability that the error has not yet been detected, $Q(e,t)$, and proportional to a function h of the magnitude of the error. According to Stewart and Melchers (7) the efficiency of the check is not constant, and therefore a function $g(t)$ is introduced. $g(t)$ describes the efficiency of the check as a function of the complexity of the problem and the qualifications of the checker. The decrement of the non detection probability is also assumed to be proportional to $g(t)$. This leads to the following equation:

$$\frac{\partial Q(e,t)}{\partial e} = -h(e)g(t)Q(e,t) \Leftrightarrow$$

$$Q(e,t) = c\exp\left[-h(e)\int_0^t g(\tau)d\tau\right] \tag{8}$$

where $c = 1$ because $Q(e,t) = 1$ for all e at $t = 0$.

5.1 *Concept errors*

The concept errors are discovered in the initial phase of the check. where the checker is trying to understand and reevaluate the design problem. It is assumed that if the checker does not detect the error in the initial phase the error will never be discovered. The efficiency with respect to concept errors is zero at $t = 0$. It increases rapidly during the first part of the check, reaches a maximum and then decreases to zero. The probability that a concept error is detected does not depend on the magnitude of the error. The function $h(e)$ is constant. The constant can be chosen as unity. This implies that the distribution of the error does not depend on the time spent in checking. If furthermore the duration of the check exceeds the initial phase where the check for concept errors is made then, the probability that an error remains in the design is constant. Therefore, it is only necessary to know the probability that the error is detected.

SCHEME 1: When the check is performed by the designer, it is assumed that the probability, of detecting a concept error is zero

SCHEME 2: In case the check is performed by an independent checker, the probability that errors in the geometry and boundary conditions are detected is assumed to be one. There exists a large number of such errors and it is unlikely that an independent checker makes the same error. The probability that the independent checker does not detect errors in the load cases and method is assumed to be equal to the probability that the checker would have made these errors.

SCHEME 3: As by scheme 2 it is assumed that errors in the geometry and boundary conditions are certain to be detected. If the designer and the checker works in the same firm, have been educated at the same university or in some way share the same tradition for selecting load cases and analysis method, the probability that such errors remain is likely to be larger than if no dependence exists.

5.2 *Design errors*

The probability that a design error is detected depends on the magnitude of the error. It is further assumed that if infinite time is spent in checking the design errors are certain to be detected. This implies that the check for design errors cannot simply be modeled by a detection probability for each error. It is assumed that large errors are easier detected than smaller errors, and that the value of $h(e)$ is the same for errors which result in a decrease as well as for errors which increase μ_R by a given factor. The last assumption is not fully correct because errors which reduce the reliability of the structure are probably detected with larger probability than errors which increase the reliability. The assumption therefore leads to a conservative estimate of the failure probability.

As the error magnitude e is defined as the change of the mean value of the resistance of the structure, then $e = 1$ implies $h(e) = 0$. Stewart and Melchers [7] state that $h(e)$ only varies little with e. The simplest possibly model is

$$h(e) = |\ln(e)|^{\alpha} \qquad (9)$$

This model is valid for all three schemes of checking, and α does not depend on the scheme.

The function $g(t)$ describes the efficiency of the check as a function of the time spent checking. It is evident, that the efficiency depends on the checking method.

SCHEME 1: When the check is performed by the designer the checking efficiency is assumed to be constant

$$g(t) = a \qquad (10)$$

where a depends of the complexity of the problem and the qualification of the checker.

SCHEME 2: The efficiency will be small in the beginning as the checker is making an attempt to understand the problem. Thereafter, $g(t)$ increases to reach a constant value. It is assumed that the increase in the efficiency will be very small in the beginning, which means that the derivative of $g(t)$ at $t = 0$ is small. It is evident that the function depends on the complexity of the problem and the qualification of the checker. If the qualification of the checker is bad or the problem very complex it can reasonable be assumed that $g(t)$ will reach the constant at a later time. Also its final value might be smaller than for simple problems or better qualifications. A simple function which can describe the efficiency of the check is:

$$g(\tau) = a(1 - (a\tau + 1)\exp[-a\tau]) \qquad (11)$$

where a for the same level of qualification takes on roughly the same value as by scheme 1.

SCHEME 3: The checking efficiency for method 3 is equal to the checking efficiency for scheme 2. The fact that the checker and designer might share the same knowledge and are prone to make the same concept errors does not effect the checking efficiency with respect to design errors.

6. CALCULATION OF FAILURE PROBABILITY

The most efficient method for the calculation of the failure probability with human errors taken into account seems to be the hazard scenario approach formalized in [6]. A hazard scenario is a more or less complex "scenario" of events. For example, an event in a scenario is a given set of human errors remaining in the design. Failure of the system due to failure in any of the hazard scenarios is failure of a series system, i.e. a system which fails if any of its links fail or in any of its hazard scenarios. Let F_i denote the failure event in scenario i, then

$$P_f = P\left[\bigcup_{\{N\}} F_i\right] \quad (12)$$

in which the union operation runs over all the events $\{N\}$. Let M_i be the set of errors in hazard scenario i, then the probability of the event F_i is the probability that only the set of errors $\{M_i\}$ remains in the design and failure occurs given the errors remain

$$P(F_i) = P(\{M_i\} \text{ remains})\, P(\text{failure} | \{M_i\} \text{ remains}) \quad (13)$$

The probability that only the errors M_i remain in the design is the probability that these errors have occurred and not been detected multiplied with the probability that no other errors remain. For each of the scenarios the probability of failure has to be calculated. This can be done by the available reliability methods such as FORM/SORM (see e.g. Madsen, Krenk and Lind [2]). Let $\boldsymbol{X}$ be a stochastic vector defining a reliability problem and let $\boldsymbol{x}$ be the outcome of this vector and $\boldsymbol{e}_i$ be the outcome of the vector $\boldsymbol{E}_i$, which describes the set of errors in $\{M_i.\}$ The probability of failure is

$$P(\text{failure} | \{M_i\}) = \iint_{g(\boldsymbol{x},\boldsymbol{e}_i)<0} f_{\boldsymbol{X}}(\boldsymbol{x}) f_{\boldsymbol{E}_i}(\boldsymbol{e}_i) d\boldsymbol{x} d\boldsymbol{e}_i \quad (14)$$

where $f_{\boldsymbol{E}_i}(\boldsymbol{e}_i)$ is the joint density function of $\boldsymbol{E}_i$, $f_{\boldsymbol{X}}(\boldsymbol{x})$ is the joint density function of $\boldsymbol{X}$ and $g(\boldsymbol{x},\boldsymbol{e}_i)$ is a limit state function defined in such a way that $g(\boldsymbol{x},\boldsymbol{e}_i) < 0$ corresponds to failure. We consider a structural component where the load can be described by the stochastic variable S and the resistance by the stochastic variable R. A representative limit state function for this structural component is

$$g = R - S \quad (15)$$

The mean value of the resistance is determined according to eq. (2). The effect of the errors is introduced as

$$\varepsilon_i = \prod_{\{M_i\}} E_j \quad (16)$$

7. EXAMPLE

For the purpose of comparing the checking schemes the following simple example is considered. The resistance of the structural component is log normally distributed with mean $\mu_R = 3.158$ and coefficient of variation $V_R = 0.2$. The load is normally distributed with mean 1.0 and coefficient of variation $V_S = 0.2$. The reliability index of the error-free structural component is $\beta = 4.066$, which corresponds to $P_f = 2.89\ 10^{-5}$. The cost of failure of the error-free structure is $C_F = 20000$.

In table 1 the occurrence probabilities and magnitudes of the errors, which are included in the analysis, are given. The occurrence probabilities are those for a designer with the qualifications of a student. The calculation errors occur with the intensity $\lambda = 30.0$. Two load case errors are included in the example. The smallest of these (magnitude 0.85) is very likely to occur. The other load case error (magnitude 0.65) has a considerably lower probability of occurrence. There is enough time available to perform the design task with normal care, and the design task is performed under suitable working conditions.

Table 1: Probability of occurrence and magnitudes (LN denotes the logarithmic normal distribution).

Error	Probability	Magnitude
Geometry	0.025	LN(1.0;0.75)
Load 1	0.15	0.85
Load 2	0.03	0.65
Calculation	0.0136	LN(1.0;2.03)

Only three different levels of qualification of both designer and checker are considered. In table 2 the parameters which depend on the qualification of the checker and designer are shown. The values of p_{ED} and p_{EC}, the probability of making a design and concept error, respectively, in table 2 are the occurrence probabilities relative to the occurrence probabilities for a student (table 1). It is estimated that the probability of making a concept error decreases more rapidly with the experience of the designer than the probability of making a design error. It is assumed that t_D, the time necessary to perform the design task with normal care, decreases with experience of the designer, and that the cost per unit time increases with the experience. The relative increase of the cost per unit time, however, is judged to be smaller than the relative decrease of the time necessary to perform the design task. This implies that the cost of the design becomes smaller as the knowledge and experience of the designer increase. The checking efficiency a is assumed to be inversely proportional to the time t_D required to perform the design task.

Table 2: Parameters

A_D/A_C	p_{ED}	p_{EC}	t_D	$c_D = c_{OA}$	a
Student	1.0	1.0	120	1.0	0.075
Normal	0.3	0.1	60	1.5	0.15
Expert	0.1	0.01	30	2.0	0.30

SCHEME 1: The results for scheme 1 are given in table 3 for all three levels of qualification of the checker/designer.

Table 3: Optimum for scheme 1.

Designer	Checker	t_{OA}	C	P_f
Student	Student	29.0	217.0	0.00342
Normal	Normal	12.0	117.8	0.000491
Expert	Expert	5.5	73.5	0.000159

Because the efficiency of the check is high throughout the checking process the optimal checking time is relatively short, about 20 - 25 % of the time spent on the design. A designer/checker with the qualification of a student is likely to make concept errors. Therefore, the failure probability at optimum is more than 100 times larger than for the error-free structure. For the normal designer/checker and the expert the failure probability at optimum is 17 and 5.6 times larger than the failure probability of the error-free structure, respectively. This demonstrates, at mentioned earlier, that in order to use this scheme certain qualifications are required. Partly because of the low failure probability and partly because an expert performs the design at the lowest cost the total cost are also smallest when the designer/checker is an expert and increase with decreasing qualification of the designer/checker.

SCHEME 2: By scheme 2 the checker is always an expert. There are, however, no special qualifications required for the designer. The example is calculated for all three different levels of qualification of the designer. The results are given in table 4.

Table 4: Optimum for scheme 2.

Designer	Checker	t_{OA}	C	P_f
Student	Expert	14.2	152.1	0.000186
Normal	Expert	12.6	118.6	0.000174
Expert	Expert	11.0	87.3	0.000160

As expected the optimal duration of the check increases as the qualification of the designer decreases. The necessary checking time when the designer has the qualification of a student, however, is only 29 % larger than when the designer is an expert. The qualification of the designer has relatively little influence on the optimal checking time.

The qualification of the designer also seems to have little influence on the optimal failure probability. Because the differences between the optimal checking times and failure probabilities are small, the differences between the optimal cost for the three levels of qualification of the designer are almost solely caused by the differences between the cost of the design.

SCHEME 3: No special qualifications of either designer or checker are required by scheme 3, but a kind of negative correlation between designer and checker exists. The checking is often planned in such a way that the inexperienced engineer checks the calculation of the more experienced engineer. The calculations performed by the inexperienced engineer are checked by an expert, who is likely to detect the errors, which have been committed by the inexperienced engineer with large probability. The following three combinations of designer and checker are investigated (see table 5).

Table 5: Combinations.

Designer	Checker
Student	Expert
Normal	Normal
Expert	Student

The probability of a concept error remaining in the design depends on the degree of dependency between the designer and checker. Three different levels of dependency are investigated.

1. No dependency: The probability that the error remains is the product of the occurrence probabilities for the designer and checker. The results are given in table 6.

2. Full dependency: The probability that the concept error remains is the minimum value of the probability of

occurrence for the checker and designer. The results are given in table 8.

3. Medium dependency: The mean value of the probabilities for fully dependent and no dependence. The results are given in table 7.

Table 6: No dependency

Designer	Checker	t_{OA}	C	P_f
Student	Expert	14.2	154.0	0.000180
Normal	Normal	24.0	130.6	0.000232
Expert	Student	40.0	105.9	0.000297

Table 7: Medium dependency.

Designer	Checker	t_{OA}	C	P_f
Student	Expert	14.2	154.1	0.000188
Normal	Normal	24.0	131.0	0.000250
Expert	Student	40.0	105.9	0.000298

Table 8: No dependency.

Designer	Checker	t_{OA}	C	P_f
Student	Expert	14.2	154.1	0.000189
Normal	Normal	24.0	131.4	0.000268
Expert	Student	40.0	106.0	0.000300

The duration of the check for concept errors and the efficiency of the check for design errors does not depend on the degree of dependency between designer and checker. This implies that the optimal duration of the check is also independent of the degree of dependency (see tables 6-8). As the dependency between designer and checker increases the probability that a concept error remains increases. Thereby, the total costs and the failure probability at optimum increase with increasing dependency (see tables 6-8).

The probability that a concept error remains is never larger than the probability that the best qualified of both checker and designer would have made the error. The designer and checker usually are chosen in such a way that either one is well qualified. This means that the probability that a concept error remains is small even though designer and checker are fully dependent. This again implies that the effect of the dependency is small. The effect of the dependency increases with increasing cost of failure. For larger cost of failure more time will be spent in order to detect design errors and the relative contribution to the total failure probability from concept errors becomes larger.

8. CONCLUSION

It might appear that the models selected are rather special and the systems are not general. Sensitivity studies not reported herein, however, showed that the numerical results and thus the conclusions are rather robust in their relative order with respect to changes in the models. In fact all numerical results should be interpreted in terms of an ordering scheme rather than by absolute values. If one accepts that judgement by such ordered values is reasonable the following main conclusions can be drawn.

The overall lowest cost are obtained by using scheme 1 with an expert designer/checker. In this case the total costs are 73.5 and the failure probability is only about 5.5 times larger than for the error-free structure. If the designer and checker were both experts and independent (scheme 2) about the same failure probability at optimum is found, but at a larger cost. In reality not all designers and checkers can be experts. If the designer has normal qualifications the total cost are about the same for scheme 1 and 2. The failure probability obtained by using scheme 1, however, is considerably larger than the failure probability obtained by scheme 2. Since society is risk adverse scheme 2 is preferable to scheme 1. If the designer has the qualifications of a student scheme 1 leads to unacceptable high failure probability and the overall largest cost. Also in this case scheme 2 is preferable.

For all levels of qualification of the designer scheme 3 leads to larger cost. Scheme 3, however, shows that the effect of dependence between designer and checker is small if only one has better than normal qualification.

From this example it can be concluded that for an expert designer the optimal choice of checking is scheme 1. For designers less qualified scheme 2 offers the best alternative. This, however, cannot be a general conclusion, because the optimum also depends of the complexity of the design task and the cost of failure. For more complex design tasks the probability of making a concept error becomes larger and, thereby, the independence between the designer and checker becomes more important. For more complex design tasks scheme 3 presumably always is to be preferred.

The above conclusions made on the basis of numerical results from a simple, yet representative, example are nevertheless conditioned in the sense that other factors such as the educational system, the system of professional qualification and the economic and juridical (civil and criminal) environment play an important role.

REFERENCES

[1] Lind, N. C., Human error in Structural Reliability, *Structural Safety*, 1, 1983, pp. 167-176.

[2] Madsen, H. O., Krenk, S., Lind, N. C., Methods of Structural Safety, Prentice-Hall, 1986, New Jersey.

[3] Matousek, M., Schneider, J., Untersuchungen zur des Sicherheitsproblem von Bauwerken, Institut für Baustatik und Konstruktion, Eidgenössische Technische Hochschule, Zürich, 1976.

[4] Melchers, R. E., Harrington, M. V., Human Error in Structural Reliability - I, Investigation of Typical Design Tasks, Civil Engineering Research Reports, Monash University, Report No. 2/1984.

[5] Nessim, M. A., Decision-Making and Analysis of Errors in Structural Reliability, The University of Calgary, 1983, Ph.D.-Thesis.

[6] Rackwitz, R., Human error in Design and Structural Failure, in: A. S. Nowak (Ed.) Modeling Human Error in Structural Design and Construction, Proc. of a Workshop Sponsored by the National Science Foundation, 1986, ASCE, New York, pp. 216-224.

[7] Stewart, M. G., Melchers, R. E., Checking Models in Structural Design, *Journ. Struc. Eng.*, 115, 1989, pp. 1309-1323.

Structural Safety & Reliability, Schuëller, Shinozuka & Yao (eds) © 1994 Balkema, Rotterdam, ISBN 90 5410 357 4

Structural vulnerability analysis

M. Hashimoto
Steel Construction and Engineering Department, NKK Corporation, Yokohama, Japan

D. I. Blockley & N. J. Woodman
Department of Civil Engineering, University of Bristol, UK

ABSTRACT: The purpose of structural vulnerability analysis is to enable the identification of the most vulnerable parts of a structure so that they may be suitably protected and monitored. A graph model is used to analyse various load paths and to define the new concepts of structural rings and clusters. A structure is represented at various hierarchical levels of definition in terms of clusters of interconnected structural rings. A failure scenario is described in terms of deteriorating events of structural rings.

1. INTRODUCTION

This paper summarises a theory, the purpose of which is to identify the most vulnerable parts of a structure so that they may be suitably protected and monitored (Wu 1991). The systems approach used is designed to take advantage of computer based pattern recognition techniques. A graph model of a structure has been developed in order to analyse load paths. The model includes some new concepts, the single most important of which is a structural ring. A structural ring is a load path which is capable of resisting an arbitrary set of applied forces. A structure then can be represented at various hierarchical levels of definition in terms of clusters of interconnected structural rings. A failure scenario is described in terms of deteriorating events and a measure of the effort required is called damage demand. In the analysis of vulnerability three types of failure scenario are defined, the minimal, the maximal and the particular.

The emphasis of this analysis of vulnerability is not the usual one of analysing a structure under some given loading condition, rather it is to examine the vulnerability of a structure to any possible loading condition. This is well-formedness of the structural rings at various levels of definition within a structure and (ii) those rings which are the most vulnerable or critical together with the actions which might cause failure.

2. HIERARCHICAL REPRESENTATION OF A STRUCTURE

2.1 *Structural Ring*

A structure can be regarded as a graph model $S=\{M, J\}$, which consists of two sets: a set of member objects $M=\{M_i|i=1,...,N_M\}$ and a set of joint objects $J=\{J_j|j=1,...,N_J\}$. Each member object is associated with at least two joint objects. A structural path is a sequence of connected members. A structural ring is a path which is either open or closed and either structurally over-stiff or just-stiff. It is therefore capable of resisting an arbitrary equilibrium set of applied forces given adequate strength.

In its most elementary form a ring consists of a series of members such as the pin-jointed triangle of Fig.1(a) and the portal frame of Fig.1(b) which includes the ground C_G as a special member linking the bases of the frame. Fig.1 also shows an abstract representation of a structural ring as a circle with joints and with arcs representing the members.

The well-formedness of a structural ring is a measure of its ability to resist loading from any arbitrary direction. The well-formedness depends on the orientation and stiffness of the members within the ring and the stiffness of the joints connecting the members in the ring. The quality of well-formedness of a joint contained in a ring is here defined as

$$q_j = det(D_{jj}) \quad (1)$$

where D_{jj} is the stiffness sub-matrix associated with members in the ring at a joint j and $det(D_{jj})$ is the determinant of the matrix D_{jj}. We then define the quality of well-formedness of a structural ring R as

$$q(R) = \Sigma_j q_j ; \ j=1,...,N(J_R) \quad (2)$$

where $N(J_R)$ is the total number of joints in the ring R. Note that the determinant of D_{jj} is equal to the product of all eigenvalues that satisfy $det(D_{jj} - \lambda I_n) = 0$. Thus q_j is independent of the co-ordinate system used. Note also that q is not dimensionless but could be made so by normalising with respect to the optimum configuration for a given joint. It is recognised that the measure may need to be improved: units will not be quoted in this paper.

2.2 *Structural Clusters*

A structural cluster C^l at a level of definition l is defined as a sub-set $C^l=\{M^l, J^l\}$ of S in which the objects are, in some sense, more

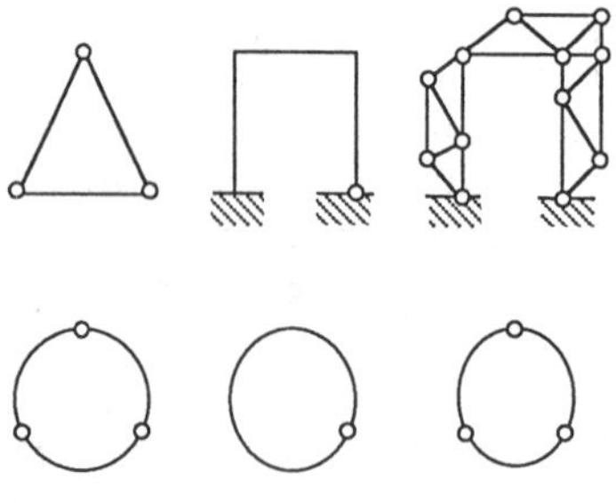

Fig 1. Examples of Structural Rings

tightly connected to each other within the cluster than to other objects outside the cluster. A primitive cluster contains only one member and two end joints. The conventional way of describing a structure uses primitive clusters and is the lowest level of definition.

If clusters of members and joints are formed and the structure is defined in terms of those clusters then that forms the second level of definition. If clusters of the second level clusters are formed then this becomes a third level of definition. This process can be repeated for even higher levels until at the top level where there is only one cluster, the whole structure. Thus the structure can be represented hierarchically as

$$S = S^0 = S^1 ... = S^l = \{ C^l_i, J^l_j \mid i=1,...,N_C^l; \ j=1,...,N_J^l \} \quad (3)$$

where N_C^l and N_J^l are the total number of clusters and joints within the structure at level of definition l.

The structural tightness of a cluster is a measure of connectivity. It depends on the number of structural rings within the cluster, the degree of overlap between them and the well-formedness of the rings. Thus the structural tightness of the cluster C^l is defined as

$$Q(C^l) = \Sigma_j det(D_{jj}) / N(J_C^l) \ ; \ j=1,...,N(J_C^l) \quad (4)$$

where $N(J_C^l)$ is the total number of joints in C^l.

2.3 *Hierarchy Formation*

The ground or foundation of the structure is effectively a very tight single cluster and must be considered specifically. In order to cluster the structure independently of the ground, the following algorithm does not include rings containing the ground until all the remaining rings not yet taken into the cluster, include the ground.

An algorithm for forming an hierarchical representation of a structure is

(1) Start at hierarchical level $l=0$, which is the lowest level of definition at which the members and joints of the basic structure are described

(2) At level l of build-up, identify all $N(R^l)$

structural rings in the structural model S^l ignoring the ground S_G unless all rings contain S_G. If no structural ring can be found then go to step (10), otherwise

(3) Calculate the well-formedness $q(R^l_i)$ for each of $N(R^l)$ rings according to Eq. 2 and rank them
(4) Choose the highest ranked ring not in a cluster as a forming cluster C^l_j and set $Q(C^l_j)=Q(R^l_i)$
(5) Assume an overlapping ring $R^l_i{}^*$ added in to the forming cluster C^l_j, and calculate the new value of structural tightness $Q^*(C^l_j)$ using equation 4
(6) Include the ring $R^l_i{}^*$ in C^l_j if $Q^*(C^l_j) > Q(C^l_j)$
(7) Repeat step (5) to (6) for all other overlapping rings
(8) Go to step (4) in order to form another cluster with $j=j+1$
(9) Go to the next hierarchical level of definition by $l=l+1$ and go to step (2)
(10) Stop

The process of cluster formation produces a hierarchical model of a structure which is also a set of structural rings at each level of definition as

$$S^l = \{ R^l_i \mid i=1,...,N(R^l) \tag{5}$$

where $N(R^l)$ is the total number of structural rings at the level of definition l.

The structure of Fig.2 is used as an example to illustrate the procedure of hierarchy formation.

At the lowest level of definition (level 0) there are 4 structural rings with well formedness values, from Equation 2, as follows:

$R^0_1 = \{M_1, J_1, M_2, J_2, M_6, J_3\}$ $q(R^0_1) = 16.0$
$R^0_2 = \{M_2, J_2, M_3, J_4, M_5, J_1\}$ $q(R^0_2) = 15.1$
$R^0_3 = \{M_1, J_3, M_4, J_4, M_5, J_1\}$ $q(R^0_3) = 15.0$
$R^0_4 = \{M_3, J_4, M_4, J_3, M_6, J_2\}$ $q(R^0_4) = 16.8$

Note that rings such as $\{ C_G, J_1, M_2, J_2 \}$ are not included. Ring 4 has the highest well formedness and is the seed for the cluster formation. Ring 1 is added to ring 4 and the structural tightness of the cluster so formed is calculated using equation 4. As Q increases the process is repeated adding in ring 2 and finally ring 3 to form cluster C^1_1. At this stage $Q(C^1_1) = 47.2$.

Since there are no more rings at this stage, we move on to the next level (level 1). Here there is only one ring $R^1_1=\{ C_G, J_1, C^1_1, J_2 \}$ including the ground cluster, it is the highest level of definition and the algorithm stops.

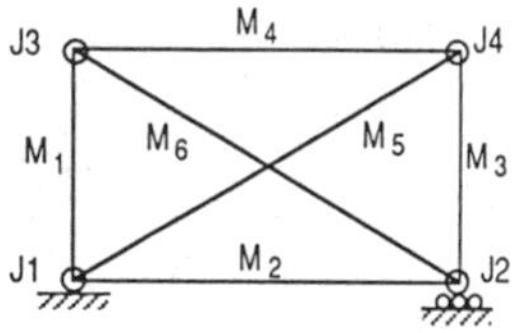

Member	A (mm)	I (mm⁴)
1	80	2000
2	90	2100
3	82.5	2050
4	92.5	2150
5	100	2500
6	120	3000

E = 200kN/mm^2

Fig 2 Example Structure

The process is illustrated in Fig.3.

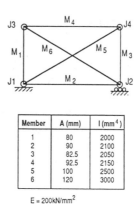

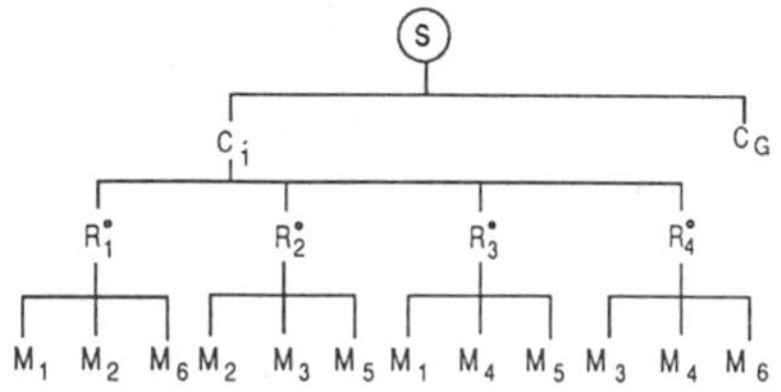

Fig 3 Cluster Formation

3. FAILURE SCENARIOS

3.1 *Deterioration Hierarchy of Structural Rings*

Fig.4 is a deterioration hierarchy for a structural ring (DHSR) and shows all of the possible ways in which a fully fixed ring can deteriorate into a mechanism. A rectangular portal frame is shown also in the figure as a practical example of a structure consisting of a single ring made of members which are primitive structural clusters and the ground.

A deteriorating event is the loss of a degree of freedom (DOF) within the ring. It may occur adjacent to a joint or inside a cluster causing it to become two separate clusters connected by a joint. If a ring deteriorates until it becomes a just-stiff ring then one more deteriorating event will cause it to become a mechanism.

3.2 *String Patterns of Structural Rings*

A structural ring can be characterised by a string pattern representing the DOF $d^l_{i,j}$ carried at each joint and by each cluster and where $d^l_{i,j} = 1$ when the force can be carried and 0 otherwise and i is a joint at level l and j is u, v, θ in the conventional x, y co-ordinate system. Thus, a fixed joint at i is $J^l_i=\{1,1,1\}$, a pin joint at i is $J^l_i=\{1,1,0\}$. For further details see Wu 1991.

3.3 *Failure Scenarios*

A path through the DHSR is a scenario $F_h(R^l)$ of a ring R^l.

$$F_h(R^l) = \{ R^l_k \mid k=1,...,m_h \} \quad (6)$$

where each R^l_k is a deteriorated ring of R^l and there are m_h such rings in the scenario. In a scenario, the R^l_k are ordered in the sense that R^l_{k+1} is more deteriorated than R^l_k. A failure scenario is a scenario in which the final element is a mechanism. One or more deteriorating events are the result of actions which would cause the loss, in a ring, of the capacity to transmit a DOF. Those actions may be natural (e.g. wind, earthquake) or human (e.g. mistake, sabotage).

Therefore for a failure scenario there are series of deteriorating events $g^l_{i,j,k}$ occurring either in a joint or in a cluster, which can be represented by

$$F_h(R^l) = \{ g^l_{i,j,k} \mid i=1,...,N(J_R^l);\ j=u,v,\theta;\ k=1,...,m_{h-1} \} \quad (7)$$

3.4 *Damage Demand*

The effort required to cause a given deteriorating event is called damage demand with respect to that event. The damage demand for a deteriorating event $g^l_{i,j,k}$ is defined as $e(g^l_{i,j,k})$ and is assumed to be directly proportional to the loss of the principal stiffness for a given DOF

The total damage demand for a failure scenario $F_h(R^l)$ is

$$E[\, F_h(R^l) \,] = \Sigma_i\, \Sigma_j\, \Sigma_k\, e(g^l_{i,j,k})\ ;\quad i=1,...,N(J_R^l);\ j=u,v,\theta;\ k=1,...,m_{h-1} \quad (8)$$

3.5 *Action Effect*

An action may cause more than one deteriorating event. We define an action effect, adjacent to a joint i (or a member i) as

$$E_i = \{ a_{i,u}, a_{i,v}, a_{i,\theta} \} \quad (9)$$

where $a_{i,j}$ ($j=u,v,\theta$) is the action effect and is 1 when the DOF remains and 0 otherwise. For instance, $E_i=\{0,0,0\}$ is the effect of complete failure. The intersection of the string patterns of the first ring R^l_1 and the first action E_1, gives a deteriorated ring R^l_2. If this is not a mechanism then a second action E_2 can be considered in a similar manner. In general for k actions

$$F_h(R^l) = \{ R^l_k \mid k=1,...,N_E \} \quad (10)$$

where $R^l_{k+1} = R^l_k \cap E_k$, and N_E is the total number of actions.

4. VULNERABILITY ANALYSIS

Among all of the possible failure scenarios for a structure three types of scenarios are of particular interest

(i) the minimal failure scenario, (ii) the maximal failure scenario, (iii) any particularly interesting failure scenario.

4.1 *The Minimal Failure Scenario*

The minimal failure scenario of a structural ring at level of definition l is the one in which the least damage demand is required to transform the structural ring into a mechanism. If a given ring can fail with little damage demand then it is a vulnerable ring. Thus minimal failure scenario at level l is

$$F_{min}(R^l) = \min_h \{E[F_h(R^l)] \mid h=1,...,p \} \quad (11)$$

where p is the total number of possible failure scenarios for a ring R^l.

The algorithm for the formation of the hierarchy and hence the levels of definition results in the least well-formed and most

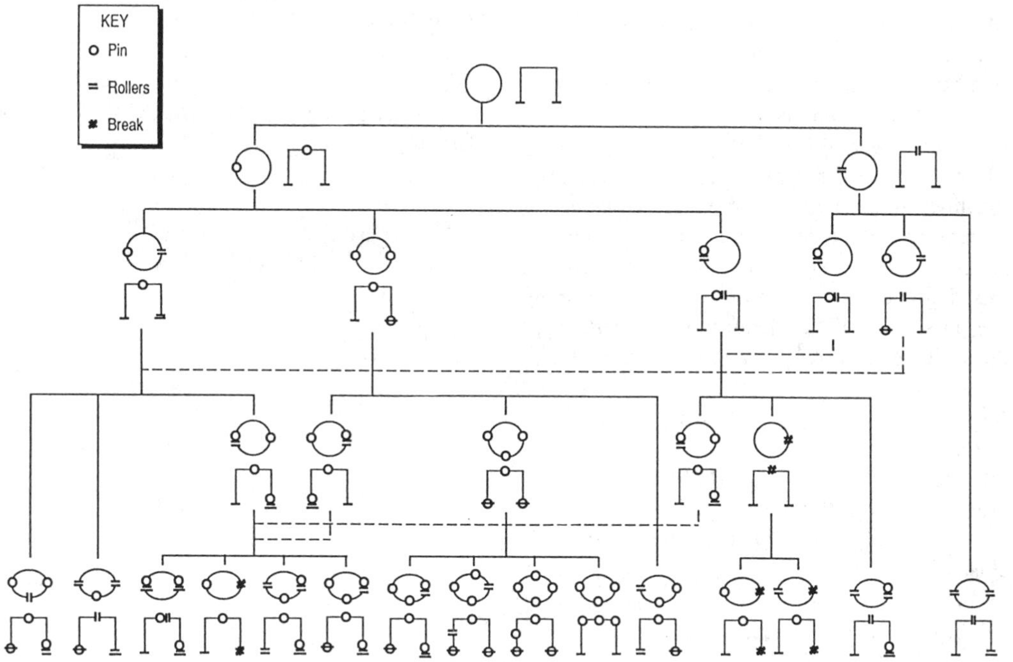

Fig. 4 Deterioration Hierachy for a Structural Ring

vulnerable rings being formed last. The analysis of structural vulnerability therefore starts at the top of level of definition and works downwards in a sort of 'unzipping' search process.

Consider again the structure of Fig.2 and limit the possible actions $E_i (i=1,..,6)$ to "cutting one member". In this case all DOFs of the member are lost.

If $R^0{}_1 = \{M_1, J_1, M_2, J_2, M_6, J_3\} =$
$\{(1,1,1),(1,1,0),(1,1,1),(1,1,0),(1,1,1),(1,1,0)$.
then if M_1 is cut
$E_1 = \{$
$(0,0,0),(1,1,1),(1,1,1),(1,1,1),(1,1,1),(1,1,1)\}$.
and the new ring $R^0{}_1$ will be obtained as
$R^0{}_1 \cap E_1 = \{$
$(0,0,0),(1,1,0),(1,1,1),(1,1,0),(1,1,1),(1,1,0)\}$.

Matching the deteriorated ring with the DHSR, we see that the ring becomes a mechanism. Thus the action E_1 triggers the failure scenario for $R^0{}_1$. The damage demand for this scenario is
$E[\ F_1(R^0{}_1)\ ] = 2160$ (Eq 8). The same procedure also can be applied to each action and the other rings. The minimal failure scenario is the one where the least damage demand is required and can be shown to be
$F_{min}(R^l) = E[\ F_5(R^0{}_1)\ ] = 1620$
i.e. cut M_5 in the ring $R^0{}_1$ and ring $R^0{}_3$.

4.2 *The Maximal Failure Scenario*

A failure scenario may or may not cause a given cluster to be separated from a reference cluster (usually the ground). The separateness of a structural ring R^l with respect to that failure scenario is defined as

$$\gamma[\ F_h(R^l)\ ] = \Sigma_i\, Q(C^l{}_i) \,/\, \Sigma_j\, Q(C^l{}_j) \qquad (12)$$

where $\Sigma_i\, Q(C^l{}_i)$ is the total tightness of all clusters which are disconnected from the reference cluster $C^l{}_r$, and $\Sigma_j\, Q(C^l{}_j)$ is the total tightness of all clusters still connected with the reference cluster $C^l{}_r$.

The effective consequence of a failure scenario at a level of definition is defined as the ratio of the separateness to the total damage demand for that scenario and is therefore

$$\xi[\ F_h(R^l)\ ] = \gamma[\ F_h(R^l)\ ] \,/\, E[\ F_h(R^l)\ ] \qquad (13)$$

The maximal failure scenario at a given level

of definition is then the one in which the smallest damage demand causes the largest number of clusters to be disconnected from a reference cluster and is therefore defined as

$F_{max}(R^l) = max_h \{\xi[F_h(R^l)] \mid h=1,...,p\}$ (14)

In order to find the maximal failure scenario, we look at the ring R^1 at level 1 where the system would loose its integrity without the reference cluster (ground cluster). Intersecting this ring and a set of actions, we may examine the possibility of the deterioration of the ring. Now the cluster $C^1{}_1$ may be represented at level *0* as

$C^1{}_1 = \{R^0{}_l \mid l=1,...,4\} = \{J_i, M_j \mid i=1,...,4, j=1,...,6 =$
$\{(1,1,0),(1,1,0),(1,1,0),(1,1,0),(1,0,0),(1,0,0),(1,0,0),(1,0,0),(1,0,0),(1,0,0)\}$

If we cut M_1, action E_1, then the deteriorated cluster is

$C^1{}_1 \cap E_1 =$
$\{(1,1,0),(1,1,0),(1,1,0),(1,1,0),(0,0,0),(1,0,0),(1,0,0),(1,0,0),(1,0,0),(1,0,0)\}$

Since this does not cause collapse a second action may be to cut M_2, action E_2

$E_1 \cap E_2 = \{$
$(0,0,0),(0,0,0),(1,1,1),(1,1,1),(1,1,1),(1,1,1)\}$

resulting in

$C^1{}_1 \cap (E_1 \cap E_2) =$
$\{(1,1,0),(1,1,0),(1,1,0),(1,1,0),(0,0,0),(0,0,0),(1,0,0),(1,0,0),(1,0,0),(1,0,0)\}$.

which is a mechanism

At level *1*, the deteriorated ring $C^1{}_1$ at this level is described as

$R^1{}_1 = \{(1,1,0),(0,0,0),(1,0,0),(1,1,1)\}$.

Now this ring is also a mechanism. The damage demand for this scenario is calculated using Eq 8 as $2160 + 1700 = 3860$

The same procedure for the other combinations of any two actions results in 15 failure scenarios.

It is clear however that unless member 2 is cut the separateness from the ground will not be a maximum. Thus the 15 failure scenarios reduce to 5 and all produce an infinite separateness (46.8/0.0 using Equation 12). The maximal failure scenario will therefore be that scenario with a cut to member 2 together with a cut to the member taken from the set of members 1, 3, 4, 5, 6 which has minimum damage demand. Thus the maximal scenario is E_2 & E_5.

5. REFERENCES

Wu X (1991) "Vulnerability Analysis of Structural Systems", *PhD Thesis*, University of Bristol, Bristol, UK.

Structural Safety & Reliability, Schuëller, Shinozuka & Yao (eds) © 1994 Balkema, Rotterdam, ISBN 90 5410 357 4

Seismic vulnerability analysis of retrofitted RC structures using extended fiber model

Kazuyuki Izuno
Ritsumeikan University, Japan

Yoshikazu Yamada
Chubu University, Japan

Hirokazu Iemura
Kyoto University, Japan

Toshiki Ohkawa
Okumura Corporation, Japan

ABSTRACT: Seismic vulnerability of retrofitted reinforced concrete structure was evaluated through comparison of ductility, hysteretic energy and damage index with the original. Retrofitted RC structures were modeled using fiber modeling technique extended to include the stress-strain relation of repair material such as grouted epoxy resin which is randomly located in the retrofitted cross section. Earthquake response analysis was then conducted using inelastic hysteretic rules. The proposed method was verified by the test results of retrofitted RC beam members. As an application of this method, earthquake response of a strengthened RC bridge pier was simulated. The bridge pier with reinforcements terminated at mid-height was modeled including retrofitting by steel jackets. Thin steel jacket showed higher capacity than thick one because of its larger ductility.

1 INTRODUCTION

The collapse of structures during the 1989 Loma Prieta earthquake in U.S.A. brought a large concern on retrofit work. The use of epoxy resin in retrofitted RC members and steel jackets in strengthening of RC structures has been found to be applicable and effective. However, a reasonable index to judge whether the structure can be retrofitted or has to be rebuilt has not been established yet. In addition, to check whether the retrofitted structures could survive in future earthquakes is not evaluated precisely either.

In this paper, seismic vulnerability for existing RC structures retrofitted with grouted epoxy resin or steel jackets was quantified using inelastic response analysis with the force-displacement relationship based on extended fiber model.

2 EXTENDED FIBER MODEL FOR RETROFITTED CROSS SECTION

Conventional fiber modeling technique for RC structure can be extended to include repair material, such as epoxy resin and steel jacket.

The conventional fiber modeling technique calculates the stress of each fiber based on the actual location of steel and concrete. The extended fiber model proposed here uses the same equation for the resin, of which real location is unknown, based on the approximated ratio of the resin to the total area of the cross section. The assumed constitutive laws are shown in Fig. 1. Fig. 1(a) shows the assumed stress-strain relation of confined and unconfined concrete (Kent and Park 1971). (The plus strain in Fig. 1 means tensile strain.) The reinforcing bars were modeled as shown in Fig. 1(b). A maximum strength plateau was added to a bilinear curve in order to describe the ultimate behavior. Steel jacket was assumed to contribute to the moment-curvature relation of each section as additional reinforcing bars. As the grouted epoxy resin usually locates randomly in each section, precise location is unpredictable from outside. Therefore, epoxy resin was defined as an area ratio of its existence in this study. Epoxy resin was assumed to behave partially linear like Fig. 1(c). The confined epoxy resin was assumed to withstand any tensile and compression force, and the unconfined epoxy resin was assumed to crush in compression.

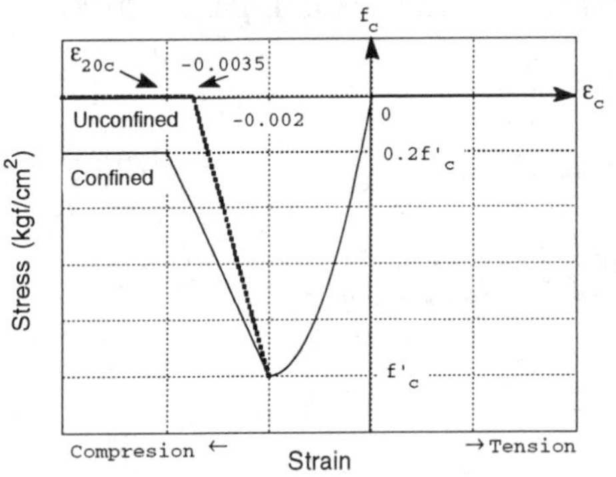

(a) Concrete

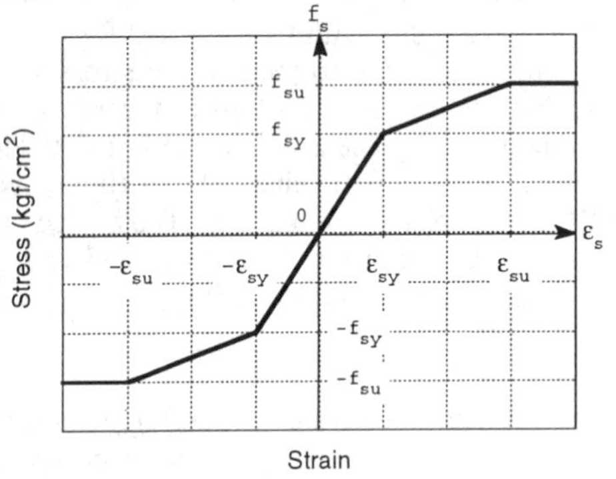

(b) Steel

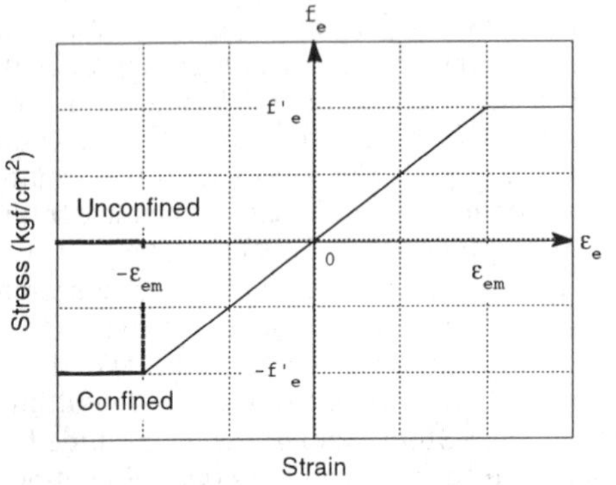

(c) Epoxy resin

Fig. 1 Assumed stress-strain relationship

Young's modulus of epoxy resin is almost 10% of concrete, but is larger than concrete in tension.

Four critical points were determined using this extended fiber model. These were the crack point where cracks of the concrete appear, the yield point where yield of the reinforcing bars occur, the maximum point where crush of the cover concrete occur, and the ultimate point where the core concrete crushes or the strength of the structure decreases to the assumed level. Four linear segments were used to connect these 4 critical points, so that the moment-curvature relation was assumed to be piecewise linear.

3 SIMPLIFIED PROCEDURE FOR SEISMIC VULNERABILITY ANALYSIS

The force-displacement relationship was derived based on the moment-curvature relationship of each section. Four points on the moment-curvature relationships were determined to approximate the relationship to piecewise linear in 4 lines.

Using the force-displacement relationship as a skeleton curve, an inelastic hysteretic model was assumed for earthquake response analysis. One of the suitable hysteretic models for RC structures, 3-parameter model (Park, Reinhorn and Kunnath 1988), was used in this study. This inelastic model achieves stiffness degradation , strength degradation and pinching behavior using 3 parameters; α, β and γ.

Inelastic earthquake response analysis was carried out for the estimated model using several earthquake records. Then, response of the retrofitted structure during a future earthquake was compared with the original structure to evaluate effectiveness of seismic retrofit. Three indices were used to evaluate quantitative effect of seismic retrofit; these were ductility, hysteretic energy and the damage index (Park and Ang 1985).

Among the many indices used to evaluate damage to the structures, ductility, energy absorbing capacity and the linear combination of these two are the basic indices.

In evaluation of retrofit effect, the maximum restoring force under static loading is the most important checking point. However, ductility and energy absorbing capacities are also important for seismic retrofit in terms of earthquake hazard mitigation.

Ductility is defined as the ratio of the maximum deformation response to the yield deformation. Hysteretic energy, which is calculated as the area of the hysteretic load-deformation responses, has strong correlation

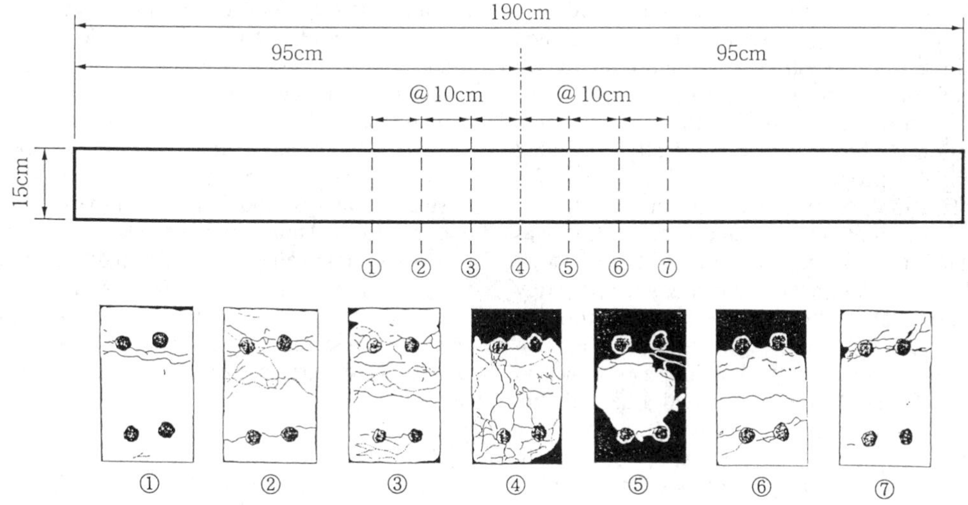

Fig. 2 Cross-sectional views of repaired specimen

with the accumulated damage. Damage index is expressed as (Park and Ang 1985)

$$D = \frac{\delta_m}{\delta_u} + \frac{\beta}{P_y \delta_u} \int dE \qquad (1)$$

in which, D is the damage index, δ_m is the maximum deformation response, δ_u is the defined ultimate deformation, P_y is the yield strength, $\int dE$ is the absorbed hysteretic energy and β is an empirical coefficient which takes the same value as β of the 3-parameter model. A damage index of more than 1 represents severe damage or collapse.

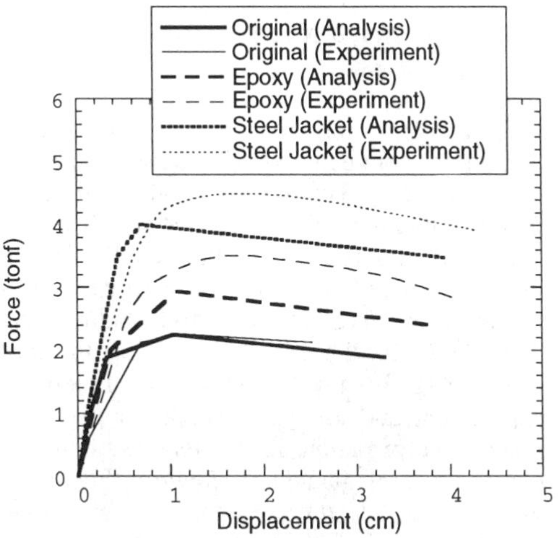

Fig. 3 Force-displacement relation of analytical and experimental results

4 VERIFICATION EXAMPLES USING TEST RESULTS

4.1 Hybrid loading tests of retrofitted RC members

Seismic behavior of repaired or strengthened RC members by epoxy resin and steel plates was tested using the hybrid testing procedure (Izuno, Yamada and Iemura 1988). Specimens used were doubly-reinforced concrete members with the dimensions of 100mm×150mm ×1900mm. Concrete was confined by stirrups every 70mm.

The specimen was simply supported at its both ends and was loaded at its center with an actuator, while the axial load was applied using high-pressure oil built up by pressurizing air. The NS component of the 1940 El Centro record was used as input earthquake motions. First, the specimens were pseudo-dynamically tested to get the damaged specimens. Second, some of them were repaired by grouted epoxy resin or covered with steel jackets.

Then, the repaired specimens and the unrepaired damaged specimens were tested again using the same input motions to see effectiveness of seismic retrofit.

The models used in the experiments mentioned above were numerically simulated using the proposed analytical procedure for verification. A model of the specimen was divided into 50 sections and each section was divided into 52 fibers. Repair had been done at the mid-span of the simple beam. Area of epoxy resin in each section was determined from the cross-sectional views of the repaired specimen cut into pieces after the loading tests shown in Fig. 2. In the figure, the shaded area and thin lines are epoxy resin used for repair, and the 4 circles are the reinforcing bars. Epoxy resin exists quite randomly in each section, which shows difficulty in modeling this section exactly using conventional fiber modeling technique or the finite element method.

4.2 Comparison between experimental and analytical results

First, force-deformation relation was compared between experiment and analysis shown in Fig. 3. The experimental result for the original member had been derived from the static loading test while the results for the retrofitted members had been estimated from the skeleton curves of the hybrid loading tests. The relation for the original specimen, the solid lines in Fig. 3, shows good agreement between experiment and analysis except for the small range of deformation. As the tensile strength of concrete was neglected as shown in Fig. 1, the yield point of the reinforcing bars was underestimated. For the retrofitted specimens (the broken lines and the dotted lines in Fig. 3), the analysis underestimated the strength compared with the experiments. Considering that the skeleton curves of the dynamic loading tests usually show higher strength, the analytically obtained force-deformation relation could give reasonable estimation.

Fig. 4 shows one of the results comparing the experimental and analytical hysteretic responses before and after repair work with epoxy resin. Analytically obtained hysteretic load-deformation responses successfully simulated the experimental results. Repaired specimens became stiffer than originals because the grouted epoxy resin has higher tensile strength than concrete. The evaluated maximum responses showed similar values as the experiments.

Fig. 5 compares analytically obtained hysteretic energy with experimentally obtained hysteretic energy and Fig. 6 compares their corresponding damage indices. The dashed line means the same values were obtained from both the experiments and the analysis. All dots scattered around this line, which shows the estimated hysteretic energy and damage index were in good agreement with the experimental results.

5 SEISMIC VULNERABILITY OF BRIDGE PIER WITH TERMINATED REINFORCEMENT

Some old bridge piers constructed under old codes have insufficient anchorage length and termination of their main reinforcement (Kawashima, Unjoh and Iida 1990). These structures might suffer a brittle shear failure at mid-height where the reinforcement terminated and need strengthening for future earthquakes. Use of steel jackets has been found applicable from experiments to prevent brittle failure. The proposed analytical method was then applied to evaluate effect of strengthening for the bridge piers with reinforcements terminated at mid-height.

5.1 Model for simulations

T-shape bridge pier of 13m high shown in Fig. 7 was modeled to have the terminated main reinforcement. The cross section has rectangle shape of 3.5m×3.0m. This pier was designed basically on the specification of highway bridge piers and 30% of the main reinforcing bars were intentionally terminated at 3.0m high to model an old type pier. Then, the pier was divided into 50 sections and the moment-curvature relationship of each section shown in Fig. 8(a) was calculated using more than 120 fibers. The ultimate state was defined as the crush of the core concrete. This figure shows that crushing may occur at 3.0m high where the reinforcement terminated.

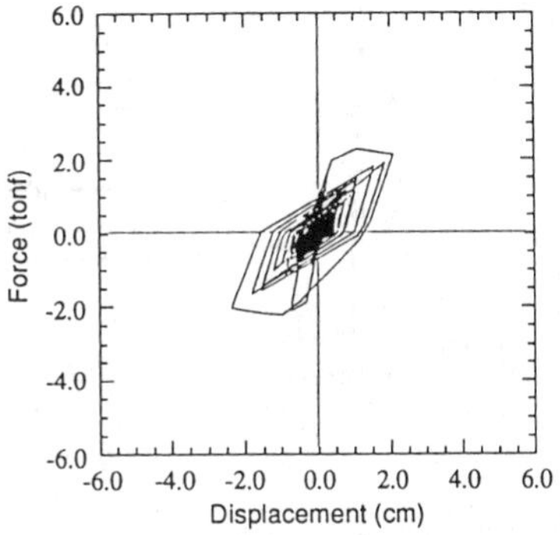

(i) Analytical result for original specimen.

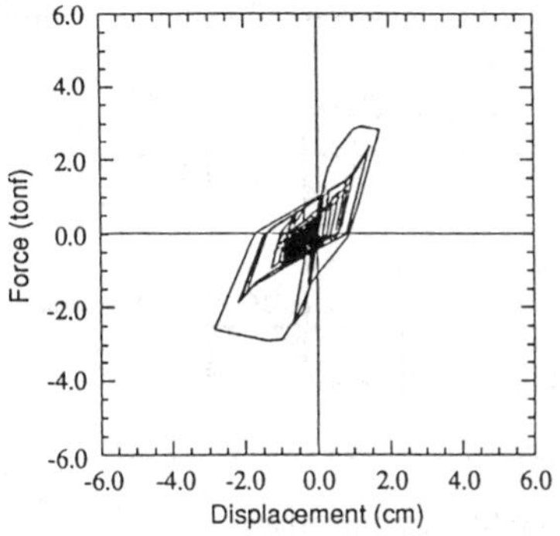

(ii) Analytical result for repaired specimen.

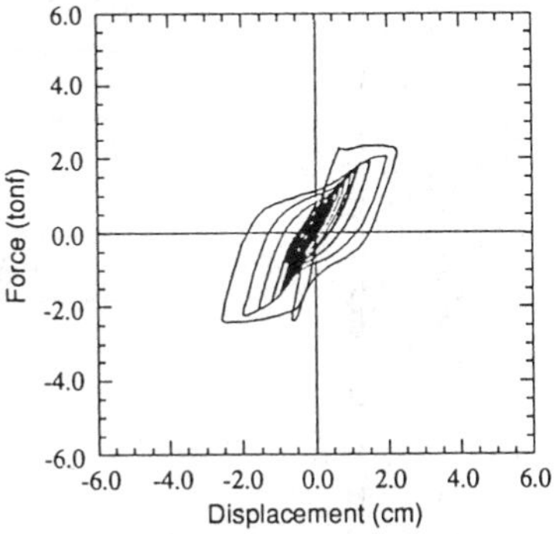

(iii) Experimental result for original specimen.

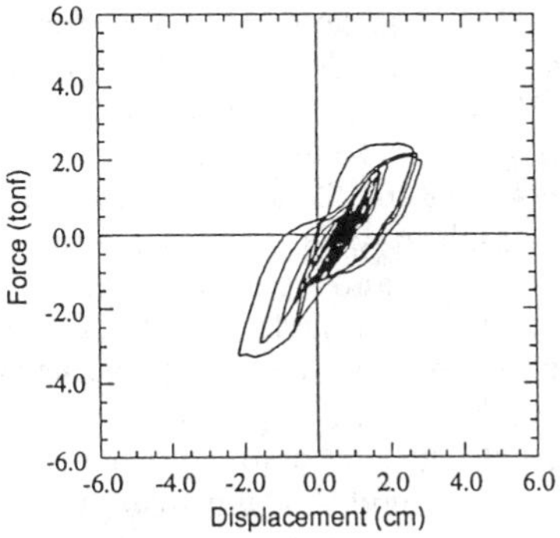

(iv) Experimental result for repaired specimen.

Fig. 4 Comparison of hysteretic responses between analysis and experiment for original and repaired specimens

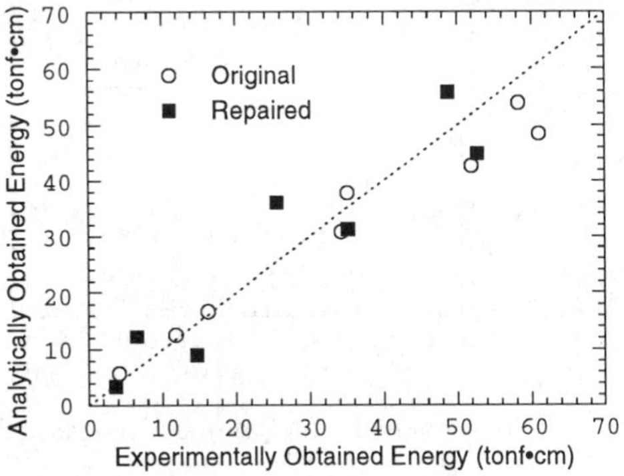

Fig. 5 Comparison between analytical and experimental absorbed hysteretic energy during experiments

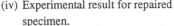

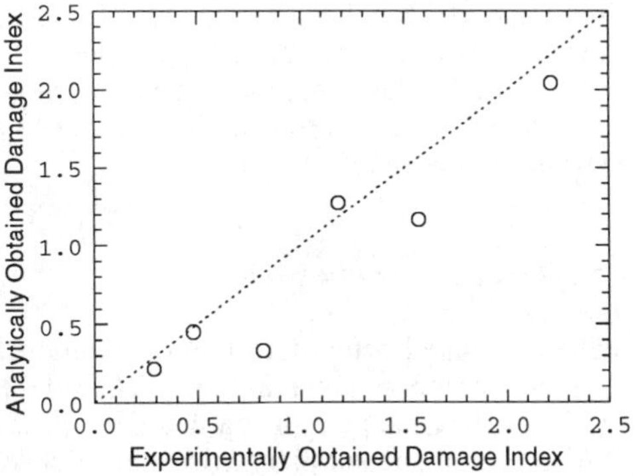

Fig. 6 Comparison of damage index obtained by analysis and experiments

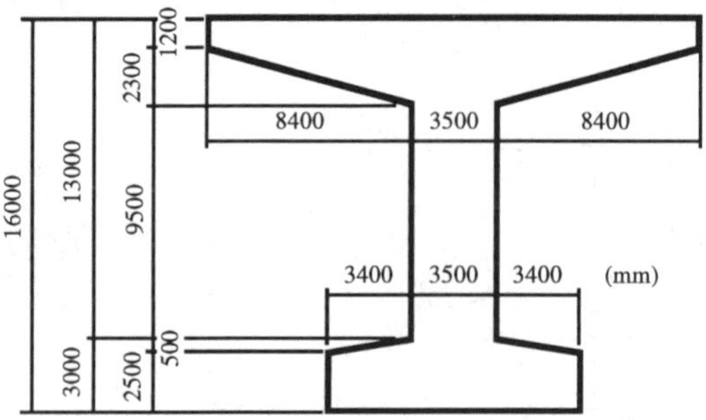

Fig. 7 T-shape bridge pier model used for simulations

Four types of steel jackets were considered: (A) 1mm thick plate which covers only at midheight; (B) 2mm thick plate which covers only at midheight; (C) 1mm thick plate which covers from bottom to midheight; and (D) 2mm thick plate which covers from bottom to midheight. For (A) and (B), the pier was covered with the steel jackets only at the midheight where the reinforcing bars had been terminated, i.e., the retrofitted pier has the steel belt at midheight and the no-strengthened pier bottom. The vertical width of the steel jackets for (A) and (B) was set to have the same length as the pier width (=3.5m). Steel jackets were assumed to bonded perfectly to the concrete surface through the seismic response regardless of epoxy resin bond. Figs. 8(b) and 8(c) show the moment-curvature relationships for the strengthened bridge pier with steel jackets for cases (A) and (B). Thicker steel jackets gave smaller curvature at the section where the reinforcement had been terminated. The critical place moved from the mid-height to the bottom, which means the strengthening had been done effectively. Note that the pier bottom of (A) and (B) had not been retrofitted.

5.2 Force-deformation relation

The calculated force-deformation relationship for the pier top is shown in Fig. 9. The solid line is for the original pier, the dashed thin line is for (A) the strengthened pier with 1mm thick steel jackets only at midheight; the dotted thin line is for (B) the strengthened pier with 2mm thick steel jackets only at midheight; and the dashed

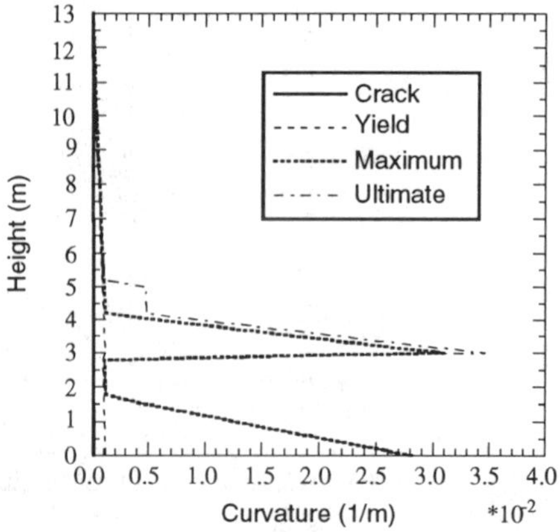

(a) Original

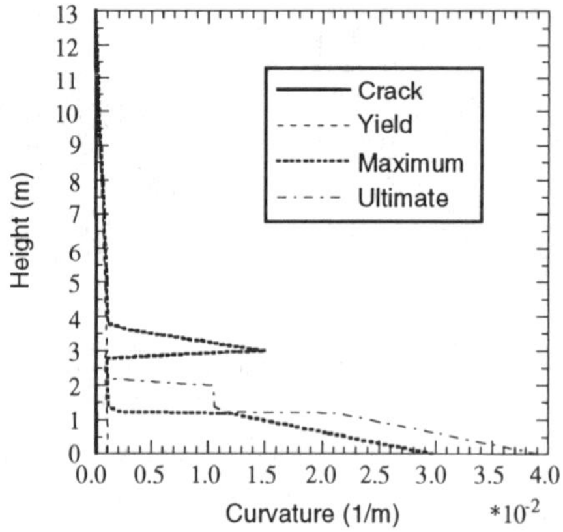

(b) Strengthened with 1mm jackets - case (A)

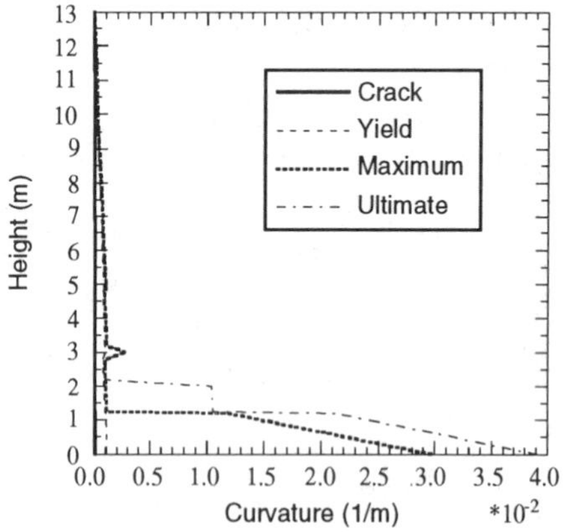

(c) Strengthened with 2mm jackets-case (B)

Fig. 8 Moment-curvature relation at each section of the bridge piers

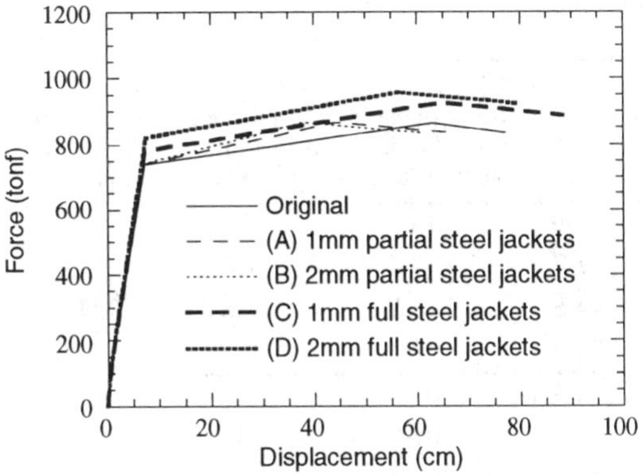

Fig. 9 Estimated force-displacement relation of original and strengthened bridge pier

thick line is for (C) the strengthened pier with 1mm thick steel jackets from bottom to midheight; the dotted thick line is for (D) the strengthened pier with 2mm thick steel jackets from bottom to midheight. The pier bottom where no strengthening had been done became the critical place for the strengthened piers for (A) and (B). Therefore, the ductility decreased because of the stiffer structure with the same ultimate strength as the original. For (C) and (D), both the ductility and the maximum restoring force increased for the retrofitted piers comparing with the original one. The pier with 2mm jackets of case (D) showed larger restoring force compared to the 1mm jackets of case (D), however, it showed smaller ductility.

5.3 Inelastic Earthquake response

The inelastic earthquake response analysis was done using the recommended earthquake record in the Japanese seismic code for the soft ground. Though the original structure may collapse in shear failure, the proposed analytical method cannot determine the failure mode nor the dynamic behavior of each section at each time step because it considers the structure as an single-degree-of-freedom system.

The absorbed hysteretic energy, the damage index of the original and strengthened bridge piers including the other indices are shown in Table 1. Some indices showed similar values for the original and retrofitted structures. All ductility factors remained about 1, and all the ratios of hysteretic energy to the total input energy were 0.6. Though the absorbed energy became smaller for the partially strengthened structures of (A) and (B), they showed larger values for the damage index because of their low ductility. Furthermore, the retrofitted pier with 2mm thick steel jackets showed larger damage index than 1mm thick steel jackets for the same reason.

The fully retrofitted piers of (C) and (D) showed smaller displacement responses and larger restoring forces, which means they became stiffer than the original. Though the retrofitted piers showed smaller damage indices, the retrofitted pier with 2mm thick steel jackets of case (D) showed larger damage index than 1mm thick steel jackets of case (C) because of its lower ductility.

Effectiveness of thinner jackets was verified, of which thickness is thick enough to change the critical place from the midheight to the pier bottom. Strengthening at the pier bottom may also be needed to satisfy ductility requirement as in cases (C) and (D). The retrofitted piers in cases (A) and (B) showed more stable hysteretic responses than the original which means less potential of seismic vulnerability, however, their damage indices became larger compared with the original. The failure mode of the structure could not be evaluated with the damage index.

6 CONCLUSIONS

Analytical methods for inelastic earthquake response and seismic vulnerability analysis of retrofitted RC structures was studied. Main conclusions obtained are as follows:

1. Analytical methods for inelastic earthquake response analysis was proposed using the skeleton curves obtained from the fiber model extended to include repair materials.
2. Analytically obtained hysteretic responses, absorbed hysteretic energy and damage indices were in good agreement with the experimental results. The hysteretic responses of both original and retrofitted RC members can be modeled by the proposed method with high accuracy.
3. Seismic vulnerability of the strengthened bridge pier was simulated, and effectiveness of thinner steel jackets was verified.

Table 1 Response of original and strengthened bridge pier

	Original	(A) Partially Strengthened with 1 mm steel jackets	(B) Partially Strengthened with 2 mm steel jackets	(C) Fully Strengthened with 1 mm steel jackets	(D) Fully Strengthened with 2 mm steel jackets
Ductility Factor; μ	1.12	1.09	1.09	1.03	0.98
Maximum Displacement (cm)	8.1	7.7	7.5	7.6	7.3
Maximum Force (tonf)	743.1	743.3	743.7	781.6	804.0
Maximum Velocity (kine)	60.7	58.2	57.5	59.7	60.3
Damage Index	0.52	0.55	0.60	0.41	0.42
1st term of Damage Index	0.11	0.11	0.12	0.09	0.09
2nd term of Damage Index	0.42	0.44	0.48	0.32	0.33
Hysteretic Energy; W_H (tonf·cm)	47573	43959	43452	44226	42853
Total Input Energy; E (tonf·cm)	79463	72789	71818	73792	71825
W_H/E	0.599	0.604	0.605	0.599	0.597

ACKNOWLEDGEMENTS

The authors thank Mr. Shinji Nakanishi of Kyoto University and Dr. William Tanzo of Saitama University for their aid in preparing this paper. We are also grateful to Mr. Susumu Inoue of Kyoto University and Mr. Tetsuo Hirose of Hanshin Highway Public Corporation who helped us in making suitable bridge pier model.

REFERENCES

Izuno, K., Y. Yamada & H. Iemura 1988. Hybrid experiments on repaired RC members considering axial-force effects, *Proc. of the 9th World Conference on Earthquake Engineering,* Vol. VII: 365-370.

Kawashima, K., S. Unjoh & H. Iida 1990. Seismic inspection and seismic strengthening of reinforced concrete bridge piers with termination of main reinforcement at mid-height, *Proc. of the 1st U.S.-Japan Workshop on Seismic Retrofit of Bridges,*: 251-279.

Kent, D. C. & R. Park 1971. Flexural members with confined concrete. *Journal of the Structural Division, ASCE,* Vol. 97, ST7: 1969-1990.

Park, Y. J. & A. H.-S. Ang 1985. Mechanistic seismic damage model for reinforced concrete. *Journal of Structural Engineering, ASCE,* Vol. 111: 722-739.

Park, Y. J., A. M. Reinhorn & S. K.Kunnath 1988. Seismic damage analysis of reinforced concrete buildings, *Proc. of the 9th World Conference on Earthquake Engineering,* Vol. VII: 211-216.

Structural Safety & Reliability, Schuëller, Shinozuka & Yao (eds) © 1994 Balkema, Rotterdam, ISBN 90 5410 357 4

Uncertainties in design and construction of actual steel structures

Tetsuro Ono
Department of Architecture, Nagoya Institute of Technology, Japan

Hideki Idota
Department of Architecture, Aichi Sangyo University, Okazaki, Japan

Toshiki Ohya
Nippon Steel Corporation, Tokyo, Japan

ABSTRACT : The purpose of this study is to evaluate the uncertainties of actual structures using the measured strain data of a column. This paper presents the statistical data of the axial strain developed on columns of actual steel structures. The uncertainties involved in the design stage are statistically evaluated by analyzing the differences between the measured values and the theoretical values. Introducing the uncertainties in the design stage and construction stage into the design method, a design factor concerned with the errors of construction and structural analysis, is presented.

1 INTRODUCTION

In order to assure the safety of structures, it is necessary to consider various uncertainties that apply to the structure, and to reflect how these uncertainties should effect structural design. Among the uncertainties that affect the safety of actual structures are those that have been rarely handled in the past, such as human error in the design stage, construction error, and load estimation error. These uncertainties involved in the design and construction stage can be understood by means of surveying a strain or deflection of actual constructed structures. Although a considerable number of studies have been made to evaluate uncertainties in design and construction by Melchers, Nowak, Ravindra, Shibata(1985), et al., no data has been colleted on the gaps between the measured strain of actually constructed structures and the calculated strain in the design stage. For the purpose of reliability evaluation for steel structures, the authors(1989) have been measuring the strain of actually constructed structures.

The purpose of this study is to evaluate the uncertainties of actual structures using the measured strain data of a column. This paper presents the statistical data of the axial strain developed in columns of actual steel structures. The uncertainties involved in the design stage are statistically evaluated by analyzing the differences between the measured values and theoretical values. Introducing the uncertainties in the design stage and construction stage into the design method, a design factor concerned with the errors of construction and structural analysis, is presented.

2 AXIAL STRAIN MEASUREMENT RESULTS FOR ACTUAL STRUCTURES

2.1 *Measured buildings*

The column axial strain measurement, using the strain gauges, was carried out for three office buildings shown in Table 1. The method of measurement is described in greater detail in references reported by the authors(1989). Strain gauge placement positions and plans of the measured buildings are shown in Fig. 1. A thermocouple was attached near each strain gauge to measure the temperature during the strain measurement and to compensate for the measured strain data with this temperature. During the construction of buildings, the dead load and live load were measured in detail at each strain measurement time and were used as load data for stress analysis.

2.2 *Results of the measurement*

Fig. 2 shows the axial strain of the three office

Table 1 Outlines of objective structures

	Building A	Building B	Building C
Location	Nakamura-ku, Nagoya	Chuo-ku, Tokyo	Chuo-ku, Tokyo
Structural Type	18 Floors above ground - Steel Structure 2 Floors under ground - SRC Rigid Frame Structure	15 Floors above ground - Steel Structure 2 Floors under ground- SRC Rigid Frame Structure	8 Floors above ground 1 Floor under ground Column of Core - SRC Column of Suspension and Beam - Steel Structure Suspended Structure
Measurement Period	Feb. 18, 1987 ~ Nov. 29, 1988	Dec. 1, 1989 ~ Dec. 17,1990	Dec. 13, 1988 ~ Cep. 12, 1989
Measured Point	6th Floor - 17 Columns 9th Floor - 11 Columns 15th Floor - 11 Columns	4th Floor - 8 Columns 6th Floor - 8 Columns 9th Floor - 8 Columns	6 Columns of SRC Columns 4 Columns of Suspension

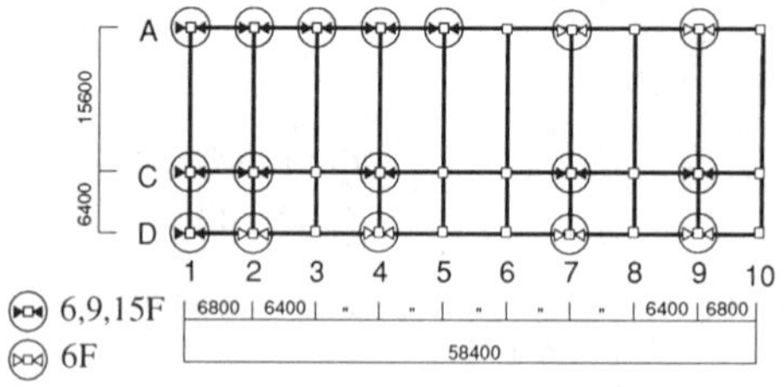

(a) Bld. A

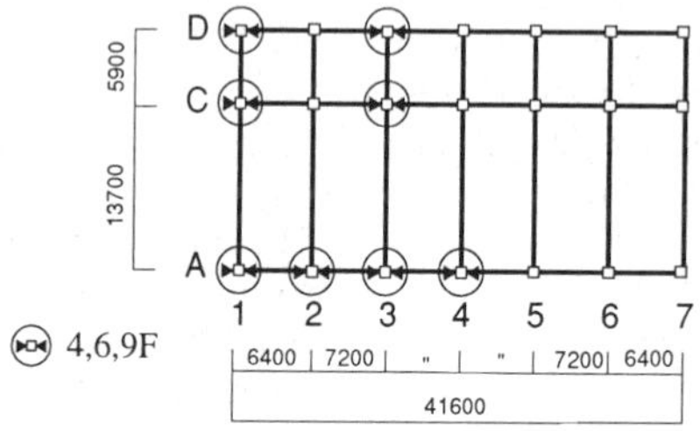

(b) Bld. B

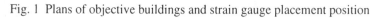

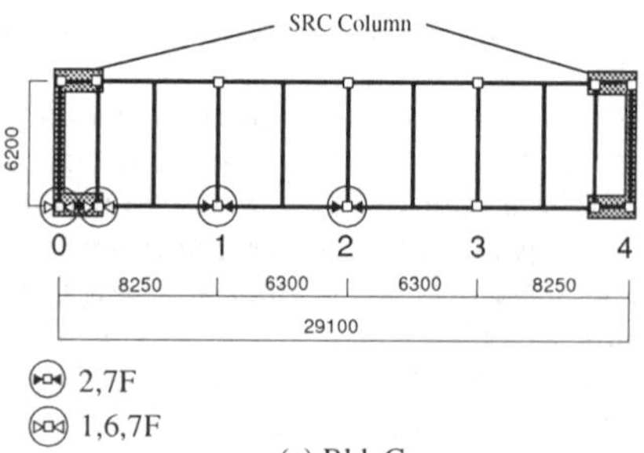

(c) Bld. C

Fig. 1 Plans of objective buildings and strain gauge placement position

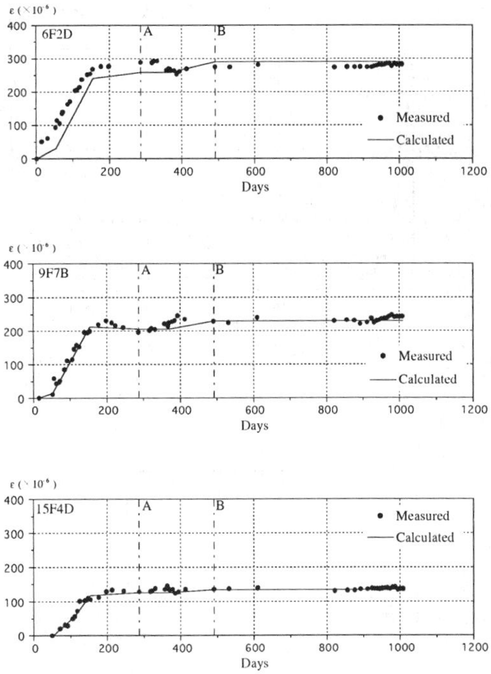

(a) Building A

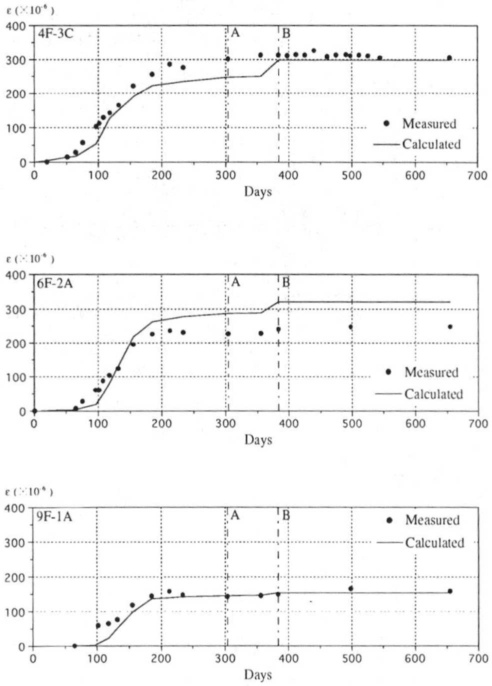

(b) Building B

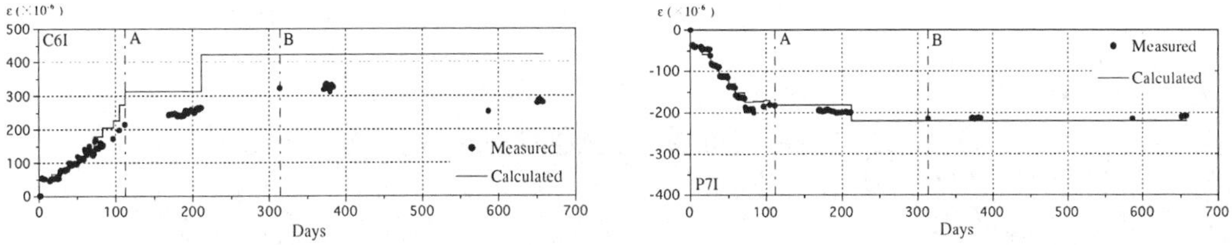

(c) Building C

Fig.2 Measured and calculated axial strain of columns

buildings. The dates of measurement and the strain ($\times 10^{-6}$) are plotted along horizontal and vertical axis, respectively. In Fig. 2, the measured strain value is plotted as dots, and the analytical value is plotted as a solid line. The analytical value of strain in the construction stage is obtained using the dead load and live load recorded during the strain measurement, and the analytical strain after completion is calculated using the design dead load and live load as is 785Pa uniformly loaded over the office area. The strain calculation considers the bending, shearing and axial deformation of members in an elastic range using the stiffness matrices method. Point A and B in Fig. 2 indicate the dates that the construction work is completed and the moving in of the tenants' furniture is finished, respectively. From the beginning of measurement through point A is defined as period I, and after point B is defined as period II. It is observed that the measured strain increases with the progression of the construction work in period I, settles down at point A, increases again due to the

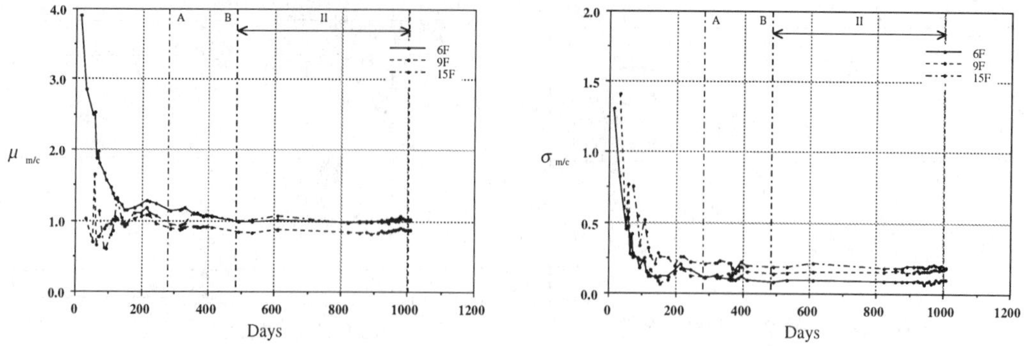

(a) Building A

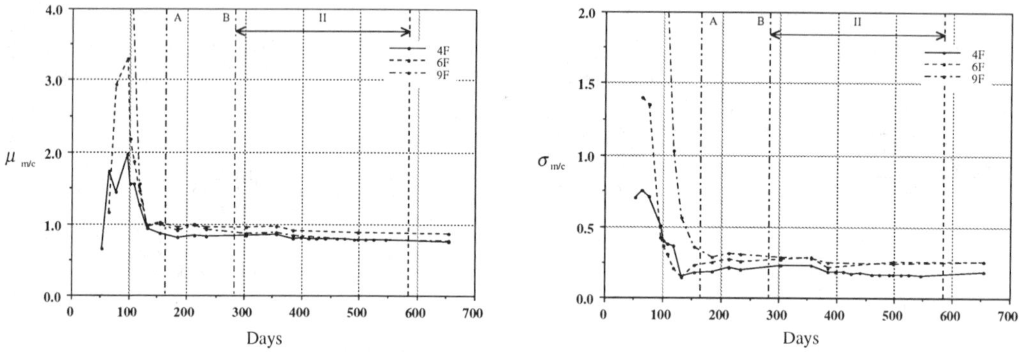

(b) Building B

Fig. 3 mean value and standard deviation of $\varepsilon_m/\varepsilon_c$

live load of installing furniture, and finally becomes a constant in period II. The measured strain values ε_m agree well with the analytical value, so the validity of the measurement method is confirmed.

In order to evaluate the difference between the measured strain and analytical strain in the structural design, it is necessary to process and arrange the difference statistically. It is assumed in this study that uncertainties in the design and construction stage can be treated as a random variable using a statistical method.

Fig. 3 shows the mean value $\mu_{m/c}$ and standard deviation $\sigma_{m/c}$ of the ratio of ε_m to ε_c for buildings A and B. Both $\mu_{m/c}$ and $\sigma_{m/c}$ change considerably during period I, but on the other hand during period II they become constant. $\sigma_{m/c}$ decreases rapidly while the construction work progresses, and it settle down at a certain level in the period II. In the period II, $\sigma_{m/c}$ becomes smaller in the order of the 15th, 9th and 6th floors.

Fig. 4 shows histograms of the $\varepsilon_m/\varepsilon_c$ ratio in each period for buildings A and B. The mean value $\mu_{m/c}$ and the standard deviation $\sigma_{m/c}$ are presented in each histogram. The standard deviation in period I is much larger than that in period II. The standard deviations observed were 0.418 for building A and 0.400 for building B in period I, and 0.196 for building A and 0.175 for building B in period II. They are closer to each other in each period.

In the structural design, the variation of axial column load in the period II is needed to evaluate its safety. Then, the histograms of $\varepsilon_m/\varepsilon_c$, averaged during period II, are shown in Fig. 5. Mean values are 0.975 and 0.839, and standard deviations are 0.175 and 0.210 for building A and building B respectively. These results are close to the results in period II presented in Fig. 4.

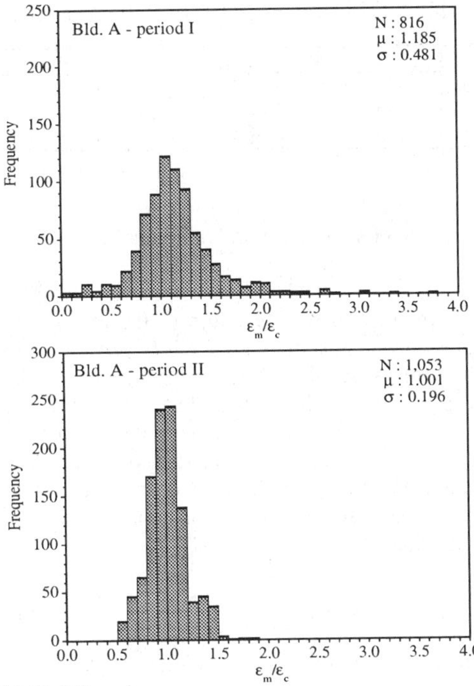

(a) Building A

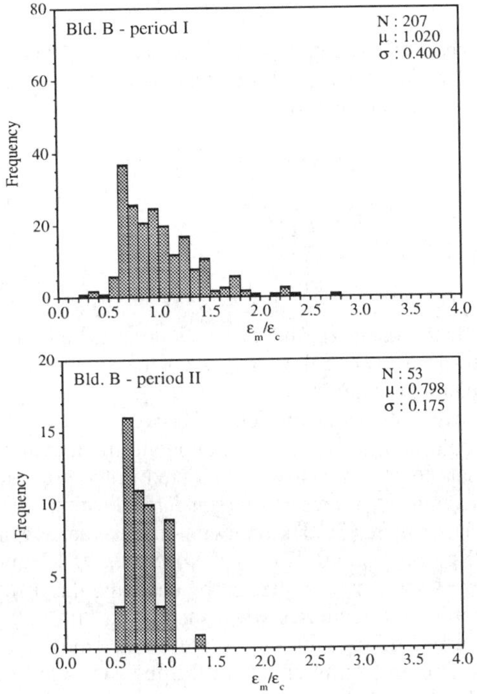

(b) Building B

Fig. 4 Histograms of $\varepsilon_m/\varepsilon_c$

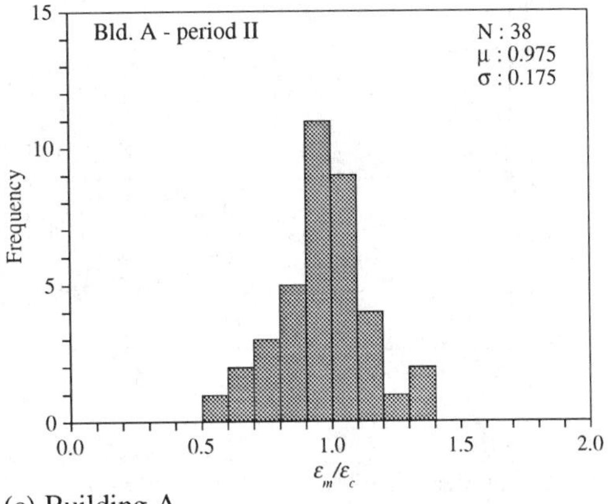

(a) Building A

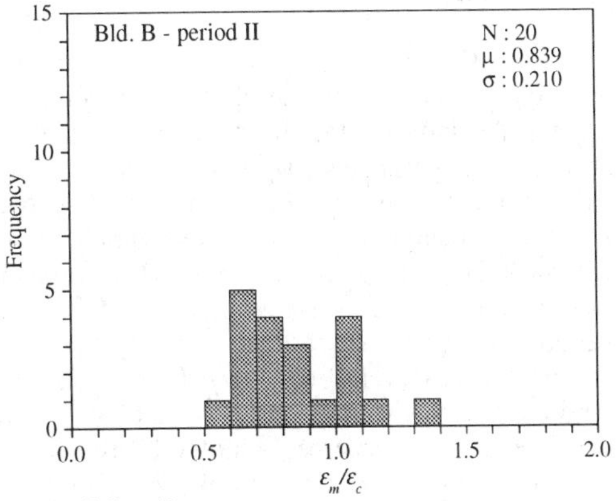

(b) Building B

Fig. 5 Histograms of $\varepsilon_m/\varepsilon_c$ averaged in period II

3 UNCERTAINTIES IN STRUCTURAL DESIGN AND CONSTRUCTION

The uncertainties that affect the reliability of actual structures are included in the member strengths, loads, design errors, construction errors and so on. The statistical data of member strengths and loads has been corrected in recent years. On the other hand, almost no measured data has been collected on uncertainties of design errors and construction errors. It is important to discuss how to treat uncertainties in the design and construction stages. In this paper, however, it is assumed that such uncertainties can be treated as random variables by means of a statistical method. Assuming that all uncertainties existing in structures is represented with a random variable U, the random variable U can be modeled by classifying them into groups of

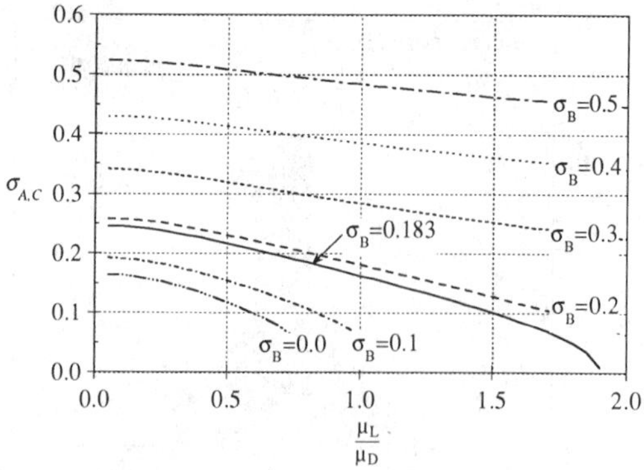

Fig. 6 Standard deviation of uncertainties $U_{A,C}$

uncertain factors.

$$U = \mu_U U_A U_C U_M U_S \tag{1}$$

Where μ_U is the mean value of U, U_A is a random variable that indicates the uncertainty existing in the structural design process, U_C is a random variable that indicates the uncertainty involved in the construction stage, U_M is a random variable that indicates the uncertainty involved in the member strengths and the material properties and U_S is a random variable that indicates the uncertainty involved in loads. In Eq.(1), U_A, U_C, U_M, U_S are random variables with the mean value of 1. μ_U is considered to be concerned with the degree of the structural design and construction works. Using the random variable U, the measured strain of structures ε_e can be related to the calculated strain ε_c based on the analytical model in the structural design stage as follows.

$$\varepsilon_e = \varepsilon_c U \tag{2}$$

Assuming that ε_e is the elastic strain caused by the dead load and live load, $\varepsilon_e/\varepsilon_c$ can be expressed with the following equation.

$$\varepsilon_e/\varepsilon_c = \mu_U U_A U_C U_{SEC} U_E (U_D + U_L) \tag{3}$$

U_{SEC}, U_E, U_D and U_L are random variables representing the randomness of the sectional area of column, young's modules, dead load and live load respectively. The mean values of U_{SEC} and U_E are 1, and μ_D and μ_L for U_D and U_L respectively (however $\mu_D+\mu_L=1$). $\varepsilon_e/\varepsilon_c$ represents the difference between the designed structures and actual structures, and corresponds to the measured uncertainties that is included in a design and construction. On the other hand, $\varepsilon_e/\varepsilon_c$ can be expressed by dividing it into two random variables; U_I representing the variation of strain in a structure, and U_B representing the variation of strain for plural buildings.

Table 2 Statistical properties of random variables

	COV	Distribution Type	3rd Moment	4th Moment
R	0.15	Lognormal	0.45	3.37
S_D	0.10	Normal	0.00	3.00
S_L	0.40	Lognormal	1.26	5.97

$$\varepsilon_e/\varepsilon_c = 1 + \mu_{U2} + U_I + U_B \tag{4}$$

Where μ_{U2} is the deviation of the mean value caused by uncertainties, and both U_I and U_B are the random variables having 0 mean values respectively. The normalized strain data $\varepsilon_e/\varepsilon_c$ obtained in this measurement represents the randomness of strain in a building, so that $\varepsilon_e/\varepsilon_c$ is equivalent to $1+U_I$ in Eq.(4). Consequently, $U_A U_C$ can be expressed by Eq.(5) substituting $1+U_I=\varepsilon_e/\varepsilon_c$ and Eq.(3) into Eq.(4).

$$U_{A,C} = U_A U_C = \frac{\mu_{U2} + \varepsilon_m/\varepsilon_c + U_B}{\mu_U U_{SEC} U_E (U_D + U_L)} \tag{5}$$

Then, $\sigma_{A,C}$, the standard deviation of $U_{A,C}$ is;

$$\sigma_{A,C}^2 = \frac{\sigma_{m/c}^2 + \sigma_B^2 - \sigma_{MAT}^2 - \sigma_S^2 - \sigma_{MAT}^2 \sigma_S^2}{1 + \sigma_{MAT}^2 + \sigma_S^2 + \sigma_{MAT}^2 \sigma_S^2} \tag{6}$$

where $\sigma_{m/c}$ and σ_B represents the standard deviation of $\varepsilon_m/\varepsilon_c$ and U_B respectively, σ_{MAT} and σ_S are expressed by the following equations.

$$\sigma_{MAT}^2 = \sigma_{SEC}^2 + \sigma_E^2 + \sigma_{SEC}^2 \sigma_E^2 \tag{7}$$

$$\sigma_S^2 = \frac{\sigma_D^2 + \sigma_L^2 \left(\frac{\mu_U}{\mu_D}\right)}{\left(1 + \frac{\mu_U}{\mu_D}\right)^2} \tag{8}$$

In this paper, μ_U and μ_{U2} are assumed to be 1 and 0 respectively because of the lack of measured data for plural structures.

Fig. 6 shows the relationship between $\sigma_{A,C}$ and the load ratio μ_L/μ_D in Eq.(6) applying the measured data to $\sigma_{A,C}$ which was presented in the previous section. σ_B represents the standard deviation of U_B shown in Eq.(6). The following statistics are used in Fig. 6; $\sigma_{SEC}=0.021$, $\sigma_E=0.048$, $\sigma_D=0.10$ and $\sigma_L=0.40$. $\sigma_{A,C}$ is affected by changing the μ_L/μ_D ratio, and decreases when increasing τηε μ_L/μ_D ratio. If measurements are carried out for a sufficient number of buildings, σ_B can be determined. Applying $\sigma_B=0.183$ obtained with three buildings presented in this paper as a matter of convenience, the standard deviation $\sigma_{A,C}$ is shown with the sold line in Fig. 6. When similar measurements are continuously carried out on more buildings, σ_B can be evaluate more clearly.

4 DESIGN FACTORS CONSIDERING UNCERTAINTIES OF DESIGN AND CONSTRUCTION

To cope with the uncertainties involved in the design and construction stage, the design method requires a new design factor which can evaluate the design and construction error. In this section, the design factor to secure safety against design and construction errors is presented and calculated using the measured data.

The uncertainties $U_{A,C}$ presented in the previous section can be introduced into the reliability evaluation equation as follows.

$$R \geq U_{A,C}S \tag{9}$$

Structures are designed to satisfy ;

$$P[R < U_{A,C}S] < p_{fd} \tag{10}$$

where p_{fd} is allowable probability of failure, and $P[E]$ represents occurrence probability of event E. On the other hand, the design formula can be defined, introducing the design factor γ_0 concerned to $U_{A,C}$, as follows:

$$\phi R_n \geq \gamma_0 \sum_{i=1}^{n} \gamma_i S_{ni} \tag{11}$$

where R_n and S_n are the nominal values of the resistance and i-th load respectively, ϕ and γ_i are the resistance factor and load factor for the i-th load respectively. γ_0 can be obtained with solving the equation $P[R^*-S^*<0]=p_{fd}$ where R^* and S^* satisfy Eq.(11). That is to say, the following equation gives γ_0.

$$p_{fd} = P[R - U_{A,C}S < 0] < P[R - \gamma_0 S < 0] \tag{12}$$

Since $U_{A,C}$ obtained in this paper represents the uncertainties concerned with the elastic strain under a dead load and live load, it can be considered that p_{fd} is the allowable probability of failure concerned with the S_D and S_L. Then S_D+S_L is substituted into S in Eq.(12). Fig. 7 shows γ_0 calculated by Eq.(12) arranging the μ_L/μ_D ratio when the reliability index β is established at 1.0, 2.0 and 3.0. Reliability index β corresponds to the allowable probability of failure through the equation $\beta=\Phi^{-1}(1-p_{fd})$, where $\Phi(\bullet)$ is the probability distribution function of the standard normal variable. The statistical properties of R, S_D and S_L are assumed as shown in Table 2. The

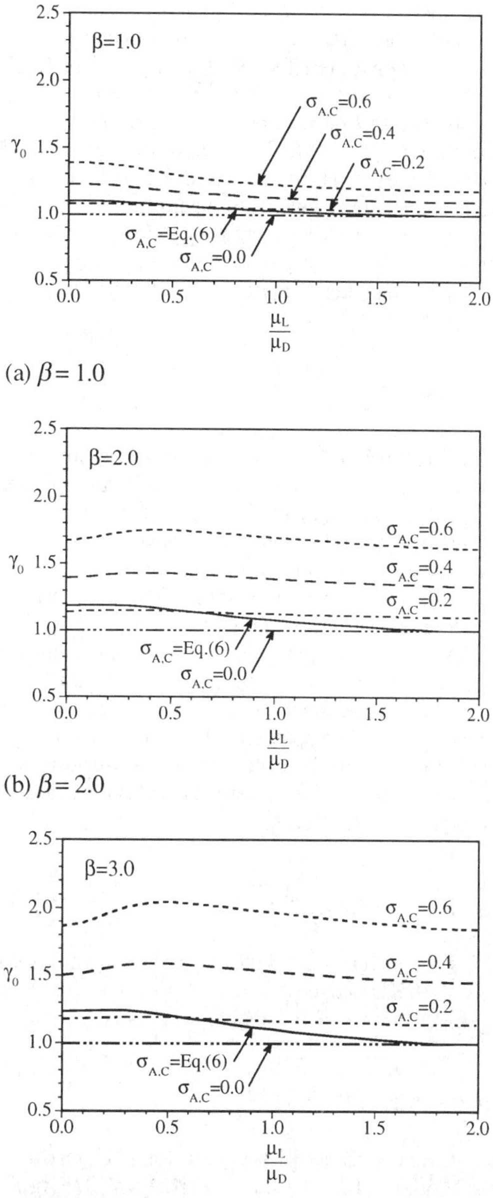

Fig. 7 Design Factor

higher-order moment standardization technique by the authors(1989) is used to evaluate $P[R-U_{A,C}S<0]$ and $P[R-\gamma_0 S<0]$ when considering the 3rd and 4th moments of R, S_D and S_L respectively. When β =1.0, γ_0 decreases continuously with μ_L/μ_D. When β=2.0 and 3.0, γ_0 takes maximum value in the neighborhood of μ_L/μ_D=0.5, but is not sensitive to changing of μ_L/μ_D. The higher the value of β is, the more sensitive the effect of $\sigma_{A,C}$ on γ_0. The thick solid line in Fig.7 represents the design factor γ_0 applied from Eq.(6) through Eq.(8) to the relation between $\sigma_{A,C}$ and μ_L/μ_D. σ_B in Eq.(6) is determined at 0.183 based on the measured data. Since $\sigma_{A,C}$ approaches zero when increasing μ_L/μ_D, γ_0 also approaches zero when increasing μ_L/μ_D. Assuming that the design factor γ_0 is defined with the maximum value in changing of μ_L/μ_D, γ_0 is 1.1 approximately when β=1.0.

5 CONCLUSIONS

The relationships between the uncertainties involved in actual steel structures and measured strain data are formulated, and a stochastic model for measured data is presented. The statistical data of uncertainties involved in design and construction stage is presented using column axial strain data measured for three actual office buildings.

Furthermore, the design factor, which can consider the uncertainties of design and construction, is calculated using the measured data. When similar measurements are continuously carried out on more buildings, it will become feasible to quantify the uncertainties and evaluate the safety of actual structures more clearly.

ACKNOWLEDGMENT

This research was supported in part by a grant from KOZAI KURABU. We wish to thank for their generous financial assistance.

REFERENCES

Melchers R. E. and M. G. Stewart, "Data-Based Models for Human Error in Design", *ICOSSAR'85, Structural Safety and Reliability*, Vol.II, pp.51-60, 1985.

Nowak A.S. and Robert I. Carr,"Errors in Structural Models", ICOSSAR'85, Structural Safety and Reliability, Vol.II, pp.41-50, 1985.

Ono T., H. Idota, Y. Nishina, T. Ohoka and T. Miyazaki,"Design and Observation of New Structural System for Office Building, *EASEC, Structural Engineering and Construction*, Vol.I, pp.817-823, 1989.

Ono T., H.Idota, T. Ohya and T. Takeuchi, "Study on Reliability of Actual Steel Structures Using Measured Data", *ICOSSAR'89, Structural Safety and Reliability*, Vol.II, pp.2087-2090, 1989.

Ono T., H.Idota and A. Totsuka, "System Reliability Using Higher-Order Moments", *ICOSSAR'89, Structural Safety and Reliability*, Vol.II, pp.959-966, 1989.

Ravindra M.K.,"Treatment of Gross Errors in Nuclear Plant Risk Studies -- A Suggestion Approach", *ICOSSAR'85, Structural Safety and Reliability*, Vol.II, pp.21-30, 1985.

Shibata H, "Role of Human Error to the Safety of Critical Structures -- Emphasized on Earthquake Failure", *ICOSSAR'85, Structural Safety and Reliability*, Vol.II, pp.11-20, 1985.

Structural Safety & Reliability, Schuëller, Shinozuka & Yao (eds) © 1994 Balkema, Rotterdam, ISBN 90 5410 357 4

A human reliability analysis of reinforced concrete slab construction

Mark G. Stewart
Department of Civil Engineering and Surveying, The University of Newcastle, Callaghan, N.S.W., Australia

ABSTRACT: A Human Reliability Analysis (HRA) has been used to simulate the effect of human error on the construction of a reinforced concrete slab in order to obtain "realistic" probabilities of structural failure. Existing code calibration statistical distributions were used to describe the effective depth of steel reinforcing and concrete compressive strength. Expert opinions were used to develop human performance models for the size of reinforcing mesh installed. It was found that "nominal" and "realistic" estimates of structural reliability were similar when the expected slab size is small (i.e., slab thickness generally less than 150 mm). However, there was a significant difference between these estimates for larger slab sizes since the likelihood of undersizing is increased.

1. INTRODUCTION

Human error is the main cause of structural failure (e.g, Brown and Yin, 1988), where human error may be defined as a human action that exceeds some limit of acceptability. For the present paper, the "limit of acceptability" may be defined as construction tolerances specified by appropriate codes of practice. It is generally accepted that discrepancies (by at least several orders of magnitude) between observed and calculated "nominal" failure rates are due mainly to human error (Allen, 1968). It has been shown (Eldukair and Ayyub, 1991) also that reinforced concrete slab construction is one of the most error prone stages of the building process, and that approximately 77% of all failures are attributed to contractor error. Brown and Yin (1988) suggest that contractor errors are dominated by ignorance, thoughtlessness, and negligence. Such an observation is not unreasonable; construction workers are generally unskilled or semi-skilled, work in unpleasant conditions, and often have poor motivation.

Consequently, engineers place a very high importance on works inspection as a measure to control construction errors (Ingles and Nawar, 1983). Accordingly, it is a regulatory requirement in some countries that construction work be inspected by a professional engineer (e.g., BCA, 1990).

The Human Reliability Analysis (HRA) approach is widely used for modelling the effect of human error in risk analysis (Swain and Guttman, 1983). This method utilises event–tree logic where the event–tree is analysed using Monte–Carlo simulation techniques. HRA models complex systems by subdividing the system into successive individual sub–tasks (or microtasks). Each microtask models a human action needed in the sequence of producing a final outcome. The principal objective of HRA is to (i) obtain "realistic" estimates of risk, and (ii) compare the effectiveness of various error control measures. HRA's have also been conducted for structural steel design, and reinforced concrete beam design and construction (Stewart and Melchers, 1989a, 1989b; Stewart, 1991, 1992a, 1993).

In the present paper, a HRA has been conducted to simulate the effect of human error in the construction of a typical cast–in–situ reinforced concrete floor slab. Microtasks incorporated in the HRA include (i) size of steel reinforcing mesh installed, (ii) effective depth of installed steel reinforcing (i.e., depth to bottom reinforcement), and (iii) compressive strength of placed concrete. These are all tasks that directly influence flexural strength. It is assumed that errors are committed by construction contractors (i.e., site staff and workmen). It is assumed also that existing code calibration statistical distributions for effective depth and concrete compressive strength include the effect of gross error; hence these distributions are used directly in the HRA. However, expert opinions have been obtained for the estimates of error rates, error magnitudes and checking efficiencies for the installation of an incorrect size of reinforcing mesh. These assessments have been used to obtain models of human performance (i.e., the distribution of error rates and error magnitudes) at the time of concrete placement; these models are described herein. The HRA provides "realistic" estimates of structural reliability. Estimates of structural reliability can also be computed from existing design code calibration methods (i.e., "nominal" failure rates). A comparison can then be made between "nominal" and "realistic" failure rates.

2. SLAB CONSTRUCTION TASK

The design and construction of a reinforced concrete simply supported one–way floor slab requires that the design dead and live loads, slab span, cover (say 20 mm, thus total slab thickness is $D = d + 30$), minimum concrete compressive strength (F'_{cn}), and reinforcing yield stress (f_{sy}) are known. It is assumed herein that the design outcome is obtained from a realistic "error–free" design conducted in accordance to the serviceability and strength provisions of the Australian Standard Concrete Structures Code (SAA, 1988). The design outcomes (i.e., construction specifications) are the effective depth to steel reinforcing (d_{nom}) and the amount of reinforcing steel ($A_{st-spec}$= size of rectangular manufactured mesh). Therefore, the construction task consists of implementing the task in accordance to the construction specifications; namely, (i) location of reinforcement (d), (ii) cross–sectional area of reinforcing mesh (A_{st}) and (iii) concrete compressive strength (F'_c).

3. DESCRIPTION OF HRA METHOD

3.1 Method

The HRA procedure for the slab construction task is as follows:

Step 1. A variable (i.e., construction outcome) is generated for d, F'_c and A_{st}.

Step 2. Values for reinforcing yield stress (f_{sy}), concrete flexural tensile strength (F'_{cf}), modelling error (ME), and dead and live load effects ($\overline{D}, \overline{L}_P$) are randomly generated from appropriate statistical distributions, see Table 1.

Step 3. An estimate of structural resistance (R) is computed (Note that $R \geq F'_c BD^2/6$).

Step 4. Check if the limit state function $G(R_U, S) = (ME \times R) - \overline{D} - \overline{L}_P \leq 0$.

The progression through Steps 1 – 4 is termed a "construction cycle" or "run". By replicating Steps 1 – 4 n_S times (i.e., n_S simulation runs) the mean probability of failure (over a 50 year period) can then be computed as:

$$p_f \approx \frac{n(G \leq 0)}{n_S} \tag{1}$$

where $n(G \leq 0)$ is the number of trials for which $G(R_U, S) \leq 0$.

The modelling of construction outcomes for Step 1. is based on (i) a "nominal" construction scenario utilising existing code calibration statistical distributions for d and F'_c or (ii) a "realistic" construction scenario incorporating existing code calibration statistical distributions for d and F'_c, and a microtask human performance model for A_{st}.

3.2 "Nominal" Construction Scenario

The "nominal" construction scenario assumes that the dimensional and material variables are randomly generated directly from the probability distributions using generally accepted code calibration statistical parameters (see Table 1 for parameters). The code calibration statistical distribution for d and F'_c were obtained from measurements made from 2805 slabs and 231 jobs respectively. These are significant sample sizes, and it is therefore likely that these samples would include within them the influence of some gross errors. However, code calibration statistical parameters for A_{st} assumes no variance due to human error; hence, this error type is currently ignored from the present method for calculating failure rates.

Table 1.Typical Code Calibration Statistical Parameters

Variable	Mean ($\bar{x}$)	C.O.V. (V)	Source
d	1.00	0.68/D	Ellingwood (1980)
A_{st}	1.00	0.00	–
F'_c	1.00	0.18	Pham (1985a)
F'_{cf}	$0.69\sqrt{F_c}$	0.22	Mirza, et al (1979)
f_{sy}	1.00	0.11	Ellingwood (1980)
ME	1.11	0.07	Ellingwood (1980)
$\bar{D}$	1.05	0.10	Pham (1985b)
$\bar{L}_p$	0.74	0.25	Pham (1985b)

3.3 "Realistic" Construction Scenario

For this scenario it is necessary to (i) develop a microtask human performance model for installation of incorrect size of reinforcing mesh (A_{st}) and (ii) develop an event-tree for this microtask for incorporation into the HRA.

3.3.1 Microtask Human Performance Models

For the present microtask, an error is defined as a departure from the construction specifications. An average error rate is defined as the number of errors encountered divided by the total number of inspections for that error type. If an error occurs, the error magnitude (or error consequence) is the size of the error as a percentage of the correct (or specified) outcome. Thus, the error magnitude (m_e) is defined as:

$$m_e = \frac{x - x_m}{x_m} \times 100\% \qquad (2)$$

where x and x_m are the incorrect (or error outcome) and correct values respectively.

A mailed survey of practising structural engineers with construction experience was conducted in order to obtain subjective estimates of the average error rate, and the average and maximum error magnitudes for each specified construction error type for A_{st}. These estimates were requested for the time of the initial engineering inspection (i.e., an assessment of the work done by the contractor before the inspection). Survey recipients were also requested to directly estimate the proportion of errors they believe are corrected at the time of concrete placement (i.e. a checking efficiency). Thus, an estimate of the proportion of errors included in the finished work could then be calculated.

The error types considered in the present are:

E1: Reduced area of tensile steel (A_{st}):
Longitudinal bars (of manufactured rectangular mesh) spanning in the wrong direction.

E2: Reduced area of tensile steel (A_{st}):
All other errors.

E3: Increased area of tensile steel (A_{st}).

Error Type E1 was included because it is not uncommon for reinforcing bars (or mesh) to be placed such that they are orientated 90 degrees to the direction specified (Ford, 1984).

A total of 21 individuals were classified as "experts" (i.e., those with more than five years professional experience). The expert judgements were statistically aggregated to obtain single point estimates of the parameters of interest; it was then shown that there was interjudge consistency or agreement of assessments (Stewart, 1992b).

Error Rates:

Error rates for individuals (for a specific task) will vary from individual to individual. This variation in error rates is represented by the lognormal distribution. The median of the lognormal distribution of error rates for errors that are included in the finished work ($\tilde{m}_i$) is calculated as:

$$\tilde{m}_i = \tilde{m}_o \times (1 - \tilde{p}_i) \qquad (3)$$

where $\tilde{m}_o$ and $\tilde{p}_i$ are the average error rate prior to engineering inspections and the checking efficiency respectively (see Table 2). The variance of the lognormal distribution is not available from this survey data. A convenient measure of variance is the "error factor" (EF) which is expressed as

$$EF = \sqrt{\frac{Pr(F_{90^{th}})}{Pr(F_{10^{th}})}} \qquad (4)$$

where $Pr(F_{10^{th}})$ and $Pr(F_{90^{th}})$ are the error rates corresponding to the 10th and 90th percentiles respectively of the distribution of error rates [Swain and Guttman, 1983]. Error factors (EF_i) applied to the tasks considered in the present study (see Table 2) are consistent with present guidelines (Swain and Guttman, 1983). Distributions of error rates (for errors in the finished work) are shown in Figure 1.

Table 2. Statistical Parameters for Distributions of Error Rates

ERROR TYPE	Initial Errors $\tilde{m}_o$	Checking Efficiency $\tilde{p}_i$	Errors in Finished Work $\tilde{m}_i$	EF_i
E1	0.0077	0.9985	0.00001	10.0
E2	0.0154	0.9905	0.00015	10.0
E3	0.0121	0.3558	0.00780	3.0

With reference to Table 2 it is observed that the checking efficiencies are highest for errors that are most detrimental to structural safety. The high checking efficiencies tend to suggest that the current engineering inspection processes may be adequate. It is also observed that the median error rates for errors that are included in the finished work are relatively low. For example, the error rate for an individual (after inspections) varies from 10^{-3} to 10^{-9}, for error type E1 (see Figure 1). However, it is the high error rates (albeit with a low probability of occurrence) that may significantly influence structural reliability.

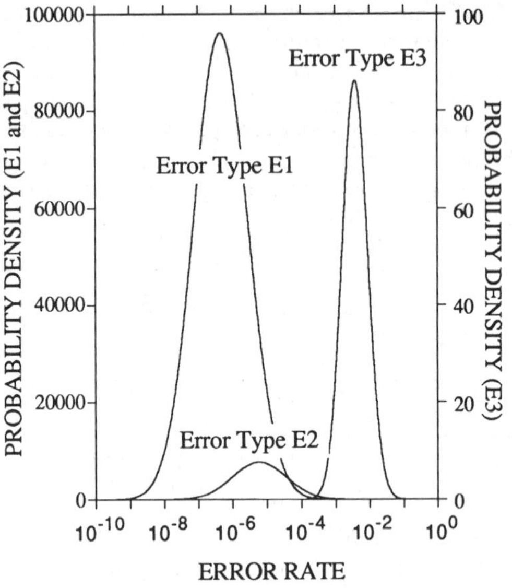

Figure 1. Distribution of Error Rates at Time of Concrete Placement

Error Magnitudes:

The distribution of error magnitudes was also modelled by the lognormal distribution, where the average estimate (λ_{BE}) and the maximum estimate (λ_{UB}) are the median and 90th percentile upper bound respectively, λ_{BE} and λ_{UB} are given in Table 3. This is a censored distribution at m_e=–100% (i.e., an error of omission) for error type E2. This is an important consideration because approximately 7% of survey respondents indicated that the "worst" error magnitude that they had witnessed was the omission of tensile steel reinforcing. A distribution of error magnitudes for error type E1 is not required since the error outcome of this error type is a known and deterministic value (i.e., A_{st} is area of cross wires).

Table 3. Statistical Parameters for Distributions of Error Magnitudes

ERROR TYPE	λ_{BE}	λ_{UB}
E1	–	–
E2	-14.6	-52.9
E3	11.3	39.8

The error outcome can not be computed directly from the error magnitude (i.e., by reference to Equation 2) because A_{st} is not a continuous variable. Therefore, it is necessary to truncate the error outcomes to the smallest and largest manufactured rectangular meshes; namely, F62 (A_{st}=156 mm^2/m) and F1218 (A_{st}=1227 mm^2/m) respectively. The determination of the incorrectly installed amount of A_{st} is limited to ten discretised values (i.e., cross–sectional area for each the ten manufactured rectangular meshes between F62 to F1218). Therefore, the error outcome (i.e., A_{st}) is based on the mesh size that has an error magnitude closest to the error magnitude generated from the appropriate distribution of error magnitudes. A typical relationship between m_e and A_{st} is shown in Figure 2, for an "error–free" mesh F818.

It may be likely that the efficiency of design checking is dependent upon the size of the error. However, in the absence of any appropriate quantitative information, it is assumed simply that the distribution of error magnitude will remain unchanged after engineering inspections.

It is of some interest to note that expert opinions were also obtained (in the same survey) for d and F'_c. It was found by Stewart (1992b) that the error rates obtained from expert opinions were significantly lower (by several orders of magnitude) than that observed from direct field measurements (i.e., existing code calibration data). In this study an error was defined as an exceedence of dimensional construction tolerances as specified by codes of practice (SAA, 1988). This suggests that it may be difficult to visually observe the exceedence of construction tolerances because the construction tolerances are generally quite low (5 – 10 mm). It is also unlikely that inspecting engineers would directly measure (e.g., by a tape measure) the effective depth to steel reinforcing. Furthermore, it is likely that laboratory results of concrete compressive strengths are known only to the contractor (who is generally the client); hence the engineer is not always in a position to make an informed assessment of concrete compressive strengths. Therefore, it is reasonable to conclude that expert opinions tend to underestimate error rates for dimensional and material measurements. This suggests that error rates obtained for reinforced concrete beam construction (Stewart, 1992a, 1993) may be somewhat optimistic (i.e., too low).

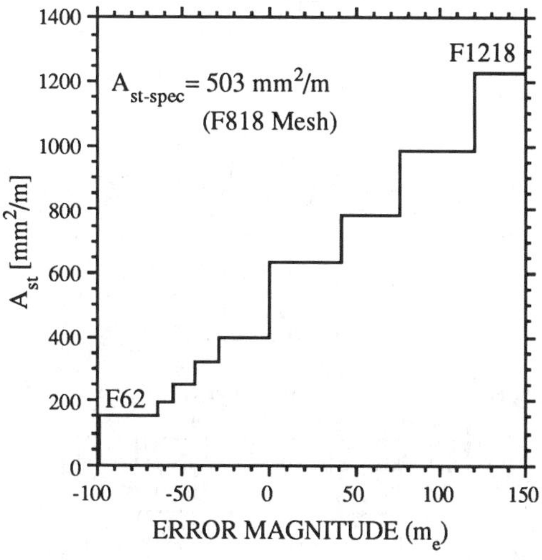

Figure 2.The Influence of Error Magnitude on Error Outcome (A_{st})

3.3.2 Event–Tree Method

For each construction cycle, a value of A_{st} was obtained from an event–tree developed for this microtask (see Figure 3). The procedure is as follows: a uniform random percentile [0 – 100] is initially generated, and is then used to obtain an error rate from the appropriate distribution of error rates (ER1). The same percentile is then applied to the distribution of error rates for other error types (ER2, ER3). Then another uniform random variable [0, 1] is generated (RN) and compared with the given microtask error rates. This enables a binary decision (either "error type E1", "error type E2", "error type E3" or "error–free") to be made. If necessary, the error magnitude is randomly generated from the appropriate distribution of error magnitude (see Table 3), and hence the value of A_{st} can be obtained. The distributions of error rates and error magnitudes are as described by the microtask human performance models (see Section 3.3.1). The generation of variables for d and F'_c are as described in Section 3.2.

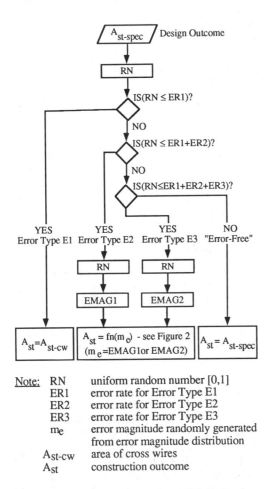

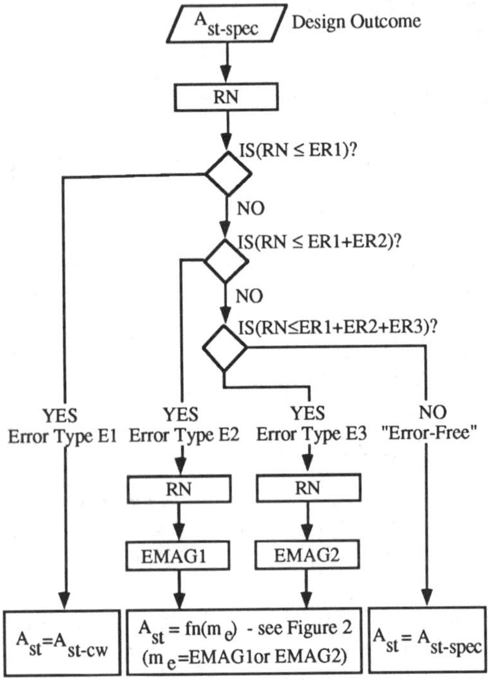

Note: RN uniform random number [0,1]
ER1 error rate for Error Type E1
ER2 error rate for Error Type E2
ER3 error rate for Error Type E3
m_e error magnitude randomly generated from error magnitude distribution
A_{st-cw} area of cross wires
A_{st} construction outcome

Figure 3. Event-Tree for Size of Reinforcing Mesh (A_{st}) Microtask.

4. HUMAN RELIABILITY ANALYSIS

A HRA was conducted for slabs with a design live load of 2.0 kPa, constant dead load of 0.25 kPa, dead load due to slab thickness, and member spans from 2.4m (D=100 mm) to 6.0 m (D=250 mm). The mean probabilities of failure (and their 95% confidence intervals) for the "nominal" ($\bar{p}_{fN}$) and "realistic" ($\bar{p}_{fE}$) construction scenarios are presented in Table 4. The proportional loss of structural safety (i.e., $\bar{p}_{fE}/\bar{p}_{fN}$) is also shown in Table 4.

Not surprisingly, it is observed that $\bar{p}_{fE}$ is higher than $\bar{p}_{fN}$ for all slab spans. The estimates of $\bar{p}_{fE}$ and $\bar{p}_{fN}$ are not significantly different when the slab thickness is relatively small; it is likely that a small reinforcing mesh size also is selected (assuming an "error–free" outcome). However, the proportional loss of structural safety tends to increase as the slab size increases. This is to be expected because the potential for undersizing (due to construction error) increases as the correctly designed slab depth and size of reinforcing mesh increases. It is the selection of undersized members that significantly increases the probability of failure. On the other hand, for "smaller slabs" it is more likely that a construction error will result in the selection of an increased slab depth or larger size of reinforcing mesh. Therefore, not all errors are detrimental to structural safety.

Thus, "nominal" probabilities of failure may be considered to be realistic only when the specified amount of steel reinforcing is very low; this situation generally occurs for slabs with a low span and/or small thickness (e.g., D≤150 mm). However, for larger slabs, the results presented herein indicate that "nominal" probabilities of failure are not particularly realistic. This is in general agreement with Stewart (1992a) who found that realistic probabilities of failure (as obtained from a HRA) were nearly an order of magnitude higher than "nominal" probabilities of failure, for large reinforced concrete floor beams (i.e., beam depths of 600 mm to 1300 mm). Finally, it should be recognised that the influence of other inspections (e.g., local government building inspectors, clerk of works) may result in a closer correlation between "realistic" and "nominal" probabilities of failure for large slab sizes.

Table 4. "Nominal" and "Realistic" Failure Probabilities as obtained from the HRA.

Slab Span (m)	Slab Thickness (mm)	$\bar{p}_{fN}$ "nominal"	$\bar{p}_{fE}$ "realistic"	$\frac{\bar{p}_{fE}}{\bar{p}_{fN}}$
2.4	100	0.41E-2	0.41E-2	1.00
2.8	110	0.23E-3	0.23E-3	1.00
3.2	130	0.26E-3	0.27E-3	1.03
3.6	140	0.10E-3	0.12E-3	1.12
4.0	160	0.30E-3	0.31E-3	1.06
4.4	180	0.26E-4	0.43E-4	1.65
4.8	190	0.26E-3	0.30E-3	1.13
5.2	210	0.23E-4	0.52E-4	2.28
5.6	230	0.13E-3	0.17E-3	1.33
6.0	250	0.15E-4	0.51E-4	3.36
n_S		10E7	10E7	
95% Confidence Interval		± 5%	± 2%	

Sensitivity analyses were conducted to assess the influence of uncertainty in the microtask human performance models on the "realistic" probabilities of failure. The results indicate that structural reliability is not particularly sensitive to variations in median error rates ($\tilde{m}_i$) which depart by ± 25% from those given in Table 2. However, the "realistic" probabilities of failure were lowered (sometimes significantly) when all error factors were set to zero. Nonetheless, the "realistic" probability of failure when EF_i=0 (for all error types) was still significantly higher than the "nominal" probability of failure, for large slab sizes.

5. CONCLUSION

A Human Reliability Analysis (HRA) has been used to simulate the effect of human error on the construction of a reinforced concrete floor slab in order to obtain "realistic" probabilities of structural failure. Existing code calibration statistical distributions were used to describe the effective depth of steel reinforcing and concrete compressive strength. Expert opinions were used to develop error rate and error magnitude distributions for the size of reinforcing mesh. It was found that "nominal" and "realistic" estimates of structural reliability were similar when the expected slab size is small (i.e., slab thickness less than 150 mm). However, for larger slab sizes the likelihood of undersizing is increased; hence the calculated "realistic" probabilities of failure (including the effects of engineering inspections) are generally substantially higher than the "nominal" probabilities of failure that are presently used to obtain measures of risk.

REFERENCES

Allen, D.E. (1968), Choice of Failure Probabilities, *Journal of Structural Engineering*, ASCE, Vol. 94, No. 9, pp. 2169–2173.

BCA (1990), Building Code of Australia, Australian Uniform Building Regulations Co-Ordinating Council, Canberra, ACT.

Brown, C.B. and Yin, X. (1988), Errors in Structural Engineering, *Journal of Structural Engineering*, ASCE, Vol. 114, No. 4, pp. 2575–2593.

Eldukair, Z.A. and Ayyub, B.M. (1991), Analysis of Recent U.S. Structural and Construction Failures, *Journal of Performance of Constructed Facilities*, ASCE, Vol. 5, No. 1, pp. 57–73.

Ellingwood, B. (1978), *Reliability Basis of Load and Resistance Factors for Reinforced Concrete Design*, NBS Building Science Series 110, National Bureau of Standards, U.S. Government Printing Office, Washington, D.C.

Ford, J.S. (1984), Construction Inspection – Can You Afford Not To Do It ?, (in) *Professional Inspection of Construction*, ASCE, S.B. Quinn (Ed.), pp. 27–41.

Ingles, O.G. and Nawar, G. (1983), Evaluation of Engineering practice in Australia, *IABSE Workshop on Quality Assurance within the Building Process*, Rigi, pp. 111–116.

Mirza, S.A., Hatzinikolas, M. and MacGregor, J.G. (1979), Statistical Descriptions of Strength of Concrete, *Journal of the Structural Division*, ASCE, Vol. 105, no. ST6, pp. 1021–1037.

Pham, L. (1985a), Reliability Analyses of Reinforced Concrete and Composite Column

Sections Under Concentric Loads, *Civil Engineering Transactions, IEAust,* Vol. CE27, No. 1, pp. 68–72.

Pham, L. (1985b), Load Combinations and Probabilistic Load Models for Limit State Codes, *Civil Engineering Transactions, IEAust,* Vol. CE27, No. 1, pp. 62–67.

Standards Association of Australia (1988), *SAA Concrete Structures Code*, AS3600–1988, Sydney, Australia.

Stewart, M.G. and Melchers, R.E. (1989a), Error Control in Member Design, *Structural Safety,* Vol. 6, No. 1, pp. 11–24.

Stewart, M.G. and Melchers, R.E. (1989b), Checking Models in Structural Design, *Journal of Structural Engineering*, ASCE, Vol. 115, No. 6, pp. 1309–1324.

Stewart, M.G. (1991), Probabilistic Risk Assessment of Quality Control and Quality Assurance Measures in Structural Design, *IEEE Transactions on Systems, Man, and Cybernetics* , Vol. 21, No. 5, pp. 1000–1007.

Stewart, M.G. (1992a), A Human Reliability Analysis of Reinforced Concrete Beam Construction, *Civil Engineering Systems* , Vol. 9, No. 3, pp. 227–247.

Stewart, M.G. (1992b), *The Occurrence and Detection of Errors in Reinforced Concrete Slab Construction,* Research Report No. 078.05.92, Department of Civil Engineering and Surveying, The University of Newcastle, Australia.

Stewart, M.G. (1993), Structural Reliability and Error Control in Reinforced Concrete Design and Construction, *Structural Safety* (in press).

Swain, A.D. and Guttman, H.E. (1983), *Handbook of Human Reliability Analysis with Emphasis on Nuclear Power Plant Applications*, NUREG/CR–1278, US Nuclear Regulatory Commission, Washington, D.C.

Offshore structures

Structural Safety & Reliability, Schuëller, Shinozuka & Yao (eds) © 1994 Balkema, Rotterdam, ISBN 90 5410 357 4

Probabilistic inspection planning for offshore structures using an inspection importance measure

Jan Inge Dalane & Ivar Langen
Statoil, Stavanger, Norway

ABSTRACT: Fatigue can be an important failure mode for offshore structures. The parameters governing the prediction of fatigue life show large uncertainties and it is therefore normally considered necessary to monitor the condition of the structures by supplementing in-service inspections. Underwater inspections are, however, very expensive and it is therefore important that the inspections are performed in such a way that they significantly increase our knowledge about the safety of the structure. In order to develop an optimal inspection plan, one should not only look at the probability of fatigue failure for the different members, but also consider the consequence of such failures. Offshore structures are normally very redundant meaning that total structural collapse will not occur before a sequence of failures has been developed. One or several fatigue failures may occur before the weakened structure fails due to overload. In order to identify the members which are most important from an inspection point of view, an inspection importance measure is suggested in this paper. This measure will tell us how an inspection of a given member will influence the reliability of the structural system.

1 INTRODUCTION

Traditionally, inspection plans for jacket type structures have been based on considerations of design fatigue lives, stress levels, some notion of criticality, inspection history and experience. Information gained from previous inspections is only used qualitatively in these traditional approaches. Therefore, probabilistic approaches within inspection planning (Madsen, et al., 1986; Lotsberg and Kirkemo, 1989) have been very successful and are now frequently used for offshore structures. With probabilistic methods it is possible to treat the large uncertainties associated with the fatigue and inspection process in a more rational way resulting in higher quality of the inspection process and also significant reduction in inspection costs.

Usually, the probabilistic inspection planning is based on a component level approach. The probabilities of fatigue failures in different members are calculated and evaluated against a specified target reliability level. This target level is established based on the importance of the considered member from a system point of view usually determined by use of nonlinear pushover analysis. These types of probabilistic approaches do not include all the probabilistic system effects in a structural system: The system reliability of a structural system exposed to fatigue loading will depend on the uncertainties in loading and strength of the different structural members, the redundancy in the system, the number of possible failure sequences, and the correlation in strength and loading between the different members. The correlation aspect is also of interest from an inspection point of view, i.e. how much will inspection of one particular member tell us about the other members which have not been inspected.

In order to include all these aspects, a system reliability formulation of the structural system is necessary. In this paper, a brief review of a system reliability formulation for fatigue is given (Karamchandani et al., 1991; Dalane, 1993). It is also discussed how simplification can be introduced in order to limit the computational effort for large structural systems. This system formulation is used in the definition of an inspection importance measure which can be used to identify the most critical members in the system from an inspection point of view.

2 SYSTEM FORMULATION

Fatigue cracks may develop in both ends of a structural member, and in the present system formulation possible locations for fatigue damage are referred to as sections. The time needed to develop a through thickness crack in a section can be derived by fracture mechanics using Paris crack growth law and linear damage accumulation. The relation between time to a through thickness crack and time to total section failure can be accounted for by a multiplicative factor R_{sec}, i.e. the time to total failure of a section is

$$T_{sec_{j_1}} = R_{sec_{j_1}} \frac{1}{C_{j_1} \nu_{0_{j_1}} E_{j_1}[S]^m} \int_{a_{0_{j_1}}}^{a_{th_{j_1}}} \frac{da}{Y_{j_1}(a)^m (\pi a)^{m/2}} \quad (1)$$

where subscript j_1 means that the variable refers to section j_1 in the intact structure; a is the crack size; C and m are the material parameters; ν_0 is the zero upcrossing rate; $E[S]^m$ is the expected value of the stress ranges raised to m; $Y(a)$ is a geometry function, and a_0 and a_{th} are initial and critical crack size, respectively.

After a fatigue failure in the structure, the stress level will change in the remaining members. The time necessary to also develop a second fatigue failure, i.e. failure of section j_2 after failure of section j_1, can be written as

$$T_{sec_{j_2|j_1}} = \frac{R_{sec_{j_2}} \int_{a_{0_{j_2}}}^{a_{th_{j_2}}} \frac{da}{Y_{j_2}(a)^m (\pi a)^{m/2}} - C_{j_2} \nu_{0_{j_2}} E_{j_2}[S]^m \; T_{sec_{j_1}}}{C_{j_2} \nu_{0_{j_2|j_1}} E_{j_2|j_1}[S]^m} + T_{sec_{j_1}} \quad (2)$$

Here, subscript $j_2|j_1$ means that the variable refers to section j_2 given that section j_1 has failed. In terms of failure sequences, the final failure of a structure will always be an overload failure. One or several fatigue failures will reduce the structural capacity so much that the structure will fail due to an extreme wave in the next storm. A failure path or a collapse sequence j in a structural system can be defined as the event

$$E_j = \{T_{sec_{j_1}} \leq T_{life}\} \cap ... \cap \{T_{sec_{j_m}|j_1 j_2 ... j_{m-1}} \leq T_{life}\} \cap \{E_{overl, \Delta T | j_1 j_2 ... j_m}\} \quad (3)$$

where $T_{sec_{j_1}}$, up to $T_{sec_{j_m}|j_1 j_2 ... j_{m-1}}$ are the points in time when the different fatigue failures in the sequence j occur and m is the total number of fatigue failures. $E_{overl, \Delta T | j_1 j_2 ... j_m}$ is the event of an subsequent overload failure in the weakened structure after sections $j_1 j_2 ... j_m$ have failed. The time of exposure for the overload failure is

$$\Delta T = T_{life} - T_{sec_{j_m}|j_1 j_2 ... j_{m-1}} \; .$$

A Branch and Bound search may be used to identify the most important failure sequences. Total system failure will occur if any of these failure sequences occur, and the system failure probability is thus obtained by taking the union of all the identified important failure sequences. Assuming n important failure sequences, the system failure probability is

$$P_{fSYS} = P(E_1 \cup E_2 \cup ... \cup E_n) \quad (4)$$

When performing an inspection, the additional information obtained may influence one or several failure sequences in the system, i.e. the inspected section may be involved in many of the failure sequences. It will also be some correlation between different sections because of common random variables. Let I be the result of one or several inspections, and let $P_{fSYS|I}$ be the conditional probability of system failure given that the inspection resulted in event I. The updated system failure probability is then

$$P_{fSYS|I} = \frac{P(\{E_1 \cup E_2 \cup ... \cup E_n\} \cap I)}{P(I)} = \frac{P(\bigcup_{j=1}^{n} \{E_j \cap I\})}{P(I)} \quad (5)$$

3 SIMPLIFIED SYSTEM FORMULATION

Current approaches for system reliability analysis of overload failure apply rather idealized mechanical models for the behaviour of the structure. This is in contrast to deterministic analysis where calculation of the ultimate capacity can be performed by using advanced nonlinear pushover analysis. Hence, a major question is then how important the probabilistic system effects are compared to the deterministic system effects. Obviously, it is not possible to state any general conclusions regarding the probabilistic characteristics of the strength of a structural system. The behaviour of the system will always be a function of the actual structural configuration and also a function of the actual loading. However, based on results from reliability analyses of different structural systems (De, 1990; Sigurdson et al., 1993; Dalane, 1993; Dalane et al., 1993), some typical probabilistic characteristics can be found for ultimate load capacity of typical jacket structures.

The capacity of a particular failure sequence in a redundant system is a function of the capacity of several individual members and an averaging of the variability for the individual member capacities will typically occur. Hence, the coefficient of variation for the strength of the failure sequence will typically be less than the coefficient of variation for the individual member capacities. This averaging effect will be large for a ductile system with a high degree of redundancy where the sequence capacity depends on many individual members. For a more brittle system or a system with low redundancy, the averaging effect will be less significant.

For offshore structures, the uncertainty in extreme wave loading is much higher than the uncertainty in the ultimate strength. Therefore, compared to the high uncertainty in loading, the uncertainty in ultimate capacity become usually less important. Due to highly correlated loads, the different failure sequence will also be highly correlated. Furthermore, offshore structures are usually not balanced, which means that there is one or a few failure sequences which dominate and contribute significantly to the failure probability. In a damaged condition, the structure will generally be more unbalanced than in the intact condition and the number of important failure sequences will be reduced even more.

All these effects result in limited probabilistic system effects for overload capacity. Hence, the ultimate load capacity of a structural system is usually well described by only one component. If $Q_{\Delta T}$ denotes the extreme environmental loading in the time period ΔT and SC the system capacity, the event describing overload failure when the sections $j_1 j_2 \ldots j_m$ have failed, can be written as

$$E_{overl,\Delta T|j_1 j_2 \ldots j_m} = \{SC_{j_1 j_2 \ldots j_m} \leq Q_{\Delta T}\} \tag{6}$$

The loading and the load capacity can be described in terms of global load characteristics as base shear or overturning moment. Calculation of the mean ultimate load capacity can be performed by using nonlinear pushover analysis.

For fatigue of structural systems, the probabilistic system characteristics are quite different. The uncertainty in environmental loading is relatively low for fatigue and most of the uncertainties are associated with the material parameters and the calculation of hotspot stresses. The correlation for both these types of uncertainties is assumed to be rather low from hot spot to hot spot and this will result in a rather low strength correlation between the different hot spots in the structure. Hence, for fatigue we will have sequences of rather independent events and then also significant probabilistic redundancy.

The correlation between the different section failures in a particular failure sequence is important for the sequence failure probability. The correlation between the different failure sequences, however, is not so important in calculation of the system failure probability. This is because the event that system failure occurs, involves union of sequence failure events, while the sequence events involves intersections of section failure events. To understand why the level of correlation has a greater impact on events involving intersections, consider two cases; a parallel and a series system. In a parallel system of k identical components, each with failure probability p, the probability of system failure (i.e. intersection of k components failure events) varies from p^k to p as the correlation between the components varies from 0 to 1. In a similar series system, for small p, the probability of system failure (i.e. union of k components failure events) varies from kp to p as the correlation varies from 0 to 1. Therefore, the correlation will have a greater impact on an event involving intersections than on an event involving unions.

The correlation between the different failure sequences must be relatively high in order to change the system failure probability significant. For fatigue there is usually a rather low correlation between the different failure sequences, and simplifications in evaluating the system failure probability can therefore be done. Consider a structural system with n important failure sequences where each failure sequence is defined by the events E_j, $j=1,2,\ldots,n$ and failure probability P_j. The system failure probability P_{fSYS} is calculated by using second order bounds which are calculated from the probabilities P_j and the intersection probabilities $P_{ij}=P(E_i \cap E_j)$. The events E_i and E_j represents parallel systems consisting of several components and the event $E_i \cap E_j$ will therefore in some cases result in a very large parallel system requiring very much computational effort to calculate P_{ij}. This will particularly be the case when inspection results introduce additional components in the parallel subsystems. For a system where correlation between the events E_i and E_j is low, a good approximation to the system failure probability can be obtain by using simple upper bounds which only makes use of the P_j's, i.e. only do a summation of the individual sequence probabilities, i.e.

$$P_{fSYS} \approx \sum_{j=1}^{n} P(E_j) \tag{7}$$

This simplification will be conservative and overestimate the system failure probability.

4 IMPORTANCE MEASURE FOR INSPECTION

A section in a structural system should only be inspected if there is a relatively large likelihood of detecting a crack. If there is a rather low chance of detecting a crack with the inspection method available, it is not worth inspecting the section. The likelihood of detecting a crack will be large only if:

I) There is a relatively high probability that a fatigue crack has been developed in that section.

II) There is a high probability that the fatigue crack can be found with the inspection method available.

These two factors can be combined to give a net probability of detecting a crack at section i, at the time of inspection T_{insp}, i.e.

$$P_{a_{det,i}} = P(T_{a_{det,i}} \leq T_{insp}) \tag{8}$$

where $T_{a_{det,i}}$ is the time necessary to develop a detectable crack $a_{det,i}$ in section i, i.e.

$$T_{a_{det,i}} = \frac{1}{C_i \nu_{0_i} E_i[S]^m} \int_{a_{0_i}}^{a_{det,i}} \frac{da}{Y_i(a)^m \ (\pi a)^{m/2}} \tag{9}$$

The detectable crack size $a_{det,i}$ is modeled as a stochastic variable reflecting the actual probability of detection (POD) curve.

Furthermore, a section is only worth inspecting if it is important for the integrity of the structure; i.e. it is not worth inspecting a section whose failure does not significantly reduce the safety of the structure. An inspection may result in two possible outcomes; a crack may either be detected or not detected. The inspection itself does not increase the reliability of the structure, but an inspection makes it possible to take the necessary corrective actions like repair etc. if a crack is detected.

Assume that an inspection defined by the event I_i is planned to be performed for section i. The effect of this inspection can then be measured by the change in system failure probability, i.e.

$$\Delta P_{fSYS|I_i} = P_{fSYS} - P_{fSYS|I_i} \tag{10}$$

where $P_{fSYS|I_i}$ is the system failure probability given an inspection has been performed for section i, and P_{fSYS} is the system failure probability without any inspection. Since fatigue is a time dependent process, the system failure probability has to be calculated for a given period of time, e.g. for the remaining service life after the inspection.

Due to correlation effects, inspection of a section may also influence the estimated reliability of sections that have not been inspected. This effect will depend on the correlation between the considered sections. The change in system failure probability given an inspection of section i can by using Eqs. (7) and (10) be written as

$$\Delta P_{fSYS|I_i} \approx \sum_{j=1}^{n} \left(P(E_j) - P(E_j|I_i) \right) \tag{11}$$

where E_j is the failure sequence j, and n is the number of failure sequences. The probability $P(I_i)$=1.0 and the inspection event I_i can be divided into two disjoint events dependent on the inspection results "no detection" and "detection of a crack", i.e.

$$P(E_j|I_i) = P(E_j \cap I_i) = P(E_j \cap I_i^0) + P(E_j \cap I_i^1) \tag{12}$$

where I_i^0 and I_i^1 denotes "no detection" and "detection of a crack" in section i, respectively. Calculation of $P(E_j \cap I_i^1)$ is very complicated for a planned inspection. If a crack is detected and the crack size is measured in an inspection, infinitely many possible outcome and repair strategies are possible and should in principle be evaluated. Herein it is assumed that if a crack is detected, repair and other corrective action will be taken to ensure that section i will not fail; i.e. using this strategy, $P(E_j \cap I_i^1) \approx 0$. This assumption may also reflect what will really happen if a crack is detected in an offshore structure. Due to the large consequences of failures of offshore structures, an observed crack will receive very much attention to ensure that no critical situation will occur. Detected cracks are more an economical problem than a safety problem. It is the undetected cracks which cause problem from a safety point of view.

Based on the above concept, $\Delta P_{fSYS|I_i}$ can be used to define an inspection importance factor IF_i for section i;

$$IF_i = \frac{\Delta P_{fSYS|I_i}}{P_{fSYS}} = 1 - \frac{P_{fSYS|I_i}}{P_{fSYS}} \approx 1 - \frac{\sum_{j=1}^{n} P(E_j \cap I_i^0)}{\sum_{j=1}^{n} P(E_j)} \tag{13}$$

The event I_i^o is defined by the event $\{T_{a_{det,i}} > T_{insp}\}$, and the probability $P(E_j \cap I_i^0)$ is then

$$P(E_j \cap I_i^0) = P(E_j \cap \{T_{a_{det,i}} > T_{insp}\}) \tag{14}$$

The factor IF_i is then a measure of how much the system failure probability can be reduced by inspection of section i. In this definition of the inspection importance factor, the probability of fatigue crack, the consequence of a fatigue failure, and the inspection reliability are included. A IF_i equal to 0 indicates that there is no effect of the inspection, while IF_i equal to 1 indicates that it is possible to eliminate system failure probability due to fatigue completely by inspection of section i only.

5 EXAMPLE JACKET STRUCTURE

As an illustration of the use of the inspection importance factor the tripod jacket structure shown in Figure 1 (left) with 30 years designlife is selected (Karamchandani et al., 1991; Dalane 1993). This structure is not very redundant and overall structural failure is assumed to occur if any two elements fail or if a leg member fails. System reliability analysis is carried out to identify important failure sequences initiated by fatigue, that lead to collapse. The identified sequences and their probabilities of occurrence are given by the failure tree in Figure 1 (right). The symbols A and B indicate the two ends of a member. In this failure tree, each branch corresponds to failure of a section in fatigue or an element under an extreme wave. Each node corresponds to a damage state of the structure, and the number in the node is the probability of reaching the corresponding damage state.

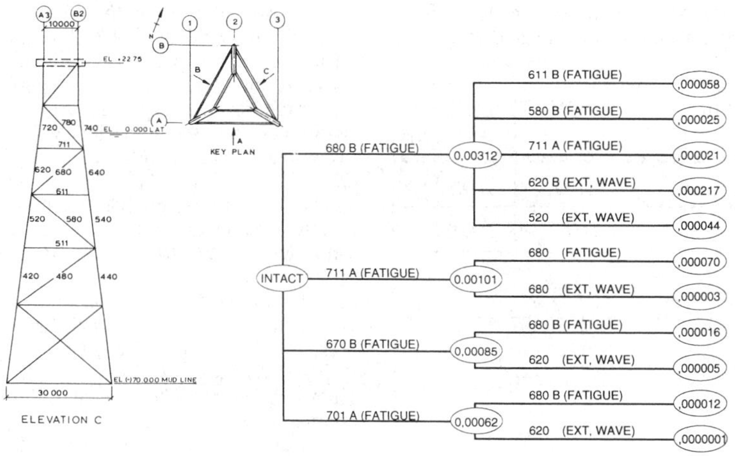

Figure 1. Schematic representation of the tripod jacket structure (left) and the failure tree (right).

The most likely to fail section is section 680B, and this section occurs in almost all the important failure sequences. The total system failure probability is $3.6 \cdot 10^{-4}$. An MPI inspection is planed to be performed after 10 years in service. The intersection probability of system failure and inspection result "no crack found" in section 680B is $6.4 \cdot 10^{-6}$ while the same result for section 711A gives a probability equal to $3.1 \cdot 10^{-4}$. Hence, the inspection importance factor for section 680B is 0.98 while it for section 711A is 0.14. For this structure, the importance of section 680B is obvious also from traditional selection criteria. This section has the lowest fatigue life and member 680 is also a highly loaded member with respect to extreme environmental loading.

6 EXAMPLE JACK-UP STRUCTURE

6.1 *Description of the jack-up structure*

For a structure with many fatigue sensitive members, the identification of the most important member from an inspection point of view is not obvious and can only be detected by detailed calculations. A jack-up platform (Dalane, 1993) is select as an example for a such structure. The finite element model of the platform is shown in Figure 2. The jack-up, which is designed to operate at a waterdepth of 108 m, consist of an almost triangular deck structure and triangular truss legs. The distance between leg centres is 62 m and the distance between chord centres in the legs is 16 m.

The foundation stiffness is a very important parameter for a jack-up platform. Traditional spudcan foundations are usually considered as pinned foundations. However, when the footings are equipped with skirts as in this case a significant rotational stiffness and moment capacity can be obtained.

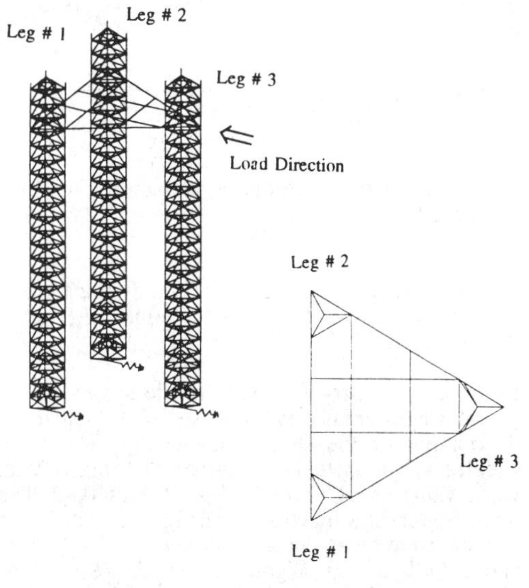

Figure 2. Element model for the jack-up structure.

6.2 *Overload failure*

The nonlinear FEM program USFOS (Søreide et al., 1988) is used to determine the Reserve Resistance Factor (*REF*-factor) for the structure. The *REF*-factor is defined as the ratio between the load level causing global failure of the structure and the design wave and current loading. The design load is defined by the 100 year wave and 10 year current and gives a base shear load equal to 15.8 MN for this structure. In calculation of the *REF*-factor, the dead weight, topside load, wind load and the buoyancy load are applied as the first load step. Then, increments of the design wave and current load are added until total collapse occurs. For the worst

Table 1. Structural strength based on 100 Monte Carlo simulations for intact and damaged structure.

Case	$E[REF]$	COV	$E[REF]/REF_{DET}$
Intact	2.66	0.10	0.82
Damaged	1.83	0.08	0.94

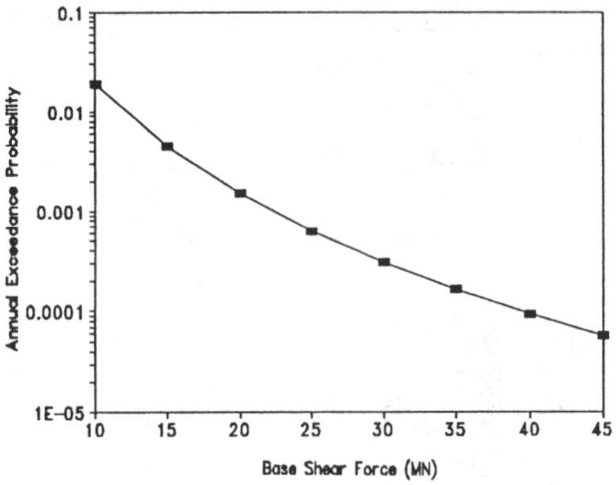

Figure 3. Annual exceedance probability for base shear load.

direction, using mean values for yield strength and no imperfections, a deterministic value, REF_{DET}, of 3.25 is found.

In order to identify the probabilistic system effect Monte Carlo simulations are used to establish the distribution of the structural capacity. The yield stresses are modelled as normal distributed with mean value of 687 MPa and a COV of 0.10. The imperfection is a function of member length L and is assumed to be lognormal with mean value $0.002L$ and a COV of 2.0 (Sigurdsson et al., 1993). The direction of the imperfections is included as a uniform distributed random variable in the range $[0, 2\pi]$. The modelling of imperfections is also assumed to reflect the uncertainties in residual stresses and eccentricities. The yield stresses, magnitude of imperfections and direction of imperfections are all assumed to be independent from member to member. Since the chord members are the most important members in this structures, only uncertainties in the capacity of chord members are included in the uncertainty modelling. For each outcome of the stochastic variables a pushover analysis is performed.

Table 1 shows the results from the simulations of the intact structure and a damaged condition. In the damaged condition, a chord member close to the deck in leg 2 is removed from the structure. For both cases, it is seen that the COV for structural capacity is less than the COV for member capacities. The structural capacity depends on the capacities of several members, and an averaging effect of the uncertainties will typically occur. Furthermore, the ratio between the expected value of the Reserve Resistance Factor, $E[REF]$ and the deterministic value REF_{DET} is relatively low for the intact case. This structure is relative sensitive for buckling of chord members and the random imperfections introduce many different failure modes.

In the damaged condition the structure is more unbalanced and the probabilistic system effects characterised by the COV and the ratio $E[REF]/REF_{DET}$, are significant smaller than for the intact structure. The typical failure mode for this damaged condition is that failures in the leg occur in the vicinity of the removed member. Hence, in the damaged condition the number of different failure sequences is significantly smaller than for the intact structure.

The system effect for overload failure seems to be more significant for jack-up structures than for "standard" jacket structures. In the jack-up structure relatively many failure sequences are possible; particularly, when random imperfections are included. A jacket structure has, usually, not so slender members and imperfections will then be less important. Further, jacket structures are usually not balanced, which means that there is only one or a few failure sequences that will dominate, i.e., $E[REF]$ and REF_{DET} will approximately have the same value. In a damaged condition, however, this will also be the case for the jack-up structure and the deterministic value REF_{DET} will then be a good estimate for $E[REF]$.

An important question is then how important the uncertainty in capacity are compared to the large uncertainty in wave load. In this study the wave model suggested by Haver (1991) together with data for the Ekofisk field are adopted. A joint description of wave and current is used. For load calculation, the Stokes 5th order wave profile is used and the drag coefficient is modelled as a random variable. The distribution of base shear load is shown in Figure 3 and is characterized with a COV of 0.52.

In the reliability analysis, model uncertainties and uncertainties in the foundation stiffness are also included. More details about the reliability formulation is given in Dalane et al. (1993). By using First Order Reliability Method (FORM) (Madsen et al.,1986), the annual failure probability for the structure is calculated. The result is shown for different values of $E[REF]$ and for different levels of uncertainties in Figure 4. It is seen that the uncertainty in capacity is negligible compared to the large uncertainty in wave loading. This will at least be the conclusion if we assume that the COV in the range of 10% as obtained from the simulation is a reasonable level for the uncertainty in capacity.

6.3 *Fatigue failure*

The primary failure mode in this structure is buckling of chord members. The shear force capacity of the X-braced legs does not seem to be a problem. Removal of a single brace member does not lead to a significant change in the ultimate capacity compared to the intact condition. The shear force capacity is still sufficient. However, if two braces in the same bay in a leg are removed, then the shear capacity will be reduced significantly. The shear forces will then introduce large bending moments in the chord members adjacent to the failed braces.

Hence, with respect to fatigue, two brace members must fail in the same bay before a significant reduction in ultimate load capacity occur, i.e. a sequence of two fatigue failures must occur. A sensitivity study of such a failure sequence is performed as an illustration. Two very important parameters then are the fatigue design life for the two brace members involved and also the residual capacity of the structure. The results are presented in terms of these two parameters. In a failure sequence of two fatigue failures, "Member *I*" and "Member *II*" are assumed to have the same fatigue design life. The service life for the structure is assumed to be 10 years.

A one-dimensional fracture mechanics model is used for the fatigue calculations. The uncertain parameters used in the study are summarized in Table 2. Uncertainties in geometry functions and calculation of hot spot stresses are accounted for by multiplying the deterministic values by the random variables ε_Y and ε_{SCF} , respectively. A shifted lognormal distribution is used to account for the fact that $R_{sec}>1.0$, i.e. R_{sec} is then written as $R_{sec}=1+R_{sec}^{\,1}$. The correlations between the material parameters lnC_I and lnC_{II}, and between the hot spot stresses ε_{SCF_I} and $\varepsilon_{SCF_{II}}$ are assumed to be 50%.

In a S-N approach, deterministic fatigue life should according to NPD (1992) be calculated by using the 2.5% fractile of the S-N curve. It is not possible to compare the S-N curve approach and the fracture mechanic approach directly. For this illustrative study, however, "design" fatigue life using fracture mechanic approach is also defined as the calculated fatigue life when the 2.5% fractile is used for the material parameter C .

The brace members transfer the shear forces in the legs as compression and tension forces in the X-braced panel. When a brace fails, only one brace is left to transfer the shear force in the considered bay, i.e. the axial force in the intact brace will increase with approximately a factor of 2. Axial forces are the dominating forces in the braces and is also associated with the largest stress concentration factors (*SCF*). Hence, in this study a factor of 2 is assumed for the increase in hot spot stress in "Member *II*" after failure of "Member *I*".

Table 2. Uncertain fatigue parameters.

Parameter	Distribution	Mean	St.dev.
a_0	exponential	0.11	
lnC	normal	-29.75	0.50
ε_Y	lognormal	1.0	0.10
ε_{SCF}	lognormal	1.0	0.15
$R_{sec}^{\,1}$	lognormal	0.2	0.20

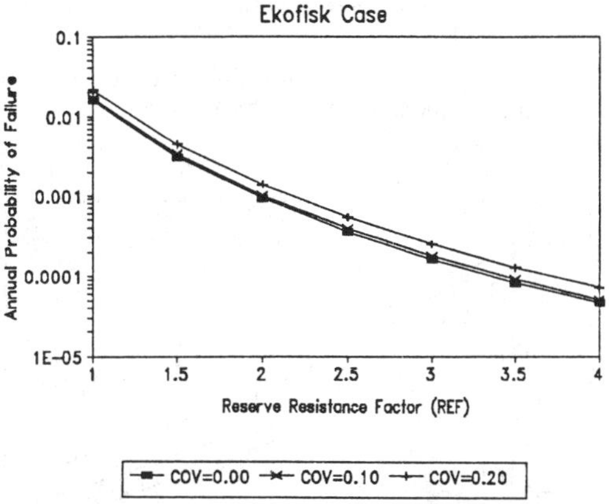

Figure 4. Annual probability of failure as function of *E*[*REF*] for different levels of uncertainty.

Total system failure probability due to fatigue is not calculated in this study. However, figure 5 displays the lifetime failure probability of a collapse sequence initiated by a fatigue failure. As expected, the failure probability will decrease with increasing fatigue design life, and also with increasing Residual Resistance Factor (*RRF*-factor). The *RRF*-factor is defined as the ratio between the load level causing global failure of the damaged structure and the design wave and current loading. It is seen that significant redundancy for fatigue failure is envisaged, even for a structure with a low *RRF*-factor.

In this study, it is assumed that the first (*MPI*) inspection is planned to be performed after five years in service. The *POD* curve is assumed exponential distributed and gives a 90 % probability of detecting a 40 mm long crack. Figure 6 shows the inspection importance factor (*IF*) for the *MPI* inspection as a function of fatigue design life and the *RRF*-factor. Since the total system failure probability is not calculated, the *IF* factor is multiplied with the system failure probability, i.e. the probabilities given in Figure 6 represent the

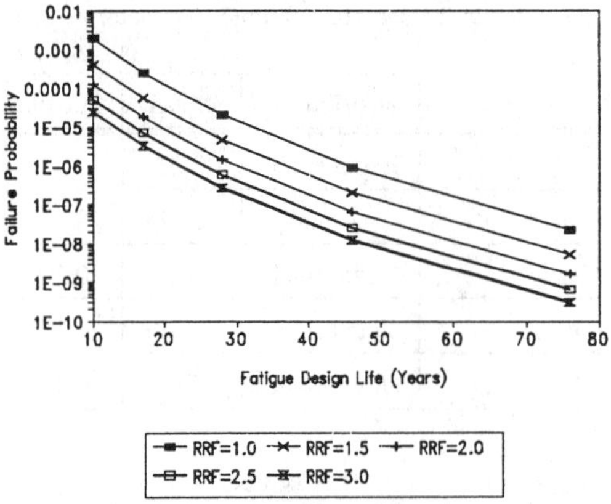

Figure 5. Lifetime (10 years) failure probability of a collapse sequence initiated by fatigue.

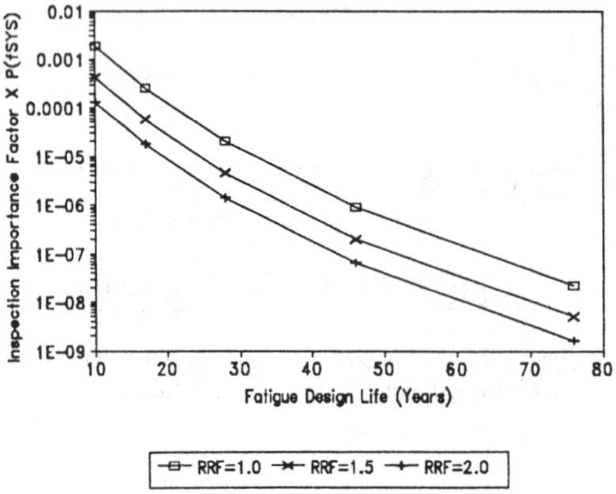

Figure 6. Inspection importance factor (*IF*) for *MPI* inspection after 5 years in service as a function of design fatigue life and the *RRF* factor.

reduction in system failure probability by inspecting "*Member I*" in the considered failure sequence. It is then also implicit assumed that the system failure probability can be found by just adding the sequence failure probabilities. As expected, the effect of inspection increases with reduced fatigue design life and with reduced *RRF*-factor.

7 CONCLUSIONS

In order to identify the most important members in a structure from an inspection point of view, it is necessary to evaluate the system effects. A complete system reliability analysis for fatigue of a large structural system, which also include additional information from the inspection history, will require large computational effort. However, simplification in the system reliability calculation can be introduced without loosing any large amount of accuracy. In order to identify the most important members from an inspection point of view, an inspection importance factor is suggested. This inspection importance factor take into account the probability of a fatigue damage, the consequence of such damage and also the inspection reliability. The applicability of the concept has been demonstrated for a jacket and a jack-up structure.

ACKNOWLEDGEMENT

Statoil is acknowledged for permission to present these results. The opinions expressed herein are those of the authors and they should not be construed as reflecting the views of Statoil.

REFERENCES

Dalane, J.I. (1993): System Reliability in Design and Maintenance of Fixed Offshore Structure, Dr.ing. thesis, Division of Marine Structures, Norwegian Institute of Technology, Trondheim, Norway.

Dalane, J.I., Haver, S., and Langen, I. (1993): Variability in Load and Strength for a Jack-up Structure, Proc. ISOPE 93, Third International offshore and Polar Engineering Conference, Vol IV, Singapore.

De, R. (1990): Offshore Structural System Reliability; Wave-load Modelling, System Behaviour, and Analysis, Report No. RMS-6, Department of Civil Engineering, Stanford University, USA.

Haver, S. (1991): Wave Climate Description for Foundation Design, Statoil, Report No. F&U-UoD-91047, Version 2, Trondheim, Norway.

Karamchandani, A., Dalane J.I. and Bjerager, P., (1991): "System Reliability of Offshore Structures Including Fatigue and Extreme Wave Loading", Marine Structures, Vol. 4, p353-379.

Lotsberg, I. and Kirkemo, F. (1989): "A Systematic Method for Planning In-service Inspection of Steel Offshore Structures", Proceedings OMAE 89, The Hague, The Netherlands.

Madsen, H.O., Krenk, S., and Lind, N.C. (1986): Methods of Structural Safety, Prentice Hall, Englewood Cliffs, New Jersey, USA.

NPD (1992): Regulations Concerning Loadbearing Structures in the Petroleum Activities", Norwegian Petroleum Directorate, Stavanger, Norway.

Sigurdson, G., Skallerud, B., Skjong, R., and Amdahl, J. (1993): Probabilistic Collapse Analysis of Jackets, Proc. ICOSSAR'93.

Søreide, T. H., Amdahl, J., Eberg, E., Holmås, T., and Helland, Ø. (1988): A computer Program for

Progressive Collapse Analysis of Steel Offshore Structures - Theory Manual, SINTEF Report No. STF71 F88038, Trondheim, Norway.

Structural Safety & Reliability, Schuëller, Shinozuka & Yao (eds) © 1994 Balkema, Rotterdam, ISBN 90 5410 357 4

Probabilistic assessment of steel jacket redundancy: The ARPEJ software

Jean Goyet
Centre Technique Industriel de la Construction Métallique, Paris, France

Christos Saouridis
Computer Control Systems, Athens, Greece

ABSTRACT: A software system is presented trough an application on a small offshore steel jacket platform. A discussion of its capabilities with regard to recent advances in reliability calculations and the respective developments in software tools are also presented.

1 INTRODUCTION

If we go through the last twenty years, we can observe that Structural Reliability has achieved a very large development. An important amount of research work has been devoted to reliability theory, reliability methods and also to the development of computer programs which are obviously essential in order to manage any reliability analysis.

For a good apprehension of the present situation we have to go back to the development mentioned above. For that, the concepts of "state of the art" and "state of the practice" are relevant.

The "state of the art" is the global methodology which allows a probabilistic evaluation of structural reliability. With this methodology, the structure is modelled as a serial system of parallel systems and the event-for which we have to compute the probability of occurrence-is a union of intersections:

$$P_f = P\left(\bigcup_{i=1}^{m}\bigcap_{j=1}^{n_j} E_{ij}\right) \qquad (1)$$

This methodology should allow to perform reliability analysis of realistic structures, meaning complex structures, in their actual environmental conditions.

*Currently at Bureau Veritas, Paris, France

Evaluation of P_f by (1) is not trivial and we generally need to assume several simplifications both in mechanical models and in the probabilistic calculations (for example by using FORM or any other simplified approach).

The "state of the practice" is constituted by all available computer programs and the experience accumulated through using these programs.

Go from the art to the practice is a challenge. Many problems dealing with loading aspects, structural analysis, probabilistic redundancy analysis and stochastic modelling have to be tackled and solved.

Otherwise, it is sometimes only practical experience on realistic structures which lead to know if our computer programs are really operational and well fitted to their assumed purposes.

There is therefore a "feedback from applications": Development of computer programs and practical applications lead to renew the methods, change the formulations and identify some new research fields for the future. Thus we are already able to speak in terms of "generations of computer programs" although the number of these programs is very limited.

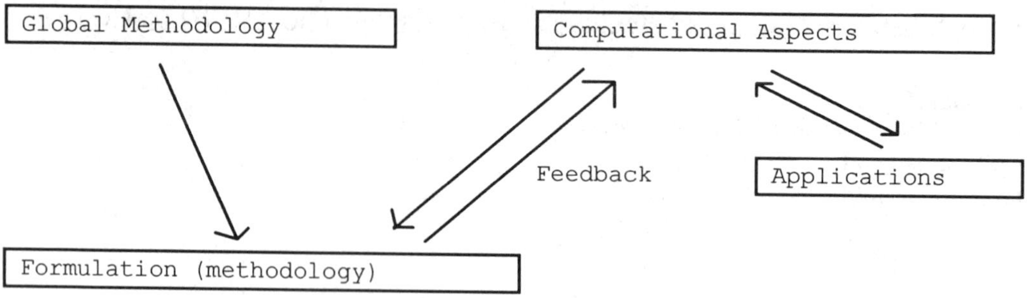

Figure 1:System Reliability-Methodology, formulation and computational aspects

Our experience in reliability indicate that results given in literature and dealing with only purely illustrative cases can not be considered as general. Simplification in structural topology and loading lead to results which are not representative of actual situations. So, only the experience obtained by analyses on realistic structures enable to give in the future precise indications and hints which can be expected from reliability analysis.

All these considerations are illustrated here within the presentation of ARPEJ (Probabilistic Redundancy Analysis for Evaluation of Jacket Structures) program.

2 FUNDAMENTAL OPTIONS OF ARPEJ

The three main options of ARPEJ concern environmental modelling, the build up and the analysis of the failure tree and the computer-machine implementation.

2.1 *Environmental modelling*

We give below the two formulations (table 1) which define the prototype version and the version 1 of ARPEJ. The key-point is the introduction of the surface response method for environmental modelling in the version 1. The prototype version was already presented in 1987/3/. In table 1, E is the set of environmental random variables (wave height, wind speed, hydrodynamic and aerodynamic coefficients - see also table 2), F is the nodal forces and moments vector applied to the structure, S is the vector of internal forces and moments in a particular member and R is the set of resistance random variables included in the limit state equation which characterises any particular component failure criteria.

As shown in table 1, the surface response method allows to perform reliability analysis in the "natural" space {E,R} without calculation at each step of the nodal forces vector corresponding to a particular realisation of the random variables included in E. Each realisation of E gives a realisation of λ_i random variables which then result to a realisation of the internal forces vector S (through the W_i). Reliability analysis uses systematically this response surface which is performed once at the beginning of the analysis.

2.2 *Build up and analysis of the failure tree*

With ARPEJ, it is possible to compute at any step t:

- $-\beta\ (FR_t)$ Where FR_t is the formal representation of the system at step t, including all complete and incomplete sequences identified until now.
- $-\beta\ (S_{it})$ Where S_{it} is any particular sub-system which can be defined from FR_t.

Table 1: Prototype and upgraded version formulations

Prototype Version	Version 1
1) Calculation of a second order probabilistic description of F: $E(F), \sum F$ 2) Calculation of a second order description of S: $E(S) = C\ E(F)$ $\sum(S) = C\ \sum(F)\ C^T$ 3) Space of basic variables: {S,R} 4) Simplifying assumptions: F, S --> Multinormal The knowledge on E is lost since it is not possible to use directly its description to deduce F.	1) Evaluation of F by a response surface model of an extreme wave loading: $F = \sum_{j=1}^{p} \lambda_j(E) F_j$ $\{F_j\}$: deterministic vectors $\{\lambda_j\}$: random scalar variables 2) Internal forces $S = CF = \sum_{j=1}^{p} \lambda_j(E) CF_j = \sum_{j=1}^{p} \lambda_j(E) W_j$ 3) Space of basic variables: {E,R} The knowledge on the environmental variables E is used 4) Simplifying assumptions Only those related to the response surface modelling.

$\{S_{it}\}$ includes the sub-system S_{1t} which consists of all mechanisms (complete sequences) already identified at step t. It is well known that:

$ß(FR_t) < ß_{system} < ß(S_{1t})$ (2)

As we advance through the failure tree, the gap between ß (FR_t) and (S_{1t}) decreases. Consequently, the evaluation of $ß_{system}$ becomes more and more precise. This process can be optimised by using the "branch and bound" method. In this method, the build up of the failure tree is achieved by selecting always the incomplete failure sequence with the smallest equivalent $ß_{eq}$ index.

The program proceeds in an interactive way. At any step of the algorithm, the user has the possibility to activate one of the four options proposed by the task manager's menu:

RUIN THEN CONTINUE (option 1), (SUB)SYSTEM SAFETY INDEX CALCULATION (option 2), GO BACK TO A PREVIOUS STAGE (option 3) and FINISH (option 4).

If option 2 is selected, the user has to prescribe the sub-system to be analysed. By this option, which is absolutely essential, we are able to define precisely the contribution of the different sub-systems in the evaluation of the probability of failure.

2.3 *Computer machine implementation*

The prototype version of ARPEJ was developed on an IBM main-frame machine. For reasons of flexibility, portability and because of the perspective to develop a graphical user interface, version 1 was developed on a Unix Workstation. A strong advantage is that one can work with some window manager (SunView, OpenLook, X-Windows etc) which allows to review results at any stage of the analysis while continue to run the program.

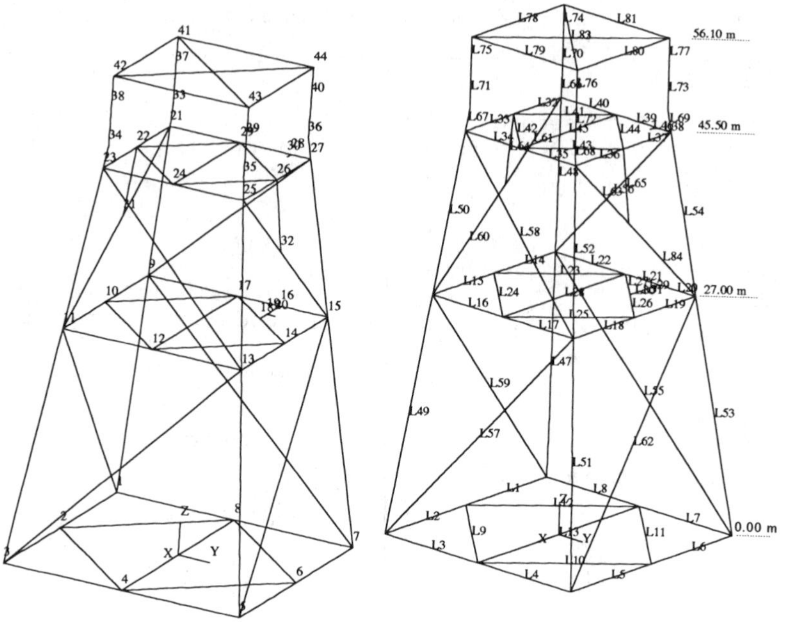

Figure 2: 3D model of the jacket (Beams and nodes)

2.4 *Other options of ARPEJ*

These options are more typical and are used in almost all general system reliability computer programs:

- Component reliability analysis: Hasofer-Lind ß index.
- System reliability analysis: multinormal integration and equivalent hyperplane.
- Sensitivity analysis: this kind of analysis may be easily performed insofar as reliability analysis is performed directly in the physical space {E,R}.
- Structural analysis: elastic global analysis with Gauss resolution for each loading case.
- Component post-failure behaviour: the internal force vector S is:

$S=C(F+F_r)$ (3)

where : C is the modified stiffness matrix and F_r is the deterministic vector of equivalent forces and moments due to residual strength of failed members.

The two limit state equations corresponding to the two component failure criteria mentioned above are described below:

a) Plastification of a tubular section (eq. 4):
If 0<Nx/Np<0.65

$$g_p = \gamma_p - \frac{\sqrt{M_y^2 + M_x^2}}{M_p} - 1.18(N_x / N_p)^2$$

If 0.65<Nx/Np

$$g_p = \gamma_p - 0.7\frac{\sqrt{M_y^2 + M_x^2}}{M_p} - 1.18(N_x / N_p)^2$$

b) Buckling of a tubular member (eq. 5)

$$g_f = \gamma_f - k_o \frac{N_x}{N_p} - k_f \frac{\sqrt{M_y^2 + M_x^2}}{M_p}$$ where

γ_p, γ_f = model random variables,
N_p= plastic axial capacity,
M_p= plastic moment capacity,
N_x, M_y, M_z= internal forces and moments,
k_o= function of the reduced slenderness ratio,
k_f= amplification coefficient.

Table 2: Probabilistic data for sensitivity analysis.

Variable name	Distribution	Mean value	Standard deviation
marine growth ϑ_{mg}	Normal	1.0	0.1
return period, T	Normal	12 years	1.2 years
wave height, H	Normal	16,8 m	1.344 m
wind intensity, λ_W	Normal	51.0 m/sec	5.1 m/sec
current intensity, λ_C	Normal	0.5256 m/sec	0.05256 m/sec
hydrodynamic coef. C_X	Normal	0.00001	0.0000035
hydrodynamic coef. C_X	Normal	0.75	0.2625
hydrodynamic coef. C_X'	Normal	0.00001	0.0000025
hydrodynamic coef. C_D'	Normal	1.5	0.375
aerodynamic coef. C_X	Normal	0.00001	0.0000000001
aerodynamic coef. C_d	Normal	5.0	0.6
elastic limit, f_y	Normal	350 N/mm^2	52.5 N/mm^2
reduced strength coef. $1/k_o$	Gumbel	**	**

**Mean value and standard deviation of 1/Ko depend on each single member and are computed by the program.

3 APPLICATION EXAMPLE WITH ARPEJ

3.1 *Structural Model*

The 3D structure shown in fig. 2 is a jacket-type platform. The model consists of 84 beam elements (six degrees of freedom per node) and 44 nodes. The structure is assumed to be perfectly fixed at the bottom of the sea (nodes 1, 3, 5 and 7 are completely blocked). The water depth is 39 m and the direction of the wave is 270 degrees (clockwise sense) with respect to the X axis of the global co-ordinate system. It is assumed that the wind and the current directions coincide with the direction of the wave action. Only the loading due to the environmental conditions is considered (self-weight and live loads, which could be modelled as user-defined random loads, are not taken into account). For each beam element, the reliability indexes are computed for plastic hinge formation in each beams's extremity and for global beam buckling.

3.2 *Basic Random Variables*

A first analysis was made with 13 random variables (see table 2): the 11 variables for the environmental conditions, the plasticity threshold f_y (one variable per component but all variables having identical probabilistic description) and the reduced strength parameter for buckling $1/K_o$.

The variables corresponding to the model uncertainties are assumed deterministic with value 1. All variables are assumed uncorrelatted. The objective of this first analysis was to investigate the sensitivity of results for each one of the random variables.

As it appears from the values of the components of vector a (the unit vector which is normal to the failure surface at the design point) for all elementary failures (3 elementary failures per beam), the results are sensitive only to the following variables: ϑ_{mg}, T H, λ_W, C_D, f_y and $1/K_o$. Thus, the complete analyses were made with a probabilistic description for only these variables while the rest of the variables are considered deterministic with value equal to their mean value. It must be noticed here that more work is needed to further improve the above assumptions on the distribution type of these random variables.

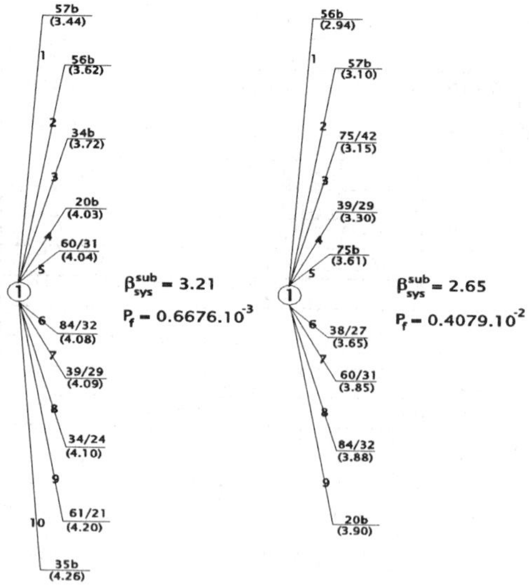

Figure 3:"Level 0" results for intact (left) and damaged (right) structure configuration.

3.3 *Reliability assessment of structural redundancy*

3.3.1 General procedure

ARPEJ first computes the nodal loads induced by the environment conditions (using the module RSEWL/1) and then proceeds to the structural analysis. At the end of this analysis, the internal forces (acting to the extremities of the beams) are computed. The software then proceeds with the computations of the Hasofer-Lind reliability indexes β_{HL} for all "survival" elementary failures of the structure at the current step. This is done by the FORM5 software library /2/ which is linked with ARPEJ. The β_{HL} values are sorted in an increasing order and a list of the first n elementary failures with the corresponding β_{HL} values for the current step is proposed to the user (n is a user-defined number and it is fixed to 10 for the following examples).

3.3.2 Constructing the failure-tree

The user may then select (based on his experience on the specific structure or on some other individual's criteria) a part or the entire set of the n elementary failures (events) to be included in the failure tree for current step. The corresponding failure sequences are then updated automatically.

3.3.3 Exploring the failure-tree

In a structural system reliability analysis it is important to identify the probabilistically most dominant complete failure sequences (mechanisms). The largest the number of the identified mechanisms the better the system failure probability approximation is. Though there is not any automatic (imposed) methodology to go through the failure tree, a consistent help is provided to the user by computing (if so requested) the reliability indexes of each one of the current failure sequences (this is done by an equivalent hyperplane method using the software library SYSREL /2/ also linked with ARPEJ).

Based on the obtained β_{eq} values of other structural criteria (excessive nodal displacement or important loss of stiffness) the user can identify or declare a failure sequence as a mechanism. By choosing to go on with the sequence presenting the lowest β_{eq} value, the most dominant mechanisms may be identified (this is a "branch-and-bound" procedure). This is possible with ARPEJ since it allows, after each step, to proceed either with a "go-back-to-a-previous-step" or a "confirm-a-failure" action . For the applications presented here a failure sequence is considered as a mechanism if its β_{eq} value is less than 1 (which corresponds to a probability of failure greater than 15.87%).

3.3.4 Performing System Reliability Computations

ARPEJ performs reliability calculations for either the entire system of failure sequences or a user specified subset of sequences. The results of these operation include the system reliability index β_{sys}^{sub} and the corresponding probability of failure (probability of a union of intersections of events) as well as the matrix of correlation between the equivalent

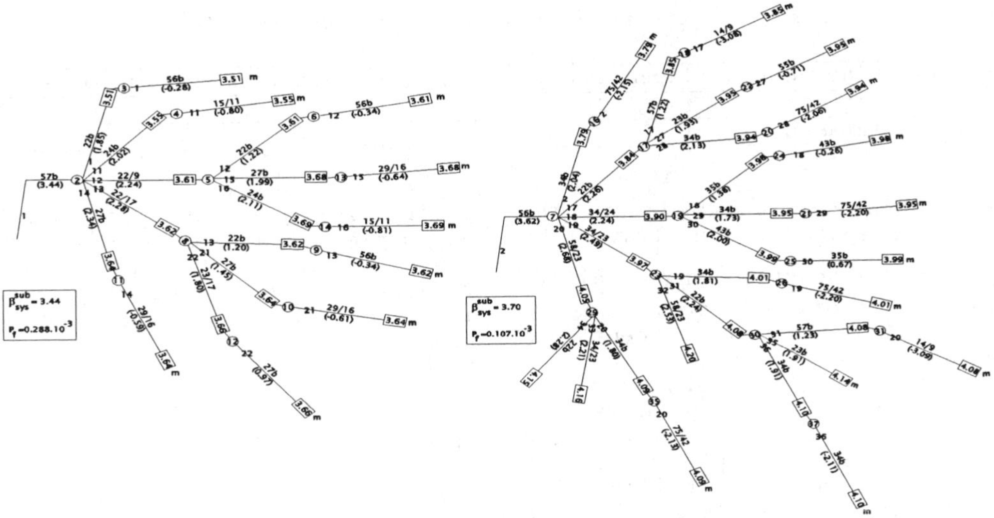

Figure 4: Failure tree of the "intact" structure configuration

safety margins. The actual probability of failure of the structure can be bounded (see equation (2)). The upper bound is the probability of the entire system composed of all the complete or incomplete failure sequences while the lower bound is the probability of the subsystem composed only of the mechanisms.

3.4 *Applications*

Two analyses were performed for the structure of fig. 2: a first one for the "intact" structure and a second for a "damaged" configuration were beam n° 34 was removed. It appears from the results below that the main failure mode is buckling rather than plastic hinge formation since there isn't any failure sequence with only plastic hinge formation modes. It is also obvious that this structure is not so much redundant since the β_{eq} values for any particular failure sequence slightly increase as events are added to this sequence. Figure 3 above shows the results obtained at "level 0" for the two analyses.

In all figures the integers on the failure tree branches represent the numbers of sequences. The numbers in circles are the step numbers. A plastic hinge formation event is described by the beam/node number pair, while buckling is described by the letter 'b' next to the beam number. Values in parenthesis are the β_{HL} values for the respective event at the current step.

The "level 1" system reliability indexes were also calculated for the two configurations in the case of 60 events selection in step 1. The corresponding β_{sys}^{sub} values computed are 3.16 for the "intact" structure and 2.62 for the "damaged" structure configuration.

In order to investigate the influence of correlation, an additional analysis for the"intact" case was performed where variables H and λ_W were fully correlated (ρ_{H,λ_W}=1.0). It appears that correlation significantly affects the results by decreasing the β_{HL} values (57b-->(3.39), 56b-->3.45) etc.) and the β_{sys}^{sub} (becomes 3.12) as well. The results presented below are obtained however with uncorrelatted variables.

A part of the total failure tree is shown on figure 4. It contains events after step 1. The framed values represent the β_{eq} values of

Table 3: Structural redundancy and system effect factors

Configuration type	Redundancy factor R_0	System effect factor S_0
"Intact"	0.665	1.76
"Damaged"	0.657	1.78

each sequence obtained at the various steps of the analysis. The β_{sys}^{sub} and the corresponding failure probability of each subsystem are also calculated. The identified mechanisms are indicated by the literal m at the ending of the respective branch. In order to obtain an approximation of the true system failure probability we have calculated the two bounds: The probability of failure for the entire system treated (45 sequences) and the one of the system composed of the mechanisms only (24 mechanisms). We therefore have,
$3.28 < \beta_{sys} < 3.31$ or,
$0.461.10^{-3} < P^f < 0.516.10^{-3}$

The same analysis was made for the damaged structure configuration. A system composed of 48 sequences (26 of them were identified as mechanisms) was treated and the resulting bounds for the system reliability are:
$2.76 < \beta_{sys} < 2.82$ or,
$0.240.10^{-2} < P^f < 0.289.10^{-2}$

The following two factors can be considered as simple indicators of the structure redundancy and the system effect: Redundancy factor R_0= (Failure probability of the entire system treated)/(Failure probability of system at "level 0")and System effect factor S_0=(Failure probability of the entire system treated)/(Most likely elementary failure at "level 0"). Table 3 summarises the values of these two factors for the two configurations analysed.

4 CONCLUSIONS

As already suggested in the introduction, the scope of this paper was not to present a new methodology but rather some improvements achieved in successive versions of ARPEJ. We also tried to show that this software allows to perform simply reliability analysis of complex structures submitted to their actual environmental conditions. This is in our opinion an important step towards a daily basis use of reliability analysis by engineers.

ACKNOWLEDGEMENTS

ARPEJ is a common property of CTICM, Elf Aquitaine and Ifremer. CCS has been largely involved in the development of Version 1 presented in this paper. The work of the association CTICM-Elf-Ifremer has been supported by the "Comité d'Etudes Petrolières Marines (CEPM)" in the framework of a general research project concerning the use of advanced procedures for a rational estimate of structural safety.

REFERENCES

/1/J. Labeyrie, Response Surface Modelling of an extreme wave loading, BRITE project P1270, Report B(1.1)5, December 1988.
/2/FORM & SYSREL manuals, RCP GmbH, S. Gollwitzer, T. Abdo and R. Rackwitz, Munich 1988.
/3/Y.Guenard, J. Goyet, B.Remy and J. Labeyrie, Structural Safety of jacket platforms, Marine Structural Reliability Symposium, pp 169-183, October 1987.
/4/S. De Rabi, A. Karamchandani and A. Cornell, Study of redundancy in near-ideal parallel structural system, ICOSSAR 1989.

Structural Safety & Reliability, Schuëller, Shinozuka & Yao (eds) © 1994 Balkema, Rotterdam, ISBN 90 5410 357 4

A reliability-based approach to seismic reassessment of offshore platforms

W.D.Iwan & G.W.Housner
California Institute of Technology, Pasadena, Calif., USA
C.C.Thiel & C.A.Cornell
Stanford University, Calif., USA

ABSTRACT: This paper presents recommendations for minimum performance levels for existing offshore platforms and procedures whereby these levels may be met. The recommendations were developed by a panel formed by the American Petroleum Institute.

INTRODUCTION

Technological innovations, as well as economic and environmental conditions, are causing an increased interest in both life extension and modification of existing offshore platforms. These factors require reassessment of existing platforms to determine their capability to withstand anticipated environmental loads. This process represents a number of challenging technical and public policy issues that are not addressed in current guidelines or regulations.

A panel, consisting of the authors of this paper, was formed by the American Petroleum Institute and charged to develop a rational basis for the seismic reassessment and requalification of offshore platforms in U.S. waters. The Panel was requested to relate recommended performance objectives and reassessment methodologies for offshore facilities to those for onshore facilities. The Panel developed recommendations for minimum acceptable levels of performance of existing offshore platforms and procedures whereby these levels may be demonstrated. These recommendations were developed independently without direction from either the petroleum industry or government regulatory agencies. This paper presents the conclusions and recommendations of the Panel (Iwan, et al. 1992).

PERFORMANCE OBJECTIVES

Four principles are used to guide the development of safety performance objectives and reassessment methodologies for existing offshore structures. These are:

1. The focus should be on limiting, to an acceptable level, the risk due to catastrophic impacts of earthquakes.
2. The seismic life-safety risk posed by an existing platform should be comparable to that posed by well-designed onshore conventional building structures.
3. The seismic environmental risk posed by an existing platform should be no greater than that posed by other major offshore petroleum release sources.
4. Offshore facilities require more rigorous site hazard and engineering behavior analyses than onshore facilities even though they may have comparable risk objectives.

These principles allow the rational development of minimum performance objectives for existing offshore platforms.

Based on the above principles, the following criteria are believed to provide acceptable levels of life and environmental safety for seismic reassessment of existing offshore platforms off the California coast:

1. The platform should have a probability of collapse in an earthquake not to exceed 0.001 per year, or, it must meet the requirements set in the 19th edition of the American Petroleum Institute Recommended Practice for Planning, Design and Constructing Fixed Offshore Platforms (RP 2A) using ground-motion levels associated with a 200 year return period for strength level analysis and a 1000 year return

period for ductility level analysis.

2. The ground motions used in life and environmental safety assessments should be determined from the median results of a probabilistic seismic hazard analysis for the site that considers all sources of uncertainty.

3. Attachments to the platform should meet the seismic requirements of the 1988 Uniform Building Code (UBC).

4. If the size of any potential earthquake-caused spill from one source is less than 2000 barrels, then the probability of the spill from that source need not be considered.

5. If the size of the spill cannot be limited to 2000 barrels, then the estimated probability of occurrence should not exceed 0.001 per year.

6. If flow shut-off devices are used to meet the environmental performance criterion, they should be shown by laboratory tests to operate at a 90% or more success rate.

The specific values recommended are based on comparisons to life-safety requirements for other structures and examinations of the thresholds for environmental safety or other sources of environmental release. The basis for these recommendations is discussed below.

Basis for life-safety performance criteria

There are well established approaches to regulation of the seismic retrofitting of existing buildings that provide useful precedents for reassessment of platforms These are appropriate precedents for setting life-safety reassessment levels since there are no standards for seismic strengthening of platforms.

The seismic design and construction of most conventional buildings are governed by building codes. Within regions of high seismic activity in the U.S., the most common regulatory standard is the Uniform Building Code (UBC). The UBC seeks to minimize, not absolutely eliminate, the loss of life in buildings resulting from major earthquakes. Its objective is to limit the probability of *substantial* loss of life. This objective is approached not through estimates of potential life loss but by adopting standards that are professionally evaluated as giving a high likelihood that structures that comply will not collapse.

The level of seismic risk in the UBC is approached by specifying the probability of the earthquake ground motion to be used. The UBC specifies that the design ground motion is one that has a 10% probability of being exceeded in 50 years, the normal lifetime for typical buildings, which is equivalent to a return period of 475 years, or an annual probability of exceedance of approximately 1/475 = 0.002. This ground motion is used in conjunction with specified response analysis and checking criteria.

Studies made for the Department of Energy estimate the expected life-safety performance of buildings designed to the UBC (Kennedy, *et al.* 1990, Murray, 1991). It is concluded that the annual probability of a *minimally* UBC code-compliant building suffering life-threatening damage during an earthquake is about 0.001, or approximately one-half the annual probability of exceedance of the ground motion used for design. Based on the collective experience and judgment of the Panel, this factor of two between design annual probability of exceedance and the probability of failure for minimally code-designed buildings is believed to be reasonable.

Offshore platforms that comply with recent editions of the RP 2A are likely to perform substantially better than typical UBC compliant land-based buildings. The RP 2A requires a specific site-hazard analysis as well as engineering field inspection and verification programs during construction and installation, and while in service. It provides for a dual-level seismic analysis with design for a "strength" level event and checking for a "ductility" level event. These requirements are expected to yield new platform structures whose seismic performance is comparable to the best of land-based conventional building structures, those that are designed and constructed to a higher standard than the code minimum.

The recommended maximum collapse probability of 0.001 is somewhat higher than that expected for new platform designs. This is consistent with current practice for land-based buildings in the U.S. where typically the ground motion level for retrofit is set at 75%, or less, of that required for new buildings (Thiel, et al. 1991). The 75% UBC ground motion level has a mean return period of under 300 years, implying a relaxation of about a factor of two in annual probability of severe damage.

The recommended approach to seismic reassessment includes independent peer review of the technical aspects of the reassessment process for both the hazard and structural performance analyses. Further, it is recommended that the reassessed platform be regularly surveyed to identify conditions that could compromise its expected seismic performance. Neither of these steps are commonly undertaken for buildings. Therefore, this should result in reassessed platforms which perform significantly better than minimally compliant UBC buildings.

Basis for environmental performance criteria

The process for determining the acceptable level of risk for seismic environmental consequences is not so well established as that for life safety. The source of a petroleum spill, whether earthquake-induced or caused by other processes, accidents, or natural events, is in some respects irrelevant. The critical issues are the size of release and the frequency of occurrence.

There are two approaches to addressing environmental risks posed by platforms. The first is to determine the probability that different levels of consequences will occur. The second approach assesses the likely worst consequences of various environmental risks and determines whether these consequences are within acceptable thresholds. A worst consequences approach to environmental performance has the advantage of being easily applied without complex analysis procedures.

The worst consequences approach is believed to be appropriate for seismic safety reassessment of offshore platforms. Using this approach, environmental issues may be addressed in much the same way as life safety has traditionally been approached for buildings. That is, the likelihood of physical performance of the structure is used as a surrogate for the environmental impacts of its performance. Earthquakes will cause spills only when there is damage to a platform or its associated structures.

The acceptability of earthquake-induced environmental hazards will be determined by: 1) the size of the spill if wells or off-production piping systems naturally flow, 2) the likelihood that a spill will occur which exceeds an acceptable size, 3) the reliability of flow-limiting devices, and 4) the size of the discharge if all on-platform storage is spilled. In order to determine an acceptable size of spill, consideration should be given to sources of spills other than earthquake, including tanker operations, other platform releases, and natural seepage.

The capacity for cleanup of spills must also be examined. In Southern California waters, tanker operations alone are expected, on a probabilistic basis, to cause at least one 2000 barrel spill per year. Since damaging earthquakes at any platform site are very rare events, this is judged to be a reasonable threshold for earthquake-induced spills in that region.

If the size of an earthquake-induced spill from an individual source cannot be limited to less than 2000 barrels from a single source, then the collapse of the platform will dominate the cause of large spills. In this case, the annual probability of collapse should be limited to less than or equal to 0.001. On this basis, large spills from platforms caused by earthquakes will be considerably less likely, by at least an order of magnitude, than similar size spills from tanker operations.

RECOMMENDED APPROACH TO SEISMIC REASSESSMENT

Because of the relatively small number of offshore platforms to which seismic reassessment might be applied, and the diverse nature of their design, construction, and condition, a specific technical standard for reassessment is not warranted. Instead, the appropriate approach to seismic safety reassessment should be a *procedural* one, based on well-articulated objectives and principles of evaluation. It is recommended that the process of seismic safety reassessment be approached as indicated in Figure 1.

In the present approach to reassessment, issues of life safety and environmental safety have been intentionally separated as they are of essentially different character. It is believed that there is no completely acceptable common measure for these very difficult consequences. This is a significant departure from other proposed reassessment methodologies which combine life safety and environmental consequences into one "basket" with a common measure such as dollars.

The reassessment process may be summarized as follows:

1. Determine whether the platform meets the seismic requirements for new platforms as described in the 19th edition of RP 2A with modest modifications. If the platform (with any proposed modifications) meets current standards for a new platform, then an acceptable level of performance is expected and there should be no further requirements other than to implement any proposed retrofit or other modifications.

2. If the platform does not meet modified RP

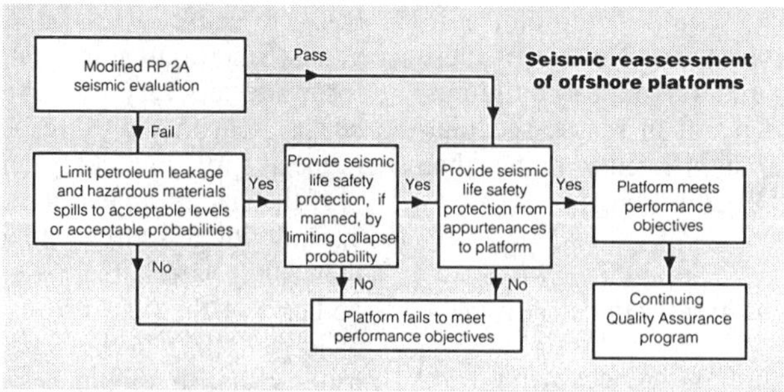

Figure 1. Schematic diagram of procedural approach to seismic safety reassessment of offshore platforms

2A requirements, then the likelihood that the platform poses an environmental hazard larger than the prescribed level should be less than or equal to an annual threshold probability of 0.001. There are two basic ways to meet this requirement: either limit the size of spills, or limit the probability of a spill occurring. If these environmental performance levels cannot be achieved, the platform should fail reassessment.

3. If the platform does not meet the modified RP 2A requirements, then the annual probability that the platform poses a life-safety risk must be less than or equal to 0.001. Two essential life-safety threats should be considered: the structural stability of the platform, and the performance of the appurtenances attached to the platform.

4. When both the life and environmental safety performance objectives have been met, a continuing quality assurance program should be implemented to insure that structural and environmental systems retain their performance integrity.

Modified RP 2A analysis

A platform that meets both the strength and ductility level seismic requirements of the 19th edition of RP 2A should pass reassessment. Alternatively, a platform that meets only the ductility level requirements should pass reassessment provided that a full dynamic time-history analysis is performed using a set of at least three time-histories. In either case, results of both the site seismic hazard determination and structural response analysis must be independently peer reviewed.

So far as specification of ground motion is concerned, it is recommended that the RP 2A analysis for seismic reassessment be performed using the following seismic hazard: for strength level performance, use a ground motion with a 200 year return period, and for ductility level performance, use a ground motion with a 1000 year return.

The alternative to passing a RP 2A reassessment entails determining whether the platform meets acceptable environmental and life-safety performance objectives as indicated above.

Environmental safety determination

For a platform to pass an environmental seismic safety reassessment, the potential pollution from all sources must be limited to within the acceptable performance level. The actions necessary to address the environmental safety assessment are shown in Figure 2. It is possible that the earthquake response of a platform and oil production systems could cause a conductor or riser to be severed at or above the mudline. If the well naturally flows, it is important that the continued flow of oil into the ocean be mitigated. It is believed that the only likely way that wells would be severed, or suffer damage severe enough to leak substantial petroleum, is when the platform collapses or its foundation fails, conditions that are expected to be highly linked. Thus, the environmental considerations should be focused on platform collapse.

If well closure systems are used to control potential releases from naturally flowing wells, the recommended closure system criterion is that the probability of platform collapse (P_f) times the probability of the shutoff system failing to function (P_s) be less than or equal to 0.001. If

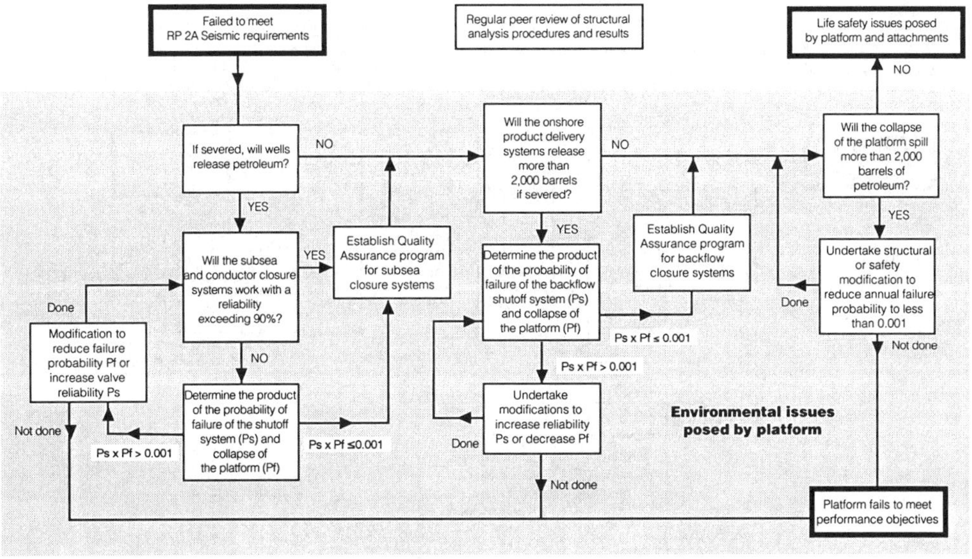

Figure 2. Schematic diagram of environmental safety determination

this performance objective is not met by an existing platform, there are several alternatives. First, the closure systems may be modified so that they meet the RP 14B reliability criteria. Then, if the annual probability of collapse of a platform is less than 0.001, nothing further need be done. Otherwise, it will be necessary to either decrease the probability of platform collapse or increase the reliability of the closure systems, or both.

Product delivery pipelines transporting petroleum and processing fluids from the platform to onshore facilities or to other platforms are another source of environmental spills. Determination of the acceptability of the seismic performance in this case parallels the approach used to limit the consequences of risers and conductors being severed. First, it must be determined whether the pipelines can backflow or not. If they can, such spills are of importance only if the possible amount of leakage from the pipelines and storage tanks to which it is connected exceeds the acceptable release level, in this case, 2000 barrels.

It is possible that storage tanks and large assemblages of processing equipment containing petroleum on the platform could remain intact if a platform were to collapse, and thereby pose no immediate environmental threat. It is also possible that all of the storage capacity could leak at damage levels below the platform collapse level. Thus, rather than focus the acceptance criteria on the probability of spillage, the focus should be on the total spill capacity of the platform. If the maximum spill that could occur from the platform is less than the acceptable release level of 2000 barrels, then possible spills from the platform may be deemed to be acceptable for purposes of reassessment.

Life-safety determination

The life-safety alternative to the modified RP 2A procedure entails determining whether the platform meets the life-safety performance objectives indicated above. There are three basic steps in determining whether the platform poses an acceptable life-safety risk: 1) determining whether the platform is manned, 2) determining the likelihood of platform collapse if manned, and 3) controlling the likelihood of injuries to failure of appurtenances. The actions necessary to address these steps are shown graphically in Figure 3.

A life-safety risk exists only if a platform is manned. While there is always some situation when a platform is occupied, the question of whether or not the platform is manned must be addressed within the context of the life-safety

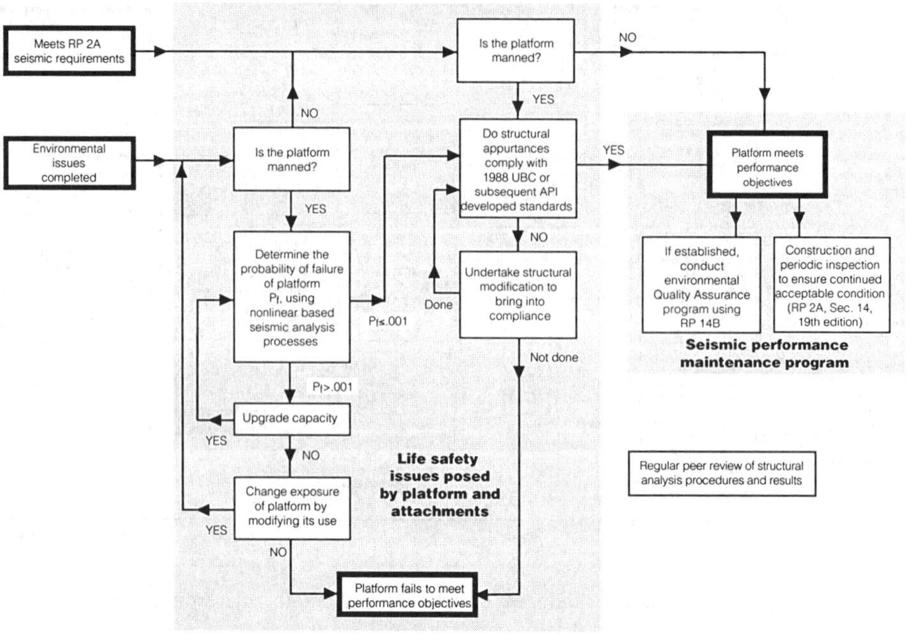

Figure 3. Schematic diagram of life-safety determination.

objective to prevent *substantial* loss of life. A "manned" platform is herein defined as one which is actually and continuously occupied by at least five persons. This definition does not depend on whether living accommodations or quarters are provided on the platform. If a platform is manned, then the life-safety issue is whether the platform has a collapse probability less than or equal to the acceptable level of 0.001 per year.

A basic performance objective is that the life-safety hazard posed by being in the proximity of an appurtenance attached to the platform should be no greater than the hazard posed if the appurtenance was on land. Therefore, all major appurtenances attached to the platform should meet the same requirements that would be applied if they were subject to the applicable sections of building code provisions. UBC provisions are keyed to seismic zones which are specified only for land sites. In this case, the appurtenances are not attached to the ground, but to the platform deck, which will respond to the ground motion input at its base. It is appropriate to use the deck's lateral response spectrum determined from the ground spectrum for the strength level earthquake (200 year return period) modified by an appropriate dynamic amplification factor to characterize the platform deck motions.

PLATFORM COLLAPSE PROBABILITY

A key element in the several parts of the proposed reassessment methodology is the probability of collapse of the platform. There are a number of approaches to assessing the probability of platform collapse including nonlinear dynamic analysis using time histories of predicted or representative ground motions, and nonlinear static analysis. While the former is preferable, it is believed that a displacement-controlled static analysis that fully takes into account the true nonlinear dynamic characteristics of the structure can, if properly performed and interpreted, provide a sufficiently reliable estimate of a platform's seismic performance. Capacity modification factors will need to be used to reflect the differences between static and dynamic behavior, as appropriate, and these should be subject to independent peer review.

There are two general sources of uncertainty that affect the probability of platform collapse: uncertainty in the future ground motion (demand) at the site, and uncertainty in the structural characteristics (capacity) of the platform. Strictly speaking, the probability of failure is the integral of the probability density of the capacity times the probability that the demand is greater than the capacity. For seismic reassessment of existing platforms, it is expected that the probability distribution of the demand will be relatively flat over the range of the capacity distribution. Therefore, the probability of failure can be approximated by determining the probability that the demand exceeds the mean or average capacity of the platform. This is equivalent to determining the probability of failure from the seismic hazard curve at the best or "unbiased" estimate of the capacity. Because the median hazard curve is used for this determination, the result is termed the *median* failure probability. To establish the unbiased estimate of capacity, care must be taken to ensure that the values and models employed do not have embedded within them unrecognized "safety factors" or other assumptions that represent conservative rather than best estimates of the behavioral properties of the platform's materials and systems.

IMPORTANCE OF PEER REVIEW

Independent peer review has become generally accepted as an indispensable element in the seismic design of major structures and facilities to insure that intended performance objectives are achieved. Independent peer review is an essential element of the reassessment process recommended herein. The numerical values assigned to life safety and environmental safety criteria are all founded on the assumption that independent peer review will be employed, beginning at the initiation of reassessment and continuing through any modifications to the platform.

INSPECTION AND SURVEY

After all the environmental and life-safety evaluations have been completed, it remains to insure that the reassessed platform is regularly inspected to determine its condition and that repairs are completed when appropriate. These actions are indicated in Figure 3 as part of the process for determining life safety. Well-reasoned inspection standards presently exist for verifying quality control of the construction process for offshore platforms. Procedures also exist for performing regular surveys to determine the physical condition of a platform. Surveys should, at a minimum, meet the frequency recommendations of RP 2A, Edition 19. If the environmental reassessment was based upon the reliability of well and pipeline closure systems, these should be regularly verified to be in operational order and likely to function in an earthquake. This requires regular inspection of the devices, the controllers, power sources and installation conditions as specified in API RP 14B and C.

CONCLUDING REMARKS

The proposed reassessment methodology was a subject of discussion at the International Workshop on Seismic Design and Reassessment of Offshore Structures which was held at the California Institute of Technology on December 7–9, 1992. The primary sponsor of the workshop was the U.S. Minerals Management Service. A majority of the Working Group on Design, Reassessment and Requalification felt that it was appropriate to separate life safety, environmental consequences, and economic decisions as recommended by the Panel but a consensus on this matter was not reached. The Working Group also endorsed the use of peer review and determined that a strength level earthquake analysis need not be performed for requalification so long as a ductility level analysis is performed. The Working Group felt that further study was needed to define manned and unmanned operations and catastrophic consequences of environmental pollution.

A separate Working Group on Public Policy concurred that peer review was an essential element in seismic design and reassessment of offshore structures. This Working Group also saw the need for additional research on what constitutes a catastrophic event, particularly as it relates to environmental consequences. Finally, this Working Group recommended that the application to other natural hazards of the Panel's approach for seismic reassessment be investigated.

In conclusion, the authors believe that implementation of the seismic reassessment approach recommended herein should meet public expectations for seismic life and environmental safety. A seismic reassessment performed according to the recommended guidelines should yield platforms whose seismic performance will be as good as commercial and industrial buildings onshore in the United States. The authors further believe that the general approach developed herein can be successfully applied to other regions of the world and to other environmental hazards.

REFERENCES

Iwan, W. D., C. C. Thiel, C. A. Cornell, G. W. Housner, 1992. Seismic safety requalification of offshore platforms, American Petroleum Institute, Dallas, Texas, May.

Kennedy, R. P., et al., 1990. Design and evaluation guidelines for department of energy facilities subjected to natural phenomena hazards, Lawrence Livermore National Laboratory Report UCRL–15910, June.

Murray, R. C., 1991. Technical basis, supporting information, and strategy for development and implementation of doe policy for natural phenomena hazards, Lawrence Livermore National Laboratory Report UCRL-ID-108242, September.

Thiel, C. C., F. J. Wilsea, J. P. Singh, 1991. Breaking the pattern: a research and development plan to improve seismic retrofitting practices for government buildings, California Seismic Safety Commission Report SSC 91-05.

ACKNOWLEDGMENT

The authors gratefully acknowledge the support of the American Petroleum Institute and the assistance provided on industry specific issues by a Technical Advisory Committee of the Institute chaired by Mr. Griff Lee. Appreciation is also expressed to Mr. Charles Smith of the U.S. Minerals Management Service.

Structural Safety & Reliability, Schuëller, Shinozuka & Yao (eds) © 1994 Balkema, Rotterdam, ISBN 90 5410 357 4

Probabilistic assessment of hydrostatic pressure test strategies for pipelines

Guoyang Jiao
SINTEF Structures and Concrete, Trondheim, Norway

ABSTRACT: Reliability analysis of hydrotesting for pipelines is performed. It is assumed that hydrotesting is reliable to eliminate weak pipe sections when subjected to a test pressure equal to 1.25 design pressure. The level 2 assessment method is applied for assessment of combined brittle fracture and plastic collapse mode. Importance sampling procedure is used to estimate and update failure probabilities. Through sensitivity analysis, it is observed that hydrotesting is effective in reducing design failure probability, provided that there is no strength degradation during pipeline operation.

1 INTRODUCTION

Pipelines are major tools for transportation of hydrocarbons from oil and gas fields. It is important that the design of pipelines is based on an adequate safety margin due to serious consequences and high cost associated with failure and repair. The pipeline design practice is thus restrictive with respect to structural design criteria and inspection and testing requirements. These design criteria limit the utilization of structural and material capacities so that probability levels of exceedance are very low. The inspection and testing requirements are to qualify the pipeline system after it is installed.

Determination of the pipeline steel wall thickness is a basic design task. The foremost consideration is to contain a usually high design pressure which induces a high hoop stress in the steel wall. The current design criterion for pressure containment design is to limit the design hoop stress well below the yield stress level. After installation, hydrostatic pressure test is further performed by filling the pipe with high pressure water which results in a hoop stress level close to the yield stress. Such hydrotesting aims at removing weak joints which do not survive the test.

As it is costly for performing hydrotesting, it is then of great interest to investigate the effects of hydrotesting on the pipeline safety. Qualitatively, this strategy is able to remove pipe section with exceptionally low strength, or with unacceptably large defects, which are not accounted for during design. Should the hydrotesting be effective in reducing the failure probability, it may be possible to achieve a cost optimal design which balances the design criterion and the testing requirement.

It is thus the aim of this paper to investigate the hydrotesting strategy for pipelines using the reliability approach. The structural failure mode is defined as fracture or plastic collapse of the pipe wall in the circumferencial direction subjected to high tensile hoop stress. Importance sampling procedure is applied to assess and update failure probability levels of pipelines before and after the hydrostatic pressure test.

2 DESIGN REQUIREMENTS

A pipeline system is to be hydrostatically pressure tested after installation. The pressure test is to prove the strength and tightness of the tested pipe sections. If burst or leak occurs, the failure is to be corrected and the section retested. The minimum test pressure may vary from land pipelines to offshore pipelines. Typically required test pressure for qualifying a pipeline for a maximum allowable operating pressure (MAOP) equal to the internal design pressure are as follows:

a) The testing pressure is to be 1.25 times design pressure, ANSI/ASME (1979), DnV (1981), or 1.5 times design pressure, ANSI/ASME (1980), BSI (1990).

b) The testing pressure should normally not result in a hoop stress exceeding 90% of the specified minimum yield strength (SMYS), BSI (1990), DnV (1981).

In the current design practice for pipelines, the design pressure is determined by limiting the hoop stress to 72% SMYS. When the test pressure is 1.25 and 1.5 times the design pressure, the resulting hoop stress is then 0.90 SMYS and 1.08 SMYS, respectively. The design requirements regarding to pressure containment are thus summarized as

$$\sigma_{Hc} \leq \eta_1 \; SMYS \qquad (1)$$

and

$$\sigma_c > \eta_2 \; SMYS \qquad (2)$$

in which σ_{Hc} is the characteristic hoop stress, σ_c is the stress at failure, η_1 is the usage factor for design, and η_2 is the usage factor for pressure test. The hoop stress may be calculated by the following equation

$$\sigma_H = \frac{(p_i - p_o)D}{2t} \qquad (3)$$

in which p_i and p_o are internal and external pressure, respectively, D is the steel pipe diameter, and t is the steel pipe wall thickness. The external pressure is only relevant for offshore pipelines.

3 FAILURE ASSESSMENT MODEL

When the pipe is subjected to high tensile hoop stress, failure is mainly related to brittle fracture

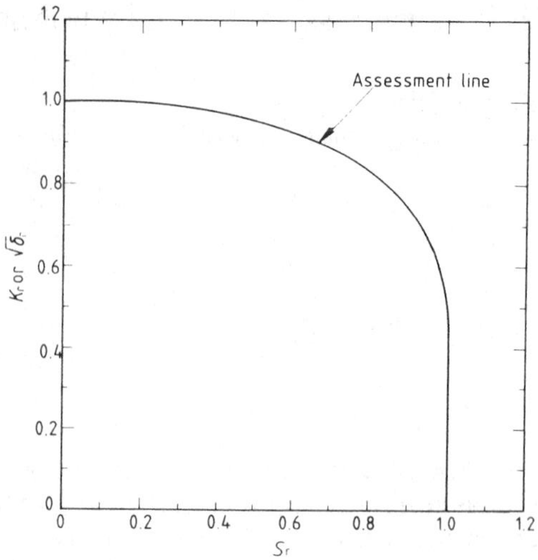

Figure 1: Level 2 failure assessment diagram

of defects, or collapse of the net section, or in a combined manner. A standard failure assessment model is then the level 2 assessment method, as specified by PD 6493 (1991).

Due to a very high hoop stress during hydrostatic pressure test, possible ductile crack growth during the test period, typically 24 hours, may be relevant. This problem is assessed by Leis and Brust (1990). Their model is concerned with ductile time-dependent growth of axial part through surface cracks in pipelines with extended "hold" times at high pressures which result in a hoop stress close or even above uniaxial yield stress. The crack driving force is assessed by J-integral analysis following the EPRI Scheme with time-dependent effects accounted for. A general conclusion from their study is that cracks are stable at hold pressures causing hoop stresses less than 0.9 SMYS, even for very deep flaws. Hence, the standard level 2 assessment method is applicable.

The level 2 method is based on the plane stress strip yield model approach. The equation of the assessment curve, as shown in Figure 1, is given by, PD 6493 (1991),

$$\sqrt{\delta_r} = S_r \left(\frac{8}{\pi^2} \ln \sec(\frac{\pi}{2} S_r) \right)^{-0.5} \qquad (4)$$

in which

$$S_r = \frac{\sigma_n}{\sigma_f} \quad (5)$$

$$\sqrt{\delta_r} = \sqrt{\frac{\delta_I}{\delta_{mat}}} + \rho \quad (6)$$

where σ_n and σ_f are net section stress and flow stress, respectively, δ_{mat} is the material toughness measured by CTOD method, δ_I is the applied CTOD, and ρ is the plasticity correction factor. The applied CTOD is further calculated by

$$\delta_I = \frac{(Y\sigma)^2 \pi a}{\sigma_Y E} \quad (7)$$

in which a is the defect size, E is the elastic modulus, σ_Y is the yield strength, σ is the applied stress, and Y is the stress intensity factor correction. Further details may be found in PD 6493 (1991).

4. RELIABILITY APPROACH

To assess implications of hydrostatic pressure test with regard to design safety, an importance sampling procedure is applied for calculating and updating reliability. Based on the selected failure assessment model, the limit state function is then defined as

$$g(x) = \sigma_c - \sigma_H(\eta_1) \quad (8)$$

in which the stress at failure σ_c is determined by the level 2 assessment curve as given by Eq. (4), and the applied hoop stress σ_H is determined by Eq. (3) with the characteristic stress level given by Eq. (1).

When the pipe survives the testing pressure, this event state can be expressed as

$$g_1(x) = \sigma_H(\eta_2) - \sigma_c < 0 \quad (9)$$

After hydrotesting without failure, or after retesting with failure correction, the failure probability is then calculated by

$$P_f = P[\,(F_1 \cap E_1) \cup (F_2 \cap \overline{E}_1 \cap E_2)\,] \quad (10)$$

in which F_1 is the failure event for the initial pipe section, E_1 is the event with no failure during testing, F_2 is the failure event for the replaced pipe section, E_2 is the event with no failure during retesting. When no failure occurs during testing, the limit state is then simply

$$P_f = P\left(g(x)<0 \cap g_1(x)<0\right) \quad (11)$$

The ISPUD program, Bourgund and Bucher (1986), is used for reliability analysis. As the limit state surface for this problem may be very complicated, extensive simulations around the design point may be necessary.

5 UNCERTAINTY MEASURES

The reliability implications of pressure testing are highly dependent upon uncertainty measures associated with strength and load (pressure) quantities. While most of the parameters involved in the limit state, Eq. (8), and event state, Eq. (9), are random variables, only several random variables may be regarded as important with respect to their contributions to the failure probability, namely, fracture toughness, flow stress, defect size, pipe wall thickness, and internal design pressure.

The CTOD based fracture toughness, or the critical CTOD, depends on the actual pipe size, the in-service temperature, as well as the location where the fracture is likely to occur. It is generally agreed that δ_{mat} is a random variable with large uncertainty. It is recommended that a lognormal distribution or a Weibull distribution be applied to model this random variable. And further, its characteristic value may be defined as the mean value minus one standard deviation, Burdekin et al. (1988). Typical characteristic values are between 0.05 - 0.3 mm. The lower bound may be more relevant for land pipelines. The coefficient of variation (cov) may vary depending on the size of test samples available. A representative cov may be between 0.5 - 0.9.

The most important characteristics of the defect is its depth a. In general, a is the sum of its initial size before testing, a_0, and possible crack increment after testing due to e.g. crack growth under pressure cycles, Δa, that is,

$$a = a_0 + \Delta a \quad \textbf{(12)}$$

It is also possible that a more critical defect is generated after the pressure test, due to e.g. accidental external damage. For axial part through defects, it is found that a_0 may follow an exponential distribution, Corder and Feanebough (1987), which implies a cov equal to 1.0. The mean value of a_0 depends on the type of defects considered. For seam weld defects, a mean value less than 1.0 mm is expected in general; while it may be up to a few mm if defects caused during installation are assumed.

Uncertainty in the steel pipe wall thickness is associated with the fabrication tolerance, corrosion protection measures such as coatings or allowance, and corrosion rate. Corrosion is also dependent upon pipeline types (land vs. offshore, gas vs. oil, etc.). Typically, a normal distribution is assumed for the thickness, Ellinas et al. (1987). A cov of 5% may be on the safe side. The characteristic value may be defined as 95% of its mean value.

Different models have been applied for calculation of the steel flow stress, Denys (1992). As specified in PD 6493 (1991), the following definition is commonly used,

$$\sigma_f = \frac{\sigma_Y + \sigma_u}{2} \quad \textbf{(13)}$$

in which σ_u is the ultimate tensile strength. With this model, uncertainty in the flow stress is then related to those in the yield strength and ultimate tensile strength. The yield strength is generally assumed as a lognormal distribution, with a cov typically varying from 3 - 8%, Ellinas et al. (1987). The SMYS is a statistical minimum of the yield strength, which may be defined as a lower 5% fractile of the distribution. The ultimate tensile strength is correlated with the yield strength. Their ratio, however, may be modelled as an independent lognormal variable with a small cov.

The internal design pressure is equal to or above MAOP. In general, it should not be exceeded except for specified transient operations when 1.1 MAOP may be the maximum allowable limit, BSI (1990). Taking into account probable exceedance of the internal design pressure, a normal distribution may be assumed for the design pressure. As the probability of exceedance is in general low, a small cov should be considered.

It should be mentioned that the model uncertainty for the level 2 assessment equation may be also important. It is generally regarded to give conservative predictions, although no sound statistics is available for estimation of the model uncertainty. It may be assumed that it is conservative without accounting for this model uncertainty.

6. CASE STUDIES

To quantify the effects of hydrostatic pressure test on the design safety, case studies are thus made. Since pipeline design can be very case specific with regard to, e.g. material selection, manufacturing process, quality control, the aim of the following case studies is thus to see the effects by performing sensitivity analysis. It is assumed that hydrotesting is reliably performed so that any failure by brittle fracture or plastic collapse is corrected.

The considered material is steel grade X65 with SMYS = 450 MPa. The elastic modulus is a constant as E = 210 GPa. The ultimate tensile strength to yield strength ratio is fixed as 1.18. The residual stress to yield strength ratio is assumed as 0.2. The pipe diameter is 500 mm. The nominal pipe wall thickness is 10 mm. The

Table 1. Uncertainty measures for case studies

Variable	distribution	mean	cov
δ_{mat} (mm)	lognormal	0.167 0.333 0.667	0.70
a_0 (mm)	exponential	0.5 1.0 1.5	1.00
a/c	lognormal	0.2	0.25
t (mm)	normal	10.5	0.04
p (MPa)	normal	12.96	0.04
σ_y (MPa)	lognormal	499.5	0.06
Δa (mm)	normal	0.5 1.0	0.40

design pressure is 12.96 Mpa. The test pressure is 16.2 Mpa. In addition, seven random variables are considered in reliability analysis. Their uncertainty measures are summarized in Table 1. Three mean values are selected for the toughness, which correspond to characteristic values of 0.05, 0.1 and 0.2 mm, respectively.

In the first case study, it is assumed that no strength degradation will occur after testing. Two mean values of the initial defect depth are selected as 1.0 and 1.5 mm. The failure probability levels before and after hydrotesting are estimated corresponding to three different thoughnesses. For each run, up to 1 million simulations are performed to ensure accuracy. As shown in Figure 2, hydrotesting is able to reduce the failure probability by between one to two orders of magnitude. This is because hydrotesting is effective in eliminating pipe sections with very low toughness and/or very large defects.

It should be mentioned that failure probability levels shown in Figure 2 are conditional. This is because the defect depths considered are large, which are caused by accidental events. When evaluating total failure probability levels, the probability levels of such accidental events, which are generally small, need to be multiplied to the conditional structural failure probabilities. Hence, generally much smaller total failure probability levels are expected.

In the second case study, it is assumed that there is crack growth during pipeline operation. This crack growth thus occurs after hydrotesting. Two mean Δa values are selected, namely 0.5 and 1.0 mm. The mean value of a_0 is assumed as 0.5 mm. The failure probability levels before and after hydrotesting are shown in Figure 3. Although hydrotesting is still able to reduce the failure probability, this reduction becomes insignificant when there is a considerable crack growth. This is because hydrotesting is no longer able to eliminate large defects which are generated after the testing.

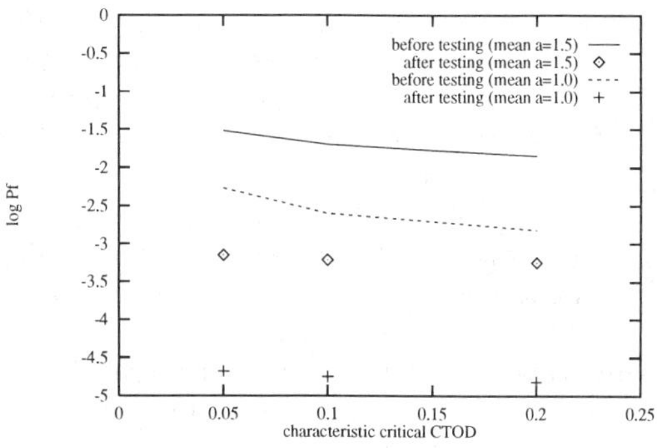

Figure 2. Case study 1

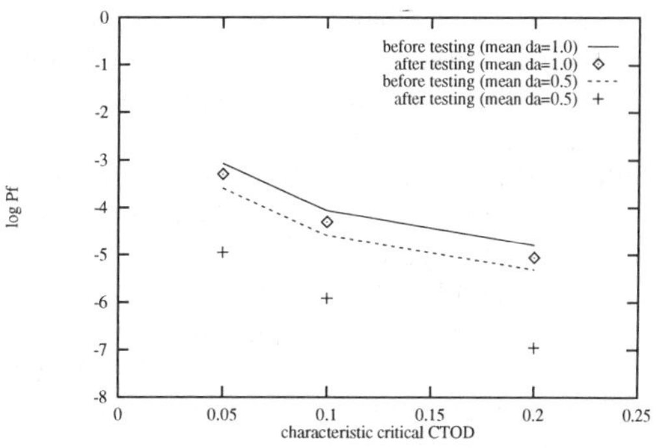

Figure 3. Case study 2

7. CONCLUSIONS

Through sensitivity analysis, it is observed that hydrotesting is effective in reducing design failure probability levels by orders of magnitude, provided that there is no strength degradation during pipeline operation. When crack growth after testing is concerned, however, the effectiveness becomes insignificant when the crack growth is considerable compared with its initial size during testing.

These conclusions are based on case studies performed in this paper only. Pipeline design may be very case and type specific. Hence, further studies are recommended in order to obtain firm conclusions.

8. REFERENCES

ANSI/ASME (1979), "B31.4 - Code for liquid petroleum transportation piping systems", American National Standards Institute.

ANSI/ASME (1980), "B31.3 - Code for chemical plant and petroleum refinery piping", American National Standards Institute.

Bourgund U and Bucher C G (1986), "Importance sampling procedure using design points - ISPUD - a user's manual", Institute of Engineering Mechanics, University of Innsbruck, Austria.

BS 8010 (1990), "Draft british standard 8010 code of practice for pipelines - part 3 pipeline subsea: design, construction and installation", British Standards Institution.

Burdekin F M et al. (1988), "The basis for the technical revisions to the fracture clauses of PD 6493", International Conference on Weld Failure, London.

Coder I and Fearnehough G D (1987), "Prediction of pipeline failure frequencies", the 2nd Int. Conf. on Pipes, Pipelines and Pipeline Systems, Utrecht, the Netherlands.

DnV (1981), "Rules for submarine pipelines", Det norske Veritas.

Ellinas C P et al. (1986), "Limit state philosophy in pipeline design", J. of Offshore Mechanics and Arctic Engineering.

Leis B N and Brust F W (1990), "Ductile crack growth model and its implications with regard to optimum hydrotest strategies", Pipeline Technology Conference, Oostende, Belgium.

PD 6493 (1991), "Guidance on methods for assessing the acceptability of flaws in fusion welded structures", British Standards Institution.

Structural Safety & Reliability, Schuëller, Shinozuka & Yao (eds) © 1994 Balkema, Rotterdam, ISBN 90 5410 357 4

Modelling of wave-induced drift forces at displaced position of compliant offshore platforms

Ahsan Kareem
University of Notre Dame, Ind., USA

Yousun Li
Shell Development Company, Houston, Tex., USA

ABSTRACT: The displacement and velocity dependent wave induced forces may introduce low frequency drift forces and wave drift damping. The compliant offshore structures, e.g., tension leg platform (TLP) are configured by design to experience large excursions under the environmental load effects, therefore, these contributions may become particularly significant. This paper focuses on the displacement dependence of wave induced loading. Computationally efficient procedures in the time and frequency domains that permit inclusion of the time-dependent drift forces, introduced by the platform displacement, in terms of linear and nonlinear feedback contributions are presented.

1. INTRODUCTION

Traditionally, the dynamic response analysis of conventional fixed jacket-type offshore structures is based on the assumption that the structural displacement is small. Although this assumption appears to be quite reasonable for fixed structures, it may have serious implications for compliant structures that are expected to experience large excursions in surge motion (Mercier, 1982; Kareem, 1985; and Kareem and Li, 1988). McIver (1976) and Rainey (1977) emphasized that the wave effects on compliant structures such as TLPs should be calculated at their instantaneous displaced position and not at their undisplaced position. It was also noted that the resulting time-dependent terms may give rise to unstable solutions. Spanos and Agarwal (1984), utilizing a simplified single-degree-of-freedom model of a TLP, demonstrated the existence of a steady offset component in the structural response for both deterministically and probabilistically described wave fields. Kareem and Li (1988) reported a procedure based on a feedback concept to model the influence of displacement induced effects on tension leg platforms. In this formulation, the wave-induced loads evaluated at the displaced position of a compliant platform result in loading contributions at frequencies outside the range of the wave excitation frequencies, i.e., at low and higher frequencies representing sum and difference frequency components. Therefore, a compliant platform like a TLP experiences steady and slowly varying forces referred here as feedback drift force. These drift forces are in addition to those resulting from second-order wave diffraction and viscous forces which are not addressed in this paper.

The hydrodynamic forces acting on moored vessels and compliant structures also depend on the body motions. In this context, the velocity dependent description of the wave drift force in surge results in providing an additional damping term that is referred to as wave drift damping. This contribution can be quantified either experimentally, or computationally (Wichers and Hijsman, 1984 and Faltinsen and Zhao, 1989).

This paper describes a computationally efficient procedure for computing the wave-induced forces on a TLP at its displaced position. Efficient numerical methods for evaluating the displacement feedback forces in the time and frequency domain approaches are introduced. The displacement-induced feedback forces consist of steady and slowly varying components resulting from both the potential and viscous load effects.

2. THEORETICAL BACKGROUND

2.1 Description of Wave Kinematics and Wave Forces Evaluated at the Platform Displaced Position

Traditionally for fixed offshore platforms, the wave kinematics and diffraction forces are described in terms of a linear transform of the reference wave-surface elevation. For example, the wave diffraction force vector with respect to the platform-fixed coordinate system may be expressed as

$$\boldsymbol{F}_S(t) = \int_{-\infty}^{+\infty} \eta(f)\, \boldsymbol{H}_S(f)\ exp\,(j2\pi ft)\, df, \qquad (1)$$

in which $H_S(f)$ is the transfer function vector which relates the wave-surface elevation and the diffraction forces, and $\eta(f)$ is the Fourier transform of water surface elevation $\eta(t)$ at the origin of the space-fixed reference frame. Any modification of the flow field due to platform motion is not included here. The preceding equation is based on the assumption that the platform-fixed coordinates coincide with the space-fixed frame. In fact, for a compliant structure, the discrepancy between these two coordinate systems may become critical. Therefore, representation of the diffraction force vector acting on a platform experiencing a displacement $X(t)$ is given by

$$F_S(t, \xi(X)) = \int_{-\infty}^{+\infty} \eta(f, \xi(X)) H_S(f) \exp(j2\pi ft)\, df \tag{2a}$$

where $\xi(X)$ represents the instantaneous location of the platform center in the direction of wave propagation, which is derived from the global displacements $X(t)$. Accordingly, the wave surface elevation at the displaced platform location is expressed by

$$\eta(f, \xi(X)) = \eta(f) \exp[-jk\xi(X)], \tag{2b}$$

where k is the wave number. Combining Eqs. (2a) and (2b) the following relationships are obtained

$$F_S(t, \xi(X)) = \int_{-\infty}^{+\infty} \eta(f) H_S(f) \exp(-jk\xi(X)) \exp(j2\pi ft)\, df \tag{3}$$

or

$$F_S(t, \xi(X)) = \int_{-\infty}^{+\infty} F_S(f, \xi(X)) \exp(j2\pi ft)\, df, \tag{4}$$

where $F_S(f, \xi(X))$ can be directly written from Eq. (3) as

$$F_S(f, \xi(X)) = F_S(f, 0) \exp(-jk\pi(X)). \tag{5}$$

The difference between $F_S(t, \xi(X))$ and $F_S(t, 0)$ is that the former describes the diffraction force vector on a platform with an instantaneous displacement $\xi(X)$ and the latter represents the force on a platform at its undisplaced position. A similar approach is appropriate for the treatment of viscous forces

2.2 Numerical Simulation by Summation of Trigonometric Functions

The summation of trigonometric functions may be utilized to compute the wave particle velocity and the associated force acting on a platform at its displaced position. For example, the surge diffraction force, $F_S(t, \xi(X))$, is given by

$$F_S(t, \xi(X)) = \sqrt{2} \sum_{p=0}^{N_f} \sqrt{G_{F_S}(f_p) \Delta f} \cos 2\pi f_p t - k\xi(X) + \Phi_p + \varepsilon_p \tag{6}$$

and the horizontal wave particle velocity, $u(t, x)$, is given by

$$u(t, x) = \sqrt{2} \sum_{p=0}^{N_f} \sqrt{G_u(f_p) \Delta f} \cos(2\pi f_p t - k\xi(x) + \Psi_p + \varepsilon_p) \tag{7}$$

in which $G_{F_S}(f_p)$ and $G_u(f_p)$ denote the spectral density functions of surge diffraction force and the horizontal water particle velocity, respectively, f_p represents $p\ \Delta f$, p is an integer number, Δf is the frequency increment, Φ_p and Ψ_p describe the phase differences, ε_p is a random number uniformly distributed between 0 and 2π, and $N_f \Delta f$ represents the total frequency range. This approach is only limited to the time domain analysis, but the large number of summations involved in the preceding equations makes it computationally inefficient.

2.3 Forces Calculated at Displaced Position in Feedback Form

Introducing a Taylor's series expansion of the global diffraction force vector $F_S(t, \xi(X))$ about $F_S(t, 0)$ (the diffraction force vector on the structure at its undisplaced position) gives

$$F_S(t, \xi(X)) = F_S(t, 0) + \left.\frac{\partial F_S(t, \xi(X))}{\partial \xi}\right|_{\xi=0} + \xi(X) + \frac{1}{2}\left.\frac{\partial^2 F_S(t, \xi(X))}{\partial \xi^2}\right|_{\xi=0} \xi^2(X) + \dots \tag{8}$$

The preceding equation may be expressed in terms of the Fourier transform of the diffraction force vector by utilizing the relationship, $\left.\frac{\partial^n}{\partial \xi^n} F_S(f, \xi(X))\right|_{\xi=0} = (-jk)^n F_S(f, 0)$.

$$F_S(t,\xi(X)) = F_S(t,0) + 2Re$$

$$\left\{\sum_{n=1}^{\infty} \frac{(-1)^n}{n!} \int_0^{+\infty} F_S(f,0)\,[jk\xi(X)]\,exp\,(j2\pi ft)\,df\right\} \quad (9)$$

The previous equation may be recast in a feedback form,

$$F_S(t,\xi(X)) = F_S(t,0) +$$

$$\theta_S^{[1]}(t)\,\xi(X) + \theta_S^{[2]}(t)\,\xi^2(X) + \ldots \quad (10)$$

where $\theta_S^{[i]}(t)$ $(i = 1, 2, \ldots)$ denotes time-variant *coefficient vector of the potential feedback* associated with the six degrees of freedom. The Fourier transform, $\theta_S^{[i]}(f)$ $(i = 1, 2, \ldots)$, of the potential feedback coefficients, according to Eqs. (9) and (10), is given by

$$\theta_S^{[n]}(f) = \frac{1}{n!} + (-jk)^n F_S(f,0)\,. \quad (11)$$

The computation of the wave-induced viscous forces involves simulation of the wave particle velocities over various locations on the wetted surface of the platform. By a procedure similar to the one described above for the diffraction forces, the wave particle velocities at any location may be expanded in terms of time-dependent coefficient vector of the velocity feedback terms. In Eq. (10) the total wave diffraction force vector is expressed as a sum of the force experienced by the platform at its undisplaced position and the contributions resulting from the platform displacement. This may be viewed as a combination of linear, quadratic and higher-order feedback terms pertaining to the wave diffraction force. The schematic representation of a nonlinear displacement feedback system for diffraction forces is shown in Fig. 1.

The velocity dependent hydrodynamic forces can also be expressed in terms of a Taylor series expansion

$$F(t,\dot{X}(t)) = F(t;0) + \frac{\partial F(t;0)}{\partial \dot{X}}\dot{X}(t) \quad (12)$$

in which $F(t;0)$ is the hydrodynamic force on a floating structure with zero velocity, and the second term is the feedback contribution that results in addition damping. $\frac{\partial F(t;0)}{\partial \dot{X}}$ represents the wave drift damping coefficient.

3. INFLUENCE OF PLATFORM DISPLACEMENT ON LOADS

In this section, the influence of displacement on wave loads is examined in light of the statistical relationship between wave loads on a fixed platform and the platform displacement. The coherence function and phase difference are utilized to express a level of statistical relationship. A compliant platform, e.g., TLP, under uni-directional waves and wind experiences response at the wave frequencies, which is fully coherent with wave loads, and low frequency response component induced by wave-drift forces and wind. The focus of this analysis is to examine interaction between fully coherent applied loads and response, and the effect of low frequency response on the applied loads.

For the sake of illustration, it is assumed that both the applied load and displacement are narrow-banded processes. The load is expressed by the following:

$$F(t,0) =$$

$$\frac{1}{\sqrt{10}}\sum_{p=0}^{19} \cos\,[2\pi\,(0.1 + 0.0015p)\,t + \varepsilon_p] \quad (13)$$

where p is an integer number and ε_p is a random number. For the sake of illustration, lets assume that the displacement response is given by

$$\xi(t) = \sqrt{3.6}\sum_{p=0}^{19}$$

$$\cos\,[2\pi\,(f_{lower} + 0.0015p)\,t + \phi_p + \varepsilon_p] \quad (14)$$

where f_{lower} denotes the lower frequency in a band-limited response given by Eq. (14), and ϕ_p denotes the phase difference between the load and displacement. The exact form of the time history of $F(t,\xi)$ is given by Eq. (6),

$$F(t,\xi) = \frac{1}{\sqrt{10}}\sum_{p=0}^{19}$$

$$\cos 2\pi\,(0.1 + 0.0015p)\,t +$$

$$\varepsilon_p - \frac{[2\pi\,(0.1 + 0.0015p)]^2}{g}\xi(t) \quad (15)$$

Its approximate form is obtained by the proposed expansion described earlier (Eqs.8 & 10). By utilizing the FFT algorithm, the spectral density functions of $F(t,0)$, and the exact and approximate descriptions of $F(t,\xi)$ are estimated (the length of the time history used here is equal to 20,000 seconds and the

Table 1: Feedback of Displacement Response - Cases 1,2,3

	Case 1		Case 2		Case 3	
	Mean	Variance	Mean	Variance	Mean	Variance
No Feedback	0.	1.	0.	1.	0.	1.
Exact	-3.09	1.08	0.	0.844	0.	1.019
Approximate	-3.02	1.06	0.	0.843	0.	1.011

Table 2: Response Computed by Different Methods

Item	Time domain	Freq. domain
Mean offset	0.560 m	0.637 m
St. dev. (< 0.02 Hz)	1.54 m	1.57 m
St. dev. (> 0.02 Hz)	2.32 m	2.30 m

time increment is 0.5 second). The response is examined for different phase angles, ϕ_p, and the low frequency cut off values, f_{lower}.

1) Case 1: Let us consider full coherence (f_{lower} = 0.1 Hz) between the loading and displacement with a phase angle of $\phi_p = \pi/2$. The low and high frequency parts resulting from the feedback to the total load are present. The response statistics are listed in Table 1. It is noted that the feedback contributes to the mean and the low and high frequency portions of the load.

2) Case 2: In this case, consider full coherence (f_{lower} = 0.1 Hz) between the loading and displacement, and a phase angle of $\phi_p = 0$. It is noted that the mean and low frequency parts of the feedback contribution vanish, but the high frequency part remains practically unchanged from Case 1 (Table 1).

3) Case 3: Consider a low frequency displacement (f_{lower} = 0.0015 Hz) by wind or wave drift loads. It is noted that the mean and low and high frequency parts of the feedback contribution have disappeared in (Table 1).

These cases can be easily explained by the first-order feedback term. If we have two narrow-banded processes $F(t)$ and $\xi(t)$ centered at ω_F and ω_ξ respectively. Let us define a function

$$F(t,\xi) = \hat{F}(t)\cos\left[\omega_F t + \phi_F + k\hat{\xi}(t)(\omega_\xi t + \phi_\xi)\right] \quad (16)$$

in which $\hat{F}(t)$ and $\hat{\xi}(t)$ are slowly varying amplitude envelopes. A Taylor series expansion of the preceding expression following Eq. 10 leads to

$$F(t,\xi) = \hat{F}(t)\cos((\omega_F + \phi_F) +)$$

$$k\hat{X}(t)\sin(\omega_F + \phi_F)\cos(\omega_\xi + \phi_\xi) =$$

$$\hat{F}(t)\cos(\omega_F + \phi_F) +$$

$$\frac{1}{2}k\hat{F}(t)\hat{X}(t)\sin[(\omega_F + \omega_\xi)t + \phi_F + \phi_\xi] +$$

$$\frac{1}{2}k\hat{F}(t)\hat{X}(t)\sin[(\omega_F - \omega_\xi)t + \phi_F - \phi_\xi] \quad (17)$$

In the right-nand side of the preceding equation, the second term is the sum frequency component and the third term is the difference frequency component. When ω_ξ is much smaller than ω_F, these two terms virtually have almost the same frequency range as the first term, $F(t,0)$, but is much smaller in magnitude. Hence, the feedback can be neglected as shown in case 3. However, if $F(t)$ and $\xi(t)$ are in the same frequency range (i.e. $\omega_F = \omega_\xi$) and fully-coherent, the second term represents high frequency component induced by the feedback which has little meaning in the study of low frequency compliant platform response. The value of the third term, which follows the envelope function of $k\hat{F}(t)\hat{\xi}(t)$, depends on $\hat{F}(t) - \hat{\xi}(t)$. Case 1 represents a completely out-of-phase condition that re-

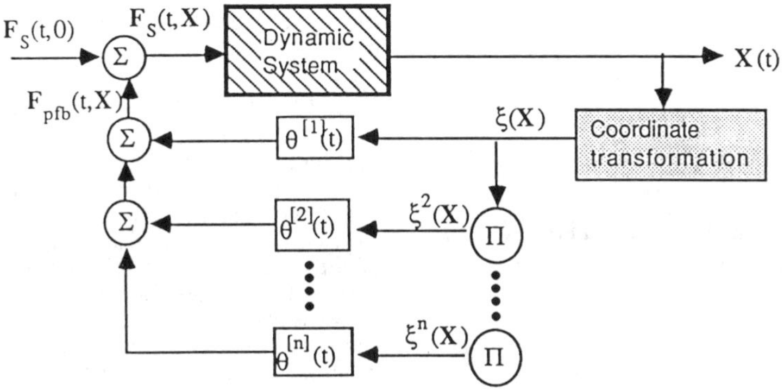

Fig. 1. Displacement Induced Feedback Force

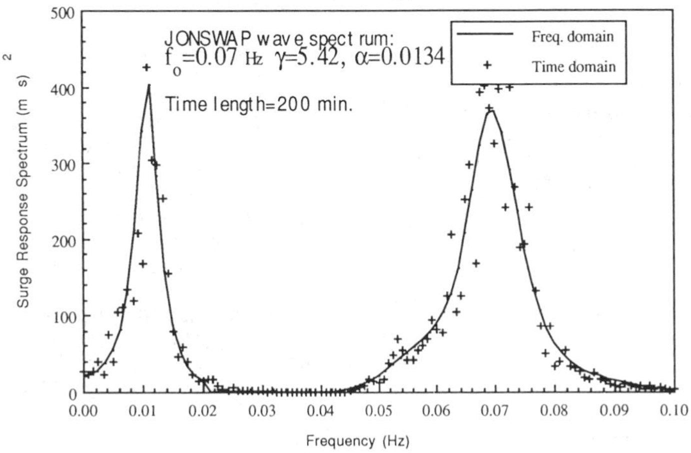

Fig. 2. Spectral Density Function of Displacement-Induced Surge Drift

sults in a low frequency feedback force, while case 2 is an example of a completely in-phase condition which contributes a negligible level of a low frequency component.

The above cases demonstrate that the feedback of the response of a compliant offshore platform to wave loads can induce a low-frequency force. This force herein is referred to as a feedback drift force. The feedback drift force largely depends on the phase differences between the displacement and wave loads. The feedback resulting from low frequency effects of the wind and drift induced response may be ignored as noted in Case 3.

4. TIME DOMAIN ANALYSIS

In the time domain simulation of wave field, ARMA and other digital filters are numerically more efficient than a direct summation of trigonometric functions. In this presentation, the simulation of the feedback forces is integrated into the overall digital filters utilized to simulate wave load effects at an instantaneous displaced position of the platform.

The time-dependent diffraction coefficients described in Eq. 10 may be obtained by the convolution of the wave diffraction forces computed at the undisplaced position of the platform with an appropriate kernal. The first-order and second-order feedback coefficients have the following transfer functions,

$$h_\theta^{[1]}(f) = -jk;\; h_\theta^{[2]}(f) = -\frac{1}{2}k^2. \tag{18}$$

Hence, the first- and second-order feedback coefficients may be written as

$$\theta_S^{[1]}(t) =$$

$$\int_0^{+\infty} [\boldsymbol{F}_S(t-\tau,0) - \boldsymbol{F}_S(t+\tau,0)]\, h_\theta^{[1]}(\tau)\, d\tau \tag{19}$$

and

$$\theta_u^{[2]}(t) = \int_0^{+\infty} [u(t-\tau,0) + u(t+\tau,0)]\, h_\theta^{[2]}(\tau)\, d\tau \qquad (20)$$

In the preceding equation $h_\theta^{[1]}(t)$ and $h_\theta^{[2]}(t)$ represent the convolution kernels that are obtained from Eqs. (21) and (22):

$$h_\theta^{[1]}(t) = 2\int_0^{+\infty} k\sin(2\pi f t)\, dt,$$

$$h_\theta^{[2]}(t) = -\int_0^{+\infty} k^2\cos(2\pi f t)\, dt. \qquad (21)$$

The higher order feedback may be computed by the procedures similar to the first and second-order feedback terms. However, if the origin of the global coordinates is established close to the estimated mean platform offset, i.e., $k\xi(X) << 1$, then terms up to the second-order are sufficient.

5. FREQUENCY DOMAIN ANALYSIS

In the frequency domain, Eq. (10) can be recast in the following form,

$$F_S[t,\xi(\tilde{X})] = F_S(t,0) + \theta_S^{[1]}(t)\,\xi(\tilde{X}) + \theta_S^{[2]}(t)\,\xi^2(\tilde{X}) + \ldots \qquad (22)$$

in which $\xi(\tilde{X})$ denotes the fluctuating displacement components along the direction of wave propagation.

Following the properties of the Gaussian random processes, it can be shown that the expected value of the third term on the right-hand side of Eq. (22) vanishes: $E[\theta^{[2]}(t)\,\xi^2(\tilde{X})] = 0$. The second term on the right-hand side now represents the potential feedback forces. This may be further decomposed into

$$F_{pfb}(t,\xi(\tilde{X})) \approx \theta_S^{[1]}(t)\,\xi(\tilde{X}) = F_{pfb}^{[0]} + F_{pfb}^{[2]}(t,\tilde{X}) \qquad (23)$$

where $F_{pfb}^{[0]}$ is the approximate mean potential drift force vector given by

$$\bar{F}_{pfb} \approx F_{pfb}^{[0]} = C_{\theta S\xi} = E[\theta_S^{[1]}(t)\,\xi(\tilde{X})], \qquad (24)$$

and the second-order potential feedback force vector is

$$F_{pfb}^{[2]}(t,\tilde{X}) = \theta_S^{[1]}(t)\,\xi(\tilde{X}) - C_{\theta S\xi}. \qquad (25)$$

Considering that the feedback coefficients in Eq. (11) are linear transforms of the diffraction forces acting on a fixed platform, the potential feedback forces may be described as the quadratic transforms of the diffraction forces, experienced by the platform at its undisplaced position and the fluctuating response component. The spectral description of potential and viscous feedback drift forces is obtained in terms of bispectra. A computationally efficient procedure to evaluate these higher-order spectral descriptions is presented by Li and Kareem (1989).

6. EXAMPLE

This example examines the response of a typical TLP to displacement-induced feedback force in random sea states. Following the procedure outlined in the previous section, the mean and standard deviation of the force and corresponding response due to the displacement feedback effects are computed. For brevity's sake all the results are not presented here. The power spectral density of the surge response under an extreme sea state condition is presented in Fig. 2 and the associated mean and standard deviation of response are presented in Table 2. The results demonstrate that the response due to the displacement feedback is comparable in magnitude to the wave-frequency response. Furthermore, the frequency domain response analysis clearly provides a good agreement with the time domain results.

7. CONCLUDING REMARKS

The contribution of the wave-induced loading resulting from computing wave loads at the displaced position of compliant offshore structures is quantified in terms of the linear and nonlinear feedback forces. The wave-induced loads on a platform at its displaced position consist of the wave forces on the platform at its undisplaced position and the feedback components in terms of the second-order forces that include mainly the mean and slowly varying drift forces. The feedback coefficients are time-dependent and linearly related to the wave-induced forces. The frequency domain response analysis provides good agreement with the time domain results.

8. ACKNOWLEDGEMENTS

The support for this research was provided in part by the National Science Foundation Grant BCS-9096274 (BCS-8352223), ONR Grant N00014-93-1-0761, Texas Advanced Technology and Research Program, and several major oil companies. The views presented here do not necessarily reflect the views of the sponsors.

9. REFERENCES

Faltinsen, M.O. and Zhao, Rong, 1989, "Slow-Drift Motions of a Moored Two-Dimensional Body in Irregular Waves," J. of Ship Res., Vol. 33, No. 2, pp. 93-108.

Kareem, A. (1985), "Wind-Induced Response Analysis of Tension Leg Platforms," Journal of Structural Engineering, Vol. 111, No. 1, pp. 37-55.

Kareem, A., and Li, Yousun (1988), "Stochastic Response of Tension Leg Platforms to Wind and Wave Fields," Department of Civil Engineering, Report No. UHCE 88-18.

Li, Yousun and Kareem, A. (1989), "On Stochastic Decomposition and Its Application in Probabilistic Dynamics," Proceedings of the 5th International Conference on Structural Safety and Reliability, ICOSSAR '89, San Francisco, California, August.

Li, Yousun and Kareem, A. (1992), "Response of Tension Leg Platforms to Wave Drift Forces," Proceedings, Offshore Technology Conference, Houston, TX.

McIver, D.B. (1976), "Parametrically-Excited Oscillations in Offshore Structures," Proceedings of BOSS '76 Conference, Trondheim, Norway, Vol. 2.

Mercier, J.A. (1982), "Evolution of Tension Leg Platform Technology," Proceeding, BOSS 82.

Rainey, R.C.T. (1977), "The Dynamics of Tethered Platforms," Trans, Royal Institute of Naval Architects, Vol. 120, pp. 59-80.

Spanos, P.D. and Agarwal, V.K. (1984), "Response of a Simple Tension Leg Platform Model to Wave Forces Calculated at the Displaced Position," Journal of Energy Resources Technology, ASME, pp. 437-443.

Structural Safety & Reliability, Schuëller, Shinozuka & Yao (eds) © 1994 Balkema, Rotterdam, ISBN 90 5410 357 4

Response statistics of tension leg platforms under wind and wave loads: A statistical quadratization approach

Ahsan Kareem
University of Notre Dame, Ind., USA

Jun Zhao
Barnett & Casbarian Inc., Houston, Tex., USA

ABSTRACT: The nonlinearities in the wind and wave loadings of compliant offshore structures result in response statistics that deviate from the Gaussian distribution. This paper focuses on the analysis of these structures to random nonlinear wind and wave load. As an extension of the commonly used linearization approach an equivalent statistical quadratization method (ESQM) is presented. The nonlinear loading is expressed in terms of an equivalent polynomila that contains terms up to quadratic. The response cumulants are based on Volterra theory and earlier work of Bedrosian and Rice (1971). A direct integration scheme and Kac-Siegert technique was utilized to evaluate the response cumulants. The equivalent quadratization technique provides accurate results for asymmetrical nonlinearities. For symmetrical nonlinearities an equivalent statistical cubicization approach is more appropriate (Kareem & Zhao, 1992). Based on the information on higher order cumulants the response pdf, crossing rates and peak value distributions are derived. The results provide a good comparison with simulation.

1. INTRODUCTION

The challenge of developing deep water oil fields has placed a growing importance on the economics and safety issues concerning drilling and production platforms. The tension leg platform (TLP) is the most promising structural concept among different structural systems being considered for deep water applications. The compliant nature of TLP motions in the horizontal plane makes these platforms sensitive to the low frequency oscillations due to wind and wave drift forces. Both the wind loads and wave loads acting on TLPs are nonlinear, e.g., wind loading in the presence of the square of the fluctuating velocity term and the wave drag forces in Morison equation. Historically, analysis of nonlinear systems in the frequency domain has been based on the statistical linearization approach. The linearization approach fails to adequately represent important features of nonlinearity, e.g., the response power spectral density function spans only in the range of excitation frequencies. The energy at the sum and difference frequency components is nonexistent and the response probability density function remains Gaussian.

Obviously, the next level of improvement involves statistical quadratization. In the present paper, the system nonlinearity is expressed in terms of an equivalent polynomial in which up to quadratic terms are included. The concept of quadratization has been used in the area of hydrodynamic loads on offshore structures, e.g., Kareem and Li, 1988, and Spanos and Donley, 1990.

This paper addresses the response of a TLP to wind and wave loads. The nonlinear loading is expressed in terms of an equivalent polynomial that contains terms up to quadratic. The response cumulants are obtained based on the work of Bedrosian and Rice (1971). We utilize the direct integration method to evaluate the first four cumulants, and also provide a scheme to obtain the cumulants by Kac-Siegert technique (Kac and Siegert, 1947), both analyses include the effects of the relative fluid structure velocity. The probability density function and the extreme value analysis are carried out based on the higher-order cumulants utilizing the Hermite moment method (Winterstein, 1985) and Maximum Entropy method (Mead and Papanicolaou, 1984, Sobczyk and Trebicki, 1990).

2. THEORETICAL BACKGROUND

Historically, there exist three fundamental approaches to solving nonlinear stochastic problems. One is based on the theory of Markov processes and the associated Fokker-Planck equation, initially applied in the analysis of Brownian motion in statistical physics. The other approach is based on spectral analysis which has been extensively applied in the field of communications and electrical engineering. The third concerns time domain Monte Carlo based simulation. In this paper the second approach is addressed which utilizes Volterra series expansion to model a nonlinear system correct to second order.

2.1 Volterra Series

A Volterra series expansion [Bedrosian and Rice, 1971] may be viewed as a regular expansion in power series, "with memory". Notice that the first-order kernel is simply the impulse response function of a linear system, while the higher-order kernels can be viewed as higher-order impulse response functions which serve to characterize the various orders of nonlinearities. The general formulation of the kernels is not available, but when the nonlinear transformation is in a polynomial form, the kernels can be evaluated. We may call the system with polynomial nonlinear transformation as a Volterra equivalent system. A general second-order Volterra equivalent system may be described as the following,

$$x(t) = a_1 \int_{-\infty}^{\infty} h_1(\tau) u(t-\tau) d\tau$$

$$+ \frac{a_2}{2} \int\int_{-\infty}^{\infty} h_2(\tau, \sigma) u(t-\tau) u(t-\sigma) d\tau d\sigma$$

$$= x_1(t) + x_2(t), \quad (1)$$

where $h_1(\tau)$ and $h_2(\tau, \sigma)$ = linear and second-order impulse response functions, respectively and $x_1(t)$ and $x_2(t)$ are linear and second-order response components.

2.2 Modelling of Wind and Wave Loads

The TLP surge response is modelled by an equivalent single-degree-of-freedom system given by

$$M\ddot{x} + C\dot{x} + Kx = F, \quad (2)$$

where M = structure mass and added mass; C = structure damping; K = structure stiffness; F = wind, wave and current forces. The wind force acting in the surge direction is expressed by [Kareem, 1985].

$$F_{wind} = K_w (w + W - \dot{x})^2, \quad (3)$$

where $K_w = \frac{1}{2}\rho_a C_a A$; w is the fluctuating wind velocity defined by the wind spectrum; and W is the mean wind velocity at a reference height. The wave force is expressed by Morison equation in terms of the relative velocity format

$$F_{wave} = K_m \dot{u} + K_d |u + U - \dot{x}| (u + U - \dot{x}), \quad (4)$$

where $K_m = \rho C_m V_e$; $K_d = \frac{1}{2}\rho C_d A_e$ and u, $\dot{u}$ = wave particle velocity and acceleration, respectively; and U is the current speed. It is important to note that although the wind and wave processes may be assumed as Gaussian, the structural velocity in the preceding equations is no longer Gaussian.

2.3. Slow Drift Approximation

To eliminate the computational difficulty imposed by the non-Guassian structural velocity, a slowly varying drift approximation is invoked. For a system with low natural frequency, the slowly varying drift motion plays an important role. This leads to a reasonable assumption, i.e., the higher-order nonlinear velocity terms can be neglected. Accordingly, the nonlinear terms in the wind and wave induced drag descriptions are given by

$$(w + W - \dot{x})^2 \cong (w + W - \dot{x}_1)^2$$

$$-2E\{w + W - \dot{x}_1\} \dot{x}_2 \quad (5)$$

$$|u + U - \dot{x}| (u + U - \dot{x}) \cong$$

$$|u + U - \dot{x}_1| (u + U - \dot{x}_1) - 2E\{|u + U - \dot{x}_1|\} \dot{x}_2 \quad (6)$$

2.4 Statistical Quadratization Method

In a quadratization approach, the nonlinear force of an arbitrary form is approximated by a polynomial expression. A quadratic expression is the obvious choice as an improvement over a linear expression. Since the wind force is already in a quadratic form, only the wave drag force is described here. Let us take the nonlinear term in the right side of Eq. 6 and evaluate the error term given below

$$\varepsilon = |u + U - \dot{x}_1| (u + U - \dot{x}_1)$$

$$-\alpha_0 - \alpha_1 (u - \dot{x}_1) - \alpha_2 (u - \dot{x}_1)^2. \quad (7)$$

Mean square minimization of the error term leads to the determination of the following coefficients (Kareem & Zhao, 1992).

$$\alpha_0 = 2U\sigma(rb_1 + b_2); \ \alpha_1 = 4\sigma(rb_1 + b_2),$$

$$\text{and } \alpha_2 = 2b_1, \quad (8)$$

where $b_1 = \frac{1}{\sqrt{2\pi}} \int_0^r exp(-\frac{y^2}{2}) dy$

$b_2 = \frac{1}{\sqrt{2\pi}} exp(-\frac{r^2}{2})$; $r = \frac{U}{\sigma}$; and

$$\sigma = E\{(u-\dot{x}_1)^2\}$$

Now the equations of motion for wave excitation can be expressed as

$$M\ddot{x}_1 + (C+a_1)\dot{x}_1 + Kx_1 = K_m\dot{u} + a_1 u,$$

$$M\ddot{x}_2 + (C+a_1)\dot{x}_2 + Kx_2 = \frac{a_2}{2}(u-\dot{x}_1)^2, \qquad (9)$$

where $a_0 = K_d\alpha_0$; $a_1 = K_d\alpha_1$; and $a_2 = 2K_d\alpha_2$.

Accordingly, the equations of motion under the wind force are given by

$$M\ddot{x}_1 + (C+a_1)\dot{x}_1 + Kx_1 = a_1 w,$$

$$M\ddot{x}_2 + (C+a_1)\dot{x}_2 + Kx_2 = \frac{a_2}{2}(w-\dot{x}_1)^2, \qquad (10)$$

where $a_0 = K_w W^2$; $a_1 = 2K_w W$; and $a_2 = 2K_w$.

Notice that the terms with $(u-\dot{x}_1)^2$ and $(w-\dot{x}_1)^2$ are squares of Gaussian processes, and include non-linear damping terms.

3. RESPONSE STATISTICS

In conventional linear analysis, the response statistics are described by the second-order statistics of response. In the present case the response distribution is no longer Gaussian, therefore, higher-order moments or cumulants are needed to describe the response statistics. The power spectrum and the first four cumulants can be obtained from the following (Bedrosian and Rice, 1971).

$$D_x(\omega) = k_1\delta(\omega) + \alpha_1^2|H_1(\omega)|^2 D(\omega) +$$

$$\frac{a_2^2}{2}|H(\omega)|^2 \int_{-\infty}^{\infty} |H_v(\theta) H_v(\omega-\theta)|^2$$

$$D(\theta) D(\omega-\theta) D(\omega-\theta)\, d\theta, \qquad (11)$$

where $H(\omega) = \dfrac{1}{K + (-\omega^2)M + i\omega(C+a_1)}$; in the wind case; $H_1(\omega) = H(\omega)$ and for the waves case, $H_1(\omega) = \left[1 + \dfrac{K_m}{a_1} i\omega\right] H(\omega)$; and $H_v(\omega) = 1 - a_1 i\omega H_1(\omega)$. $H_v(\omega)$ represents the feedback filter involving relative fluid structure velocity; $D_x(\omega)$ is the two-sided spectrum of $x(t)$; and $D(\omega)$ represents the two-sided spectrum of $u(t)$.

For brevity's sake, let $H(1)$, $H(1+2)$, $D(1)$ are used to represent $H_1(\omega_1)$, $H(\omega_1+\omega_2)$ and $D(\omega_1)\ d\omega_1$, respectively. The associated cumulants are given below

$$k_1 = \frac{a_2}{2}\sigma_v^2 H(0),$$

$$k_2 = a_1^2\sigma_x^2 + \frac{a_2^2}{2}\int\int_{-\infty}^{\infty} |H(1+2) H_v(1) H_v(2)|^2$$

$$D(1) D(2)$$

$$k_3 = 3a_1^2 a_2 \int\int_{-\infty}^{\infty} H_1(1) H_1(2) H(-1-2)$$

$$H_v(-1) H_v(-2) D(1) D(2),$$

$$+ a_2^3 \int\int\int_{-\infty}^{\infty} H(1+2) H(-1+3) H(-2-3)$$

$$|H_v(1) H_v(2) H_v(3)|^2 D(1) D(2) D(3).$$

$$k_4 = 12 a_a^2 a_2^2 \int\int\int_{-\infty}^{\infty} H_1(1) H_1(2)$$

$$H(-1+3) H(-1-3)$$

$$H_v(-1) H_v(-2) |H_v(3)|^2 D(1) D(2) D(3), \qquad (12)$$

$$+ 3a_2^4 \int\int\int\int_{-\infty}^{\infty} H(1+2)$$

$$H(-1+3) H(-2+4) H(-3-4)$$

$$|H_v(1) H_v(2) H_v(3) H_v(4)|^2$$

$$D(1) D(2)\ (D3)\ (D4)$$

where $\sigma_v^2 = \int_{-\infty}^{\infty} |H_v(\omega)|^2 D(\omega)\, d\omega$, and

$$\sigma_x^2 = \int_{-\infty}^{\infty} |H_1(\omega)|^2 D(\omega)\, d(\omega).$$

The skewness and kurtosis representing a departure from Gaussianity are given by

$$\gamma_3 = \frac{k_3}{k_2^{3/2}};\ \text{and}\ \gamma_4 = \frac{k_4}{k_2^2}. \tag{13}$$

3.1 Direct Integration Method

The calculation of the fourth-order cumulant was considered as prohibitive, not only because of the behavior of the integrand, but also due to very extensive computational effort needed in evaluating the multi-fold integrals. Bedrosian and Rice (1971) stated that the four-fold integral in the above equations can not be carried out because of its complexity. Recently, Spanos and Donley (1990) reported a similar difficulty. In this study the four-fold integral in the above equations was evaluated after some mathematical manipulations by reducing it to a three-fold integral. Equation 12 is recast in the following form.

$$k_1 = \frac{a_2}{2}\sigma_v^2 H(0)$$

$$k_2 = a_1^2\sigma_x^2 + \frac{a_2^2}{4}\int_0^{\infty}$$

$$[C_{22}(\omega_1, \omega_1) + C_{22}(-\omega_1, -\omega_1)]\, S(\omega_1)\, d\omega_1$$

$$k_3 = \frac{3}{2}a_1^2 a_2 \int_0^{\infty} Re\,(H(\omega_1)$$

$$[C_{11}(-\omega_1) + C^*{}_{11}(\omega_1)]\, S(\omega_1)\, d\omega_1$$

$$+\frac{a_2^3}{4}\iint_0^{\infty} Re\,\{H(\omega_1+\omega_2) H_v(\omega_1) H_v(\omega_2)$$

$$[C_{22}(\omega_1, -\omega_2) + C_{22}(\omega_2, -\omega_1)]$$

$$+H(\omega_1-\omega_2) H_v(\omega_1) H_v(-\omega_2)$$

$$[C_{22}(\omega_1, \omega_2) + C_{22}(-\omega_2, -\omega_1)]$$

$$S(\omega_1)\, S(\omega_2)\, d\omega_1 d\omega_2$$

$$k_4 = 3a_1^2 a_2^2 \int_0^{\infty} 2Re C_{11}(\omega_1)\, C_{11}(-\omega_1) +$$

$$[|C_{11}(\omega_1)|^2 + |C_{11}(-\omega_1)|^2]\, S(\omega_1)\, d\omega_1$$

$$+\frac{3}{8}a_2^4 \iint_0^{\infty} \{2Re\,[C_{22}(-\omega_1, \omega_2)\, C_{22}(\omega_1, -\omega_2)] +$$

$$[C_{22}(-\omega_1, \omega_2)\, C_{22}(\omega_1, \omega_2)]$$

$$+2|C_{22}(-\omega_1, \omega_2)|^2 + |C_{22}(-\omega_1, -\omega_2)|^2 +$$

$$|C_{22}(\omega_1, \omega_2)|^2\}\, S(\omega_1)\, S(\omega_2)\, d\omega_1 d\omega_2$$

$$C_{11}(\omega) = H_v(\omega) \int_0^{\infty} H_1(\alpha)$$

$$H(-\alpha+\omega)\, H_v(-\alpha)\, S(\alpha)\, d\alpha$$

and

$$C_{22}(\omega_1, \omega_2) = H_v(-\omega_1)\, H_v(\omega_2) \int_0^{\infty}$$

$$H(\alpha-\omega_1)\, H(-\alpha+\omega_2)\, |H_v(\alpha)|^2 S(\alpha)\, d\alpha \tag{14}$$

In this manner, the solution of the fourth-order cumulant involves an effort equal to that needed for solving the third-order cumulant without any compromise on the accuracy.

3.2 Kac-Siegert Technique

The other approach is named after Kac and Siegert who first applied it to the theory of noise in radio receivers with square law detectors in 1947. In fact, it may be taken as the generalized Fourier series representation method (Cochran, 1972), based on the theory of linear integral equations. Its application to ocean engineering was first given by Neal (1974). This approach has been used extensively by Naess (1985) for offshore problems. The generalized Fourier series expansion can be written in uni- and two-dimensional forms.

$$Q_1(x) \cong \sum_{i=1}^{\infty} \alpha_i \phi_i(x);\ \text{and}$$

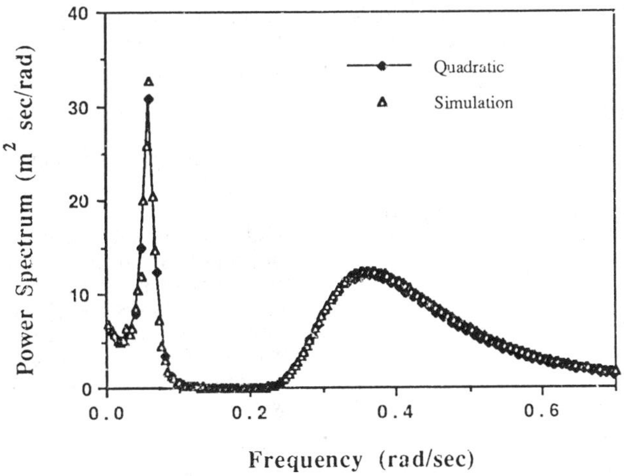

Fig. 1. Comparison of Linearization, Quadratization, and Simulation Results

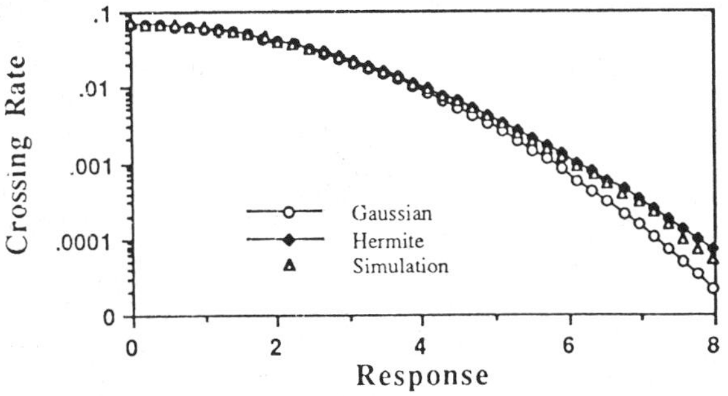

Fig. 2. Comparison of Crossing Rates

$$Q_2(x,y) \equiv \sum_{i=1}^{\infty} \sum_{j=1}^{\infty} \beta_{ij} \phi_i(x) \phi^*_j(y) . \qquad (15)$$

The above expression is sometimes called degenerate kernel or separable kernel. In the case when $Q_2(x,y)$ is Hermitian, $\phi_i(.)$'s are chosen as the characteristic functions of the following Fredholm homogeneous integral equation of the second kind.

$$\int_a^b Q_2(x,y)\phi(y)\,dy = \lambda\phi(x) . \qquad (16)$$

For the present application

$$Q_2(x,y) =$$

$$G(x) H_v(x) H(x-y) H_v(-y) G(y) \qquad (17)$$

where $G(x) = D(x)^{1/2}$, and $Q_2(x,y)$ is Hermitian. Both x and y represent frequencies, which are discretized as $\{\omega_i, i=1,2,\ldots,N\}$, with equal length Δ.

The discrete form of the homogeneous integral equation constitutes a linear algebraic eigenvalue problem as the following,

$$\sum_{j=1}^{N} Q_2(\omega_i, \omega_j) W_j \phi(\omega_j) \Delta +$$

$$\sum_{j=1}^{N} Q_2(\omega_i, -\omega_j) W_j \phi(-\omega_j) \Delta = \lambda\phi(\omega_i) ,$$

$$\sum_{j=1}^{N} Q_2(-\omega_i, \omega_j) W_j \phi(\omega_j) \Delta +$$

$$\sum_{j=1}^{N} Q_2(-\omega_i, -\omega_j) W_j \phi(-\omega_j) \Delta = \lambda\phi(-\omega_i) \qquad (18)$$

where W_j is the weighting factors determined by the numerical method used to evaluate these equations.

For slowly vary drift response applications, off-diagonal terms can be ignored. The preceding equations can be solved numerically to obtain the eigenvalues and the corresponding eigenvectors. Then, the system transfer functions can be obtained as,

$$G(\omega_1) H_v(\omega_1) H(\omega_1 - \omega_2) H_v(-\omega_2) G(\omega_2) =$$

$$\sum_{i=1}^{N} \lambda_i \phi_i(\omega_1) \phi^*_i(\omega_2)$$

$$H_1(\omega) G(\omega) = \sum_{j=1}^{N} \alpha_j \phi_j(\omega), \qquad (19)$$

where, $\alpha_j = \int_{-\infty}^{\infty} H_1(\omega) G(\omega) \phi^*_j(\omega) \, d\omega,$

Utilizing these expressions leads to the following description of the cumulants:

$$k_1 = \frac{a_2}{2} \sum_{i=1}^{M} \lambda_i;\; k_2 = a_1^2 \sum_{i=1}^{M} \alpha_i^2 + \frac{a_2^2}{2} \sum_{i=1}^{M} \lambda_i^2;$$

$$k_3 = 3a_1^2 a_2 \sum_{i=1}^{M} \alpha_i^2 \lambda_i + a_2^3 \sum_{i=1}^{M} \lambda_i^3, \text{ and}$$

$$k_4 = 12 a_1^2 a_2^2 \sum_{i=1}^{M} \alpha_i^2 \lambda_i^2 + 3a_2^4 \sum_{i=1}^{M} \lambda_i^4. \qquad (20)$$

4. COMBINED WIND, WAVE AND CURRENT EFFECTS

The equation of motion due to combined wind, wave and current loadings after introducing some assumptions and mathematical manipulations are given by

$$M\ddot{x}^{wind} + (C + C_{wave}) \dot{x}^{wind} + Kx^{wind} =$$

$$K_w (w + W - \dot{x}^{wind})^2$$

$$M\ddot{x}^{wave} + (C + C_{wind}) \dot{x}^{wave} + Kx^{wave} =$$

$$K_m \dot{u} + K_d |u + U - \dot{x}^{wave}| (u + U - \dot{x}^{wave}) \qquad (21)$$

The solution of the preceding equations is obtained following the two procedures outlined earlier in the text. Based on the assumptions made earlier the total response cumulants can be obtained by a simple summation of the response cumulants due to all loadings (Pugachev, 1984).

$$k_m^{total} = k_m^{wind} + k_m^{wave}, \; m = 1, 2, 3, 4, \ldots \qquad (22)$$

5. PROBABILITY DISTRIBUTION OF RESPONSE

Following the evaluation of the first four moments or cumulants of response, the non-Gaussian distribution of response processes can be obtained with a subsequent estimation of the extreme value distribution. In this study, Gram-Charlier Series, Hermite Moment Approach and Maximum Entropy method are utilized. Details are omitted here for brevity's sake.

6. EXAMPLE

To illustrate the nonlinear effects introduced by wind and wave loads, an idealized TLP model is utilized. Details are omitted here due to space limitations. Only very limited results are presented here. In Fig. 1, a comparison of the linearization, quadratization and simulation procedures is made by examining the response power spectral density functions. It is noted that for the cases with the currents, quadratization and simulation are almost coincident. This is not true in the absence of currents. Indeed this is the limitation of the quadratization method which degenerates to a linearization case in the absence of currents. This feature is captured by a cubicization approach (Kareem and Zhao, 1992). The crossing rates are plotted in Fig. 2, the results based on quadratization show a good agreement with the simulation results and exhibit departure from Gaussian.

7. CONCLUDING REMARKS

The quadratization approach presented in this paper addresses the treatment of nonlinearities in the frequency domain analysis that result from wind and wave loadings on TLPs. The results obtained in terms of TLP response spectra and cumulants are in good agreement with simulation results. The higher-order cumulants are used to determine the response probability distribution and extreme value distribution. The derived response distributions are in good agreement with simulated results.

8. ACKNOWLEDGMENTS

The support for this study was provided in part by NSF grant BCS-9096274, ONR grant N00014-93-1-0761, Texas Advanced Research Program and several major oil companies. The views presented here do not necessarily reflect the view of the sponsors.

9. REFERENCES

Bedrosian, E. and Rice, S., 1971, "The Output Properties of Volterra Systems (Nonlinear Systems with Memory) Driven by Harmonic and Gaussian Inputs", Proc. of IEEE, Vol. 59, No. 12.

Cochran, J.A., 1972, The Analysis of Linear Integral Equations. McGraw-Hill.

Kac, M. and Siegert, A., 1947, "On the Theory of Noise in Radio Receivers with Square Law Detectors", Journal of Applied Physics, Vol. 18.

Kareem, A. and Li, Yousun, 1988, "Stochastic Response of Tension Leg Platforms to Wind and Wave Fields," Tech. Rep. No. UHCE88-18, Dept. of Civil Engrg., University of Houston, TX.

Kareem, A. and Zhao, Jun, 1992, "Response Statistics of Tension Leg Platforms to Wind and Wave Loadings," Technical Report, Dept. of CE/GEOS, University of Notre Dame.

Mead, L.R. and Papanicolaou, N., 1984, "Maximum Entropy in the Problem of Moments", Journal of Math. Phys., Vol. 25, No. 8.

Naess, A., 1985, "Statistical Analysis of Second-Order response of Marine Structures", Journal of Ship Research, Vol. 29, No. 4.

Neal, E., 1972, "The Statistical Distribution of Second-Order Slowly Varying Forces and Motions," Applied Ocean Research, 8(2).

Pugachev, V.S., 1984, Probability Theory and Mathematical Statistics for Engineers, Pergamon Press.

Sobczyk, K. and Trebicki, J., 1990, "Maximum Entropy Principle in Stochastic Dynamics", Probabilistic Engineering Mechanics, Vol. 5, No. 3.

Spanos, P.D. and Donley, M.G., 1990, "Stochastic Response of a Tension Leg Platform to Viscous Drift Forces", Proceedings of the 9th OMAE, Conference, ASME, Houston, TX.

Winterstein, S., 1985, "Non-Normal Responses and Fatigue Damage", Journal of Engineering Mechanics, Vol. 111, No. 10.

Structural Safety & Reliability, Schuëller, Shinozuka & Yao (eds) © 1994 Balkema, Rotterdam, ISBN 90 5410 357 4

Reliability assessment of marine structures by long-term response analysis and importance sampling

B.J. Leira & D. Karunakaran
SINTEF Structures and Concrete, Trondheim, Norway

T. Moan
The Norwegian Institute of Technology, Trondheim, Norway

ABSTRACT: A method for reliability analysis of dynamically responding offshore structures subjected to nonlinear wave loading is outlined. The response analysis is carried out in the time doamin by a stepwise integration procedure. Samples of the response maxima constitute the basis for calculation of long term response distributions. Subsequently, importance sampling is introduced to estimate failure probabilities. The procedure is illustrated by application to a jack-up platform.

1 INTRODUCTION

For slender offshore platforms constructed for operation in relatively deep waters, nonlinear drag loading and dynamic amplification of response are features that must be accounted for in design against extreme forces. The extremes of the internal load effects are most accurately predicted by means of a nonlinear stochastic dynamic response analysis. Furthermore, the interplay between natural frequency, damping level and wave environment may in some cases imply that the extreme response most likely will occur at relatively low sea states. This is accounted for most consistently by a so-called long-term response analysis.

In order to perform a complete reliability analysis of such structures, probability distributions are introduced for the basic physical parameters describing the structure, the wave environment and the foundation properties. Selecting different values of these variables, the main contribution to the extreme response will come from different parts of the wave-height/peak-period domain.

In this paper, a procedure is outlined that will take these effects properly into account. It is based on a combination of long-term analysis supplemented by an efficient simulation scheme for the time-invariant basic random variables. A jack-up platform is analysed to illustrate the approach.

2 PROCEDURE FOR NONLINEAR LONG TERM RESPONSE ANALYSIS

2.1 *Long Term Response Distribution*

The long term distribution of structural response is obtained by combining a joint distribution of significant wave height and peak period (wave scatter diagram) with the short term distribution of response, Battjes (1972) and Inglis et al. (1985). The probability of exceeding a fixed response level ζ is given by

$$Q_r(\zeta) = \int_{H_s}\int_{T_p} w(h_s, t_p)\, Q_{r\,|H_s, T_p}(\zeta|h_s, t_p)\, f_{H_s, T_p}(h_s, t_p)\, dt_p\, dh_s \quad (1)$$

where $Q_{r\,|H_s, T_p}(\zeta|h_s, t_p)$ is the conditional probability distribution of response maxima for a short term sea state with given values of significant wave height h_s and peak period t_p. The procedure to obtain this density function will be described in Section 2.3; $f_{H_s, T_p}(h_s, t_p)$ is the joint distribution of significant wave height, H_s and peak period, T_p; $w(h_s, t_p)$ is a weight function accounting for changes in characteristic response frequency from one sea state to another. This change implies that the number of maxima per time unit will vary, and hence most weight is put on sea states with the relatively largest number of maxima per unit time.

For structures with significant nonlinear loading and/or response behaviour, the short term distribution needs in principle to be obtained for all sea states in the wave scatter diagram. Without significant loss of accuracy, however, a simplification based on blocking of the scatter diagram is utilized. It is assumed that the short term response distribution does not change significantly within each block. Each block comprises several sea states and within each of these nonlinear dynamic response analysis is performed for one representative sea state. From the dynamic response analysis the short term response distribution is established.

In the subsequent sections the basic input required to obtain the long term distribution of response are discussed.

2.2 *Environmental Description*

The following information regarding the environmental conditions are needed as input to a long term response analysis:

- a long term joint distribution of the primary sea state characteristics: significant wave height, H_s, spectral peak period, T_p, and possibly, the mean wave direction.
- a procedure making it possible to select and define a proper wave spectrum model for all possible combinations of the primary sea state characteristics.

Some details of these aspects are given below.

2.2.1 *Long term modelling of wave characteristics*

It is often assumed that the omni-directional wave conditions are reasonably well characterized by the significant wave height, H_s and the spectral peak period, T_p. Thus, the long term climate is described by a joint probability density function for these variables, i.e.:

$$f_{H_sT_p}(h_s, t_p) = f_{H_s}(h_s)\, f_{T_p|H_s}(t_p|h_s) \qquad (2)$$

where $f_{H_s}(h_s)$ is the marginal distribution of H_s and $f_{T_p|H_s}(t_p|h_s)$ is the conditional distribution for T_p given H_s.

A joint model according to Eq. (2) for the northern North Sea is presented by Andersen et al. (1987). The marginal distribution for H_s is modelled by:

$$f_{H_s}(h_s) = \begin{cases} \dfrac{1}{\sqrt{2\pi}\,\kappa\, h_s} \exp\left(-\dfrac{1}{2}\dfrac{(\ln h_s - \lambda)^2}{\kappa^2}\right) & for\ h_s \le \eta \\[2ex] \dfrac{\beta}{\rho}\left(\dfrac{h_s}{\rho}\right)^{\beta-1} \exp\left(-\left(\dfrac{h_s}{\rho}\right)^{\beta}\right) & for\ h_s > \eta \end{cases} \qquad (3)$$

where η, λ, κ^2, ρ and β are selected to fit the observations at a given site.

A log-normal distribution is adopted for the spectral peak period, i.e.:

$$f_{T_p|H_s}(t_p|h_s) = \frac{1}{\sqrt{\pi}\varphi t} \exp\left(-\frac{1}{2}\frac{(\ln t_p - \mu)^2}{\varphi^2}\right) \qquad (4)$$

where the parameters, $\mu = E(\ln T_p)$ and $\phi^2 = Var(\ln T_p)$ are functions of h_s, see Andersen et al. (1987).

2.2.2 *Short term description of sea states*

Assuming that the sea surface for short time periods can be modelled as a stationary Gaussian process, and, furthermore, that the waves are long crested, the sea surface is in a statistical sense completely characterized by the one-dimensional frequency spectrum. A rather general model for the spectrum is given by, Torsethaugen (1987). A typical range of the peakedness parameter γ entering this model for storm spectra in the North Sea is 2 - 3.5.

An artificial lower boundary for the spectral peak period has been introduced to avoid unphysical sea states, i.e.:

$$T_p \geq C_L \sqrt{H_s} \qquad (5)$$

In this study a single-moded spectral model with a lower boundary corresponding to $C_L = 3.2$ is introduced, see Haver (1990a) and Karunakaran et al. (1992).

2.3 Non-linear Dynamic Response Analysis

2.3.1 Response analysis method

Each short term sea-state is specified by a model spectrum with the significant wave height (H_s) and peak period (T_p) as parameters. The relationship between the load and response stochastic processes is given by the dynamic equilibrium equation in incremental form:

$$\boldsymbol{M}\Delta\ddot{\boldsymbol{r}}(t) + \boldsymbol{C}\Delta\dot{\boldsymbol{r}}(t) + \boldsymbol{K}\Delta\boldsymbol{r}(t) = \Delta\boldsymbol{F}(t) \qquad (6)$$

where

- **M** - Mass matrix
- C - Damping matrix
- **K** - Stiffness matrix
- **r** - Structural response vector
- **F**(t) - External load vector

Δ denotes increments, and dot superscript designates differentiation with respect to time.

The mass and stiffness matrices are based on a finite element model. The latter matrix also contains soil stiffness contributions. The damping matrix is determined from the Rayleigh model, i.e. as a linear combination of the stiffness and mass matrices.

Sample time series of wave kinematics are simulated using independent random phase angles uniformly distributed between 0 and 2π, generated by Monte-Carlo simulation, refer Karunakaran (1991). The hydrodynamic force vector at each discrete time step is determined from the Morison equation containing the relative velocity between water particles and the structure, in addition to the particle acceleration.

The response process samples are obtained by a step-wise numerical integration procedure. A Newmark-ß method with $ß = 0.25$ and $\gamma = 0.5$ is employed, corresponding to the so-called constant average acceleration algorithm. Since the force process depends on the response an iteration is required at each step.

2.3.2 Short term response distributions

Based on broad experience, it has been found that Weibull 2- and 3-parameter distributions are able to represent a wide range of response types accurately. The long term distribution of response maxima is obtained by amalgamating Eq. (1) and the Weibull 3-parameter short term distribution of maxima presented below:

$$Q_{r_{|H_{mo},T_p}}(\zeta, h, t) = e^{-\left[\frac{\zeta - \mu}{\sigma}\right]^{\beta}} \qquad (7)$$

where

- μ - Weibull location parameter
- σ - Weibull scale parameter
- ß - Weibull shape factor

These parameters are estimated from sample time series for each response quantity to be studied.

3 RELIABILITY ANALYSIS METHOD

3.1 Basic Formulation

Based on the long-term response distribution, the probability of exceeding a given critical response level during the specified life-time can be computed. This probability corresponds to a fixed set of values for the parameters defining the ocean environment and the structural properties. Hence, it must be regarded as a conditional quantity that will change if these parameter values are different. Collecting the parameters in a vector **x**, we introduce the notation $p_{f|x}$ for the conditional failure probability. Generally, the vector **x** also contains strength parameters defining the critical response level (e.g. yield stress for stress response).

In a reliability formulation, probability density functions are introduced for each of the components in the vector **x** instead of fixed values. The failure probability for this case is found by integration over the random variables, i.e.

$$p_f = \int_x p_{f|x} \, f_x(x) \; dx \qquad (8)$$

where the integration domain is the definition range for all the variables.

3.2 Calculation Procedure

The expression in Eq. (8) could in principle be evaluated by generating random samples of the vector **x** and calculating $p_{f|x}$ from a long-term analysis in each case. Obviously, this will require a large number of such analyses if e.g. pure Monte Carlo simulation is employed.

To reduce this number, a more efficient scheme based on importance sampling is applied. The failure proability is then expressed as

$$p_f = \int_x p_{f|x} \left(f_x(x)/h_x(x)\right) \, h_x(x) \; dx \qquad (9)$$

where $h_x(\mathbf{x})$ is the new sampling density and the expression in parenthesis plays the role of a weight function.

By selecting the sampling density properly, the convergence rate as a function of sample size is greatly improved. In our context, this new density function is obtained by shifting the peak of the original density function $f_x(\mathbf{x})$ to the region with the highest relative contribution to the failure probability.

4. EXAMPLE STRUCTURE

4.1 Structural Description

The jack-up platform considered is designed as a production facility to operate at a water depth of 108 m. The legs are spaced 62 m apart. Each leg consists of a triangular truss leg structure supported by a spud - can foundation. The typical member diameters in the leg truss work range from 0.2 m to 0.8 m.

A three dimensional FEM computer model is prepared, which is shown in Figure 1. The leg truss work is idealized as a string of beam elements with equivalent stiffness properties. The computer model consists of 31 nodes and 34 elements. The leg-soil interaction is modelled by linear springs.

The first natural period of the structure is 7.5 sec.

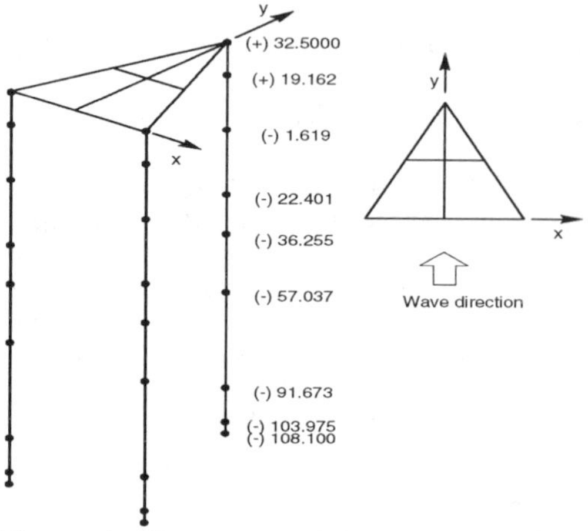

Figure 1 Computer model.

The base case soil and structural damping together is assumed to be 2% and modelled as proportional damping (Rayleigh damping) at the first and third natural modes. The nonlinear hydrodynamic drag damping of the structure is included by using relative velocities in the Morison equation.

The hydrodynamic coefficients used in this analysis are:

Mass coefficient C_M: 2.0
Drag coefficient C_D: 1.0

The value 1.0 is taken on the basis of earlier experience and results from experimental studies carried out by a variety of investigators, refer the review study by Spidsøe (1990).

4.1.1 Long term design condition

The long term response is calculated using the joint probability of significant wave height and spectral peak period (wave scatter diagram), given in Eq. (3). In the generated scatter diagram the peak period discretization interval is 0.25 sec, and for the significant wave height it is 0.5 m. The small interval for the peak period is needed to identify properly the contributions to long term response from sea states giving resonant response. A wave spectrum of the JONSWAP type is employed.

The scatter diagram is divided into five blocks. The response functions are calculated at five representative sea states corresponding to these.

4.1.2 Current velocity

The following current profile, which corresponds to a 10 year return period is used in all analyses.

at MWL	1.00 ms^{-1}
- 50 m	1.00 ms^{-1}
- 100 m	0.90 ms^{-1}
at mud line	0.75 ms^{-1}

This current profile is employed for all sea states.

5. RESULTS AND DISCUSSION

Only two response quantities are considered for presentation, which are base shear and overturning moment. All other response quantities exhibit a dynamic behaviour similar to either of these.

A summary of the statistical properties of the basic random variables which are input to the analysis are given in Table 1. The importance sampling distribution is obtained by shifting the mean values of each variable.

Two different sets of shifted values were investigated. In the first case, the failure probabilities were calculated for two different sample sizes. These results are presented in Table 2 for a reference period of 100 years. It is observed that as few as 200 samples give accurate results for both choices of sampling point. However, the first choice gives slightly better estimates than the second one. This is due to a more correct estimate of the true "design point" for the most important random variables.

Table 1 List of random variables

Variable name	Distribution type	c.o.v.
Drag coefficient, C_D	Normal	0.2
Inertia coefficient, C_M	Normal	0.1
Wave model correction factor	Normal	0.1
Soil stiffness, k_{soil}	Normal	0.3
Deck mass, M	Lognormal	0.1
Damping ratio, λ	Lognormal	0.25
Significant wave height, H_S	Eq.(3)	0.1
Peak period, T_P	Eq.(4)	0.1
Current	Lognormal	0.20
Base shear capacity	Normal	0.15
Overturning moment capacity	Normal	0.15
Allowable deck displacement	Normal	0.15

Mean value of capacity levels:
Base shear : 32 MN
Overturning moment : 3600 MNm
Deck displacement : 2.1 m

Table 2 Annual probabilities of failure

Response quantity	Sample density number 1		2
	200 samples	400 samples	200 samples
Base shear	$6.9.10^{-4}$	$6.5.10^{-4}$	$6.1.10^{-4}$
Overturning moment	$6.7.10^{-4}$	$6.2.10^{-4}$	$5.2.10^{-4}$
Deck displacement	$3.1.10^{-4}$	$3.1.10^{-4}$	$2.4.10^{-4}$

6 CONCLUSIONS

A method for reliability analysis of marine structures was outlined. The procedure is based on a combination of nonlinear long-term response analysis and importance sampling with respect to the basic variables. Contributions to the response exceedance probability from all parts of the scatter diagram can then be accounted for. The approach was illustrated by application to a jack-up platform in a North Sea environment. It was found that as few as 200 samples gave quite accurate results. Furthermore, the sensitivity to choice of shifted sampling point seems to be quite low. This indicates that the scheme is robust.

7 REFERENCES

1. Battjes, T.A.: "Long-term Wave Height Distribution at Seven Stations around the British Isles", Deutschen Hydrographischen Zeitschr., Vol. 25, 1972, pp. 179-189.
2. Inglis, R.B., Pijfers, J.G.L. and Vugts, J.H.: "A Unified Probabilistic Approach to Predicting the Response of Offshore Structures, Including the Extreme Response", Proc. of the 4th Intl. Conf. on Behaviour of Offshore Structure, BOSS'85, The Hague, 1985.
3. Andersen, O.T., Førland, E., Haver, S. and Straas, P.: "Design Basis - Environmental Conditions - Veslefrikk", STATOIL Report No. 87004A, 1987.
4. Torsethaugen, K.: "On the Range of Validity for the JONSWAP Spectrum", (in Norwegian), Oceanographic Center, Rep. no. 02.0788.00/09/87, Trondheim, 1987.
5. Haver, S.: "On a Possible Lower Limit for the Spectral Peak Period", Statoil F&U-MT 90009, Stavanger, 1990.
6. Karunakaran, D., Leira, B.J., Haver, S. and Moan, T.: "Parametric Influence on Extreme Dynamic Response of Drag-dominated Platforms", Proc. Second Int. Offshore and Polar Engineering Conference, San Fransisco, 1992.
7. Karunakaran, D.: "Procedure for Nonlinear Dynamic Response Analysis of Offshore Structures - Both for Extreme and Fatigue Response", SINTEF Report STF71 F91016, Trondheim, 1991.
8. Spidsøe, N.: "Summary Report - Effect of Surface Elevation", SINTEF Report STF71 A89027, Trondheim, 1990.

Structural Safety & Reliability, Schuëller, Shinozuka & Yao (eds) © 1994 Balkema, Rotterdam, ISBN 90 5410 357 4

Failure modes and reliability analysis of flexible risers

B.J. Leira
SINTEF Structures and Concrete, Trondheim, Norway

A. Engseth
Kværner Engineering Lysaker, Norway

ABSTRACT: Main types of failure modes for flexible pipes are described. A procedure for assessment of failure probabilities related to specified failure modes of a flexible riser system is outlined. As an example of application, the probabilities of having a tensile failure or axial compression failure are evaluated for a flexible riser of a floating production system. The relationships between safety factors and failure probabilities are also assessed.

1 INTRODUCTION

High pressure flexible pipes are increasingly being used for offshore production of oil and gas. Such pipes are employed as flowlines, dynamic risers, kill and choke lines, topside jumpers and spool pieces. Most of the experience with flexible pipes has been gained in mild environments such as offshore Brazil, where more than 900 km of flexible pipe have been installed since the late 1970's /1/. The use of flexible pipes as dynamic risers is now also considered to be accepted technology in the harsh environments of the North Sea.

The majority of the flowlines and risers installed in Brazil are in relatively shallow waters (less than 240 m), they have small diameters (2-6 inches inner diameter), operate at relatively low pressure (40-200 bar) and at low temperature (typically below 70°C). By contrast, for North Sea applications there is a need for flexible pipes satisfying the following demands

- inner diameter up to 20 inches (508 mm) (for relatively low pressure)
- design pressure up to 1000 bar
- design temperature up to 110-130°C
- pipes for transportation of large quantities of gas

This development calls for an improved understanding of the failure mechanisms and failure modes of flexible pipes and flexible riser systems.

In this paper, main types of failure modes are described. A procedure for assessment of failure probabilities related to specified failure modes of a flexible riser system is outlined. The analysis approach is illustrated by application to a floating production system.

2 DESCRIPTION OF PIPE

Flexible pipes are today grouped into two fundamentally different categories: the nonbonded and bonded types. We focus here on the nonbonded type. Such pipes consist of separate steel and plastic layers which are free to move relative to each other, and thereby constitute a flexible structure. The layers interact only through contact pressure and friction.

There are two basic designs of nonbonded flexible risers: rough bore and smooth bore. The rough bore structures are used whenever gas may be present in the transported fluid. An example of such a structure is shown in Fig 1 where each layer has the following functions:

- The inner interlocked steel carcass prevents

the collapse of the internal thermoplastic sheath. The fluid is allowed to leak through the carcass, which is then only subjected to external compressive forces.

- The internal thermoplastic sheath makes the pipe leakproof.
- The main function of the Zeta spiral is to sustain loads due to internal pressure, but it will also resist external pressure and crushing effects.
- An intermediate thermoplastic sheath is applied on dynamic risers to reduce friction between the layer which is resisting pressure and the tensile armours.
- The double crosswound armour layers provide good resistance to axial loads and torsion. These layers are usually made of flat steel wires with a lay angle of about 35°.
- The external thermoplastic layer protects the metallic layers against corrosion and abrasion, and binds the underlying armours.

3 FAILURE MODES AND FAILURE CAUSES

The statistical basis for identification of the most probable failure modes is poor, due to the limited long-term experience with flexible pipes so far. AME /2/ has compiled a database of incidents involving North Sea offshore pipelines in which containment was lost or jeopardized. The reported failures may not necessarily be relevant for future applications of flexible pipes involving larger water depths, increased pressures and temperatures. As a consequence of the lack of failure data, the study of failure modes and failure causes for flexible pipes is usually based on a careful evaluation of pipe structure, function of each layer, interface between pipe and end fitting, materials, documentation of pipe properties, operational experience and type of application.

In principle, there exist only two main failure modes that can impede the fluid transportation through a flexible pipe:

- Leakage
- Reduction of internal cross section

As illustrated by the "failure tree" in Fig 2, leakage can be caused by a hole through the pipe wall, excessive diffusion and separation of pipe/nipple.

The factors that need to be considered when designing flexible pipes may be grouped into two categories: (i) initial strength parameters and (ii) service life parameters. The latter aspect can further be divided into: (a) mechanical deterioration, and (b) material degradation.

4 FATIGUE MECHANISMS OF NONBONDED PIPES

4.1 Wear and Fatigue of Tensile Armour

In oscillatory bending, armour wires in different layers of a nonbonded pipe will be sliding against each other. Laboratory tests have shown that failure of the pipe is controlled by wear, fatigue and fretting of the armour wires, cf Feret et al. /4/, Kastelein et al. /5/ and Sævik et al. /6/. Analytical models require the following type of input information for a given loading condition:

(a) stresses in each wire
(b) contact pressure between sliding layers
(c) relative slip between two surfaces in contact as a function of curvature change
(d) friction level and wear characteristics

Information concerning (a), (b) and (c) above, may be obtained by using simple expressions for the mechanical behaviour, or by use of more refined computer models. Friction and wear parameters may be obtained by tribological experiments.

4.2 End Terminations

The end termination is probably the most critical part of a pipe construction. A well designed transition zone is required, such that all the wall components converge into one flange or connector piece that carries all the pipe wall forces, see Fig 3. The termination is designed to provide a comprehensive sealing system both radially and axially and also to achieve continuity of loads.

According to Mallen et al. /7/ the validity of the design of Coflexip and terminations have been confirmed by observations made during burst tests. No failure has occurred in the end terminations, not even in burst tests carried out with an internal temperature of 100°C. However, a systematic description and discussion of the different problem areas related to design of end terminations, have not yet been presented anywhere.

5 DESIGN FORMATS FOR ULTIMATE LIMIT STATE

There are several standards and guidelines for the design and operation of high pressure pipes, for example, NPD /8/, API /9/ and Veritec /10/. Only the API and Veritec guidelines address flexible pipes explicitly. In addition, various oil companies have issued design specifications for internal use.

5.1 Permissible Stress Format

The design formats in these documents are generally of the form

$$S_c \leq R_c/SF \qquad (1)$$

where S_c is the characteristic load effect for the actual loading, R_c is the characteristic resistance for the considered failure mode, and SF is the corresponding safety factor accounting for all uncertainties in load, resistance, analysis methods, etc.

The safety factors are specified as minimum values and due consideration should be given to aspects related to consequences of failure and the degree of novelty for the given application.

Minimum safety factors specified in various design documents are listed in Table 1. (Tech Spec A, B and C contain confidential information and the sources are hence not referred to explicitly.) Incorporation of all associated uncertainties in a single safety factor (SF) is obviously questionable. As a result, the permissible stress method may give inconsistent safety levels and in some cases even a nonconservative design.

5.2 Partial Coefficient Design Format

This design format is based on degree of utilization for the individual pipe layers under combined loads. The safety format for the ultimate limit state is expressed as

$$\gamma_E S_{CE} + \gamma_F S_{CF} \leq R_C/\gamma_m \qquad (2)$$

where subscript C denotes characteristic values. Subscripts E and F refer respectively to environmental and functional load effects (e g stresses) in a particular pipe layer. The partial coefficients (γ_E, γ_F and γ_m) refer to a given failure mode, and may generally be tailored to each separate mode.

6 RELIABILITY ANALYSIS

In order to determine the reliability level corresponding to a given design format, the design criteria are rewritten in terms of basic variables, contained in a vector **x**. These should be chosen such that they reflect all the inherent uncertainties properly. To exemplify, we consider failure due to tensile rupture of the steel tendons in the armour layer. The failure function can be expressed on the form

$$g(\mathbf{x},t) = \sigma_C(\mathbf{x}) - \sigma_S(\mathbf{x}) - \sigma_D(\mathbf{x},t) - \sigma_p(\mathbf{x}) \quad (3)$$

where $\sigma_C(x)$, $\sigma_S(x)$, $\sigma_D(x,t)$ and $\sigma_p(x)$ represent respectively the random tensile stress capacity, the random static stress, the time-varying stochastic dynamic stress, and the axial stress due to internal pressure. Failure is now defined by the function $g(\mathbf{x})$ being negative. For the axial compression failure criterion, the failure function is the same as in Eq (3), but with the negative value of the compressive capacity replacing the tensile capacity $\sigma_C(\mathbf{x})$ and with a positive sign in front of the static contribution.

All the random variables entering into the vector **X** in Eq (3) are listed in Ref /15/. The main steps of the reliability analysis procedure for a given riser system can be summarized as:

i) Specification of limit states.
ii) Uncertainty analysis with respect to the basic random variables.

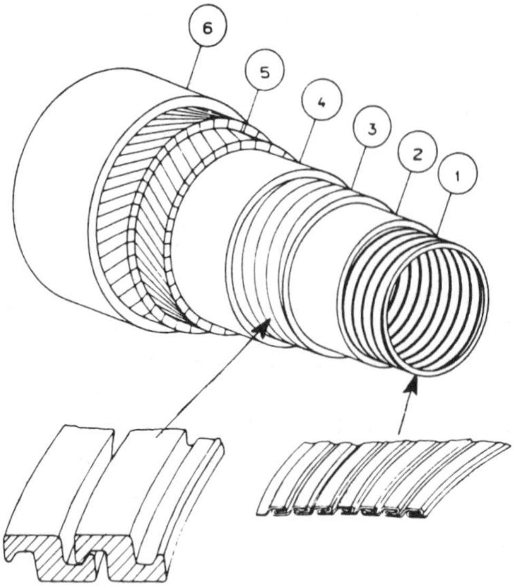

1 Inner interlocked steel caracass
2 Internal thermoplastic sheath
3 Zeta spiral
4 Intermediate thermoplastic sheath
5 Double crosswound armours
6 External thermoplastic sheath

Fig 1. Typical Coflexip rough bore structure

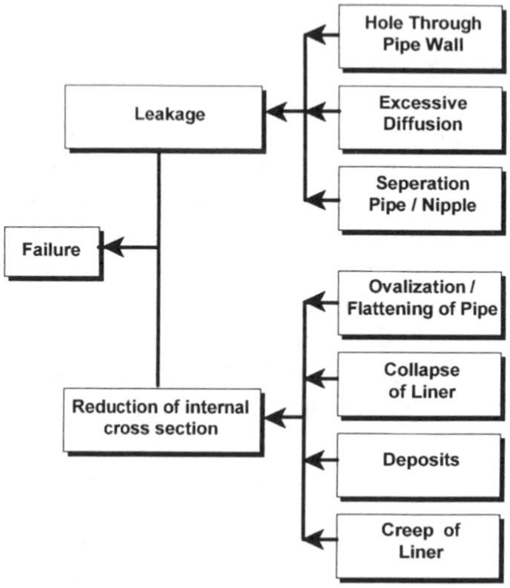

Fig 2. Main failure tree for flexible pipes

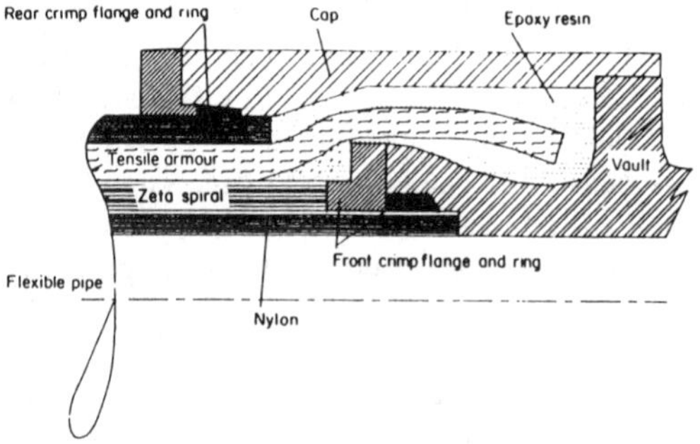

Fig 3. Termination for Coflexip, nonbonded pipe

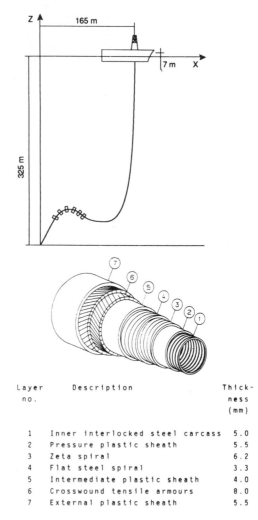

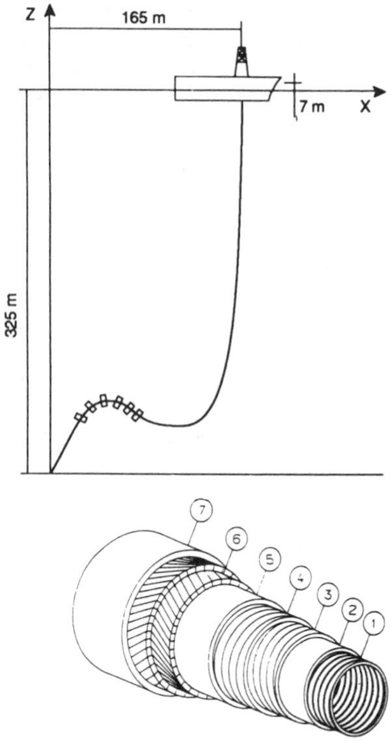

Layer no.	Description	Thickness (mm)
1	Inner interlocked steel carcass	5.0
2	Pressure plastic sheath	5.5
3	Zeta spiral	6.2
4	Flat steel spiral	3.3
5	Intermediate plastic sheath	4.0
6	Crosswound tensile armours	8.0
7	External plastic sheath	5.5

Fig 4. Riser system 3

iii) Establishment of failure functions based on functional relations between basic variables and selected load effects).
iv) Identification of statistical properties of selected load effects, accounting for all sources of statistical uncertainty.
v) Evaluation of annual and lifetime failure probability e g by FORM/SORM algorithms.

Details of the mathematical formulation are found in Ref /15/. Some details of steps, (iii) and (iv), will be given here.

6.1 Failure Functions

For reliability applications, the interest lies in the probabilistic properties of the response maxima and extremes, rather than that of the process itself. Three-parameter Weibull distributions are fitted to samples of response maxima obtained from simulated response time series. The location (1), scale (s) and shape (η) parameters are then expressed as functions of the random vector **X**. The probability of an arbitrary response maximum exceeding a fixed level can now be expressed as a function of the basic variables. Formally, we get

$$P(\sigma_{max} > \zeta) = 1 - W_3(1(\mathbf{x}),s(\mathbf{x}),\eta(\mathbf{x})) \quad (4)$$

where $W_3(\ )$ is the three parameter Weibull cumulative distribution function; ζ is a given value of the dynamic stress maxima. This expression is then utilized in combination with Eq (3).

6.2 Calculation of Failure Probability

The probability of failure (P_f), is calculated as:

$$P_f = \bar{\nu}^+(0) \cdot T \int_{g(u,x)<0} f_x(x)\ \phi(u)\ dxdu \quad (6)$$

where u is an auxilary Gaussian random variable and $\phi(u)$ is the standard normal density function;

T is a specified time period; $\bar{\nu}^+(0)$ is an average zero-upcrossing frequency; and g(u,**x**) is a modified failure function, /7/.

7 CASE STUDY

7.1 General

A general view of a flexible riser configuration is shown in Fig 4. The waves and current are assumed to act along the axis of symmetry of the system. This modelling will give response levels that generally are higher than if a fully three-dimensional environmental representation was employed. The reported design values of dynamic load effects are calculated by a regular wave analysis. These design values tend to be higher than from the long term analysis, but with some statistical spread /14/.

The computer program PROBAN /2/ is employed for the numerical calculations. A summary of the main results concerning statistical properties of static, dynamic and total load effects is given in Table 2. Two specific failure modes are considered here. Also the importance factors for the dominating random variables are given in each case.

7.1.1 Tensile failure mode

Weibull parameters for the static, dynamic and total load effects are given in Table 2. The sum of the static and extreme dynamic tension corresponding to a regular wave design analysis is 294 kN. The design pressure is 405 bar, which produces a helix tension of 840 kN. Applying a safety factor of 2.0, the required tensile capacity becomes 2268 kN. The calculated annual probability of failure for a pressure c.o.v. of 0.3 is $9 \cdot 10^{-7}$ (FORM) ($8.8 \cdot 10^{-7}$ SORM). The most important random variables are those pertaining to the pressure and capacity. The pressure uncertainty variable has an importance factor of 50 %. The model uncertainty related to stress calculation has a factor of 20 %, and for the helix thickness variable it is 10 %. The safety factor/failure probability relations are shown in Fig 6.

7.1.2 Axial compression failure mode, sag bend

The worst combination for this limit state is

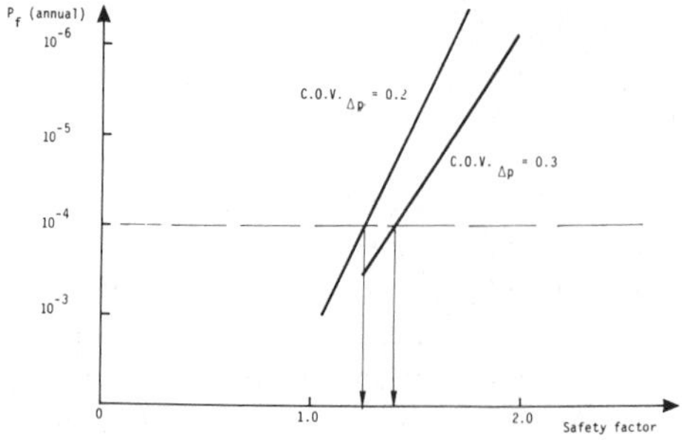

Fig 5. Relation between annual failure probability and safety factor. Tensile failure mode of riser system 3, top end

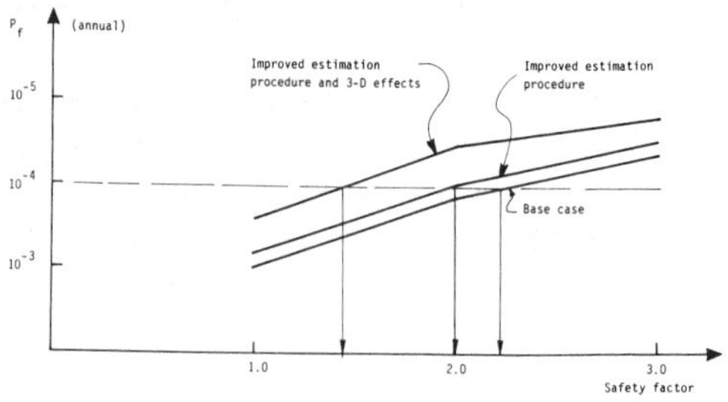

Fig 6. Relation between annual failure probability and safety factor. Axial compression failure mode of riser system 3, sag bend

Table 1. Comparison of minimum safety factors

	Veritec /3/ (1987)	API /2/ (1988)	Tech. Spec. A	Tech. Spec. B I II III	Techn. Spec. C
Burst failure	2.0	2.0	2.0(3.0[2])	2.25 2.0 1.3	2.25
Tensile failure	2.0	2.0	2.0	2.75 2.0 1.3	2.0
Axial compression failure	2.0	Manufac-turer's value	2.0	2.75 2.0 1.3	-[3]
Hydrostatic collapse	1.5	1.5[1]	1.5	2.25 2.0 1.3	2.0[4]
Torsional failure	2.0	Manufac-turer's value	2.0	2.75 2.0 1.3	2.0
Bending failure (MBR)	Storage 1.1 Static 1.25 Dynamic 1.5	Manufac-turer's value	Storage 1.1 Temporary 1.5 Operation 1.5	Storage 1.1 Dynamic 1.5	Manufac-turer's value

1) Atmospheric internal pressure
2) Gas carrying inside "zone 2" (i e risers)
3) Not allowed, unless specifically designed
4) With max design external loads (max axial and curvature)

I: Operation phase - functional loads only
II: Operation and temporary phase - functional and environmental loads
III: Operation and temporary phase - accidental, functional, and environmental loads

when the waves are moving in the positive x-direction, with opposite current direction. Weibull parameters are shown in Table 2 also for this limit state. The 100-year extreme dynamic compression obtained by a regular wave analysis is 14 kN. Using this as the design value (instead of the difference load effect), the required axial compressive capacity becomes 28 kN for a safety factor of 2.0. The (annual) probability of failure is computed to be $2 \cdot 10^{-4}$. The relation between failure probability and safety factor for this limit state is shown in Fig 7. A safety factor of 2.3 or higher is required in order to provide an annual failure probability smaller than a target value of 10^{-4}. The most important random variables influencing the

Table 2 Statistical properties of static, dynamic and total load effects. All sources of uncertainty included

Limit state	Weibull parameters and c.o.v									Importance factors for main variables		
	Static			Dynamic			Total					
	Shape	Scale	c.o.v	Shape	Scale	c.o.v	Shape	Scale	c.o.v	Static	Dynamic	Total
Top end tension	3.26	197 kN	0.32	2.5	52kN	0.43	4.2	296kN	0.25	M/C (90%) S.A.(10%)	D.A./D.R.(60%) M/C (20%)	M/C (60%) D.A./D.R.(20%)
Sag bend compression	2.8	9kN	0.35	0.75	15kN	1.3	0.63	8.5kN	1.8	M/C (70%) V.O.(25%)	Sc/Sh (40%) T_p (10%)	Sc/Sh (45%) T_p (10%)

M/C = measured to calculated response ratio
D.A./D.R. = correlated pair of dynamic analysis uncertainty and damping ratio
V.O. = surface vessel static offset
Sc/Sh = correlated pair of Weibull scale and shape parameter
T_p = peak period

probability of failure are those related to the dynamic load effect. The two primary ones are the shape/scale uncertainty variables (importance factor 35%), the peak period (importance factor 15%) and the significant wave height (importance factor 10%).

8 CONCLUSIONS

In this paper, main types of failure modes of flexible pipes were discussed. A method for case-by-case reliability analysis of flexible riser systems was outlined. Its application was illustrated by a case study.

The results which are presented need to be supplemented by additional studies in order to serve as a basis for a complete code calibration. However, the presented methodology may serve as a tool for reliability analysis of flexible riser systems also for new applications.

9 ACKNOWLEDGEMENT

The results presented in this paper are based on work performed within the research project FPS2000 Flexible Risers and Pipes, sponsored by BP Norge, Elf Aquitaine Norge, Esso Norge, Norsk Hydro, The Royal Norwegian Council for Scientific and Technical Research, SAGA Petroleum, Statoil and Total Norge.

10 REFERENCES

/1/ Neffgen, J.M.: "Optimization of Subsea and Floating Production Systems Utilizing Flexible Pipe Technology", Proc. 5th International Conference on Floating Production Systems, London, December 1989.

/2/ Advanced Mechanics & Engineering Ltd (AME): "The Update of Loss of Containment Data for Offshore Pipelines", Final Report prepared for United Kingdom Offshore Operators Association (UKOOA), Job No. 226.1, August 1990.

/3/ Eriksen, M.: "Inspection Philosophy and Recommendations for Further Work", FPS2000 Report 2.3-3 prepared by Robit, March, 1990.

/4/ Feret, J.J., Bournazel, C.L. and Rigaud, J.; "Evaluation of Flexible Pipes Life Expectancy under Dynamic Conditions", OTC Paper 5230, Offshore Technology Conference, Houston, 1986.

/5/ Kastelein, H.J., Out, J.M.N. and Birch, A.D.: "Shell's Research Efforts in the Field of High-Pressure Flexible Pipe", Deepwater Offshore Technology Conference, Monaco, October, 1987.

/6/ Sævik, S. and Berge, S.: "Fatigue Testing and Theoretical Studies of Two 8-inch

Flexible Pipes", Proc. MARINFLEX'92, London, 1992.

/7/ Mallen, J., Estrier, P. and Amilhau, S.G.: "Quality and Reliability of Flexible Steel Pipes", Trans. Institute of Maritime Engineering, Vol. 101, pp. 253-268, 1989.

/8/ Norwegian Petroleum Directorate (NPD), Regulations Concerning Pipeline Systems in The Petroleum Activities, 1990.

/9/ API RP 17B Recommended Practice for Flexible Pipe, American Petroleum Institute, Washington, 1988.

/10/ Veritec, Guidelines for Flexible Pipes, Rev 2, Det norske Veritas, Høvik, 1987.

/11/ Sødahl, N.R.: "Methods for Design and Analysis of Flexible Riser Systems", Dr ing Thesis, Div of Marine Structures, NTH, Trondheim, 1991.

/12/ PROBAN, "Computer Program for Probabilistic Reliability and Sensitivity Analysis", Program Manuals, Version 2, Veritas Research, Høvik, 1990.

/13/ Madsen, H.O., Krenk, S. and Lind, N.: "Methods of Structural Safety", Prentice-Hall, Inc., New Jersey, 1986.

/14/ Olufsen, A., Larsen, C.M., Leira, B.J. and Passano, E.: "Extreme Response Estimation for Flexible Risers", SINTEF Structural Engineering, Report STF71 F91015. (FPS-Report 2.1-17), Trondheim, 1991.

/15/ Leira, B.J., Olufsen, A. and Jiao, G.: "Reliability Analysis of Flexible Riser Systems", Proc, ISOPE, San Fransisco, 1992.

Structural Safety & Reliability, Schuëller, Shinozuka & Yao (eds) © 1994 Balkema, Rotterdam, ISBN 90 5410 357 4

Fatigue reliability analysis of offshore structures considering the effect of inspection and repair

Torgeir Moan & Geir Olav Hovde
Department of Marine Structures, The Norwegian Institute of Technology, Trondheim, Norway

Guoyang Jiao
SINTEF Structures and Concrete, Trondheim, Norway

ABSTRACT: Current fatigue design criteria for offshore structures are based on SN-data and the Miner-Palmgren approach. The acceptable cumulative damage in some codes is made dependent on a vaguely defined feature of accessibility for inspection. This paper presents a reliability based fracture mechanics (FM) calibration of the allowable cumulative damage which depends on the quality of inspection in terms of probability of crack detection curves. Target safety levels are taken to correspond to those for a cumulative damage of 0.1 and 0.3 with no effect of inspection.

1 INTRODUCTION

Fatigue and fracture are important failure modes of welded offshore structures. Significant progress has recently been made in developing rational design methods to handle these failure modes, especially by using reliability methods, see e.g. Moan et al. (1991).

Fatigue design criteria are commonly based on the SN-Miner-Palmgren approach where the allowable cumulative damage, $D = \sum n_i/N_i$, is made dependent upon the consequences of failure and access for inspection as specified e.g. by the Norwegian Petroleum Directorate (1991). However, existing criteria are qualitative.

The purpose of the present paper is to quantify the effect of inspection depending upon its quality for a given inspection strategy.

Welded joints in TLP tethers and tubular joints of jackets are considered for illustration purposes. This is done by determining the allowable cumulative damage in the SN-approach as a function of the quality of inspection. The calibration of this simple design approach is based on a probabilistic fracture mechanics model of fatigue crack growth.

2 CRACK GROWTH MODEL

The crack growth is described by a generalization of the Paris-Erdogans equation,

$$\frac{da}{dN} = C(\Delta K)^m \qquad (1)$$

as introduced by Newman & Raju (1981). In this approach the surface crack is assumed to be semielliptical with semiaxes a and c at any stage during crack growth, where a is the crack depth, c is the half crack length, N is the number of stress cycles, ΔK is the stress intensity factor range, and C and m are material parameters.

Analysis of the distribution of the material parameter, C, for steels shows C to approximately follow a lognormal distribution. The crack growth parameters may be reasonably modelled as: C lognormally distributed with mean of $1.1 \cdot 10^{-13}$ MPa$\sqrt{\text{mm}}$ (DNV 1984) and a coefficient of variation of 0.55. Instead of using C as a random variable, lnC is used herein, see Tables 1 and 2. m is assumed fixed and equal to 3.1 (DNV 1984).

The stress intensity factor range at any point of the crack front for surface cracks in plane plates subjected to tension and bending loads is expressed as (Newman & Raju 1981)

$$\Delta K = (S_t + S_b)\frac{1 + H\dfrac{S_b}{S_t}}{1 + \dfrac{S_b}{S_t}}\sqrt{\frac{\pi a}{Q}}\, F\left(\frac{a}{c},\frac{a}{t},\frac{c}{W},\varphi\right)$$

$$= S\sqrt{\pi a}\, Y_{plate}\left(\frac{a}{c},\frac{a}{t},\frac{c}{W},\varphi,\frac{S_b}{S_t}\right) \qquad (2)$$

where S_t and S_b are tension and bending stress ranges, respectively. S is the total stress range, t is the plate thickness and W is the width of the plate. φ is the parametric angle to identify the position along an elliptical flaw front. In welded joints, consideration also has to be given to the non-linear stress fields arising from local stress concentrations at the weld toe. These have been modelled using a local stress intensity magnification factor, M_k. The stress concentrations due to local weld geometry are especially present in tubular joints. Ideally, both crack growth in depth a ($\varphi = \pi/2$) and growth in length 2c ($\varphi = 0$) should be calculated, but due to coalescence of cracks the two-dimensional model proposed in (Newman & Raju 1984) may give the wrong aspect ratio evolution. For this reason a constant crack aspect ratio is used for TLP tethers; a/c = 0.83 (Jiao et al. 1990) and for tubular joints the aspect ratio a/c is assumed to be a function of the crack depth a (forcing function). In the general case the stress intensity factor range can then be stated as

$$\Delta K = S\sqrt{\pi a}\, Y \tag{3}$$

Proposed fabrication specifications for TLP tethers call for grinding of the butt welds on both faces. In such cases the influence of the weld, expressed in the local stress intensity magnification factor M_k, is negligible, and with no bending stresses the geometry function is taken as

$$Y = Y_{plate}\ (\frac{a}{c}, \frac{a}{t}, \frac{c}{\pi D}, \frac{\pi}{2}, 0) \tag{4}$$

where D is the diameter of the tether.

On the other hand, M_k is of great importance to tubular joints due to the T-butt welds. Weld toe geometry is often characterized by the weld toe radius ρ and the local weld toe angle θ among other parameters. Thus the local stress intensity magnification factor can be formulated as (MTD Ltd 1992)

$$M_k = 1 + M_{red}(\rho, \theta) \cdot \left[1.24\exp\left(\frac{-22.1a}{t}\right) + 3.17\exp\left(\frac{-357.0a}{t}\right)\right] \tag{5}$$

where M_{red} is introduced in order to include effects due to local weld toe smoothing by for example grinding. In addition to the weld toe stress concentration and combined bending and tension stresses, the stress distribution in the circumferential direction is nonlinear. Therefore, the stress field at the tip of a surface crack changes as the crack propagates. The beneficial effect of the nonlinear stress variations in the circumferential direction is neglected and a constant hot spot stress over the cracked region is applied as a conservative assumption. Hence, the geometry function used for tubular joints is

$$Y = Y_{plate}\ (\gamma_{a/c}\frac{a}{c}, \frac{a}{t}, 0, \frac{\pi}{2}, \frac{S_b}{S_t})\, M_k \tag{6}$$

where the crack aspect ratio is randomized by multiplying by a random variable $\gamma_{a/c}$ which is assumed lognormal distributed with a mean of 1.0 and a COV of 10%. However, the present method may be extended to account for nonlinear stress variations in the circumferential direction.

Neutrally bouyant tethers with dimensions of D = 1100 mm and t = 37 mm are adopted, and a representative bending/tension stress ratio of 4.0 and a thickness of 50 mm are used for the tubular joints.

3 CRACK ASPECT RATIO FOR A TUBULAR JOINT

To account for coalescence of cracks in tubular joints a forcing function, which makes the crack aspect ratio dependent on the crack depth, is used.

As a result of micro-crack coalescence and growth, it is assumed that a small semi-elliptical crack has developed with an aspect ratio of about 1, not immediately next to another small crack. These cracks develop independently until their crack tips meet. Due to this coalescence one long shallow crack evolves, which then will grow through the wall of the tubular.

During crack growth two different stages are considered; one stage before coalescence and one stage during and after coalescence. Assuming that for small crack sizes the tubular joint can be modelled as a T-plate, the crack aspect ratio before coalescence will develop as follows (Vosikovsky et al. 1985):

$$\frac{a}{c} = e^{-ka} \quad ; \quad for\ \frac{a}{t} \le \left(\frac{a}{t}\right)_{coalescence} \tag{7}$$

where $k = 0.2\ mm^{-1}$ (Moan et al. 1993). As coalescence occurs the crack aspect ratio will develop as suggested by Burns et al. (1987), in which the shape of the forcing function is dependent upon the bending-tension stress ratio. The shape of the forcing function for $a/t > (a/t)_{coal.}$ is described by the following relationship:

$$\frac{a}{c} = \frac{1}{\alpha_1\left(\frac{a}{t} - \left(\frac{a}{t}\right)_{coal.}\right) + 1} + \alpha_2 \frac{\frac{S_b}{S_t} - 1}{\frac{S_b}{S_t} + 1}\left(\frac{a}{t} - 1\right) + \left(\frac{a}{c}\right)_{failure} \tag{8}$$

where α_1 and α_2 are parameters. The combined forcing function, Eq. (7) and Eq. (8), is shown in Fig. 1.

4 FATIGUE LOADING

The longterm distribution of stress ranges due to waves is modelled by a two parameter Weibull distribution

$$F_S(s) = 1 - \exp\left[-\left(\frac{s}{A}\right)^B\right] \tag{9}$$

where A is the scale parameter and B is the dimensionless shape parameter.

5 INSPECTION RELIABILITY

The inspection reliability is characterized by the relationship between the probability of detection, $P_D(a_D)$,

Table 1: Probabilistic data for welded joints in TLP tethers. Stresses and geometrical sizes are given in MPa and mm, respectively.

Variable	Distribution	Mean μ	Std.Dev. σ
$\log_{10}K$	Normal	14.0342	0.2041
Δ	Lognormal	1.0	0.3
a_0	Exponential	0.11	0.11
a_R	Exponential	0.11	0.11
a_D	Exponential	1.0 / 1.5	1.0 / 1.5
ν_0	Normal	$5.0 \cdot 10^6$	$3.5 \cdot 10^5$
lnC	Normal	-29.97	0.514
m	Fixed	3.5[1)] 3.1[2)]	
lnA	Normal	2.486[3)]	0.198
1/B	Normal	0.877	0.044
γ_Y	Normal	1.0	0.05
thickness	Normal	37.0	1.48
W	Fixed	3450.0	
a/c	Fixed	0.83	

[1] Used in SN approach
[2] Used in fracture mechanics (FM) approach
[3] Corresponds to $\Delta_d = 0.1$

Table 2: Probabilistic data for welded joints in jackets. Stresses and geometrical sizes are given in MPa and mm, respectively.

Variable	Distribution	Mean μ	Std.Dev. σ
$\log_{10}K$	Normal	12.515	0.2484
Δ	Lognormal	1.0	0.3
a_0	Exponential	0.11	0.11
a_R	Exponential	0.11	0.11
a_D	Exponential	1.0 / 1.5	1.0 / 1.5
ν_0	Normal	$5.0 \cdot 10^6$	$3.5 \cdot 10^5$
lnC	Normal	-29.97	0.514
m	Fixed	3.0[1)] 3.1[2)]	
lnA	Normal	1.336[3)]	0.294
1/B	Normal	1.25	0.125
γ_Y	Normal	1.0	0.10
thickness	Normal	50.0	2.0
W	Fixed	$1.0 \cdot 10^5$	
a/c	Fixed	Eq. (7)+(8)	
$\gamma_{a/c}$	Lognormal	1.0	0.1
S_b/S_t	Fixed	4.0	
M_{red}	Fixed	0.8	

See fotnotes of Table 1

and the crack depth, a_D. The relationship is of the following form

$$P_D(a_D) = 1 - \exp\left(-\frac{a_D}{\lambda}\right) \qquad a_D > 0 \tag{10}$$

This is the analytical expression for the Probability Of Detection, or POD, curve assumed herein.

6 STRATEGY FOR INSPECTION AND REPAIR

It is assumed that inspections basically are performed every fourth year during a life time of 20 years, i.e. after $T_1 = 4$, $T_2 = 8$, $T_3 = 12$ and $T_4 = 16$ years. Only one repair strategy is considered; all detected cracks are repaired by welding and grinding. At each inspection a crack may either be missed or detected and then repaired. Thus 16 different repair courses are possible. The event tree is illustrated in Fig. 2, in which 0 denotes no detection and 1 denotes repair. In addition a simpler inspection scheme with inspection after 4 and 12 years is used to perform sensitivity studies.

7 SAFETY AND EVENT MARGINS

The safety margin for failure before time t can be formulated as (Madsen et al 1990)

$$M(t) = \int_{a_0}^{a_C} \frac{da}{(\gamma_Y Y\sqrt{\pi a})^m} - C\nu_0(t - T_0)A^m\Gamma(1 + \frac{m}{B}) \tag{11}$$

where the initiation period, T_0, is neglected. ν_0 is number of stress cycles a year. The geometry function is randomized by multiplying by a random variable γ_Y which is normally distributed with a mean of 1.0 and a COV of 5% and 10% for tethers and tubular joints respectively. A lower uncertainty is used for the TLP tethers because the stress intensity factor range is close to the flat plate solution, while the uncertainties in the magnification factor imply a greater COV value. The critical crack length a_C is taken as the plate thickness.

The first inspection at time T_1 leads either to crack detection or no crack detection, and the event margin is defined as

$$H = \int_{a_0}^{a_D} \frac{da}{(\gamma_Y Y\sqrt{\pi a})^m} - C\nu_0(T_1 - T_0)A^m\Gamma(1 + \frac{m}{B}) \tag{12}$$

H is negative when a crack is detected and positive otherwise. a_D is the detectable crack size which is assumed given by the POD curve. If a crack is detected and repaired at time T_2 and no crack was detected during inspection number 3 the safety margin for failure before t, where $T_3 < t \le T_4$, is

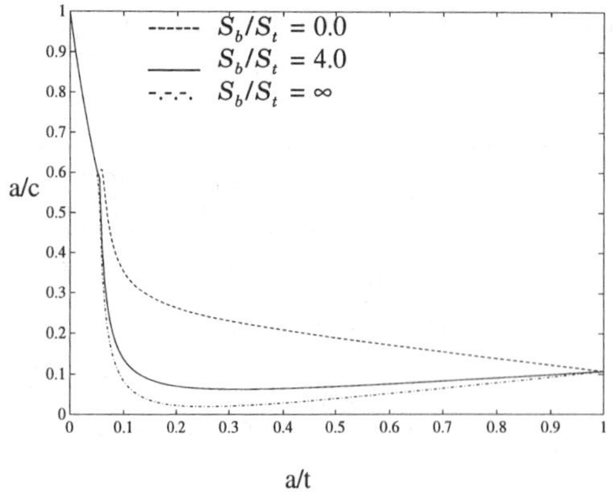

Fig. 1: Aspect ratio, a/c, of cracks in tubular joints as a function of a/t (forcing function). The aspect ratios are calculated by using $\alpha_1 = 150$, $\alpha_2 = 0.15$, $(a/t)_{coal.} = 0.05$, $(a/c)_{fail} = 0.1$ and $t = 50$ mm.

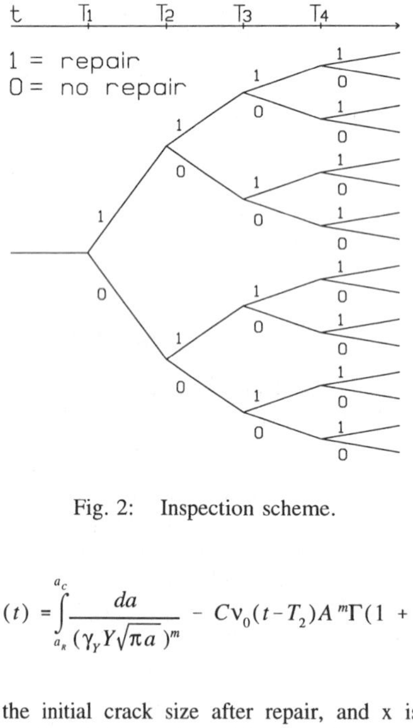

Fig. 2: Inspection scheme.

$$M^{x10}(t) = \int_{a_R}^{a_C} \frac{da}{(\gamma_Y Y\sqrt{\pi a}\,)^m} - C\nu_0(t-T_2)A^m\Gamma(1 + \frac{m}{B}) \tag{13}$$

a_R is the initial crack size after repair, and x is any outcome after first inspection, either 1 or 0.

The event margin for crack detection at time T_4 is

$$H^{x10} = \int_{a_R}^{a_D} \frac{da}{(\gamma_Y Y\sqrt{\pi a}\,)^m} - C\nu_0(T_4 - T_2)A^m\Gamma(1 + \frac{m}{B}) \tag{14}$$

Safety and event margins are defined similarly for the other branches.

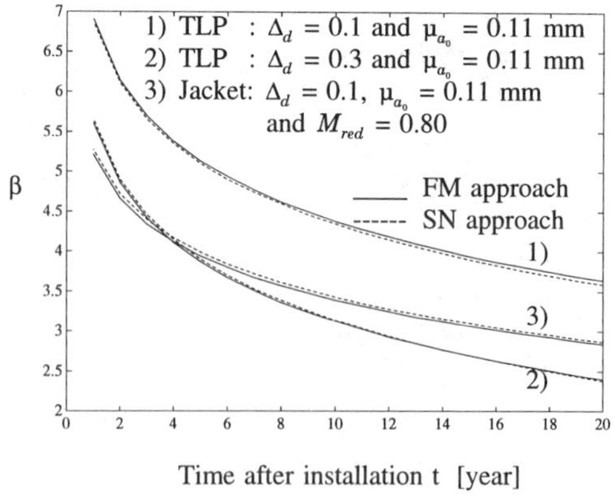

Fig. 3: Reliability index for welded joints in TLP tethers and jackets as a function of time. No inspection and repair.

8 FAILURE PROBABILITIES

The probability of failure before time t is $P_F(t)$, with a corresponding reliability index given by

$$\beta(t) = -\Phi^{-1}(P_F(t)) \tag{15}$$

The following three expressions for the failure probability are given as presented by Madsen et al. (1990). Each expression corresponds to a specific interval of the time parameter, t.

For $T_0 \le t \le T_1$:

$$P_F(t) = P[M(t) \le 0] \tag{16}$$

For $T_1 < t \le T_2$:

$$\begin{aligned} P_F(t) &= P_F(T_1) + \Delta P_F^0(T_1,t) + \Delta P_F^1(T_1,t) \\ &= P_F(T_1) + P[M(T_1)>0 \cap H>0 \cap M^0(t)\le 0] \\ &\quad + P[M(T_1)>0 \cap H\le 0 \cap M^1(t)\le 0] \end{aligned} \tag{17}$$

For $T_2 < t \le T_3$:

$$\begin{aligned} P_F(t) &= P_F(T_2) + \Delta P_F^{00}(T_2,t) \\ &\quad + \Delta P_F^{01}(T_2,t) + \Delta P_F^{10}(T_2,t) + \Delta P_F^{11}(T_2,t) \\ &= P_F(T_2) \end{aligned}$$

$$+ P[M(T_1)>0 \cap H>0 \cap M^0(T_2)>0 \cap H^0>0 \cap M^{00}(t) \le 0]$$

$$+ P[M(T_1)>0 \cap H>0 \cap M^0(T_2)>0 \cap H^0 \le 0 \cap M^{01}(t) \le 0]$$

$$+ P[M(T_1)>0 \cap H \le 0 \cap M^1(T_2)>0 \cap H^1>0 \cap M^{10}(t) \le 0]$$

$$+ P[M(T_1)>0 \cap H \le 0 \cap M^1(T_2)>0 \cap H^1 \le 0 \cap M^{11}(t) \le 0] \tag{18}$$

Similar equations can be established for time intervals $T_3 < t \le T_4$ and $T_4 < t$. $P_F(t)$ is calculated by FORM using PROBAN (Olesen 1992) and importance sampling using ISPUD (Bourgund and Bucher 1986). The simulations are performed around the design point, calculated by FORM.

It is difficult to calculate these intersections by FORM, especially if some of the events are covered by other events. Such problems have mainly been solved by neglecting unnecessary events, which are verified by simulations.

Event margins of a format of the type g = R - S, confer e.g. Eq. 11, may give inconsistent failure probabilities, e.g. the calculated probability of failure after 6 years may be larger than the calculated failure probability after 7 years. The experience is also that using event margins formulated as g = lnR - lnS give the correct FORM solution, but in this case the design point in a very limited number of the intersections is found. Thus, the failure margins are given as

$$g = sign(R)\,|R|^{\zeta} - sign(S)\,|S|^{\zeta} \tag{19}$$

where $\zeta = 0.25$ is found as the most favourable exponent.

9 CALIBRATION OF THE FRACTURE MECHANICS MODEL

The purpose is to calibrate a design check: $D = \sum n_i/N_i \le \Delta_d$ so that it implies a consistent life time failure probability by determining Δ_d as a function of the consequences of failure and reliability of inspection and repair. For instance the Norwegian Petroleum Directorate (1991) prescribes $\Delta_d = 0.1$ for structural components of major importance for the structural integrity and $\Delta_d = 0.3$ for components of minor importance, when no inspections and repairs are performed. When an inspection and repair strategy is implemented, the allowable cumulative damage Δ_d can be relaxed.

To account for the effect of inspection and repair a fracture mechanics approach needs to be used, in which a_0 and M_{red} are important parameters. The initial crack size, a_0, and M_{red} are estimated by requiring that the reliability level for the fracture mechanics approach should be the same as that implied by $\Delta_d = 0.1$ or 0.3 for the SN approach when no inspection and repair are considered.

The SN curve is of the form

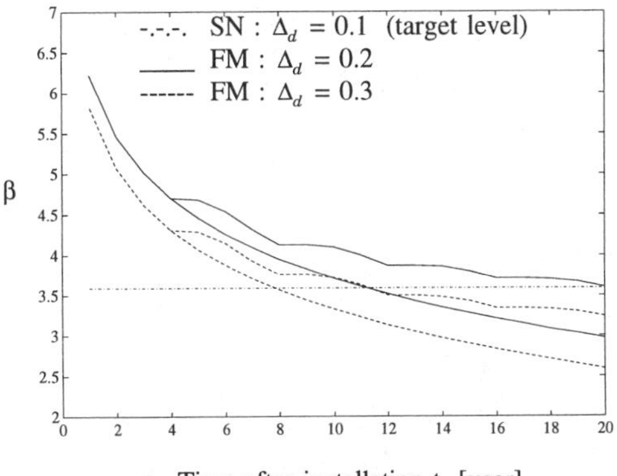

Fig. 4: Reliability index for welded joints in TLP tethers as a function of time. The target level is given by $\Delta_d = 0.1$ and no use of inspection and repair, corresponding to $\mu_{a_0} = 0.11$ mm. The inspection and repair scheme is characterized by 4 inspections, $\mu_{a_D} = 1.50$ mm and $\mu_{a_R} = 0.11$ mm.

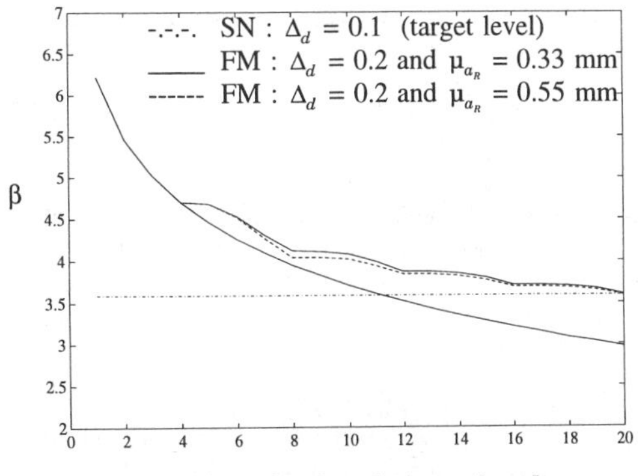

Fig. 5: Reliability index for welded joints in TLP tethers as a function of time. The target level is given by $\Delta_d = 0.1$ and no use of inspection and repair, corresponding to $\mu_{a_0} = 0.11$ mm. The inspection and repair scheme is characterized by 4 inspections, $\mu_{a_D} = 1.50$ mm and $\mu_{a_R} = 0.33/0.55$ mm.

$$N = KS^{-m} \tag{20}$$

where K is a parameter relating to the location of the SN curve, and m is the inverse slope of the curve. The DNV C-curve is used for the TLP tethers and the T-curve corrected for a thickness of 50 mm is used for the tubular joints. No correction is performed for the tethers, because of the butt welds (DNV 1984). Probabilistic data are given in Tables 1 and 2.

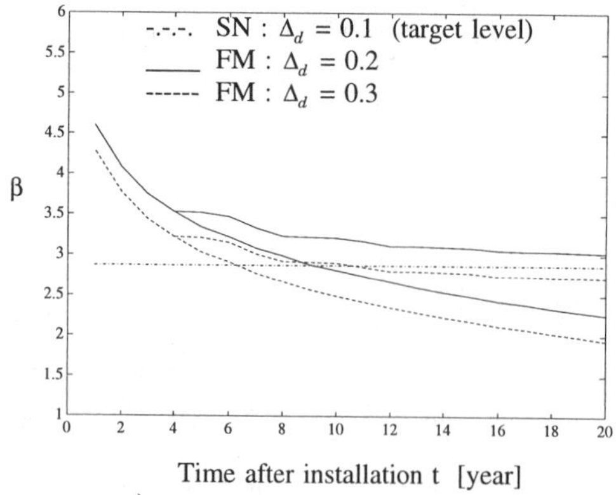

Fig. 6: Reliability index for welded joints in jackets tethers as a function of time. The target level is given by $\Delta_d = 0.1$ and no use of inspection and repair, corresponding to $\mu_{a_0} = 0.11$ mm and $M_{red} = 0.80$. The inspection and repair scheme is characterized by 4 inspections, $\mu_{a_D} = 1.50$ mm and $\mu_{a_R} = 0.11$ mm.

The safety margin for failure before time t according to the SN approach is

$$M(t) = \Delta - \frac{\nu_0 t}{K} A^m \Gamma(1 + \frac{m}{B}) \tag{21}$$

where Δ is the Miner's sum at failure. Δ is taken to have a lognormal distribution with mean of 1.0 and a COV of 30% (Engesvik et al 1987). The scale parameter A is determined from

$$D = \frac{\nu_0 t}{K_c} A^m \Gamma(1 + \frac{m}{B}) = \Delta_d \tag{22}$$

where K_c is a characteristic value, Δ_d is the design Miner's sum equal to 0.1, 1/B is 0.877 for TLP tethers (Jiao et al. 1990) and 1.25 for tubular joints, and t is the design life taken as 20 years.

By assuming that the distribution of initial crack size is exponential with a mean value of 0.11 mm and M_{red} equal to 0.8 it has been found that the fracture mechanics approach, Eq. 11, and the SN approach, Eq. 21, result in the same probability of failure. The mean depth is identical to that found by Bokalrud and Karlsen (1982) based on measured undercuts in production buttwelds of plates with thickness from 10 to 25 mm. The tether analysis is also performed with a target level of Δ_d equal to 0.3. The mean value of a_0 is then calibrated to be 0.15 mm.

The SN data for the C- and T-curves are based on welds subjected to NDT. Thus, the mean crack depth obtained from these calibrations are estimates of the initial depths after the fabrication inspections. But, due to nonequal m values in the fracture mechanics approach and the SN approach, the initial crack size becomes load dependent, which is physically unrealistic. The target levels are presented in Fig. 3, and the input data are given in Tables 1 and 2.

10 CALIBRATION OF FATIGUE DESIGN CRITERIA

The purpose is to calculate a relaxed Δ_d, when inspection and repair are implemented, such that it corresponds to the failure probability for $\Delta_d = 0.1$ and $\Delta_d = 0.3$, when no inspection and repair are implemented. The calculations are performed with the parameter λ in the POD curve, Eq. 10, equal to 1.0 and 1.5 mm. This implies a 90% probability of detecting a 2.3 mm and a 3.5 mm deep crack, respectively. In case of tethers, the allowable Δ_d can be relaxed from 0.1 to 0.25 when the inspection and repair are characterised by a λ of 1.0 mm and a mean crack size after repair of $\mu_{a_R} = 0.11$ mm. By reducing the inspection quality to λ = 1.5 mm the allowable Miner's sum is 0.2, Fig. 4. These inspection qualities are very good. However, they are of the same order as those determined from experimental work performed by UCL, NDE Centre (1990) for the magnetic particle inspection method (MPI). In contrast, mean detectable crack size by visual inspection ranges from about 6.5 mm to 13 mm (MTD Ltd 1992). According to the above results it is assumed that visual (VI) qualities hardly give any effect on the reliability for tethers. NPD (1991) prescribes a design fatigue damage of 0.3 when inspections are performed. This level can either be reached by using an even better inspection quality than above or more frequent inspections. The latter is more realistic.

A mean initial crack size of 0.11 mm after repair is

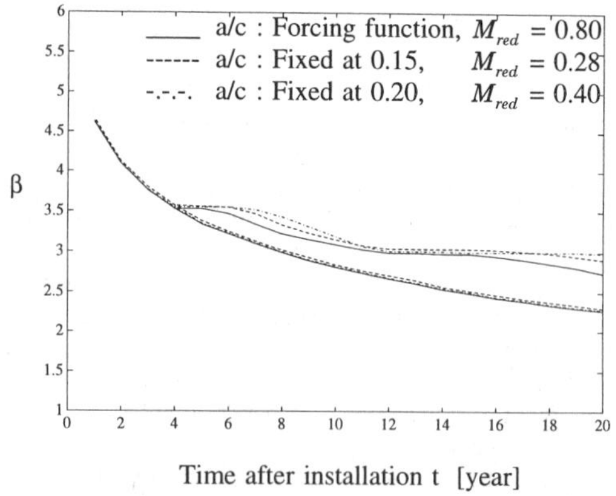

Fig. 7: Reliability index for welded joints in jackets as a function of time, when the crack aspect ratio is kept constant and is made dependent on a/t. The target level is given by $\Delta_d = 0.1$ and no use of inspection and repair, corresponding to μ_{a_0} = 0.11 mm. The inspection and repair scheme is characterized by 2 inspections, $\mu_{a_D} = 1.50$ mm and $\mu_{a_R} = 0.11$ mm.

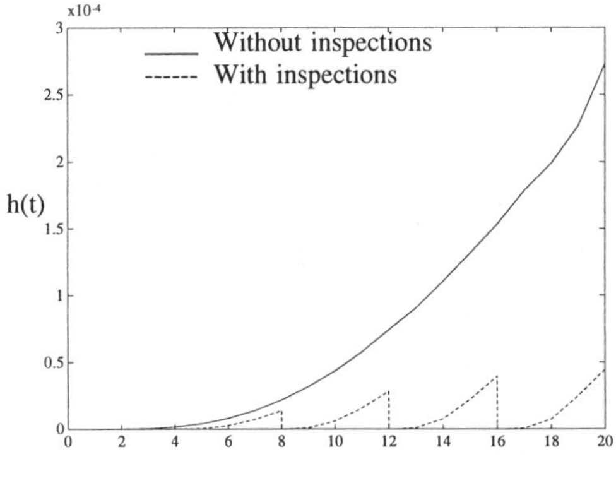

Fig. 8: Hazard rate function for welded joints in TLP tethers. The load corresponds to $\Delta_d = 0.2$, and the initial crack depth is $\mu_{a_0} = 0.11$ mm. The inspection and repair scheme is characterized by 4 inspections, $\mu_{a_D} = 1.50$ mm and $\mu_{a_R} = 0.11$ mm.

rather good. A more realistic value may be 0.33 mm or 0.55 mm. While there is a small difference in the failure probability in short time periods, t, for cases with different initial cracksizes after repair, the failure probability after 20 years is almost the same as for 0.11 mm, see Fig. 5.

If the structural component is of minor importance for the structural integrity a larger failure probability can be allowed, e.g. by using a target level corresponding to a Miner's sum equal to 0.3. This gives a design fatigue damage of 0.7 if a mean detectable crack size of 1.5 mm is used, compared to 0.2 for the target level corresponding to $\Delta_d = 0.1$.

Importance sampling using ISPUD (Bourgund et al. 1986) around the FORM design point gave approximately the same reliability index as FORM. For a period of 20 years with 4 inspections the reliability index achieved by importance sampling was 4% higher (Moan et al. 1993) than the FORM solution. Hence, FORM is conservative.

The behaviour of tubular joints is different from welded joints in TLP tethers. Most important, the stress level through the plate thickness is not constant, such as for the tethers. Due to the bending stresses the stress level in tubular joints is at its maximum at the plate surface, which explains why the time history of β in Fig. 6 ($S_b/S_t = 4.0$) is not as steep as in Fig. 4 ($S_b/S_t = 0.0$). When inspections are carried out every fourth year and the mean detectable crack size is 1.0 mm and 1.5 mm, respectively, a design fatigue damage of 0.3 and 0.25 can be used to achieve the target failure probability. The effect of a given inspection scheme on the allowable cumulative damage is hence more significant for tubular joints than for TLP tethers. Results for the tubular joint case with $\mu_{a_D} = 1.5$ m is given in Fig. 6.

Fig. 7 shows the time history of β for the the tubular joint case for different assumptions of the crack aspect ratio development. Three cases are analysed, i.e. a/c = 0.15, a/c = 0.20 and a/c given by Eqs. 7 and 8. It is seen that the crack aspect ratio assumption is of major importance, and it is also seen that the assumed forcing function gives the smallest reliability.

The hazard rate function,

$$h(t) = \frac{dP_F(t)/dt}{1 - P_F(t)} = -\frac{\phi(-\beta(t))}{1 - \Phi(-\beta(t))}\frac{d\beta(t)}{dt} \qquad (23)$$

expresses the likelihood of failure in the time interval t to t+dt as dt→0, given that failure has not occured prior to time t. Fig. 8 shows $h(t)$ for welded joints in TLP tethers, and it is clearly seen that the hazard rate is significantly reduced due to inspections. The hazard rate functions for the tubular joints have approximately the same shape as for the TLP tether joints, except for the absolute level ,which is higher due to a lower target level, and the hazard rate tops (inspection case), which decrease instead of increasing with time t.

11 CONCLUDING REMARKS

In this paper the effect of inspection depending upon its quality for a given inspection strategy for two different types of welded joints in offshore structures has been quantified using probabilistic methods. The conclusions are based on the following assumptions: welded joints in North Sea structures, 4 year inspection interval, mean crack size after repair $\mu_{a_R} \leq 0.55$ mm and mean detectable crack size $\mu_{a_D} \leq 1.5$ mm.

For TLP tethers, loaded by tensile stresses only, the allowable cumulative damage, when no inspection and repair is implemented, is assumed to be 0.1. This requirement may be relaxed to 0.2-0.25, when inspection scheme as described above is considered. For components of minor importance loaded in tension, an allowable cumulative damage of 0.3 can be relaxed to 0.7.

The allowable damage criterion, for tubular joints, can also be relaxed in case of inspection and repair. For a case with a bending-tension stress ratio of 4.0 the criterion can be relaxed from 0.1 to 0.25-0.3.

The inspection reliability is rather ambitious, especially for tubular joints in jackets, therefore, the relaxation in design criteria shown represents a maximum effect of inspection on design criteria.

REFERENCES

Bokalrud, T. and A. Karlsen 1982. Control of Fatigue Failure in Ship Hulls by Ultrasonic Inspection. *Norwegian Maritime Research*: No. 1, 9-15.

Bourgund, U. and C.G. Bucher 1986. Importance Sampling Procedure Using Design Points - A user's Manual. Inst. Of Engineering Mechanics, University of Innsbruck, Austria.

Burns, D.J., S.B. Lambert and U.H. Mohaupt 1987. Crack

Growth Behaviour and Fracture Mechanics Approach. *Proc. of Conference on Steel in Marine Structures*: Paper PS 6. Delft: Delft University of Technology.
DNV 1984. Fatigue Strength Analysis for Mobile Offshore Units - Classification Notes No. 30.2. Det Norske Veritas, Høvik, Norway.
Engesvik, K. et al. 1987. Literature survey on parameter uncertainties related to structural analysis of marine structures. SINTEF Report STF71 A87008, Trondheim, Norway.
Jiao, G., T. Moan and M.J. Marley 1990. Reliability Analysis of TLP Tether Systems. *Proc. 9th OMAE Conference*: Vol. II, 65-71. New York: The American Society of Mechanical Engineering.
Madsen, H.O. et al. 1986. *Methods of Structural Safety*, New Jersey: Prentice-Hall, Inc., Englewood Cliffs.
Madsen, H.O. and J.D. Sorensen 1990. Probability-based optimization of fatigue design, inspection and maintenance. *Proc. 4th Integrity of Offshore Structures*: 421-438. London: Elsevier Applied Science.
Marine Technology Directorate Limited 1992. Probability-based fatigue inspection planning. MTD Ltd Publication 92/100, London, England.
Moan, T. et al. 1991. Report of ISSC Committee IV.1 Design Philosophy. *Proc. of 11th ISSC*: 575-661. London: Elsevier Applied Science.
Moan T., G.O. Hovde and A.M. Blanker 1993. Reliability-Based Fatigue Design Criteria for Offshore Structures Considering the Effect of Inspection and Repair. *Proc. 25th OTC*: Vol. II, 591-599, OTC 7189. Houston: Offshore Technology Conference.
NDE Centre 1990. Tentative POD curves. Private Communication, University College, London, England.
Newman, J.C. and I.S. Raju 1981. An empirical stress-intensity factor equation for the surface crack. *Engineering Fracture Mechanics*: Vol. 15, No. 1-2, 185-192.
NPD 1991. Regulations for Structural Design of Loadbearing Structures intended for Exploration of Petroleum Resources, Norwegian Petroleum Directorate, Stavanger, Norway.
Olesen, R. 1992. PROBAN General Purpose Probabilistic Analysis Program User's Manual. Veritas Sesam Systems, Report No. 92-7049, Høvik, Norway.
Vosikovsky et al. 1985. Fracture mechanics Assessment of Fatigue Life of Welded Plate T-Joints Including Thickness Effect. *Proc. of 4th BOSS*: 453-464. Amsterdam: Elsevier, Amsterdam.

Structural Safety & Reliability, Schuëller, Shinozuka & Yao (eds) © 1994 Balkema, Rotterdam, ISBN 90 5410 357 4

Reliability method for offshore structures by linearization of mechanical behaviour

A. M. Mohamed & J. P. Muzeau
Laboratory of Civil Engineering, Blaise Pascal University, Clermont-Ferrand, France
M. Lemaire
French Institute of Advanced Mechanics, Clermont-Ferrand, France

ABSTRACT : An efficient reliability method is proposed for offshore structures. The method is based on the linearization of the mechanical behaviour. The elastic-plastic behaviour of material is piecewise linearized by a plastic criterion taking into account the interaction between internal forces. The buckling of members is represented by using an equivalent negative strain-hardening. The structural reliability is estimated by the safety margin concept. Numerical example shows a good performance of the model.

1. INTRODUCTION

In the last decade, a great attention has been given to the reliability analysis of offshore structures. Steel tubular space frames are widely used in such constructions. Due to high degree of redundancy, these structures can resist to external loadings, even in a deteriorated state. The analysis of the post-elastic behaviour is of great importance because it affects the redistribution of internal forces, and consequently, the determination of dominant failure paths.

The reliability analysis of structures implies interdependence between the mechanical and the reliability models. Knowing that mechanical behaviour is generally nonlinear, the reliability analysis is possible by simulation process, but calculation cost becomes very expensive. A great efficiency can be obtained by decoupling the two models ; this can be done by linearization of the "real" mechanical behaviour.

An example of linearized mechanical models is given by Thoft-Christensen and Murotsu (1986). Generally, the linearization process leads to a significant loss of model accuracy. This affects the reliability evaluation. In the present work, a linearized method is proposed, with an improved modelling of yielding of cross-sections and buckling of members. The model can also take into account the unloading of cross-sections.

The elastic-plastic behaviour is represented by the well-known plastic flow theory (Lemaitre et Chaboche 1985). The interaction between normal force N and bending moments M_y, M_z is taken into account by a linearized plastic criterion. The unloading of cross-sections is integrated in the general formulation. During member buckling, the behaviour is modelled by using an equivalent negative strain-hardening representing the loss of normal force capacity. This negative strain-hardening allows the linearization of the buckling phenomenon. In this manner, the structural nonlinear response is represented by a set of linear sequences.

In structural systems, the random variables are divided into external variables as applied loads, and into internal variables as cross-section strengths. In case of linearized structural systems, the internal forces are linear functions of random variables. In order to separate the effect of contribution of each random variable, we have to integrate analytically the plastic deformations all over the loading history. The failure probability of each structural component is calculated by describing its safety margin in function of random variables and influence coefficients. The system reliability is obtained by combination in series and in parallels of component failure modes.

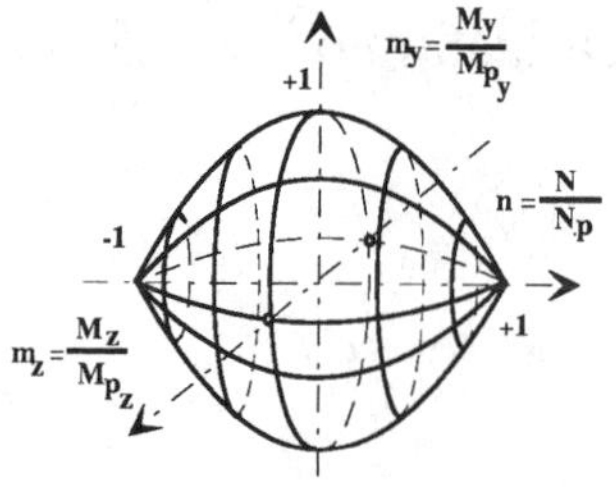

Figure 1. Criterion of tubular cross-sections.

2. STRUCTURAL NONLINEAR BEHAVIOUR

There are two types of non-linearities known as : material nonlinearity and geometrical nonlinearity. In our model, the former type is represented by describing a plastic criterion and a plastic flow law. The second type is represented by an equivalent negative strain-hardening.

2.1. *Tubular section plastic criterion*

For tubular cross-sections of space frames, we shall consider the interaction between normal force "N" and bending moments "M_y, M_z" about the transversal axes "y" and "z" respectively. The effect of shearing forces and torsional moment on the plastic behaviour has been neglected. For this kind of cross-sections, the real plastic criterion is shown in figure 1.

In order to linearize the material behaviour, a plastic criterion formed of 24 plane surfaces is proposed (Mohamed 1993). The criterion is identified by three parameters γ, ξ and η. The choice of these parameters allows to represent different shapes of cross-sections. In the case of tubular cross-sections, suitable values are : $\gamma = 0.25$, $\xi = 0.03$ and $\eta = 0.40$.

The general criterion equation is given by :

$$f = \left(\frac{1-\gamma\mathcal{J}_n}{1-\xi+\eta\mathcal{J}_y\mathcal{J}_z}\right)\left[\mathcal{J}_y\left|\frac{M_y}{M_{py}}\right| + \mathcal{J}_z\left|\frac{M_z}{M_{pz}}\right|\right] + \mathcal{J}_n\left|\frac{N}{N_p}\right| - (1+a) \quad (1)$$

where N_p, M_{p_y}, M_{p_z} are respectively the cross-section plastic capacities under pure normal force and bending moments ; $\mathcal{J}_y$, $\mathcal{J}_z$ and $\mathcal{J}_n$ are indexes which define the equation of the considered plane

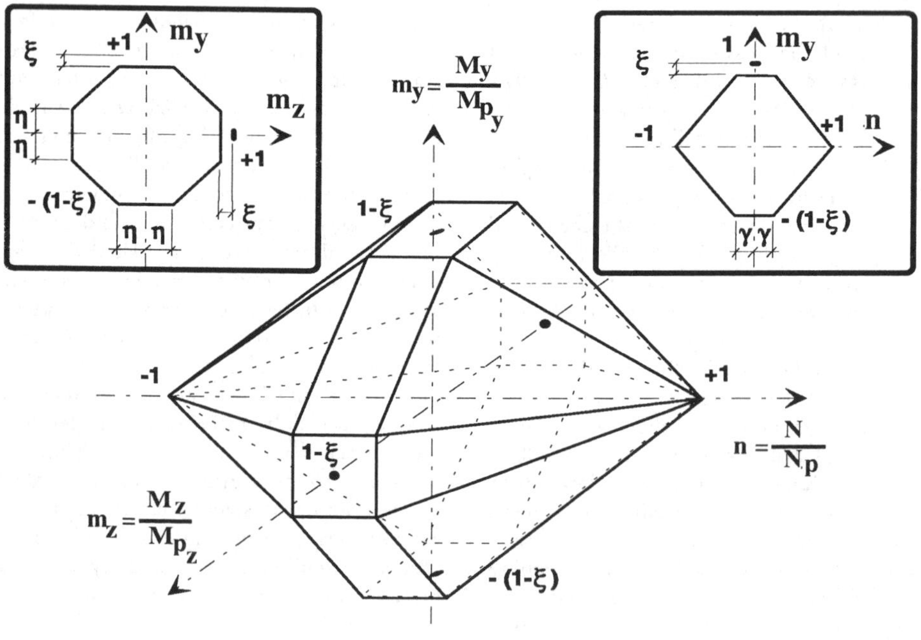

Figure 2. Linearized plastic criterion.

surface (their values are "0" or "1"); a is the strain-hardening magnitude. The criterion is illustrated in figure 2.

2.2. *Elastic-plastic formulation*

For a linearized plastic criterion (case of standard plasticity), the tangent stiffness matrix $[\overline{K}_e]$ is constant during each analysis increment as long as the strain-hardening parameter is unchanged. The incremental form of element equilibrium is given by :

$$\{dF_e\} = [\overline{K}_e] \cdot \{dq_e\} \tag{2}$$

where $\{dF_e\}$, $\{dq_e\}$ are respectively vectors of incremental nodal forces and displacements. Knowing that platform loadings can be represented by nodal forces, the yieldings are expected only at member ends. In order to make the inelastic behaviour fully linearized, element internal yieldings are reported at the ends. This means that numerical integrations, which are time consuming, are completely eliminated. In this way, the nonlinear behaviour is piecewise linearized.

In classical incremental methods, it is not possible to distinguish between internal forces due to each random variable. The integration of plastic deformations all over the loading history allows to separate the contribution of different random variables (loads and strengths). The integration constants are determined by initial conditions. They are evaluated as "equivalent forces" $\{\overline{F}_e\}$ corresponding to a vanishing nodal displacement vector.

The integrated equilibrium equation becomes as following (Mohamed 1993) :

$$\{F_e\} = [\overline{K}_e] \cdot \{q_e\} + \{\overline{F}_e\} \tag{3}$$

with

$$[\overline{K}_e] = [K_e] - \frac{[K_e] \cdot \{\mathcal{N}_k\} \cdot \{\mathcal{N}_k\}^t \cdot [K_e]}{L\ N_p^2 + \{\mathcal{N}_k\}^t \cdot [K_e] \cdot \{\mathcal{N}_k\}}$$

$$\{\overline{F}_e\} = N_p\ \frac{[K_e] \cdot \{\mathcal{N}_k\}}{L\ N_p^2 + \{\mathcal{N}_k\}^t \cdot [K_e] \cdot \{\mathcal{N}_k\}}$$

where $[K_e]$, $[\overline{K}_e]$ are respectively the elastic and the elastic-plastic stiffness matrix; $\{\mathcal{N}_k\}$ is the vector normal to the plastic criterion; L is the strain-hardening parameter. This expression

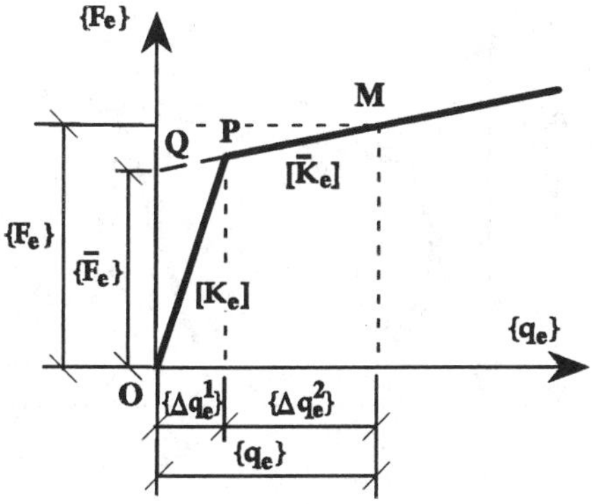

Figure 3. Elastic-plastic element behaviour.

represents a linear relation between element total forces and displacements. The equivalent forces are not function of external loading, but only of internal strengths of yielded cross-sections (they vanish in case of pure elastic behaviour).
The elastic-plastic behaviour of elements is shown in figure 3. For classical incremental methods (equation 2), the analysis is carried out according to the bi-linear behaviour OPM. For integrated method (equation 3), the behaviour is represented by only one linear relation QM, the distance OQ being the initial condition.

2.3. *Unloading of yielded cross-sections*

In offshore structures, unloading may occur during the redistribution of internal forces. In many cases, this is due to buckling of bracing elements. A realistic model must take into account the effect of such phenomenon. According to the plastic flow theory, unloading can be detected by the observation of negative plastic multipliers.

The behaviour after unloading follows an elastic relationship with conservation of plastic deformations obtained just before unloading (figure 4). The incremental element equilibrium is given by :

$$\{dF_e\} = [K_e] \cdot \{dq_e^e\} \tag{4}$$

where $\{dq_e^e\}$ is the vector of incremental elastic displacements. The integration of this equation gives the following form :

$$\{F_e\} = [K_e] \cdot (\{q_e\} - \{q_e^p\}^*) \tag{5}$$

where $\{q_e\}$ is the vector of total element dis-

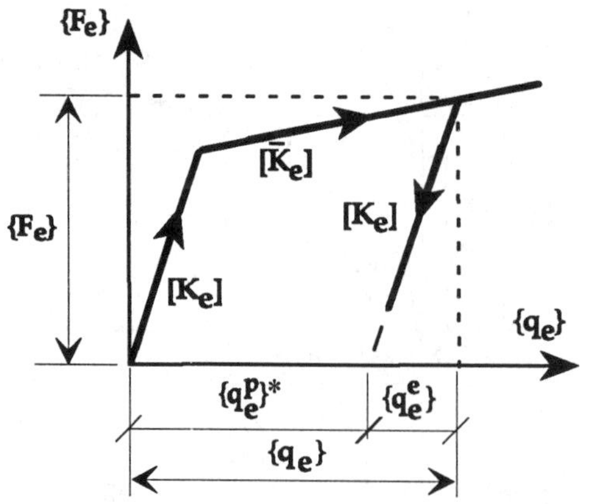

Figure 4. Element behaviour during unloading.

placements and $\{dq_e^p\}^*$ is the vector of plastic displacements at the moment of unloading (this vector is constant in the rest of the analysis).

Equation (5) can be written as following :

$$\{F_e\} = [K_e] \cdot \{q_e\} + \{F_e^*\} \tag{6}$$

where $\{F_e^*\}$ represents the vector of equivalent forces in case of unloading; it is given by : $\{F_e^*\} = -[K_e] \cdot \{q_e^p\}^*$. It can be seen that equation (6) takes the same form of the element equilibrium as in case of loading (equation 3). This allows us to keep the same formulation for the two cases of loading and unloading, using the appropriate stiffness matrix and vector of equivalent forces.

2.4. *Negative strain-hardening*

The elastic-plastic buckling is characterized by the reduction of ultimate axial strength and by the decreasing of member capacity during the post-critical behaviour. The critical force can be described by introducing a reduction factor to the axial plastic force N_p. The increasing loss of axial force capacity is represented by a negative strain-hardening introduced to the elastic-plastic model (Lemaire et al. 1991). The parameter of this strain-hardening is evaluated by the analysis of the member buckling curve, having the remaining structure as boundary conditions.

For a deformed member, the element shape must satisfy the internal and external forces equilibrium, as well as the boundary conditions. In order to find the so-called "exact" element shape, we have elaborated a numerical method based on a shape function (finite element context) formed by a high degree polynomial.

The member axial shortening is obtained by the sum of axial elastic-plastic deformations and the shortening due to element curvature. The calculation of a few points on the buckling curve allows to evaluate the corresponding negative strain-hardening parameter. This one is injected into the elastic-plastic model in order to perform linearized mechanical analysis.

In case of axially loaded elements, the strain-hardening parameter L is given by :

$$L = \frac{k_e}{N_p^2(\Delta P/\Delta\delta - 1)} \tag{7}$$

where k_e is the initial axial rigidity ; ΔP is the rate of decrease of axial force and $\Delta\delta$ is the rate of member shortening.

3. STRUCTURAL RELIABILITY

The previous mechanical model allows to obtain the structure response in a piecewise linearized form. The contribution of each random variable is also evaluated. The output of this model is linear function of its input, this is written as :

$$N = \sum_{i=1}^{ngb} c_{n_i} \; P_{r_i} + \sum_{j=1}^{nfe} \bar{c}_{n_j} \; N_{p_j} \qquad (8.a)$$

$$M_y = \sum_{i=1}^{ngb} c_{m_{yi}} \; P_{r_i} + \sum_{j=1}^{nfe} \bar{c}_{m_{yj}} \; N_{p_j} \qquad (8.b)$$

$$M_z = \sum_{i=1}^{ngb} c_{m_{zi}} \; P_{r_i} + \sum_{j=1}^{nfe} \bar{c}_{m_{zj}} \; N_{p_j} \qquad (8.c)$$

where P_{r_i} are the loading random variables, N_{p_j} are the strength random variables, c_{n_i}, $\bar{c}_{n_j}$, $c_{m_{yi}}$, $\bar{c}_{m_{yj}}$, $c_{m_{zi}}$, $\bar{c}_{m_{zj}}$ are the coefficients of the linear function (their values depend on the mechanical behaviour) and ngb, nfe are respectively the number of loading random variables and the number of equivalent forces (i.e. number of yielded elements).

The linearized plastic criterion defines the failure condition of cross-sections (structural components). Hence, the safety margin M_s is written

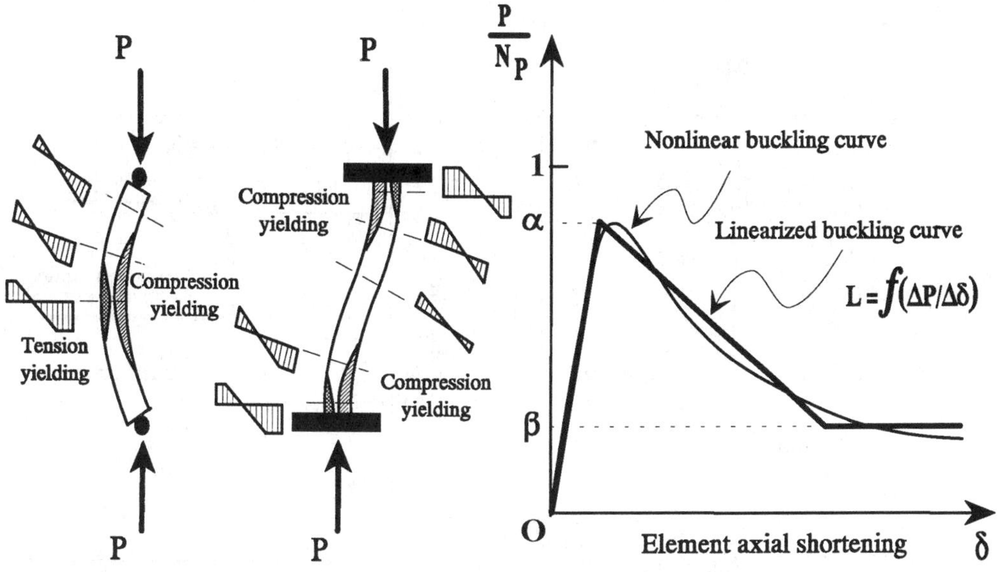

Figure 5. Elastic-plastic buckling modelling by negative strain-hardening.

as a linear function of internal and external random variables. Using equations (1) and (8), we can write the safety margin M_{s_k} for the cross-section "k".

$$M_{s_k} = N_{p_k} + \sum_{j=1}^{nfe} \left(\overline{a}_j + \frac{\overline{b}_{y_j}}{Z_{p_{yk}}} + \frac{\overline{b}_{z_j}}{Z_{p_{zk}}} \right) N_{p_j} - \sum_{i=1}^{ngb} \left(a_i + \frac{b_{y_i}}{Z_{p_{yk}}} + \frac{b_{z_i}}{Z_{p_{zk}}} \right) P_{r_i} \tag{9}$$

which is equivalent to :

$$M_{s_k} = \text{Strength at } k - \text{ Stress at } k$$

where $\overline{a}_j, \overline{b}_{y_j}, \overline{b}_{z_j}, a_i, b_{y_i}, b_{z_i}$ are respectively the influence coefficients of axial force and bending moments due to internal and external random variables ; $Z_{p_{yk}}, Z_{p_{zk}}$ are plastic coefficients giving a dimensional characteristic defined by $Z_{py} = M_{py}/N_p, Z_{pz} = M_{pz}/N_p$ (they may be considered as random in this formula). The influence coefficients are calculated by the mechanical behaviour analysis programme. The failure probability P_f of this safety margin is given by :

$$P_f = P(M_{s_k} \leq 0) = \Phi(-\beta_{HL_k}) \tag{10}$$

where $\Phi(\cdot)$ is the standard probability distribution function and β_{HL_k} is the Hasofer and Lind reliability index (1974) evaluated for the cross-section "k".

For different deteriorated states of the structure, safety margins of surviving components can be described by the mechanical model. Knowing these safety margins, the structural system reliability can be estimated by identification of the dominant failure paths (Thoft-Christensen and Murotsu 1986).

4. NUMERICAL EXAMPLE

The proposed model is applied to the offshore structure shown in figure 6.a (Maroini et al. 1992). The structure is exposed to an equivalent storm loading. The mechanical response is shown in figure 6.b by the relation between loading factor and lateral top displacement in "y" direction (node "42"). Our results present a very good agreement with nonlinear analysis performed by the software "USFOS" (SINTEF 1991). The mechanical response is characterized by the buckling of bracing element number "57".

The reliability analysis of the structure is done for the design loading assumed as normally dis-

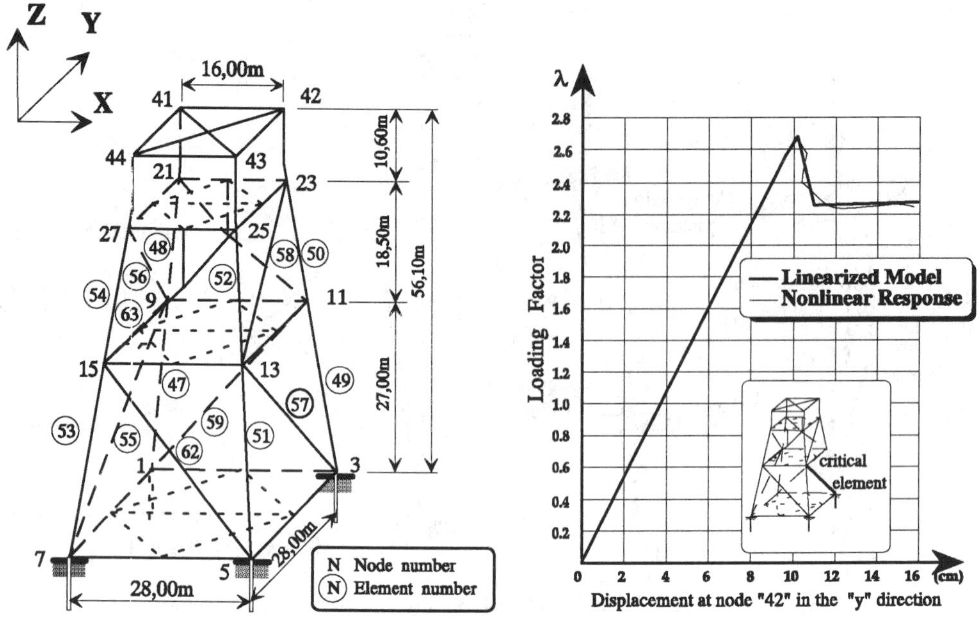

(a) (b)

Figure 6. Mechanical behaviour of the structure.
a) Platform configuration. b) Mechanical response.

tributed with a variation coefficient of 40%, where the cross-section strengths are considered as log-normally distributed with a variation coefficient of 8% (Maroini et al. 1992). The software "STRUREL" (RCP GmbH and RCP ApS 1991) has been used to calculate system reliability. At the intact state the reliability of the structure is found to be :

$$\beta_{sys} = 2.66 \qquad (P_f \approx 7.00 \times 10^{-2})$$

Figure 7 shows the reliability indexes of critical components (i.e. member ends) for the undamaged configuration of the structure. Between the shown component indexes, the most critical failure modes are identified by the yieldings at sections S_{20} ($\beta = 2.70$) and S_{18} ($\beta = 2.91$), and the buckling of the lower bracing element E_{57} ($\beta = 3.91$).

The partial failure tree considering the most critical failure paths is shown in figure 8. The analysis is carried out for four deterioration levels. Under the design loading conditions, the

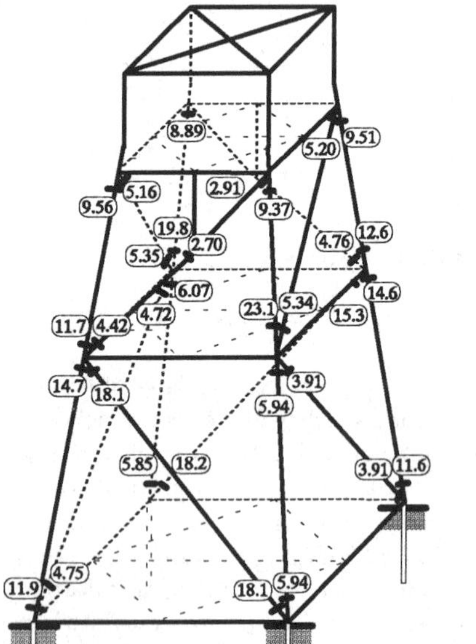

Figure 7. Cross-section reliability indexes.

reliability of the structural system at different levels of damage is given in the following table .

5. CONCLUSION

A linearized method allowing mechanical and reliability analysis of structures is proposed. The mechanical behaviour is fully linearized ; it takes into account the interaction between internal forces and the elastic-plastic buckling. The integration of plastic deformations allows to separate the influence of different internal and external random variables. As the mechanical model

Table 1. Reliability indexes of the structural system.

Level	Reliability index	Most probable damage
Level 1	$\beta_{sys} = 2.66$	yielding at S_{20}
Level 2	$\beta_{sys} = 2.87$	yielding at S_{18}
Level 3	$\beta_{sys} = 3.79$	Buckling of element E_{57}
Level 4	$\beta_{sys} = 3.94$	Failure of element E_{56}

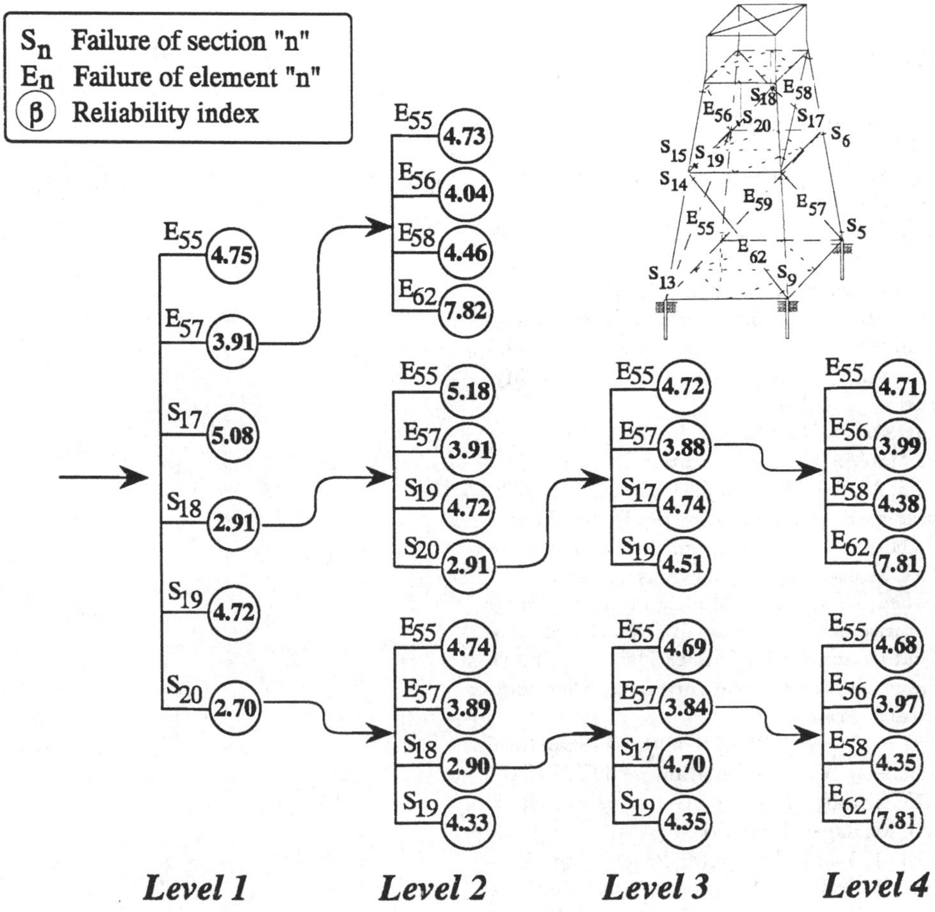

Figure 8. Reliability analysis of the platform : Partial failure tree.

describes linear safety margins, the reliability analysis is simply done by classical procedures. Numerical example has been presented to show the good performance of the model. With reference to nonlinear mechanical models, our model proved a very high efficiency and a good accuracy. With reference to nonlinear reliability analysis (i.e. simulations), our model needs a very little time consumption (to compare accuracy, a large scale examples have to be treated). In both cases, the accuracy is satisfactory for practical use and the computing time is considerably reduced.

ACKNOWLEDGEMENTS

The authors wish to acknowledge the french company ELF AQUITAINE PRODUCTION for the help that it gave to financially support this research.

REFERENCES

Hasofer, A. M. and Lind, N. C. 1974. Exact and Invariant Second-Moment Code Format. Journal of the Engineering Mechanics Division 100 : 111-121.

Lemaire, M., Chung, J. F., Mohamed, A. M. and Muzeau, J. P. 1991 Model of Negative Strain-Hardening to Evaluate the Reliability of Beam-Columns. Sixth International Conference on Applications of Statistics and Probability in Civil Engineering, Mexico City, Mexico : 201-208.

Lemaitre, J. et Chaboche, J. L. 1985. Mécanique des Matériaux Solides. Edition Dunod, Paris.

Maroini, A., Ben Mariem, J. et Birades, M. 1992. Stratégie de Requalification des Plates-Formes Offshore. Journée Clarom du 12 Novembre 1992, Saint-Rémy-lès-Chevreuse, France.

Mohamed, A. M. 1993. Modèle Mécanofiabiliste Linéarisé pour l'Analyse des Structures. Application aux Plates-Formes Marines. Ph. D. Thesis, Blaise Pascal University, Clermont-Ferrand, France.

RCP GmbH and RCP ApS 1991. A Structural Reliability Analysis Program "STRUREL". RCP GmbH, München, Germany and RCP ApS, Mariager, Denmark.

SINTEF 1991. A Computer Program for Progressive Collapse Analysis of Steel Offshore Structures "USFOS". User's and Theory manuals, SINTEF, Division of Structural Engineering, Trondheim, Norway.

Thoft-Christensen, P. and Murotsu, Y. 1986. Application of Structural Systems Reliability Theory. Springer-Verlag.

Structural Safety & Reliability, Schuëller, Shinozuka & Yao (eds) © 1994 Balkema, Rotterdam, ISBN 90 5410 357 4

Probabilistic collapse analysis of jackets

G. Sigurdsson & R. Skjong
DNV Research AS, Høvik, Norway

B. Skallerud & J. Amdahl
SINTEF, Trondheim, Norway

ABSTRACT

For realistic structures the structural behavior near collapse can be very complex to model and the computation time can be exhaustive. This complexity can be due to the non-linear mechanical behavior of the structure, the applied loading and load distribution near failure. The structural behavior beyond the first member-failure, depends not only on the degree of static indeterminacy, but also on the ability of the structure to redistribute the load and on the post-failure behavior, e.g. the ductility of the individual members. In addition, the probability of system failure depends on the uncertainty of the load, uncertainty of the member capacity and the correlation between the uncertain parameters. In a reliability analysis of fixed offshore structures, the problem is generally load driven, i.e. the uncertainties in load are much greater than in the system capacity. The following questions naturally arise in the reliability formulation :
The environmental parameters are certainly probabilistic in nature, but what about the transfer from the environmental conditions (current speed and direction, wind speed and direction and wave direction, height and period) to loading ? Can the loading be treated as a deterministic function of the environmental parameters, or should a probabilistic description be used ? Is a deterministic model of the structure adequate to quantify the probability of collapse or is a probabilistic description required ?

The purpose of this paper is to gain insight into the above questions by performing studies on randomness of the ultimate capacity of different types of actual jackets. The uncertainties in the structural capacity model are assumed to be due to the yield stresses and the member imperfections (magnitude and direction) The uncertainties in the structural loading model are assumed to be due to the wave height, the thickness of the marine growth and the drag and inertia coefficients in Morison's equation. The distribution for the ultimate capacity of the structures are determined by means of Monte-Carlo simulation. A state-of-the-art non-linear structural collapse analysis program USFOS [2]-[4] are used for the progressive collapse analysis, and a general PROBabilistic ANalysis program PROBAN [1] is used for the reliability calculation and to generate outcome of the stochastic parameters in the simulation studies of the ultimate capacity of the structures.

1. INTRODUCTION

The first offshore platform was installed in the late forties in the Gulf of Mexico. Since then over 6000 platforms have been installed worldwide, mostly in water depths of less than 200 m. Given the unfamiliar and harsh environment of the oceans, much research has been carried out over the past four decades, and is still being undertaken, to ensure that existing and new platforms are safe. The objective of this past and ongoing research has been and still is to demonstrate that the chance of structural failure either by high cycle fatigue or by overloading (caused by extreme storms, ship impact etc.) is acceptably small. In the present paper, we aim to focus only on the probability of structural failure owing to overloading and in particular we will address the extreme storm situation.

In the 1990's, we have now reached the point where existing design procedures certainly achieve highly reliable structures. However, the design procedures are mostly based on 'experience' and do not quantify the probability of failure. If we are to make use of our improved knowledge in each individual discipline (reliability theory, structural mechanics, hydrodynamics etc.) to try to estimate the probability of structural failure, we must adopt a 'probabilistic framework'. Into this framework we must provide the components of the overall model and the statistical data. Only then will it be possible to quantify the risk levels and take decisions with improved confidence.

The following problems (questions) will be discussed :

Problem 1: Can the randomness in the system capacity be estimated independent of the randomness in the loading, i.e. can the system capacity be related directly to the total base-shear on the structure ?

Problem 2: Is a deterministic model of the structure adequate to quantify the probability of collapse or is a probabilistic description required ?

Problem 3: The environmental parameters are certainly probabilistic in nature, but what about the link between the environmental situation (e.g. current speed and direction, wind speed and direction and wave direction, height and period) and the loading ? i.e can the randomness in the loading be related directly to the randomness in the environmental parameters $\mathbf{Z}_{sea\ state}$, or should a probabilistic model for prediction of the wave forces be applied ?

The purpose of this paper is to gain insight into the above questions by performing studies on randomness of the ultimate capacity of different types of actual jackets compared to the randomness of the environ-

mental loading acting on the structures. The uncertainties in the structural capacity model are assumed to be due to the yield stresses and the member imperfections (magnitude and direction). The uncertainties in the structural loading model are assumed to be due to the wave height, the thickness of the marine growth and the drag and inertia coefficients in Morison's equation. The wave period and the current speed are assumed to be deterministic functions of the significant wave height.
The distribution for the ultimate capacity of the structures, i.e. distribution shape and parameters, are determined by means of Monte-Carlo simulation. A state-of-the-art non-linear structural collapse analysis program USFOS [2]-[4] are used for the progressive collapse analysis. The basic principle behind USFOS is to represent each individual member in the structure by one finite element. This is allowed for by using the exact solution to the differential equations for a beam subjected to end forces as shape function to the elastic displacements, which take into account large lateral rotations at the element. Closed form solutions are obtained for the elastic total and incremental stiffness matrices which contain all information needed to identify buckling of members or subsystems. Nonlinear material behavior is modeled by means of plastic hinge theory where the yield criterion is expressed in terms of two plastic interaction functions, one representing first fiber yield, the other full plastification of the cross-section. A general PROBabilistic ANalysis program PROBAN [1] is used for the reliability calculation and to generate outcome of the stochastic parameters in the simulation studies of the ultimate capacity of the structures.

2. PROBABILISTIC MODELING OF OVERLOAD FAILURE

For realistic structures the structural behavior near collapse failure can be very complex and numerically expensive to assess. This complexity can be due to the non-linear mechanical behavior of the structure and the applied loading and load distribution near failure. The structural behavior beyond the first post-failure, depend not only on the degree of static indeterminacy, but also on the ability of the structure to redistribute the load which depends upon the ductility of the individual members. In addition, the probability of system failure depends on the uncertainty of the load, uncertainty of the member capacity and the correlation between the uncertain parameters. For a perfectly balanced structure (i.e. the first member failure has the same probability of occurrence in all members in a linear analysis) the system effects for overload capacity beyond the first member failure are strictly due to the randomness in the member capacities. In contrast, in a more realistic structure, i.e more unbalanced structure, the system effects are both due to deterministic and probabilistic effects. In an unbalanced structure, the mean member capacity to mean member force is different for the members. Hence, deterministic system effects are present because the remaining members in the structure can still carry the load after one or several members have failed. In addition, randomness in the member capacities gives a probabilistic contribution to the system capacity. By relating the system capacity directly to the total base-shear or the total over-turning moment and assuming that the load pattern has minor effect on the system capacity, the system capacity and the loading (i.e. the base-shear or the over turning moment) can be treated separately. This means that the system capacity can be represented by a single random variable SC, and the loading can also be represented by a single variable L. The annual system failure probability can be calculated as the annual probability that the load exceeds the system capacity, i.e.

$$P_{f_{sys,annual}} = P\{L_{annual} \geq SC\} = \int_0^\infty [1 - F_{L_{annual}}(x)] f_{SC}(x)\,dx$$

$$= \int_0^\infty F_{SC}(x) f_{L_{annual}}(x)\,dx \qquad (1)$$

where $F_{L_{annual}}(\cdot)$ and $F_{SC}(\cdot)$ are the cumulative annual probability distributions of the load and the system capacity, respectively, and $f_{L_{annual}}(\cdot)$ and $f_{SC}(\cdot)$ are the corresponding probability density functions. For a small $P_{f_{sys,annual}}$, the total system failure probability $P_{f_{sys}}$ for a given time T_{life} can be estimated as:

$$P_{f_{sys}} \approx T_{life} \cdot P_{f_{sys,annual}} \qquad (2)$$

In general for offshore structures, the uncertainties in prediction of wave forces are greater than the variability in prediction of the system capacity. Uncertainties in wave loads are caused by inherent randomness in the wave process, uncertainties in the sea state parameters and uncertainties in prediction of wave forces for the given sea state. Therefore, it is necessary to represent the uncertainties in loading as good as possible, in order to assure an accurate reliability assessment. The annual distribution of the extreme load on the structure can be written as :

$$F_{L_{annual}} = F_{L_{annual}}(\mathbf{Z}_{sea-state}, \mathbf{Z}_{wave-forces}) \qquad (3)$$

where $\mathbf{Z}_{sea-state}$ is a vector of random variables modeling the uncertainties in the sea state description and $\mathbf{Z}_{wave-forces}$ is a vector of random variables modeling the uncertainties in prediction of wave forces for a given sea state.
The distribution of the system capacity of the structure can be written as :

$$F_{SC} = F_{SC}(\mathbf{Z}_{structure}, \mathbf{Z}_{sea-state}, \mathbf{Z}_{wave-forces}) \qquad (4)$$

where $\mathbf{Z}_{structure}$ is a vector of random variables modeling the uncertainties in the structural model.

3. SIMULATION AND SENSITIVITY STUDY

3.1 Introduction

When the design-point at failure is not very far away from the mean value (which will usually be the case for the system capacity) and not in the tail of the distribution (which will usually be the case for the load variable), it is important to establish a good estimate of the central part of the distribution. To establish the distribution of the system capacity, SC, Monte Carlo simulation will be used. For each stochastic realization i of the stochastic variables the system capacity sc_i is calculated. Estimation of the expected value $E[SC]$ and the coefficient of variation $COV[SC]$ in the following studies are based on 100 stochastic realization. All results are checked carefully, e.g. load-displacement curves for at least one node are plotted for every collapse analysis. The basis for the simulation and the sensitivity studies, are two existing jacket structures,

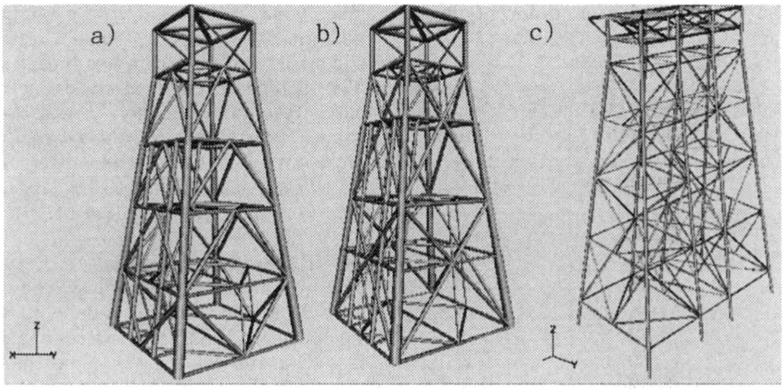

Figure 1 Sketches of a): $Model-A_{original}$ b): $Model-A_{x-brace}$ c): $Model-B$

which will be called $Model-A$ and $Model-B$. $Model-A$ is a four legged jacket structure, and $Model-B$ is a eight legged jacket structure. The study is carried out for one configuration of structural $Model-B$ and for three different configurations of $Model-A$, namely : $Model-A_{original}$: A slightly simplified model of the original structure. $Model-A_{optimized}$: A more optimized model of $Model-A_{original}$. $Model-A_{x-brace}$: A modification of $Model-A_{original}$ to get an x-brace structure. The SESAM program WAJAC [5] is used to calculate the wave loading for a given outcome of $\mathbf{Z}_{sea-state}$ and $\mathbf{Z}_{wave-forces}$ (only used for $Model-A$).

3.2 Brief description of the structural models

$Model-A$ is a model of a four legged riser platform located in the North Sea. The water depth is 70. m. A simplified model of the original structure have been used, i.e. the structural elements which do not contribute to the load-carrying capacity are not included in the computer model. Topside loads and wind loading are applied as concentrated forces at the four topside corners.

The original structure was designed for the following environmental situation : Wave (100 year return period): height H = 29.0 m ; period T = 17.5 sec. Current speed (10 year return period): at level 0.0 m (SWL) u_c = 1.25 m/sec. ; at level -25 m u_c = 0.65 m/sec. ; at level -70 m (at bottom) u_c = 0.35 m/sec. Marine growth Mg = 0.06 m

$Model-B$ is a model of an eight legged platform. The water depth is 77 m. The environmental design loads on the structure are defined by the following environmental criteria: Wave : height : H = 12.20 m ; period : T = 11.7 sec. Current speed : u_c = 1.0 m/sec. at level 0.0 m (SWL) and 1/7 power decay law is assumed. The marine growth is not included.

3.3 Problem 1

To investigate whether the system capacity can be related directly to the total base-shear force, a deterministic model of the structure is applied, i.e. all uncertainties are assumed to be on the load side. First we look at the importance of modeling the uncertainty in prediction of the wave and current loading. For a given sea state the wave height and period and the current profile are kept fixed. $Model-A$ will be applied for this purpose, and only one wave direction is considered. The next step will be to look at the influence of the wave height and period, defining the load-pattern. $Model-A_{x-brace}$ and $Model-B$ will be applied for this purposes. Only limited number of wave heights and periods are considered for this purpose.

Uncertainty in the structural loading model :

The loads due to the structural weight, buoyancy and wind are assumed to be deterministic.

Stochastic loading due to wave and current ($\mathbf{Z}_{wave-force}$):

The total wave force per unit length perpendicular to a structural member of diameter D_o at time t is estimated, using Morison's equation :

$$f(t) = f_D(t) + f_I(t) \tag{5}$$

where the drag force f_D and the inertia force f_I can be written as :

$$f_D(t) = C_D\ D\ \frac{\rho}{2}\ (u(t)+u_c)\ |u(t)+u_c| = C_D\ D\ f_{D,basic}(t) \tag{6}$$

$$f_I(t) = C_M\ D^2\ \frac{\rho\pi}{4}\ \dot{u}(t) = C_I\ D^2\ f_{I,basic}(t) \tag{7}$$

$$D = D_o + 2\ Mg \tag{8}$$

where C_D is the drag coefficient, C_M is the inertia coefficient, u and $\dot{u}$ are the water particle velocity and the water particle acceleration perpendicular to the structural member, u_c is the current velocity perpendicular to the structural member (assumed to be time independent), ρ is the mass density of the water and Mg is the thickness of the marine growth.

The stochastic variables :

C_D: Log-Normal distributed with expected value $E[C_D] = 0.75$ and coeff. of variation $COV[C_D] = 0.25$

C_M: Log-Normal distributed with $E[C_M] = 1.7$ and $COV[C_M] = 0.10$

Mg: Log-Normal distributed with $E[Mg] = 0.050\ m$ for level +2 m to -40 m, and $E[Mg] = 0.025\ m$ for level -40 m down to the bottom, and $COV[Mg] = 0.50$

C_D, C_M and Mg are assumed to be uncorrelated, but individual fully correlated between members.

Table 1 The system capacity for different load-pattern, for $Model-A_{x-brace}$ and $Model-B$.

Case	Base-Shear for load-factor=1 MN	Load-factor at collapse	Base-Shear at collapse (=SC) MN
a1	9.20	20.99	193.2
a2	10.41	19.59	203.8
a3	15.74	12.95	203.8
a4	20.27	10.06	203.8
a5	30.01	6.93	208.0
a6	40.58	5.11	207.4
b1	5.88	3.33	19.6
b2	2.71	7.10	19.2
b3	3.43	5.62	19.3
b4	8.45	2.34	19.8
b5	15.02	1.30	19.5
b6	9.50	1.99	18.9

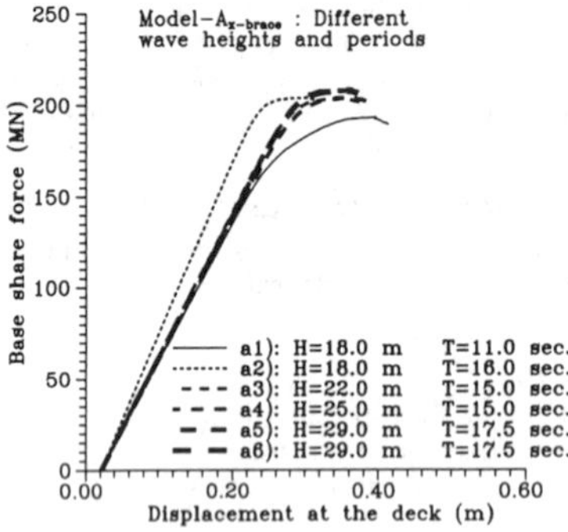

Figure 2 Different wave situation for $Model-A_{x-braces}$

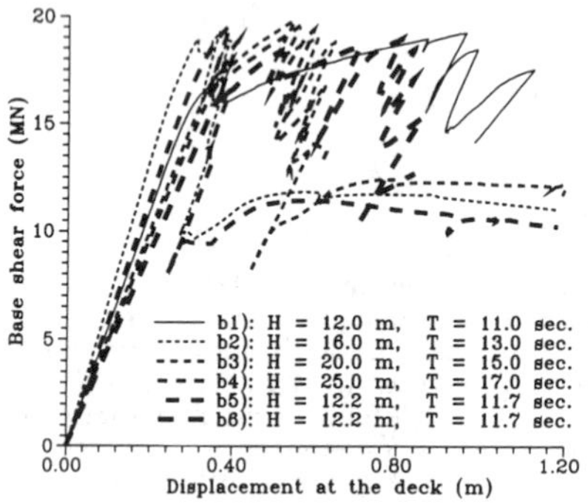

Figure 3 Different wave situation for $Model-B$

For a given time t (i.e. wave position) the *basic* drag force $f_{D,basic}$ and the *basic* inertia force $f_{I,basic}$ is calculated by WAJAC as two separate load cases, i.e. setting all member diameters equal to one and: *load case* 1 : C_M = 0.0 and C_D = 1.0 (gives $f_{D,basic}$) ; *load case* 2 : C_M = 1.0 and C_D = 0.0 (gives $f_{I,basic}$). For a given outcome of the stochastic variables C_D, C_M and Mg, the overall load profile can be calculated, using equations (5)-(8).

The yield stresses for the structural members is assumed to be 386.0 N/mm^2, and no imperfections of the structural members are modeled.
The deterministic system capacity, i.e. the system capacity obtained using the expected values of the stochastic variables, for $Model-A$ was obtained as :
$Model-A_{original}$: SC = 168.3 MN (load-factor = 4.38) ;
$Model-A_{optimized}$: SC = 185.9 MN (load-factor = 4.88) ;
$Model-A_{x-brace}$: SC = 207.4 MN (load-factor = 5.11).

Simulation procedure :
The automated simulation procedure of collapse, using USFOS and PROBAN, is done as follows:

1) Choose the number of realizations, *NSIM*
2) Sample *NSIM* sets of outcomes of the stochastic variables, i.e. C_D, C_M and Mg, using PROBAN.
3) Establish the *basic* drag and inertia load profiles $f_{D,basic}$ and $f_{I,basic}$ for the chosen wave height and period, e.g. the 100 year wave, using WAJAC.
4) For each set of outcome of the stochastic variables
   4.1) Establish the overall wave/current load profile.
   4.2) Perform a static push-over analysis by scaling up the load profile, established in 3).
   4.3) Save the results.

The following cases are considered for $Model-A_{original}$ and $Model-A_{x-brace}$:
Case 1: C_D and C_M stochastic and Mg deterministic ;
Case 2: C_D and C_M deterministic and Mg stochastic ;
Case 3: C_D, C_M and Mg stochastic.
All analysis gave successful results. In all cases the $COV[SC]$ became less than 0.4% and the $E[SC]$ approximately equal to the deterministic SC. These results indicate that the SC can be estimated without accounting for the randomness in C_D, C_M and Mg. But what about the sensitivity of the chosen load-pattern ? To investigate this effect, we consider $Model-A_{x-brace}$ and $Model-B$ for different load-patterns. The load-pattern for different wave heights H and periods T are calculated using WAJAC. 5th order Stokes waves are applied. For $Model-A$ the expected values are used for C_D and C_M, but C_D = 0.7 and C_M = 2.0 are used for $Model-B$. The yield stress for the structural members in $Model-B$ is assumed to be 283.0 N/mm^2, and no imperfections of the structural members are modeled.
The following cases are included :

$Model-A_{x-brace}$:
a1): H=18 m, T=11.0 sec. ; a2): H=18 m, T=16.0 sec. a3): H=22 m, T=15.0 sec. ; a4): H=25 m, T=15.0 sec. ; a5): H=29 m, T=17.5 sec. ; a6): H=29 m, T=17.5 sec. ;
In cases a1)-a5) the current and marine growth are not included, but in case a6) both current and marine growth are included (the same input as in the deterministic case).

$Model-B$:
b1): H=12.2 m, T=11.7 sec. (including the design current profile) ; b2): H=12 m, T = 11.0 sec. ; b3): H=16 m, T = 13.0 sec. ; b4): H=20 m, T = 15.0 sec. ; b5): H=25 m, T = 17.0 sec. ; b6): The load pattern used in the design (H = 12.2 m, T = 11.7 sec. including the design current profile)

The results for $Model-A$ are shown in table 1 and in figure 2 and for $Model-B$ in table 1 and in figure 3.
As we can see from these results, the system capacity is not very sensitive for changes in the wave height and period, even though the structural response can be

Table 2 The system capacity for different studies of *Model –A* and *Model –B*.
* : $COV[Y]$=0.10, ** : $COV[Y]$=0.15.

Case	Model	Stoch. variables	Num. of failed analysis	$E[SC]$ MN	$COV[SC]$	Skew-ness	Kur-tosis	Deter-ministic SC (MN)
c1	$Model-A_{original}$	Y, ϵ, θ	0	165.1	0.027	0.35	3.48	168.3
c2	$Model-A_{original}$	Y	0	166.7	0.025	0.37	3.23	168.3
c3	$Model-A_{original}$	ϵ, θ	0	166.9	0.009	-1.65	6.39	168.3
c4	$Model-A_{optimized}$	Y	0	181.4	0.038	0.02	2.66	185.9
c5	$Model-A_{x-brace}$	Y, ϵ, θ	0	207.2	0.024	0.19	2.87	207.4
c6	$Model-A_{x-brace_{mod}}$	Y^{*}, ϵ, θ	1	207.7	0.051	0.15	2.54	214.5
c7	$Model-A_{x-brace_{mod}}$	Y^{**}, ϵ, θ	1	200.8	0.069	-0.04	2.96	214.5
c8	$Model-B$	Y, ϵ	15	19.6	0.035	0.62	3.93	18.9

very sensitive for the changes, e.g. for Model-A where different wave periods, but the same wave height H = 18.0 m, gives significant different response, see figure 2 (case a1 and case a2). The same results are obtained for the current, i.e. different responses are obtained for cases with and without current but the SC almost the same in both cases, see figure 2 (case a5 and case a6).

The results obtained in this section indicate:

- The SC can be related directly to the base-shear force on the structure.
- The SC can be estimated without taking into account the uncertainty of describing the load-pattern.

3.4 Problem 2

In this section we will investigate if a deterministic description of the structural resistance is suitable or not in order to quantify the probability of collapse.

To estimate the probability of collapse failure, we need to establish the probability distribution of SC as well as for the loading L, see eqs. (2.1) and (2.2). It is important to establish a realistic distribution for both SC and L. Assuming that the results obtained in section 3.3 can be applied, the SC of the structure can be related directly to the base-shear force on the structure, and the distribution of SC, F_{SC}, can be established independent of the loading. This mean that the system capacity can be represented by a single variable SC and the loading (i.e. the total base-shear) can be represented by a single variable L.

The distribution of SC, F_{SC}, becomes only function of the random variables modeling the uncertainty in the structural model $\mathbf{Z}_{structure}$. The distribution of the annual maximum base-shear on the structure L, $F_{L_{annual}}$ is established using a random model of the environmental description $\mathbf{Z}_{sea-state}$, given by Haver (1991) [7], and the uncertainty model for the structural loading $\mathbf{Z}_{wave-force}$ for a given sea-state described in section 3.3.

For a given outcome of $\mathbf{Z}_{sea-state}$ and $\mathbf{Z}_{wave-force}$ WAJAC is applied to calculate the total base-shear on the structure.

3.4.1 Distribution of the system capacity F_{SC}

Uncertainty in the structural model ($\mathbf{Z}_{structure}$) :

The yield stresses, Y :

Distribution type : Normal, with $E[Y_i] = 386\ N/mm^2$, i=1,..num. of struct. members for *Model –A* and $E[Y_i] = 283\ N/mm^2$, i=1,..num. of struct. members for *Model –B*.

The coeff. of variation : $COV[Y_i] = 0.06$ for both models. The yield stress within the same structural member is assumed to be fully correlated but uncorrelated between different structural members.

The imperfections, ϵ and θ :

The magnitude, ϵ, is assumed to be Log-Normal distributed with $E[\epsilon_i] = 0.0019 \times L_i$, where L_i is length of the struct. member i, and with $COV[\epsilon_i] = 1.9$.

The direction, θ, is assumed to be Uniform distributed within $[0\ ;\ 2\pi]$. ϵ and θ, for all members, are assumed to be uncorrelated.

The $E[Y_i]$ for *Model –A* and *Model –B* are obtained by assuming that the yield stress, used in design is 1% fractile value with $COV[Y_i] = 0.06$. The evaluation of $E[\epsilon]$ and $COV[\epsilon]$ are discussed in [11].

Simulation procedure :

The automated collapse simulation, using USFOS and PROBAN, is carried out as follows:

1) **Establish wave/current load pattern for a fixed wave, e.g. the 100 year wave height and the 10 years current pattern.**
2) Choose the number of realizations, *NSIM* (here *NSIM* = 100 is used)
3) Sample *NSIM* sets of outcomes of the stochastic variables, i.e. Y_i, ϵ_i and θ_i for each structural member i, using PROBAN.
4) For each set of outcome of the stochastic variables:
   - 4.1) Perform a static push-over analysis by scaling up the load profile, established in 1).
   - 4.2) Save the results

The load-pattern for *Model –A* is established using WAJAC with : H = 29.0 m, T = 17.5 sec., C_D = 0.75, C_M = 1.7, Mg = 0.050 m for level +2 m to -40 m and Mg = 0.025 m for level -40 m down to the bottom. The current profile given in section 3.2

The load-pattern applied for *Model –B* is the same as the design load, using : H = 12.2 m, T = 11.7 sec., C_D = 0.7, C_M = 2.0, without accounting for the marine growth. The current profile given in section 3.2.

The eight cases shown in table 2 are considered. In case c6 and case c7, a more optimized model of the $Model-A_{x-brace}$ is used, named $Model-A_{x-brace_{mod}}$, and $COV[Y]$ = 0.10 for case c6 and $COV[Y]$ = 0.15 for case c7. The deterministic SC for $Model-A_{x-brace_{mod}}$ is 214.5

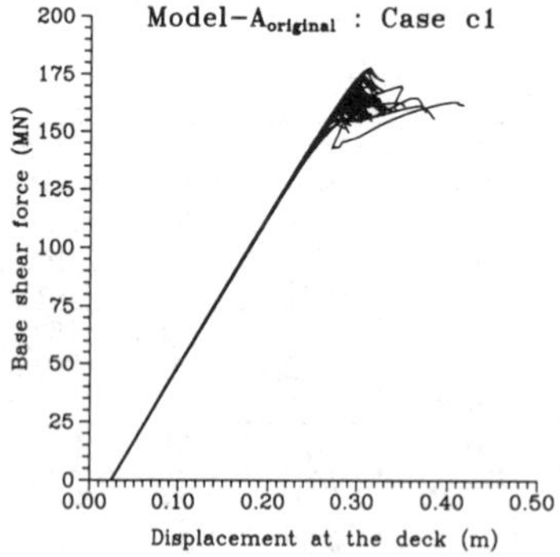

Figure 4 Simulation results for case c1.

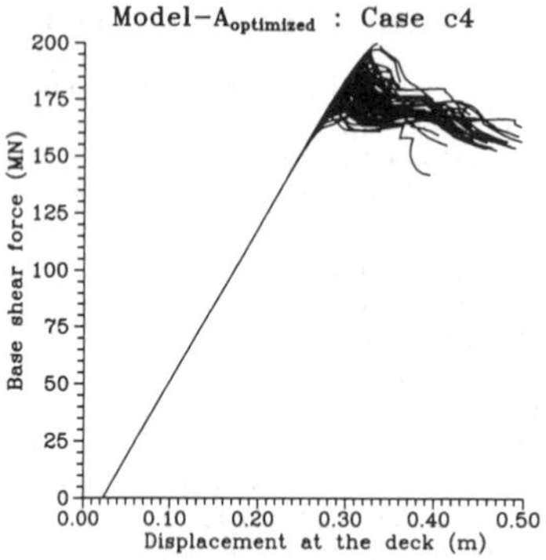

Figure 5 Simulation results for case c4.

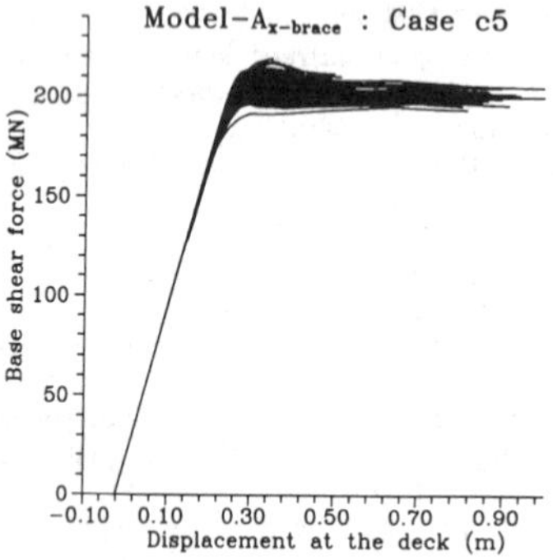

Figure 6 Simulation results for case c5.

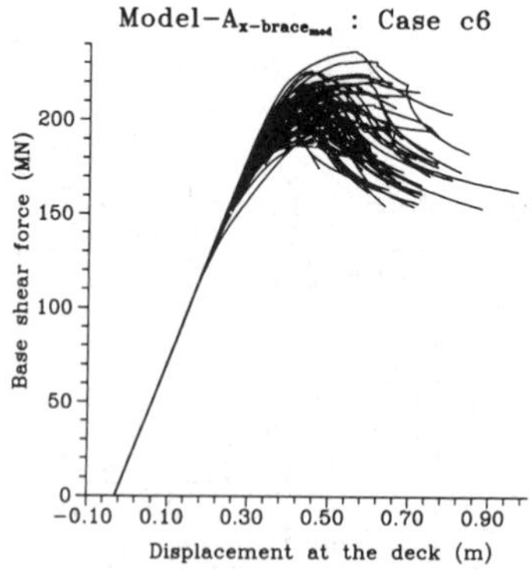

Figure 7 Simulation results for case c6.

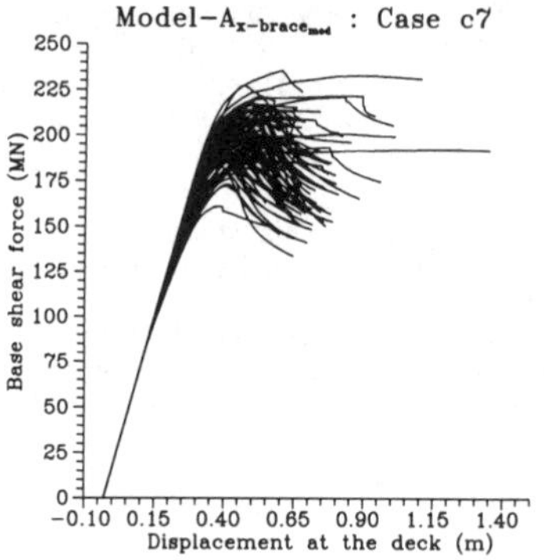

Figure 8 Simulation results for case c7.

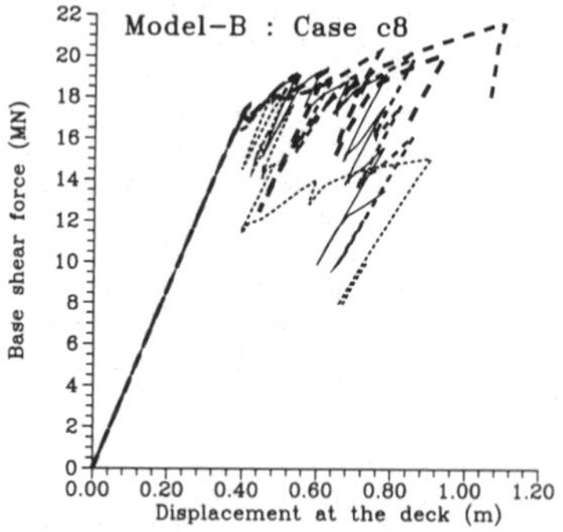

Figure 9 Simulation results for case c8.

MN. Load-displacement curves for cases c1, c4-c8 are shown in figure 4-9. For case c8 results from only a few of the analysis are shown.

Example of density functions for *SC* (only shown for case c7) is shown in figure 10. All analysis which failed are not included.

The results obtained for cases c1-c8 indicate :

- USFOS seems to be robust enough for automated analysis. However, the results must be checked carefully in each case.
- Different collapse mechanisms are identified.
- The imperfections have insignificant influence on the *SC* .
- The $COV|SC|$ is much less than the $COV|Y|$ (approximately 50%).
- The normal distribution gives an acceptable good fit to the F_{SC}, for all cases.

3.4.2 Distribution of the maximum annual base-shear

To establish the probability distribution of the annual maximum base-shear loading acting on a given structure, we need a realistic description of the environmental situation, i.e. joint distribution of the wave height, wave period and the current. For this purpose a model described in [7] is used. The uncertainty model for the structural loading for a given environmental situation described in section 3.3 is applied. The $Model-A_{original}$ is applied for this purpose, and the environmental parameters available for the Ekofisk area are applied, where the water depth is approximately 70 meters.
In the following a brief description of the environmental model is given.

Environmental model

The distribution of the maximum significant wave height H_S exceeding a certain threshold h_{th} in a given storm, is modeled by a truncated Weibull distribution:

$$F_{H_S}(h_s)=1-\exp\left\{-\left(\frac{h_s}{\theta}\right)^{\gamma}+\left(\frac{h_{th}}{\theta}\right)^{\gamma}\right\}; \quad h_s>h_{th} \tag{9}$$

where the scale parameter θ and the shape parameter γ are estimated from available storm data collected at a given site. The expected value $E[T_p]$ and the coefficient of variation $COV[T_p]$ for the spectral peak period are given as:

$$E[T_p]=a_0+a_1 h_s^{a_2} \tag{10}$$

$$COV[T_p]=0.009 \tag{11}$$

where a_0, a_1 and a_2 are estimated from data obtained at a given site. In here T_p will be assumed deterministic, and equal to the expected value given by eq. (10). The number of annual storms N_{th} where the significant wave height exceed the threshold h_{th} is assumed to be Poisson distributed, and is given by

$$p_{N_{th}}(n_{th}) = \frac{(\phi\, t)^{n_{th}}}{n_{th}!} e^{\phi t} \tag{12}$$

Assuming that the annual largest significant wave height appear in the annual largest storm and by using eq. (9) and eq. (12) the distribution function of the annual largest significant wave height $H_{s,max}$ can be written as:

$$F_{H_{s,max}}(h_{s,max})=\exp\left\{-\phi \exp\left[\left(\frac{h_{s,max}}{\theta}\right)^{\gamma}+\left(\frac{h_{th}}{\theta}\right)^{\gamma}\right]\right\} \tag{13}$$

However in the load calculation, the distribution of the largest crest height C_{max} and the corresponding wave period $T_{C_{max}}$ are needed. It is assumed that the $T_{C_{max}}$ can be related to the spectral peak period as : $T_{C_{max}} = 0.9\, T_p$. Assuming that the sea surface, for a short period Δt, can be modeled by a Gaussian process, the conditional distribution function for the largest crest height C_{max} can be given by, [8]:

$$F_{C_{max}|H_{s,max}T_z}(c_{max}|h_{s,max},t_z) = \exp\left\{-\frac{\Delta t}{t_z}\exp\left\{-8\left(\frac{c_{max}}{h_{s,max}}\right)^2\right\}\right\} \tag{14}$$

where T_z is the zero up-crossing wave period approximated by : $T_z = 0.74\, T_p$. The duration Δt = 6 hrs. = 21600 sec. is applied herein. The relation between the crest height and the wave height for a 5th order Stokes wave is given by: $H_{max} = \alpha_1 + \alpha_2 C_{max}$, where α_1 = 3.55 and α_2 = 1.42 are used. The current speed at the SWL is assumed to be related to the significant wave height as: $u_{c,SWL} = (0.08H_s - 0.24)\ m/sec.$, and the current speed profile is assumed to be similar to the profile used for design i.e. level -25 m : $u_{c,-25m}$ = 0.52 $u_{c,SWL}$ and level -70 m : $u_{c,-70m}$ = 0.28 $u_{c,SWL}$

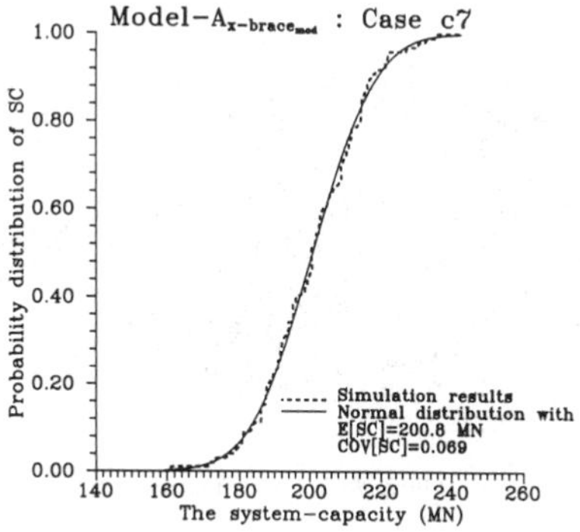

Figure 10 F_{SC} in case c7.

Table 3 Parameters for the applied storm wave climate description (available for Ekofisk)

Storm threshold h_{th}	6.5 m
Expected number of annual storms ϕ	5.8
Weibull parameters for $H_{s,max}$	γ = 1.25 θ = 2.03
Parameters for spectral peak period	a_0 = -2.60 a_1 = 6.59 a_2 = 0.382

The storm wave climate parameters for describing the crest height available for Ekofisk are given in table 3, Haver (1991) [7].

The expected maximum annual wave height is estimated to 17.2 m, the COV is estimated to 0.11 and the 100-years wave height, i.e the expected value of the maximum wave height in 100 years period, is estimated to 24.0 m. The 100-years crest height can be calculated to 14.4 m. This result predict approximately the same 100-years value as obtained in [6] which gave 14.2 m.
For a given outcome of the stochastic variables $\mathbf{Z}_{wave-force}$ i.e. C_D, C_M, and Mg, and $\mathbf{Z}_{sea-state}$ i.e. H_{max} (which depends on $H_{s,max}$ and $C_{max}|H_{s,max}$) the base-shear force on the structure is calculated using WAJAC.

3.4.3 Estimation of probability of collapse failure

To investigate whether a deterministic description of the structural resistance is suitable (or unsuitable) for quantifying the probability of collapse failure, a probability analysis is required.
Using the results obtained in section 3.3, the limit state for the annual collapse failure can be written as:

$$g(\mathbf{z}_{structure},\mathbf{z}_{sea-state},\mathbf{z}_{wave-force}) =$$

$$sc(\mathbf{z}_{structure}) - l_{annual}(\mathbf{z}_{sea-state},\mathbf{z}_{wave-force}) \tag{15}$$

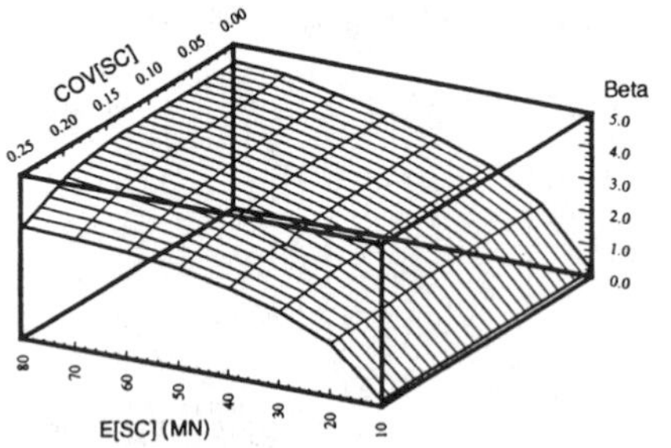

Figure 11 Reliability index as function of $E[SC]$ and $COV[SC]$.

In section 3.4.1 it was shown that the distribution of SC can be estimated by the Normal distribution, at least for the central part of the distribution. Applying the model for the maximum annual base-shear loading described in section 3.4.2, the annual probability of collapse failure can easily be calculated using e.q. (15). The values for $E[SC]$ obtained in section 3.4.1 are not interesting for this purpose, because they will depend on how over designed the structures are. Therefore a parametric study on $E[SC]$ and $COV[SC]$ is performed. The $E[SC]$ takes values from 10 MN to 80 MN with step of 10 MN and $COV[SC]$ take values from 0.00 to 0.25 with step of 0.01. $COV[SC] = 0.00$ is equivalent to assume SC as deterministic. The values for $E[SC]$ are chosen due to the distribution of the maximum annual base-shear force given in section 3.4.2, The expected maximum annual base-shear force is estimated to 10.9 MN, the COV is estimated to 0.35 and the 100-year base-shear force is estimated to 23.9 MN. The values for $COV[SC]$ are chosen due to the results obtained in section 3.4.1, where the calculated $COV[SC]$ was less than 10% in all cases. However the study is extended for $COV[SC]$ up to 0.25, to investigate the effect of very large randomness in SC. The values chosen in this study will be suitable for most structures, which have been designed based on deterministic requirements. The results are shown in figure 11.

The following conclusions are obtained:

- The reliability index is almost independent of the $COV[SC]$, for $COV[SC]$ less than approximately 0.15. This means that a deterministic description of the structural resistance is suitable to quantify the probability of collapse failure for $COV[SC] < 0.15$.
- The design point for SC in case of $COV[SC] <$ 0.15 is less than one standard deviation, $SD[SC]$, from the $E[SC]$. This indicate that the Normal assumption for SC is acceptable good, due to the results obtained in section 3.4.1.
- The results obtained for $COV[SC] > 0.15$ are not reliable, because the design points for SC are in tail of the distribution and the Normal assumption, which is based on only 100 simulations is not sufficient to describe the tail.
- For reliability indices greater than four, the design points for the wave height are greater than 30 m. However, the model describing the maximum annual wave height is not verified for so large waves.
- Assuming deterministic description of the SC, the over estimation in the reliability index calculated by applying the deterministic system capacity SC_{det} in stead of $E[SC]$ is approximately 2 $(SC_{det} - E[SC])/SC_{det}$. In section 3.4.1 the $(SC_{det} - E[SC])/SC_{det}$ is obtained to be less than the $COV[SC]$, which indicates that by using SC_{det} instead of $E[SC]$ give acceptable results in most cases.

3.5 Problem 3

In this section we will investigate whether it is necessary to include a probabilistic model for prediction of the wave force, i.e whether it is sufficient to establish the maximum annual base-shear force only by including the randomness in the sea state description.
This can easily be done by comparing the sensitivity factors for the stochastic variables applied in the sea state description $\mathbf{Z}_{sea-state}$ and in the wave loading description $\mathbf{Z}_{wave-force}$.
The stochastic variables applied in the study are: $\mathbf{Z}_{wave-force}$: C_D, C_M and Mg ; $\mathbf{Z}_{sea-state}$: $H_{s,max}$ and $C_{max}|H_{s,max}$. More details are given in Sigurdsson et al. [11].
The following results/conclusions are obtained:

- The marine growth Mg and the inertia coefficient C_M in Morison's equation can be modeled as deterministic.
- The results are dominated by $\mathbf{Z}_{sea-state}$. Especially the uncertainty in the annual largest significant wave height $H_{s,max}$ is important. However, the importance of including the uncertainty in C_D can not be ignored.

4. CONCLUSIONS

Questions about whether it is necessary to perform a probabilistic collapse analysis to estimate the probability of collapse failure of jackets, have been discussed in this paper. All results obtained in this paper are based on study on a limited number of structures, and for a given environmental description and can not be generalized to be available for all jacket structures. In the following a resume of the main conclusions/remarks are given. More detail conclusions can be found in each section.

- The SC can be related directly to the base-shear force on the structure and can be estimated without taking into account the uncertainty of describing the load-pattern.
- The imperfections have insignificant influence on the SC.
- The $COV[SC]$ is much less than the $COV[Y]$ (approximately 50%).
- The Normal distribution give a acceptable good fit to the F_{SC}.
- A deterministic description of the structural resistance (i.e the system capacity SC) is suitable to quantify the probability of collapse failure.
- Assuming a deterministic description of the SC, using SC_{det} instead of $E[SC]$ give acceptable results in most cases.
- The marine growth Mg and the inertia coefficient C_M in Morison's equation can be modeled as deterministic.
- The reliability is dominated by uncertainty in $\mathbf{Z}_{sea-state}$. Especially the uncertainty in the annual

largest significant wave height $H_{s,max}$ is important. However, the importance of including the uncertainty in C_D can not be ignored.

ACKNOWLEDGMENT

This study has been carried out under the project "Probabilistic Collapse Analysis of Jackets" in DNV-Research AS and SINTEF, Norway. The project is a joint effort by Shell, Conoco, Elf and Statoil. The companies permission to publish the results are gratefully acknowledged.

5. REFERENCES

|1| Olesen R., "PROBAN - User Manual", DNV SESAM report No.:92-7049/Rev. 0, 17-Jan-1992

|2| Soreide T. et al., "USFOS Theory manual", SINTEF report STF71 F88038, rev. 1990-07-01, 1990.

|3| Soreide T., Amdahl J., Rembar H., "The idealized structural unit method on space tubular frames", Proc. Int. Conf. Steel and alum. struct., Cardif 1987.

|4| Hopperstad O., Eberg E., Skallerud B., "Plasticity models for cyclic behavior of steel frames", 3. Int. Conf. Computational Plasticity, Barcelona 1992.

|5| "WAJAC - User Manual", DNV SESAM report No.:92-7052/Rev. 0, 25-May-1992.

|6| Haver S., Stass P., Bjarke P.E., and Staveland L. (1985) "Design Basis Environmental Conditions for Tommeliten", Report 68 F&U, Statoil, Stavanger, Norway.

|7| Haver S. (1991), "Wave Climate Description for Foundation Design", Statoil, R&D-dep., Rep.no. 91047, Version 2, Trondheim, January 1992.

|8| Madsen H.O., Krenk S., and Lind N.C., (1986) "Methods of structural safety", Prentice Hall Inc., Englewood Cliffs, New Jersey.

|9| Moan T., "The inherent safety of structures designed according to the NPD regulations", SINTEF Report STF71 F88043, 1988.

|10| Bjærhovde R., "Deterministic and probabilistic approaches to the strength of beam-columns" Ph.D. Thesis, Lehigh university, 1972.

|11| Sigurdsson G. et al.,"Probabilistic Collapse Analysis of Jackets", DNV-Research report No.:92-2045/Rev. 0, 6-Oct-1992.

Structural Safety & Reliability, Schuëller, Shinozuka & Yao (eds) © *1994 Balkema, Rotterdam, ISBN 90 5410 357 4*

Current induced skewness of dynamic response for offshore structures

Jin Wang
Bechtel Corporation, Houston, Tex., USA

Loren D. Lutes
Texas A&M University, College Station, Tex., USA

ABSTRACT: Adding current to a Gaussian sea model causes a number of changes in the nonlinear hydrodynamic force exerted on an offshore structure in the random seaway. This study focuses on one of those changes, namely the introduction of skewness to the wave force and to the dynamic response of the offshore structure. The response skewness is calculated from simulated time histories of the dynamic response of the structure for various Morison equation excitations. For the nonlinear Morison equation models, it is shown that the skewness may be substantial and the maximum values are achieved when the current velocity is approximately equal to the standard deviation of the wave induced velocity. An approximate analytical approach for predicting the response skewness, in which the unsymmetric hydrodynamic force is modeled as a filtered non-Gaussian white noise, is also discussed.

INTRODUCTION

The primary loading exerted on an offshore structure is the environmental force induced by the ocean wave and current. It is known that ocean current and wave-current interaction have significant effects on the hydrodynamic loading and wave induced motions of offshore structures. In general, the current effects on the nonlinear drag force lead to unsymmetric hydrodynamic loading on an offshore structure. Consequently, the unsymmetric loading causes unsymmetric response stress time histories in the structural members. In particular, the unsymmetric nonlinear drag may be modeled by the commonly used Morison equation with the flow velocity being a non-zero mean random process due to the presence of a current. This gives an unsymmetric non-Gaussian loading time history which, in turn, results in unsymmetric non-Gaussian dynamic response time histories. The effects of current induced skewness on the structural dynamic response will be the focus of this paper. This study of skewness is part of a larger study of the non-Gaussian aspects of the dynamic structural response performed in order to obtain an improved assessment of structural reliability related to first-passage and fatigue failure.

The inclusion of the structural response velocity and current in the expression for the drag force in the Morison equation makes the equation of motion nonlinear and the response unsymmetric. Stochastic descriptions of the hydrodynamic forces due to the combination of ocean waves and a current, and of the dynamic responses of offshore structures can be found in the studies of Spanos and Chen (1981), and Gudmestad and Connor (1983). While previous studies of the current effects have been focused on the mean and standard deviation of the response, existing information on the current effect on the skewness of the dynamic response induced by the wave-current loading is quite limited due to the complexity of the problem. There is still no exact analytical solution available for determining the response statistics such as the response skewness and kurtosis for this problem. Time domain simulation, however, is generally not restricted by the nonlinearities in the structural system. Therefore, numerical simulation is first applied to obtain nonlinear unsymmetric response time histories and the response skewness values for various Morison equation models. The response skewness may also be approximated by using analytical procedures. One approach considered here involves modeling the unsymmetric hydrodynamic force as a filtered non-Gaussian white noise, since this approach has previously been shown to give satisfactory modeling of the response kurtosis for the symmetric situation with no current (Wang and Lutes 1992).

RESPONSE TO WAVE-CURRENT EXCITATION

For simplicity, the offshore structure is idealized as a single degree of freedom (SDOF) system. The equation of motion then may be written as

$$m_0\ddot{x} + c_0\dot{x} + k_0 x = F(t) \tag{1}$$

in which x, $\dot{x}$ and $\ddot{x}$ are the displacement, velocity and acceleration of the system response and m_0, c_0, and k_0 are the structural mass, damping coefficient, and stiffness, respectively. The force $F(t)$ is the hydrodynamic wave loading acting on the structure. For a flexible slender structure, this force may be expressed by the Morison equation. The generalized Morison equation including the current velocity expresses F as

$$F = K_m\dot{u} - m_a\ddot{x} + K_d|u_w + u_c - \dot{x}|(u_w + u_c - \dot{x}) \tag{2}$$

in which u_w denotes the zero-mean symmetric process of the water particle velocity induced by waves, u_c denotes the current velocity, and K_m, K_d, and m_a are constants related to the inertia, drag, and added mass. It should be mentioned here that the values of K_m and K_d in the Morison equation are generally influenced by the presence of the current superimposed on the wave (Sarpkaya and Isaacson 1981). To avoid unnecessary complication, the values of K_m and K_d for a non-zero current situation are considered to be constant and equal to the K_m and K_d values for the zero current situation in this paper.

Since the unsymmetric relative velocity and nonlinear drag term pose severe difficulties for an exact analytical solution, simplified approximations of the full interactive nonlinear Morison equation given by Eq. (2) are often used. To investigate the accuracy of different approximations, three commonly used simplified versions of the Morison equation models, as well as the nonlinear fluid-structure interaction model, are used to describe the wave-current induced hydrodynamic force. These models are:

Model 1,

$$(m_0 + m_a)\ddot{x} + (c_0 + K_d c_a)\dot{x} + k_0 x = K_m\dot{u} + K_d(a + bu) \tag{3}$$

Model 2,

$$(m_0 + m_a)\ddot{x} + c_0\dot{x} + k_0 x = K_m\dot{u} + K_d|u|u \tag{4}$$

Model 3,

$$(m_0 + m_a)\ddot{x} + (c_0 + K_d c_a)\dot{x} + k_0 x = K_m\dot{u} + K_d|u|u \tag{5}$$

Model 4,

$$(m_0 + m_a)\ddot{x} + c_0\dot{x} + k_0 x = K_m\dot{u} + K_d|u - \dot{x}|(u - \dot{x}) \tag{6}$$

in which $u = u_w + u_c$, $a = E(|u|u)$ and $b = c_a = 2E(|u|)$. For the non-zero mean Gaussian process $u(t)$ with mean μ and standard deviation σ, it can be shown that minimum mean squared error linearization gives:

$$a = E(|u|u) = (\mu^2 + \sigma^2)\left[1 - 2\Phi\left(-\frac{\mu}{\sigma}\right)\right] + 2\mu\sigma\frac{\exp[-\mu^2/(2\sigma^2)]}{\sqrt{2\pi}} \tag{7}$$

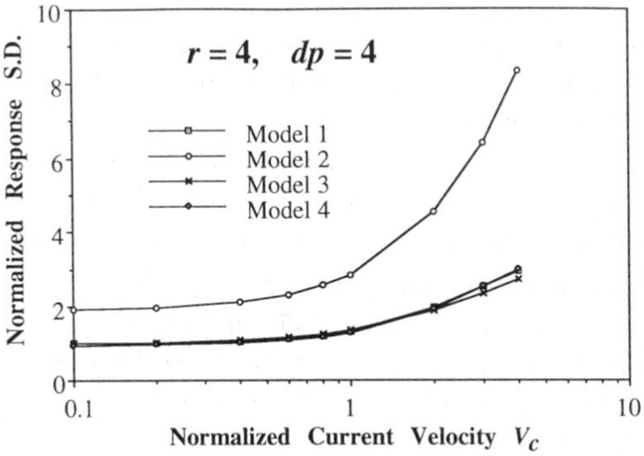

Fig. 1. Response standard deviation

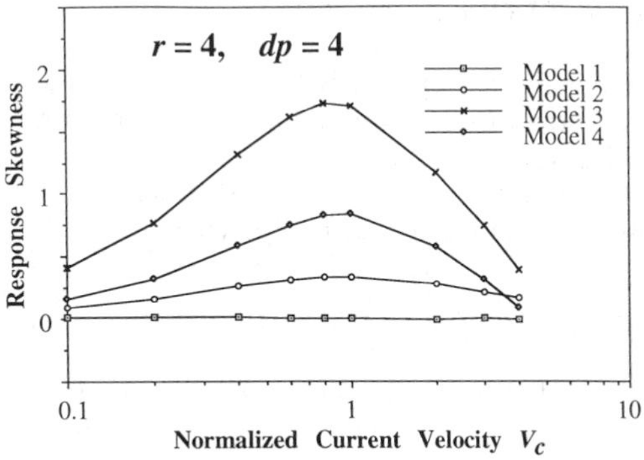

Fig. 2. Response skewness

$$b = c_a = 2E(|u|) = 2\mu\left[1 - 2\Phi\left(-\frac{\mu}{\sigma}\right)\right] + 4\sigma\frac{\exp[-\mu^2/(2\sigma^2)]}{\sqrt{2\pi}} \quad (8)$$

in which $\Phi(x) = \int_{-\infty}^{x} \frac{\exp(-t^2/2)}{\sqrt{2\pi}}dt$.

It should be noted that Models 1, 2, and 3 all represent linear structures inasmuch as they are linear in $\dot{x}$. Only Model 1 is also linear in u. The response statistical properties of the three nonlinear unsymmetric Morison equation models (Models 2, 3 and 4) described above can be obtained by using numerical simulation, while the linear model (Model 1) can be readily solved analytically.

SIMULATION RESULTS

Numerical simulation studies have been conducted for one particular fixed type structure. The frequency ratio, defined as the structure natural frequency over the peak wave frequency, has been taken as r = 4.0 and the zero current drag parameter, defined as the rms drag force over the rms inertia force, as dp = 4.0. Thus, the situation studied represents a relatively rigid structure with a drag-dominated hydrodynamic loading.

Simulation results of the response standard deviation and skewness are presented in Figs. 1 and 2 as a function of the normalized current velocity, which is defined as the actual current velocity normalized by the standard deviation of the wave fluid velocity, i. e., $V_c = u_c / \sigma_u$. Note that the results of Model 1 are presented along with simulation results for the three nonlinear models in the figures so that one can also examine how Model 1 compares with other simplified models in approximating Model 4 for the non-zero current situation. It can be seen that the response standard deviation of Model 4 is best approximated by Model 1 and Model 3, and its magnitude grows as the current velocity increases for all the four models presented in Fig. 1. This is expected since the presence of a current increases the standard deviation of the wave-current force due to the nonlinearity in the drag term, even though the standard deviation for the fluid particle velocity remains unchanged. In Fig. 2, it is shown that the response skewness tends to increase as the current velocity increases for V_c less than unity, then it starts to decrease as V_c becomes greater than one. It should be kept in mind that skewness is a form of the third cumulant normalized by the standard deviation cubed. Thus, the decrease in skewness for very large current values may be attributed to the large growth in standard deviation for this situation. The maximum skewness for Model 4 is nearly equal to 1.0 indicating a significant departure from the symmetric Gaussian distribution. It is also noted that the skewness curve of Model 4 is between those of Models 2 and 3. Model 1, of course, gives zero skewness since it includes no nonlinearity.

ANALYTICAL APPROXIMATION

The analytical approach considered is a filter method which models the unsymmetric hydrodynamic force as a filtered non-Gaussian white noise. A simple second order filter is used here and it may be expressed as

$$\ddot{Y}+2\beta_g\omega_g\dot{Y}+\omega_g^2 Y=W(t) \tag{9}$$

in which $W(t)$ is unsymmetric δ-correlated white noise with a power spectral density (PSD) of S_0 and bispectrum of B_0. The filtered derivative process $\dot{Y}$, instead of the process Y, is then used to approximate the Morison wave force on the right-hand-side of Eqs. (3) - (6), since the PSD of $\dot{Y}$ has a shape much closer to that of the PSD of the Morison equation wave force. This approach gives a simple analysis procedure for Models 1, 2, and 3, since they are linear in $\dot{x}$, although it is of no particular value for Model 4.

The filter equation Eq. (9) and a linear structural motion equation may be combined into a single linear system described by four first order differential equations. In matrix form this is

$$\dot{\mathbf{X}}+\mathbf{A}\mathbf{X}=\mathbf{f}(t) \tag{10}$$

in which the transpose of $\mathbf{X}$ is $\{Y,\dot{Y},X,\dot{X}\}$. For the third response cumulants of this system, the number of simultaneous state space equations is equal to 20, and they may be written as

$$\dot{\mathbf{C}}+\mathbf{Z}\mathbf{C}=\mathbf{I}(t) \tag{11}$$

where $\mathbf{C}$ is the vector containing all the third cumulants of $\mathbf{X}$. The $\mathbf{Z}$ matrix, which depends only on the $\mathbf{A}$ matrix in Eq. (10), has a dimension of 20 by 20 and can be obtained using an algorithm developed by Lutes and Chen (1992). For stationary response $\dot{\mathbf{C}}=0$ so that Eq. (11) becomes a set of linear algebraic equations, for which the solutions are not difficult to find using a linear equation solver.

However, the parameters of the linear filter must be selected before one can solve for the third cumulants using the above method. It can be seen that there are four parameters to be determined, namely the damping β_g and natural frequency ω_g of the filter, and the PSD S_0 and bispectrum B_0 of the white noise excitation. The values of β_g, ω_g , S_0 and B_0 in Eq. (9) should be selected in such a way that both the ordinary PSD and bispectrum of $\dot{Y}$ match those of F as closely as possible. Four equations are required to determine these four unknown parameters which define the filter system.

The first criterion which will be used is matching the area under the PSD curves of $\dot{Y}$ and F. This is most important since this area represents the energy of the process. Similarly, for the bispectrum, the volume enclosed by the bispectral density function represents the third cumulant ξ, which is important in characterizing the degree of nonnormality. Thus, the following two equations are imposed:

$$\lambda_{0_{\dot{Y}}}=\lambda_{0_F} \quad \text{and} \quad \xi_{\dot{Y}}=\xi_F \tag{12}$$

where $\lambda_{0_{\dot{Y}}}$ and λ_{0_F} are the areas under the PSD curves of $\dot{Y}$ and F; and $\xi_{\dot{Y}}$ and ξ_F are the skewness of $\dot{Y}$ and F, respectively.

Next, it is important that the frequencies of the processes are matched. Ideally, the frequency components of both the ordinary PSD and the bispectrum should be matched at every frequency interval. However, this would be very difficult and would require very complicated filters with numerous undetermined parameters. For the second order linear filter used here, one should not expect to achieve exact matching for the Morison equation wave force PSD. The following two equations are used as an approximation. In particular, the ordinary PSD is matched at two frequency points which have an identical preselected PSD value between 0 and the peak value, with one point being on each side of the peak frequency ω_{F_p}:

$$G_{\dot{Y}}(\omega_1)=G_F(\omega_1)=\alpha G_{F_p} \tag{13}$$

$$G_{\dot{Y}}(\omega_2)=G_F(\omega_2)=\alpha G_{F_p} \tag{14}$$

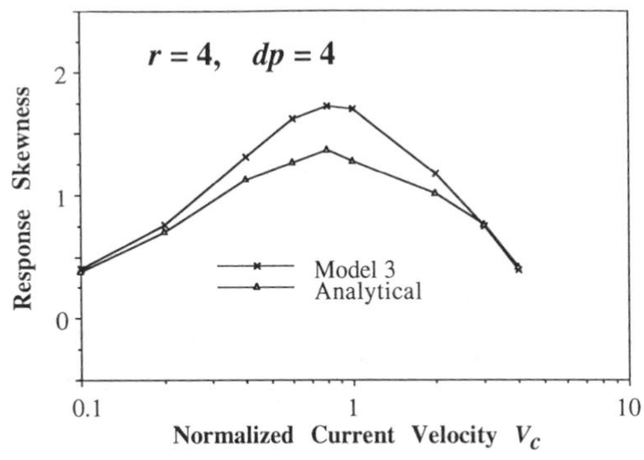

Fig. 3. Results of analytical model

in which α is a constant between 0 and 1, and G_{F_p} is the maximum PSD value of the Morison wave force spectrum which occurs at frequency ω_{F_p}.

Using Model 3 as a simplified approximation for Model 4, the response skewness has been calculated using the filtered white noise approach. The values of α for this relatively stiff system have been chosen quite near unity (greater than 0.9). More general rules for choosing α have been formulated in the study of kurtosis (Wang and Lutes 1992). The response skewness results are shown in Fig. 3. It can be seen that the filter method gives skewness values which compare fairly well with simulation results.

CONCLUSIONS

1. The current induced response skewness can be substantial.

2. The maximum response skewness occurs when the current velocity is approximately equal to the standard deviation of the wave fluid velocity.

3. The filtered white noise approach can be a useful analytical tool for evaluating the response skewness of an offshore structure under non-Gaussian wave-current loading.

ACKNOWLEDGEMENT

Portions of this research were sponsored by the Offshore Technology Research Center at Texas A&M University, including support from the NSF Engineering Research Centers Program Grant No. CDR-8721512.

REFERENCES

Gudmestad, O.T. and Connor, J. J. 1983. "Linearization methods and the influence of current on the nonlinear hydrodynamic drag force," *Applied Ocean Research*, No. 12, pp.14-21.}

Lutes, L. D. and Chen, D. C. K. 1992. "Stochastic response moments for linear systems," *Probabilistic Engineering Mechanics*, Vol. 7, No. 3, pp.165-173.

Sarpkaya, T. and Isaacson, M. 1981. *Mechanics of wave forces on offshore structures*, Litton Education Publishing, New York, NY.

Spanos, P-T. D. and Chen, T. W. 1981. "Random response to flow-induced forces," *Journal of the Engineering Mechanics Division,* ASCE, Vol. 107, No. 12, pp.1173-1190.

Wang, J. and Lutes, L.D. 1992. "Effects of Morison equation nonlinearity on stochastic dynamics and fatigue of offshore structures," *Report, Offshore Technology Research Center*, Texas A&M University, College Station, Texas.

Structural Safety & Reliability, Schuëller, Shinozuka & Yao (eds) © 1994 Balkema, Rotterdam, ISBN 90 5410 357 4

Environmental parameters for extreme response: Inverse form with omission factors

S. R. Winterstein, T.C. Ude & C.A. Cornell
Civil Engineering Department, Stanford University, Calif., USA

P. Bjerager
Det Norske Veritas Sesam AS, Høvik, Norway

S. Haver
Statoil, Postuttak, Trondheim, Norway

ABSTRACT

In structural reliability problems there is generally uncertainty both in the gross load environment, and in the extreme response given the loading. We show here how these uncertainties can be approximately decoupled. We find contours of environmental parameters along which specified extreme fractiles lie (e.g., 100-year values of any structural response quantity). These contours are independent of the structure, making them a practical way to display a two (or higher) dimensional environmental hazard at a site. Based on the first-order reliability method (FORM), the inverse-FORM method is introduced. This searches a hypersphere of constant radius β to find the maximum response. FORM omission factors are used to permit correct results based on only the *median* response, which may be estimated either analytically or by simulation.

Applications to various offshore structure problems are shown, including prediction of extreme wave crests and the base shear of a shallow-water jacket structure. Results are found to compare well with full FORM analysis.

INTRODUCTION

In practical structural reliability problems there is generally uncertainty both in the gross load environment, and in the extreme dynamic response given the loading. Denoting the environmental variables by $\boldsymbol{X}=[X_1 \dots X_n]$ and the response by Y, the failure probability p_F can be written formally as

$$p_F = \int_{all\,\boldsymbol{x}} P[Y > y_{cap}|\boldsymbol{X} = \boldsymbol{x}]f(\boldsymbol{x})d\boldsymbol{x} \quad (1)$$

In principle Eq. 1 can be estimated with FORM/SORM or simulation. Several practical difficulties may arise, however, which motivate this paper. First, Eq. 1 requires a full, coupled environment-response model; i.e., the joint description $f(\boldsymbol{x})$ of all environmental variables and the conditional failure probability, $P[Y > y_{cap}|\boldsymbol{X} = \boldsymbol{x}]$, for all $\boldsymbol{x}$.

For the structural analyst, however, it is simpler to require that only a limited set of environmental conditions (values of $\boldsymbol{x}$) be checked to ensure adequate capacity y_{cap}. For example, many shallow-water ocean structures are most sensitive to the significant wave height, $H_S=4\sigma_\eta$, in which σ_η is the wave elevation rms over a stationary "seastate." To estimate 100-year responses, one may then simply apply a seastate with 100-year H_S level, and representative values of other variables such as wave period, current, etc.:

$$\begin{aligned} X_1 &= H_S = H_{100}\,; \\ X_i &= \text{Median}[X_i|X_1 = H_{100}]\,,\ i = 2,3,\dots \end{aligned} \quad (2)$$

This also simplifies the task of the environmental analyst, who need only report parameters of this H_S-driven 100-year seastate.

More generally, however, the critical environment will be structure-dependent. For example, deeper-water structures may be affected by seastates with smaller H_S but larger current U, or with resonant values of peak spectral period T_P. This leads to interest in defining joint contours (e.g., H_S–T_P or H_S-U contours) along which 100-year levels of any deterministic response quantity will lie (e.g., Haver, 1987). This again decouples the environmental description from the specific structural design concept. By introducing the "inverse FORM" method, we show here how such contours can be directly generated. These contours, unlike some others proposed, contain Eq. 2 as a special case.

A further complication is that the foregoing approach—Eq. 2 and its generalizations—ignore uncertainty in the response Y given the environment $\boldsymbol{X}$. By using a representative Y, such as its median $y(\boldsymbol{X})$ given $\boldsymbol{X}$, extreme response fractiles may be underestimated. Again we may resort to Eq. 1, but we lose the benefit of the decoupled, environmental contour. Also, for nonlinear stochastic response, it may be expensive to accurately estimate the full distribution of Y given $\boldsymbol{X}$.

To overcome these obstacles, we introduce "inflated" environmental contours. These are based on FORM omission factors (Madsen, 1988). These determine how much the environmental contour return period should be inflated (e.g., from 100 to 200–300 years or more) to compensate for omitting uncertainty in the 100-year response given $\boldsymbol{X}$.

Contours for Deterministic Response

We first consider the response Y as a deterministic function of the seastate variables $\boldsymbol{X}$. For FORM purposes we transform $\boldsymbol{X}$ to a set of standard normal variables $\boldsymbol{U}$ (e.g., Madsen et al, 1986), so that

$$Y = y(\boldsymbol{U}) \tag{3}$$

In this case uncertainty in the environment (randomness in $\boldsymbol{X}$) is assumed to dominate, so that conditional uncertainty in Y given $\boldsymbol{X}$ (or given $\boldsymbol{U}$) is negligible. We generalize these results in the next section to include conditional uncertainty in Y as well.

The "Forward" FORM problem typically seeks the failure probability, p_F, associated with exceeding a known response capacity, y_{cap}. The FORM estimate of p_F, and associated reliability index $\beta=\Phi^{-1}(1-p_f)$, is formally found through the optimization problem:

$$\text{Given } y_{cap}: \quad \begin{array}{l} \text{Find } \beta = \min|\boldsymbol{U}|; \\ \text{Subject to } g(\boldsymbol{U}) = y_{cap} - y(\boldsymbol{U}) = 0 \end{array} \tag{4}$$

In probabilistic design, however, the capacity y_{cap} is often not given but rather sought, with the goal that a desired reliability β be achieved. This may be solved iteratively with Forward FORM; i.e., assume a capacity y_{cap}, find β through Eq. 4, and iterate with variable y_{cap} until the desired β is found. Alternatively, we may find the capacity y_{cap} which provides a given reliability β, as estimated by Eq. 4, through the "Inverse FORM" method:

$$\text{Given } \beta: \quad \begin{array}{l} \text{Find } y_{cap} = \max y(\boldsymbol{U}); \\ \text{Subject to } |\boldsymbol{U}| = \beta \end{array} \tag{5}$$

Physically, the Forward FORM method minimizes distance from a known failure surface, finding the most likely failure point in $\boldsymbol{U}$-space. In contrast, in Inverse FORM we specify the exceedance probability p_F and hence minimal distance β to the failure surface. To set the capacity y_{cap} we then search all possible FORM design points with given return period (a hypersphere with radius β), to find the maximum response $y(\boldsymbol{U})$ we must withstand. (We assume throughout that the failure surface is star-shaped with respect to the origin, so that the maximum $y(\boldsymbol{U})$ for $|\boldsymbol{U}| \leq \beta$ occurs along the bounding hypersphere $|\boldsymbol{U}|=\beta$.)

Advantages of Inverse FORM. The formulation of Inverse FORM, Eq. 5, carries several advantages. Perhaps its main benefit lies in decoupling the description of the environmental variables $\boldsymbol{X}$ and the response Y. For desired β the environmental analyst need only report a contour of critical values $\boldsymbol{X}$, corresponding to the sphere $|\boldsymbol{U}|=\beta$ in $\boldsymbol{U}$-space. These contours may then be used to find the specified fractile of any structural response quantity.

Eq. 5 may also confer some computational benefits. It yields y_{cap} without iteration, and it optimizes in one less dimension; e.g., the $n-1$ directions $\phi_i=\cos^{-1}(U_i/\beta)$. Thus, the minimal distance constraint $|\boldsymbol{U}| = \beta$ is used directly to simplify the problem, rather than imposed by trial and error after Forward FORM is done. Also, by recasting the problem in terms of angles ϕ_i, the constraints are simplified into "box-like" regions (e.g., $|\phi_i| \leq \pi$). A greater number of optimization routines are available for this problem, as opposed to those needed in Forward FORM with the nonlinear constraint $g(U)=0$. Finally, some experience with the method suggests that it may be better suited than Forward FORM to numerically noisy g-functions.

Example 1: Extreme Wave Crest Heights

Throughout our offshore examples we shall model the wave elevation $\eta(t)$ as a Gaussian process, over a series of stationary seastates. These seastates are then defined by the power spectrum of $\eta(t)$, parametrized here by the significant wave height $H_S=4\sigma_\eta=X_1$, and the peak spectral period $T_P=X_2$. We consider here a Northern North Sea wave climate, for which a Weibull distribution has been fit to H_S (Haver and Nyhus, 1986):

$$P[H_S < h] = F_{H_S}(h) = 1 - \exp[-(h/2.822)^{1.547}] \quad (6)$$

Conditional on H_S, T_P is assumed lognormally distributed with parameters

$$E[\ln T_P|H_S] = 1.59 + 0.42\ln(H_S+2);$$
$$Var[\ln T_P|H_S] = .005 + .085\exp(-0.13{H_S}^{1.34}) \quad (7)$$

Following this reference an alternate lognormal distribution is used for H_S values < 3.27 m. This has little impact, however, on the extreme response calculations done here.

Figure 1 shows H_S–T_P contours for response return periods of 10, 100, and 1000 years. These have been found by relating H_S and T_P to standard normal variables U_1 and U_2:

$$H_S = F^{-1}_{H_S}(\Phi(U_1)); \quad T_P = F^{-1}_{T_P|H_S}(\Phi(U_2)) \quad (8)$$

From Eq. 5, Eq. 8 gives a contour with return period T_r by varying U_1 and U_2 along the circle $\sqrt{U_1^2+U_2^2}=\beta$, where

$$\beta = \Phi^{-1}(1-p_f) = \Phi^{-1}(1 - \frac{T_{SS}}{365 \times 24 \times T_r}) \quad (9)$$

The factor of 365 × 24 converts the units of seastate duration T_{SS} [hrs] into those of T_r [yrs]. For example, the T_r=100-year contour follows by setting p_f=.01 per year or 3.43 $\times 10^{-6}$ per 3-hour seastate, so that $\beta=\Phi^{-1}(1-p_f)=4.5$. Note that by including the point $U_1=\beta$, $U_2=0$, this contour contains the wave-dominated seastate from Eq. 2: $H_S=H_{100}$=14.5m and associated median T_P=15.9s. Similarly, the 10- and 1000-year contours in Figure 1 correspond to circles with radius β=4.0 and 5.0, respectively.

The response in this example is the extreme crest height $Y=\eta_{max}$, of interest in setting the deck level to avoid wave impact loads. Y is readily modelled by assuming Poisson upcrossings of level y (e.g., Madsen et al, 1986):

$$P[Y > y] = \exp\{-(T_{SS}/T_Z)\exp[-8(y/H_S)^2]\} \quad (10)$$

Assuming uncertainty here to be dominated by H_S, we estimate Y by its median value, found by setting Eq. 10 to 0.5:

$$Y = y(H_S, T_P) = 0.25H_S\sqrt{2\ln(1.44T_{SS}/T_Z)};$$
$$T_Z \approx T_P(1-0.29\gamma^{-0.22}) \quad (11)$$

The latter approximation to T_Z is empirical, found to fit fairly well for the JONSWAP wave spectrum with peak factor γ between 1 and 5. Results are shown here for γ=3.3.

Following Eq. 5, the extreme 10-, 100-, and 1000-year wave crests are found by searching the appropriate contour for the maximum response (Eq. 11). Figure 1 shows these extreme crest values to be η_{max}=12.1, 13.7, and 15.2 [m]. Also shown are contours of constant $\eta_{max}(H_S,T_P)$. As might be expected, the (median) extreme wave crest is essentially produced by the largest possible H_S; e.g., the 100-year crest height is produced by the seastate with 100-year H_S (Eq. 2). Note, however, that the same contours remain valid for any structural problem, i.e., for any function $y(H_S,T_P)$, and that other responses may not be dominated by H_S alone. For example, extreme heave motions of tension-leg platforms may be governed by seastates along these contours for which T_P is twice the structural period, due to resonance with second-order load effects (Winterstein et al, 1992).

Note also that these contours are not simply contours of the joint probability density function of H_S and T_P, selected to enclosed area $1-p_F$. In fact, they will generally enclose somewhat less area. They are instead constructed so that the area inside the failure region, $Y \geq y_{cap}$, is estimated by FORM to be p_F.

Contours for Stochastic Response

The foregoing analysis assumes deterministic response: in the example, crest height Y is essentially proportional to H_S (Eq. 11). It follows that the seastate with maximum H_S produces the maximum Y. Typically, this assumption is unconservative. For example, it ignores the chance that the largest Y can be produced in a seastate with less-than-maximum H_S. Equivalently, it underestimates Y by neglecting its uncertainty given the seastate parameters; i.e., the difference between the random Y in Eq. 10 and its median estimate in Eq. 11.

To include this uncertainty, we supplement Eq. 3 by adding a random error term ϵ, reflecting conditional uncertainty in Y given $\boldsymbol{U}$:

$$Y = y(\boldsymbol{U}) + \epsilon \qquad (12)$$

We define ϵ to have zero median value, so that $y(\boldsymbol{U})$ remains the median response given $\boldsymbol{U}$. In a full analysis, ϵ should be included as an additional random variable. If ϵ is normally distributed, for example, an additional standard normal variable V is introduced into the inverse FORM problem:

$$\begin{array}{l}\text{Find } y_{cap} = \max Y(\boldsymbol{U}, V) = \max y(\boldsymbol{U}) + \sigma_\epsilon V; \\ \text{Subject to } |\boldsymbol{U}|^2 + V^2 = \beta^2\end{array} \qquad (13)$$

To avoid explicit inclusion of this additional variable, we seek here a new, inflated contour, along which the *median response* $y(\boldsymbol{U})$ yields the correct capacity:

$$\begin{array}{l}\text{Find } y_{cap} = \max Y(\boldsymbol{U}, V=0) = \max y(\boldsymbol{U}); \\ \text{Subject to } |\boldsymbol{U}| = \beta^*\end{array} \qquad (14)$$

Because this result ignores the conditional uncertainty in Y, to compensate we must choose a contour with larger radius; i.e., $\beta^* \geq \beta$. The value of β^* depends on both σ_ϵ and the precise form of $y(\boldsymbol{U})$. Consider first a simple linear variation of $y(\boldsymbol{U})$ with each U_i:

$$\begin{aligned} Y(\boldsymbol{U}, V) &= m_Y + \sum_i c_i U_i + \sigma_\epsilon V \\ &= m_Y + \sigma_Y (\sum_i \alpha_i U_i + \alpha_o V) \end{aligned} \qquad (15)$$

The latter form is in terms of the total variance of Y, $\sigma_Y^2 = \sum_i c_i^2 + \sigma_\epsilon^2$, and the relative variance contributions $\alpha_i^2 = c_i^2/\sigma_Y^2$ and $\alpha_o^2 = \sigma_\epsilon^2/\sigma_Y^2$ due to U_i and V. Combining Eqs. 13 and 15, the exact Inverse FORM method gives the familiar result

$$y_{cap} = m_Y + \sigma_Y \beta \qquad (16)$$

Thus $\beta = (y_{cap} - m_Y)/\sigma_Y$, the ratio of mean safety margin, $E[M]$, to its standard deviation, σ_M. For the linear/normal model of Eq. 15, this gives the exact p_F value. In the reduced Inverse FORM problem (Eqs. 14–15), the exact y_{cap} value requires the inflated contour radius

$$\beta^* = \beta / \sqrt{1 - \alpha_o^2} \qquad (17)$$

Eqs. 14 and 17 form the basis for "inflated" environmental contours, to compensate for approximating the true stochastic response by its median value. Eq. 17 is best motivated for linear/Gaussian safety margins: replacing a factor by its mean preserves the mean margin $E[M]$, but reduces σ_M^2 to $\sigma_M^2(1 - \alpha_o^2)$. (Here, as previously, α_o^2 is the contribution of the omitted variable to σ_M^2.) The reliability index, $\beta = E[M]/\sigma_M$, is then increased by a factor of $1/\sqrt{1 - \alpha_o^2}$. Eq. 17 states that it is this artificially inflated β we must seek if we set the omitted variable to its mean value. More generally, if we replace V by an arbitrary value v_o the altered reliability index, β^*, is

$$\beta^*(V = v_o) = (\beta - \alpha_o v_o)/\sqrt{1 - \alpha_o^2} \qquad (18)$$

This is the FORM omission sensitivity factor (Madsen, 1988). That reference suggests the fixed value $v_o = \beta\alpha_o/2$ so that $\beta^* \approx \beta$. Here we instead retain the median response ($v_o = 0$), and hence inflate the contour through Eq. 17.

In general, $y(\boldsymbol{U})$ will be nonlinear and ϵ non-Gaussian. However, Eq. 15 will apply locally near the design point; indeed it is the basis of the FORM approximation. For relatively small α_o^2, this local linearization may not change significantly after ϵ is omitted, and hence Eq. 17 may remain accurate. This is studied in the examples to follow.

Given β^*, Eq. 9 can be inverted to find an inflated return period T_r^+. Figure 2 shows this inflated return period, for a target return period of T_r=100 years, versus α_o^2. Note that while the reliability ratio β^*/β depends only on α_o^2, the return period ratio T_r^+/T_r depends also on the seastate duration T_{SS} in Eq. 9. Figure 2 shows two cases: (1) all 3-hour seastates are modelled (T_{SS}=3 [hrs]);

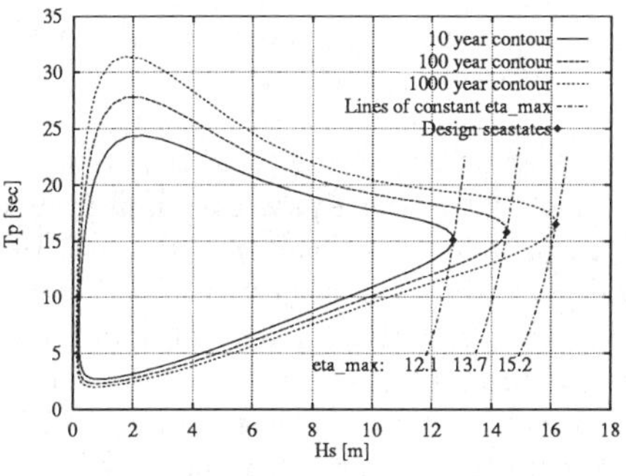

Figure 1: H_S–T_P contours for deterministic response.

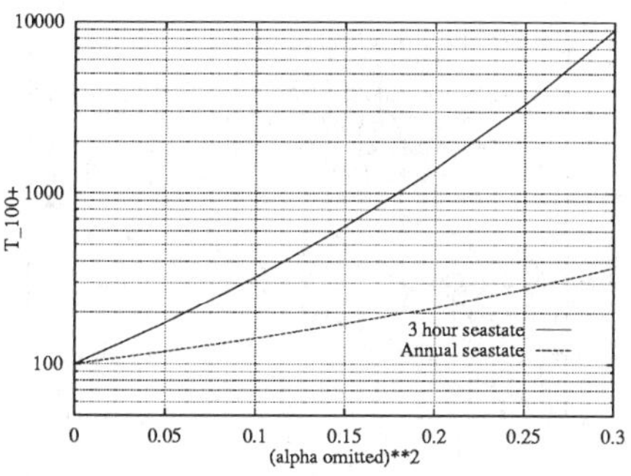

Figure 2: Inflated 100-year return period, T_r^+, for various α_o^2.

and (2) only the annual extreme storm is modelled, reflected by taking T_{SS}=1 [yr]=365×24 [hrs] in Eq. 9. For example, to find the 100-year response if α_o^2=.10, we should search the T_r^+=140-year contour of annually occurring seastate parameters, or the T_r^+=320-year contour of 3-hour seastates. If the omitted importance α_o^2 increases to .20, to compensate these return periods must be increased to 215 years (annual seastates) and 1390 years (3-hour seastates).

Of course such results are in a sense circular: the inflated contour radius β^* and return period T_r^+ use α_o^2, whose precise value requires solution of the full FORM or Inverse FORM problem. We hope that growing experience—including results shown here—will suggest a reasonable range for α_o^2. For extreme response of offshore structures we find α_o^2 generally between .05–.25, and most commonly .10–.20. Hence the return period values cited above for α_o^2=.10 and .20 may be useful in estimating likely ranges of response variation.

Example 2: Extremes of Random Waves

To continue our first example, we return to the extreme wave crest problem. Figure 3 shows previous "median extreme crest" results, from the median response in Eq. 11 and the basic contour with return period T_r (Eq. 9). Also shown are "exact" results, which include the actual extreme response distribution (Eq. 10) and solve the resulting 3-variable FORM problem. As expected these are larger: the actual 100-year extreme crest is found to be 14.9m, as opposed to 13.7m if only the median crest is considered.

We seek here to predict the extreme crest from only a median crest model (Eq. 11), but with an inflated contour. Figure 3 shows that for return periods from 10 to 1000 years, the exact result is bracketed by using inflated contours from Eq. 17, with α_o^2 between .10 and .20. For example, this gives the range 14.5m–15.5m for the 100-year extreme crest, which includes the exact result 14.9m. Figure 3 also shows the result if the inflated contour is used with exact α_o^2, as found from the 3-variable FORM analysis. (Of course this value will not generally be available; however, the comparison serves to test the validity of the theory.) If the exact α_o^2 is used, the inflated contour is found to give rather accurate extreme crest estimates. It is somewhat conservative with respect to the exact FORM result in this case. This reflects that the actual failure surface tends to curve toward the origin, and hence the actual contour should be inflated less than the linear model implies. In principle, curvature (SORM) information could be used to correct for this error.

Example 3: Extreme Base Shear

As a final example, we consider the extreme base shear on a shallow-water jacket structure. From simultaneous hindcast of wind, waves, and current in the Southern North Sea (DHI, 1989), we include six environmental parameters: $\boldsymbol{X}=[H_S, T_Z, U, \Delta D, W, G]$. Here H_S is the annual maximum significant wave height in a T_{SS}=6-hour storm. This storm with annual maximum H_S is also characterized by its mean zero-upcrossing wave period T_Z, current U, surge level ΔD, mean wind speed W and gust factor G.

A correlated model of these variables has

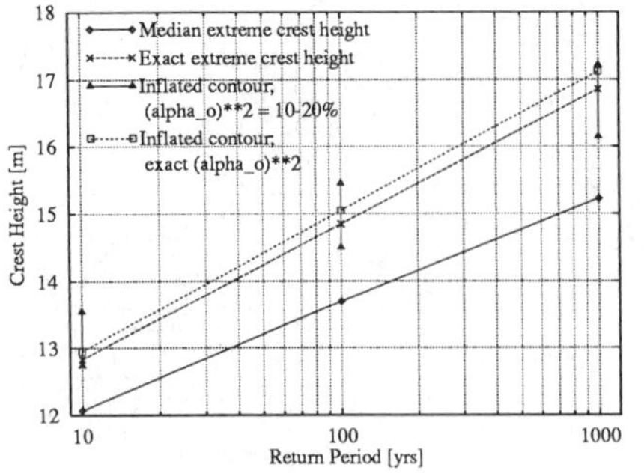

Figure 3: Extreme wave crest with various return periods.

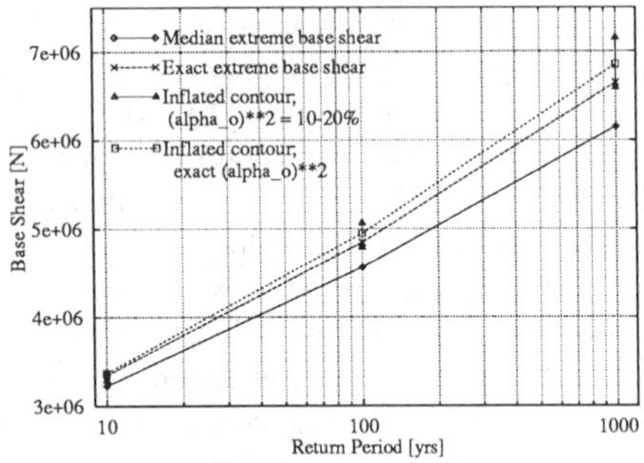

Figure 4: Extreme base shear with various return periods.

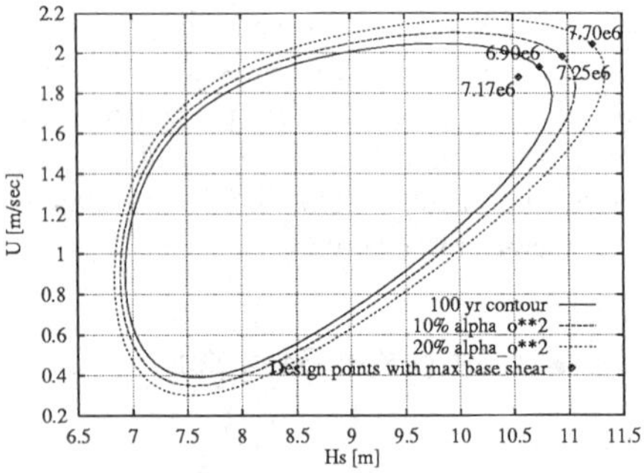

Figure 5: 100-year base shear from various wave height–current contours.

been fit to the hindcast data (Haver and Winterstein, 1990; Winterstein and Haver, 1991). These references also establish an analytical estimate of the extreme base shear Y in a seastate, in terms of its corresponding extreme crest height η_{max}:

$$\begin{aligned} Y &= n_{eq}\rho r C_D[\lambda_0 + \lambda_1\eta_{max} + \lambda_2\eta_{max}^2 + \lambda_3\eta_{max}^3] \\ &\quad + C_W C_Z^2 A_P W^2(1+2G) \end{aligned} \tag{19}$$

Here C_D and C_W are drag coefficients for waves and wind, n_{eq} is the equivalent number of legs with radius r for wave loads, and C_Z and A_P are height correction and projected area factors for wind loads. The factors λ_n depend in turn on the wave parameters T_Z, U, and ΔD. Nonlinear dependence of Y on η_{max}, reflected by λ_2 and λ_3, is due to both the nonlinear Morison drag force and the integration of distributed wave forces to the exact water surface.

Figure 4 shows base shear results versus return period, analogous to Figure 3 for the extreme wave crest. The lowest result considers Y a deterministic (median) function of the environmental variables, solving the 6-variable FORM problem by substituting the median crest height (Eq. 11) into Eq. 19. The exact result, larger as expected, follows by solving the 7-variable problem with the full extreme crest distribution (Eq. 10). As in Figure 3, these exact results are bracketed by using the median crest (Eq. 11) with an inflated contour, assuming an omitted importance factor α_o^2 ranging from .10–.20. (In fact, from the exact results we find that α_o^2 varies from .15–.19 in this case.)

Finally, we seek to alter this North Sea example to better reflect Gulf of Mexico storm conditions. First, based on their unimportance in the North Sea case, we set T_Z, ΔD, W and G to their conditional mean values based on H_S. We then revise the two variable (H_S–U) model, preserving the observed correlation ρ=.52 but rescaling the marginal mean current $E[U]$ from 0.4 m/s (North Sea case) to 1.2 m/s. Figure 5 shows resulting H_S–U contours, both including and excluding response variability (in crest height and hence base shear). Excluding this variability, the 100-year H_S–U contour yields the extreme base shear Y=6.90 $\times 10^6$ [N]. If this variability is kept, the exact result Y=7.17 $\times 10^6$ [N] is produced by searching H_S, U, and η_{max} (from its full distribution in Eq. 10). This maximum Y occurs at somewhat smaller H_S and U as shown, but somewhat larger-than-median Y given H_S and U). Our proposed method uses the median Y but searches the inflated contours shown in Figure 5, for which α_o^2=.10 and .20 (and, from Figure 2, T_r^+=140 and 215 years). The resulting Y estimates are shown to be fairly accurate and somewhat conservative, due to failure surface curvature and the slight overestimation in this case of α_o^2 (exact value=.094). Note that in all cases, the design point is at slightly less-than-maximum H_S for the given return period, to accommodate a slightly larger-than-median current given H_S.

CONCLUSIONS

- A method has been shown to construct contours of environmental parameters, along which extreme responses with given return period should lie. For any deterministic response these are found by transforming a hypersphere in standard normal space, with radius β (Eq. 9), to the physical space of environmental variables. Environmental contours can thus be produced which, when searched for the maximum response, yield response levels with the desired return periods (Figure 1).

- For stochastic response, the foregoing method will tend to underestimate extreme response levels because it neglects response variability. The degree of error is reflected by α_o^2, the contribution to uncertainty due to the response *given* the load environmental parameters. Across a range of offshore structural problems we typically find α_o^2 values from .05 to .25, and most often from .10 to .20.

- From an assumed α_o^2 value we can determine how to inflate the return period (Figure 2) and hence the environmental contours (Figure 5) along which the median response has the desired return period. The exact result is found often to be well-approximated, and usually bracketed, by choosing contours for which α_o^2=.10 and .20 (e.g., Figures 3–5). Thus, to estimate 100-year levels from only the median response, we should search environmental contours with return periods T_r^+ ranging from about 140–215 years (annual extreme seastates), and about 320–1400 years if all 3-hour seastates are considered.

ACKNOWLEDGMENTS

This work has grown out of the joint efforts and discussions with a number of the authors' colleagues, both at their respective institutions and among the other sponsors of the Reliability of Marine Structures Program of Stanford University. The Stanford authors also gratefully acknowledge the financial support from this program, as well as from the Office of Naval Research through Contract No. N00014-87-K-0475.

REFERENCES

DHI, (1989). *Environmental design conditions and design procedures for offshore structures,* Danish Hydraulic Institute, Copenhagen.

Haver, S. (1987). On the joint distribution of heights and periods of sea waves. *Ocean Eng.,* **14**(5), 359–376.

Haver, S. and K.A. Nyhus (1986). A wave climate description for long term response calculations. *Proc., 5th OMAE Symp.,* ASME, **IV**, 27–34.

Haver, S. and S.R. Winterstein (1990). The effects of a joint description of environmental data on design loads and reliability. *Proc., 9th Int. Offshore Mech. Arc. Eng. Sym.,* ASME, **II**, 7–14.

Madsen, H.O. (1988). Omission sensitivity factors. *Struc. Safety,* **5**, 35–45.

Madsen, H.O., S. Krenk, and N.C. Lind (1986). *Methods of structural safety,* Prentice-Hall, Inc., New Jersey.

Winterstein, S.R. and S. Haver (1991). Statistical uncertainty in extreme waves and structural response. *J. Offshore Mech. Arc. Eng.,* ASME, **113**, 156–161.

Winterstein, S.R., T. Marthinsen, and T.C. Ude (1992). Second-order springing effects on TLP extremes and fatigue. *J. Engrg. Mech.,* ASCE, submitted for possible publication.

Structural Safety & Reliability, Schuëller, Shinozuka & Yao (eds) © *1994 Balkema, Rotterdam, ISBN 90 5410 357 4*

Dynamic response of compliant offshore platforms to non-collinear wave, wind and current loading

O. Yılmaz & A. Incecik
University of Glasgow, UK

ABSTRACT: A non-linear time domain analysis procedure and calculation tools based on these procedures to predict the motion and structural response of various types of compliant offshore platforms under wave, wind and current loading was developed. These tools have been validated with experimental measurements. The compliant platform types considered by the authors were a coupled ship and catenary moored buoy, a coupled ship and articulated tower and a tension leg platform. In the first part of the paper a brief summary of the non-linear analysis procedures is given, together with some results obtained from predictions and experimental measurements. In the second part of the paper the results of parametric studies investigating the effects of variations in external wave, wind and current force magnitudes and directions on the dynamic motion and structural response of three different compliant platforms is illustrated. These results show the sensitivity of certain types of response values in different types of compliant structures as the external forces vary in magnitude and in direction. Such results provide essential information for probability based design procedures.

1 INTRODUCTION

Increases in oil prices over the last ten years have made the small offshore oil reservoirs in deep waters more attractive. Compliant offshore platforms are increasingly used in these oil fields due to their lower construction cost ,(compared with fixed platforms) and the fact that they can be towed to other oil fields to be used again. Design of compliant offshore platforms operating in severe weather conditions has become a popular research topic both in the universities and in the offshore industry. Tension leg platforms and single point mooring systems are some of the examples of compliant offshore production platforms.

Compliant offshore structures can sustain severe weather conditions by moving under the environmental forces and adopting the position with minimum mooring loads. Compliant offshore structures are particularly sensitive to directionality and there are several advantages to including the directionality and taking into account the probability of joint occurrence of various environmental events in design calculations (Incecik 1993). These advantages are:

. It serves us as a sound basis for combining environmental events in a rational manner
. It provides a framework for incorporating improved knowledge of the environment
. It provides greater accuracy in design calculations
. It allows a more consistent design approach in moving from one area to another
. It avoids unintentional and implicit conservatism.

In this paper having briefly described non-linear time-domain analysis techniques to predict dynamic motion response and mooring load characteristics of Tanker-Buoy , Articulated Tower-Tanker and a Tension Leg Platform systems under wave , wind and current forces as they act from arbitrary directions the effects of variations in environmental force directions and magnitudes on motion response and dynamic mooring load characteristics of these compliant platforms are summarised.

2 TIME DOMAIN ANALYSIS TECHNIQUES FOR COMPLIANT OFFSHORE STRUCTURES

2.1 Tanker-Buoy System

In modelling the tanker-buoy system surge, sway and yaw motions of the tanker and surge and sway motions of the buoy only were considered(See Fig. 1) (Yılmaz & Incecik 1992). Fluid reactive forces acting on the system were taken into account by using Cummins'

TABLE 1: GEOMETRICAL CHARACTERISTICS OF THE TLP

HULL LENGTH	510 m
NUMBER OF HULLS	4
HULL DIAMETER	13.6 m
COLUMN DIAMETER	25.0 m
COLUMN SPACING	76.0 m
COLUMN HEIGHT	63.0 m
DRAUGHT	37.5 m
WATER DEPTH	310.0 m
TOTAL WEIGHT	590, 080.0 kN
TENDON PRETENSION	448, 281.0 kN
TENDON AXIAL RIGIDITY	1, 014 MN/m
LONG. DECK AREA EXP. TO WIND	7,000 m^2
TRANS. DECK AREA EXP. TO WIND	12000 m^2

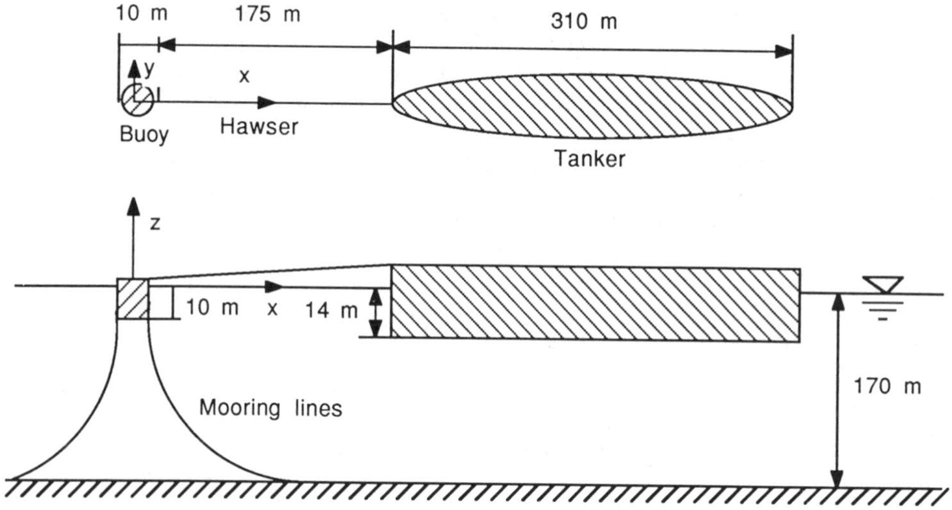

Fig. 1 CALM System

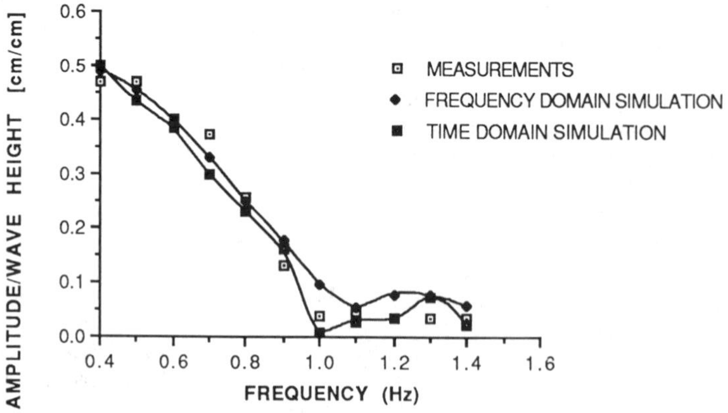

Fig. 2 Surge response of the ship (Current Force=0.06 N.)

technique (Cummins 1962). Since the tanker was modelled as an elliptical cylinder all hydrodynamic forces acting on the tanker were calculated using Mathieu Functions. Morison's equation was used in calculating the wave forces on the buoy. Wave drift forces were calculated according to Faltinsen's asymptotic theory (Faltinsen 1979). Wind and current forces were evaluated using semi-empirical formulas. ITTC friction resistance formula and the cross-flow principle were used to calculate current forces and moment. The Munk moment was also included in the current yaw moment calculations (Faltinsen 1979). Mooring forces were calculated using catenary equations and the non-linear load-elongation characteristics of hawser were taken into account in the prediction model (Yılmaz & Incecik 1991).

The Motion responses of the tanker-buoy system were compared with experimental measurements carried out at Hydrodynamics Laboratory of Glasgow University. The results of predictions agree well with those of experimental measurements. (See Figs. 2 and 3).

2.2 Articulated Tower-Tanker System

In order to predict the motion and structural response values of the coupled articulated tower-tanker system in waves, the system was modelled as a single degree of freedom system by assuming the tanker to be a rigid extension of the tower and that the coupled system rotates about the articulating joint (See Fig. 4). The heave, pitch and roll motions of the tanker were neglected. In formulating the wave forces and hydrodynamic properties.the tanker was modelled as a rectangular box. The wave forces and hydrodynamic properties of the articulated tower were formulated using Morison's equation. The detailed derivations of the motion and structural response equations are given in (Helvacıoglu 1990). Wave drift, steady current and wind forces and moments were calculated according to the methods described in section 2.1.

2.3 Tension Leg Platform

In order to predict the dynamic motion response characteristics of the TLP described in Table 1 due to combined wave, wind and current forces, a simple mathematical model was developed which takes into account the large amplitude displacements of the platform due to steady wave, wind and current drift. In the model first-order motions of the TLP due to waves and steady motions due to wind and current were predicted using the Morison formulation (Incecik 1993). The steady and slowly varying displacements of a TLP due to waves were obtained following the procedure described by (Pinkster 1979). The slowly varying oscillations of a TLP due to wind were predicted by combining a wind spectrum which represents the air turbulence, aerodynamic admittance and the dynamic response characteristics of the platform to a unit wind gusting velocity. Tables 2 and 3 show the effect of variation in wave direction and sea state on the dynamic response characteristics of the TLP described in Table 1.

3 PARAMETRIC STUDIES AND DISCUSSION OF RESULTS

A group of simulation studies for the tanker-buoy system were carried out using a non-linear time domain simulation computer program based on the prediction method described in section 2.1. At the beginning of each simulation the tanker was placed along the x axis with a zero yaw angle and the hawser was unstretched. Results of the parametric study are tabulated using the steady and oscillatory motion responses of the buoy and the tanker, which were obtained through a F.F.T analysis of the time domain simulations. During the simulations the effects of directionality of wave, wind and current force were investigated and the results of these simulations are given in Table 4. These indicate that maximum steady and oscillatory sway and yaw motions occur when wave and current forces make a 90 degree angle with the wind forces. Similarly maximum sway motions of the buoy occur when wave and current forces make a 90 degree angle with the wind forces. Another conclusion which can be drawn from Table 4 is that wind direction does not significantly affect the motions and that mean sway displacement and yaw angle increase as the current direction changes from 0 to 90 degrees. However maximum oscillatory sway motion of the buoy occurs when wave force direction makes a 0 degree and wind and current directions make a 45 degree angle with the horizontal axis. Maximum steady and oscillatory surge motions of the buoy and ship occur when wave, wind and current forces act co-linearly.

A group of simulation studies was carried out for the articulated tower-tanker system. In Table 5 the yaw angle (in degrees) corresponds to a steady angle after the articulated tower-ship system reaches an equilibrium state under wave, wind and current forces. Surge and sway values given in Table 5 indicate the amplitude of oscillations about the equilibrium state. Similarly, oscillatory components of the axial and trans-

TABLE 2: Motion Response of the TLP due to Co-linear Wave, Wind and Current

		FIRST ORDER RESPONSE						SECOND ORDER RESPONSE			TOT. RESP.
Wind speed [m/s]	Damp. Coef.	Hs Wave [m]	Surge As [m]	Sway As [m]	Yaw As [Deg]	Offset x [m]	Offset z [m]	Dynamic Wind [m]	S.o. waves [m]	Steady drift [m]	Total Surge [m]
10.00	0.05	2.01	0.36	0.00	0.00	3.36	0.00	0.42	0.01	0.01	3.92
15.00	0.10	4.74	0.53	0.00	0.00	3.70	0.00	0.73	0.18	0.05	4.66
20.00	0.15	8.49	0.85	0.00	0.00	4.18	0.00	1.05	0.98	0.12	5.97
30.00	0.20	19.05	2.26	0.00	0.00	5.55	0.00	1.89	6.64	0.35	13.16

TABLE 3: Effect of Wave Direction on the Motion Response of the TLP

15 deg

		FIRST ORDER RESPONSE						SECOND ORDER RESPONSE			TOT. RESP.
Wind speed [m/s]	Damp. Coef.	Hs Wave [m]	Surge As [m]	Sway As [m]	Yaw As [Deg]	Offset x [m]	Offset z [m]	Dynamic Wind [m]	S.o. waves [m]	Steady drift [m]	Total Surge [m]
10.00	0.05	2.01	0.20	0.20	4.67	3.36	0.00	0.42	0.01	0.01	3.83
15.00	0.10	4.74	0.38	0.09	6.55	3.70	0.00	0.73	0.18	0.05	4.58
20.00	0.15	8.49	0.74	0.22	7.07	4.18	0.00	1.05	0.98	0.12	5.91
30.00	0.20	19.05	2.16	0.62	7.31	5.55	0.00	1.89	6.64	0.35	13.13

30 deg

		FIRST ORDER RESPONSE						SECOND ORDER RESPONSE			TOT. RESP.
Wind speed [m/s]	Damp. Coef.	Hs Wave [m]	Surge As [m]	Sway As [m]	Yaw As [Deg]	Offset x [m]	Offset z [m]	Dynamic Wind [m]	S.o. waves [m]	Steady drift [m]	Total Surge [m]
10.00	0.05	2.01	0.09	0.05	4.95	3.36	0.00	0.42	0.01	0.01	3.79
15.00	0.10	4.74	0.24	0.17	7.38	3.70	0.00	0.73	0.18	0.05	4.53
20.00	0.15	8.49	0.63	0.41	8.00	4.18	0.00	1.05	0.98	0.12	5.87
30.00	0.20	19.05	1.93	1.67	8.28	5.55	0.00	1.89	6.64	0.35	13.06

45 deg

		FIRST ORDER RESPONSE						SECOND ORDER RESPONSE			TOT. RESP.
Wind speed [m/s]	Damp. Coef.	Hs Wave [m]	Surge As [m]	Sway As [m]	Yaw As [Deg]	Offset x [m]	Offset z [m]	Dynamic Wind [m]	S.o. waves [m]	Steady drift [m]	Total Surge [m]
10.00	0.05	2.01	0.03	0.06	2.20	3.36	0.00	0.42	0.01	0.01	3.79
15.00	0.10	4.74	0.19	0.20	2.76	3.70	0.00	0.73	0.18	0.05	4.52
20.00	0.15	8.49	0.53	0.54	3.06	4.18	0.00	1.05	0.98	0.12	5.83
30.00	0.20	19.05	1.60	1.61	3.31	5.55	0.00	1.89	6.64	0.35	12.98

60 deg

		FIRST ORDER RESPONSE						SECOND ORDER RESPONSE			TOT. RESP.
Wind speed [m/s]	Damp. Coef.	Hs Wave [m]	Surge As [m]	Sway As [m]	Yaw As [Deg]	Offset x [m]	Offset z [m]	Dynamic Wind [m]	S.o. waves [m]	Steady drift [m]	Total Surge [m]
10.00	0.05	2.01	0.02	0.13	4.82	3.36	0.00	0.42	0.01	0.01	3.79
15.00	0.10	4.74	0.15	0.28	7.50	3.70	0.00	0.73	0.18	0.05	4.51
20.00	0.15	8.49	0.48	0.65	8.53	4.18	0.00	1.05	0.98	0.12	5.81
30.00	0.20	19.05	1.16	1.94	9.12	5.55	0.00	1.89	6.64	0.35	12.89

75 deg

		FIRST ORDER RESPONSE						SECOND ORDER RESPONSE			TOT. RESP.
Wind speed [m/s]	Damp. Coef.	Hs Wave [m]	Surge As [m]	Sway As [m]	Yaw As [Deg]	Offset x [m]	Offset z [m]	Dynamic Wind [m]	S.o. waves [m]	Steady drift [m]	Total Surge [m]
10.00	0.05	2.01	0.01	0.24	4.92	3.36	0.00	0.42	0.01	0.01	3.79
15.00	0.10	4.74	0.08	0.40	7.15	3.70	0.00	0.73	0.18	0.05	4.50
20.00	0.15	8.49	0.22	0.76	8.03	4.18	0.00	1.05	0.98	0.12	5.75
30.00	0.20	19.05	0.61	2.17	8.52	5.55	0.00	1.89	6.64	0.35	12.82

90 deg

		FIRST ORDER RESPONSE						SECOND ORDER RESPONSE			TOT. RESP.
Wind speed [m/s]	Damp. Coef.	Hs Wave [m]	Surge As [m]	Sway As [m]	Yaw As [Deg]	Offset x [m]	Offset z [m]	Dynamic Wind [m]	S.o. waves [m]	Steady drift [m]	Total Surge [m]
10.00	0.05	2.01	0.00	0.36	0.00	3.36	0.00	0.42	0.01	0.01	3.79
15.00	0.10	4.74	0.00	0.54	0.00	3.70	0.00	0.73	0.18	0.05	4.49
20.00	0.15	8.49	0.00	0.85	0.00	4.18	0.00	1.05	0.98	0.12	5.73
30.00	0.20	19.05	0.00	2.26	0.00	5.55	0.00	1.89	6.64	0.35	12.80

Constant values: Current speed 1.5 m/s, wind angle 0 deg and current angle 0 deg

Table 4: Effect of Directionality of Environmental Forces on the Motion Response of the CALM System

Simulation No	Direction of Variables			MEAN SHIP MOTION RESPONSE			Mean Hawser Tension(kN)
	Wave	Current	Wind	Surge(m)	Sway(m)	Yaw(deg)	
1	0	0	0	342.80	0.00	0.00	2674
2	0	90	0	331.80	73.06	16.70	2585
3	15	90	0	312.50	128.70	25.10	2802
4	30	90	0	283.00	185.30	29.72	2617
5	45	90	0	254.17	224.50	33.85	3874
6	60	90	0	219.60	259.57	43.50	2518
7	75	90	0	183.40	288.90	52.33	5571
8	90	90	0	146.20	311.10	62.30	6522
9	0	45	0	323.90	63.89	2.24	2838
10	0	45	15	320.80	78.77	1.76	2947
11	0	45	30	317.10	91.35	2.31	3039
12	0	45	45	314.60	104.20	3.91	3099
13	0	45	60	315.60	105.50	5.59	3036
14	0	45	75	317.30	104.10	7.07	2953
15	0	45	90	322.30	91.82	8.36	2783
16	0	15	0	290.10	21.69	2.06	2990
17	0	30	0	313.70	45.98	1.10	2857
18	0	45	0	323.90	63.89	2.24	2838
19	0	60	0	325.53	73.11	7.49	2786
20	0	75	0	327.90	69.73	12.08	2681

Table 5: Effect of Directionality of Environmental Forces on the Motion Response of the SALM System

Simulation No	Direction of Variables			Yawing Angle(deg)	Axial Yoke F.(MN)	Transverse Yoke F.(MN)	Steady Joint F.(MN)
	Wave	Current	Wind				
1	0	0	0	0.00	6.60	0.00	2.60
2	0	0	90	6.97	6.60	1.00	2.50
3	30	0	90	28.16	6.60	0.30	2.30
4	45	0	90	37.93	6.50	1.10	1.70
5	60	0	90	49.25	6.50	1.60	1.40
6	90	0	90	73.40	6.70	2.40	2.20
7	0	90	0	17.20	6.80	2.40	2.50
8	0	90	30	18.06	6.80	2.50	2.50
9	0	90	45	18.28	6.90	2.60	2.50
10	0	90	60	18.76	6.90	2.60	2.50
11	0	90	90	19.49	7.00	2.70	2.50
12	0	30	90	9.70	6.50	1.40	2.70
13	0	45	90	12.48	6.50	1.80	2.50
14	0	60	90	15.18	6.70	2.20	2.50

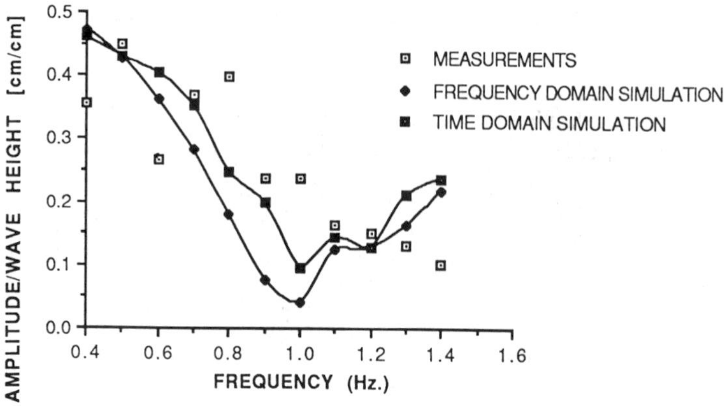

Fig. 3 Surge response of the buoy (Current Force=0.06 N.)

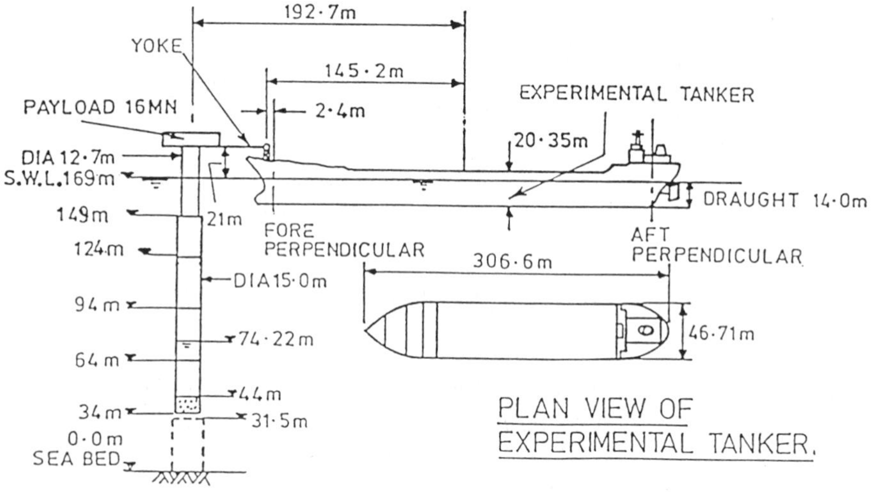

Fig.4 Articulated tower and tanker geometry

verse yoke forces, as well as the steady forces at the base joint, are given in Table 5.

Table 5 shows that wave excitation is the dominant environmental force in the prediction of steady yaw angle and surge and sway motion amplitudes. Examination of Table 5 also reveals that maximum axial and transverse yoke forces occur when the direction of wave approach makes a 90 degree angle with wind and current directions.

Parametric studies carried out with the TLP Platform shows that with an increasing wave angle from 0 to 90 degrees total surge response decreases by 3-4% and that maximum yaw response occurs when the wave angle of attack is 60 degrees (Tables 2-3).

4 CONCLUSIONS

1. Maximum steady and oscillatory sway and yaw motions of the tanker in the CALM system occur when wave and current forces make a 90 degree angle with the wind forces. Similarly maximum sway motions of the buoy occur when wave and current forces make a 90 degree angle with the wind forces. It was found that the mean mooring line forces were not generally sensitive to changes in current and wind loading. and that there was not a linear relationship between the wave height and the response values or the mooring forces of the CALM system.
2. Maximum axial and transverse yoke forces in the

SALM system occur when the direction of wave approach makes a 90 degree angle with wind and current directions. It was also found that neither the wind and current speed nor their directions had a significant effect on axial and transverse yoke force predictions.

3. Since the motion displacements due to steady wind and current as well as second-order wave and dynamic wind forces contribute significantly to the total surge response the effect of variation of wave direction on total surge displacements is not significant. However as the direction of wind and current varies from 0 to 90 degrees the total surge response decreases by about 25%.

REFERENCES

Yılmaz, O. & Incecik A. 1992. Identification of Non-linear Effects in Predicting the Motion Response of a CALM System. *Proc. of the Eleventh International Conference on Offshore Mechanics and Arctic Engineering*, Calgary.

Cummins, W.E. 1962. The Impulse Response Function and Ship Motions *Schiffstechnik* 9. No. 47.

Faltinsen, O.M. , Kjaerland, O. , Liapis, N. and Walderhaug, H. 1979. Hydrodynamic Analysis of Tankers at Single Point Mooring Systems. *Proceedings of the Second International Conference on Behaviour of Offshore Structures*, No. 59, London.

Yılmaz, O. & Incecik A. 1991. Non-Linear Dynamic Time Domain Simulation of Moored Floating Systems. *Proc. of the Tenth International Conference on Offshore Mechanics and Arctic Engineering*, Stavanger.

Helvacıoglu, I.H.H. 1990. Dynamic Analysis of Coupled Articulated Tower and Floating Production Systems. Department of Naval Architecture and Ocean Engineering, University of Glasgow.

Incecik, A. 1993. The Effect of Joint Occurrence of Wave, Wind and Current Loading on Dynamic Response of a Tension Leg Platform. To be presented at the *International Conference on OMAE '93*, Glasgow.

Pinkster, J.A. 1979. Mean and Low Frequency Wave Drifting Forces on FloatingStructures. *Ocean Engineering* 6.

Wood structures

Structural Safety & Reliability, Schuëller, Shinozuka & Yao (eds) © 1994 Balkema, Rotterdam, ISBN 90 5410 357 4

Lengthwise variability of bending strength of timber beams

I.Czmoch
Chalmers University of Technology, Göteborg, Sweden

ABSTRACT: The weak-zone model of timber beam is employed in the analysis. It is assumed that timber is composed of short weak zones (knot or group of knots) connected by strong sections of clear wood. The failure can only occur in the middle of the weak zones. The spatial distribution of bending strength is modelled by means of the composite random point series: random series of bending strength is assigned to randomly distributed weak zones. Stationarity of both point series is assumed. The paper presents the application of the filtered Poisson process, which is the composite point process with zero correlation between the strengths of weak zones and zero correlation between the distances between weak zones. The distribution functions of the load carrying capacity of the weak-zone model of timber beam have been evaluated for three load cases.

1. INTRODUCTION

Timber as a building material can be considered as a composition of clear wood and defects. Clear wood is an anisotropic material whose properties along grains are different than in direction perpendicular to grain. Particularly, the tensile strength parallel to grain is several times greater than tensile strength perpendicular to grain. Therefore knots and other defects, which change the angle between the grains and the stress vector, are important source of the lengthwise variability of bending strength of timber. Furthermore, since the strength of cross-section with knot can be several times lower than the strength of clear wood, the failure of timber beam generally occurs in the vicinity of a group of knots or is initiated in such region. In general, the character of failure is brittle. Those facts are recognised by the present codes of practice, which demand that in the tests for determination of the timber bending strength, the length with constant moment should contain "a strength reducing grade determining defect". However, the lengthwise variability of timber properties is not taken into account in design of timber structures. In design practice timber is considered as a homogeneous material, whose design value of bending strength is based on the 0.05-quantile of the extreme (minimum) value distribution.

The lengthwise variability of bending strength timber beams is investigated in this paper with the help of the weak-zone model of timber beam. The spatial distribution of bending strength is modelled by means of the composite random point series: random series of bending strength is assigned to randomly distributed weak zones. Stationarity of both point series is assumed.

The paper presents the application of the filtered Poisson process, which is the composite point process with zero correlation between the strengths of weak zones and zero correlation between the distances between weak zones. The distribution functions of the load carrying capacity of the weak-zone model of simply supported timber beam have been evaluated for three load cases.

2. WEAK-ZONE MODEL OF TIMBER BEAM

The bending strength of a timber beam cross-section with knot is considerably reduced due to:

- grain distortions around knot resulting in grains perpendicular to tensile stress vector,
- stress concentrations caused by knot holes and encased knot,
- differences between mechanical properties of knot itself and surrounding wood.

Furthermore, the bending strength of the cross-section with knot depends on many factors, such as:

- knot size and shape
- knot position within the cross-section (tension or compression zone)
- inclination of grains in the vicinity of knot in relation to the stress vector
- intersection of grains around knot by beam surface (specially in tension zone)

The quantitative knowledge about the influence of knots on mechanical properties is scarce. Therefore a simply mechanical model of the timber beam subjected to bending stresses is assumed in this study.

The weak-zone model of timber beam is based on the following assumptions:

- timber beam is modelled as a composition of short weak zones connected by longer strong sections of clear wood,
- weak zones correspond to knots or group of knots and are randomly distributed,
- failure can only occur in the middle of weak zones, i.e. a weak zone means actually a weak cross-section,
- bending strength of weak zones is random

Hence, the spatial distribution of the bending strength of timber beam is represented by means of the composite random point series: random series of bending strength $\{R_i\}$ is assigned to randomly distributed weak zones $\{\Delta_i\}$, Fig. 1.

It is assumed that both series $\{R_i\}$ and $\{\Delta_i\}$ may be modelled as stationary random series. The statistical description of the stationary random series $\{R_i\}$, $\{\Delta_i\}$ is limited to the marginal probability distribution functions $F_R(r)$, $F_\Delta(\delta)$ and the correlation functions

$$\rho_R(k) = \rho_R(R_i, R_{i+k}), \qquad \rho_\Delta(k) = \rho_\Delta(\Delta_i, \Delta_{i+k}).$$

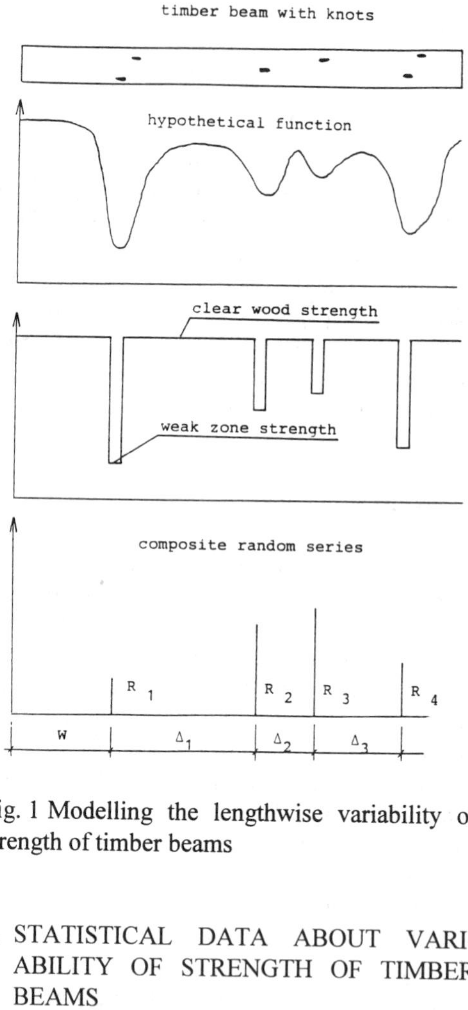

Fig. 1 Modelling the lengthwise variability of strength of timber beams

3. STATISTICAL DATA ABOUT VARIABILITY OF STRENGTH OF TIMBER BEAMS

Statistical data about lengthwise distribution of knots and strength of weak-zones with knots is very scarce.
Colling & Dinort (1987) investigated the Knot-Area-Ratio (KAR) values for 456 boards and KAR lengthwise distribution. The boards were taken from three different growing areas. It

turned out that the knot distribution does not depend on the growth area and timber grade. Colling & Dinort report a mean distance of 0.5 meters between major knots or knot groups. Riberholt & Madsen (1979) fitted the exponential distribution to experimental data concerning distances between knots or groups of knots. The mean distance was about 0.3-0.5 meters.

Very limited experimental data is available about the bending strength of weak zones (cross-sections with knots) or spatial distribution of bending strength of structural lumber. Lam & Varoglu (1991) investigated the within member variation of tensile strength parallel to grain in 38x89 No 2 Spruce-Pine-Fir lumber. The tests indicate a considerable strength correlation between cross-sections 0.5-1.0 metre apart. The strength correlation decreases to zero for distance greater than 1.8 metre.

High correlation between the strength of timber and the localized modulus of elasticity has been shown in many experimental tests. Therefore the degree of correlation between the strength of consecutive weak zones can be also estimated on the basis of the degree of correlation for the modulus of elasticity for timber. Czmoch (1991) presents the statistical analysis of randomness of bending stiffness and reports that the correlation distance for the modulus of elasticity decreases from about 1.3 metre for the serviceability load level down to 0.5 metre for a load level equal to about 80% of the ultimate load.

It is worth noticing that above results about the degree of correlation for strength concern structural lumber, which is a composition of weak and strong zones. The degree of correlation for strength of weak zones might be expected to decrease even faster. Geometrical features of consecutive weak zones (i.e. size of knots, their position, inclination of grains) suggest that the degree of correlation for strength of consecutive weak zones can be assumed to be equal to zero.

4. STATISTICAL MODEL OF VARIABILITY OF STRENGTH OF TIMBER BEAM

In this paper the lengthwise variability of bending strength of timber is modelled by means of the filtered Poisson process (Parzen 1962)

$$R(x) = \sum_{i=1}^{N(x)} R_i \, \delta(x - x_i) \tag{1}$$

where $\delta(x)$ is Dirac function, $N(x)$ is number of weak zones in the interval $(0,x]$ following the Poisson process with intensity $\nu(x)$ and R_i (weak zone strength) is a sequence of independent random variables, identically distributed.

5. LOAD CARRYING CAPACITY OF WEAK-ZONE MODEL OF TIMBER BEAM AS THE FIRST DOWNCROSSING PROBLEM

The load carrying capacity generally denotes the maximum external load which can be imposed to a member without causing its failure. The load carrying capacity depends on the shape and sizes of a structure as well as on the load configuration. In this paper the load carrying capacity is expressed in terms of stresses and is equal to the maximum stress allowed by the assumed failure criterion.

Since structural timber is a brittle material, the failure of statically determined timber beam occurs when the allowable stress level is attained at any cross-section. The stress function $S(x)$ can be expressed as the product of the maximal stress Z and the normalized bending moment function $m(x)$

$$S(x) = \frac{M(x)}{W} = \frac{M_{max}}{W} \frac{M(x)}{M_{max}} = Z\, m(x) \tag{2}$$

where $M(x)$ is the bending moment function and W is the section modulus.

The first downcrossing of the stress function $S(x) = Z\, m(x)$ by the allowable stress R_i at the i-th weak zone denotes the failure of timber beam. Taking into account equation (1) the load carrying capacity is defined as the maximal value of parameter Z for which the condition

$$R(x_i) = R_i \geq S(x_i) = Z\,m(x_i) \tag{3}$$

is fulfilled for every weak zone within beam span L; $i = 1, ..., N(L)$.

6. DISTRIBUTION FUNCTION OF THE LOAD CARRYING CAPACITY OF TIMBER BEAMS

According to the assumed model of bending strength of timber beam the random variables $\{R_i\}$ are assigned to weak zones whose lengthwise distribution is described by the Poisson process with the intensity $\nu(x)$. Therefore the sequence of events defined by the condition that the random variables $\{R_i\}$ downcross the stress function $S(x) = Z\,m(x)$ constitutes a new Poisson process with intensity

$$\nu_z(x) = \nu(x)\,F_R(S(x)) = \nu(x)\,F_R(Z\,m(x)) \tag{4}$$

where $F_R(r)$ is the marginal distribution function for random variables $\{R_i\}$.

Hence, the distribution function of the load carrying capacity $F_Z(z)$ is equal to the probability of the resistance function $R(x)$ being above the stress function $S(x)$ within the length L

$$\begin{aligned} F_Z(z) &= P\big(R(x) > S(x) = z\,m(x);\ 0 \leq x \leq L\big) \\ &= 1 - \exp[-I(z)] \end{aligned} \tag{5}$$

where

$$I(z) = \int_0^L \nu(x)\,F_R(z\,m(x))\,dx \tag{6}$$

The integral (6) depends on the intensity function $\nu(x)$ describing the distribution of weak zones, on the distribution function for bending strength of weak zone $F_R(r)$ and on the normalized bending moment function $m(x)$. In most cases the integral (6) must be evaluated by numerical methods for specific data.

However, the integral (6) can be derived analytically, if the Weibull distribution is assumed as a model of the variability of bending strength of weak zones R_i. The cumulative distribution function of the Weibull model is

$$F(r) = 1 - \exp\left[-\left(\frac{r-\mu}{\sigma}\right)^{\lambda}\right] \tag{7}$$

where μ is the location parameter, σ is the scale parameter and λ is the shape parameter. The Weibull distribution is a very flexible model. It becomes the exponential model for $\lambda = 1$ and it can as well approximate the Normal model for $\lambda \cong 3.6$.

7. DISTRIBUTION FUNCTION OF THE LOAD CARRYING CAPACITY OF SIMPLY SUPPORTED TIMBER BEAMS

The load carrying capacity depends on the load configuration. Typical load cases, considered in the design of simply supported timber beams, are

a) the middle point load

$$m_a(x) = \begin{cases} 2\dfrac{x}{L} & \text{if } 0 \leq x \leq \dfrac{L}{2} \\ 2\left(1-\dfrac{x}{L}\right) & \text{if } \dfrac{L}{2} \leq x \leq L \end{cases} \tag{8}$$

b) the one-third point load (uniformly distributed load produces similar bending moment function)

$$m_b(x) = \begin{cases} 3\dfrac{x}{L} & \text{if } 0 \leq x \leq \dfrac{L}{3} \\ 1 & \text{if } \dfrac{L}{3} \leq x \leq \dfrac{2}{3}L \\ 3\left(1-\dfrac{x}{L}\right) & \text{if } \dfrac{2}{3}L \leq x \leq L \end{cases} \tag{9}$$

c) pure bending

$$m_c(x) = 1 \qquad \text{if } 0 \le x \le L \tag{10}$$

Assuming the homogenous Poisson process for the distribution of weak zones $\nu(x) = \nu$, the integral (6) has been evaluated for the normalized bending moment functions (8), (9), (10), and for the Weibull model of the weak zone strength (7). The corresponding distribution functions of load carrying capacity have the following forms

a) the middle point load

$$F_Z(z) = 1 - \exp[-I_1(z)] \tag{11}$$

b) the one-third point load

$$F_Z(z) = 1 - \exp\left[-\frac{2}{3} I_1(z) - \frac{1}{3} I_2(z)\right] \tag{12}$$

c) pure bending

$$F_Z(z) = 1 - \exp[-I_2(z)] \tag{13}$$

where

$$I_1(z) = -\nu L\left(1 - \frac{\mu}{z}\right) \sum_{n=1}^{\infty} \frac{(-1)^n}{n!\,(1+\lambda n)} \left(\frac{z-\mu}{\sigma}\right)^{\lambda n} \tag{14}$$

$$I_2(z) = \nu L \left\{1 - \exp\left[-\left(\frac{z-\mu}{\sigma}\right)^{\lambda}\right]\right\}$$

In design practice the most important parameter is the characteristic value of the load carrying capacity Z_k defined as the p-th quantile of the distribution F_Z, i.e.

$$F_Z(Z_k) = 1 - \exp[-I(Z_k)] = p \tag{15}$$

where p is usually equal to 0.05 for the resistance variables.

The characteristic values of the load carrying capacity for the first two load cases can be obtained by solving the equations

a) middle point load

$$I_1(Z_k) = -\ln(1-p) \tag{16}$$

b) the one-third point load

$$\frac{2}{3} I_1(Z_k) + \frac{1}{3} I_2(Z_k) = -\ln(1-p) \tag{17}$$

For the timber beam under pure bending the characteristic value is given by the formula

$$Z_k = \mu + \sigma\left[-\ln\left(1 + \frac{1}{\nu L}\ln(1-p)\right)\right]^{\frac{1}{\lambda}} \tag{18}$$

or by the simplified approximate formula

$$Z_k = \mu + \sigma\left(\frac{-\ln(1-p)}{\nu L}\right)^{\frac{1}{\lambda}} \tag{19}$$

The lengthwise variability of the timber beam strength is observed in the experimental tests (Madsen & Buchanan 1986) and is referred to as the size and load case effects, i.e. the characteristic value of bending strength of timber depends on the sizes of tested beams (mainly length and depth) and on the type of applied load (from pure bending up to the middle point load).

Taking into account the formulae (16-19) the relationships between the characteristic values Z_{k1}, Z_{k2} corresponding to the beam spans L_1, L_2 can be obtained. In the case of a beam under pure bending, the derived relationship is similar to the formula for the length effect derived on the basis of the Weibull weakest link theory (Madsen & Buchanan 1986),

$$\frac{Z_{k1} - \mu}{Z_{k2} - \mu} = \left(\frac{L_2}{L_1}\right)^{\frac{1}{\lambda}} \tag{20}$$

The formula (20) shows that the shape parameter λ in the Weibull distribution describing the randomness of the bending strength of weak zones determines the degree of length effect. Thus the experimental data about the variation of characteristic value of timber strength with the test beam span could be used in order to determine the parameters μ and λ of the Weibull distribution modelling the variability of the strength of weak zones.

The characteristic value of load carrying capacity of timber beam also depends on the type of applied load. The characteristic value attains the minimum for a beam subjected to the pure bending and the maximum for a beam subjected to the middle point load. The formula for the load configuration effect has the general form

$$F_Z(Z_{k1}) = F_Z(Z_{k2}) \tag{21}$$

where Z_{k1}, Z_{k2} are the characteristic values of the load carrying capacities for two load cases. Taking into account equations (11-15) the formulae for the load effect can be derived, in terms of the parameters of the Weibull distribution describing the variability of the strength of weak zones. Thus the load configuration relationships could be used in order to determine the parameters of the Weibull distribution μ, σ, λ. The accuracy of this estimation should be a subject of further research.

For the considered solution the length effect and the load configuration effect do not depend on the intensity of weak zones ν, because the distribution of weak zones $\{x_i\}$ is modelled by the homogeneous Poisson process, $\nu(x) = \nu$.

8. NUMERICAL EXAMPLE

Calculations have been carried out for the artificial but realistic data. The intensity of weak zones was chosen according to the results of measurement presented by Colling & Dinort (1987) and Riberholt & Madsen (1979)

$$\nu = \frac{2 \text{ weak zones}}{1 \text{ metre beam length}} \tag{22}$$

Three statistical models of the variability of the bending strength of weak zones have been considered by assuming three values of the shape parameter λ in the Weibull model (Table 1). Fig. 1 presents the corresponding probability density functions.

Table 1 Parameters of the Weibull distribution used in the example

Data set	μ	σ	λ	mean value	standard deviation	coefficient of variation
A1	10.	60.	1.5	63.5	35.7	0.56
A2	10.	60.	3.0	63.3	19.2	0.30
A3	10.	60.	4.5	64.6	13.7	0.21

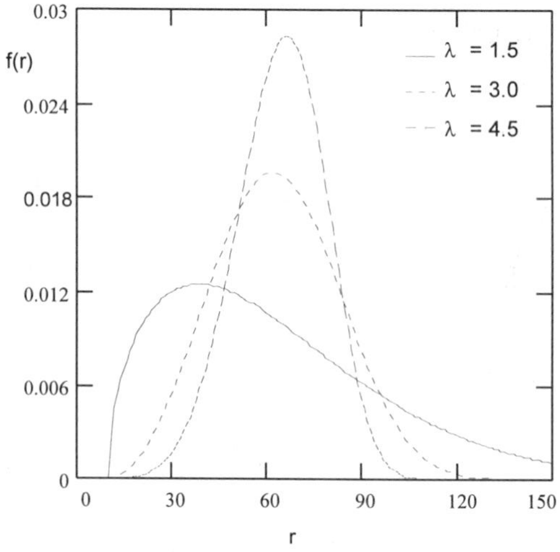

Fig. 2 Probability density functions of the Weibull distribution used in the example

Table 2 presents the characteristic values (the 0.05-th quantiles) of the load carrying capacity for timber beams under pure bending (eqn 18) and for timber beams under the middle point load (eqn 16). Those values are compared with the characteristic values of the load carrying capacity estimated by means of the Monte Carlo simulation, (Czmoch et al. 1992).

The model assumed for the purpose of the simulation is more general than the model considered in this paper. The sequence of distances between weak zones $\{\Delta_i\}$ has been assumed as the stationary random series with the marginal distribution function $F_\Delta(\delta)$ and the correlation function

$$\rho_\Delta(k) = \rho_\Delta(\Delta_i, \Delta_{i+k}). \tag{23}$$

The sequence of weak zones strength $\{R_i\}$ has been assumed as the stationary random series with the marginal distribution function $F_R(r)$ and the correlation function

$$\rho_R(k) = \rho_R(R_i, R_{i+k}). \tag{24}$$

The Monte Carlo simulations have been carried out for the following assumptions:

- the distance between weak zones has been modelled by the exponential distribution with the mean value of distance equal to 0.5 m.,
- the bending strength of weak zones has been modelled by the Weibull distribution with statistical parameters as in Tab.1,
- non-zero degree of correlation has been represented by the Gaussian type correlation functions,

$$\rho_\Delta(k) = \exp\left[-\left(\frac{k}{\theta_\Delta + 1}\right)^2\right] \quad \text{for} \quad \theta_\Delta > 0$$

$$\rho_R(k) = \exp\left[-\left(\frac{k}{\theta_R + 1}\right)^2\right] \quad \text{for} \quad \theta_R > 0$$

- zero degree of correlation has been represented by the Dirac function,

$$\rho_\Delta(k) = \rho_R(k) = \delta(k) \qquad \text{for } \theta_\Delta = \theta_R = 0$$

where θ_Δ, θ_R are the scales of fluctuations. The common value of the scale of fluctuation $\theta = \theta_\Delta = \theta_R$ was assumed in the simulation.

- 2000 sample beams were generated for each data set.

The load carrying capacity Z has been calculated for each sample beam from relation

$$Z = \min\left(\frac{R_i}{m(x_i)}; \;\; i = 1, \ldots, N(L)\right) \tag{25}$$

where R_i is allowable stress at the i-th weak zone and $m(x_i)$ is the normalized bending moment function at the i-th weak zone.

The non-parametric estimation of the characteristic Z_k value for the set of 2000 sample beams is equal to

$$Z_k = Z_{(100)} \tag{26}$$

where $Z_{(100)}$ denotes the 100-th ordered observation.
The approximate (distribution free) 95% confidence interval for the 0.05-th quantile estimated on the basis of the set of 2000 samples is given as follows

$$P\left(Z_{(81)} \le Z_k \le Z_{(119)}\right) = 0.95 \tag{27}$$

Table 2 contains both point estimates and the 95% confidence interval for the characteristic values calculated by means of simulation for the scale of fluctuation $\theta = 0$. Differences between the point estimates of the characteristic values by means of the Monte Carlo simulation and the values derived according to the analytical expressions are small. The characteristic values determined from the formulae (16) and (18) are between the limits of the corresponding confidence intervals in all considered cases.

If the scale of fluctuation $\theta = 4$, than the simulation results in higher characteristic values of load carrying capacity. Thus the analytical model, which does not take into account any correlation, could be used in practice, since it gives the lower (on the safe side) estimation the characteristic values of the load carrying capacity of timber beams.

Table 2 Characteristic values of load carrying capacity

Data set	Beam span [m]	the middle point load characteristic value according to					pure bending characteristic value according to				
		equation (16)	simulation				equation (18)	simulation			
			θ = 0			θ = 4		θ = 0			θ = 4
			point estimate	95% confidence interval		point estimate		point estimate	95% confidence interval		point estimate
A1	1.8	20.3	19.8	18.8	20.5	20.7	13.5	13.3	12.9	13.9	15.0
	3.6	17.3	17.7	17.1	18.3	19.0	12.2	12.2	12.0	12.6	13.5
	5.4	16.0	16.2	15.9	16.9	17.9	11.7	11.8	11.6	12.0	13.0
	7.2	15.3	15.4	15.0	15.9	17.5	11.4	11.5	11.2	11.7	12.5
A2	1.8	35.9	35.0	33.8	36.3	36.5	24.6	24.0	23.3	25.3	27.3
	3.6	31.0	31.7	30.3	32.4	33.7	21.6	21.6	20.9	22.4	24.4
	5.4	28.5	29.3	28.4	30.5	32.1	20.1	20.3	19.7	21.0	23.5
	7.2	27.0	27.5	26.7	28.3	31.4	19.2	19.5	18.5	20.1	22.2
A3	1.8	46.2	45.5	43.8	46.5	46.0	33.4	32.7	31.9	34.1	36.2
	3.6	41.2	41.6	40.4	42.7	42.7	30.0	30.1	29.3	31.0	33.2
	5.4	38.6	39.4	38.4	40.8	41.7	28.3	28.6	27.8	29.4	32.2
	7.2	36.9	37.2	36.6	38.4	40.9	27.2	27.6	26.3	28.3	30.8

9. CONCLUSIONS

The distribution functions of the load carrying capacity of the weak-zone model of simply supported timber beams have been evaluated analytically for a special case:

- the lengthwise distribution of weak zones is modelled as a homogeneous Poisson process,
- the strength of weak zones is considered as a sequence of independent random variables, with identical Weibull distributions.

Practical application of the results depends on the extensive experimental verification of the assumptions.

Project dealing with experimental analysis of the distribution of weak zones along 380 timber beams, 5.8 meters long, is being carried out currently at Chalmers University of Technology.

10. ACKNOWLEDGEMENTS

This project was carried out under supervision of Dr Kamal Handa, with the financial support from the Swedish Council for Building Research (BFR).

11. REFERENCES

Colling, F., Dinort, R. 1987. Ästigkeit des in den Leimbaubetrieben verwendeten Schnittholzes. *Holz als Roh- und Werkst.* 45:23-26

Czmoch, I. 1991. Lengthwise Variability of Bending Stiffness of Timber Beams, *1991 International Timber Engineering Conference*. London. United Kingdom.

Czmoch, I., Thelandersson, S., Larsen, H.J. 1992. Simulation of Lengthwise Variability of Bending Strength of Structural Timber, *Chalmers University of Technology*, Göteborg.

Lam, F., Vargolu, E. 1991. Variation of tensile strength along length of timber, *Wood Sci. Tech.* 25(5):351-360.

Madsen, B., Buchanan, A.H. 1986. Size effect in timber explained by a modified weakest link theory, *Can. J. Civ. Eng.* 13(2):218-232

Parzen, E. 1962. *Stochastic Processes*. Holden-Day. Inc. San Francisco

Riberholt, H., Madsen, P.H. 1979. Strength distribution of timber structures. Measured variation of the cross sectional strength of structural lumber, *Struct. Research Lab., Technical University of Denmark, Report No R 114*

Structural Safety & Reliability, Schuëller, Shinozuka & Yao (eds) © 1994 Balkema, Rotterdam, ISBN 90 5410 357 4

Stochastic stiffness properties and buckling of laminated wood columns

Bryan Folz & Ricardo O. Foschi
The University of British Columbia, Vancouver, B.C., Canada

ABSTRACT: A characterization of the within specimen variability in the modulus of elasticity (MOE) of wood as a one-dimensional stochastic field is presented. Simulation of MOE-profiles is achieved via a spectral approach. Central to this study is the use of test data, from lumber specimens, to calibrate the parameters which define the spectral density function. This material model is incorporated in a simple one-dimensional stochastic finite element model for predicting the elastic buckling load of laminated wood columns. Monte Carlo simulation is then used to evaluate the resulting variability in the elastic buckling load as influenced by the stochasticity in the elastic modulus.

1 INTRODUCTION

Wood as an organic material can exhibit substantial variation in stiffness properties both *within* and *between* specimens; a specimen herein refers to a lumber piece of structural size. This variability can arise from the presence of naturally occuring defects (knots) and fiber (grain) deviation. Quantifying the variability between specimens is a key objective of most test programs and is requisite for reliability-based code development. To this end, simple standardized test procedures have been adopted. For example, a nominal modulus of elasticity (MOE) value for a specimen, over a given gauge length, can be obtained from a conventional tension or bending test.

Considerable research interest is now focusing on the within specimen variability in material properties and its contributing influence to the response of various structural systems. This study examines the lengthwise variability in the elastic modulus of wood using experimental data which approximates, with reasonable accuracy, the actual underlying MOE process. A simple procedure for generating MOE profiles, based on a spectral approach, is presented and the results are compared against the test data. An application is then made to the elastic buckling of laminated wood columns using a simple one-dimensional stochastic finite element model.

2 STOCHASTIC ELASTIC MODULUS

The lengthwise variability in the elastic modulus of a wood specimen can be determined, in an approximate manner, by means of a stiffness-based grading machine; which measures, at a number of discrete points along the length of the member, the force required to produce a prescribed midspan deflection over a short test span. For each position of the specimen, as it is fed through the machine, the nominal MOE value over the test span is obtained and can be taken as an approximation to the *true* localized MOE value about the point of loading. When the specimen has passed through the machine a record, or profile, of the within specimen lengthwise variability in the elastic modulus is provided by the machine's data acquisition system. A method to enhance the accuracy of the generated MOE profiles has been proposed, but the scheme is not numerically robust (Foschi 1987).

It is noted that the commercial application of this test procedure is to stress rate the material through the statistical correlation between the stiffness of the specimen and the strength over the test length. Obtaining the MOE profile is only a by-product of this test procedure, albeit of key importance to the problem at hand.

A data set consisting of 179 MOE records from Spruce and Pine specimens was used in this study. The specimens were of rectangular cross-section,

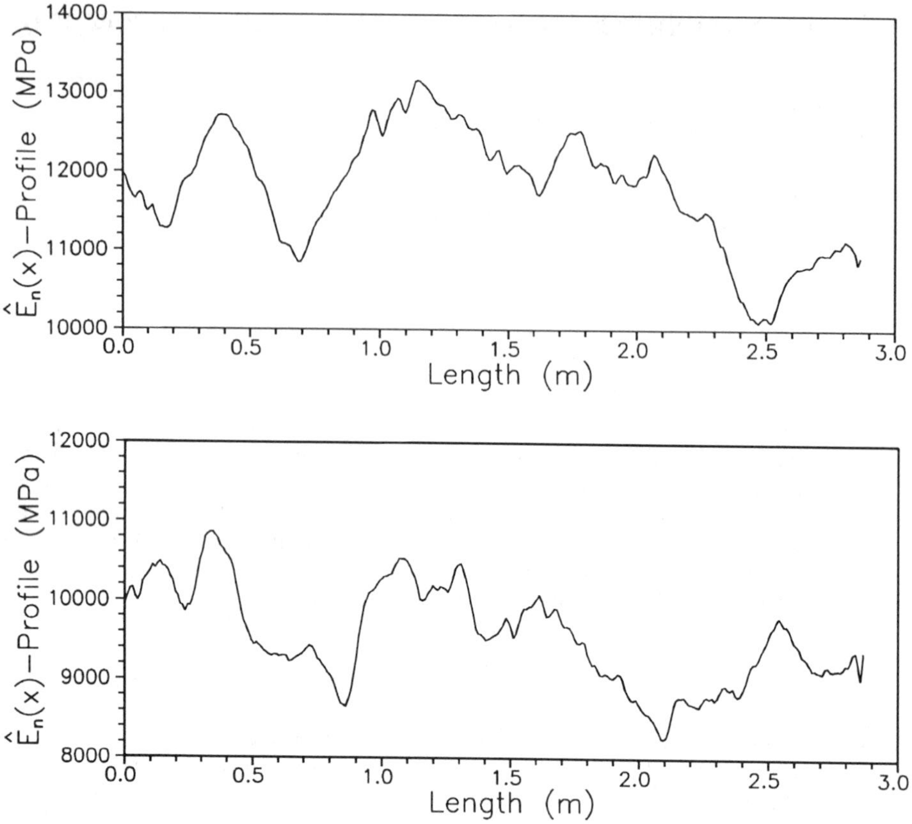

Figure 1: Experimental $\hat{E}_n(x)$–profiles.

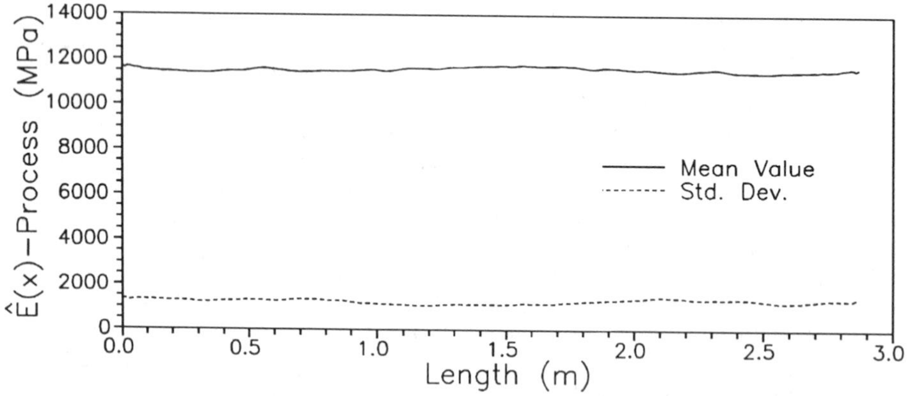

Figure 2: Ensemble statistics for $\hat{E}(x)$.

dimensioned either 38 mm × 89 mm (102 pcs.) or 38 mm × 140 mm (77 pcs.) and having a test length $L_t = 2865$ mm. Note, any variation in the elastic modulus along the width of the specimen is ignored. The particular grading machine used to produce the MOE profiles had a test span of 910 mm and enforced a center point deflection of 4.45 mm. The machine was set up so that bending of the specimens took place about the weak axis and a total of 291 MOE point–estimates, spaced 9.88 mm apart, were obtained from each specimen.

Let $\hat{E}(x)$ denote the ensemble MOE corresponding to the test data and $\hat{E}_n(x)$ a generic test record, two of which are shown in Fig. 1. Figure 2 shows the ensemble mean and standard deviation of $\hat{E}(x)$, from which it is apparent that these statistics do not show substantial variation with respect to sampling location. The analysis thus far suggests that the MOE process can be considered, at the least, a weakly homogeneous stochastic field.

For subsequent analysis each record is decomposed as follows:

$$\hat{E}_n(x) = \hat{E}_n^* + \hat{e}_n(x) \tag{1}$$

where $\hat{E}_n^*$ is the specimen mean over the test length L_t and $\hat{e}_n(x)$ is the zero–mean fluctuating part. From specimen to specimen $\hat{E}_n^*$ varies, resulting in a non–parametric mean of 11.56 GPa and a standard deviation of 1.04 GPa over the ensemble. Thus the MOE process is non–ergodic, when examined over this test length. The specimen means are assumed, without loss of generality, to obey a 2–P Weibull distribution:

$$F(\hat{E}_n^*) = 1 - \exp\left[-(\hat{E}_n^*/m)^k\right] \tag{2}$$

where $m = 12.0$ GPa and $k = 13.8$ are, respectively, the fitted scale and shape parameters. Figure 3 shows the ensemble mean and standard deviation of the zero–mean fluctuating part $\hat{e}(x)$. Further evaluation of the $\hat{e}(x)$–process will be carried out via its spectral representation. To this end, the finite length one–sided spectral density function $G_{\hat{e}_n}(\kappa)$, corresponding to the $\hat{e}_n(x)$ record, can be evaluated as

$$G_{\hat{e}_n}(\kappa) = \frac{1}{\pi L_t} \cdot \mid \mathcal{F}_{\hat{e}_n}(\kappa) \mid^2 \tag{3}$$

where $\mathcal{F}_{\hat{e}_n}(\kappa)$ is the finite Fourier transform expressed in terms of wave number κ. Employing a FFT algorithm allows for an efficient evaluation of Eq. (3) at discrete wave numbers $\kappa_n = n\Delta\kappa$. The resolution of the individual spectrums, however, is relatively poor, with $\Delta\kappa = 2.19$ rads/m; this being a direct consequence of the chosen sampling rate of the MOE records in the spatial domain. The ensemble one–sided spectral density function for the $\hat{e}(x)$–process, as shown in Fig. 4, is obtained from

$$G_{\hat{e}}(\kappa) = \frac{1}{\pi L_t} \cdot \mathcal{E}\left[\mid \mathcal{F}_{\hat{e}_n}(\kappa) \mid^2\right] \tag{4}$$

with $\mathcal{E}[\cdot]$ denoting the operation of ensemble averaging. From Fig. 4 it is noted that there exists, over the ensemble, a dominate spectral amplitude at $\kappa_1 = 2.19$ rads/m. In fact the dominate spectral amplitude could occur at a wave number less than κ_1, but can not be captured from the data due to the coarse spectral resolution. This long wave length component may simply be a feature of the MOE process. An alternate explanation, however, may be that a linear or higher order trend exists in the process, arising possibly from the growth characteristics of the wood, with one end of a lumber piece consisting of more mature (stiffer) wood compared to the other end. This of course would render the MOE process non–homogeneous. Unfortunately, the chosen test length for the specimens is not long enough and the procedure of specimen collection was inadequate to provide a judgement on this matter. Consequently, it will be assumed herein that the MOE process is homogeneous up to second order moments.

3 SIMULATING MOE RECORDS

Simulation of one–dimensional univariate homogeneous, and ergodic, zero–mean stochastic fields can be achieved by a summation of sinusoids with random phase (Shinozuka 1987):

$$e(x) = \sum_{n=1}^{N} \sqrt{2G_e(\kappa_n)\Delta\kappa}\cos(\kappa_n x + \phi_n) \tag{5}$$

where $G_e(\kappa)$ is the ensemble one–sided spectral density function, ϕ_n are independent random phase angles uniformly distributed over the interval 0 to 2π and N is the number of participating sinusoids. Equation (5) results in a simulated process which is periodic with a wave length equal to $2\pi/\Delta\kappa$.

In order to relate $G_{\hat{e}}(\kappa)$ to $G_e(\kappa)$ in Eq. (5), it is advantageous to fit $G_{\hat{e}}(\kappa)$ to an analytical expression. From examination of Fig. 4 an appropriate form is

$$G_e(\kappa) = \kappa^{c_1}\left\{c_2\exp(-c_3\kappa^2) + c_4\exp(-c_5\kappa)\right\} \tag{6}$$

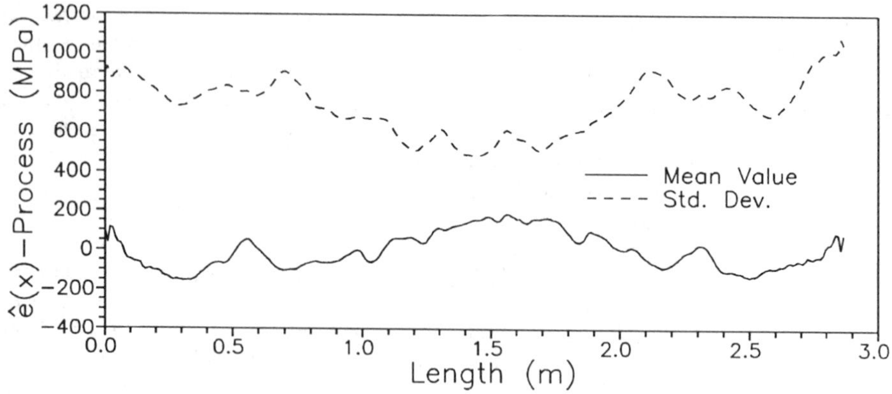

Figure 3: Ensemble statistics for $\hat{e}(x)$.

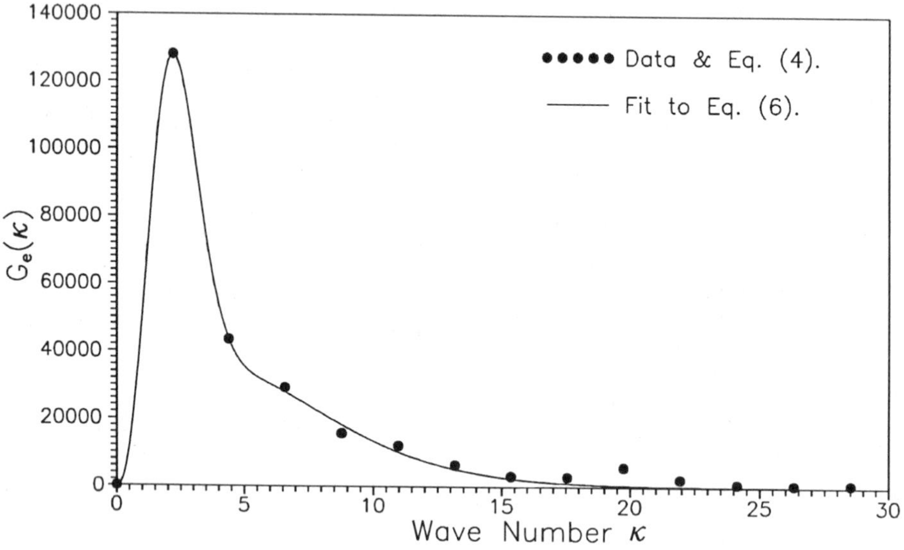

Figure 4: Ensemble spectral density $G_e(\kappa)$.

where the five non–negative constants c_1, c_2, c_3, c_4 and c_5 can be determined by regression analysis of the data, as obtained from Eq. (4), with Eq. (6). The regression procedure based on nonlinear function minimization, in a least squares sense, takes the form:

$$\Psi_1 = \sum_{n=1}^{N} \{G_{\hat{e}}(\kappa_n) - G_e(\kappa_n)\}^2 \rightarrow \min \tag{7}$$

Given the value of $\Delta\kappa$ obtained from the data (2.9 rad/m) and from examining Fig. 4, N was set equal to 15. The regression analysis can be complemented with two constraint conditions: first, that the mean sample variance from the data must be met, that is the integral of Eq. (6) must equal $\mathcal{E}[\sigma^2_{\hat{e}_n}]$, and second, that $G_e(\kappa)$ be unimodal with a maximum value at κ_1. Thus only 3 free parameters remain to be fitted. The results of the fit are given in Table 1. For subsequent reference let this approach be designated as Method A.

Two simulated $e_n(x)$ records, based on Method A, are shown in Fig. 5. Figure 6 presents a comparison of the cumulative distribution of sample root mean square (rms) $\sigma_{\hat{e}_n}$ and σ_{e_n} as obtained, respectively, from the data and simulated using

Table 1: Spectral parameter values for $G_e(\kappa)$.

	c_1	c_2	c_3	c_4	c_5
Method A	2.39	61700	0.271	8660	0.507
Method B					
Mean	2.39	61700	0.271	8660	0.507
Std. dev.	0.26	57000	0.020	7300	0.051

Eq. (5), with the fitted parameters. The feature that Eq. (5) produces ergodic samples over the record length L_t is readily apparent.

One means of achieving better agreement for the sample variance is to let the five spectral parameters be random. It is assumed that each is an independent log-normal variate. Fitting of the mean and standard deviation of each c_i can, again, be achieved using non-linear function minimization, which in this case takes the form:

$$\Psi_2 = \sum_{j=1}^{19} \left\{ \sigma_{\hat{e}}^2|_{q_j} - \sigma_e^2|_{q_j} \right\}^2 \rightarrow \min \tag{8}$$

with the variance evaluated at the quantiles $q_j = 0.05j$. The fitted parameter values are given in Table 1 and the resulting cumulative distribution of sample rms σ_{e_n} is shown in Fig. 6, both under the designation of Method B. Note that within the accuracy of the regression analysis the mean parameter values for the c_i equal those obtained by Method A.

4 STOCHASTIC FINITE ELEMENT MODEL

As an application for the stochastic MOE information, given in the previous section, the elastic buckling analysis of laminated wood columns is considered. The structural analysis can be performed in a straight forward manner via the finite element method. A simple one-dimensional beam-column element is used, which has two nodes and eight degrees-of-freedom, with the element formulation based on a higher order shear deformation theory (Heyliger and Reddy 1988).

For each lamina within an element the assignment of an elastic modulus value is obtained by sampling the stochastic field at the midpoint along the element. As an example, for the j-th lamina in the i-th element

$$E(i,j) = E_j^* + e_j(x_i) \tag{9}$$

where x_i locates the midpoint of the i-th element and $e_j(x_i)$ is determined according to Eq. (5). Note that if the overall length of the column exceeds L_t each lamina must be constructed from more than one MOE record.

The governing equation for the elastic buckling analysis can be posed as a linear eigenvalue problem:

$$(\mathbf{K} - \lambda_{cr}\mathbf{K}_\sigma)\mathbf{\Phi} = \mathbf{0} \tag{10}$$

where $\mathbf{K}$ and $\mathbf{K}_\sigma$ are, respectively, the linear and geometric global stiffness matrices, λ_{cr} is an eigenvalue and $\mathbf{\Phi}$ the corresponding eigenvector. The lowest λ_{cr} can be extracted from Eq. (10) using an inverse iteration technique (Bathe 1982).

5 RESPONSE VARIABILITY ANALYSIS

The next step in the analysis is to determine the response variability in the elastic buckling load of laminated wood columns as a function of their stochastic stiffness properties. Monte Carlo simulation is used in conjunction with the stochastic finite element model described previously.

Attention is restricted to columns under the action of a concentric end load and simple support conditions. The columns are of rectangular cross-section, each having a width of 130 mm and depths of either 38 mm, 76 m, 152 mm or 304 mm, corresponding to 1, 2, 4 and 8 laminae (each 38 mm thick), with perfect bond existing between the laminae. Each column has a span-to-depth ratio of 40 and buckling is constrained to take place parallel to the depth dimension of the cross-section. The length of available lamination material was taken to be random and uniformly distributed between 1.0 and 2.75 m, so that in all cases the wave length of the simulated elastic modulus for the lamination piece is longer than its length. The ratio of the mean elastic modulus to the shear rigidity was set equal to 15, which is a typical value for softwoods. Note that this resulted in only a 1% reduction in the mean buckling load when compared with the classical Euler load.

In the analysis which follows three material models are considered:

Model 1: Stochastic $E(x) = E^* + e(x)$;

Model 2: Random $E(x) = E^*$;

Model 3: Stochastic $E(x) = \mathcal{E}[E^*] + e(x)$.

Model 1 utilizes the full information available for the elastic modulus, as presented in the previous

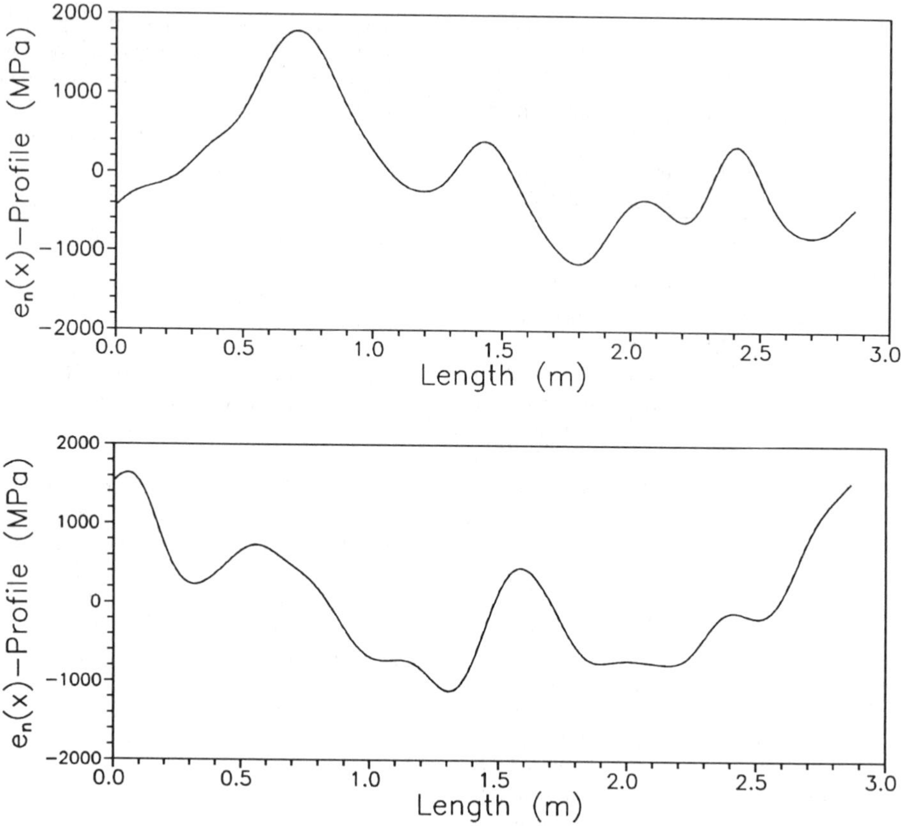

Figure 5: Simulated $e_n(x)$–profiles.

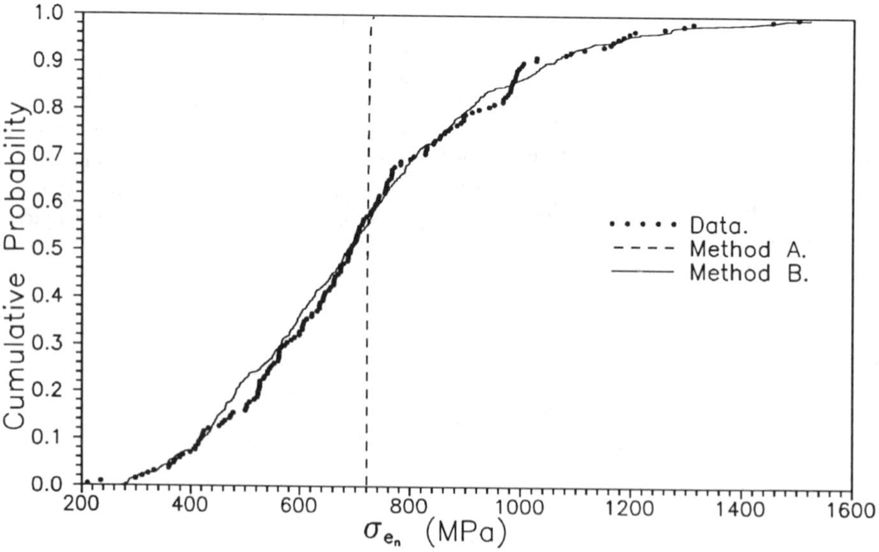

Figure 6: Cumulative distribution of $\sigma_{\hat{e}}$ and σ_e.

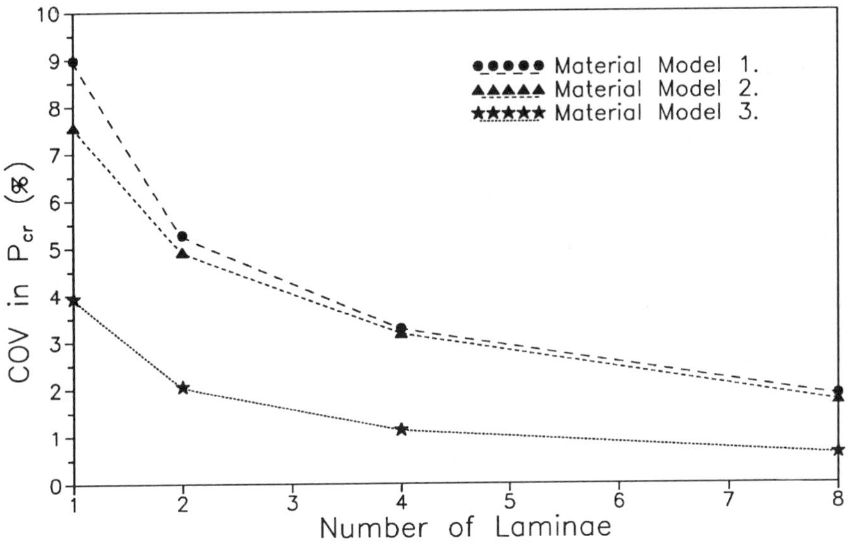

Figure 7: COV in P_{cr}.

sections. Model 2, on the other hand, ignores totally the fluctuating $e(x)$–process and only considers the mean elastic modulus as a random quantity. With Model 3 the mean elastic modulus is taken to be constant between laminae (equal to the mean of the sample means), with variability only in the $e(x)$–process. It is noted that stochastic material modeling equivalent to Model 3 has been adopted in many studies which do not rely directly on supporting experimental data.

The simulation results for each of the material models are presented in Fig. 7, with the quantity of interest being the coefficient of variation (COV) in the elastic buckling load P_{cr}. All of the results are based on a simulation size of 250 and a column discretization of $10 \times n_l$ finite elements, where n_l equals the number of laminae in the member.

First, it is noted that the Model 1 and 3 results were insensitive to the method by which the MOE profiles were generated (*i.e.* using either Method A or B). From Fig. 7, it is observed that when the number of laminae exceeds one, Models 1 and 2 yield results which are in close agreement with each other. Thus, for this structural application it is sufficient to use the simpler approach offered by Model 2. From the Model 3 results it is observed that the stochastic fluctuations in the MOE process have a less pronounced influence on the response than that produced by the variability between laminae. As expected for all of the material models the COV in P_{cr} decreases with an increase in the number of laminae. In particular, when the number of laminae reaches 4 the COV in P_{cr} is reduced to less than 4%, which suggests that at this point a deterministic analysis may be adequate.

6 CONCLUSIONS

A straight forward procedure, based on a spectral approach, has been presented for simulating, in a realistic manner, the lengthwise variability in the elastic modulus of wood specimens. As an application, Monte Carlo simulation in combination with a finite element formulation was used to analyze the variability in the elastic buckling load of laminated wood columns with stochastic stiffness properties. It was observed, however, that consideration of only the variability in the elastic modulus between laminae (*i.e.* a random variable approach) was sufficient in providing an accurate estimate of the response variability. Furthermore, as the number of laminae in a column exceeds four, a deterministic analysis may be deemed adequate.

ACKNOWLEDGEMENT

This work was supported by Forestry Canada, the Natural Science and Engineering Research Council of Canada and the Canadian Wood Council under a cooperative research partnership.

REFERENCES

Bathe, K.J. 1982. *Finite element procedures in engineering analysis,* Prentice Hall, Englewood Cliffs.

Foschi, R.O. 1987. A procedure for the determination of localized modulus of elasticity. *Holz als Roh-und Werkstoff* 45: 257-260.

Heyliger, P.R. and J.N. Reddy 1988. A higher order beam finite element for bending and vibration problems. *Journal of Sound and Vibration* 126(2): 309-326.

Shinozuka, M. 1987. Stochastic fields and their digital simulation. In *Stochastic methods in structural dynamics.* eds. G.I. Schueller and M. Shinozuka: 93-133. Martinus Nijhoff, Dordrecht.

Structural Safety & Reliability, Schuëller, Shinozuka & Yao (eds) © 1994 Balkema, Rotterdam, ISBN 90 5410 357 4

Stochastic finite element analysis of progressive failure in a laminated wood beam

Bryan Folz & Ricardo O. Foschi
The University of British Columbia, Vancouver, B.C., Canada

ABSTRACT: A stochastic finite element analysis of the load–carrying capacity of a laminated wood beam, under progressive tensile fracture, is presented. The stochasticity in the problem arises from the spatial randomness of the elastic modulus and tensile strength along each lamina. Monte Carlo simulation is used to evaluate the variability in the load–carrying capacity statistics as a function of the stochastic material modeling assumptions. A statistical size effect, arising from the stochastic tensile strength model, is also demonstrated.

1 INTRODUCTION

In this study, the stochastic finite element method (SFEM) is employed to examine the effect of spatially varying strength and stiffness properties on the load–carrying capacity of a glued–laminated wood beam. This problem is a particularly appropriate application of the SFEM given the inherent variability in these mechanical properties in wood; resulting, for example, from knots and grain deviation. The failure of wood beams in flexure is typically governed by brittle fracture of the material in tension. In this presentation the stochastic material modeling is limited to only the elastic modulus and the tensile strength of the lamination material.

At present, test data are readily available to facilitate modeling of the elastic modulus along a lamina as a one–dimensional stochastic field (Wang and Foschi 1992). Similar tensile strength information, on the other hand, is essentially non–existent. In turn, the within lamina spatial correlation between the two processes is also unknown. Before undertaking a program of data collection some important questions to first consider are: How important are the fluctuations in the elastic modulus in redistributing the stresses? Is it necessary to model this property as a stochastic field, or is knowing the mean value sufficient? At what level of within lamina variability in the tensile strength does it become expedient to model it as a random field? Does the cross–correlation structure between the elastic modulus and the tensile strength have to be quantified? To address these questions three material models are considered in this study (listed with decreasing level of complexity):

<u>Material Model 1</u> stochastic elastic modulus and tensile strength;

<u>Material Model 2</u> stochastic tensile strength only;

<u>Material Model 3</u> random tensile strength only.

A response variability study, using Monte Carlo simulation, is conducted to evaluate the influence of each of these material modeling assumptions on a glued–laminated beam's load–carrying capacity.

The particular glued–laminated beam under consideration has a rectangular cross–section, is simply supported and subjected to a monotonically increasing uniformly distributed load to failure. The beam has a span of 6080 mm, a cross–sectional breadth and depth of 130 mm and 304 mm respectively, with a horizontal layup of 8 laminae, each 38 mm thick. To simplify the analysis it is assumed that the laminae are perfectly bonded one to another and free of end–joints. Thus, the accumulation of damage under increasing load is completely confined to the lamination material.

Table 1: Material Model 3 tensile strength statistics.

b_s	$\delta_S = 0.10$		$\delta_S = 0.20$		$\delta_S = 0.30$	
(m)	$\bar{S}_{min}$ (MPa)	$\delta_{S_{min}}$ (%)	$\bar{S}_{min}$ (MPa)	$\delta_{S_{min}}$ (%)	$\bar{S}_{min}$ (MPa)	$\delta_{S_{min}}$ (%)
0.5	38.8	6.1	27.7	17.0	16.9	38.2
1.0	40.4	6.6	30.9	17.4	21.5	35.4
2.0	42.2	7.2	34.4	17.7	26.8	32.9

Simulation size = 1000 laminations.

Table 2: Load–carrying capacity statistics obtained from the various material models.

Material Model	$\delta_S = 0.10$		$\delta_S = 0.20$		$\delta_S = 0.30$	
	$\bar{P}_u$ (kN/m)	δ_{P_u} (%)	$\bar{P}_u$ (kN/m)	δ_{P_u} (%)	$\bar{P}_u$ (kN/m)	δ_{P_u} (%)
	$b_e = b_s = 0.5$ m			($n_e = 75$)		
Stochastic $E(x)$ and $S(x)$, $\gamma_{es}^2 = 0$	21.4	5.6	19.1	10.0	16.6	14.8
Stochastic $E(x)$ and $S(x)$, $\gamma_{es}^2 = 1$	23.1	3.3	19.8	8.9	17.1	14.6
Stochastic $S(x)$	21.7	5.3	19.2	9.9	16.3	16.2
Random S_{min}	19.6	5.9	14.5	13.6	8.8	34.7
	$b_e = b_s = 1.0$ m			($n_e = 50$)		
Stochastic $E(x)$ and $S(x)$, $\gamma_{es}^2 = 0$	22.2	7.3	20.1	11.7	18.0	17.1
Stochastic $E(x)$ and $S(x)$, $\gamma_{es}^2 = 1$	23.7	3.9	20.9	10.2	18.6	15.2
Stochastic $S(x)$	22.5	6.9	20.2	11.4	18.0	17.1
Random S_{min}	20.0	6.4	15.3	14.4	10.0	35.4
	$b_e = b_s = 2.0$ m			($n_e = 25$)		
Stochastic $E(x)$ and $S(x)$, $\gamma_{es}^2 = 0$	23.4	8.9	21.8	13.7	20.0	18.4
Stochastic $E(x)$ and $S(x)$, $\gamma_{es}^2 = 1$	24.3	4.7	22.3	12.2	20.8	17.0
Stochastic $S(x)$	23.4	8.4	21.8	13.8	20.3	19.1
Random S_{min}	20.9	6.8	17.1	14.7	12.2	31.6

Simulation size = 500 beams.
n_e = number of finite elements used to discretize the beam.

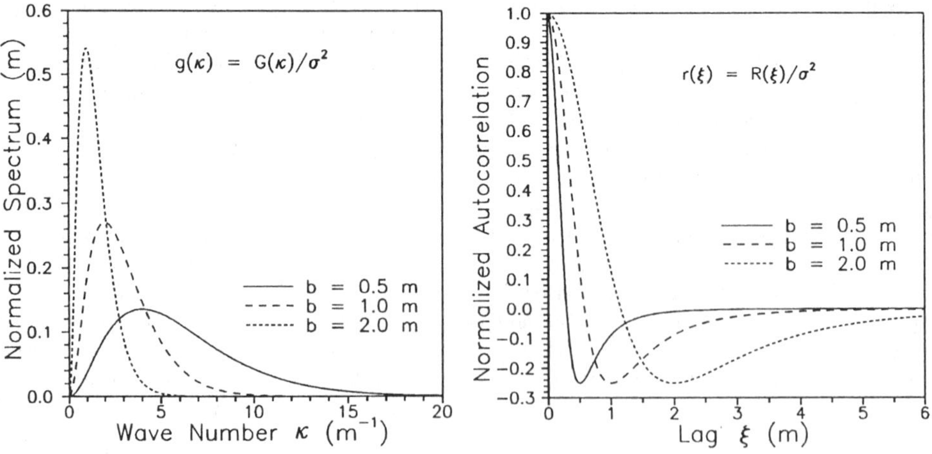

Figure 1: Normalized spectral density $g(\kappa)$ and autocorrelation $r(\xi)$.

2 STOCHASTIC MATERIAL MODELING

In this section the underlying theory of Material Model 1 is presented in detail, followed by the simplifying assumptions which produce Material Models 2 and 3.

Spanwise along each lamina, the modulus of elasticity $E(x)$ and the tensile strength $S(x)$ are modeled as homogeneous one-dimensional stochastic fields:

$$E(x) = E^* + e(x) \qquad (1)$$

$$S(x) = S^* + s(x) \qquad (2)$$

where E^* and S^* are the corresponding ensemble mean values and $e(x)$ and $s(x)$ are stochastic fields representing the fluctuations of each material property about its mean value. Note that the constraint conditions $e(x) > -E^*$ and $s(x) > -S^*$ must be enforced in order to avoid physically unrealizable material behaviour.

Adopting a spectral approach, the zero-mean stochastic fields $e(x)$ and $s(x)$ are characterized by their one-sided spectral density functions, having the plausible form,

$$G_e(\kappa) = \frac{1}{2}\sigma_e^2 b_e^3 \kappa^2 \exp(-b_e\kappa) \qquad (3)$$

$$G_s(\kappa) = \frac{1}{2}\sigma_s^2 b_s^3 \kappa^2 \exp(-b_s\kappa) \qquad (4)$$

with κ denoting the wave number, b_e and b_s spectral constants and σ_e and σ_s denoting the standard deviation of $e(x)$ and $s(x)$, respectively. As shown in Fig. 1, the parameters b_e and b_s determine the shape of each spectrum, as well as the scale of correlation of each field; with a larger parameter value implying a longer distance over which a significant level of correlation will persist in the random field. Note, to facilitate numerical implementation of this spectral approach it is assumed that there exists an upper cut-off wave number κ_u, such that $G_e(\kappa) = G_s(\kappa) = 0$ for $\kappa \geq \kappa_u$.

In general, some statistical dependence will exist between the two processes $E(x)$ and $S(x)$, which can be expressed in terms of the cross-correlation function $R_{ES}(\xi)$:

$$R_{ES}(\xi) = E^*S^* + R_{es}(\xi) \qquad (5)$$

where ξ is the lag between sampling points along the x-axis and $R_{es}(\xi)$ is the cross-correlation function for $e(x)$ and $s(x)$. Alternatively, this cross-correlation structure can be represented by the one-sided cross-spectral density function $G_{es}(\kappa)$. In turn, the interdependence between $G_e(\kappa)$, $G_s(\kappa)$ and $G_{es}(\kappa)$ can be expressed in terms of the coherence function γ_{es}^2, with

$$\gamma_{es}^2(\kappa) = \frac{|\, G_{es}(\kappa)\,|^2}{G_e(\kappa)G_s(\kappa)}, \qquad 0 \leq \gamma_{es}^2(\kappa) \leq 1 \qquad (6)$$

Realizations of the $e(x)$ and $s(x)$ processes can then be generated in accordance with the following formulae (Shinozuka 1987):

$$e(x) = \sum_{i=1}^{N}[2G_e(\kappa_i)\Delta\kappa]^{1/2} \cdot \cos(\kappa_i x + \phi_{1i}) \qquad (7)$$

$$s(x) = \sum_{i=1}^{N}[2G_s(\kappa_i)\Delta\kappa]^{1/2} \cdot \{\gamma_{es}\cos(\kappa_i x + \phi_{1i}) + (1-\gamma_{es}^2)^{1/2}\cos(\kappa_i x + \phi_{2i})\} \qquad (8)$$

in which $\Delta\kappa = \kappa_u/N$ and $\kappa_i = (2i-1)\Delta\kappa/2$. The variables ϕ_{1i} and ϕ_{2i} are independent random phase angles uniformly distributed on the interval 0 to 2π. Note that by virtue of the central limit theorem the simulated fields given by Eqs. (7) and (8) are asymptotically Gaussian as $N \to \infty$. Further, from these equations, it is seen how $\gamma_{es}^2(\kappa) = 1$ implies total coherence and $\gamma_{es}^2(\kappa) = 0$ total incoherence in the simulation of the random fields. Lacking knowledge about an appropriate form for $\gamma_{es}^2(\kappa)$, the two extreme values of the coherence function will be applied in the subsequent application of Material Model 1.

With Material Model 2 only the tensile strength is a stochastic field, which is obtained using Eqs. (2) and (8), with $\gamma_{es}^2(\kappa) = 0$. Material Model 3 is obtained by assigning to each lamina the minimum strength value obtained over the length of the lamina from Model 2: $S_{min} = S^* + s_{min}$. Thus, for each lamina, S_{min} is a random variable. For both of these models the elastic modulus is simply assigned the value $E(x) = E^*$.

3 FINITE ELEMENT MODEL

State-of-the-art deterministic finite element analysis has produced very sophisticated solution strategies to model the progressive failure in laminated composites. However, with stochastic finite element modeling the level of sophistication of the structural analysis model must be balanced against the knowledge one has, or can feasibly obtain, regarding the governing stochastic material properties in the problem. Seeking such a balance, a simple one-dimensional finite element formulation

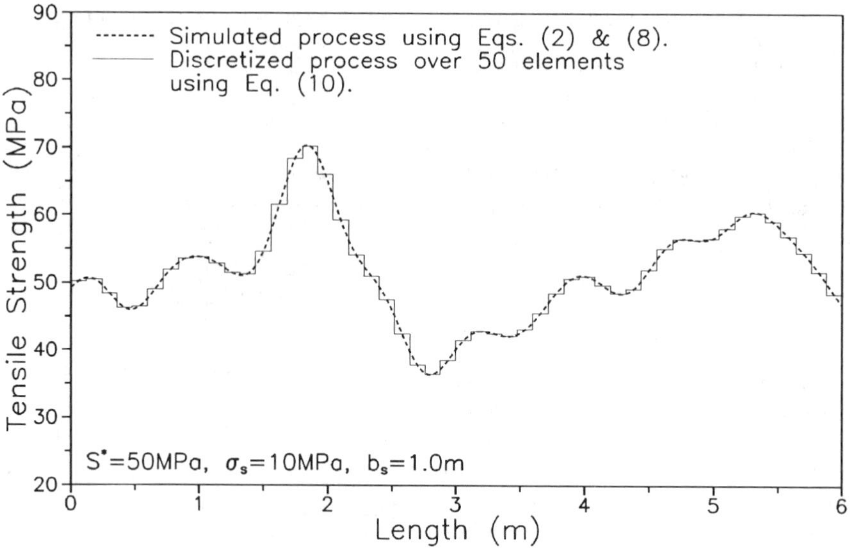

Figure 2: Realization of tensile strength process.

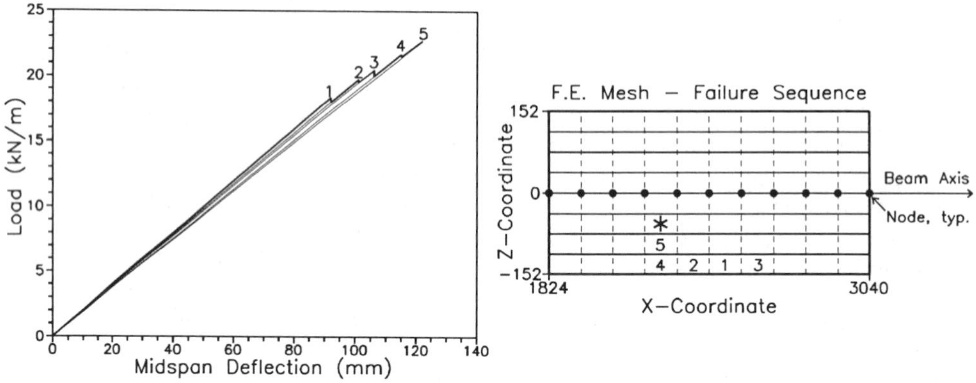

Figure 3: Progressive failure analysis to determine P_u.

which is based on a higher order shear deformation beam theory (Heyliger and Reddy 1988) has been adopted for this study. The resulting element has two nodes, each with four degrees–of–freedom.

For each lamina within an element the assignment of an elastic modulus and tensile strength value is based on a simple collocation scheme where by the stochastic fields are sampled at the midpoint along the element. As an example, for the j–th lamina in the i–th element, material properties are assigned as follows

$$E(i,j) = E_j^* + e_j(x_i) \qquad (9)$$
$$S(i,j) = S_j^* + s_j(x_i) \qquad (10)$$

where x_i locates the midpoint of the i–th element and $e_j(x_i)$ and $s_j(x_i)$ are determined according to Eqs. (7) and (8), respectively. Figure 2 provides a comparison between the tensile strength process generated using Eqs. (2) and (8) and a piecewise constant approximation to it using Eq. (10).

Material strength is assumed to be governed by a maximum tensile stress failure criterion under which brittle fracture occurs where ever the tensile strength is exceeded. Under a given load level, say P_o, the finite element model determines the tensile stresses induced in the beam. Let $\sigma_o(i,j)$ denote the resulting average tensile stress in the j–th lamina of the i–th element. Assuming that

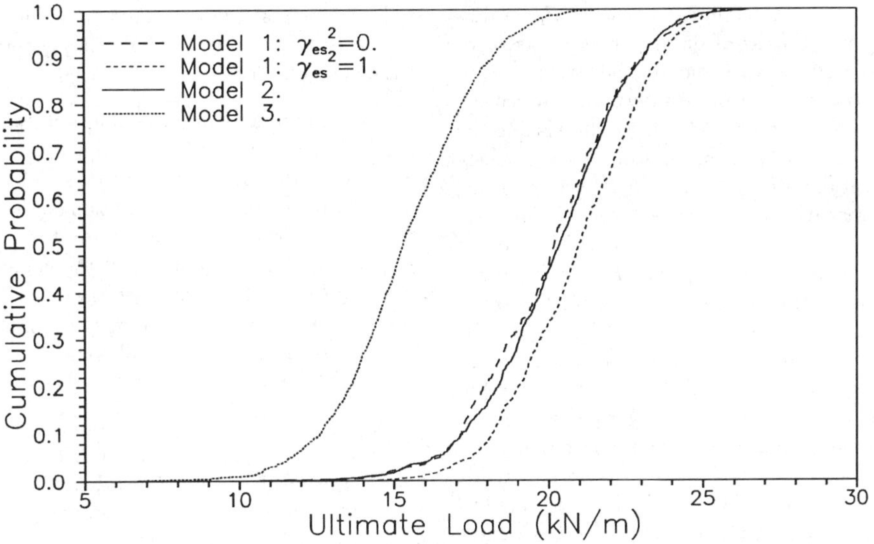

Figure 4: Distribution of load–carrying capacity ($b_s = 1.0$ m, $\delta_S = 0.20$).

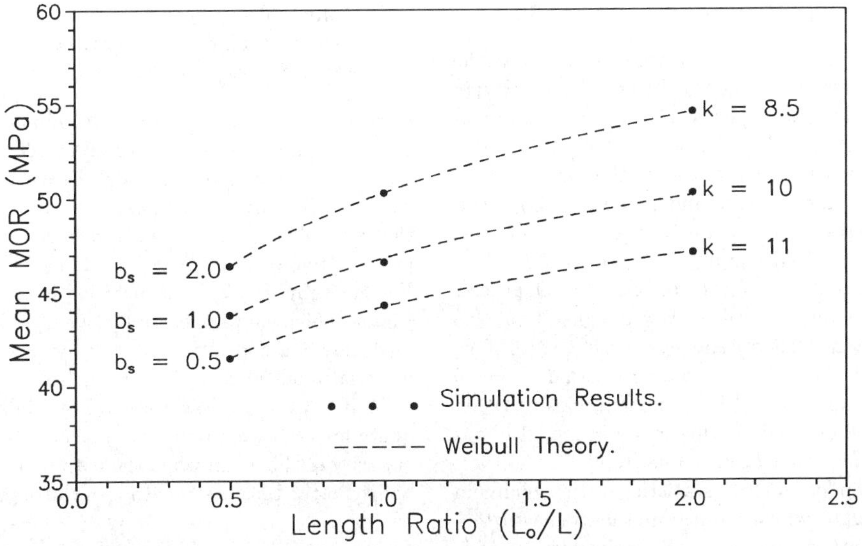

Figure 5: Mean MOR versus normalized beam span ($\delta_S = 0.20$).

the beam response is linear to failure, the load to produce failure at this location is given by

$$P(i,j) = P_o \cdot \frac{S(i,j)}{\sigma_o(i,j)} \tag{11}$$

with $S(i,j)$ determined according to Eq. (10). In turn, Eq. (11) can be evaluated over the entire finite element mesh. The load to produce *first failure* within a lamina is then given by

$$P_1 = \min\{P(1,1),\ldots,P(1,n_l),\ldots,P(k,l), \ldots,P(n_e,1),\ldots,P(n_e,n_l)\} \tag{12}$$

where n_l and n_e denote, respectively, the number of laminae and finite elements in the beam. Assume, for example, $P_1 = P(k,l)$. In general, the load–carrying capacity of the beam will not be exhausted at the first failure load level P_1. Additional localized areas of damage may be accomodated through redistribution of the stresses. Subsequent failure loads can be determined by repeating the above procedure after first eliminating the bending stiffness contribution from the previously failed location; *i.e.* setting $E(k,l) = 0.0$. Thus, through repeated linear analysis steps the *progressive failure* of the beam can be modeled and the ultimate load–carrying capacity P_u can be determined as illustrated in Fig. 3. Note, the load–displacement envelope shown in Fig. 3 can be likened to the results from a displacement controlled test. One is reminded that this simple linear finite element model is only valid as long as the material response in compression remains linear elastic and the deformations before impending failure remain small. When axial loads are applied or when the compression zone undergoes ductile yielding a full non-linear finite element model is required (Folz 1993).

4 RESPONSE VARIABILITY ANALYSIS

Monte Carlo simulation was used to evaluate the response variability in the beam's load–carrying capacity as a consequence of the chosen material model. To simplify the analysis each laminate in the beam was selected from the same population (*i.e.* same species and grade). Using available experimental data for Spruce lamination material $E^* = 12$ GPa and $\delta_E = \sigma_e/E^* = 0.10$ (Folz 1993). Lacking data on the tensile strength process a plausible range of values were assigned to b_s (0.5 m, 1.0 m and 2.0 m) and $\delta_S = \sigma_s/S^*$ (0.10, 0.20 and 0.30). Further it was assumed that $S^* = 50$ MPa. Material Model 1 was simulated using either $\gamma^2_{es}(\kappa) = 0$ or 1. With this material model it was also assumed that b_e was equal to b_s.

The non–parametric statistics of the minimum strength S_{min} which result from the stochastic tensile strength model, using the various values of b_s and δ_S, are presented in Table 1. Note that obtaining these statistics is a principal objective of a conventional test program. Available data on Spruce lamination material gives $\bar{S}_{min} = 26$ MPa and $\delta_{S_{min}} = 0.24$ for a test length of 3.0 m (Folz 1993). This suggests that the assumed parameter values for the tensile strength process used in this study are not unrealistic.

The load–carrying capacity statistics (mean and coefficient of variation) for the beam, resulting from the chosen material model and the assigned material parameters, are presented in Table 2. From this sensitivity study the following general observations can be made:

1. The difference in the response statistics using Model 1 with $\gamma^2_{es}(\kappa) = 0$ and with $\gamma^2_{es}(\kappa) = 1$ is most pronounced when $\delta_S = \delta_E = 0.10$; however, under this condition δ_{P_u} is relatively small, in all cases being less than 9%. For $\delta_S \geq 0.2$, the influence of the coherence between the two processes is not as large; for example, by assuming $\gamma^2_{es}(\kappa) = 0$ the maximum error in the $\bar{P}_u$ and δ_{P_u} will be less than 1 kN/m and 2%, respectively.

2. The response statistics obtained using Model 1, with $\gamma^2_{es}(\kappa) = 0$, and Model 2 are in close agreement one with the other; the maximum difference in $\bar{P}_u$ and δ_{P_u} was only 0.3 kN/m and 1.4%, respectively.

3. The response statistics obtained with Model 3 are significiantly different than those obtained with the other material models, especially when $\delta_S \geq 0.2$; the maximum difference in $\bar{P}_u$ and δ_{P_u} was greater than 8 kN/m and 20%, respectively.

Observations 1 and 2, taken in combination with the fact that within lamina cross–correlation data between the elastic modulus and the tensile strength would be difficult to obtain experimentally, suggest that a viable approximation is to consider only the tensile strength as a stochastic field in the problem. Further support for this conclusion is provided by Fig. 4, which shows the cumulative distribution of load–carrying capacity predicted by each of the 3 material models, using $b_s = 1.0$ m and $\delta_S = 0.20$.

Model 3 can be dismissed as woefully inadequate for accurate prediction of the load–carrying capacity of the beam when the within lamina variability in the tensile strength is of the order of 20% or more.

Also presented in Table 2 are the finite element discretization and simulation sizes which were required to produce stable response statistics. The dependence of the finite element mesh on the value of b_e and b_s is duly noted.

An investigation was also undertaken to determine what, if any, statistical size (length) effect would result from the stochastic tensile strength model. To this end, additional simulations were performed with the beam span halved and doubled from the reference value $L_o = 6080$ mm. To facilitate comparison, the response quantity of interest is now the modulus of rupture (MOR), as determined by the elastic flexure formula from elementary beam theory. The simulation results confirm a length effect as shown in Fig. 5. Also presented in Fig. 5 is a fit of these results to Weibull's brittle fracture theory (Bolotin 1969):

$$\bar{R} = \bar{R}_o \left(\frac{L_o}{L}\right)^{\frac{1}{k}} \tag{13}$$

where $\bar{R}$ is the mean MOR at length L, $\bar{R}_o$ is the mean reference MOR at length L_o and k is the fitted size effect parameter.

Regarding the computational performance of this stochastic finite element approach it is noted that a simulation of 500 beams (with $n_e = 50$) was performed in less than 1500 seconds on a 33-MHz 486 microcomputer, which is not prohibitively time consuming.

5 CONCLUSIONS

A stochastic finite element analysis of the load–carrying of a glued–laminated beam, under progressive tensile fracture, has been presented. Lacking sufficient experimental data, numerous assumptions were made with respect to the stochastic material modeling of the elastic modulus and tensile strength. A response variability study, based on Monte Carlo simulation, showed that, for the particular beam under consideration, sufficient accuracy was achieved in predicting the response statistics when only the tensile strength was modeled as a stochastic field. This tensile strength model also manifested a statistical size effect.

ACKNOWLEDGEMENT

This work was supported by Forestry Canada, the Natural Science and Engineering Research Council of Canada and the Canadian Wood Council under a cooperative research partnership.

REFERENCES

Bolotin V.V. 1969. *Statistical Methods in Structural Mechanics.* Holden Day, San Francisco.

Folz, B. 1993. Stochastic finite element analysis of the load–carrying capacity of laminated wood beam–columns. Ph.D. Thesis, The University of British Columbia, Canada.

Heyliger, P.R. and J.N. Reddy 1988. A higher order beam finite element for bending and vibration problems. *Journal of Sound and Vibration* 126(2): 309–326.

Shinozuka, M. 1987. Stochastic fields and their digital simulation. In *Stochastic methods in structural dynamics.* Eds. G.I. Schueller and M. Shinozuka: 93–133. Martinus Nijhoff, Dordrecht.

Wang, Y.T. and R.O. Foschi 1992. Random field stiffness properties and reliability of laminated wood beams. *Structural Safety* 11: 191–202.

Structural Safety & Reliability, Schuëller, Shinozuka & Yao (eds) © *1994 Balkema, Rotterdam, ISBN 90 5410 357 4*

Reliability analysis of wood including viscoelastic effects

K.J. Fridley
The University of Oklahoma, Okla., USA

D.V. Rosowsky
Purdue University, West Lafayette, Ind., USA

T. Philpot
Murray State University, Ky., USA

ABSTRACT: Current deflection serviceability reliability analyses for wood members are based on an elastic deflection analysis and an essentially time-independent approach. In this paper, the effects of the viscoelastic creep mechanism on the deflection characteristics of wood members and the associated reliabilities are evaluated for stochastic load combinations involving occupancy live load and snow load. The development of the mathematical model for the creep response of the beam is reported, and the extension to parallel-member wood systems is examined. Implications for probability-based design for wood is examined in the context of the new LRFD specification for wood construction in the United States.

1 INTRODUCTION

Current design specifications for wood construction such as the *National Design Specification of Wood Construction* (NDS) (*National,* 1991) provide only limited provisions to account for creep deflections. Typically, a creep factor is provided which, when multiplied by the immediate elastic deflection, is assumed to predict the long-term deflection of the member. In the NDS, two creep factors are offered, one for glued laminated timber or seasoned lumber and one for unseasoned lumber. However, no consideration is apparently given to the type or duration of the loads. These same factors are being considered for adoption into the load and resistance factor design (LRFD) specification for wood construction which is currently being written in the United States. To offer a more rational and appropriate design procedure, the effects of the viscoelastic creep mechanism on the deflection characteristics of wood members and the associated reliabilities are evaluated for stochastic load combinations involving occupancy live load and snow load, and implications for probability-based design for wood is examined in the context of the new LRFD specification for wood construction in the United States.

2 VISCOELASTIC RESPONSE OF WOOD

Wood is a viscoelastic material and as such possesses the following characteristics: (1) elastic deformation, (2) primary (decelerating) creep, (3) secondary (steady-state) creep, (4) zero stress creep recovery, and (5) stress relaxation. Various models are available for predicting the stress-strain-time relationship for wood, including empirical, semi-empirical and phenomenological models. The four-element Burger model, a phenomenological model which consists of a Maxwell element added serially to a Kelvin element, has been used in the past to model the viscoelastic response of wood. The Burger model predicts the five characteristics listed above with the following single expression:

$$\sigma(t) = K_e\varepsilon_e(t) + \mu_v\frac{d\varepsilon_v(t)}{dt} + \mu_k\frac{d\varepsilon_k(t)}{dt} + K_k\varepsilon_k(t) \quad (1)$$

where $\sigma(t)$ = the stress in the material, K_e = the Maxwell element spring parameter, μ_v = the Maxwell element viscous damper parameter, K_k = the Kelvin element elastic spring parameter, and μ_k = the Kelvin element viscous parameter. The strains $\varepsilon_e(t)$, $\varepsilon_v(t)$, and $\varepsilon_k(t)$ = those portions of the total strain owing to the Maxwell element elastic

Table 1. Creep Model Parameter Distributions for Structural Lumber (from [5]).

Parameter	15% MC		19% MC		28% MC	
	Mean	COV	Mean	COV	Mean	COV
K_e (10^6 psi)	1.68	0.16	1.56	0.16	1.31	0.16
K_k (10^6 psi)	2.77	0.41	2.17	0.41	1.63	0.41
μ_k (10^{10} psi-min)	4.56	0.42	3.36	0.42	1.79	0.42
μ_v (10^{14} psi-min)	5.56	0.39	4.44	0.39	3.24	0.39

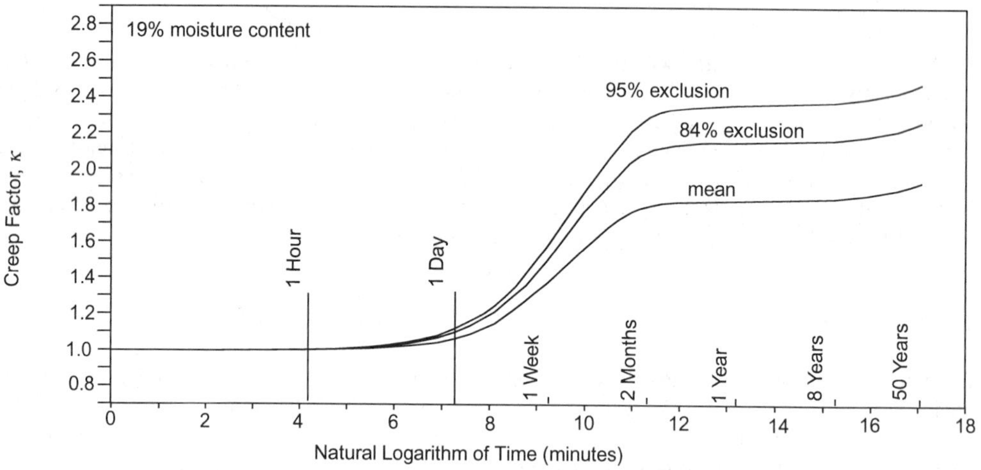

Fig. 1. Relationship Between κ and Duration of Constant Load.

spring (elastic strain), the Maxwell element viscous damper (visco-plastic strain), and the Kelvin element (visco-elastic strain), respectively.

Solving (1) for the case of constant applied stress (i.e., $\sigma(t) = \sigma$) yields

$$\varepsilon(t) = \frac{\sigma}{K_e} + \frac{\sigma t}{\mu_v} + \frac{\sigma}{K_k}\left[1 - \exp\left(-\frac{K_k t}{\mu_k}\right)\right] \tag{2}$$

where $\varepsilon(t)$ = the total, time-dependent strain at time t beyond the application of the stress σ.

The apparent time-dependent stiffness of wood, often referred to as the relaxation modulus, may be expressed in terms of its elastic stiffness, K_e, or

$$K(t) = \frac{K_e}{\kappa(t)} \tag{3}$$

where $K(t)$ = relaxation modulus and $\kappa(t)$ = a time-dependent creep factor. Assuming the Burger model with a constant applied stress, $\kappa(t)$ is derived from (2) as

$$\kappa(t) = 1 + \frac{K_e t}{\mu_v} + \frac{K_e}{K_k}\left[1 - \exp\left(-\frac{K_k t}{\mu_k}\right)\right] \tag{4}$$

Fridley et al. (1992) estimated the four Burger model parameters from a large population of select structural grade Douglas-fir lumber. The four model parameters were found to follow lognormal distributions with mean and COV values as defined in Table 1. The general relationship between the creep factor and the duration of a constant applied load for structural lumber (Eq. 1) is illustrated in Fig. 1.

Table 2. Dead and Occupancy Live Load Process Statistics.

	Intensity			Arrival	
Load Component	Mean	COV	CDF	Mean rate/year	Duration
Dead	$1.05D_n$	0.10	LN	n/a	50 years
Sustained Live					
200 sq. ft.	$0.24L_n$	0.90	Gamma	0.125	8 years
800 sq. ft.	$0.30L_n$	0.60	Gamma	0.125	8 years
Extraordinary Live					
200 sq. ft.	$0.16L_n$	0.84	Gamma	1	1 week
800 sq. ft.	$0.19L_n$	0.66	Gamma	1	1 week
Snow	$0.20S_n$	0.87	LN	13	2 weeks

3 STRUCTURAL LOAD MODELS

Considerable work has been done in the area of probabilistic modeling of structural loads including dead load, occupancy live load, and roof snow load (e.g., Chalk and Corotis, 1980; Ellingwood and Culver, 1977; Ellingwood et al., 1980). Complete load histories are required for reliability analyses of structures constructed from materials with time-dependent resistances or which are subject to cumulative damage. A renewal pulse process model which includes information on load duration and arrival rate in addition to magnitude can then be used to represent structural loads for the specified reference period. Such pulse process models can be combined to represent various load combinations. Consideration is given in this paper to the dead plus live and dead plus snow load combinations taken to represent typical floor and roof loadings.

3.1 *Dead Load Modeling*

Although the self weight of a structure is well defined, the dead load is usually assumed to be underestimated by a small amount due to uncertainty in other dead load components such as permanent equipment, partitions, floor coverings, etc. (Ellingwood et al., 1980). Dead load statistics are summarized in Table 2.

3.2 *Occupancy Live Load Modeling*

Occupancy live load is typically modeled as having two components: a sustained component and an extraordinary component (Chalk and Corotis, 1980). The sustained component includes items normally associated with the intended use of the structure. For example, furnishings and occupants in typical office and residential space are included as sustained live load. The extraordinary live component accounts for atypical use of a structuresuch as crowding of people during special events or the temporary use of the space for storage during renovation (Ellingwood and Culver, 1977). Statistics for both components of occupancy live load are summarized in Table 2.

3.3 *Snow Load Modeling*

A Bernoulli pulse model has been used for roof snow loading based on collected climatological data (Ellingwood and Redfield, 1983). Given the distribution parameters for the annual maximum roof snow load and the temporal parameters for the pulse model (i.e., arrival rate, probability of "non-zero" load magnitude), the individual snow pulse intensities can be simulated (Rosowsky and Ellingwood, 1990). In recent reliability studies, a composite set of snow statistics corresponding to a general site in the northeast U.S. has been adopted. These statistics are adopted herein and are presented in Table 2 (statistics shown are for the annual extreme roof snow load).

4 LIMIT STATE FUNCTION INCLUDING CREEP

A limit state function can be derived to include creep effects from a specific design equation. Considering a deflection serviceability check where the actual deflection of a member, δ_{actual}, must be less than some allowable value, $\delta_{allowable}$, the limit state function including creep effects will take the form

$$g(x) = \delta_{allowable} - \delta_{actual} \tag{5}$$

The actual deflection is typically assumed as the elastic deflection. For a viscoelastic material, however, the actual deflection must include creep deflection. Using (3), the actual deflection of a viscoelastic member can be written as

$$\delta_{actual} = \kappa(t) \cdot \delta_e \tag{6}$$

where δ_e = the elastic deflection. For a load pulse process, both $\kappa(t)$ and δ_e vary with each pulse. In a reliability analysis, the maximum product of $\kappa(t)$ and δ_e which occurs during serviceability reference period, or

$$\delta_{actual} = \max\{\kappa(t) \cdot \delta_e\} \tag{7}$$

must be determined and compared with the allowable deflection. The elastic deflection is a function of the actual modulus of elasticity, actual loading, and the creep factor, each of which is assumed to be a random variable, and the design span and beam moment of inertia, both of which are assumed to be deterministic.

By setting the nominal beam deflection equal to the deflection limit, the allowable deflection can be written as a function of the nominal loading and the nominal elastic modulus. By substituting this information into (5) and reducing, the following limit state equation is obtained for the deflection limit state including creep effects:

$$g(x) = \frac{E}{\phi_E E_n}\left(\frac{\gamma_D}{L_n / D_n} + \gamma_L\right) - \max\left\{\kappa_i \cdot \left(\frac{D+L}{L_n}\right)\right\} \tag{8}$$

in which ϕ_E = the resistance factor, E and E_n = the random and nominal elastic moduli, γ_D and γ_L = dead and live load factors, D and L = the random dead and live loads, D_n and L_n = the nominal dead and live loads, and κ_i = the creep factor for each load i in the reference period.

Typically, serviceability analyses utilize unfactored loads, i.e., $\gamma_D = \gamma_L = 1$; however, some design specifications and building codes suggest other load combinations when considering creep. For example, the *National Design Specification of Wood Construction* (1991) implies a $1.5D_n + L_n$ load combination be used when considering creep effects in wood structures comprised of seasoned lumber by recommending deflections owing to "permanent" loads be increased by a factor of 50%.

5 RELIABILITY ANALYSIS

Reliability analyses including creep effects must consider the in-time behavior of the structure throughout the specified reference period. Therefore, first-order second-moment techniques are not applicable and Monte Carlo simulation procedures must be utilized. For serviceability, reduced references periods, such as one year and eight years, are often considered in the reliability analysis. Obviously, the longer the reference period, the lower the associated reliability index. A target reliability index also must be established in order to propose appropriate creep factors for use in design. A value of $\beta = 2$ has been identified in the past as an appropriate index for serviceability (e.g., Galambos and Ellingwood, 1986). This value is generally lower than the target reliability for a strength analysis owing to the reduced consequence of failure.

Using the general viscoelastic material response presented in Table 1 and Fig. 1 for structural lumber, the load statistics presented in Table 2, and load combinations of $D_n + L_n$ and $D_n + S_n$, the relationship between the reliability index, β, and the resistance factor, ϕ_E, was determined and is presented in Fig. 2 for one and eight year reference periods. For the $D_n + L_n$ load combination, a resistance factor, ϕ_E, of 0.45 is required to attain the target reliability index for a one year reference period and 0.35 for an eight year reference period. Likewise, for D_n + S_n the load combination, a resistance factor of 0.60 is required for a one year reference period and 0.40 for an eight year reference period.

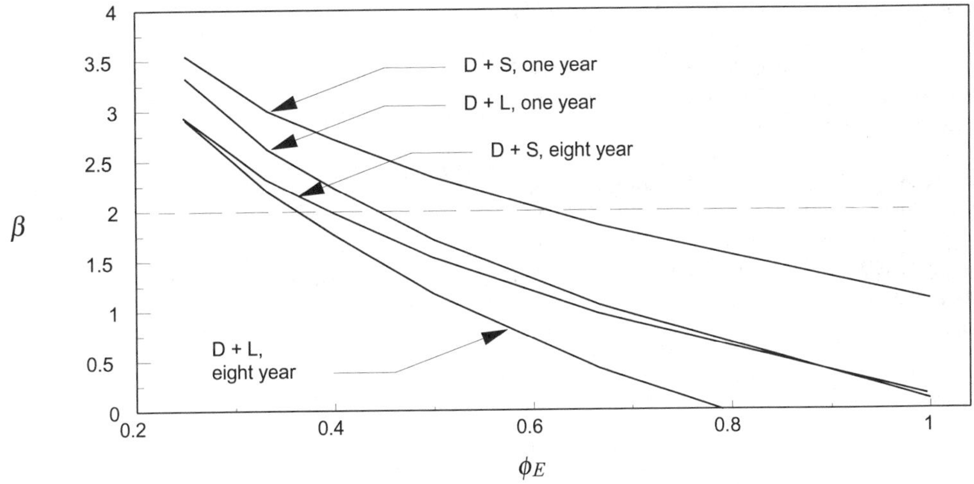

Fig. 2. $\beta - \phi_E$ Relationship for D + L and D + S Load Combinations and One and Eight Year Reference Periods.

6 PARALLEL-MEMBER SYSTEMS

The analysis presented above can be extended to multi-member (parallel) systems in order to investigate the effect of differential creep and the interaction of system effects with creep behavior. Of particular interest is the role that load sharing plays in reducing the potentially negative effects of creep. This can have implications for strength design as well as serviceability (deflection), as load sharing in redundant systems is assumed to be a function of relative member stiffnesses. Therefore, system reliability analyses which include creep effects are required in order to evaluate system factors for use in LRFD. The extension of this analysis to parallel-member systems is currently being performed and will be reported on by the authors in the near future.

7 CONCLUSIONS

The serviceability reliability of wood including creep effects has been discussed. The viscoelastic response of wood has been defined, and information on the load duration and arrival rate as well as load magnitude has been used in the analysis. The deflection limit state function was written as a function of a resistance factor, the random and nominal elastic moduli, the random dead and live loads, the nominal dead and live loads, and a creep factor for each load pulse in the reference period. Resistance factors for use in LRFD were generated for the dead plus live and dead plus snow load combinations and one and eight year reference periods. The analysis presented can be extended to multi-member (parallel) systems in order to investigate the effect of differential creep and the interaction of system effects with creep behavior in order to evaluate systems factors for use in LRFD.

REFERENCES

Chalk, P., and Corotis, R.B. (1980). "A Probability Model for Design Live Loads," *J. of the Struct. Div.*, ASCE, 106(ST10):2017-2033.

Ellingwood, B., and Culver, C.G. (1977). "Analysis of Live Loads in Office Buildings," *J. of the Struct. Div.*, ASCE, 103(ST8):1551-1560.

Ellingwood, B., Galambos, T.V., MacGragor, J.B., and Cornell, C.A. (1980). "Development of a Probability Based Load Criterion for American National Standard A58," Special Pub. No. 577, National Bureau of Standards, USA.

Ellingwood, B., and Redfield, R.K. (1983). "Ground Snow Loads for Structural Design." *J. of Struct. Engrg.*, ASCE, 109(4):950-964.

Fridley, K.J. (1992). Designing for Creep in Wood Structures. *Forest Products J.*, 42(3):23-28.

Fridley, K.J., R.C. Tang, and L.A. Soltis. 1992. Creep Behavior Model for Structural Lumber. *J. of Struct. Engrg.*, ASCE, 118(8):2261-2277.

Galambos, T.V., and Ellingwood, B. (1986).

Serviceability Limit States: Deflection. *J. of Struct. Engrg.*, ASCE, 112(1):67-84.

National Design Specification for Wood Construction. (1991). National Forest Products Association, Washington, D.C., USA.

Rosowsky, D.V., and Ellingwood, B.R. (1990). Stochastic Damage Accumulation and Probabilistic Codified Design for Wood. Civil Engrg. Report No. 1990-02-02, Dept. of Civil Engrg., The Johns Hopkins University, Baltimore, MD, USA.

Structural Safety & Reliability, Schuëller, Shinozuka & Yao (eds) © 1994 Balkema, Rotterdam, ISBN 90 5410 357 4

Duration of load and reliability of wood structures

David V. Rosowsky
Purdue University, West Lafayette, Ind., USA

Bruce R. Ellingwood
The Johns Hopkins University, Baltimore, Md., USA

Timothy A. Philpot
Murray State University, Ky., USA

ABSTRACT: The wood industries in the United States and Canada recently have completed the development of limit states design specifications for engineered wood construction using structural reliability concepts. Unlike steel and concrete, where time-dependent material behavior can be ignored, wood is a natural material with highly variable mechanical properties and strength that depends on both the rate of application and duration of loads. This dependence of wood strength on loads implies that complete stochastic load histories are required to evaluate reliability over a specified period of time. During the past decade, new duration of load models have been developed and have been calibrated to the results of tests on dimension lumber specimens loaded in flexure, compression and tension. These models represent damage by a state variable, and can be used in a probabilistic cumulative damage analysis similar to that often used in studies of low-cycle or static fatigue of metals. However, the state variable approach cannot be related to the physical condition of the structure or its residual strength at a given time. Accordingly, it cannot be used in performing a condition assessment of an existing structure. A failure criterion based on a cumulative energy of deformation offers an alternative approach. This paper addresses these two approaches to reliability analysis of wood structures and structural systems.

1 DURATION OF LOAD MODELING AND RELIABILITY ANALYSIS

Failure of wood structural members occurs by a creep-rupture phenomenon in which voids are propagated in the microstructure of the wood at a stress level that is lower than the short-term static strength when the member is subject to a sustained load (Barrett and Foschi, 1978; Gerhards, 1979). Existing models that describe the time-dependent behavior of wood members represent damage accumulation leading to creep-rupture by means of a state variable, $\alpha(t)$. The rate of damage accumulation is defined as a function of the intensity of the applied stress and the current damage state, $d\alpha/dt = f(\sigma,\alpha)$, in which $\sigma(t)$ is the applied stress process and $\alpha(t)$ is the current damage state at time t. The state variable is a stochastic process because of the inherent variability in material properties, randomness in loads, and uncertainties in the damage accumulation model itself. In the initial undamaged state, $\alpha(t) = 0$; $\alpha(t)$ increases monotonically thereafter due to applied load. The creep-rupture limit state probability is the probability that $\alpha(t)$ increases to above 1.0 during a specified service period, T:

$$P_f = P[\alpha(t) > 1.0] \qquad 0 < t < T \tag{1}$$

The reliability analysis thus is in the form of a first-passage problem. Closed-form approximate solutions are available for Eqn 1 in certain idealized cases (Ditlevsen, 1990). Monte Carlo simulation also may be used to evaluate limit state probabilities. The associated reliability index, β, is computed as $\beta = \Phi^{-1}(1 - P_f)$ in which $\Phi^{-1}(\cdot)$ is the inverse standard normal cumulative distribution function.

Several models for the rate of damage accumulation in wood have been developed at Forintek Canada (e.g., Barrett and Foschi, 1978; Foschi et al., 1989) and at the U.S. Forest Products Laboratory, FPL

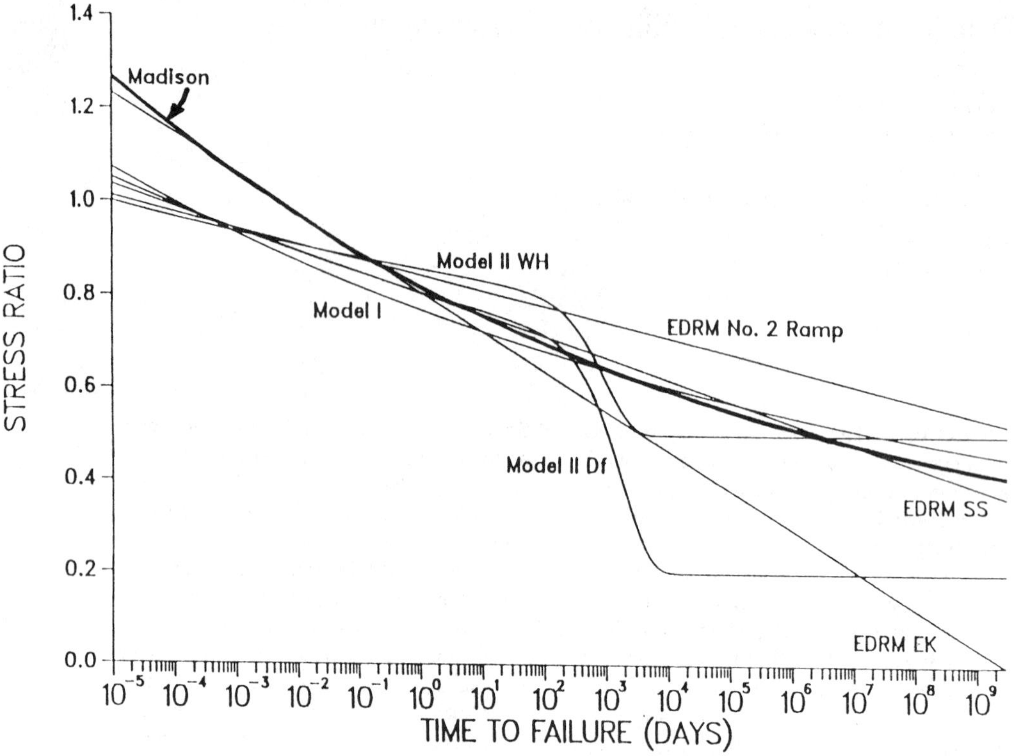

Figure 1: Comparison of DOL models

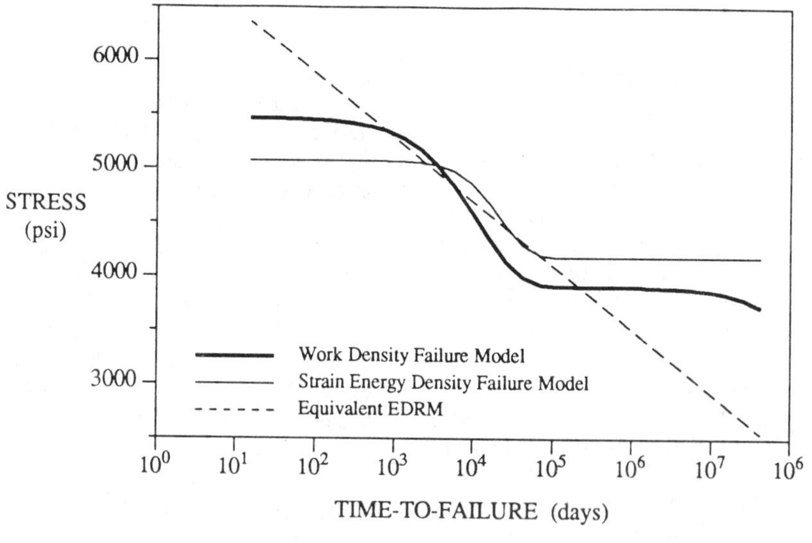

Figure 2: Typical relationship between stress and time-to-failure for energy-based failure criteria

(Gerhards and Link, 1986). These models are summarized below:

$$\frac{d\alpha}{dt} = A(\sigma - \sigma_o)^B + C\alpha \qquad (2)$$

Barrett and Foschi Model II (1978)

$$\frac{d\alpha}{dt} = \exp(-A + B\sigma)$$

Exponential Damage Rate Model, FPL (1986) (3)

$$\frac{d\alpha}{dt} = A[\tau(t) - \sigma_o\tau_s]^B + C[\tau(t) - \sigma_o\tau_s]^n\alpha \qquad (4)$$

Foschi et al. (1989)

in which σ = stress ratio (applied stress divided by short-term strength), σ_O = stress threshold, τ = applied stress, τ_S = short-term strength, and A,B,C and n are model parameters. The state variable, as defined above, is an abstract quantity. In some engineering applications, $\alpha(t)$ can be related to a physical quantity, e.g., the ratio of the crack depth to component thickness, fraction of section lost to corrosion, etc. Such relations have not been attempted for wood, however. Instead, the customary approach has been to: (1) postulate a functional form for the rate of damage accumulation $d\alpha/dt$ based on a physical (i.e., Caulfield, 1985) or heuristic argument; (2) integrate this equation with the assumption that the applied stress history is known (usually, stress that increases linearly to test level and then is held constant to failure); and (3) determine the constants in the equation for the damage model from available test data obtained under similar circumstances (Gerhards and Link, 1986; Foschi and Barrett, 1982).

These models are compared in Figure 1, where they have been fitted to the results of duration of load tests of dimension lumber in flexure. The two models developed at Forintek include a stress threshold below which no damage accumulates, comparable to a fatigue endurance limit. The models are similar at stress levels above about 70% of the modulus of rupture (MOR) determined from a ramp load test of approximately 5 minutes duration, but diverge at lower stress levels.

To the writers' knowledge, only limited variable amplitude tests have been performed to validate these damage rate models (e.g., Gerhards, 1988). Moreover, only recently have limited tests been conducted in which specimens were held at a constant stress level for an extended period of time, were unloaded, and subsequently tested monotonically to failure (Fridley et al., 1991). Since the relationship between $\alpha(t)$ and residual strength is unknown, one cannot utilize Eqns 2 - 4 to evaluate explicitly the residual capacity of a structural member or system following the occurrence of some extreme event or to perform an in-service condition assessment of an existing structure.

1.1 Load models

Structural loads must be modeled as random processes because of the time-dependent nature of the creep-rupture failure mechanism in wood. Fortunately, simple pulse process load models are sufficiently accurate to model the variation in time of static gravity loads such as dead, live and snow loads. Fluctuating forces due to wind and earthquake do not appear to lead to creep-rupture failure in the sense described above. Details of the stochastic load models used in structural reliability studies for wood construction may be found in (Ellingwood and Rosowsky, 1991).

1.2 Single Member Reliability Analysis and Time Effects Factors

Target reliabilities for development of resistance and duration of load (DOL) factors for an LRFD code are based on current working stress design (NFPA, 1986) reliability levels. The details of the calibration procedure may be found in (Ellingwood and Rosowsky, 1991). Current WSD procedures for combinations involving live load and snow load were found to lead to inconsistent measures of reliability. A single target reliability for use in LRFD of β = 2.3 for a 50-year service period was subsequently adopted.

The proposed LRFD format for engineered wood construction in the U.S. contains the following set of equations:

$$\lambda\phi R = \sum_i \gamma_i Q_i \qquad (5)$$

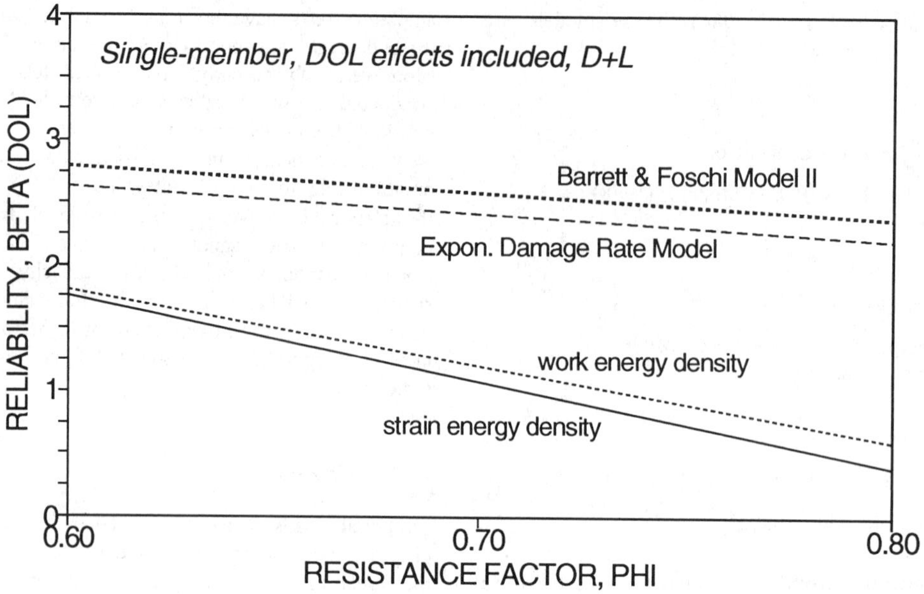

Figure 3: Comparison of DOL-included BETA-PHI curves (dead + live load)

Table 1: Time Effect Factors for LRFD

Limit State	ϕ	**Load Combination**	λ US-FPL	λ Forintek
Flexure	0.85	D	0.59	0.50
Flexure	0.85	D + L	0.76	0.84
Flexure	0.85	D + S	0.75	0.83
Compression	0.90	D + S	0.75	-
Tension	0.80	D + S	0.75	-
Flexure	0.85	D + L(roof)	0.78	-
Flexure	0.85	D + L(storage)	0.67	-

Table 2: System Factors for LRFD

Lumber:	ψ US-FPL	ψ Forintek
DFL, 2 x 10, No. 2	1.25	1.30
SP, 2 x 10, No. 2-KD	1.35	1.40
HF, 2 x 8, No. 2	1.46	1.52

in which R and Q are the nominal (code-specified) resistance and load, λ is a DOL factor, ϕ is a resistance factor, and γ_i are load factors. The right-hand side of Eqn 5 corresponds to the load combination requirements given in the ASCE Standard 7-88 (ASCE, 1990). The resistance factor, ϕ, accounts for the inherent variability in short-term strength and the mode and consequence of failure. The time effect factor, λ, adjusts the design strength so that essentially the same reliability is obtained for combinations involving loads with different temporal characteristics. For example, the nominal strength in flexure is $R = F_n S_x$ in which F_n is the nominal strength (typically, the 5th-percentile of the short-term strength adjusted for end-use conditions), and S_x is the section modulus in bending. The factors λ and ϕ are determined to ensure the target reliability is achieved. These values can be obtained for different load combinations by performing the reliability analysis twice, once with load duration effects included and once with load duration effects ignored, and determining the value of λ needed to maintain the target reliability.

A summary of some of the results are shown in Table 1 for two damage models: one developed at the U.S. Forest Products Laboratory (Gerhards and Link, 1986) and the other developed at Forintek Canada (Barrett and Foschi, 1978). The values for λ were selected to achieve a target reliability of β = 2.3. While values of λ depend on the temporal characteristics of the load process, they are essentially independent of limit state. Values of ϕ, however, are dependent on the limit state.

1.3 System Analysis and Load Sharing

Light frame construction using dimension lumber exhibits significant load sharing capabilities. The *National Design Specification* (NDS) allows a 15% increase in allowable stress for design calculations involving an assembly of repetitive framing members such as joists or rafters (NFPA, 1986). This increase attempts to distinguish between the behavior of the system and that of a single member. Such system behavior arises from partial composite action between the sheathing and the joists and two-way action or load sharing. Since failure in wood occurs by progressive accumulation of damage, the system reliability analysis must account for duration of load effects.

A system behavior model consists of (1) a mechanism for load sharing and load redistribution following member failure, and (2) a system limit state definition. A review of previous studies of wood floor systems, including behavioral model development and experimental validation (Rosowsky and Ellingwood, 1991), led to the model used in this study in which the sheathing was assumed to be rigid. This assumption is supported by full-scale tests of floor systems (Criswell, 1981), which suggested that for models in which both composite T-beam action and two-way action are included, the effect is to produce approximately equal midspan deflections of the inner (non-edge) joists. Assuming one-way flexural action and that no rigid body rotation occurs due to non-symmetrical joist resistances, the joists in the system with the rigid sheathing have the same midspan deflection, and the load is distributed to the individual joists in proportion to their flexural rigidities. Following member failure, the load once carried by the failed member is distributed to all of the remaining members in proportion to their flexural rigidities.

The following lumber types, representative of those most commonly used in the U.S., were considered: (1) Douglas-fir Larch (DFL), 2 x 10, No.2, (2) Southern Pine (SP), 2 x 10, No.2-KD, and (3) Hem-fir (HF), 2 x 8, No.2. The mechanical properties were taken from the results of the In-Grade Test Program (Green and Evans, 1987). Consistent with the IGTP values, a correlation of 70% was assumed between the MOR and MOE. In order to take system effects into account within the context of a single-member safety check, Eqn 5 is modified to include a system factor, ψ, on the left-hand side. Simulation is used to develop a relationship between system reliability and percent increase in nominal (short-term) strength. The results are used to obtain values of ψ (percent increase) such that the system reliability and single-member reliability are comparable. A comparison of LRFD system factors for both the FPL and Forintek damage models (with ϕ = 0.85 and λ = 0.8 in Eqn 5) is shown in Table 2. From these results, a system factor of about 1.25 appears acceptable for all three common species. The current value of 1.15, which corresponds to the current NDS repetitive-use factor, appears somewhat conservative for all three lumber types. While Table 2 suggests that the system factors are relatively insensitive to choice of damage accumulation model, the load sharing (system) effects and load duration effects appear to be coupled. System factors which neglect these time effects, therefore, may be unduly high.

2 AN ENERGY-BASED FAILURE CRITERIA

Fridley et al. (1992) have proposed a failure criteria which postulates the existence of a critical energy density which can be used as a predictor for the failure strength of wood. Energy density (i.e., the energy per unit volume at any point in the member) rather than total energy (i.e., the sum of all energy throughout the complete member) was selected to allow the criteria to be applied to members of various sizes and spans and subject to different load combinations. This critical energy density is based on tests of dimension lumber including natural and processing defects rather than on a theoretical value based on the wood strength at a microscopic level. The critical energy density, μ_{cr}, was calibrated to the point of impending collapse as marked by the proportional limit or the end of secondary creep (Fridley et al., 1992). It is interesting to point out that μ_{cr} was found to be essentially invariant with respect to load history, environment history, and material quality factors such as strength, stiffness, and grade.

In order to evaluate the energy-based failure model for use in a reliability analysis, a study was performed to compare the limit state probabilities of flexural members subject to combinations of dead, live, and snow load. Consideration was given to the energy density failure model[1] and two commonly used damage accumulation models (the EDRM-SS and the Barrett and Foschi Model II). To avoid uncertainties associated with grade, species and size, only Select Structural grade Douglas-fir nominal 2 × 4 in. dimension lumber was considered. A "typical use" moisture condition (surface dry, 19% m.c.) was further assumed. Relevant statistics for the MOR were taken from the results of the In-Grade Test Program (Green and Evans, 1987).

Assuming a constant critical energy density, typical examples of the relationship between failure stress and time to failure are shown on Figure 2. The energy models are shown along with the EDRM for comparison. While Caulfield's model based on chemical kinetics (1985) suggests a model with an exponential form, the energy models considered herein suggest a model more closely resembling the Barrett and Foschi model (cf. Figures 1 and 2). The results of the reliability analysis for a service life of 50 years of combined dead plus occupancy live load are shown in Figure 3, which shows the member reliability as a function of the resistance factor. Similar results were found for combinations involving snow load. As can be seen, the reliability indices predicted by the energy-based models differ markedly from those obtained using the damage accumulation models. This large discrepancy may be attributed to the different failure definitions for the two model types (i.e., start of tertiary creep vs. ultimate collapse).

[1] Actually, two distinct critical energy density models were developed and compared with the existing damage models. The details of these two models, and their differences, may be found in (Philpot, 1992).

3 SUMMARY

There is a lack of experimental data, particularly on the response of wood flexural members subjected to repeated and arbitrary load histories, necessary to make meaningful evaluations of failure models. Although energy-based models, such as those discussed herein, cannot be recommended at the present time as an alternative to the current damage accumulation models for use in reliability analyses, they appear to offer the potential for an improved understanding of the duration of load phenomenon by offering a more physical descriptor of damage. Further exploration of these models, particularly with regard to additional experimental verification and the improved characterization of creep recovery effects, is warranted.

REFERENCES

American Society of Civil Engineers 1990. Minimum design loads for buildings and other structures. ASCE. New York, NY.

Barrett, J.D. and Foschi, R.O. 1978. Duration of load and failure probability in wood. Part I. modeling creep rupture. *Canadian Journal of Civil Engineering*. 5(4):505-514.

Caulfield, D.F. 1985. A chemical kinetics approach to the duration-of-load problem in wood. *Wood and Fiber Science*. 17(4):504-521.

Criswell, M.E. 1981. New floor design procedures. *Proceedings:* Wall and floor systems: design and performance of light-frame systems. Forest Products Research Society, Madison, WI.

Ditlevsen, O. 1990. Asymptotic first-passage time distributions in compound poisson processes. *Structural Safety*. 8(1):327-336.

Ellingwood, B. and Rosowsky, D. 1991. Duration

of load effects in LRFD for wood construction. ASCE *Journal of Structural Engineering*. 117(2):584-599.

Foschi, R.O. and Barrett, J.D. 1982. Load-duration effects in western hemlock lumber. ASCE *Journal of the Structural Division*. 108(7):1494-1510.

Foschi, R.O., Folz, B.R. and Yao, F.Z. 1989. Reliability-based design of wood structures. SRS Report 34, Dept. of Civil Engineering, Univ. of British Columbia, Vancouver, 282 pp.

Fridley, K.J., Tang, R.C. and Soltis, L.A. 1991. Environmental effects on the load-duration behavior of structural lumber. *Proceedings:* 1991 International timber engineering conference, TRADA, London, U.K., pp. 4.180-4.187.

Fridley, K.J., Tang, R.C. and Soltis, L.A. 1992. Load duration effects in structural lumber: strain energy approach. ASCE *Journal of Structural Engineering*. 118(9):2351-2369.

Gerhards, C.C. 1988. A note on load duration of Douglas-fir 2 by 4's under repeated loads. *Wood and Fiber Science*. 20(3):365-369.

Gerhards, C.C. 1979. Time-related effects of loading on wood strength: a linear cumulative damage theory. *Wood Science*. 11(3):139-144.

Gerhards, C.C. and Link, C.L. 1986. Effect of loading rate on bending strength of Douglas-fir 2 by 4's. *Forest Products Journal*. 36(2):63-66.

Green, D.W. and Evans, J.W. 1987. Mechanical properties of visually graded lumber. Vols. 1-5, USDA, Forest Products Laboratory, Madison, WI.

National Forest Products Association 1986. *National Design Specification for Wood Construction*. NFPA, Washington, D.C.

Philpot, T.A. 1992. The effect of creep on the performance and reliability of wood structures. *Ph.D. dissertation*. School of Civil Engineering, Purdue University, West Lafayette, IN.

Rosowsky, D. and Ellingwood, B. 1991. System reliability and load sharing effects in light frame wood construction. ASCE *Journal of Structural Engineering*. 117(4):1096-1114.

Materials

Structural Safety & Reliability, Schuëller, Shinozuka & Yao (eds) © 1994 Balkema, Rotterdam, ISBN 90 5410 357 4

Fracture of random quasibrittle materials: Markov process and Weibull-type models

Zdeněk P. Bažant & Yunping Xi
Northwestern University, Evanston, Ill., USA

ABSTRACT: Quasibrittle materials, for example concretes, rocks, certain composites, toughened ceramics and ice, are materials that fail by fracture with a large fracture process zone. They have a heterogeneous random microstructure, which causes significant scatter in material strength. The classical statistical strength models of Weibull-type, as well as some recent random process models for crack growth, are not applicable because they ignore the stress redistributions and energy release caused by large stable crack growth before failure, and the variability of critical energy release rate (the R-curve). This paper presents, in the first part, a new random process model for crack growth which takes the R-curve into account. The second part focuses on another related problem — the Weibull-type model. It reviews a recent nonlocal generalization of Weibull theory which can take into account the existence of a large fracture process zone and stable crack growth prior to maximum load.

INTRODUCTION

Probabilistic fracture modeling of quasibrittle materials such as concrete is a problem of formidable complexity which must be simplified to be tractable. The simplifications required should obviously be patterned after the probablistic theories for the fracture of metals, although with certain salient special features which reflect the fact that the size of fracture process zone in quasibrittle materials is normally comparable to the cross section dimensions and that quasibrittle structures exhibit a significant stable fracture growth before the maximum load is reached. In contrast to metals, the fractures at maximum load of concrete structures typically occupy 50% to 90% of the cross section.

The present brief conference paper will expound two simplification, one describing the probabilistic nature of fracture from the static viewpoint and another from the evolutionary viewpoint. The former simplification represents an adaptation of Weibull theory in which the failure probability is estimated on the basis of the stress state of the structure before failure. This approach makes it possible to deduce a simple law for the effect of the structure size. The latter simplification treats fracture propagation as a random process, for which the Markov process is a natural choice. This approach can yield the probabilities of fracture growth at various stages. First we will consider the evolutionary aspect of random process and then will briefy outline the essential results in Weibull-type modeling.

1 MARKOV CHAIN MODEL FOR RANDOM CRACK GROWTH

The previous investigations of fracture in concrete structures concentrated mainly on the deterministic and statistical behaviors at the peak load (Bažant and Kazemi, 1990; Gettu et al., 1990; Bažant and Xi, 1991a, 1991b). A realistic theory is needed to consider the probabilistic nature of the steps in the crack growth process, especially, the question how the survival probability of one elementary volume is influenced by the preceding failure of an adjacent elementary volume. This requires following the incremental jumps of the fracture process in a probabilistic manner. To determine the probabilities of crack growth in each loading step, one must consider the probabilities of the lengths of the jumps of the crack tip for a given load increment when the major crack extends to a certain point. Furthermore, if a structure with a crack of a certain length is surviving at a given load level, one must decide what is the failure probability for a given load increment.

To answer these questions, a stochastic model for crack growth is required. We will now present a probabilistic model that can capture the randomness of progressive crack growth in a quasibrittle material such as concrete. The present method

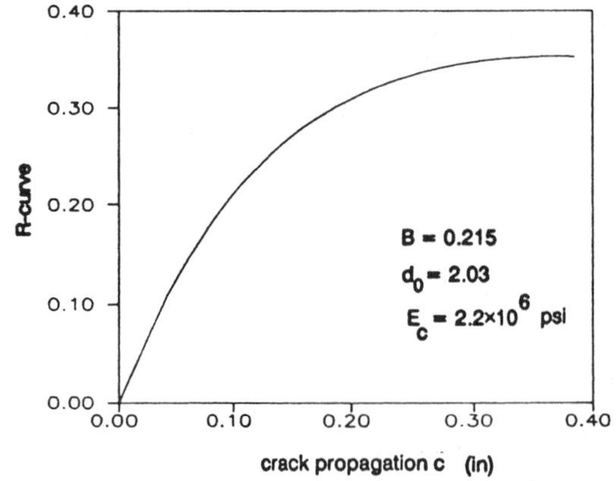

Fig. 1 R-curve

will be based on the Markov chain model and R-curve behavior which is derived from a new generalization of the size effect law combining Weibull statistical theory and nonlocal concept (Bažant and Xi, 1991a, b). The standard deviation of the peak load will be the only statistical information needed for the model. The parameter estimation method will be formulated, and some applications will be illustrated.

1.1 General Formulation

The Markov process (or Markov chain) is a general model commonly used to characterize and simulate many kinds of accumulative damage processes (Bogdanoff and Kozin, 1985). The well-known basic evolution equation for the Markov process is

$$\boldsymbol{p}_x = \boldsymbol{p}_0 \boldsymbol{P}^X \tag{1}$$

where X is the loading level, $\boldsymbol{p}_0$ is the initial state probability (a vector), $\boldsymbol{p}_0 = (\pi_1, \pi_2, \ldots \pi_{B-1}, 0)$, with $\sum \pi_j = 1$, in which π_j = Prob(damage state j is initially occupied) and $\boldsymbol{P}$ is the probability transition matrix which characterizes the material properties. Eq. 1 means that the probability of crack advance depends only on the current state, i.e. is independent of the preceding states (the history). In the present study, we assume that always $\pi_1 = 1$, with all other $\pi_j = 0$, which means the crack or damage always starts from state 1; $\boldsymbol{p}_x$ is the damage state probability, $\boldsymbol{p}_x = \boldsymbol{p}_x(1), \boldsymbol{p}_x(2), \ldots \boldsymbol{p}_x(B)$, in which $\boldsymbol{p}_x(j)$ = Prob (damage state j is occupied at stress level X); and B denotes the failure state. Here we assume the crack can propagate only one unit at one loading level, which means a unit jump model is called for; then

$$\boldsymbol{P} = \begin{bmatrix} p_1 & q_1 & 0 & \ldots & 0 \\ 0 & p_2 & q_2 & \ldots & 0 \\ 0 & 0 & p_3 & \ldots & 0 \\ \ldots & \ldots & \ldots & \ldots & \\ 0 & 0 & 0 & p_{B-1} & q_{B-1} \\ 0 & 0 & 0 & 0 & 1 \end{bmatrix} \tag{2}$$

where p_1 = probability of remaining in the state i during one loading step, and q_1 = probability that in one loading step the damage moves from state i to state $i+1$.

Let the random variable $X_{1,B}$ denote the load at failure reached by starting in damage state 1 at $X = 0$. Then the first two central moments of $X_{1,B}$ are found to be (Bogdanoff and Kozin, 1985):

$$E(X_{1,B}) = \sum_{j=1}^{B-1} (1 + r_j), \tag{3a}$$

$$\text{Var}\,(X_{1,B}) = \sum_{j=1}^{B-1} r_j (1 + r_j) \tag{3b}$$

where

$$r_j = \frac{p_j}{q_j}, \quad p_j = \frac{r_j}{1 + r_j}, \quad q_j = \frac{1}{1 + r_j} \tag{4}$$

We will focus on the statistical scatter of the failure load, which actually represents the macroscopic reflection of the microscopic randomness within the crack process zone. In other words, the failure load will be considered as a fictitious source of randomness, while the real source of randomness of course is the heterogeneity of material properties.

In the case of real engineering problems, the failure load and its standard deviation are the data most likely to be available, especially the failure loads obtained from small specimens. So, a suitable equation has to be introduced as a mean curve. One must realize that this mean equation is not just the equation obtained from fracture mechanics handbooks; it has to be calibrated from the peak load test results and averaged over all the specimens of different sizes. For quasi-brittle materials which show a very strong dependence of the nominal strength on the specimen size, the size effect law proposed by Bažant and Xi (1991a,b) might be one of the available method satisfying such a requirement.

1.2 Mean R-Curve

The deterministic equation for the nominal stress may generally be written in the form:

$$\bar{X} = \frac{\sqrt{R(a - a_0)\ E_c}}{\sqrt{\pi a}\ F(a/d)} \tag{5}$$

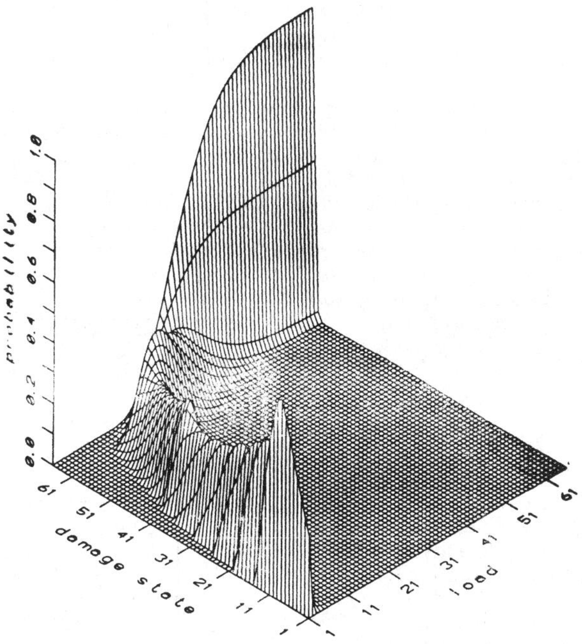

Fig. 2 Probability - loading level - damage state

where $\bar{X}$ represents the mean nominal stress (which is proportional to the applied load), E_c is the initial elastic modulus, $R(a - a_0)$ is the R–curve, $F(a/d)$ is a geometry-dependent function, available in fracture handbooks (e.g. Tada, 1983), a_0 is the current crack length, and a_0 is the initial crack (or notch) length.

The R–curve can be deduced from the statistical generalization of the size effect law proposed by Bažant and Xi (1991a,b). By fitting this law to the test results on maximum load values for geometrically similar fracture specimens of sufficiently different sizes, one can determine the fracture energy G_f of the material (defined as the energy required to propagate a crack in an infinitely large specimen), and the effective length of the process zone, c_f (defined for an infinitely large specimen). Then, according to the method of Bažant and Kazemi (1990), one can obtain the R-curve as follows:

$$R(a - a_0) = G_f \frac{g'(\alpha)c}{g'(\alpha_0)c_f} \frac{1}{1 - \frac{2n}{m}} \tag{6a}$$

$$\frac{c}{c_f} = \frac{g'(\alpha_0)}{g(\alpha_0)} \left(\frac{g(\alpha)}{g'(\alpha)} - \alpha + \alpha_0 \right) \tag{6b}$$

$$G_f = \frac{(bf_u)^2}{c_n^2 E} d_0 g(\alpha_0) \tag{7a}$$

$$c_f = \frac{d_0 g(\alpha_0)}{g'(\alpha_0)} \left(\frac{d}{d_0} \right)^{2n/m} \tag{7b}$$

in which $c = a - a_0, \alpha_0 = a_0/d, \alpha = a/d, g(\alpha)$ is the nondimensionalized energy release rate obtained from handbooks, and c_n is a factor in the size effect law chosen arbitrarily for convenience (see Bažant and Kazemi, 1990), b and d_0 are two constants (see Eq. 11). When m approaches infinity, Eqs. 6-7 degenerate to the same form as Bažant and Kazemi (1990). By choosing a series of α values, the corresponding c-values are obtained from Eq. 6b, and then, substituting each c into Eq. 6a, the corresponding R–curve values can be calculated. Then Eq. 5 will represent the mean curve of nominal stress as a function of the crack length.

1.3 Variance of Nominal Stress before Failure

As a simplifying approximation, we may assume the variance σ_j^2, at state j, to be proportional to the length of the fracture process zone. This assumption seems reasonable in view of the experience from testing. The length of the process zone at the initial state is 0, and the standard deviation of the load is also 0. At the failure state, the process zone is fully developed and the standard deviation of the failure load must, therefore, also reach its maximum value at the same time. Since the fracture process zone size depends upon the crack length a, the variance at state j may be expressed approximately as a linear function of the crack length a;

$$\sigma_j^2 = \frac{(a_j - a_0)}{(a_{\max} - a_0)} \sigma_{\max}^2 \quad (8)$$

where $a_{\max}$ can be obtained from Eq. 5. $\sigma_{\max}^2$, representing the variance of peak load, may be considered to be size independent, since, as we already explained, the random scatter is mainly related the size of the fracture process zone during the loading process, and at ultimate state the fracture process zone size is almost independent of the structure size. This has been shown by test results of laser speckle interferometry (Ansari, 1989) and by size effect analysis (Bažant and Kazemi, 1991).

1.4 Markov Chain Model Combined with R–Curve

Based on Eqs. 5 and 8, we can derive the expression for the parameters in Eq. 2 and damage state B_j. First, we divide the damage states $j = 1, \ldots B-1$ into j groups as follows: $1, \ldots B_1 - 1; B_1, \ldots B_2 - 1; \ldots$; and $B_j, \ldots B-1$. Then we assume r_1 for $1, \ldots B_1 - 1; r_2$ for $B_1, \ldots B_2 - 1; \ldots$, and r_j for $B_j, \ldots B-1$. The following recursive equations are obtained

$$B_j = \frac{(\bar{X}_j - \bar{X}_{j-1})^2}{(\bar{X}_j - \bar{X}_{j-1}) + (\sigma_j^2 - \sigma_{j-1}^2)} + B_{j-1}, \quad (9a)$$

$$r_j = \frac{\bar{X}_j - \bar{X}_{j-1}}{B_j - B_{j-1}} - 1 \quad (9b)$$

1.5 Numerical Example

Consider a notched three-point-bend beam specimen of high strength concrete (Gettu et al. 1990) as an example. The R–curve obtained from the peak loads by the size effect method is shown in Fig. 1. Fig. 2 shows the probability of each damage state and nominal stress. One can see that, for example, at loading level 61 (almost the peak load, which is 66) the probability for the occurrence of the damage state 61 (almost the failure state, which is 66) is very high, more than 90%. On the other hand, the probabilities for the occurrence of the lower damage states, 1 - 50, at the same loading level are almost 0; this is true in reality, because at such a high loading level the probabilities for a very low damage state should be very small. The ridge of the probability surface represents the mean path of damage evolution.

An advantage of the present model is that the sample curve can be easily simulated by the computer. Fig. 3 shows the relation between crack extension and loading level. One can clearly see that the generated sample curves represent the observed test curves quite well. This means that the present model can characterize the probabilistic structure for the entire loading history from the initial state up to the failure load.

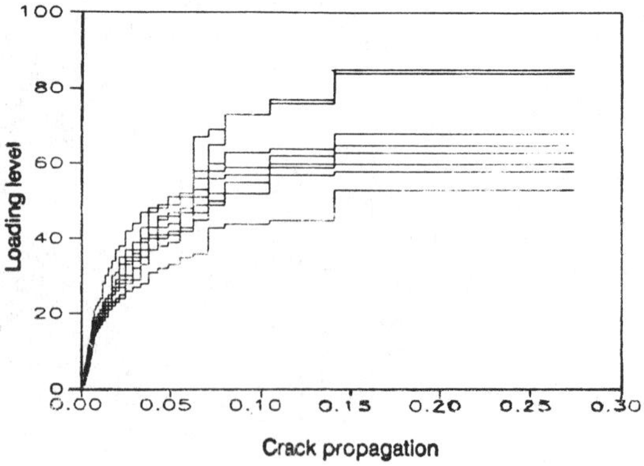

Fig. 3 Sample of load - crack propagation

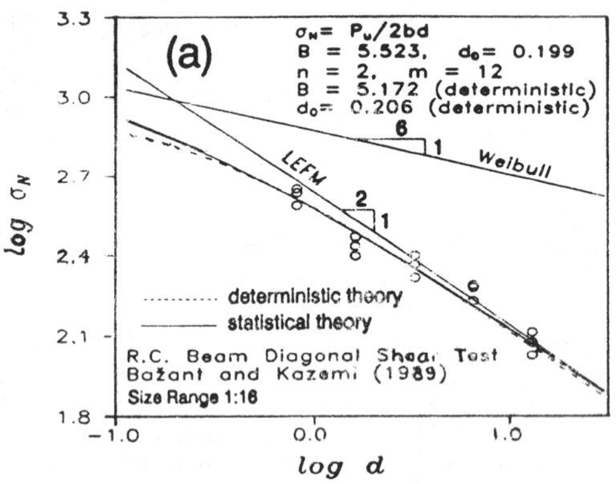

Fig. 4 Test result on diagonal shear failure of reinforced concrete beam without stirrups, and optimum fit by (11)

1.6 Conclusions on Markov Chain Model

1. The present discrete Markov chain model can be used to determine the probabilistic structure of progressive cracking under monotonic loading conditions. The crack propagation probability for an existing crack at any loading step before the peak load and the failure probability from any damage state can be calculated.

2. The determination of model parameters requires a large number of sample curves. When only the peak load data are available and the number of samples for the peak load does not suffice to obtain the statistical parameters required in the model, the mean curve based on fracture mechanics may be used as a substitute.

3. The standard deviation of the mean curve in the entire process of crack propagation may

be predicted from the standard deviation of the peak loads of specimens. The standard deviation of the peak load is affected by the crack process zone. In turn, the crack process zone is affected by the length of crack propagation. Hence, the peak-load deviation can be assumed as a function of the crack propagation length.

4. The R-curve obtained by size effect analysis of a series of experimental results on peak loads for different sizes is employed as the mean curve. The peak load deviation is assumed to be a linear function of the length of crack propagation. The obtained probabilistic structure for the progressive damage process agrees with the general observation of experimental results. (Note: In detail, the Markov process model will be presented in a journal article; Bažant and Xi, 1993).

2 NONLOCAL GENERALIZATION OF WEIBULL STATISTICAL THEORY OF RANDOM STRENGTH

The second part of the conference presentation briefly reviews a recent development of nonlocal Weibull-type theory for concrete and other quasibrittle materials (Bažant and Xi, 1991a,b) and discusses some new consequences of this model. The classical Weibull theory applies only to those case where the maximum load (failure load) is attained at the initiation of the macroscopic crack propagation. Applications have been made to concrete structures, in which large crack growth with large stress redistributions that occur prior to the maximum load are ignored. But such application fs are not very realistic. They do not yield the correct size effect.

One might think of remedying the problem by substituting the stress distribution according to linear elastic fracture mechanics into the Weibull-type probability integral, but the integral then diverges. The root of the problem is that the probability of material failure is, in the classical Weibull approach, assumed to depend on the local stress at any given point of the material. This is not realistic, as clarified by the recent deterministic nonlocal continuum models for strain softening damage. In similarity to those models, it is proposed that the material failure probability be considered as a function of the average deformation of a certain neighborhood of a given point. Thus, generalizing the basic results of the classical Weibull theory, the basic hypothesis is that

$$\ln(1-P_f) = \int_v \sum_{i=1}^{n} \left(\frac{E\bar{\epsilon}_i(x)}{\sigma_0}\right)^m \frac{dV(x)}{V_r} \qquad (10)$$

where P_f is failure probability, x is coordinate vector, E is elastic modulus, σ_0 and V_r are constants, m is Weibull modulus, n is the number of dimension, and $\bar{\epsilon}_i(x)$ are the averaged principal strains.

The most important consequence of the nonlocality is a change in the size effect on the nominal strength, σ_N, of geometrically similar structures of different sizes. It can be shown that (Bažant and Xi, 1991b):

$$\sigma_N = \frac{bf_u}{\left[\left(\frac{d}{d_0}\right)^{2n/m} + \left(\frac{d}{d_0}\right)\right]^{1/2}} \qquad (11)$$

where f_u is the tensile strength, d is the characteristic dimension, and b and d_0 are two constants which can be identified by linear regression of the test results. This law has as the deterministic limit ($m \to \infty$) the size effect law proposed by Bažant and Kazemi (1990).

Eq. 11 has been considered in dealing with the test results by Bažant and Kazemi (1991) on the diagonal shear strength of longitudinally reinforced concrete beans without stirrups, which had a particularly broad size range, 1:16. These results are shown in Fig. 4, in comparison to the optimum fits by Eq. 11, as well as by the deterministic size effect law. Both fits are good, but the difference between them is minor. It has been concluded that, in this type of failure, the deterministic size effect due to energy release dominates, and the statistical contribution to the size effect is unimportant. This may be explained by the fact that, due to localization of cracking, major contribution to the probability integral in Eq. 10 comes only from the fracture process zone, the size of which is almost the same even for specimens of different sizes. However, in other quasibrittle structures, in which such localization of damage prior to the maximum load does not take place, the nonlocal generalization of Weibull statistical theory could be more important.

ACKNOWLEDGMENT — Partial financial supports from AFOSR under grant 91-0140 to Northwestern University, and from NSF Science and Technology Center for Advanced Cement-Based Materials at Northwestern University are gratefully acknowledged.

References

Ansari, F. 1989. Mechanism of microcrack formation in concrete. *ACI Materials Journal*: Sept.–Oct., 459-464.

Bažant, Z.P., and Kazemi, M.T. 1991. Size effect on diagonal shear failure of beams without stirrups. *ACI Structural Journal*: 88(3), 268-276.

Bažant, Z.P., and Kazemi, M.T. 1990. Size effect in fracture of ceramics and its use to determine fracture energy and effective process zone length. *Journal Am. Ceramic Society*: 73(7), 1841-1953.

Bažant, Z.P., Xi, Y., and Reid, S.G. 1991a. Statistical si: effect in quasibrittle structures Part I. Is Weibull theory applicable? *Journal of Engineering Mechanics*, ASCE: 117(11), 2609-2622.

Bažant, Z.P., and Xi, Y. 1991b. Statistical size effect in quasibrittle structures Part II. Nonlocal theory. *Journal of Engineering Mechanics*, ASCE: 117(11), 2623-2640.

Bažant, Z.P., and Xi, Y. 1993. Random growth of a crack with R-curve: Markov process model. *Journal of Engineering Mechanics*, in press.

Bogdanoff, J.L. and Kozin, F. 1985. *Probabilistic models of cumulative damage*. New York: John Wiley & Sons.

Gettu, R., Bažant, Z.P., and Karr, M.E. 1990. Fracture properties and brittleness of high-strength concrete. *ACI Material Journal*: Nov.–Dec., 87, 608-618.

Tada, H. 1983. *The stress analysis of cracks handbook*. St. Louis, MO: Del Research Corp.

Structural Safety & Reliability, Schuëller, Shinozuka & Yao (eds) © 1994 Balkema, Rotterdam, ISBN 90 5410 357 4

Stochastic analysis of viscoelastic composite delaminations*

Harry H. Hilton & Sung Yi
University of Illinois at Urbana-Champaign, Ill., USA

ABSTRACT: The previously formulated deterministic viscoelastic quadratic time dependent delamination onset criterion is generalized to complete three dimensional stochastic environments. The analysis includes stochastic processes due to combined random loads and random delamination failure stresses as well as random anisotropic viscoelastic material properties including the influence of stochastic temperature fields, moisture contents and boundary conditions. It is shown that times for delamination onset occurrences in composites can be predicted probabilistically depending on any one or all of the above conditions. Illustrative examples are presented showing the relationship in terms of parametric variations between times to delamination and corresponding probabilities that such events will occur. Since uniaxial tension, compression and shear viscoelastic delamination failure stresses decrease in time, the loading history is of significant importance. For cases where deterministic criteria predict no delamination failures, the present stochastic failure theory indicates high probabilities of failure at either early or long times depending on the load-time relations. The early time high probabilities of delamination onset predict short life times and occur in conditions where the composite internal stresses relax at faster rates than the failure stresses are degrading. The effects of fiber orientation and of number of plies on delamination probabilities are also examined.

INTRODUCTION

Delamination of composite plies due to interlaminar stresses at or near free edges is not necessarily to be considered as an ultimate structural failure, but rather an initiation failure process which may ultimately lead to entire structural disintegration. Delamination may under go either stable or unstable growth, with the latter leading to eventual complete failures. However its initiation at a flight vehicle outer surface may severely damage its aerodynamics without necessarily totally destroying its flight capabilities. In this paper, the word failure is used in a generic sense interchangeably with the occurrence of delamination onset.

Stochastic failure analyses must be treated on two distinct levels: (1) the analysis of stresses and displacements due to random loads, creep and relaxation functions (moduli and compliances), temperatures, moisture contents and boundary conditions; and (2) the formulation of appropriate failure criteria including proper random failure property descriptions and prescriptions. Composite materials, whether polymeric or rubber like, inherently exhibit time dependent viscoelastic responses to loads, temperatures and moisture contents and equally viscoelastic failure properties which degrade with time, temperatures and moistures at comparable rates, but wholly independently, both in value and mechanisms, of the degradations associated with their viscoelastic moduli.

In Reference 1 an extensive literature survey was presented regarding the effects of random loads, temperatures, humidity and

---

[1] The complete version of this paper has been published in the *Journal of Composite Materials* (Ref. 22).

[2] Professor Emeritus of Aeronautical and Astronautical Engineering and of Supercomputing Applications.

[3] Post Doctoral Research Associate NCSA.

Table I. Values for Weibull distributions.

Case	Fig. No. (Ref.12)	Mean $\tilde{v}$	σ/v	$\tilde{v}'_{max}$	α	$\tilde{\beta}$	β
1	7	8,460	.0760	10,879	15.78	8,730	.8024
2	7	4,740	.118	7,117	9.67	4,970	.6983
3	6	5,380	.230	11,319	5.26	5,850	.5168
4	8 (M=0%)	15,630	.0980	23,083	10.05	16,340	.7079
5	9 (M=1.6%)	14,820	.154	24,497	7.86	15,750	.6429

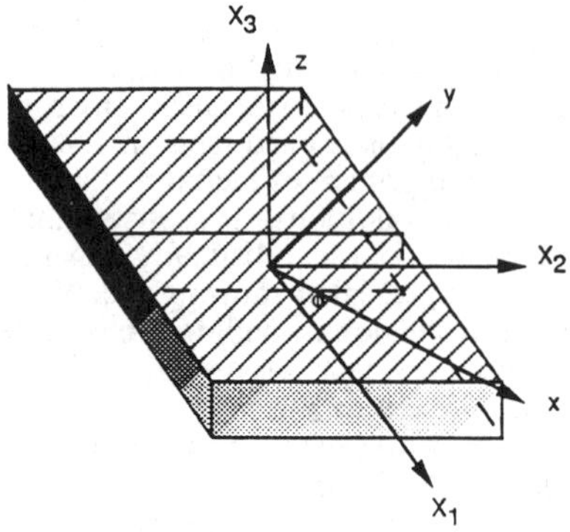

Fig.1 Composite plate geometry

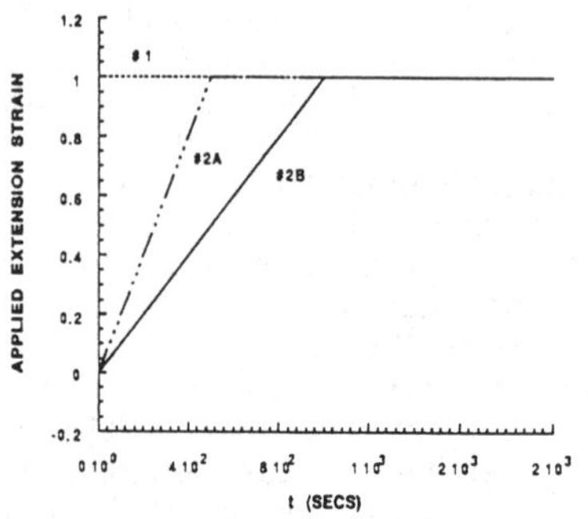

Fig. 2 Applied strains in x_1-direction

material properties on viscoelastic stress analysis. Additionally, that paper also extends the traditional deterministic elastic-viscoelastic analogies [2,3] to analyses with stochastic material properties.

Analytical or numerical solutions of viscoelastic problems are always obtainable by analogy if corresponding elastic solutions exist [1-3,10]. In more complicated problems, finite element methods (FEM) must be used. A large number of deterministic FEM are available and were compared by the present authors [7-9] who also have developed more computationally efficient anisotropic viscoelastic FEM for quasi-static and dynamic deterministic viscoelasticity and were used to determine stresses in the composite.

The quadratic delamination onset criterion was first introduced by Brewer and Lagace [4] and has recently been extended to cover time dependent creep strengths and stresses by Yi [5]. Hilton [6] has formulated general stochastic failure criteria under combined loads and strengths. In the present paper, the Yi delamination criterion has been generalized to cover 3D stress states due to random loads, temperatures, humidity, moduli and time dependent failure stresses using the stochastic combined load failure criteria of Ref. 6. This formulation results in analytical relations between probabilities of failure and times at which delaminations may occur [22].

RESULTS AND DISCUSSION

There is a veritable pauperage of statistical experimental delamination failure data. Hiel *et al* [12] present experimental probability distributions for interlaminar tensile strengths of curved beams and in one case examine the effects of 0% and 1.6% moisture contents. Weibull distributions are used to curve fit the data, although other distribution functions, such as Gaussian, beta, etc. could equally readily be used. Other unidirectional composite time dependent failure data has been reported by Lifshitz and Rotem [13], Phoenix and Tierney [14], Phoenix [15], Watson and Smith [16], Farquhar *et al* [17] and Larder and Beadle [18]. In the absence of other experimental data, one can consider that

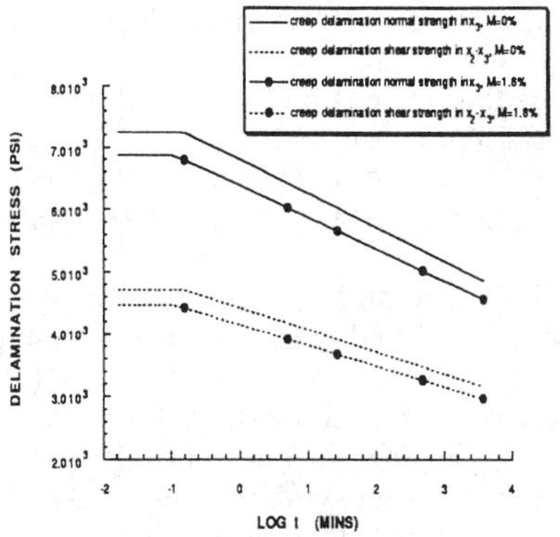

Fig. 3 Uniaxial mean delamination stresses (Ref.11)

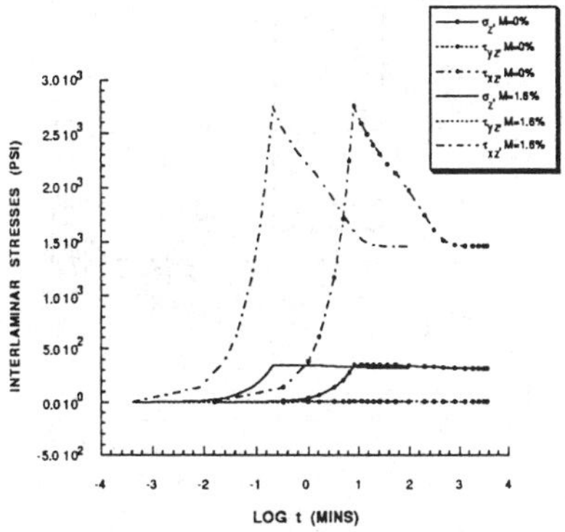

Fig. 5 Stresses for condition D15PL2A

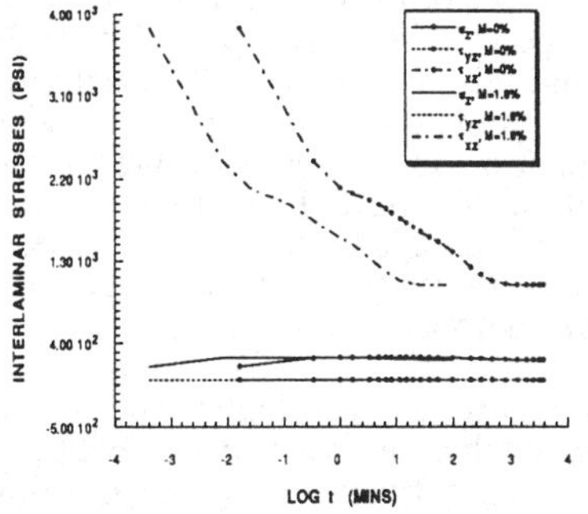

Fig. 4 Stresses for condition D15P5L1

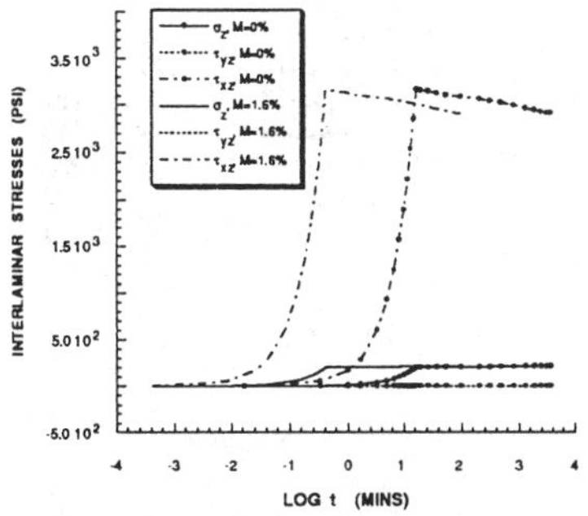

Fig. 6 Stresses for condition D15PL2B

the density function is representable by any of the five examples given by Hiel *et al* [12] for deterministic temperatures and moistures.

A limited number of examples have been constructed in order to illustrate the stochastic delamination criterion with some parametric studies. Consider the flat plate shown in Fig. 1 with one side free at $x_2 = b$ and three sides clamped. The plate dimensions are b = .2", a thickness of .005" per ply and with a > > b to achieve a plane strain condition in the x_1 direction. One to five plies and fiber orientations of 15, 30 and 45 degrees are considered. The plate is loaded by uniform quasi-static strains e_{11} = constant and $e_{11}(t)$. Due to the linear viscoelastic analysis the shear stresses t_{23} develop a singularity along the free edge (x_1, b, x_3) and have been artificially cut off at a level equal to their values at $x_2 = x_{2co} = .875b$ as suggested in Ref. 4 based on experimental observations.

The Weibull distributions [19,20] reported by Hiel *et al* [12] and summarized in Table I were used in these examples. Since the variable v' in Eq. (1) varies from 0 to 1, the $\tilde{v}'$ and $\tilde{\beta}$ values of Ref. 12 (Table I) must be nondimensionalized to conform to the present pattern. The Weibull probability distribution is given by [19,20]

$$P = 1 - \exp\left[-(v'/\beta)^{\alpha}\right] \qquad (0 \le v' \le 1) \qquad (1)$$

For the purposes of the present analysis, the largest value $\tilde{v}'_{max}$ is determined as the lowest $\tilde{v}'$ at which P = 1. From a computational point of view, for most mainframes and work stations, a unity value in Eq. (1) occurs when the exponential function reaches values of the order of 10^{-14} (double precision). These $\tilde{v}'_{max}$ values, of course, depend on a and $\tilde{b}$ and smaller values than 10^{-14}, i.e. 10^{-26}, could be used in quadruple precision. However, due to the highly nonlinear nature of the

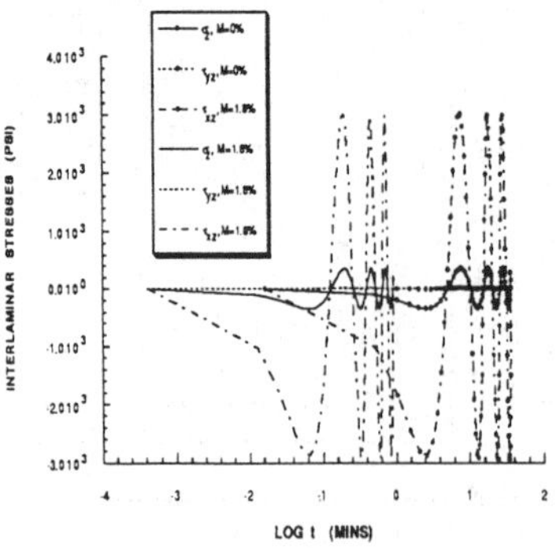

Fig. 7 Stresses for condition D15PL3A

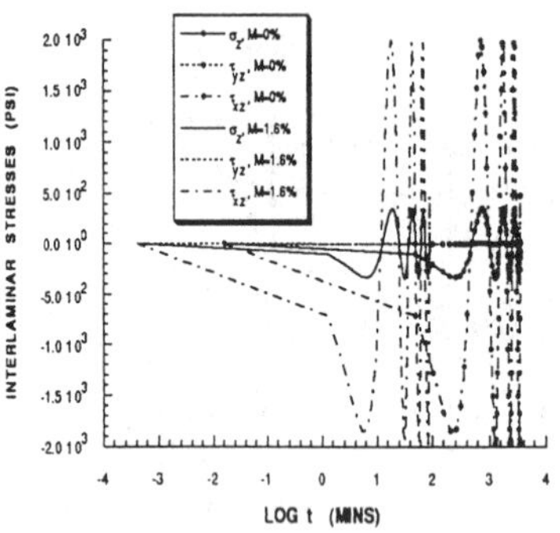

Fig. 8 Stresses for condition D15PL3B

exponentional functions and due to the high degree of experimental data scatter and lack of sufficiently large number of experiments, i.e. uncertain α and β values reported by Hiel *et al* [12], such other $\tilde{v}'_{max}$ definitions do not contribute significally to the probability predictions with any greater certainty and/or confidence. Therefore, $\tilde{v}'_{max}$ is defined here as

$$\left(\tilde{v}'_{max}/\tilde{\beta}\right)^{\alpha} = 32.2 \qquad (2)$$

At the low end of the $\tilde{v}'$ scale where $P = 1 - (1\text{-}10^{-14})$, or approaching 0 with same number of significant digits, the corresponding value of $\tilde{v}'_{min}$ is 10 to 15 psi depending on α and β values. For all practical purposes, $\tilde{v}'_{min}$ is taken as zero here when compared to $\tilde{v}'_{min}$.

The plate stresses were obtained from analytical anisotropic viscoelastic analyses [10] for prescribed uniform strain loadings (Fig. 2 and 3) on $x_1 = a$.

Loading 1: $e_{11} = C_1 \quad (t \geq 0) \quad$ (3a)

Loading 2: $e_{11} = C_2\, t/t_1 \quad (0 \leq t \leq t_1)$

$e_{11} = C_2 \quad (t \geq t_1) \quad$ (3b)

Loading 3: $e_{11} = C_3 \sin wt \quad (t \geq 0) \quad$ (3c)

Loadings 2A and 2B are for respective t_1 values of 500 and 1000 secs and Loadings 3A and 3B are for $w = 10^{-2}$ and 10^{-4} Hz respectively. For all loadings $C_1 = C_2 = .0014$ and $C_3 = .001$. The various plates are designated by the symbols such as D15P5L2A, where D15 refers to the fiber orientation, P5 to the number of plies and L2A to the loading condition.

Figs. 4 to 12 depict results for the various loadings 1, 2A, 2B, 3A and 3B, for distinct fiber orientations, different number of plies and for the five probability density distributions of Ref. 12. The deterministic uniaxial tension, compression and shear delamination failure stresses are taken from Dillard and Brinson [11]. It can be seen that the delamination stresses increase indefinitely, but elastic values ($t = 0$) are available [4] which provide an upper limit occuring at $t = .149$ min for this material, T300/934. These curves are shown in Fig. 3. The higher and lower pairs are the tension and shear delamination failure stresses respectively. The larger values for each pair are for 0% moisture and smaller ones are for 1.6% moisture. The procedure used to arrive at the non zero percent moisture curves will be discussed subsequently.

In all cases, deterministic predictions indicate the absence of delamination since all v' values are comfortably less than unity for all service times of $0 \leq t \leq 10^4$ mins. at room temperature. (This time range will significantly decrease by several orders of magnitude as the temperature increases). However, stochastic analysis results with only random failure conditions and deterministic applied loads, material properties, etc. shown in Figs. 4 to 13, indicate that for all 5 probability distributions of Table I, $P < 1$. It is of further interest to note that inloading conditions 1 and 2A, Figs. 4 and 5, the viscoelastic internal stresses relax at much faster time rates than those at which corresponding uniaxial failure stresses degrade. The very much higher probabilities of failure for very small times, i.e., $P \cong .7$ at $t \cong 0$, as opposed to $P \cong 0$ at $t \cong 200$ to 1000 minutes, forces one to conclude for loading Condition 1 that delaminations are more likely to occur very early or not at all if the composite survives the

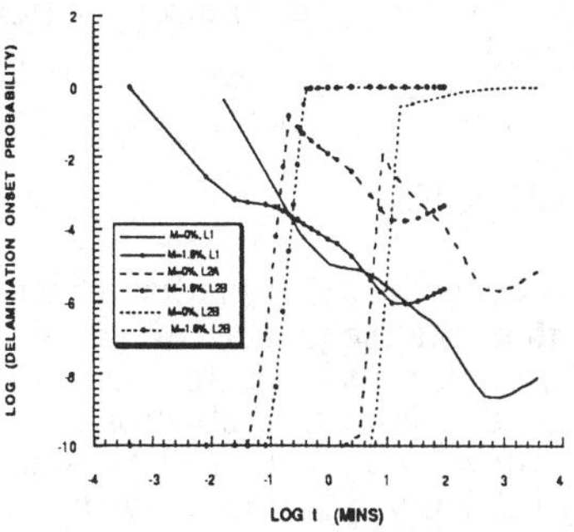

Fig. 9 Loading effects for conditions D15P5L1/L2A/L2B

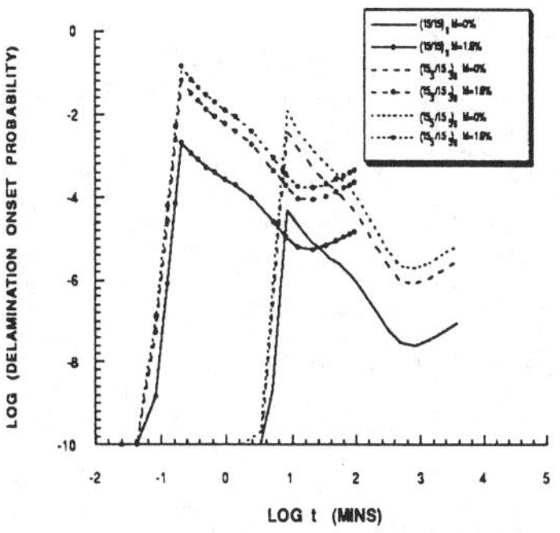

Fig. 11 Thickness effects for conditions D15P1L2A/P3/P5

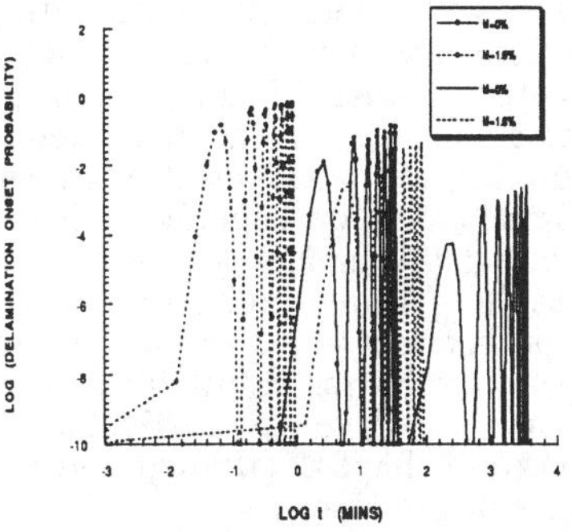

Fig. 10 Loading effects for conditions D15P5L3A/L3B

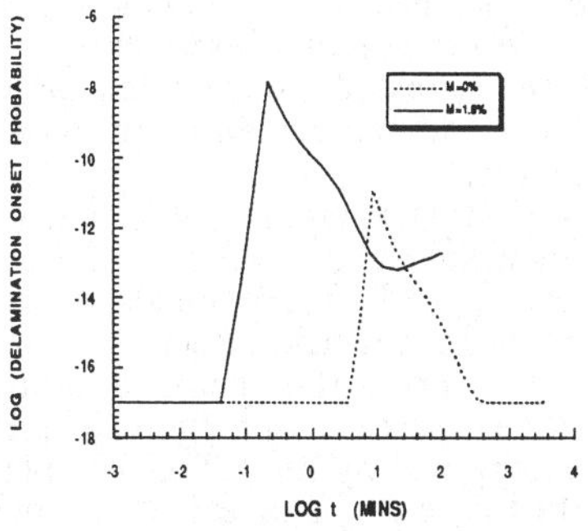

Fig. 12 Fiber orientation effects for conditions D15P5L2A/D30/D45

early time histories (Fig. 10).

As seen in Table I, cases 4 and 5 represent stochastic failure data for the same composite material (T300/934) with moisture contents of 0 and 1.6% as reported by Hiel *et al* [12]. Unfortunately Dillard and Brinson [11] indicate deterministic uniaxial failure stress history only at 0% moisture. In order to obtain some feel for possible moisture effects on failure the Dillard parameters were multiplied by the ratio of the mean v's. The shift factor [2,3,7-9] for 1.6% moisture was found in the work of Crossman *et al* [21] to be $10^{1.6}$. Figs. 9 to 13 show the probability of delamination onset curves for 0 and 1.6% moisture contents against real time in minutes plotted on a log scale. The increased moisture content according to this however imperfectly conjectured moisture model produces higher delamination probabilities at earlier times. This is not an unexpected trend when one considers that the influence of moisture which serves to decrease relaxation times and causes viscoelastic actions to occur in shorter earlier real times. Increases in temperature will produce similar shifts to the left.

Figs. 7 and 8 show the stresses for the sinusoidal loading conditions D15P5L3A and D15P5L3B for forcing frequencies 10^{-2} and 10^{-4} Hz respectively. The effect of the smaller frequency is to delay the occurance of the initial peak stresses by about two orders of magnitude on the time scale. The smaller forcing frequencies also produce smaller stress and delamination onset probabilities due to their frequency dependence.

Fig. 9 examines the effects of loading conditions and moisture content on delamination probabilities for Conditions 1, 2A and 2B of Fig. 2. For the loadings of Conditions 1 and 2A, the largest delamination probabilities occur very early and then decrease as stress relaxation out paces delamination stress degradation with time. The responses for 2A and 2B are dissimilar and the shallower strain loading produces higher

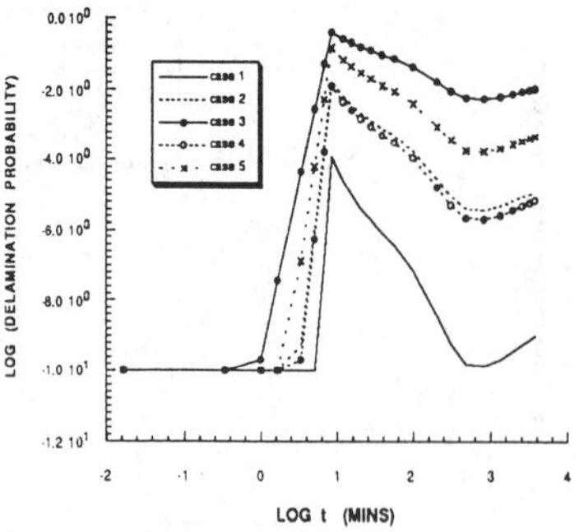

Fig. 13 Probability distribution effects for condition D15P5L2A

probabilities since full deformation is reached in longer times at less relaxed stresses. In all three cases increased moisture content leads to lager delamination probabilities in significantly shorter times.

Fig. 10 is a similar plot to Fig. 9 except that the sinusoidal strains 3A and 3B of Fig. 3 have been applied. These are also quasi-static analyses without inertia effects. It should be noted that now the peak delamination probabilities increase with time since in each cycle energy lost by relaxation is replenished at a greater rate. The effect of increased moisture is to again increase probabilities of failure in shorter times. The effect of increased forcing frequency from 10^{-5} to 10^{-4} Hz is to decrease peak delamination probabilities and to delay their occurrence, i.e., shift them to longer times.

The thickness effects due to 1, 3 and 5 plies for loading 2A are examined in Fig. 11 for an orientation angle of 15°. It is readily seen that the largest delamination probabilities occur for the largest (thickness) number of plies when the same strain loading is imposed on these plates.

The results for fiber orientation of 15, 30 and 45° for 5 plies and loading 2A are shown in Fig. 12. Increases in fiber angles strengthen the plate and result in lower probabilities of delamination onset at later times. Moisture effects in these three examples are the same as before, i.e. larger probabilities of failure in shorter times with increased moisture content for each ply orientation.

Finally, in Fig. 13 the delamination probabilities for the five Weibull distributions of Table I are shown for Condition 2A. It is clear that the predicted delamination probabilities are extremely sensitive to the a and b combinations of these distributions, although no clear pattern emerges from these limited studies.

CONCLUSIONS

Stochastic delamination onset studies show that the sequence of larger failure probabilities is important. In steady state strain loadings the largest probabilities occur in shorter times, when stress relaxation has not yet significantly manifested itself and when failure stresses have not had sufficient time to degrade. For sinusoidal strain loadings delamination probabilities increase with time as relaxation energies are replenished in each cycle while delamination onset stresses decrease with time. In these case failure predictions are delayed as opposed to the steady state loadings. In all cases the effect of increased moisture content and/or temperature is to predict delaminations at times decreased by several orders of magnitude. Increases in fiber orientation angles strengthen the plate and result in lower probabilities, while the effect of increases in the number of plies leads to larger delamination probabilities. The results are very sensitive to particular Weibull distribution shapes, although no apparent pattern is discernable.

REFERENCES

1. Hilton, H. H. , J. Hsu and J. S. Kirby. "Linear Viscoelastic Analysis with Random Material Properties", Journal of Probabilistic Engineering Mechanics, 6:57-69 (1991).
2. Christensen, R. M. Theory of Viscoelasticity: An Introduction, 2nd Ed., Academic Press, New York (1982).
3. Hilton, H. H. "Viscoelastic Analysis", Engineering Design for Plastics, Eric Baer, Ed., Reinhold Publishing Corp., New York, 199 - 276 (1964).
4. Brewer, J. C. and P. A. Lagace. "Quadratic Stress Criterion for Delamination", Journal of Composite Materials, 22:1141-1155 (1988).
5. Yi, S. "Thermoviscoelastic Analysis of Delamination Onset and Free Edge Response in Epoxy Matrix Composite Laminates", AIAA - 91 - 0962 - CP (1991).
6. Hilton, H. H. "Structural Reliability and Minimum Weight Analysis for Combined Random Loads and Strengths", Submitted to Journal of Probabilistic Engineering Mechanics (1992).
7. Hilton, H. H. and S. Yi. "Finite Element

Formulation of Thermoviscoelastic Analyses of Composite Structures Subjected to Mechanical and Hygrothermal Loadings," Proceedings First NCSA Conference on Finite Element Applications in Computational Mechanics, University of Illinois U-C, 1-14 (1990).
8. Hilton, H. H. and S. Yi. "Bending and Stretching Finite Element Analysis of Anistropic Viscoelastic Composite Plates," Proceedings of Third Air Force/NASA Symposium on Recent Advances in Multidisciplinary Analysis and Optimization, 488 - 494 (1990).
9. Hilton, H. H. and S. Yi. "Dynamic Finite Element Analysis of Viscoelastically Damped Composite Structures," Proceedings Second International Conference on Applications of Supercomputers in Engineering, 495 - 511 (1991).
10. Hilton, H. H. and S. B. Dong. "An Analogy for Anistropic, Nonhomogeneous Linear Viscoelasticity Including Thermal Stresses," Proceedings Eighth Midwestern Mechanics Conference, 58 - 73 (1964).
11. Dillard, D. A. and H. F. Brinson. "A Numerical Procedure for Predicting Creep and Delayed Failures in Laminated Composites," Long Term Behavior of Composites, ASTM STP 813, T. K. O'Brien, Ed., 23 - 37 (1983).
12. Hiel, C. C., M. Sumich and D. P. Chappell. "A Curved Beam Test Specimen for Determining the Interlaminar Tensile Strength of a Laminated Composite," Journal of Composite Materials, 25: 854 - 868 (1991).
13. Lifshitz, J. M. and A. Rotem. "Time-Dependent Longitudinal Strength of Unidirectional Fibrous Composites," Fibre Science and Technology, 3:1-20 (1970).
14. Phoenix, S. L. and L-J. Tierney. "A Statistical Model for the Time Dependent Failure of Unidirectional Composite Materials Under Local Elastic Load-Sharing Among Fibers," Engineering Fracture Mechanics, 18:193-215 (1982).
15. Phoenix, S. L. "Statistical Aspects of Failure of Fibrous Materials," Composite Materials: Testing and Design, ASTM STP 674, S. W. Tsai, Ed., 455-483 (1979).
16. Watson, A. S. and R. L. Smith. "An Examination of Statistical Theories for Fibrous Materials in the Light of Experimental Data," Journal of Materials Science, 20: 3260-3270 (1985).
17. Farquhar, D. S., F. M. Mutrelle, S. L. Phoenix and R. L. Smith. "Lifetime Statistics for Single Graphite Fibres in Creep Rupture," Journal of Material Science, 28:2151-2164 (1989).
18. Larder, R. A. and C. W. Beadle. "The Stochastic Finite Element Simulation of Parallel Fiber Composites," Journal of Composite Materials, 10:21-31 (1976).
19. Lin, Y. K. Probabilistic Theories of Structural Dynamics, McGraw-Hill Book Co., New York, 340 (1967).
20. Law, A. M. and W. D. Kelton. Simulation Modeling and Analysis, 2nd Ed., McGraw-Hill, New York, 333-335 (1990).
21. Crossman, F. W., R. E. Mauri and W. J. Warren. "Moisture Altered Viscoelastic Response of Graphite/Epoxy Composites," Advanced Composites Materials - Environmental Effects, ASTM STP 658, J. R. Vinson, Ed., 200-205 (1978).
22. Hilton, H. H. and S. Yi. "Stochastic Viscoelastic Delamination Onset Failure Analysis of Composites", University of Illinois U-C T.R. UILU ENG 92-0501 (1992) Journal of Composite Materials, 27:1097-1113 (1993).

Structural Safety & Reliability, Schuëller, Shinozuka & Yao (eds) © 1994 Balkema, Rotterdam, ISBN 90 5410 357 4

Reliability of glass windows

A. Nilsson & L. Sentler
Lund Institute of Technology, Department of Structural Engineering, Sweden

ABSTRACT: A general methodology to evaluate the reliability of window glass is presented. The strength properties of glass are considered in the strain domain where both an elastic and a viscous strain response are included. Failure criteria are derived within a stochastic viscoelastic material characterization which provides failure surfaces in the space - time domain. This makes the failure stress both size and time dependent. Impact, sustained and long term fluctuating wind loads can be considered in a rational manner as well as the influence of the environment.

1 INTRODUCTION

The reliability of a structural system and its components can be defined as satisfactory performance over time when subject to loads and the environment in which it is operating. The environment can often be of significant importance since this introduces conditions in which a structure has to operate in reality. For this reason it is desirable to consider reliability in a wider framework than is normally the case today.

It is necessary to base an extended evaluation of the reliability on an advanced material characterization where a time dependent response can be accounted for. This will be the case if the strain response is considered. For most materials there exist two types of strain responses, an elastic response or a viscous response, which are completely different in nature. The first response is typical for low stress levels and for direct loadings to failure. The latter response occurs after a transition in time if the stress level exceeds a certain value. Both response forms may occur under operational conditions and as a consequence of this there will be two failure forms. It is necessary to consider this in an advanced reliability analysis.

The material glass, when used in windows and structural glazing, is a building component which will be subject to actions in terms of wind loads and the environment. This makes the time dependent properties of importance and it is necessary to evaluate the reliability within the framework of an advanced material characterization.

2 GLASS CHARACTERISTICS

Glass has been produced with approximately the same composition for a long period of time. This well defined nature of glass in terms of its composition is not reflected in an equally well defined structure. The silica network which makes up the basic structure of glass is far from perfect. This is in particular the case in the surface regions there will be defects in this structure.

Glass plates are covered with minute flaws of varying geometries and orientation. Such flaws are initially introduced during the production. During the use new flaws can be added or existing flaws can be amplified. This will be the result of mechanical wear and chemical reactions. For instance, the water vapour in the air can react with the silica network and create new breaks.

When surface flaws are exposed to tensile stresses they will grow in size. In the presence of water this process will be magnified which reflects a form of stress corrosion. For sustained stresses the time to failure will be reduced. This is a process which is often referred to as static fatigue by glass researchers like Baker and Preston (1946) and Charles (1958).

When a glass plate is loaded, the surface flaws interact with the surface tensile stresses and cause large local stress concentrations. Glass plate failure occurs when the local stress associated with one of these surface flaws becomes large enough to initiate a fracture (Allen and Dalgliesh 1973, Beason and Morgan 1984, Reed and Simiu 1984). But since the number of flaws on a glass surface is extremely large and their size and orientation can vary considerably it is not possible to consider these flaws individually. Instead a stochastic continuum mechanical approach is preferable.

3 VISCOELASTIC CHARACTERIZATION OF GLASS STRENGTH

Glass is sometimes referred to as a super cooled liquid. This is a characteristic which has a very direct influence on the material behaviour. This is illustrated in Figure 1 where the centre deflection of a glass plate is compared with the deflection predicted with the most advanced plate theory, the Levy theory for large deflections (Timoshenko and Woinowsky-Krieger 1959). It is only for small deflections, of the same size as the thickness of the glass plate, where a reasonable agreement is obtained. For increasing deflections the plate theory is completely inaccurate. Measured elastic strains also deviate significantly from that predicted by the elastic theory (Bowles and Sugerman 1962).

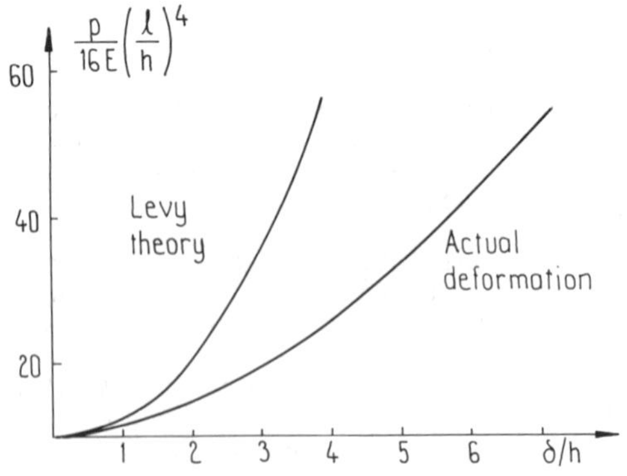

Fig. 1 Actual and predicted glass plate centre deformations.

The deformations which can be obtained for glass plates is a typical manifestation of a viscoelastic material response. For an increasing stress there will be a shift in time from an elastic response towards a viscous response. This shift in time is very fast at room temperature and much more pronounced than for most other materials.

The viscoelastic nature of glass makes it appropriate to consider the strain response ε(t) in time. For a one dimensional constant stress this can be expressed as (Penny 1971)

$$\epsilon(t) = \epsilon_e t^{\alpha} + \epsilon_v t \tag{1}$$

where ε_e is an elastic response and ε_v is a viscous response. The elastic response can be expressed as

$$\epsilon_e t^{\alpha} = \frac{\sigma}{E}\left(1 + \sum_{k=1}^{\infty} \frac{\alpha^k \ln^k(t)}{k!}\right) \tag{2}$$

where σ is the applied stress and E is the elastic modulus. The series expansion of the time term reveals that the elastic strain can be split up in an initial time independent response and a time dependent part which normally is associated with primary creep.

The viscous strain response is often represented with the Norton creep law. For a constant stress this response can be expressed as

$$\epsilon_v t = \frac{(\sigma - \sigma_c)^n}{\eta} t \tag{3}$$

where σ is the applied stress which has to exceed a temperature dependent threshold stress σ_c, and n and η are material dependent parameters in the same way as the elastic modulus. The viscous strain, which is manifested in terms of secondary creep, reflects a damage accumulation in a material. For glass it will reflect the growth of flaws which will increase the flaw density in time. It is here assumed that this response can be considered as a strain softening and not as a time softening.

For other stress conditions than a constant stress equations 2 and 3 have to be differentiated.

The elastic strain and the viscous strain capacities are not unlimited. This is illustrated in Figure 2 for small circular specimens subject to sustained stresses in bending at different temperatures (Charles 1958b). It is mainly a viscous strain responses which are obtained except for very low temperatures where elastic strain responses occur. The pronounced time dependence, which is typical for a viscous response, will not occur for an elastic response.

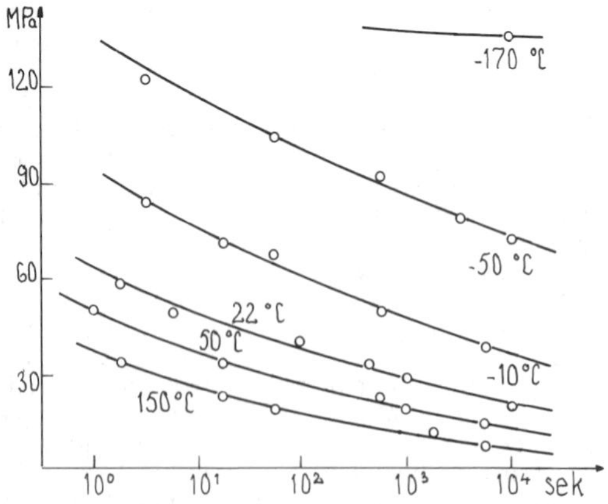

Fig. 2 Time dependent failure boundaries for glass at different temperatures.

In addition to the time dependence there will also be a size dependence. This size dependence is illustrated in Figure 3 which shows the failure stress for small specimens of window glass of different lengths and thicknesses subject to a direct loading to failure (Nilsson 1993). All specimens had the same width. The rupture stresses are normalized with respect to the surface area subject to tensile stresses. The increasing failure stress for increasing thicknesses reflects the influence of residual stresses which are introduced on purpose during the production. Otherwise the rupture stress should be of the same magnitude if other dimensions are the same since it can be expected that the surface characteristics are similar.

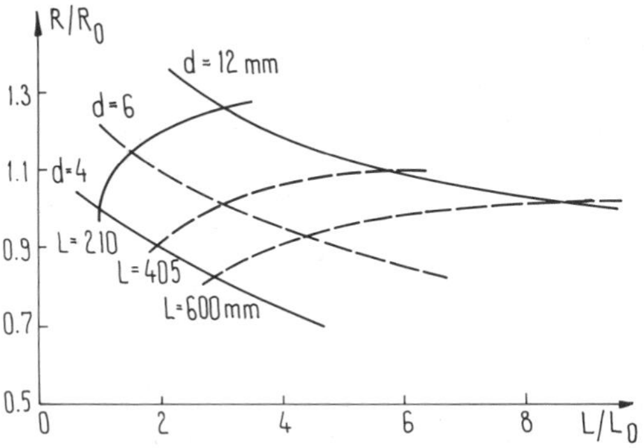

Fig. 3 The size dependence of glass plates.

Similar results as those presented in Figure 3 have also been obtained for large glass plates (Bowles and Sugarman 1962). Often the Weibull theory is used to represent this size dependence. For new glass it is found that the size parameter will be around 8.

4 STOCHASTIC STRAIN FAILURE CHARACTERIZATION

The strength properties of glass will depend on the strain limitations in terms of elastic strain or viscous strain. These strain limitations will be influenced by the presence of existing flaws or the growth of flaws in time. For this reason the failure strain ε_u will not only depend on the stress system which is acting but it will also be influenced by an effective surface area and an effective duration associated with this stress (Sentler 1987). It is here assumed that glass can be considered as an isotropic media but the flaw characteristics will differ between the surface and the interior.

For a body subject to a two dimensional stress system, which can be defined in volumetric stress components σ_x and σ_y or a distortion stress $\sigma_d = (\sigma^2_x + \sigma_x\sigma_y + \sigma^2_y + 3\tau^2_{xy})^{1/2}$, the failure strain responses can be expressed as

$$\varepsilon_u = g_e(\sigma_x; A_x, D_x) + g_e(\sigma_y; A_y, D_y) + g_v(\sigma_d; V_d, D_d) \tag{4}$$

where g_e and g_v are strain functions for an elastic or a viscous strain response respectively, A_i reflects an effective area subject to tensile stress, and D_i reflects an effective duration of this stress.

A failure may occur because of an elastic strain response or because of a viscous strain response. In the first case the strain capacity will mainly depend on existing flaws in the material which will emphasize a size dependence. This is the background to the well established Weibull (1939) theory of a size dependent rupture strength. In the second case the strain capacity will mainly depend on the growth of existing flaws which will emphasize a time dependence.

An elastic strain response and a viscous strain response are two completely different response forms. An elastic strain is a natural response form where the distance between the constituents increase or decrease. A viscous strain is a consequence of flaws in a material and this strain results in a damage accumulation which is not recoverable. For this reason the two response forms can be considered as independent of each other.

The ultimate strain at failure is not determined by the average material behaviour. Instead, due to the influence of flaws, the ultimate strain will depend on the weakest part of the material. This behaviour is considered in the most rational way within the statistical theory of extreme values. For an independence between the strain responses the statistical distribution of the minimum value, or no failure, is obtained from the theorem of multiplication of probabilities which can be expressed as

$$F(\varepsilon_u) = 1 - \left(1 - F_e(\sigma_x; D_x)\right)^{\Sigma A_i} \left(1 - F_e(\sigma_y; D_y)\right)^{\Sigma A_j} \left(1 - F_v(\sigma_d; A_d)\right)^{\Sigma D_s} \tag{5}$$

where $F_e(.)$ and $F_v(.)$ are the cumulative statistical distribution functions of elastic strain for a small surface area A_i and A_j or viscous strain during a short duration D_s. For i, j and s large in the summations of equation 5 this expression can be written in terms of the asymptotical extreme value distribution. The initial distributions $F_e(.)$ and $F_v(.)$ are limited to the left since only positive arguments are possible, and the behaviour of the arguments can be expressed in terms of power laws. Thus equation 5 can be written in terms of the asymptotical extreme value distribution of type 3, the Weibull distribution, as distribution, as

$$F(\epsilon_u(\sigma;A,D)) = 1-\exp\left(-\frac{A_x}{A_0}\left(\varphi(D_x)\frac{\sigma_x}{c_e}\right)^k - \frac{A_y}{A_0}\left(\varphi(D_y)\frac{\sigma_y}{c_e}\right)^k - \frac{D_d}{D_0}\left(\psi(A_d)\frac{\sigma_d}{c_v}\right)^h\right) \tag{6}$$

where the size and time dependence are expressed in normalized form, k and h are size and time parameters respectively, $\varphi(.)$ and $\psi(.)$ are functions which have to meet certain continuity requirements, and c_e and c_v are normalizing constants. Also $F(\varepsilon_u)$ is expressed with the arguments which determine the strain capacity.

The smooth transition from primary creep to secondary creep, which reflects a change from an elastic response to a viscous response, requires continuity in time and indirectly in space too. These continuity requirements can be evaluated in detail for the mean value behaviour. Details of this can be found in (Nilsson 1993). This results in the following expressions

$$\varphi(D.) = \left(\frac{D.}{D_0}\right)^{1/h} \tag{7}$$

$$\psi(A_d) = \left(\frac{A_d}{A_0}\right)^{1/k} \tag{8}$$

where . stands for x or y.

With equations 7 and 8 introduced in equation 6 and the statistical distribution of a strain response failure is obtained. Because a viscous strain response only can occur for stresses which exceeds a threshold stress σ_c this is also introduced. The result is

$$F(\epsilon_u(\sigma;A,D)) = 1-\exp\left(-\frac{A_x}{A_0}\left(\frac{D_x}{D_0}\right)^{k/h}\left(\frac{\sigma_x}{c_e}\right)^k - \frac{A_y}{A_0}\left(\frac{D_y}{D_0}\right)^{k/h}\left(\frac{\sigma_y}{c_e}\right)^k - \frac{D_d}{D_0}\left(\frac{A_d}{A_0}\right)^{h/k}\left(\frac{\sigma_d-\sigma_c}{c_v}\right)^h\right) \tag{9}$$

The first two parts, which reflect an elastic strain response, is an extension of the original Weibull theory where time also is included. The last part, which reflects a viscous strain response, is similar to the Norton creep law but where a size dependence also is introduced. The Norton creep law, which is empirical, can thus be derived from the extreme value theory.

Equation 9 provides two failure surfaces in space and in time. These failure surfaces may have the form presented in Figure 4. The exact location of the two failure surfaces will depend on the type of material but also on the loading conditions. A comparison with Figure 2 indicates that one of the two failure surfaces will dominate for glass subject to normal loading conditions.

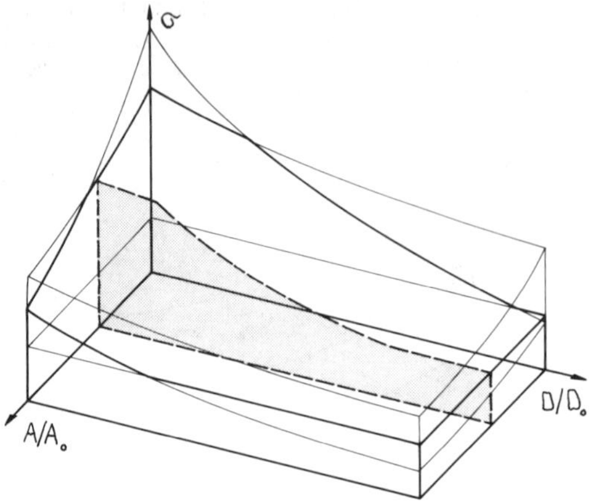

Fig. 4 The failure surfaces which are given by equation 9.

5 THE RUPTURE STRESS

The rupture stress can only be expressed in terms of the mean value and the variance of equation 9. Because there are two possible strain responses the mean value can be derived in terms of an elastic response or a viscous response. For the special case of a uniform sustained stress this is obtained with the help of variable transformations like $u_e = (A_x/A_0)^{1/k}(D_x/D_0)^{1/h}\sigma_x/c_e$, $\mu = (A_y/A_x)^{1/k}(D_y/D_x)^{1/h}\sigma_y/\sigma_x$. The mean value elastic failure stress can be expressed in terms of the maximum stress σ_x as

$$E[\sigma_x] = \frac{c_e}{(1+\mu)^{1/k}}\left(\frac{A_0}{A_x}\right)^{1/k}\left(\frac{D_0}{D_x}\right)^{1/h}\Gamma\left(1+\frac{1}{k}\right) \tag{10}$$

where $\Gamma(.)$ is the gamma function.

Table 1. Reduction of the failure stress as a function of μ in equation 10.

μ	0.0	0.2	0.4	0.6	0.8	1.0
k=8	1.00	0.98	0.96	0.94	0.93	0.92

An evaluation of the influence μ is presented in Table 1 for k = 8. It can be seen that it is only when μ is close to 1, which reflects square windows, where there will be a significant contribution from the minimum stress. In other cases the minimum stress σ_y will only have a marginal influence on the rupture stress.

For a pure viscous response the mean value failure stress is obtained in a similar way as

$$E[\sigma_d] = \sigma_c + c_v\left(\frac{A_0}{A_d}\right)^{1/k}\left(\frac{D_0}{D_d}\right)^{1/h}\Gamma\left(1+\frac{1}{h}\right) \qquad (11)$$

It is possible to introduce a reference strength σ_0 which corresponds to a reference area and a reference duration. This makes it possible to express both mean values in a similar form as

$$\bar{\sigma} = \bar{\sigma}_0\left(a + b\left(\frac{A_0}{A_x}\right)^{1/k}\left(\frac{D_0}{D_x}\right)^{1/h}\right) \qquad (12)$$

where a and b are empirical constants. Similar expressions can also be derived for the variances.

6 GENERAL STRESS CONDITIONS

The stress system which is acting will be of importance and it has to be considered appropriately. For the special case of uniform stress conditions the effective volume is given by the total volume and for a sustained stress the effective duration is given by the total time. For stresses which vary in space and in time the effective volume and effective duration have to be evaluated from

$$F(\epsilon_u(\sigma;A,D)) = 1-\exp\left(-\iiint_{\sigma_x} t_x^{k/h}\left(\frac{\sigma_x(x,y,t)}{c_e}\right)^k dxdydt - \iiint_{\sigma_y} t_y^{k/h}\left(\frac{\sigma_y(x,y,t)}{c_e}\right)^k dxdydt - \iiint_{\sigma_d}(xy)^{h/k}\left(\frac{\sigma_d(x,y,t)-\sigma_c}{c_v}\right)^h dxdydt\right) \qquad (13)$$

where $\sigma_x(x,y,t)$, $\sigma_y(x,y,t)$and $\sigma_d(x,y,t)$ are stress variations over an area and in time. For certain regular stress variations in space and in time explicit solutions are available. This is the case for stresses which vary linearly in space or in time. For non linear stress variations a numerical procedure is normally necessary.

A viscous response will be of particular importance for glass since it tends to be a dominating response form. Davenport (1983) concludes that the pressure fluctuations during a two hour wind storm will cause about as much damage as the peak pressure during 60 seconds.

A stationary stress time history as that presented in Figure 5 can be expressed as (Box and Jenkins 1976)

$$\sigma_d(t) = \bar{\sigma}(1+\zeta\Omega(t)) \qquad (14)$$

where $\sigma_d(t)$ is the stress variation in time, $\bar{\sigma}$ is a reference value or mean value, $\zeta = \sigma_{max}/\bar{\sigma}$ and $0 \le \zeta \le 1$, and $\Omega(t)$ is a stationary random process with the condition $|\Omega(t)|_{max} = 1$. In addition the autocovariance function $R_{\sigma\sigma}$ of $\sigma_d(t)$ has to behave as $1/\lambda \int |R_{\sigma\sigma}| dt < \infty$ where λ is the duration.

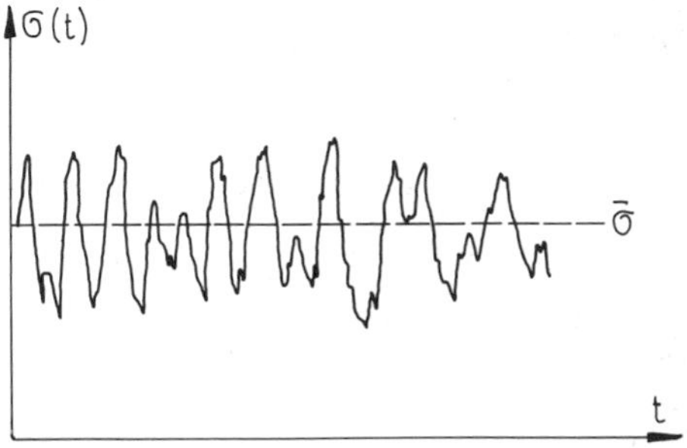

Fig. 5 Stationary stress time history.

A constant stress level can be dealt with directly within equation 9. The random process can be analyzed by expressing the time integral in equation 13 in terms of a discrete summation as

$$g(h,\zeta) \simeq \frac{1}{\lambda}\sum_{t=0}^{\lambda}(1+\zeta\Omega_t)^h\Delta\lambda \qquad (15)$$

where Ω_t is the discrete correspondence to $\Omega(t)$. Here the threshold stress σ_0 is put equal to zero for mathematical reasons. If $\Delta\lambda = 1$ the expression $(1 + \zeta\Omega_t)^h$ can be replaced by a binomial series and expressed as

$$g(h,\zeta) \simeq \frac{1}{\lambda}\sum_{t=0}^{\lambda}\sum_{r=0}^{h}\binom{h}{r}(\zeta\Sigma\Omega_t)^h \qquad (16)$$

where the summations can be interchanged and the series expressed as

$$g(h,\zeta) \approx \frac{1}{\lambda}\sum_{t=0}^{\lambda} 1 + \frac{h}{1!}\frac{\zeta}{\lambda}\sum_{t=0}^{\lambda}\Omega_t + \frac{h(h-1)}{2!}\frac{\zeta^2}{\lambda}\sum_{t=0}^{\lambda}\Omega_t^2 + \dots \qquad (17)$$

which can be identified as the sum of the statistical moments of the distribution of $\zeta\Omega_t$. From measurements of the stress variation which occurs during a wind loading and a determination of the statistical moments the possible damage accumulation can be accounted for in a rational manner. A similar methodology can be found in (Beason and Morgan 1984).

Wind loads are often characterized with a Weibull distribution. Normally it seems as if only the first four moments are needed to represent the stress variation in time for this statistical distribution. Results for three stress variability forms of ζ and a time response parameter of h = 16 are given in Table 2. Similar results can also be derived for other statistical distributions.

Table 2. Increase of the damage accumulation expressed as g(h,ζ) in equation 15.

ζ	0.1	0.2	0.3
h = 16	1.064	1.247	1.512

7 THE INFLUENCE OF THE ENVIRONMENT

Glass is a material which exhibits a viscous response for stresses which exceeds a threshold stress of 6 - 8 MPa. This will increase the flaw density under the continuous action of loads. It can also easily can be affected by the environment. Results available in the literature indicate that in particular the surface area dependence will increase in outdoor environments (Baker and Preston 1946). This can be expressed as $k = k_0$ - f(environment) where k_0 is an initial value. The functional relationship is not known but the size parameter will often be reduced from an initial value of $k_0 \approx 8$ to 4 - 6 (Beason and Morgan 1984). The time parameter which initially is $h \approx 16$ will also be affected but only marginally (Wiederhorn 1967).

8 CONCLUSIONS

It is possible to evaluate the reliability of glass plates in an advanced way. This is in particular the case for term fluctuating loading conditions. Both the variability of wind loads and material properties are accounted for as well as the influence of the environment.

ACKNOWLEDGEMENT

This research is supported by the Swedish Council for Building Research that is gratefully acknowledged.

REFERENCES

Allen, D.E. and Dalgliesh 1973. Dynamic wind loads and cladding design, IABSE Symp., Lisbon, Portugal, pp. 279-285.

Baker, T.C. and Preston, F.W. 1946. The effect of water on the strength of glass, Journal of applied physics, Vol. 17, No. 3.

Bowles, R. and Sugerman, B. 1962. The strength and deflection characteristics of large rectangular glass panels under uniform pressure, Glass Technology, Vol. 3, No. 5, pp. 156-170.

Beason, W.L. and Morgan, J.R. 1984. Glass failure prediction model, Journal of structural division, ASCE, Vol. 110, No.2.

Box, G.E. and Jenkins, G.M. 1976. Time series analysis: forecasting and control, Holden-Day, San Francisco.

Charles, R.J. 1958. Static fatigue of glass I, Journal of applied physics, Vol. 29, No. 11.

Charles, R.J. 1958. Static fatigue of glass II, Journal of applied physics, Vol. 29, No. 11

Davenport, A.G. 1983. The relationship of reliability to wind loading, Jounal of Engineering and Industrial Aerodynamics, Vol. 13, pp. 3-13.

Minor, J.E. 1981. Window glass design practices: a review, Journal ofthe structural division, ASCE, Vol. 107, No. 1.

Nilsson, A. 1993. Stochastic characterization of glass strength properties, D4:93, Swedish Council for Building Research, Stockholm.

Penny, R.K. 1971. Design for creep, McGraw-Hill.

Reed, D.A. and Simiu, E. 1984. Wind loading and strength of glass cladding, Journal of structural Division, ASCE, Vol. 110, No.4.

Sentler, L. 1987. A strength theory for viscoelastic materials, D9:1987, Swedish Council for Building Research, Stockholm.

Sentler, L. 1988. A stability theory for orthotropic viscoelastic structural members, D5:1988, Swedish Concil for Building Research, Stockholm.

Timoshenko, S.P. and Wionowsky-Krieger S. 1959. Theory of plates and shells, McGraw-Hill.

Weibull, W. 1939. A statistical theory of strength of materials, Proc. R. Sci. Soc. 151, 1939.

Wiederhorn, S.M. 1967. Influence of water vapor on crack propagation in sodalime glass, Journal of the American ceramic society, Vol. 50, No.8.

Structural Safety & Reliability, Schuëller, Shinozuka & Yao (eds) © 1994 Balkema, Rotterdam, ISBN 90 5410 357 4

Micromechanically based random fields

Martin Ostoja-Starzewski
Department of Materials Science and Mechanics, Michigan State University, East Lansing, Mich., USA

ABSTRACT: A procedure for determination of random fields of stochastic constitutive laws is outlined on the basis of linear elastic polycrystalline and matrix-inclusion microstructures. Central to the procedure is the concept of a scale-dependent representative volume element, which leads to two response tensors convergent from above and below to the effective moduli of an infinitely large microstructure. In a given boundary value problem these response tensors are to be used to bound the effective solution at a given scale of resolution. This micromechanical approach also forms basis for studies of spatially random plastic and damaging materials.

1. INTRODUCTION

Discreteness of materials is the key cause of their non-deterministic response. Thus, there arises a need for development of continuum type stochastic constitutive laws, which we discuss using the paradigms of elastic polycrystalline and matrix-inclusion materials. Setting up the analysis in a terms of discrete microstructure, allows one to grasp the geometrical and physical micro-scale material randomness, as well as to assess the strength of fluctuations as a function of the scale of observation. The latter observation leads to the concept of a scale-dependent Representative Volume Element, which forms the basis of a strategy for determination of effective constitutive laws of a class of linear elastic random microstructures, continuum random field characterizations of such microstructures, and methods involved in solution of stochastic boundary value problems. Thus, following (Ostoja-Starzewski 1992a, 1993a, b, c), we review a strategy for determination of stochastic constitutive laws of linear elastic random microstructures, continuum random field characterizations of such microstructures, and methods involved in solution of stochastic boundary value problems. Additionally, we also discuss extensions of this methodology to inelastic materials (plastic and damaging).

2. MODELS OF RANDOM MICROSTRUCTURE

Fundamental role in our formulation is played by the concept of a *random microstructure*, which, as is commonly done in mechanics of random media (Ostoja-Starzewski 1993b), is taken as a family $\boldsymbol{B} = \{\mathbf{B}(\omega);\ \omega \in \Omega\}$ of deterministic media $\mathbf{B}(\omega)$, where ω indicates one specimen (realization), and Ω is an underlying sample (probability) space. Formally, Ω is equipped with a σ-algebra F and a probability distribution P. In an experimental setting P may be specified by a set of stereological measurements, while in a theoretical setting P is usually specified by a chosen model of a micro- structure. For the purpose of further discussion, we take every specimen $\mathbf{B}(\omega)$ to be modeled by a realization of a *Voronoi tessellation* (Fig. 1 a), which corresponds to one realization of a space-homogeneous planar Poisson process of a given density. Each cell of this tessellation, centered at x, is assumed to be occupied by a homogeneous continuum governed by a stiffness tensor $\underline{C}(x, \omega)$ following the same space-homogeneous probability distribution $P(\underline{C})$ and satisfying the so-called *ellipticity conditions*: $\exists \alpha, \beta > 0$ such that for any $\underset{\sim}{\varepsilon}$ the following inequalities hold

$$\alpha \underset{\sim}{\varepsilon}\underset{\sim}{\varepsilon} \le \underset{\sim}{\varepsilon}\underline{C}\underset{\sim}{\varepsilon} \le \beta \underset{\sim}{\varepsilon}\underset{\sim}{\varepsilon} \qquad (2.1)$$

Thus, we have a realistic model of an ergodic polycrystalline medium without holes and rigid inclusions described by a random field $\underline{C} = \{\underline{C}(x, \omega);\ x \in \boldsymbol{B},\ \omega \in \Omega\}$ with piecewise-constant realizations. We note here that:

i) there are many kinds of random elastic microstructures, other than of Voronoi geometry, described by

random fields with piecewise-constant realizations - for example, matrix-inclusion composites with randomly located circular inclusions that have random diameters, Fig. 1 b);
ii) this formulation is sufficiently general to deal with other linear transport problems, such as, for example, conductivity, as well as the inelastic materials (though in this case (2.1) may not hold);
iii) while for simplicity and clarity of presentation, the discussion is conducted in two dimensions (2-D) here, a generalization to 3-D would follow same lines.

3. RANDOM FIELDS FOR LINEAR ELASTIC MATERIALS

3.1 Two scale-dependent meso-continuum approximations

First, with the help of Fig. 1 (Ostoja-Starzewski 1993a), we introduce a square-shaped *window* of scale

$$\delta = \frac{L}{d} \tag{3.1}$$

Equation (3.1) defines a nondimensional parameter $\delta \geq 1$ specifying the scale L of observation (and/or measurement) relative to a typical microscale d (i.e. grain size) of the material structure. $\delta = 1$ is the smallest scale we consider: scale of a crystal. In view of the fact that the Voronoi tessellation is a random medium, the window bounds a random microstructure $B_\delta = \{B_\delta(\omega);\ \omega \in \Omega\}$, where $B_\delta(\omega)$ is a single realization from a given specimen $\mathbf{B}(\omega)$.

In order to define the effective moduli of $B_\delta(\omega)$, we introduce two types of uniform boundary conditions on its boundary ∂B_δ :

- displacement-controlled (essential) boundary conditions

$$u_i = \bar{\varepsilon}_{ij} x_j \tag{3.2}$$

- stress-controlled (natural) boundary conditions

$$t_i = \bar{\sigma}_{ij} n_j \tag{3.3}$$

Using (3.2) (or (3.3)) and postulating the existence of an effective homogeneous continuum of the same volume (i.e. area in 2-D), whose potential energy $U(\omega)$ (or complementary energy $U^*(\omega)$) under given uniform boundary conditions equals that of a microstructure $B_\delta(\omega)$ under the same boundary conditions, we obtain an effective stiffness tensor

$$\underline{C}^e_\delta = \underline{C}^e_\delta(\omega) \tag{3.4}$$

or

$$\underline{C}^n_\delta = (\underline{S}^n_\delta)^{-1} = (\underline{S}^n_\delta(\omega))^{-1} \tag{3.5}$$

In (3.4) ^{e} and ^{n} stand for essential and natural boundary conditions, respectively.

We make here the following observations:
i) Due to the heterogeneity of the microstructure $B_\delta(\omega)$, the inverse

$$\underline{C}^n_\delta(\omega) = [\underline{S}^n_\delta(\omega)]^{-1} \tag{3.6}$$

is for any finite δ, in general, different from $\underline{C}^e_\delta$ of (3.4).
ii) In view of the spatial homogeneity of microstructure's statistics, $\underline{C}^n_\delta(\omega)$ and $\underline{C}^e_\delta(\omega)$ converge as δ tends to infinity; this defines a deterministic continuum $\mathbf{B}_{det}$ for a single specimen $\mathbf{B}(\omega)$

$$\underline{C}^{det}(\omega) = \underline{C}^n_\infty(\omega) = \underline{C}^e_\infty(\omega) \tag{3.7}$$

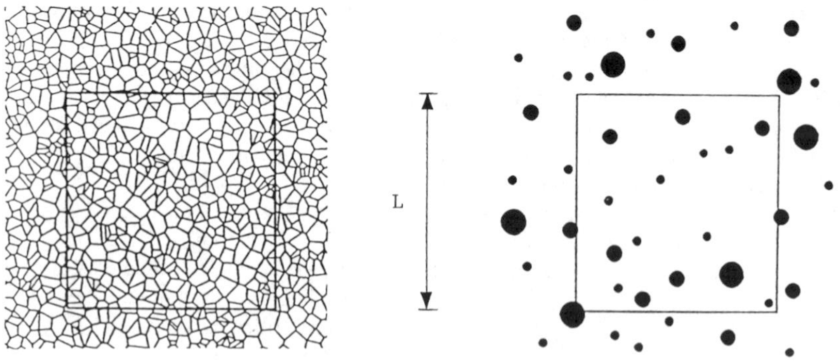

a) b)

Fig. 1 a) A Voronoi tessellation with an average cell size d; b) a matrix-inclusion composite with inclusions of average diameter d; in both cases a window of size L is indicated.

whereby the window of infinite extent plays the role of an RVE of deterministic elasticity theory; in other words, it is at $\delta \rightarrow \infty$ that the invertibility of the constitutive law is obtained.
iii) Ergodicity of the microstructure implies that

$$\underline{C}^{det}(\omega) = \underline{C}^{eff} \qquad \forall \omega \in \Omega \tag{3.8}$$

where $\underline{C}^{eff}$ is the effective response tensor (independent of ω) of a homogeneous medium.
iv) At any finite δ both response tensors are, in general, anisotropic, with the nature of anisotropy dependent on any specific $B_\delta(\omega)$; isotropy of the material is approached only asymptotically as $\delta \rightarrow \infty$, and simultaneously with the coefficient of variation tending to zero.
v) since the window may be located arbitrarily in the domain of $\mathbf{B}(\omega)$, the essential and natural boundary conditions (3.3) and (3.5) define two different inhomogeneous tensor fields at the scale δ with continuous realizations, which lead to two basic *random continuum approximations*: $\boldsymbol{B}^e_\delta = \{\mathbf{B}^e_\delta(\omega)\,;\ \omega \in \Omega\}$ and $\boldsymbol{B}^n_\delta = \{\mathbf{B}^n_\delta(\omega)\,,\ \omega \in \Omega\}$, respectively; accordingly, a window of size δ may be considered as an RVE of these two continuum models; the fact that there are two different, generally anisotropic, response laws at any finite δ, calls into question the 'unique' stochastic constitutive law commonly assumed in other studies (see e.g. Benaroya and Rehak 1988)

$$\sigma_{ij} = \lambda(x, \omega)\,\delta_{ij}\varepsilon_{kk} + 2\mu(x, \omega)\,\varepsilon_{ij} \tag{3.9}$$

vi) This definition of two inhomogeneous tensor fields is conceptually similar - but not the same(!) - to the procedure of *local averaging* in the theory of random fields applied to a single realization $\underline{C}(\omega)$; $\omega \in \Omega$; it becomes the same in case of a 1-D model only. In 2-D and 3-D, computational mechanics methods, such as conjugate gradient, have to be used to find the tensors in (3.4) and (3.5).

3.2 Scale-dependent bounds on the elastic response

Observe that the principles of minimum potential energy and of minimum complementary energy imply a hierarchy of bounds on the effective stiffness tensor $\underline{C}^{eff}$ expressed by the following inequalities

$$\begin{aligned}(\underline{S}^R)^{-1} \equiv \langle \underline{S}^n_1\rangle^{-1} \le \langle \underline{S}^n_{\delta'}\rangle^{-1} \le \langle \underline{S}^n_\delta\rangle^{-1} \le \underline{C}^{eff} \le \\ \langle \underline{C}^e_{\delta'}\rangle \le \langle \underline{C}^e_{\delta'}\rangle \le \langle \underline{C}^e_1\rangle \equiv \underline{C}^V \qquad \forall \delta' < \delta\end{aligned} \tag{3.10}$$

Hereinafter, for two fourth-rank tensors $\underline{A}$ and $\underline{B}$, an order relation $\underline{B} \le \underline{A}$ means

$$t_{ij}B_{ijkl}t_{kl} \le t_{ij}A_{ijkl}t_{kl} \qquad \forall \underline{t} \ne 0 \tag{3.11}$$

An inversion of (3.9) gives a hierarchy of bounds on the effective compliance tensor $\underline{S}^{eff} = (\underline{C}^{eff})^{-1}$.

Hierarchies of that type have been first derived by Huet (1990 and 1991). Ostoja-Starzewski and Wang (1989 and 1990) obtained numerically such hierarchies on $\underline{C}^e_\delta$ for so-called Delaunay networks, which, by virtue of duality to Voronoi tessellations, are generic models of granular and fibrous media. A numerical specification of these hierarchies for random chessboard (checkerboard) composites has been given by Ostoja-Starzewski (1993b).

3.3 Higher-order moments and correlation functions

It is clear from the preceding section that at any $\delta < \infty$ we deal with two random continuum fields - $\underline{C}^e_\delta$ of the approximation $\boldsymbol{B}^e_\delta$ and $\underline{S}^n_\delta$ of the approximation $\boldsymbol{B}^n_\delta$ - bounding the response of an actual heterogeneous material. It is seen from (3.9) that the fluctuations (i.e. scatter) in these two tensor fields become zero at $\delta \rightarrow \infty$. On the other hand, they must be strongest at $\delta = 1$. Thus, some monotonic properties resembling (3.10) are expected for higher-order moments. In fact, by using the same procedure as in Section 3.2, it can be shown that the following inequalities hold between second-order moments of successively smaller windows

$$\begin{aligned}(C^{eff})^2 = \langle (C^e_\infty)^2\rangle \le \langle (C^e_\delta)^2\rangle \le \\ \langle (C^e_{\delta'})^2\rangle \le \langle (C^e_1)^2\rangle\end{aligned} \tag{3.12}$$

where the equality on the left-hand side expresses the fact that there is no scatter at the infinite scale. Similarly, we have for the second-order moments of com pliances

$$\begin{aligned}(S^{eff})^2 = \langle (S^n_\infty)^2\rangle \ge \langle (S^n_\delta)^2\rangle \ge \\ \langle (S^n_{\delta'})\rangle \ge \langle (S^n_1)\rangle^2\end{aligned} \tag{3.13}$$

These results may be generalized to higher powers n > 2.

With the help of the above we immediately confirm the heuristic result that the scatter is strongest at the microscale $\delta = 1$, while zero at infinity, which is the deterministic continuum limit.

As noted in point vi) of Section 3.1, definitions of (3.2) and (3.3) are analogous to a moving locally averaged random field, although no direct straightforward averaging is possible, but, rather, computations must be carried out. It follows that the normalized *autocorrelation* (or *autocovariance) functions* of C_{ijkl}'s are to be obtained from micromechanics too. These are defined by

$$\rho_{ijkl}(\underset{\sim}{x}_1, \underset{\sim}{x}_2) = \frac{B_{ijkl}(x_1, x_2)}{\sigma_{ij}\sigma_{kl}} \tag{3.14}$$

where

$$B_{ijkl}(\underset{\sim}{x}_1, \underset{\sim}{x}_2) = \langle C_{ij}(\underset{\sim}{x}_1) C_{kl}(\underset{\sim}{x}_2)\rangle - \langle C_{ij}(\underset{\sim}{x}_1)\rangle\langle C_{kl}(\underset{\sim}{x}_2)\rangle \tag{3.15}$$

Henceforth we will use the terms auto-covariance function for $\rho_{1111}, \rho_{2222}, \ldots$, and cross-covariance function for $\rho_{1112}, \rho_{1112}, \ldots$.

Calculation of ρ_{ijkl}'s has to be done in the Monte Carlo sense by conducting computations of effective constitutive moduli at two different spatial positions for sufficiently many samples $B(\omega)$ from the Ω space. Since there are basically two possible definitions of effective moduli, two different autocorrelation functions will result. In Fig. 2 a) through d) we show a set of correlation functions ρ_{1111} and ρ_{1212}, ρ_{1112} and ρ_{1122} of $\underset{\sim}{C}^e_\delta$ for a matrix-inclusion composite at volume (i.e. area) fraction 22% of disks, which are ten times stiffer than the matrix. In view of the statistical homogeneity of the microstructure, all the ρ_{ijkl}'s are parametrized by $\underset{\sim}{\xi} = |\underset{\sim}{x}_1 - \underset{\sim}{x}_2|$. In Fig. 3 a) through d) we show the same functions for $\underset{\sim}{S}^n_\delta$ for the same composite. Calculations were performed at $\delta = 2.5$ over a 20 x 20 mesh in units of d/4, d = disk diameter. It is evident that the character of the correlation functions of both response tensors is the same.

Above analyses together with those carried out in the same vein by Ostoja-Starzewski & Wang (1989 and 1990) for Delaunay networks and by Ostoja-Starzewski (1993c) for random chessboards and random composites, lead to the following principal conclusions, which may be argued to hold for a number of other microstructures:

i) the auto-covariance functions of shear moduli are isotropic, while those of other moduli are anisotropic;

ii) some cross-covariance functions display a rather unexpected character: for example, ρ_{1112} in Figs. 2 c) shows that C^e_{11} and C^e_{11} are practically uncorrelated even for completely overlapping windows - as the distance $|\underset{\sim}{\xi}|$ increases beyond δ, the near zero level tends to reflect the actual independence of two separate windows;

iii) as a result of such observations as in point ii) above, the exponential correlation function, which is usually assumed in conventional stochastic finite element studies, has to be replaced by functions fitting the plots of Figs. 2 and 3;

iv) as the scale of resolution δ increases to the large and very large values, the ρ_{ijkl}'s of plots a), b), and d) in Figs. 2 and 3 become cone-shaped, and a simplest possible approximation can be set up in terms of a moving average over a homogeneous uncorrelated random field;

v) the uniform strain approximation results in practically identical autocorrelation functions as those obtained by the exact method; this suggests a very inexpensive computational method for finding the second moments.

4. RANDOM FIELDS FOR INELASTIC MATERIALS

Continuum thermomechanics of solids may be formulated on the basis of two functionals: free energy ψ and dissipation function ϕ, see e.g. (Zigler & Wehrli, 1987). As pointed out by Maugin (1992), damage and fracture enter the same framework, and hence there was the so-called *continuum damage mechanics* - a development of the eighties. However, the deterministic character of this formulation has to be contrasted with the probabilistic nature of fracture and damage phenomena at the microscale, see e.g. (Duxbury, 1990). This distinction has already been noted in (Ostoja-Starzewski, 1989), and hence, a continuum constitutive law has to be a stochastic one. Such a law is defined with the help of a window of (meso-) scale δ, recall equation (3.1). Also here, in principle, two types of boundary conditions - essential and natural - may be considered, leading in each case to two random scale-dependent functionals ψ_δ and ϕ_δ; these provide upper and lower estimates on elastic/damage response. Considering the fact that the window may be placed arbitrarily in the body domain, it plays the role of a Representative Volume Element of a continuum random field of either type, i.e. there are two approximations - upper and lower - of *stochastic continuum damage mechanics*, which becomes the conventional deterministic continuum damage mechanics in the limit of δ going to infinity.

A similar formulation may also be developed in case where the dissipative function models plasticity alone. (Ostoja-Starzewski, 1992b) generalized the theory of rigid-perfectly-plastic solids to a situation of yield limit being a (weakly) random field of space coordinates. Here the choice of δ automatically determines the mesoscale in solution of a given boundary value problem.

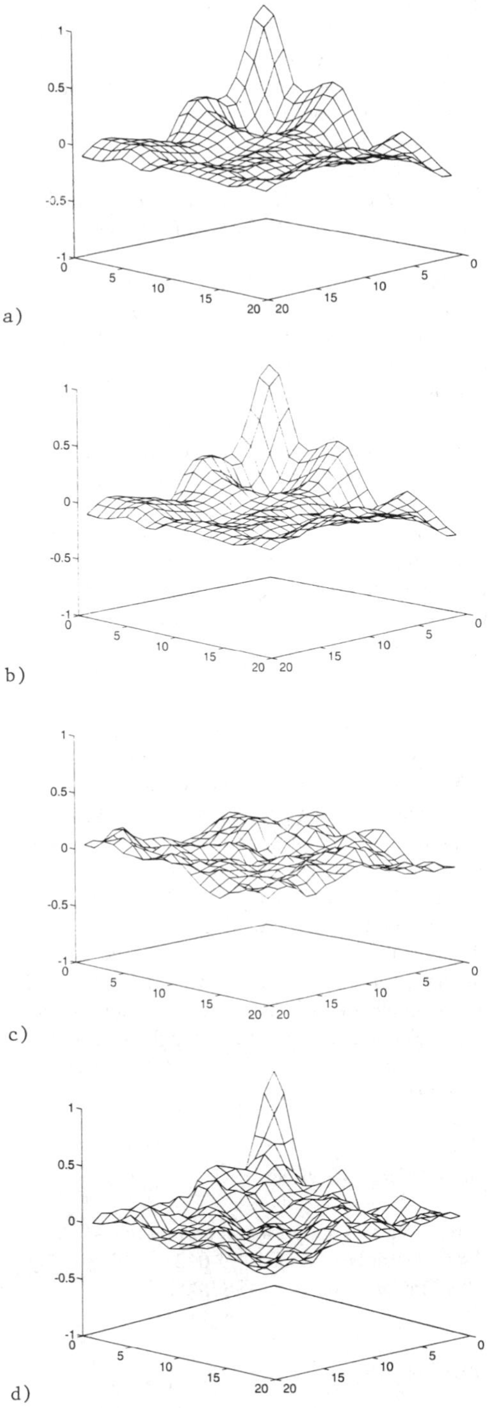

Fig 2. Correlation coefficient $\rho_{ijkl}(\underset{\sim}{\xi})$ of the components of $\underline{C}^e_\delta$; a) ρ_{1111}; b) ρ_{1212}; c) ρ_{1112}; d) ρ_{1122}; ξ_1-axis goes to the right, ξ_2-axis goes to the left.

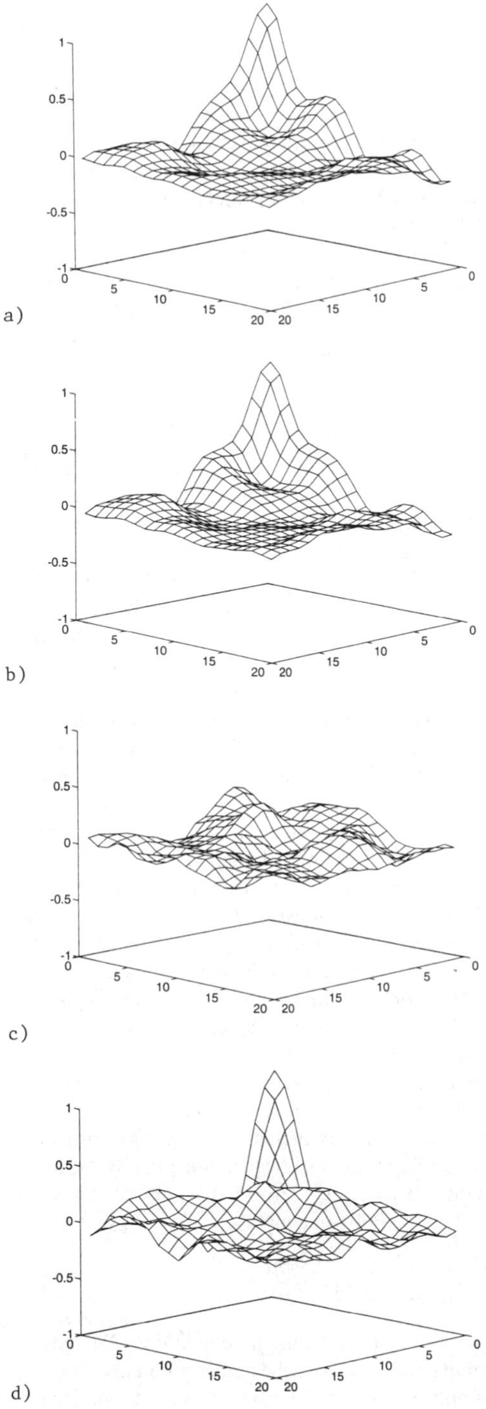

Fig 3. Correlation coefficient $\rho_{ijkl}(\underset{\sim}{\xi})$ of the components of $\underline{S}^n_\delta$; a) ρ_{1111}; b) ρ_{1212}; c) ρ_{1112}; d) ρ_{1122}; ξ_1-axis goes to the right, ξ_2-axis goes to the left.

5. SOLUTION OF STOCHASTIC BOUNDARY VALUE PROBLEMS

In this section we briefly discuss the methodology of solution of boundary value problems in elasticity. Three measuring levels may be introduced: microscale $\delta = 1$, mesoscale δ_{meso}, and macroscale δ_M, where δ_M = macroscopic dimension of the body $\boldsymbol{B}$. At this point we note that the scatter in both tensors $\underset{\sim}{S}^n_{\bar{\delta}}$ and $\underset{\sim}{C}^e_{\bar{\delta}}$ becomes eventually negligible at some large, or very large, $\bar{\delta}$ so that we arrive at an invertible constitutive law with very small scatter

$$\underset{\sim}{C}^{eff} \cong \underset{\sim}{C}^n_{\bar{\delta}} \cong \underset{\sim}{C}^e_{\bar{\delta}} \tag{5.1}$$

$\bar{\delta}$ may then be taken as the scale of an RVE of a deterministic medium $\mathbf{B}_{det}$. However, in case $\bar{\delta}$ of equation (5.1) is on the order of, or even greater than, the macroscale δ_M, one is forced to deal with spatial random fluctuations on the macroscale. Thus, a statistical rather than a deterministic continuum model has to be used. Accordingly, a solution of the problem at hand is then conducted with a certain choice of a mesoscale δ_{meso} of an RVE of the statistical continuum, which corresponds directly to the size of any given finite element. However, since two alternative definitions of boundary conditions are possible - displacement-controlled (3.2) and stress-controlled (3.3) - two different random anisotropic continua result. It follows that a given boundary value problem may then be solved analytically or numerically (by e.g. stochastic finite elements) to find the upper and lower bounds on response according as $\underset{\sim}{C}^e_\delta$ and $\underset{\sim}{C}^n_\delta$ are used. The above features, dictated by micromechanics, distinguish our approach from that of the conventional stochastic finite element methods; see e.g. (Liu, Belytschko & Mani 1986), (Benaroya & Rehak 1988), (Ghanem & Spanos 1991).

An example problem has recently been treated by Alzebdeh and Ostoja-Starzewski (1993). It concerns the elastostatics of a membrane having the microstructure of a matrix-inclusion composite with both phases being locally isotropic. Specifically, we studied the (out-of-plane) deformations $u(x_1, x_2)$ governed by the Poisson equation

$$\frac{\partial}{\partial x_i}\left(C_{ij}(\underset{\sim}{x}, \omega)\frac{\partial u}{\partial x_j}\right) = f(\underset{\sim}{x}) \tag{5.2}$$

under the Dirichlet boundary conditions. We note here that this also describes other problems - for example, torsion of elastic composites or conductivity in presence of heat sources. $C_{ij}(x, \omega)$ in (5.2) correspond to either a random continuum approximation $\boldsymbol{B}^e_\delta$ or $\boldsymbol{B}^n_\delta$ at some scale δ - that is, they are components, and realizations, of the random tensor fields $\underset{\sim}{C}^e_\delta$ or $\underset{\sim}{C}^n_\delta$. We use here the same type of notation for conductivity and out-of-plane elasticity, whereby $\underset{\sim}{\varepsilon}$ and $\underset{\sim}{\sigma}$ are vectors, C_{ij} is conductivity, and its inverse S_{ij} is resistivity.

The above membrane problem was solved in the following steps:

a) making a choice of a finite difference (or finite element) net for solution of (5.2), which immediately specifies δ;

b) generation, in a Monte Carlo sense, of each realization $\mathbf{B}(\omega)$ of a matrix-inclusion composite and calculation of $\underset{\sim}{C}^e_\delta(\omega)$ under (3.2), and $\underset{\sim}{C}^n_\delta(\omega)$ under (3.3), for each one of the windows;

c) solution of (5.2) by a finite element method according as $\underset{\sim}{C}^e_\delta$ or $\underset{\sim}{C}^n_\delta$ are used for each specific ω to get bounds on response; by repeating it a number of times, in a Monte Carlo sense, a statistical range of reponses is obtained.

It is interesting to observe here that other analytical methods for stochastic finite elements (such as Neumann series) could be used in place of a Monte Carlo approach, to obtain the upper and lower bounds on response. Also, a finer net could be used for determination of the effective moduli of windows. However, the computational expense in the micromechanical characterization of two continuum random fields of material properties goes up with the window size. On the other hand, the computational expense in the finite element solution goes down with the coarseness of the mesh. Thus, it appears that in the micromechanically-based stochastic boundary value problems there will be an optimum δ_{meso} leading to the lowest total computational costs.

Finally, a discussion of other issues relating to a particular choice of a stochastic variational formulation, the choice of a triangular versus quadrilateral mesh, and choices of linear versus higher-order interpolation functions are discussed in (Ostoja-Starzewski, 1993c).

ACKNOWLEDGEMENT

This research was supported in part by the AFOSR under Contract No. AFOSR-89-0423 and in part by the NSF under Grant No. MSS 9202772.

REFERENCES

Alzebdeh K & Ostoja-Starzewski M 1993. Micromechanically based stochastic finite elements, Finite Elem. Anal. Design, in press.

Benaroya H. & M. Rehak 1988. Finite element methods in probabilistic structural analysis: a selective review. Appl. Mech. Rev. 41(5): 201-213.

Duxbury, P.M. 1990. Breakdown of diluted and hierarchical systems, in Statistical Models for the Fracture of Disordered Media, Elsevier (North-Holland).

Ghanem, R. & P.D. Spanos 1991. Stochastic finite elements: a spectral approach, Springer-Verlag.
Huet, C. 1990. Application of variational concepts to size effects in elastic heterogeneous bodies. J. Mech. Phys. Solids 38(6): 813-841.

Huet, C. 1991. Hierarchies and Bounds for Size Effects in Heterogenous Bodies, in Continuum Models and Discrete Systems - Vol. 2 (G.A. Maugin, Ed.): 203-210, LONGMAN Scientific and Technical.

Liu, W.K., T. Belytschko & A. Mani 1986. Random field finite elements. Int. J. Num. Meth. Engng. 23: 1831-1845.

Maugin, G.A. 1992. The Thermomechanics of Plasticity and Fracture, Cambridge University Press.

Ostoja-Starzewski, M. 1989. Damage in a Random Microstructure: Size Effects, Fractals, and Entropy Maximization. Appl. Mech. Rev. 42 (11, Part 2): S202-S212.

Ostoja-Starzewski, M. 1992a. Random fields and processes in mechanics of granular materials, in Advances in Micromechanics of Granular Materials (Proc. 2nd US-Japan Seminar on Micromechanics of Granular Materials; H. Shen et al, Eds.); Stud. Appl. Mech. 31: 71-80, Elsevier; extended version in Mech. Mater. in press.

Ostoja-Starzewski, M. 1992b. Plastic Flow of Random Media: Micromechanics, Markov Property and Slip-Lines, Appl. Mech. Rev. (Special Issue: Material Instabilities) 45 (3, Part 2): S75-S82.

Ostoja-Starzewski, M. 1993a. Micromechanics as a basis of random elastic continuum approximtions. Probabilistic Engng. Mech. 8(2): 107-114.

Ostoja-Starzewski, M. 1993b. Micromechanics as a basis of continuum random fields. Appl. Mech. Rev. (Special Issue: Micromechanics of Random Media), in press.

Ostoja-Starzewski, M. 1993c. Micromechanics as a basis of stochastic finite elements and differences - an overview. Appl. Mech. Rev. (Special Issue: Mechanics Pan-America 1993), in press.

Ostoja-Starzewski, M. & C. Wang 1989. Linear elasticity of planar Delaunay networks: random field characterization of effective moduli. Acta Mech. 80: 61-80.

Ostoja-Starzewski, M. & C. Wang 1990. Linear elasticity of planar Delaunay networks - II: Voigt and Reuss bounds, and modification for centroids. Acta Mech. 84: 47-61.

Shinozuka, M. & F. Yamazaki 1988. Stochastic finite element analysis: an introduction. Stochastic Structural Dynamics: Progress in Theory and Applications (S.T. Ariaratnam, G.I. Schuëller & I. Elishakoff, Eds.), 241-291, Elsevier Applied Sci., London.

Ziegler, H. & Ch. Wehrli 1987. The derivation of constitutive relations from the free energy and the dissipation functions, Adv. Appl. Mech. 25: 182-238, Academic Press.

Structural Safety & Reliability, Schuëller, Shinozuka & Yao (eds) © 1994 Balkema, Rotterdam, ISBN 90 5410 357 4

A study on temperature dependence of flexural strength distribution of alumina ceramics

Tatsuo Sakai
Faculty of Science and Engineering, Ritsumeikan University, Kita-ku, Kyoto, Japan

Yoshinobu Nakama
School of Science and Engineering, Ritsumeikan University, Kita-ku, Kyoto, Japan

Richard C. Bradt
Mackay School of Mines, University of Nevada-Reno, Nev., USA

ABSTRACT: Many kinds of engineering ceramics have been developed and used as parts of mechanical structures. These ceramics are often used as machine parts which are exposed to elevated temperatures. The strength distribution of those materials significantly depends upon the temperature. In this study, distribution characteristics of the flexural strength for alumina ceramics were examined at the temperatures of room temperature (RT), 800℃, 1000℃ and 1200℃, respectively. Based on the temperature dependence of distribution parameters, the Probability–Strength–Temperature characteristics were quantitatively analysed and an analytical model was proposed on the temperature dependence of the strength distribution.

1 INTRODUCTION

Many kinds of engineering ceramics have been developed and used as parts of mechanical structure. When these ceramics are applied to the mechanical use, it is the most critical probrem that they are brittle and the strength has a large scatter (Sakai and Fujitani (1989)). These ceramics are often used as machine parts which are exposed to elevated temperatures. In such a case, the strength distribution at the high temperature governs the reliability of the fabricated ceramic components, and it is important to clarify the temperature dependence of the strength distribution.

In this study, distribution characteristics of the flexural strength for alumina ceramics were first examined at room temperature (RT), 800℃, 1000℃ and 1200℃, respectively. Based on the temperature dependence of distribution parameters, the strength distribution at any temperature was given quantitatively as Probability–Strength–Temperature characteristics. An analytical model for the temperature dependence of the strength distribution was finally constructed by combining the crack size distribution and the temperature dependences of the defect sensitivity and the matrix strength. Analytical results thus obtained were in good agreement with the experimental aspect of the strength distributions in a wide range of the testing temperature.

Table 1. Typical properties of material

Alumina content	99.98[%]
Specific gravity	3.95
Vickers hardness	20.6[GPa]
Young's modulus	386[GPa]
Coefficient of thermal expansion	0.84×10^{-5} [1/K]
Thermal conductivity	37[W/m K]
Specific heat	0.78[J/g K]

2 SPECIMEN AND EXPERIMENTAL PROCEDURE

The material used in this study is engineering alumina ceramics for the mechanical use which are supplied by Kyocera Corporation. Typical properties of this material are listed in Table 1. Specimens were machined into the configuration specified in JIS R1601 [width (b=3mm) x height (h=4mm) x length (l_0=40mm)]. Three point flexural test in this standard were carried out by means of the loading system as shown in Fig.1. The flexural strength σ was calculated from the ultimate load for the fracture and geometries of the specimen and loading device as follows;

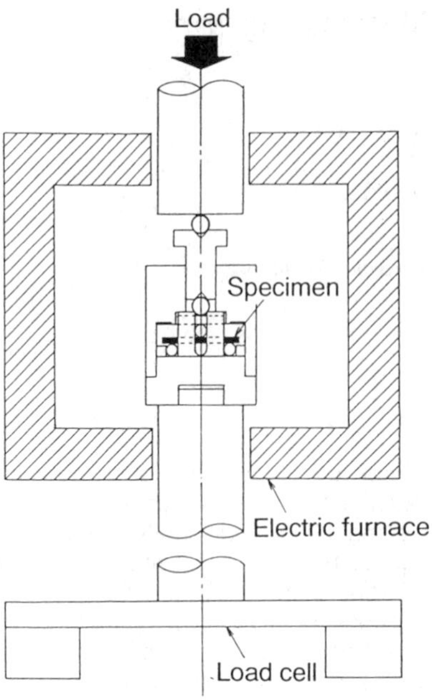

Fig.1 Loading system for 3-point flexural test

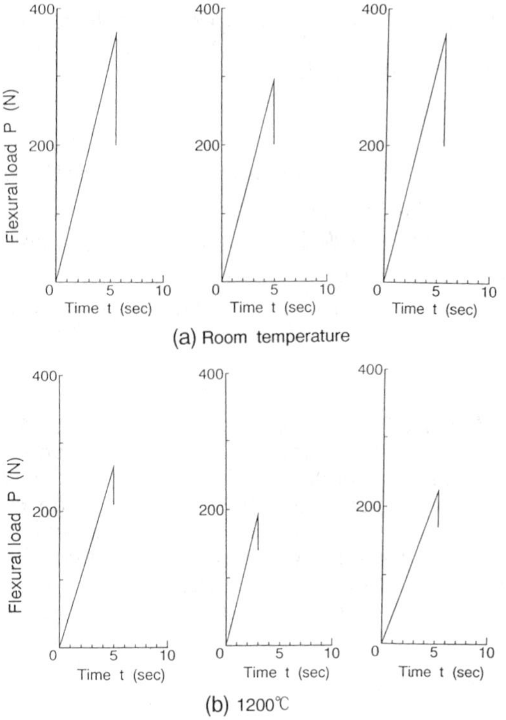

Fig.2 Typical examples of flexural load vs. time diagrams

$$\sigma = \frac{3Pl}{2bh^2} \qquad (1)$$

where P is the ultimate load and 1 the span of supports. The flexural load was plotted on X-Y recorder as a function of the time. Typical examples are shown in Fig.2. The ultimate load P was easily obtained from the peak of each diagram. In order to clarify distribution characteristics of the strength, tests were repeatedly performed by using 20 specimens at each temperature of RT, 800℃, 1000℃ and 1200℃, respectively.

3 EXPERIMENTAL RESULTS

Experimental distributions of the flexural strengths at the respective temperatures are shown as Weibull plots in Fig.3. The value of $F(\sigma_i)$ for the i'th smallest strength σ_i is calculated by $F(\sigma_i)= (i-0.5)/n$, where n is total number of specimens tested at each temperature (Sakai and Tanaka (1982)). The excellent linearity is confirmed on the results at RT and 800℃, but the distribution characteristics at 1000℃ and 1200℃ are no longer represented by any straight line in the figure. This fact means that the distribution can be well approximated by two-parameter Weibull distribution at RT and 800℃, whereas it should be represented by other type of distribution such as three-paramter Weibull distribution in the higher temperature region. Probability density and distribution functions in Weibull type are written as follows;

$$f(\sigma)=\frac{m}{\beta}\left(\frac{\sigma-\gamma}{\beta}\right)^{m-1}\exp\left\{-\left(\frac{\sigma-\gamma}{\beta}\right)^{m}\right\} \qquad (2)$$

$$F(\sigma)=1-\exp\left\{-\left(\frac{\sigma-\gamma}{\beta}\right)^{m}\right\} \qquad (3)$$

where m, β and γ are shape,scale and location parameters, respectively. These parameters estimated by the correlation coefficient method proposed by the authors(Sakai and Tanaka (1984)) are listed in Fig.3.

Temperature dependence of these parameters is plotted in Fig.4. Shape parameter m and scale parameter β decrease with an increase of the temperature. The marked decreasing is found around T=900℃ for both parameters. However, another parameter γ keeps almost zero in the region of T<900℃ and it increases distinctly as the temperature exceeds the critical level T_c.

Based on the temperature dependence of the respective parameters, one can calculate

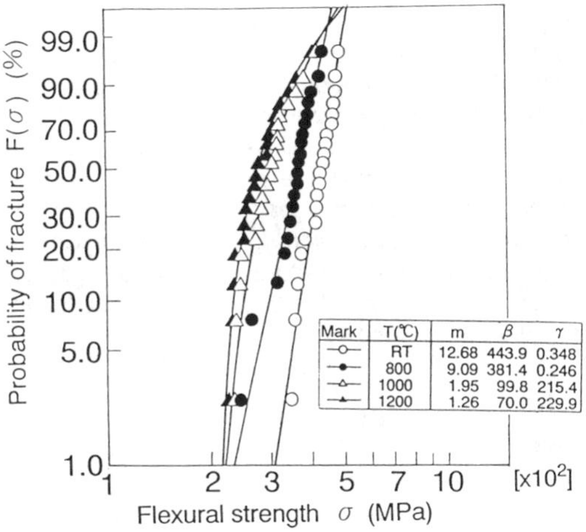

Fig.3 Weibull plots of flexural strength
(Experimental results and fitted distributions)

Mark	T(℃)	m	β	γ
—○—	RT	12.68	443.9	0.348
—●—	800	9.09	381.4	0.246
—△—	1000	1.95	99.8	215.4
—▲—	1200	1.26	70.0	229.9

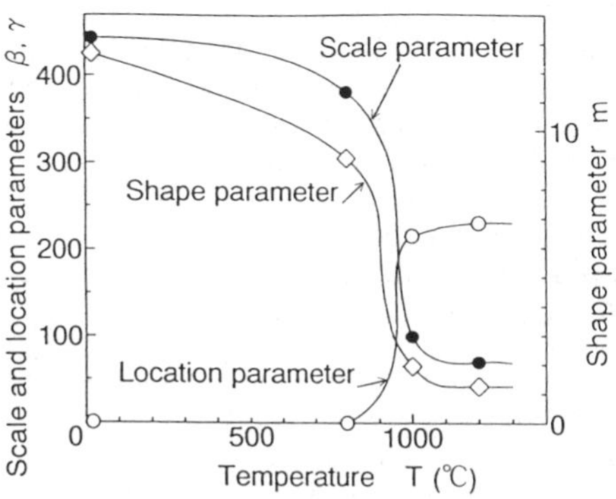

Fig.4 Temperature dependence of
Weibull parameters

the strength distribution at any temperature.
Therefore, the relationship of the strength vs.
the temperature can be obtained for any level
of fracture probability by substituting parameter
values at any temperature and given level of
fracture probability into Equation (2) or (3).
Probability–Strength–Temperature
characteristics thus obtanied are depicted in
Fig.5. A dashed curve indicates the
temperature dependence of the location
parameter γ. This curve provides the lower
bound for the occurrence of the specimen
fracture. Typical aspect on this parameter is
the fact that the value of γ is negligible in
$T<T_c$ and keeps a definite level in $T>T_c$. This
feature would be attributed to the occurrence
of plastic flow in the specimen in the region of

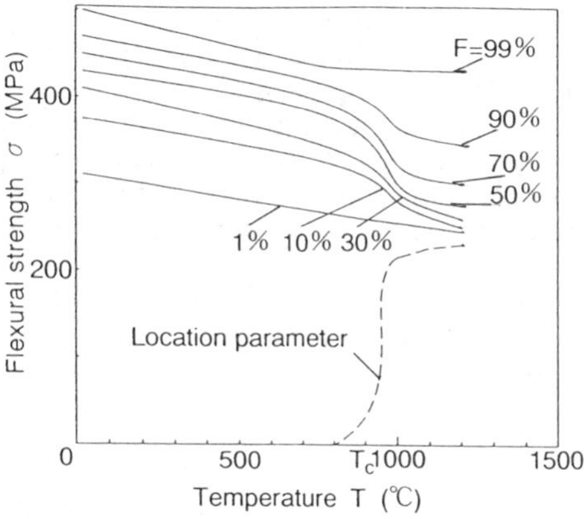

Fig.5 Probability–Strength–Temperature
characteristics fitted to experimental results

$T>T_c$ as discussd later.

4 ANALYTICAL DISCUSSION

4.1 Flexural Strength Distribution at Room Temperature

Flexural strength distribution of the present
material at room temperature is well
represented by two–parameter Weibull
distribution as shown in the previous section.
Thus we have its probability density and
distribution functions by putting $γ=0$ in Eqs. (2)
and (3) as follows;

$$f(\sigma)=\frac{m}{\beta}\left(\frac{\sigma}{\beta}\right)^{m-1}\exp\left\{-\left(\frac{\sigma}{\beta}\right)^m\right\} \quad (4)$$

$$F(\sigma)=1-\exp\left\{-\left(\frac{\sigma}{\beta}\right)^m\right\} \quad (5)$$

4.2 Crack Size Distribution

The maximum stress is induced at the
bottom surface of the specimen under the
flexural load in this work. Suppose that the
fracture strength is governed by the size of the
largest microcrack on the specimen surface.
Then one can derive the distribution
characteristics of the crack size from the
fracture strength distribution. When the crack
shape is assumed by a semi–ellipse with the
diameter of 2a, the flexural strength of the
specimen is given by

$$\sigma=\frac{K_{IC}}{\sqrt{\pi a}}=\xi a^{-1/2} \quad (6)$$

where K_{IC} is fracture toughness and $\xi = K_{IC}/\pi^{1/2}$. The value of K_{IC} is put as K_{IC}=4 (MPa•$m^{-1/2}$) making reference to some reports (Wakai, Sakuramoto, Sakaguchi and Matsuno (1986), Kobayashi, Nakamura, Kitayama and Ohote (1986)). Since K_{IC} is the material constant, ξ is also a constant peculiar to the material.

Probability density function of the crack size is now derived transforming the random variable σ into a;

$$f_c(a)=f(\sigma)\left|\frac{d\sigma}{da}\right| \tag{7}$$

Differentiating σ of Eq.(6) with respect to a, $d\sigma/da$ is easily obtained and $f(\sigma)$ is already provided by Eq.(4). Thus we have the probability density function of the crack size as follows;

$$f_c(a)=\frac{\xi}{2a^{3/2}}\frac{m}{\beta}\left(\frac{\xi}{\beta\sqrt{a}}\right)^{m-1}\exp\left\{-\left(\frac{\xi}{\beta\sqrt{a}}\right)^{m}\right\} \tag{8}$$

Then the distribution fuction is given by

$$F_c(a)=\int_0^a f_c(a)da \tag{9}$$

Probability density function of the crack size calculated by Eq.(8) is depicted in Fig.6. Since the fracture strength of specimen is governed by the largest crack of the specimen surface, the distribution in Fig.6 provides the distribution characteristics of the size for the largest crack.

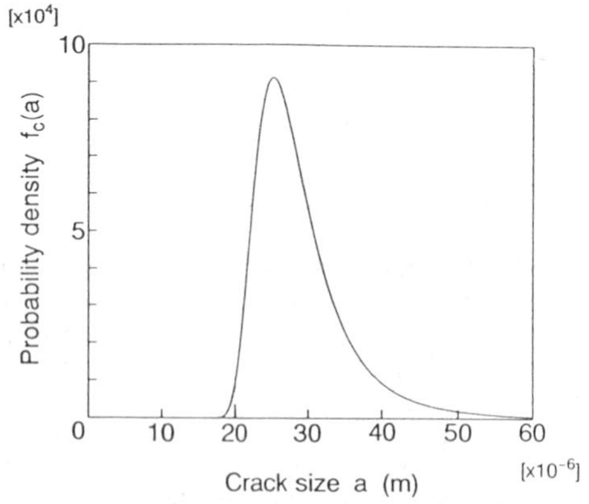

Fig.6 Probability density of crack size

4.3 Defect Sensitivity

Structural ceramics have many defects such as microcracks, fine pores and grain boundaries, which are inevibably introduced in their fabrication processes. Every defect of these different types on the specimen surface are replaced by semi-ellipse in this work as an attempt. The larger the crack size "a" is, the smaller the critical load to the fracture becomes. In other words, the crack becomes sensitive to the fracture as the size is increased. Thus, sensitivity of a defect "s" is introduced as follows;

$$\sigma=\frac{K_{IC}}{\sqrt{\pi a}}=\frac{K_{IC}}{s} \tag{10}$$

where

$$s=\sqrt{\pi a} \quad (T<T_c) \tag{11}$$

If the temperature exceeds the critical level T_c, distribution pattern of the flexural strength changes drastically on the Weibull plot as mentioned previously. On this aspect, an analytical model is constructed by considering the temperature dependence of the defect sensitivity. In the case of alumina specimens, plastic flow begins to occur at 800~1000℃(Davidge(1979)). Such a plastic flow is facilitated at the crack tip due to the stress concentration under a given load. When the flow takes place at the crack tip, the defect sensitivity tends to decrease due to the stress relaxation. It is noted that the lower bound (location parameter) appears on the strength distribution as the temperature exceeds the critical level T_c. This feature suggests that the defect sensitivity tends to saturate to a certain value even if the crack size is increased. In order to represent all of

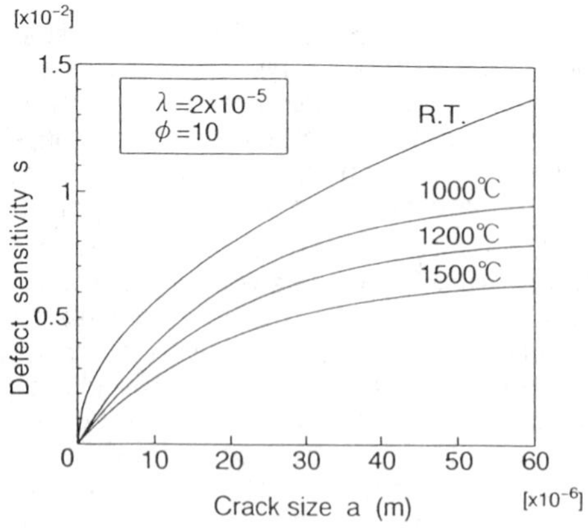

Fig.7 Relationship between defect sensitivity and crack size

these aspects, the following expression is assumed on the temperature dependence of the sensitivity;

$$s=\frac{\phi}{T}\left\{1-\exp\left(-\frac{a}{\lambda}\right)\right\} \quad (T \geqq T_c) \qquad (12)$$

where ϕ and λ are material constants. In the case of $T<T_c$, it is also assumed that the defect sensitivity is constant and is given by Eq.(11).

By putting $\phi=10$ and $\lambda=20\times10^{-6}$, the defect sensitivity was calculated at T=1000℃,1200℃ and 1500℃ by Eq.(12) and plotted in Fig.7. The defect sensitivity in $T<T_c$ given by Eq.(11) is also plotted by a curve for RT.

4.4 Flexural Strength Distribution at Elevated Temperature

In general, the strength of ceramics tends to decrease with an increase of the testing temperature(Lange(1974)). This is supposed to be true even if the material has no defect. The strength of the material without any defect is hereafter referred to matrix strength of the material and denoted by σ_M. On the temperature dependence of this matrix strength, the following expression is employed from a viewpoint of the reaction rate process(Glasstone,Laidler and Eyring (1941));

$$\sigma_M=\sigma_0\left[1-\exp\left\{\frac{-\mu(T_L-T)}{BT}\right\}\right] \qquad (13)$$

where T_L is melting temperature, B is Boltzman constant and σ_0 is matrix strength at absolute zero.

Similar expression was used on the temperature dependence of the fracture toughness for the matrix K_M. Thus we have

$$K_M=K_0\left[1-\exp\left\{-\frac{\mu(T_L-T)}{BT}\right\}\right] \qquad (14)$$

where K_0 is fracture toughness at absolute zero.

Replacing K_{IC} in Eq.(10) by above K_M, we obtain the flexural strength at any temperature as follows;

$$\sigma=\frac{K_0}{\sqrt{\pi a}}\left[1-\exp\left\{\frac{-\mu(T_L-T)}{BT}\right\}\right] \quad (T<T_c) \qquad (15)$$

$$\sigma=\frac{TK_0\left[1-\exp\left\{\frac{-\mu(T_L-T)}{BT}\right\}\right]}{\phi\left\{1-\exp\left(-\frac{a}{\lambda}\right)\right\}} \quad (T \geqq T_c) \qquad (15)'$$

Now, by using the following formula on transformation of random variables;

$$f(\sigma)=f_c(a)\left|\frac{da}{d\sigma}\right| \qquad (16)$$

one can derive the strength distribution at any temperature as follows;

$$f(\sigma)=\frac{m}{\beta_0}\left(\frac{\sigma}{\beta_0}\right)^{m-1}\exp\left\{-\left(\frac{\sigma}{\beta_0}\right)^m\right\} \quad (T<T_c) \qquad (17)$$

$$f(\sigma)=\frac{m}{2\beta_1}\frac{r}{\sigma(\sigma-r)}\frac{1}{\left(\log\frac{\sigma}{\sigma-r}\right)^{3/2}} \times\left\{\frac{1}{\beta_1\left(\log\frac{\sigma}{\sigma-r}\right)^{1/2}}\right\}^{m-1}\exp\left[-\left\{\frac{1}{\beta_1\left(\log\frac{\sigma}{\sigma-r}\right)^{1/2}}\right\}^{m}\right] \quad (T \geqq T_c) \qquad (17)'$$

where

$$\left.\begin{aligned} \beta_0&=\frac{\beta K_0}{\sqrt{\pi}\,\xi}\left[1-\exp\left\{-\frac{\mu(T_L-T)}{BT}\right\}\right] \\ \beta_1&=\frac{\sqrt{\lambda}\,\beta}{\xi} \\ r&=\frac{TK_0}{\phi}\left[1-\exp\left\{\frac{-\mu(T_L-T)}{BT}\right\}\right] \end{aligned}\right\} \qquad (18)$$

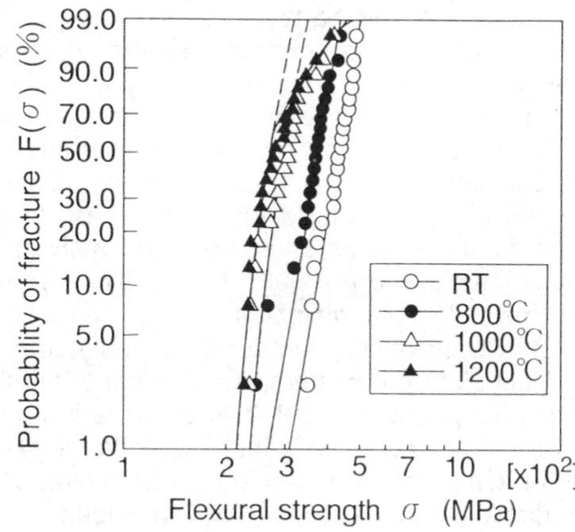

Fig.8 Weibull plots of flexural strength (Experimental results and analytical distributions)

Therefore, the cumulative distribution of the flexural strength is given by integration of Eq.(17) or (17)'. By putting K_0=4.0, μ=1.0x10^{-3} and T_L=1600, the cumulative distribution was calculated at each temperature of RT, 800℃, 1000℃ and 1200℃. These analytical distributions are all indicated by solid curves in Fig.8, where the experimental results at the respective temperatures are also replotted for the sake of comparison. Analytical results at 1000℃ and 1200℃ show a disagreement from the experimental ones in higher strength region of the distribution. This fact suggests that the defect sensitivity of initial cracks smaller than the critical size is decreased at such high temperatures. With respect to this aspect, the sensitivity was modified as to provide the weakened level in small cracks of a<24μm. Analytical distribution thus obtained are plotted by solid curves in Fig.8, where dashed and solid curves are almost overlapped at RT and 800℃. The experimental distribution pattern at every temperature is well explained by the present analytical model.

It should be noted that solid curves in Fig.3 indicate fitted results by correlation coefficient method whereas the curves in Fig.8 give analytical strength distributions derived by the present model.

5 CONCLUDING REMARKS

Temperature dependence of the flexural strength distribution for alumina was examined experimentally and an analytical model was constructed. Main conclusions obtained in this study are summarized as follows;

1) Distribution pattern of the flexural strength changes drastically at the critical temperature T_C. The strength distribution in $T<T_C$ is well represented by two-parameter Weibull distribution, but lower bound of the distribution distinctly increases as the temperature exceeds T_C so that the distribution characteristics are well fitted by three-parameter Weibull distribution.

2) Based on the temperature dependence of distribution parameters, the flexural strength at any temperature is obtained under any given level of fracture probability. The entire aspect on the temperature dependence of the strength distribution is successfully provided as Probability-Strength-Temperature characteristics.

3) The distribution of surface crack size is first derived from the strength distribution at room temperature. Then the defect sensitivity and its temperature dependence are formulated togther with the temperature dependence of the matrix strength. By combining the crack size distribution and the temperature dependences of the defect sensitivity and the matrix strength, an analytical model was finally constructed on the temperature dependence of the strength distribution. Analytical results thus obtained were in good agreement with the experimental aspects in a wide range of the testing temperature.

REFERENCES

SAKAI,T. and FUJITANI,K. 1989. A statistical aspect on fatigue behavior of alumina ceramics in rotating bending. Engng. Fract. Mech., 32:653-664.

SAKAI,T. and TANAKA,T. 1982. Reliability design of machines and structures. J. Soc. Mat. Sci., Japan, 31:941-947.

SAKAI,T. and TANAKA,T. 1984. Parameter estimation of Weibull-type-fatigue life distributions including non-failure probability. Proc. FATIGUE84, Vol.2, 1125-1137.

WAKAI,F. SAKURAMOTO,H. SAKAGUCHI,S. and MATSUNO,Y. 1986. Evaluation of crack propagation in ceramics by double-torsion. J. Soc. Mat. Sci., Japan, 35:898-903.

KOBAYASHI,H. NAKAMURA,H. KITAYAMA,A. and OHOTE,S. 1986. Evaluation of fracture toughness for ceramics by chevron-notched specimen. J. Soc. Mat. Sci., Japan, 35:892-897.

DAVIDGE,R.W. Mechanical Behaviour of Ceramics, Chapter 5, (1979), Cambridge University Press.

Lange,F.F. J. Am. Ceram. Soc., 57-2(1974), 84

Glasstone,S. Laidler,K.J. and Eyring,H. The Theory of Rate Processes, 1(1941), McGraw-Hill Book Co., Imc.

Structural Safety & Reliability, Schuëller, Shinozuka & Yao (eds) © 1994 Balkema, Rotterdam, ISBN 90 5410 357 4

Strength characterization of high performance fiber composites

L. Sentler
Lund Institute of Technology, Department of Structural Engineering, Sweden

ABSTRACT: High performance fiber composites will in addition to an elastic response in a direct loading to failure easily exhibit a viscous response when subject to high sustained or variable stresses. Therefore a strength characterization has to be done in the strain domain. The elastic and the viscous strain capacities will be influenced by existing flaws subject to stress or by the growth of flaws under the action of a sustained stress respectively. Based on an extreme value approach two failure surfaces in the space - time domain are derived for each form of response. The result is a stochastic viscoelastic strength characterization of fiber composites where also the highly anisotropic nature of these materials is considered.

1 INTRODUCTION

Fiber composites are materials with several interesting properties. In particular, if the fibers are oriented in one direction extremely high strength characteristics can be obtained. Such fiber composites are often termed high performance fiber composites. At the same time many of these materials exhibit extremely good durability properties as well as other desirable qualities. This make fiber composites of interest in many structural applications.

The commercial use of fiber composites has made it necessary with a trustworthy rupture characterization. This is in particular the case for the behaviour of fiber composites subject to high sustained stresses. In such applications it is necessary that the time dependence of the rupture strength is accounted for properly in addition to the size dependence. This makes presently used reliability methods less appropriate and an extended reliability evaluation methodology is necessary.

1.2 Background

A theoretical strength theory for loose bundles of fibers based on the law of large numbers was proposed by Daniels (1945). The analytical treatment of unidirectional fibrous composites based on actual test information indicated that the weakest link theory developed by Weibull (1939) made more sense. This was the conclusion in the papers of Rosen (1964) and Zweben (1968). A particular model of this type was presented by Harlow and Phoenix (1981).

The statistical justification of these and other simplified models was examined by Smith (1980) and later by Batdorf (1982). Bader and Priest (1982) provided the first major experimental investigation. Based on this information Watson and Smith (1985) have made an extensive evaluation of possible models where in particular the weakest link theory was evaluated.

The mechanical aspects of the weakest link theory were examined by Sentler (1987). Based on the material response, which can reflect either an elastic strain or a viscous strain, an extension and generalization of the weakest link theory was derived for isotropic materials. In this theory both a size and a time dependence are addressed. An extension of this theory where anisotropic material characteristics are considered is given by Sentler (1988).

2 FIBER COMPOSITES

A high performance fiber composite is made up of parallel fibers in a matrix. Fibers of commercial interest are glass fibers, aramid fibers and carbon fibers. Matrixes normally used are thermosettings like vinyl ester resins or epoxy resins if a high durability is pursued. In other cases polyester resins are used.

The matrix which surrounds the fibers has two important functions, it protects the fibers against the surrounding environment and it makes it possible to redistribute stresses between fibers.

Typical for fibers are that they are far from perfect. The diameter varies between fibers as well as along the length of each fiber. There are also flaws or defects on the surface of a fiber. For a fiber composite as a whole there will be defects in the matrix itself in terms of air voids and bonding defects in the interphase between fibers and the matrix.

2.1 Fiber composite behaviour

The strength properties of a fiber composite will depend on the type of fibers, the relative fiber content, the type of matrix and the production method. This makes it possible to obtain strength characteristics within a wide range. These properties are not in general a geometrical average of the properties of the constituents. This is in particular the case for high performance fiber composites where an interaction between fibers and the matrix takes place. The rupture strength of a fiber composite with a high relative fiber content can be considerably higher than that of the fibers alone. This is because a relatively high size dependence of individual fibers in the length direction will be reduced by the matrix. This is illustrated with the result given in Table 1 for single carbon fibers, bundles of carbon fibers alone and carbon fibers in an epoxy matrix (Bader and Priest, 1982).

Table 1 Summary of test data from (Bader and Priest, 1982).

Length mm	Number of obs.	Mean GPa	St.dev. GPa	COV
Single fibers				
1	57	4.24	0.85	0.20
10	64	3.05	0.62	0.20
20	70	2.45	0.49	0.20
50	66	2.25	0.41	0.18
Bundles of fibers				
5	28	1.92	0.07	0.036
20	25	1.68	0.10	0.059
100	29	1.58	0.13	0.082
200	27	1.38	0.11	0.080
Impregnated bundles				
20	28	2.82	0.16	0.056
50	30	2.81	0.17	0.060
150	32	2.68	0.19	0.071
300	29	2.50	0.23	0.092

The obvious size dependence of the rupture strength of fibrous materials has been one reason to use a weakest link theory. The original Weibull theory also explains the size dependence of the mean values for all test series in Table 1 reasonably well. But a maximum likelihood analysis methodology will not confirm a weakest link theory (Watson and Smith, 1985). Not even for single fibers where the coefficient of variation is reasonably consistent with the theory. For fiber bundles, impregnated or not, there is a change of the coefficient of variation for increasing lengths which indicates that something else also could be taking place. The fiber bundle theory (Daniels, 1945) can be shown to be less appropriate, especially to predict the length dependence (Watson and Smith, 1985).

New test information has revealed that a fiber composite can rupture in two different ways. For a relatively fast increase of a stress a failure will most likely occur because of an existing defect in the fiber composite. For a slow increase of a stress or a high constant stress a failure will occur because of a growth of small micro cracks in the interphase between fibers and the matrix. This brings back the size dependence of individual fibers and it triggers off local fiber breaks which eventually will result in a total failure. This makes the time dependence more pronounced.

The failure stress will be reduced in relation to the duration of the stress in the way shown in Figure 1 (Overbeck, 1987). Two types of time dependencies can be identified which reflect two different failure modes. The importance of the type of matrix is also evident.

An interesting feature with fiber composites is that both failure forms can occur in the duration range of ordinary tests. Since the final failures often are explosive in nature it can be difficult to differentiate them from each other. A close investigation of the failure surface can sometimes reveal the nature of a failure. Instead the failure form has to established from the difference in time dependence in the way shown in Figure 1. There will also be a difference in the size dependence which could be used.

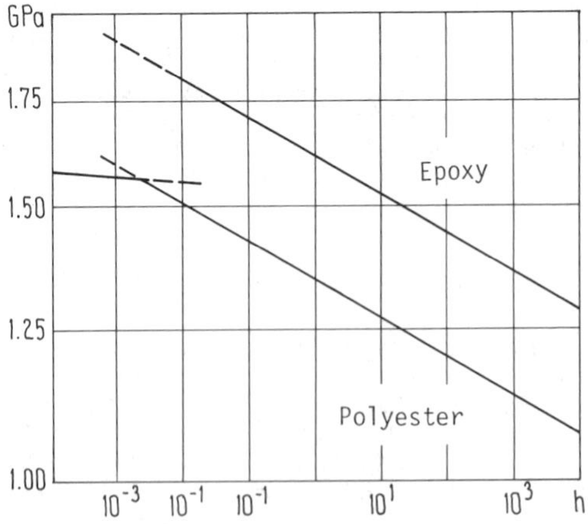

Fig.1 The rupture stress as a function of the duration of a stress (Overbeck, 1987).

3 VISCOELASTIC STRAIN RESPONSE

The strain response of a fiber composite as well as most other materials subject to a tensile stress is not limited to an initial strain which reflects an applied stress. This is especially pronounced for a sustained stress where the initial strain will be followed be creep. For low stress levels only primary creep will occur, but for an initial stress above a certain threshold stress, primary creep will be followed by secondary creep and eventually by tertiary creep under certain conditions. These responses are illustrated in Figure 2 for sustained stresses of different magnitudes and for a continuously increasing stress.

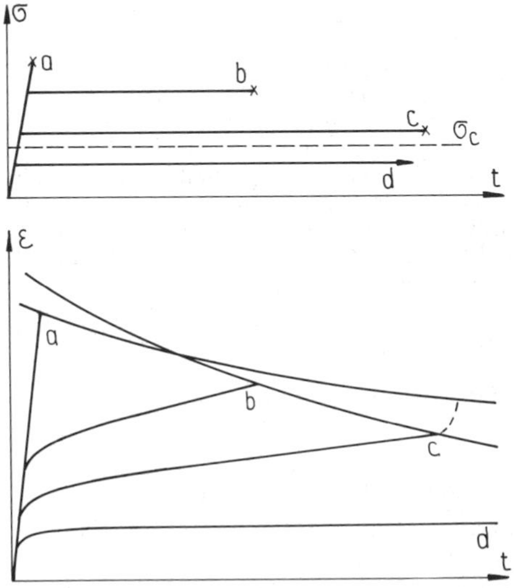

Fig. 2 Time dependent material responses.

The strain response can permanently alter the material properties. In this context it is assumed that such changes need not be considered out of mathematical reasons. This is a common assumption if tertiary creep, which is of little engineering interest, is not considered (Penny, 1971). This makes it possible to separate otherwise dependent variables as

$$\epsilon = g(\sigma,t,T) = g_1(\sigma)g_2(t)g_3(T) \tag{1}$$

where ε is the strain response as a function of the applied stress σ, the time t and the temperature T. In this context temperature is not included even if it easily can be accounted for.

Dependent on the type of material and stress conditions equation 1 will reflect either a strain hardening or a strain softening. For fiber composites a strain softening can be expected.

The strain responses presented in Figure 2 can be represented in the following form

$$\epsilon(t) = \epsilon_e t^{\alpha} + \epsilon_v t \tag{2}$$

where ε_e is an elastic response and ε_v is a viscous response.

For low stress levels the response is elastic. But there will also be a delayed time dependent strain response. For a constant stress this elastic response is often expressed as an extension to Hook´s law as

$$\epsilon_e t^{\alpha} = \frac{\sigma}{E}\left(1 + \frac{\alpha \ln(t)}{1!} + \dots\right) \tag{3}$$

where σ is the applied stress, E is the elastic modulus and the parameter $\alpha < 1$. The series expansion of the time term reveals that there is an initial time independent response and a time dependent creep response. This primary creep is normally recoverable at unloading but delayed in time. Also in a fast loading to failure the response can be mainly elastic of the type given by equation 3.

Secondary creep which takes place above a certain stress level without any increase of the stress is a viscous strain which often is represented with the Norton creep law. The strain rate for a constant stress can be expressed as

$$\epsilon_v = \frac{(\sigma - \sigma_c)^n}{\eta} \tag{4}$$

where σ is the applied stress which has to exceed a threshold stress σ_c and the parameters n and η are material dependent parameters. This strain is not recoverable after unloading and it is therefore a damage accumulation in the material. For a fiber composite this damage accumulation reflects mainly the growth of cracks in the interphase of fibers and the matrix. For a slow increase of a stress there may also be a transition to a viscous response which was demonstrated in Figure 1.

The strain capacities in terms of an elastic strain and a viscous strain are not unlimited which are shown as failure bounds in the strain domain in Figure 2. These limitations also show up as rupture stress bounds in the time domain in Figure 1. These bounds in the time domain will be conditional on an initial size dependence.

4 STRAIN FAILURE CHARACTERIZATION

On an atomic level only elastic strains are possible. These strains are a consequence of an increase or decrease of the distance between atoms. In actual materials the presence of a multitude of different types of flaws will make the internal structure complicated and it is in principle made up of conglomerates of atoms or molecules which modifies the strain behaviour. A consequence of this is that a viscous strain response is introduced. A viscous strain response reflects a shear like permanent deformation between such groups. In strain hardening, which is typical for metallic materials, broken bonds can be recreated at another location. In strain softening, which is the manifestation of fiber composites, broken bonds cannot be recreated which results in a crack growth. In both cases the flaw density will increase. Also the elastic strain response will be affected because primary creep is introduced.

Locally it is reasonable to assume that there will be a coupling between these strain responses since they are influenced by the same flaws. On a material level there will be a mixture of influences of different types of flaws which reduces this dependence to a level where it will have a marginal influence. Besides, the two responses do not occur naturally at the same time. A consequence of this material behaviour is that the elastic strain response and the viscous strain response can be considered as independent of each other.

The strength properties of a material will depend on the strain limitations in terms of elastic strain or viscous strain. These strain limitations will be influenced by the presence of existing flaws or the growth of flaws in time. The elastic strain capacity will depend directly on initially existing flaws in the material which will emphasize a size dependence. But there will also be a time dependence as a consequence of a delayed elastic response. The viscous strain capacity will depend directly on the growth of existing flaws which will emphasize a time dependence. But there will also be an initial size dependence. For this reason the failure strain ε_u will not only depend on the magnitude of the tensile stress which is acting but it will also be influenced by an effective volume subject to this stress and an effective duration associated with this stress. For a fiber composite the anisotropic nature also has to be considered by splitting up the size dependence appropriately.

For a circular bar subject to a tensile stress σ the failure bounds can be expressed in a functional form as

$$\varepsilon_u = g_e(\sigma;L,A,D) + g_v(\sigma;L,A,D) \tag{5}$$

where g_e and g_v are strain functions of the form given by equations 3 and 4 for an elastic or a viscous strain response respectively, L is an effective length, A is an effective cross section, and D is an effective duration. Occasionally surface flaws are more important than internal flaws which results in a circumferential dependence instead of a cross section dependence.

The ultimate strain at failure is not given by the average material behaviour if a size and a time dependence should be considered in a rational way. Instead, due to the influence of flaws, the ultimate strain will reflect the weakest part of the material. This behaviour is considered in the most rational way within the statistical theory of extreme values. The statistical distribution of the minimum value, or no failure, is obtained from the theorem of multiplication of probabilities which can be expressed as

$$F(\varepsilon_u) = 1 - \left(1 - F_e(\sigma;A,D)\right)^{\Sigma L_i} \left(1 - F_v(\sigma;L,A)\right)^{\Sigma D_j} \tag{6}$$

where $F_e(.)$ and $F_v(.)$ are the cumulative statistical distribution functions of elastic strain for a small length L_i or viscous strain during a short duration D_j. For i and j large in the summations of the volume and duration in equation 6 this expression can be written in terms of an asymptotical extreme value distribution. The initial distributions $F_e(.)$ and $F_v(.)$ will reflect the behaviour on an atomic or molecule level. For this reason these distributions are limited to the left since only positive arguments are possible. The behaviour of the arguments can be expected to follow a power law based on the equations normally used to characterize the relations between atoms. Thus equation 6 can be written in terms of the asymptotical extreme value distribution of type 3, the Weibull distribution as

$$F(\varepsilon_u(\sigma;L,A,D)) = 1 - \exp\left(-\frac{L}{L_0}\left(\varphi(A,D)\frac{\sigma}{c_e}\right)^{k_L} - \frac{D}{D_0}\left(\psi(L,A)\frac{\sigma}{c_v}\right)^{h_L}\right) \tag{7}$$

where the size and the time dependence in the length direction are expressed in normalized form, k_L and h_L are size and time parameters in the length direction respectively, $\varphi(.)$ and $\psi(.)$ are functions which have to meet certain continuity requirements, and c_e and c_v are normalizing constants. Here the arguments which determine the ultimate strain ε_u also are introduced.

The smooth transition from primary creep to secondary creep which reflects a change from an elastic response to a viscous response requires continuity in time and indirectly in space too. This continuum requirement can only be met if the two functions have the form

$$\varphi(.) = \left(\frac{A}{A_0}\right)^{1/k_A}\left(\frac{D}{D_0}\right)^{1/h_L} \tag{8}$$

$$\psi(.) = \left(\frac{L}{L_0}\right)^{1/k_L}\left(\frac{A}{A_0}\right)^{1/k_A} \tag{9}$$

where the cross area dependence is expressed in normalized form and k_A is a size parameter over a cross section. A more detailed analyzes based on the mean value behaviour is given in (Sentler, 1988). With these expressions introduced in equation 7 and the introduction of a threshold stress over which a transition to a viscous strain response can take place the following expression is obtained

$$F(\epsilon_u(\sigma;L,A,D)) = 1 - \exp\left(-\frac{L}{L_0}\left(\frac{A}{A_0}\right)^{k_L/k_A}\left(\frac{D}{D_0}\right)^{k_L/h_L}\left(\frac{\sigma}{c_e}\right)^{k_L} - \frac{D}{D_0}\left(\frac{L}{L_0}\right)^{h_L/k_L}\left(\frac{A}{A_0}\right)^{h_L/k_A}\left(\frac{\sigma-\sigma_c}{c_v}\right)^{h_L}\right) \quad (10)$$

which is the statistical distribution of strain response failures. If the two size parameters k_L and k_A are equal obvious simplifications can be made and the strain failure response for a material with isotropic strength properties is obtained. For this simplification the original Weibull theory can be identified as the first part in equation 10 but where time also is included. The second part reflects the Norton creep law but where a size dependence also is introduced.

Equation 10 can be visualized as two failure surfaces in time and in space which are shown in Figure 3. The locations of the two failure surfaces are not uniquely defined. The initial relation between them reflects a specific material but the actual location will also depend on the loading conditions. The shaded area reflects the time response for a given volume subject to stress and it corresponds in this case to the time dependence given in Figure 1.

Dependent on the location of the failure surfaces the strain path to failure may result in a different failure mode than the strain response. This will be the case for the failure surfaces which are presented in Figure 3 and this seems to be typical for many structural materials. For a fast increase of a stress an elastic strain response can normally be expected but the viscous strain failure surface is outcrossed first. For a sustained stress which results in a viscous strain response the elastic strain failure surface is outcrossed first. The result is a failure which is brittle in character. Such failure behaviours would normally be expected for mild steel at normal temperatures. At low temperatures a viscous response is suppressed and all failures tend to be brittle.

Fiber composites seem to behave in a similar manner as ordinary steel even if both failure forms appear to be brittle like. The transition to a viscous response will take place more easily though.

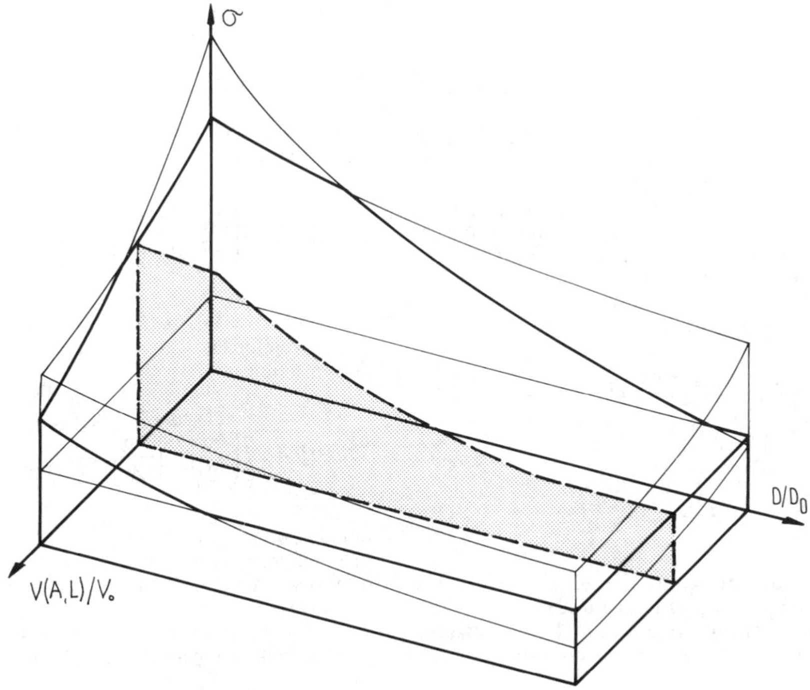

Fig. 3 The failure surfaces in space and in time.

4.1 Effective volume and duration

For uniform stress conditions, which is normally the case for a bar, the effective length is given by L and the effective cross section by A. For a sustained stress the effective duration is given by the total time. It is only for a time varying stress where the effective duration has to be evaluated separately.

For a continuously increasing stress the effective duration can be evaluated from equation 10 as $D = t/(k_L + 1)$ for an elastic failure response and $D \approx t/(h_L + 1)$ for a viscous failure response where t is the time to failure. For other stress variations a numerical evaluation is often necessary.

5 THE RUPTURE STRESS

The rupture stress can only be expressed in terms of the statistical moments for a given volume subject to stress and the corresponding duration. Because there are two strain failure surfaces the rupture stress can be derived both as an elastic strain failure or as a viscous strain failure. Normally the lowest failure surface will be decisive for a failure of engineering interest.

The mean value is derived from

$$E[\sigma] = \int_r^{\infty} f(\sigma) d\sigma = \int_r^{\infty} (1 - F(\sigma)) d\sigma \tag{11}$$

where r is a threshold stress. Often it is desirable to assume that this threshold stress is equal to zero for mathematical reasons. The physical consequences of this simplification is marginal if the actual threshold stress is low. This is the case in most applications.

For uniform stress conditions in space and in time the mean value is derived with the help of variable transformations like $u_e = (L/L_0)^{1/kL}(A/A_0)^{1/kA}(D/D_0)^{1/hL}\sigma/c_e$, $u_v = (L/L_0)^{1/kL}(A/A_0)^{1/kA}(D/D_0)^{1/hL}(\sigma-\sigma_c)/c_v$, and $\mu = c_v/c_e$. For an elastic response the mean value is obtained as

$$E[\sigma] = c_e \left(\frac{L_0}{L}\right)^{1/k_L} \left(\frac{A_0}{A}\right)^{1/k_A} \left(\frac{D_0}{D}\right)^{1/h_L} \int_0^{\infty} \exp\left(-u_e^{k_L} - \mu^{h_L} u_e^{h_L}\right) du_e \tag{12}$$

where the integral equals $\Gamma(1 + 1/k_L)$ for $\mu = 0$, which corresponds to an elastic response, where $\Gamma(.)$ is the gamma function. For a viscous response the corresponding mean value is

$$E[\sigma] = \sigma_c + c_v \left(\frac{L_0}{L}\right)^{1/k_L} \left(\frac{A_0}{A}\right)^{1/k_A} \left(\frac{D_0}{D}\right)^{1/h_L} \int_0^{\infty} \exp\left(-u_v^{h_L} - \frac{u_v^{k_L}}{\mu^{k_L}}\right) du_v \tag{13}$$

where the integral equals $\Gamma(1 + 1/h_L)$ for $\mu = \infty$, which corresponds to a viscous response. The behaviour of the integrals in equations 12 and 13 are visualized in Figure 4 for $0 < \mu < \infty$. Initially the integral in equation 12 will dominate completely up to the point where the integral in equation 13 takes over and dominates completely. The transition zone depends on the parameters k_L and h_L and it is decreasing for increasing values of these parameters.

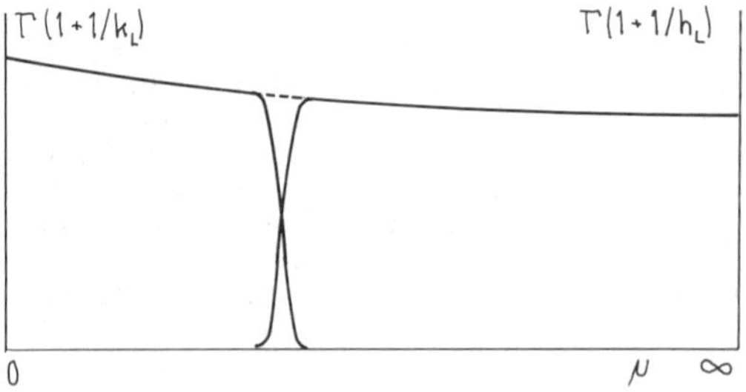

Fig. 4 The relation between the integrals in equation 11 and equation 12.

The corresponding expressions for the variances can also be derived with the help of variable transformations. For an elastic strain response the variance is

$$Var[\sigma] = c_e^2 \left(\frac{L_0}{L}\right)^{2/k_L} \left(\frac{A_0}{A}\right)^{2/k_A} \left(\frac{D_0}{D}\right)^{2/h_L} G_e(k_L, h_L, \mu) \tag{14}$$

where $G_e(.) = \Gamma(1 + 2/k_L) - \Gamma^2(1 + 1/k_L)$ for an elastic response. For a viscous strain response the variance is

$$Var[\sigma] = c_v^2 \left(\frac{L_0}{L}\right)^{2/k_L} \left(\frac{A_0}{A}\right)^{2/k_A} \left(\frac{D_0}{D}\right)^{2/h_L} G_v(k_L, h_L, \mu) \tag{15}$$

where $G_v(.) = \Gamma(1 + 2/h_L) - \Gamma^2(1 + 1/h_L)$ for a viscous response. For mixed responses the integrals will contain incomplete gamma functions.

From a variable transformation both an elastic strength and a viscous strength can be expressed as

$$\sigma = \sigma_0 \left(a + b \left(\frac{L_0}{L}\right)^{1/k_L} \left(\frac{A_0}{A}\right)^{1/k_A} \left(\frac{D_0}{D}\right)^{1/h_L} \right) \quad (16)$$

where a reference strength σ_0 has been introduced, and a and b empirical constants. The coefficient of variation can be expressed as

$$COV = \frac{b \left(\frac{L_0}{L}\right)^{1/k_L} \left(\frac{A_0}{A}\right)^{1/k_A} \left(\frac{D_0}{D}\right)^{1/h_L}}{a + b \left(\frac{L_0}{L}\right)^{1/k_L} \left(\frac{A_0}{A}\right)^{1/k_A} \left(\frac{D_0}{D}\right)^{1/h_L}} f(\lambda) \quad (17)$$

where f(λ) contains gamma functions. For a pure elastic or viscous response this function can be expressed as

$$f(\lambda) = \left(\frac{\Gamma(1+2/\lambda)}{\Gamma^2(1+1/\lambda)} - 1 \right)^{1/2} \simeq \frac{\pi}{\lambda\sqrt{6}} \quad (18)$$

where also the first term in a series expansion is given. For an elastic response λ equals k_L and and for a viscous response λ equals h_L.

Equations 16 and 17 provides a means to estimate parameters from experimental information. The limitation in this procedure is that only one material response can be considered at the time. If there is a mixture between an elastic and a viscous strain response failure these equations will not be appropriate for a direct use. For this reason it is necessary that each failure mode is treated separately.

6 APPLICATIONS

The methodology presented to analyze the strain response of fiber composites is applicable not only to explain experimental findings but in particular to predict the behaviour under real conditions. This makes this methodology useful in many ways.

The results given in Table 1 can now be given a wider interpretation. If the two failure surfaces crosses each other, as is shown in Figure 3, this will alter the failure mode both for increasing durations and increasing lengths of a fiber composite. A change from a viscous strain failure to an elastic strain failure will increase the time dependence as well as the size dependence. This will be reflected in an increased coefficient of variation. This is especially the case for fiber bundles and impregnated fiber bundles. A plot of the information for impregnated bundles of fibers reveal that such a change seem to take place. This may explain why a maximum likelihood estimate will not confirm a Weibull distribution.

The result given in Figure 1 reflects the test information obtained for a commercially available prestressing tendon. This fiber composite has been tested for a large number of loading conditions including fatigue. The size dependence has been determined for a large variation of lengths and the time dependence in sustained load tests for up to more than one year. All results presented fits well into the model given above.

7 CONCLUSIONS

The failure behaviour and thus the reliability of fiber composites can be analyzed satisfactory with this approach. This include both the short term strength which is obtained in a direct loading to failure and the long term strength for more or less constant stress conditions. This makes it possible to understand experimental results in a satisfactory manner and predict the behaviour in real applications. At the same time it is possible to evaluate the reliability in a rational manner.

ACKNOWLEDGEMENT

This research was supported by the Swedish Council for Building Research which is gratefully acknowledged..

REFERENCES

Bader M.G. and Priest, A.M. 1982. in "Progress in Science and Engineering of Composites", ed. T. Hayashi, K. Kawata and S. Umekaw (Japanese Society for Composite Materials, Tokyo) p. 1129.

Batdorf, S.B. 1982. Tensile strength of unidirectionally reinforced composites, J. Reinf. Plas. Compos. p.153.

Daniels, H.E. 1954. The statistical theory of the strength of bundles of threads, Proc. R. Soc. A. 183 p.404.

Harlow, D.G. and Phoenix, S.L. 1981. Probability distributions for the strength of composite materials, Int. J. Fracture 17 p. 347.

Overbeck, E. 1987. Zur Bruchfestigkeit und Zeitfestigkeit von Glasfasern und unidirektionalen GFK-stäben, VDI-Fortschrittsberichte, Reihe 5, Nr. 127, VDI-Verlag.

Penny, R.K. 1971. Design for creep, McGraw-Hill, 1971.

Rosen, B.W. 1964. Tensile failure of fibrous composites, AIAA J. 2 p.1985.

Sentler, L. 1987. A strength theory for viscoelastic materials, D9:1987, Swedish Council for Building Research.

Sentler, L. 1988. A stability theory for orthotropic viscoelastic structural members, D5:1988, Swedish Concil for Building Research.

Smith, R.L. 1980. A probability model for fibrous composites with local load sharing, Proc. R. Soc. A 372 p.539.
Watson, A.S. and Smith, R.L. 1985. An examination of statistical theories for fibrous materials in the light of experimental data, J.Mater. Sci. 20.
Weibull, W. 1939. A statistical theory of strength of materials, Proc. R. Sci. Soc. 151.
Zweben, C. 1968. Tensile failure of fiber composites, AIAA J. 6 p.2325.

Optimum design

Structural Safety & Reliability, Schuëller, Shinozuka & Yao (eds) © 1994 Balkema, Rotterdam, ISBN 90 5410 357 4

Optimum decisions related to design and replacement of seismic energy dissipators

L. Esteva & O. Díaz
Institute of Engineering, National University of Mexico, D.F. Mexico

ABSTRACT: A decision approach to the earthquake resistant design of structures provided with energy dissipators is developed. It includes the selection of optimum seismic design coefficients, as well as the formulation of optimum policies for the repair of damage on the conventional structural members and for the preventive replacement of the energy-dissipating devices. The process of damage accumulation in structural members and energy dissipators is modeled as a Markov process with transition probabilities which account for the joint distribution of the waiting times and intensities of earthquakes, as well as for the initial damage conditions of the structural system prior to the occurrence of each earthquake.

1 INTRODUCTION

Modern criteria and methods of structural design to resist earthquakes have succeeded at permitting the balance of safety and economy, albeit at the cost of accepting that structures may undergo substantial damage during high-intensity earthquakes, while remaining standing, thus offering protection to human lives. However, the high costs and long times tied to the repair of structural damage produced by earthquakes points at the need for alternate approaches to earthquake resistant design. Particularly promising among those approaches is the use of energy dissipating devices, at least for an important class of structural types.

Practically speaking, energy dissipating devices may be dealt with as structural members with special properties, among which are particularly relevant: a) the stability of their hysteretic behavior cycles under the action of a large number of alternating load applications, and b) their possibility of being replaced in case of damage, by means of simple and reliable operations, not requiring the reconstruction of the elements anchoring them to the members of the conventional structure. These advantages are partly overshadowed by the cost increments tied to the fabrication and installation of the mentioned devices; hence the importance of studying the economics of possible applications.

In this paper, where the process of seismic damage accumulation is formulated as a Markov process, with the states of the system described by the level and spatial distribution of local damage at the end of each earthquake and by the resulting failure probabilities as functions of earthquake intensities. The transition probabilities between successive states of the system depend on both natural environment (seismicity) and human actions (seismic design decisions; policies and rules for repair of damage and preventive replacement of energy-dissipating devices). The framework for making those decisions and establishing those policies and rules is provided by a formal decision approach based on initial construction costs and expected present values of benefits and damage, the latter including failure consequences, damage repair and device replacement. This framework is developed in the sequel, which also includes the formulation of models to describe the accumulation of damage, the transition probabilities between successive states of a system, and therefore the evolution of system reliability with seismic history.

The paper presents a general formulation of the problem, as well as a detailed description of the components of the mathematical framework. Quantitative results are preliminary

2 GENERAL FORMULATION

Seismic design criteria for conventional structures are expressed by means of a design response spectrum, the ordinates of which are reduced in order to account for nonlinear behavior. When energy-dissipating devices (EDD) are used, these criteria are complemented by a statement concerning the fractions of the lateral force and of the lateral stiffness which are to be taken by each system: the conventional frame (CF) and the EDD. These fractions may vary along the system's height. For the particular case of a simple (single-bay, single-story) system, the design variables are fully determined through the initial tangent stiffnesses, k_F and k_D, of CF and EDD, respectively, and by the corresponding lateral strengths, R_F and R_D (Fig 1). Typical idealized load-deformation curves for these elements are shown in Figs 1a and b, which distinguish between the deterioration of stiffness and strength typical of reinforced concrete members, and the stability of the hysteretic cycles of the EDD.

Every time that a moderate or high intensity earthquake occurs, damage increments take place on both elements. These increments are functions of the amplitudes and numbers of the deformation cycles, and are additive to those accumulated as a consequence of previous events, provided the CF is not repaired and the EDD is not replaced. It is assumed that damage on the CF is visually identified, thus leading to repair actions when it reaches a preestablished level; and that the EDD is replaced when it breaks or, in accordance with a preventive strategy, when it has been subjected to a number of high intensity earthquakes such that the probability of failure in the event of the next potential earthquake is unacceptably high. An optimum strategy for the design, construction and maintenance of such a system implies selecting the structural design parameters, as well as the threshold levels for repair of the CF and the criteria for preventive replacement of the EDD, in such a way as to minimize the sum of initial, maintenance and damage costs, all these updated to a common reference time.

Denoting by T_i, $i = 1, \dots \infty$ the times of occurrence of earthquakes, by C the initial cost and by L_i, $i = 1, \dots \infty$ the expenditures associated with failure, damage, repair and maintenance costs immediatly following each earthquake, then the quantity to be minimized is

$$U = C + E\left(\sum_{i=1}^{\infty} L_i \, e^{-\gamma T_i} \right) \tag{1}$$

where E stands for expected value, and γ is an adequate discount rate.

The damage accumulated on the CF and the EDD up to and including the j-th event are D_{Fj} and D_{Dj}, respectively. If a repair or replacement operation has taken place after the i-th earthquake, D_{Fi} and D_{Di} are transformed into their updated values, D'_{Fi} and D'_{Di}. The increments corresponding to the (j+1)th earthquake are $\delta_{F(j+1)}$ and $\delta_{D(j+1)}$; hence,

$$D_{F(j+1)} = D'_{Fj} + \delta_{F(j+1)} ,$$

$$D_{D(j+1)} = D'_{Dj} + \delta_{D(j+1)} \tag{2a, b}$$

In the case of collapse of the structure during the i-th event, followed by immediate reconstruction, L_i equals C + A, where C is the construction cost and A includes all other related costs, including direct and indirect material costs, as well as those arising from human losses, social impact, and so forth. If collapse is not reached, L_i includes the repair costs of structural and infill elements, the replacement of

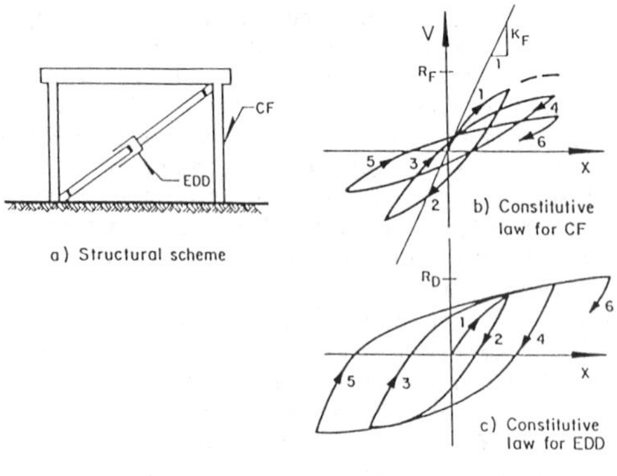

Fig 1 SDOF system with EDD

those EDD which have failed or which are estimated to have attained significant fatigue damage, and the losses inflicted by the eventual interruption of the normal operation of the system.

Every time that the damage accumulated on the CF exceeds a preestablished value D_{rF}, the repair must eliminate the accumulated damage; that is, it must restore the initial properties R_F and K_F of the structural frame. Here it is assumed that the damage is visually apparent in the members of a structural frame; but that in the energy dissipators it remains hidden, in the form of partial fatigue, until it reaches so high values that the risk of their failure during the next high-intensity earthquake is excessively high. This justifies the adoption of a preventive policy consisting in replacing the dissipators when they have been subjected to a given number of earthquakes, on the basis of an index of calculated damage, D_{CD}, to be defined in the sequel. Consequently,

$$D'_{Fi} = D_{Fi} \quad \text{if} \quad D_{Fi} < D_{rF}$$

$$= 0 \quad \text{otherwise}$$

$$D'_{Di} = D_{Di} \text{ if } D_{Di} < 1$$

and no preventive replacement takes place

$$= 0 \text{ otherwise}$$

Because of eqs 2a, b, and because $\delta_{F(j+1)}$ and $\delta_{D(j+1)}$ depend on D_{Fj} and D_{Dj}, the process of damage accumulation is a Markov process, regardless of whether earthquakes occur in accordance with a Poisson, a renewal or a Markov process.

The transition probability matrices between consecutive events are obtained from the probability density functions of $\delta_{F(j+1)}$ and $\delta_{D(j+1)}$ conditional to D_{Fj} and D_{Dj}, as explained in the following paragraph.

Inmediately after the i-th earthquake, the state of the system is expressed in terms of two sets of variables, describing respectively the state of damage on the system and the seismic process. D_{Fi} and D_{Di} belong to the first set. The second is assumed to be described in terms of the vector S_i of seismological variables, if the characteristics of consecutive earthquakes are not independent; otherwise, their probability distribution does not depend on i. In the general case, in order to determine the conditional probability distributions of $D_{F(i+1)}$, $D_{D(i+1)}$ and S_{i+1} given the values corresponding to instant i, it is required, on one hand, to determine the joint p.d.f. of the waiting time and the intensity (T_{i+1}, Y_{i+1}, respectively) of the next earthquake, and on the other, to determine the states of damage D'_{Fi} and D'_{Di} of the system's components, just after the repair of the frame and/or the replacement of some or all the energy dissipators have taken place, if required. The decision process is summarized in the left portion of Fig 2: a) if the damage accumulated in the structural frame exceeds level D_{rF}, the frame is repaired so as to recover the properties it had right after its construction was finished; b) if the energy dissipator fails, or if its calculated damage index exceeds D_{rD}, it is replaced by a new one. The joint p.d.f. of $\delta_{F(i+1)}$ and $\delta_{D(i+1)}$ will be

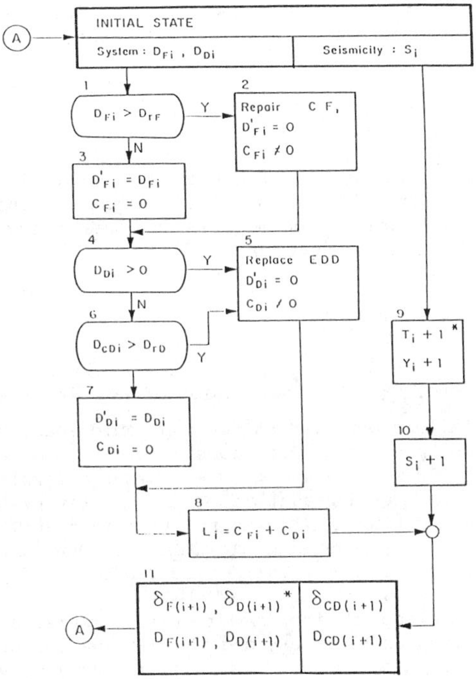

Fig 2 Flow diagram for response- repair cycles

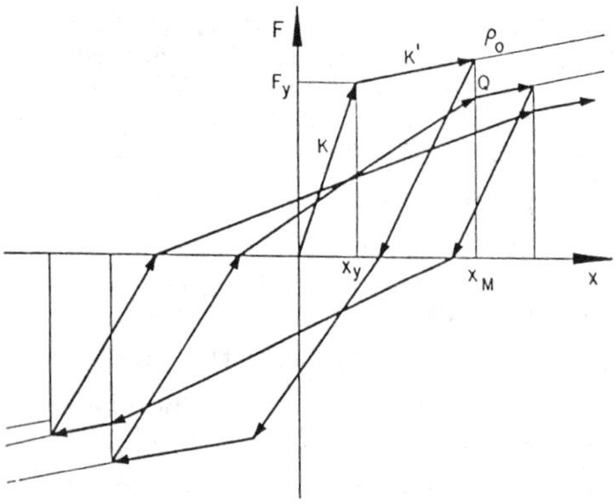

Fig 3 Shah and Wang's model

obtained as a function of δ_i, and of D'_{Fi} and D'_{Di}. L_i includes among other costs those associated to the repair and replacement operations mentioned above.

The transition probabilities obtained in accordance with the foregoing paragraphs relate the states of the system at the end of two consecutive earthquakes. These probabilities depend on the seismic activity as well as on the adopted seismic design criteria, and the repair and replacement strategies. Their determination and use in the decision making process are described later.

3 CUMULATIVE DAMAGE AND STRUCTURAL RESPONSE

The following definition of cumulative damage is adopted for both, the structural frame members and the energy dissipating devices (1, 2).

$$D = k \sum_{j=1}^{n} \xi_j^m \tag{3}$$

Here, ξ_j is the ratio of x_j to the failure deformation for monotonic load; k and m are empirical constants.

D is a measure of the damage determining the degradation of both, stiffness and strength. This degradation is quantitatively expressed by means of Shah and Wang's model (1987), illustrated in Fig 3. In accordance with this figure, the envelope of the cyclic load-deformation curve is bilinear, and the degradation of mechanical properties is represented by means of the evolution of that envelope in each loading direction, as a function of D, and of the peak amplitude reached in any cycle in the direction of interest. The updated envelope is defined by point Q in the figure, having an abscisa equal to x_M, the peak deformation, and an ordinate equal to the product of $(1-\varepsilon)$ by $P_o(x_M)$, where the latter is the load corresponding to x_M in the original envelope and ε is a degradation coefficient related with D in accordance with Eq. 4 of Ref 2, or with and improved version, Eq 6 of Ref 1.

Figure 4 shows means and variation coefficients of δ_F and δ_D in terms of intensities, for an initially undamaged system similar to that of Fig 1a, with the properties summarized in Table 1. Fig 5 is typical of many obtained, reporting values of final damage for a given intensity, and different combinations of initial damage in CF and EDD. The excitation was a collection of recorded and simulated accelerograms, typical of the soft soil area in Mexico City, normalized at different intensitites. Typical ground motion time histories and response spectra are shown in Fig 6. The responses were calculated by means of nonlinear step-by-step studies. The intensities are expressed in terms of the maximum ordinate of the linear pseudo acceleration spectrum for damping equal to 0.05 of critical.

For multistory systems, neither the damage on the CF elements, nor that on the EDD's can be expressed in terms of a scalar quantity; instead, a description is required, at least, of the variability of the corresponding damage measure, as well as of its average value throughout the system. Assume, for instance, that the system being studied is an n-story building, having one energy dissipating element at the m lowest stories ($m \le n$). Then, the damage state just after the occurrence of the n-th earthquake may be described by the average values $\hat{D}_{Di}$ and $\hat{D}_{Di}$, and by the sample dispersions $\hat{S}_{Fi}$ and $\hat{S}_{Di}$, given by

$$\hat{D}_{Fi} = \frac{1}{n} \sum_{j=1}^{n} D_{Fji} ,$$

$$\hat{D}_{Di} = \frac{1}{m} \sum_{j=1}^{m} D_{Dji} \tag{4a}$$

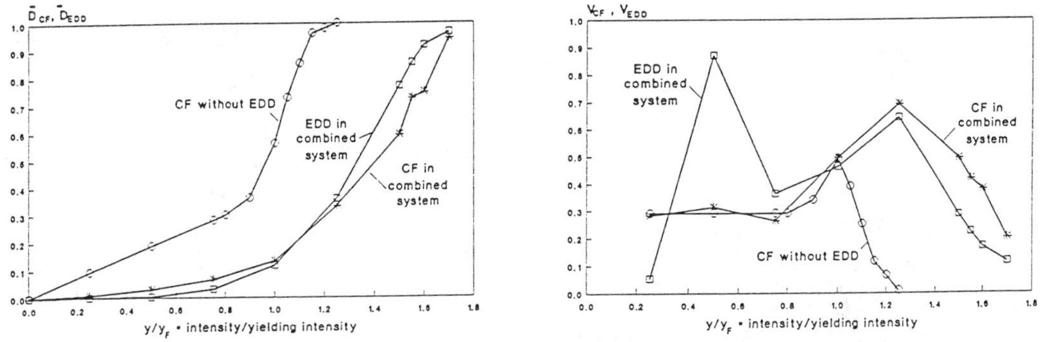

a) Mean value

b) Variation coefficient

Fig 4 Variation of damage with the normalized intensity y/y$_F$, for the case of zero initial damage

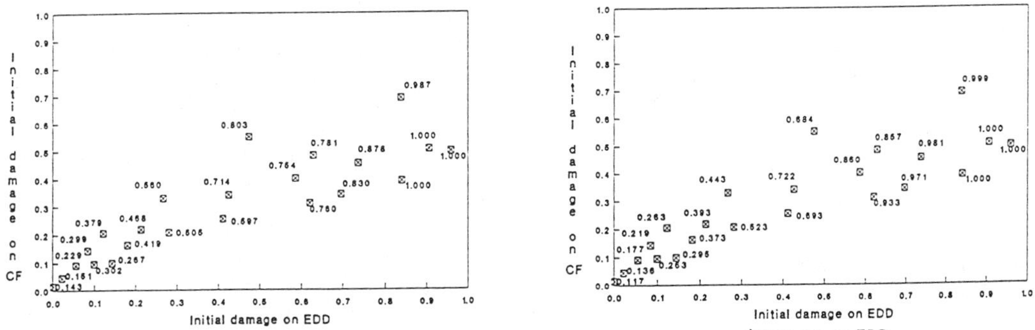

a) Damage on CF

a) Damage on EDD

Fig 5 Mean values of accumulated damage for y/y$_F$ = 1.0

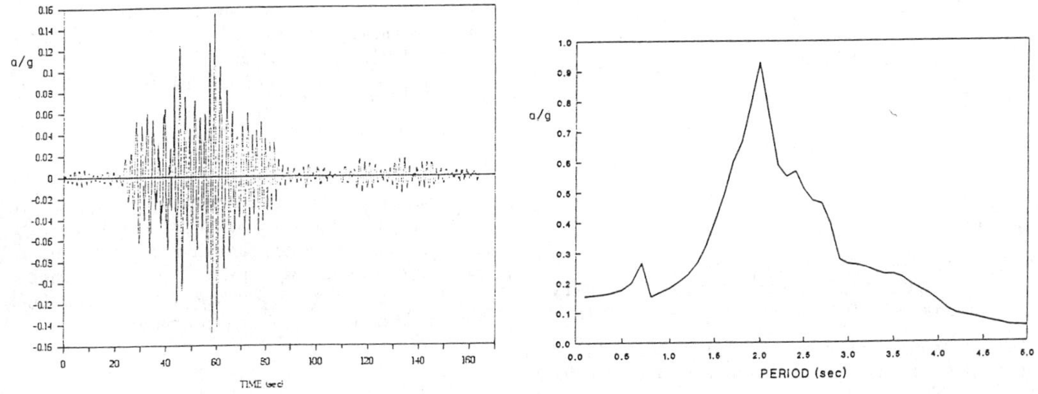

a) accelerogram

b) Response spectra

Fig 6 Typical strong motion record

$$\hat{S}^2_{Fi} = \frac{1}{n} \sum_{j=1}^{n} (D_{Fji} - \hat{D}_{Fi})^2,$$

$$\hat{S}^2_{Di} = \frac{1}{m} \sum_{j=i}^{m} (D_{Dji} - \hat{D}_{Di})^2 \quad (4b)$$

where D_{Fji} and D_{Dji} are respectively the damage values at the j-th member of the CF and at the j-th EDD, just after the occurrence of the i-th earthquake. After taking the actions derived from decisions related to repair of CF elements and/or replacement of EDD's, the foregoing values are transformed into those to exist at the beginning of the (n+1)th event: $\hat{D}'_{Fi}$, $\hat{D}'_{Di}$, $\hat{S}'_{Fi}$ and $\hat{S}'_{Di}$. Cumulative damage studies would now be oriented to generating from these sample values the probability distributions of those to be expected at the end of the (i+1)th earthquake, as a function of its intensity. Repair and replacement decisions would be taken individually for each damaged element.

4 TRANSITION PROBABILITIES BETWEEN DAMAGE STATES

Let $f_{FD(i+1)}(u,v|D'_{Fi},D'_{Di})$ designate the joint p.d.f. of $D_{F(i+1)}$ and $D_{D(i+1)}$ given D'_{Fi} and D'_{Di}. In order to account for the uncertainty associated to Y_{i+1}, the intensity of the (i+1)th earthquake, $F_{FD(i+1)}$ is calculated as follows:

$$f_{FD(i+1)}(u, v|D'_{Fi}, D'_{Di}) = \quad (5)$$

$$\int F_{FD(i+1)}(u,v|D'_{Fi}, D'_{Di}, y)F_{Y(i+1)}(y)dy$$

The marginal probability density functions of D_{Fi} and D_{Di} are obtained recursively from the initial condition of the system. If the actions of preventive replacement of EDD are neglected, after some algebraic transformations the recursive relation is the following (1).

$$F_{FD(i+1)}(u,v) = \int_0^1 \int_0^{D_{rF}}$$

$$f_{FD(i+1)}(u,v|u',v')f_{FDi}(u',v')du'dv'$$

$$+ \int_0^1 f_{FD(i+1)}(u,v|0, v')f_{Di}(v')$$

$$P(D_{Fi}>D_{rF}|v')dv'$$

$$+ \int_0^{DrF} f_{FD(i+1)}(u,v|u', 0)f_{Fi}(u')$$

$$P(D_{Di}>1|u')du' \quad (6)$$

$$+ f_{FD(i+1)}(u,v|0,0)P(D_F>D_{rF}, D_{Di}>1)$$

The recursive computation of F_{FDi} can be done by numerical integration in Eq 6 or by means of Monte Carlo simulation. However, both approaches may turn out to be too cumbersome, in order to achieve sufficiently accurate estimates of U in Eq 1. Hence the convenience of resorting to simplications or alternate approaches. One of them consists in simulating values of D_{Fi} and D_{Di} in each event by means of Monte Carlo methods, and using those values to estimate their probability distributions. Another approach consists in assuming the forms of those distributions and orienting the simulations to determining their moments. This has the advantage of demanding a much smaller number of simulations than the first approach, but requires, instead, a very careful selection of the forms of the mentioned distributions.

Another drawback of Eq 6 is that it does not account for the possibility of preventive replacement of the EDD's. Incorporating it into the second approach would imply handling a third variable in the mentioned equation and working with the corresponding joint p.d.f. These difficulties are circunvented through the use of the Monte Carlo approach.

5 PREVENTIVE REPLACEMENT OF EDD'S

For s.d.o.f. systems, D_{cD} (calculated damage index) after a given earthquake can be estimated on the basis of two sources of information: the seismic history and the relation between intensity and damage increment. The former is expressed in terms of intensities and times of occurrence of earthquakes, while the latter is contained in Figs 4 and 5. Given this information, D_{cD} equals, by definition, the sum of the expected values of the damage increments δ_{Di} during the earthquakes reported. For the purpose of this analysis, the expected influence of the repair actions on the CF may be accounted for or neglected, depending on the curves used from Figs 4 and 5. The determination of D_{cD} for complex systems is left for future publications.

6 SEISMICITY MODELS

Monte Carlo simulation of values of T_i to be used to obtain the expected value appearing in eq 1 is straightforward when the occurrence of earthquakes is represented by a renewal process, including the Poisson process as a particular case. Accounting for the influence of seismic history on the probabilistic description of future activity can be handled by means of Markov models. In the general case, the joint distribution of the waiting time to the next (large) earthquake and to its intensity is a function of the seismic history (magnitudes, locations and times) covering an interval of the order of one or two centuries, or perhaps smaller. Incorporating this influence on the Monte Carlo simulation of the values of T_i and Y_i offers no problem.

7 MULTISTORY FRAMES

Two parallel studies are underway, trying to understand the process of damage accumulation in multistory frames. One is a comparison of the responses of frames characterized by Takeda's and Shah and Wang's behavior models. The other aims at obtaining incremental damage functions similar to Figs 4 and 5. For this purpose, five three-bay ten-story frames were designed in accordance with Mexico City seismic code of 1987, and subjected to sets of simulated ground motion records with statistical properties similar to those of the SCT (Mexico City) EW accelerogram of 19 September 1985. Mean values of acting loads and mechanical properties of the structural members were derived from the nominal design values, and adopted as deterministic parameters of the systems studied. Systems I to III were conventional frames, having different periods. Systems IV and V were frames provided with EDD's, where the contribution of the latter to the total lateral stiffness was 0.75 and 0.5, respectively, in all stories. D-ε curves defined by the equation shown in Fig 3 were used, with α = 1 for the CF and 20 for the EDD. Design response spectra for linear systems were divided by a factor Q, in order to account for nonlinear behavior. The values of Q adopted for CF's and EDD's are summarized in the following table.

Table 1

CASE	I	II	III	IV	V
CF	4.0	1.5	4.0	1.0	4.0
EDE	4.0	1.5	4.0	4.0	4.0

Detailed time histories and maximum values of story deformations are reported in Ref 3. Briefly, ductility demands at stories of frames with Takeda's behavior occasionally reached values as high as 10. Under the assumption of Shah and Wang's behavior, systems II and IV survived under the action of most of the assumed excitations; the other systems collapsed induced by total loss of strength at one or more members of the CF. For this reason, it was not possible to pass to the next stage in the study, that is, calculating the response of systems having initial damage. The resulting rate of the number of failures to the number of cases studied was considered to be excessively high, and therefore not representative of reality. The authors believe that the very large values of the calculated responses are associated to the inadequacy of the D-ε curves adopted (Fig 3), and to the assumption of instantaneous loss of strength of members of both the CF and the EDD elements when D reaches a value of unity. A new stage in the study is currently in progress, aiming at eliminating the mentioned drawbacks.

8 CONCLUDING REMARKS

The use of energy-dissipating devices is an efficient approach for the reduction of repair costs of structures damaged by earthquakes. Many types of structural members and connections, as well as infill elements, may play the role of those devices, but the success in their use is attached to an approach to seismic design which envisages the reduction of structural (and non-structural) damage and the easiness of its repair as an essential part of the design process, in addition to the determination of the member strengths and stiffnesses required to comply with conventional design rules.

Energy dissipating elements can be dealt with as structural members characterized by very stable load-deformation curves when subjected to alternating loads. Therefore, a very promising approach to their use consists in assigning them the function of preventing the possibility of development of large stresses and strains in other members and hence reducing their vulnerability. This is achieved by adequately selecting both their location in the structural system and the relative values of their strengths, stiffnesses and safety factors with respect to those of the conventional structural members, in such a manner that demands for nonlinear dynamic behavior are concentrated on the former. As a complement to this strategy, their fastening to the conventional system must facilitate their replacement in case they are damaged or when a preventive strategy determines it so.

In order to implement the criteria described in the foregoing paragraphs, it will be necessary to count with tools for determining optimum locations and properties of energy dissipators, as well as required stiffness and strengths of the integrated system, and relative contributions of conventional structural members and energy dissipators to those properties. The models proposed in this paper provide a rational framework for reaching the corresponding decisions, as well as for establishing optimum repair/replacement policies. With the aid of those models, simple guidelines and recommendations must be prepared for preliminary design in typical practical cases.

9 REFERENCES

Esteva, L. and Díaz, O. "Influence of cumulative damage on the seismic reliability of multistory frames", Proc. IFIP 5th WG 7.5 Working Conference on Reliability and Optimization of Structural Systems, Takamatsu, Japan (1993)

Shah, S. P. and Wang, M. L. "Reinforced concrete hysteresis model based on the damage concept", Earthquake Engineering and Structural Dynamics, 15 (1987), 993-1003

Rodríguez, J. F. F. "Respuesta sísmica de edificios con disipadores de energía", Thesis for the degree of Civil Engineer, School of Engineering, National University of Mexico (1993)

Structural Safety & Reliability, Schuëller, Shinozuka & Yao (eds) © 1994 Balkema, Rotterdam, ISBN 90 5410 357 4

Reliability-based optimization of composite-hybrid plate girders

Samer Hendawi & Dan M. Frangopol
Department of Civil Engineering, University of Colorado, Boulder, Colo., USA

ABSTRACT: In the United States the design of plate girders for highway bridges is made in agreement with the AASHTO-Standard Specifications for Highway Bridges. Primarily, these specifications set forth minimum requirements which are consistent with current bridge design practice. A major limitation of the AASHTO Specifications is that reliability and optimization are not explicitly included in the design procedure. This paper presents the results of a reliability-based optimization approach carried out to design both unstiffened and stiffened composite-hybrid plate girders. Reliability-based optimization examples using reserve and residual reliability requirements are also presented.

1 INTRODUCTION

Plate girders are actually deep beams used generally for spans of about 70 to $150ft$. They are built up from plate elements to achieve a more efficient arrangement of material than is possible with rolled beams (Salmon and Johnson 1990). A special case of plate girder is the hybrid plate girder which is particularly efficient in composite construction. Due to the large compression capacity of the concrete slab the tension flange of the composite-hybrid plate girder controls the strength. For this reason, as shown by Salmon and Johnson (1990), there are obvious economical advantages to use a higher strength grade steel for the tension flange as compared to that of compression flange and the web.

The early work of Goble and Desantis (1966) and the subsequent developments by Burns and Ramamurthy (1983) and Dhillon and Kuo (1991), provided basic approaches for the optimum design of plate girders based on total cost and total weight. Unfortunately, these optimization approaches used the deterministic design philosophy. Recently, Frangopol and Moses (1993) have shown that in some instances deterministic optimization promotes structures with smaller ultimate overload margins than obtained with more conventional design procedures. Consequently, deterministic optimized structures will usually have higher failure probabilities than unoptimized structures. For this and other reasons it is much more rational to use reliability-based design concepts in structural optimization (Enevoldsen 1991, Thoft-Christensen 1991, Hendawi 1993).

In this paper, a reliability-based structural optimization approach for the design of both unstiffened and stiffened composite-hybrid plate girders is presented. Nondeterministic composite-hybrid plate girders under random loads are automatically sized for least-weight in full conformance with the AASHTO-Standard Specifications for Highway Bridges (1983). Reliability-based optimization examples are also presented.

2 OPTIMUM DESIGN PROBLEM

The contribution of Dhillon and Kuo (1991) deserves special attention in view of the formulation of the deterministic optimum design problem for composite-hybrid plate girders. Although this approach does not consider reliability-based constraints, the formulation presented by Dhillon and Kuo (1991) is very general. It is based on the load factor design method of the AASHTO-Standard Specifications for Highway Bridges (1983). In this paper, the formulation of Dhillon and Kuo (1991) is

Table 1. **Design Variables (X_1 *to* X_{22}) and Parameters (Y_1 *to* Y_{13}) as Considered by Dhillon and Kuo(1991)**

Notation	Variable	Units
X_1	Width of top flange	in
X_2	Thickness of top flange	in
X_3	Thickness of web	in
X_4	Depth of web	in
X_5	Width of bottom flange	in
X_6	Thickness of bottom flange	in
X_7	Distance from top of concrete slab to centroid of compression flange	in
X_8	Distance from top of concrete slab to centroid of web	in
X_9	Distance from top of concrete slab to centroid of tension flange	in
X_{10}	Distance from top of concrete slab to neutral axis of composite section	in
X_{11}	Moment of inertia of composite section	in^4
X_{12}	Distance from top of concrete slab to neutral axis of plate girder section alone	in
X_{13}	Moment of inertia of plate girder alone	in^4
X_{14}	Dead-load moment	kips-ft
X_{15}	Ratio of the neutral axis distance from the outer edge of the tension flange to the depth of the steel section	-
X_{16}	Strength reduction factor	-
X_{17}	Top flange compression stress caused by noncomposite dead load	ksi
X_{18}	Web buckling coefficient for stiffened plate girder	-
X_{19}	Width of stiffener	in
X_{20}	Thickness of stiffener	in
X_{21}	Stiffener spacing	in
X_{22}	Variable $J = [2.5(X_4/X_{21})^2 - 2]$ for transverse stiffener stiffness	-
Y_1	Deck slab thickness	in
Y_2	Yield strength of top flange steel	ksi
Y_3	Girder span length	ft
Y_4	Modular ratio	-
Y_5	Dead load excluding girder weight	kips/ft
Y_6	Max. live load moment including impact	kips-ft
Y_7	Max. live load shear including impact	kips
Y_8	Yield strength of bottom flange steel	ksi
Y_9	Yield strength of web steel	ksi
Y_{10}	Compressive strength of concrete	ksi
Y_{11}	Superimposed dead load moment	kips-ft
Y_{12}	Effective width of deck slab	in
Y_{13}	Unit weight of steel	$kips/ft^3$

enhanced for reliability-based structural optimization by including the consideration of uncertainties in loads and strengths.

2.1 *Deterministic optimization formulation of Dhillon and Kuo (1991)*

This subsection summarizes the deterministic optimization formulation of simply supported, unstiffened and stiffened, composite, welded, hybrid plate girders of uniform cross section presented by Dhillon and Kuo(1991). In this formulation, concrete deck slab, bearing stiffeners, welded connections and shear connectors are not accounted for. The top flange and web are assumed to be of the same steel and the bottom tension flange is assumed to be of higher-strength steel. In the case of the stiffened plate girder, the stiffeners are assumed to be placed on one side only and equally spaced.

The design variables X_1 to X_{22} are summarized in Table 1. It is interesting to note that only nine of the twenty-two design variables in Table 1 are independent (X_1 to X_6 and X_{19} to X_{21}). Once the values of these variables are determined, the structure is completely defined. The dependent variables X_7 to X_{18} and X_{22} may be expressed in function of a combination of independent variables only or by a combination of independent variables and design parameters Y_1 to Y_{13} listed also in Table 1. Both the strengths and geometric constraints (see Table 2) are based on the requirements imposed by the load factor design method

Table 2. Constraints on Strengths and Dimensions.

Type	Constraint
Ultimate Flexural Strength	
Stress in bottom flange	$g_1(X_6, X_9, X_{10}, X_{11}, X_{12}, X_{13}, X_{14}, X_{16}, Y_6, Y_8, Y_{11}) \leq 0$
Stress in top flange	$g_2(X_{10}, X_{11}, X_{12}, X_{13}, X_{14}, Y_1, Y_2, Y_6, Y_{11}) \leq 0$
Ultimate Shear Strength	
Unstiffened case	$g_3(X_3, X_4, X_{14}, Y_3, Y_7, Y_{11}) \leq 0$
	$g_4(X_3, X_4, X_{14}, Y_3, Y_7, Y_9, Y_{11}) \leq 0$
Stiffened case	$g_5(X_3, X_4, X_{14}, X_{18}, Y_3, Y_7, Y_9, Y_{11}) \leq 0$
Concrete Compression Stress	$g_6(X_{10}, X_{11}, Y_4, Y_6, Y_{10}, Y_{11}) \leq 0$
Aspect Ratio	
Unstiffened Case	$g_7(X_3, X_4) \leq 0$
Stiffened Case	$g_8(X_3, X_4, Y_9) \leq 0$
	$g_9(X_{19}, X_{20}, Y_9) \leq 0$
	$g_{10}(X_3, X_{19}, X_{20}, X_{21}, X_{22}) \leq 0$
Flange Local Buckling	$g_{11}(X_1, X_2, X_{17}) \leq 0$
Geometric Constraints	$g_{12}(X_4, X_5) \leq 0$

Note: The expressions of the functions $g_1(.)$ to $g_{12}(.)$ are indicated in Dhillon and Kuo(1991)

Table 3. AASHTO (1983) Constraints for Composite-Hybrid Plate Girders.

(a) Behavioral Constraints

1. *Ultimate Flexural Strength (§10.53.2)*

The maximum flexural strength shall be the moment at first yielding of the flanges multiplied by the reduction factor R, i.e.

$F_{yf}SR = M_u$

M_u = ultimate moment;

F_{yf} = minimum specified yield stress of tension flange;

S = section modulus of the composite section;

R = reduction factor(for unsymmetrical sections).

2. *Ultimate Shear Strength*

∗ For unstiffened plate girder(§10.48.2.1.e)

The maximum factored shear force is limited to:

$1.015 \times 10^8 \frac{t_w^3}{D} \geq V_u$

and

$0.58 F_{yw} D t_w \geq V_u$

∗ For stiffened plate girder(§10.48.2.e)

$0.58 F_{yw} C D t_w \geq V_u$

V_u = ultimate shear force(lb);

$C = \text{shear factor} = 18,000(\frac{t_w}{D})\sqrt{\frac{1+(\frac{D}{d_o})^2}{F_{yw}}} - 0.3 \leq 1.0$

F_{yw} = yield strength of web steel(psi);

D = unsupported distance between flange components(in);

t_w = thickness of web(in);

d_o = spacing of stiffeners(in)

3. *Concrete Compressive Stress (§10.45.3)*

The maximum compressive stress in the concrete slab shall not exceed $0.85 f'_c$, where f'_c is the compressive strength of concrete.

(b) Constraints on Dimensions

1. *Aspect Ratio*

∗ For unstiffened plate girder(§10.48.2.1)

$\frac{D}{t_w} \leq 150$

∗For stiffened plate girder(§10.48.5.1)

$\frac{D}{t_w} \leq \frac{36,500}{\sqrt{F_{yw(psi)}}}$ or $\frac{1,154.23}{\sqrt{F_{yw(ksi)}}}$

∗ The width-to-thickness ratio of transverse stiffeners(§10.48.5.5) shall not exceed

$\frac{2600}{\sqrt{F_{yw(psi)}}}$ or $\frac{82.22}{\sqrt{F_{yw(ksi)}}}$

∗ The moment of inertia of transverse stiffeners(§10.48.5.5) is $I_s \geq d_o t_w^3 J$

where $J = 2.5(\frac{D}{d_o})^2 - 2 \geq 0.5$

2. *Flange Local Buckling(§10.50.c)*

The ratio of top compression flange width to thickness must not exceed $\frac{2200}{\sqrt{1.3 F_{dl}(psi)}}$ or $\frac{61.02}{\sqrt{F_{dl}(ksi)}}$ where F_{dl} is the top flange compressive stress caused by noncomposite dead load.

of the AASHTO-Standard Specifications for Highway Bridges (1983) summarized in Table 3.

For a fixed deck-slab thickness ($Y_1 = ct$), the objective function to be minimized is the weight of the steel plate girder alone expressed for the unstiffened case as

$$W_1 = f_1(X_1 \ to \ X_6, Y_3, Y_{13}) \tag{1}$$

and for the stiffened case as

$$W_2 = f_2(X_1 \ to \ X_6, X_{19} \ to \ X_{21}, Y_3, Y_{13}) \tag{2}$$

In Dhillon and Kuo (1991) the solution to the

Table 4. Deterministic Parameters and Random Variables

Notation	Parameter or Random Variable	Units
Y_1	7.0	in
Y_2	(37.8,0.10)	ksi
Y_3	80.0	ft
Y_4	8.0	-
Y_5	(0.865,0.04)	kips/ft
Y_6	(1119.,0.214)	kips-ft
Y_7	(59.7,0.214)	kips
Y_8	(105.,0.11)	ksi
Y_9	(37.8,0.10)	ksi
Y_{10}	(3.4,0.18)	ksi
Y_{11}	(137.3,0.09)	kips-ft
Y_{12}	84.0	in
Y_{13}	0.49	kips/ft^3

Note: Random variables are indicated in parenthesis by their mean values and coefficients of variation

Table 5. Reliability-Based Optimization Results

	Unstiffened Case		Stiffened Case	
Variables	Optimum Values	Discrete Values	Optimum Values	Discrete Values
X_1	9.5571	9.5	9.7260	10.0
X_2	0.76835	0.750	0.42556	0.5
X_3	0.39984	0.4375	0.31189	0.375
X_4	43.112	43.25	58.2630	58.5
X_5	12.262	12.25	14.2560	14.5
X_6	0.5695	0.75	0.32563	0.375
X_{19}			3.96750	4.
X_{20}			0.10	0.25
X_{21}			50.092	50.
W_1	8.5924	9.59		
W_2			7.4639	9.13

Note: X_1 to X_6 and X_{19} to X_{21} are in inches, and W_1 and W_2 are in kips.

above formulation was given by a generalized geometric programming technique.

2.2 *Reliability-based optimization formulation*

This subsection summarizes the reliability-based optimization formulation of simply supported, unstiffened and stiffened, composite, welded, hybrid plate girders of uniform cross section proposed by Hendawi and Frangopol (1993). In order to formulate the optimization problem in a reliability-based format, eight of the design parameters Y_1 to Y_{13} indicated in Table 1 are treated as random variables. They represent the yield strength of the steel (Y_2, Y_8, Y_9), the dead load (Y_5), the maximum live load moment (Y_6) and shear (Y_7) including impact, the compressive strength of concrete (Y_{10}), and the superimposed dead load moment (Y_{11}). Using this reliability-based context, the unstiffened and stiffened minimum weight formulations are as follows.

1. Unstiffened Case

Minimize W_1

such that :

$$\beta_1 = \beta_{bot.fl} \geq \beta_1^* \quad (3)$$
$$\beta_2 = \beta_{top.fl} \geq \beta_2^* \quad (4)$$
$$\beta_3 = \beta_{shear1,unstiff.} \geq \beta_3^* \quad (5)$$
$$\beta_4 = \beta_{shear2,unstiff.} \geq \beta_4^* \quad (6)$$
$$\beta_6 = \beta_{conc.} \geq \beta_6^* \quad (7)$$
$$g_7 \leq 0 \quad (8)$$
$$g_{11} \leq 0 \quad (9)$$
$$g_{12} \leq 0 \quad (10)$$

where β_1 and β_2 are reliability indices with respect to maximum stress in bottom and top flanges, respectively, β_3 and β_4 are reliability indices with respect to maximum shear forces, β_6 is the reliability index with respect to maximum compressive

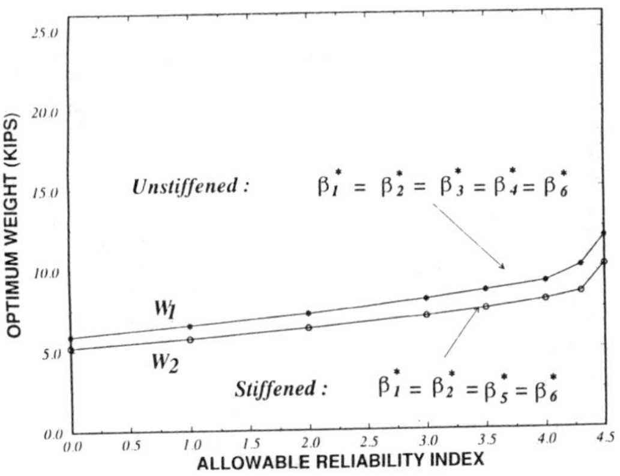

Figure 1. Effect of Allowable Reliability on Optimum Weight

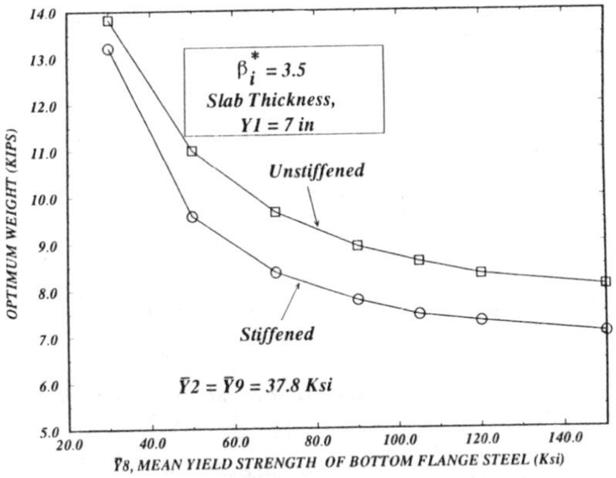

Figure 2. Effect of Hybrid Behavior on Optimum Weight

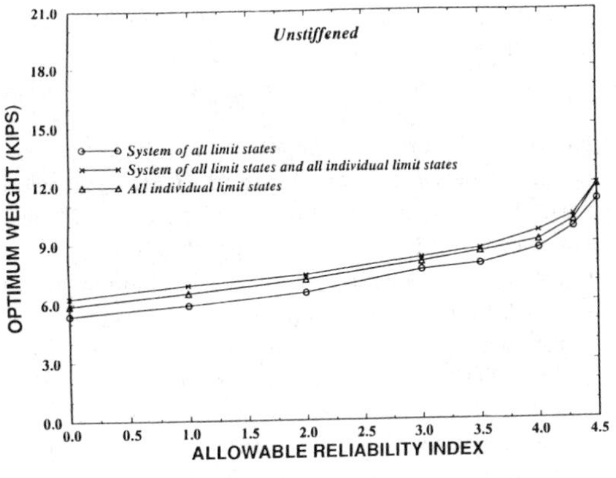

Figure 3. Limit States and System Reliability Effects on Optimum Weight

stress in the concrete slab, g_7, g_{11} and g_{12} are defined in Table 2, and β_i^* are allowable reliability indices.

2. Stiffened Case

Minimize W_2

such that :

$$\beta_1 = \beta_{bot.fl} \geq \beta_1^* \quad (11)$$
$$\beta_2 = \beta_{top.fl} \geq \beta_2^* \quad (12)$$
$$\beta_5 = \beta_{shear,stiff.} \geq \beta_5^* \quad (13)$$
$$\beta_6 = \beta_{conc.} \geq \beta_6^* \quad (14)$$
$$g_8 \leq 0 \quad (15)$$
$$g_9 \leq 0 \quad (16)$$
$$g_{10} \leq 0 \quad (17)$$
$$g_{11} \leq 0 \quad (18)$$
$$g_{12} \leq 0 \quad (19)$$

where β_5 is the reliability index with respect to maximum shear force, and g_8 to g_{12} are constraints defined in Table 2. For practical purposes, the yield strength of web steel is considered deterministic when the aspect ratio requirements for the stiffened case are checked (see constraints (15) and (16)). It is also noted that randomness in loads and resistances is explicitly considered in the above formulations. Therefore, loads and/or resistances are not affected by the factors used in AASHTO-Standard Specifications for Highway Bridges (1983).

For solving the above formulations the general purpose deterministic optimization program ADS (Vanderplaats 1986) was linked to the structural reliability analysis program RELTRAN (Lee et al 1993). The failure probabilities are estimated in RELTRAN by first order reliability methods. RELTRAN is used for both component and system reliability problems. In the case of a system problem defined as a union of components, RELTRAN uses Ditlevsen (1979) bounds. Also, sensitivity information is provided by RELTRAN.

3 NUMERICAL EXAMPLES

For comparison purposes, the deterministic numerical example considered by Dhillon and Kuo (1991) is treated herein in a reliability-based context. It consists of an 80 ft simple span, composite-hybrid plate girder. Results in this section complement information presented in Hendawi and Frangopol (1993) and Frangopol et al(1993).

Table 4 summarizes both the deterministic parameters and random variables considered in computations. The random variables are all considered independent and normal distributed. Furthermore, the allowable reliability indices are all considered to be the same $\beta_1^* = \beta_2^* = ... = \beta_6^* = 3.5$.

Table 5 shows the results of the optimization process for both the unstiffened and stiffened cases. The optimum solutions were obtained using as strategy the sequential linear programming, as optimizer the modified method of feasible directions, and as one-dimensional search bounds followed by polynomial interpolation. The optimum solution for unstiffened case was obtained in 2.85 *sec* (CPU time) after 6 iterations and 49 function evaluations, and that of stiffened case in 3.11 *sec* after 6 iterations and 50 function evaluations. The computer used was a VAX 6000 – 510.

The effect of allowable reliability index, $\beta_i^* = \beta_1^* = \beta_2^* = ... = \beta_6^*$, on the optimum weight is shown in Figure 1 for both unstiffened and stiffened cases. As expected (a) the stiffening effect reduces the optimum weight, and (b) the optimum weight increases with β_i^*.

The effect of hybrid behavior on the optimum weight of both unstiffened and stiffened plate girders is shown in Figure 2. In this figure, the mean yield strength of bottom flange steel $\bar{Y}_8$ is varied from 37.8 *ksi* to 150 *ksi*. As expected, this effect is extremely important. Use of a stronger steel for tension flange will reduce significantly the weight of the girder under the same reliability requirements. The rate of decrease in optimum weight with increasing the mean yield strength of bottom flange steel is similar for the unstiffened and stiffened cases.

For the case of the unstiffened plate girder, Figure 3 shows the effect of allowable reliability index $\beta_i^* = \beta^*$ on the optimum weight for three cases of reliability computations: (a) when all individual limit states have to satisfy the same level of reliability requirement (i.e., $\beta_i^* \geq \beta^*$); (b) when all individual limit states have to satisfy the same level of reliability requirement (i.e., $\beta_i^* \geq \beta^*$) and, additionally, the system made of all individual limit states has also to satisfy the same level of reliability requirement (i.e., $\beta^{system} \geq \beta_i^* = \beta^*$); and (c) when only the system made of all individual limit states has to satisfy the reliability requirement $\beta^{system} \geq \beta^*$. Evidently, the least weight is associated with case (c), and the maximum weight is associated with case (b).

Optimum reliability-based design under both initial and damaged states is also reported in this paper by considering steel corrosion of the plate girder

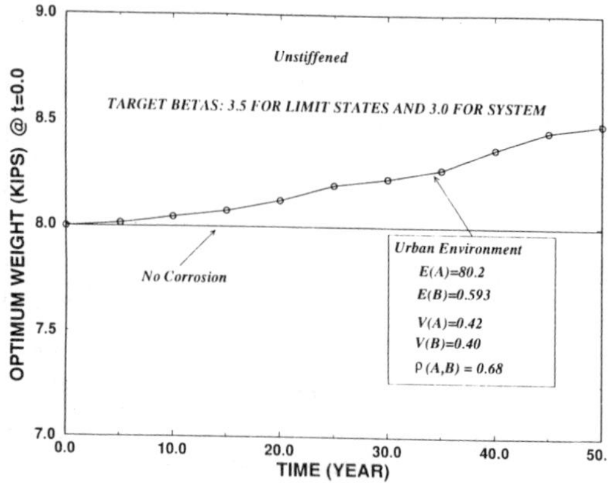

Figure 4. Time of Exposure Effect on Optimum Weight

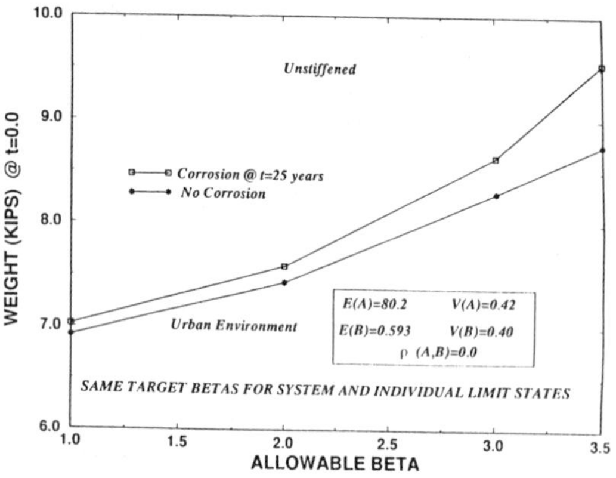

Figure 5. Allowable Residual Reliability Effects on Optimum Weight

according to the exponential kinetic equation given in Townsend and Zoccola (1982)

$$C = At^B \tag{20}$$

where C = corrosion penetration (μm), t = exposure time (*years*), and A and B are regression variables with mean values, standard deviations, and correlation coefficients depending on the environment (e.g., urban, rural, marine) and type of steel (Albrecht and Naeemi 1984).

Following the assumptions stated out by Hendawi (1993), and considering the unstiffened corroded plate girder case, Figure 4 shows the sensitivity of optimum weight with respect to the time of exposure to an urban environment, and Figure 5 compares the optimum weight associated with various allowable reliability levels for the case of no corrosion (i.e., only reserve reliability requirements) and the case of exposure to an urban environment (i.e., considering both reserve and residual reliability requirements) during 25 years. The mean values, $E(A)$, $E(B)$, the coefficients of variation, $V(A)$, $V(B)$, and the correlation coefficient, $\rho(A, B)$, of the variables A and B considered in computations are also indicated in Figures 4 and 5.

4 CONCLUDING REMARKS

From the theoretical considerations and numerical results, the following concluding remarks can be made.

1. Using previous results (Hendawi and Frangopol 1993, Frangopol et al. 1993, Hendawi 1993), a general approach to design unstiffened and stiffened composite-hybrid plate girders for highway bridges based on both reliability and optimization has been presented. This approach represents a reliability-based extension of the deterministic optimum design formulation of composite-hybrid plate girders proposed by Dhillon and Kuo(1991). The proposed reliability-based optimization formulation takes into consideration both behavior and side constraints according to the AASHTO-Standard Specifications for Highway Bridges (1983). The solution is obtained by linking a general purpose optimization program to a reliability analysis program.

2. The numerical examples demonstrate the feasibility of considering multiple limit states and/or system reliability requirements in reliability-based design of composite-hybrid plate girders. Both reserve and residual reliability requirements should be considered in the design of an optimal structure exposed to damage conditions during its expected life cycle.

3. Modern structural reliability-based optimization methodology has reached the stage where it is possible to design structural systems based on reliability constraints with respect to all limit states required by code specifications. As the structural reliability-based optimization technology moves from the research to practical applications, more emphasis should be placed on (a) selecting more realistic objective functions and using multicriteria optimization, (b) including both system reliability and consequences of failure in the optimization process, (c) providing sensitivity information, and (d) moving upward in the design hierarchy towards global life cycle optimization.

ACKNOWLEDGEMENTS

The research was supported in part by the United States National Science Foundation under Grant MSM-9013017 and by the Jordan University of Science and Technology for graduate studies of the first author. These supports are gratefully acknowledged.

REFERENCES

AASHTO - Standard Specifications for Highway Bridges. 1983. 13th Edition. Washington, D.C.

Albrecht, P. and Naeemi, A.H. 1984. *Performance of weathering steel in bridges.* National Cooperation Highway Research Program. Report 272. National Research Council. Washington, D.C.

Burns, S. and Ramamurthy, S. 1983. Mathematical programming in structural design. *Journal of Structural Engineering.* ASCE. 109(7):1669-1679.

Chong, K.P. 1976. Optimization of unstiffened hybrid beams. *Journal of the Structural Division.* ASCE. 102(2): 401-409.

Dhillon, B.S. and Kuo, C-H. 1991. Optimum design of composite hybrid plate girders. *Journal of Structural Engineering.* ASCE. 117(7): 2088-2098.

Ditlevsen, O. 1979. Narrow reliability bounds for structural systems. *Journal of Structural Mechanics.* 7:453-472.

Enevoldsen, I. 1991. *Reliability-Based Structural Optimization.* Ph.D. Thesis. Department of Building Technology and Structural Engineering. Aalborg University. Aalborg. Denmark.

Frangopol, D.M., Hendawi, S. and Tudor, M. 1993. Sensitivity information in reliability-based optimization of plate girders for highway bridges. *Proceedings of the 5th IFIP WG 7.5 Conference on Reliability and Optimization of Structural Systems.* Takamatsu-shi. Kagawa. Japan. in print.

Frangopol, D.M. and Moses, F. 1993. Reliability-based structural optimization. Chapter 10 in *Advances in Design Optimization.* ed. H. Adeli. Chapman and Hall Ltd. London. in print.

Goble, G.G. and Desantis, P.V. 1966. Optimum design of mixed steel composite girders. *Journal of the Structural Division.* ASCE. 92(ST6): 25-43.

Goble, G.G. and Moses, F. 1971. Automated optimum design of unstiffened girder cross section. *Engineering Journal.* AISC. 8(2):43-47.

Hendawi, S. and Frangopol, D.M. 1993. Design of composite hybrid plate girders based on reliability and optimization. submitted for publication.

Hendawi, S. 1993. *Structural Design Based on Reliability and Optimization.* Ph.D. Thesis. Department of Civil Engineering. University of Colorado. Boulder. Colorado. in preparation.

Lee, Y-H., Hendawi, S. and Frangopol, D.M., 1993. *RELTRAN- A Structural Reliability Analysis Program: Version 2.0.* Report No. CU/SR-93/6. Structural Engineering and Structural Mechanics Research Series. Dept. of Civil Eng. University of Colorado. Boulder. Colorado.

Salmon, C.G. and Johnson, J.E. 1990. *Steel Structures.* Third Edition. Harper & Row Publ. New York.

Thoft-Christensen, P. 1991. On reliability based structural optimization. *Lecture Notes in Engineering.* eds. A.Der Kiureghian and P. Thoft-Christensen. Springer Verlag. Berlin. 61:387-402.

Townsend, H.E. and Zoccola, J.C. 1982. Eight year atmospheric corrosion performance of weathering steel in industrial, rural and marine environments. ASTM. STP 767:45-59.

Vanderplaats, G.N. 1986. *ADS. Version 1.10.* Eng. Design Opt. Inc. Santa Barbara. California.

Structural Safety & Reliability, Schuëller, Shinozuka & Yao (eds) © 1994 Balkema, Rotterdam, ISBN 90 5410 357 4

Reliability-based optimization method of structural system using information integration method

Shigeyuki Matsuho & Wataru Shiraki
Tottori University, Japan

ABSTRACT: In this study, an efficient method of reliability-based optimization of structural system is proposed using IIM (Information Integration Method). We must discuss safety of structural system from the viewpoint of probabilistic theory because structural strength and loads acting on structure have stochastic properties. So, in this study, the safety evaluation of structural system is based on reliability theory. On the other hand, structural design should be performed so that the structural system satisfied various requirements besides the safety point, for example, economic and social requirements etc. Therefore, general design problem of structural system should be formulated as multi-objective optimum design problem with multi-constraint conditions. In such design problem, the evaluation of the system with respect to various evaluation items is necessary. However, the formulation and calculation of this evaluation are difficult. In this study, to overcome these difficulties, IIM is used to evaluate at what level the system satisfies all the evaluation items. IIM is efficient calculation method of system evaluation (Matsuho et al. 1990) and it is also easy to optimize the system by making use of IIM. In this study, simple mathematical programming technique is used as optimization method. Moreover, the genetic algorithm is also applicable to this proposed method as optimization method. In the appendix of this paper, more efficient optimization method than the original genetic algorithm is also proposed and is applied to the design method based on the IIM. For a numerical example, optimum design problem of highway bridge girder is considered. Results obtained by this optimum design method are compared with those of the present design method, and then efficiency of this proposed method is demonstrated.

1. INTRODUCTION

In this study, an efficient method of reliability-based optimization of structural system is proposed using IIM (Information Integration Method).

The structural design is performed economically so that safety margin of the system becomes as large as possible. However, structures should be designed considering not only these safety and economic points but also social requirements, because structures are generally public facilities. Social requirements include conditions for city planning, economic effect owing to construction of the structure and so on. Recently, aesthetics of landscape is also becoming one of important evaluation items. This implies that we have to keep in our mind that the most economical solution is not always optimum one. Therefore, general design problem of structural system should be formulated as multi-objective optimum design problem with multi-constraint conditions. Moreover, we must discuss safety of structural system from the viewpoint of probabilistic theory because structural strength and loads acting on structure have stochastic properties. So, the safety evaluation of structural system should be based on reliability theory.

In optimization design problem with the multi-objective functions, the evaluation of the system with respect to various evaluation items, which correspond to the objective functions, is necessary. However, the formulation and calculation of this evaluation are difficult. In this study, to overcome these difficulties, IIM is used to evaluate at what level the system satisfies all the evaluation items. Because IIM is very efficient and effective method of system evaluation (Matsuho et al. 1990), it is also easy to optimize the system by making use of the IIM. In this study, simple mathematical programming technique is used as optimization method. Moreover, the genetic algorithm is also applicable to this proposed method as optimization method. In the appendix of this paper, more efficient optimization method than the original genetic algorithm is also proposed and hybrid design method of this optimization algorithm and the IIM is presented.

For a numerical example, optimum design problem of highway bridge girder is considered. In this example, only two evaluation items such as economic and safety requirements are considered for simplicity. In evaluation of safety, reliability analysis concerning failure due to bending is carried out by using actual observed data (HDL Committee 1986). Results obtained by this optimum design method are compared with those of the current design method, and then efficiency of this proposed method is demonstrated.

2. OVERALL EVALUATION OF BRIDGE STRUCTURE USING INFORMATION INTEGRATION METHOD

The IIM is aimed at evaluating alternative of system concerning many evaluation items efficiently and rationally (Suh et al. 1978, Nakazawa 1987). Evaluation result with respect to each evaluation item generally has different dimension, so synthesis of evaluation results about various evaluation items is very difficult without scaling of each evaluation result. In traditional methods, this scaling was performed by weighting each evaluation result subjectively. In the IIM, synthesis of evaluation results is carried out by systematically converting each evaluation result into non-dimensional measure of information without subjective.

2.1 *Definition of Measure of Information*

In the theory of information, a measure of information I(A) concerning a certain event A is defined as follows (Shannon and Weaner 1949):

$$I(A)=\ln\{1/P(A)\} \quad \text{(Unit: nat)} \qquad (1)$$

in which P(A) is a probability that event A takes place, and 'ln' indicates natural logarithm. Eq.(1) shows that event A with small probability P(A) results in large measure of information I(A). In the IIM, measure of information I(A) is interpreted (Nakazawa 1987) as information, material and energy necessary for causing event A. It should be noted that measure of information is essentially different from the entropy which has different equation of definition.

2.2 *Application of Information Integration Method to system evaluation*

Civil engineering system consists of four stages, i.e. planning, design, construction and administration (including control). In this section, an example applying IIM to one of problems about planning stage is considered in order to illustrate calculation method of Eq.(1). This problem is to choose best one from four alternatives A, B, C and D of highway route which connects a certain city with another city.

First, three items, i.e. ① number of improved routes for traffic congestion, ② improvement (%) of traffic connection to other area, ③ initial construction cost (yen) are considered as evaluation items of this planning system (see Table 1). These items are also system parameters, because these are variables characterizing the state of system for evaluation items.

Second, let us define technical terms necessary for

Table 1 Effect for each evaluation item

Alternative	A	B	C	D
① Number of Improved Routes for Traffic Congestion	20	30	40	50
② Improvement of Traffic Connection to Other Area (%)	10-20	50-70	25-40	20-35
③ Construction Cost (billion yen)	120	100	130	140

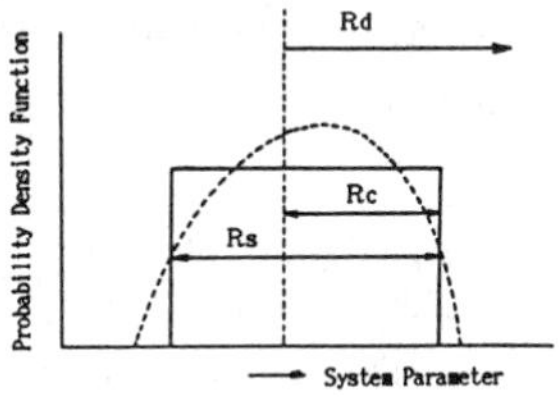

Fig.1 Illustration of measure of information

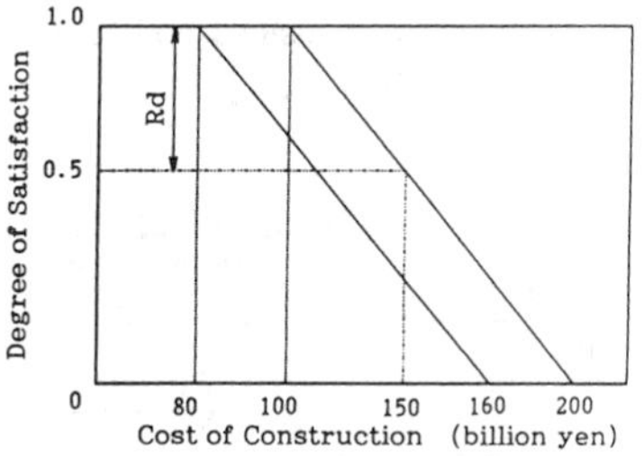

Fig.3 Function of satisfaction for construction cost

calculating Eq.(1). When system parameter is random variable, p.d.f. (probability density function) can be considered as shown by broken line in Fig.1. In most cases, random system parameter can be approximated to be uniformly distributed as shown in Fig.1, although precise p.d.f. should be used to make a strict discussion. Therefore, hypothesis of uniformly distributed system parameter is also used for simplicity in this study. In this figure, the system range R_S is a possible range which the system parameter can take and the design range R_d is a desirable range demanded for the design. Furthermore, R_C is a common range of both R_S and R_d. In the case that system parameter is deterministic variable and the case that design range is assumed to have a fuzziness, a function of satisfaction (Nakazawa 1987) may be introduced. The value of the function of satisfaction takes 1 for the design with perfect satisfaction, and 0 for the design to be rejected. Fig.2 represents function of satisfaction for number of improved routes, i.e. system parameter ①. In this figure, the ordinate y is a degree of satisfaction, and the abscissa x is a system parameter. The degree of satisfaction is considered to be a variable with fuzziness. The upper and lower functions of satisfaction correspond

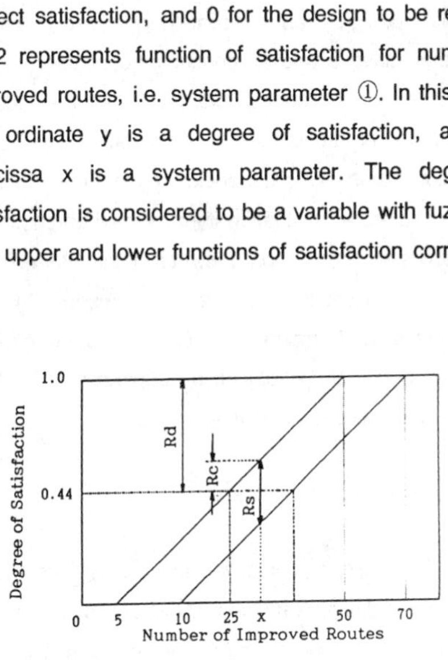

Fig.2 Function of satisfaction for number of improved routes

to the upper and lower bounds of the degree of satisfaction (see Fig.2).

In this example, system range of each system parameter shown in Table 1 is used. Functions of satisfaction concerning evaluation items ① and ③ are shown in Figs.2 and 3, respectively. Moreover, design range R_d of each system parameter is $R_d \geqq 25$ [routes] for system parameter ①, $R_d \geqq 30$ [%] for system parameter ②, and $R_d \geqq 150$ [billion yen] for system parameter ③. Therefore, common range R_C of each system parameter is as follows: namely, $0.44 \leqq R_C \leqq 1.0$ [routes] for system parameter ①, $R_C \geqq 30$ [%] for system parameter ②, and $0.5 \leqq R_C \leqq 1.0$ [billion yen] for system parameter ③. Upper function of satisfaction was used to convert to design range represented by degree of satisfaction (see Figs.2 and 3).

After the above-mentioned preparation, measure of information of Eq.(1) is given by

I = (Information necessary for causing P_2)
 -(Information necessary for causing P_1)
 $= \ln (1/P_2) - \ln (1/P_1) = \ln (P_1/P_2) = \ln (R_S/R_C)$ (2)

where P_1 is probability that system parameter takes a value within system range and P_2 is probability that system parameter takes a value within common range. Table 2 shows calculation results about each measure of information.

Finally, summation of measures of information over all the evaluation items, which is called integrated information measure, is calculated. This is because integrated information measure indicates measure of information for the system (i.e. measure of information when all the system parameters satisfy the

Table 2 Information measures and integrated information measure

(Unit: nat)

Alternative	A	B	C	D
Information Measure for Evaluation Item ①	∞	0.6506	0	0
Information Measure for Evaluation Item ②	∞	0	0.4055	1.0986
Information Measure for Evaluation Item ③	0	0	0.4855	1.2528
Integrated Information Measure	∞	0.6506	1.1911	2.3514

corresponding design ranges at the same time). This is called theorem on additivity of measure of information. The system, whose integrated information measure is smallest, is decided as optimum solution because of interpretation of Eq.(1) (see Section 2.1). Table 2 also gives integrated information measure of each alternative, and shows that alternative B is optimum solution.

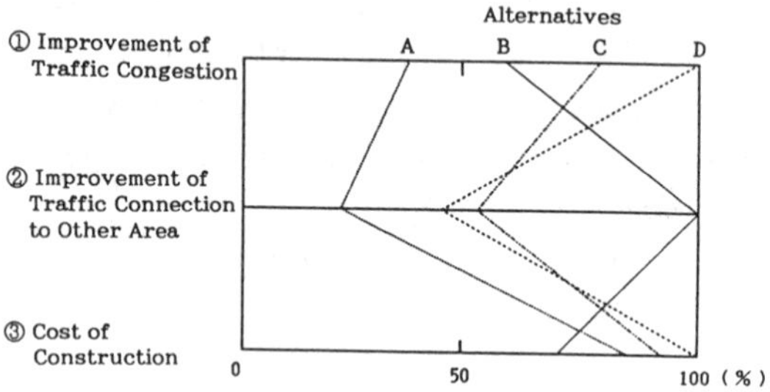

Fig.4 Results using factor-profile method

On the other hand, Fig.4 shows results by the factor-profile method (Yoshikawa 1985) which is used usually. In the factor-profile method, evaluation result for each evaluation item is normalized so that maximum value of evaluation result is 100[%], and normalized results are expressed graphically to compare results each other. But, it is difficult to evaluate the system concerning many evaluation items and not able to choose best alternative rationally and systematically.

3. RELIABILITY-BASED OPTIMIZATION DESIGN METHOD USING INFORMATION INTEGRATION METHOD

In chapter 2, IIM was shown to be powerful evaluation method of system. In this chapter, a new reliability-based optimization design method is proposed using IIM and mathematical programming technique. Moreover, it is shown that this proposed method is effective, through numerical example.

3.1 *Optimization design method*

As stated above, general design problem should be formulated by multi-objective optimization problem. And the safety of structural system, which is most important evaluation item, should be evaluated based on reliability theory. In the design method proposed by this study, integrated information measure is used as objective function. Then, multi-objective function corresponding each evaluation item can be represented as single-objective function by means of the IIM as stated in chapter 2. Moreover, the optimization problem with constraint conditions can be also converted to optimization problem without constraint condition because design range concerning each system parameter plays a role as constraint condition.

But, it should be kept in our mind that the theorem on additivity of measure of information is valid when evaluation items are independent each other, to state strictly. Therefore, evaluation items should be chosen so that stochastic correlation between these items is as small as possible. If IIM is used in the above-mentioned manner, design problem can be formulated by single-objective optimization problem without constraint conditions as follows:

$$F(\underline{X}) = I_1+I_2+\cdots+I_n \rightarrow \text{minimize} \tag{3}$$

in which $F(\underline{X})$ is objective function, $\underline{X}$ is vector consisting of design variables, I_i $(i=1,2,\cdots,n)$ is measure of information concerning each evaluation item i, and n is total number of evaluation items. Mathematical programming technique may be used for optimization of

Eq.(3). In this chapter, method of steepest decent, which is one of simple mathematical programming techniques, is used. In the appendix, application of the genetic algorithm (Goldberg 1989) to this optimization problem is also explained.

Integrated information measure evaluated in the above-mentioned manner can be optimized by the following iterative procedure:

Step①: To give initial values $\underline{X}^{(0)}$ to vector $\underline{X}$ consisting of design variables.

Step②: To decide direction vector $S^{(k)}$ for searching optimization solution $\underline{X}^{(k)}$ in the design space as follows:

$$S^{(k)}=-\nabla F(\underline{X}^{(k)}) \quad (4)$$

in which ∇F is gradient vector of objective function.

Step③: To decide next vector $\underline{X}^{(k+1)}$ using present vector $\underline{X}^{(k)}$ and direction vector $S^{(k)}$ of Eq.(4) as follows:

$$\underline{X}^{(k+1)}=\underline{X}^{(k)}+\alpha S^{(k)} \quad (5)$$

where step size α is decided so that right side of Eq.(5) is smallest, namely,

$$\frac{\partial F(X^{(k)}+\alpha S^{(k)})}{\partial \alpha}=0 \quad (6)$$

Step④: Steps ② and ③ are repeated until $\nabla F(\underline{X}^{(k+1)})$ and norm of $\underline{X}^{(k+1)}-\underline{X}^{(k)}$ converge into zeros.

3.2 *Numerical example*

As a numerical example of proposed design method, the design of bridge girder subjected to vehicle live loads is considered. This girder bridge (HDL Committee 1986), which is used on the Hanshin (Osaka-Kobe area in Japan) Expressway, has four vehicle lanes, seven main girders and one cross beam. Each girder of the bridge has I-beam section and is simply supported. In the design, it is assumed that web width t of girder is equal to thickness of flange and design variable is this parameter t, for simplicity. Numerical values using in the design are as follows: namely, total height of I-beam section h=2m, total width of I-beam section B=32cm, moment of inertia about outside girder $I=106.18\times10^5 cm^4$, moment of inertia about cross beam $I_Q=14.62\times10^5 cm^4$, allowable bending stress σ_a =24000 tf/m², unit weight of steel ρ=7.85 tf/m³, span length of girder l=39.2m, interval between the main girders a=2.85m.

Vehicle live load acting on each girder is modeled based on statistical processing of simulated values using the observed data on traffic streams obtained by the Hanshin Expressway Authority of Japan (HDL Committee 1986). Through the simulation, it is shown that annual maximum bending moment can be modeled by the type I extreme value distribution. In this simulation, effect of cross beams is also considered. For example, in the case of outside girder into which large bending moment is expected to induced, mean and standard deviation of the bending moment are 312.8 tf•m and 28.6 tf•m, respectively.

Based on optimization design method of section 3.1 and the above-mentioned conditions, main girders of the bridge are designed. In the design, failure probability and weight of each girder per unit length are considered as system parameters. Failure probability is assumed to be the probability when the maximum bending moment of each girder during lifetime period (50 years) exceeds the allowable moment calculated according to the Japanese current design code (Japanese Society of Highway 1990). In calculation of the allowable moment, safety factor 1.7 is multiplied by the bending moment calculated considering impact coefficient. Moreover, functions of satisfaction of Figs. 5 and 6 are used for failure probability and steel weight, respectively. Calculation results in the design process is shown in Fig.7. This figure shows that optimum solution is t_{opt}=0.0262m (optimum weight per unit length w_{opt}=532.9kgf/m, optimum failure probability $P_{f,opt}=2.12\times10^{-5}$). Although this solution t_{opt} is a little

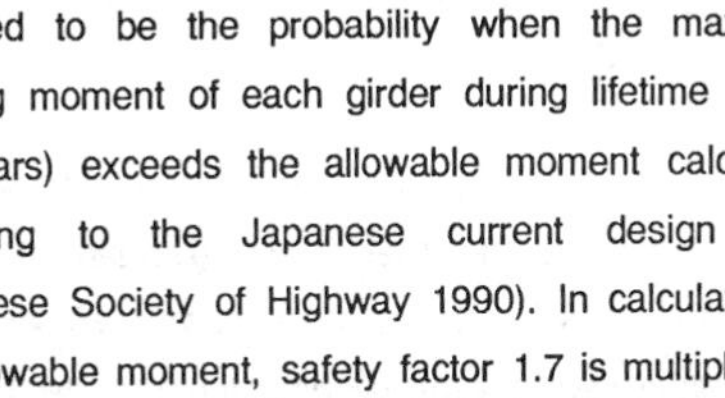

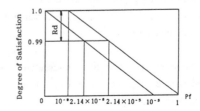

Fig.5 Function of satisfaction for failure probability

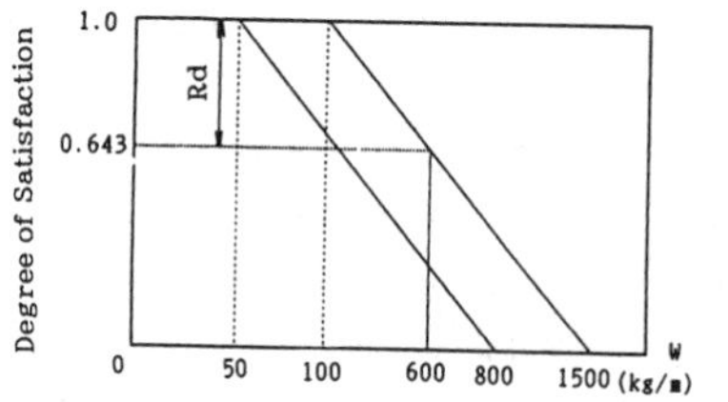

Fig.6 Function of satisfaction for steel weight

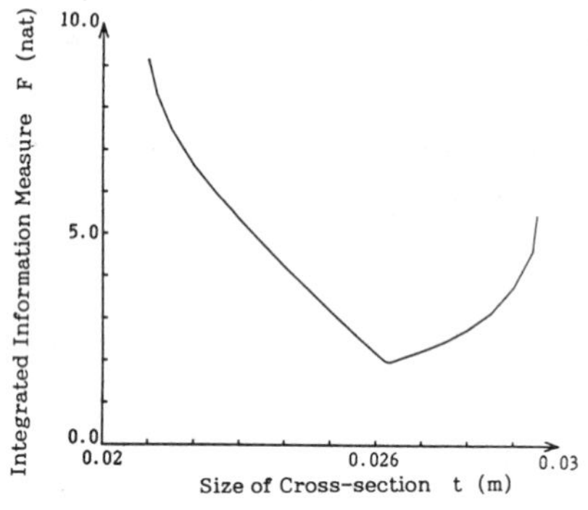

Fig.7 Objective function versus design variable

larger than the section of actual bridge (HDL Committee1986), the reliability improvement of 10^{-3} was realized by the proposed method. Moreover, by considering degree of slope near this solution t_{opt} in this figure, it is shown that this optimum solution is not sensitive to accuracy for execution of work. It should be noted that the solution (of Pareto optimum problem) can be decided uniquely.

4. CONCLUDING REMARKS

In this study, a method of reliability-based optimization was proposed using information integration method. And the efficiency was demonstrated by numerical examples. Merits of this method are as follows:

1. Alternatives can be evaluated easily and rationally for the many different types of evaluation items by using the logarithmic scaling.
2. The optimization design problem can be converted to the single-objective problem without constraint conditions. Therefore, optimization and sensitive analysis on design variable can be easily performed.
3. Fuzziness of the constraint conditions demanded for the design can be incorporated by means of function of satisfaction.
4. Optimization considering importance of each evaluation items can be carried out by setting of design range or function of satisfaction. For example, it may set design range narrowly for an important item.

Last, it should be noted that the information integration method can be applied to various fields of evaluation problems (for example, various types of expert-system, civil engineering system, the whole structural system, structural damage and so on). This study was performed by a Grant-in-Aid for Scientific Research provided by the Japanese Ministry of Education in 1989.

APPENDIX - APPLICATION OF THE GENETIC ALGORITHM

The genetic algorithm approach (Goldberg 1989), which is one of effective optimization methods drawing much attention, has a lot of advantages. One of them is of good efficiency. Applicability to the optimization problem of the objective function with the discretized design parameters is also one of the advantages. This is justified by considering that optimal design result should be selected out of standardized materials in many cases, for example (Matsuho 1990). Therefore, development of the optimization method combined with this approach is significant. In this appendix, more efficient optimization method than the original genetic algorithm is proposed.

The genetic algorithm consists of computational components of natural selection, reproduction, cross-over, mutation and so on. Flow chart of the algorithm is shown in Fig.A1 (Goldberg 1989) and can be easily applied to the optimization problem of Eq.(3). However, this approach do not always lead to global optimal result (minimum value or maximum value), because this

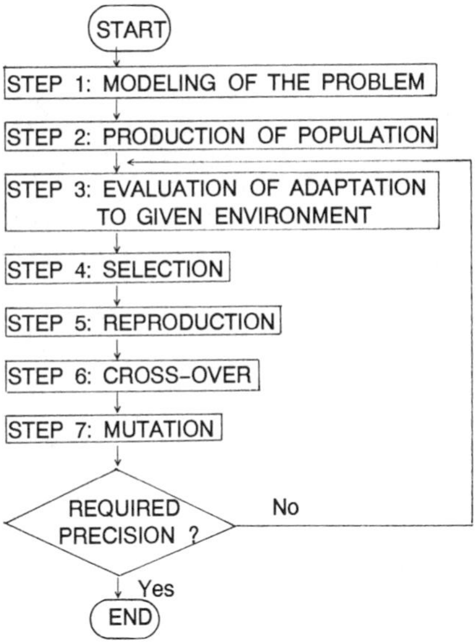

Fig.A1 Flow chart of the genetic algorithm

approach uses probabilistic transition rules, not deterministic rules. This is also accelerated when the objective function has a number of local minimum values (or local maximum values). In the original genetic algorithm, operations of cross-over and mutation prevent a choice of local minimum solution. However, more efficiency of this optimization can be realized by transformation of the objective function. In this study, the original objective function $f(\underline{x})$ with many local minimum values (or local maximum values) is transformed into the one $f^*(\underline{x})$ with only single global minimum value (or global maximum value). $\underline{x}$ in $f(\underline{x})$ or $f^*(\underline{x})$ is vector whose elements are design parameters. This transformation is based on the principle of relative peaking, which can be expressed as follows (Tsuda 1977):

$$h(\underline{x},t) \rightarrow \delta(\underline{x}-\underline{x}_M) \quad \text{as t goes to infinity} \qquad \text{(A1)}$$

where $\delta(\cdot)$ is Dirac's delta function and $\underline{x}_M$ is the vector consisting of optimal design parameters. Moreover, $h(\underline{x},t)$ is the objective function transformed by relative peaking and is defined as follows:

$$h(\underline{x},t) = [F(\underline{x})]^t / \int_D F(\xi)^t d\xi, \qquad F(\underline{x}) = \exp[f(\underline{x})] \qquad \text{(A2)}$$

This proposed method was applied to the same optimization problem as in the section 3.2. In calculation, suitable large value of t was used. Finally, same result was gotten, but it was shown that this algorithm was very efficiently.

REFERENCES

Goldberg, David E. 1989. *Genetic Algorithms in Search, Optimization & Machine Learning:* Addison-Wesley Publishing Company Inc.

HDL Committee 1986. *Report on Investigation of Design Load Systems on Hanshin Expressway Bridges.* Hanshin Expressway Public Corporation. (in Japanese)

Japanese Society of Highway 1990-2. *Japanese Steel Highway Bridge Specification and Its Comments:* Vol.I. (in Japanese)

Matsuho,S. and Shiraki,W. 1990. Design Method Satisfying Safety Requirements for Various Limit - States Based on Information Integration Method: *Proc. of ICOSSAR'89:* Vol.3, 2243-2246. ASCE.

Nakazawa, H. 1987. *Information Integration Method:* Corona Publishing Co. Ltd. (in Japanese)

Shannon, C.E. and Weaner, W. 1949. *The Mathematical Theory of Communication:* University of Illinois Press.

Suh, N.P. et al. 1978. On an Axiomatic Approach to Manufacturing Systems: *Jour. of Eng. Indus. Trans. ASME:* Vol.100, 127-130.

Tsuda, T. 1977. *Monte-Carlo Method and Simulation:* Baifukan Publishing Co. Ltd. (in Japanese)

Yoshikawa, K.(Editor) 1985. *Exercises in Systems Planning for Civil Engineering:* Morikita Publishing Co. Ltd. (in Japanese)

Structural Safety & Reliability, Schuëller, Shinozuka & Yao (eds) © *1994 Balkema, Rotterdam, ISBN 90 5410 357 4*

Reliability-based minimum weight design of space truss: A case under constraint of system reliability

T. Takada & Y. Kohama
Department of Architecture, Mie University, Tsu, Japan

A. Miyamura
Nagoya City Women's College, Chikusa, Japan

ABSTRACT: The present paper deals with the reliability-based minimum weight design of space truss systems under the constraint of system failure probability given by Ditlevesen bounds. Ditlevesen bounds demand predominant combination between enormous failure modes. The present searching technique can pursue predominant combinations effectively. A numerical design example by the present design criteria is compared with the relaxed criteria.

1 INTRODUCTION

There are many aspects of reliability-based optimal design problem mostly accompanying reliability analysis which require large numerical burden for sizable structural systems. Jendo[1991] discusses the multicriteria optimization of bar structures with regard to the cost of material and reliability criteria on the assumption of two states failure or failure mechanism and safe state. The failure tree approach is applied to a dome truss with 52 bar-members. Vanderplaats[1972] deals with an optimum geometry design problem in terms of practical systems such as transmission and offshore towers or long span roof trusses. Frangopol[1985a] proposes the practical design solution by combination of reliability-based design procedures and optimization techniques to hold an adequate margin of safety against occurrence of both serviceability and ultimate limit states. Frangopol[1985b] extends this method to the reliability-based optimal design sensitivity of redundant ductile structures. Frangopol also developed a reliability-based optimization technique to design optimal plastic frame structures. In this procedure, both load and strength are assumed to be random variable. Load correlations and strength correlations are also considered. The procedure to obtain the best design is based on minimization of the total expected weight for a specified probability of collapse for the structure. Levy[1987] discusses the sequential linear programming, dual methods. Optimality criteria is applied to truss systems on the emphasis of convergence of optimality criteria and mathematical programming methods. Relating to these reliability-based design problems Ditlevsen[1991] proposes the effectivity factor that reduces the random maximal yield moment capacities of rigid-plastic hinges. This effectivity factor is a function of random load and strength variable. It is approximated by the zero order or first order Taylor expansion with expansion point equal to the most central limit state point corresponding to a plastic collapse mode. Enevoldsen[1991] deals with reliability sensitivity analysis of a series system of parallel systems. Direct and sequential optimization procedures to solve the optimal problems are discussed by the bounds iteration method(DIM). Shao[1991] proposes a systematic technique to find multiple check points on the limit state surface with linearized limit state functions at the selected points, where approximation by AFOSM method is used. Frangopol[1985a] deals with structural systems subject to probability constraints imposed at both serviceability and ultimate limit state and the sensitivities of the optimal solutions to changes in parameters. Dobbs[1969] discusses that such problems or design parameters are naturally classified into sizing of cross section, configuration of members and structural topology in terms of truss system.

This paper deals with the minimum weight design of space trusses under the constraint of the system failure probability estimated by Ditlevsen bounds which requires effective search for predominant modes between enormous combinations of failure mode. This can be, herein, accomplished by the generate-and-test technique with heuristic rule, and a practical design criteria is proposed with implementation on a numerical example.

2 OPTIMALITY CRITERIA

The present reliability-based minimum weight design criteria can be described by the following:

$$W_T \rightarrow \min \qquad \text{subject to} \quad P_{f,sys} \leq P_{fa} \tag{1}$$

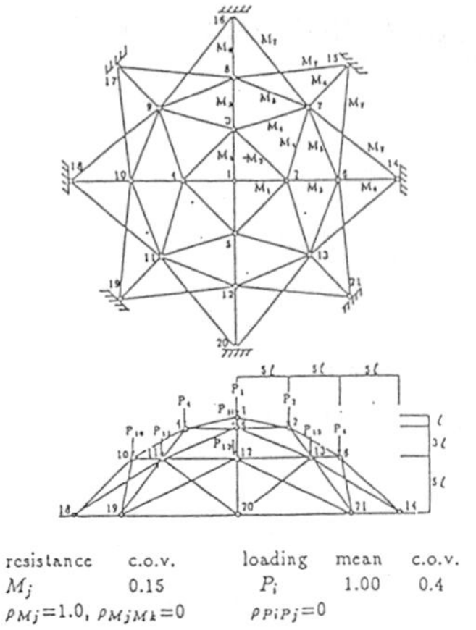

Fig. 1 Example system

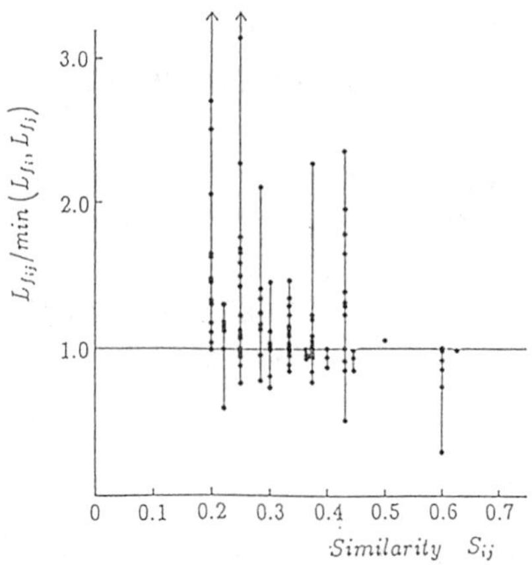

Fig. 2 $L_{fij}/min(L_{fi}, L_{fj}) - S_{ij}$ relationship

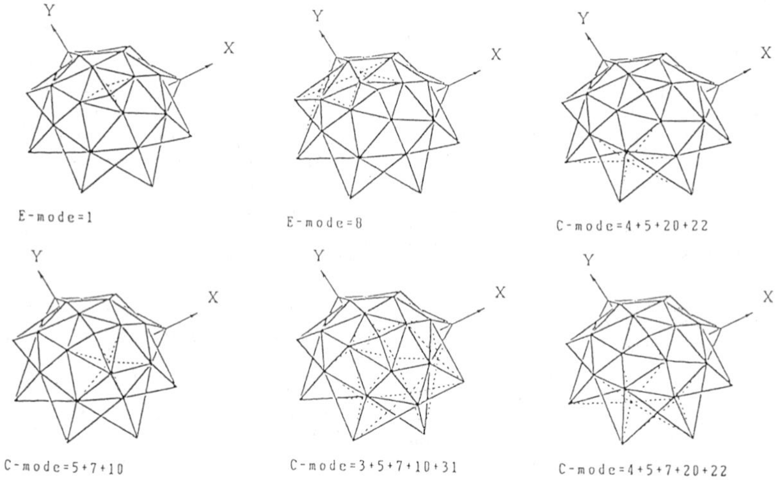

Fig. 3 Predominant failure modes

Table 1 Elemetary failure modes

No.	1	2	3	4	5	6	7	8	9	10	11	12	13	14	15	16	17	18	19	20
n_f	1	1	1	2	2	2	2	4	4	2	2	1	1	1	1	1	1	1	1	5
n_y	2	2	4	4	4	4	4	6	6	4	4	3	3	3	3	3	3	3	3	7
No.	21	22	23	24	25	26	27	28	29	30	31	32	33	34	35	36	37	38	39	
n_f	5	4	8	4	6	2	1	1	2	8	8	4	1	1	1	1	8	8	1	
n_y	5	6	9	6	7	4	4	4	4	6	6	6	4	4	4	4	6	6	2	

n_f : the number of failed nodes, n_y : the number of yielding members where $P_{f,sys}$ means the system reliability, which is a function of mode failure probabilities in terms of both failure probabilities for possible failure modes and their joint probabilities.

Since the above equation contains large redundancy, a large discrepancy is expected between individual mode failure probabilities. Consequently, regarding those individual mode probabilities it is desirable to accomplish optimal design within certain tolerance. Hence, the present criteria is rewritten by adding constraints of the mode failure probabilities as follows:

$$W_T \rightarrow \min \quad \text{subject to} \quad P_{f,sys} \leq P_{fa} \quad \text{and} \quad P_{f,k} \leq P_{m,a} \; ; \quad k = 1, \ldots, K \tag{2}$$

where $P_{f,k}$ represents the k-th mode failure probability and K, the number of possible failure modes, respectively. $P_{m,a}$ is a tolerance of mode failure probability that is determined to satisfy the constraint of system failure probability.

It is difficult to evaluate the rigorous system failure probability whose upper and lower bounds are estimated in various ways. Herein, Ditlevsen bounds[1979] is assumed for the upper bound of $P_{f,sys}$. Thus Eq.(2) become:

$$\begin{aligned} &W_T \rightarrow \min \\ &\text{subject to} \\ &P_{f,sys} \leq \sum_{i=1}^{K} P(E_i) - \sum_{i=2}^{K} \max_{j<i} P_L(E_i E_j) \leq P_{fa} \\ &P(E_i) \leq P_{ma} \; ; \quad k = 1, \ldots, K \end{aligned} \tag{3}$$

where E_i means the i-th mode failure event and $P(E_i)$, its failure probability, respectively.

When assumed normally distributed, each probability can be expressed by:

$$\begin{aligned} P(E_i) &= \Phi(-\beta_i) \\ P_L(E_i E_j) &= \max(P_{ij}, P_{ji}) \\ P_{ij} &= \Phi(-\beta_i)\Phi(-\frac{\beta_j - \rho_{ij}\beta_i}{\sqrt{1-\rho_{ij}^2}}) \end{aligned} \tag{4}$$

where β_i means the i-th mode reliability index and ρ_{ij}, the correlation coefficient of safety margin between the i- and j-th modes, respectively.

K is the number of all possible failure modes for $P_{f,sys}$, which is not so large even for a sizable system because of smaller number of active failure mode to design variables. Furthermore, since the failure probability of a non-active mode is negligible compared to an active mode, the number, K, can be replaced by the number of active modes, K_a. For these active modes $P(E_i) = P_{m,a}$ and Eq.(3) provide:

$$\Phi(-\beta_{ma}) \left\{ K_a - \sum_{i=2}^{K_a} \max_{j<i} \Phi(-\sqrt{\frac{1-\rho_{ij}}{1+\rho_{ij}}}\beta_{ma}) \right\} = \Phi(-\beta_{fa}) \tag{5}$$

where $P_{m,a} = \Phi(-\beta_{ma})$ and $P_{fa} = \Phi(-\beta_{fa})$, respectively. This equation provides the relationship between the system failure probability and a tolerance for failure probabilities of active mode.

By Eq.(5) this optimality criteria becomes:

$$\begin{aligned} &W_T = \{L\}^t\{\overline{N}_y\} \rightarrow \min \\ &\text{subject to} \\ &\beta_k = \frac{\overline{M}_k(\{\overline{N}_y\})}{\sigma_{Mk}(\{\overline{N}_y\})} \geq \beta_{ma} \; ; \quad k = 1, \ldots, K_a \end{aligned} \tag{6}$$

where $\{\overline{N}_y\}$ and $\{L\}$ are design variables regarding mean axial yield resistance and member length, respectively. $\overline{M}_k(\cdot)$ represents mean of the safety margin. β_{ma} is provided by Eq.(5). The mean and variance of $M_k(\cdot)$ are obtained by the virtual work equation:

$$\begin{aligned} &\overline{M}_k(\{\overline{N}_y\}) = \overline{U}_k - \overline{W}_k = \{e_{pk}\}^t\{\overline{N}_y\} - \{d_k\}^t\{\overline{P}\} \\ &C_{M_j M_k}(\{\overline{N}_y\}) = C_{U_j U_k} + C_{W_j W_k} \\ &\quad = \{\overline{N}_y\} di.[e_{pj}] di.[V_{N_y}][\rho_{N_y}] di.[V_{N_y}] di.[e_{pk}]\{\overline{N}_y\} \\ &\qquad + \{d_j\}^t[C_P]\{d_k\} \\ &\sigma_{M_k}^2(\{\overline{N}_y\}) = C_{M_k M_k}(\{\overline{N}_y\}) \\ &\rho_{ij} = \frac{C_{M_i M_j}(\{\overline{N}_y\})}{\sigma_{M_i}(\{\overline{N}_y\})\sigma_{M_j}(\{\overline{N}_y\})} \end{aligned} \tag{7}$$

where $\{e_{pk}\}$ represents the axial plastic displacement vector and $\{d_k\}$, the node displacement vector at the k-th failure mode, respectively. $\{\overline{P}\}$ means a vector of nodal loads and $[C_P]$, a covariance matrix, respectively. $di.[V_{N_y}]$ means a diagonal matrix of coefficients of variance of member resistance and $[\rho_{N_y}]$, its correlation matrix, respectively.

To solve Eqs.(5) to (7) there appears difficulty due to both non-linearity of constraints and laborious generation of possible failure modes. Hence, β_k is linearized by Taylor expansion as follows:

$$\beta_k \approx \widetilde{\beta_k} = \beta_k^* \left[1 + \frac{\{e_{pk}\}^t\{\overline{N}_y\} - \overline{W}_k}{\overline{M}_k^*} - \frac{\{\overline{N}_y^*\}[G_k]\{\overline{N}_y\} + \sigma_{Wk}^2}{\sigma_{Mk}^{*2}} \right] \tag{8}$$

where

$$\begin{aligned} &\beta_k^* = \overline{M}_k^* / \sigma_{Mk}^* \\ &\overline{M}_k^* = \{e_{pk}\}^t\{\overline{N}_y^*\} - \overline{W}_k \\ &[G_k] = di.[e_{pk}] di.[V_{N_y}][\rho_{N_y}] di.[V_{N_y}] di.[e_{pk}] \\ &\sigma_{Mk}^{*2} = \{\overline{N}_y^*\}[G_k]\{\overline{N}_y^*\} + \sigma_{Wk}^2 \end{aligned} \tag{9}$$

Thus Eq.(6) can be transformed into the linear programming problem as follows:

$$\begin{aligned} &W_T = \{L\}^t\{\overline{N}_y\} \rightarrow \min \\ &\text{subject to} \\ &\widetilde{\beta_k} \geq \beta_{ma}, \quad k = 1, \ldots, K_a \end{aligned} \tag{10}$$

Table 2 Generated modes for design

System constraint with $\beta_{fa} = 2.0$					Mode constraint with $\beta_{ma} = 3.0$				
	combination	β_k	n_f	n_y		combination	β_k	n_f	n_y
	e-mode 1	2.85*	1	2	1	15+33	3.00	1	4
	e-mode 2	2.85*	1	2	2	8+32	3.10	4	8
	e-mode 8	2.85*	4	6	3	10+20	3.76	5	7
	e-mode 9	2.85*	4	6	4	8+26+32	3.07	4	7
1	15+33	2.85*	1	4	5	10+20+25	6.19	8	9
2	4+22	6.55	4	6	6	3+28	3.00	1	4
3	4+5+22	2.86	4	6	7	2+39	3.00	1	2
4	6+11	3.46	2	4	8	26+29	3.00	2	4
5	4+5+10+22	2.85*	4	4	9	4+22	7.90	4	6
6	4+5+7+22	2.85*	4	4	10	4+5+22	3.01	4	6
7	4+5+20+22	2.85*	4	6	11	8+26+29+32	5.29	4	7
8	4+5+22+25	2.86	4	5	12	4+5+10+22	3.00	4	4
9	4+5+22+23+25	2.85*	4	5	13	4+5+7+22	3.00	4	4
10	4+5+22+23	2.86	4	5	14	4+5+20+22	3.00	4	6
11	3+28	3.12	1	4	15	4+29	3.00	1	4
12	5+10	3.12	1	4	16	4+5+22+25	3.01	4	5
13	5+7+10	3.12	1	4	17	4+5+22+23+25	3.00	4	4
14	5+7+10+31	4.08	8	7	18	5+10	3.00	1	4
15	5+7+10+11+31	2.89	8	8	19	4+5+22+23	3.01	4	5
16	3+5+7+10+31	2.85*	8	8	20	8+26+32+38	3.04	8	9
17	4+5+10+20+22	7.03	5	5	21	4+8+26+32+38	3.11	8	9
18	5+7+10+30+31	3.34	11	9	22	1+39	3.00	1	2
19	5+7+10+27+31	3.44	7	8	23	13+36	3.00	1	4
20	5+7+10+27+31	4.01	8	8	24	5+7+10	3.00	1	4
21	8+32	2.85*	4	8	25	5+7+10+31	5.36	8	7
22	26+29	2.85*	2	4	26	5+7+10+11+31	3.86	8	8
23	4+5+7+20+22	2.86	5	5	27	3+5+7+10+31	3.00	8	8
24	12+35	2.85*	1	4	28	5+7+10+30+31	3.67	7	8
					29	5+7+10+30+31	4.64	11	9
					30	5+7+10+27+31	4.36	8	8
					31	8+21+26+32+38	3.00	7	9
$2.02 \leq \beta_{sys} \leq 2.29$					$2.26 \leq \beta_{sys} \leq 2.51$				

n_f : the number of displaced nodes, n_y : the number of yielded members

Table 3 Design results

design result case	W_f	mean strengths : $\overline{N}_{0j}$								β_{sys}
		M_1	M_2	M_3	M_4	M_5	M_6	M_7	M_8	
$\beta_{fa} = 2.5$	3870	11.2	5.5	24.1	6.0	17.0	29.0	0.1	23.4	$2.54 \leq \beta_{sys} \leq 2.72$
$\beta_{fa} = 2.0$	3522	9.9	4.5	20.5	5.8	14.9	27.2	0.1	22.7	$2.02 \leq \beta_{sys} \leq 2.29$
$\beta_{ma} = 3.0$	3303	8.6	4.9	16.4	6.1	13.0	27.0	0.1	22.5	$2.26 \leq \beta_{sys} \leq 2.51$

Thus, the present design process can be summarized as follows:

1. Given both tolerance level of β_{fa}, the system reliability index, and K_a, the order of $\{\overline{N}_y\}$, β_{ma} can be calculated approximately by Eq.(5) assuming $\rho_{ij} = 0$. By the present optimality procedure the initial design variable vector, $\{\overline{N}_y^*\}$, is obtained based on elementary failure modes.
2. An approximate design variable vector, $\{\overline{N}_y\}$, is calculated by the sequential linear programming(SLP). By Eq.(5) β_{ma} is improved at each calculating step.
3. By reliability analysis of any designed system with $\{\overline{N}_y\}$, then, a predominant failure mode can be obtained, and β_c and $\{e_p\}$ for this predominant mode can be calculated.
4. If a new predominant mode under $\beta_c < \beta_{ma}$ is obtained, then it should be added to a set of design modes and returned to the previous step 2.
5. When there exists no more predominant failure mode, the present optimal design process is completed. And $\{\overline{N}_y\}$ becomes the optimal design.

3 NUMERICAL EXAMPLE

A numerical example space truss with 13 nodes and 52 members shown in Fig. 1 is designed by the present optimality criteria. For simplicity eight classes are assumed of the design variables, $\{\overline{N}_y\}$, entirely composed of pipe sections whose compressive buckling resistance can be expressed as:

$$N_y \approx \frac{\pi^2}{8}\left(\frac{D_j}{l_j}\right)^2 EA_j = \frac{\pi^2}{8}\frac{E_j}{\sigma_{yj}}\left(\frac{D_j}{l_j}\right)^2 N_{0j} \tag{11}$$

where E_j means Young's modulus of the j-th member, D_j, the diameter assumed constant, l_j, the length of the j-th member and $N_{0j} = \sigma_{yj}A_j$, respectively. Young's modulus and the yield point are assumed deterministic.

Table 1 shows the number of the yielded members and the displaced nodes for 39 elementary failure modes, among which almost half of them collapse partially with both less than two displaced nodes and 4 yielded members. This is due to characteristics of circuit matrix from topology of the truss system.

Table 2 shows reliability indices and numbers of displaced node with both yielded member for predominant modes with smaller β-value to design modes subject to $\beta_{fa} = 2.0$ and 2.5 and system reliabilities, $P_{sys} = \Phi(-\beta_{sys})$, by Ditlevsen bounds with orders of less than 5 combined modes. Herein, asterisk shows an active mode. The discrepancy of lower bound between the target reliability index, β_{fa}, and the lower bound, β_{sys} is negligible. This is dependent on discrepancy between numbers of active modes for design and actual mode. The relationship between the system reliability index, β_{sys}, and a mode reliability index, β_k, depends upon the number of active modes. The present example takes 12 which implies the difference; $\beta_k - \beta_{sys} = 0.7 \sim 1.0$. The present minimum weight can be finally attained under the target mode reliability index, β_k, which is the sum of the target system reliability index, β_{fa} and the difference.

Table 3 shows the resulting total weight and mean resistance, $\{\overline{N}_y\}$, compared with the results subjected to constraints of mode failure reliability by means of the sequential linear programing technique(Kohama[1991]). There is few discrepancy between them.

4 CONCLUDING REMARKS

Regarding the present reliability-based minimum weight design the following remarks are obtained:

a. Homogeneous qualified system can be optimally designed by adding the constraints of the mode failure probability to the system reliability constraint.
b. Although the present problem contains nonlinearity, the optimality can be attained by alternative implementation of both SLP and failure analysis of superimposed modes.
c. Search for predominant modes can be accomplished by recursive generation of the failure modes and pruning futile alternatives based upon the similarity index.
d. For the present example 5-mode combinations provides satisfactory evaluation of the system reliability. After twenty or more design stages the total weight becomes constant, and there appear no more modes with less than 80% active β.
e. The optimally designed system satisfies the prescribed system reliability resulting in the relationship, $\beta_k - \beta_{sys} = 0.7 \sim 1.0$.

APPENDIX 1: FAILURE MODE GENERATION

The present reliability-based design process is subjected to the constraint of the system reliability whose bound is evaluated by the mode approach. Thus, systematic generation of failure modes is indispensable and its formulation can be accomplished by the graph theory. The compatibility condition at any failure mode is equivalent to the closing condition of a failure circuit which is described as follows:

$$\begin{aligned}
&\sum_{circuit} L_{ij}\left\{\cos\phi_{ij}\omega_{y_{ij}} - \sin\theta_{ij}\sin\phi_{ij}\omega_{z_{ij}}\right\} = 0 \\
&\sum_{circuit} L_{ij}\left\{-\cos\phi_{ij}\omega_{x_{ij}} + \cos\theta_{ij}\sin\phi_{ij}\omega_{z_{ij}}\right\} = 0 \\
&\sum_{circuit} L_{ij}\left\{\sin\theta_{ij}\sin\phi_{ij}\omega_{x_{ij}} - \cos\theta_{ij}\sin\phi_{ij}\omega_{y_{ij}}\right\} = 0
\end{aligned} \tag{a}$$

where L_{ij} means a length of the ij-th member and $\omega_{x_{ij}}, \omega_{y_{ij}}, \omega_{z_{ij}}$, the rotational components subjected to plastic rigid rotation at collapse, respectively. ϕ_{ij} and θ_{ij} represent Euler angles.

Applying topology of space truss system the final expression of nodal displacement vector for the virtual work equation is summarized as follows:

$$\left\{\begin{array}{c}\{r\}\\ \{e_D\}\\ \{e_I\}\end{array}\right\} = \left[\begin{array}{c}[C_I]\\ [E]\end{array}\right]\{e_I\} = [C]\{e_I\} \qquad \text{(b)}$$

where $\{r\} = \{\{v_t\}^t, \{v_{\bar{t}}\}^t\}^t$ and $\{e\} = \{\{e_D\}^t, \{e_I\}^t\}^t$. $\{v_t\}^t$ means the rigid rotational displacement vector of tree members, $\{e_D\}$ and $\{e_I\}$, the corresponding dependent and independent axial displacement vectors, respectively. $[E]$ means the unit matrix.

The k-th column vector, $\{C_k\}$, from $[C]$ is produced by a displacement mode for $e_{Ik} = 1$. Thus, $[C]$ is called an elementary failure mode matrix because of independence of each element of $\{C_k\}$. The above equation implies that non-zero elements of the axial vector, $\{e\}$, for each elementary failure mode or the column vector of $[C]$ corresponds to a yielded member.

The displacement vector of the corresponding node, $\{D\}(= \{\{D_x\}^t, \{D_y\}^t, \{D_z\}^t\}^t)$, can be easily obtained by the path matrix(Kohama[1989]) The limit state function for the k-th failure mode is the virtual work equation:

$$M_k = \{N_y\}^t\{e_k\} - \{P\}^t\{D_k\} = 0 \qquad \text{(c)}$$

where $\{P\}$ means the nodal load vector and $\{N_y\}$, the axial yield resistance vector, respectively. The axial yield resistance of the j-th member, an element of $\{N_y\}$, is assigned to either tensile yield strength or compressive buckling resistance.

Eqs.(6) and (8) are necessarily implemented for the entire possible failure modes whose generation becomes practically burdensome. Consequently, the most possible predominant failure mode due to a set of approximate design variables, $\{\overline{N}_y\}$, at each design stage is introduced.

This can be accomplished by searching a failure mode with minimum reliability index from elementary failure modes and their linear combinations. For these combination there are two techniques; the minimum problem under constraints by the sequential linear programing(Kohama[1991]) and combinatorial searching for the goal mode with minimum reliability index by the generate-and-test procedure. Practically for sizable systems the former requires a large burden of computation. The latter that is algorithmically quite simple results also in explosive increase of combinations which can be, however, drastically narrowed by application of effective heuristic rule.

After generation of a sufficient number of elementary failure modes the combined modes are expressed by their linear combination. When both yield elongation vector and the nodal displacement vector of the i-th elementary mode is expressed by $\{Y_i\} = \{\{e_p\}_i^t, \{D\}_i^t\}$, the combination of any two modes $\{Y_i\}$ and $\{Y_j\}$, which may be either elementary or combined, becomes:

$$C_{i+j}(\{Y_i\}\{Y_j\}) = A_i\{Y_i\} + A_j\{Y_j\} \qquad \text{(d)}$$

subject to

$$\begin{aligned} A_i e_{pi,s} + A_j e_{pj,s} &= 0\\ A_i \overline{W}_i + A_j \overline{W}_j &= 1\\ s \in H_i \cap H_j & \\ H_i \cap H_{dj} &= \phi\\ H_j \cap H_{di} &= \phi \end{aligned} \qquad \text{(e)}$$

where $e_{pi,s}$ and $e_{pj,s}$ mean the plastic deformations at a critical section, s, for i-th and j-th modes. H_i and H_j represent a set of yielded members for i-th and j-th modes, H_{di} and H_{dj}, a set of yielded members that are eliminated to generate both i-th and j-th modes, and ϕ, an empty set, respectively.

Although Eqs.(d) and (e) can progressively produce any combined mode, it is difficult to implement effective search of predominant modes by the exhaustive enumeration method with the generate-and-test. If an approximate estimation of effective combination is implemented before actual combination procedure the amount of calculation decreases drastically even for a large sized structural systems.

Although the predominance of a failure mode is closely related to its failure load factor or its central safety factor. This suggests that a combination of the smaller internal virtual work to the external one becomes predominant. Hence, when yield elongations decrease by the combination of appropriate modes the corresponding internal work decreases. This can be more accomplished for two candidate modes whose common yield members become larger in number, in other words, whose yield allocation becomes more similar. Thus pruning of the futile searching space is attained with less burden if it is possible to evaluate an extent of similarity of yield distribution. Such evaluation is established by enumeration of both common and non-common yield members between the i- and j-th modes, and satisfied to some extent by means of the following similarity index, S_{ij}:

$$S_{ij} = \frac{\sum_s \sum_a \min(h_{is}, h_{js})}{\sum_s \sum_a \max(h_{is}, h_{js})}; \quad i,j = 1,\ldots,K \qquad \text{(f)}$$

where h_{is} is a binary parameter as follows:

$$h_{is} = \begin{cases} 0, & \text{if } e_{pis} = 0;\\ 1, & \text{otherwise.}\end{cases} \qquad (g)$$

Eq.(f) provides the ratio of the number of common yielded members and the number of total yielded members between any two failure modes. Thus, $h_{is} = 1$ corresponds to a yield member at the s section in the i-th failure mode. Furthermore, $0 \leq S_{ij} \leq 1$ is valid.

Fig. 2 shows the relationship between the similarity index and the combination load factor ratio, $L_{fij}/\min(L_{fi}, L_{fj})$, for 39 elementary failure mode of the present example space truss, where L_{fij} means the failure load factor from combination of i and j modes and L_{fi}, L_{fj}, those of each elementary modes, respectively. It is apparently understood that the load factor ratio tends to decrease with larger L_{fij}, nevertheless whose discrepancy is not uncommitted. This is applied for pruning of the searching space for predominance. Consequently, the present heuristics becomes: before implementation of the conventional generate-and-test for the exhaustive enumeration the similarity index, S_{ij}, by Eq.(f) should be firstly calculated and pruned if the similarity index is smaller. This is extended to further multi-combination modes whose similarity indices are easily calculated by Eq.(f) with substitution of both $i = n$ and $j = n - 1$.

As stated previously, the predominant failure mode that has the minimum mode reliability index can be searched by the generated-and-test technique from elementary modes and their linear combinations. For these combined modes progressively generated by Eqs.(d) and (e) their similarity index by Eq.(f) can narrow effectively the searching space where the corresponding reliability indices are calculated. Thus, searching should accompany balance between tolerance of similarity index and combination. For smaller S_{ij} and larger combination the corresponding searching space becomes explosively large with heavy computing burden. While, for larger S_{ij} and smaller combination the searching space decreases with difficulty of mode generation with smaller β-value.

Herein, two-mode combinations are completed from only elementary modes. Then three-mode combinations are generated from combination of the two-modes with the elementary modes. Similarly, the higher modes can be obtained by Eqs.(d) and (e).

For these combinations two searching techniques can be applied:

(i) the breadth-first search that requires a set of the same order modes,

(ii) the depth-first search that provides a series of the successively-increasing order modes.

Herein, the latter is applied, because, first, empirically five-mode combinations in maximum can satisfy sufficient accuracy, second, survey of two-mode combinations shows $S_{ij} \geq 0.3$ as in Fig. 2 and, third, some different modes from elementary modes, that is, modes with smaller similarity index are preferable for the present design.

APPENDIX 2: REFERENCES

1. O.Ditlevsen and T.Arubjerg-Nielsen[1991], Effectivity factor method in structural reliability, Proc. of the 4th IFIP-WG7.5 Conf., pp.171-179, Munich, Sept., Springer-Verlag
2. I.Enevoldsen and J.D.Sorensen[1991], Strategies for optimal design of structural systems, Proc. of the 4th IFIP-WG7.5 Conf., pp.191-206, Munich, Sept., Springer-Verlag
3. S.Jendo and J.Putresza[1991], Multicriteria optimization of elasto-plastic bar structures, Proc. of the 4th IFIP-WG7.5 Conf., pp.239-249, Munich, Sept., Springer-Verlag
4. S.Shao and Y.Murotsu[1991], Reliability evaluation methods for systems, Proc. of the 4th IFIP-WG7.5 Conf., pp.325-338, Munich, Sept., Springer-Verlag
5. F.Moses and J.D.Stevenson[1970], Reliability-based structural design, Proc. of ASCE, ST2, Feb., pp.221-244.
6. G.N.Vanderplaats and F.Moses[1972], Automated design of trusses for optimum geometry, Proc. of ASCE, ST3, March, pp.671-690
7. D.M.Frangopol[1985a], Structural optimization using reliability concepts, Proc. of ASCE, SE11, Nov., pp.2288-2301
8. D.M.Frangopol[1985b], Sensitivity of reliability-based optimum design, Proc. of ASCE, SE8, August, pp.1703-1721
9. R.Levy and O.E.Lev[1987], Recent developments in structural optimization, Proc. of ASCE, SE9, Sept., pp.1939-1962
10. M.Zhou and R.Xia[1990], An efficient method of truss design for optimum geometry, Computers and Structures, Vol.35, No.2, pp.115-119
11. A.Kaufmann and M.M.Gupta[1988], Fuzzy mathematical models in engineering and management science, North-Holland
12. Y.Kohama and A.Miyamura[1989], Predicated minimal path to failure reliability of space truss, Proc. of International Conference on Computer and Optimal Design of Structures, Computer Mechanics Pub., June, pp.275-284
13. Y.Kohama and A.Miyamura[1991], Reliability-based minimum weight design of space truss, Proc. of ICASP6, pp.362-369
14. O.Ditlevsen[1979], Narrow reliability-bounds for structural systems, J. Struct. Mech., ASCE, Vol.7, pp.435-451

Structural Safety & Reliability, Schuëller, Shinozuka & Yao (eds) © 1994 Balkema, Rotterdam, ISBN 90 5410 357 4

Reliability-based life cycle costing in structural design

Zongwei Tao, J. Hugh Ellis & Ross B. Corotis
Johns Hopkins University, Baltimore, Md, USA

ABSTRACT: The objective of this research is to provide a framework for structural life cycle cost analysis with the application of Markov decision process (MDP) and structural reliability theory. An MDP is a stochastic sequential decision process for a system that develops dynamically under the control of a decision maker. This paper assumes that owners and users represent two distinct groups involved in the design and lifetime process of a structure. The decision maker, separate from owners and users, pursues an optimal criterion in the initial design stage, which combines the goals of both owners and users. Several multiobjective solution techniques are applied to assist in considering various relationships between owners and users in the decision model.

1 INTRODUCTION

Life cycle cost is defined as the present value of all anticipated costs to be incurred during the lifetime of a structure. The objective of life cycle cost analysis is to assess and control these costs, and also to assist in deciding design and operating alternatives of a structure. The principles underlying life cycle cost analysis imply that these alternatives generate distinctive trade-offs among initial costs, recurring operations and maintenance cost, and structural performance [5].

This paper, with applications of structural reliability theory and Markov decision processes (MDP), provides a systematic approach for life cycle cost analysis. Structural reliability theory is applied to tackle the uncertainty associated with anticipating future structural performance, which is the fundamental phase in life cycle cost analysis. Markov decision processes provide a unique and workable framework to systematically characterize and aggregate human remedial actions and costs during the lifetime of a structure.

When cost is included in decision making, it is important to address by whom the cost is incurred. The life cycle cost is aggregated from costs incurred by several parties, who can hold conflicting views of loss and benefit with respect to the performance and operation of a structure. Failure to distinguish different party's costs in life cycle cost analysis may produce misleading results. Based on this fact, this paper assumes that owners and users represent two different types of people involved in the design and lifetime process of a structure. Their life cycle costs are then aggregated separately. The decision maker, separate from owners and users, pursues an optimal criterion in the initial design stage, which combines the goals of owners and users. The MDP program must then incorporate multiobjective considerations.

2 Markov Decision Processes

A Markov decision process (MDP) is a stochastic sequential decision process for a dynamic system under the control of a decision maker. At time t, a decision maker finds the system in state i, with a choice of available actions. Choosing a particular action k has two related effects:

- the decision maker incurs an immediate cost C_{ik};
- the system moves to state j at time $t+1$ with transition probability P_{ij}^k.

The objective of a decision maker is to find an optimal policy, under some prespecified optimality criteria, that prescribes rules for making decisions. Minimizing expected total discounted cost is a typical optimality criterion in an MDP. The elements of an MDP include a state space of the system, transition probabilities, cost classification and aggregration, policy modelling, and a mathematical program. These elements will be introduced subsequently, with their interpretations in structural engineering. A complete description of the MDP approach and associated references are provided in [6].

3 Transition Probability and Lifetime Reliability

3.1 Self-transition Probability of a Structure

In the conceptualization of the state space of a structure, each state represents a certain level of performance. Taking into account the randomness of both loading and resistance, a structure can be either safe or have failed for a particular deterioration of its resistance. In order to describe both the serviceability and safety of a structure, a reliability-based state space of a structure is proposed here:

$$\mathbf{S} = \{1, ..., i, ..., M\} \tag{1}$$

$$s_t = i \Leftrightarrow \begin{cases} R_t = \gamma_i R_o \cap \text{safe at time t}, \\ \quad \text{for } i = 1, \ldots, M-1. \\ \text{failed at time t}, \\ \quad \text{for } i = M. \end{cases} \tag{2}$$

in which $\mathbf{S}$ is the state space with M states, s_t the state variable at time t, i the state index R_t the remaining resistance at time t, and γ_i the ratio of the remaining resistance R_t to the initial resistance R_0. Here, "safe at time t" or "failed at time t" should be interpreted as the structure remaining safe or failing during the inspection interval $(t-1, t)$.

To consider whether the structure will remain safe until the next inspection, a decision maker is interested in the maximum loading effect and minimum resistance during the next inspection interval. A typical realization of a loading effect $S(t)$ and resistance $R(t)$ during the lifetime of a structure is shown in Figure 1. A discrete lifetime structural resistance and loading model is introduced in Figure 2. The structural resistance is discretized so that R_t is time-independent within each inspection interval. The loading effect S_t represents the maximum of $S(t)$ during the interval $(t-1, t)$. It is assumed that the loadings in each interval are independent and identically distributed.

According to structural reliability analysis, the safety or failure of a structure during interval $(t-1, t)$ can be modelled by limit state functions as

$$\text{safe at time } t \Leftrightarrow S_t < R_t \tag{3}$$

$$\text{failed at time } t \Leftrightarrow S_t \geq R_t \tag{4}$$

A self-transition matrix is defined to describe the changes of states of a structure under the probabilistic laws of loads and resistances without any human interaction during its lifetime. Let $\mathbf{P^S}$ be the self-transition matrix of a structure. Then

$$\mathbf{P^S} = \{P^S_{ij}\} \tag{5}$$

$$P^S_{ij} = P[s_{t+1} = j | s_t = i], \quad \text{for } i, j = 1, \ldots, M. \tag{6}$$

It is a requirement for a transition matrix that

$$\sum_{j=1}^{M} P^S_{ij} = 1, \quad \text{for } i = 1, \ldots, M. \tag{7}$$

For $i, j = 1, \ldots, M-1$,

$$\begin{aligned} P^S_{ij} &= P[\text{safe at time } t+1 \cap R_{t+1} = \gamma_j R_0 \\ &\qquad |\text{safe at time } t \cap R_t = \gamma_i R_0] \quad (8) \\ &= P[S_{t+1} < R_{t+1} \cap R_{t+1} = \gamma_j R_0 \\ &\qquad |S_t < R_t \cap R_t = \gamma_i R_0] \quad (9) \\ &= \frac{\int_0^\infty F_{S_t}(\gamma_i r) F_{S_{t+1}}(\gamma_j r) f_{R_0}(r) dr}{\int_0^\infty F_{S_t}(\gamma_i r) f_{R_0}(r) dr} \times \\ &\qquad P[R_{t+1} = \gamma_j R_0 | R_t = \gamma_i R_0] \quad (10) \end{aligned}$$

in which F_{S_t} and $F_{S_{t+1}}$ are the CDF's of loading effects S_t and S_{t+1}, respectively, and f_{R_0} is the PDF of the initial resistance R_0. $P[R_{t+1} = \gamma_j R_0 | R_t = \gamma_i R_0]$ is an element of a resistance deterioration matrix. P^S_{MM} is equal to 1 since M represents the failure state.

3.2 Decision Effect Probability of a Structure

A homogenous decision space $\mathbf{\Phi}$ is defined to be a space comprising all practically applicable actions for a structure. Each action in the space has an unambiguous interpretation of its cost and effect. The effect of an action is defined as the amount of resistance added to the structure, as a result of the implementation of the action [9].

A policy in MDP is defined as the rule of decision making. With MDP, maintenance operations can be modelled as a long-run maintenance policy A, represented by a policy matrix $\mathbf{D}$, where elements D_{ik} are interpreted as the probability that the decision k is made, given the structure is in state i:

$$D_{ik} = \text{P[decision=}k|\text{state=}i] \tag{11}$$

$$\sum_{k=1}^{K} D_{ik} = 1, \quad \text{for } i = 1, ..., M \tag{12}$$

A decision effect matrix $\mathbf{P^E}(A)$ describes changes to a structure under policy A. Its elements $P^{E,k}_{il}(A)$ are interpreted as the conditional probability of a structure being in state l, following implementation of the decision, $d_i(A) = k$, given that it is in state i and decision $d_i(A) = k$ is made at time t. Mathematically,

$$\mathbf{P^E}(A) = \{P_{il}^{E,k}(A)\} \quad (13)$$

$$P_{il}^{E,k}(A) = P[s_{c,t+\delta_k} = l | s_t = i \cap d_i(A) = k], \text{ for } i, l = 1, \ldots, M; \; k \in \Phi. \quad (14)$$

in which δ_k is the time needed for completing action k.

It is assumed that the self-transition of the structure does not occur during the intervals δ_k. Therefore, $\mathbf{P^E}(A)$ describes completely the probabilistic effects of decisions to the structure. Its evaluation depends on the effects of actions.

3.3 Joint-transition Probability of a Structure

At time t in the lifetime of a structure, a decision maker inspects a structure, finds it in state i, and then decides to take action k, under a certain policy. The structure then moves to state j at the next inspection time, $t+1$. This performance of a structure under both probabilistic laws, governing its resistances and loads, and effects of human interaction, given by policy A, can be described by a joint transition matrix $\mathbf{P^J}(A)$,

$$\mathbf{P^J}(A) = \{P_{ij}^{J,k}(A)\} \quad (15)$$

$$P_{ij}^{J,k} = P[s_{t+1} = j | s_t = i \cap d_i(A) = k], \text{ for } i, j = 1, \ldots, M; \; k \in \Phi. \quad (16)$$

Noting the one-step-memory property of Markov chains, the probability of a structure being in state j at time $t+1$, depends only on its state at time $t + \delta_k$, and is independent of its state at time t, this joint probability can be decomposed as

$$P_{ij}^{J,k}(A) = \sum_{l=1}^{M} P[s_{t+1} = j | s_{t+\delta_k} = l] \times P[s_{t+\delta_k} = l | s_t = i \cap d_i(A) = k] \quad (17)$$

If the δ_k is assumed to be very small compared to the inspection interval, $(t, t+1)$, then $t + \delta_k$ can be approximated as t. According to the loading and resistance model defined early, the likelihood that the structure fails during $(t, t+1)$ is larger than that during $(t + \delta_k, t+1)$. Therefore, this approximation is conservative from the point of view of structural safety. With this approximation, Equation (17) becomes

$$P_{ij}^{J,k}(A) = \sum_{l=1}^{M} P[s_{t+1} = j | s_t = l] \times P[s_t = l | s_t = i \cap d_i(A) = k] \quad (18)$$

$$= \sum_{l=1}^{M} P_{il}^{E,k}(A) P_{lj}^{S}, \quad (19)$$

A matrix form of Equation (19) is

$$\mathbf{P^J}(A) = \mathbf{P^E}(A) \cdot \mathbf{P^S} \quad (20)$$

3.4 Lifetime Failure Probability

Structural reliability can be evaluated by the so-called first-passage-time analysis. Let $f_{ij}^{(n)}$ denote the probability that the first passage time from state i to state j is equal to n. It can be shown that these probabilities satisfy the following recursive equations [4],

$$f_{ij}^{n} = P_{ij}^{(n)} - f_{ij}^{(1)} P_{jj}^{(n-1)} \cdots - f_{ij}^{(n-1)} P_{jj}. \quad (21)$$

in which P_{ij}^{n} is the transition probability from state i to state j in n steps. This transition probability can be evaluated for both the self-transition and joint transition. Let T_L be the design lifetime of the structure, M the failure state, and q_i the initial state reliability. The failure probability P_f is then evaluated as

$$P_f = F_T(T_L) = P[T \leq T_L] = \sum_{i=1}^{M} \sum_{n=1}^{T_L} f_{1M}^{(n)} q_i \quad (22)$$

4 Costs Characterization and Accounting

Generally, the costs incurred during the lifetime of a structure include different types, such as failure cost, maintenance cost and replacement cost. These costs are closely related to both structural performance and human interaction. Here, the following guideline for classifying these costs is developed as the first step of life cycle cost analysis:

1. costs of structural performance;
2. costs of maintenance;
3. cost of lost benefit due to maintenance.

In addition, it is also important to classify the parties by whom the costs are incurred. In this life cycle cost research, all costs are assumed to be incurred by either owners or users. Owners are responsible for initial investment, and lifetime operation and maintenance of a structure. Users represents all persons or organizations who make use of the structure directly or indirectly. The distinction between their costs is shown in the following sections.

4.1 Costs with Structural Performance

A structure can either be in operating states, perhaps with deteriorated load-carrying capacity, or the failure state during its lifetime. The costs associated with structural performance depend on whether a structure is capable of providing satisfactory service when it is in a given state.

The costs incurred when a structure operates with a deteriorated load-carrying capacity are usually not accounted for in current research of life cycle costs. Users, however, may incur costs due to the unsatisfactory service of a deteriorated structure. Owners may lose part of the benefit due to the fact that a deteriorated structure is usually not fully utilized.

The cost associated with the failure of a structure is the so-called "failure cost". The owners' loss includes the initial investment and the remaining revenue, if it is assumed that the owners receive a steady revenue stream during the lifetime of the structure. The users' loss relates to human life, environmental, socio and economic factors, for which monetary values are usually difficult to assign. Although in reality different costs may be associated with various modes of structural failure, this distinction has not been incorporated in the research. A socio-economic criterion has been developed to assess the users' loss by Flint and Baker [3], which is based on the level of risk at which society would not be prepared to pay for the cost of additional safety,

$$C = \frac{\nu n_r^2}{K_s} \qquad (23)$$

in which n_r is the average number of people within or near the structure during the period of the risk, ν a constant, and K_s a social criterion factor for various types of structures, given in the following table,

Table 1. Social Criterion Factor for Different Types of Structures

Nature of structure	K_s
Places of public assembly, dam	0.005
Domestic, office or trade and industry	0.05
Bridges	0.5
Towers, masts, offshore structures	5.0

(Flint and Baker 1976)

4.2 Costs of Maintenance

Various levels of maintenance action are possible, such as "no action", "repair" and "replacement". Since in this research owners are assumed to be responsible for the maintenance of a structure, the costs of lifetime maintenance are incurred to them.

With the tool of microeconomics, the space defining all maintenance actions is modelled as a production activity [9], in which a certain action in the space is defined to be the production activity at a given output level. The cost of an action is the expense of resources efficiently used to achieve that output level. The output of such a production activity can be quantified by the ratio of the amount of added resistance to the initial resistance of the structure, while its input resources are usually structural material, labor and equipment.

4.3 Costs of Lost Benefit Due to Maintenance

This cost is incurred when the normal operation of a structure is affected by the implementation of a maintenance action. When a structure is being replaced, owners will either lose revenue if the life of the structure is assumed to be the same as that before the replacement, or incur a loss due to the deferral of the revenue if the life of the structure is extended after the replacement. Some costs are also certainly incurred by users due to the unserviceability of the structure under replacement.

The above guideline shows how costs incurred during the lifetime of a structure can be classified. The advantage of this guideline relies on the fact that it relates costs with structural performance and maintenance operations, and serves as a basis for cost aggregation and decision making with the MDP, as will be shown subsequently.

5 MDP-based Cost Aggregation

The MDP-based cost aggregation includes the inspection-interval aggregation, and the lifetime aggregation. The inspection-interval aggregation provides a cost matrix to aggregate the costs incurred during each inspection interval, and assumes these aggregated costs to incur at time $t = 1, 2, \ldots$. The lifetime aggregation provides a recursive equation, with an assumed long-run maintenance policy, to further aggregate these costs to time $t = 0$.

5.1 Inspection-interval Aggregation

At time t, a decision maker inspects a structure

with a choice of available actions. Choosing a particular action results in some expected costs before the next inspection. These costs are aggregated by means of an MDP cost matrix, and assumed to occur immediately after the current inspection. Mathematically, this cost matrix is expressed as

$$\mathbf{C} = \{C_{ik}\} \text{ for } i = 1, \ldots, M; \ k = 1, \ldots, K. \quad (24)$$

in which i is the state index of a structure, k the decision index, and M and K are, respectively, the number of states and decisions. C_{ik} is the expected cost incurred before the next inspection if the structure is in state i and decision k is made at the current inspection. Mathematically, cost C_{ik} can be evaluated as

$$C_{ik} = C_M(k) + \sum_{j=1}^{M-1} P_{ij}^{J,k} C_S(j) + P_{iM} C_F \quad (25)$$

in which $C_M(k)$ is the cost of implementation of action k, $C_S(j)$ is the cost due to the structure being in state j, and C_F is the failure cost.

As mentioned previously, owners and users can have different points of views with respect to the costs C_{ik}. The owners' costs and users' costs are then aggregated respectively, by means of matrices $\mathbf{C}_o$ and $\mathbf{C}_u$, where

$$\mathbf{C}_o = \{C_{ik,o}\} \quad (26)$$
$$\mathbf{C}_u = \{C_{ik,u}\} \quad (27)$$

5.2 Lifetime Aggregation

The lifetime aggregation provides a recursive cost equation to aggregate the costs to time $t = 0$, under an assumed long-run maintenance policy A, as

$$V_{i,o}^n(A) = C_{ik,o} + \alpha \sum_{j=1}^{M} P_{ij}^{J,k} V_{j,o}^{n-1}(A) \quad (28)$$
$$V_{i,u}^n(A) = C_{ik,u} + \alpha \sum_{j=1}^{M} P_{ij}^{J,k} V_{j,u}^{n-1}(A) \quad (29)$$

in which $V_{i,o}^n(A)$ and $V_{i,u}^n(A)$ are the expected discounted life cycle costs incurred by owners and users, respectively, for a structure starting in state i at time $t = 0$ under policy A and evolving n time periods, α is a discount factor, and k is the decision at state i given by policy A. Although not included here, it is possible for owners and users to have different discount factors.

As n approaches infinity, equation(28) and equation(29) can be shown [4] to converge to

$$V_{i,o}(A) = C_{ik,o} + \alpha \sum_{j=1}^{M} P_{ij}^{k} V_{j,o}(A) \quad (30)$$
$$V_{i,u}(A) = C_{ik,u} + \alpha \sum_{j=1}^{M} P_{ij}^{k} V_{j,u}(A) \quad (31)$$

where $V_{i,o}(A)$ and $V_{i,u}(A)$ are the expected long-run discounted life cycle costs of owners and users for a structure starting in state i and continuing indefinitely.

6 Decision Making Analysis

The decision making analysis is the final step of the life cycle cost analysis procedure. A decision maker must evaluate alternatives with regard to objectives at the initial design stage. The MDP provides an optimization program which will be applied as the main solution procedure in this life cycle cost research.

6.1 A Decision Maker's Alternatives

Traditionally, a decision maker's alternatives at the initial design state are limited to the structural design variables $\underline{X}$, describing the mechanical properties and geometric configuration of the structure. The performance of a structure in its lifetime, however, is not only controlled by these properties but also by lifetime maintenance operations as well. It is necessary for a decision maker to consider the maintenance operations along with the structural design variables at the initial design stage.

6.2 A Decision Maker's Objectives

A decision maker's objectives at the initial design stage depend on his or her status with the parties involved with the design and lifetime process of a structure. With the assumption of independence of owners and users, the decision maker's objectives should include the goals of the owners and users, related to initial cost, life cycle cost, structural serviceability, and structural safety.

Initial cost can be modelled as a function of structural design variables $\underline{X}$ [7], denoted as $C_I(\underline{X})$.

Life cycle cost is a function of both structural design variables $\underline{X}$ and the long-run maintenance policy A. Mathematically, the owners' and users' expected life cycle costs, represented by $EC_{L,o}$

and $EC_{L,u}$ respectively, can be expressed as

$$EC_{L,o} = \sum_{i=1}^{M} V_{i,o}(A)q_i \quad (32)$$

$$EC_{L,u} = \sum_{i=1}^{M} V_{i,u}(A)q_i \quad (33)$$

in which $V_{i,o}(A)$ and $V_{i,u}(A)$ were defined previously by Equations (30) and (31)

The structural serviceability can be evaluated by the users' life cycle cost $EC_{L,u}$. The higher the $EC_{L,user}$, the less satisfying service the structure provides.

The structural safety can be evaluated based on lifetime failure probability as shown by Equation (22).

6.3 Mathematical Programs

The mathematical programs applied in life cycle cost analysis here are based on the optimization program provided by the MDP [4]. The MDP program can be applied directly to find the optimal long-run maintenance policy with respect to the minimum life cycle cost, for a given initial design,

$$\text{Min:} \quad Z = \sum_{i=1}^{M}\sum_{k=1}^{K} C_{ik}y_{ik} \quad (34)$$

$$\text{s.t:} \quad \sum_{k=1}^{K} y_{jk} - \alpha \sum_{i=1}^{M}\sum_{k=1}^{K} P_{ij}^{k}(\underline{X})y_{ik} = q_j, \quad \text{for } j = 1, \ldots, M. \quad (35)$$

$$y_{ik} \geq 0, \quad \text{for } i = 1, \ldots, M; \; k = 1, \ldots, K. \quad (36)$$

in which the structural design variables $\underline{X}$ are predetermined, and the policy variables y_{ik} are the only decision variables in the program. The objective function $Z = \sum_{i=1}^{M}\sum_{k=1}^{K} C_{ik}y_{ik}$ represents the expected life cycle cost, which is equal to EC_L The solution y_{ik} defines a long-run maintenance policy. The matrix form of this policy is evaluated as

$$D_{ik} = \frac{y_{ik}}{\sum_{k=1}^{K} y_{ik}}, \quad \text{for } i = 1, ..., M; \; k = 1, ..., K. \quad (37)$$

It is also possible to include the decision maker's comprehensive alternatives and objectives in a single program. The MDP program is then extended to the following multiobjective program,

$$\text{Min:} \quad \{C_I(\underline{X}), EC_{L,o}, EC_{L,u}, P_f\} \quad (38)$$

$$\text{s.t:} \quad \sum_{k=1}^{K} y_{jk} - \alpha \sum_{i=1}^{M}\sum_{k=1}^{K} P_{ij}^{k}(\underline{X})y_{ik} = q_j, \quad \text{for } j = 1, \ldots, M. \quad (39)$$

$$y_{ik} \geq 0, \quad \text{for } i = 1, \ldots, M; \; k = 1, \ldots, K. \quad (40)$$

where the structural design variables $\underline{X}$ become decision variables along with the decision variables y_{ik} for maintenance policy.

It is further assumed that owners and users have the same criterion for the safety of the structure, represented by an upper bound on failure probability P_f, and that initial cost is constrained to be no greater than C_I^*. The multiobjective program then becomes

$$\text{Min:} \quad \{EC_{L,o}, EC_{L,u}\} \quad (41)$$

$$\text{s.t:} \quad C_I(\underline{X}) \leq C_I^* \quad (42)$$

$$P_f \leq P_f^* \quad (43)$$

$$\sum_{k=1}^{K} y_{jk} - \alpha \sum_{i=1}^{M}\sum_{k=1}^{K} P_{ij}^{k}(\underline{X})y_{ik} = q_j, \quad \text{for } j = 1, \ldots, M. \quad (44)$$

$$y_{ik} \geq 0, \quad \text{for } i = 1, \ldots, M; \; k = 1, \ldots, K. \quad (45)$$

7 Multiobjective Solution Techniques

Multiobjective solution techniques provide approaches for a decision maker to aggregate the multiple objectives. While there are several existing techniques [2], their underlying interpretations are very different. The weighting method and the ϵ-constraint method [1] are the most commonly used techniques in multiobjective programming.

The MDP multiobjective program is based on the decision maker's consideration of owners' and users' goals, represented by their life cycle costs. The decision maker, independent of owners and users, pursues an optimal criterion which combines their points of views.

The relationship between owners and users is usually complex during the lifetime of a structure. They are not extremely opposed to each other, while they can hold conflicting views of loss and benefit with respect to the performance and operation of a structure. Several multiobjective solution techniques are applied here to assist in considering such a relationship in the decision making model. The detailed background of these techniques can be found in [1], [2], and [10].

7.1 Compromise Programming

This program defines the best solution as that which minimizes the distance from an infeasible goal point to the set of noninferior solutions. The objective function Z is expressed as

$$\text{Min:} \quad Z = \{w_o^p \left|\frac{EC_{L,o} - EC_{L,o}^{min}}{EC_{L,o}^{max} - EC_{L,o}^{min}}\right|^p + w_u^p \left|\frac{EC_{L,u} - EC_{L,u}^{min}}{EC_{L,u}^{max} - EC_{L,u}^{min}}\right|^p\}^{\frac{1}{p}} \quad (46)$$

where w_o and w_u are the weights, $EC_{L,o}^{min}$ and $EC_{L,o}^{max}$ are the minimal (goal value) and maximum (worst value) life cycle costs of owners, and $EC_{L,u}^{min}$ and $EC_{L,u}^{max}$ for users. The minimax method is a special case of compromise programming when the value of p is ∞.

The assumption implied in compromise programming is that owners and users are two competing parties in the decision maker's model. It is impossible for a decision maker to find the design and maintenance alternatives which minimize owners' and users' life cycle costs at the same time. Instead, the decision maker pursues an objective which balances their goals at an intermediate level.

7.2 Goal Programming

Instead of comparing with ideal, but unachievable, costs of owners and users, goal programming uses predetermined target costs. The objective is to minimize the sum of the absolute values of the differences between these target costs and the actually achieved costs, as

$$\text{Min:} \quad Z = |EC_{L,o} - T_o| + |EC_{L,u} - T_u| \quad (47)$$

where T_o and T_u are the predetermined goal or target values of life cycle costs of owners and users.

By assigning priority factors to the deviational slack variables associated with goals, goal programming allows the decision maker to have the flexibility needed to deal with conflicting goals. In addition, goal programming allows the decision maker to have preference for under- and over-achievement of a goal [4].

7.3 Cooperative Game Programming

In contrast to compromise programming, the best solution here maximizes the distance between the worst values of objectives and noninferior solutions. The worst values of objectives here refer to the maximum losses of owners and users. With some assumptions [2], cooperative game programming provides the following objective function

$$\text{Max:} \quad Z = |EC_{L,o} - EC_{L,o}^{max}|^{p_1} \times |EC_{L,u} - EC_{L,u}^{max}|^{p_2} \quad (48)$$

where $EC_{L,o}^{max}$ and $EC_{L,u}^{max}$ are the maximum life cycle costs of owners and users respectively.

This approach implies that owner and users are cooperating together to achieve a mutually satisfactory solution instead of competing against each other. It is then possible for a decision maker to find the design and maintenance alternatives which satisfy the goals of owners and users at their maximum cooperative level. Besides, the optimization starts at a feasible and unsatisfactory point, and then proceeds to improve one or more objectives.

In real life, owners and users can also have some points of views in common, along with some conflicting views, with regard to the loss and benefit due to the performance and operation of a structure. Hence, cooperative game programming is worth further investigation for life cycle cost research.

8 Conclusion

With structural reliability theory and Markov Decision processes, the costs incurred by owners and users during the lifetime of a structure can be systematically characterized and aggregated to the initial design stage. Initial design and maintenance policy are two important considerations for a decision maker. The MDP optimization program is extended as a multiobjective program in accordance with the decision maker's multiple goals in the initial design stage. Several multiobjective solution techniques are introduced to assist in decision making under different situations. This paper presents a systematic approach for decision making incorporating structural life cycle costs.

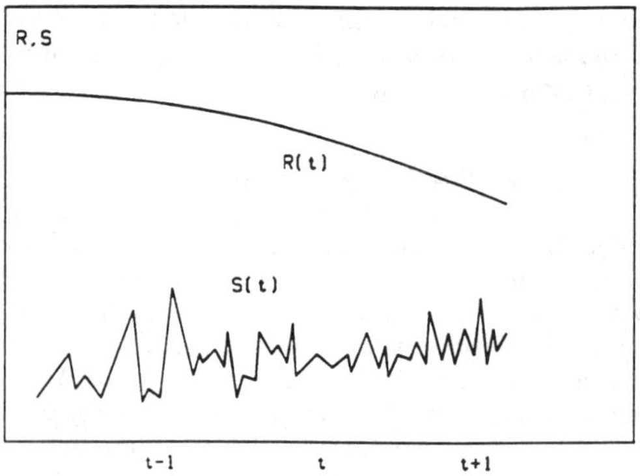

Fig 1. A Realization of Lifetime Load and Resistance

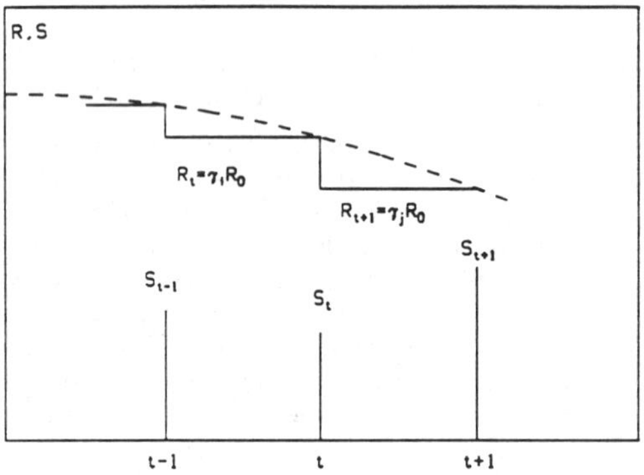

Fig 2. Lifetime Resistance and Loading Model

REFERENCES

[1] Cohon, J. (1978). *Multiobjective Programming and Planning*, Academic Press, New York, NY.

[2] Duckstein, L. (1983). "Multiobjective Optimization in Structural Design: The Model Choice Problem." *New Directions in Optimum Structural Design*, edited by E. Atrek, et.al., John Wiley & Sons, New York, N.Y.

[3] Flint, A. R. and M. J. Baker (1976). "Rationalization of Safety and Serviceability Factors in Structural Codes." *Construction Industry Research and Information Association, Report 63.* London, UK.

[4] Hiller, F. S., and Lieberman, G. J. (1986). *Introduction to Operation Research.* McGraw-Hill, New York, N.Y.

[5] National Research Council (1991). *Pay Now or Pay Later: Controlling Cost of Ownership from Design through the Service Life of Public Buildings.* Building Research Board, Committee on Setting Federal Construction Standards to Control Building Life Cycle Costs, National Academy Press, Washington, D.C.

[6] Puterman, M. L., "Markov Decision Processes", Stochastic Models: Handbooks in Operations Research and Management Science, Vol 2, D.P. Heyman and M.S. Soble eds., North Holland, 1990.

[7] Soltani, M. and Corotis, R. B. (1988). "Failure Cost Design of Structural Systems." *Structural Safety*, 5(1988).

[8] Tao, Z., J. H. Ellis, and R. B. Corotis.(1992) "Markov Decision Processes in Structural Optimization." *Probabilistic Mechanics and Structural and Geotechnical Reliability.* edited by Y. K. Lin, ASCE, pp. 539-542.

[9] Tao, Z., R. B. Corotis, and J. H. Ellis.(1992) "Reliability-based Structural Optimization with Markov Decision Processes." *Technical Report*, Department of Civil Engineering, The Johns Hopkins University, Baltimore, MD.

[10] Zeleny, M. (1982). *Multiple Criteria Decision Making*, McGraw-Hill, New York, N.Y.

Design codes

Structural Safety & Reliability, Schuëller, Shinozuka & Yao (eds) © 1994 Balkema, Rotterdam, ISBN 90 5410 357 4

A new reliability-based method to determine allowable stresses

M. Ichikawa
University of Electro-Communications, Tokyo, Japan

S. I. Bae
Chang-Won National University, Korea

ABSTRACT: In the conventional reliability–based design method, the reliability is used together with the confidence when distribution parameters are unknown. This method of using a two–dimensional measure is rather complicated, and comparison of allowable stresses based on different confidences is not easy. In the present paper, a new reliability–based design method is proposed to overcome these shortcomings. It uses a one–dimensional measure of the mean reliability. The method is applied to the determination of the design allowable stress for the cases when the material strength follows a normal distribution and a two–parameter Weibull distribution. A method for specifying the design goal of the mean reliability is also presented. Furthermore, it is shown that the product of the reliability and the confidence used in the conventional method gives an upper bound of the mean reliability.

1 INTRODUCTION

In the conventional reliability–based method to determine a design allowable stress, the reliability or the probability of survival is used together with the confidence level to take account of uncertainty in estimation of the distribution parameters. For example, the A–value and B–value in MIL(1976) standard are specified by a reliability of 99% with a confidence of 95%, and by a reliability of 90% with a confidence of 95%, respectively. A reliability used together with a confidence is a lower confidence limit of reliability for a given confidence level C. To make this point clear, we express a reliability in this sense by R_c.

Combined use of a reliability and a confidence has some shortcomings. First, it is rather complicated for engineers because two kinds of probabilistic concepts are used at a time. Second, it is not easy to compare reliabilities with different confidences. This is recognized when one tries to answer the question: which is a higher reliability between R_c=99% with C=95% and R_c=95% with C=99%.

In this paper, a new method is proposed to overcome the above shortcomings, and it is applied to determination of a design allowable stress for the cases when the material strength follows a normal distribution and a 2–parameter Weibull distribution. In this method, a one dimensional measure of "mean reliability" is used instead of a two–dimensional measure consisting of a reliability and a confidence.

2 THE CASE OF NORMAL DISTRIBUTION

2.1 *The case when only the mean is unknown*

Suppose that the material strength x follows a normal distribution $N(\mu,\sigma^2)$ with the unknown mean μ and the known variance σ^2. Let $\hat{\mu}$ be the estimate of μ obtained from testing of n specimens. Let the design allowable stress S_a be expressed as

$$S_a=\hat{\mu}-k\sigma \qquad (1)$$

where k is a coefficient to be determined below. Since $\hat{\mu}\sim N(\mu,\sigma^2/n)$, it follows $S_a\sim N(\mu-k\sigma,\sigma^2/n)$ as shown in Fig.1. When S_a takes a certain value S_a^*, the probability of failure is given by

$Pr(x<S_a^*)$, where Pr( ) denotes the probability of occurrence of the event in the parenthesis. Since S_a is a random variable, the overall probability of failure is given by $Pr(x<S_a)$, which is the mean of $Pr(x<S_a^*)$. In order to distinguish $Pr(x<S_a)$ from $Pr(x<S_a^*)$, we call $Pr(x<S_a)$ the mean probability of failure and express it by $\bar{P}_f$. Similarly, we call $\bar{R}\equiv 1-\bar{P}_f$ the mean reliability.

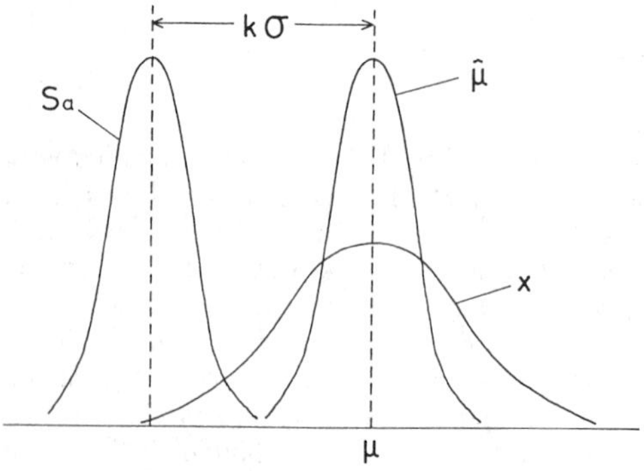

Fig.1 Distributions of the strength x and the design allowable stress S_a.

$\bar{P}_f$ can be calculated as follows. Since $x-S_a \sim N(\mu-(\mu-k\sigma),\sigma^2+\sigma^2/n)$ or $N(k\sigma,\sigma^2+\sigma^2/n)$, it follows

$$\bar{P}_f = Pr(x-S_a<0)$$

$$=\frac{1}{\sqrt{2\pi(\sigma^2+\frac{\sigma^2}{n})}}\int_{-\infty}^{0}\exp[-\frac{(\xi-k\sigma)^2}{2(\sigma^2+\frac{\sigma^2}{n})}]d\xi$$

$$=\Phi(-\frac{k\sigma}{\sqrt{\sigma^2+\frac{\sigma^2}{n}}})=\Phi(-\frac{k}{\sqrt{1+\frac{1}{n}}}) \qquad (2)$$

where $\Phi(\cdot)$ is the standard normal distribution function defined by

$$\Phi(u)\equiv\frac{1}{\sqrt{2\pi}}\int_{-\infty}^{u}\exp(-\frac{u^2}{2})du \qquad (3)$$

We determine the coefficient k so that $\bar{P}_f$ does not exceed a certain specified value P_0. Let u_{P0} be the upper $100P_0$ percentile of the standard normal distribution. Then, the above condition is expressed as

$$\frac{k}{\sqrt{1+\frac{1}{n}}}=u_{P0} \qquad (4)$$

which gives

$$k=u_{P0}\sqrt{1+\frac{1}{n}} \qquad (5)$$

The solid curves in Fig.2 show Eq.(5) for the cases of $\bar{R}=1-P_0=90\%$ and 99%.

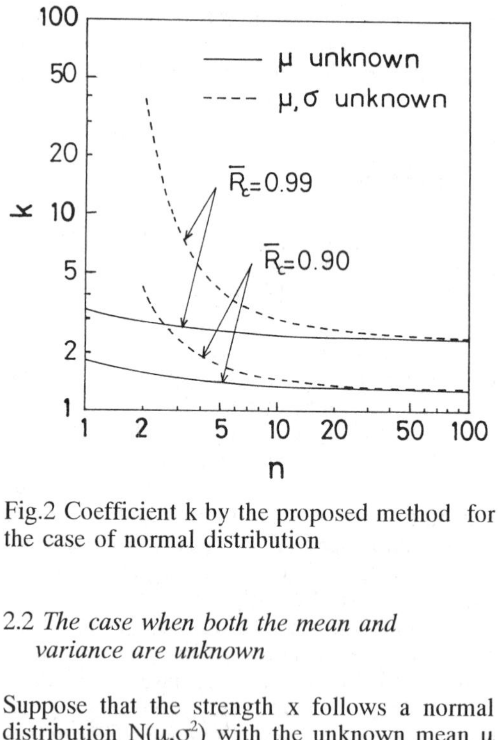

Fig.2 Coefficient k by the proposed method for the case of normal distribution

2.2 *The case when both the mean and variance are unknown*

Suppose that the strength x follows a normal distribution $N(\mu,\sigma^2)$ with the unknown mean μ and the unknown variance σ^2. Let $\hat{\mu}$ and $\hat{\sigma}$ be the estimates of μ and σ, respectively, obtained from testing of n specimens. $\hat{\sigma}$ is the unbiased estimate. Let the design allowable stress S_a be expressed as

$$S_a=\hat{\mu}-k\hat{\sigma} \qquad (6)$$

where k is the coefficient to be determined below. The mean probability of failure $\bar{P}_f$ is expressed as

$$\bar{P}_f = Pr[x<S_a]$$
$$= Pr[\mu + u\sigma < \hat{\mu} - k\hat{\sigma}] \quad (7)$$

where u is the random variable defined by u=(x−μ)/σ. Nothing that u follows the standard normal distribution, Eq.(7) can be rewritten as

$$\bar{P}_f = Pr[\frac{\frac{1}{\sqrt{1+n}}(\frac{\hat{\mu}-\mu}{\sigma/\sqrt{n}} - \sqrt{n}\mu)}{\sqrt{\frac{(n-1)\hat{\sigma}^2/\sigma^2}{n-1}}} > \frac{k}{\sqrt{1+\frac{1}{n}}}] \quad (8)$$

The numerator and the denominator of the left side of the inequality in the right side of Eq.(8) follows a standard normal distribution N(0,1) and the χ^2 distribution with the degrees of freedom n−1, respectively. Hence, the left side of the inequality follows the student's t distribution with the degrees of freedom n−1. Let $t_{P0}(n-1)$ be the upper $100P_0$ percentile of the student's t distribution with the degrees of freedom n−1. Then, the condition that $\bar{P}_f$ does not exceed P_0 is expressed as

$$\frac{k}{\sqrt{1+\frac{1}{n}}} = t_{P0}(n-1) \quad (9)$$

which gives

$$k = t_{P0}(n-1)\sqrt{1+\frac{1}{n}} \quad (10)$$

The dotted curves in Fig.2 show Eq.(10) for the cases of $\bar{R}=1-P_0$=90% and 99%. As in the conventional method, a larger k is required when both the mean and variance are unknown than when only the mean is unknown.

3 THE CASE OF 2−PARAMETER WEIBULL DISTRIBUTION

3.1 *The case when only the scale parameter is unknown*

Suppose that the material strength x follows a two−parameter Weibull distribution W(α,β). Its distribution function is expressed as

$$H(x) = 1 - \exp[-(\frac{x}{\beta})^\alpha] \quad (11)$$

We first consider the case when only the scale parameter β is unknown and the shape parameter α is known from past data. Let the design allowable stress S_a be expressed in the following form:

$$S_a = \frac{\hat{\beta}}{k^{1/\alpha}} \quad (12)$$

where $\hat{\beta}$ is the most likelihood estimate of β obtained from testing of n specimens. The mean probability of failure $\bar{P}_f$ is given by

$$\bar{P}_f = Pr[x<S_a] = Pr[x<\frac{\hat{\beta}}{k^{1/\alpha}}] \quad (13)$$

which is rewritten as

$$\bar{P}_f = Pr[2nk(\frac{x}{\beta})^\alpha < 2n(\frac{\hat{\beta}}{\beta})^\alpha] \quad (14)$$

When x~W(α,β), $2nk(x/\beta)^\alpha$ follows the exponential distribution with the mean 2nk, and $2n(\hat{\beta}/\beta)^\alpha$ follows the χ^2 distribution with the degrees of freedom 2n. Introduce new variables ξ and η as

$$\xi = 2nk(\frac{x}{\beta})^\alpha, \quad \eta = 2n(\frac{\hat{\beta}}{\beta})^\alpha \quad (15)$$

Then, the distribution function of ξ is given by

$$F(\xi) = 1 - \exp(-\frac{\xi}{2nk}) \quad (16)$$

and the probability density function of η is given by

$$g(\eta)=\frac{1}{2^n\Gamma(n)}\eta^{n-1}\exp(-\frac{\eta}{2}) \quad (17)$$

Hence, $\bar{P}_f$ is calculated as

$$\bar{P}_f=Pr[\xi<\eta]$$
$$=\int_0^\infty[1-\exp(-\frac{x}{2nk})]\frac{1}{2^n\Gamma(n)}x^{n-1}\exp(-\frac{x}{2})dx$$
$$=1-(1+\frac{1}{nk})^{-n} \quad (18)$$

From the condition that $\bar{P}_f$ does not exceed a certain specified value P_0, k is determined as

$$k=[n\{(1-P_0)^{-\frac{1}{n}}-1\}]^{-1} \quad (19)$$

The solid curves in Fig.3 show Eq.(19) for the cases of $\bar{R}=1-P_0=90\%$ and 99%. It is to be noted that the ordinate of Fig.3 is not $k^{1/\alpha}$ but k. For a given value of α, S_a is obtained from Eq.(12).

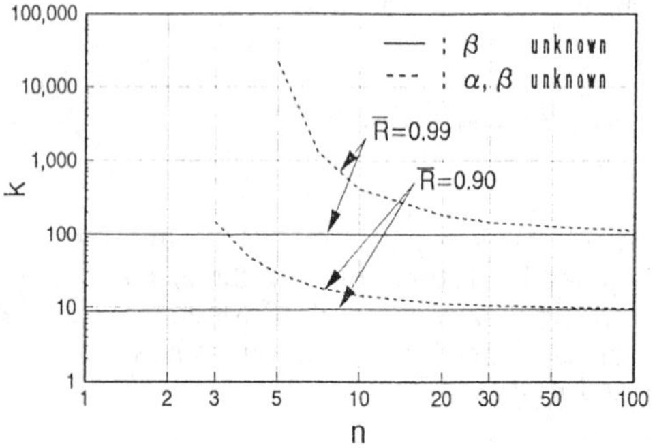

Fig.3 Coefficient k by the proposed method for the case of 2-parameter Weibull distribution.

3.2 *The case when both the shape and scale parameters are unknown*

Let $\hat{\alpha}$ and $\hat{\beta}$ be the most likelihood estimate of the shape parameter α and the scale parameter β. In this case, we express the design allowable stress S_a as follows.

$$S_a=\frac{\hat{\beta}}{k^{1/\hat{\alpha}}} \quad (20)$$

The mean probability of failure $\bar{P}_f$ is expressed as

$$\bar{P}_f=Pr[x<\frac{\hat{\beta}}{k^{1/\hat{\alpha}}}] \quad (21)$$

In this case, $\bar{P}_f$ cannot be calculated analytically, and Monte Carlo simulation is employed in the following way(Ishikawa et al., 1991).

The most likelihood estimates $\hat{\alpha}$ and $\hat{\beta}$ are the solutions of the following equations.

$$\frac{1}{n}\sum_{i=1}^{n}(\frac{x_i}{\hat{\beta}})^{\hat{\alpha}}=1 \quad (22a)$$

$$\frac{n}{\hat{\alpha}}=\sum_{i=1}^{n}\left\{(\frac{x_i}{\hat{\beta}})^{\hat{\alpha}}-1\right\}\ln\frac{x_i}{\hat{\beta}} \quad (22b)$$

Introducing Y_i, U and V as

$$Y_i=(\frac{x_i}{\beta})^{\alpha},\ U=\frac{\hat{\alpha}}{\alpha},\ V=(\frac{\hat{\beta}}{\beta})^{\hat{\alpha}} \quad (23)$$

Eqs.(22a) and (22b) are rewritten as

$$V=\frac{1}{n}\sum_{i=1}^{n}Y_i^U \quad (24a)$$

$$\frac{\sum_{i=1}^{n}Y_i^U\ln Y_i}{\sum_{i=1}^{n}Y_i^u}-\frac{1}{U}=\frac{1}{n}\sum_{i=1}^{n}\ln Y_i \quad (24b)$$

As Y_i follows an exponential distribution with the mean of unity, U and V are independent of α and β. Using U and V, Eq.(21) is rewritten as

$$\bar{P}_f=Pr[(\frac{x}{\beta})^{\alpha}<\frac{1}{k^{\alpha/\hat{\alpha}}}(\frac{\hat{\beta}}{\beta})^{\alpha}]$$

$$=Pr[Y<(\frac{V}{k})^{\frac{1}{U}}] \quad (25)$$

Generating n random variables which follow the exponential distribution with the mean of unity, and obtaining a set (U,V) by solving Eqs.(24a) and (24b), $(V/k)^{1/U}$ in Eq.(25) is calculated for a given value of k. Y in Eq.(25) is obtained by generating a random variable which follows the exponential distribution with the mean of unity. Then, $(V/k)^{1/U}$ and Y thus obtained are compared with each other. By replicating the above procedure many times, $\bar{P}_f$ is obtained from Eq.(25). Carrying out the above simulation for various values of k, $\bar{P}_f$ can be obtained as a function of k.

Newton–Raphson method was employed in solving Eqs.(24a) and (24b). The limit of error was chosen as 10^{-5}. The number of replication was chosen as 1×10^6~5×10^6 depending on the level of $\bar{P}_f$. Fig.4 shows the result of Monte Carlo simulation. From Fig.4, the relation between k and n was obtained for $\bar{R}=1-P_0=90\%$ and 99%, and shown in Fig.3 by the dotted curves. It is to be noted the ordinate of Fig.4 is not $k^{1/\hat{\alpha}}$ but k. For any $\hat{\alpha}$, S_a is calculated from Eq.(20).

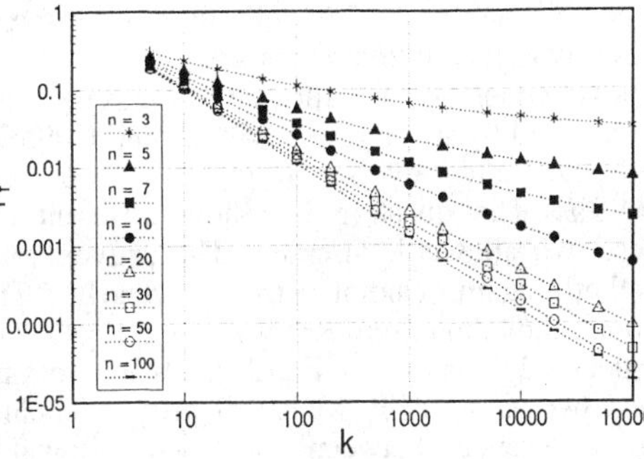

Fig.4 Result of Monte Carlo simulation.

4 DETERMINATION OF REQUIRED VALUE OF MEAN RELIABILITY

In the proposed method, the allowable stress S_a was determined so that the mean reliability exceeds a certain value $1-P_0$. The next problem is how to determine $1-P_0$. It seems reasonable to employ $1-P_0$ which results in the same level of $\bar{P}$ as that attained when S_a is determined by the conventional method.

Now, taking the case of normal distribution with the unknown mean and the known variance as an example, we determine $1-P_0$ based on the above idea. In the conventional method, the coefficient k in $S_a=\hat{\mu}-k\sigma$ for a given reliability of $R_C=1-P$ and a given confidence of $C=1-\gamma$ is expressed by (Ichikawa, 1988)

$$k=u_P+\frac{u_\gamma}{\sqrt{n}} \quad (26)$$

where u_P and u_γ are the upper 100P and 100γ percentiles of the standard normal distribution, respectively. Substituting Eq.(26) into Eq.(2), we obtain

$$\bar{R}=\Phi\left(\frac{u_P+\frac{u_\gamma}{\sqrt{n}}}{\sqrt{1+\frac{1}{n}}}\right) \quad (27)$$

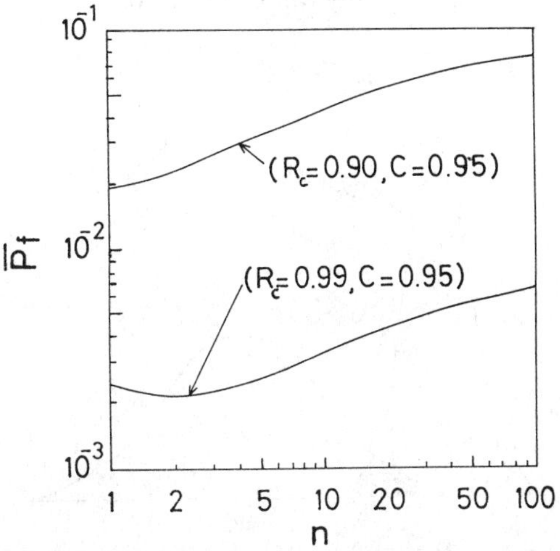

Fig.5 Mean reliability attained by the conventional method.

Eq.(27) gives $\bar{R}$ which is attained by the conventional method. Fig.5 shows Eq.(27) for $R_C=90\%$ with C=95% and for $R_C=99\%$ with C=95%. It is seen that $\bar{R}$ by Eq.(27) depends on the sample size n. For the present purpose of determining $1-P_0$, it is reasonable to take the greatest value of $\bar{R}$, which is about 99.8% for R_C=99% with C=95%, and about 98% for R_C=90% with C=95%. Hence, in order to attain the same level of $\bar{R}$ as attained by R_C=99% with C=95%, $1-P_0$=99.8% is required. Similarly, in order to attain the same level of $\bar{R}$ as attained by R_C=90% with C=95%, $1-P_0$=98% is required. In the similar way, $1-P_0$ can be determined for any combination of R_C and C. The above method can be applied to the case of normal distribution with unknown μ and σ, and also to the case of 2-parameter Weibull distribution.

5 LOWER BOUND OF MEAN RELIABILITY

When the design allowable stress S_a is determined using the conventional method so that the lower confidence limit of reliability R_C exceeds a certain value 1−P with the confidence level C=1−γ, the relative positions of the distributions of the strength x and S_a are as shown in Fig.6. Applying the upper and lower bounds formula concerning the stress-strength model (Okamura et al., 1979) to Fig.6, the following inequality can be obtained.

$$R_C C < \bar{R} < R_C + C - R_C C \qquad (28)$$

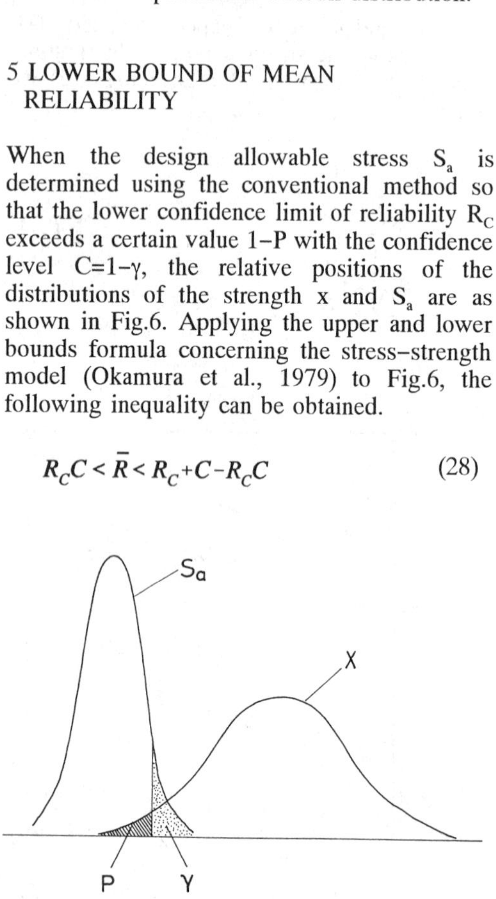

Fig.6 Illustration to derive inequality (28).

It is seen that the product of a lower confidence limit of reliability R_C and a confidence C gives the lower bound of the mean reliability $\bar{R}$. Therefore, it follows that $\bar{R}$>0.99×0.95=0.9405 for the A-value, and $\bar{R}$>0.90×0.95=0.8550 for the B-value in MIL standard. It is important to note that inequality (28) holds irrespective of the distribution form of the material strength x.

6 DISCUSSION

In order to compare the new reliability-based design method proposed in this paper with the conventional one, we consider some illustrative problems below. For simplicity, assume that the material strength follows a normal distribution with unknown mean and known variance. It is also assumed that the mean is estimated by testing 10 specimens.

Suppose that an allowable stress was determined using the conventional method by specifying R_C=99% and C=95%. The question is : what is the overall reliability in this case? The conventional method gives some feeling but cannot answer clearly. On the other hand, the proposed method gives an answer based on Eq.(27) that the overall reliability is $\bar{R}$=99.7%. This value of $\bar{R}$=99.7% tells us what we actually did by specifying $\bar{R}_C$=99% and C=95%.

Next, consider two allowable stresses : one determined by specifying $\bar{R}_C$=99% and C=95%, and the other by specifying $\bar{R}$=90% and C=95%. The question is : how different are these two allowable stresses? The conventional method cannot answer clearly. The proposed method gives an answer based on Eg.(27) that the overall reliabilities are $\bar{R}$=99.7% and $\bar{R}$=95.7%, respectively. Comparison of $\bar{R}$=99.7% and $\bar{R}$=95.7% tells us how different these two allowable stresses are.

Next, consider two allowable stresses ; one determined by specifying R_C=99% and C=95%, and the other by specifying R_C=95% and C=99%. The question is : how different are these two allowable stresses? The conventional method again cannot answer clearly. The proposed method gives $\bar{R}$=99.7% and $\bar{R}$=98.8%, respectively, based on Eq.(27). By comparing these two values of $\bar{R}$, we can clearly understand the difference between the two allowable stresses.

As may be understood from the above examples, the most outstanding advantage of the proposed method is that it can give an overall reliability quite clearly in terms of one measure of the mean reliability. In contrast, the conventional method needs to use two measures of R_C and C, hence it cannot give an overall reliability clearly. Furthermore, the proposed method can give a lower bound of overall reliability based on Eq.(28) if the distribution forms are unknown.

7 CONCLUSIONS

(1) A new reliability–based design method was proposed. The method uses a one–dimensional measure of the mean reliability $\bar{R}$ in contrast with the conventional method of using the lower confidence limit of reliability R_C and the confidence C. Hence, the proposed method can give an overall reliability very clearly and overcome some shortcomings of the conventional method.

(2) Applying the above method, the design allowable stress was determined for the case when the material strength follows a normal distribution, and for the case when the material strength follows a two–parameter Weibull distribution.

(3) A method was given to specify a design goal of the mean reliability.

(4) It was shown that the product of the lower confidence limit of reliability R_C and the confidence C gives a lower bound of the mean reliability $\bar{R}$, that is, $R_C C<\bar{R}$. This holds irrespective of distribution forms of the material strength.

REFERENCES

Ichikawa,M. 1988. *Structural reliability engineering,* Tokyo : Kaibundou.

Ishikawa,H., Iomo,H., Utsumi,A. and Ishikawa,H. 1991. Reliability–based fatigue–proof design of machines and structural components based on statistical fatigue failure probability. *J. Soc. Mater. Sci, Japan.* 40 : 276–282.

MIL–HDBK–5C, Metallic Materials and Elements for Aerospace Vehicle Structures. 1976. *Department of Defense,* USA.

Okamura,H and Itagaki,H. 1979. *Statistical treatment of strength,* Tokyo : Baifukan.

Structural Safety & Reliability, Schuëller, Shinozuka & Yao (eds) © *1994 Balkema, Rotterdam, ISBN 90 5410 357 4*

Reliability-based design for steel-concrete beam-columns

G.F. Kogut
Department of Civil and Environmental Engineering, Syracuse University, N.Y., USA
K.C.Chou
Department of Civil Engineering, The University of Tennessee, Knoxville, Tenn., USA

ABSTRACT: In load and resistance factor design (LRFD) format, a single resistance factor is used for each type of structural components. This factor would be sufficient if the structural component is made of one type of material. For composite members where the properties for each material are significantly different from the other, a single resistance factor may not adequately reflect the uncertainty of each material property. In this case, partial resistance factors are desired. In this paper, the focus is on concrete-filled beam-columns. Reliability indices using partial resistance factors for steel and concrete strengths are compared with that using a single resistance factor.

1. INTRODUCTION

Load and resistance factor design format is widely adopted by engineering design codes. However, probability based LRFD is not quite as universal. While all the design codes in use in Europe and Canada are probability based, the only probability- based design code in use in U.S. is the American Institute of Steel Construction's (AISC) LRFD manual (AISC, 1986). There are, however, three design codes, LRFD AASHTO, engineering wood and masonry, which are being developed with the same philosophy.

In the AISC LRFD manual, the design format is given as

$$\phi \Sigma R_n \geq \Sigma \gamma Q \qquad (1)$$

in which R_n = nominal resistance; Q = load effect; ϕ = resistance factor; and γ = load factor. The nominal resistance can be moment capacity for flexure members; axial capacity for short columns or tensile members; load transfer capacity for connections; stability capacity for slender columns or any combination of these. Most structural components such as beams, columns, connections are primarily made of steel. Thus, a single resistance factor is sufficient to reflect the uncertainty in steel. The ϕ factor varies depending on the limit states. For example, ϕ = 0.90 for flexure members while ϕ = 0.85 for compression members.

In the case of composite sections, such as concrete-encased or concrete-filled steel beams or columns, the current practice of an overall single resistance factor may not be adequate. Studies have shown that variabilities in steel, concrete and reinforcement strengths are different (Ellingwood et al., 1980), so are their respective probability density functions. As the height and number of skyscrapers continue to increase, steel-concrete composite structural members will be vital to future constructions. Thus, it is necessary to examine the adequacy of a single resistance factor.

In this paper, concrete-filled square beam-columns are considered. The reliability indices, β, using partial resistance factors for steel strength, F_y, and compressive concrete strength, f'_c, are compared with that using a single resistance factor. The bending moment, M, and the axial load, P, are assumed to be perfectly independent, that is, M=Pe, with e being a random variable.

2. LRFD BEAM-COLUMNS DESIGN

A beam-column is a member that is designed to withstand compressive axial loads as well as bending moments. The design is achieved through the use of an interaction equation. The design criterion is that the sum of the ratios between the required strength (factored load effect) and design strength (reduced nominal strength) for both flexure and compression must not exceed 1. This criterion in general is very conservative. The AISC LRFD Manual suggests two interaction equations according to the load effect (axial or flexure) that has the greater impact. The equations depend on the ratio of factored compressive load to reduced nominal axial strength. For $P_u/\phi_a P_n \geq 0.2$:

$$\frac{P_u}{\phi_a P_n} + \frac{8}{9}\left[\frac{M_{ux}}{\phi_b M_{nx}} + \frac{M_{uy}}{\phi_b M_{ny}}\right] \leq 1 \tag{2}$$

For $P_u/\phi_a P_n < 0.2$:

$$\frac{P_u}{2\phi_a P_n} + \left[\frac{M_{ux}}{\phi_b M_{nx}} + \frac{M_{uy}}{\phi_b M_{ny}}\right] \leq 1 \tag{3}$$

where P_u = the factored compressive axial load; P_n = the nominal compressive axial strength; M_u = the required flexural strength; M_n = the nominal flexural strength; ϕ_a = the axial resistance factor, 0.85; ϕ_b = the bending resistance factor, 0.90.

The required axial strength, P_u, is determined from:

$$P_u = \gamma_{DL} DL + \gamma_{LL} LL \tag{4}$$

where LL = live load; DL = dead load; γ_{DL} and γ_{LL} = load factors for DL and LL respectively.

For a composite member, the nominal compressive strength P_n is the product of the cross-sectional area of steel of the member A_s, and the critical stress F_{cr}. The critical stress is dependent upon the slenderness parameter, λ_c. For $\lambda_c \leq 1.5$:

$$F_{cr} = \left(0.658^{\lambda_c^2}\right) F_m \tag{5}$$

For $\lambda_c > 1.5$:

$$F_{cr} = \left[\frac{0.877}{\lambda_c^2}\right] F_m \tag{6}$$

where λ_c is given by:

$$\lambda_c = \frac{kl}{\pi r}\sqrt{\frac{F_m}{E_m}} \tag{7}$$

and F_m is given by:

$$F_m = F_y + c_2 f'_c (A_c/A_s) \tag{8}$$

in which F_y, f'_c = the yield strengths of steel and concrete, respectively, ksi; c_2 = a numerical constant equal to 0.85 for concrete-filled cross-sections; A_s and A_c = the areas of steel and concrete, respectively, in^2.

In addition, the modulus of elasticity is adjusted as follows:

$$E_m = E + c_3 E_c (A_c/A_s) \quad (9)$$

where E, E_c = the moduli of elasticity for steel and concrete, respectively, ksi. c_3 = a numerical constant equal to 0.4 for concrete-filled cross-sections.

If a frame is sufficiently braced such that no lateral translation exists, the factored flexural moment about the strong axis is given by:

$$M_u = \frac{P_e P_u e}{P_e - P_u} = \frac{\pi^2 r^2 (A_s E_m)(1.6LL + 1.2DL)e}{\pi^2 r^2 (A_s E_m) - k^2 l^2 (1.6LL + 1.2DL)} \quad (10)$$

in which P_e = the Euler buckling load.

For a concrete-filled cross-section, the nominal flexural capacity is determined from:

$$M_n = M_p = ZF_y \quad (11)$$

where Z = the plastic section modulus of the steel section.

Substituting Eqs. 5 to 11 into Eqs. 2 and 3 yields the following performance (limit state) functions. For $\lambda_c \leq 1.5$

$$G = 1 - C_A \frac{P_u}{\phi_a A_s F_m \, 0.658^{\lambda_c^2}} - C_F \left[\frac{\pi^2 r^2 A_s E_m P_u e}{\phi_b M_p (\pi^2 r^2 A_s E_m - k^2 l^2 P_u)} \right] = 0 \quad (12)$$

For $\lambda_c > 1.5$

$$G = 1 - C_A \frac{1.140251 k^2 l^2 P_u}{\phi_a \pi^2 r^2 A_s E_m} - C_F \left[\frac{\pi^2 r^2 A_s E_m P_u e}{\phi_b M_p (\pi^2 r^2 A_s E_m - k^2 l^2 P_u)} \right] = 0 \quad (13)$$

where for $P_u/\phi_a P_n \geq 0.2$: $C_A=1$ and $C_F=8/9$; for $P_u/\phi_a P_n < 0.2$: $C_A=1/2$ and $C_F=1$.

3. RELIABILITY INDEX USING PARTIAL RESISTANCE FACTORS

During the study, concrete-filled square beam-columns were designed using a single overall resistance factor as discussed in AISC LRFD Manual. The design load parameters are summarized in Table 1. To ensure that small as well as large loads would be sustained by the designs, the total axial load was multiplied by factors from 1 through 5 noted as load multiplier in Table 1.

Next, the reliability indices for each of the 17 designs for a total of 100 load combinations were computed using the performance functions defined by Eqs. 12 and 13. The design variables that are considered as random variates and their corresponding statistics and probability models are summarized in Table 2. The eccentricity, e, was considered to be a random variable with a value of 10 percent of the width of the square tubing. A "target" or "reference" reliability index, β_0, using the current design criteria is determined using a weighted average function, (Israel et al., 1987)

$$\beta_o = \frac{1}{n} \sum_{i=1}^{n} \omega_i \, \beta_i^{LRFD} \quad (14)$$

Table 1. Beam-Column Design Load Parameters

Nominal Live Load	125 psf, 250 psf
Influence Area	100 ft^2, 320 ft^2
Live Load / Dead Load	0.50, 0.75, 1.00, 1.25, 1.50
Load Multiplier	1, 2, 3, 4, 5

Table 2. Load and Resistance Statistics

Variable	Distribution	Mean	C.O.V.
LL	Gumbel	$1.15LL_n$	0.25
DL	Normal	$1.05DL_n$	0.10
A_s	Normal	nominal	0.05
A_c	Normal	nominal	$0.35/\sqrt{\mu_{Ac}}$
F_y	Lognormal	1.10Fy	0.11
f_c'	Normal	3.39 ksi	0.18
l	Normal	nominal	0.05
e	Normal	nominal	0.05
Z	Normal	nominal	0.05

where n = the number of designs or load combinations considered; ω_i = the weight associated with the ith design or load combination; β_i^{LRFD} = the reliability of the ith design or load combination designed with LRFD. Since no one particular design was assigned a greater importance than any other, all values of ω_i are equal to 1.

The beam-columns were then redesigned using partial ϕ factors. These partial resistance factors were applied to the yield strength of concrete and steel as follows:

$$F_m = \phi_s F_y + \phi_c c_2 f_c^{/}(A_c/A_s) \qquad (15)$$

The performance functions, Eqs. 12 & 13, remained the same with the exception that the single resistance factors, ϕ_a and ϕ_b, were eliminated and F_m was replaced by Eq. 15.

The resistance factors used for steel strength ϕ_s were 0.85 and 0.9. The factors used for concrete strength ϕ_c ranged from 0.6 to 0.75 with an increment of 0.05. The same loading combinations as those for single overall resistance factor were used for the redesigns. The reliability indices were then evaluated for these designs. The optimal partial resistance factor combination is determined so as to minimize the following function, (Israel et al., 1987):

$$I(\phi_s,\phi_c) = \sum_{i=1}^{n} \omega_i(\beta_i(\phi_s,\phi_c) - \beta_o)^2 \qquad (16)$$

where $\beta_i(\phi_s, \phi_c)$ = the ith beam-column designed using the partial factors ϕ_s and ϕ_c.

The "target" reliability β_o from the 17 single ϕ factor designs was determined to be 4.03.

Table 3. Results of Reliability Indices Minimization

Case	ϕ_s	ϕ_c	Avg. β	I
1	0.85	0.60	3.98	1.47
2	0.85	0.65	3.98	1.41
3	0.85	0.70	3.98	1.50
4	0.85	0.75	3.97	1.57
5	0.90	0.60	3.99	1.48
6	0.90	0.65	4.00	1.48
7	0.90	0.70	4.01	1.54
8	0.90	0.75	3.98	1.33

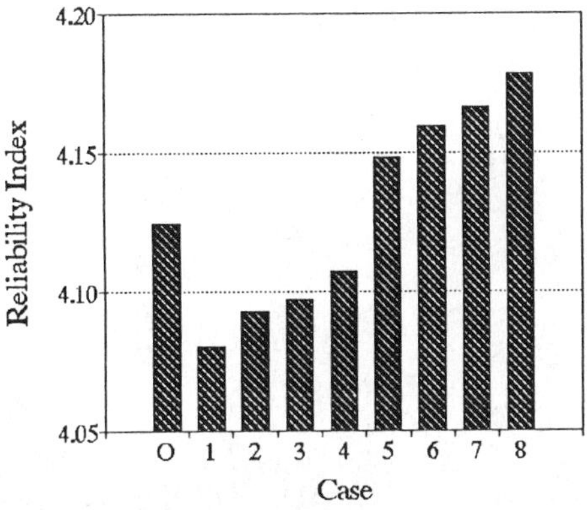

Figure 1. Reliability index versus resistance factor combination

The results of the minimization process (Eq. 16) for each pair of (ϕ_s, ϕ_c) are summarized in Table 3. Among the 8 combinations considered, the best set of combinations is when $\phi_s = 0.9$ and $\phi_c = 0.75$ (Case 8 in Table 3). However, one may observe that the variations among the average β and the optimal function, I, are small. In addition, these average β's are not significantly different from the β_0 based on single overall ϕ factor designs.

The sections that were designed using current practice of a single overall resistance factor were re-evaluated using the partial resistance factors considered here. Comparisons of the reliability indices for designs using the various partial resistance factors and a single resistance factor indicated that the optimal partial resistance factors developed here yields the highest reliability index. A typical comparison of reliability indices for a square concrete-filled beam-column section is shown in Figure 1. In this figure, Case O represents the overall resistance factor design.

4. CONCLUSION

Concrete-filled steel beam-columns were designed using a range of resistance factors for steel and concrete strengths. An optimal set of partial resistance factors was determined. The target reliability index used in this study was the average reliability index among seventeen designs for 100 load combinations using a single overall resistance factor. A comparison of the reliability indices among various partial factors designs and overall resistance factor design for sections designed under current practice was also performed. It was found that sections using the optimal partial resistance factor combination developed here yields the highest reliability index. Thus, the partial ϕ factor design concept may yield more economical beam-column sections for similar reliability levels as designs using the single ϕ factor method.

REFERENCES

AISC (1986) *Manual of Steel Construction Load & Resistance Factor Design*, American Institute of Steel Construction, Inc.

Ang, A. H.-S. and Tang, W.H. (1984) *Probability Concepts in Engineering Planning and Design*, Volume II, John Wiley & Sons, New York.

Ellingwood, B.R., Galambos, T.V., MacGregor, J.G., and Cornell. C.A., (1980) "Development of a probability-based load criterion for American national standard A58," *National Bureau of Standards Special Publication No. 577*, U.S. Department of Commerce.

Israel, M., Ellingwood, B.R. & Corotis, R.B. (1987) "Reliability-Based Code Formulations for Reinforced Concrete Buildings," *Journal of Structural Engineering*, ASCE Vol. 113, No. 10, pp. 2235-2252.

Structural Safety & Reliability, Schuëller, Shinozuka & Yao (eds) © 1994 Balkema, Rotterdam, ISBN 90 5410 357 4

Multi-stage treatment of nonlinear functions and selectable-type design code

Naotsugu Sato
Kanto-Gakuin University, Yokohama, Japan

Fumio Nishino
University of Tokyo, Japan

ABSTRACT: Multi-stage treatment is an alternative in reliability analysis scheme, which decomposes nonlinear limit state functions into simpler functions such as summation, multiplication, and some other primary transformations of variables in a hierarchical structure. This is designated to show the implicit background in the safety checking of the conventional design method. Furthermore, this paper offers selectable-type(menu) design code which makes it possible for designers to use fractile probabilities as a mesure to "select" design values, which can be understood as a limit state design of level I and a half.

1 INTRODUCTION

The terms "closed design"and "open design" are used to express the types of design methods for structures. The former is understood as the design method in which details are specified by rules, such as codes, and little room is left for the designers' discretion, whereas the latter refers to the method in which ample freedom is given to the designers, and specification by rules is minimized. From the structural mechanics aspect, every detail including geographical shapes is specified for each portion in the conventional design of ships. This is an example of the "closed" design in which no mechanics computation is required of the designers. On the other hand, it is common practice in structural design in construction field(i.e. buidings, bridges, etc.) for designers to check safety by mechanics computation. This can be called a relatively "open" design procedure, in which there is ample freedom in the selection of member sizes.

Meanwhile,the design values of loads and material strength are usually specifiedin the design codes for construction field, and the designer's discretion is limited. Designers are not usually provided with sufficient materials and knowledge to decide the loads and material strengths by themselves as in structural analysis. These design values are specified by the code writers. They are makingdecisions not only with the engineering judgement but also the statistical analysis of collected data. The authors have described them as a "design by large load values and small strength values"[1].

In Japan, several design codes based on limit state method have been offered. The Standard Specification for Concrete ; 1986 [2] and the Design code for Steel Structures ; 1987 [3] were published by the JSCE. With regard to the design values of loads and strength, the former presents five safety factors, each of which has a certain range, whereas the latter describes that the selection of loads and (unique) safety factor is specified as "in accordance with the standard of similar structures", and not giving many kind of safety factors. Differences in both method may reflect the difference in the practices of the fields. At the same time, an aspect of openness is found in the selection of design values, when compared with the closed conventional design procedures.

As a minimum restriction to open design, the safety levels to be satisfied should be indicated. In other words, in an extreme case of the open design, only the required (or target) probabilities of failure (or safety indices, β) may be presented, and the designers may be allowed to design structures of any shape through any procedure, so long as the fulfillment of the probabilities or indicesis confirmed by any objective means. The concepts of the Levels II and III limit state design are close to this. However, it is not very likely in reality that such a form of design could come into practice in near future, but at the same time, sticking to the idea that the design conditions should be given from an authority (even if it could be a level I design) can be a constraint on the development of design engineering. It can also spoil the opportunities for enhancing designers' abilities. In current design methods, designers have many degree of freedom not only in structural analysis, but in many other areas including selection of materials, structural forms, and construction methods. It is though obvious that the ranges of selection are not limitless. In this sense, itis worthwhile from an engineering point of view to provide the designers with a menu (selection table) of design values of loads and strength that is not bestowed from an authority nor excessively open, but allows selection from a range of alternatives based on the information appropriately provided. This may be positioned as a level I and a half method ; between levels I and II.

Some existing method such that the safety factors are specified in range can be understood as a continuous type of a menu. Unfortunately, it is not clearly indicated how to select the values. This could cause confusion on the designers' side. This is not the fault of the menu method itself, but rather is due to information deficiency,i.e.the incompleteness of the form. If, for example, the fractile (exceeding) probabilities or the return periods of the loads are indicated with the design loads or its alternatives (this corresponds to the "price list" of a menu), they will be useful as information for the designers.

The authors propose here a reliability design method which helps designers who are accustomed to conventional design methods to understand the statistical meaning of the design values, utilizes the menu-like(selectable) characteristics of current design code. The involvement of the fractile probabilities of design values is understood in relation to the procedure of the design computation, by setting them as a kind of safety measure, using the fractile-based reliability theory proposed by the authors[1].

2 MULTI-STAGE TREATMENT OF NONLINEAR LIMIT STATE FUNCTIONS

Nonlinear functions that appear in design problems often contain combinations of the following terms.

① Sums of variables (X1 + X2 + ... + Xn, including differences)

② Products of variables (X1•X2• ... •Xn, including quotients)

③ Mapping (f(X), including power)

Taking function Z of Eq.(1), for example, its decomposition into the forms of ①,② and ③ above is considered.

$$Z = (X_1+X_2)\cdot X_3 + \exp[X_4\cdot X_5] \quad (1)$$

When Y1=X1+X2 and Y2=X4•X5, Eq.(1) is expressed as :

$$Z = Y_1\cdot X_3 + \exp[Y_2] \quad (2)$$

Further, when U1=Y1•X3 and U2=exp[Y2], it is expressed as :

$$Z = U_1 + U_2 \quad (3)$$

and it finally becomes a "linear" function. Fig. 1 (next page) shows the flow of the above deformation of the equations.

The deformation of the equations as shown in Fig. 1 is called here "multi-stage analysis". By this analysis each stage of the functions in Eq.(1) belongs to one of the above ①,② or ③. Logically there could be a case where this analysis is unapplicable, but most of the functions used in the design of civil engineering are considered to be processable in this way. Consequently, the argument proceeds within these limits.

The application of the fractile probability method to each stage, ①,② and ③ is considered.

① Sums of variables:
Assume

$$Y = X_1+X_2+ \dots +X_n \tag{4}$$

In design process, X_i^* is taken as the design value for each X_i, determined so as to have fractile probability of e_{xi}, i.e.

$$e_{Xi} = \text{Prob.}[X_i \geqq X_i^*] \tag{5}$$

and Y calculated from them is given as

$$Y^* = X_1^*+X_2^*+ \dots +X_n^* \tag{6}$$

If Xi's are mutually independent and are normally distributed,

$$e_{Xi} = \Phi[-(X_i^*-\mu_{Xi})/\sigma_{Xi}] \tag{7}$$

is satisfied. $\Phi[\cdot]$is cumulative function of standard normal distribution and μ_{Xi} and σ_{Xi} are mean and standard deviation of Xi respectively. Then

$$X_i^* = \mu_{Xi} - \sigma_{Xi}\cdot\Phi^{-1}[e_{Xi}] \tag{8}$$

By substituting Eq.(8) into Eq.(6), and considering that Y also distributes normally,the probability of Y to exceed Y^*, eY, is given by

$$e_Y = \Phi[-(Y^*-\mu_Y)/\sigma_Y] = \Phi[\Sigma\{\sigma_{Xi}\cdot\Phi^{-1}[e_{Xi}]\}/\{\Sigma\,\sigma_{Xi}^2\}^{1/2}] \tag{9}$$

Eq.(9) gives the correct relationship of the fractile probabilities onlywhen all of Xi are mutually independent and are normally distributed, but even when they have other distribution forms, Eq.(9) should be utilized as an approximate equation.

(Note: Here the fractile probability is related to exceeding probability in all cases,but when considering the application to actual problems, it should be regarded correspondingly as nonexceeding probability for the variables on the strength side and exceeding probability for those on the load side. e_{Xi}'s should be taken "very small" in both side.)

② Products of variables :
Assume

$$Y = X_1\cdot X_2\cdot \dots \cdot X_n \tag{10}$$

and Y is calculated from design values of Xi's which satisfy Eq.(5) as above.

$$Y^* = X_1^*\cdot X_2^*\cdot \dots \cdot X_n^* \tag{11}$$

It should be noted that the logarithms of both sides of Eq.(10) have a "linear" relationship. In other words, by assuming that all of Xi are mutually independent and are log-normally distributed, same development as above is possible for the logarithm of both sides of Eq.(10).In this case, considering following relationship between variance of (lnXi) and coefficient of variation of Xi;

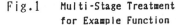

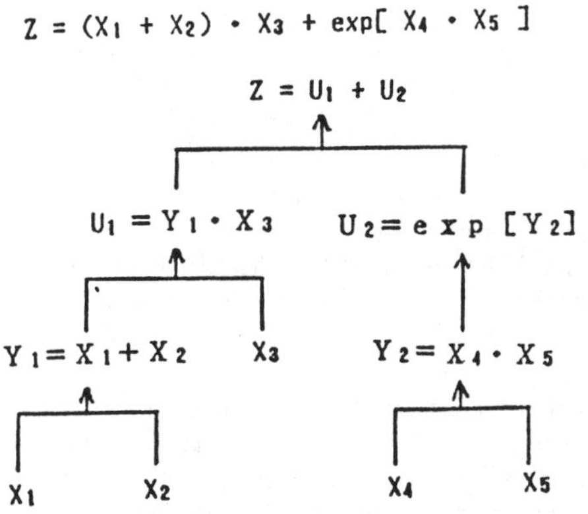

Fig.1 **Multi-Stage Treatment for Example Function**

$$\sigma_{\ln Xi}^2 = \ln[1+V_{Xi}^2] \tag{12}$$

fractile probability of Y^* of Eq.(11) is given by

$$e_Y = \Phi[\{\Sigma(\ln[1+V_{Xi}^2])^{1/2}\cdot\Phi^{-1}[e_{Xi}]\} / \{\Sigma(\ln[1+V_{Xi}^2])\}^{1/2}] \tag{13}$$

When Vxi is small (< 0.2), the approximation $\ln[1+V_{Xi}^2]\fallingdotseq V_{Xi}^2$ may be employed. In the case where the distribution of each Xi is not log-normal, this relationship is similarly applied as in ①.

③ Mapping :
In this case, the relationship of the fractile probabilities depends on the properties of the mapping. In

$$Y = f(X) \tag{14}$$

Y computed from the design value of X is given by

$$Y^* = f(X^*) \tag{15}$$

when f is a one-to-one map of a monotonic increase, such as logarithmic and exponential functions,

$$e_Y = e_X \tag{16}$$

holds. In the case of periodic functions, such as trigonometric functions, the inverse mapping of Eq.(14) is considered, and partial region of X where $Y > Y^*$ is obtained. In stability and vibration problems, multivalued functions could appear, in which case this treatment is so comlicated. However, these problems are not usually dealt with in the field of practical design in which this study is involved. It is considered that there is practically no problem in neglecting these cases.

Using the above relational expressions,

fractile probability of variables on the upper level is estimated from the design value of variables on the lower level one after another by applying them to, e.g., the relationship shown in Fig.1 In the design problems, the uppermost function expresses the limit state,and the fractile probability obtained in the end is the probability of failure corresponding to the limit state. As seen from Eqs.(9) and (13), in order to proceed to the upper level after estimating a fractile probability of a level,information such as variance, or coefficients of variation, is necessary. Therefore, they should be estimated by other means. In the case of sums, for example,

$$\sigma_Y^2 = \sigma_{X1}^2+\sigma_{X2}^2+ \dots +\sigma_{Xn}^2 \qquad (17)$$

holds regardless of the distributions. In the case of products, if a log-normal distribution is assumed,

$$1+V_Y^2 = (1+V_{X1}^2)\cdot(1+V_{X2}^2)\cdot \dots \cdot(1+V_{Xn}^2) \qquad (18)$$

follows, and this may be applied.

Conversely, the fractile probabilities can be determined downward by setting the target probability of failure, i.e. the fractile probability of the uppermost level. This corresponds to the process of code writing. In this case, since the degree of freedom increases from one stage to another,it can happen that the decision cannot be made uniquely, thus engineering judgment is required, in consideration of such conditions as the number of data.

3 PRACTICAL EXAMPLES

Taking the design of a simple steel beam of an uniform cross-section as shown in Fig. 2 as an example, possible menus are presented in this chapter from both aspects of load(load effect) and strength.

3.1 Loads/load effects

Suppose the following conditions are given to two loading patterns. Investigation is conducted for each pattern.

① Pattern A (wind load type): Uniformly distributed load. The load values (the maximum values of the years) are supposed to have the extreme I type distribution with the mean μq=1000kg (9.81kN/m) and the coefficient of variation, Vq = 0.2.

② Pattern B (live load type): Five mutually independent concentrated loads (each has the extreme I type distribution with the mean, μp=2000kg(19.62kN) and the coefficient of variation, Vp=0.2) are supposed to be applied at the points as shown in Fig. 2 (c).

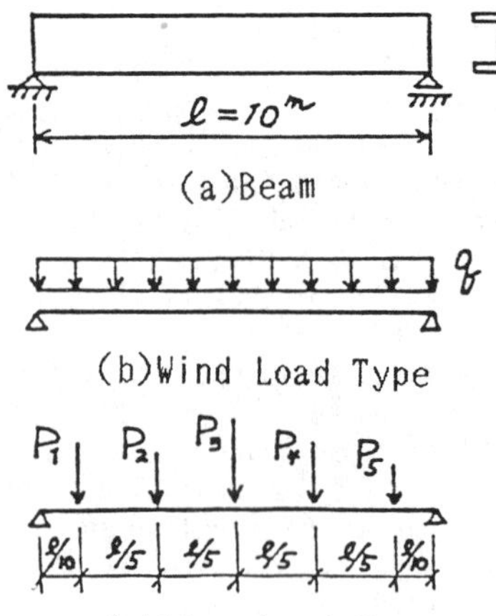

Fig.2 Beam and Loading Conditions

The terms"wind load type" and "live load type" here are used for convenience, and refer to the models with no direct relation to actual loads. They are used to illustrate that the number of degrees of freedom to be dealt with independently is important in the load problems.

The following two ways are possible for presenting load menus, in relation to the frame of design codes (i.e. who handle the transformation of fractile probabilities? designer or code writer?)

① To present menus of several design values of load with fractile probabilities in the code, regardless of structures to design. Designers select the values and conduct a structural analysis to compute load effects. They also compute the fractile probabilities on the load effect level as well.

② To present menus of several design values of load with fractile probabilities in the code, but the fractile probabilities on the load effect level are still estimated by code writers. The offered probabilities are essentially related to load effect level but formally to load level. Designers make selections and conduct a structural analysis similarly to ① above,but they do not have to transform the probabilities.

The service period of the beam is assumed to be 50 years, and the following three are proposed as the candidates for the design values.

① The value with which the exceeding probability becomes 10% in 50 years: This corresponds to the value considered equivalent to the maximum in 50 years of service of roadway bridges [4]. Incidentally, the yearly exceeding probability of this load value is $1-(0.9)^{1/50} = 0.0021$, and the return period is 475 years.

② The expected value of the maximum value in 50 years: This follows the regulation of characteristics values in the Standard Specification for Concrete [2]. The exceeding probability corresponding to this is 0.4296, the annual exceeding probability is 0.0112, and the return period is 89.6 years.

③ The expected value of the annual maximum value: This is considered for expedience as the load of "relatively frequently occurring magnitude" against service ability limit state. The annual exceeding probability for this is 0.4296; the probability for 50 years is $1-6.42*10^{-13}$, which can be practically regarded as 1. Therefore, it is meaningless to discuss the limit state probability for the design by this load value in 50 years. This is positioned as secondary to the loads of ① and ②. The limit state probability in 50 years for the load value of ② equals the annual limit state probability for this load value.

3.1.1 The designers compute the exceeding probabilities on the load effect level from the load menu and the fractile probabilities.

[1-1-1] Load values

According to the aforementioned rules, the following menus are presented

(A) Wind load type:

A-1: q_1 = 1870 kg/m (18.34 kN/m)
e_q = 10% [50years]

A-2: q_2 = 1610 kg/m (15.79 kN/m)
e_q = 42.96% [50years]

A-3: q_3 = 1000 kg/m (9.81 kN/m)
e_q = 42.96% [1year]

(B) Live load type:

B-1: P_{i1} = 3740 kg (36.69 kN)
e_P = 10% [50years]

B-2: P_{i2} = 3220 kg (31.59 kN)
e_P = 42.96% [50years]

B-3: P_{i3} = 2000 kg (19.62 kN)
e_P = 42.96% [1year]

Here means and coefficients of variation of q and P should be given to the designers as a premise for carrying out the operation of [1-1-2], but the information of the distribution forms are not necessary.

[1-1-2] Computation of load effects

Since this is the design of a beam of uniform section, a bending moment at the midspan is adopted as the load effect. Depending on the load values designers may select, the results vary as follows.

(A)Wind load type: M (l/2) = 12.5q. Since this is simple multiplication by constants, the load effects have no exceeding probability change.

S_{A1} = 23375 kg•m (229.3 kN•m)
e_S = 10% [50years]

S_{A2} = 20125 kg•m (197.4 kN•m)
e_S = 42.96% [50years]

S_{A3} = 12500 kg•m (122.6 kN•m)
e_S = 42.96% [1year]

In the next step to obtain the limit state probability in 50 years in combination with R, μs = 20125kgm and Vs = 0.2 are employed.

(B)Live load type: M (l/2) = $0.5(P_1+P_5)+1.5(P_2+P_4)+2.5P_3$. In this case, the five load values should not necessarily be the same. First, the result of the same load values is indicated.

S_{B1} = 24310 kg•m (238.5 kN•m)
e_S = 0.65% [50years]

S_{B2} = 20930 kg•m (205.3 kN•m)
e_S = 36.55% [50years]

S_{B3} = 13000 kg•m (127.5 kN•m)
e_S = 35.55% [1year]

These e_S's are obtained by applying Eq.(9) to the variables assumed as X1 = 0.5P1, X2=1.5P2, and so on. On the other hand, since P3 has the greatest influence on the load effect, it is possible to select 10-percentile value only for P3 and

42.96% for others as a load combination. In this case,

S_{B1+2} = 22230 kg•m (218.1 kN•m)
e_S = 4.84% [50years]

is obtained. Thus, it is one of the advantages of the menu method that not only is there freedom of selection for each value, but load values of different fractiles can be combined depending on the degrees of freedom of the loading method. In this case, μs = 20930 kgm, and Vs = 0.064.

3.1.2 The exceeding probabilities on the load effect level are considered by code writers.

(A)Wind load type: Since S = 12.5 q in this case,es = eq is realized.Therefore, consideration of the fractile probabilities on the load level is sufficient even on the code writing, and the menu loads in A-1 - A-3 above can be used as they are. In other words, when the load has one degree of freedom and only static effect is considered, 3.1.1 and 3.1.2 are equivalent.

(B)Live load type: In this case 3.1.1 and 3.1.2 are different. For code writers it is likely that sufficient data and time for investigation is available to specify the design values. There is room for adopting more precise techniques, such as probability density integration of the five variables or Monte Carlo simulation. Here the distribution form of S was estimated by the simulation (average of 2 iterations of 10^5 times was adopted). As a result, the value for 50 years 10% was 22680kgm (222.5kNm), and that for 50 years 42.96% was 21020kgm (206.2kNm). There may be a variety of ways to express these load values in design codes. If uniformly distributed load are adopted, the values are expressed, by dividing them by 12.5, as

B'-1: q'_1 = 1815 kg/m (17.81 kN/m)
e_q = 10% [50years]

B'-2: q'_2 = 1680 kg/m (16.48 kN/m)
e_q = 42.96% [50years]

Needless to say, eq in this case means es.

3.2 Strength

Here a concept of application of the menu method is exemplified for strength under simplified conditions as in the case of loads. In contrast to the load effect conditions adopted in 3.1, the rolled H shaped steel member shown in Fig. 3 is adopted as the cross-section of the beam. The problem of buckling is the predominant factor in steel structure designs. For members treated here decreases in strength due to buckling have to be considered, unless considerableshift constraint is provided in transverse directions.

Therefore, in the following argument , the beam strength is expressed as

$$R = Mu = \sigma y \cdot f(\lambda) \cdot W \qquad (19)$$

where σy is material yield strength, $f(\lambda)$ is nondimensionalized load carrying capacity curve of the beam, λ is parameter of slenderness ratio and W is elastic section modulus. In this study,a geometric quantity W is not treated statistically, and only the first two quantities are paidattention to. According to [5], nondimensional material factor $M=\sigma y \div$(nominal yield strength ; JIS value) conforms to a log-normal distribution of μM=1.16 and VM=0.11. Since in reality trancation of the probability distribution is carried out by inspection, its effects should be taken into account, but the natural distribution is adopted here for convenience. Fractile (non-exceeding) probability of adopting the nominal strength as the design value (M^* = 1) is estimated as eM = 9.71%.

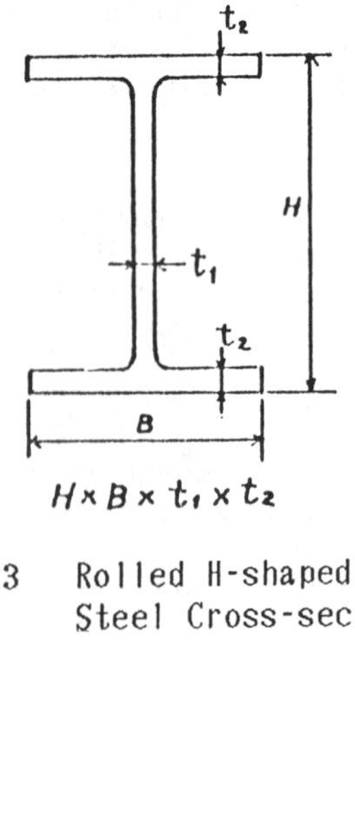

Fig.3 Rolled H-shaped Steel Cross-section

According to ECCS curve for rolled H shaped steel [6], $f(\lambda)$ is expressed as (20).

$$f(\lambda) = (1+\lambda^{2n})^{-1/n} \quad (20)$$

On rolled beams, n = 2.5 and n = 1.5 agree well with the mean curve and the (mean−2*standard deviation) curve, respectively. It is known empirically that the results of these sorts of load carrying capacity tests distribute analogously to normal distributions. In this study λ = 0.7 is assumed and then $f(\lambda)$ = 0.94 is estimated to lead to the mean ; ep = 50% ; $f(\lambda)$ = 0.822 is estimated to lead to ep =2.28% ; and $f(\lambda)$ = 0.881, which is in the middle of these two, is estimated to lead to ep = 15.9%. The coefficient of variation, Vp = 0.0628.
By incorporating the above investigations, the above three alternatives for the selection of $f(\lambda)$ are considered as the menu of resistance R, without altering the adoption of the nominal material strength as the design value (eM = 9.71%). By substituting the data into the log-normal type Eq.(13), the following menu of three design strengths and fractile probabilities is presented.

R^*_1 = 2400*0.94*W
= 2256W kg·m (22.13W kN·m)
e_R = 13.0%

R^*_2 = 2400*0.881*W
= 2114W kg·m (20.74W kN·m)
e_R = 5.22%

R^*_3 = 2400*0.822*W
= 1973W kg·m (19.36W kN·m)
e_R = 1.70%

The coefficient of variation on the level of R is estimated as 0.127.

3.3 Safety Check and Limit State Probability

Table 1 shows the required section moduli and the limit state probabilities corresponding to them for some of the combinations of the load effect and strength menus indicated in 3.1 and 3.2. The probabilities in the table are obtained by estimating Prob.[S/R>1] as a 2-variable problem of R and S. This computation is based on the log-normal type Eq.(13).

4 CONCLUDING REMARKS

The points of this study are summed up as follows:

① Multi-stage treatment of nonlinear limit sytate functions is developed and proposed. This method is applied to reliability theory of fractile probability and is interpreted in relation with the frameworks of actual design codes.

② The menu method in selecting the design

Table 1 Cross-section moduli and Limit State Probabilities for each Resistance(R)-Load(s) Combination

R\s		A−1	A−2	A−3	B−1	B−2	B−3	B−1+2
R*1	upper	1036	892	554	1078	928	576	985
	lower	4.59E-2	2.25E-1	←(1yr.)	1.67E-2	1.23E-1	←(1yr.)	3.97E-2
R*2	upper	1106	952	591	1150	990	615	1052
	lower	2.54E-2	1.53E-1	←(1yr.)	5.08E-3	5.43E-2	←(1yr.)	1.40E-2
R*3	upper	1185	1020	634	1232	1061	659	1128
	lower	1.32E-2	9.86E-2	←(1yr.)	1.29E-3	2.03E-2	←(1yr.)	4.13E-3

upper: Cross-section modulus(cm^3)   lower: Limit State Probabilities values was proposed, and fractile probabilities were given as the measure for selecting the menu.

③ Examples of application of this menu method was presented. In addition, with regard to the loads,the importance of the influence of the number of degrees of independent freedom was systematized as well as the difference resulting from considering fractile probabilities on the load level and on the load effect level. With regard to the strength, the handling of the variability of material strength and of the non-dimensional load carrying capacity curve was systematized, taking the steel structure as an example.

References

[1] SATO, N., A.HASEGAWA, F.NISHINO: A Design Format Based on Fractile-Based Reliability, Proc. of ICOSSAR'85, pp.II-127-136, May, 1985.

[2] JSCE: Standard Specification for Concrete <Design>, Japan Society of Civil Engineers, Oct. 1986 (in Japanese).

[3] JSCE: Design Code for Steel Structures <Structures in General>,Japan Society of Civil Engineers, Nov. 1987 (in Japanese).

[4] JRA: 1st Report of The Working Group on Loads, Subcomittee on Limiti State Design, Bridge Comittee, the Japan Road Association, Nov.1986(in japanese).

[5] Tokai Steel Structure Group:Estimation of Ultimate Strength of Steel Structural Members and Application to Reliability Design, Bridges and Foundations, Vol.14, No.11,12 Nov., Dec. 1980 (in Japanese).

[6] JSCE: Guidelines for Stability Design of Steel Strucures, Japan Society of Civil Engineers, Oct. 1987 (in Japanese).

Structural Safety & Reliability, Schuëller, Shinozuka & Yao (eds) © 1994 Balkema, Rotterdam, ISBN 90 5410 357 4

Code calibration as a decision problem

J.D. Sørensen, I.B. Kroon & M.H. Faber
Department of Building Technology and Structural Engineering, University of Aalborg, Denmark

ABSTRACT:

Calibration of partial coefficients for a class of structures where no code exists is considered. The partial coefficients are determined such that the difference between the reliability for the different structures in the class considered and a target reliability level is minimized. Code calibration on a decision theoretical basis is discussed. Results from code calibration for rubble mound breakwater designs are shown.

1. Introduction

During the last decades calibration of partial coefficients in level 1 codes for structural systems and civil engineering structures has been performed in a number of codes of practice.

The calibration is generally performed for a given class of structure, materials or loads in such a way that the reliability measured by the first order reliability index β estimated on the basis of structures designed using the new calibrated partial coefficients are as close as possible to the reliability indices estimated using existing design methods. Procedures to perform this type of calibration of partial coefficients are described in for example Thoft-Christensen & Baker [1].

In this paper the problem of determining partial coefficients is considered for a class of structures where no code exists but where some experience with the behaviour of the structures is reported. For the class of structures considered it is assumed that no tradition and well proven design methodology exists. A calibration procedure is described which includes the common basic steps in code calibration, see e.g. Nowak [2], namely definition of scope of the code, definition of the code objective, selection of code format, selection of target reliability index levels, calculation of calibrated partial coefficients and verification of the system of partial coefficients.

A first guess of the partial coefficients is obtained by solving an optimization problem where the objective is to minimize the difference between the reliability for the different structures in the class considered and a target reliability level. In order to ensure that all the structures in the class considered have a satisfactory reliability, constraints are imposed on the reliability for the whole range of structures considered. It is shown how this optimization problem can be solved in an efficient way. Next, the partial coefficients determined in this way are adjusted taking into account current engineering judgement and tradition.

As an example of this type of code calibration the results from a recently performed code calibration for rubble mound breakwater designs are shown. The reliability of these structures are known to vary considerably, reflecting that the structures are used under widely different conditions - from interim structures where failure has very low economic and environmental consequences to structures where failure results in significant economic losses and loss of human lives.

Next the above formulation for calibration of partial coefficients for a given class of structures to specified reliability levels is extended such that the partial coefficients can be calibrated on a decision theoretical basis. The objective is to minimize the total expected consequences for the society during the expected lifetime of the structures. Therefore, the rates of application of the structures considered are taken into account. The consequences include among others structural costs due to failure, maintenance costs, costs due to loss of human lives, environmental costs and benefits obtained from the structures. The parameters in the code calibration process may be partial coefficients if a level 1 code is calibrated or they may be reliability indices if a level 2 code is calibrated.

Generally the objective of structural engineering design may be taken as a maximization of the total expected utility of the structure, given a prescribed reference data set and availability of materials and labour. This results in two complementary aspects of the design task that must be considered, see Melchers [3]:

1) optimization of the total expected utility of a given structure by the designer - with the design parameters as optimization variables.
2) optimization of the structural design code by the code writing committee - with the partial coefficients as optimization variables.

Task 1 is considered in section 2. Task 2, which also as an important step requires design optimization is considered in the following sections.

2. Design Optimization

A design optimization problem can be formulated as

$$\min \quad C(\overline{z}) \tag{1}$$

$$s.t. \quad c_i(\overline{z}) = 0 \quad , i = 1, ..., m_e \tag{2}$$

$$c_i(\overline{z}) \geq 0 \quad , i = m_e + 1, ..., m \tag{3}$$

$$z_i^l \leq z_i \leq z_i^u \quad , i = 1, ..., N \tag{4}$$

where $\overline{z} = (z_1, \ldots, z_N)$ are the design/optimization variables. C is the objective function and c_i , $i = 1, 2, ..., m$ are the constraints. The optimization variables are assumed to be related to parameters defining the geometry of the structure (for example diameters and thicknesses of tubular elements) and coordinates (or related parameters) defining the geometry (shape) of the structural system.

The objective function C is often chosen as the weight of the structure. The m_e equality constraints in (2) can be used to model design requirements (e.g. constraints on the geometrical quantities) and to relate the load on the structure to the response (e.g. finite element equations). Often equality constraints can be avoided because the structural analysis is incorporated directly in the formulation of the inequality constraints.

The inequality constraints in (3) ensure that response characteristics such as displacements and stresses do not exceed codified critical values. Determination of the inequality constraints usually includes finite element analyses of the structural system. The inequality constraints can also include general design requirements for the design variables. The constraints in (4) are so-called simple bounds. z_i^l and z_i^u are lower and upper bounds to z_i. Generally the optimization problem (1) - (4) is non-linear and non-convex.

The objective function and the constraints in (1)-(3) are formulated using partial coefficients as specified in the code being used. The optimization problem can be solved using a number of different techniques, for example fully stressed design methods, see Vanderplaats [4], optimality criteria methods, see e.g. Arora [5], sequential quadratic programming methods, e.g. the NLPQL algorithm, Schittkowski [6], the GRG (Generalized Reduced Gradient) method, see Vanderplaats [4], sequential linear programming methods, e.g. the CONLIN algorithm, Fleury & Braibant [7] and Decomposition techniques - Large optimization problems, e.g. Sobieszczanski-Sobieski & Barthelemy [8].

An important part of the above optimization methods is to be able to perform a sensitivity analysis of the structural system giving the derivatives of the stress effects or the constraints with respect to the design variables.

The following techniques can be used to calculate these derivatives, see Vanderplaats [4] and Haug et al. [9]: overall differentation, semianalytical differentation with pseudo loads or with adjoint loads and the continuum approach with direct differentation or with adjoint design sensitivities.

3. Probabilistic Code Calibration

Code calibration can be performed by judgement, fitting, optimization or a combination of these, see Madsen et al. [10]. Calibration by **judgement** has been the main method until the last decades. **Fitting** of codes is used when a new code format is introduced and the parameters in this code are determined such that the same level of safety is obtained as in the old code. In **code optimization** the following steps are generally performed :

1) Definition of the scope of the code, i.e. the class of structures to be considered is defined.

2) Definition of the code objective. In a level 1 reliability method (which uses a single characteristic value of each uncertain quantity - for example the method of partial safety factors) the objective may be to obtain on average the same reliability as obtained by a reliability method on a higher level.
In a level 2 reliability method the objective can be a target reliability index.
In a level 3 reliability method the objective can be to obtain a constant reliability level for the whole class of structures considered.
In a level 4 reliability method the objective may be that the total expected utility of the whole class of structures considered is maximized from a society point of view, see section 5.

3) Determination of the frequency at which each type of safety check is performed.

4) Definition of a measure of closeness between code realizations and the code objective.

5) Determination of the "best" code format, i.e. the code format which is closest to the objective measured by the closeness criteria.

Calibration of partial coefficients is generally performed for a given class of structures, materials or loads in such a way that the reliability measured by the first order reliability index β estimated on the basis of structures designed using the new calibrated partial coefficients is as close as possible to the reliability indices estimated using existing design methods, see Thoft-Christensen & Baker [1], Ditlevsen & Madsen [11], Östlund [12], Shinozuka et al. [13], Vrouwenvelder [14] and Hauge et al. [15]. Procedures to perform this type of calibration of partial coefficients are described in e.g. Thoft-Christensen & Baker [1].

In the following this procedure is described and extended in some directions. For each failure mode the **limit state function** (failure function) is written

$$g(\overline{x}, \overline{p}, \overline{z}) = 0 \tag{5}$$

where $\overline{x} = (x_1, \ldots, x_n)$ is a realization of $\overline{X} = (X_1, \ldots, X_n)$ modelling n stochastic variables describing the uncertain quantities. $\overline{p} = (p_1, \ldots, p_m)$ are m parameters (deterministic) and $\overline{z} = (z_1, \ldots, z_N)$ are N design variables (which are used to design the actual structure). Realizations $\overline{x}$ of $\overline{X}$ where $g(\overline{x}, \overline{p}, \overline{z}) \leq 0$ correspond to failure states, while $g(\overline{x}, \overline{p}, \overline{z}) > 0$ correspond to safe states.

Using FORM (First Order Reliability Methods), see Madsen et al. [13] the reliability index β can be determined. The corresponding estimate of the probability of failure is

$$P_f = \Phi(-\beta) \tag{6}$$

where Φ is the standard normal distribution function.

The design equation

$$G(\overline{x}^c, \overline{p}, \overline{z}, \overline{\gamma}) \geq 0 \tag{7}$$

is used by the designer to design the structures. $\overline{x}^c = (x_1^c, \ldots, x_n^c)$ are characteristic values corresponding to the stochastic variables $\overline{X}$. $\overline{\gamma} = (\gamma_1, \ldots, \gamma_M)$ are M partial coefficients ($\gamma_i \geq 1$, $i = 1, \ldots, M$). The design equation is closely connected to the failure function (5). The only difference is that the state variables $\overline{x}$ are exchanged by the characteristic values $\overline{x}^c$ and the partial coefficients $\overline{\gamma}$.

The application area for the code is described by the set I of L different vectors $\overline{p}_i, i = 1, \ldots, L$. The set I may e.g. contain different geometrical forms of the structure, different parameters for the stochastic variables and different statistical models for the stochastic variables.

The partial coefficients $\overline{\gamma}$ are calibrated such that the reliability indices corresponding to the L vectors $\overline{p}$ are as close as possible to a target probability of failure P_{ft} or equivalently a target reliability index $\beta_t = -\Phi^{-1}(P_{ft})$. This is formulated by the following optimization problem

$$\min_{\overline{\gamma}} \quad W(\overline{\gamma}) = \sum_{j=1}^{L} w_j (\beta_j(\overline{\gamma}) - \beta_t)^2 \tag{8}$$

where $w_j, j = 1, \ldots, L$ are weighting factors ($\sum_{j=1}^{L} w_j = 1$) indicating the relative frequency of appearance of the different design situations. $\beta_j(\overline{\gamma})$ is the reliability index for combination j obtained as described below. In (8) the deviation from the target reliability index is measured by the squared distance.

The reliability index $\beta_j(\overline{\gamma})$ for combination j is obtained as follows. First, for given $\overline{\gamma}$ the design equation (7) is used. If the number of design variables is one then (7) is simply solved giving z_1. If $m > 1$ then a design optimization problem is formulated (often related to the cost of the structure) and solved, giving $\overline{z}$, see section 1. Next, the reliability index $\beta_j(\overline{\gamma})$ is estimated.

It should be noted that, following the procedure described above for estimating the partial coefficients then the two (or more) partial coefficients are not always uniquely determined. They can be functionally dependent, in the simplest case as a product, which has to be equal to a constant.

In the above procedure there is no lower limit on the reliability. An improved procedure which has a constraint on the reliability and which takes the non-uniqeness problem into account can be formulated by the optimization problem

$$\min_{\overline{\gamma}} \quad W(\overline{\gamma}) = \sum_{j=1}^{L} [w_j(\beta_j(\overline{\gamma}) - \beta_t)^2 + \delta_j \sum_{i=1}^{M} (\gamma_i - \gamma_{ji}^*)^2] \qquad (9)$$

$$s.t. \quad \beta_i(\overline{\gamma}) \geq \beta_t^{\min} \quad , i = 1, \ldots, L \qquad (10)$$

$$\gamma_i^l \leq \gamma_i \leq \gamma_i^u \quad , i = 1, \ldots, M \qquad (11)$$

where $w_j, j = 1, \ldots, L$ and $\delta_j, j = 1, \ldots, L$ are weighting factors ($\sum_{j=1}^{L} w_j = 1$). $\beta_j(\overline{\gamma})$ is the reliability index for combination j obtained as described above. γ_{ji}^* is an estimate of the partial coefficient obtained by considering combination j in isolation. The second term in the objective function (9) is added due to the non-uniqueness-problem and has the effect that the partial coefficients are forced in the direction of the "simple" definition of partial coefficients. For load variables : $\gamma = \frac{x^*}{x^c}$. If only one combination is considered then $\gamma_{ji}^* = \frac{x_{ji}^*}{x_{ji}^c}$ where x_{ji}^* is the design point.

The constraints (10) have the effect that no combination has a reliability index smaller than $\beta_t^{\min}$ and the constraints in (11) are simple bounds on the partial coefficients.

This type of code calibration has been used in Burcharth [16] for code calibration of rubble mound breakwater designs. These structures are known to have reliabilities which vary considerably. The reason is that the structures are used under widely different conditions.

A first guess of the partial coefficients is obtained by solving these optimization problems. Next, the final partial coefficients are determined taking into account current engineering judgement and tradition.

4. Example - Calibration of Partial Coefficients for Dolos Armours

Dolos armours for rubble mound breakwaters are widely used. The reliability requirements highly depend on the environment where they are used. In this example partial coefficients are calibrated for different levels of reliability requirements and different expected lifetimes. Hydraulic stability failure for the Dolos armours is the failure mode considered. The following failure function can be used to model this failure mode, see Burcharth & Liu [17]:

$$g = Z(47 - 72r)\varphi D^{1/3} N_z^{-0.1} - \frac{H_S Z_{H_S}}{\Delta D_n} \qquad (12)$$

where Z is the model uncertainty connected with eq. (12), D_n is the size of the Dolos blocks, r is the Dolos waist ratio, φ is the packing density, D is the relative number of Dolos units within levels SWL $\pm 6.5 D_n$, N_z is the number of waves, H_S is the significant wave height, Z_{H_S} is a stochastic variable modelling the uncertainty in the statistical description of the significant wave height and Δ is the ratio of block and water densities - 1.

The parameters are modelled as follows:

Table 1. Stochastic models for parameters. N(μ, σ) indicates a normal distribution with expected value μ and standard deviation σ.

par.	distribution	variation of $\overline{p}$
Z	N(1,0.22)	
r	N(p_1, 0.01 p_1)	p_1 = (0.32, 0.42)
φ	N(p_2, 0.1 p_2)	p_2 = (0.6, 1.0)
D	0.05	
N_z	N(5000,1000)	
H_S	*(p_3)	p_3 =(...)
Z_{H_S}	N(1,0.10)	
Δ	N(1.4, 0.03)	
D_n	N(z, 0.01 z)	

p_1, p_2 and p_3 define the application area for the code. $*(p_3)$ indicates that 12 different stochastic models for H_S are used, see Burcharth [16]. Three distribution types (exponential, Gumbel and Weibull) are used to fit sample data from 4 different geographic locations. The application area is thus described by $L = 2 \times 2 \times 12 = 48$ different parameters sets.

Two partial coefficients are used, namely γ_{H_S} multiplied by the charateristic value of the significant wave height and γ_R divided into the resistance term. In this example there is only one design variable, namely D_n. The optimal design is determined from

$$D_n = \frac{\gamma_R \gamma_{H_S} \hat{H}_S^T}{\mu_\Delta(47 - 72\mu_r)\mu_\varphi D^{1/3} \mu_{N_z}^{-0.1}} \quad (13)$$

where μ_X indicates the expected value of the stochastic variable X. $\hat{H}_S^T$ is the central estimate of the significant wave height which on average is exceeded once every T years.

Dolos blocks are used under quite different conditions. The expected lifetime can be from 5 years (interim structure) to 100 years (permanent structure) and the accepted level of probability of failure can vary from a very small number, e.g. 10^{-4} if failure of the breakwater results in significant damage to large probabilities, e.g. 0.4 if the consequences are insignificant. Therefore, the partial coefficients are calibrated for three levels of the expected lifetime $T = 20$, 50 and 100 years and three levels of the target reliability index $\beta_t =$ 2.33, 1.28 and 0.25 corresponding to target probability of failures $P_f^t = \Phi(-\beta_t)$=0.01, 0.1 and 0.4

The optimization problem (9)-(11) is solved with $\beta_t^{\min} = 0$. The results are shown in table 2. It is seen that the partial coefficients for the significant wave height are smaller than those for the resistance term. This reflects that the uncertainty connected with the wave height is smaller than the total resistance uncertainty. Further, the partial coefficients for the wave height decrease with longer lifetime and larger target probability of failure.

Table 2. Partial coefficients for different expected lifetimes T and different target probabilities P_f^t.

T	P_f^t	γ_R	γ_{H_S}
20	0.01	1.93	1.21
50	0.01	1.96	1.18
100	0.01	1.97	1.16
20	0.1	1.36	1.15
50	0.1	1.37	1.13
100	0.1	1.37	1.12
20	0.4	1.05	1.06
50	0.4	1.06	1.05
100	0.4	1.06	1.05

5. Application of Decision Theory in Code Calibration

Application of decision theory in code calibration has been considered in [11] and [18]. In this section code calibration using level 4 reliability methods and decision theoretical methods is discussed. One possibility is to determine target reliability indices to be used in a code based on level 2 reliability methods, i.e. using reliability indices as design criteria. In principle the same steps as for calibration of partial coefficients have to be taken. It is assumed that M target reliability indices are used to define the code format. The optimization problem is written

$$\max_{\overline{\beta}_t} \quad W(\overline{\beta}_t) = \sum_{j=1}^{L} w_j (B_j - C_{Ij}(\overline{\beta}_t) - C_{Fj} P_{fj}(\overline{\beta}_t)) \quad (14)$$

$$s.t. \quad \beta_{ti}^l \le \beta_{ti} \le \beta_{ti}^u \quad ,i = 1, \ldots, M \quad (15)$$

where w_j are weigting factors. The index j indicates which of the L different points in the application area is considered. B_j is the expected benefits (for the society), C_{Ij} is the initial (or construction) costs and C_{Fj} is the cost of failure. The probability of failure of the whole structure P_{Fj} is dependent on the target reliabilities. Often the probability of failure of the system can be determined assuming system failure if one of the single failure modes fails (series system model). The cost of failure C_{Fj} is assumed to be independent of the target reliabilities. The objective is thus to maximize the total expected consequences for the society during the expected lifetime of the structures.

$C_{Ij}(\overline{\beta}_t)$ and $P_{fj}(\overline{\beta}_t)$ are determined on the basis of the solution of the following reliability-based optimization problem

$$\min_{\overline{z}} \quad C_{Ij}(\overline{z}) \tag{16}$$

$$s.t. \quad \beta_{ij}(\overline{z}) \geq \beta_{ti} \quad , i = 1, \ldots, M \tag{17}$$

where $\overline{z}$ are the design variables. The objective function is the initial cost, and the constraints are related to reliability requirements. $\beta_{ij}(\overline{z})$ is the reliability index for combination j and failure mode type i given the design $\overline{z}$. The probability of failure of the structure P_{fj} is determined on the basis of the reliabilities of the single failure modes at the optimum design point $\overline{z}^*$. Alternatively, if the code is to be a level 1 code, then the optimization variables in (14)-(15) would be the partial coefficients. C_{Ij} and β_{ij} are determined as functions of the partial coefficients $\overline{\gamma}$.

6. Example - Calibration of Partial Coefficients for Dolos Armours (continued)

The same structure as in section 4 is considered. The application area is described by the same $L =$ 48 combinations. Since only one failure mode is considered, $M = 1$. Further, it is assumed that

$$B_j = 10.000$$
$$C_{Ij} = 2.000 + 200D_n$$
$$C_{Fj} = 10.000 \quad \text{or} \quad 2.000$$

The calibration problem (14)-(15) is solved for two different values of the failure costs and for three values of the expected lifetime, namely 20, 50 and 100 years. The results are

Table 3. Target reliability indices for different expected lifetimes T = 20, 50 or 100 years.

T	$\beta_t(C_F = 10.000)$	$\beta_t(C_F = 2.000)$
20	2.19	1.44
50	2.17	1.41
100	2.15	1.38

This example shows that the target reliability index is almost independent of the expected lifetime of the structure but is highly dependent on the cost of failure, i.e. the relative distribution of costs between construction and failure.

7. Conclusions

Code calibration is considered for a class of structures where no code exists but where some experience with the behaviour of the structures is reported. A calibration procedure is described which includes formulation of an optimization problem from which the 'best' partial coefficients can be determined.

The formulation for calibration of partial coefficients for a given class of structures to specified reliability levels is extended such that the calibration is performed on a decision theoretical basis. The objective is to minimize the total expected consequences for the society during the expected lifetime of the structures. As an illustrative example results from code calibration for rubble mound breakwater designs are shown.

8. Acknowledgements

Part of this paper is supported by the research project "Risk Analysis and Economic Decision Theory for Structural Systems" sponsored by the Danish Technical Research Council which is greatly acknowledged.

9. References

[1] Thoft-Christensen, P. & M.B. Baker: Structural Reliability Theory and Its Applications. Springer Verlag, 1982.

[2] Nowak, A.S. : Probabilistic basis for bridge design codes. Proc. ICOSSAR'89, pp. 2019-2026, 1989.

[3] Melchers, R.E.: Structural reliability, analysis and prediction. John Wiley & Sons, 1987.

[4] Vanderplaats, G.N.: Numerical Optimization Techniques for Engineering Design. McGraw-Hill, New York, 1984.

[5] Arora, J.S.: Introduction to Optimum Design. McGraw-Hill, 1989.

[6] Schittkowski, K.: NLPQL: A FORTRAN Subroutine Solving Constrained Non-Linear Programming Problems. Annals of Operations Research, 1986.

[7] Fleury, C. & Braibant, V.: Structural Optimization - a new Dual Method using Mixed Variables. International Journal of Numer-

ical Methods in Engineering, Vol. 23, pp. 409-428, 1986.

[8] Sobieszczanski-Sobieski, J. & Barthelemy, J.-F.: Improving Engineering System Design by Formal Decomposition, Sensitivity Analysis, and Decomposition. NASA Technical Memorandum 86377, 1985.

[9] Haug, E.J., K.K. Choi & V. Komkov: Design sensitivity analysis of structural systems. Academic press, 1986.

[10] Madsen, H.O., S. Krenk & N.C. Lind: Methods of Structural Safety. Prentice-Hall, 1986.

[11] Ditlevsen, O. & H.O. Madsen: Bærende Konstruktioners sikkerhed. SBI-rapport 211, Statens Byggeforskningsinstitut, 1990 (in Danish).

[12] Östlund, L.: General European principles of codes concerning reliability. Proc. ICOSSAR'89, pp. 1943-1948, 1989.

[13] Shinozuka, M. & H. Furuta & S. Emi: Reliability-based LRFD for bridges : theoretical basis. Proc. ICOSSAR'89, pp. 1981-1986, 1989.

[14] Vrouwenvelder, A.C.W.M. & A.J.M. Siemes: Probabilistic calibration procedure for the derivation of partial safety factors for the Netherlands building codes. HERON, Vol. 32, No. 4, pp. 9-29, 1987.

[15] Hauge, L.H., R. Loseth & R. Skjong: Optimal Code Calibration and Probabilistic Design. Proc. OMAE'92, Vol. II, pp. 191-199, 1992.

[16] Burcharth, H.F.: Development of a partial coefficients system for the design of rubble mound breakwaters. PIANC Working Group 12, December 1991.

[17] Burcharth, H.F. & Z. Liu: Stresses in Dolos Breakwater armour units. PIANC-AIPCN bulletin no. 72, pp. 158-166, 1991.

[18] Jordaan, I.J.: Decision analysis applied to problems of structural code formulation. Proc. NATO ASI Series E - No. 70, Martinus Nijhoff Publishers, pp. 594-604, 1983.

Design (ongoing research)

Structural Safety & Reliability, Schuëller, Shinozuka & Yao (eds) © 1994 Balkema, Rotterdam, ISBN 90 5410 357 4

Multicriterion reliability-based optimization of structural systems

S. Jendo & J. Putresza
Institute of Fundamental Technological Research, Poland

ABSTRACT: Cost and reliability of structures are two conflicting criteria. Usually, both of them are considered separately during a design process. In this paper we consider them simultaneously. Systematic investigation of the feasible set leads us to the set of non inferior solutions. Next, one preferable solution is chosen from the non inferior set. Calculation of the reliability of the structure is carried out using the failure tree method, which allows a reduction of the number of the possible collapse modes considered.

1 PROBLEM STATEMENT

Multicriterion optimization problem can be formulated in the following way: find a vector of design variables x satisfying the following conditions:

$$f(x^*) = \min_{x \varepsilon \Omega} \{-\beta_{sys}(x), W(x)\}, \tag{1}$$

$$\Omega = \{x: x_{min} \leq x \leq x_{max}\ q = N\lambda,\ \lambda \geq 0,\ Cw = q,\ Pw = 1\},$$

where $\beta_{sys}(x)$ is the reliability index of the structural system, N is the matrix of yield surfaces gradients, λ is the vector of plasticity coefficients, W is the weight of the structure, q is the vector of increments of the plastic deformations, C is a compatibility matrix. A safety index is calculated as a minimal distance from the origin of the random variables coordinates system to the failure region F which is defined by the safety margin g_f:

$$F: g_f = q^T s - P^T w \leq 0 \tag{2}$$

A negative value of the safety margin means a failure state. It is defined according to an ultimate states theory in plasticity. Kinematically compatible strains and displacements define a mechanism (mode) of failure. A great number of possible mechanisms must be considered in the case of redundant structures.

2 MULTICRITERIA OPTIMIZATION METHODS

Two criteria considered in formulation (1) are conflicting therefore an unique solution of this problem does not exist. However, it is possible to find a compromise solution by solving one of the following alternative optimization problems presented below.

- Constrained objectives method: find a vector of design variables x satisfying the following conditions:

$$f(x^*) = \min_{x \varepsilon \Omega_1} \{-\beta_{sys}(x), W(x)\}, \tag{3}$$

$$\Omega_1 = \{x: x_{min} \leq x \leq x_{max}\ q = N\lambda,\ \lambda \leq 0,\ Cw = q,\ Pw = 1,\ W(x) \leq W^0\},$$

where W^0 is an arbitrarily set upper value of the weight of the structure;

-Min- max method:

find design variables vector x, for which:

$$f(x^*) = \min_{x \varepsilon \Omega_2} \left\{ \alpha_1 \frac{\beta_{sys}(x) - \beta^{id}}{\beta^{max} - \beta^{id}} + \alpha_2 \frac{W(x) - W^{id}}{W^{max} - W^{id}} \right\}, \tag{4}$$

$$\Omega_2 = \{x: x_{min} \leq x \leq x_{max}\ q = N\lambda,\ \lambda \geq 0,\ Cw = q,\ Pw = 1\},$$

where W^{id} and W^{max} are minimal and maximal

value of the objective function respectively.

- Metric function method:

find design variables vector x, for which:

$$f(x^*) = \min_{x \varepsilon \Omega_3} \sqrt[p]{\left(\alpha_1 \frac{\beta_{sys}(x) - \beta^{id}}{\beta^{max} - \beta^{id}}\right)^p + \left(\alpha_2 \frac{W(x) - W^{id}}{W^{max} - W^{id}}\right)^p},$$

$$\Omega_3 = \{x: x_{min} \leq x \leq x_{max}\, q = N\lambda, \quad \lambda \geq 0,\ Cw = q,\ Pw = 1\}, \tag{5}$$

where p is user set parameter.

3 FAILURE TREE METHOD

Y. Murotsu and P. Thoft- Christensen (1986) proposed to identify only the most probable collapse modes using a failure tree method. Critical plastic hinges are selected by systematic simulation of the incremental collapse. An intact structure is analyzed in the first step of the simulation. Internal forces in elements depend on random loads P only:

$$s^0 = Z^0 P, \tag{6}$$

where Z^0 is an influence matrix, index'0' means that all structural elements are in the safe state. Accordingly to (2) a failure domain F^k is formulated for k-th element in the following way:

$$F^k: g_f^{k0} = - Y(s^{k0}), \tag{7}$$

where $Y(s^{k0})$ is the yield potential of the k-th element on the 0-th stage of the structural deterioration. Then, a probability of the k-th element's failure $P(g_f^{k0} \leq 0)$ can be calculated then by integrating the probability density function of random variables over the failure area F^k. From all elements are chosen only those with the smallest value of the safety index β:

$$\beta = -\Phi^{-1}(\, P(g_f^{k0} \leq 0)\,), \tag{8}$$

where Φ is the Gaussian probability distribution function. Selection is held according to the following criterion:

$$\beta_i^0 < \beta^{k0} < \beta_i^0 + \Delta\beta^0, \tag{9}$$

where and β_i^0 and $\beta_i^0 + \Delta\beta^0$ are arbitrarily chosen limits. On the m-th stage of the degradation, i.e. after removing m-1 elements , internal forces in remaining elements are computed as follows:

$$s^m = Z^m (P + P^m), \tag{10}$$

where Z^m is a modified influence matrix, P^m is the additional load equivalent to plastic modules of removed elements. Next failure functions of the remaining elements are formulated: $g_f^{km} = -Y(s^m)$. The fact that elements $k_1, \ldots, k_{m-1}$ have already failed is taken into account, therefore a conditional probability is calculated in the following way:

$$\beta_f^{km/k1,\ldots,km-1} = -\Phi^{-1}(P(g_f^{k0} \leq 0) \cap \cap(g_f^{k1/k0} \leq 0) \cap \ldots \cap (g_f^{km/k0,\ldots,km-1} \leq 0)). \tag{11}$$

Selection is continued until the structure becomes a mechanism. Fundamental mechanisms are automatically generated accordingly to the chosen set of critical elements. It can be done using V. B. Watwood's (1979) algorithm. For each fundamental mechanism the safety margin is defined as a balance of internal and external work:

$$g_f^i = q^T{}_s - w^T P,\ i=1,\ldots,nm, \tag{12}$$

where nm is the number of fundamental mechanisms. Linear combination of fundamental mechanisms can be considered as a complex mechanism. The search for the most probable (dominant) mechanism is done by solving the following optimization problem: find a vector c* of a linear combination of collapse mechanisms, for which:

$$\beta_{sys}(c^{*T} g_f) = \min \beta_{sys}(c^T g_f), \tag{13}$$

provided that:

$$Pw = 1. \tag{14}$$

This problem can be solved using a suitable nonlinear programming procedure. Due to the fact that the optimized function is differentiable only locally, the problem (1) was reformulated by adding additional constraints on change of design variables x and the increment of the safety index β_{sys} during each iteration step. Finally the sequence of the following optimization sub problems has to be solved:

$$f(x^*) = \min_{x \varepsilon \Omega'} \{-\beta_{sys}(x), W(x)\}, \tag{15}$$

where:

$$\Omega' = \{x: x^r - \varepsilon^1 \leq x \leq x^r + \varepsilon^1, q=Cw, Pw=1, q=N\lambda, \quad \lambda \geq 0, -\beta_{sys}(x^r) - \varepsilon \leq \beta_{sys}(x^r) \leq -\beta_{sys}(x^r) + \varepsilon\}, \tag{16}$$

where ε, ε^1 are arbitrarily assumed parameters, x^r is a reference value of design variables x.

4 EXAMPLES

1) The 15 bar truss structure (Fig. 1) is optimized with regard to a weight of the material and a reliability. Loads and plastic modules are normally distributed random variables. Loads are fully independent, yield stresses are correlated with a correlation coefficient 0.9. The cross sectional areas of elements are taken as the design variables, A_i, i=1,....,9. Each design variable represents one of the following groups of elements:

group number	element number
I	1, 2
II	3, 4
III	5
IV	6, 7
V	8, 9
VI	10
VII	11, 12
VIII	13, 14
IX	15

Table 1. Structural data

random variable	mean value	standard deviation
$Y_{1,2,3,4,5,6,7}$	320.0MPa	48.0MPa
P_1	45.0MN	2.25MN
P_2	1.0MN	0.3MN
P_3	0.667MN	0.2MN
P_4	0.126MN	0.0378MN

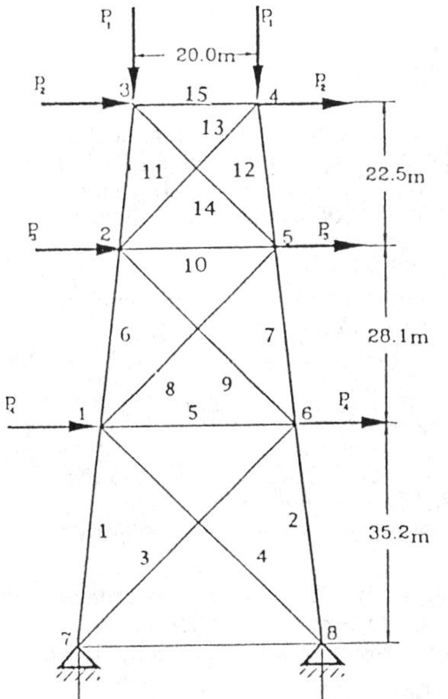

Fig. 1. Layout and a loading scheme of the structure

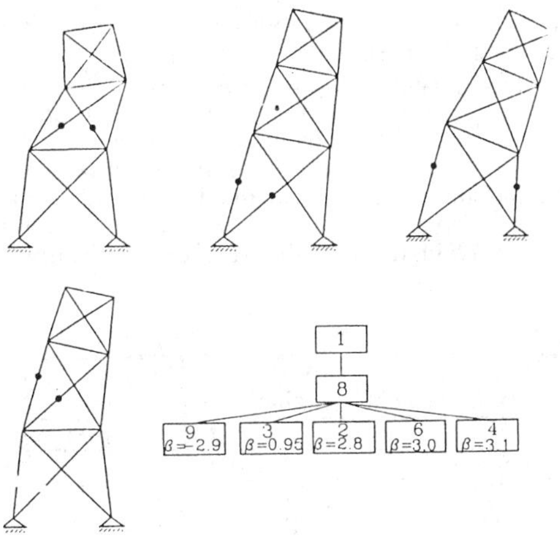

Fig. 2. Failure tree and dominant mechanisms of the reference structure

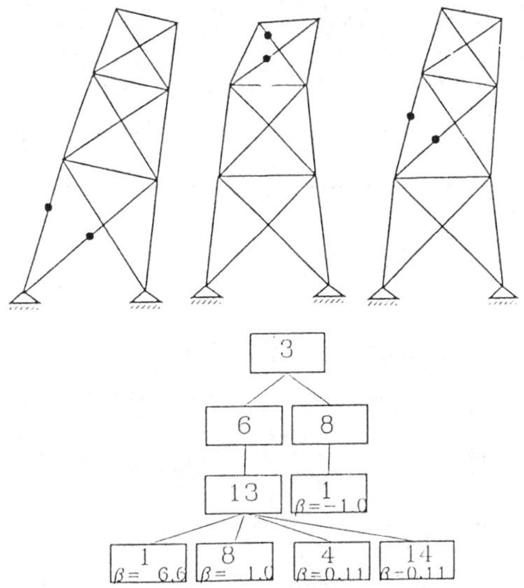

Fig. 3. Failure tree and dominant mechanisms of the optimal structure

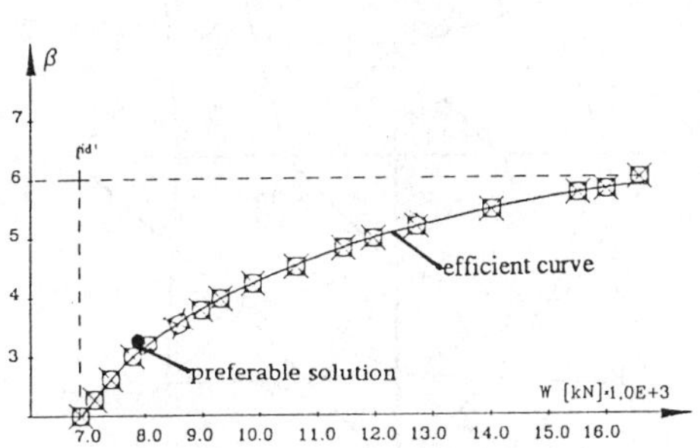

Fig. 4. The non inferior set of solutions

The set of feasible solutions is defined by the technological constraints:
$0.02m^2=A_{imin}\leq A_i\leq A_{imax}=1.0m^2$. Pareto- optimal solutions were generated using a constrained objectives method. A non inferior solution curve is shown on Fig. 4. The preferable solution was distinguished using a metric function method.
2) The 52 bar structure shown on Fig. 5 was optimized with regard to the weight and reliability.

Table 2. Structural data

Yield stress in groups of elements	Mean value	Standard deviation
1-4	200 MPa	20 MPa
5-8, 21-28	200 MPa	20 MPa
17-20	200 MPa	10 MPa
9-16	200 MPa	10 MPa
29-36	200 MPa	10 MPa
37-52	200 MPa	20 MPa
Loads	Mean value	Standard deviation
$P_{3Z}, P_{5Z}, P_{8Z}, P_{12Z}$	-18 kN	3.6 kN
P_{1X}, P_{2X}, P_{4X}	19 kN	4.0 kN
P_{2Z}, P_{4Z}	-25 kN	5.0 kN
$P_{6X}, P_{7X}, P_{9X}, P_{10X}, P_{11X}, P_{13X}$	10 kN	4.0 kN
$P_{6Z}, P_{9Z}, P_{10Z}, P_{11Z}, P_{13Z}$	-60 kN	1.5 kN

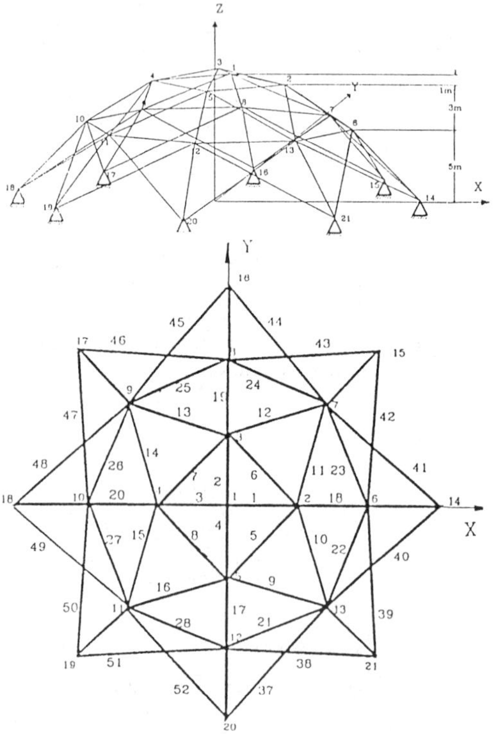

Fig. 5 A layout of the 52- bar structure

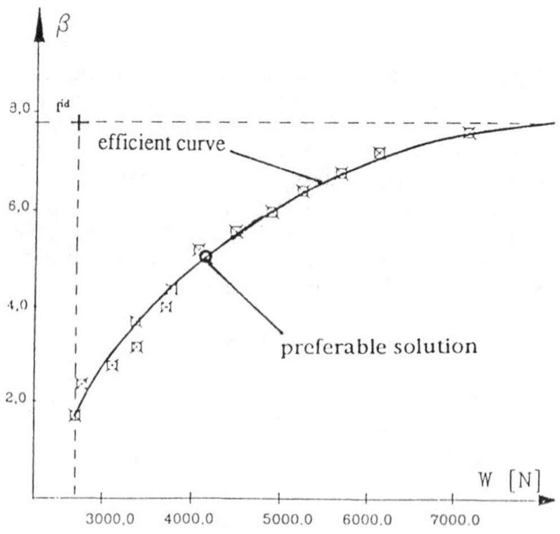

Fig. 6. Efficient solutions curve and the preferable solution

The correlation coefficients are 0.5 and 0.7 between loads and yield stresses respectively. Cross- sectional areas of bars are taken as design variables. The whole structure consists of 7 types of elements. Each design variable represents one element type. The range of each variable is constrained: $1.0E\text{-}3m^2 < A_i < 1.0E\text{-}2m^2$, $i=1,...7$. The set of Pareto- optimal solutions has been generated using a min-max method. The set of efficient solutions is shown on Fig. 6. A preferable solution was distinguished using metric function method.

5. CONCLUSIONS

1) The number of selected dominant mechanisms increases during optimization process when the optimal solution is approached. It means that increasing number of fundamental mechanisms must be taken into account.
2) A sequence of failed elements in the optimal structure is longer than an analogous sequence in the reference structure.

REFERENCES

1. Thoft-Christensen, P., Murotsu, Y., *Application of Structural Systems Reliability Theory,* Springer Verlag, Berlin Heidelberg, New York, Tokyo, 1986.
2. Watwood, V. B., *Mechanism generation for limit analysis of frames,* ASCE, journal of the Structural Division, Vol. 109, NO. ST1, 1979.

Structural Safety & Reliability, Schuëller, Shinozuka & Yao (eds) © 1994 Balkema, Rotterdam, ISBN 90 5410 357 4

Application of dynamic reliability theory to probabilistic structural design

T. Mochio
Mitsubishi Heavy Industries, Ltd, Nagasaki, Japan

ABSTRACT : Recent researches related to probability-based structural design become more active, especially in the field of the load and resistance factor design (LRFD) technique. This paper also describes some contributions to the improvement of LRFD method.
The required code format and load factors were determined experimentally in many former LRFD studies. A rational and analytical method in order to determine those design parameters is proposed on the basis of reliability theory for a dynamic problem with load combination theory, considering the dynamic characteristics of a structure.

1 INTRODUCTION

Load and resistance factor design (hereinafter, abbreviated as LRFD) method has attracted attention as a more rational structural design method than the conventional allowable stress design method.

In this paper, a technique is discussed, which can estimate the selection of design code formats and the setting-up of optimal load factors, taking the dynamic characteristics of structures in consideration.

2 APPLICATION OF DYNAMIC RELIABILITY THEORY TO SELECTION OF REQUIRED CODE FORMAT

2.1 Necessity of selecting code format

LRFD method becomes the following expression.

$$\sum_i \eta_{ji} G_i \leqq \phi R \quad (j=1, 2, 3, \cdots) \qquad (1)$$

In order to express the above equation more concretely, a dead load and two dynamic loads as external forces are considered as follows,

$$\zeta_1 D + \gamma_{11} Q_1 + \gamma_{12} Q_2 \leqq \phi R \qquad (2)$$

$$\zeta_2 D + \gamma_{21} Q_1 \leqq \phi R \qquad (3)$$

$$\zeta_3 D + \gamma_{32} Q_2 \leqq \phi R \qquad (4)$$

where,

- R : resistance value (stress, displacement, etc.) of a structure
- D : load effect as the structural response due to a dead load
- Q_1, Q_2 : load effects due to dynamic loads F_1 and F_2
- ϕ : resistance factor
- ζ_i : load factor for D
- γ_{ij} : load (combination) factors for Q_1 and Q_2

As a rule, it is necessary to execute all the checkings of safety according to equations (2) to (4). However, when the case that two dynamic loads F_1 and F_2 occur incidentally and intermittently is supposed, if the influence that the event due to the simultaneous occurrence of both dynamic loads exerts to the whole failure probability is small, namely negligible, it becomes unnecessary to consider equation (2) as the code format.

Dead load is a static load and acts always on a structure. Suppose variability of the dead load is negligibly small, then it is required only to execute the comparison and evaluation of the failure probability due to two dynamic loads only, for the discussion on whether equation (2) should be taken in consideration or not.

2.2 Selection of code formats with the aid of dynamic reliability theory

The single-degree-of-freedom linear model as shown in Fig. 1 is taken up as a typical model.

For relative displacement x_r $(=x-x_g)$, the following equation comes out,

$$m\ddot{x}_r + c\dot{x}_r + kx_r = F(t) - m\ddot{x}_g \quad (5)$$

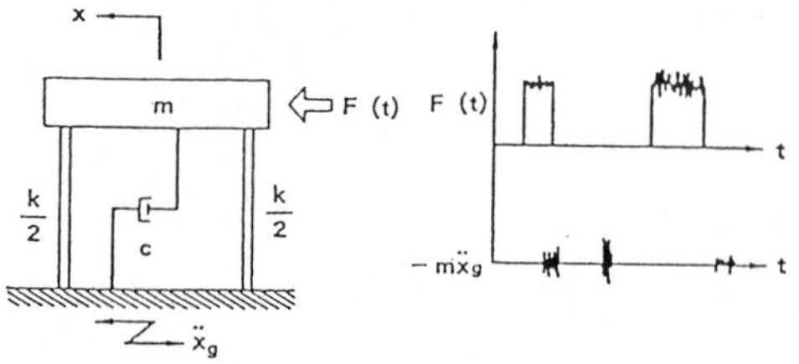

Fig. 1 Single-degree-of-freedom model with dynamic loads

where, $F(t) = F_o + F'(t)$ is assumed. Further $F'(t)$ as the random process and $\ddot{x}_g$ are considered to be mutually independent stationary random processes, and are assumed as follows,

$$\frac{F'(t)}{m} = n_f(t), \quad -\ddot{x}_g = n_g(t) \quad (6)$$

where, $n_f(t)$ and $n_g(t)$ are assumed to be the Gaussian white noises with power spectral intensity S_{of} and S_{og}, respectively.

On the other hand, the first passage failure probability in the system in Fig.1 is approximately expressed by utilizing Wen's model (Wen 1981) as follows,

$$P_f(T) = T\nu_f\mu'_{df}\lambda_f + T\nu_g\mu'_{dg}\lambda_g + T\nu_{fg}\mu'_{dfg}\lambda_f\lambda_g(\mu_{df}+\mu_{dg}) \quad (7)$$

where,

- ν_I (I=f, g) : conditional mean rate of upcrossing for one load
- ν_{fg} : conditional mean rate of upcrossing when two loads exist
- λ_I (I=f, g) : mean occurrence rate of load I
- μ_{dI} (I=f, g) : mean duration time of load I (year)
- μ'_{dI} (I=f, g), μ'_{dfg} : mean duration time expressed by seconds (sec)
- T : service life (year)

Therefore, when the value of the third term is sufficiently small as compared with that of the first and second terms on the right side of equation (7), the effect that it exerted to the whole failure probability is small, consequently, the checking of safety according to equation (2) can be omitted.

ν_I (I=f, g) and ν_{fg} in equation (7) can be determined as follows,

$$\nu_f = \frac{\omega_o}{\pi}\exp\left\{-\frac{\xi\omega_o^3(d_{max}-d_o)^2}{\pi S_{of}}\right\} \quad (8)$$

$$\nu_g = \frac{\omega_o}{\pi}\exp\left(-\frac{\xi\omega_o^3 d^2_{max}}{\pi S_{og}}\right) \quad (9)$$

$$\nu_{fg} = \frac{\omega_o}{\pi}\exp\left\{-\frac{\xi\omega_o^3(d_{max}-d_o)^2}{\pi(S_{of}+S_{og})}\right\} \quad (10)$$

where,

$$\omega_o = \sqrt{\frac{k}{m}}, \quad \xi = \frac{c}{2\sqrt{mk}}$$

$$d_o = \frac{F_o}{k}, \quad d_{max} \text{ ; threshold level}$$

2.3 Example of application

Substituting equations (8) to (10) into equation (7), the following expression is obtained.

$$P_f(T) = T(\nu_f\mu'_{df}\lambda_f + \nu_g\mu'_{dg}\lambda_g) \times\left\{1 + \frac{\nu_{fg}\mu'_{dfg}\lambda_f\lambda_g(\mu_{df}+\mu_{dg})}{(\nu_f\mu'_{df}\lambda_f + \nu_g\mu'_{dg}\lambda_g)}\right\}$$

$$= T(\nu_f\mu'_{df}\lambda_f + \nu_g\mu'_{dg}\lambda_g) \times\left[1 + \underbrace{\mu'_{dfg}\lambda_f\lambda_g(\mu_{df}+\mu_{dg}) \times\exp\left\{-\frac{\xi\omega_o^3(d_{max}-d_o)^2}{\pi(S_{of}+S_{og})}\right\} \Big/ \left[\mu'_{df}\lambda_f\exp\left\{-\frac{\xi\omega_o^3(d_{max}-d_o)^2}{\pi S_{of}}\right\} + \mu'_{dg}\lambda_g\exp\left(-\frac{\xi\omega_o^3 d^2_{max}}{\pi S_{og}}\right)\right]}_{①}\right] \quad (11)$$

① in the above equation shows the proportion of the failure probability due to the simultaneous occurrence to the case in which the failure probability by neglecting the effect of the simultaneous occurrence of loads is assumed to be 1. The results of ① in equation (11) with $\xi = 0.01$ are shown in Fig.2 and Fig.3. For example, when the load characteristics are in the region of $10^{-2} \leq \lambda_f (= \lambda_g) \leq 10^1$, and

$2.0\times10^{-6}\leqq \mu_{df}(=\mu_{dg})\leqq 2.0\times10^{-4}$, if the result of investigation becomes the form in Fig. 3, at this time, it becomes necessary to consider the simultaneous occurrence phenomenon, and it becomes possible to judge quantitatively that equation (2) cannot be omitted.

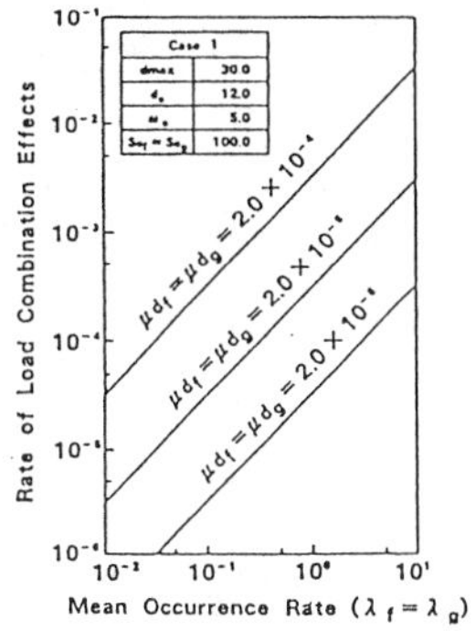

Fig. 2 Load combination effect (Case 1)

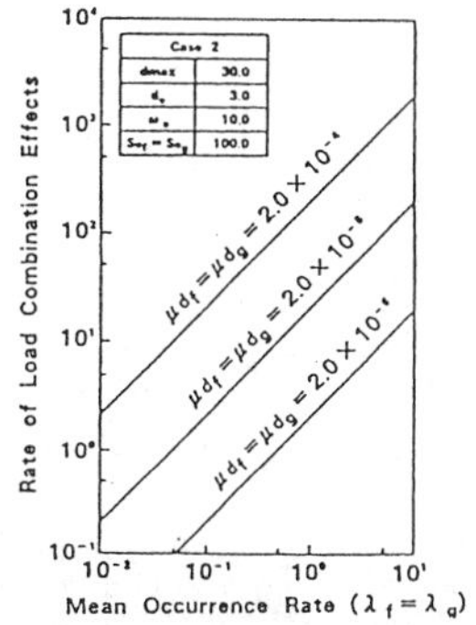

Fig. 3 Load combination effect (Case 2)

3 APPLICATION OF DYNAMIC RELIABILITY THEORY TO SETTING-UP OF LOAD FACTOR

3.1 Setting-up of load factor with the aid of dynamic reliability theory

As a typical example to show setting up the optimal load factor, the following equation is specified.

$$\gamma_f Q_f + \gamma_g Q_g \leqq R \tag{12}$$

where resistance R is assumed to be deterministic value because of its small variability, therefore the dead load effect can be included in the expression of R. Q_f and Q_g in equation (12) are the characteristic values of response (load effect), and approximately represented by (mean value+standard deviation). Moreover, it is assumed that the structure can be modeled with a cantilever beam, and its length, Young's modulus, area moment of inertia, modulus of section and allowable stress are represented with ℓ, E, I, Z and a_b, respectively. Further it is decided to adopt the equal sign in equation (12) so as to become the reasonable (economical) design, then the following equation is obtained.

$$\frac{\ell^3}{3EI}\gamma_f(F_o+\sigma_{F'})+\frac{\ell^3}{3EI}\gamma_g m\sigma_{\ddot{X}_g} = R \equiv \frac{a_b Z\ell^2}{3EI} \tag{13}$$

$\sigma_{F'}$ and $\sigma_{\ddot{X}_g}$ in equation (13) are approximated as follows,

$$\sigma_{F'} = m\sqrt{2\Omega_f S_{of}}\quad,\quad \sigma_{\ddot{X}_g} = \sqrt{2\Omega_g S_{og}} \tag{14}$$

where Ω_f and Ω_g show the maximum frequencies included in respective loads. Then, the natural circular frequency of the structure $\omega_o\ (=\sqrt{k/m}\ )$ is obtained as follows.

$$\omega_o = \sqrt{\left(\frac{3EI}{a_b Z\ell^2}\right)} \times\sqrt{\left\{\gamma_f\left(\frac{F_o}{m}+\sqrt{2\Omega_f S_{of}}\right)+\gamma_g\sqrt{2\Omega_g S_{og}}\right\}} \tag{15}$$

On the other hand, the first passage failure probability P(T) is determined as follows by utilizing equations (7) and (10).

$$P(T) = \frac{T\mu'_{dfg}\lambda_f\lambda_g(\mu_{df}+\mu_{dg})\omega_o}{\pi} \times\exp\left\{-\frac{\xi\,\omega_o^3(d_{max}-d_o)^2}{\pi(S_{of}+S_{og})}\right\} \tag{16}$$

where,

$$\left.\begin{aligned} d_{max} &= R = \frac{a_b Z\ell^2}{3EI} \\ d_o &= \frac{F_o}{k} = \frac{1}{\omega_o^2}\left(\frac{F_o}{m}\right)\end{aligned}\right\} \tag{17}$$

Now the following objective function is introduced.

$$J(\gamma_f,\ \gamma_g) = \sum_{i=1}^{N}\sum_{j=1}^{N}\left\{\frac{\log_{10}P^*-\log_{10}P_{ij}(T)}{\log_{10}P^*}\right\}^2 \tag{18}$$

Therefore, if γ_f and γ_g that make equation (18) into the minimum can be determined, these values become the optimal load factors γ_f^o and γ_g^o.

3.2 Example of application

The example of exploring the optimal load factors with the parameters in equation (19) is shown in Fig. 4. The result in Fig. 4 shows the exisitence of optimal load factors by which limit state probabilities of many structures are expected to be close to the target limit state probability.

$$\left.\begin{aligned} &30 \leqq \left(\frac{a_b Z \ell^2}{3EI}\right) \leqq 50 \\ &1 \leqq \left(\frac{F_o}{m}\right) \leqq 100 \\ &T = 20, \quad \xi = 0.01 \\ &\Omega_f = \Omega_g = 3140 \end{aligned}\right\} \qquad (19)$$

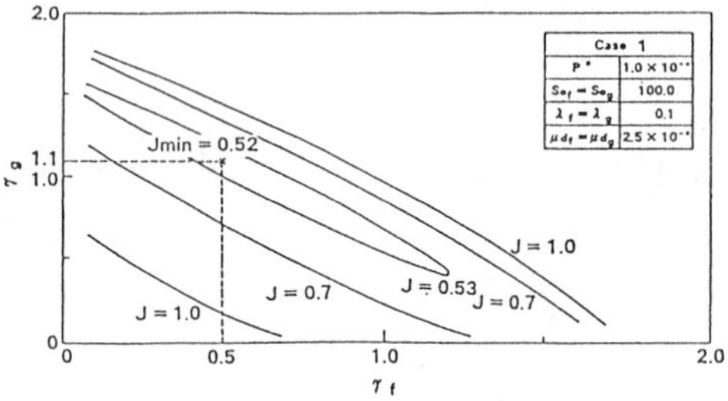

Fig. 4 Example of optimum factors

4 CONCLUSIONS

It is shown that with the aid of reliability theory for a dynamic problem, the structural reliability design taking the vibration characteristics of structures in consideration can be carried out more reasonably and quantitatively.

REFERENCE

Wen, Y.K. 1981. Stochastic dependencies in load combination. Proc. 3rd ICOSSAR : 89 −102.

Structural Safety & Reliability, Schuëller, Shinozuka & Yao (eds) © 1994 Balkema, Rotterdam, ISBN 90 5410 357 4

On the application of reliability based methods to the structural design assessment of submersibles and other externally pressurised vessels

A.C. Morandi, P.K. Das & D. Faulkner
Department of Naval Architecture and Ocean Engineering, University of Glasgow, UK

ABSTRACT: A summary is given of the work carried out on reliability based design and assessment of externally pressurised vessels in recent years. Four methods were used for the evaluation of the failure probability: Advanced First Order Second Moment, Second Order Reliability Method, Monte Carlo Simulation and Importance Sampling Using Design Points; and comparative results for sample design cases are given, for the individual failure modes as well as for the total failure probability.

1 INTRODUCTION

Submersibles and other externally pressurised vessels can be designed according to codes of practice, like the BS5500 (1991), where a deterministic approach is used. Representative buckling modes are selected, for which collapse pressures are calculated with fairly simple engineering models "calibrated" by experimental results and related to the desired operational pressure by suitable safety factors. Since excessive conservativeness may be implied in such procedures, the application of reliability based methods is suggested. In this, all relevant parameters (material properties, shape imperfections, geometry, etc.) would be treated as statistical variables, and appropriate partial safety factors for direct use in design could be derived, in principle, for any desired operational pressure, material, geometry or code of practice used. The most relevant parameters, as well as the safety of in-service structures, could also be readily assessed.

Initial studies by Faulkner et al (1988, 1991a, 1991b) showed the potential application of reliability concepts in connection with improved strength formulations to the design and assessment of externally pressurised vessels. They were followed by an investigation by Morandi, Das and Faulkner (1992a, 1992b) in which theory and finite element results were combined to verify the effect of mode interaction between frame tripping and interframe shell collapse in the elastic range. Also the Advanced First Order Second Moment (AFOSM) method (Hasofer and Lind, 1974, Rackwitz and Fiessler, 1978) and crude Monte Carlo Simulation (MCS) (Rubinstein, 1981) were applied to three design cases. Finally, a study of the failure modes associated with general instability was carried out (Morandi, Das, Faulkner, 1993) combining theory, finite element and experimental results. Also, using reliability concepts, possible weight savings were identified.

2 MODES OF FAILURE AND RELIABILITY METHODS CONSIDERED

This paper is concerned with the application of further reliability theory developments, such as the Second Order Reliability Method (SORM), using a curvature fitted approach (Fiessler, Neumann and Rackwitz, 1979) with the Breitung (1984) approximation for the failure probability and the Importance Sampling Procedure Using Design Points (ISPUD) (Schueller and Stix, 1986), to the sample design cases of Morandi, Das, Faulkner, 1992a. Four modes of failure were considered: interframe collapse, frame flange yielding, plate yielding and frame tripping, and the strength models used for the critical pressure corresponding to each of them are shown in detail in Morandi, Das, Faulkner, 1992a. The failure surfaces were then defined as $g_i = X_{mi}\, p_i - p_d$, where:

p_i = Critical pressure for each mode of failure
p_d = Design operational pressure (deterministic)
X_{mi} = Modelling parameter for each mode of failure

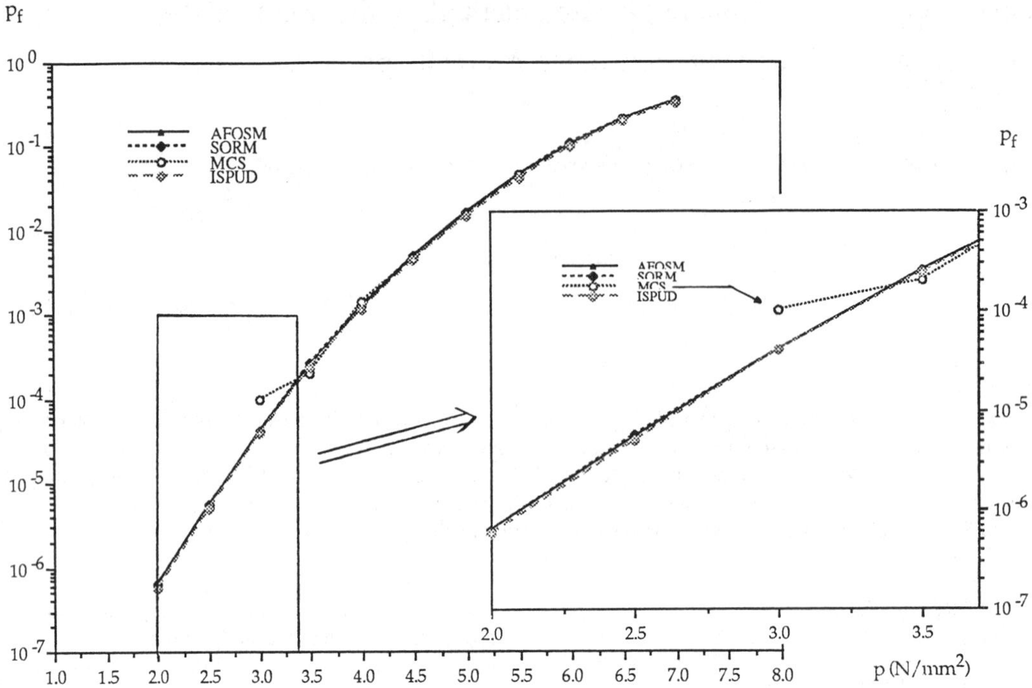

Figure 1: Failure Probability, Frame Tripping, SUB3 of Morandi, Das, Faulkner, 1992b

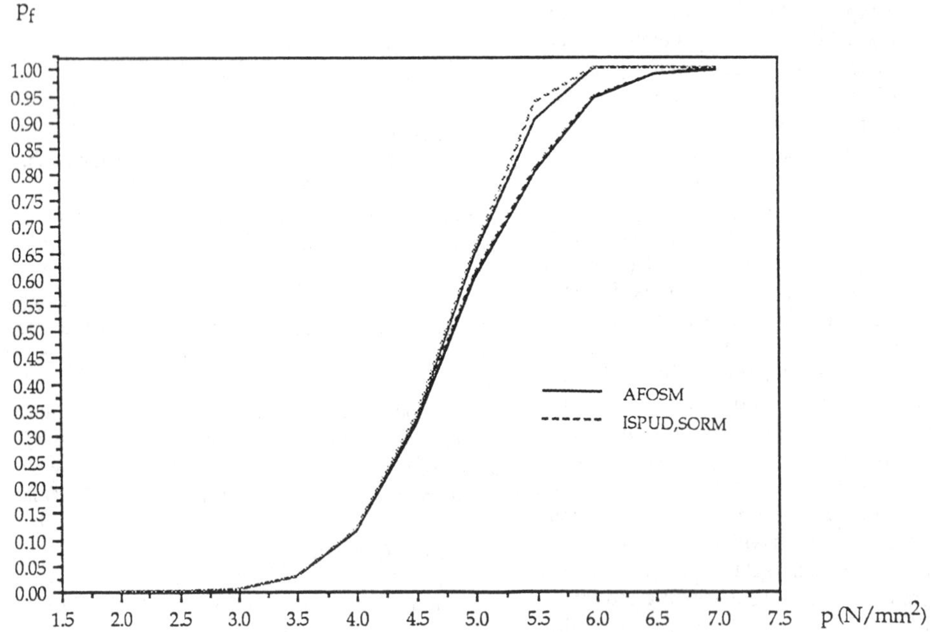

Figure 2: Ditlevsen Bounds for the Total Failure Probability, SUB3 of Morandi, Das, Faulkner, 1992b

3 NUMERICAL RESULTS

As far as computer time is concerned, analytical methods such as AFOSM and SORM are quite attractive but, on the other hand, they involve an idealisation of the failure surface at the design point, as a hyperplane and as a hyperparaboloid, respectively. Such idealisation may not be suitable if the failure surface has higher order terms or rapid local changes of curvature around the design point. Both methods may also give inadequate results if there are multiple design points with similar probability content. To overcome these shortcomings, simulation methods may be used. MCS is a rather straightforward method that converges toward exact results as the number of simulation increases but, if small probabilities of failure p_f have to be evaluated, a great number of simulations is necessary ($100/p_f$ as a rule of thumb, Torhaug et al, 1991) and excessive computer time may be required. ISPUD, in which simulation is made with the sampling density located in or near the design point obtained from a first order approximation, is a computationally efficient method but also has problems with multiple design points. Finally, if precise estimations of the failure probability can be obtained for individual modes, the total failure probability may be estimated with Ditlevsen Bounds (Ditlevsen, 1979).

These methods were applied to the sample design cases of Morandi, Das, Faulkner, 1992a and it was verified that the modelling parameters X_{mi} had by far the major influence on the results, so that for the ISPUD, the sampling densities corresponding to the other variables were kept the same as the original ones. Figure 1 illustrates the results obtained for the individual failure modes, and Figure 2 illustrates results obtained for the total failure probability. MCS and ISPUD results are based on 10 000 and 500 simulations respectively.

4 CONCLUSIONS

- For the design cases considered, the four methods gave somewhat similar results as far as the failure probabilities of the individual modes were concerned. The only exception was MCS for values of p_f lower than .01 due to the number of simulations used. ISPUD gave reasonable results with only 500 simulations.

- MCS allows a precise estimation of larger values of the total failure probability. According to Torhaug et al (1991) this is the recommended method for p_f larger than 0.05. For values smaller than that, Ditlevsen Bounds may be calculated based on the failure probabilities of the individual failure modes obtained with one of the other reliability methods. These bounds are very narrow for such small values and, for the cases analysed, the first order approximation was still adequate.

5 REFERENCES

Breitung, K., 1984. Asymptotic approximations for multinormal integrals.*J.Engg.Mech.Div*, ASCE, 110(3), pp.357-366.

British Standards Institution, 1991. Specification for unfired fusion welded pressure vessels, BS5500, HMSO, London.

Ditlevsen, O., 1979. Narrow reliability bounds for structural systems.*J. Strc. Mech.*,7(4), pp. 453-472.

Faulkner, D., Das, P.K. and Garside, J.F.,1988. 'Pilot Study into the application of reliability analysis to a structural design code for externally pressurised vessels. *MoD contract with BMT Ltd.*, Wallsend.

Faulkner, D.,1991a. Application of reliability theory in submarine design. *Intl. Conf. Advances in Marine Structures*, ARE, Dunfermline, Scotland, May 1991.

Faulkner, D. and Das, P.K., 1991b. Application of reliability theory to structural design and assessment of submarines and other externally pressurised cylindrical structures. *Integrity of Offshore Structures-3*, D.Faulkner et al, Elsevier Applied Science Publishers.

Fiessler, B., Neumann, H.J. and Rackwitz, R., 1979. Quadratic limit states in structural reliability. *J. Engg. Mech. Div.*, ASCE, 105(4), pp. 661-676.

Hasofer, A.M., and Lind, N.C., 1974. An exact invariant second-moment code format. *J.Engg. Mech. Div.*, ASCE, 100(1), pp.111-121.

Morandi, A.C., Das, P.K. and Faulkner, D., 1992a. An outline of the application of reliability based techniques to structural design and assessment of submarines and other externally pressurised cylindrical structures. Dept. of Naval Arch. and Ocean Eng., *Report NAOE-92-22, University of Glasgow.*

Morandi, A.C., Das, P.K. and Faulkner, D., 1992b. An outline of the application of reliability based techniques to structural design and assessment of submarines and other externally pressurised cylindrical structures. *Charles Smith Memorial Conference*, DRA, Dunfermline, July 1992.

Morandi, A.C., Das, P.K. and Faulkner, D., 1993. Reliability based design of submersibles: an investigation on the general instability of externally pressurised vessels. *Proc. 12th. OMAE*, vol. II, pp.297-304, ASME.

Rackwitz, R., Fiessler, B., 1978. Structural reliability under combined random load sequences. *Computers and Structures*, 9, pp.489-494.

Rubinstein, R.Y., 1981. *Simulation and the Monte Carlo Method.* John Wiley and Sons., New York.

Schuëler, G.I., and Stix, R., 1986. A critical appraisal of methods to determine failure probabilities. *Structural Safety*, 4(4), pp.293-309.

Torhaug, R., Løseth, R., Crocker, A., Ronold, K. and Gran, S., 1991. Draft classification note on use of structural reliability methods. *Report 91-2014, A.S. Veritas Research*, Oslo.

ACKNOWLEDGEMENTS

The first author wishes to acknowledge the sponsorship of the National Research Council of Brazil (CNPq) and the support of the Naval Architecture and Ocean Engineering Department of the University of Glasgow

Structural Safety & Reliability, Schuëller, Shinozuka & Yao (eds) © 1994 Balkema, Rotterdam, ISBN 90 5410 357 4

The relative safety of lattice towers and buildings designed to Australian standards

L. Pham & J. D. Holmes
CSIRO Division of Building, Construction and Engineering, Australia

ABSTRACT: Australian communication tower construction industries have, over the years, developed their own design and construction procedures. This practice has been made possible because almost all communication towers belong to a government-owned statutory authority. The procedure, in working stress format, includes both the wind load determination and the strength assessment. Both are at variance with relevant Australian standards for general building structures. With the deregulation of the communication industries, it has been found necessary to develop a new national Standard for this type of structure. The major concern is to prove that the safety of towers is comparable to that of buildings built in the same type of environment. Reliability analysis has been used for this purpose and is described in this paper.

Probabilistic models for wind loads on towers and the resistance of major components of the tower, such as legs and bracings, were developed and compared with those for buildings. The models are generally similar except for the factor for structural analysis, which is placed in the resistance model for towers and in the wind loads model for buildings. Their magnitudes were also different to account for the reserve of strength in buildings through system effects which is not available in towers. The resulting safety indices for towers are generally lower than those for buildings. Safety indices for type I towers (for populated areas) are comparable to those for general buildings, while that for type II towers (for open areas) are close to those for farm buildings.

1 INTRODUCTION

The Australian communication industries have, over the years, developed their own design and construction procedures and specifications. This practice was possible because almost all communication towers belong to a single government authority. With the deregulation of the Australian communication industries, it has been found necessary to develop an Australian Standard for this type of structure. Since the design procedure for towers is at variance with that for buildings, the major concern is to prove to the regulatory authority that the safety of towers is comparable to that of buildings in the same type of environment. Reliability analysis has been used for this purpose and is described in this paper.

2 PROBABILISTIC MODELS

2.1 Wind load effects models

2.1.1 Towers. The wind load effects, W, on exposed structural members of a tower can be written as:

$$W = C_d K_s K_i A_z q_z \qquad (1)$$

where C_d is the drag force coefficient for the tower and ancillaries; K_s is a factor for the shielding of downwind members or ancillaries by upwind members or ancillaries; K_i is a factor for the interference from components in the same plane; A_z is the projected area of the component in the plane normal to the wind direction; and q_z is the freestream gust dynamic wind pressure at height z and is given by:

$$q_z = \rho V_z^2 \quad (2)$$

where ρ is the air density; and V_z is the wind velocity at height z and is given by:

$$V_z = M_z M_t M_d V \quad (3)$$

where V is the basic wind velocity for a particular geographical region; M_z is the terrain and height multiplier; M_t is the topographic multiplier; and M_d is the wind directionality multiplier.

Statistical estimates of the variables are given in terms of the ratio of mean to nominal in Table 1. The nominal values are computed in accordance with the Australian Standard AS 3995–1991 for wind loads on lattice towers (Standards Australia 1991a).

2.1.2 Buildings. The probabilistic wind load effects, W, on buildings can be written as:

$$W = H B q_z \quad (4)$$

where H is a factor to convert wind loads into wind load effects; B is a factor covering all components of the wind loads such as pressure coefficients and area reduction factors; and q_z is the freestream gust dynamic wind pressure at height z, and has the same definition as in the case of towers (i.e. as given by Eqs (2) and (3)).

Statistical estimates of the variables are given in terms of the ratio of mean to nominal in Table 1. The nominal values are computed in accordance with the Australian Standard AS 1170.2–1989 for wind loads in buildings (Standards Australia 1989).

The differences between the models are (a) slight differences in the estimates of parameters concerning wind speeds because of the different procedures used in the tower and building standards; and (b) the factor B in the building model is the product of a number of variables for which separate estimates can be made (Pham *et al.* 1983); however it is likely that these variables may be correlated and have therefore not been treated separately.

Table 1. Statistical estimates of parameters of wind loads on towers and buildings.

Parameter	Mean	Cov
Towers		
M_z/M_{zn}	1.00	0.08
M_t/M_{tn}	1.00	0.10
M_d/M_{dn}	0.95	0.02
V/V_n non-cyclonic	0.86	0.12
V/V_n cyclonic	0.91	0.28
K_s/K_{sn}	0.95	0.10
K_i/K_{in}	0.95	0.10
C_d/C_{dn}	1.00	0.05
W/W_n non-cyclonic	0.53	0.39
W/W_n cyclonic	0.60	0.63
Buildings		
M_z/M_{zn}	1.00	0.06
M_t/M_{tn}	1.05	0.06
M_d/M_{dn}	0.90	0.05
V/V_n non-cyclonic	0.86	0.12
V/V_n cyclonic	0.91	0.28
H/H_n	0.80	0.10
B/B_n	0.90	0.25
W/W_n non-cyclonic	0.47	0.40
W/W_n cyclonic	0.53	0.65

2.2 Strength models

2.2.1 Towers. The probabilistic strength model for a member in a tower, R, is written as:

$$R = k_s k_t k_m k_f R_n \quad (5)$$

where k_s is a factor for the accuracy of the structural analysis method for the whole tower; k_t is a factor for the accuracy of the strength prediction method for a structural member; k_m is a factor for the variability of the material properties; k_f is a factor for the variability of the fabrication and erection procedure; and R_n is the nominal strength of the member in accordance with the tower Standard AS 3995–1991.

Quantitative estimates of the components of the model are given for angle members in compression in Table 2 from the data of Pham *et al.* (1992).

2.2.2 Buildings. The probabilistic strength model for a member in a building, R, is written as:

$$R = k_t k_m k_f R_n \quad (6)$$

where k_t is a factor for the accuracy of the strength prediction method for a structural member; k_m is a factor for the variability of the material properties; k_f is a factor for the variability of the fabrication and erection procedure; and R_n is the nominal strength of the member in accordance with steel building Standard AS 4100–1991 (Standards Australia 1991b).

Quantitative estimates of the components of the model for a typical member in a building are given in Table 2 from the data of Pham *et al.* (1986).

Table 2. Statistical estimates of parameters of member strength in towers and buildings.

Parameter	Mean	Cov
Towers		
k_s	1.05	0.07
k_t	0.93	0.09
k_m	1.14	0.06
k_f	1.00	0.10
R/R_n	1.11	0.16
Buildings		
k_t	1.05	0.06
k_m	1.14	0.06
k_f	1.00	0.10
R/R_n	1.20	0.13

The two models are similar except for the factor for structural analysis which is placed in the resistance model for towers and in the wind load model for buildings. The reason for this was that the building code was structured for member-by-member design and the effect of structural analysis was not accounted for in the resistance model formulated originally for code calibration purpose.

3 SAFETY INDICES

Reliability calculation has been carried out for cases involving wind loads only for simplicity. The basic strength design equation can be written as:

$$\phi R_n > \gamma W_n \qquad (7)$$

The capacity factor $\phi = 0.9$ is the same for both tower and building design. For building design, the load factor γ is effectively equal to 1.2 for important buildings, 1.0 for general buildings and 0.8 for farm structures. For tower design, the load factor is equal to 1.0, 0.85 and 0.7 for three standard types of towers depending on the remoteness of the location.

The safety index β is defined by:

$$p_f = \Phi(-\beta) \qquad (8)$$

where p_f is the probability of failure; and Φ is the cumulative distribution function of a unit normal variate. For the purpose of this paper, β is approximated by:

$$\beta = \frac{\ell n(\overline{R}/\overline{W}) \sqrt{(V_W^2 + 1)/(V_R^2 + 1)}}{\sqrt{\ell n[(V_W^2 + 1)/(V_R^2 + 1)]}} \qquad (9)$$

where $\overline{R}$ and $\overline{W}$ are the mean values, and V_R and V_W are the coefficients of variation of R and W respectively.

The 50-year lifetime safety indices for both towers and buildings are summarised in Table 3.

Table 3. Safety indices for towers and buildings for designs to Australian Standards.

Type	Non-cyclonic	Cyclonic
Towers		
1	2.3	1.4
2	1.9	1.1
3	1.4	0.7
Buildings		
Important	3.2	2.1
General	2.7	1.8
Farm	2.1	1.3

The notional safety indices are lower for towers than for buildings, confirming the general perception that the design for towers is much 'tighter' than for buildings. There are reserves of strength in buildings that are not taken account for in design such as the system effects, and this has been accounted for in the model by the factor H. From the above analysis, safety indices for type

I towers (for populated areas) are comparable to those for general building structures, while that for type II towers (open, less populated areas) are close to those for farm buildings.

4 CONCLUSIONS

The paper has considered the relative safety of lattice towers and buildings designed to Australian Standards. It has been shown that the reliability of most towers is comparable to those of buildings for the same type of environment.

REFERENCES

Pham, L., R.Q. Bridge and M. Bradford 1986. Calibration of the proposed limit states design rules for steel beams and columns. *Civil Engineering Transactions, Institution of Engineers Australia* CE28(3): 286–274.

Pham, L., J.D. Holmes and R.H. Leicester 1983. Safety indices for wind loading in australia. *Journal of Wind Engineering and Industrial Aerodynamics* 14: 3–14.

Pham, L., J.D. Holmes and J. Yang 1992. Reliability analysis of Australian communication lattice towers. *Journal of Construction Steel Research* 23: 255–272.

Standards Australia 1989. *SAA loading code. Part 2: Wind loads: AS 1170.2–1989.* Standards Australia, North Sydney.

Standards Australia 1991a. *Interim standard for the design of steel lattice towers and masts: AS 3995–1991.* Standards Australia, North Sydney.

Standards Australia 1991b. *Steel structures: AS 4100–1991.* Standards Australia, North Sydney.